Medical and Veterinary Entomology

Medical and Veterinary Entomology

Second Edition

Edited by

Gary R. Mullen

Department of Entomology and Plant Pathology,
Auburn University,
Auburn, AL 36849-5413, USA

Lance A. Durden

Department of Biology,
Georgia Southern University,
Statesboro, GA 30460-8042, USA

AMSTERDAM • BOSTON • HEIDELBERG • LONDON
NEW YORK • OXFORD • PARIS • SAN DIEGO
SAN FRANCISCO • SINGAPORE • SYDNEY • TOKYO
Academic Press is an imprint of Elsevier

Front-cover image: Reproduced with permission of the American Mosquito Control Association. Egg of *Aedes caspius*, from Linley et al. 1993. The egg of *Aedes caspius* and *Aedes africanus* (Diptera: Culicidae). Mosquito Systematics 25(1): 25–34.

Academic Press is an imprint of Elsevier
30 Corporate Drive, Suite 400, Burlington, MA 01803, USA
525 B Street, Suite 1900, San Diego, California 92101-4495, USA
84 Theobald's Road, London WC1X 8RR, UK

Library of Congress Cataloging-in-Publication Data

APPLICATION SUBMITTED

British Library Cataloguing-in-Publication Data

A catalogue record for this book is available from the British Library.

ISBN: 978-0-12-372500-4

For information on all Academic Press publications
visit our Web site at www.elsevierdirect.com

Companion website available:
http://books.elsevier.com/companions/9780123725004

Printed in China
09 10 9 8 7 6 5 4 3 2

Contents

Contributing Authors xi

Preface xiii

Acknowledgments xv

Chapter 1 Introduction 1

Lance A. Durden, and Gary R. Mullen

General Entomology 1
Medical-Veterinary Entomology Literature 1
A Brief History of Medical-Veterinary Entomology 2
Identification and Systematics of Arthropods of Medical-Veterinary Importance 3
Types of Problems Caused by Arthropods 3
  Annoyance 3
  Toxins and Venoms 3
  Allergic Reactions 4
  Invasion of Host Tissues 4
Arthropod-Borne Diseases 4
Food Contaminants 5
Fear of Arthropods 5
Delusional Disorders 5
Formicophilia 6
Host Defenses 6
Minor Arthropod Problems of Medical-Veterinary Interest 6

Chapter 2 Morphological Adaptations of Parasitic Arthropods 13

Nathan D. Burkett-Cadena

Body Shape and Wings 13
Mouthparts 13
Legs 15
Sensory Structures 17

Chapter 3 Epidemiology of Vector-Borne Diseases 19

William K. Reisen

Components of Transmission Cycles 20
  Host Immunity 20
  The Vertebrate Host 21
  The Arthropod Vector 22
Modes of Transmission 23
  Vertical Transmission 23
  Horizontal Transmission 24
Transmission Cycles 25
Interseasonal Maintenance 26
  Continued Transmission by Vectors 27
  Infected Vectors 27
  Infected Vertebrate Hosts 27
Vector Incrimination 27
  Infection Rates 27
  Vector Competence 28
  Vectorial Capacity 28
Surveillance 29
  Environmental Conditions 30
  Vector Abundance 30
  Enzootic Transmission Rates 30
  Clinical Cases 32
Emerging Vector-Borne Diseases 32

Chapter 4 Forensic Entomology 35

William L. Krinsky

History 35
Legal Cases Involving Liability 36
  Structural Entomology 36
  Stored-Products Entomology 36
  Occupational Hazards Associated with Arthropods 36
  Veterinary and Wildlife Entomology 36
Legal Cases Involving Homicides, Suspicious and Accidental Deaths, and Abuse and Neglect 36
  Sudden Death with Arthropod Association 37
  Automobile-Accident Death 37
  Arthropods as Signs of Neglect or Abuse, or as Agents of Murder 37
  Illicit Drug Transport, Use, and Overdose 37
  Suspicious Deaths 38
Stages of Decomposition 38
Insect Succession and Postmortem Interval 39

Chapter 5 Cockroaches (Blattaria) 43

Richard D. Kramer, and Richard J. Brenner

Taxonomy 43
Morphology 43
Life History 44
Behavior and Ecology 45
Common Cockroach Species 46
  Oriental Cockroach (*Blatta orientalis*) 46
  Turkestan Cockroach (*Blatta lateralis*) 46
  American Cockroach (*Periplaneta americana*) 47
  Australian Cockroach (*Periplaneta australasiae*) 47
  Brown Cockroach (*Periplaneta brunnea*) 48
  Smokybrown Cockroach (*Periplaneta fuliginosa*) 48
  Florida Woods Cockroach (*Eurycotis floridana*) 48
  Brownbanded Cockroach (*Supella longipalpa*) 49
  German Cockroach (*Blattella germanica*) 49

Asian Cockroach (*Blattella asahinai*) 50
Surinam Cockroach (*Pycnoscelus surinamensis*) 50
Public Health Importance 50
Pathogenic Agents 51
Intermediate Hosts 53
Cockroach Allergies 53
Veterinary Importance 54
Prevention and Control 55
Sanitation 55
Harborage Elimination 55
Physical Control 55
Biological Control 55
Insect Growth Regulators (IGRs) 56

Chapter
6 Lice (Phthiraptera) 59
Lance A. Durden, and John E. Lloyd
Taxonomy 59
Morphology 60
Life History 63
Behavior and Ecology 64
Lice of Medical Importance 66
Human Body Louse (*Pediculus humanus humanus*) 66
Human Head Louse (*Pediculus humanus capitis*) 67
Human Crab Louse (*Pthirus pubis*) 67
Lice of Veterinary Importance 68
Lice of Cattle 68
Lice of Other Livestock Animals 70
Lice of Cats and Dogs 71
Lice of Laboratory Animals 71
Lice of Poultry and Other Birds 71
Public Health Importance 72
Epidemic Typhus 72
Louse-Borne Relapsing Fever 73
Trench Fever 74
Other Pathogens Transmitted by Human Body Lice 74
Lice as Intermediate Hosts of Tapeworms 75
Veterinary Importance 75
Lice of Livestock 76
Lice of Wildlife 78
Lice of Cats and Dogs 78
Lice of Laboratory Animals 79
Lice of Poultry and Other Birds 79
Prevention and Control 79

Chapter
7 True Bugs (Hemiptera) 83
William L. Krinsky
Kissing Bugs (Reduviidae) 84
Taxonomy 85
Morphology 85
Life History 86
Behavior and Ecology 86
Public Health Importance 88
Chagas Disease (American Trypanosomiasis) 88
Other Human Parasites Associated with Kissing Bugs 92
Veterinary Importance 92
Prevention and Control 92
Bed Bugs (Cimicidae) 93
Taxonomy 93
Morphology 94
Life History 95
Behavior and Ecology 95
Public Health Importance 96
Other Cimicids that Occasionally Attack Humans 97
Veterinary Importance 97
Prevention and Control 97

Chapter
8 Beetles (Coleoptera) 101
William L. Krinsky
Taxonomy 101
Morphology 101
Life History 102
Behavior and Ecology 102
Public Health Importance 103
Meloidae (Blister beetles) 104
Oedemeridae (False blister beetles) 105
Staphylinidae (Rove beetles) 106
Tenebrionidae (Darkling beetles) 107
Dermestidae (Larder beetles) 107
Scarabaeidae (Scarab beetles) 108
Coccinellidae (Lady beetles) 109
Veterinary Importance 109
Ingestion of Toxic Beetles 109
Transmission of Pathogens 109
Intermediate Hosts of Parasites 110
Nest Associates and Ectoparasites 111
Dung Beetles and Biocontrol 111
Prevention and Control 112

Chapter
9 Fleas (Siphonaptera) 115
Lance A. Durden, and Nancy C. Hinkle
Taxonomy 115
Morphology 116
Life History 118
Behavior and Ecology 119
Fleas of Medical-Veterinary Importance 120
Human Flea (*Pulex irritans*) 120
Cat Flea (*Ctenocephalides felis*) 121
Dog Flea (*Ctenocephalides canis*) 121
Oriental Rat Flea (*Xenopsylla cheopis*) 121
European Rabbit Flea (*Spilopsyllus cuniculi*) 121
Sticktight Flea (*Echidnophaga gallinacea*) 122
Chigoe (*Tunga penetrans*) 122
Northern Rat Flea (*Nosopsyllus fasciatus*) 122
European Chicken Flea (Hen Flea in Europe) (*Ceratophyllus gallinae*) 123
European Mouse Flea (*Leptopsylla segnis*) 123
Public Health Importance 123
Flea-Associated Allergies 124
Plague 124
Murine Typhus 127
Other Flea-Borne Rickettsial Agents 128
Other Flea-Borne Pathogens 129
Tungiasis 129
Fleas as Intermediate Hosts of Helminths 130
Veterinary Importance 130
Flea-Bite Dermatitis 131
Tungiasis 131
Myxomatosis 131
Murine Trypanosomiasis 131
Other Flea-Borne Pathogens and Parasites 132
Fleas as Intermediate Hosts of Helminths 132
Prevention and Control 132

Chapter
10 Flies (Diptera) 137
Robert D. Hall, and Reid R. Gerhardt
Taxonomy 137
Morphology 138
Life History 141
Behavior and Ecology 142
Families of Minor Medical or Veterinary Interest 142
Tipulidae (Crane flies) 142
Bibionidae (March flies) 143
Sciaridae (Darkwinged fungus gnats) 143
Chaoboridae (Phantom midges) 144
Chironomidae (Chironomid midges) 144
Rhagionidae (Snipe flies) 145
Athericidae (Athericid flies) 145
Stratiomyidae (Soldier flies, Latrine flies) 146
Phoridae (Humpbacked flies, Scuttle flies) 146
Syrphidae (Flower flies, Hover flies) 147
Piophilidae (Skipper flies) 147
Drosophilidae (Small fruit flies) 148
Chloropidae (Grass flies, Eye gnats) 148
Public Health Importance 149
Veterinary Importance 150
Prevention and Control 150

Chapter
11 Moth Flies and Sand Flies (Psychodidae) 153
Louis C. Rutledge, and Raj K. Gupta
Taxonomy 153
Sycoracinae 153
Psychodinae 153
Phlebotominae 154
Morphology 154
Psychodinae 154
Phlebotominae 155
Life History 156
Psychodinae 156
Phlebotominae 156
Behavior and Ecology 156
Psychodinae 156
Phlebotominae 156
Public Health Importance 157
Psychodinae 157
Phlebotominae 158
Vesicular Stomatitis Virus Disease 158
Chandipura Virus Disease 160
Sand Fly Fever 160
Changuinola Virus Disease 160
Bartonellosis 161
Leishmaniasis 162
Cutaneous Leishmaniasis 163
Visceral Leishmaniasis 165
Veterinary Importance 165
Leishmaniasis 165
Vesicular Stomatitis Virus Disease 165
Prevention and Control 166
Psychodinae 166
Phlebotominae 166

Chapter
12 Biting Midges (Ceratopogonidae) 169
Gary R. Mullen
Taxonomy 169
Morphology 170
Life History 171
Behavior and Ecology 172
Public Health Importance 174
Oropouche Fever 176
Other Viral Agents 176
Mansonellosis 176
Veterinary Importance 178
Bluetongue Disease 178
Epizootic Hemorrhagic Disease 180
African Horsesickness 181
Other Viral Agents 183
Blood Protozoans 183
Equine Onchocerciasis 184
Other Filarial Nematodes 185
Equine Allergic Dermatitits 185
Prevention and Control 186

Chapter
13 Black Flies (Simuliidae) 189
Peter H. Adler, and John W. McCreadie
Taxonomy 189
Morphology 190
Life History 191
Behavior and Ecology 192
Public Health Importance 194
Biting and Nuisance Problems 195
Human Onchocerciasis 196
Mansonellosis 199
Other Diseases Related to Black Flies 199
Veterinary Importance 200
Bovine Onchocerciasis 201
Leucocytozoonosis 201
Other Parasites and Pathogens of Veterinary Importance 202
Simuliotoxicosis 202
Prevention and Control 203
Onchocerciasis Control 204

Chapter
14 Mosquitoes (Culicidae) 207
Woodbridge A. Foster, and Edward D. Walker
Taxonomy 207
Morphology 209
Life History 215
Behavior and Ecology 217
Public Health Importance 223
Mosquito Bites 223
Mosquito-Borne Viruses 223
Togaviridae (*Alphavirus*) 225
Other Alphaviruses 229
Flaviviridae (*Flavivirus*) 229
Yellow Fever 230
Dengue 232
Japanese Encephalitis Virus Complex 234
Other Flaviviruses 237
Bunyaviridae (*Orthobunyavirus* and *Phlebovirus*) 237
Malaria 239
Filariasis 243
Veterinary Importance 248
Mosquito-Borne Viruses of Animals 248
Nonhuman Malarias 250
Dog Heartworm 251
Other Filarial Nematodes of Animals 252
Prevention and Control 252
Control of Pathogen Transmission 254

Chapter
15 Horse Flies and Deer Flies (Tabanidae) 261
Bradley A. Mullens
Taxonomy 261
Morphology 263
Life History 264
Behavior and Ecology 266
Public Health Importance 268
Loiasis 269
Tularemia 270
Other Tabanid-Transmitted Human Pathogens 270
Veterinary Importance 270
Surra and Related Trypanosomiases 271
Equine Infectious Anemia 271
Anaplasmosis 271
Elaeophorosis 272
Other Pathogens of Veterinary Importance 272
Prevention and Control 272

Chapter
16 Muscid Flies (Muscidae) 275
Roger D. Moon
Taxonomy 275
Morphology 275
Life History 279
Behavior and Ecology 280
Species of Medical-Veterinary Importance 282
House Fly (*Musca domestica*) 282
Bazaar Fly (*Musca sorbens*) 283
Bush Fly (*Musca vetustissima*) 283
Face Fly (*Musca autumnalis*) 283
Cluster Fly (*Pollenia rudis*) 283
Stable Fly (*Stomoxys calcitrans*) 283
Horn Fly (*Haematobia irritans irritans*) and Buffalo Fly (*Haematobia irritans exigua*) 284
False Stable Fly (*Muscina stabulans*) and Its Relatives 284
Little House Fly (*Fannia canicularis*) and Its Relatives 284
Garbage Flies (*Hydrotaea* spp.) 285
Sweat Flies (*Hydrotaea* spp.) 285
Public Health Importance 285
House Fly (*Musca domestica*) 285
Bazaar Fly (*Musca sorbens*) 286
Bush Fly (*Musca vetustissima*) 286
Face Fly (*Musca autumnalis*) and Cluster Fly (*Pollenia rudis*) 286
Stable Fly (*Stomoxys calcitrans*) 287
False Stable Fly (*Muscina stabulans*) and Its Relatives 287
Little House Fly (*Fannia canicularis*) and Its Relatives 287
Garbage Flies (*Hydrotaea* spp.) 287
Sweat Flies (*Hydrotaea* spp.) 287
Veterinary Importance 288
House Fly (*Musca domestica*) 288
Bush Fly (*Musca vetustissima*) 289
Face Fly (*Musca autumnalis*) 289
Stable Fly (*Stomoxys calcitrans*) 290
Horn Fly (*Haematobia irritans irritans*) and Buffalo Fly (*Haematobia irritans exigua*) 291
Sweat Flies (*Hydrotaea* spp.) 292
Prevention and Control 292

Chapter
17 Tsetse Flies (Glossinidae) 297
William L. Krinsky
Taxonomy 297
Morphology 298
Life History 299
Behavior and Ecology 300
Public Health Importance 302
African Sleeping Sickness 302
West African Trypanosomiasis 303
East African Trypanosomiasis 303
Life Cycle of Trypanosomes 303
Veterinary Importance 305
Nagana 305
Prevention and Control 306

Chapter
18 Myiasis (Muscoidea, Oestroidea) 309
Philip J. Scholl, E. Paul Catts, and Gary R. Mullen
Taxonomy 310
Morphology 313
Life History 314
Ecology and Behavior 315
Myths 317
Flies Involved in Myiasis 317
Stratiomyidae (Soldier Flies) 317
Syrphidae (Flower Flies, Hover Flies, Rat-tailed Maggots) 317
Piophilidae (Skipper Flies) 318
Neottiophilidae (Nest Skipper Flies) 318
Drosophilidae (Pomace Flies, Vinegar Flies, Fruit Flies, and Wine Flies) 318
Chloropidae (Grass Flies and Australian Frog Flies) 319
Anthomyiidae (Root Maggots) 319
Fanniidae (Faniid Flies) 319
Muscidae (Dung Flies) 319
Tropical Nest Flies 320
Calliphoridae (Blow Flies, Carrion Flies, Floor Maggots, Nest Maggots, Screwworms) 320
Sarcophagidae (Flesh Flies) 323
Oestridae (Bot Flies) 324
New World Skin Bot Flies (Cuterebrinae) 325
Old World Skin Bot Flies (Hypodermatinae) 327
Nose Bot Flies (Oestrinae) 329
Stomach Bot Flies (Gasterophilinae) 331
Public Health Importance 332
Clinical Use of Maggots 333
Veterinary Importance 333
Prevention and Control 335
Screwworm Eradication Program 336
Cattle Grub Control 336

Chapter
19 Louse Flies, Keds, and Related Flies (Hippoboscoidea) 339
John E. Lloyd
Taxonomy 339
Morphology 340
Hippoboscidae 340
Streblidae 341
Nycteribiidae 341

Life History 342
Behavior and Ecology 342
Common Species of Hippoboscids 343
Sheep Ked (*Melophagus ovinus*) 343
Dog Fly (*Hippobosca longipennis*) 344
Hippobosca equina 344
Hippobosca variegata 345
Deer Keds (*Lipoptena and Neolipoptena* spp.) 345
Pigeon fly (*Pseudolynchia canariensis*) 346
Public Health Importance 346
Veterinary Importance 347
Prevention and Control 350

Chapter
20 Moths and Butterflies (Lepidoptera) 353
Gary R. Mullen
Taxonomy 353
Morphology 354
Spicule Hairs 355
Spine Hairs 356
Life History 357
Behavior and Ecology 357
Urticating Caterpillars 358
Megalopygidae 358
Limacodidae (Cochlidiidae, Eucleidae) 358
Saturniidae 360
Lymantriidae 361
Arctiidae 362
Lasiocampidae 362
Noctuidae 363
Nolidae 363
Thaumetopoeidae 363
Nymphalidae 363
Morphoidae 363
Lachryphagous Moths 364
Geometridae 364
Pyralidae 364
Notodontidae 364
Noctuidae 364
Sphingidae 365
Thyatiridae 365
Wound-Feeding and Skin-Piercing Moths 365
Public Health Importance 366
Veterinary Importance 367
Caterpillar-induced Equine Abortion 368
Prevention and Control 368

Chapter
21 Ants, Wasps, and Bees (Hymenoptera) 371
Hal C. Reed, and Peter J. Landolt
Taxonomy 371
Morphology 373
Life History 374
Behavior and Ecology 375
Hymenoptera Venoms 376
Ant Venoms 376
Vespid Venoms 377
Honey Bee Venom 377
Ants 377
Fire Ants (*Solenopsis* species) 378
Harvester Ants (*Pogonomyrmex* species) 380
Pavement Ant (*Tetramorium caespitum*) 380
Pharaoh's Ant (*Monomorium pharaonis*) 381
Wasps 381
Solitary Wasps 381
Social Wasps (Vespidae) 382
Bees 386
Solitary Bees 386
Social Bees 387
Public Health Importance 389
Veterinary Importance 392
Prevention and Control 392

Chapter
22 Scorpions (Scorpiones) 397
Gary R. Mullen, and Scott A. Stockwell
Taxonomy 397
Buthidae 397
Microcharmidae 398
Pseudochactidae 398
Chaerlidae 398
Chactidae 399
Euscorpiidae 399
Superstitioniidae 399
Troglotayosicidae 399
Iuridae 399
Vaejovidae 399
Bothriuridae 400
Liochelidae 400
Heteroscorpionidae 400
Hemiscorpiidae 400
Urodacidae 400
Diplocentridae 400
Scorpionidae 401
Morphology 401
Life History 403
Behavior and Ecology 404
Public Health Importance 405
Scorpions of Medical Importance 406
Veterinary Importance 407
Prevention and Control 407

Chapter
23 Solpugids (Solifugae) 411
Gary R. Mullen

Chapter
24 Spiders (Araneae) 413
Gary R. Mullen, and Richard S. Vetter
Taxonomy 413
Mygalomorph Spiders 413
Araneomorph Spiders 414
Morphology 416
Life History 417
Behavior and Ecology 418
Public Health Importance 418
Tarantism 419
Tarantulism 419
Atraxism 421
Phoneutriism 422
Cheiracanthism 422
Tegenarism 422
Loxoscelism 423
Latrodectism 426

Veterinary Importance 430
Prevention and Control 430

Chapter
25 Mites (Acari) 433
Gary R. Mullen, and Barry M. OConnor
Taxonomy 433
Morphology 433
Life History 435
Behavior and Ecology 436
Public Health Importance 436
Mite-Induced Dermatitis 436
Stored-Products Mites 442
Skin-Invading Mites 445
Mite-Induced Allergies 449
Internal Acariasis 452
Mite-Borne Diseases of Humans 453
Rickettsialpox 453
Tsutsugamushi Disease 454
Intermediate Hosts of Human Parasites 455
Delusory Acariasis and Acarophobia 456
Veterinary Importance 456
Mite-Induced Dermatitis 456
Laelapidae 458
Trombiculidae 459
Fur Mites 460
Feather Mites 463
Mange Mites 464
Other Sarcoptid Genera 471
Notoedres Species 472
Mite-Induced Allergies 478
Ear Mites 478
Respiratory Mites 481
Mite-Borne Diseases 484
Mites as Intermediate Hosts of Tapeworms 485

Chapter
26 Ticks (Ixodida) 493
William L. Nicholson, Daniel E. Sonenshine, Robert S. Lane, and Gerrit Uilenberg
Taxonomy 493
Family Ixodidae (Hard Ticks) 493
Family Argasidae (Soft Ticks) 495
Family Nuttalliellidae 495
Morphology 495
External Anatomy 495
Ixodidae 496
Argasidae 498
Internal Anatomy 498
Life History 499
Ixodid Life Cycles 499
Argasid Life Cycles 500
Behavior and Ecology 501
Tick Species of Medical-Veterinary Importance 505
Public Health Importance 511
Human Babesiosis 512
Tick-Borne Encephalitis Complex 513
Colorado Tick Fever 514
Rocky Mountain Spotted Fever 515
Boutonneuse Fever 517
Other Spotted Fever Group Rickettsiae 517
Human Ehrlichiosis 518
Human Granulocytic Anaplasmosis 519
Q Fever 519
Lyme Disease 520
Tick-Borne Relapsing Fever 522
Tularemia 524
Tick Paralysis 525
Tick-Bite Allergies 526
Veterinary Importance 526
Piroplasmoses 526
Louping Ill 528
African Swine Fever 529
Diseases Caused by Members of the Family Anaplasmataceae 529
Borrelioses 532
Tularemia 532
Q Fever 533
Dermatophilosis 533
Tick Paralysis 534
Tick Toxicoses 534
Prevention and Control 535
Personal Protection 535
Acaricides 535
Pheromone-Assisted Control 536
Passive Treatment 536
Hormone-Assisted Control 537
Vaccines 537
Management 537
Eradication 538

Chapter
27 Molecular Tools Used in Medical and Veterinary Entomology 543
Dana Nayduch
Cloning Genes and Genomics 543
Cloning Genes: Recombinant DNA Technology 543
Genomics: Cataloguing an organism's complete genetic sequence 544
Library Construction and Genome Assembly 544
Bioinformatics and databases 546
Genomes of Vectors and Vector-borne Pathogens 546
Polymerase Chain Reaction (PCR) 546
Applications of PCR 546
Analyzing Gene Expression 549
RNA Analysis 550
Diagnostic Techniques 553
Rapid Detection and Quantification of Pathogens in Hosts and Vectors 553
Visualization of Pathogens in Hosts and Vectors 553
Immune-based Diagnosis of Host Infection 553
Conclusions 554

Appendix: Arthropod-Related Viruses of Medical and Veterinary Importance 557
Michael J. Turell

Glossary 565

Taxonomic Index 611

Subject Index 625

Contributing Authors

Peter H. Adler
Department of Entomology, Soils and Plant Sciences, Clemson University, Clemson, SC 29634-0315, USA

Richard J. Brenner
Beltsville Agricultural Research Center, ARS, USDA, Office of Technology Transfer, Beltsville, MD 20705, USA

Nathan D. Burkett-Cadena
Department of Entomology and Plant Pathology, Auburn University, Auburn, AL 36849-5413, USA

E. Paul Catts
Deceased, formerly, Department of Entomology, Washington State University, Pullman, WA 99164-6382, USA

Lance A. Durden
Department of Biology, Georgia Southern University, Statesboro, GA 30460-8042, USA

Woodbridge A. Foster
Department of Entomology, The Ohio State University, Columbus, OH 43210-1242, USA

Reid R. Gerhardt
Department of Entomology and Plant Pathology, University of Tennessee, Knoxville, TN 37996-4560, USA

Raj K. Gupta
Walter Reed Army Institute of Research, Office of the Science Director, Silver Spring, MD 20910, USA

Robert D. Hall
Office of Research, University of Missouri, Columbia, MO 65211, USA

Nancy C. Hinkle
Department of Entomology, University of Georgia, Athens, GA 30602-2603, USA

Richard D. Kramer
Innovative Pest Management, Inc., Brookeville, MD 20833-1912, USA

William L. Krinsky
Division of Entomology, Peabody Museum of Natural History, Yale University, New Haven, CT 06520-8118, USA

Peter J. Landolt
Yakima Agricultural Research Laboratory, USDA, ARS, Wapato, WA, 98951-9651, USA

Robert S. Lane
University of California, Division of Organisms and Environment, Department of Environmental Science, Policy and Management, Berkeley, CA 94720, USA

John E. Lloyd
Emeritus in Entomology, University of Wyoming, Laramie, WY 82071, USA

John W. McCreadie
Department of Biology, University of South Alabama, Mobile, AL 36688-0002, USA

Roger D. Moon
Department of Entomology, University of Minnesota, St. Paul, MN 55108, USA

Gary R. Mullen
Department of Entomology and Plant Pathology, Auburn University, Auburn, AL 36849-5413, USA

Bradley A. Mullens
Department of Entomology, University of California, Riverside, CA 92521, USA

Dana Nayduch
Department of Biology, Georgia Southern University, Statesboro, GA 30460-8042, USA

William L. Nicholson
Disease Assessment Team, Rickettsial Zoonoses Branch, National Center for Zoonotic, Vector-borne, and Enteric Diseases, Centers for Disease Control and Prevention, Atlanta, GA 30333, USA

Barry M. OConnor
Museum of Zoology, University of Michigan, Ann Arbor, MI 48109-1079, USA

Hal C. Reed
Biology Department, Oral Roberts University, Tulsa, OK 74171, USA

William K. Reisen
Department of Pathology, Microbiology and Immunology, School of Veterinary Medicine, University of California, Davis, CA 95616, USA

Louis C. Rutledge
[retired, U.S. Army], Mill Valley, CA 94941-3420, USA

Philip J. Scholl
United States Department of Agriculture/Agricultural Research Service (retired), Richland Center, WI 53581, USA

Daniel E. Sonenshine
Department of Biological Sciences, Old Dominion University, Norfolk, VA 23529-0266, USA

Scott A. Stockwell
[retired, U.S. Army], Lubbock, TX 79423, USA

Michael J. Turell
Virology Division, U.S. Army Medical Research Institute of Infectious Diseases, Fort Detrick, MD 21702, USA

Gerrit Uilenberg
"A Surgente", route du Port, (Corsica), FRANCE

Richard S. Vetter
Department of Entomology, University of California, Riverside, CA 92521, USA

Edward D. Walker
Department of Microbiology and Molecular Genetics, Michigan State University, East Lansing, MI 48824, USA

Preface

It has been seven years since the publication of the first edition of this book. During this time, significant advances have been made in our knowledge of certain arthropod-borne diseases, the geographic movement of arthropod vectors from one part of the world to another, expansion of the range of many vector species into both temperate and tropical regions, and the emergence of new or previously unrecognized arthropod-related diseases of medical and veterinary concern. New pathogens and parasites continue to be discovered and characterized utilizing the latest genetic and molecular tools. The latter have improved dramatically, not only the ability to distinguish closely related arthropods and pathogens and to recognize new species, but also to provide rapid laboratory identification of disease agents for earlier diagnosis and treatment of cases of arthropod-related diseases. This edition is intended to provide you with as current a picture as possible of the insects and related arthropods of importance to human and animal health.

One of the primary objectives of the first edition was to provide a textbook suitable for teaching courses in medical and veterinary entomology at the college and university level. In keeping with that goal and the format of the first edition, the book is organized from an entomological perspective, with each chapter devoted to a particular taxonomic group of insects or related arthropods (including spiders, scorpions, mites, and ticks). As in the first edition, each chapter includes the following subheadings: Taxonomy, Morphology, Life History, Behavior and Ecology, Public Health Importance, Veterinary Importance, Prevention and Control, and References and Further Reading. The separate sections on public health and veterinary entomology are designed to assist instructors in using this book to teach courses in either medical or veterinary entomology, or courses combining these two closely related disciplines.

In addition to its value to students as a textbook, this volume should appeal to a much broader audience as a comprehensive reference source for biologists in general, entomologists, zoologists, parasitologists, physicians, public-health personnel, veterinarians, wildlife biologists, vector biologists, military entomologists, the general public, and others looking for a readable, authoritative book on this important topic.

Several new features have been added. These include the following three new chapters: Morphological Adaptations of Parasitic Arthropods; Forensic Entomology; and Molecular Tools in Medical and Veterinary Entomology. In addition, there is an Appendix titled Arthropod-related Viruses of Medical and Veterinary Importance, and a Glossary of approximately 1700 terms. The latter, together with the new chapter on morphological adaptations, is intended to help you understand entomological, medical, and other terminology used in the book, with which you may not be familiar. Hopefully this will facilitate use of the book by the widest possible range of readers, specialists and nonspecialists alike, in diverse disciplines relating either directly or indirectly to the subject matter. The text is illustrated with 481 figures, including 109 new color images and 17 revised, or new, full-color maps.

We welcome as new contributors to the book the following nine individuals: Nathan D. Burkett-Cadena (Chapter 2), Nancy C. Hinkle (Chapter 9), Richard D. Kramer (Chapter 5), Peter J. Landolt (Chapter 21), Dana Nayduch (Chapter 27), Philip J. Scholl (Chapter 18), Michael J. Turell (Appendix), Gerrit Uilenberg (Chapter 26), and Richard S. Vetter (Chapter 24). Together with 23 contributors to the first edition, 32 in all, they have helped significantly in revising and updating the text, adding new chapters and topics, strengthening the veterinary aspects of several chapters, and achieving the desired balance between medical and veterinary entomology as two closely related disciplines.

Given the success of the first edition, as reflected by the widespread adoption of this book at colleges and universities throughout the United States and other parts of the world, we hope the second edition will be equally successful in helping to educate the next generation of medical and veterinary entomologists.

Gary R. Mullen
Lance A. Durden

Acknowledgments

As with any undertaking of this magnitude, there are many individuals to whom the editors are indebted. Foremost are the contributing authors, whose combined expertise and commitment to promoting medical and veterinary entomology have made the second edition of this book possible. We also are grateful to the many other individuals who contributed so generously of their time in reviewing chapters or select parts of the text, offering suggestions for improving and updating the first edition, preparing illustrations, and providing original photographs and other previously unpublished color images for the figures.

We particularly want to recognize the following people who have shared their talents in preparing the figures, maps, and other illustrations throughout the book: Rebecca L. Nims (Social Circle, GA), for the outstanding work she did in redrawing from various sources more than half the black-and-white figures for the first edition, notably for the mite chapter, virtually all of which have been retained in the second edition; Margo A. Duncan (Gainesville, FL), for her original illustrations, particularly those in the Lepidoptera chapter (originals 20.1, 20.4), and redrawn figures (6.6, 20.2, 20.3, 20.5) that appear in both the first and second editions; Nathan D. Burkett-Cadena (Auburn University, AL), for providing the new, original color figures that accompany his chapter on morphological adaptations of parasitic arthropods (2.1, 2.2, 2.3, 2.4), preparing the illustrations for the epidemiology chapter (3.2, 3.3) and the majority of revised, or new, color maps that appear throughout the book (7.10, 11.7, 14.21, 14.22, 14.24, 14.27, 14.28, 14.30, 14.31, 14.34, 14.35, 17.9, 21.7, 26.21, 26.25). We are grateful to the late E. Paul Catts (Washington State University, Pullman, WA) for being able to include in this edition the fine original figures of myiasis-causing flies that he prepared for the first edition, prior to his untimely death in 1996 (18.1, 18.2, 18.3, 18.4, 18.5, 18.6, 18.7, 18.8, 18.9, 18.11, 18.13, 18.14, 18.15, 18.17, 18.20, 18.23, 18.26, 18.31, 18.32, 18.35, 18.36, 18.39). In addition, we wish to recognize the following individuals for their original illustrations that appear within these pages: Woodbridge A. Foster (The Ohio State University, Columbus, OH; 14.14), Susan J. M. Hope (Mebane, NC; 15.1), Takumasa Kondo (Palmira, Valle, Colombia; 10.9), William L. Krinsky (Yale University, New Haven, CT; 4.3, 7.6), Dana Nayduch (Georgia Southern University, Statesboro, GA; 27.1, 27.2, 27.3, 27.4, 27.5, 27.6, 27.7), and Blair Sampson (US Department of Agriculture, Agricultural Research Service, Poplarville, MS; 21.5).

The following individuals have kindly provided original photographs, slides, or digital images for reproduction as figures in this edition: W. V. Adams, Jr. (Louisiana State University, Baton Rouge, LA; 15.14); Peter H. Adler (Clemson University, Clemson, SC; 13.5); the late Roger D. Akre (Washington State University, Pullman, WA; 21.9, 21.13, 21.14, 21.23, 21.30); Hans Bänziger (Chaing Mai University, Thailand; 20.19, 20.20, 20.21, 20.22, 20.23, 20.24); Yehuda Braverman (Kimron Veterinary Institute, Israel; 12.9); Alberto B. Broce (Kansas State University, Manhattan, KS); Nathan D. Burkett-Cadena (Auburn University, Auburn, AL; 4.1, 4.4, 5.4, 9.5, 9.6, 9.7, 9.8, 19.5, 24.6, 25.11); João P. Burini, Pontifica Universidade Católica de São Paulo, Brazil; 22.2, 22.10, 24.4, 24.5, 24.10); Lyle Buss (University of Florida, Gainesville, FL; 20.25); Jerry F. Butler (University of Florida, Gainesville, FL; 8.3, 19.4, 20.13, 25.25, 25.40, 25.43, 26.10, 26.19); Bonnie Buxton (Suwanee, GA; 14.41); James Castner (University of Florida, Gainesville, FL; 5.3, 5.5, 5.6, 5.7, 5.8, 5.9, 5.10, 5.11, 5.12, 20.26, 24.21); Ronald D. Cave (Indian River Research and Education Center, University of Florida, Fort Pierce, FL; 18.25); Valerie J. Cervenka (University of Minnesota, St. Paul, MN); Dr. Chan Chee Keong (Panang Island, Malaysia; 8.6); Eddie W. Cupp (Owensboro, KY; 13.7, 13.9); Neil K. Dawe (Canadian Wildlife Service; 13.12); Aaron T. Dossey (Gainesville, FL; 1.1); Lance A. Durden (Georgia Southern University, Statesboro, GA; 9.3); the late S. Allen Edgar (Auburn University, Auburn, AL; 14.36, 14.37); Marc E. Epstein (California Department of Food and Agriculture, Sacramento, CA; 20.12); Debbie R. Folkerts (Auburn University, Auburn, AL; 23.1); Woodbridge A. Foster (The Ohio State University, Columbus, OH; 14.11, 14.17, 14.19, 14.26, 14.29, 14.33); James Gathany (Centers for Disease Control and Prevention, USA; 11.2, 26.16, 26.17, 26.18); Carolyn Grissom (Shelbyville, TN; 24.17); Joyce Gross (San Leandro, CA; 9.1); Duane J. Gubler (Duke-National University of Singapore Graduate Medical School, Singapore; 14.25); Robert G. Hancock (Metropolitan State College, Denver, CO; 14.23); the late Elton J. Hansens (Asheville, NC; 6.9, 6.10, 9.14, 15.7, 15.10, 16.12); Nancy C. Hinkle (University of Georgia, Athens, GA; 6.11, 8.11); Mac C. Horton (Clemson University, Clemson, SC; 21.27); Kevin Humphreys (Huntsville, AL; 24.16); the late Lacy L. Hyche (Auburn University, Auburn, AL; 20.15); Gregory D. Johnson (Montana State University, Bozeman, MT; 6.19); Phillip E. Kaufman (University of Florida, Gainesville, FL; 9.10); Takumasa Kondo (Palmira, Valle,

Colombia; 21.22); William L. Krinsky (Yale University, New Haven, CT; 4.2, 4.5); Dwight R. Kuhn (Dwight Kuhn Photography, Dexter, ME; 13.2); Richard C. Lancaster (Agriculture and Agri-Food Canada, Lethbridge, AB; 12.2); Lloyd L. Lauerman (Alabama State Veterinary Diagnostic Lab, Auburn, AL; 12.9, 12.10, 12.11); John E. Lloyd, Sr. (University of Wyoming, Laramie, WY); Stephen A. Marshall (University of Guelph, Guelph, ON, Canada; 13.1, 13.3, 13.4); Sturgis McKeever (Georgia Southern University, Statesboro, GA; 15.5, 18.18, 20.6, 20.7, 20.8, 20.9, 20.10, 20.14, 20.18, 24.15, 24.20); Hendrick J. Meyer (North Dakota State University, Fargo, ND; 16.23, 16.24); Roger D. Moon (University of Minnesota, St. Paul, MN; 16.16, 16.19); Gary R. Mullen (Auburn University, Auburn, AL; 7.14, 8.3, 12.4,12.5, 12.12, 12.13, 16.18, 16.20, 18.19, 20.11, 21.8, 21.19, 21.20, 21.21, 21.28, 21.29, 24.13, 26.9, 26.13, 26.14, 26.15); Bradley A. Mullens (University of California, Riverside, CA; 15.2, 15.8, 15.11); Harold D. Newson (Michigan State University, East Lansing, MI; 14.38, 14.40); Yoshiro Ohara (Tohoku University School of Medicine, Sendai, Japan; 15.13); Jonathan D. Patterson (Michigan State University, East Lansing, MI; 14.39); the late Laverne L. Pechuman (Cornell University, Ithaca, NY; 15.6); Phil Pellitteri (University of Wisconsin, Madison, WI; 13.13); Eric Poggenphol (Amherst, MA; 13.8); Robert J. Raven (Queensland Museum, South Brisbane, Australia; 24.9); Hal C. Reed (Oral Roberts University, Tulsa, OK; 21.17); Will K. Reeves (USDA-ARS Arthropod-borne Animal Disease Research Laboratory, Laramie, WY; 19.4); Mary Elizabeth Rogers (Waukegan, IL; 12.17); Philip J. Scholl (US Department of Agriculture/Agricultural Research Service, Richland Center, WI; 18.34); Justin O. Schmidt (Southwestern Biological Institute, Tucson, AZ; 21.4, 21.26); the late Joseph A. Shemanchuk (Department of Agriculture and Agri-Food Canada, Lethbridge, AB, Canada; 13.10; 13.11); Scott A. Stockwell (Lubbock, TX; 22.1, 22.8, 22.9); Daniel R. Suiter (University of Georgia, Griffin, GA; 5.2); Robert B. Tesh (University of Texas Medical Branch, Galveston, TX; 7.4); Gerrit Uilenberg (Corsica, France; 26.27, 26.28, 26.29, 26.30); P. Kirk Visscher (University of California, Riverside, CA; 12.6); Jan Votýpka (Czech Academy of Science, Czech Republic; 13.14); Laurel L. Walters (Lieen-Follican Research, Bishop, CA; 11.8); D. Wesley Watson (University of North Carolina, Raleigh, NC; 8.10); the late J. Weintraub (Agriculture and Agri-Food Canada, Lethbridge, AB; 18.27, 18.30); Julian White (Toxicology Department, Women's & Children's Hospital, North Adelaide, Australia; 24.8, 24.22); Ralph E. Williams (Purdue University, West Lafayette, IN; 16.17); and Germano Woehl, Jr. (Instituto Rã-bugio para Conservação da Biodiversidade, Jaraguá do Sul, Brazil; 20.16).

We also express our appreciation to the following persons for the multiple services they have provided us in preparing this edition: Arthur G. Appel (Auburn University, Auburn, AL) for his expertise on cockroaches and assistance in photographing select species; Byron Blagburn (College of Veterinary Medicine, Auburn University, Auburn, AL) for providing fleas for photographing the developmental stages; Art Borkent (Salmon Arm, British Columbia) for providing taxonomic literature on the Ceratopogonidae and helpful contact information; Paulo Bretanha Ribeiro (Universidade Federale de Pelotas, Pelotas, RS, Brazil) for expertise regarding *Dermatobia*, the Calliphoridae, and parasitic wasps that attack cockroaches; Charles H. Calisher, Colorado State University, Ft. Collins, Co.; Anne-Marie A. Callcott (US Department of Agriculture, APHIS, Gulfport, MS) for assistance in providing latest information on distribution of imported fire ants in the United States; Simon Carpenter (Institute for Animal Health, Pirbright, UK) for information on laboatory colonies of *Culicoides* species and control techniques for ceratopogonids; Douglas D. Colwell (Agriculture and Agri-Food Canada, Lethbridge, AB) for assistance with the myiasis chapter; Nancy C. Hinkle (University of Georgia, Athens, GA) for her helpful input in preparing the Glossary and Acknowledgments; Lawrence J. Hribar (Florida Keys Mosquito Control District, Gulf Marathon, FL) for reviewing the forensic chapter and providing literature sources on formicophilia; Claudine Jenda (Auburn University Libraries, Auburn, AL) for her help in literature and related searches; Marcelo B. Labruna (University of São Paulo, Brazil) for input and assistance with images for the flea chapter; Phil Lounibos (Florida Medical Entomology Laboratory, University of Florida, Vero Beach, FL) for his help in making available SEM photographs of mosquito eggs by the late John Linley for the cover design of this book; Timothy J. Lysyk (Agriculture and Agri-Food Canada, Lethbridge, AB) for providing images and assistance in obtaining permission to reproduce others; C. Steven Murphree (Belmont University, Nashville, TN) for reviewing the forensic chapter; Richard G. Robbins (Walter Reed Army Institute of Research, Washington, DC) for help in tracking down literature and locating people we needed to contact; John F. Roberts (Alabama Veterinary Diagnostic Lab, Auburn, AL) for input regarding caterpillar-induced equine abortions; Michael W. Service (Liverpool, UK) for providing a review and corrections to be made to the first edition, David E. Stallknecht (Southeastern Cooperative Wildlife Disease Study, The University of Georgia, Athens, GA) for information on serotypes of bluetongue and epizootic hemorrhagic disease viruses; David Taylor (USDA-ARS, Midwest Livestock Insects Laboratory, Lincoln, NE); Richard Wall (School of Biological Sciences, The University of Bristol, UK) for his helpful taxonomic input, comments, and suggestions regarding myiasis-causing flies; Scott C. Weaver, University of Texas Medical Branch, Galveston, Tx; Yu Yi-Xin (Institute of Microbiology and Epidemiology, Beijing, China) for information on ceratopogonid midges in China; and the following three individuals for pertinent literature and helpful suggestions regarding the tick chapter—Anne M. Kjemtrup (Vector-borne Disease Section, California Department of Public Health, Sacramento, CA), Sarah E. Randolph (University of Oxford, Oxford, UK), and Tom G. Schwan (National Institute of Allergy and Infectious Diseases, Rocky Mountain Laboratories, Hamilton, MT).

Appreciation is also extended to the following agencies and institutions for providing information,

images, and other forms of assistance: American Mosquito Control Association (Mount Laurel, NJ), American Museum of Natural History (New York), Entomological Society of America, Centers for Disease Control and Prevention (USA), Auburn University College of Veterinary Medicine (Auburn, AL), Department of Agriculture and Agri-Food Canada, Food and Agricultural Organization of the United Nations, National Geographic Society, The Rockefeller Foundation, The Natural History Museum (London), US Armed Forces Institute of Pathology, US Department of Agriculture, including the Animal and Plant Health Inspection Service (APHIS), Animal Research Service (ARS) laboratories at Laramie, WY, and Kerrville, TX, the Foreign Animal Disease Diagnostic Laboratory (Plum Island, NY) and Forest Service; US Public Health Service, US National Tick Collection (Georgia Southern University, Statesboro, GA), and World Health Organization (WHO) Vector Control and Prevention Programme.

It is our special privilege to recognize three contributors to the first edition of this book who are no longer with us: Roger D. Akre and E. Paul Catts (Department of Entomology, Washington State University, Pullman, WA) and Robert Traub (Bethesda, MD). Because of the major contributions that Paul Catts made as lead author of the myiasis chapter, and in recognition of his 22 original figures that once again appear in that chapter in this second edition, Chapter 18 still bears his name. We dedicate this book to their memory and to other medical and veterinary entomologists who have devoted their careers to protecting humans and animals from injurious arthropods and vector-borne diseases.

And, finally, we would like to acknowledge the Elsevier staff, without whom this book could not have come to fruition. We thank our Developmental Editors Pat Gonzalez, Christine Minihane, and Kelly Sonnack (San Diego); and Senior Acquisition Editor, Life Science Books, Andy Richford (London); and especially Project Manager Christie Jozwiak (Burlington, MA).

Chapter

1

Introduction

Lance A. Durden . Gary R. Mullen

Medical entomology is the study of insects, insect-borne diseases, and other associated problems that affect humans and public health. **Veterinary entomology** is the study of insects and insect-related problems that affect domestic animals, particularly livestock and companion animals (dogs, cats, horses, caged birds, etc.). In addition, veterinary entomology includes insect-associated problems affecting captive animals in zoological parks and wildlife in general. **Medical-veterinary entomology** combines these two disciplines.

Traditionally the fields of medical and veterinary entomology have included health-related problems involving arachnids (particularly mites, ticks, spiders, and scorpions). This broad approach encompassing insects and arachnids is followed in this text. Alternatively, the study of health-related problems involving arachnids is called **medical-veterinary arachnology** or, if just mites and ticks are considered, **medical-veterinary acarology**.

Historically, both medical and veterinary entomology have played major roles in the development of human civilization and animal husbandry. Outbreaks of insect-borne diseases of humans have profoundly influenced human history; these include such diseases as yellow fever, plague, louse-borne typhus, malaria, African trypanosomiasis, Chagas disease, and lymphatic filariasis. Likewise, livestock scourges such as bovine babesiosis, bovine theileriosis, scabies, pediculosis, and botfly infestations, all of which are caused or transmitted by arthropods, have greatly influenced animal production and husbandry practices. Arthropod-related disorders continue to cause significant health problems in humans, domestic animals, and wildlife. At the same time new strains of known pathogens, as well as previously unrecognized disease agents transmitted by arthropods, are causing newly recognized diseases (e.g., Lyme disease and human granulocytic anaplasmosis) and the resurgence of diseases that had been suppressed for many years (e.g., malaria). In fact, emerging and resurging arthropod-borne diseases are recognized as a growing health concern by public-health and veterinary officials (Wilson and Spielman, 1994; Walker et al. 1996; Gubler, 1998; Winch, 1998; Gratz, 1999) and are addressed in Chapter 3 of this book.

GENERAL ENTOMOLOGY

Basic concepts of entomology such as morphology, taxonomy and systematics, developmental biology, and ecology provide important background information for medical and veterinary entomologists. General entomology books that are helpful in this regard include Gillot (1995), Elzinga (2000), Chapman (1998), Romoser and Stoffolano (1998), Gullan and Cranston (2005), Triplehorn and Johnson (2005), and Pedigo and Rice (2006). References that provide a more taxonomic or biodiversity-oriented approach to general entomology include works by Arnett (2000), Richards and Davies (1994), Bosik (1997), and Daly et al. (1998). General insect morphology is detailed in Snodgrass (1993), and Torre-Bueno (1962) contains a useful glossary of general entomology. An encyclopedia of entomology (Capinera, 2008) and a dictionary of entomology (Gordh and Headrick, 2001) are also available. Texts on **urban entomology**, the study of insect pests in houses, buildings, and urban areas, which also has relevance to medical-veterinary entomology, have been prepared by Ebeling (1975), Hickin (1985), Mallis et al. (2004), and Robinson (1996). General texts on **acarology** include works by Woolley (1987), Evans (1992), and Krantz and Walter (2009).

MEDICAL-VETERINARY ENTOMOLOGY LITERATURE

Textbooks or monographs pertaining to medical entomology, veterinary entomology, or the combined discipline of medical-veterinary entomology are listed under these headings at the end of this chapter. Most of these publications emphasize arthropod morphology, biology, systematics, and disease relationships, whereas some of the more recent texts emphasize molecular aspects of medical-veterinary entomology, such as Crampton et al. (1997) and Marquardt et al. (2005). Other works are helpful regarding common names of arthropods of medical-veterinary importance (Pittaway, 1992), surveillance techniques (Bram, 1978), control measures (Drummond et al., 1988), repellents (Debboun et al., 2007), or ectoparasites (Andrews, 1977; Marshall, 1981; Kim, 1985; Uilenberg, 1994; Barnard and Durden, 1999). Publications that devote substantial sections to arthropods associated with wildlife and the pathogens they transmit include Davidson et al. (1981), Fowler (1986), Davidson and Nettles (1997), and Samuel et al. (2001).

Several journals and periodicals are devoted primarily to medical or veterinary entomology. These include the following:

- *Journal of Medical Entomology*, published by the Entomological Society of America
- *Medical and Veterinary Entomology*, published by the Royal Entomological Society (UK)
- *Journal of Vector Ecology*, published by the Society for Vector Ecology
- *Review of Medical and Veterinary Entomology*, published by CAB International
- *The Annals of Medical Entomology*, published in India

Journals specializing in parasitology, tropical medicine, or wildlife diseases that also include papers on medical-veterinary entomology include:

- *Parasitology*, published by the British Society for Parasitology
- *Journal of Parasitology*, published by the American Society of Parasitologists
- *Parasite-Journal de la Société Française de Parasitologie*, published in France
- *Advances in Disease Vector Research*, published by Springer-Verlag
- *Bulletin of the World Health Organization*, published by the World Health Organization
- *Journal of Wildlife Diseases*, published by the Wildlife Disease Association
- *Emerging Infectious Diseases*, published by the Centers for Disease Control and Prevention
- *American Journal of Tropical Medicine and Hygiene*, published by the American Society of Tropical Medicine and Hygiene
- *Memorias Do Instituto Oswaldo Cruz*, published in Brazil

Various Internet web sites pertaining to medical-veterinary entomology can also be accessed for useful information.

A BRIEF HISTORY OF MEDICAL-VETERINARY ENTOMOLOGY

Problems caused by biting and annoying arthropods and the pathogens they transmit have been the subject of writers since antiquity (Service, 1978). Homer (mid-eighth century BC), Aristophanes (ca. 448–380 BC), Aristotle (384–322 BC), Plautus (ca. 254–184 BC), Columella (5 BC–AD 65), and Pliny (AD 23–79) all wrote about the nuisance caused by flies, mosquitoes, lice, or bedbugs. However, the study of modern medical-veterinary entomology usually is recognized as beginning in the late nineteenth century, when blood-sucking arthropods were first proven to be vectors of human and animal pathogens.

Englishman **Patrick Manson** (1844–1922) was the first to demonstrate pathogen transmission by a blood-feeding arthropod. Working in China in 1877, he showed that the mosquito *Culex pipiens fatigans* is a vector of *Wuchereria bancrofti*, the causative agent of Bancroftian filariasis. Following this landmark discovery, the role of various blood-feeding arthropods in transmitting pathogens was recognized in relatively rapid succession.

In 1891, Americans **Theobald Smith** (1859–1934) and **F. L. Kilbourne** (1858–1936) implicated the cattle tick, *Boophilus annulatus*, as a vector of *Babesia bigemina*, the causative agent of Texas cattle fever (bovine babesiosis). This paved the way for a highly successful *B. annulatus*-eradication program in the United States directed by the US Department of Agriculture (USDA). The eradication of this tick resulted in the projected goal—elimination of indigenous cases of Texas cattle fever throughout the southern United States.

In 1898, Englishman Sir **Ronald Ross** (1857–1932), working in India, demonstrated the role of mosquitoes as vectors of avian malarial parasites from diseased to healthy sparrows. Also in 1898, the cyclical development of malarial parasites in anopheline mosquitoes was described by Italian **Giovanni Grassi** (1854–1925). In the same year, Frenchman **Paul Louis Simond** (1858–1947), working in Pakistan (then part of India), showed that fleas are vectors of the bacterium that causes plague.

In 1848, American physician **Josiah Nott** (1804–1873) of Mobile, Alabama, published circumstantial evidence that led him to believe that mosquitoes were involved in the transmission of yellow fever virus to humans. In 1881, the Cuban-born Scottish physician **Carlos Finlay** (1833–1915) presented persuasive evidence for his theory that what we know today as the yellow fever mosquito, *Aedes aegypti*, was the vector of this virus. However, it was not until 1900 that American **Walter Reed** (1851–1902) led the US Yellow Fever Commission at Havana, Cuba, which proved *Aedes aegypti* to be the principal vector of yellow fever virus.

In 1903, Englishman **David Bruce** (1855–1931) demonstrated the ability of the tsetse fly *Glossina palpalis* to transmit, during blood feeding, the trypanosomes that cause African trypanosomiasis.

Other important discoveries continued well into the twentieth century. In 1906, American **Howard Taylor Ricketts** (1871–1910) proved that the Rocky Mountain wood tick, *Dermacentor andersoni*, is a vector of *Rickettsia rickettsii*, the causative agent of Rocky Mountain spotted fever. In 1907, **F. P. Mackie** (1875–1944) showed that human body lice are vectors of

Borrelia recurrentis, the spirochete that causes louse-borne (epidemic) relapsing fever. In 1908, Brazilian **Carlos Chagas** (1879–1934) demonstrated transmission of the agent that causes American trypanosomiasis (later named Chagas' disease in his honor) by the conenose bug *Panstrongylus megistus*. In 1909, Frenchman **Charles Nicolle** (1866–1936), working in Tunis, showed that human body lice are vectors of *Rickettsia prowazekii*, the agent of louse-borne (epidemic) typhus.

These important discoveries, as well as others of historical relevance to medical-veterinary entomology, are discussed in more detail in the references listed at the end of this chapter. Because of the chronology of many major discoveries relevant to this topic in the 50-year period starting in 1877, this time has been called the "golden age of medical-veterinary entomology" (Philip and Rozeboom, 1973).

IDENTIFICATION AND SYSTEMATICS OF ARTHROPODS OF MEDICAL-VETERINARY IMPORTANCE

Table 1.1 provides a list of the eight orders of insects and four orders of arachnids that are of particular interest to medical-veterinary entomologists. Accurate identification of these arthropods is an important first step in determining the types of problems they can cause and, subsequently, in implementing control programs.

Table 1.1 Principal Orders of Insects and Arachnids of Medical-Veterinary Interest

Order	Common Names
Class Insecta	
Order Blattaria	Cockroaches
Order Phthiraptera	Lice
Order Hemiptera	True bugs: bedbugs, kissing bugs, assassin bugs
Order Coleoptera	Beetles
Order Siphonaptera	Fleas
Order Diptera	Flies: mosquitoes, black flies, no-see-ums, horse flies, deer flies, sand flies, tsetse flies, house flies, stable flies, horn flies, bot flies, blow flies, flesh flies, louse flies, keds, etc.
Order Lepidoptera	Moths and butterflies
Order Hymenoptera	Wasps, hornets, velvet ants, ants, bees
Class Arachnida	
Order Scorpionida	Scorpions
Order Solpugida	Solpugids, sun spiders, camel spiders, barrel spiders
Order Acari	Mites, ticks
Order Araneae	Spiders

Although taxonomy and identification are discussed in more detail with respect to arthropod groups treated in the chapters that follow, some publications provide a broader perspective on the classification, taxonomy, and identification of a range of arthropods of medical-veterinary importance. These include two works published by the US Centers for Disease Control and Prevention (CDC) (1979, 1994), as well as Service (1988), Hopla et al. (1994), Lago and Goddard (1994), and Davis (1995). Also, some medical-veterinary entomology books are very taxonomically oriented, with emphasis on identification; for example, Baker et al. (1956), Smith (1973), Lane and Crosskey (1993), and Walker (1995).

TYPES OF PROBLEMS CAUSED BY ARTHROPODS

Annoyance

Irrespective of their role as blood-feeders (hematophages), parasites, or vectors of pathogens, certain arthropods cause severe annoyance to humans or other animals because of their biting behavior. These include lice, bedbugs, fleas, deer flies, horse flies, tsetse flies, stable flies, mosquitoes, black flies, biting midges, sand flies, chiggers, and ticks. Some, however, do not bite but instead are annoying because of their abundance, small size, or habit of flying into or around the eyes, ears, and nose. Nonbiting arthropods that cause annoyance include the house fly, chironomid midges, and eye gnats. Large populations of household or filth-associated arthropods such as houseflies and cockroaches also can be annoying. Nuisance arthropods are commonly problems for humans at outdoor recreational areas including parks, lakes, and beaches.

Toxins and Venoms

Several terms are used when discussing chemical substances that have adverse effects on humans and other animals. A **poison** is any substance that when taken into the body interferes with normal physiological functions. A **toxin** is a poison of plant or animal origin that can result in a pathological condition called **toxicosis**. A **venom** is a poisonous mixture of compounds containing one or more toxins, which is produced in venom glands and injected into animal tissues via specialized morphological structures (e.g., stings, modified spines, and chelicerae in arthropods). The act of injecting venom into animal tissues is called **envenomation**.

Toxins produced by arthropods represent a wide range of chemical substances, from simple inorganic or organic compounds to complex alkaloids and heterocyclic compounds. Venoms often contain various pharmacologically active compounds that facilitate the spread and effectiveness of the toxic components. They commonly include amines (e.g., histamine, catecholamines, serotonin), peptides, polypeptides (e.g., kinins), specific proteins, and enzymes (e.g., phospholipase, hyaluronidase, esterases) that vary significantly among different arthropod taxa. Depending on what types of cells or tissues they affect, toxins and venoms can be characterized, for example, as neurotoxins, cytotoxins, or hemotoxins. Frequently they cause such symptoms as pain, itching, swelling, redness, hemorrhaging, or blisters, the severity of which is largely dependent on the particular types and amounts of toxin involved.

Further information on arthropod toxins and venoms is provided by Beard (1960), Roth and Eisner (1962), Bücherl and Buckley (1971), Bettini (1978), Schmidt (1982), Tu (1984), Meier and White (1995), and Eisner et al. (2005).

Allergic Reactions

A relatively wide spectrum of allergic reactions can occur in humans or animals exposed to certain arthropods. Bites or stings by arthropods such as lice, bedbugs, fleas, bees, ants, wasps, mosquitoes, and chiggers all can result in allergic host reactions. Contact allergies can occur when certain beetles or caterpillars touch the skin. Respiratory allergies can result from inhaling allergenic airborne particles from cockroaches, fleas, dust mites, and other arthropods. The recirculation of air by modern air-handling systems in buildings tends to exacerbate inhalation of insect allergens. For a review of arthropod allergens, see Arlian (2002).

Humans and animals usually react to repeated exposure to bites or stings from the same or antigenically related arthropods in two ways, depending on the nature of the antigen or toxin inoculated and the sensitivity of the host: (1) desensitization to the bites or stings with repeated exposure, and (2) allergic reactions that, in extreme cases, can develop into life-threatening anaphylactic shock. However, a distinct five-stage sequence of reactions typically occurs in most humans when repeatedly bitten or stung by the same, or related, species of arthropod over time. Stage 1 involves no skin reaction but leads to development of **hypersensitivity**. Stage 2 is a **delayed-hypersensitivity** reaction. Stage 3 is an **immediate-sensitivity** reaction followed by a delayed-hypersensitivity reaction. Stage 4 is immediate reaction only, whereas Stage 5 again involves no reaction (i.e., the victim becomes **desensitized**). These changes reflect the changing host immune response to prolonged and frequent exposure to the same arthropod or to cross-reactive allergens or toxins.

Invasion of Host Tissues

Some arthropods invade the body tissues of their host. Various degrees of invasion occur, ranging from subcutaneous infestations to invasion of organs such as the lungs and intestines. Invasion of tissues allows arthropods to exploit different host niches and usually involves the immature stages of parasitic arthropods.

The invasion of host tissues by fly larvae, called **myiasis**, is the most widespread form of host invasion by arthropods. Larvae of many myiasis-causing flies move extensively through the host tissues. As they mature, they select characteristic host sites (e.g., stomach, throat, nasal passages, or various subdermal sites) in which to complete the parasitic phase of their development.

Certain mites also invade the skin or associated hair follicles and dermal glands. Others infest nasal passages, lungs and air sacs, or cloaca, stomach, intestines, and other parts of the alimentary tract of their hosts. Examples include scabies mites, follicle mites, nasal mites, lung mites, and a variety of other mites that infest both domestic and wild birds and mammals.

ARTHROPOD-BORNE DISEASES

Table 1.2 lists the principal groups of insects and arachnids involved in arthropod-borne diseases and the associated types of pathogens. Among the wide variety of arthropods that transmit pathogens to humans and other animals, mosquitoes are the most important, followed by ticks. Viruses and bacteria (including rickettsiae) are the most diverse groups of pathogens transmitted by arthropods, followed by protozoa and filarial nematodes. A standardized nomenclature has been proposed for parasitic diseases of animals including those with arthropod vectors (Kassai et al., 1988).

All the viruses listed in Table 1.2 are arthropod-borne viruses, usually referred to as **arboviruses**, indicating that they are transmitted typically by insects or other arthropod hosts. The study of arboviruses is termed **arbovirology**. These and related terms are discussed in more detail in Chapter 3 on the epidemiology of vector-borne diseases and in the appendix devoted to arboviruses.

Pathogens are transmitted by arthropods in two basic ways, either biologically or mechanically. In **biological transmission**, pathogens undergo development or reproduction in the arthropod host. Examples of diseases that involve biological transmission are malaria, African trypanosomiasis, Chagas' disease, leishmaniasis, and lymphatic filariasis. In **mechanical transmission**, pathogens are transmitted by arthropods via contaminated appendages (usually mouthparts) or regurgitation of an infectious blood meal. Examples of diseases that involve mechanical transmission are equine infectious anemia and myxomatosis. Biological transmission is by far the more common and efficient mechanism for pathogen maintenance and transmission.

A wide range of life-cycle patterns and degrees of host associations is characterized by arthropod vectors. Some ectoparasites, such as sucking lice, remain on their host for life. Others, such as mosquitoes and most biting flies, have a more fleeting association with the host, some being associated with it only during the brief acts of host location and blood-feeding. Between these two extremes is a wide range of host associations exhibited by different arthropod groups.

Table 1.2 Examples of Arthropod-Borne Diseases of Medical-Veterinary Importance (for more Comprehensive Coverage, see the Individual Chapters Devoted to each Arthropod Group)

Arthropod Vectors	Diseases Grouped by Causative Agents
Mosquitoes	VIRUSES: Yellow fever, dengue, Rift Valley fever, myxomatosis; eastern equine encephalomyelitis, western equine encephalomyelitis, Venezuelan equine encephalomyelitis, St. Louis encephalitis, LaCrosse encephalitis, Japanese encephalitis, Murray Valley encephalitis, Chikungunya fever, O'nyong nyong fever, Ross River fever, West Nile fever PROTOZOAN: Malaria FILARIAL NEMATODES: Brugian filariasis, Bancroftian filariasis, Timorian filariasis, dog heartworm
Black flies	FILARIAL NEMATODES: Human onchocerciasis (river blindness), bovine onchocerciasis
Biting midges	VIRUSES: Bluetongue disease, epizootic hemorrhagic disease, African horse sickness, leucocytozoonosis, Oropouche fever FILARIAL NEMATODES: Equine onchocerciasis, mansonellosis
Sand flies	VIRUSES: Sand fly fever, vesicular stomatitis BACTERIA: Oroya fever (Veruga Peruana) PROTOZOANS: Leishmaniasis
Horse flies and deer flies	VIRUSES: Equine infectious anemia, hog cholera BACTERIA: Tularemia PROTOZOAN: Surra (livestock trypanosomiasis) FILARIAL NEMATODES: Loiasis, elaeophorosis
Tsetse flies	PROTOZOANS: African trypanosomiasis, nagana
Triatomine bugs	PROTOZOAN: American trypanosomiasis (Chagas disease)
Lice	VIRUS: Swine pox BACTERIA: Epidemic typhus, trench fever, louse-borne relapsing fever
Fleas	VIRUS: Myxomatosis BACTERIA: Plague, murine (endemic) typhus, tularemia
Ticks	VIRUSES: Tick-borne encephalitis, Powassan encephalitis, Colorado tick fever, Crimean-Congo hemorrhagic fever, African swine fever BACTERIA: Lyme disease, Rocky Mountain spotted fever, Boutonneuse fever, tick-borne ehrlichiosis, Q fever, heartwater fever (cowdriosis), anaplasmosis, tick-borne relapsing fever, avian spirochetosis, theileriosis (East Coast fever), bovine dermatophilosus PROTOZOAN: Babesiosis
Mites	BACTERIA: Tsutsugamushi fever (scrub typhus), rickettsialpox

Literature references on vector-borne diseases, together with their epidemiology and ecology, are provided in the section, "Arthropod-Borne Diseases," at the end of this chapter.

FOOD CONTAMINANTS

Many arthropods can contaminate or spoil food materials. In addition to causing direct damage to food resources, arthropods or their parts (e.g., setae, scales, shed cuticles, or body fragments) may be ingested accidentally. This can lead to toxic or allergic reactions, gastrointestinal myiasis, and other disorders. At least one case of millipedes (*Nopoiulus kochii*) infesting human intestines, for several years, has been documented (Ertek et al., 2004).

Insects such as the house fly may alight on food and regurgitate pathogen-contaminated fluids prior to or during feeding. While feeding they also may defecate, contaminating the food with potential pathogens. Because the alimentary tract of arthropods may harbor pathogenic microorganisms, subsequent consumption of the contaminated food can lead to the transmission of these pathogens to humans or other animals. Similarly, the integument of household pests such as flies and cockroaches (particularly their legs and tarsi) can serve as a contact source of pathogens that may be readily transferred to food items. Some of these arthropods previously may have visited fecal matter, garbage heaps, animal secretions, or other potential sources of pathogens, thereby further contributing to health risks.

Additional information on insects and other arthropods that can contaminate food is provided by Olsen et al. (1996) and in reviews by Terbush (1972), Hughes (1976), and Gorham (1975, 1991a, b).

FEAR OF ARTHROPODS

Some people detest arthropods, or infestation by them, to such a degree that they suffer from **entomophobia**, the fear of insects, **arachnophobia**, the fear of spiders and other arachnids, or **acarophobia**, the fear of mites (including ticks). Showing concern or disapproval toward the presence of potentially injurious arthropods is probably a prudent and healthy reaction, but phobic behaviors reflect an unusually severe psychological response. Such persons exhibit more-than-normal fear when they encounter an arthropod, often resorting to excessive or obsessive measures to control the problem (e.g., overtreatment of themselves or their homes with insecticides and other chemical compounds).

DELUSIONAL DISORDERS

A psychological state occurs in which an individual mistakenly believes that he or she is being bitten by, or infested with, parasites. This is variously known as **delusory parasitosis**, **delusional parasitosis**, **delusions of parasitosis**, **Ekbom syndrome**, and **Elliot's disease** (also Elliott's disease). This condition is distinct from simply a fear, or phobia, of insects or other arthropods, and represents a more deeply rooted psychological problem. This delusional condition is most frequently experienced by middle-aged or elderly persons, particularly women, and is one of the more difficult situations in which entomologists may become involved.

Remarkable behavioral traits are sometimes attributed to the parasites by victims. These include descriptions of tiny animals jumping into the eyes when a room is entered or when a lamp is switched on. Some victims have failing eyesight; others may have real symptoms from other conditions such as psoriasis, which are attributed to imagined parasites. Victims become convinced that the parasites are real, and they often consult a succession of physicians in a futile attempt to secure a diagnosis and satisfactory treatment to resolve the problem. Patients typically produce skin scrapings or samples of such materials as vacuumed debris from carpets, draperies, and window sills that they believe contain the elusive parasites.

Victims of delusory disorders often turn to extension entomologists or medical entomologists as a last resort, out of frustration with being unable to resolve their condition through family physicians, allergists, and other medical specialists. Because patients are convinced that arthropods are present, they are usually reluctant to seek counseling or other psychiatric help. Dealing with these cases requires careful examination of submitted specimens, tact, and professional discretion on the part of the entomologist. Additional information on delusory parasitosis is provided by Driscoll et al. (1993), Koblenzer (1993), Kushon et al. (1993), Poorbaugh (1993), Webb (1993a,b), Goddard (1995), and Hinkle (2000).

Morgellons disease is a term, coined in 2002, for a condition generally regarded by the medical community as delusional parasitosis. The name refers to a medical case in 1674. Also referred to as **unexplained dermopathy** ("unexplained skin disease"), Morgellons can manifest as a range of skin conditions, including crawling, biting, and stinging sensations; granules, fibers, or dark specks in, or emerging from, the skin; and rashes or sores. Some patients diagnosed with Morgellons disease also may experience fatigue, joint pain, visual changes, short-term memory loss, or mental confusion. Although published reports and anecdotal accounts have suggested possible involvement, there is no firm evidence to date implicating insects, or other arthropods, as a direct cause. Morgellons disease has been reported worldwide, with a particular focus of attention in certain parts of the United States (e.g., California, Florida, and Texas). For further information about this disorder, see Koblenzer et al. (2006), Murase et al. (2006), and Savely et al. (2006).

FORMICOPHILIA

An unusual human psychosexual disorder, called formicophilia ("ant-loving"), can involve insects. In such cases an individual experiences self-induced sexuoerotic arousal and orgasm when ants, cockroaches, or other small creatures (e.g., snails) are allowed to crawl, creep, or nibble on the body, notably the genitalia, perianal area, or nipples (Dewaraja and Money, 1986; Dewaraja, 1987).

HOST DEFENSES

Humans and other animals have developed elaborate means to defend themselves against infestation by arthropods and infection by pathogens they may transmit. Both behavioral or immunological responses are used to resist infestation by arthropods. **Behavioral defenses** include evasive, offensive, or defensive action against biting flies such as mosquitoes, black flies, ceratopogonids, stable flies, and horse flies. Grooming and preening by animals (e.g., biting, scratching, or licking) are defensive behaviors used to reduce or prevent infestations by ectoparasites and other potentially harmful arthropods. Host **immunological defenses** against arthropods vary with different arthropods and with respect to previous exposure to the same or antigenically related taxa. Details concerning such host immune responses are beyond the scope of this book, but some general trends are noteworthy. Repeated feeding attempts by the same or antigenically cross-reactive arthropods often lead to fewer arthropods being able to feed successfully, reduced engorgement weights, greater mortality, and decreased fecundity of female arthropods. Widespread arthropod mortality rarely results. For more information concerning the types of host immune responses and cell types involved against various ectoparasites, see Wikel (1996b) and other works listed at the end of this chapter.

Many blood-feeding arthropods partially or completely counteract the host immune response by inoculating immunomodulators or immunosuppressive compounds into the bite site. In fact, a wide range of pharmacologically active compounds is known to be released at the bite site by various arthropods (Ribeiro, 1995). These compounds range from **anticoagulants** to prevent the blood from clotting, local **analgesics** to reduce host pain, **apyrase** to prevent platelet aggregation and promote capillary location, and various enzymes and other factors for promoting blood or tissue digestion. Some of these compounds are perceived by the host as antigens and may elicit an immune response whereas others can cause localized or systemic toxic responses and itching.

MINOR ARTHROPOD PROBLEMS OF MEDICAL-VETERINARY INTEREST

In addition to arthropod groups detailed in the chapters that follow, a few arthropods in other groups may have minor, incidental, or occasional significance to human and animal health. These include springtails (Order Collembola), mayflies (Order Ephemeroptera), locusts (Order Orthoptera), walkingsticks (Order Phasmatodea), earwigs (Order Dermaptera), thrips (Order Thysanoptera), bark lice and book lice (Order Pscoptera), caddisflies (Order Trichoptera), millipedes (Class Diplopoda), and centipedes (Class Chilopoda).

Some **walkingsticks**, or stick insects, possess glands that are used to spray defensive fluids at potential predators, such as ants, beetles, rodents, and birds, or when otherwise threatened. A pair of large, elongate glands is located in the anterior part of the thorax, where they open on the anterolateral margins of the pronotum, just behind the head. Two species in the United States that produce and spray defensive secretions are *Anisomorpha buprestoides* (Figure 1.1) and *A. ferruginea*, called two-striped walkingsticks. They forcefully discharge an emulsion of malodorous, milky fluid, as a fine mist containing a terpene dialdehyde (anisomorphal) as the active ingredient. A Madagascar species, *Parectatosoma mocquerysi*, produces a similar defensive monoterpene compound (parectadial).

A number of cases have been documented in which animals, particularly humans and dogs, have been sprayed in the eyes by these and other walkingstick species, from as far away as 30 cm. The result in severe cases is immediate, sometimes excruciating, pain, with burning and dull aching of the eye(s) for several hours, and impaired vision that may persist for one to several days. Recommended treatment is immediate and thorough irrigation of the affected eye(s) with cool water, followed by administration of an analgesic.

For more information on the chemical aspects of defensive secretions in walkingsticks, see Meinwald et al. (1962), Happ et al. (1966), Carlberg (1985), and Eisner (1965, 2005); for human cases, see Stewart (1937), Albert (1947), Hatch et al. (1993), and Paysse et al. (2001); and for a dog case, see Dziedzyc (1992).

Figure 1.1 Two-striped walkingstick (*Anisomorpha buprestoides*); a pair, with smaller male atop larger female; a common species in the southeastern United States that sprays a defensive secretion from a pair of thoracic glands, which can severely irritate the eyes of humans, dogs, and other animals. (Photo by Aaron T. Dossey)

Springtails, or collembolans, have been reported infesting human skin (Scott et al., 1962; Scott, 1966). However, in most cases in which springtails are found on, or adhering to, skin of humans and household pets (e.g., dogs and cats), they are believed to be incidental associations, most likely due to contact of the skin with soil, compost, decomposing plant material, or moist ground debris where springtails typically live. It is not surprising, therefore, to find them as contaminants of moist dermal lesions and other skin problems, where they can survive for at least short periods of time. This stated, the association of collembolans with cases of delusory parasitosis (Altschuler et al., 2004) has led to considerable controversy over the interpretation of this association.

Some **bark lice** (psocids) are known to cause allergies or dermatitis in humans (Li and Li, 1995; Baz and Monserrat, 1999); adult **mayflies** and **caddisflies** can cause inhalational allergies, especially when they emerge in large numbers from lakes, rivers, or streams (Seshadri, 1955). **Locusts** can cause allergic reactions among personnel working in insectaries where locusts are reared, and during outbreaks of migratory locusts in Sudan and other parts of Africa. Symptoms include rhinitis, bronchitis, and difficulty breathing, due to inhalation of microscopic particles of locusts or of their dried feces (Hunter-Jones, 1966; Wirtz, 1980).

Thrips, which have tubular mouthparts adapted for sucking plant fluids, occasionally pierce the skin and have been known to imbibe blood (Williams, 1921; Hood, 1927; Bailey, 1936; Arnaud, 1970). When local populations are particularly high, as in the case of the flower thrips (*Frankliniella tritici*), they have been documented causing significant discomfort at outdoor gatherings in the southeastern United States. On rare occasions, **earwigs** also have been recorded as imbibing blood (Bishopp, 1961). Bishopp further noted that some earwigs have been known to pierce human skin with their pair of caudal pincers (cerci) and may stay attached for an extended period.

Some miscellaneous arthropods inhabit the feathers of birds or the fur of mammals. The exact nutritional requirements of some of these arthropods remain unknown; most of them, however, do not appear to be true ectoparasites. Representatives of two of the three suborders of earwigs (Suborders Arixeniina and Hemimerina) live in mammal fur. Members of the Arixeniina are associated with Old World bats whereas members of the Hemimerina are found on African cricetine rodents (Nakata and Maa, 1974). These earwigs may feed on skin secretions or sloughed cells but their effect on the health of their hosts is poorly understood. Some other occasional inhabitants of host pelage, such as various beetles, cheyletid mites, and pseudoscorpions, are predators of ectoparasites and are therefore beneficial to their hosts (Durden, 1987).

A few arthropods that are not mentioned in the following chapters can occasionally serve as intermediate hosts of parasites that adversely affect domestic and wild animals. These include certain springtails and psocids (bark lice) as intermediate hosts of tapeworms (Baz and Monserrat, 1999).

Occasionally, entomologists are asked questions about millipedes and centipedes. Defensive sprays of some millipedes contain hydrochloric acid that can chemically burn the skin and can cause long-term skin discoloration (Radford, 1975). Centipedes, especially some of the larger tropical species, can cause envenomation when they "bite" with their claws ("forcipules," which are modifications of the first pair of legs), onto which they open the ducts of their venom glands (Remington, 1950).

REFERENCES AND FURTHER READING

GENERAL ENTOMOLOGY

Arnett, R. H., Jr. (2000). *American Insects: A Handbook of the Insects of America North of Mexico* (2nd ed.). Boca Raton: CRC Press.

Bosik, J. J. (1997). *Common Names of Insects and Related Organisms.* Lanham, MD: Entomological Society of America. [This list is now updated electronically and can be accessed through the web site for the Entomological Society of America.]

Capinera, J. L. (2008). *Encyclopedia of Entomology* (2nd ed.). New York: Springer.

Chapman, R. F. (1998). *The Insects: Structure and function* (4th ed.). Cambridge University Press.

Daly, H. V., Doyen, J. T., & Purcell III, A. H. (1998). *Introduction to Insect Biology and Diversity* (2nd ed.). Oxford University Press.

Ebeling, W. (1975). *Urban Entomology.* Berkeley: University of California Press.

Elzinga, R. J. (2000). *Fundamentals of Entomology* (5th ed.). Upper Saddle River, NJ: Prentice Hall.

Gillott, C. (1995). *Entomology* (2nd ed.). Plenum Press.

Gordh, G., & Headrick, D. H. (2001). *A Dictionary of Entomology.* New York: CABI Publishing.

Grimaldi, D., & Engel, M. S. (2005). *Evolution of the Insects.* Cambridge University Press.

Gullan, P. J., & Cranston, P. S. (2005). *The Insects: An Outline of Entomology* (3rd ed.). Oxford: Blackwell Publishing.

Mallis, A., Hedges, S. A., & Moreland, D. (2004). *Handbook of Pest control: The Behavior, Life History and Control of Household Pests* (9th ed.). Mallis Handbook & Technical Training Co.

Pedigo, L. P., & Rice, M. E. (2006). *Entomology and Pest Management* (5th ed.). Upper Saddle River, NJ: Pearson, Prentice Hall.

Richards, O. W., & Davies, R. G. (1994). *Imm's General Textbook of Entomology* (Vol. 1, 10th ed.). Structure, Physiology and Development, Vol. 2. Classification and biology. Chapman & Hall.

Robinson, W. H. (1996). *Urban Entomology: Insect and Mite Pests in the Human Environment.* Chapman & Hall.

Romoser, W. S., & Stoffolano, J. G., Jr. (1998). *The Science of Entomology* (4th ed.). Boston: WCB/McGraw Hill.

Snodgrass, R. (1993). *Principles of Insect Morphology.* Cornell University Press.

Torre-Bueno, J. R. de la (1962). *A Glossary of Entomology.* Brooklyn: Brooklyn Entomological Society.

Triplehorn, C. A., & Johnson, N. F. (2005). *Borror and DeLong's Introduction to the Study of Insects* (7th ed.). Belmont, CA: Thomson, Brooks/Cole.

GENERAL ACAROLOGY

Evans, G. O. (1992). *Principles of Acarology.* Wallingford, UK: CAB International.

Krantz, G. W., Walter, D. E. (Eds.). (2009). *A Manual of Acarology* (3rd ed.). Texas Tech University Press, Lubbock.

Woolley, T. A. (1987). *Acarology: Mites and Human Welfare.* New York: Wiley.

MEDICAL-VETERINARY ENTOMOLOGY

Baker, E. W., Evans, T. M., Gould, D. J., Hull, W. B., & Keegan, H. L. (1956). *A Manual of Parasitic Mites of Medical or Economic Importance.* New York: National Pest Control Assoc.

Baker, J. R., Apperson, C. S., & Arends, J. J. (1986). *Insect and Other Pests of Man and Animals.* North Carolina State University.

Crampton, J. M., Beard, C. B., & Louis, C. (1997). *Molecular Biology of Insect Disease Vectors: A Methods Manual.* Chapman & Hall.

Debboun, M., Frances, S. P., & Strickman, D. (2007). *Insect Repellents: Principles, Methods and Uses.* Boca Raton: CRC Press.

Eldridge, B. F., & Edman, J. D. (2003). *Medical Entomology: A Textbook on Public Health and Veterinary Problems Caused by Arthropods* (2nd ed.). Norwell, MA: Kluwer Academic.

Harwood, R. F., & James, M. T. (1979). *Entomology in Human and Animal Health* (7th ed.). New York: Macmillan.

Hickin, N. E. (1985). *Pest Animals in Buildings: A World Review.* London: G. Godwin; New York: Longman.

Jobling, B. (1987). *Anatomical Drawings of Biting Flies.* London: British Museum (Natural History) & Wellcome Trust.

Kettle, D. S. (1995). *Medical and Veterinary Entomology* (2nd ed.). Wallingford, UK: CAB International.

Kim, K. C. (Ed.). (1985). *Coevolution of Parasitic Arthropods and Mammals.* Wiley.

Lehane, M. (2005). *Biology of Blood-Sucking Insects* (2nd ed.). Cambridge University Press.

Mario Vargas, V. (2001). *Los acaros en la salud humana y animal.* San Juan: Universidad de Costa Rica.

Marquardt, W. C. (Ed.). (2005). *Biology of Disease Vectors* (2nd ed.). Burlington, MA: Elsevier.

Marquardt, W. C., Demaree, R. S., & Grieve, R. B. (1999). *Parasitology and Vector Biology* (2nd ed.). San Diego: Harcourt Academic Press.

Marshall, A. G. (1981). *Ecology of Ectoparasitic Insects.* London: Academic Press.

Nutting, W. B. (Ed.). (1994). *Mammalian Diseases and Arachnids, Vol. 1. Pathogen Biology and Clinical Management, Vol. 2. Medico-Veterinary, Laboratory, and Wildlife Diseases, and Control.* Boca Raton: CRC Press.

Parish, L. C., Nutting, W. B., & Schwartzman, R. M. (1983). *Cutaneous Infestations in Man and Animals.* New York: Praeger Press.

Patton, W. S. (1931). *Insects, Ticks, Mites and Venomous Animals of Medical and Veterinary Importance. Part II: Public health.* Liverpool University Press.

Patton, W. S., & Evans, A. M. (1929). *Insects, Ticks, Mites and Venomous Animals of Medical and Veterinary Importance. Part I: Medical.* Liverpool University Press.

Pittaway, A. R. (1992). *Arthropods of Medical and Veterinary Importance: A Checklist of Preferred Names and Allied Terms.* Tucson: University of Arizona Press.

Walker, A. R. (1995). *Arthropods of Humans and Domestic Animals: A Guide to Preliminary Identification.* New York: Chapman & Hall.

MEDICAL ENTOMOLOGY

Alexander, J. O. (1984). *Arthropods and Human Skin.* Berlin: Springer-Verlag.

Andrews, M. L. A. (1977). *The Life that Lives on Man.* New York: Taplinger Publishing.

Burgess, N. H. R., & Cowan, G. L. O. (1993). *A Colour Atlas of Medical Entomology.* Chapman & Hall.

Busvine, J. M. (1980). *Insects and Hygiene: The Biology and Control of Insect Pests of Medical and Domestic Importance* (3rd ed.). London: Chapman & Hall.

Daniel, M., Stramova, H., Absolonova, V., Dedicova, D., Lhotova, H., Maskova, L., et al. (1992). Arthropods in a hospital and their potential significance in the epidemiology of hospital infections. *Folia Parasitologica, 39,* 159–170.

Furman, D. P., & Catts, E. P. (1982). *Manual of Medical Entomology.* Cambridge University Press.

Goddard, J. (2007). *Physician's Guide to Arthropods of Medical Importance* (5th ed.). Boca Raton: CRC Press.

Goddard, J. (1998). Arthropods and medicine. *Journal of Agromedicine, 5,* 55–83.

Gordon, R. M., & Lavoipierre, M. M. J. (1962). *Entomology for Students of Medicine.* Oxford: Blackwell Scientific.

Gratz, N. G. (1999). Emerging and resurging vector-borne diseases. *Annual Review of Entomology, 44,* 51–75.

Gubler, D. J. (1998). Resurgent vector-borne diseases as a global health problem. *Emerging Infectious Diseases, 4,* 442–450.

Harwood, R. F., & James, M. T. (1979). *Entomology in Human and Animal Health* (7th ed.). New York: Macmillan.

Herms, W. B. (1961). *Medical Entomology* (5th ed.). New York: Macmillan.

Horsfall, W. R. (1962). *Medical Entomology: Arthropods and Human Disease.* New York: Ronald Press.

James, M. T., & Harwood, R. F. (1969). *Herms' Medical Entomology* (6th ed.). New York: Macmillan.

Lane, R. P., & Crosskey, R. W. (Eds.). (1993). *Medical Insects and Arachnids.* Chapman & Hall.

Leclercq, M. (1969). *Entomological Parasitology: The Relations Between Entomology and the Medical Sciences.* Oxford: Pergamon Press.

Marples, M. J. (1965). *The Ecology of the Human Skin.* Springfield, IL: Charles C. Thomas.

Matheson, R. (1950). *Medical Entomology.* Ithaca: Comstock Publishing Co.

McClelland, G. A. H. (1992). *Medical Entomology: An Ecological Perspective* (12th ed.). Davis: University of California Press.

Orkin, M., & Maibach, H. I. (Eds.). (1985). *Cutaneous Infestations and Insect Bites.* New York: Marcel Dekker.

Peters, W. (1992). *A Colour Atlas of Arthropods in Clinical Medicine.* London: Wolfe Publishing Ltd.

Riley, W. A., & Johannsen, O. A. (1938). *Medical Entomology: A Survey of Insects and Allied Forms Which Affect the Health of Man and Animals* (2nd ed.). New York: McGraw-Hill.

Service, M. W. (1980). *A Guide to Medical Entomology.* London: Macmillan.

Service, M. W. (2005). *Medical Entomology for Students* (3rd ed.). Cambridge University Press.

Smith, K. G. V. (1973). *Insects and Other Arthropods of Medical Importance.* London: British Museum (Natural History).

Walker, D. H., Barbour, A. G., Oliver, J. H., Jr., Lane, R. S., Dumler, J. S., Dennis, D. T., et al. (1996). Emerging bacterial zoonotic and vector-borne diseases. *Journal of the American Medical Association, 275,* 463–469.

Wilson, M. E., & Spielman, A. (Eds.). (1994). Vector-borne terrestrial diseases. In *Disease in evolution: Global changes and emergence of infectious diseases* (pp. 123–224). *Annals of the New York Academy of Sciences, 740,* 1–503.

Winch, P. (1998). Social and cultural responses to emerging vector-borne diseases. *Journal of Vector Ecology, 23,* 47–53.

VETERINARY ENTOMOLOGY

Axtell, R. C., & Arends, J. J. (1990). Ecology and management of arthropod pests of poultry. *Annual Review of Entomology, 35,* 101–126.

Barker, B. (1999). *Livestock Entomology Laboratory Manual* (2nd ed.). Dubuque, IA: Kendall/Hunt.

Barnard, S. M., & Durden, L. A. (1999). *A Veterinary Guide to the Parasites of Reptiles, Vol. 2. Arthropods (excluding mites).* Melbourne, FL: Krieger Press.

Bay, D. E., & Harris, R. L. (1988). *Introduction to Veterinary Entomology.* Bryan, TX: Stonefly Publishing.

Bowman, D. D. (1999). *Georgi's Parasitology for Veterinarians* (7th ed.). Philadelphia: W. B. Saunders.

Bram, R. A. (1978). Surveillance and collection of arthropods of veterinary importance. U.S. Dep. Agric, Agriculture Hdbk. No. 518.

Drummond, R. O., George, J. E., & Kunz, S. E. (1988). *Control of Arthropod Pests of Livestock: A Review of Technology.* Boca Raton: CRC Press.

Flynn, R. J. (1973). *Parasites of Laboratory Animals.* Ames: Iowa State University Press.

Foil, L. D., & Foil, C. S. (1990). Arthropod pests of horses. *Compendium for Continuing Education for the Practicing Veterinarian, 12,* 723–731.

Georgi, J. R. (1990). *Parasitology for Veterinarians* (5th ed.). Philadelphia: W. B. Saunders.

Guimarães, J. H., Tucci, E. C., & Barros-Battesti, D. M . (2001). *Ectoparasitos de Importância Veterinária.* Fundação de Amparo á Pesquisa do Estado de São Paulo.

Jones, C. J., & DiPietro, J. A. (1996). Biology and control of arthropod parasites of horses. *Compendium for Continuing Education for the Practicing Veterinarian, 18,* 551–558.

Lancaster, J. L., & Meisch, M. V. (1986). *Arthropods in Livestock and Poultry Production.* Wiley: Halstead Press.

Soulsby, E. J. L. (1982). *Helminths, Arthropods and Protozoa of Domesticated Animals* (7th ed.). London: Bailliere, Tindall and Cassell.

Uilenberg, G. (Coordinator). (1994). Ectoparasites of animals and control methods. *Rev Sci Tech Off Int Epizoot, 13,* 979–1387.

Wall, R., & Shearer, D. (1997). *Veterinary Entomology.* Chapman and Hall.

Wall, R., & Shearer, D. (2001). *Veterinary Ectoparasites: Biology, Pathology and Control.* Wiley Blackwell.

Williams, R. E., Hall, R. D., Broce, A. B., & Scholl, P. J. (Eds.). (1985). *Livestock Entomology.* Wiley.

WILDLIFE ENTOMOLOGY

Davidson, W. R., Hayes, F. A., Nettles, V. F., & Kellogg, F. E. (1981). *Diseases and Parasites of White-Tailed Deer.* Misc. Pub. No. 7. Tall Timbers Res. Stn., Tallahassee, FL.

Davidson, W. R., & Nettles, V. F. (1997). *Field Manual of Wildlife Diseases in the Southeastern United States* (2nd ed.). Athens, GA: Southeastern Cooperative Wildlife Disease Study.

Fowler, M. E. (Ed.). (1986). *Zoo and Wild Animal Medicine* (2nd ed.). Philadelphia: W. B. Saunders.

Fowler, M. E., & Miller, R. E. (Eds.). (1999). *Zoo and Wild Animal Medicine: Current Therapy* (4th ed.). Philadelphia: W. B. Saunders.

Samuel, W. M., Pybus, M. J., & Kocan, A. A. (2001). *Parasitic Diseases of Wild Mammals* (2nd ed.). Ames: Iowa State University Press.

HISTORY OF MEDICAL-VETERINARY ENTOMOLOGY

Anon. (1996). History of CDC [CDC's 50th anniversary]. *Morbidity and Mortality Weekly Report, 45,* 526–530.

Augustin, G. (1909). *History of Yellow Fever.* New Orleans: Searcy & Pfaff.

Bayne-Jones, S. (1964). *Preventive Medicine in World War II, Vol. VII, Communicable Diseases, Arthropod-Borne Diseases Other than Malaria.* Office of the Surgeon General, Dept. of the Army, Washington DC.

Bean, W. B. (1982). *Walter Reed—A Biography.* Charlottesville: University of Virginia Press.

Bockarie, M. J., Gbakima, A. A., & Barnish, G. (1999). It all began with Ronald Ross: 100 years of malaria research and control in Sierra Leone (1899–1999). *Annals of Tropical Medicine and Parasitology, 93,* 213–224.

Busvine, J. R. (1976). *Insects, Hygiene and History.* London: Athlone Press.

Busvine, J. R. (1993). *Disease Transmission by Insects: Its Discovery and 90 Years of Effort to Prevent It.* Berlin: Springer-Verlag.

Calisher, C. H. (1996). From mouse to sequence and back to mouse: Peregrinations of an arbovirologist. *Journal of Vector Ecology, 21,* 192–200.

Cartwright, F. F. (1972). *Disease and History; the Influence of disease in Shaping the Great Events in History.* London: Hart-Davis.

Chernin, E. (1983). Sir Patrick Manson: An annotated bibliography and a note on a collected set of his writings. *Review of Infectious Diseases, 5,* 353–386.

Chernin, E. (1987). A unique tribute to Theobald Smith, 1915. *Review of Infectious Diseases, 9,* 625–635.

Collins, W. E. (1976). Fifty years of parasitology: Some fulgent personalities in arthropodology. *The Journal of Parasitology, 62,* 504–509.

Cook, G. C. (1992). *From the Greenwich Hulks to Old St. Pancras: A History of Tropical Disease in London.* London: Athlone Press.

Cox, F. E. G. (Ed.). (1996). *The Wellcome Trust Illustrated History of Tropical Diseases.* London: Trustees of the Wellcome Trust.

Cushing, E. C. (1957). *History of Entomology in World War II.* Washington, DC: Smithsonian Inst.

Delaporte, F. (1991). *The History of Yellow Fever: An Essay on the Birth of Tropical Medicine.* MIT Press.

Desowitz, R. S. (1991). *The Malaria Capers: More Tales of Parasites and People, Research and Reality.* New York: W. W. Norton & Co.

Desowitz, R. S. (1997). *Who Gave Pinta to the Santa Maria? Tracing the Devastating Spread of Lethal Tropical Diseases.* New York: W. W. Norton & Co.

Dolman, C. E. (1982). Theobald Smith (1859–1934), pioneer American microbiologist. *Perspectives in Biology and Medicine, 25,* 417–427.

Eldridge, B. F. (1992). Patrick Manson and the discovery age of vector biology. *Journal of the American Mosquito Control Association, 8,* 215–222.

Ellis, J. H. (1992). *Yellow Fever and Public Health in the New South.* Lexington: University of Kentucky Press.

Gillett, J. D. (1979). Vitamin C, yellow fever and plague; the near misses. *Antenna, Bulletin of the Royal Entomological Society of London, 3,* 64–70.

Gillett, J. D. (1985). Medical entomology, past, present and future: A personal view. *Antenna, Bulletin of the Royal Entomological Society of London, 9,* 63–70.

Gorgas, M. D., & Hendrick, B. J. (1924). *William Crawford Gorgas: His Life and Work.* Garden City, NY: Garden City Publishing Co.

Harwood, R. F., & James, M. T. (1979). Historical review. In *Entomology in human and animal health* (7th ed., pp. 3–10). Macmillan Publishing Co.

Horsman, R. (1987). *Josiah Nott of Mobile: Southerner, Physician, and Racial Theorist.* Baton Rouge: Louisiana State University Press.

Laurence, B. R. (1989). The discovery of insect-borne disease. *Biologist, 36,* 65–71.

Lewis, D. J. (1984). Reminiscences of medical entomology in the last fifty years. *Antenna, Bulletin of the Royal Entomological Society of London, 8,* 117–122.

Lockwood, J. A. (1987). Entomological warfare: History of the use of insects as weapons of war. *American Entomologist, 33,* 76–82.

Miller, G. L. (1997). Historical natural history: Insects and the Civil War. *American Entomologist, 43,* 227–245.

Nye, E. R., & Gibson, M. E. (1997). *Ronald Ross: Malariologist and Polymath—A Biography.* New York: Macmillan/St. Martin's.

Oldstone, M. B. A. (1998). *Viruses, Plagues, and History.* Oxford: Oxford University Press.

Peterson, R. K. D. (1995). Insects, disease, and military history. *American Entomologist, 41,* 147–160.

Philip, C. B. (1948). Tsutsugamushi disease (scrub typhus) in World War II. *The Journal of Parasitology, 34,* 169–191.

Philip, C. B., & Rozeboom, L. E. (1973). Medico-Veterinary Entomology: A Generation of Progress. In R. F. Smith, T. E. Mittler, C. N. Smith (Eds.), *History of Entomology* (pp. 333–360). College Park: Entomological Society of America.

Schmidt, C. H., & Fluno, J. A. (1973). Brief history of medical and veterinary entomology in the USDA. *Journal of the Washington Academy of Sciences, 63,* 54–60.

Service, M. W. (1978). A short history of medical entomology. *Journal of Medical Entomology, 14,* 603–626.

Sosa, O., Jr. (1989). Carlos J. Finlay and yellow fever: A discovery. *Bulletin of the Entomological Society of America, 35,* 23–25.

Woodward, T. E. (1973). A historical account of the rickettsial diseases with a discussion of unsolved problems. *Journal of Infectious Diseases, 127,* 583–594.

Young, M. D. (1966). Scientific exploration and achievement in the field of malaria. *The Journal of Parasitology, 52,* 2–8.

Zinsser, H. (1935). *Rats, Lice and History.* Boston: Little, Brown & Co.

Zinsser, H. (1936). Biographical memoir of Theobald Smith, 1859–1934. *National Academy of Sciences, Biographical Memoir, 17,* 261–303 [reprinted in Rev. Inf. Dis. 9, 636-654].

IDENTIFICATION AND SYSTEMATICS OF TAXA OF MEDICAL-VETERINARY IMPORTANCE

CDC. (1979). *Introduction to Arthropods of Public Health Importance.* Atlanta: Centers for Disease Control.

CDC. (1994). *Pictorial Keys: Arthropods, Reptiles, Birds and Mammals of Public Health Importance.* Atlanta: Centers for Disease Control and Prevention.

Davis, G. M. (1995). Systematics and public health. *Bioscience, 45,* 705–714.

Hopla, C. E., Durden, L. A., & Keirans, J. E. (1994). Ectoparasites and classification. *Revue Scientifique et Technique Office International des Epizooties, 13,* 985–1017.

Lago, P. K., & Goddard, J. (1994). Identification of medically important arthropods. *Laboratory Medicine, 25,* 298–305.

Service, M. W. (Ed.). (1988). *Biosystematics of Haematophagous Insects.* Oxford University Press.

TYPES OF PROBLEMS CAUSED BY ARTHROPODS

ANNOYANCE

Burns, D. A. (1987). The investigation and management of arthropod bite reactions acquired in the home. *Clinical and Experimental Dermatology, 12,* 114–120.

Frazier, C. D. (1973). Biting insects. *Archives of Dermatology, 107,* 400–402.

Newson, H. D. (1977). Arthropod problems in recreation areas. *Annual Review of Entomology, 22,* 333–353.

ALLERGIC REACTIONS

Arlian, L. G. (2002). Arthropod allergens and human health. *Annual Review of Entomology, 47,* 395–433.

Bellas, T. E. (1982). *Insects as a Cause of Inhalant Allergies: A Bibliography* (2nd ed.). CSIRO Aust. Div. Entomol. Rep. No. 25. CSIRO, Canberra.

Carlsen, K. C., Carlsen, L. K. H., Buchmann, M. S., Wikstrom, J., & Mehl, R. (2002). Cockroach sensitivity in Norway: A previously unidentified problem? *Allergy, 57,* 529–533.

Feinberg, A. R., Feinberg, S. M., & Benaim-Pinto, C. (1956). Asthma and rhinitis from insect allergens. *The Journal of Allergy, 27,* 436–444.

Feingold, B. F., Benjamini, E., & Michaeli, D. (1968). The allergic responses to insect bites. *Annual Review of Entomology, 13,* 137–158.

Frazier, C. A., & Brown, F. K. (1980). *Insects and Allergy and What to do About Them.* Norman: University of Oklahoma Press.

Henson, E. B. (1966). Aquatic insects as inhalant allergens; a review of American literature. *Ohio Journal of Science, 66,* 529–532.

Heyworth, M. F. (1999). Importance of insects in asthma. *Journal of Medical Entomology, 36,* 131–132.

Levine, M. I., & Lockley, R. F. (1981). *Monograph on Insect Allergy.* Pittsburgh: Typecraft.

Musken, H., Franz, J. T., Fernandez-Caldas, E., Maranon, F., Masuch, G., & Bergmann, K. C. (1998a). Psocoptera spp. (dust lice): A new source of indoor allergies in Germany. *The Journal of Allergy and Clinical Immunology, 101,* 121.

Musken, H., Franz, J. T., Fernandez-Caldas, E., Masuch, G., Maranon, F., & Bergmann, K. C. (1998b). Psocoptera (dust lice): New indoor allergens? *Allergologie, 21,* 381–382.

Schulman, S. (1967). Allergic responses to insects. *Annual Review of Entomology, 12,* 323–346.

Wirtz, R. A. (1980). Occupational allergies to arthropods—Documentation and prevention. *Bulletin of the Entomological Society of America, 26,* 356–360.

Wirtz, R. A. (1984). Allergic and toxic reactions to non-stinging arthropods. *Annual Review of Entomology, 29,* 47–69.

ARTHROPOD-BORNE DISEASES

Beck, J. W., & Davies, J. E. (1981). *Medical Parasitology.* C. V. Mosby Co.

Benenson, A. S. (Ed.). (1995). *Control of Communicable Diseases Manual* (16th ed.). American Public Health Assoc.

Beran, G. W., & Steele, J. (1994). *Handbook of Zoonoses: Section A. Bacterial, Rickettsial, Chlamydial, and Mycotic* (2nd ed.). Boca Raton: CRC Press.

Busvine, J. R. (1979). *Arthropod Vectors of Disease. The Institute of Biology's Studies in Biology No. 55.* London: Edward Arnold.

Cook, G. C. (Ed.). (1996). *Manson's Tropical Diseases* (20th ed.). Orlando: W. B. Saunders.

Dye, C. (1992). The analysis of parasite transmission by bloodsucking insects. *Annual Review of Entomology, 37,* 1–19.

Ewald, P. W. (1983). Host-parasite relations, vectors and the evolution of disease severity. *Annual Review of Ecology and Systematics, 14,* 465–485.

Faust, E. C., Beaver, B. C., & Jung, R. C. (1975). *Animal Agents and Vectors of Human Disease* (4th ed.). Philadelphia: Lea & Febiger.

Goddard, J. (1999). *Infectious Diseases and Arthropods.* Humana Press.

Gratz, N. (2006). *Vector- and Rodent-Borne Diseases in Europe and North America: Distribution, Public Health Burden, and Control.* Cambridge: Cambridge University Press.

Horsfall, F. L., & Tamm, I. (Eds.). (1965). *Viral and Rickettsial Infections of Man.* Philadelphia: J. B. Lippincott.

Hubbert, W. T., McCullough, W. F., & Schnurrenberger, P. R. (Eds.). (1975). *Diseases Transmitted from Animals to Man* (6th ed.). Charles Thomas.

Jeffrey, H. C., Leach, R. M., & Cowan, G. O. (1991). *Atlas of Medical Helminthology and Protozoology.* New York: Churchill Livingstone.

Kassai, T., Cordero del Campillo, M., Euzeby, J., Gafaar, S., Hiepe, T., & Himonas, C. A. (1988). Standardized nomenclature of animal parasitic diseases (SNOAPAD). *Veterinary Parasitology, 29,* 299–326.

McHugh, C. P. (1994). Arthropods: vectors of disease agents. *Lab Medicine, 25,* 429–437.

Mills, J. N., Childs, J. E., Ksiazek, T. G., Peters, C. J., & Velleca, W. M. (1995). *Methods for Trapping and Sampling Small Mammals for Virologic Testing.* Atlanta: Centers for Disease Control & Prevention.

Monath, T. P. (Ed.). (1988). *The Arboviruses: Epidemiology and Ecology. Vol. I. General Principles. Vol. II. African Horse Sickness to Dengue. Vol. III. Eastern Equine Encephalomyelitis to O'nyong Virus Disease. Vol. IV. Oropouche Fever to Venezuelan Equine Encephalomyelitis. Vol. V. Vesicular Stomatitis to Yellow Fever.* Boca Raton: CRC Press.

Moore, C. G., McLean, R. G., Mitchell, C. J., Nasci, R. S., Calisher, C. H., Marfin, A. A., Moore, P. S., Gubler, D. J. (1993). *Guidelines for Arbovirus Surveillance Programs in the United States.* Fort Collins: Centers for Disease Control & Prevention.

Service, M. W. (1986). *Blood-Sucking Insects: Vectors of Disease.* London: Edward Arnold.

Service, M. W. (Ed.). (1989). *Demography and Vector-Borne Diseases.* Boca Raton: CRC Press.

Snow, K. R. (1974). *Insects and Disease.* New York: Wiley.

Strickland, G. T. (Ed.). (1991). *Hunter's Tropical Medicine* (7th ed.). W. B. Saunders Company.

Theiler, M., & Downs, W. G. (1973). *The Arthropod-Borne Viruses of Vertebrates: An Account of the Rockefeller Foundation Virus Program, 1951–1970.* New Haven: Yale University Press.

WHO. (1989). *Geographical Distribution of Arthropod-Borne Diseases and Their Principal Vectors.* WHO/VBC/89.967.

WHO. (1995). *Vector Control for Malaria and Other Mosquito-Borne Diseases* (p. 720). WHO Tech. Rep. Ser. No. 857.

FOOD CONTAMINANTS

Gorham, J. R. (1975). Filth in foods: Implications for health. *Journal of Milk and Food Technology, 38,* 409–418.

Gorham, J. R. (1991a). *Insect and mite pests in food: An illustrated key* (Vols. 1 & 2). USDA Agric. Handbook No. 655.

Gorham, J. R. (Ed.). (1991b). *Ecology and Management of Food-Industry Pests.* FDA Tech. Bull. No. 4. Arlington, VA: AOAC International.

Hughes, A. M. (1976). *The Mites of Stored Food and Houses.* Ministry of Agriculture, Fisheries and Food, Her Majesty's Stationery Office, London.

Olsen, A. R., Sidebottom, T. H., & Knight, S. A. (Eds.). (1996). *Fundamentals of Microanalytical Entomology; A Practical Guide to Detecting and Identifying Filth in Foods.* Boca Raton: CRC Press.

Terbush, L. E. (1972). The medical significance of mites of stored food. *FDA By-Lines, 3,* 57–70.

DELUSIONAL DISORDERS, PHOBIAS, AND FORMICOPHILIA

Altschuler, D. Z., Crutcher, M., Dulceanu, N., Cervantes, B. A., Terinte, C., & Sorkin, L. N. (2004). Collembola (springtails) (Arthropoda: Hexapoda: Entgnatha) found in scrapings from individuals diagnosed with delusory parasitosis. *Journal of the New York Entomological Society, 112,* 87–95.

Beerman, H., & Nutting, W. B. (1984). Arachnid-related phobias: Symbiphobia, preventions, and treatments. In W. B. Nutting (Ed.), *Mammalian diseases and arachnids, Vol. II. Medico-veterinary, laboratory, and wildlife diseases, and control* (pp. 103–112). Boca Raton: CRC Press.

Dewarja, R., & Money, J. (1986). Transcultural sexology; formicophilia, a newly named paraphilia in a young Buddhist male. *Journal of Sex and Marital Therapy, 12,* 139–145.

Dewarja, R. (1987). Formicophilia, an unusual paraphilia treated with counseling and behavior therapy. *American Journal of Psychotherapy, 41,* 593–597.

Driscoll, M. S., Rothe, M. J., Grant-Kels, J. M., & Hale, M. S. (1993). Delusional parasitosis—A dermatological, psychiatric, and pharmacological approach. *Journal of the American Academy of Dermatology, 29,* 1023–1033.

Goddard, J. (1995). Analysis of 11 cases of delusions of parasitosis reported to the Mississippi Department of Health. *Southern Medical Journal, 88,* 837–839.

Hinkle, N. C. (2000). Delusory parasitosis. *Am Entomol, 46,* 17–25.

Koblenzer, C. S. (2006). The challenge of Morgellons disease. *Journal of the American Academy of Dermatology, 55,* 920–922.

Koblenzer, C. S. (1993). The clinical presentation, diagnosis and treatment of delusions of parasitosis—A dermatologic perspective. *Bulletin of the Society for Vector Ecology, 18,* 6–10.

Kushon, D. J., Helz, J. W., Williams, J. M., Lau, K. M. K., Pinto, L., & St. Aubin, F. E. (1993). Delusions of parasitosis: A survey of entomologists from a psychiatric perspective. *Bulletin of the Society for Vector Ecology, 18,* 11–15.

Murase, J. E., Wu, J. J., & Koo, J. (2006). Morgellons disease: A rapport-enhancing term for delusions of parasitosis. *Journal of the American Academy of Dermatology, 55,* 913–914.

Poorbaugh, J. H. (1993). Cryptic arthropod infestations: separating fact from fiction. *Bulletin of the Society for Vector Ecology, 18,* 3–5.

Robles, D. T., Romm, S., Combs, H., Olson, J., & Kirby, P. (2008). Delusional disorders in dermatology: a brief review. *Dermatology Online Journal, 14*(6), 2.

Savely, V. R., Leitao, M. M., & Stricker, R. B. (2006). The mystery of Morgellons disease: Infection or delusion? *American Journal of Clinical Dermatology, 7,* 1–5.

Szepietowski, J. C., Salomon, J., Hrehorow, E., Pacan, P., Zalewska, A., & Sysa-Jedrzejowska, A. (2007). Delusional parasitosis in dermatological practice. *Journal of the European Academy of Dermatology and Venereology, 21,* 462–465.

Vloten, W. A. (1998). Delusions of parasitosis. A psychiatric disorder to be treated by dermatologists? An analysis of 33 patients. *British Journal of Dermatology, 138,* 1030–1032.

Walling, H. W., & Swick, B. L. (2007). Psychocutaneous syndromes: A call for revised nomenclature. *Clinical and Experimental Dermatology, 32,* 317–319.

Webb, J. P., Jr. (1993a). Delusions of parasitosis: A symposium; coordination among entomologists, dermatologists, and psychiatrists. *Bulletin of the Society for Vector Ecology, 18,* 1–2.

Webb, J. P., Jr. (1993b). Case histories of individuals with delusions of parasitosis in southern California and a proposed protocol for initiating effective medical assistance. *Bulletin of the Society for Vector Ecology, 18,* 16–25.

TOXINS AND VENOMS

Beard, R. L. (1960). Insect toxins and venoms. *Annual Review of Entomology, 8,* 1–18.

Bettini, S. (1978). *Arthropod Venoms.* Berlin: Springer-Verlag.

Bücherl, W., & Buckley, E. E. (1971). *Venomous Animals and Their Venoms. Vol. 3. Venomous Invertebrates.* Academic Press.

Camazine, S. (1988). Hymenopteran stings: reactions, mechanisms and medical treatment. *Bulletin of the Entomological Society of America, 34,* 17–21.

Cloudsley-Thompson, J. (1995). On being bitten and stung. Antenna. *Bulletin of the Royal Entomological Society of London, 19,* 177–180.

Eisner, T., Eisner, M., & Siegler, M. (2005). *Secret Weapons: Defense of Insects, Spiders, Scorpions, and Other Many-legged Creatures.* Cambridge, MA: Harvard University Press.

Goddard, J. (1994). Direct injury from arthropods. *Laboratory Medicine, 25,* 365–371.

Keegan, H. L. (1969). Some medical problems from direct injury by arthropods. *International Journal of Pathology, 10,* 35–45.

Meier, J., & White, J. (1995). *Handbook of Clinical Toxicology of Animal Venoms and Poisons.* Boca Raton: CRC Press.

Nichol, J. (1989). *Bites and Stings: The World of Venomous Animals.* New York: Facts on File.

Papp, C. S., & Swan, L. A. (1983). *A Guide to Biting and Stinging Insects and Other Arthropods* (2nd ed.). Entomography Publications.

Roth, L. M., & Eisner, T. (1962). Chemical defenses of arthropods. *Annual Review of Entomology, 7,* 107–136.

Schmidt, J. O. (1982). Biochemistry of insect venoms. *Annual Review of Entomology, 27,* 339–368.

Tu, A. T. (1984). *Handbook of Natural Toxins. Vol. 2. Insect Poisons, Allergens, and Other Invertebrate Venoms.* Marcel Dekker.

HOST DEFENSES

Barriga, O. O. (1981). Immune reactions to arthropods. In O. O. Barriga (Ed.), *The Immunology of Parasitic Infections: A Handbook for Physicians, Veterinarians, and Biologists* (pp. 283–317). Baltimore: University Park Press.

Nelson, W. A., Bell, J. F., Clifford, C. M., & Keirans, J. E. (1977). Interaction of ectoparasites and their hosts. *Journal of Medical Entomology, 13,* 389–428.

Nelson, W. A., Keirans, J. E., Bell, J. F., & Clifford, C. M. (1975). Host-ectoparasite relationships. *Journal of Medical Entomology, 12,* 143–166.

Ribeiro, J. M. C. (1995). Blood-feeding arthropods: live syringes or invertebrate pharmacologists? *Infectious Agents and Diseases, 4,* 143–152.

Wikel, S. K. (1982). Immune responses to arthropods and their products. *Annual Review of Entomology, 27,* 21–48.

Wikel, S. K. (1996a). Host immunology to ticks. *Annual Review of Entomology, 41,* 1–22.

Wikel, S. K. (1996b). *The Immunology of Host-Ectoparasitic Arthropod Relationships.* Wallingford, UK: CAB International.

Wikel, S. K. (1999). Modulating the host immune system by ectoparasitic arthropods. *BioScience, 49,* 311–320.

WALKINGSTICKS

Albert, R. O. (1947). Another case of injury to the human eye by the walking stick, *Anisomorpha* (Phasmidae). *Entomological News, 58,* 57–59.

Carlberg, U. (1985). Chemical defense in *Anisomorpha buprestoides* (Houttuyn in Stoll') (Insecta: Phasmida). *Zool Anz Jena, 215,* 177–188.

Dziedzyc, J. (1992). Insect defensive spray-induced keratitis in a dog. *Journal of the American Veterinary Association, 200,* 1969.

Eisner, T. (1965). Defensive spray of a phasmid insect. *Science, 148,* 966–968.

Happ, G. M., Strandberg, J. D., & Happ, C. M. (1966). The terpene-producing glands of a phasmid insect. *Journal of Morphology, 219,* 143–160.

Hatch, R. L., Lamsens, S. D., & Perchalski, J. E. (1993). Chemical conjunctivitis caused by spray of *A. buprestoides,* two-striped walking stick. *Journal of the Florida Medical Association, 80,* 758–759.

Meinwald, J., Chadha, M. S., Hurst, J. J., & Eisner, T. (1962). Defense methods of arthropods – IX. Anisomorphal, the secretion of a phasmid insect. *Tetrahedron Letters, 1,* 29–33.

Paysse, E. A., Holder, S., & Coats, D. K. (2001). Ocular injury from the venom of the southern walkingstick. *Ophthalmology, 108,* 190–191.

Stewart, M. A. (1937). Phasmid injury to the human eye. *Canadian Entomologist, 69,* 84–86.

MINOR ARTHROPOD PROBLEMS OF MEDICAL-VETERINARY INTEREST

Arnaud, P. H., Jr. (1970). Thrips "biting" man. *Pan-Pacific Entomologist, 46,* 76.

Bailey, S. F. (1936). Thrips attacking man. *Canadian Entomologist, 68,* 95–98.

Baz, A., & Monserrat, J. (1999). Distribution of domestic Psocoptera in Madrid apartments. *Medical and Veterinary Entomology, 13,* 259–264.

Bishopp, F. C. (1961). Injury to man by earwigs. *Proceedings of the Entomological Society of Washington, 63,* 114.

Durden, L. A. (1987). Predator-prey interactions between ectoparasites. *Parasitology Today, 3,* 306–308.

Ertek, M., Aslan, I., Yazgi, H., Torun, H. C., Ayyildiz, A., & Tasyaran, M. A. (2004). Infestation of the human intestine by the millipede, *Nopoiulus kochii. Medical and Veterinary Entomology, 18,* 306–307.

Hood, J. D. (1927). A blood-sucking thrips. *Entomologist, 60,* 201.

Hunter-Jones, P. (1966). Allergy to animals: A zoological hazard. *New Scientist,* 615–616.

Li, D. N., & Li, J. C. (1995). [Report on human dermatitis caused by *Liposcelis divinatorius* (Psocoptera).]. *Chinese Journal of Parasitology and Parasite Diseases, 13,* 283 [In Chinese].

Nakata, S., & Maa, T. C. (1974). A review of the parasitic earwigs (Dermaptera: Arixeniina; Hemimerina). *Pacific Insects, 16,* 307–374.

Radford, A. J. (1975). Millipede burns in man. *Tropical and Geographical Medicine, 27,* 279–287.

Remington, C. L. (1950). The bite and habits of a giant centipede (*Scolopendra subspinipes*) in the Philippine islands. *American Journal of Tropical Medicine and Hygiene, 30,* 453–455.

Scott, H. G. (1966). Insect pests. Part I. Springtails. *Modern Maintenance Management, 18,* 19–21.

Scott, H. G., Wiseman, J. S., & Stojanovich, C. J. (1962). Collembola infesting man. *Annals of the Entomological Society of America, 55,* 428–430.

Seshadri, A. R. (1955). An extraordinary outbreak of caddis-flies (Trichoptera) in the Meltrudam township area of Salem district, South India. *Indian Journal of Entomology, 3,* 337–340.

Williams, C. B. (1921). A blood-sucking thrips. *Entomologist, 54,* 163–164.

Writz, R. A. (1980). Occupational allergies to arthropods-documentation and prevention. *Bulletin of the Entomological Society of America, 23,* 356–360.

Chapter

2

Morphological Adaptations of Parasitic Arthropods

Nathan D. Burkett-Cadena

Among the Insecta and Arachnida are several independent lineages of arthropods that are parasitic on vertebrates in one or more stages of their life cycles. Many morphological features of these parasitic arthropods have been modified in a number of ways as adaptations for parasitic relationships. Common adaptations include those for feeding on vertebrate blood and other body fluids, attaching to and clinging to hosts, and dispersal to new hosts. Morphological structures that play a particularly important role in host associations are body shape, mouthparts, and legs. Other features that often are modified in parasitic arthropods include the wings, eyes, and various sensory organs.

BODY SHAPE AND WINGS

A recurring theme among ectoparasites is modification of the general body shape to facilitate movement on the host and to enable them to hide in tight spaces when off the host. This usually involves being dorso-ventrally or laterally flattened. Dorso-ventral flattening is characteristic of bed bugs and bat bugs, lice, louse flies and keds, and ticks. In contrast, lateral flattening is best exemplified by fleas.

Wings are indispensable features of many parasitic insects, without which they would not be able to reach their hosts. In other parasitic insects, however, wings have become secondarily reduced, or even completely lost, as in fleas, lice, bed bugs, and the sheep ked. Still others, such as hippoboscids of the genus *Lipoptena* that parasitize deer, may have fully functional wings as adults but shed them after reaching a suitable host. In the latter case, the wings break off at a specific location near their base, leaving a small stub.

MOUTHPARTS

Mouthparts of parasitic arthropods typically are adapted for feeding on host body fluids, particularly blood, but also lymph, skin secretions, and tears. They also may be adapted for feeding externally on skin, sloughed skin scales, hair, or feathers. Those arthropods that feed directly on host tissues generally retain the chewing-type mouthparts, like those of cockroaches (Figure 2.1A). Fluid-feeding parasites, on the other hand, have mouthparts adapted for piercing host skin to reach and feed on internal fluids. It is among this group that the mouthparts have become the most modified and specialized. Arthopods that use their mouthparts to lacerate host skin and feed on blood that pools at the bite site as a result of damage to the surrounding blood vessels are called **telmophages**. Examples of telmophages include black flies, biting midges, horse flies, and deer flies. They typically cause more immediate pain and discomfort at the bite site due to open puncture wounds, damage to surrounding skin and related tissues, and the greater risk of secondary infections. Hematophagous arthropods with highly specialized piercing-sucking mouthparts that penetrate individual capillaries and then feed directly on host blood are called **solenophages**. Examples of solenophages are mosquitoes, bed bugs, kissing bugs, and sucking lice. They represent the more refined blood feeders, with highly modified, styletiform mouthparts that typically leave little or no evidence of an actual puncture of the skin at the bite site.

Biting midges are excellent examples of telmophages. The labrum, mandibles, maxillae, and hypopharynx are bladelike, pointed or serrate apically, and function in piercing host skin (Figure 2.1C). The labium serves as a sheath for protecting the other, more delicate, mouthparts.

Mosquitoes are classic examples of solenophages. The labrum, maxillae, mandibles, hypopharynx, and labium are very elongate, forming a feeding apparatus called the proboscis (Figure 2.1B). The mosquito labium serves as a protective sheath and a guide for the other mouthparts. In insects with chewing mouthparts, however, the labium is analogous to a "lower lip" (Figure 2.1A). In mosquitoes the labrum, maxillae,

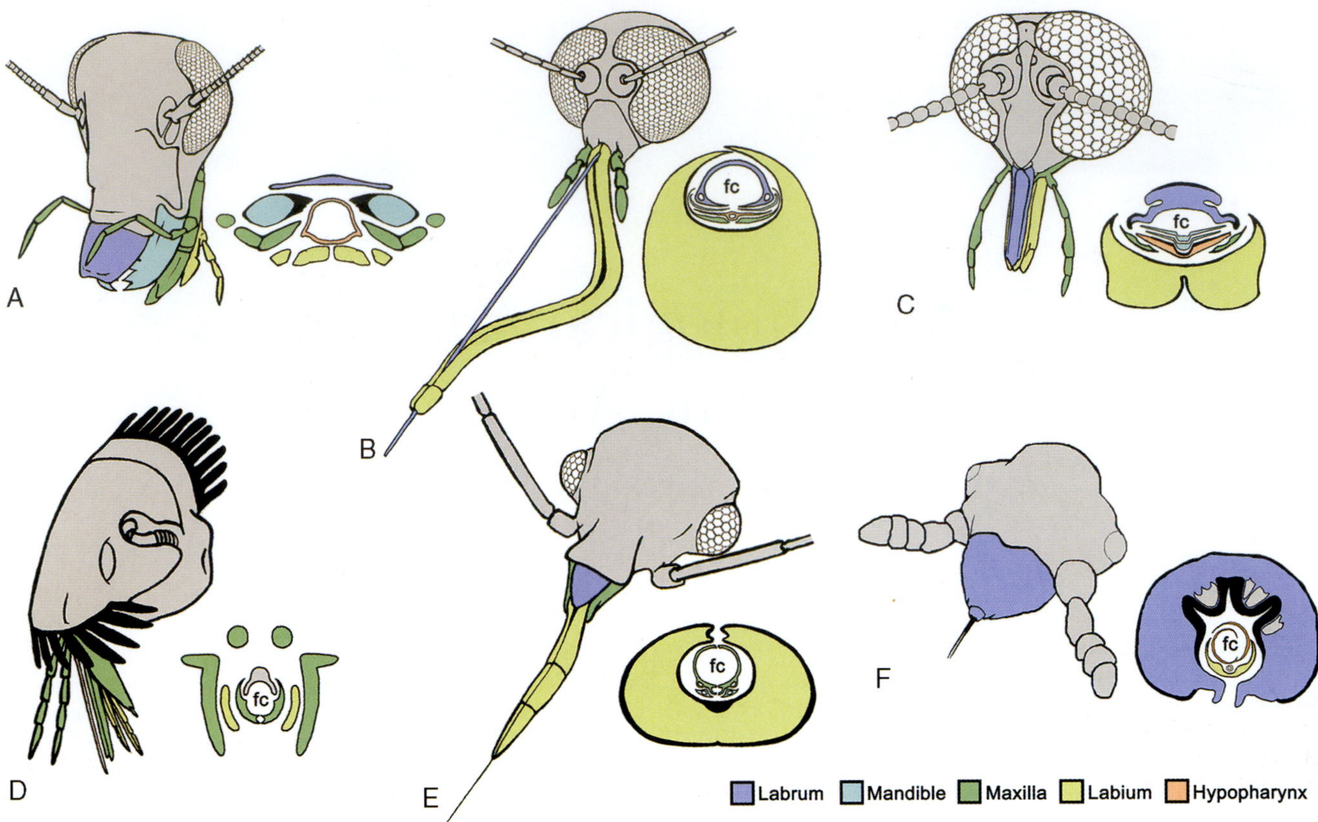

Figure 2.1 Head and mouthparts of medically important insects, with cross-section of mouthparts. (A) Cockroach, *Periplaneta* (Blattidae); (B) Mosquito, *Aedes* (Culicidae) with labium reflexed to show styletiform mouthparts; (C) Biting midge, *Culicoides* (Ceratopogonidae); (D) Cat flea, *Ctenocephalides felis* (Pulicidae); (E) Bed bug, *Cimex* (Cimicidae); (F) Human body louse, *Pediculus humanus* (Pediculidae). (Original by N. D. Burkett-Cadena)

mandibles, and hypopharynx are in the form of fine stylets and together form a fascicle, or bundle, which penetrates the host skin and then serves to deliver blood to the digestive system. The labrum, which functions as the "upper lip" of insects with chewing mouthparts (Figure 2.1A), forms the blood-feeding tube in mosquitoes. The mandibles and maxillae, typically used for manipulating and masticating food by arthropods with chewing mouthparts (Figure 2.1A), are modified in mosquitoes for piercing the host epidermis. The hypopharynx, a tongue-like structure in insects with chewing mouthparts (Figure 2.1A), is also styletiform in mosquitoes and is used to pierce host tissue. Running the length of the hypopharynx is a channel that delivers saliva to the apical portion of the mouthparts during feeding.

In fleas, the epipharynx (an outgrowth of the body wall unique to fleas) and the maxillae are in the form of stylets and are used to pierce skin (Figure 2.1D). During blood feeding the tip of the epipharynx is inserted into a capillary. The pair of maxillae, together with the epipharynx, forms the feeding tube. In fleas the labium is reduced, with only the palps visible externally. The palps help to guide the blood-feeding stylets.

In blood-feeding hemipterans, such as bed bugs and kissing bugs, the maxillae and mandibles are styletiform and held within a sheath-like, segmented labium (Figure 2.1E). Each of the paired maxillae is curved medially and interlocks with the other on their dorsal and ventral surfaces to form the food canal. The maxillae, mandibles, and labium form the rostrum, which is directed posteriorly and held under the head and thorax of the insect when not in use. The labrum is relatively unmodified and resembles the labrum of insects with chewing mouthparts.

In sucking lice the feeding apparatus is quite different from other parasitic insects. The labrum of sucking lice is highly modified to form a snoutlike structure, called the **haustellum**, which surrounds the other mouthparts (Figure 2.1F). At the tip of the haustellum are prestomal, or haustellar, "teeth," which are used to anchor the mouthparts to the host. The maxillae, hypopharynx, and labium are modified as stylets for piercing host tissues. The hypopharynx also serves as a salivary canal, and the maxillae form the food canal.

Among the arachnids, ticks represent a highly specialized group of mites that are obligate parasites of terrestrial vertebrates. Ticks feed on host blood during each of their developmental stages (i.e., as larvae, nymphs, and adults). The mouthparts are modified for piercing host tissue, anchoring the tick to its host and drawing blood into the alimentary tract. In most groups of mites, the tip of each chelicera is pincerlike, or **chelate**, with a fixed digit and opposable

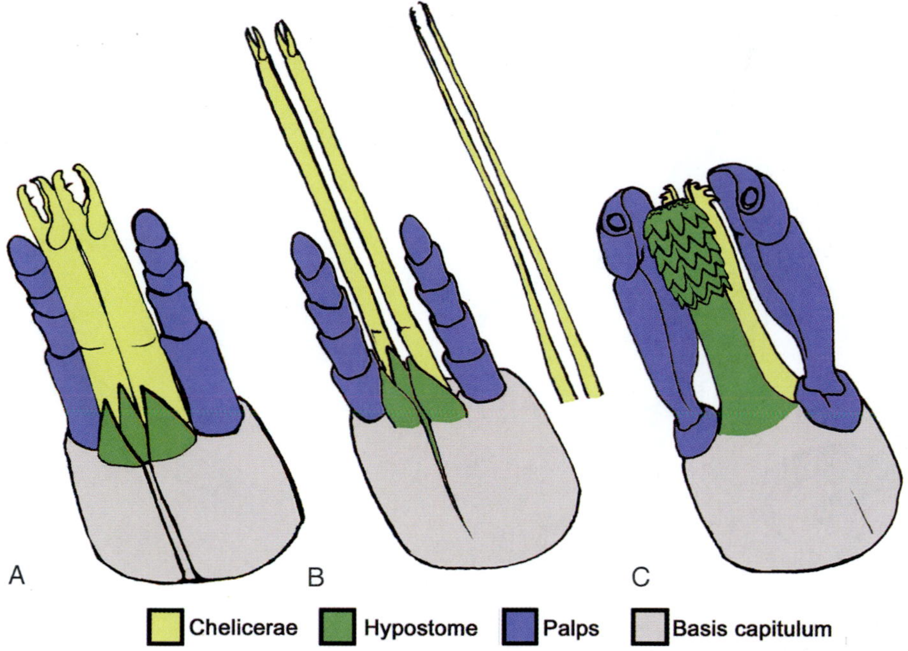

Figure 2.2 Mouthparts and palps (gnathosoma) of parasitic mites, ventral views. (A) General morphology of mesostigmatid mite, with well-developed, chelate chelicerae (Laelapidae, *Laelaps*); (B) Chelicerae long and slender, with reduced chelae (Macronyssidae, *Ornithonyssus*); inset of nonchelate chelicerae, adapted for blood-feeding (Dermanyssidae, *Dermanyssus*); (C) Hard tick (Ixodidae, *Amblyomma*), showing hypostome highly modified as an attachment organ. (Original by N. D. Burkett-Cadena)

movable digit for grasping and manipulating food (Figure 2.2A). In ticks, however, the chelicerae have lost the terminal chelae and have become modified as short blade-like structures, with serrate tips, adapted for piercing host skin (Figure 2.2C). The hypostome is greatly enlarged and serves as an attachment organ. Backward projecting teeth on the hypostome serve to anchor the tick securely to its host after it has cut a hole through the skin with its chelicerae. The palps of ticks are relatively unmodified and do not penetrate the skin of the host. In other parasitic mites, including members of the families Dermanyssidae and Macronyssidae, the chelicerae may be long, slender, and retractable, and are adapted for piercing host skin. The terminal portion of the chelicera may be chelate, as in *Ornithonyssus* (Macronyssidae) (Figure 2.2B), or serrate, lacking the movable element, as in *Dermanyssus* (Dermanyssidae) (Figure 2.2B inset).

LEGS

In both insects and arachnids, those taxa that live for extended periods of time on their hosts often have specialized structures on their legs to facilitate attachment to the host and movement amidst the host hair or feathers. Structures for grasping often are coupled with stout, heavily sclerotized and generously muscled appendages. This combination of characters allows ectoparasites not only to obtain access to new hosts, but also to avoid being displaced or removed by host grooming.

The insect leg typically consists of five segments (Figure 2.3A). The basal segment is the **coxa**, followed by the **trochanter**, **femur**, **tibia**, and **tarsus**. The tarsus is divided into subsegments or **tarsomeres**, providing flexibility. Claws and other structures, when present, are typically borne on the apical tarsomere. In many ectoparasitic insects the typical leg has become modified to facilitate host attachment and dispersal to new hosts. Adult fleas exhibit perhaps the most widely recognized modification of the legs for getting on and off hosts. The hind legs of fleas are particularly modified to enable them to jump remarkable distances to reach a host or evade removal by host grooming (Figure 2.3B). Modifications of the hind leg includes an enlarged, muscular femur (as in other jumping insects) and an elastic protein in the integument called **resilin**. Resilin, an important structural component of the flight mechanism in flying insects, helps to store energy and significantly increase the efficiency of the hind legs in the jumping ability of fleas. Contraction of muscles in the hind legs compresses pads of resilin located at the bases of the hind coxae. A release mechanism causes rapid expansion of the resilin pads, propelling the flea forward and upward during the jump.

Lice are noted for their ability to cling tenaciously to their hosts. The legs of sucking lice are particularly well adapted for grasping host hair. The claws are formed by modifications of the tibia and tarsus, called tibiotarsal claws (Figure 2.3C, D). The tarsus generally is reduced to form a stout, curved, often sickle-shaped, movable element. This element articulates with a

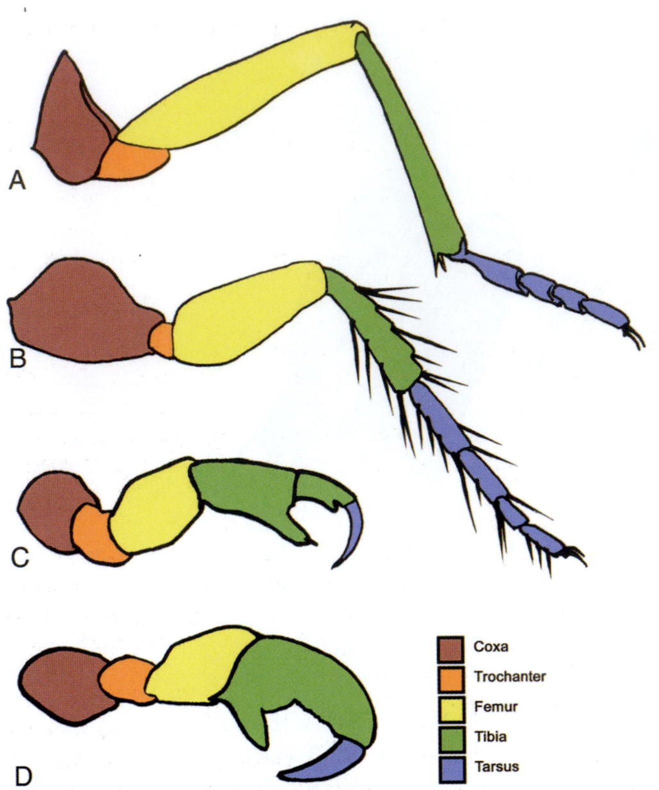

Figure 2.3 Legs of medically important insects. (A) Hind leg of blister beetle (Meloidae, *Epicauta*); (B) Hind leg of cat flea, *Ctenocephalides felis* (Pulicidae); (C) Foreleg of human body louse, *Pediculus humanus* (Pediculidae); (D) Hind leg of human pubic louse, *Pthirus pubis* (Pthiridae).(Original by N. D. Burkett-Cadena)

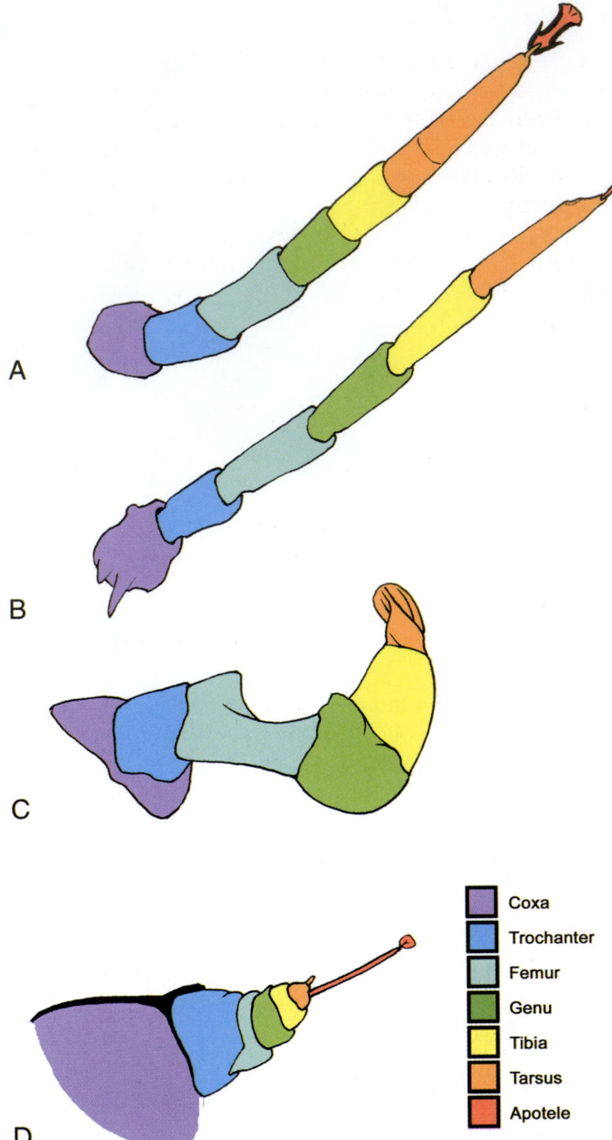

Figure 2.4 Legs of medically important mites. (A) Leg I, with membranous apotele and lacking claws (Laelapidae, *Laelaps*); (B) Leg I of hard tick, with well-developed pair of claws and padlike empodium (Ixodidae, *Amblyomma*); (C) Leg III of fur mite, with highly modified tarsus for grasping host hair (Myocoptidae, *Trichoecius*); (D) Leg I of scabies mite, with highly reduced leg segments and long, sucker-tipped apotele (Sarcoptidae, *Sarcoptes scabiei*). (Original by N. D. Burkett-Cadena)

sclerotized projection of the tibia to form an effective grasping structure.

Among the Diptera, hippoboscids have legs that are specialized for host attachment. Unlike other blood-feeding dipterans, for which host contact is usually brief, hippoboscids may spend their entire lives on their hosts. The legs are therefore stout and usually spinose, with enlarged tarsal claws. These features enable hippoboscids to hold onto their hosts and to move about quickly and efficiently amidst host pelage.

In mites, each leg typically consists of seven segments (Figure 2.4A). The basal segment is the coxa, followed by the trochanter, femur, **genu**, tibia, tarsus, and the terminal **apotele**. The apotele may bear claws, setae, and an empodium, a padlike structure arising between the bases of the claws. In parasitic mites, various modifications of the legs enable these arthropods to locate and attach to their hosts. In ticks, for example, the forelegs have enlarged claws (Figure 2.4B) that enable them to quickly grasp passing hosts and facilitate holding onto host skin or pelage during feeding and mating. Certain feather mites (e.g., Analgidae) have suckerlike empodia and large spurs on their legs for securing themselves to their avian hosts. Many of the mites that parasitize snakes (e.g., Ixodorhynchidae and Ophionyssidae) also have suckerlike empodia, which facilitate movement on their host and holding onto the smooth surfaces of the body scales. In many other parasitic mites, hind legs, rather than forelegs, are modified for host attachment. Mites in the family Mycoptidae, for example, have legs III and IV enlarged with opposable digits for clasping the fur of their rodent hosts (Figure 2.4C). In scabies mites (Sarcoptidae) all four pairs of legs are reduced and have elongate apoteles with terminal suckers (Figure 2.4D). These structures allow the mites to move about quickly on the surface of the skin and to hold tightly to the epidermis.

SENSORY STRUCTURES

Sensory structures play an integral role in host location and recognition, and have thus become highly specialized in many parasitic arthropods. Various sensory structures of parasitic arthropods function to detect motion, vibrations, temperature, moisture, carbon dioxide, and a plethora of chemical substances produced by potential hosts. In combination with one another these environmental and host-associated cues are often specific for a single host species or group of closely related host animals.

In insects, the antennae and eyes are the primary sensory organs. The antennae of blood-feeding insects, particularly hematophagous dipterans, have receptors that detect molecules emanating from the skin and present in the exhaled breath of potential hosts. Host substances known to attract mosquitoes include carbon dioxide, lactic acid, octenol, estrogen, fatty acids, and amino acids. In mosquitoes, sensory receptors in the basal segment of the antenna are highly developed to form **Johnston's organ**, which is specialized for detecting airborne vibrations. Host-seeking female mosquitoes may cue in on vibrations produced by host movements and even vocalizations by hosts such as birds and frogs. Fleas exhibit an interesting modification of the antennae, in which the antenna is short, flattened, and fits into a protective groove on the side of the head (Figure 2.1D). This allows the antennae to be retracted so as not to become damaged or impede movement as the flea maneuvers amidst host hair or feathers.

In fleas and lice, the eyes generally are reduced in size (Figure 2.1D, F) or may be altogether absent. In some cases, such modifications of the eyes help to prevent damage to the sense organs, whereas in other cases the reduction of eyes reflects the relative unimportance of vision in the life of the parasite. In many other insects, such as mosquitoes (Figure 2.1B), biting midges (Figure 2.1C), and horse flies, the eyes are greatly enlarged, reflecting the more significant role that light perception and vision plays among these groups in locating, or orienting toward, potential hosts.

Sensory structures are often particularly numerous or well developed on the mouthparts of parasitic arthropods. In many solenophages, for example, these receptors are concentrated near the tip of the proboscis or rostrum (e.g., mosquitoes and bed bugs, respectively) and are used to detect the precise location of capillaries beneath the surface of the skin. In biting midges, sensory receptors for detecting environmental and host cues are concentrated in a specialized pit, in the form of a sensory organ on the enlarged third segment of the maxillary palp (Figure 2.1C).

In fleas, the dorsal portions of the terminal abdominal segments are modified as a sensory organ, called the **sensilium**. The associated sensory structures are specialized for detecting host-associated cues such as vibrations and temperature gradients.

In mites, chemical and tactile cues are perceived by sensory structures on the pedipalps, legs, and various other parts of the body. Specialized sensory setae with an associated socket-like base, called **trichobothria**, are common in many groups of mites and other arachnids for detecting airborne and substrate vibrations, and other tactile cues. In certain groups of mites, the first pair of legs may be unusually long and slender, with numerous receptors, serving as a sensory organ, much like the antennae of insects. In ticks, a complex sensory structure, called **Haller's organ**, is located on the dorsal aspect of the tarsus of the first pair of legs (Figure 26.6) and functions in detection of temperature, air movements, host odors, and other host and environmental cues.

In summary, a number of advantageous morphological modifications are evident in ectoparasitic and blood-feeding arthropods. Recurring adaptations include modifications of the body shape, feeding apparatus, locomotary appendages, and sensory structures. Each modification has allowed parasitic arthropods to more efficiently exploit their vertebrate hosts. A number of excellent works have been produced, especially in some of the older literature by R. E. Snodgrass, that examines in detail the morphological adaptations in parasitic arthropods. For further information on general insect and arachnid morphology, see Snodgrass (1935), Chapman (2008), and Krantz and Walter (2009).

REFERENCES AND FURTHER READING

Chapman, R. F. (2008). *The Insects: Structure and Function* (4th ed.). Cambridge, UK: Cambridge University Press.

Harbach, R. E., & Knight, K. L. (1980). *Taxonomist's Glossary of Mosquito Anatomy.* Biological Research Institute of America, Plexus Publishing, Inc.

Krantz, G. W., & Walter, D. E. (Eds.). (2009). *A Manual of Acarology,* (3rd ed.). Lubbock, TX: Texas Tech University Press.

Labrzycka, A. (2006). A perfect clasp—Adaptation of mites to parasitize mammalian fur. *Biological Letters, 43,* 109–118.

Marshall, A. G. (1981). *The Ecology of Ectoparasitic Insects.* London: Academic Press.

Snodgrass, R. E. (1935). *Principles of Insect Morphology.* McGraw-Hill.

Snodgrass, R. E. (1943). The feeding apparatus of the biting and disease-carrying flies: A wartime contribution to medical entomology. *Smithsonian Miscellaneous Collections, 104,* 1–51.

Snodgrass, R. E. (1944). The feeding apparatus of the biting and sucking insects affecting man and animals. *Smithsonian Miscellaneous Collections, 104,* 1–113.

Snodgrass, R. E. (1946). The skeletal anatomy of fleas (Siphonaptera). *Smithsonian Miscellaneous Collections, 104,* 1–110.

Stojanovich, C. J. (1945). The head and mouthparts of the sucking lice (Insecta: Anoplura). *Microentomology, 10,* 1–46.

Chapter

3

Epidemiology of Vector-Borne Diseases

William K. Reisen

Medical-veterinary entomologists play a pivotal role in understanding the epidemiology of vector-borne diseases and are a key component of multidisciplinary programs that research, monitor, and control vector-borne parasites. Medical entomology especially comes to the forefront in public health during periods of war, famine, or natural disasters that disrupt public health programs, displace populations, and increase exposure to vectors. Historically, the large-scale movement of populations (such as military troops) into areas endemic for vector-borne diseases has had devastating effects on both the invading army and the local population, because neither was immune to the new parasites to which they were exposed. The recent globalization of commerce, improved and readily available rapid transportation, and the relaxing of international travel health regulations have combined to increase the movement of parasites of humans and domestic animals and their vectors into new geographical areas, placing previously unexposed populations at risk of infection. Globally, outbreaks of many pathogens have been on the increase, and collectively these have been called **emerging infectious diseases**. Among these pathogens are a large number of vector-borne parasites that are emerging, because the primary vectors or parasites have been able to expand their distributions in either time or space due to anthropogenic factors or because the parasites have evolved into more virulent or drug-resistant forms.

Although specific methods of investigation vary considerably among the vast array of vector-borne parasites, basic concepts unify the pattern of information necessary to understand the epidemiology of vector-borne disease. Information progresses from discovery of the parasite as the causative agent of a disease, to identifying its mode of transmission among vectors and vertebrate hosts, to monitoring, forecasting, and control. During the discovery period, clinical case definition and diagnosis are established, enabling the tracking of cases in time and space, and the causative agent is identified, perhaps indicating that an arthropod may be responsible for transmission. The incrimination of the vector(s) requires a combination of field and laboratory investigation that determines vector abundance in time and space, host selection patterns, field infection rates, and vector competence. Although short-term studies rapidly may determine the mode(s) of transmission, delineating transmission cycles and interseasonal maintenance mechanisms typically requires years of careful, often frustrating, ecological investigation and laboratory experimentation. Effective surveillance and control programs are best implemented after maintenance, amplification, and epidemic transmission patterns have been described. Unfortunately, discovery rarely progresses in the orderly fashion outlined earlier. Frequently, monitoring and management of cases progresses more rapidly than the discovery of the pathogen or the mode(s) of transmission.

This chapter provides an introduction to concepts needed to understand the epidemiology and emergence of vector-borne diseases. Epidemiology developed as a science through the investigation of outbreaks of infectious diseases. As a modern discipline, **epidemiology** (etymology: *epi* = upon, *demos* = people, *logos* = study) deals with the natural history and spread of diseases within human and animal populations. Vector-borne diseases consist minimally of a triad that includes an arthropod vector, a vertebrate host, and a parasite. The spread of pathogens by arthropods is especially complex, because in addition to interactions between the vertebrate host and the parasite, an arthropod is required for transmission of the parasite to uninfected hosts. Environmental factors such as temperature and rainfall impact these processes by affecting the rate of parasite maturation within the arthropod host as well as arthropod and vertebrate host abundance in time and space.

In medical entomology, a **vector** is an arthropod responsible for the transmission of parasites (not diseases) among vertebrate hosts. Disease is the response of the host to invasion by or infection with a parasite. A parasite is any organism, including viruses, bacteria, protozoa, helminths, and arthropods, that is dependent upon the host for its survival. Parasites may or may not cause disease. When a parasite injures its host and causes disease, it is referred to as a **pathogen** or **disease agent**. A vector-borne disease, therefore, is an illness caused by a pathogen that is transmitted by an arthropod. **Facultative parasites** have both free-living and parasitic forms, whereas **obligate parasites** are totally dependent upon their host(s) to provide their requirements for life. **Ectoparasites** live on or outside the host, whereas **endoparasites** live inside the host. When interacting with their hosts, ectoparasites produce an infestation that typically remains topical or peripheral, whereas endoparasites produce an infection upon invasion of host tissues or cells. The occurrence and severity of disease depends upon the host–parasite interaction after infection. A host carrying a parasite is infected, whereas an infected host capable of transmitting a parasite is infective. A host capable of parasite maintenance without clinical symptoms is a **carrier**.

A complete understanding of the epidemiology of arthropod-borne disease requires knowledge of the ecology, physiology, immunology, and genetics of parasite, arthropod, and vertebrate host populations and how they interact in their environment. The degree of contact between the vertebrate host and vector ranges from intermittent (e.g., mosquitoes) to intimate and continuous (e.g., sucking lice). Frequently the host provides the vector not only food in the form of blood or other tissues, but also a habitat or place in which to live. Blood feeding by the vector typically brings parasite, vector, and vertebrate host together in time and space, and ultimately is responsible for the transmission of parasites from infected to susceptible vertebrate hosts. A vector usually must take at least two blood meals during its lifetime to transmit a parasite, the first to acquire the infection and the second to transmit it. Blood meals are taken to provide the arthropod with nutrients necessary for metabolism, metamorphosis, and reproduction.

The **gonotrophic cycle**, or reproductive cycle of the arthropod, includes the sequence of questing or searching for a host, blood feeding, blood meal digestion, egg maturation, and oviposition. **Parous** females have completed one or more gonotrophic cycles and have a greater probability of being infected with parasites than **nulliparous** females that have not reproduced and are feeding for the first time. Unlike parasites that are transmitted directly from host to host, parasites transmitted by arthropods generally have replaced free-living or environmentally resistant stages with those that can multiply and develop within the arthropod and be transmitted during the blood feeding process.

COMPONENTS OF TRANSMISSION CYCLES

The components of a transmission cycle of an arthropod-borne disease include:

- A parasite that can multiply within both vertebrate and invertebrate host tissues
- A vertebrate host (or hosts) that develops a level of infection with the parasite that is infectious to a vector
- An arthropod-host or vector that acquires the parasite from the infected host and is capable of transmission (Figure 3.1)

Vector-borne parasites have evolved mechanisms for tolerating high constant body temperatures and evading the complex immune systems of the vertebrate hosts as well as for tolerating variable body temperatures and avoiding the very different defensive mechanisms of the arthropod vectors. Asexual parasites such as viruses and bacteria employ the same life form to infect both vertebrate and arthropod hosts, whereas more highly evolved heterosexual parasites such as protozoa and helminths have very different life stages in their vertebrate and arthropod hosts. Some asexual parasites such as the plague bacillus may bypass the arthropod host and be transmitted directly from one vertebrate host to another.

Among sexually reproducing parasites, the host in which gametocyte union occurs is called the **definitive host**, whereas the host in which asexual reproduction occurs is called the **intermediate host**. Vertebrates or arthropods can serve as either definitive or intermediate host, depending upon the life cycle of the parasite. For example, humans are the definitive host for the filarial worm, *Wuchereria bancrofti*, because adult male and female worms mate within the human lymphatic system, whereas the mosquito vector, *Culex quinquefasciatus*, is the intermediate host where larval worm development occurs without reproduction. In contrast, humans are the intermediate host of the *Plasmodium* protozoan that causes malaria, because only asexual reproduction occurs in the human host; gametocytes produced in the human host unite only in the gut of the definitive host, the Anopheles mosquito.

Host Immunity

A disease is the response of the host to infection with the parasite and can occur in either vertebrate or arthropod hosts. Immunity includes all properties of the host that confer resistance to infection and plays an important role in determining host suitability and the extent of disease or illness. Some species or individuals within species populations have natural (or innate) immunity and are refractory to infection. Natural immunity does not require that the host have previous contact with the parasite, but may be age dependent. For example, humans do not become infected with avian malaria parasites such as *Plasmodium relictum*, even though infective *Culex* mosquito vectors feed frequently on humans.

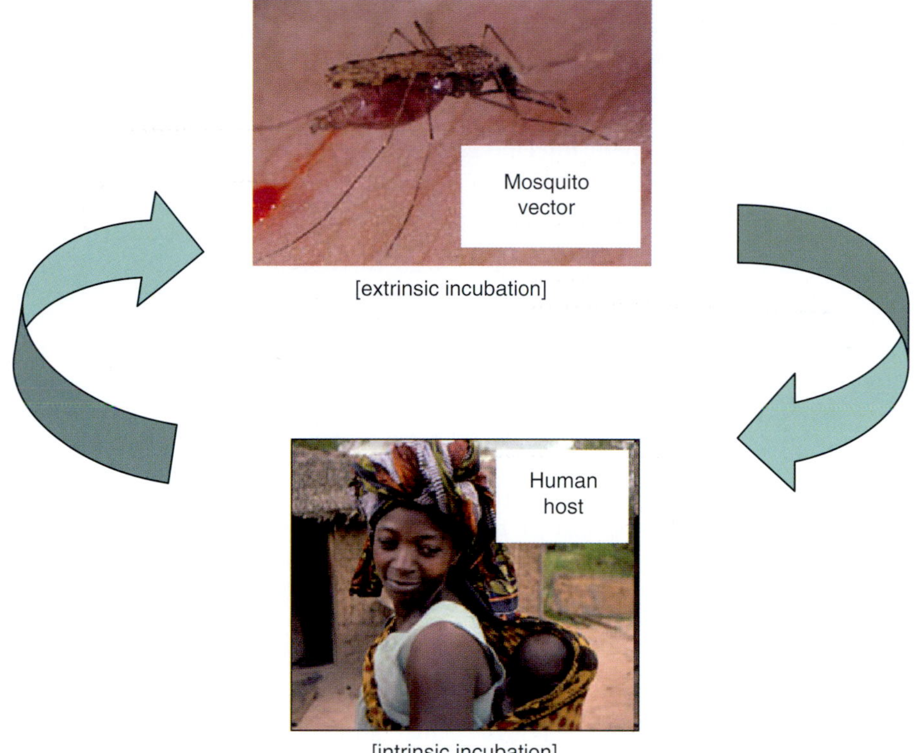

Figure 3.1 Components of the transmission cycle of an anthroponosis (e.g., malaria and dengue), which does not involve nonhuman vertebrate hosts; with intrinsic incubation of the pathogen in humans and extrinsic incubation in mosquitoes.

Conversely, mosquitoes do not become infected with measles or polio viruses that infect humans, even though these viruses undoubtedly are ingested by mosquitoes blood-feeding on viremic human hosts.

Individuals within populations often become infected with parasites, recover, and in the process actively acquire immunity. This acquired immunity to the parasite ranges from transient to life-long, and may provide partial to permanent protection. A partial immune response may permit continued infection, but may reduce the severity of disease, whereas complete protection results in a cure and usually prevents immediate reinfection. Acquired immunity may be humeral and result in the rapid formation of antibodies, or may be cellular and result in the activation of T-cells and macrophages. Antibodies consist of five classes of proteins called **immunoglobulins** that have specific functions in host immunity. **Immunoglobulin G** (abbreviated **IgG**) is most common, comprising over 85% of the immunoglobulins present in the sera of normal individuals. The IgGs are relatively small proteins, typically develop to high concentration several weeks after infection, and may persist at detectable and protective levels for years. Therefore, parasites such as western equine encephalomyelitis virus that induce long-lasting immunity are good candidates for vaccine development. In contrast, **immunoglobulin M** (**IgM**) is a large macroglobulin that appears shortly after infection, but decays rapidly relative to IgG. For the laboratory diagnosis of many diseases, serum samples typically are tested during periods of acute illness and convalescence, two to four weeks later. A fourfold increase in parasite-specific IgG antibody concentration in these paired sera provides diagnostic serological evidence of infection. The presence of elevated concentrations of IgM presumptively implies current or recent infection. **T-cells** and **macrophages** are several classes of cells that are responsible for the recognition and elimination of parasites. In long-lived vertebrate hosts, such as humans, acquired immunity may decline over time, eventually allowing reinfection.

Clinically, the host response to infection ranges from inapparent or asymptomatic to mildly symptomatic to acute. Generally it is beneficial for the parasite if the host tolerates infection and permits parasite reproduction and development without becoming severely ill and dying before vectors are infected. However, the fitness of many parasites is contingent upon elevated virulence that causes high parasite concentrations and frequent mortality in the vertebrate host but facilitates transmission by increased infection of the vector. Generally, the susceptibility of the vector to infection dictates the concentration levels or virulence of parasites in the vertebrate host necessary to complete the transmission cycle.

The Vertebrate Host

One or more primary vertebrate hosts are essential for the maintenance of parasite transmission, whereas secondary or incidental hosts are not essential to maintain transmission, and may or may not contribute to parasite amplification. **Amplification** refers to the general increase in the number of parasites present in a given area. An **amplifying host** increases the number of parasites and theoretically also the number of infected vectors. Amplifying hosts typically do not remain infected for long periods of time and may develop disease of varying severity. A **reservoir host** supports parasite development, remains infected for long periods, and serves as a source of vector infection, but usually does not develop acute disease. Attributes of a primary vertebrate host include accessibility, susceptibility, and transmissibility.

Accessibility The vertebrate host must be abundant and fed upon frequently by vectors. Host seasonality, daily activity, and habitat selection determine availability in time and space to host-seeking vectors. For example, the avian hosts of eastern equine encephalomyelitis (EEE) virus generally begin nesting in swamps coincidentally with the emergence of the first spring generation of the mosquito vector, *Culiseta melanura*, thereby bringing EEE virus, susceptible avian hosts, and mosquitoes together in time and space. Diel activity patterns also may be critical. For example, larvae (microfilariae) of the filarial worm *W. bancrofti* move to the peripheral circulatory system of the human host during specific hours of the night that coincide with the biting rhythm of the mosquito vector, *Culex quinquefasciatus*. Historically, epidemics of vector-borne diseases have been associated with increases in human accessibility to vectors during wars, natural disasters, environmental changes, or human migrations.

Susceptibility Once exposed, a primary host must be susceptible to infection and permit the development and reproduction of the parasite. Dead-end hosts either do not support a level of infection sufficient to infect vectors or become extremely ill and die before the parasite can complete development, enter the peripheral circulatory system or other tissues, and infect additional vectors. Ideal reservoir hosts permit parasites to survive in the peripheral circulatory system (or other suitable tissues) in sufficient numbers for sufficiently long time periods to be an effective source for vector infection. Asexual parasites such as viruses and bacteria typically produce intensive infections that produce large numbers of infectious organisms for relatively short periods during which the host either succumbs to infection or develops protective immunity. In the case of EEE virus, for example, one ml of blood from an infected bird may contain as many as 10^{10} virus particles during both day and night for a two- to five-day period; birds that survive such infections typically develop long-lasting, protective immunity. In contrast, highly evolved parasites produce comparatively few individuals during a longer period. *Wuchereria bancrofti*, for example, maintains comparatively few microfilaria in the blood stream (usually <10 microfilaria per mm^3 of blood), which circulate most abundantly in the peripheral blood during periods of the day when the mosquito vectors blood-feed. However, because both the worms and the human host are long-lived, transmission is enhanced by repeated exposure rather than by an intense parasite presentation over a period of a few days. Infection with >100 microfilaria per female mosquito may prove fatal for the vector; therefore, in this case, limiting the number of parasites that infect the vector actually may increase the probability of transmission.

Transmissibility Suitable numbers of susceptible vertebrate hosts must be available to become infected and thereby maintain the parasite population. Transmission rates typically decrease concurrently with a reduction in the number of susceptible (i.e., nonimmune) individuals remaining in the host population. The **epidemic threshold** refers to the number of susceptible individuals required for epidemic transmission to occur, whereas the **endemic threshold** refers to the number of susceptibles required for parasite persistence. These numeric thresholds vary depending on the immunology and dynamics of infection in the host population. Therefore, suitable hosts must be abundant and either not develop lasting immunity or have a relatively rapid reproductive rate to ensure the rapid recruitment of susceptibles into the population. In the case of malaria, for example, the parasite elicits an immune response that rarely is completely protective, and the host remains parasitemic and susceptible to reinfection. In contrast, encephalitis virus infections of passerine birds typically produce life-long protection, but bird life expectancy is short and the population replacement rate is rapid, thereby ensuring the rapid renewal of susceptible hosts.

The Arthropod Vector

Literally, a **vector** is a "carrier" of a parasite from one host to another. An effective vector generally exhibits characteristics that compliment those listed earlier for the vertebrate hosts and include host selection, infection, and transmission.

Host Selection A suitable vector must be abundant and feed frequently upon infective vertebrate hosts during periods when stages of the parasite are circulating in the peripheral blood or other tissues accessible to the vector. Host-seeking or biting activity during the wrong time or at the wrong place on the wrong host will reduce contact with infective hosts and reduce the efficiency or force of transmission. Patterns of host-selection determine the types of parasites to which vectors are exposed. **Anthropophagic** vectors feed selectively on humans and are important in the transmission of human parasites. Anthropophagic vectors that readily enter houses to feed on humans or to rest on the interior surfaces are termed **endophilic** (literally, "inside loving"). Vectors that rarely enter houses are termed **exophilic** ("outside loving"). **Zoophagic** vectors feed primarily on vertebrates other than humans. **Mammalophagic** vectors blood-feed primarily on mammals and are important in the maintenance of mammalian parasites, whereas **ornithophagic** vectors feed primarily on avian hosts and are important in the maintenance of avian parasites. There is a distinction between vectors attracted to a host and those which successfully blood-feed on the host. Mammalophagic vectors therefore represent the successful subset of blood-feeding vectors among those mammalophilic vectors that are attracted to mammalian hosts.

Infection The vector must be susceptible to infection and survive long enough for the parasite to complete multiplication or development. Not all arthropods that ingest parasites support parasite maturation, dissemination, and transmission. For example, the mosquito *Cx. quinquefasciatus* occasionally becomes infected with western equine encephalomyelitis (WEE) virus; however, because this virus rarely escapes the midgut, this species rarely transmits WEE virus. Some arthropods are susceptible to infection under laboratory conditions, but in nature seldom feed on infected vertebrate hosts or survive long enough to allow parasite development. *Aedes aegypti*, for example, readily becomes infected with the filarial worm, *Brugia malayi*, in the laboratory and has been used as a model research system, but this mosquito is not considered a vector in nature.

The **transmission rate** is the number of new infections per unit of time and is dependent upon the rate of parasite development to the infective stage and the frequency of blood feeding by the vector. Because arthropod vectors are poikilothermic and contact their homeothermic vertebrate hosts intermittently, parasite transmission rates frequently are dependent upon ambient temperature. Therefore, transmission rates for many parasites are more rapid at tropical than temperate latitudes and at temperate latitudes progress most rapidly during summer. The frequency of host contact and therefore the transmission rate also depends upon the life history of the vector. For example, epidemics of malaria in the tropics transmitted by a mosquito that feeds at two-day intervals progress faster than epidemics of Lyme disease at temperate latitudes where the spirochetes are transmitted to humans principally by the nymphal stage of a hard tick vector that may have one generation and one blood meal per life stage per year.

Transmission Once infected, the vector must exhibit a high probability of refeeding on one or more susceptible hosts to ensure the transmission of the parasite. Diversion of vectors to nonsusceptible or dead-end hosts dampens transmission effectiveness. The term **zooprophylaxis** (literally, "animal protection") arose to describe the diversion of host-seeking *Anopheles* infected with human malaria parasites from humans to cattle, a dead-end host for the malaria parasites. With zooprophylaxis the dead-end host typically exhibits natural immunity in which host tissues are unacceptable to parasites and do not permit growth or reproduction. Alternatively, transmission to a dead-end host may result in serious illness, because the host–parasite relationship has not coevolved to the point of tolerance by the dead-end host. WEE virus, for example, can cause serious illness in humans, which are considered to be dead-end hosts, because they rarely produce a viremia sufficient to infect mosquitoes. In zoonoses such as West Nile virus with complex transmission cycles, *Culex* vector blood-feeding on a variety of avian hosts in diverse or complex ecosystems may dampen or dilute amplification transmission, whereas feeding on a reduced variety of highly or moderately competent hosts may enable transmission in simple suburban/urban ecosystems.

MODES OF TRANSMISSION

The transmission of parasites by vectors may be vertical or horizontal. **Vertical** transmission is the passage of parasites directly to subsequent life stages or generations within vector populations. **Horizontal** transmission describes the passage of parasites between vector and vertebrate hosts.

Vertical Transmission

Three types of vertical transmission are possible within vector populations: transstadial, transgenerational, and venereal.

Transstadial transmission is the sequential passage of parasites acquired during one life stage or stadium through the molt to the next stage(s) or stadium. Transstadial transmission is essential for the survival of parasites transmitted by mites and hard ticks that blood-feed once during each life stage and die after oviposition. Lyme disease spirochetes, for example, that are acquired by larval ticks must be passed transstadially to the nymphal stage before they are transmitted to vertebrates.

Transgenerational transmission is defined as the vertical passage of parasites by an infected parent to its offspring (the next generation). Some parasites may be maintained transgenerationally for multiple generations, whereas others require horizontal transmission for amplification. Transgenerational transmission normally occurs transovarially (through the ovary) after the parasites infect the ovarian germinal tissue and then transstadially to the next reproductive or blood-feeding stage. In true transovarial transmission most of the progeny are infected. Other parasites do not actually infect the ovary and, although they are passed on to their progeny, transmission is not truly transovarial. This situation is usually less efficient and only a small percentage of the progeny are infected. Transgenerational transmission in vectors such as mosquitoes also must include transstadial transmission, because the immature life stages do not blood-feed.

Venereal transmission is the passage of parasites between male and female vectors during mating and is relatively rare. Venereal transmission usually is limited to transovarially infected males who infect females during insemination, which, in turn, infect their progeny during fertilization.

La Crosse virus (Figure 3.2) is an example of a vertically maintained parasite where the arthropod host serves as the reservoir. This virus is maintained vertically by transgenerational transmission within clones of infected *Aedes (Ochlerotatus) triseriatus* mosquitoes and is amplified by horizontal transmission among squirrels and chipmunks. Because this temperate

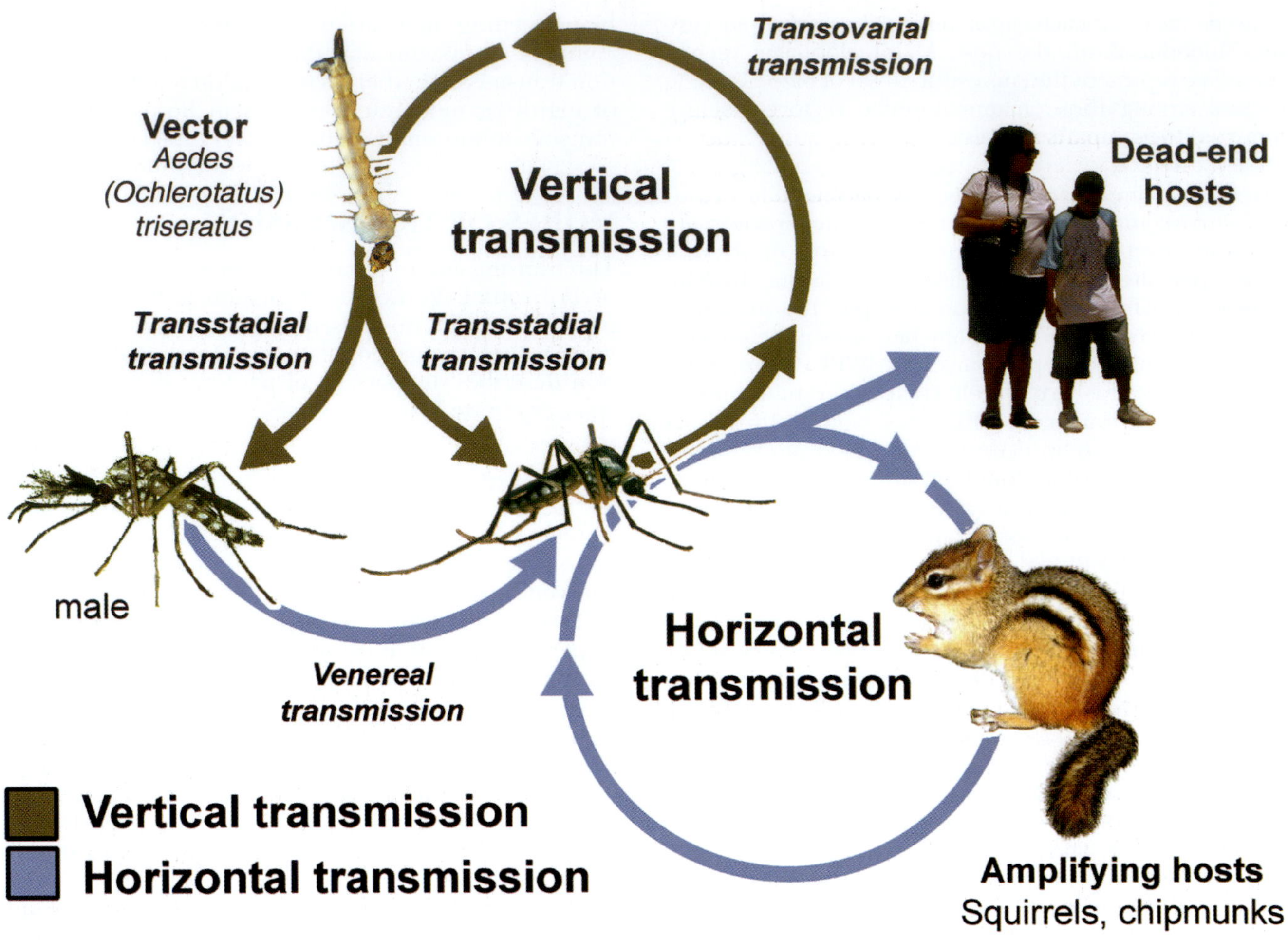

Figure 3.2 Modes of transmission of La Crosse encephalitis virus, involving both vertical transmission among mosquitoes (including transovarial, transstadial, and venereal transmission) and horizontal transmission to other host species. Humans are dead-end hosts for this virus. (Illustration by Nathan D. Burkett-Cadena)

mosquito rarely has more than two generations per year, La Crosse virus spends long periods in infected vectors and relatively short periods in infected vertebrate hosts. Females infected vertically or horizontally transmit their infection transovarially to first-instar larvae. These larvae transmit virus transstadially through the four larval stadia and the pupal stage to the adults. These transgenerationally infected females then take a blood meal and oviposit infected eggs, often in the same tree hole from which they emerged. Some blood meal hosts such as chipmunks become highly viremic and amplify the number of infected *Ae. triseriatus* when uninfected females feeding on these rodents become infected. Venereal transmission of virus from transgenerationally infected males to uninfected females has been demonstrated in the laboratory and may serve to establish new clones of infected females in nature.

Horizontal Transmission

Horizontal transmission is essential for the maintenance of almost all vector-borne parasites and is accomplished by either anterior (biting) or posterior (defecation) routes. **Anterior-station** transmission occurs when parasites are liberated from the mouth parts or salivary glands during blood feeding (e.g., malaria parasites, encephalitis viruses, filarial worms). **Posterior-station** (or **stercorarian**) transmission occurs when parasites remain within the gut and are transmitted via contaminated feces. The trypanosome that causes Chagas' disease, for example, develops to the infective stage within the hindgut and is discharged onto the host skin when the triatomid vector defecates during feeding. Irritation resulting from salivary proteins introduced into the host during feeding causes the host to scratch the bite and rub the parasite into the wound. Louse-borne relapsing fever and typhus fever rickettsia also employ posterior-station modes of transmission.

There are four types of horizontal transmission, depending upon the role of the arthropod in the life cycle of the parasite: mechanical, multiplicative, developmental, and cyclodevelopmental.

Mechanical transmission occurs when the parasite is transmitted among vertebrate hosts without amplification or development within the vector, usually by contaminated mouthparts. Arthropods that are associated intimately with their vertebrate hosts and feed at frequent intervals have a greater probability of transmitting parasites mechanically. The role of the arthropod may be little more than an extension of contact transmission between vertebrate hosts. Eye gnats, for example, have rasping, sponging mouth parts and feed repeatedly at the mucous membranes of a variety of vertebrate hosts, making them an effective mechanical vector of the bacteria and viruses that cause conjunctivitis or "pink-eye." Pink-eye also may be transmitted from infected to susceptible hosts by contact. Mechanical transmission also may be accomplished by contaminated mouthparts if the vector is interrupted while blood-feeding and then immediately refeeds on a second host in an attempt to complete the blood meal.

In contrast, **nonviremic** transmission is a special form of nonpropagative transmission where the infectious vectors are able to transmit viruses through the host directly to concurrently feeding uninfected vectors without or prior to host infection. With ticks, this apparently occurs through the skin in viruses such as in Russian spring summer encephalitis virus. With mosquitoes, this has been demonstrated experimentally for West Nile virus when infectious *Culex* vectors inject virus directly into the circulatory system of small vertebrate hosts such as mice or House finches.

Multiplicative (or **propagative**) transmission occurs when the parasite multiplies asexually within the vector and is transmitted only after a suitable incubation period is completed. In this case, the parasite does not undergo metamorphosis (or development) and the form transmitted is indistinguishable from the form ingested with the blood meal. Arboviruses such as St Louis encephalitis (SLE) virus for example, are not transmitted until the virus replicates within and passes through the mosquito vector midgut, is disseminated throughout the hemocoel, and enters and replicates within the salivary glands. The number, but not the form of the virus, changes during this process. The number of parasites expectorated by the vector during transmission frequently may be much less than the number ingested with the blood meal.

Developmental transmission occurs when the parasite develops and metamorphoses, but does not multiply, within the vector. Microfilariae of *Wuchereria bancrofti*, for example, are ingested with the blood meal, penetrate the mosquito gut, move to the flight muscles where they molt twice, and then move to the mouthparts where they remain until deposited during blood feeding. These filarial worms do not reproduce asexually within the mosquito vector; the number of worms available for transmission is always equal to or less than the number ingested.

Cyclodevelopmental transmission occurs when the parasite metamorphoses and reproduces asexually within the arthropod vector. In the life cycle of the malaria parasite, for example, gametocytes that are ingested with the blood meal unite within the mosquito gut and then change to an invasive form (ookinete) that penetrates the gut and forms an asexually reproducing stage (oocyst) on the outside of the gut wall. Following asexual reproduction, this stage ruptures and liberates infective forms (sporozoites) that move to the salivary glands from where they are transmitted during the next blood meal. Malaria parasites also reproduce asexually with the liver and blood cells of the intermediate human host.

The **extrinsic** incubation period is the time interval between vector infection and parasite transmission or when the parasite is away from the vertebrate host. The **intrinsic** incubation period is the time from infection to the onset of symptoms (or infectiousness) in the vertebrate host. Repeated lag periods of consistent duration between clusters of new cases at the onset of epidemics were first noticed by early epidemiologists who coined the term "extrinsic incubation." These intervals actually represent the combined duration of extrinsic incubation in the vector and intrinsic incubation periods.

The duration of the extrinsic incubation period is typically temperature dependent. The rate of parasite development normally increases as a linear degree-day function of ambient temperature between upper and lower thresholds. After being ingested by the mosquito vector, arboviruses such as SLE virus, for example, must enter and multiply in cells of the midgut, escape the gut, be disseminated throughout the hemocoel, and then infect the salivary glands, after which the virus may be transmitted by bite. Under hot summer conditions, this process may be completed within six or seven days, and the vector mosquito, *Culex tarsalis*, is capable of transmitting virus during the next blood meal. In contrast, under cooler spring conditions transmission may be delayed for more than two weeks, until the third blood meal. Some parasites may alter vector behavior by increasing the frequency of vector blood-feeding and thereby enhance transmission. The plague bacillus, for example, remains within and eventually blocks the gut of the flea vector, *Xenopsylla cheopis*. Regurgitation occurs during blood-feeding, causing vector starvation, more frequent blood-feeding attempts, and therefore transmission at progressively more closely spaced intervals before the vector succumbs to starvation.

TRANSMISSION CYCLES

Transmission cycles vary considerably depending upon their complexity and the role of humans as hosts for the parasite. A vector-borne **anthroponosis** is a disease resulting from a parasite that normally infects only humans and one or more anthropophagic vectors (Figure 3.1). Malaria, dengue, some forms of filariasis, and louse-borne typhus are examples of anthroponoses. Humans serve as reservoir hosts for these parasites, which may persist for years as chronic infections. Vectors of anthroponoses selectively blood-feed upon humans and are associated with domestic or peridomestic environments. Widespread transmission of an anthroponosis with an increase in the number of diagnosed human cases during a specified period of time is called an **epidemic**. When human cases reappear consistently in time and space, transmission is said to be **endemic**.

Zoonoses are diseases of animals that occasionally infect humans. Likewise, **ornithonoses** are diseases of wild birds that are transmitted occasionally to humans.

In most vector-borne zoonoses, humans are not an essential component of the transmission cycle, but rather become infected when bitten by a vector that fed previously on an infected animal host. Although humans frequently become ill, they rarely circulate sufficient numbers of parasites to infect vectors and thus are termed dead-end hosts. The **enzootic** transmission cycle is the basic or primary animal cycle (literally "in animals"). When levels of enzootic transmission escalate, transmission may become **epizootic** (an outbreak of disease among animals). Transmission from the enzootic cycle to dead-end hosts is called **tangential** transmission (i.e., at a tangent from the basic transmission cycle). Often different vectors are responsible for enzootic, epizootic, and tangential transmission. **Bridge vectors** transmit parasites tangentially between different enzootic and dead-end host species. Human involvement in zoonoses may depend on the establishment of a secondary **amplification cycle** among vertebrate hosts inhabiting the peridomestic environment.

Lyme disease, caused by infection with the spirochete *Borrelia burgdorferi*, is an example of a tick-borne zoonosis (Figure 3.3) that is now epidemic in eastern North America. If left untreated, the spirchete causes serious chronic disease in humans presenting a variety of symptoms that may include meningoencephalitis, myocarditis, frank arthritis, and fatigue. The vectors are principally ticks in the *Ixodes ricinus* complex including *scapularis* in eastern and *pacificus* in western North America. Hard ticks require blood meals for both molting and reproduction. In the eastern United States, larval ticks acquire *Borrelia* blood-feeding on mice during summer that have infectious spirochetemias (elevated numbers of spirochetes in the blood), maintain infections during winter, and then pass this infection transstadially to the nymphal stage the following spring (Figure 3.3). Nymphal ticks subsequently transmit their infection to a variety of hosts including rodents, squirrels, lizards, birds, and humans, but, if uninfected, may acquire *Borrelia* during blood feeding. Lizards, some birds, and humans are refractory or dead-end hosts and their infection may actually reduce the rate of *Borrelia* amplification. Infected nymphs also pass their infection transstadially to the adult stage and adults may transmit to large mammals during blood-feeding, although deer seem to be refractory to infection.

There is minimal evidence to support vertical transmission of *Borrelia* to the eggs, and therefore larvae and mice seem to be the reservoirs of infection. The changing landscape and reforestation of eastern North America, accompanied by large increases in Whitetail deer and *Peromyscus* mouse populations, and the construction of housing adjacent to or within wooded areas have combined to create perfect epidemiological situations for the large-scale outbreak of Lyme disease. Infected immature tick populations residing suburban gardens and lawns greatly increase the risk of transmission to humans.

INTERSEASONAL MAINTENANCE

An important aspect of the ecology of vector-borne parasites is the mechanism(s) by which they persist between transmission seasons or outbreaks. Parasite transmission typically is most efficient when weather conditions are suitable for vector activity and

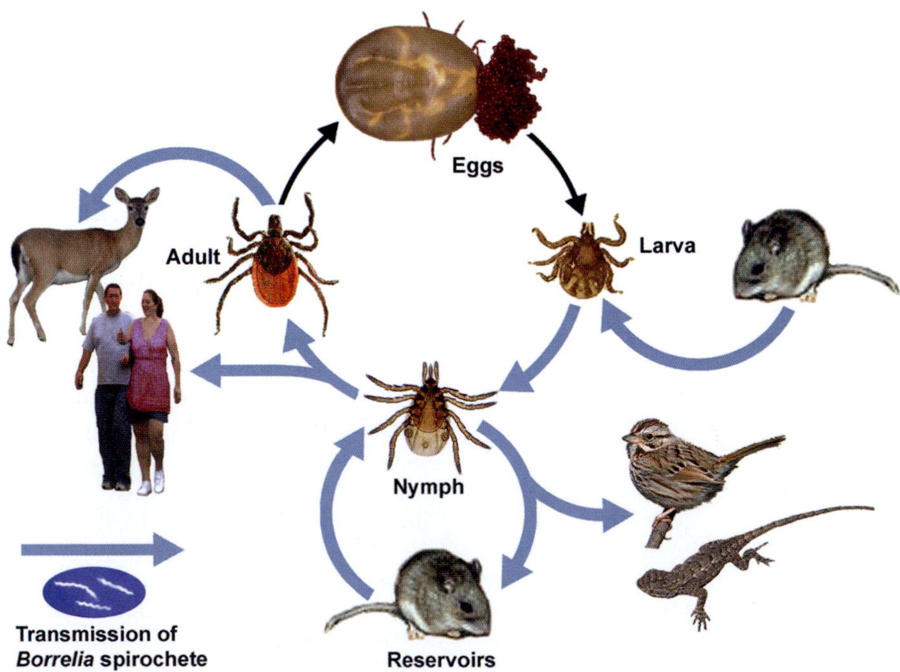

Figure 3.3 Components of the transmission cycle of a zoonosis, such as Lyme disease. In this case the pathogen, a spirochete of the genus *Borrelia*, is maintained by transmission among nonhuman hosts (e.g., woodland mice), but also can be transmitted to humans, deer, and other vertebrates. (Illustration by Nathan D. Burkett-Cadena)

population growth, and warm temperatures expedite parasite replication and dissemination within the vector. In temperate latitudes, overwintering of parasites becomes problematic when vertebrate or arthropod hosts either enter winter dormancy or migrate. Similar problems face tropical parasites, when transmission is interrupted by prolonged dry or wet seasons. The apparent seasonality that is characteristic of most vector-borne parasites may be due to either the periodic amplification of a constantly present parasite or to the consistent reintroduction of parasites following focal extinction.

There are several mechanisms of parasite maintenance during or reintroduction after periods of unfavorable weather.

Continued Transmission by Vectors

During periods of unfavorable weather, vectors may remain active and continue to transmit parasites, although transmission rates may be slowed by cold temperature or low vector abundance. In temperate latitudes with cold winters, transmission may continue at a slow rate, because the frequency of blood feeding and rate of parasite maturation in the vector is diminished. In tropical latitudes, widespread transmission may be terminated during extended dry seasons that reduce vector abundance and survival. In both instances, transmission may be restricted spatially and involve only a small portion of the vertebrate host population. Human infections during adverse periods may be highly clumped and restricted to members of the same household, because vector dispersal is limited at this time.

Infected Vectors

Many vectors enter a state of dormancy as non-blood-feeding immatures or adults. Vertically infected vectors typically remain infected for life and therefore may maintain parasites during periods when horizontal transmission is interrupted. California encephalitis virus, for example, is maintained during winter and drought periods within the transovarially infected eggs of its vector, *Aedes (Ochlerotatus) melanimon*. Infected eggs of this floodwater mosquito may remain dormant and infected for up to several years and are able to withstand winter cold, summer heat, and extended dry periods. Inundation of eggs during spring or summer produces broods of adult mosquitoes that are infected at emergence. Tick-borne parasites, such as Lyme disease, persist through winter within infected immature stages of the tick vector. Similarly, vectors that inhabit the nests of migratory hosts such as cliff swallows often remain alive and infected for extended periods until their hosts return.

Infected Vertebrate Hosts

Parasite maintenance may be accomplished by infected reservoir hosts that either continue to produce stages infective for vectors or harbor inactive stages of the parasite that relapse or recrudesce during the season when vectors are blood-feeding. Adult filarial worms, for example, continue to produce microfilariae throughout their lifetime, regardless of the population dynamics or seasonality of the mosquito vector. In contrast, some Korean strains of *Plasmodium vivax* malaria overwinter as dormant stages in the liver of the human host and then relapse in spring concurrent with the termination of diapause by the mosquito vector(s).

Alternatively, parasites may become regionally extinct during unfavorable weather periods and then are reintroduced from distant refugia. Two possible mechanisms may allow the reintroduction of parasite:

Migratory vertebrate hosts. Many bird species overwinter in the tropics and return to temperate or subarctic breeding sites each spring, potentially bringing with them infections acquired at tropical or southern latitudes. It also is possible that the stress of long flights and ensuing reproduction triggers relapses of chronic infections. In addition, many large herbivores migrate annually between summer (or wet) and winter (or dry) pastures bringing with them an array of parasites. Rapid long-range human or commercial transportation is another possible mode for vector and parasite introduction. The seasonal transport of agricultural products and the movements of migratory agricultural workers may result in the appearance of seasonality.

Weather fronts. Infected vectors may be carried long distances by prevailing weather fronts. Consistent weather patterns such as the sweep of the southeastern monsoon from the India Ocean across the Indian subcontinent may passively transport infected vectors over hundreds of kilometers. The onset of Western equine encephalitis activity in the north-central United States and Canada has been attributed to the passive dispersal of infected mosquitoes by storm fronts as has the repeated introduction of Japanese encephalitis virus into northern Australia.

VECTOR INCRIMINATION

To understand the epidemiology and control vector-borne disease, it is essential to establish which arthropod(s) is the primary vector(s) responsible for parasite transmission. Partial or incomplete vector incrimination has resulted in the misdirection of control efforts at arthropod species that do not play a substantial role in either enzootic maintenance or epidemic transmission. Vector incrimination combines field and laboratory data that measure field infection rates, vector competence, and vectorial capacity.

Infection Rates

The collection of infected arthropods in nature is an important first step in identifying potential vectors, because it indicates that the candidate species feeds on vertebrate hosts carrying the parasite. Infection data may be expressed as a percentage at one point in time or as an **infection prevalence** (i.e., number of vectors infected/number examined × 100). The more commonly employed infection rate refers to infection incidence, and includes the number infected per unit of population over a specified time period. When the infection prevalence is low and arthropods are tested in groups or pools, data are referred to as a minimum infection rate (number of pools of vectors positive/total specimens tested/unit of time × 100 or 1000). Minimum infection rates are relative values with ranges delineated by pool size. For example, minimum infection rates of vectors tested in pools consisting of 50 individuals each must range from 0 to 20 per 1000 females tested.

It is important to distinguish between infected hosts harboring a parasite and infective hosts capable of transmission. In developmental and cyclodevelopmental vectors, the infective stages may be distinguished by location in the vector, morphology, or biochemical properties. Distinguishing infective from noninfective vectors is difficult, if not impossible, with viral or bacterial infections, because the parasite form does not change, although the location of recovery within the vector can be informative. The ability to transmit may be implied by testing selective body parts such as the cephalothorax, salivary glands, or head. With some tick pathogens, however, parasite movement to the mouthparts does not occur until several hours after attachment. As mentioned previously, the transmission rate is the number of new infections per time period. The annual parasite incidence frequently used in malaria control programs is the number of new infections detected per year per 1000 population (or other unit of population size).

The **entomological inoculation rate** is the number of potentially infective bites per unit of time. This frequently is determined from the human or host biting rate and the proportion of vectors that are infective and is calculated as bites per human per time period × infectivity prevalence. In malaria, for example, this is the number of *Anopheles* positive for sporozoites per human time period.

Vector Competence

Vector competence is defined as the susceptibility of an arthropod species to infection with a parasite and its ability to transmit this acquired infection. Vector competence typically is determined experimentally by feeding the candidate arthropod vector on a vertebrate host circulating the infective stage of the parasite, incubating the blood-fed arthropod under suitable ambient conditions, refeeding the arthropod on a noninfected susceptible vertebrate host, and then examining this host to determine if it became infected. Because it often is difficult to maintain natural vertebrate hosts in the laboratory and control the concentration of parasites in the peripheral circulatory system, laboratory hosts or artificial feeding systems frequently are used to expose the vector to the parasites. Susceptibility to infection may be expressed as the percentage of arthropods that became infected among those blood feeding. When the arthropod is fed on a range of parasite concentrations, susceptibility may be expressed as the **median infectious dose** or the concentration required to infect 50% of blood-fed arthropods (**ID_{50}**). The ability to transmit may be expressed either as the percentage of blood fed or infected females that transmitted or the percentage of recipient hosts that became infected.

Failure of a blood-fed arthropod to become infected with or transmit a parasite may be attributed to the presence of one or more barriers to infection. For most parasites, the arthropod midgut provides the most important barrier to infection. Often parasites will grow in a nonvector species if they are inoculated into the hemocoel, thereby bypassing this gut barrier. After penetrating and escaping from the midgut, the parasite then must multiply or mature and be disseminated to the salivary glands or mouthparts. Arthropod cellular or humoral immunity may clear the infection at this point, creating a dissemination barrier. Even after dissemination to the salivary glands, the parasite may not be able to infect or be transmitted from the salivary glands due to the presence of salivary gland infection or salivary gland escape barriers, respectively. For parasites transmitted at the posterior station, vector competence may be expressed as the percentage of infected vectors passing infective stages of the parasite in their feces.

Vectorial Capacity

The concept of **vectorial capacity** summarizes quantitatively the basic ecological attributes of the vector relative to parasite transmission. Although developed for the mosquito vectors of malaria parasites and most easily applied to anthroponoses, the components of the model provide a framework to conceptualize how the ecological components of the transmission cycle of many vector-borne parasites interact.

Vectorial capacity is expressed by the formula:

$$C = ma^2(P^n)V/(-\ln P)$$

where C = vectorial capacity as new infections per infection per day, ma = bites per human per day, a = human biting habit, P = probability of daily survival, n = extrinsic incubation period in days, and V = vector competence or innate transmission efficiency.

The **biting rate**, *ma*, frequently is estimated by collecting vectors as they attempt to blood-feed and is expressed as bites per human per day or night (e.g., 10 mosquitoes per human per night). The human biting habit, *a*, combines vector feeding frequency and host selection. **Feeding frequency** is the length of time between blood meals and frequently is expressed as the inverse of the length of the gonotrophic cycle. Host selection patterns are determined by testing blood-fed vectors to determine what percentage fed on humans or the primary reservoir. Therefore, if the blood-feeding frequency is two days and 50% of host-seeking vectors feed on humans, a = (1/2 days) × (0.5) = 0.25. In this example, ma^2 = 10 bites/human/night × 0.25 = 2.5. a is repeated, because infected vectors must refeed to transmit.

The probability of the vector surviving through the extrinsic incubation period of the parasite, P^n, requires information on the probability of vector survival, P, and the duration of the extrinsic incubation period, n. P is estimated either vertically by age-grading the vector population, or horizontally by marking cohorts and monitoring their death rate over time. In Diptera, P may be estimated vertically from the parity rate (proportion of parous females/number examined). In practice, $P =$ (parity rate)$^{1/g}$, where g is the length of the gonotrophic cycle. The extrinsic incubation period may be estimated from ambient temperature from data gathered during vector competence experiments by testing the time from infection to transmission for infected vectors incubated at different temperatures. Continuing our example, if $P = 0.8$ and $n = 10$ days, then the duration of infective life, $P^n/(-\ln P) = 0.8^{10}/(-\ln 0.8) = 0.48$. In addition it is useful to also account for vector competence, V. For this example, we will assume that 90% of vectors become infected and 90% of infected females are capable of transmission, so $V = 0.9 \times 0.9 = 0.81$. Therefore, vectorial capacity, $C = 2.5 \times 0.48 \times 0.81$ or 0.97 parasite transmissions per infective host per day.

SURVEILLANCE

The number of cases of most vector-borne diseases typically varies over both time and space. Surveillance (from the French "watching over") programs typically monitor diseases to measure their impact on public or veterinary health. Information on the number of cases can be gathered from morbidity and mortality records maintained by state or national governmental agencies for the human population. **Morbidity** data are records of illness, whereas **mortality** data are records of the cause of death. These data vary greatly in their quality and timeliness, depending upon the accuracy of determining the cause of illness or death and the rapidity of reporting. In the United States, the occurrence of confirmed cases of many vector-borne diseases including yellow fever, plague, malaria, and encephalitis by law must be reported to municipal health authorities. However, infections with many arthropod-borne parasites such as Lyme disease and the mosquito-borne encephalitides frequently are asymptomatic or present variable clinical symptoms and therefore remain largely undiagnosed and under-reported.

The frequency of case detection and accuracy of reporting systems are dependent on the type of surveillance employed and the ability of the medical or veterinary community to recognize suggestive symptoms and request appropriate confirmatory laboratory tests. In addition, some laboratory tests vary in their specificity and sensitivity complicating the interpretation of laboratory results. Cases may be classified as suspect or presumptive based on the physician's clinical diagnosis, or confirmed based on a diagnostic rise in specific antibodies or the direct observation (or isolation) of the parasite from the case. Surveillance for clinical cases may be active or passive.

Active surveillance involves active case detection, in which health workers visit communities, seek out, and test suspect cases that fit a predetermined case definition. In malaria control programs, for example, a field worker visits every household biweekly or monthly and collects blood films from all persons with a current or recent fever. Fever patients are treated with antimalarial drugs presumptively, and these suspect cases are confirmed by detection of malaria parasites in a blood smear. Confirmed cases are revisited and additional medication administered, if necessary. This surveillance provides population infection rates regardless of case classification criteria.

Most surveillance programs rely on passive surveillance, which utilizes passive case detection to identify clinical human or veterinary cases. In this system individuals seeking medical attention at primary health care organizations such as physicians' offices, hospitals, and clinics are diagnosed by an attending physician who requests appropriate confirmatory laboratory tests. However, because many arthropod-borne diseases present a variety of nonspecific symptoms (e.g., headache, fever, general malaise, arthralgia), cases frequently may be missed or not specifically diagnosed. In mosquito-borne viral infections the patient often spontaneously recovers, and cases frequently are listed under fevers of unknown origin or aseptic (or viral) meningitis without a specific diagnosis. In a passive case detection system, it is the responsibility of the attending physician to request laboratory confirmation of suspect clinical cases and then to notify the regional public health epidemiologist that a case of a vector-borne disease has been documented.

The reporting system for clinical cases of vector-borne diseases must be evaluated carefully when interpreting surveillance data. This evaluation should take into account the disease, its frequency of producing clinically recognizable symptoms, the official case definition, the sensitivity and specificity of confirmatory laboratory tests, and the type and extent of the reporting system. Usually programs that focus on the surveillance of a specific disease and employ active case detection provide the most reliable epidemiological information. In contrast, broad-based community health care systems that rely on passive case detection typically produce the least reliable information, especially for relatively rare vector-borne diseases with nonspecific symptoms. Unfortunately, in the modern era, the extent of diagnosis and therefore the sensitivity of surveillance frequently are becoming dependent upon the extent of medical insurance coverage and the physician's decision to provide a cost-effective definitive diagnosis.

Diseases that are always present or reappear consistently at a similar level during a specific transmission season are classified as endemic. The number of cases in a population is expressed as incidence or prevalence. Population is defined as the number of individuals at risk from infection in a given geographical area at a given time. Incidence is the number of new cases per unit of population per unit of time. Incidence data is derived from two or more successive samples spaced over time (longitudinal survey). Prevalence is the frequency of both old and new infections among members of a population. Prevalence typically is determined by a single point in time estimate (cross sectional survey) and frequently is expressed as the percentage of the population tested that was found to have been infected.

The level of parasite endemicity in a population may be graded as **hypoendemic** (low), **mesoendemic** (medium), or **hyperendemic** (high), depending upon the incidence of infection or the immune status of the population. In malaria surveys, for example, the percentage of children with palpable spleens and the annual parasite incidence are used to characterize the level of endemicity. In endemic disease, the percentage of individuals with sera positive for IgG class antibodies typically increases as a linear function of age or residence history, whereas in hypoendemic disease with intermittent transmission, this function is disjunct with certain age groups expressing elevated positivity rates. The occurrence of an above-normal number of human infections or cases is termed an epidemic. Health agencies, such as the World Health Organization (WHO), typically monitor incidence data to establish criteria necessary to classify the level of endemicity and to decide when an epidemic is underway. A geographically widespread epidemic on a continental or global scale is called a **pandemic**.

Serological surveys (or serosurveys) are a useful epidemiological tool for determining the cumulative infection experience of a population with one or more parasites, host-related factors affecting the efficiency or risk of transmission, and reinfection rates. When coupled with morbidity data, serosurveys provide information on the ratio of apparent to unapparent infections. Random sampling representatively collects data on the entire population and may provide ecological information retrospectively by analysis of data collected concurrently with each serum sample. This information may assign risk factors for infection such as sex, occupation, and residence history or may help in ascertaining age-related differences in susceptibility to disease. Stratified sampling is not random and targets a specific cohort or subpopulation. Although stratified samples may have greater sensitivity in detecting rare or contiguously distributed parasites, the data is not readily extrapolated to infection or disease trends in the entire population. Repeated serological testing of the same individuals within a population can determine the time and place of infection by determining when individuals first become **seropositive** (i.e., serologically positive with circulating antibodies against a specific parasite). This change from seronegative to seropositive is called a **seroconversion**.

Forecasting the risk of human infection usually is accomplished by monitoring environmental factors, vector abundance, the level of transmission within the primary and/or amplification cycles, and the numbers of human or domestic animal cases. As a general rule, the accuracy of forecasting is related inversely to the time and distance of the predictive parameter from the detection of human cases. Surveillance activities typically include the time series monitoring of environmental conditions, vector abundance, enzootic transmission rates, and clinical cases.

Environmental Conditions

Unusually wet or warm weather may indicate favorable conditions for vector activity or population increases, concurrently increasing the risk of parasite transmission. Parameters frequently monitored include temperature, rainfall, snow pack (predictive of vernal flooding), and agricultural irrigation schedules. Recently, global circulation models and the real time monitoring of ocean temperatures have been used to skillfully predict climate variation. In the Pacific Ocean, for example, warming of temperatures near the South American coast (termed the El Niño–southern oscillation index) can forecast rainfall patterns in North America and Africa. El Niño patterns leading to increased rainfall in east Africa have been used to successfully predict outbreaks of Rift Valley fever virus several months in advance.

Vector Abundance

Standardized vector monitoring at fixed sites and time intervals can be used to compare relative changes in temporal and spatial vector abundance that are useful in detecting an increased risk of parasite transmission. Trap type and placement are critical in effectively sampling vectors such as mosquitoes for surveillance purposes. Devices are deployed in or adjacent to habitats that maximize their effectiveness in collecting the target mosquito species. Sampling in a systematic fashion over time can produce historical data useful in determining anomalous increases or decreases in abundance. Extraordinary increases in vector abundance and survival may accurately forecast increased enzootic transmission and, in turn, epidemics.

Enzootic Transmission Rates

Systematically monitoring the level of parasite infection in vector or vertebrate populations provides direct evidence that the parasite is present and being actively transmitted. The level of enzootic transmission usually is directly predictive of the risk of human or domestic animal involvement. Enzootic transmission activity may be monitored by vector infection rates, vertebrate-host infection rates, sentinel seroconversion rates, and clinical cases.

Vector Infection Rates Sampling vectors and testing them for parasites determines the level of infection in the vector population in various habitats. When vectors are tested individually, prevalence data are expressed as percentages (e.g., 10 females infected per 50 tested is a 20% infection rate). When combined with abundance estimates, infection rates also may be expressed as infected vectors per sampling unit per time interval; 100 bites per human per night × 0.2 infection rate = 20 infective bites per human per night. These data provide an index of the transmission rate or the **entomological inoculation rate**. When infection rates are low, vector populations large, and sampling independent of vector age, vectors usually are tested in lots or pools. It is statistically advantageous to keep pool size constant and therefore the chance of detecting infection the same. Because there may be >1 infected vector per pool, infection rates are expressed as a minimum infection rate = positive pools/total individuals tested × 100 or 1,000. There are several new mathematical approaches to calculating these estimates that rely on maximum likelihood estimation methods.

Vertebrate Host Infection Rates Introduced zoonoses, such as sylvatic plague in North American rodents, frequently produce elevated mortality that may be used to monitor epizootics of these parasites over time and space. Large numbers of dead American crows have been a hallmark of the ongoing West Nile virus epidemic in North America (Figure 3.4E) and counts of reports by the public have been used for surveillance purposes to indicate recent transmission as well as to forecast human risk using predictive spatial models. In contrast, endemic zoonoses rarely result in vertebrate host mortality. Testing reservoir or amplifying hosts for infection is necessary to monitor the level of enzootic parasite transmission. Stratified sampling for these parasites (directly by parasite isolation or indirectly by seroprevalence) usually focuses on the young of the year to determine ongoing or recent transmission. For example, examining nestling birds for viremia can provide information on the level of enzootic encephalitis virus transmission.

Monitoring the incidence of newly infected individuals in a population over time is necessary to detect increased transmission activity. Because many parasites are difficult to detect or are present only for a limited time period, sampling frequently emphasizes the monitoring of seropositivity or seroprevalence. Figure 3.4A, B shows methods of collecting and bleeding free ranging birds to detect infection status by monitoring seroprevalence. Testing for IgM antibody, which rises rapidly after infection, is parasite-specific and decays relatively quickly, can indicate the level of recent infection, whereas monitoring IgG antibody documents the population's historical experience with the parasite. Sampling, marking, releasing, recapturing, and resampling wild animals is most useful in providing information on the time and place of infection in free-roaming animal populations.

Sentinel Seroconversion Rates Sentinels typically are animals that can be monitored over time to quantify the prevalence of a parasite. Trapping wild animals or birds is labor intensive and determining seroprevalence may provide little information on the time and place of infection, especially if the host is a bird species that has a large home range. To circumvent this problem, caged or tethered natural hosts or suitable domestic animals of known infection history are placed in sensitive habitat and repeatedly bled to detect infection. A suitable sentinel should be fed upon frequently by the primary vector species, easy to diagnose when infected, unable to infect additional vectors (i.e., not serve as an amplifying host), not succumb to infection, and be inexpensive to maintain and easy to bleed or

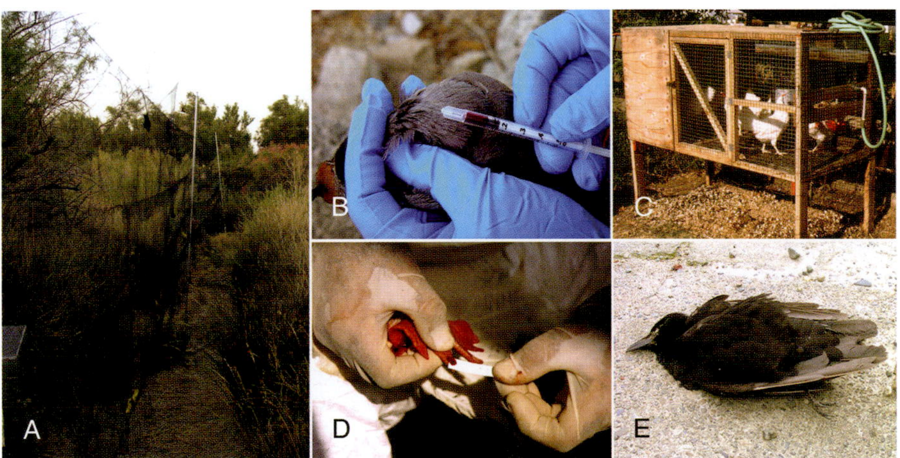

Figure 3.4 Surveillance methods for encephalitis viruses. (A) A mist net used to capture birds to monitor for seroprevalence as an indication of previous viral infection. (B) Taking blood sample from a bird caught in a mist net. (C) A coop used to house sentinel chickens. (D) Collection of blood droplets on a strip of filter paper after pricking the comb of a chicken with a lancet; individual birds can be bled systematically over time to detect seroconversion. (E) A dead American crow; reports of dead birds and confirmation of the cause of death by testing specimens for virus provide a means of monitoring acute infections in vertebrate reservoir populations.

otherwise sample for infection. Chickens, for example, are useful sentinels in mosquito-borne encephalitis virus surveillance programs (Figure 3.4C, D). Flocks of seronegative chickens are placed at farm houses or other suitable localities and housed in standard coops, and then bled weekly or biweekly to determine seroconversions to viruses such as WEE, SLE, or WN viruses. Small blood samples taken on filter paper (Figure 3.4D) can be tested by a semi-automated enzyme-linked immunoassay (ELISA) to detect seroconversions. Because the chickens are confined and the date of seroconversion known, the time and place of infection is determined and the number seroconverting estimates the intensity of transmission.

Clinical Cases

Detecting infection among domestic animals may be an important indication that an epizootic is under way and that the risk of human infection has become elevated. Domestic animals often are more exposed to vectors than humans and therefore provide a more sensitive indication of parasite transmission, unless they are protected by widespread vaccination campaigns. Clinical human cases in rural areas in close association with primary transmission cycles may be predictive of future epidemic transmission in urban settings. For most arboviruses such as West Nile virus, most human infections are unapparent. Symptoms for clinical cases are varied and range from mild fever through high fever with debilitating sequellae. For those patients with a neuroinvasive infection (the virus breaks through the blood–brain barrier), the resulting disease is often severe with devastating sequellae throughout life or the patient succumbs.

Vector-borne diseases frequently affect only a small percentage of the human population or do not impart long-lasting protective immunity, thereby making vector control the intervention method of choice. Control programs attempt to maintain vector abundance below thresholds necessary for the transmission of parasites to man or domestic animals. When these programs fail, personal protection by repellents or insecticide-impregnated clothing, bed nets, or curtains is often the only recourse. Vaccination may be a viable alternative method of control for specific vector-borne diseases, if the vaccine imparts lasting immunity as in the case of yellow fever virus. However, many parasites such as malaria have evolved to the point where infection elicits a weak immune response that provides only short-term and marginal protection. The need for continued revaccination at short intervals severely limits their global usefulness, especially in developing countries where delivery systems are rudimentary. Although breakthroughs in chemotherapy have been useful in case management, it remains the mandate of the medical/veterinary entomologist to devise strategies that combine epidemiological and ecological information to effectively reduce or eliminate the risk of vector-borne diseases.

EMERGING VECTOR-BORNE DISEASES

An expanding and rapidly growing human population, increased and rapid travel, the globalization of commerce, and a variety of anthropogenic factors, including global warming and urbanization, have produced conditions conducive for the emergence and/or resurgence of infectious diseases, including several transmitted by vectors. By definition an **emerging disease** has shown a significant increase in incidence, severity and/or distribution within recent history and threatens to continue to increase in the future. In many emerging diseases, a cascade of historical events has altered human demography, vector or pathogen distribution, caused anthropogenic environmental change and led to the evolution of enhanced virulence in pathogens. These changes enable the emergence or introduction of new pathogens leading the major human, domestic animal or wildlife health problems.

1. **Demography**. Although the per capita population growth rate has slowed in recent years, the earth's human population continues to increase at an alarming rate. More humans living in closer approximation linked by rapid travel enable the rapid transit of pathogens, especially anthroponoses utilizing humans as a reservoir or amplifying host. Historically, the large scale displacement of ethnic groups has created patterns of frequent travel that may be exploited by pathogens.
2. **Globalization of commerce**. The globalization of commerce, originally by the sailing ships of the colonial European empires, and recently by the rapid international exchange of goods such as in the used tire trade, established conditions suitable for the inadvertent transport of both vectors and the pathogens they transmit. *Aedes aegypti* and *Ae. albopictus* both lay drought resistant eggs in dark areas such as water barrels or tires that collect rainwater, enabling the transport of the immature stages and the circumglobal establishment of both effective arbovirus vectors. Of special concern is the recent wide scale, and often illegal, trade in exotic pets that may bring with them infections that escape into new geographic areas. The arrival of West Nile virus in New York may have due to wild bird importation.
3. **Anthropogenic changes**. A rapidly expanding human population has markedly altered the environment in ways conducive to the increase and spread of vectorborne pathogens.
    a. *Urbanization*. The large scale movement of the rural poor to urban centers to find jobs in developing countries and an unchecked human population growth has caused large scale unplanned urbanization in previously agrarian societies, the overwhelming of municipal infrastructure, and the creation of inadequate housing, a lack of potable water that frequently must be stored for domestic use, and an absence of adequate waste systems. This

has created conditions suitable for the explosion of peridomestic mosquitoes such as *Aedes aegypti* and the *Culex pipiens* complex that preferentially rest in houses and feed on humans. Increase in *Ae. aegypti* populations have been closely linked to the circumglobal resurgence and increased severity of Dengue virus epidemics. In addition, the simplification of species diversity and the increase in peridomestic commensals such as crows and House sparrows has enabled the highly efficient urban transmission of zoonoses such as West Nile virus by *Culex pipiens* complex mosquitoes. Ecosystem simplification by both agriculture and urbanization has made amplification more rapid and transmission more efficient than in complex natural ecosystems, because there is less diversion of pathogens to non-competent vertebrate or invertebrate hosts.

b. *Ecotones.* An ecotone is the transition area between landscapes dominated by differing ecosystems and may appear as a gradual blending or a sharply delineated boundary. The alteration of landscapes and the spatial expansion of human and associated domestic animal populations into or adjacent to natural areas have created suitable disease ecotones enabling the expansion of zoonoses. Movement of upscale housing into wooded areas of the northeast USA, for example, has increased the forest ecotone, expanded browse increasing deer and associated tick populations, and given rise the on-going Lyme disease epidemic.

c. *Climate.* Global warming and associated climate change has contributed to the receptivity of new areas for tropical pathogens. Warming of northern latitudes and higher elevations have resulted in milder and shorter winters and therefore longer and warmer transmission seasons, range expansions by vector species, and outbreaks in highly susceptible populations. Examples include the invasion of Canada and the northern USA by West Nile virus and the African highlands by *falciparum* malaria, both pathogens of tropical African origin now creating outbreaks at new latitudes and elevations, respectively.

5. **Surveillance**. Detection of introduced pathogens provides a challenge for existing health programs and surveillance systems. Syndromic surveillance and associated specific confirmational testing may preclude the identification of new invading pathogens, especially if there are cross reactions in assays with endemic pathogens. Passive case detection systems rely of the ability of health care providers to suspect exotic pathogens, take appropriate samples and send these to appropriate laboratories for testing. Typically, only a few facilities with a wide range of diagnostic capacity are capable of detecting newly invading pathogens.
6. **Pathogen evolution**. Outbreaks frequently are caused by genetic changes in the pathogen that enhances their virulence, fitness and ability to cause disease. The invading genotype of West Nile virus, for example, contains a minor mutation of the helicase gene leading to increased virulence in crows and a subsequent minor change to the new North American genotype enables earlier transmission by *Culex* mosquitoes, permitting efficient virus amplification at northern latitudes.

In summary, Medical Entomologists provide a key element in preventive medicine programs charged with detecting and investigating the emergence of new vectorborne pathogens and developing strategies for their control. Urbanization and the invasion of previous wilderness areas will undoubtedly continue to be accompanied by veterinary and public health problems. Devising methods of vector and parasite containment and the eradication of invasions will provide endless challenges for public health and preventive medicine into the foreseeable future.

REFERENCES AND FURTHER READING

Beaglehole, R., Bonita, R., & Kjellstrom, T. (1993). *Basic Epidemiology*. Geneva: World Health Organization.

Bruce-Chwatt, L. J. (1980). *Essential Malariology*. London: William Heinemann Medical Books Ltd.

Davis, J. R., & Lederberg, J. (2001). *Emerging Infectious Diseases from the Global to the Local Perspective*. Workshop Summary. Washington, DC: National Academies Press.

Garrett-Jones, C. (1970). Problems of epidemiological entomology as applied to malariology. *Miscellaneous Publications of the Entomological Society of America, 7*, 168–178.

Gregg, M. B. (1988). Epidemiological principles applied to arbovirus diseases. In Monath, T. P. (Ed.), *The Arboviruses: Epidemiology and Ecology* (Vol. 1, pp. 292–309). Boca Raton, FL: CRC Press.

Herms, W. B., & James, M. T. (1961). How arthropods cause and carry disease. In Herms, W. B. (Ed.), *Medical Entomology* (5th ed., pp. 15–26). New York: Macmillan & Co.

Last, J. M. (Ed.). (1995). *A Dictionary of Epidemiology* (3th ed.). Oxford University Press.

Jawetz, E., Melnick, J. L., Adleberg, E. A. (1972). Host-parasite relationships. In: *Review of Medical Microbiology* (pp. 128–135). Los Altos, CA: Lange Medical Publications.

Macdonald, G. (1957). *The Epidemiology and Control of Malaria*. Oxford University Press.

Moore, C. G., McLean, R. G., Mitchell, C. J., Nasci, R. S., Tsai, T. F., Calisher, C. H., et al (1993). *Guidelines for Arbovirus Surveillance Programs in the United States*. Ft. Collins, CO: Centers for Disease Control and Prevention, Division of Vector Borne Infectious Diseases, US Dept Hlth Human Svcs.

Rice, P. L., & Pratt, H. D. (1992). Epidemiology and Control of Vectorborne Diseases. U. S. Dept. Hlth, Educ., & Welfare, Publ. Hlth Serv. Pub. No. (HMS0) 72–8245.

Chapter

4 Forensic Entomology

William L. Krinsky

Forensic entomology is the study of insect biology as it relates to societal problems that come to the attention of the legal profession and that often must be resolved by legal proceedings. The term "forensic" (from the Greek *forum*) refers to the public forum or courts of law. Although the forensic cases involving insects that receive the most publicity are those involving unnatural deaths, in which insects may be used to help date the time of death or determine whether a corpse has been moved after death, forensic entomology encompasses many other aspects of entomology.

Because insects and their arthropod relatives are found in every environment inhabited by humans, the forensic entomologist may be consulted about problems that concern structural entomology; stored product entomology; occupational hazards involving arthropods; veterinary and wildlife entomology; as well as those in which arthropods are associated with injuries, death, or criminal activity (medico-legal entomology). Because of the nature of this textbook, the latter aspects of forensic entomology will be emphasized in this chapter. An entomologist who is a specialist in any of these areas, medical or not, may at times be consulted and act in the capacity of a forensic entomologist.

Traditionally, forensic entomology has not been a part of medical or veterinary entomology, but a small specialty of forensic medicine. However, within the last 50 years, some medical entomologists have become specialists in forensic entomology because carrion insects feeding on corpses seen in many forensic cases are the same or similar to species that feed on blood and other tissues of living humans that are typically studied by medical entomologists.

A person trained as a forensic entomologist must obtain broad, intensive training in all aspects of insect biology including anatomy, physiology, morphology, taxonomy, ecology, and life cycles. Such a background is essential because virtually any insect may at some time or place provide useful information for the investigation of a problem requiring legal intervention. Therefore, a forensic entomologist should be trained first as an entomologist and then as a forensic specialist. Because forensic cases require the accurate identification of the arthropods involved as well as a clear understanding of their developmental stages and life cycles, a forensic entomologist must have the knowledge to pursue such analyses and the background to understand the work of other entomological specialists. Familiarity with the species and habits of insects found in a given geographic area is essential for an analysis of a case in that region. A forensic entomologist who provides information in medico-legal cases involving unnatural deaths must have expertise in the taxonomy, developmental biology, and ecology of Diptera.

For many years in the United States, and more recently in Europe, entomological testimony has been admissible in criminal cases. The forensic entomologist must limit his or her analysis and testimony to the entomological evidence and base conclusions solely on observations of insect biology. Other considerations related to suspects, motives, nonarthropod organisms, and other physical evidence are matters for experts in those areas of study. The rationale for the use of insects in forensic investigations is that these organisms can provide more objective evidence than that provided by witness testimony, decompositional changes occurring in corpses greater than 24 to 48 hours old, and other forms of evidence lacking a scientific basis.

HISTORY

The earliest crime in which insects are known to have provided evidence occurred between AD 907 and 960 in China. Flies settling on the head of a dead man indicated where a deadly blow had been made (Greenberg and Kunich, 2002). The first known case in which insects associated a murderer with his victim was described by Sung Tz'u, also in China, in 1247 (McKnight, 1981). A Chinese farmer was murdered by being slashed with a sickle. At an inquest, all the farmers of the village were asked to place their sickles on the ground. Within a short time, flies, apparently attracted to residual blood, landed on only one of the 70 to 80 sickles assembled, incriminating its owner as the murderer.

Most of the history of forensic entomology deals with medico-criminal cases, especially homicides, rather than civil law proceedings such as liability cases in which remuneration is sought for damages to structures, commodities, animals, or humans. The forensic entomologist usually is asked to analyze the arthropods associated with a corpse as a means of approximating the time since death (**postmortem interval**) or often more accurately, the postinfestation interval. The first recognition that insect development on a corpse could be useful in determining the interval from the time of death until discovery of the body occurred in Europe in the mid-nineteenth century, and the first book to describe various arthropods infesting a dead body was published by Mégnin (1894).

One of the most famous English murder cases in which insects were used to help date the time of death involved the investigation of Dr. Buck Ruxton in 1935. The dismembered remains of Mrs. Ruxton and her nursemaid were found in a stream bed. The neat disarticulation of the bodies suggested that the murderer had some anatomical knowledge. Staging of the blow fly maggots (*Calliphora* species) found on the remains indicated that they must have been placed in the ravine 12 to 14 days before discovery, a time consistent with the disappearance of the victims from Dr. Ruxton's house. The fly evidence helped to convict Dr. Ruxton, who was found guilty and hanged.

Recent studies of carcasses and corpses have refined the analysis of the progression of arthropods associated with different stages of decomposition (Nuorteva, 1977; Smith, 1986; Campobasso et al., 2001). In 1990, a step-by-step procedural guide was published as a handbook for medico-criminal forensic entomologists (Catts and Haskell, 1990). Besides murder investigations, medico-criminal entomological proceedings may involve investigations of suicide, abuse or neglect, rape, and illicit drug trafficking (Catts and Goff, 1992; Benecke and Lessig, 2001).

LEGAL CASES INVOLVING LIABILITY

Structural Entomology

Forensic entomologists may be consulted when property evaluated by pest control personnel to be pest-free, either before or after pest control treatment, begins to deteriorate as the result of the activity of wood-boring insects. The forensic entomologist often is asked to evaluate whether inspection of the now damaged property or treatment was carried out properly.

Stored-Products Entomology

Most legal cases involving stored product infestations rely on a forensic entomologist to provide information about the identification, origin, and destructive habits of the infesting arthropods.

Occupational Hazards Associated with Arthropods

In warehouses, factories, arthropod-rearing laboratories, and other facilities where insects and other arthropods or their parts, secretions, or excretions may be present, employees may become sensitized to these organisms or their associated materials and develop dermatological or respiratory problems. The latter may be life threatening. In all these cases, where there is a question of liability, the potentially causative arthropods must be identified and a forensic entomologist may be called upon to explain the relationship between the arthropods and the discomfort experienced by the employees. Specific allergic and toxic problems that have arisen in the past have been associated with wasps (Vespidae) and bees (Apidae) attracted to food storage and processing facilities; paederine staphylinid beetles attracted to lights on oil rigs and buildings; dermestid, silvanid, and curculionid beetle infestations of stored products; and with various insects, such as bees, wasps, moths, mosquitoes, grasshoppers, locusts, cockroaches, bed bugs, and lady beetles in rearing facilities.

Veterinary and Wildlife Entomology

Legal cases involving insects and domestic animals often result from poor animal husbandry practices leading to uncontrolled populations of nuisance or pestiferous insects, such as house flies and stable flies. A forensic entomologist being consulted about such cases should have a thorough knowledge of arthropods of veterinary importance. One type of criminolegal case involving wildlife entomology is poaching. Arthropods feeding on carcasses of animals that have been shot or poisoned can be analyzed to help determine when and where the animals were killed, providing evidence that may be helpful in identifying a person guilty of illegal hunting or criminal mischief. Knowledge of the arthropods that infest living wild animals as well as those species that feed on carrion is necessary to discern which organisms may have been present when the animal was alive as opposed to those that colonized the body after death. Confusing these organisms can lead to erroneous estimates of the postmortem interval.

LEGAL CASES INVOLVING HOMICIDES, SUSPICIOUS AND ACCIDENTAL DEATHS, AND ABUSE AND NEGLECT

Medico-legal entomology is the area of forensic entomology that involves the interpretation of events surrounding various kinds of injuries and deaths in which arthropods are associated premortem or postmortem with a body. Specimens and data collected by a forensic entomologist can be used to help determine the time since death, the geographical location at which death occurred, the season of the year when death occurred, movement or storage of a body following death, specific sites of trauma on a body, time of dismemberment, duration and intervals of submersion of a body, presence of illicit drugs or other pharmaceuticals; and to link suspects with the scene of a crime, or with victims of murder, child neglect, or elderly or sexual abuse (Catts and Goff, 1992; Benecke and Lessig, 2001; Campobasso and Introna, 2001).

Identification of arthropods and knowledge of their defensive, feeding, and reproductive behaviors can help differentiate between pre- and postmortem wounds, and can help to interpret the streaking, tracking, and deposition of blood, tissues, and bodily fluids present at a crime scene. Ant bites and roach feeding sites on corpses have been misinterpreted as premortem chemical burns. Blood regurgitated by flies, and blood spots through which flies and roaches have crawled may confuse blood spatter patterns that can provide evidence of how a victim was assaulted or injured during an attack or fall.

Sudden Death with Arthropod Association

In cases where sudden death has occurred and an arthropod is found associated with the body or in the vicinity of the body, the possibility that death may have been caused by hypersensitivity to an arthropod bite or sting, or as the result of an anaphylactic response to such an injury must be considered. In such cases, the forensic/medical entomologist must identify the arthropod and evaluate its potential to cause such a reaction. A definitive determination of the cause of death will rely on a careful analysis of the available entomological and medical information with particular attention to evidence of allergic or unusual immune responses recorded in the medical history of the deceased.

Automobile-Accident Death

In some cases, a stinging insect (wasp or bee) or insect that resembles such an insect (wasp- or bee-mimicking fly) may be recovered from a vehicle in which a person has died. In these instances, the possibility must be raised that the driver panicked or was distracted by such an insect, resulting in an accident. Determining the responsible party in a fatal or injurious accident will have direct bearing on the legal outcome of such an occurrence.

Arthropods as Signs of Neglect or Abuse, or as Agents of Murder

Several families of flies (Muscidae, Calliphoridae, Sarcophagidae) include carrion-feeding species that have larvae that may also invade living vertebrate tissue causing myiasis. Myiasis seen in patients in hospitals and nursing facilities may indicate poor nursing care or neglect. Surgical dressings or bandages that are not frequently changed, and untreated wounds or bedsores, may attract female flies that oviposit or larviposit at these sites. Similarly, myiasis may occur in cases where flies are attracted to diapers of infants or small children who have not been changed regularly. In any of these situations, myiasis may be cited as a sign of neglect or abuse during legal proceedings. An entomologist in consultation with the medical personnel involved must investigate and determine the identity of the flies, and analyze the circumstances under which the flies gained access to the infested individuals. Testimony by a forensic entomologist is often crucial in providing the background necessary to incriminate the party guilty of neglect or abuse.

Reports of the purposeful use of venomous arthropods as agents of abuse or murder are quite rare. However, individuals are known to have placed infants in rooms with stinging insects, and biting arthropods have been placed in proximity to intended murder victims. As in other cases of abuse or death, the finding of a venomous arthropod at the scene should at least suggest the possibility that the arthropod was directly related to the cause of injury or death.

Illicit Drug Transport, Use, and Overdose

Insects and mites of various kinds have been found associated with *Cannabis sativa* (marijuana or hemp). Identification of the arthropods found within shipments or seizures of marijuana recovered by authorities can sometimes help pinpoint the original source of this drug. Insects recovered have included species of Diptera, Lepidoptera, Coleoptera, and Hymenoptera. Although many of the species found are cosmopolitan stored product pests, species from some marijuana samples are known only from specific geographic regions within Asia and from particular ecological settings. In these cases, arthropods provide evidence that helps incriminate drug traffickers. More than 270 species of mites and insects have been recorded from hemp around the world (Batra, 1976).

Human remains often are found in an advanced state of decay in which the tissues have largely decomposed. In these cases, where there may not be any available material for toxicological testing, the carrion-infesting arthropods or their remains (e.g., fly larval and puparial skins, dermestid exuvuiae, and frass) may provide material that can be assayed for various pharmaceuticals or toxins. Gas and liquid chromatography and mass spectrometry have been used successfully to discern the presence of various barbiturates as well as cocaine, heroin, and other controlled substances, and heavy metals, such as arsenic and mercury, in insect tissues associated with corpses. The results of qualitative analyses of the insects compared in some cases to analyses of tissues from the corpses have been consistent, but quantitative analyses have not been reliable. Until further studies are done, toxicological information from carrion insects as evidence of drug use, overdoses, or poisoning must be used with caution in legal proceedings (Introna et al., 2001).

Suspicious Deaths

Arthropod evidence may be helpful in discerning whether transport of a corpse has occurred from the site where a suicide, abduction, attack, or murder has taken place. Careful attention to the species of arthropods found associated with a body might indicate that the species or types of arthropods are not consistent with the ecological setting where the body was found, or that the species are not known from the geographic area where the body was discovered. Recovery of arthropods and arthropod parts from clothing, debris, and any containers or vehicles in which the body may have been transported can be helpful in discerning the initial location of the body. Arthropod remains found on the outside of a vehicle (e.g., on the windshield, radiator grill, or surface of a vehicle) can prove useful in this regard and may also provide evidence of where a suspect has traveled. In a recent multiple homicide case in California, the prime suspect had rented a car in Ohio and claimed he had never driven it outside of the state. Insects that were recovered from the car radiator and air filter included the remains of a grasshopper, a paper wasp, and two true bugs, all species found only west of Kansas or only in far western states. The insect evidence helped to convict the murderer (Kimsey, 2007).

Arthropod evidence can be used to link a suspect to a victim. Finding the same kind of arthropod at the site of a body as one associated with the clothing or vehicle of a potential suspect has helped to incriminate individuals in several murder cases. As in the case of flies pinpointing the murder weapon (sickle) in thirteenth century China leading to a confession by the owner (see earlier), in recent years, arthropods as diverse as grasshoppers and chigger mites have brought murderers to justice. In the first case, part of a grasshopper leg was found on the corpse. Subsequently, the rest of the same grasshopper (matched by viewing the alignment of the broken leg with the damaged whole insect under a microscope) was found in the cuff of the pants of a potential suspect, placing him at the scene of the crime.

During a murder investigation in Ventura County, California, chiggers (trombiculid mites) provided evidence to help convict a murderer. Investigators at the scene where the victim was found developed very itchy skin rashes, with spots on the ankles, waist, and buttocks. Similar lesions were seen on a suspect in the case. Further analysis and field study indicated that these lesions were caused by the bites of a chigger, *Eutrombicula belkini*, a species found in very limited geographic areas, which included where the murder victim was discovered. The association of the chigger bites on the investigating team and the suspect and the presence of these rare mites at the scene in southern California helped incriminate the suspect. He was found guilty of first-degree murder and sentenced to life imprisonment (Webb et al., 1983).

In some cases, accumulations of maggots or other insects at particular anatomical sites have been used as indicators of trauma at those sites. Finding large numbers of maggots feeding on surfaces of the body other than at the common oviposition sites at body orifices may suggest the presence of wounds. However, insect evidence alone is not sufficient to conclude that such sites attractive to insects were created by trauma to the victim. The insect distribution may only be used to corroborate other evidence, such as pathological or physical evidence of stab wounds, cuts, gunshot wounds, or abrasions.

STAGES OF DECOMPOSITION

A body begins to decay at the moment of death and continues to decompose in a progression that will vary depending upon the temperature and conditions to which the body is exposed. Cold or hot temperatures, low or high humidity, submergence in water or other fluids, burial, and access of insects, scavengers, and other animals to a corpse will affect the degree of decomposition at any given time and the rate of decay. Putrefaction starts at about 50° F and is most active between 70° and 100° F.

Mummification and **adipocere formation** are two somewhat unusual forms of decomposition occurring under specific environmental conditions. Mummification occurs when dehydration of the body begins before putrefactive changes in very dry, hot environments. The decay process is prevented or slowed by temperatures between 100° and 212° F, at which fluids rapidly evaporate and mummification can occur. Mummified bodies have darkened, leathery skin that is not readily attractive to carrion insects. In cases where partially mummified corpses have been found that are infested with fly larvae, it has sometimes been assumed that vertebrate scavengers have disrupted or dismembered the bodies, allowing fly colonization to occur (Westerfield trial, 2002).

Saponification, the conversion of tissues to a yellowish-white waxy substance called adipocere, may occur if a body has been continuously exposed to moisture, either submerged or on a damp substrate. Bodies decay about twice as quickly in air as in water, and about eight times more quickly in air than when buried in earth. Although the decay process (excluding mummification and saponification) is a continuum, for the purpose of comparing different corpses in different environmental situations and as a means of standardizing postmortem analyses, four or five stages of decomposition have been defined that reflect the condition of the body and the arthropods most often found during those periods in the decay process. These stages are characterized by the progression: discoloration, bloating, liquefaction, and skeletonization.

The fresh (first) stage of decay shows little or no signs of decomposition, but external color changes begin to appear. Internal changes are occurring as a result of bacterial and protozoal activity. Bloating occurs with the breakdown of the intestinal tract releasing gas-forming, anaerobic bacteria that cause swelling of the corpse and the odors of decaying tissues. Active decay begins when the gas escapes and the body collapses. Active decomposition results from **autolysis** of tissues following the liberation of enzymes from cells and the action of bacteria and fungi growing on the remains. The skin begins to liquefy and has a darkened appearance and the odor of decay is very strong. Advanced decay (skeletonization) is the period when the body is drying out; some flesh and hair may remain and fermentation occurs in anaerobic pockets of the body. The dry remains stage includes the slow decay of the remaining desiccated tissues, hair, teeth, and bones over the course of months or years. The duration of each of these stages varies with the environmental conditions to which a corpse is exposed. Body temperature can give an indication of the postmortem interval within about the first 24 hours. Some bodies reach a state of putrefaction within 48 to 72 hours, which under other conditions, would require 10 to 14 days to attain. Because of this variability, estimates of the postmortem interval based on the degree of decomposition in cases where a body has been found more than 24 hours since death are fraught with difficulty. Observations of insects, for which developmental times are known for given temperatures, provide more objective evidence for making such estimates.

INSECT SUCCESSION AND POSTMORTEM INTERVAL

The stages of decomposition can also be characterized by the **succession** of insects that inhabit the body at any given stage. Most of the insects of forensic importance in staging the degree of decomposition are species of *Diptera*, *Coleoptera*, and *Lepidoptera* (Table 4.1). Just as geographic and environmental conditions affect the rate of decomposition, the species and numbers of arthropods infesting a corpse also vary with climate and geography. Greater diversity is seen when a corpse is exposed in warm months or rainy seasons than in colder or dry seasons. In addition, any given species may be present at more than one stage of decomposition, depending on the condition of different parts of the body, and the presence of microhabitats on the body suitable for different developmental stages of a particular arthropod species. The rate of decomposition and the progression of different stages are also somewhat dependent on the changes brought about by the feeding and activity of arthropods at any given stage.

Typically, the first arthropods to arrive at a corpse, often within minutes, are the green and blue blow flies (Calliphoridae) (Figures 4.1 and 4.2). Gravid female flies feed on body secretions, especially around the eyes, nose, mouth, exposed anal and urogenital

Table 4.1 Insect Species Most Commonly Associated with Different Stages of Decomposition

Stage of Decomposition	Order	Family	Common Genera and Species
Fresh	Diptera	Calliphoridae (blow flies)	*Phaenicia sericata, P. coeruleiviridis*, Phormia regina, Cochliomyia macellaria*, Calliphora vicina, C. vomitoria*
Bloated	Diptera	Calliphoridae (as above) plus: Sarcophagidae (Flesh flies)	*Sarcophaga haemorrhoidalis, S. bullata**
		Muscidae (House, latrine & dump flies)	*Musca domestica, Fannia scalaris, Hydrotaea leucostoma*
	Coleoptera	Staphylinidae (rove beetles)	*Creophilus maxillosus, Platydracus* spp.
Active Decay	Diptera	Calliphoridae, Sarcophagidae & Muscidae (as above)	
	Coleoptera	Staphylinidae (as above) plus: Silphidae (Carrion beetles)	*Necrophila americana*, Nicrophorus* spp., *Oiceoptoma* spp.
Advanced Decay	Coleoptera	Staphylinidae & Silphidae (as above) plus: Histeridae (Hister beetles)	*Hister* spp., *Saprinus* spp.
	Diptera	Sepsidae (Black Scavenger flies)	*Sepsis* spp.
		Sphaeroceridae (Small Dung flies)	
		Scathophagidae (Dung flies)	*Scathophaga* spp.
		Stratiomyidae (Soldier flies)	*Hermetia illucens*
		Phoridae (Scuttle flies)	
Dry Remains	Diptera	Piophilidae (Skipper flies)	*Piophila casei*
	Coleoptera	Cleridae (Checkered beetles)	*Necrobia rufipes*
		Nitidulidae (Sap beetles)	*Omosita* spp.
		Dermestidae (Larder & Carpet beetles)	*Dermestes* spp., *Anthrenus* spp., *Attagenus* spp.
		Trogidae (Hide beetles)	*Trox* spp.
	Lepidoptera	Pyralidae (Pyralid moths)	*Aglossa* spp.
		Tineidae (Clothes moths)	*Tinea pellionella, Tineola bisselliella, Trichophaga tapetzella*

All familes listed are found worldwide (except Antarctica); species marked with an asterisk (*) are found only in the Western Hemisphere; all others have a wide global distribution.

Figure 4.1 Green blow fly, *Phaenicia sericata* (Calliphoridae), adult. (Photo by Nathan D. Burkett-Cadena)

orifices, or open wounds, and then lay eggs at these moist sites (Figure 4.2). Individual females may deposit as many as 300 eggs. The eggs hatch within about 24 hours under optimal conditions, and the first instar larvae (maggots) begin to feed. The duration of each successive stage of the fly life cycle varies with the species of fly, ambient temperature, and other environmental conditions. The first instar larvae molt into second instars, these will feed and molt, and the third instar larvae will feed for the longest of the three stages, often leaving the corpse within 7 to 10 days (after the eggs were deposited) to seek dry places in which to transform into puparia, and eventually emerge as adult flies (Figure 4.3). Because blow flies are usually the first insects to arrive at a corpse, they often provide the most useful information about the time elapsed since a corpse was present in an environment in which insects had access. The durations of the developmental cycles of the forensically important blow flies have been determined under different laboratory conditions, allowing the estimation of time elapsed from fly arrival until the recovery of a given stage of development on a corpse.

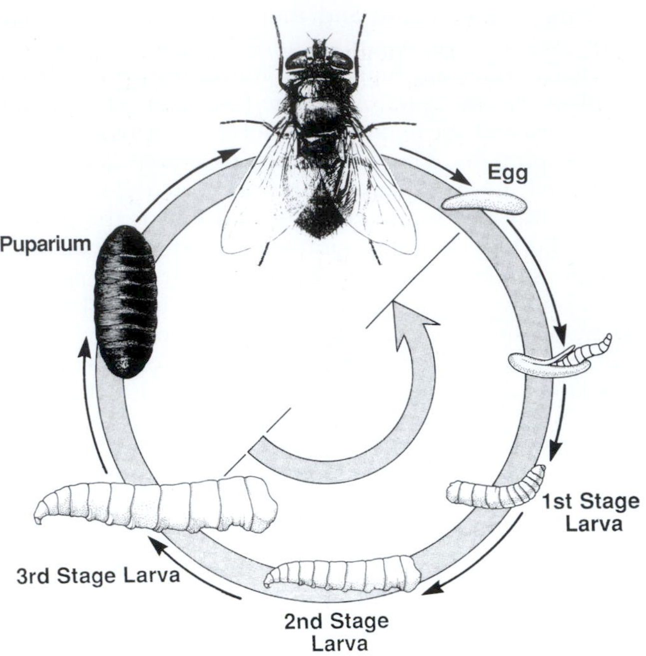

Figure 4.3 Blow fly development. Life cycle showing developmental stages: egg, larva, pupa, and adult. (Original by W. L. Krinsky)

Various other flies (house and latrine flies) that are attracted to feces or urine that may be associated with a decaying corpse arrive within the first few days. Predaceous insects (staphylinid, histerid, and silphid beetles (Figure 4.4); ants, vespid wasps, some muscid larvae, such as *Hydrotaea* – *Ophyra* species) and other arthropods (spiders and mites) that feed on blow fly larvae may appear after the fly eggs hatch. These predaceous species persist, feeding on fly larvae and puparia as well

Figure 4.2 Green blow fly, *Phaenicia sericata* (Calliphoridae), female with freshly deposited eggs on dead animal tissue; blue blowfly, *Calliphora vicina*, feeding at upper left. (Photo by W. L. Krinsky)

Figure 4.4 Carrion beetle, *Necrophila americana* (Silphidae), adult. (Photo by Nathan D. Burkett-Cadena)

as on the immature stages of each other in some cases. As the corpse dries out and the period of blow fly larval feeding ends, cheese skipper flies (Piophilidae), clerid beetles (*Necrobia rufipes*), sap beetles (Nitidulidae), and various larder (Dermestidae) and hide beetles (Trogidae) may arrive to feed on the dried skin, hair, and connective tissues. A few species of moths (some representatives of the families Pyralidae and Tineidae) can breed in the completely dried remains.

An entomological analysis of a crime scene is useful only if it can be determined why each of the diverse arthropods recovered was present at the given location, and whether its presence was consistent with the environment in which it was found. Arthropods collected should be assigned to one of four species group designations: necrophagous, predaceous or parasitic, omnivorous, and adventitious. An understanding of the interactions of these species with a corpse and with each other enables the forensic entomologist to make an estimate of the postmortem interval (Campobasso et al., 2001).

Ideally, a forensic entomologist should be the one who collects any arthropod material at the scene of an investigation. This both insures that the material will be properly preserved and maintained and allows the entomologist to make his or her own observations of the circumstances involved in the case. Often an entomologist must rely on specimens, records, and photographs provided by others. In such cases, the conclusions drawn must be tempered by the nature of the materials provided and the level of experience of the personnel who collected the evidence and recorded the information.

The most accurate estimate of the postmortem interval based on entomological data can be made when the interval between death and discovery of a body is within two to four weeks. Blow flies are the insects most often used for determining an estimate of the time interval. The interval between the arrival of flies on a body and the discovery of the insects on a body is the postinfestation interval (PII). When an uninfested person dies in an environment where insects do not have immediate access to the body, the postmortem interval (PMI) will be longer than the postinfestation interval. Alternatively, if the victim is infested with flies (or other insects) before death (myiasis in a conscious, comatose, or paralyzed individual), the postinfestation interval will be longer than the postmortem interval. In the absence of both pathological evidence of myiasis and physical or environmental evidence of insects being excluded from the corpse for some period postmortem, the postinfestation interval is considered to be approximately equal to the postmortem interval.

If a body is found on which blow fly development has ceased, insects active at this later stage of decomposition can be identified and compared with the succession of insects known to infest corpses. Living insects as well as insect remains (exuviae, shed puparial or pupal cases, frass, or dead insects) that can be identified and staged can help to develop a time line of insect infestation.

Establishing how long insects have been developing on a body requires climatic data from the site at which a body was discovered as well as the identification of the infesting insects and their developmental stages. The development of blow flies, as for most plants and other insects, is dependent upon temperature, with increased temperature (below lethal values) directly related to increased growth. In blow flies, as in plants, assuming other conditions (humidity, light, etc.) are suitable for growth, by summing the number of degrees of temperature over time, it is possible to correlate a total (accumulated degree hours) with a given stage of development. Accumulated degree hours (degrees × hours = ADH) are calculated to determine the minimal PMI (or PII). The ADH values needed for development of the egg and development to each larval stage, puparium, and the adult of most of the forensically important blow flies have been determined by rearing the flies at different temperatures in the laboratory. Identification of the species and developmental stages of blow flies found on a corpse enables a comparison of the most mature stages found with known information about how much time had to have elapsed at different temperatures for these stages (egg, particular larval instar, or puparium) to have developed. By obtaining hourly temperature readings that most closely represent the temperatures that occurred over a period of days at the site where a body is found, we can add the degree hours (=ADH) that occurred from the collection of specimens backward in time (Figure 4.3, large counterclockwise arrow) to determine how many hours must have elapsed since the first fly eggs were laid on the body.

The PMI has been estimated in numerous murder cases based on the determination of the ages of eggs, larvae, or puparia found associated with a victim. Larvae have been used most often for this determination (Figure 4.5). In one case, over 4,000 maggots of *Phormia regina* were recovered from a body wrapped in a piece of carpet and dumped along an interstate highway. By correlating climatic conditions with the ages

Figure 4.5 Partially mummified corpse infested with blow fly maggots. This homicide victim was found and placed in the body bag within seven days of the murder.

of the larvae and the timing of emergence of adult flies from the oldest larvae collected, a PMI estimate of about seven days was made, which was within a day of when the victim was last seen alive (National Library of Medicine; entomology case study).

The ADH method is a means of reaching a minimal estimate of the PMI. Many factors may interfere with the time of arrival of insects on a body, the nature of the insect stages found on a body, and with the assumed linear correlation of temperature and insect development. Factors affecting the microclimate on a corpse, such as insulating materials (clothing, wrappings, position of a body in a building, container, or vehicle), and position of a body in relation to vegetation or overshadowing materials must be considered. In cases where there is evidence that the victim was feverish, chilled, or even frozen before death or before being placed in a location where insects would have access to the body, estimates of the time required for development of the insects found must be adjusted to accommodate these differences. The developing maggots themselves when aggregated into a maggot mass at one site on a body may markedly raise the temperature at which they are feeding and growing compared to the ambient temperature.

Disruption or dismemberment of a body, and predation of necrophagous insects by vertebrate scavengers, changes in the rate of insect development caused by insect ingestion of illicit drugs from tissues of the corpse, and changes in microclimatic conditions at the discovery site (e.g., wind, precipitation) or on the corpse itself related to the number and location of maggots (Slone and Gruner, 2007) must all be considered when an estimate of PMI or PII is made based on insect development.

Molecular techniques have begun to be applied to forensic entomology. Nucleic acid sequencing has been used experimentally to identify larvae and puparia of different fly species, as well as to differentiate among immature stages arising from different female flies. In the future, refinement of the techniques may allow for their use in rapid identification of species collected at crime scenes and other sites of forensic interest. Additional potential uses of DNA analysis include specific identification of blood meals from hematophagous insects collected at a crime scene or of semen in the alimentary tract of necrophagous insects that might link a victim or a scene to a specific suspect.

As the methodology in forensic entomology becomes more sophisticated and the interpretation of entomological events following death becomes more precise, entomology may become even more important in criminal investigations. Forensic entomology will continue to provide a different and valuable approach to the investigative process that is a collaborative effort of law enforcement personnel, forensic pathologists, and other forensic specialists.

REFERENCES AND FURTHER READING

Batra, S. W. T. (1976). Some insects associated with hemp or marijuana (*Cannabis sativa* L.) in northern India. *Journal of the Kansas Entomological Society, 49,* 385–388.

Benecke, M. (2001). A brief history of forensic entomology. *Forensic Science International, 120,* 2–14.

Benecke, M., & Lessig, R. (2001). Child neglect and forensic entomology. *Forensic Science International, 120,* 155–159.

Byrd, J. H., & Castner, J. L. (2001). *Forensic Entomology: Utility of Arthropods in Legal Investigations.* Boca Raton: CRC Press.

Campobasso, C. P., Di Vella, G., & Introna, F. (2001). Factors affecting decomposition and Diptera colonization. *Forensic Science International, 120,* 18–27.

Campobasso, C. P., & Introna, F. (2001). The forensic entomologist in the context of the forensic pathologist's role. *Forensic Science International, 120,* 132–139.

Catts, E. P., & Goff, M. L. (1992). Forensic entomology in criminal investigations. *Annual Review of Entomology, 37,* 253–272.

Catts, E. P., & Haskell, N. H. (Eds.).(1990). *Entomology and Death: A Procedural Guide.* Clemson, SC: Joyce's Print Shop.

Erzinçlioğlu, Y. Z. (2000). *Maggots, Murder and Men: Memories and Reflections of a Forensic Entomologist.* Colchester, UK: Harley.

Greenberg, B. (1991). Flies as forensic indicators. *Journal of Medical Entomology, 28,* 565–577.

Greenberg, B., & Kunich, J. C. (2002). *Entomology and the Law: Flies as Forensic Indicators.* Cambridge Univ. Press.

Hart, A. J., & Whitaker, A. P. (2006). Forensic entomology: Insect activity and its role in the decomposition of human cadavers. *Antenna, Bulletin of the Royal Entomological Society, 30,* 159–164.

Introna, F., Campobasso, C. P., & Goff, M. L. (2001). Entomotoxicology. *Forensic Science International, 120,* 42–47.

Kimsey, L. (2007). The Case of the Red-Shanked Grasshopper. *Los Angeles Times,* Wednesday, 4 July, p. A21.

McKnight, B. E. (1981). *The Washing Away of Wrongs: Forensic Medicine in Thirteenth Century China* (S. Tz'u, Trans.).Ann Arbor: Center for Chinese Studies, University of Michigan.

Mégnin, P. (1894). *La Faune des Cadavres.* Encyclopédie scientifique des Aide-Mémoire No. 101B, Gautier-Villars *et fils,* Paris.

National Library of Medicine. *Forensic views of the body—Entomology case study.* http://www.nlm.nih.gov/exhibition/visibleproofs/galleries/cases/insect.html

Nuorteva, P. (1977). Sarcosaprophagous insects as forensic indicators. In C. G. Tedeschi, W. G. Eckert, L. G. Tedeschi (Eds.), *Forensic Medicine: A Study in Trauma and Environmental Hazards* (Vol. 27, pp. 1072–1095). Philadelphia: W. B. Saunders Co.

Sachs, J. S. (2001). *Corpse—Nature, Forensics, and the Struggle to Pinpoint Time of Death.* Cambridge, MA: Perseus Publishing.

Slone, D. H., & Gruner, S. V. (2007). Thermoregulation in larval aggregations of carrion-feeding blow flies (Diptera: Calliphoridae). *Journal of Medical Entomology, 44,* 516–523.

Smith, K. G. V. (1986). *A Manual of Forensic Entomology.* London: British Museum (Natural History), and Cornell Univ. Press.

Vincent, C., Kevan, D. K. McE., Leclercq, M., & Meek, C. L. (1985). A bibliography of forensic entomology. *Journal of Medical Entomology, 22,* 212–219.

Webb, J. P., Jr., Loomis, R. B., Madon, M. B., Bennett, S. G., & Greene, G. E. (1983). The chigger species *Eutrombicula belkini* Gould (Acari: Trombiculidae) as a forensic tool in a homicide investigation in Ventura County, California. *Bulletin of the Society of Vector Ecologists, 8,* 141–146.

Westerfield trial. (2002). *Transcripts.* www.signonsandiego.com/news/metro/danielle/transcripts/20020710-9999-pm1.html

Whitworth, T. (2006). Keys to the genera and species of blow flies (Diptera: Calliphoridae) of America north of Mexico. *Proceedings of the Entomological Society of Washington, 108,* 689–725.

Chapter

5

Cockroaches (Blattaria)

Richard D. Kramer ▪ Richard J. Brenner

Cockroaches are among the oldest and most primitive of insects. They evolved about 350 million years ago during the Silurian Period, diverging together with the mantids from an ancestral stock that also gave rise to termites (Boudreaux, 1979). Cockroaches are recognized as the order Blattaria. Although the majority of species are feral and not directly associated with people, a few species have evolved in proximity to human habitations where they have adapted to indoor environments. Their omnivorous feeding behavior, facilitated by their unspecialized chewing mouthparts, has contributed to a close physical relationship between cockroach populations and humans, with resultant chronic exposure of humans to these pests.

The presence of some species in the home and commercial kitchens (e.g., German and brownbanded cockroaches) often is an indicator of poor sanitation or substandard housekeeping. Although they are primarily nuisance pests, their presence can have important health implications. Cockroaches are generalists that feed on virtually any organic substance grown, manufactured, stored, excreted, or discarded by humans. Consequently, food supplies and preparation surfaces are at risk of contamination by pathogens associated with cockroaches. Because species that infest structures typically have high reproductive rates, humans commonly are exposed to high levels of potentially allergenic proteins associated with cockroaches that can lead to significant respiratory ailments. Cockroaches also can serve as intermediate hosts of parasites that debilitate domestic animals.

TAXONOMY

There are about 4,000 species of cockroaches worldwide. About 70 species occur in the United States, 24 of which have been introduced from other parts of the world. According to Atkinson et al. (1991), 17 of these species are pests of varying degrees. There are five cockroach families, three of which include most of the pest species: *Blattidae*, *Blattellidae*, and *Blaberidae*. Species in the *Cryptocercidae* are unusual in that they have gut symbionts similar to those found in termites, and they live in family groups in decaying logs. Members of the *Polyphagidae* include those dwelling in arid regions where they are capable of moving rapidly through sand. Species in these two families are rarely pests. The family Blattidae includes relatively large cockroaches that are the most common peridomestic pests throughout much of the world. Blattellid cockroaches range in length from less than 25 mm (e.g., *Supella* and *Blattella*) to 35 to 40 mm (e.g., *Periplaneta* and *Parcoblatta* spp.). *Parcoblatta* species are feral, occasionally invading homes but seldom reproducing indoors. Blaberid cockroaches range greatly in size and include some of the more unusual species such as the Cuban cockroach that is green as an adult and the Surinam cockroach that is parthenogenetic in North America. Nearly all the Blaberids that occur in the United States are restricted to subtropical regions and have minor medical or veterinary significance. Taxonomic keys for adults are provided by McKittrick (1964), Cornwell (1968), Roth (1985), and Helfer (1987). A pictorial key for identifying the egg cases of common cockroaches is provided by Scott and Borom (1964).

MORPHOLOGY

Cockroaches have retained their basic ancestral form. The Blattaria are distinguished from other insect orders by morphological characters associated with wing size and venation, biting-chewing mouthparts, and prominent cerci. They differ from other orthopteroid insects by having hind femora that are not enlarged, cerci typically with eight or more segments, a body that is dorsoventrally flattened and generally ovoid, and a head that is largely concealed from above by a relatively large pronotum.

An indicator of cockroach infestations is their egg cases, or **oothecae** (sing. *ootheca*), which can be useful in differentiating species infesting buildings. Most cockroach oothecae persist in the environment after the nymphs have emerged, providing a history of infestation.

These purse-shaped capsules typically contain five to 40 embryos (Figure 5.1) and their coloration ranges from

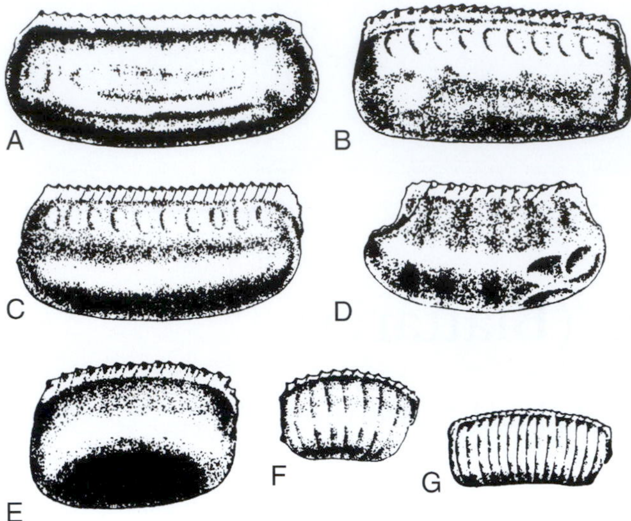

Figure 5.1 Cockroach oothecae (egg cases). (A) Australian cockroach (*Periplaneta australasiae*); (B) brown cockroach (*P. brunnea*); (C) smokybrown cockroach (*P. fuliginosa*); (D) Oriental cockroach (*Blatta orientalis*); (E) American cockroach (*P. americana*); (F) brownbanded cockroach (*Supella longipalpa*); (G) German cockroach (*Blatella germanica*). (Courtesy of U.S. Public Health Service)

light brown to chestnut brown, depending on the degree of sclerotization. A keel that runs the anterior length of the ootheca permits transport of water and air to the developing embryos. Each embryo is contained in a separate compartment that may or may not be obvious externally. In some species (e.g., German and brownbanded cockroaches) lateral, anterior-to-posterior indentations denote the individual developing embryos. Others have only weak lateral indentations (e.g., brown and smokybrown cockroaches), and still others have no lateral indentations but differ in their symmetry (e.g., Oriental, American, and Australian cockroaches).

The mouthparts of cockroach nymphs and adults are characterized by strongly toothed mandibles for biting and chewing. Maxillary and labial palps are well developed with five and three segments, respectively. Antennae are long and whip-like, originate directly below the middle of the compound eyes, and consist of numerous small segments. The arrangement of three ocelli near the antennal sockets is variable: they are well developed in winged species (*macropterous*) but rudimentary or lacking in species with reduced wings (*brachypterous*) or those lacking wings altogether (*apterous*).

Adults generally have two pairs of wings that are folded fan-wise at rest. The front wings, called **tegmina** (sing. *tegmen*), typically are hardened and translucent with well-defined veins. The hindwings are membranous and larger. In some species, such as the wood cockroaches (e.g., *Parcoblatta* species), females are brachypterous and incapable of flight, whereas males are macropterous. Other species, such as the Florida woods cockroach (*Eurycotis floridana*), have only vestigial wing buds as adult males and females. In cockroaches, all three pairs of legs are well developed, with large coxae and slender, long segments that aid in the rapid running characteristic of these insects. Each femur has two longitudinal keels that typically are armed with spines. The tibiae are often heavily spined and are used for defense against predators. Each tarsus consists of five segments with a pair of claws and may bear a pad-like **arolium** that aids in walking on smooth surfaces. Ventral pads, or **pulvilli**, are present on tarsomeres 1–4. A pair of caudal **cerci** have small ventral hairs that are sensitive to vibrations caused by low-frequency sound and air movement; their stimulation initiates an escape response.

The posterior end of the abdomen of some nymphs and all males bears a pair of **styli** (sing. *stylus*) between the cerci, arising from the sternum of the ninth abdominal segment. In winged species, the styli may be apparent only when viewed ventrally. The structure of the styli serve to distinguish males from females. Generally the males also can be recognized by their more slender bodies, with laterally tapered and dorsally flattened external genitalia (**terminalia**). The terminalia of the more robust females are notably broader than in males, and bear a conspicuous subgenital plate that is rounded or keel-like when viewed ventrally. Associated with this plate is a relatively large genital chamber (genital pouch) in which the ootheca develops. For a more detailed description of cockroach genitalia, see McKittrick (1964) or Cornwell (1968). Nymphal stages are similar in appearance to adults, but lack wings, have incompletely developed genitalia, and may vary markedly in color from the adult.

LIFE HISTORY

Cockroaches are paurometabolous insects. The immatures generally are similar in appearance to the adults except for their undeveloped sexual organs and lack of fully developed wings (Figure 5.2). Reproduction in cockroaches is typically sexual, although parthenogenesis is reported in a few species. Comparative life-history data

Figure 5.2 Developmental stages of cockroaches, represented by *Periplaneta brunnea*. Left to right: first, second, and third nymphal instars; adult female, adult male. (Courtesy of Daniel R. Suiter)

Table 5.1 Life Histories of Selected Common Species of Cockroaches, Showing the High Degree of Variability within Species Due to Environmental Temperatures and Nutritional Availability

Cockroach	Number of Eggs/Ootheca	Number of Nymphal Instars	Developmental Time (Days)	Embryonic Development
German	30–40	5–7	103	Internal/extruded
Asian	38–44	5–7	52–80	Internal/extruded
Brownbanded	14–18	6–8	90–276	External
American	12–16	10–13	168–700	External
Smokybrown	20	9–12	160–716	External
Australian	24	10–12	238–405	External
Oriental	16	7–10	206–800	External
Surinam	26	8–10	127–184	Internal

for some of the more common cockroach pests are provided in Table 5.1.

In cockroaches, **embryogenesis** and **oviposition** occur in one of three ways. Most species are **oviparous**, including all *Periplaneta* species and the Oriental and brownbanded cockroaches. Eggs of oviparous species are protected inside a thick-walled, impermeable ootheca that is deposited soon after it is formed. Embryonic development occurs external to the female. The German cockroach is oviparous, but the female carries the ootheca protruding from the genital chamber until just hours before hatching occurs. The ootheca is softer than in *Periplaneta* species, allowing uptake of water and nutrients from the genital pouch. Females of a few cockroaches, such as *Blaberus* species and the Surinam cockroach, produce an ootheca that is extruded, rotated, and then retracted into the genital pouch. The eggs are incubated internally until hatching. *Diploptera punctata,* a pest species in Hawaii, is the only cockroach in which the eggs hatch while still in the genital pouch. Embryogenesis takes one to eight weeks, depending on the species.

The number of nymphal instars varies from five to 13, depending on the species, nutritional sources, and microclimate. Development of pestiferous species through the nymphal stadia requires six or seven weeks for German cockroaches, to well over a year for *Periplaneta* species and other larger cockroaches. Typically, the nymphs exhibit strong aggregation tendencies, governed largely by **aggregation pheromones**. These pheromones act as locomotory inhibitors; when cockroaches perceive the pheromone they become relatively stationary. Studies of various species have shown that development to the adult stage is quicker when nymphs are reared in groups rather than in isolation. However, aggregation does have a biological cost; those reared in groups typically are smaller in size, and cannibalism may occur. Longevity of cockroaches varies from several weeks to over a year.

BEHAVIOR AND ECOLOGY

Mating in cockroaches generally is preceded by courtship behavior facilitated by sex pheromones. In some species a blend of volatile compounds is produced by virgin females to attract and orient males (e.g., *Periplaneta* species and the brownbanded cockroach). In the German cockroach, the sex pheromone is a blend of nonvolatile and volatile cuticular components that elicits courtship by males following palpation of the female's integument by the male's antennae. Once courtship is initiated in the male, he turns away from the female and raises his wings to expose dorsal tergal glands; the female feeds on pheromones from these glands as the male grasps her genitalia with his pair of caudal claspers. Most species copulate in an end-to-end position. During the hour or so that follows, a spermatophore is formed and passed from the male into the genital chamber of the female. Only about 20% of females mate again after the first gonotrophic cycle.

Cockroaches can be categorized ecologically as domestic, peridomestic, or feral. **Domestic** species live almost exclusively indoors and are largely dependent on humans for resources (food, water, and harborage) for survival. They rarely are able to maintain themselves outdoors. Although this group contains the smallest number of species, it presents the greatest concern to human health. Domestic species include the German and brownbanded cockroaches. **Peridomestic** species are those that survive in or around human habitation. Although they do not require humans for their survival, they are adept at exploiting the amenities of civilization. This group is represented by American, Australian, brown, and smokybrown cockroaches (all *Periplaneta* species), the oriental cockroach, and the Florida woods cockroach. **Feral** species are those in which survival is independent of humans. This group includes more than 95% of all species in the world. Only a few occur indoors as occasional and inadvertent invaders that typically do not survive in a domestic environment. They are of little or no medical importance.

Cockroach behavior and survival are strongly influenced by their need for food, water, and safe harborage from potential predators and detrimental microclimates. They are omnivorous, and will consume virtually any organic matter, including fresh and processed foods, stored products, and even book bindings and pastes on stamps and wallpaper when more typical foodstuffs are not available. Cockroaches have the same general problems with water balance as do other terrestrial arthropods. Their relatively small size results in a high surface area-to-volume ratio, and a high risk of losing water through respiration, oral and anal routes, or the cuticle. Temperature, air flow, relative humidity, and availability of liquid water greatly affect water regulation.

As a result of these physiological considerations, physical constraints of the environment usually determine habitat preferences of cockroaches in and around structures. Oriental, Turkestan, and American cockroaches, for example, require high moisture and occur in damp terrestrial environments such as septic tanks and municipal sewer systems. Brown, smokybrown, and Florida woods cockroaches occur in a wider range of habitats associated with trees, wood and leaf piles, wall voids, and foundation blocks of buildings. Brownbanded cockroaches are more tolerant of drier conditions and commonly occur in kitchens, pantries, and bedrooms. German cockroaches occupy harborages near food and water. Consequently, they are found primarily in kitchens and pantries, and secondarily in bathrooms when their populations are high. In mixed populations of German and brownbanded cockroaches, the German cockroach tends to out-compete the brownbanded cockroach within nine months.

Cockroaches are adept crawlers, and are capable of rapid movement even across windows and ceilings. Flight ability varies with species. Some are incapable of flight except for crude, downward gliding used as an escape behavior. Others are weak fliers, occasionally seen flying indoors when disturbed. Still others are relatively strong fliers that are particularly active at sunset when they may be attracted indoors by lights and brightly lit surfaces. Attraction to light is especially common in the Asian, Surinam, and Cuban cockroaches, and in many of the wood cockroaches (*Parcoblatta* species).

Pestiferous cockroaches that occur indoors are typically nocturnal, and tend to avoid lighted areas. This enables them to increase their numbers and become established in structures before human occupants even become aware of their presence.

COMMON COCKROACH SPECIES

The following cockroach species are commonly encountered in and around structures in the United States and are the ones most frequently brought to the attention of medical entomologists.

Oriental Cockroach *(Blatta orientalis)*

This peridomestic cockroach (Figure 5.3) is believed to have originated in northern Africa and from there spread to Europe and western Asia, South America, and North America. It is a relatively lethargic species that prefers cooler temperatures than the German cockroach, and is primarily a concern in temperate regions of the world. Adults are black and 25 to 33 mm long. Males are short-winged but do not fly, and females are brachypterous. Their tarsi lack aroliar pads, precluding this cockroach from climbing on smooth vertical surfaces. Oothecae are 8 to 10 mm long, each typically containing 16 eggs. Also commonly known as waterbug and shad roach, this species usually is associated with damp or wet conditions, such as those found in decaying wood, heavy ground cover (e.g., ivy), water-meter boxes, floor drains, sump pumps, and the lower levels of structures. It infests garbage chutes of apartment complexes, sometimes reaching upper floors. Development is slow compared to that of most other species, typically requiring about a year or more depending on temperature conditions. Adults may live for many months. Mobility is fairly restricted, making control easier than for most other species. This species is rarely seen during the daytime.

Figure 5.3 Oriental cockroach (*Blatta orientalis*), female. (Courtesy of University of Florida/IFAS)

Turkestan Cockroach *(Blatta lateralis)*

This peridomestic cockroach (Figure 5.4) originated in North Africa, the Middle East, and Asia. It was introduced to the United States in the late 1970s by military personnel returning from the Middle East and is now

Figure 5.4 Turkestan cockroach (*Blatta lateralis*), female. (Photo by Nathan D. Burkett-Cadena)

found in California, Arizona, and Texas. Adults are 14 to 25 mm long. Females have very short, nonfunctional wings and are black in color, with light bands on the anterior margin of the thorax. Males are fully winged and light brown in color. Oothecae are 9 to 12 mm long, each typically containing 18 eggs. The biology of this species is similar to the Oriental cockroach, although their developmental time (3–4 months from egg to adult) is much shorter. The adults can live for almost one year. Despite being considered a desert species in the United States, this cockroach is typically found infesting warehouses, steam tunnels, and sewers.

American Cockroach (*Periplaneta americana*)

The American cockroach (Figure 5.5) is a large species with adults 34 to 53 mm in length. It is reddish brown, with substantial variation in light and dark patterns on the pronotum. Adults are winged and capable of flight. Nymphs typically complete development in 1.5 to 2 years while undergoing 10 to 13 molts. Adults live an average of seven months, but longevity may exceed two years. Females drop or glue their oothecae (8 mm long) to substrates within a few hours or days of formation. Each ootheca has 12 to 16 embryos. A female generally produces nine to 10 egg cases during her life.

The American cockroach is perhaps the most cosmopolitan peridomestic pest species. Together with other closely related *Periplaneta* species, *P. americana* is believed to have spread from tropical Africa to North America and the Caribbean on ships engaged in slave trading. Today this species infests most of the lower latitudes of both hemispheres and extends significantly into the more temperate regions of the world.

The habitats of this species are quite variable. American cockroaches commonly are found in commercial buildings, landfills, municipal sewage systems, storm drainage systems, septic tanks, crawl spaces beneath buildings, attics, treeholes, canopies of palm trees, voids in walls, ships, caves, and mines. Studies conducted in Arizona indicated movement by a number of individuals several hundred meters through sewer systems and into neighboring homes. This species often can be seen at night on roofs and in air stacks or vents of sewage systems through which they enter homes and commercial buildings. Entrance also is gained to homes through laundry vent pipes and unscreened or unfiltered attic ventilation systems. This cockroach is known to move from crawl spaces of hospitals via pipe chases into operating theaters, patients' rooms, storage facilities, and food-preparation areas. Consequently, the potential of this cockroach for disseminating pathogenic microorganisms can be a significant concern for health-care personnel.

Australian Cockroach (*Periplaneta australasiae*)

Adult body coloration is similar to that of the American cockroach, but with paler lateral markings on the anterior-lateral edges of the tegmina (Figure 5.6). The pronotum is ringed with similar coloration. Adults are slightly smaller than American cockroaches, measuring 32 to 35 mm in length. Developmental time is about one year, and females typically live for another four to six months. A female can produce 20 to 30 oothecae during her lifetime; the ootheca is about 11 mm long and contains about 24 embryos. Embryonic development requires about 40 days. Nymphs are strikingly mottled, distinguishing them from nymphs of other *Periplaneta* species.

This peridomestic species requires somewhat warmer temperatures than the American cockroach and does not occur in temperate areas other than in greenhouses and other pseudo-tropical environs. In the United States, outdoor populations are well established in Florida and along the coastal areas of Louisiana, Mississippi, Alabama, and Georgia. It commonly is found in environments similar to those inhabited by the smokybrown cockroach. In situations where both species occur (e.g., treeholes, attics), the Australian cockroach tends to displace the smokybrown. It can be a serious pest in greenhouses and other tropical environments in more

Figure 5.5 American cockroach (*Periplaneta americana*), female. (Courtesy of University of Florida/IFAS)

Figure 5.6 Australian cockroach (*Periplaneta australasiae*), female. (Courtesy of University of Florida/IFAS)

temperate latitudes (e.g., atriums and interior plantscapes), where it can cause feeding damage to plants, notably seedlings. Once in structures they occupy habitats similar to American cockroaches.

Brown Cockroach (*Periplaneta brunnea*)

The brown cockroach (Figure 5.7) is smaller than the American cockroach (33–38 mm), and its pronotal markings are more muted. The most apparent diagnostic character for separating these two species is the shape of the last segment of the cercus; in the brown cockroach, the length is about equal to the width, whereas in the American cockroach the length is about three times the width. The ootheca of the brown cockroach usually is larger (7–13 mm) and contains more embryos (24). The brown cockroach affixes its oothecae to substrates using salivary secretions. They give the ootheca a grayish hue not typical of other *Periplaneta* species that attach their oothecae with salivary secretions. This species is more subtropical than the American cockroach, occurring throughout the southeastern United States, where it infests homes and outbuildings. It is less frequently associated with sewage than is the American cockroach. Because of its similar appearance to the American cockroach, it is often misidentified and may be more widely distributed than is commonly recognized. In Florida, *P. brunnea* is commonly found in canopies of palm trees and attics. It also readily infests natural cavities, buildings, and other structures, like the American cockroach.

Smokybrown Cockroach (*Periplaneta fuliginosa*)

The smokybrown cockroach (Figure 5.8) has become a major peridomestic pest throughout the southern United States, including southern California, and extends as far north as the midwestern states. It can be differentiated from the American cockroach by its slightly smaller size (25–33 mm) and uniform dark coloration. Although developmental times are quite variable, individuals mature in 1.5 to 2 years. Adults may live for about seven months. Females produce several ootheca that are 10 to 11 mm in length with 20 embryos, at 11-day intervals.

Figure 5.7 Brown cockroach (*Periplaneta brunnea*), female. (Courtesy of University of Florida/IFAS)

Figure 5.8 Smokybrown cockroach (*Periplaneta fuliginosa*), female. (Courtesy of University of Florida/IFAS)

Primary foci for this peridomestic species in the southeastern United States are treeholes, canopies of palm trees, loose mulches such as pine straw or pine bark, and firewood piles. Within structures, *P. fuliginosa* seeks the ecological equivalent of treeholes, areas characterized as dark, warm, protective, and moist, with little air flow and near food resources. These include the soffit (eaves) of underventilated attics, behind wall panels, the interstices of block walls, false ceilings, pantries, and storage areas. From these harborages, individuals forage for food and water, generally returning to the same refugia. Mark-release-recapture studies using baited live traps have shown that the median distance traveled between successive recaptures is less than one meter, but that some adults may forage at distances of more than 30 meters.

Florida Woods Cockroach (*Eurycotis floridana*)

This cockroach is restricted to a relatively small area of the United States along the Gulf of Mexico from eastern Louisiana to southeastern Georgia. It is mentioned here only because of its defensive capabilities. It is a large, dark-reddish-brown to black cockroach (Figure 5.9), 30 to 40 mm long. Although small wing pads are evident, adults are apterous and are relatively slow moving. Oothecae are 13 to 16 mm long and contain about 22 embryos. *Eurycotis floridana* occurs in firewood piles, mulches, treeholes, attics, wall voids, and outbuildings. Last-instar nymphs and adults, if alarmed, can spray a noxious mix of aliphatic compounds that are both odoriferous and caustic. If sprayed into the eyes or onto soft tissues, a temporary burning sensation is experienced. Domestic dogs and cats quickly learn to avoid this species. Among its common names are the Florida cockroach, the Florida woods roach, the Florida stinkroach, and palmetto bug. The latter term also is commonly used for other *Periplaneta* species.

Figure 5.9 Florida woods cockroach (*Eurycotis floridana*), female. (Courtesy of University of Florida/IFAS)

Brownbanded Cockroach (*Supella longipalpa*)

Like the German cockroach, this domestic species (Figure 5.10) probably originated in tropical Africa where it occurs both indoors and outdoors. In North America and Europe it is confined almost exclusively to indoor environments of heated structures. In warm climates, infestations occur particularly in apartments without air conditioning, in commercial establishments with relatively high ambient temperatures, such as pet stores and animal-care facilities, office buildings, and schools. Adults are similar in size to the those of the German cockroach (13–14.5 mm long) but lack pronotal stripes. Adults have two dark bands of horizontal stripes on the wings, whereas nymphs have two prominent bands running across the mesonotum and first abdominal segment. The brownbanded cockroach derives its name from these bands. Populations tend to occur in the nonfood areas of homes such as bedrooms, living rooms, and closets; also in electronics equipment, such as CPUs and telephones, and storage cabinets in commercial office buildings. Male brownbanded cockroaches occasionally fly, and are attracted to lights. Members of this species seek harborage higher within rooms than does the German cockroach. Developmental time from egg to adult averages five to six months. The ootheca is small, only 5 mm long, with an averages of 18 embryos, and an incubation time of 35 to 80 days. Females affix their oothecae to furniture, in closets, on or behind picture frames, and in bedding. Transporting *S. longipalpa* with furniture to new locales is common. Although this species occurs with other cockroaches in homes, the German cockroach often out-competes it within a few months.

German Cockroach (*Blattella germanica*)

This cockroach also is known as the steamfly in Great Britain. It is believed to have originated in northern or eastern Africa, or Asia, and has spread from there via commerce. The German cockroach is considered to be the most important domestic pest species throughout the developed world. Adults are about 16 mm long with two dark, longitudinal bands on the pronotum (Figure 5.11). It requires warm (optimally 30–33° C), moist conditions near adequate food resources. It primarily inhabits kitchens and pantries, with secondary foci in bathrooms, bedrooms, and other living spaces in heavily infested structures. Although this species is nocturnal like most other cockroaches, some individuals may be seen moving about on walls and in cupboards during the daylight hours where infestations are heavy. Their wing musculature is vestigial, making them unable to fly except for short, gliding, downward movements. *Blattella germanica* does not readily move between buildings; however, it does occur in garbage collection containers and outbuildings near heavily infested structures.

The German cockroach has a high reproductive potential. Females produce an ootheca (6–9 mm) containing about 30 to 40 embryos within seven to 10 days after molting to the adult, or about two or three days after mating. The female carries the egg case until a few hours before hatching of the nymphs, preventing access of any oothecal parasitoids or predators. Oothecae are produced at intervals of 20 to 25 days, with a female producing four to eight oothecae during her

Figure 5.10 Brownbanded cockroach (*Supella longipalpa*), female. (Courtesy of University of Florida/IFAS)

Figure 5.11 German cockroach (*Blattella germanica*), female. (Courtesy of University of Florida/IFAS)

lifetime. Nymphs complete their development in seven to 12 weeks.

This species is the main cockroach pest in most households and apartment complexes. Control is difficult, in part because of their movement between apartments through plumbing chases in shared or adjacent walls. Researchers studying over 1,000 apartments in Florida concluded that the median number of cockroaches per apartment was more than 13,000. This high biotic potential makes this species a major nuisance, as well as a pest with implications for human health.

Asian Cockroach (*Blattella asahinai*)

The Asian cockroach is closely related to the German cockroach from which it is difficult to distinguish morphologically. In fact, Asian and German cockroaches are capable of hybridizing and producing fertile offspring, which further complicates their identifications. Techniques have been developed to differentiate these two species and their hybrids based on cuticular hydrocarbons in the waxy layer of the integument.

Despite their morphological similarity, *B. asahinai* differs from *B. germanica* in several aspects of its behavior and ecology. It is both a feral and a peridomestic species. Nymphs of the Asian cockroach commonly occur, sometimes in large numbers, in leaf litter and in areas of rich ground cover or well-maintained lawns. Unlike the German cockroach, the adults fly readily and are most active beginning at sunset when they fly to light-colored walls or brightly lit areas. This behavior can make invasion a nightly occurrence in homes near heavily infested areas. Flight does not occur when temperatures at sunset are below 21°C.

Like those of the German cockroach, Asian cockroach females carry their oothecae until shortly before they are ready to hatch. The ootheca is similar in size and contains a similar number of embryos as the German cockroach (30–40). Nymphs are smaller than their *B. germanica* counterparts and are somewhat paler in appearance. Development from egg to adult requires about 65 days, with females producing up to six oothecae during their life span. Adults are slightly smaller than those of *B. germanica* (average of 13 mm).

The Asian cockroach was first described in 1981 from specimens collected in sugar cane fields on the Japanese island of Okinawa. When it was first discovered in the United States in 1986, the Asian cockroach was found only locally in three counties in Florida from Tampa to Lakeland; populations already had become established with densities as high as 250,000 per hectare. By 1993 this species had spread to at least 30 Florida counties and had infested citrus groves throughout the central part of the state. This species is now also established in Alabama and Georgia (Snoddy and Appel, 2008). It feeds on succulent early growth of citrus nursery stock, tassels of sweet corn, strawberries, cabbage, tomatoes, and other agricultural products, although there has been no evidence of significant economic damage.

Infestations of apartments by *B. ashinai* have become common in central Florida. This cockroach also has become an increasing problem in warehouses, department stores, hotels, fast-food establishments, automobile dealerships, and other businesses with hours of operation that extend beyond dusk.

Surinam Cockroach (*Pycnoscelus surinamensis*)

This species is believed to have originated in the Indo-Malayan region. It commonly occurs outdoors in the southeastern United States from North Carolina to Texas. The adults are fairly stout, 18 to 25 mm in length, with shiny brown wings and a black body (Figure 5.12). Nymphs characteristically have shiny black anterior abdominal segments, whereas the posterior segments are dull black and roughened. In North America this species is unusual in that it is parthenogenetic, producing only female offspring; elsewhere both males and females are found. The ootheca is 12 to 15 mm long, is poorly sclerotized, and contains about 26 embryos. The ootheca is retained inside the genital chamber from which the nymphs emerge in about 35 days. Females produce an average of three oothecae and live about 10 months in the laboratory. This cockroach commonly burrows into compost piles and the thatch of lawns. Transfer of fresh mulch into the home for potting plants can result in household infestations. Adult females fly and are attracted to light. They are most likely to be noticed by homeowners at night when they fly into brightly lit television screens. This species commonly is transported in commercial mulch and plant material to more temperate areas of the United States where it has been known to infest greenhouses, indoor plantings in shopping malls, atriums in office buildings, and zoos.

PUBLIC HEALTH IMPORTANCE

Cockroaches infesting human dwellings and workplaces represent a more intimate and chronic association than do most other pests of medical-veterinary importance. High populations of any cockroach species may adversely affect human health in several ways. These include contamination of food with their

Figure 5.12 Surinam cockroach (Pycnoscelus surinamensis). (Courtesy of University of Florida/IFAS)

excrement, mechanical dissemination of pathogens, induced allergies, psychological stress, and bites. Although documentation of bites is limited, there are reports of cockroaches feeding on fingernails, eyelashes, skin calluses of hands and feet, and food residues about the faces of sleeping humans, causing blisters and small wounds (Roth and Willis, 1957, 1960). There are other accounts of bites around the mouths of infants in heavily infested homes, and even in hospitals. American and Australian cockroaches are the more often implicated species. Bites by the Oriental cockroach have resulted in inflammation of the skin, degeneration of epithelial cells, and subsequent necrosis of the involved tissues.

Although many individuals develop a tolerance for cockroach infestations, others may experience psychological stress. The level of stress tends to be proportional to the size of the cockroaches and the magnitude of the infestation. An aversion to cockroaches may be so strong that some people become irrational in their behavior, imagining a severe infestation even when there is none. This illusion of abundant cockroaches has caused some families to move out of their homes. High cockroach populations also produce a characteristic odor that can be unpleasant or even nauseating to some people. Food stuffs may become contaminated with the excrement of cockroaches that, on subsequent ingestion, may cause vomiting and diarrhea.

The presence of cockroaches in homes does not necessarily imply poor housekeeping. Peridomestic species such as the American and the Oriental cockroach commonly infest municipal sewage systems or septic tanks and may move into homes through sewage lines. Any of the *Periplaneta* species may develop high outdoor populations, inducing individuals to seek less crowded environments. At such times, they often enter homes through attic vents, breaches in construction joints, or through crawl spaces. This tends to occur in early fall. The smokybrown cockroach, Asian cockroach, and feral wood roaches (*Parcoblatta* species) are active at night, and often find their way into even the best-kept homes. Adults frequently alight on doors illuminated by entrance lights, or on window screens of lighted rooms. Entrance is gained once the door is opened, or by squeezing past window-screen frames.

Poor housekeeping and unsanitary conditions contribute significantly to cockroach infestations. The German cockroach, and to a lesser degree the brown-banded cockroach, are the principal bane of apartment dwellers. Their survival is enhanced by crowded living quarters, associated clutter, and the accumulated organic debris associated with food preparation. Construction practices used to build apartment complexes (e.g., common wiring ducts, sewage lines, and refuse areas) can contribute to the spread of cockroaches in multi-unit dwellings.

Pathogenic Agents

The significance of cockroaches in public health remains controversial despite the logical assumption that they play a role in transmitting pathogenic agents.

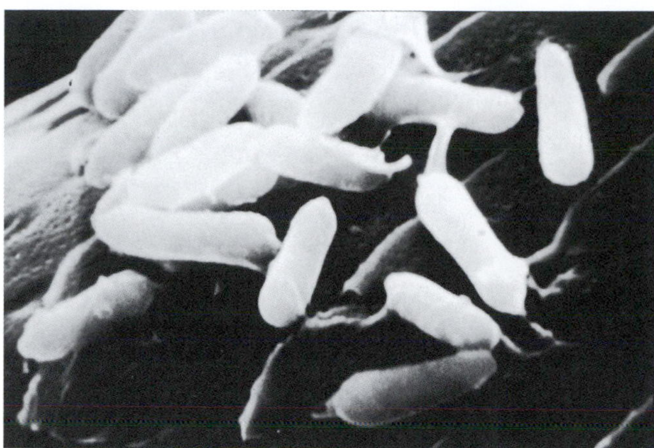

Figure 5.13 Bacteria adhering to tarsus of German cockroach (*Blatella germanica*). (From Gazivoda & Fish, 1985; permission of New York Entomological Society)

Given that cockroaches are so closely associated with humans and poor sanitation, the potential for acquiring and mechanically transmitting disease agents is very real. They are capable of transmitting microorganisms (Figure 5.13) and other disease agents indirectly by contaminating foods or food-preparation surfaces.

Table 5.2 lists pathogenic organisms that have been isolated from cockroaches in domestic and peridomestic environments. At least 32 species of bacteria in 16 genera are represented. These include such pathogens as *Bacillus subtilis*, a causative agent of conjunctivitis; *Escherichia coli* and nine strains of *Salmonella*, causative agents of diarrhea, gastroenteritis, and food poisoning; *Salmonella typhi*, the causative agent of typhoid; and four *Proteus* species that commonly infect wounds. These isolations primarily have involved American, German, and Oriental cockroaches. Cockroaches also have been found harboring the eggs of seven species of helminths, at least 17 fungal species, three protozoan species, and two strains of poliomyelitic virus (Brenner et al., 1987; Koehler et al., 1990; Brenner, 1995). Researchers in Costa Rica have shown that Australian, American, and Madeira cockroaches become infected with the protozoan *Toxoplasma gondii* after eating feces of infected cats. This suggests the possibility of cockroach involvement in the maintenance and dissemination of this parasite, which causes toxoplasmosis in humans, cats, and other animals.

Although many pathogens have been recovered from natural populations of cockroaches, this does not necessarily mean that cockroaches serve as their vectors. Isolation of pathogens from cockroaches simply may be indicative of the natural microbial fauna and flora in our domestic environment. Under certain circumstances, however, cockroaches have the potential for serving as secondary vectors of agents that normally are transmitted by other means.

Anecdotal accounts associating diseases in humans with the occurrence of cockroaches and microbes lend some credence to the hypothesis that these pests can serve as vectors. Burgess (1982) reported the isolation

Table 5.2 Bacteria Pathogenic to Humans that Have Been Isolated from Field-Collected Cockroaches

Pathogen	Associated Disease	Cockroach Species
Acinetobacter sp.	Nosocomial infection	*Blattella germanica* *Periplaneta americana*
Aeromonas sp.	Wound and other infections; diarrhea	*Blattella germanica* *Diploptera punctata*
Alcaligenes faecalis	Gastroenteritis, secondary infections, urinary tract infections	*Blatta orientalis* *Periplaneta americana*
Bacillus subtilis	Conjunctivitis, food poisoning	*Blaberus craniifer* *Blatta orientalis* *Blattella germanica* *Periplaneta americana*
Bacillus cereus	Food poisoning	*Blaberus craniifer*
Campylobacter jejuni	Enteritis	*Blatta orientalis* *Periplaneta americana*
Clostridium perfringens	Food poisoning, gas gangrene	*Blatta orientalis*; Other species
Enterobacter sp.	Bacteremia	*Blattella germanica* *Diploptera punctata,* *Periplaneta americana*
Enterococcus sp.	Urinary-tract and wound infections	*Blattella germanica* *Periplaneta am ericana*
Escherichia coli	Diarrhea, wound infection	*Blatta orientalis* *Blattella germanica* *Diploptera punctata* *Periplaneta americana*
Hafnia alvei	Diarrhea	*Blattella germanica* *Periplaneta americana*
Klebsiella sp.	Pneumonia, urinary-tract infections	*Blattella germanica* *Diploptera punctata,* *Periplaneta americana*
Mycobacterium leprae	Leprosy	*Blattella germanica,* *Periplaneta americana,* *Periplaneta australasiae*
Nocardia sp.	Actinomycetoma	*Periplantea americana*
Morganella morganii	Wound infection	*Blattella germanica* *Periplaneta americana*
Oligella urethralis		*Periplaneta americana*
Pantoea sp.	Wound infection	*Blattella germanica*
Proteus rettgeri	Wound infection	*Periplaneta americana*
Proteus vulgaris	Wound infection	*Blaberus craniifer* *Blatta orientalis,* *Diploptera punctata* *Periplaneta americana*
Proteus mirabilis	Gastroenteritis, wound infection	*Periplaneta americana*
Pseudomonas sp.	Respiratory infections, gastroenteritis	*Diploptera punctata* *Blaberus craniifer* *Blatta orientalis* *Blattella germanica* *Periplaneta americana*
Salmonella sp.	Food poisoning, gastroenteritis	*Diploptera punctata*
Salmonella bredeny	Food poisoning, gastroenteritis	*Periplaneta americana*
Salmonella newport	Food poisoning, gastroenteritis	*Periplaneta americana*
Salmonella oranienburg	Fod poisoning, gastroenteritis	*Periplaneta americana*
Salmonella panama	Food poisoning, gastroenteritis	*Periplaneta americana*
Salmonella paratyphi-B	Food poisoning, gastroenteritis	*Periplaneta americana*
Salmonella pyogenes	Pneumonia	*Blatta orientalis*
Salmonella typhi	Typhoid	*Blatta orientalis*
Salmonella typhimurium	Food poisoning, gastroenteritis	*Blattella germanica* *Nauphoeta cinerea*
Salmonella bovis-morbificans	Food poisoning, gastroenteritis	*Periplantea americana*
Salmonella bareilly	Food poisoning, gastroenteritis	*Periplaneta americana*
Sphingobacterium sp.	Sepsis	*Blattella germanica* *Peiplaneta americana*
Serretia sp.	Food poisoning	*Blatta orientalis* *Blattella germanica* *Diploptera punctata* *Periplaneta americana*

Table 5.2 Bacteria Pathogenic to Humans that Have Been Isolated from Field-Collected Cockroaches—Cont'd

Pathogen	Associated Disease	Cockroach Species
Shigella dysenteriae	Dysentery	*Blattella germanica* *Blatta lateralis*
Sphingobacterium mizutae		*Periplaneta americana*
Staphylococcus aureus	Wound infection, skin infection, infection of internal organs	*Blaberus craniifer* *Blatta orientalis* *Blattella germanica* *Diploptera punctata* *Periplaneta americana*
Staphylococcus epidermidis	Wound infection	*Blattella germanica* *Periplaneta americana*
Streptococcus faecalis and other spp.	Pneumonia	*Blatta orientalis* *Blattella germanica* *Periplaneta americana*
Vibrio spp.	Not applicable	*Blatta orientalis*
Yersinia pestis	Plague	*Blatta orientalis*

from German cockroaches of a serotype of *S. dysenteriae* that was responsible for an outbreak of dysentery in Northern Ireland. Mackerras and Mackerras (1948) isolated *Salmonella bovis-morbificans* and *Salmonella typhimurium* from cockroaches captured in a hospital ward where gastroenteritis, attributed to the former organism, was common. In subsequent experimental studies, *Salmonella* organisms remained viable in the feces of cockroaches for as long as 40 days postinfection (Mackerras and Mackerras, 1949).

Some of the most compelling circumstantial evidence suggesting that cockroaches may be vectors was noted in a correlation between cases of infectious hepatitis and cockroach control at a housing project during 1956–1962 in southern California (Tarshis, 1962). The study area involved more than 580 apartments and 2,800 persons; 95% of the apartments had German cockroaches and a lesser infestation of brownbanded and Oriental cockroaches. After pest control measures were initiated, the incidence of endemic infectious hepatitis decreased for one year. When treatments were discontinued during the following year because the insecticide was offensive to apartment dwellers, the cockroach population increased, accompanied by a corresponding increase in the incidence of hepatitis. Effective control measures were applied the following two years, and cockroach populations and cases of infectious hepatitis dropped dramatically while hepatitis rates remained high in nearby housing projects where no pest control measures were conducted.

Intermediate Hosts

Cockroaches can serve as intermediate hosts for animal parasites (Table 5.3). Roth and Willis (1960) published an extensive list of biotic associations between cockroaches and parasitic organisms that potentially infest humans. The eggs of seven species of helminths have been found naturally associated with cockroaches. These include hookworms (*Ancylostoma duodenale* and *Necator americanus*), giant human roundworm (*Ascaris lumbricoides*), other *Ascaris* species, pinworm (*Enterobius vermicularis*), tapeworms (*Hymenolepis* species), and the whipworm *Trichuris trichuria.* Development of these helminths in cockroaches has not been observed. These relationships probably represent incidental associations with the omnivorous feeding behavior of cockroaches. However, cockroaches may serve as potential reservoirs and possible vectors through mechanical transfer in areas where a high incidence of these pathogens in humans is accompanied by substantial cockroach infestations. Human infestations by spirurid nematodes associated with cockroaches are known only for the cattle gullet worm (*Gongylonema pulchrum*) in the United States, Europe, Asia, and Africa; and for the stomach worm *Abbreviata caucasia* in Africa, Israel, Colombia, and Chile. Human cases involving these parasites are rare and cause no pathology.

Cockroach Allergies

The importance of cockroach allergies and their implications on childhood asthma has been well documented in recent years. Allergic reactions result after initial sensitization to antigens following inhalation, ingestion, dermal abrasion, or injection. Allergens produced by cockroaches now are recognized as one of the more significant indoor allergens of modernized societies. Among asthmatics, about half are allergic to cockroaches. This rate is exceeded only by allergies to house-dust mites. Sensitivity to cockroaches also affects about 10% of nonallergic individuals, suggesting a subclinical level of allergy.

Symptoms exhibited by persons allergic to cockroaches are similar to those described by Wirtz (1980), who reported on occupational allergies in entomologists. They include sneezing and a runny nose, skin reactions, and eye irritation in about two-thirds of the cases. In the more severe cases, individuals may experience difficulty breathing or, even

Table 5.3 Cockroaches as Intermediate Hosts of Parasites of Veterinary Importance

Phylum and Parasite	Scientific Name	Definitive Host	Cockroach Intermediate Host
ACANTHOCEPHALA (thorny-headed worms)	*Moniliformis moniliformis*	Rat, mice, dog, cat (primates)	*Blatta orientalis* *Blattella germanica*
	Moniliformis dubius	Rat	*Blattella germanica* *Periplaneta americana* *Periplaneta brunneus*
	Prosthenorchis elegans *Prosthenorchis spirula*	Captive primates	*Blattella germanica* *Leucophaea maderae*, others
PENTASTOMIDA (tongue worms)	*Raillietiella hemidactyli*	Reptiles	*Periplaneta americana*
NEMATODA (round worms)			
Esophageal and gastrointestinal worm	*Abbreviata caucasica*	Primates (humans)	*Blattella germanica*
Stomach worm	*Cyrnea colini*	Prairie chicken, turkey, bobwhite, quail	*Blattella germanica* *Periplaneta americana*
Esophagus worm	*Gongylonema neoplasticum*	Rodents, rabbit	*Blatta orientalis* *Periplaneta americana*
Gullet worm	*Gongylonema pulchrum*	Cattle (humans)	*Blattella germanica*
Gullet worm	*Gongylonema* sp.	Marmosets and Tamarins	*Periplaneta americana*
Stomach worm	*Mastophorus muris*	Rodents, cat	*Leucophaea maderae* *Periplaneta americana*
Eye worm	*Oxyspirura mansoni*	Chicken, turkey	*Pycnoscelus surinamensis*
Eye worm	*Oxyspirura parvorum*	Chicken, turkey	*Pycnoscelus surinamensis*
Esophageal worm	*Physaloptera rara*	Dog, cat, raccoon, coyote, wolf, fox	*Blattella germanica*
Esophageal worm	*Physaloptera praeputialis*	Dog, cat, coyote, fox	*Blattella germanica*
Round worms	*Protospirura bonnei* *Protospirura muricola*	Monkeys	*Blattella germanica* *Supella longipalpa*
Stomach worm	*Spirura rytipleurites*	Cat, rat	*Blatta orientalis*
Stomach worm	*Tetrameres americana*	Chicken, bobwhite, ruffed grouse	*Blattella germanica*
Stomach worm	*Tetrameres fissipina*	Ducks, geese, waterfowl; chicken, turkey, pigeon, quail	Various species

Rare definitive hosts are listed in parentheses.

more alarming, anaphylactic shock following exposure to cockroaches. Such allergic reactions can be life-threatening (Brenner et al., 1991).

In recent years, research has focused on determining the specific components of cockroaches that cause allergy. Laboratory technicians exhibit strong allergies to cast skins and excrement of German cockroaches, whereas most patients seen at allergy clinics react primarily to cast skins and whole-body extracts of German cockroaches. Once individuals are hypersensitized, they may experience severe respiratory distress simply by entering a room where cockroaches are held.

Several proteins that can cause human allergies have been identified in the German cockroach. Different exposure histories are likely to result in allergies to different proteins. Cast skins, excrement, and partially consumed food of cockroaches, in addition to living cockroaches, all produce allergenic proteins. Some are extremely persistent and can survive boiling water, ultraviolet light, and harsh pH changes, remaining allergenically potent for decades. Traditionally, whole-body extracts have been used to screen for allergens in skin tests and in bronchial challenges for diagnosing cockroach allergies (Figure 5.13). However, use of more specific antigens that become aerosolized in cockroach-infested homes may be more appropriate, as this is likely to be the sensitizing material. Studies with laboratory colonies have shown that a population of several thousand German cockroaches produced several micrograms of aerosolized proteins in 48 hours. Consequently, presence of cockroaches may have profound respiratory implications for asthmatic occupants of infested structures. For a general discussion on aerosolized arthropod allergens, see Solomon and Mathews (1988).

Development of an allergy to one insect species can result in broad cross-reactivity to other arthropods including shrimp, lobster, crab, crawfish, sowbugs (isopods), and house-dust mites. Chronic indoor exposure to cockroach allergens, therefore, may have significant and widespread effects on human health. Studies are currently under way to determine the clinical impact of cockroach mitigation and allergen reduction on allergies and asthma in affected children.

VETERINARY IMPORTANCE

Cockroaches serve as intermediate hosts for a number of parasitic worms of animals (Table 5.3). Most of these relationships are of no economic importance. The majority of the parasites are nematodes in the order Spirurida, all members of which use arthropods as intermediate hosts. Species infesting dogs and cats, among other hosts, attach to the mucosa of the gastrointestinal tract where erosion of tissue may occur at the points of attachment. Although serious damage seldom occurs, anemia and slow growth may result. Several cockroach-associated nematodes occur in Europe and North America. The esophageal worms *Physaloptera rara* and *P. praeputialis* are the most widespread species in the United States. They develop in the German cockroach, field crickets, and several species of beetles.

Poultry also are parasitized by nematodes that undergo development in cockroaches. The Surinam cockroach is the intermediate host for the poultry eye worms *Oxyspirura mansoni* and *O. parvorum*. Both occur in many parts of the world. In the United States, their distribution is limited to Florida and Louisiana. The German cockroach has been incriminated as the intermediate host for chicken and turkey parasites, including the stomach worms *Tetrameres americana*, *T. fissispina*, and *Cyrnea colini*; the latter also develops in the American cockroach. *Cyrnea colini* apparently causes no significant damage to poultry, but *Oxyspirura* species can cause pathology ranging from mild conjunctivitis to severe ophthalmia with seriously impaired vision. *Tetrameres fissispina* can cause severe damage to the proventriculus of infested birds.

Several nematode parasites of primates, rats, and cattle utilize cockroaches as intermediate hosts (Table 5.3). These include *Gongylonema neoplasticum* and *Mastophorus muris* in rodents. Both genera occur widely in the United States where they cause no known pathological problems. The gullet worm of cattle, *Gongylonema pulchrum*, has been shown experimentally to undergo development in the German cockroach. A *Gongylonema* species has been isolated from tongue scrapings of primates in zoos, where cockroaches and dung beetles have been found to be the intermediate hosts.

Exotic zoo animals also can become infested with parasitic nematodes for which cockroaches serve as possible intermediate hosts. *Protospirura bonnei* and *P. muricola*, for example, have been found in cockroaches collected in cages of monkeys. In a case of "wasting disease" in a colony of common marmosets, more than 50% of German and brownbanded cockroaches captured in the animal room in which they were housed contained the coiled larvae of *Trichospirura leptostoma* in muscle cells (Beglinger et al., 1988).

Acanthocephalans (thorny-headed worms) commonly infest primates in zoos and research facilities. *Prosthenorchis elegans* and *P. spirula* occur naturally in South and Central America. Their natural intermediate hosts are unknown. In captivity, primates become infected after eating any of several cockroach species in which the intermediate stages of the parasite have completed development. Heavily infested primates frequently die within a few days. The proboscis of acanthocephalan adults commonly penetrates the intestines of the primate host, causing secondary infections, perforation of the gut wall, and peritonitis.

One pentastomid (tongue worm), *Raillietiella hemidactyli*, develops in cockroaches and reptilian hosts. In Singapore, infested geckos are a common occurrence in houses where heavy infestations of *R. hemidactyli* larvae have been found in American cockroaches. Remnants of cockroaches are found commonly in the guts of these lizards.

For additional information on the veterinary importance of cockroaches, see Chitwood and Chitwood (1950), Roth and Willis (1957), Levine (1968), and Noble and Noble (1976).

PREVENTION AND CONTROL

Integrated Pest Management (IPM) strategies commonly are used to manage cockroach infestations. IPM is a multifaceted approach that attempts to eliminate the habitat and conditions that sustain cockroach populations, utilizing mechanical, biological, physical, or chemical means.

Sanitation

Cleanup to reduce cockroaches should focus mainly on the food residue in and around coffee machines, sinks, stoves, microwave ovens, refrigerators, trash cans, and furniture where food residues accumulate. Removal of clutter, such as corrugated cardboard, is especially important since it provides excellent harborage for cockroaches.

Harborage Elimination

Permanent reduction of cockroach populations can be achieved by eliminating harborage through caulking and sealing. If not performed properly this can make a bad problem worse by creating additional inaccessible harborage. Building designs and construction techniques can significantly influence cockroach survival. By manipulating microclimates in areas of structures frequented by cockroaches, homes and other buildings can be rendered less hospitable to pest species, while greatly reducing aerosolized allergens. Nontoxic repellents can be used to deter cockroaches from entering specific areas.

Physical Control

This includes a variety of mechanical techniques such as vacuuming, sticky traps, and pitfall traps, which are used to reduce cockroach numbers by removing them from the environment. Heat, cold, anoxia, and steam also can be used to kill cockroaches.

Biological Control

This has drawn increased attention in recent years. Among the natural agents that have been investigated are parasitic wasps, nematodes, and sporulating fungi. Females of the eulophid wasp *Aprostocetus hagenowii* and the encyrtid wasp *Comperia merceti* deposit their eggs in the oothecae of certain peridomestic cockroaches. Major shortcomings in utilizing these wasps are difficulties involved in their mass production and the fact that they do not completely eliminate cockroach infestations. However, *A. hagenowii* has been shown to reduce populations of the peridomestic *Periplaneta* species following inundative or augmentative releases of this wasp. *Comperia merceti* parasitizes oothecae of the brownbanded cockroach and is the only known parasitoid of a domestic species. The use of parasitic nematodes (e.g., *Steinernema carpocapsae*) and several fungal pathogens that have been isolated from cockroaches have not yet proved to be effective as practical management tools. Another drawback to their use is the allergenic nature of several components of nematodes and many sporulating fungi that can become airborne and, upon inhalation, cause asthmatic responses in humans.

Traditionally, cockroaches were controlled using residual pesticides, such as organophosphates and carbamates (most are no longer registered for use) applied to harborage sites and areas frequented by foraging individuals (Ebling, 1975; Rust et al., 1995). Other widely used products for these purposes include pyrethroids and botanicals such as pyrethrins, as well as several new classes of insecticides that are metabolic inhibitors or disruptors (e.g., chlornicotinyls, pyrroles, macrocyclic lactones, amidinohydrazone, and phenylpyrazoles). These active ingredients are formulated into a variety of products, such as wettable powders, emulsifiable concentrates, aerosols, dusts, microencapsulates, and baits. Other materials with different modes of action can be used. For instance, boric acid is delivered as a fine powder or a dilute solution, which when ingested damages the gut epithelium of cockroaches and kills them by interfering with nutrient absorption. Inorganic silica dust is absorptive, reducing cuticular lipids and causing desiccation.

The use of baits containing several of the active ingredients just mentioned currently are used extensively to control cockroaches. These baits are used indoors in the form of child-resistant bait stations or gels that are applied in cracks and crevices, making them inaccessible to children and pets. Scatter baits commonly are used outdoors to treat mulches and other landscaping materials that harbor cockroaches.

Insect Growth Regulators (IGRs)

These can be used to prevent cockroaches from reaching maturity. Two commonly used IGRs are juvenile hormone analogs and chitin synthesis inhibitors. Juvenile hormone analogs regulate morphological maturation and reproductive processes. They are highly specific to arthropods, have very low mammalian toxicity, and are effective at exceptionally low rates of application. Such compounds include hydroprene and fenoxycarb. Chitin synthesis inhibitors prevent normal formation of chitin during molting. These compounds cause many of the affected nymphs to die during the molting process. Males that survive to the adult stage often have reduced life expectancies, whereas females tend to abort their oothecae.

REFERENCES AND FURTHER READING

Atkinson, T. H. P., Koehler, G., & Patterson, R. S. (1991). Catalog and atlas of the cockroaches (Dictyoptera) of North America north of Mexico. *Miscellaneous Publications of the Entomological Society of America, 78*, 1–86.

Beglinger, R., Illgen, B., Pfister, R., & Heider, K. (1988). The parasite *Trichospirura leptostoma* associated with wasting disease in a colony of common marmosets, *Callithrix jacchus. Folia Primatologica, 51*, 45–51.

Bell, W. J., Roth, L. M., & Nalepa, C. A. (2007). *Cockroaches: Ecology, Behavior, and Natural History.* Baltimore: The Johns Hopkins University Press.

Boudreaux, H. B. (1979). *Arthropod Phylogeny with Special Reference to Insects.* New York: John Wiley & Sons.

Brenner, R. J. (1988). Focality and mobility of some peridomestic cockroaches in Florida. *Annals of the Entomological Society of America, 81*, 581–592.

Brenner, R. J. (1991). Asian cockroaches: Implications to the food industry and complexities of management strategies. In J. R. Gorham (Ed.), *Ecology and Management of Food-industry Pests* (pp. 121–130). U.S. Food & Drug Admin. Tech. Bull. No. 4

Brenner, R. J. (1995). Economics and medical importance of German cockroaches. In M. K. Rust, J. M. Owens, D. A. Reierson (Eds.), *Understanding and Controlling the German Cockroach.* Oxford University Press.

Brenner, R. J., Barnes, K. C., Helm, R. M., & Williams, L. W. (1991). Modernized society and allergies to arthropods: Risks and challenges to entomologists. *American Entomologist, 37*, 143–155.

Brenner, R. J., Koehler, P. G., & Patterson, R. S. (1987). *Infections in Medicine, 4*, 349–355, 358, 359, 393.

Brenner, R. J., Patterson, R. S., & Koehler, P. G. (1988). Ecology, behavior, and distribution of *Blattella asahinai* (Orthoptera: Blattellidae) in central Florida. *Annals of the Entomological Society of America, 81*, 432–436.

Burgess, N. R. (1982). Biological features of cockroaches and their sanitary importance. In D. Bajomi, G. Erdos (Eds.), *The Modern Defensive Approach of Cockroach Control. Internat. Sympos* (pp. 45–50). Bologna.

Chitwood, B. G., & Chitwood, M. B. (1950). In B. G. Chitwood(Eds.), *An Introduction to Nematology. Section I. Anatomy.* Baltimore.

Cornwell, P. B. (1968). *The Cockroach* (Vol. 1). London: Hutchinson & Co. Ltd.

Ebling, W. (1975). *Urban Entomology.* Berkeley: University of California., Div. Agr. Sci.

Elgderi, R. M., Ghenghesh, K. S., & Berbash, N. (2006). Carriage by the German cockroach (*Blattella germanica*) of multiple-antibiotic-resistant bacteria that are potentially pathogenic to humans, in hospitals and households in Tripoli, Libya. *Annals of Tropical Medicine and Parasitology, 100*, 55–62.

Gazivoda, P., & Fish, D. (1985). Scanning electron microscope demonstration of bacteria on tarsi of *Blattella germanica. Journal of the New York Entomological Society, 93*, 1064–1067.

Helfer, J. R. (1987). *How to Know the Grasshoppers, Cockroaches and their Allies.* Dubuque: W. C. Brown Co.

Koehler, P. G., Patterson, R. S., & Brenner, R. J. (1990). Cockroaches, Chapter 3. In K. Story (Ed.), *Mallis Handbook of Pest Control* (7th ed.). Chapter 3. Cleveland: Franzak and Foster Co.

Lemos, A. A., Lemos, J. A., Prado, M. A., Pimenta, F. C., Gir, E., Silva, H. M., et al. (2006). Cockroaches as carriers of fungi of medical importance. *Mycoses, 49*, 23–25.

Levine, N. D. (1968). *Nematode Parasites of Domestic Animals and of Man.* Minneapolis: Burgess Pub. Co.

Mackerras, M. J., & Mackerras, I. M. (1948). *Salmonella* infections in Australian cockroaches. *Australian Journal of Science, 10*, 115.

Mackerras, I. M., & Mackerras, M. J. (1949). An epidemic of infantile gastroenteritis in Queensland caused by *Salmonella bovis-morbicans* (Basenau). *Journal of Hygiene, 47*, 166–181.

Massicot, J. G., & Cohen, S. G. (1986). Epidemiologic and socioeconomic aspects of allergic diseases. *The Journal of Allergy and Clinical Immunology, 78*, 954–958.

McKittrick, F. A. (1964). *Evolutionary Studies of Cockroaches.* Cornell Univ. Agric. Exp. Sta Memoir No. 389.

Mpuchane, S., Allotey, J., Matsheka, I., Simpanya, M., Coetzee, S., Jordaan, A., et al. (2006). Carriage of micro-organisms by domestic cockroaches and implications on food safety. *International Journal of Tropical Insect Science, 26,* 166–175.

Noble, E. R., & Noble, G. A. (1976). *Parasitology* (4th ed.). Philadelphia: Lea & Febiger.

Pai, H. H., Chen, W. C., & Peng, C. F. (2004). Cockroaches as potential vectors of nosocomial infections. *Infection Control and Hospital Epidemiology, 25,* 979–984.

Peterson, R. K. D., & Shurdut, B. A. (1999). Human health risk from cockroaches and cockroach management: a risk analysis approach. *American Entomologist, 45,* 142–148.

Roth, L. M. (1985). A taxonomic revision of the genus *Blattella* Caudell (Dictyoptera, Blattaria: Blattellidae). *Entomology Scandinavian,* Suppl. 22.

Roth, L. M., & Willis, E. R. (1957). The medical and veterinary importance of cockroaches. *Smithsonian Miscellaneous Collection, 134,* 1–147.

Roth, L. M., & Willis, E. R. (1960). The biotic associations of cockroaches. *Smithsonian Miscellaneous Collection, 141.*

Rust, M. K., Owens, J. M., & Reierson, D. A. (Eds.). (1995). *Understanding and Controlling the German Cockroach.* Oxford University Press.

Scott, H. G., & Borom, M. R. (1964). *Cockroaches: Key to Egg Cases of Common Domestic Species.* Atlanta, GA: U. S. Dept. of Health, Education and Welfare, Public Health Service, Communicable Disease Center.

Smith, E. H., & Whitman, R. C. (2000). *NPCA Field Guide to Structural Pests.* Fairfax, VA: National Pest Management Association.

Snoddy, E. T., & Appel, A. G. (2008). *Distribution of Blatella asahinai* (Dictyoptera: Blattellidae) in southern Alabama and Georgia. *Annals of the Entomological Society of America, 101,* 397–401.

Solomon, W. R., & Mathews, K. P. (1988). Aerobiology and inhalant allergens. In E. , Middleton Jr. C. E. Reed, E. F. Ellis, N. F. Adkinson Jr., J. W. Yunginger (Eds.), *Allergy Principles and Practice* (3rd ed.). Washington, D.C: C. V. Mosby Company.

Tarshis, I. B. (1962). The cockroach—A new suspect in the spread of infectious hepatitis. *The American Journal of Tropical Medicine and Hygiene, 11,* 705–711.

Wirtz, R. A. (1980). Occupational allergies to arthropods—Documentation and prevention. *Bulletin of the Entomological Society of America, 26,* 356–360.

Chapter

6

Lice (Phthiraptera)

Lance A. Durden . John E. Lloyd

Lice can be a menace to humans, pets, and livestock, not only through their blood-feeding or chewing habits, but also because of their ability to transmit pathogens. The human body louse has been indirectly responsible for influencing human history through its ability to transmit the causative agent of epidemic typhus. However, most of the approximately 5,000 known species of lice are ectoparasites of wild birds or mammals and have no known medical or veterinary importance.

The order Phthiraptera is divided into two main groups, sucking lice and chewing lice. All sucking lice are obligate, hematophagous ectoparasites of placental mammals, whereas the more diverse chewing lice include species that are obligate associates of birds, marsupials, or placental mammals. Although certain chewing lice imbibe blood, most species ingest host feathers, fur, skin, or skin products. Because of the different feeding strategies of the two groups, the blood-feeding sucking lice are far more important than the chewing lice in transmitting pathogens to their hosts.

TAXONOMY

The order Phthiraptera is divided into four suborders (Figure 6.1; Tables 6.1, 6.2), the Anoplura (sucking lice), and the Amblycera, Ischnocera, and Rhynchophthirina (collectively known as chewing lice or biting lice). Previous classifications treated the Anoplura and Mallophaga as separate orders with the Amblycera, Ischnocera, and Rhynchophthirina all included in the Mallophaga. However, phylogenetic analyses have shown that the chewing lice do not represent a monophyletic group and that some are more closely related to members of the Anoplura than to other chewing lice. Further, sucking and chewing lice originated from a common nonparasitic ancestral group closely related to, or within, the order Psocoptera (book lice and bark lice). These two groups diverged in the late Jurassic or early Cretaceous, 100 to 150 million years ago. Unfortunately, the fossil record for Phthiraptera is sparse with, at present, only one proven specimen, an amblyceran menoponid chewing louse from Germany dated at 44.3 ± 0.4 million years old (Dalgleish et al., 2006).

About 550 species of sucking lice have been described (Durden and Musser, 1994a), all of which parasitize placental mammals. These lice are currently assigned to 50 genera and 15 families. Price et al. (2003) recognize 4,464 valid species and subspecies of chewing lice; most of these taxa are associated with birds, but 553 of them (12.4%) parasitize mammals. The chewing lice are divided into three suborders (Table 6.1), 11 families, and 205 genera. According to Price et al. (2003), within the chewing lice, the Amblycera includes seven families, about 76 genera, and about 850 species; Ischnocera includes three families, about 130 genera, about 1800 species and Rhynchophthirina includes one family, one genus, and three species.

Major taxonomic syntheses for the sucking lice include a series of eight volumes by Ferris (1919–1935) that remains the most comprehensive treatment of this group on a worldwide basis. Ferris (1951) updated much of his earlier work in a shorter overview of the group. Kim et al. (1986) have compiled an authoritative manual and identification guide for the sucking lice of North America. Durden and Musser (1994a) provide a taxonomic checklist for the sucking lice of the world with host records and geographical distribution for each species.

The chewing lice are taxonomically less well known than are the sucking lice, and fewer authoritative identification guides are available. These include a synopsis of the lice associated with laboratory animals (Kim et al., 1973), guides to the lice of domestic animals (Tuff, 1977; Price and Graham 1997), and an identification guide to the lice of sub-Saharan Africa (Ledger, 1980). These three publications provide information on both sucking lice and chewing lice. Checklists of the chewing lice of the world (Price et al., 2003) and of North America (Emerson, 1972) are useful taxonomic references for this group.

Because of the relatively high degree of host specificity exhibited by both chewing and sucking lice, several host-parasite checklists have been prepared.

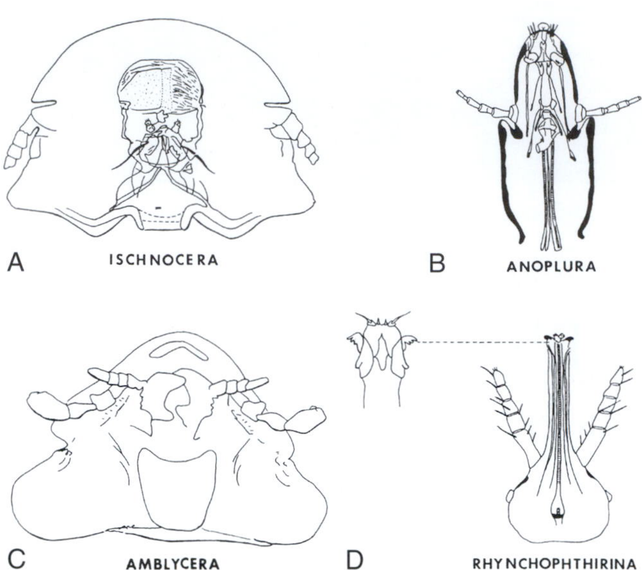

Figure 6.1 Head and mouthparts of representatives of each of the four principal groups of lice. (A) Ishnocera (Clay, 1938); (B) Anoplura (Ferris, 1931); (C) Amblycera (Bedford, 1932); (D) Rhynchophthirina (Ferris, 1931).

These include a detailed list of both sucking and chewing lice associated with mammals (Hopkins, 1949), a host-parasite list for North American chewing lice (Emerson, 1972), a world host-parasite list for the chewing lice (Price et al., 2003), and a host-parasite checklist for the Anoplura of the world (Durden and Musser 1994b).

Sucking lice of medical importance are assigned to two families, the Pediculidae and Pthiridae, whereas sucking lice of veterinary importance are assigned to five families, the Haematopinidae, Hoplopleuridae, Linognathidae, Pedicinidae, and Polyplacidae (Table 6.2). Only one species of chewing louse, the dog biting louse, in the family Trichodectidae, has public health importance. Chewing lice of veterinary significance typically are placed in five families, the Boopiidae, Gyropidae, Menoponidae, Philopteridae, and Trichodectidae (Table 6.1).

Table 6.1 Classification and Hosts of Chewing Lice of Medical and Veterinary Importance

Lice	Hosts
Suborder Amblycera	
Family Boopiidae:	
Heterodoxus spiniger	Dog, other carnivores
Family Gyropidae:	
Slender guineapig louse, *Gliricola porcelli*	Guinea pig
Oval guinea pig louse, *Gyropus ovalis*	Guinea pig
Family Menoponidae:	
Chicken body louse, *Menacanthus stramineus*	Domestic fowl
Shaft louse, *Menopon gallinae*	Domestic fowl
Goose body louse, *Trinoton anserinum*	Geese
Large duck louse, *Trinoton querquedulae*	Ducks
Suborder Ischnocera	
Family Philopteridae:	
Slender goose louse, *Anaticola anseris*	Geese
Slender duck louse, *Anaticola crassicornis*	Ducks
Large turkey louse, *Chelopistes meleagridis*	Turkey
Chicken head louse, *Cuclotogaster heterographus*	Domestic fowl
Fluff louse, *Goniocotes gallinae*	Domestic fowl
Brown chicken louse, *Goniodes dissimilis*	Chicken
Large chicken louse, *Goniodes gigas*	Domestic fowl
Wing louse, *Lipeurus caponis*	Domestic fowl
Slender turkey louse, *Oxylipeurus polytrapezius*	Turkey
Family Trichodectidae:	
Cattle biting louse, *Bovicola bovis*	Cattle
Goat biting louse, *Bovicola caprae*	Goat
Angora goat biting louse, *Bovicola crassipes*	Goat
Horse biting louse, *Bovicola equi*	Horse
Bovicola limbata	Goat
Donkey biting louse, *Bovicola ocellata*	Donkey
Sheep biting louse, *Bovicola ovis*	Sheep
Cat biting louse, *Felicola subrostrata*	Cats
Dog biting louse, *Trichodectes canis*	Dog, other canids
Suborder Rhynchophthirina	
Family Haematomyzidae:	
Elephant louse, *Haematomyzus elephantis*	Elephants

MORPHOLOGY

Lice are small (0.4–10 mm in the adult stage), wingless, dorso-ventrally flattened insects. The elongate abdomen possesses sclerotized dorsal, ventral, or lateral plates in many lice (Figure 6.2); these provide some rigidity to the abdomen when it is distended by a blood meal or other food source. In adult lice the abdomen is 11-segmented and terminates in genitalia and associated sclerotized plates. In females, the genitalia are accompanied by two pairs of finger-like gonopods, which serve to guide, manipulate, and glue eggs onto host hair or feathers. The abdomen is adorned with numerous setae in most lice. Immature lice closely resemble adults but are smaller, have fewer setae, and lack genitalia. After each nymphal molt, the abdomen is beset with progressively more setae, and the overall size of the louse increases.

The male genitalia in lice (Figure 6.3) are relatively large and conspicuous, sometimes occupying almost half the length of the abdomen. The terminal, extrusable, sclerotized, pseudopenis (aedeagus) is supported anteriorly by a basal apodeme. Laterally, it is bordered by a pair of chitinized parameres. Two or four testes are connected to the vas deferens, which coalesce posteriorly to form the vesicula seminalis. In the female, the vagina leads to a large uterus to which several

Table 6.2 Classification and Hosts of Sucking Lice (Anoplura) of Medical and Veterinary Importance

Lice	Hosts
Family Echinophthiriidae:	
Echinophthirius horridus	Harbor seals
Family Haematopinidae:	
Horse sucking louse, *Haematopinus asini*	Horse, donkey
Shortnosed cattle louse, *Haematopinus eurysternus*	Cattle
Cattle tail louse, *Haematopinus quadripertusus*	Cattle
Hog louse, *Haematopinus suis*	Swine
Buffalo louse, *Haematopinus tuberculatus*	Asiatic buffalo, cattle
Family Hoplopleuridae:	
Hoplopleura captiosa	House mouse
Tropical rat louse, *Hoplopleura pacifica*	Domestic rats
Family Linognathidae:	
African blue louse, *Linognathus africanus*	Goat, sheep, deer
Sheep face louse, *Linognathus ovillus*	Sheep
Sheep foot louse, *Linognathus pedalis*	Sheep
Dog sucking louse, *Linognathus setosus*	Dog, other canids
Goat sucking louse, *Linognathus stenopsis*	Goat
Longnosed cattle louse, *Linognathus vituli*	Cattle
Little blue cattle louse, *Solenopotes capillatus*	Cattle
Family Pedicinidae:	
Pedicinus spp.	Old World Primates
Family Pediculidae:	
Head louse, *Pediculus humanus capitis*	Human
Body louse, *Pediculus humanus humanus*	Human
Family Polyplacidae:	
Rabbit louse, *Haemodipsus ventricosus*	Domestic rabbit
Mouse louse, *Polyplax serrata*	House mouse
Spined rat louse, *Polyplax spinulosa*	Domestic rats
Family Pthiridae:	
Crab louse, *Pthirus pubis*	Human

ovarioles supporting eggs in various stages of development are connected by the oviducts. Two or more large accessory glands, which secrete materials to coat the eggs, and a single spermatheca, in which sperm is stored, are situated posteriorly in the abdomen. Except for the human body louse, all lice cement their eggs, called **nits,** onto the hair or feathers of their host. Eggs are usually subcylindrical with rounded ends and a terminal cap, the operculum (Figure 6.4). On the top of the operculum is a patch of holes or areas with thin cuticle, called aeropyles, through which the developing embryo respires. Most of the egg is heavily chitinized, which helps to protect the embryo from mechanical damage and desiccation (and from insecticides in many cases). A suture of thin cuticle encircles the base of the operculum. At the time of hatching, the first-instar nymph emerges from the egg by cracking this suture and pushing off the

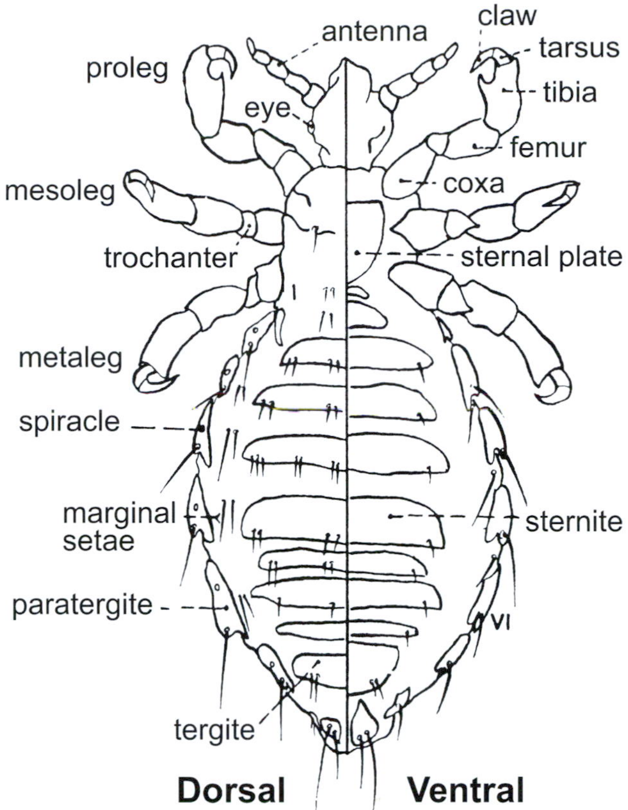

Figure 6.2 A generalized sucking louse (Anoplura), showing dorsal (left) and ventral (right) morphology (Ignoffo, 1959).

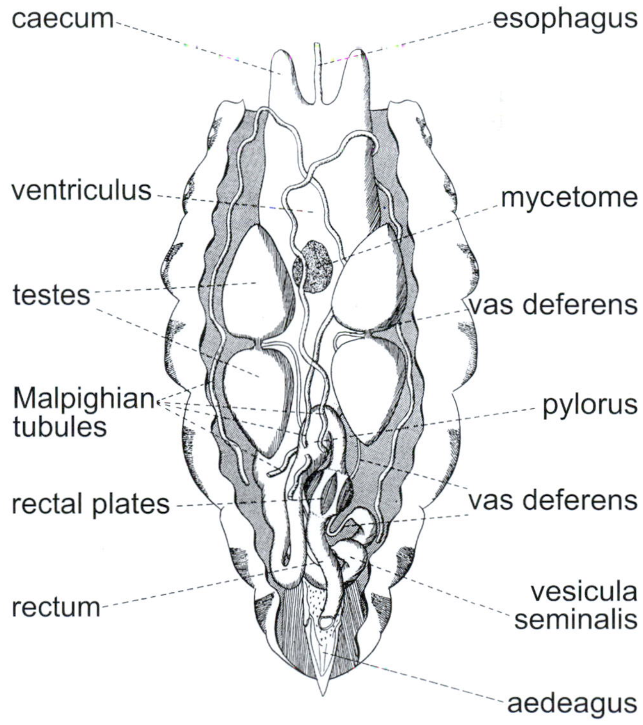

Figure 6.3 Internal abdominal anatomy of a male human body louse (*Pediculus humanus humanus*) (Ferris, 1951).

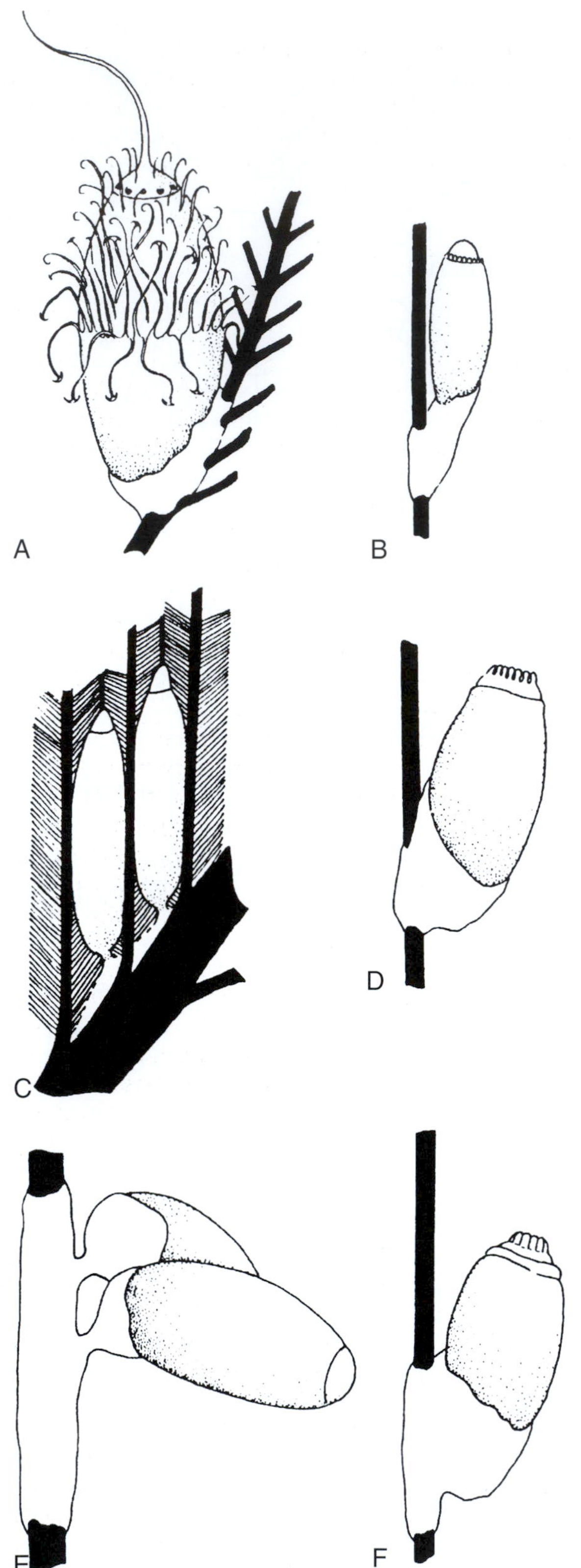

Figure 6.4 Eggs (nits) of representative lice. (A) *Menacanthus stramineus* (Amblycera); (B) *Gyropus ovalis* (Amblycera); (C) *Columbicola columbae* (Ischnocera); (D) *Bovicola bovis* (Ischnocera); (E) *Haematomyzus elephantis* (Rhynchophthirina); (F) *Pediculus humanus capitis* (Anoplura). (Modified from Marshall, 1981)

operculum. In chewing lice, the head is broader than the thorax (Figure 6.5).

Amblyceran chewing lice have four-segmented antennae, and have retained the maxillary palps characteristic of their psocopteran ancestor. However, ischnoceran chewing lice have three to five antennal segments and lack maxillary palps. In the Amblycera, the antennae are concealed in lateral grooves, whereas in the Ischnocera and Rhynchophthirina, the antennae are free from the head (Figures 6.1, 6.5).

There is a gradation in the specialization of the mouthparts, and of the internal skeleton of the head, or *tentorium*, from the psocopteran ancestor of the lice through the Amblycera, Ischnocera, Rhynchophthirina, and Anoplura. Although all chewing lice possess chewing mouthparts (Figure 6.6), the components and mechanics of these mouthparts differ for different groups. For example, members of the Rhynchophthirina possess tiny mandibles that are situated at the tip of an elongated rostrum (Figures 6.1D, 6.5F). Also, through extreme modifications, members of the chewing louse genus *Trochilocoetes* (parasites of humming birds) have evolved mouthparts that can function as sucking organs.

The thorax in chewing lice usually appears dorsally as two, occasionally three segments. Chewing lice

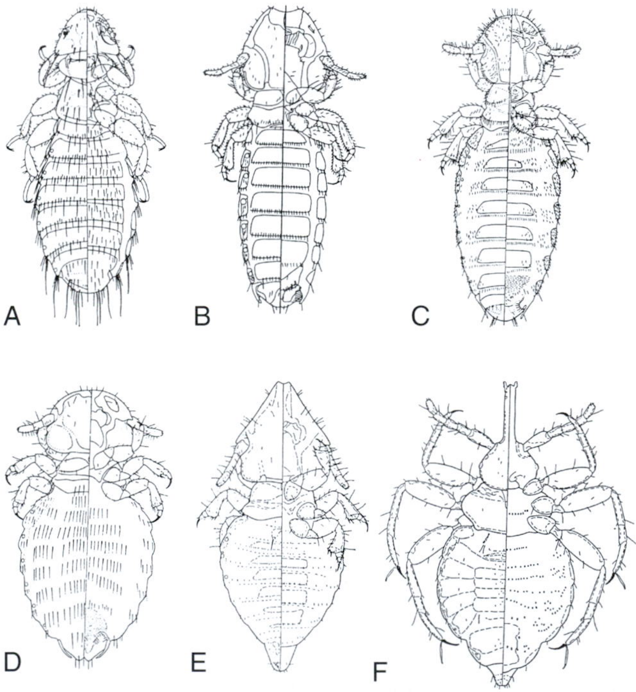

Figure 6.5 Chewing lice of veterinary importance, showing dorsal morphology (left) and ventral morphology (right) in each case. Not drawn to scale. (A) *Heterodoxus spiniger*, female, from carnivores; (B) *Tricholipeurus parallelus*, male, from New World deer; (C) Sheep biting louse (*Bovicola ovis*), female; (D) Dog biting louse (*Trichodectes canis*), female; (E) Cat biting louse (*Felicola subrostrata*), male; (F) Elephant louse (*Haematomyzus elephantis*), male. (A-E, from Emerson and Price, 1975; F, from Werneck, 1950)

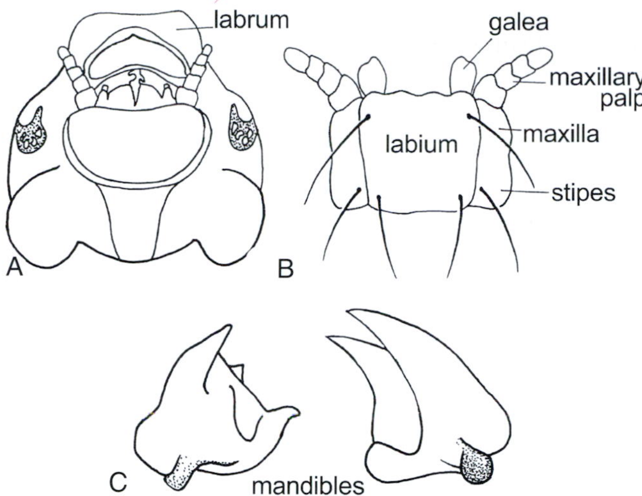

Figure 6.6 Generalized mouthparts of a chewing louse. (A) Ventral view of head; (B) Labium and associated structures; (C) Mandibles. (Drawn by Margo Duncan)

possess one or two simple claws on each leg; species that parasitize highly mobile hosts, especially birds, typically have two claws.

In sucking lice (Figures 6.2, 6.7), the head is slender and narrower than the thorax. Anoplura have three- to five-segmented antennae and lack maxillary palps.

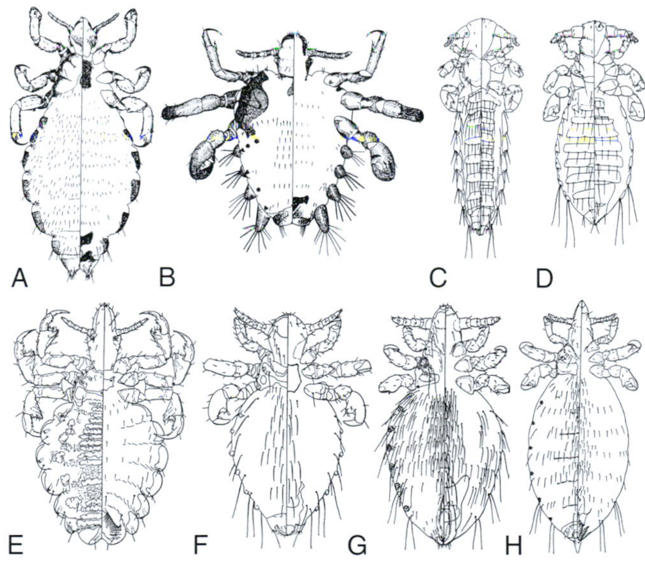

Figure 6.7 Sucking lice (Anoplura) of medical and veterinary importance, showing dorsal morphology (left) and ventral morphology (right) in each case. Not drawn to scale. (A) Human body louse (*Pediculus humanus humanus*), female; (B) Human crab louse (*Pthirus pubis*), female; (C) Flying squirrel louse (*Neohaematopinus sciuropteri*), male; (D) Spined rat louse (*Polyplax spinulosa*), male; (E) Hog louse (*Haematopinus suis*), female; (F) Little blue cattle louse (*Solenopotes capillatus*), male; (G) Dog sucking louse (*Linognathus setosus*), male; (H) Longnosed cattle louse (*Linognathus vituli*), female (Ferris, 1923–1935).

Noncompound eyes, which represent groups of ocelli, are reduced or absent in most sucking lice but are well developed in members of the medically important genera *Pediculus* and *Pthirus* (Figure 6.7A, B). **Ocular points**, or eyeless projections posterior to the antennae, are characteristic of sucking lice in the genus *Haematopinus* (Figure 6.7E).

As indicated by their name, anopluran mouthparts function as sucking devices during blood-feeding (Figure 2.1F). At rest, the mouthparts are withdrawn into the head and are protected by the snout-like haustellum, representing the highly modified labrum. The haustellum is armed with tiny recurved teeth that hook into the host skin during feeding. The stylets, consisting of a serrated labium, the hypopharynx, and two maxillae then puncture a small blood vessel (Figure 2.1F). The hypopharynx is a hollow tube through which saliva (containing anticoagulants and enzymes) is secreted. The maxillae oppose each other and are curved to form a food canal through which host blood is imbibed (Figure 2.1F).

In sucking lice, all three thoracic segments are fused and appear as one segment dorsally. In most species, the second and third pairs of legs terminate in highly specialized claws for grasping the host pelage. These **tibio-tarsal claws** consist of a curved tarsal element that opposes a tibial spur (Figure 2.3C, D) to enclose a space that typically corresponds to the diameter of the host hair.

The internal anatomy of lice (Figure 6.3) is best known for the human body louse. As in most hematophagous insects, strong cibarial and esophageal muscles produce a sucking action during blood-feeding. The esophagus leads to a spacious midgut composed primarily of the ventriculus. The posterior region of the midgut is narrow and forms a connection between the ventriculus and the hindgut. Ventrally, **mycetomes** (also called bacteriomes or stomach discs by some authors) containing symbiotic microorganisms connect to the ventriculus. These symbiotes synthesize B vitamins that are lacking in the blood meal.

LIFE HISTORY

Lice are hemimetabolous insects. Following the egg stage, there are three nymphal instars, the last of which molts to an adult. Although there is wide variation between species, the egg stage typically lasts for four to 15 days, each nymphal instar for three to eight days, and adults live for up to 35 days. Under optimal conditions many species of lice can complete 10 to 12 generations per year, but this is rarely achieved in nature. Host grooming, resistance, molting and feather loss, hibernation, hormonal changes, as well as predators (especially insectivorous birds on large ungulates), parasites and parasitoids, and unfavorable weather conditions can reduce the number of louse generations.

Fecundities for fertilized female lice vary from 0.2 to 10 eggs per day. Males are unknown in some parthenogenetic species such as the *Damalinia* sp. louse that causes hair-loss syndrome in deer, and typically constitute less than 5% of adults in the cattle biting louse, and less than 1% in the horse biting louse.

BEHAVIOR AND ECOLOGY

Blood from the host is essential for the successful development and survival of all sucking lice. Anoplura are vessel feeders, or **solenophages**, that imbibe blood through a hollow dorsal stylet derived from the hypopharynx (Figure 2.1F). Contraction of powerful cibarial and pharyngeal muscles create a sucking reaction for imbibing blood.

Chewing lice feed by the biting or scraping action of the mandibles. Bird-infesting chewing lice typically use their mandibles to sever small pieces of feather, which drop onto the labrum and are then forced into the mouth. Chewing lice that infest mammals use their mandibles in a similar manner to feed on host fur. Many chewing lice that infest birds and mammals can also feed on other integumental products such as skin debris and secretions. Some species of chewing lice are obligate, or more frequently facultative, hematophages. Even those species of chewing lice that imbibe blood scrape the host integument until it bleeds. The rhynchophthirinan *Haematomyzus elephantis*, which parasitizes both African and Asian elephants, feeds in this manner.

Symbionts are thought to be present in all lice that imbibe blood. Symbionts in the mycetomes (Figure 6.3) (also called bacteriomes or stomach discs) synthesize vitamins essential to growth and reproduction, and lice deprived of them die after a few days; female lice lacking symbionts also become sterile. In female human body lice, some symbionts migrate to the ovary where they are transferred transovarially to the next generation of lice.

Many lice exhibit host specificity, some to such a degree that they parasitize only one species of host. The hog louse, slender guinea pig louse, large turkey louse, and several additional species listed in Tables 6.1 and 6.2 all are typical parasites of a single host species. Host specificity is less evident in some lice. Certain lice of veterinary importance parasitize two or more closely related hosts. Examples include the three species that parasitize domestic dogs: *Linognathus setosus* (Figure 6.7G), *Trichodectes canis* (Figure 6.5D), and *Heterodoxus spiniger* (Figure 6.5A). These lice also parasitize foxes, wolves, coyotes, and occasionally other carnivores. Similarly, the horse sucking louse (*Haematopinus asini*) parasitizes horses, donkeys, asses, mules, and zebras, whereas the African blue louse, *Linognathus africanus*, parasitizes both sheep and goats. At least six species of chewing lice are found on domestic fowl, most of them parasitizing chickens, but some also feed on turkeys, guinea fowl, pea fowl, or pheasants (Table 6.1). Lice found on atypical hosts are termed **stragglers**.

Some sucking lice, such as the three forms that parasitize humans, the sheep foot louse, and sheep face louse, are not only host specific but also infest specific body areas from which they can spread in severe infestations. Many chewing lice, particularly species that parasitize birds, also exhibit both host specificity and site specificity; examples include several species that are found on domestic fowl, and species confined to turkeys, geese, and ducks (Table 6.1). Lice inhabiting different body regions on the same host typically have evolved morphological adaptations in response to specific attributes of the host site. These include characteristics such as morphological differences of the pelage, thickness of the skin, availability of blood vessels, and grooming or preening activities of the host.

Site specificity in chewing lice is most prevalent in the more sedentary, specialized, ischnocerans than in the mostly mobile, morphologically unspecialized amblycerans. For example, on many bird hosts, round-bodied ischnocerans with large heads and mandibles are predominately found on the head and neck. Elongate forms with narrow heads and small mandibles tend to inhabit the wing feathers, whereas morphologically intermediate forms occur on the back and other parts of the body.

Some chewing lice inhabit highly specialized host sites. These include members of the amblyceran genus *Piagetiella*, which are found inside the oral pouches of pelicans, and members of several amblyceran genera, including *Actornithophilus* and *Colpocephalum*, which live inside feather quills. Several bird species are parasitized by five or more different species of site-specific chewing lice, and up to 12 species may be found on the neotropical bird *Crypturellus soui* (a tinamou).

Site specificity is less well documented for sucking lice. However, domestic cattle may be parasitized by as many as five anopluran species, each predominating on particular parts of the body. Similarly, some Old World squirrels and rats can support up to six species of sucking lice, often on different parts of the body.

Because of the importance of maintaining a permanent or close association with the host, lice have evolved specialized host-attachment mechanisms to resist grooming activities of the host. The robust tibio-tarsal claws of sucking lice (Figure 2.3C, D) are very important in securing them to their hosts. Various arrangements of hooks and spines, especially on the head of lice that parasitize arboreal or flying hosts, such as squirrels and birds, also aid in host attachment. Mandibles are important attachment appendages in ischnoceran and rhynchophthirinan chewing lice. In some species of *Bovicola*, a notch in the first antennal segment encircles a host hair to facilitate attachment. A few lice even possess ctenidia (combs) that are convergently similar in morphology to those characteristic of many fleas. They occur most notably among lice that parasitize coarse-furred, arboreal, or flying hosts. Additionally, chewing lice that parasitize arboreal or flying hosts often have larger, more robust claws than do their counterparts that parasitize terrestrial hosts.

Because of their reliance on host availability, lice are subjected to special problems with respect to their long-term survival. All sucking lice are obligate blood-feeders; even a few hours away from the host can prove fatal to some species. Some chewing lice also are hematophages and similarly cannot survive prolonged periods off the host. However, many chewing lice, particularly those that subsist on feathers, fur, or other skin products, can survive for several days away from the host. For example, the cattle biting louse can survive for up to 11 days (this species will feed on host skin scrapings) and *Menacanthus* spp. of poultry for up to three days off the host. Off-host survival is generally greater at low temperatures and high humidities. At 26° C and 65% relative humidity (RH), 4% of human body lice die within 24 hours, 20% within 40 hours, and 84% within 48 hours. At 75% RH, a small proportion of the sheep foot louse survive for 17 days at 2° C, whereas most die within seven days at 22° C. Recently fed lice generally survive longer than unfed lice away from the host. Although most lice are morphologically adapted for host attachment and are disadvantaged when dislodged, the generalist nature of some amblyceran chewing lice better equips them for locating another host by crawling across the substrate. Amblycerans are more likely than other lice to be encountered away from the host, accounting for observations of these lice on bird eggs or in unoccupied nests and roosts.

Host grooming is an important cause of louse mortality. Laboratory mice infested by the mouse louse, for example, usually limit their louse populations to 10 or less individuals per mouse by regular grooming. Prevention of self or mutual grooming by impaired preening action of the teeth or limbs of mice can result in heavy infestations of more than 100 lice. Similarly, impaired preening due to beak injuries in birds can result in tremendous increases of louse populations. Biting, scratching, and licking also reduce louse populations on several domestic animals.

Whereas most species of lice on small and medium-sized mammals exhibit only minor seasonal differences in population levels, some lice associated with larger animals show clear seasonal trends. Some of these population changes have been attributed to host molting, fur density and length, hormone levels in the blood meal, or climatological factors such as intense summer heat, sunlight, or desiccation. On domestic ungulates in temperate regions, louse populations typically peak during the winter or early spring and decline during the summer. An exception to this trend is the cattle tail louse, in which populations peak during the summer.

Another important aspect of louse behavior is the mode of transfer between hosts. Direct host contact appears to be the primary mechanism for louse exchange. Transfer of lice from an infested mother to her offspring during suckling (in mammals) or during nest sharing (in birds and mammals) is an important mode of transfer. Several species of lice that parasitize livestock transfer during suckling, including the sheep face louse and the sheep biting louse, both of which move from infested ewes to their lambs at this time. Lice can also transfer during other forms of physical contact between hosts such as mating or fighting. Transfer of lice between hosts also can occur between hosts that are not in contact. The sheep foot louse, for example, can survive for several days off the host and reach a new host by crawling across pasture land. Nests of birds and mammals can act as foci for louse transfer but these are infrequent sites of transfer.

Dispersal of some lice occurs via **phoresy**, in which they temporarily attach to other arthropods and are carried from one host to another (Figure 6.8). During phoresy, most lice attach to larger, more mobile blood-feeding arthropods, usually a fly such as a hippoboscid or muscoid. Phoresy is particularly common among ischnoceran chewing lice and occurs when the louse mandibles grasp fly setae or appendages. Phoresy is relatively rare among sucking lice. This is probably because attachment to the fly is achieved by the less efficient mechanism of grasping with the tibio-tarsal claws.

Mating in lice occurs on the host. It is initiated by the male pushing his body beneath that of the female and curling the tip of his abdomen upward. In the human body louse, the male and female assume a vertical orientation along a hair shaft with the female supporting the weight of the male as he grasps her with his anterior claws. Other lice appear to exhibit similar orientation behavior during mating. Notable exceptions include the crab louse of humans in which both sexes continue to clasp with their claws a host hair, rather than each other, during mating; and the hog louse in which the male strokes the head of the female during copulation. Some male ischnoceran chewing lice possess modified hook-like antennal segments with which they grasp the female during copulation.

Oviposition behavior by female lice involves crawling to the base of a host hair or feather and cementing one egg at a time close to the skin surface. Two pairs of finger-like gonopods direct the egg into a precise

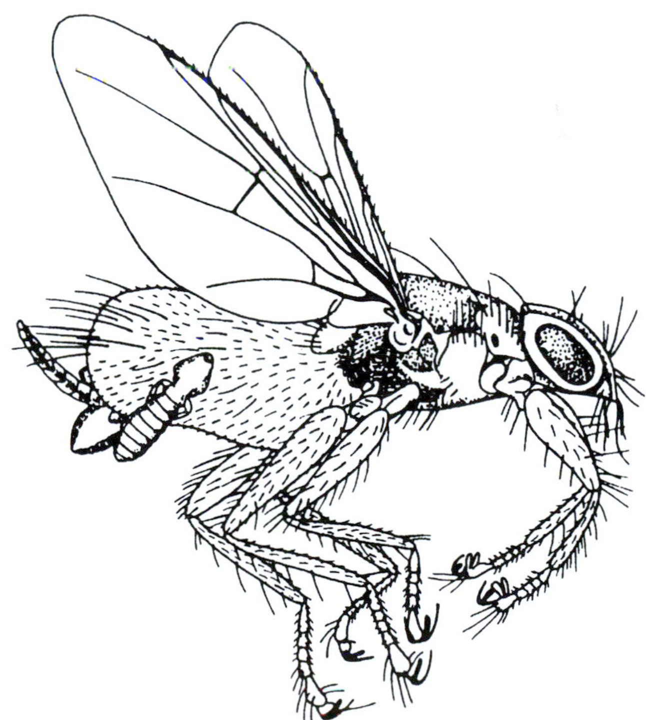

Figure 6.8 Two ischnoceran chewing lice phoretic on a hippoboscid fly attached by their mandibles to the posterior abdomen (Rothschild and Clay, 1952).

location and orientation as a cement substance is secreted around the egg and hair base. Optimal temperature requirements for developing louse embryos inside eggs are very narrow, usually within a fraction of a degree, such as may occur on a precise area on the host body. For this reason, female lice typically oviposit preferentially on an area of the host that meets these requirements.

LICE OF MEDICAL IMPORTANCE

Three taxa of sucking lice parasitize humans throughout the world: the body louse, head louse, and crab louse (pubic louse). All are specific ectoparasites of humans; rarely, dogs or other companion animals may have temporary, self-limiting infestations.

Human head and body lice are closely related and can interbreed to produce fertile offspring in the laboratory. For this reason, often they are treated as separate subspecies of *Pediculus humanus*, as they are in this chapter. Nevertheless, they rarely interbreed in nature, which has prompted some epidemiologists to treat them as separate species, *Pediculus humanus* (body louse) and *P. capitis* (head louse). Recent publications based on gene sequences have provided conflicting evidence for the recognition of human head and body lice as either separate species, subspecies, or strains of a single species.

Recent genetic analyses of human lice have produced intriguing indirect evidence for important events during human history (Reed et al., 2004). For example, analysis of two separate head louse genetic lineages suggests that *Homo sapiens* and *Homo erectus* physically interacted at some time during prehistory and that both louse lineages infested *H. sapiens* when *H. erectus* became extinct. Similarly, genetic analysis of crab lice show that humans acquired their pubic lice from gorillas about 3.3 million years ago; gorilla and human crab lice have since evolved into distinct species. Further, because human body lice infest clothing, the origin of body lice on humans might correspond with the time when humans first starting wearing clothes; one study along these lines suggests that this occurred 42,000 to 72,000 years ago.

Human Body Louse (*Pediculus humanus humanus*)

Infestations with the human body louse are sometimes referred to as *pediculosis corporis*. The human body louse (Figures 6.7A, 6.9), or *cootie*, was once an almost ubiquitous companion of humans. Today it is less common, especially in developed nations. Body lice persist as a significant problem in less developed nations in parts of Africa, Asia, and Central and South America, and on populations of some homeless people worldwide. This is significant because *P. h. humanus* is the only louse of humans that is known to naturally transmit pathogens. The large-scale reduction in body louse infestations worldwide has led to a concomitant decrease in the prevalence of human louse-borne diseases. However, situations that result in human overcrowding and unsanitary conditions (e.g., wars, famines, natural disasters, homelessness) can lead to a resurgence of body louse infestations, often accompanied by one or more louse-borne diseases.

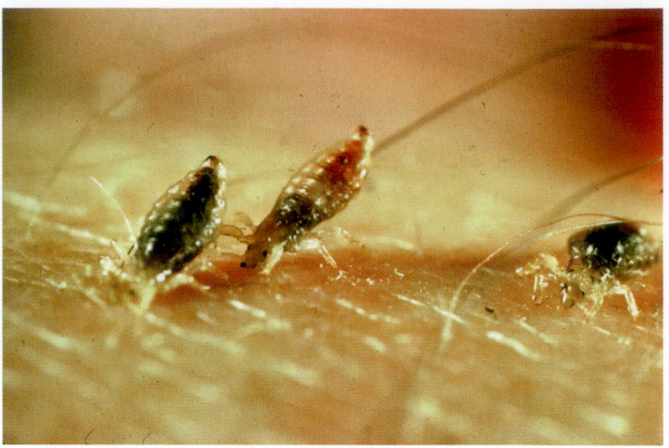

Figure 6.9 Human body lice (*Pediculus humanus humanus*) feeding on a human. (Photo by Elton J. Hansens)

Adult human body lice (Figures 6.7A, 6.9) are 2.3 to 3.6 mm long. Under optimal conditions their populations can multiply dramatically if unchecked; for example, if clothes of infested individuals are not changed and washed in hot water at regular intervals. In unusually severe infestations, populations of more than 30,000 body lice on one person have been recorded. Body lice typically infest articles of clothing and only crawl onto the body to feed. Females lay an average of four or five eggs per day, and these typically hatch after eight days. Unique among lice, females oviposit not on hair but on clothing (Figure 6.10), especially along seams and creases. Each nymphal instar lasts for three to five days, and adults can live for up to 30 days.

Figure 6.10 Eggs of human body louse (*Pediculus humanus humanus*) attached to clothing. (Photo by Elton J. Hansens)

Biting by body lice often causes intense irritation, with each bite site typically developing into a small red papule with a tiny central clot. The bites usually itch for several days but occasionally for a week or more. Persons exposed to numerous bites over long periods often become desensitized and show little or no reaction to subsequent bites. Persons with chronic body louse infestations may develop a generalized skin thickening and discoloration called **Vagabond's disease** or **Hobo's disease**, names depicting a lifestyle than can promote infestation by body lice. Several additional symptoms may accompany chronic infestations. These include lymphadenitis (swollen lymph nodes), edema, increased body temperature often accompanied by fever, a diffuse rash, headache, joint pain, and muscle stiffness.

Some people develop allergies to body lice. Occasionally, patients experience a generalized dermatitis in response to one or a small numbers of bites. A form of asthmatic bronchitis has similarly been recorded in response to louse infestation allergies. Secondary infections such as impetigo or (rarely) blood-poisoning can also result from body louse infestations.

Body lice tend to leave persons with elevated body temperatures and may crawl across the substrate to infest a nearby person. This has epidemiological significance because high body temperatures of lousy persons often result from fever caused by infection with louse-borne pathogens.

Human Head Louse (*Pediculus humanus capitis*)

Infestations with head lice can be referred to as *pediculosis capitis*. The human head louse is virtually indistinguishable from the human body louse on the basis of morphological characters and its life cycle. Unless a series of specimens is available for analysis it is often impossible to separate these two lice. Generally, adult head lice are slightly smaller (2.1–3.3 mm in length) than body lice.

As indicated by their name, human head lice typically infest the scalp and head region. Females attach their eggs to the base of individual hairs. As the hair grows, the eggs become further displaced from the scalp. An indication of how long a patient has been infested can be gleaned by measuring the farthest distance of eggs from the scalp and comparing this to the growth rate of hair.

Today, head lice are far more frequently encountered than body lice, especially in developed countries. Transmission occurs by person-to-person contact and via shared objects such as combs, brushes, headphones, and caps. School-age children are at high risk because they are often more likely to share such items. About 8% of all children three to 12 years old in the United States are infested with head lice, but in some school districts in the United States and Britain, infestation prevalences approach 50%. It has been estimated that six to 12 million people, principally children, are infested with head lice annually in the United States. Some ethnic groups, such as persons of African origin, have coarser head hairs and are less prone to head louse infestations. The reason for this is simply that the tibio-tarsal claws of these lice cannot efficiently grip the thicker hairs.

Although head lice are not typically important in transmitting pathogens, they can mechanically transmit the bacteria *Staphylococcus aureus* and *Streptococcus pyogenes*, both of which can cause skin/tissue infections. Under optimal laboratory conditions, head lice can transmit the causative agent of epidemic typhus, *Rickettsia prowazekii*, but only human body lice are implicated as vectors of this agent in nature. Although *Bartonella quintana*, the causative agent of trench fever, has been reported to have been molecularly detected in head lice from Nepal, there is no evidence of transmission of this pathogen by this louse. However, heavy infestations of head lice can cause severe irritation. As is the case with human body lice, the resultant scratching often leads to secondary infections such as impetigo, pyoderma, or (rarely) blood poisoning (septicemia). Severe head louse infestations occasionally result in the formation of scabby crusts and matted hair beneath which the lice tend to aggregate, a condition referred to as *plica polonica*. Enlarged lymph nodes in the neck region may accompany such infestations.

Human Crab Louse (*Pthirus pubis*)

Infestations with crab lice can be referred to as *pediculosis inguinalis* or *pthiriasis*. The crab louse, or pubic louse, is a medium-sized (1.1–1.8 mm long), squat louse (Figure 6.7B), with robust tibio-tarsal claws used for grasping thick hairs, especially those in the pubic region. It also may infest coarse hairs on other parts of the body such as the eyebrows, eyelashes, chest hairs, beards, moustaches, and armpits. This louse typically transfers between human partners during sexual intercourse and other intimate contact; in France, crab lice are sometimes described as *papillons d'amour* (butterflies of love). Transfer via infested bed linen, sofas, or toilet seats can also occur. This is uncommon, however, because crab lice can survive for only a few hours off the host.

Female crab lice lay an average of three eggs per day. Eggs hatch after seven or eight days; the three nymphal instars together last for 13 to 17 days. Under optimal conditions the generation time is 20 to 25 days. The intense itching caused by these lice often is accompanied by purplish lesions at bite sites and by small blood spots from squashed lice or louse feces on underwear. Crab lice are widely distributed and relatively common throughout the world. They are not known to transmit any pathogens.

Note: The generic name *Pthirus* is the taxonomically correct spelling as placed on the "Official List of Generic Names in Zoology" (1958, Opinion 104) with both *Phthirus* and *Phthirius* officially treated as invalid emendations of the original spelling *Pthirus* (see Kim et al., 1986). The correct family name is Pthiridae (not Phthiridae), following the valid generic spelling.

LICE OF VETERINARY IMPORTANCE

A variety of lice infests domestic livestock, poultry, pets, and laboratory animals (Tables 6.1, 6.2). Small rodents usually support few if any lice whereas larger hosts such as livestock animals, including poultry, may be parasitized by extremely large numbers of lice. For example, fewer than 10 mouse lice (*Polyplax serrata*) on a house mouse is a typical burden but 0.5 to1.0 million sheep biting lice (*Bovicola ovis*) may be present on heavily infested sheep. Although many species of wildlife have their own species of lice, seldom are they a problem. Relatively few lice species of veterinary concern are vectors of pathogens.

Lice of Cattle

Cattle lice are a major problem worldwide. Both dairy and beef breeds are affected. Domestic cattle can be parasitized by five species of lice: two species of *Haematopinus*, one *Linognathus*, one *Solenopotes*, and one *Bovicola*. Domestic Asiatic buffalo typically are parasitized by *Haematopinus tuberculatus* (Tables 6.1, 6.2), which has also successfully transferred to cattle in tropical climates.

The cosmopolitan cattle biting louse (*Bovicola bovis*) (Figure 6.11) is the only species of chewing louse to infest cattle. The species is primarily parthenogenic. Males are occasionally seen. The adult female is about 1.7 mm in length. Females lay an average of 0.7 eggs per day, which hatch seven to 10 days later. Nymphal instars last five to seven days each, and adult longevity can be as long as 10 weeks. The preferred host site for this louse is the top line of the back, especially the withers area from which it spreads to the rump and poll area (Watson et al., 1997). In heavy infestations, these lice spread to other body regions. In the most severe infestations, lice may be found beneath heavily encrusted scurf.

The longnosed cattle louse (*Linognathus vituli*) (Figure 6.12) is also a worldwide pest. Adult females and males are about 2.4 and 1.8 mm in length, respectively. Females deposit one egg per day, and the life cycle is completed in approximately 21 days. This louse is noticed in greater numbers on calves than on mature cattle. The species is widely distributed over the body of the host but preferred infestation sites are the shoulder, back, neck, and dewlap.

The little blue cattle louse (*Solenopotes capillatus*) (Figures 6.13, 6.14) is also worldwide in distribution. It is a common species on cattle but, because of its small size (adult female 1.5 mm; adult male 1.1 mm) it is commonly mistaken for nymphs of the longnosed cattle louse, which are also blue in color. Females lay one or two eggs per day. Eggs hatch after about 12 days and the time from egg to egg is about 28 days. Infestation by *S. capillatus* usually is noticed when dark blue

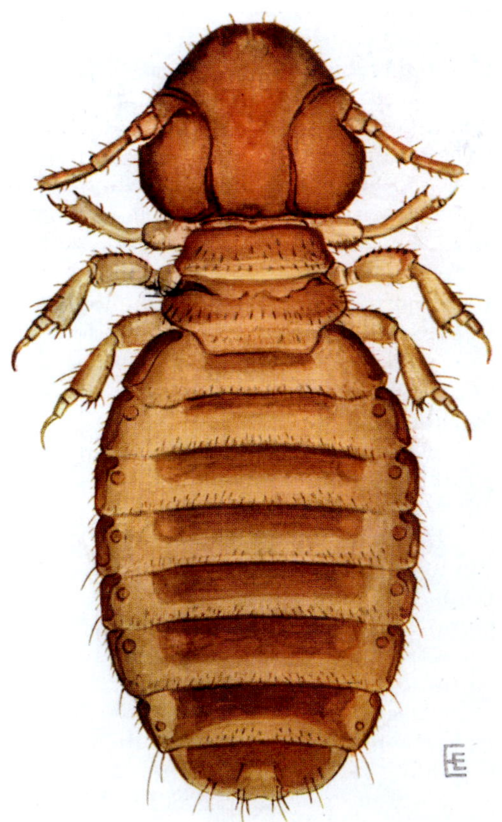

Figure 6.11 The cattle biting louse (*Bovicola bovis*), female. (Matthysse 1946; original illustration by Ellen Edmonson)

Figure 6.12 The longnosed cattle louse (*Linognathus vituli*), female. (Matthysse, 1946; original illustration by Ellen Edmonson)

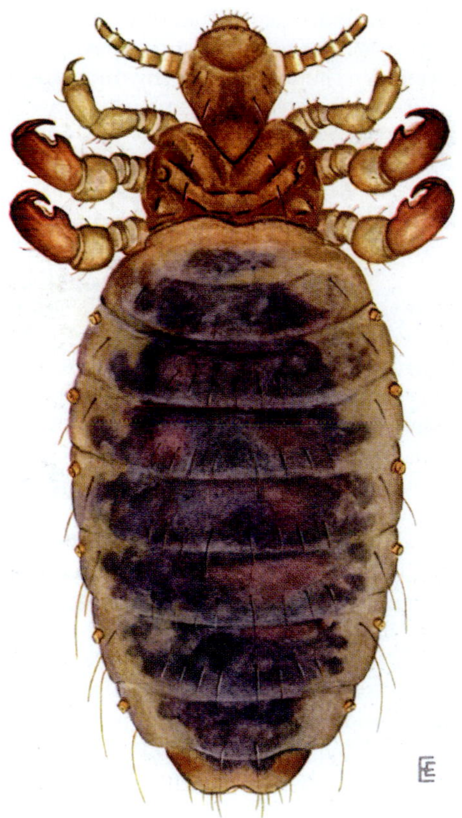

Figure 6.13 The little blue cattle louse (*Solenopotes capillatus*), female. (Matthysse, 1946; original illustration by Ellen Edmonson)

Figure 6.14 Life stages of the little blue cattle louse (*Solenopotes capillatus*). From left to right: egg (0.7 mm), 3 nymphal instars (0.69, 0.82, and 1.06 mm), male (1.08 mm) and female (1.5 mm). (Photograph by Maureen Grubbs; courtesy of the Entomological Society of America)

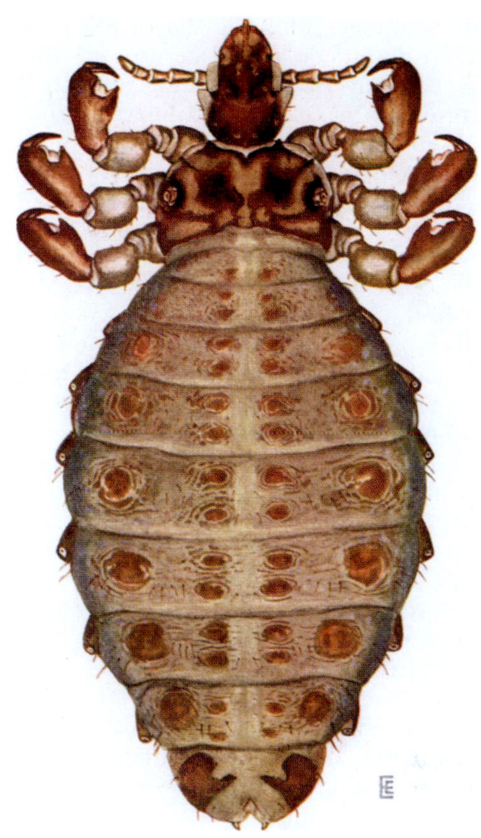

Figure 6.15 The shortnosed cattle louse (*Haematopinus eurysternus*), female. (Matthysse, 1946; original illustration by Ellen Edmonson)

patches, representing aggregations of this louse, appear on the face of the host. Clusters of *S. capillatus* typically occur on the facial area (i.e., muzzle, cheeks, and around the eyes). Occasionally, the longnosed cattle louse may also be seen within the cluster. As spring approaches, heavier infestations of the little blue cattle louse may extend to the neck and dewlap.

The cosmopolitan shortnosed cattle louse (*Haematopinus eurysternus*) (Figure 6.15) is the largest louse found on cattle in the northern United States. Adult females and males measure 2.9 and 2.3 mm in length, respectively. The female lays 1.4 eggs per day for about two weeks, nymphs reach adulthood in about 14 days, and adult longevity is 10 days for males and 15 days for females. The cycle from egg to egg normally requires 28 days. Preferred infestation sites are the top of the neck, the dewlap, and brisket. In severe infestations, the entire region from the base of the horns to the base of the tail can be infested. In warmer weather this species can also be found abundantly inside and on the tips of the ears. Although heavy infestations occasionally are encountered, the shortnosed cattle louse is the least common species on cattle in Wyoming, Nebraska, and probably neighboring Rocky Mountain and Great Plains states. At one time it was thought to be a common species on western range cattle.

The cattle tail louse (*Hematopinus quadripertusus*) is a tropical sucking louse that was inadvertently introduced into Florida in 1945. It has since spread to warmer regions of the United States, including Florida, the Gulf Coast states, and Southern California. The cattle tail louse is larger than the closely related shortnosed cattle louse. Adult females are 4.0 mm and adult males are 3.2 mm. Unlike other cattle lice in North America, *H. quadripertusus* is most abundant during the summer. Adult females of this louse, which normally are found on the distal area of the tail, oviposit on the tail hairs. Hatching may be delayed 40 or more days in January and February due to cool weather. Consequently, the tail brush becomes matted with eggs that hatch when temperatures begin to rise in the spring. In severe infestations, hair may be shed. Eggs hatch after nine days under optimal conditions and the entire life cycle can be as short as 25 days. Nymphs migrate over the host body surface and may be found in the areas of the face, neck, vulva, and anus, but adults typically are confined to the tail. Although the cattle tail louse is spread by direct contact or contaminated facilities and equipment, phoresy appears to be common. Third instar nymphs may migrate to the backs and shoulders of the host where they attach to flies and are carried to a new host. In a sample of 5000 horn flies collected in Florida, 100 were carrying louse nymphs (Kaufman et al., 2005).

Except for *H. quadripertusus*, cattle lice increase in numbers during the winter and early spring in temperate regions. Chronically infested cattle, often referred to as **carriers**, may carry heavy burdens of lice, even in the summer. Cattle producers customarily cull these animals from a herd as they are considered a source of lice for the other animals. Thorough examination of cattle with no clinical signs of lice infestation in the summer, however, has shown that these animals, too, may harbor all four species of cattle lice (Hanlin, 1994).

Lice of Other Livestock Animals

Horses, donkeys, hogs, goats, and sheep are each parasitized by one or more species of louse (Tables 6.1, 6.2). Except for hogs, all these animals are parasitized by at least one species of sucking lice and one of chewing lice. The horse biting louse (*Bovicola equi*) is the most important louse of equids worldwide, occurring even in warmer climates. Adult females and males average about 1.9 mm and 1.3 mm, respectively. Females of this louse oviposit on fine hairs near the skin, usually singly, avoiding the coarse hairs of the mane and tail. This louse typically infests the side of the neck, the flanks, and tail base but can spread to most of the body with the exception of the mane, tail, ears, and lower legs. Long-haired horse breeds are more prone to infestation by *B. equi*. *Haematopinus asini*, the horse sucking louse, is worldwide in distribution, but more common in areas of cooler climate. Although commonly occurring on horses, donkeys, and mules, it has also been reported on zebras. The adult females and males are 3.0 mm and 2.3 mm, respectively. Generally, it is found in the areas of coarse hair avoided by the horse biting louse: the forelock, mane, base of the tail, and above the hooves.

Infestations by both species of horse lice are heavier during the winter months. As with other lice species, advanced infestations will spread to additional regions of the body. Lice are normally spread by direct contact; however, the use of contaminated grooming equipment and blankets can contribute to their spread.

Domestic swine are parasitized by one louse species, the hog louse (*Haematopinus suis*) (Figure 6.7E). This is a large species of sucking louse in which adult females measure 5 to 6 mm in length, and the males measure over 4.1 mm. Geographically, the hog louse is found wherever hogs are raised but is more common in cooler climates. Hog lice normally frequent skin folds of the neck, the ears (often deep within the canal), the tender skin behind the ears, inside of the legs, and inner flanks of swine. Hog lice tend to favor upper regions of their predilection sites in summer and lower regions in winter. The heaviest infestations of hog lice occur in winter, usually December to March in North America, Europe, and northern Asia. Eggs are deposited singly on hairs (except when infestations become very heavy) along the lower parts of the body, in skin folds on the neck, and on and in the ears. Under optimum conditions the female deposits three to six eggs per day. The egg and nymphal stages last approximately two weeks each. Adults are thought to live about a month and there are six to 12 generations per year.

Domestic sheep and goats are parasitized by several species of sucking lice and chewing lice (Tables 6.1, 6.2). Worldwide, the sheep biting louse, *Bovicola bovis*, is the number one louse problem on domestic sheep. It is also one of the most studied. In Australia, a major sheep producing country, Roberts (1952) noted that the sheep biting louse occurred throughout the country in sheep raising areas, but was less frequent in, and sometimes absent from, the drier inland districts. This is probably true in the United States as well. Although the sheep biting louse is known to occur in the United States, its distribution has not been documented. In 40 years it has not been seen in Wyoming, one of the United States' major sheep producing states, and veterinary entomologists in neighboring states of Montana and Nebraska report that the species is uncommon, if not absent, in those states as well.

Females of the sheep biting louse are about 1.8 mm long and males are around 1.0 mm. Females attach eggs primarily to wool fibers close to the skin. The incubation period is about 10 days, and the three nymphal stages together last about three weeks. The female lays eggs at the relatively slow rate of one egg every two to four days and can live for up to 30 days. In spite of the low reproductive rate, infestations may reach hundreds of thousands or even a million. *Bovicola ovis* feeds on epidermal scales, scurf, and dermal secretions. It is doubtful that this louse actually consumes wool. In the winter, when louse populations are high, most *B. ovis* are found on the back and mid-sides of the sheep. Lighter summer populations are on the ribs, lower flank, and abdomen.

The African blue louse, *Linognathus africanus*, is a parasite of both sheep and goats. Originally described from sheep in Africa, as its name implies, the species now appears to be distributed almost worldwide. In the United States, it has been reported from the southern and southwestern states. Recently, it appears to have established in sheep producing areas of several western states where it has become a major pest of sheep. The species also has been reported from mule deer, Columbian black-tailed deer, and white-tailed deer. Females are 2.2 mm long and males, 1.7 mm. In the winter, when infestations are heaviest, *L. africanus* is found most abundantly on the loin, back, rib and shoulder areas of sheep. Populations may reach several thousand. On goats, the distribution is different with lice occurring on the upper neck, base of the ears, poll, and ventral surface of the jaw.

Lice of Cats and Dogs

Domestic cats are parasitized by one species of chewing louse whereas dogs are parasitized by two species of chewing lice and one species of sucking louse. All four species appear to be distributed worldwide but none of them are common associates of healthy cats or dogs in North America or Europe.

The cat biting louse (*Felicola subrostrata*) (Figure 6.5E) parasitizes both domestic and wild cats. It may occur almost anywhere on the body.

Both the dog biting louse (*Trichodectes canis*) (Figure 6.5D) and the dog sucking louse (*Linognathus setosus*) (Figure 6.7G) parasitize dogs and closely related wild canids. For example, *T. canis* also parasitizes coyotes, foxes, and wolves. A second species of chewing louse of dogs is *Heterodoxus spiniger* (Figure 6.5A), which evolved in Australia from marsupial-infesting lice and apparently switched to dingo hosts. It now parasitizes various canids and other carnivores throughout the world. *Trichodectes canis* usually infests the head, neck, and tail region of dogs where it attaches to the base of a hair using its claws or mandibles. *Linognathus setosus* occurs primarily on the head and neck and may be especially common beneath collars. *Heterodoxus spiniger* can typically be found anywhere on its host.

Lice of Laboratory Animals

The principal species of lice that parasitize laboratory mammals have been treated by Kim et al. (1973). These lice also parasitize feral populations of their respective hosts.

The house mouse (*Mus musculus*) often is parasitized by the mouse louse (*Polyplax serrata*). Populations of this louse are typically low with 10 or fewer lice per infested mouse, unless self or mutual host grooming is compromised. Eggs of this louse typically hatch seven days after oviposition. Together the three nymphal instars last only six days under optimal conditions, which can result in a generation time as short as 13 days.

Domestic rats often are parasitized by the spined rat louse (*Polyplax spinulosa*) (Figure 6.7D) and the tropical rat louse (*Hoplopleura pacifica*). Common hosts include the black rat (*Rattus rattus*) and the Norway rat (*R. norvegicus*). The spined rat louse parasitizes these hosts throughout the world, whereas the tropical rat louse is confined to tropical, subtropical, or warm temperate regions including the southern United States.

Laboratory rabbits are parasitized by the rabbit louse (*Haemodipsus ventricosis*). This louse originated in Europe but has accompanied its host wherever it has been introduced throughout the world.

Lice of Poultry and Other Birds

At least nine species of chewing lice commonly infest poultry (Table 6.1) in various parts of the world. Individual birds can be parasitized by multiple species, each of which often occupies a preferred host site.

The chicken body louse (*Menacanthus stramineus*) (Figure 6.16) is the most common and destructive louse of domestic chickens. It is thought originally to be a pest of wild turkeys that transferred to domestic poultry and is now common on both chickens and turkeys. The chicken body louse has a worldwide distribution and often reaches pest proportions. Unlike other chicken lice, it is found on the host's skin rather than the feathers. It may be detected by parting the feathers, especially in the vent area of the bird. This louse is most abundant on the sparsely feathered vent, breast, and thigh regions; however, in heavily infested poultry it may be found on any part of the body. Adults measure 3 to 3.5 mm in length. Females lay one or two eggs per day, cementing them in clusters at the bases of feathers especially around the vent. Eggs typically hatch after four or five days. Each nymphal instar lasts about three days, and the generation time typically is 13 to 14 days.

Several other chewing lice are pests of poultry more or less throughout the world (Table 6.1). Adults of the shaft louse (*Menopon gallinae*) measure about 2 mm in length, and may be seen in a line along the shaft of a feather. Although these lice do not normally rest on the skin, they quickly disperse to the skin if disturbed. Females deposit eggs singly at the base of the shaft on thigh and breast feathers. Eggs of the wing louse (*Lipeurus caponis*) hatch four to seven days after the female has cemented them to the base of a feather. Nymphal stages of this species each last five to 18 days, generation time typically is 18 to 27 days, and females can live up to 36 days. Females of the chicken head louse (*Cuclutogaster heterographus*) attach their eggs to the bases of downy feathers. Eggs hatch after five to seven days, each nymphal instar lasts six to 14 days, and average generation time is 35 days.

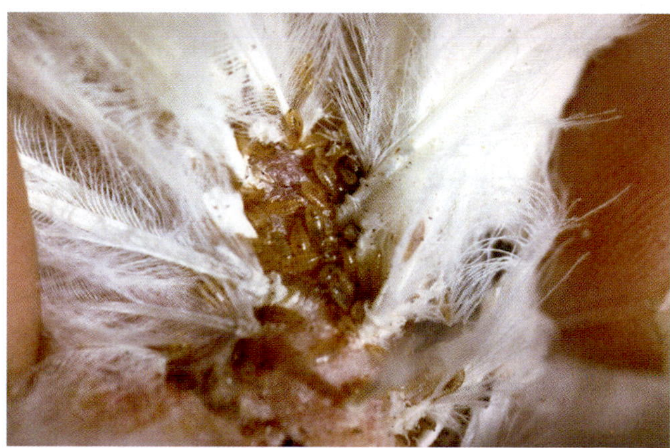

Figure 6.16 Chicken body lice (*Menacanthus stramineus*) feeding on a chicken. (Courtesy of Nancy C. Hinkle)

Poultry lice typically transfer to new birds by direct host contact. However, because most species can survive for several hours or days off the host, they also can infest new hosts during transportation in inadequately disinfected cages or vehicles.

PUBLIC HEALTH IMPORTANCE

Three important pathogens are transmitted to humans by body lice. These are the agents of epidemic typhus, trench fever, and louse-borne relapsing fever. Today, the prevalence and importance of all three of these louse-borne diseases is low compared to times when human body lice were an integral part of our lives. However, trench fever has reemerged as an opportunistic disease of immunocompromised individuals, including persons who are positive for human immunodeficiency virus (HIV). A few other pathogens can also be transmitted by body lice, particularly under optimal laboratory conditions.

Epidemic Typhus

Epidemic typhus is a rickettsial disease caused by infection with *Rickettsia prowazekii.* The entire genome sequence of *R. prowazekii* was reported by Andersson et al. (1998). The disease is also known as louse-borne fever, jail fever, and exanthematic typhus. Epidemic typhus persists in several parts of the world, most notably in Burundi, Democratic Republic of Congo, Ethiopia, Nigeria, Rwanda, areas of northeastern and central Africa, Russia, Central and South America, and northern China. Epidemic typhus is largely a disease of cool climates, including higher elevations in the tropics. It thrives in conditions of widespread body louse infestations, overcrowding, and poor sanitary conditions. This disease apparently was absent from the New World until the 1500s when the colonizing Spanish inadvertently introduced it. One resulting epidemic during 1576–1577 killed 2 million Indians in the Mexican highlands alone.

The vector of *R. prowazekii* is the human body louse. Lice become infected when they feed on a person with circulating *R. prowazekii* in their blood. Infective rickettsiae invade cells that line the louse gut, and multiply there, eventually causing the cells to rupture. Liberated rickettsiae either reinvade gut cells or are voided in louse feces. Other louse tissues typically do not become infected. Because salivary glands and ovaries are not invaded, anterior-station and transovarial transmission do not occur. Infection of susceptible humans occurs via louse feces (posterior-station) when infectious rickettsiae are scratched into the skin in response to louse bites. *Rickettsia prowazekii* can remain viable in dried louse feces for 60 days. Infection by inhalation of dried louse feces or by crushed lice are less frequent means of contracting the disease.

Transmission of *R. prowazekii* by body lice was first demonstrated by Charles Nicolle, working at the Institut Pasteur in Tunis in 1909. During these studies, Nicolle accidentally became infected with epidemic typhus from which he fortunately recovered. He was awarded the Nobel prize in 1928 for his groundbreaking work on typhus. Several other typhus workers also were infected with *R. prowazekii* during laboratory experiments. The American researcher Howard T. Ricketts working in Mexico and Czech scientist Stanislaus von Prowazek working in Europe both died from their infections and were recognized posthumously when the etiologic agent was named.

Infection with *R. prowazekii* is ultimately fatal to body lice as progressively more and more infected gut cells are ruptured. Infective rickettsiae are first excreted in louse feces three to five days after the infective blood meal. Lice usually succumb to infection seven to 14 days after the infectious blood meal although some may survive to 20 days.

The disease caused by infection with *R. prowazekii* and transmitted by body lice is called **classic epidemic typhus** because it was the first form of the disease to be recognized. Disease onset occurs 10 to 14 days after infection by a body louse in classic epidemic typhus. Abrupt onset of fever accompanied by malaise, muscle and head aches, cough, and general weakness usually occur at this time. A blotchy, often reddish-blue rash spreads from the abdomen to the chest and then often across most of the body, typically within four to seven days following the initial symptoms. The rash rarely spreads to the face, palms, and soles, and then only in severe cases. Headache, rash, prostration, and delirium intensify as the infection progresses. Coma and very low blood pressure often signal fatal cases. A case fatality rate of 10 to 20% is characteristic of most untreated epidemics, although figures approaching 50% have been recorded.

Diagnosis of epidemic typhus involves the demonstration of positive serology, usually by microimmunofluorescence. DNA primers specific to *R. prowazekii* can also be amplified by polymerase chain reaction (PCR) from infected persons or lice. Antibiotic treatment, especially with doxycycline, tetracycline, or chloramphenicol, usually results in rapid and complete recovery. Vaccines are available but are not considered to be sufficiently effective for widespread use.

Persons who recover from epidemic typhus typically harbor *R. prowazekii* in lymph nodes or other tissues for months or years. This enables the pathogen to reinvade other body tissues to cause disease later. This form of the disease is called **recrudescent typhus** or **Brill-Zinsser disease**. The latter name recognizes two pioneers in the study of epidemic typhus, Nathan Brill, who first recognized and described recrudescent typhus in 1910, and **Hans Zinsser**, who demonstrated in 1934 that it is a form of epidemic typhus. Zinsser's (1935) book, *Rats, Lice and History,* is a pioneering account of the study of epidemic typhus in general.

Recrudescent typhus was widespread during the nineteenth and early twentieth centuries in some of the larger cities along the east coast of the United States (e.g., Boston, New York, and Philadelphia). At that time, immigrants from regions that were rampant with epidemic typhus, such as eastern Europe, presented with Brill-Zinsser disease after being infected initially in their country of origin. Some of these patients experienced relapses more than 30 years after their initial exposure, with no overt signs of infection with *R. prowazekii* between the two disease episodes. Because infestation with body lice was still a relatively common occurrence during that period, the lice further disseminated the infection to other humans, causing regional outbreaks. The last outbreak of epidemic typhus in North America occurred in Philadelphia in 1877. Today, even recrudescent typhus is a rare occurrence in North America. However, this form of typhus is still common in parts of Africa, Asia, South America, and occasionally in eastern Europe. Further, some travelers returning to Europe or North America from endemic areas have presented with classic epidemic typhus.

The southern flying squirrel (*Glaucomys volans*) has been identified as a reservoir of *R. prowazekii* in the United States where it was first found to be infected in Virginia during vertebrate serosurveys for Rocky Mountain spotted fever. Since the initial isolations from flying squirrels in 1963, *R. prowazekii* has been recorded in flying squirrels and their ectoparasites in several states, especially eastern and southern states. Peak seroprevalence (about 90%) in the squirrels occurs during late autumn and winter when fleas and sucking lice are also most abundant on these hosts. Although several ectoparasites can imbibe *R. prowazekii* when feeding on infected flying squirrels, only the sucking louse *Neohaematopinus sciuropteri* is known to maintain the infection and transmit the pathogen to uninfected squirrels; nevertheless, a squirrel flea, *Orchopeas howardi*, is also a likely vector.

Several cases of human infection have been documented in which the patients recalled having contact with flying squirrels (Reynolds et al., 2003), especially during the winter months when these rodents commonly occupy attics of houses. To distinguish this form of the disease from classic and recrudescent typhus, it is called **sporadic epidemic typhus** or sylvatic epidemic typhus. Many details such as the prevalence and mode of human infection remain unresolved. Because the louse *N. sciuropteri* does not feed on humans, it is speculated that human disease may occur when infectious, aerosolized particles of infected louse feces are inhaled from attics or other sites occupied by infected flying squirrels.

In addition to flying squirrels in North America, and humans in various parts of the world, some livestock animals may be reservoirs of *R. prowazekii.* Reports published in the 1950s to 1970s that various species of ticks and livestock animals harbored *R. prowazekii* were later considered to be inaccurate but recent evidence suggests that some of these data may be valid.

Historically, epidemic typhus has been the most widespread and devastating of the louse-borne diseases. Zinsser (1935) and Snyder (1966) have documented the history of this disease, and have highlighted how major epidemics have influenced human history. For example, Napoleon's vast army of 1812 was arguably defeated more by epidemic typhus than by opposing Russian forces. Soon thereafter (ca. 1816–1819) 700,000 cases of epidemic typhus occurred in Ireland. Combined with the potato famine of that period, this encouraged many people to emigrate to North America; some of these people carried infected lice or latent infections with them. During World War II, several military operations in North Africa and the Mediterranean region were hampered by outbreaks of epidemic typhus. One epidemic in Naples in 1943 resulted in over 1400 cases and 200 deaths. This outbreak is particularly noteworthy because it was the first epidemic of the disease to be interrupted by human intervention through widespread application of the insecticide DDT to louse-infested persons.

Today, epidemic typhus is much less of a health threat than it once was. This is largely because few people, especially in developed countries, are currently infested by body lice.Higher sanitary standards, less overcrowding, regular laundering, frequent changes of clothes, effective pesticides and antibiotics, and medical advances have contributed to the demise of this disease. Nevertheless, epidemic typhus has the potential to reemerge. This is evidenced by the largest outbreak of epidemic typhus since World War II that affected about 50,000 people living in refugee camps in Burundi in 1997 to 1998. Further, more than 5,600 cases were recorded in China during 1999, recent cases have been recorded in parts of Russia, and some people in Mexico and Texas have antibodies to *R. prowazekii* (Reeves et al., 2008). *Rickettsia prowazekii* currently is listed as a select agent (category B) by the U.S. Centers for Disease Control and Prevention because of its potential to cause epidemics with high mortality and its bioweapon potential. Additional information about epidemic typhus is provided by PAHO/WHO (1973), Azad (1988), and Reynolds et al. (2003).

Louse-Borne Relapsing Fever

Also known as epidemic relapsing fever, this disease is caused by the spirochete bacterium *Borrelia recurrentis.* This pathogen is transmitted to humans by the human body louse, as first demonstrated by Sergent and Foley in 1910. Clinical symptoms include the sudden onset of fever, headache, muscle ache, anorexia, dizziness, nausea, coughing, and vomiting. Thrombocytopenia (a decrease in blood platelets) also can occur, which can result in bleeding, a symptom that may initially be confused with a hemorrhagic fever. Episodes of fever last two to 12 days (average, four days), typically followed by periods of two to eight days (average, four days) without fever, with two to five relapses being most common. As the disease progresses, the liver and spleen enlarge, leading to abdominal discomfort and labored, painful breathing as the lungs and diaphragm are compressed. At this stage, most patients remain quietly prostrate with a glazed expression, often shivering and taking shallow breaths. Case fatality rates for untreated outbreaks range from 5 to 40%. Antibiotic treatment is with penicillin or tetracycline. Humans are the sole known reservoir of *B. recurrentis.*

Body lice become infected when they feed on an infected person with circulating spirochetes. Most of the spirochetes perish when they reach the louse gut, but a few survive to penetrate the gut wall where they multiply to massive populations in the louse hemolymph, nerves, and muscle tissue. Spirochetes do not invade the salivary glands or ovarian tissues and are not voided in louse feces. Therefore, transmission to humans occurs only when infected lice are crushed during scratching, which allows the spirochetes in infectious hemolymph to invade the body through abrasions and other skin lesions. However, *B. recurrentis* is also capable of penetrating intact skin. As with *R. prowazekii* infections, body lice are killed as a result of infection with *B. recurrentis*.

An intriguing history of human epidemics of louse-borne relapsing fever is provided by Bryceson et al. (1970). Hippocrates described an epidemic of "caucus" or "ardent fever" in Thasos, Greece, which can clearly be identified by its clinical symptoms as this malady. During 1727 to 1729, an outbreak in England killed all inhabitants of many villages. An epidemic that spread from eastern Europe into Russia during 1919 to 1923 resulted in 13 million cases and 5 million deaths. Millions also were infected during an epidemic that swept across North Africa in the 1920s. Several major epidemics subsequently have occurred in Africa, with up to 100,000 fatalities being recorded for some of them. During and immediately after World War II, more than a million persons were infected in Europe alone.

The only current epidemic of louse-borne relapsing fever is in Ethiopia where 1,000 to 5,000 cases are reported annually accounting for about 95% of the world's recorded infections. Other smaller foci occur intermittently in other regions such as Burundi, Rwanda, Sudan, Uganda, People's Republic of China, Russia, Central America, and the Peruvian Andes. Resurgence of this disease under conditions of warfare or famine is a possibility.

Trench Fever

Also known as five-day fever and wolhynia, trench fever is caused by infection with the bacterium *Bartonella* (formerly *Rochalimaea*) *quintana*. Like the two preceding diseases, the agent is transmitted by the human body louse. Human infections range from asymptomatic through mild to severe, although fatal cases are rare. Clinical symptoms are nonspecific and include headache, muscle aches, fever, and nausea. The disease can be cyclic with several relapses often occurring. Previously infected persons often maintain a cryptic infection that can cause relapses years later with the potential for spread to other persons if they are infested with body lice. Effective antibiotic treatment of patients involves administering drugs such as doxycycline or tetracycline. *Bartonella quintana* DNA has been molecularly detected in a 4,000-year-old human tooth (Drancourt et al., 2005).

Lice become infected with *B. quintana* after feeding on the blood of an infected person. The pathogen multiplies in the lumen of the louse midgut and in the cuticular margins of the midgut epithelial cells. Viable bacteria are voided in louse feces, and transmission to humans occurs by the posterior-station route when louse bites are scratched. *Bartonella quintana* can remain infective in dried louse feces for several months, contributing to aerosol transmission as a rare but alternative route of transmission. Transovarial transmission does not occur in the louse vector. Infection is not detrimental to lice and does not affect their longevity.

Trench fever was first recognized as a clinical entity in 1916 as an infection of European troops engaging in trench warfare during World War I. At that time, more than 200,000 cases were recorded in British troops alone. Between the two world wars, trench fever declined in importance but reemerged in epidemic proportions in troops stationed in Europe during World War II. Because of the presence of asymptomatic human infections, the current distribution of trench fever is difficult to determine. However, since World War II, infections have been recorded in several European and African nations, Japan, the People's Republic of China, Mexico, Bolivia, and Canada.

Until recently, *B. quintana* was considered to be a rare disease of humans. However, several inner-city homeless or immunocompromised people, including HIV-positive individuals, particularly in North America, Europe, and Asia, have presented with *B. quintana* infections. This is manifested not as trench fever but as vascular tissue lesions (bacillary angiomatosis), liver pathology, chronically swollen lymph nodes, and/or inflammation of the heart valves. Because this disease typically occurs in inner cities, it has been given the name **urban trench fever**. The disease is probably widespread worldwide having been recorded, for example, in the cities of New York, Seattle, Marseille, Tokyo, and Moscow.

Other Pathogens Transmitted by Human Body Lice

Some additional bacterial pathogens can be transmitted by body lice under certain conditions. For example, *Salmonella typhi*, the agent of typhoid (salmonellosis), can be louse-borne during outbreaks of this disease. However, other modes of transmission such as through contaminated food account for most human cases. Another bacterium, *Acinetobacter baumannii*, which can cause infections of human skin, wounds, urinary tract, lungs, and meninges (and is often drug resistant), has been molecularly detected in body lice removed from humans. In laboratory trials, body lice can successfully imbibe *A. baumannii* while feeding on experimentally infected rabbits. These lice cannot transmit *A. baumannii* by bite but they excrete viable bacteria in their feces, which could represent a source of human infection. Human body lice can also transmit *Rickettsia rickettsii* and *Rickettsia conorii* to rabbits under laboratory conditions; these pathogens cause Rocky Mountain spotted fever and Boutenneuse fever, respectively, and typically are transmitted by ticks.

Lice as Intermediate Hosts of Tapeworms

Occasionally humans become infested with the double-pored tapeworm (*Dipylidium caninum*). Although carnivores are the normal definitive hosts for this parasite, humans can be infested if they accidentally ingest infested dog biting lice (*Trichodectes canis*), which serve as intermediate hosts. Although this would appear to be an unlikely event, infants, especially babies playing on carpets or other areas frequented by a family dog, may touch an infested louse with sticky fingers, which may then be put into their mouth to initiate an infestation.

VETERINARY IMPORTANCE

A variety of chewing lice and sucking lice parasitize domestic animals (Tables 6.1, 6.2). The potential effects of the various species of lice on livestock production are many. Even infestations that are considered relatively light may, in some way, have a measurable negative impact. Infestation by lice, presumably because of both their feeding and movement on the host, may be extremely annoying. Host animals become restless. The host may develop a dermatitis, allergic response, or secondary infection contributing to the pruritus. The host response to the irritation is licking and rubbing the affected body parts, which may produce a number of negative consequences. Hair, wool, mohair, or feather loss might hinder thermoregulation and result in unsightly animals with a lower market value. Scratches reduce the value of hides from slaughtered animals because defects must be trimmed away. Licking can lead to the formation of hair balls in the stomach, seen especially in cats and calves. Large animals, especially, cause considerable damage to gates, fences, and other livestock equipment against which they rub to relieve irritation. Feeding by lice may directly, or through an immunological response by the host, result in blemishes in the hide.

The effect of blood-feeding parasites such as lice is more than merely robbing nutrients necessary for normal growth of the host. Feeding and salivary secretions of parasites can stimulate immunologic and nonspecific defense mechanisms ultimately influencing behavior and physiology of a host, and preventing the host from reaching its full growth potential. This effect may not only reduce weight gains but also production of byproducts such as milk and eggs. Although lice may be barely detectable, they can multiply to extremely high numbers, particularly on young, old, sick, or stressed animals. Often this is because hosts are unable to effectively groom themselves or they are immunocompromised. Sucking lice infestations, especially when they become severe, may affect the vitality of the host, producing anemia, toxic anemia, abortion, and death, and likely, as suggested by Campbell (1988), increasing susceptibility to contagious diseases.

Few pathogens are known to be transmitted to domestic animals by lice (Table 6.3). The most

Table 6.3 Pathogens Transmitted by Lice

Disease	Disease Agent	Vector(s)	Host(s)	Geographic Distribution
VIRAL:				
Swinepox	Pox virus	*Haematopinus suis*	Hogs	Widespread
BACTERIAL:				
Epidemic typhus	*Rickettsia prowazekii*	*Pediculus h. humanus*	Humans	Global (focal)
Louse-borne relapsing fever	*Borrelia recurrentis*	*Pediculus h. humanus*	Humans	Global (focal)
Trench fever	*Bartonella quintana*	*Pediculus h. humanus*	Humans	Global
Tularemia	*Francisella tularensis*	Rodent & lagomorph lice	Rodents, rabbits	Global
Murine Mycoplasma Infection	*Mycoplasma muris*	*Polyplax spinulosa*	Domestic rats	Global
Murine Mycoplasma Infection	*Mycoplasma coccoides*	*Polyplax serrata*	Domestic mice	Global
FUNGAL:				
Bovine dermatomycosis (ringworm)	*Trichophyton verrucosum*	Cattle lice	Cattle	Global
HELMINTHIC:				
Seal heartworm	*Dipetalonema spirocauda*	*Echinophthirius horridus*	Harbor seal	Northern Hemisphere
Avian filariasis	*Eulimdana* spp.	Bird chewing lice	Charadriiform birds	Widespread
Avian filariasis	*Pelecitus fulicaeatrae*	*Pseudomenopon pilosum*	Aquatic birds	Holarctic region
Avian filariasis	*Sarconema eurycerca*	*Trinoton anserinum*	Geese, swans	Holarctic region
Double-pored tapeworm*	*Dipylidium caninum*	*Trichodectes canis*	Dogs, humans	Global

*Lice are intermediate hosts, not vectors, of this tapeworm.

important of these are the viral agent of swinepox and the bacterial agents of murine mycoplasma infection (formerly murine haemobartonellosis of rats and murine eperythrozoonosis of mice), caused by *Mycoplasma muris* (formerly *Haemobartonella muris*) and *Mycoplasma coccoides* (formerly *Eperytrozoon coccoides*), respectively (Table 6.3). Additionally, *Mycoplasma* (formerly *Eperythrozoon*) *suis*, which causes an acute febrile disease in feeder pigs, has been reported to be transmitted by the hog louse by some researchers. Other pathogens have been detected in various species of lice (Reeves et al., 2006), but there is no current evidence that lice are vectors of these organisms.

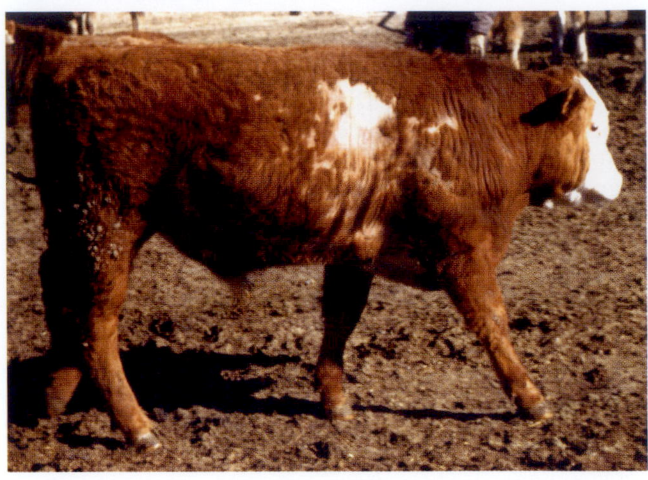

Figure 6.17 Steer infested with the cattle biting louse (*Bovicola bovis*); note areas where hair has been rubbed off by the host. (Photo by John E. Lloyd)

LICE OF LIVESTOCK

Lice cause major economic losses because of reduced livestock productivity and diminished health. These losses include the cost of treatment, which are considerable. Lice infestation may lead to pruritus and its side effects; reduced productivity; and diminished general health. During colder months of the year, lice infestations can reach into the thousands, hundred of thousands, or even a million per animal. Under these conditions, the most serious detrimental effects to the host occur, including death. Animals most heavily infested are often the young, ill, nutritionally deprived, and immunocompromised.

Cattle sucking lice can decrease cattle weight gains and milk production and can necessitate additional feed to maintain lice populations and additional time on feed (Jones 1996). The shortnosed cattle louse (Figure 6.15) in North America can be a cause of severe, and terminal, anemia in range cattle. The cattle tail louse is considered the most damaging cattle louse in Florida. Results of studies of the impact of the longnosed cattle louse (Figure 6.12), little blue cattle louse (Figures 6.13, 6.14), and cattle biting louse (Figure 6.11) on weight gains have been mixed. However, Gibney et al. (1985) found significant differences between weight gains of cattle heavily infested with this complex of lice and weight gains of uninfested cattle. This lice complex is typical on yearling cattle throughout the Great Basin and Rocky Mountain region of the United States. Weight gains are typically lower in stressed cattle and in those receiving inadequate nutrition. Nelson (1984) identified a "synergistic relationship, whereby a coexisting malnutrition and parasitic infection are more deleterious to the host than is either one alone." Campbell (1988) further suggested a synergistic relationship between lice infestation and cold winter weather.

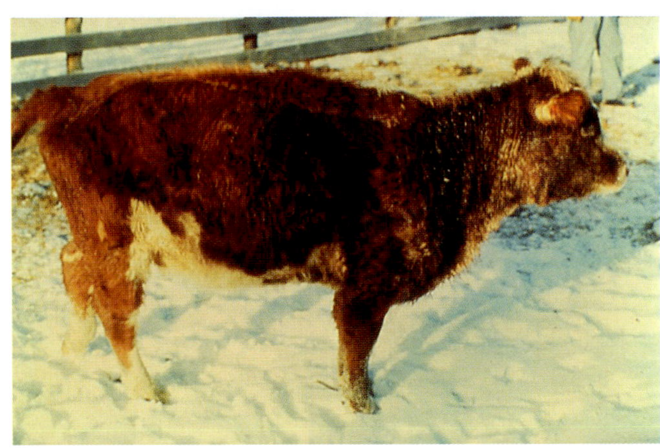

Figure 6.18 Hereford yearling heavily infested with the shortnosed cattle louse (*Haematopinus eurysternus*). Note areas of hair loss. (Photo by John E. Lloyd)

Damage to leather caused by cattle lice is costly. Irritation, leading to rubbing and subsequent hide damage (Figures 6.17, 6.18), may even be caused by small numbers of lice in sensitive cattle. The cattle biting louse not only is responsible for increased scratches in cattle hide due to irritation (Figure 6.17), but it is the major cause of the defect known as **light spot**, 1–3 mm lesions that result from erosion of grain enamel of the hide. Light spot is responsible for annual losses in the United Kingdom of £15 to £20 million (Coles et al., 2003). This rubbing also damages livestock facilities, a major concern in cattle feed lots.

Under laboratory or confined conditions, at least three pathogens can be transmitted by cattle sucking lice: the causative agents of bovine anaplasmosis, dermatomycosis (ringworm) (Table 6.3), and rarely, theileriosis. The importance of cattle lice in transmitting any of these pathogens in nature is unknown but presumed to be low.

Lice of horses and other equids typically do not greatly debilitate their hosts except when they are present in large numbers. As with other host species, horses that are malnourished or in poor health become heavily infested. Horses in poor heath may be infested by several thousand individuals of the horse biting louse. Pruritus, hair loss, and coat deterioration may occur in severely infested animals. Horses with severe louse infestations, or horses that are extremely sensitive to infestation, are nervous and irritable. They stamp their hooves, lick, and rub in response to lice. They bruise and scratch themselves. Hair can be rubbed from the neck, shoulders, flanks, and tail base, resulting in an unthrifty appearance that may affect the market value of the horse. No pathogens are known to be transmitted by equid lice.

Hog lice are extremely irritating to the host, and frequent feeding by the lice causes hogs to rub on objects to the point where blood is drawn. Hair is lost and the skin becomes rough and scaly with lesions. Heavily infested hogs are restless and eat less, which interferes with their growth. Hog lice can imbibe significant volumes of blood, especially from piglets, which often have larger infestations than adult pigs. Anemia may occur. Hog lice are thought to affect the vitality of swine (Cobbett and Bushland, 1956), and are potential vectors of several disease agents of swine. *Haematopinus suis* is a vector of the virus that causes **swinepox** (Table 6.3), a serious and potentially fatal disease characterized by large pock-mark lesions mainly on the belly of infected animals. Some studies have also implicated this louse as a vector of *Mycoplasma* (formerly *Eperythrozoon*) *suis* and *Mycoplasma* (formerly *Eperythrozoon*) *parvum*, causative agents of swine mycoplasma infection (formerly swine eperythrozoonosis), and of African swine fever virus. However, transmission of these pathogens by lice appears to be rare, if it occurs at all, in nature.

Lice that parasitize sheep and goats (Tables 6.1, 6.2) can cause serious losses, even when present in relatively small numbers, because of damage to wool and mohair (Figures 6.19, 6.20). Biting and rubbing in response to irritation damages the skin and devalues wool and mohair. The fleece becomes pulled and ragged with broken fibers and it may slip from or be pulled from the skin creating bare areas (Figure 6.20). The fleece becomes contaminated with lice, cast exoskeletons of lice, ova, and feces. Sucking lice can stain the wool with blood (Figure 6.19), presumably due to undigested blood in their feces; this wool will not scour.

Significant losses due to the sheep biting louse (Figures 6.5C, 6.21) are incurred in the major sheep producing countries (e.g., Australia, New Zealand, South Africa, etc.). In Western Australia, louse infestation reduces production of clean wool by 0.3 to 0.8 kg/animal (Wilkinson, 1982). In addition to causing fleece devaluation worldwide, the sheep biting louse is the major cause of cockle in pelts. Some sheep develop hypersensitivity to the sheep biting louse. **Cockle**, a nodular condition of the skin, arises in response to infestation by *B. ovis* and a hypersensitivity to louse antigens (Heath, 1996). The term **scatter cockle** has been used to describe the distribution of this defect and to distinguish it from **rib cockle**, a pelt blemish caused by sheep ked (*Melophagus ovinus*).

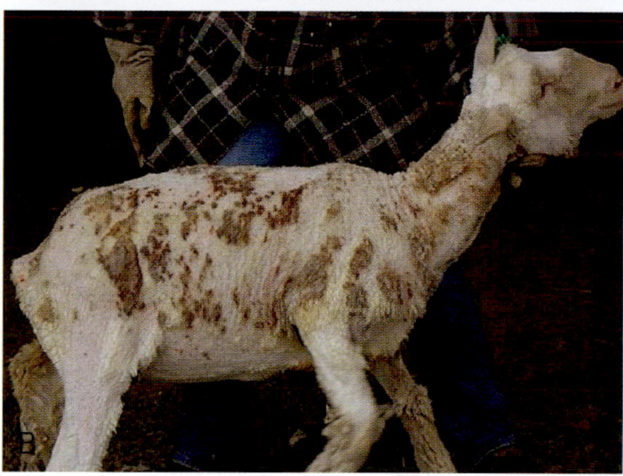

Figure 6.19 A lamb infested with the African blue louse (*Linognathus africanus*); (A) in full fleece, with large patch of bloody frass (fecal matter) from the lice; (B) the same lamb after shearing, with bloody patches of skin (brownish) where the lice are clustered. (Courtesy of Greg Johnson)

Figure 6.20 Fleece damage (wool slippage) in a ewe, caused by a severe infestation with the African blue louse (*Linognathus africanus*). (Photo by J. E. Lloyd)

Sheep heavily infested with the African blue louse develop bare spots along the sides of the body (Figure 6.19). Wool slips from the infested areas leaving a bare area surrounded by a circle of densely infested fleece. Additional effects that have been reported

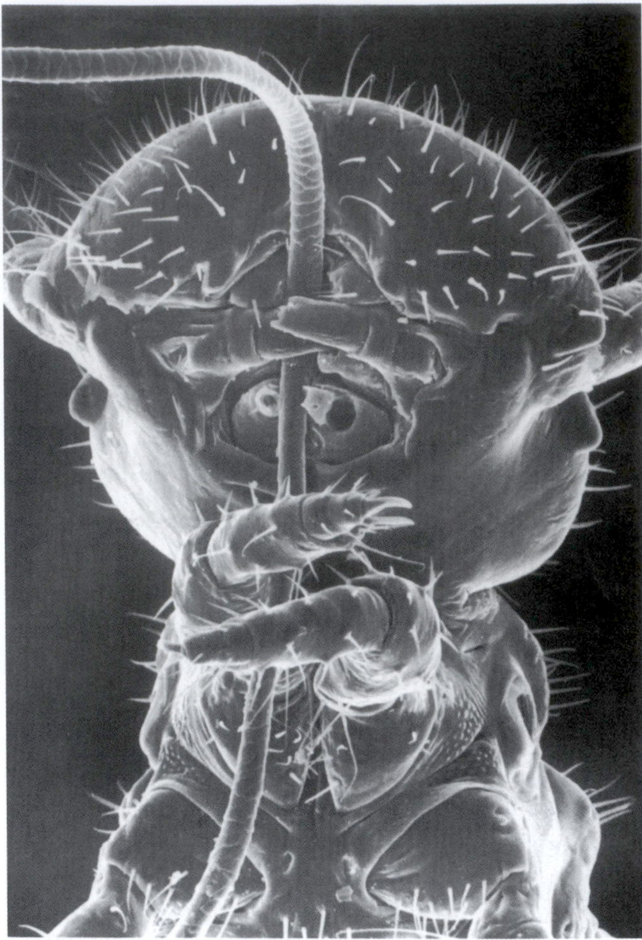

Figure 6.21 Sheep biting louse (*Bovicola ovis*), showing prothorax and head with mandibles characteristically grasping a host hair; scanning electron micrograph. (Price and Graham, 1997)

include anemia and death in goats, especially kids, and skin irritation and fleece damage in Angora goats.

Wild and working elephants (Indian and African) often are infested by the elephant louse (*Haematomyzus elephantis*) (Figures 6.1D, 6.5F). In large infestations, this louse can cause intense irritation in its hosts, which can result in rubbing and skin loss.

Lice of Wildlife

Typically, lice of wildlife cause little or no apparent health problems unless their hosts are immunosuppressed, stressed, or unable to groom efficiently perhaps because of injury. However, a chewing louse of North American wild deer (*Odocoileus* spp.) has been reported to multiply to huge numbers and cause severe hair loss (hair loss syndrome) and possibly death from hypothermia in winter. Although the louse appears to be widespread on deer in the United States, most pathological cases have been reported from the Pacific Northwestern states of Oregon and Washington from Columbian black-tailed deer (*Odocoileus hemionus columbianus*) (Bildfell et al., 2004). The causative chewing louse has been identified as an unknown species of *Damalinia*, a louse genus that is not thought to be native to North America. Massive infestations of the African blue louse, *Linognathus africanus*, have been associated with mule deer deaths in California due to exsanguination, anemia, and winter stress.

Lice of Cats and Dogs

Louse infestations of cats and dogs are most noticeable on sick, malnourished hosts, often unable to groom themselves, or on very young or old animals. Under these conditions, louse populations can increase dramatically. Apparently healthy pets will readily pick up lice when exposed to infested animals. Severe infestations by any of the four species of lice cause host restlessness, scratching, skin inflammation, a ruffled or matted coat, and hair loss. Heavy infestations of sucking lice may produce anemia.

The dog biting louse (*Trichodectes canis*) (Figure 6.5D) is distributed worldwide but apparently is becoming less common in the United States (Kim et al., 1973). The species may be found on the head, neck, and tail of the host, and will aggregate around wounds and body openings. Infestations appear to be heaviest on puppies and old dogs in poor condition. The dog biting louse is an intermediate host of the double-pored tapeworm (*Dipylidium caninum*) (Table 6.3). Lice become infected when they ingest viable *D. caninum* eggs from dried host feces. The tapeworm develops into a cysticercoid stage in the louse where it remains quiescent unless the louse is ingested by a dog, usually during grooming. In the dog gut, the cysticercoid is liberated and metamorphoses into an adult tapeworm.

The dog sucking louse (*Linognathus setosus*) (Figure 6.7G) occurs worldwide but is encountered only occasionally. The lice are found primarily on the neck, shoulders, and under the collar, mainly on long-haired breeds of dog. The dog sucking louse may cause anemia due to blood loss. The lice may cause irritation resulting in sleeplessness, nervousness, biting, and scratching, resulting in secondary bacterial infection and hair loss. This species has been shown to harbor immatures of the filarial nematode *Dipetalonema reconditum* which parasitizes dogs, but whether or not these lice are efficient vectors remains unknown.

The cat biting louse (*Felicola subrostrata*) (Figure 6.5E) is worldwide in distribution but relatively uncommon. Infestations occur mainly on unhealthy, older, or longhaired cats, or cats unable to groom. The intense irritation caused by this species may cause severe scratching, dermatitis, and hair loss on the back of the cat. Secondary bacterial infections may develop as a result of the scratching.

Lice of Laboratory Animals

Some lice that parasitize laboratory animals initiate serious health problems by causing pruritus, skin lesions, scab formation, anemia, and hair loss. Others are vectors of pathogens that can cause severe problems in animal colonies (Table 6.3). The mouse louse (*Polyplax serrata*) is a vector of the bacterium *Mycoplasma* (formerly *Eperythrozoon*) *coccoides*, which causes murine mycoplasma infection (formerly murine eperythrozoonosis), a potentially lethal infection of mice that occurs worldwide. Infection of this blood parasite in mice can either be inapparent or result in severe anemia. Transmission of this pathogen in louse-infested mouse colonies is usually rapid. The spined rat louse (*Polyplax spinulosa*) (Figure 6.7D) is a vector of *Mycoplasma* (formerly *Haemobartonella*) *muris*, which causes murine mycoplasma infection (formerly murine haemobartonellosis) (Table 6.3), another potentially fatal blood infection that can cause severe anemia in laboratory rats.

Laboratory and wild guinea pigs are parasitized by two species of chewing lice, the slender guinea pig louse (*Gliricola porcelli*) and the oval guineapig louse (*Gyropus ovalis*). Small numbers of these lice cause no noticeable harm, whereas large populations can cause host unthriftiness, scratching (especially behind the ears), hair loss, and a ruffled coat.

Large infestations of the rabbit louse (*Haemodipsus ventricosis*) can cause severe itching and scratching that results in the host rubbing against its cage, often resulting in hair loss. Young rabbits are more adversely affected than are adults and may experience retarded growth as a consequence of infestation by *H. ventricosis*. The rabbit louse is also a vector of the causative agent of tularemia among wild rabbit populations (Table 6.3).

Lice of Poultry and Other Birds

Although louse populations may be very large on domestic fowl including domestic chickens, turkeys, guinea fowl, pea fowl, and pheasants, no pathogens are known to be transmitted by these lice. The chicken body louse (*Menacanthus stramineus*) often causes significant skin irritation and reddening through its persistent feeding (Figure 6.16). Occasionally the skin or soft quills pushing through the skin bleed from the gnawing and scraping action of lice with the lice readily imbibing the resultant blood. Populations of chicken body lice are influenced by the host's ability to groom. Debeaked birds tend to become more heavily infested. In general, louse-infested chickens do not gain as much weight or produce as many eggs as do lice-free chickens, and heavily infested young chicks may die.

The shaft louse (*Menopon gallinae*) also causes significant losses to the poultry industry, including deaths of young birds with heavy infestations. Large infestations of chicken body lice, shaft lice, and other poultry lice may be injurious to the host by causing feather loss, lameness, low weight gains, inferior laying capacity, or even death.

The vast majority of chewing lice are parasites of wild or peridomestic birds. Several of these lice are suspected vectors of avian pathogens. Some chewing lice of aquatic birds, including geese and swans, are vectors of filarial nematodes (Table 6.3). Pet parrots, parakeets, budgerigars, and other birds also are subject to infestation by chewing lice, which is usually noticed mainly by the associated host scratching and ruffled or lost feathers. Large populations of these lice can debilitate their hosts. Ranch birds such as ostriches, emus, and rheas are prone to similar adverse effects caused by their associated chewing lice.

PREVENTION AND CONTROL

Several techniques have been used in attempts to rid humans and animals of lice and louse-borne diseases. Preventing physical contact between lousy persons or animals and the items they contact, as well as various chemical, hormonal, and biological control mechanisms comprise the current arsenal of techniques. Chemicals used to kill lice are called **pediculicides**.

Clothes of persons with body lice should be changed frequently, preferably daily, and washed in very hot, soapy water to kill lice and nits. Washing associated bed linen in this manner is also advisable. Infested people should also receive a concurrent whole-body treatment with a pediculicide. Overcrowded and unsanitary conditions should be avoided whenever possible during outbreaks of human body lice and louse-borne diseases because it is under these situations that both can thrive.

Crab lice can often be avoided by refraining from multiple sexual partners and changing or laundering bed linen slept on by such persons. Pediculicides should be applied to the pubic area and to any other infested body regions.

To reduce the spread of head lice, the sharing of combs, hats, earphones, and blankets, especially by children, should be discouraged. Often, parents of children with head lice are notified to keep youngsters away from school or other gatherings until the infestation has been eliminated. If the parents are also infested, this can further involve ridding lice from the entire family to prevent reinfestations. Various pediculicidal shampoos, lotions, and gels are widely available for controlling head lice. In the United States, topically applied 1% permethrin or 0.33% pyrethrins are typically used to kill human lice and 0.5% malathion lotions are often used in cases of initial treatment failures. Topically applied pediculicides are not very effective against body lice, which spend most of their lives in the clothing. These treatments typically kill all nymphal and adult head lice but only a proportion of viable louse eggs. Therefore, treatments should be repeated at weekly intervals for two to four weeks in order to kill any recently hatched lice. Hatched or dead nits that remain glued to hair may be unsightly or embarrassing, and can be removed with a fine-toothed louse comb. Louse combs have been used, in various forms, since antiquity to remove head lice (Mumcuoglu, 1996). Because lice are highly susceptible to desiccation, one recently reported effective technique for killing head lice and their eggs is to expose infested human scalps to a stream of hot air (Goates et al., 2006). An increasingly popular method for avoiding louse infestations is to impregnate clothing with permethrin; this action prevents, for example, health care workers or soldiers who are interacting with louse-infested refugees, from becoming infested.

A wide range of pediculicides is commercially available. Although its use is now banned in many developed countries, the organochlorine DDT is widely used, especially in less developed countries, for controlling human and animal lice. Several alternative pediculicides such as lindane, chlorpyrifos, diazinon, malathion, permethrin, or pyrethrins currently are used throughout the world. Pediculicides can be used in powders, fogs, or sprays to treat furniture or premises for lice. Several general parasiticides show promise as pediculicides. Avermectins such as abamectin, doramectin, and ivermectin can kill human body lice and livestock lice. Prescribed doses of these compounds can be administered orally, by injection, or as topical applications of powders, dusts, and pour-ons. However, many of these compounds have not been approved yet for use on humans. The development of novel control agents for lice is a constant process because resistance to various pediculicides has developed in lice in many parts of the world (Mumcuoglu, 1996; Burgess 2004, 2008).

Healthy, well-groomed, and well-nourished animals are the least likely to develop infestations of lice requiring treatment. Since lice are spread mainly by contact, introduction of infested animals should be avoided. Exposure of uninfested animals to facilities shortly after removal of infested animals, or exposure to contaminated equipment are also common means of infestation. Nevertheless, infestations still occur in spite of best efforts. For this reason, an almost bewildering variety of insecticide products and treatment methods are available. The following classes of insecticide have at least one compound registered for use against lice of pets and livestock: organochlorines, organophosphates, carbamates, pyrethrins, synthetic pyrethroids, macrocyclic lactones (avermectins and milbemycins), formamadines, chloronicotinyls and spinosyns, insect growth regulators, and synergists. For a complete presentation of current products, see the *Merck Veterinary Manual*, 9th edition (2005) and the *Merck/Merial Manual for Pet Health* (2007).

Small animals like dogs and cats may be treated topically with shampoos, powders, aerosols, rinses, spot-ons, and others, of low mammalian toxicity. Care must be taken to follow the label directions to insure that a product is indicated for cats as they tend to lick themselves. Lice may be controlled by clipping away matted hair, and then bathing the host, followed by the use of a mild insecticidal shampoo, powder, spray, or other treatment according to label directions. A second treatment in two weeks may be indicated on the product label to control lice that were present as eggs at the time of initial treatment.

Because of the importance of lice control to the livestock industry, a large number of insecticidal products and treatment methods are available. Withdrawal times following treatment must be observed to avoid objectionable residues in meat or milk. A number of products are approved only for nonlactating cattle or goats. Caution must be taken in treating young, very old, or debilitated animals, as these may be more susceptible to toxicity. Some products may be labeled for several species of livestock; however, in the United States, the specific use must be approved by the EPA (Environmental Protection Agency) or FDA (Food and Drug Administration) and indicated on the product label. The label should always be read and understood prior to treatment. As with pets, treatment of livestock may require a follow-up treatment to control lice that were present as ova at the time of the initial treatment.

Beef cattle, dairy cattle, sheep, goats, swine, and equines may be treated effectively with a whole body mist or spray for lice control, although easier and less stressful methods may be available. Animal systemic insecticides, which have gained wide acceptance among livestock producers, are those that enter the circulatory system and eventually reach and control blood-feeding parasites like sucking lice. These insecticides may be administered topically/dermally, orally, or as a subcutaneous injection. The most popular method of application is the pour-on, a single line of liquid along the midline of the back (a variation is a single spot on the midline). Pour-on formulations of systemics also control chewing lice, mainly through absorption and spread through the skin and hair coat of the treated animal. Several nonsystemic insecticides control lice in a similar manner—they spread through the skin and the hair coat from the site of application.

In the United States, more cattle are treated for lice than any other species of livestock. Cattle may be sprayed, but most commonly they are treated with the various pour-on formulations, both systemic and nonsystemic. A preventive fall treatment with a systemic insecticide is a common practice among cattlemen for control of both cattle lice and cattle grubs (*Hypoderma bovis* and *H. lineatum*). Fall treatment for lice control prevents the increase of lice infestations to potentially damaging levels.

Even when cattle are treated in the fall, it may be difficult for the cattle producer to keep his or her animals lice-free. The producer will not usually see clinical manifestations of a lice infestation until wintertime. At this time, animals normally are seen licking and rubbing in response to the lice. Bare patches may be noted, and if the hair is parted, lice may be visible. When clinical infestations are observed, cattle may be treated with most of the products approved for lice control. Most methods can be used in cold weather; however, some, like the pour-ons, have an advantage in that they do not require elaborate equipment for spraying and the animals do not become completely wet and subject to chilling. Caution must be taken to avoid treatment of grub-infested cattle with a systemic in the winter because of the possibility of the "host parasite reaction" occasionally seen in cattle when grubs are killed in the esophageal area or in the area of the spinal canal.

Various treatment methods can be used in the winter to treat the entire herd, or to treat only the animals that exhibit signs of infestation. Animals restrained in a stanchion may be individually dusted by hand. Because louse-infested animals are prone to rub, self-treatment devices like dust bags, oilers, and back rubbers can aid in the control of lice. Insecticidal ear tags, applied in the summer for fly control, may aid in the control of lice since they are also present on the cattle in the summer.

Lice infestations on sheep, goats, and swine may be treated with whole body or pour-on treatments of insecticide. Horses may be treated with a whole-body spray, but since horses may react to sprayer noise, liquid insecticide often is applied as a wipe. For control of severe infestations of swine an insecticide dust may be used in the bedding.

Contact of domestic poultry with potentially infested wild birds should be avoided. When poultry houses are vacated for a new group of birds, it is important to remove all feathers that may be infested with nits. Poultry can be treated with whole body pediculicidal sprays or dusts and repeated in seven to 10 days. Although host treatment is most efficacious, bedding materials and cages can also be treated to aid in lice control.

With respect to louse-borne diseases, vaccines have been developed only against epidemic typhus, and none are completely safe or currently approved for widespread use. Live attenuated vaccines have been administered to humans, particularly in certain African nations, in attempts to quell epidemic typhus outbreaks but have not been highly effective.

REFERENCES AND FURTHER READING

Andersson, S. G. E., Zomorodipour, A., Andersson, J. O., Sicheritz-Pontén, T., Alsmark, U. C. M., Podowski, R. M., et al. (1998). The genome sequence of *Rickettsia prowazekii* and the origin of mitochondria. *Nature, 396,* 133–140.

Azad, A. F. (1988). Relationship of vector biology and epidemiology of louse- and flea-borne rickettsioses. In D. H. Walker (Ed.), *Biology of rickettsial diseases* (Vol. I, pp. 51–61). Boca Raton, FL: CRC Press.

Bedford, G. A. H. (1932). Trichodectidae (Mallophaga) found on South African Carnivora. *Parasitology, 24,* 350–364.

Bildfell, R. J., Mertens, J. W., Mortenson, J. A., & Cottam, D. F. (2004). Hair-loss syndrome in black-tailed deer of the Pacific Northwest. *Journal of Wildlife Diseases, 40,* 670–681.

Bryceson, A. D. M., Parry, E. H. O., Perine, P. L., Warrell, D. A., Vukotich, D., & Leithead, C. S. (1970). Louse-borne relapsing fever. A clinical and laboratory study of 62 cases in Ethiopia and a reconsideration of the literature. *Quarterly Journal of Medicine, New Series, 39,* 129–170.

Burgess, I. F. (2004). Human lice and their control. *Annual Review of Entomology, 49,* 457–481.

Burgess, I. F. (2008). 9. Human body lice. In X. Bonnefoy, H. Kampen, & K. Sweeney (Eds.), *Public Health Significance of Urban Pests* (pp. 289–301). Geneva: World Health Organization. (also available electronically on the WHO website).

Butler, J. F. (1985). Lice affecting livestock. In R. E. Williams, R. D. Hall, A. B. Broce, & P. J. Scholl (Eds.), *Livestock entomology* (pp. 101–127). New York: Wiley.

Campbell, J. B. (1988). Arthropod induced stress in livestock. *Veterinary Clinics of North America: Food Animal Practice, 4,* 551–555.

Clay, T. (1938). New species of Mallophaga from *Afroparvo congensis* Chapin. *American Museum Novitates,* No. 1008.

Coles, G. C., Hadley, P. J., Milnes, A. S., Gren, L. E., Stosic, P. J., & Garnsworthy, P. C. (2003). Relationship between lice infestation and leather damage in cattle. *The Veterinary Record, 153,* 255–259.

Cobbett, N. G., & Bushland, R. C. (1956). The hog louse. In A. Stefferud (Ed.), *Animal Diseases. The Yearbook of Agriculture* (pp. 345–347.

Dalgleish, R. C., Palma, R. L., Price, R. D., & Smith, V. S. (2006). Fossil lice (Insecta: Phthiraptera) reconsidered. *Systematic Entomology, 31,* 648–651.

Drancourt, M., Tran-Hung, L., Courtin, J., De Lumley, H., & Raoult, D. (2005). *Bartonella quintana* in a 4000-year-old human tooth. *Journal of Infectious Diseases, 191,* 607–611.

Durden, L. A., & Musser, G. G. (1994a). The sucking lice (Insecta, Anoplura) of the world: A taxonomic checklist with records of mammalian hosts and geographical distributions. *Bulletin of the American Museum of Natural History, 218,* 1–90.

Durden, L. A., & Musser, G. G. (1994b). The mammalian hosts of the sucking lice (Insecta, Anoplura) of the world: A host-parasite checklist. *Bulletin of the Society for Vector Ecology, 19,* 130–168.

Emerson, K. C. (1972). *Checklist of the Mallophaga of North America (North of Mexico). Parts I–IV.* Dugway, Utah: Desert Test Center.

Emerson, K. C., & Price, R. D. (1975). Mallophaga of Venezuelan mammals. *Brigham Young University Science Bulletin, Biological Series, 20*(3), 1–77.

Emerson, K. C., & Price, R. D. (1985). Evolution of Mallophaga and mammals. In K. C. Kim (Ed.), *Coevolution of Parasitic Arthropods and Mammals* (pp. 233–255). New York: Wiley.

Ferris, G. F. (1919–1935). Contributions toward a monograph of the sucking lice. Parts I–VIII. *Stanford University Publications, University Series, Biological Sciences, 2,* 1–634.

Ferris, G. F. (1931). The louse of elephants *Haematomyzus elephantis* (Mallophaga: Haematomyzidae). *Parasitology, 23,* 112–127.

Ferris, G. F. (1951). The sucking lice. *Memoirs Pacific Coast Entomological Society, 1,* 1–320.

Gibney, V. G., Campbell, J. B., Boxler, D. J., Clanton, D. C., & Deutscher, G. H. (1985). Effects of various infestation levels of cattle lice (Mallophaga: Trichodectidae and Anoplura: Haematopinidae) on feed efficiency and weight gains of beef heifers. *Journal of Economic Entomology, 78,* 1304–1307.

Goates, B. M., Atkin, J. S., Wilding, K. G., Birch, K. G., Cottam, M. R., Bush, S. E., et al. (2006). An effective nonchemical treatment for head lice: A lot of hot air. *Pediatrics, 118,* 1962–1970.

Gratz, N. G. (1997). *Human Lice: Their Prevalence, Control and Resistance to Insecticides. A Review 1985–1997.* World Health Organization/CTD/WHOPES/97.8.

Hanlin, S. J. (1994). *The summer distribution of cattle lice in Wyoming.* Masters Thesis, University of Wyoming.

Heath, A. C. G., Bishop, D. M., Cole, D. J. W., & Pfeffer, A. T. (1996). The development of cockle, a sheep pelt defect, in relation to size of infestation and time of exposure to *Bovicola ovis,* the sheep-biting louse. *Veterinary Parasitology, 67,* 259–267.

Hopkins, G. H. E. (1949). The host-associations of the lice of mammals. *Proceedings of the Zoological Society of London, 119,* 387–604.

Ignoffo, C. M. (1959). Keys and notes to the Anoplura of Minnesota. *American Midland Naturalist, 61,* 470–479.

Jones, C. J. (1996). Immune responses to fleas, bugs and sucking lice. In S. K. Wikel, (Ed.), *The Immunology of Host-Ectoparasitic Arthropod Relationships* (pp. 150–174). Wallingford: CAB International.

Kaufman, P., Koehler, P. G., & Butler, J. F. (2005). *Cattle Tail Lice. ENY-271 (IG127), Entomology and Nematology Department, Florida Cooperative Extension Service.* Institute of Food and Agricultural Sciences, University of Florida.

Kim, K. C. (1985). Evolution and host associations of Anoplura. In K. C. Kim (Ed.), *Coevolution of Parasitic Arthropods and Mammals* (pp. 197–231). New York: Wiley.

Kim, K. C., Emerson, K. C., & Price, R. D. (1973). Lice. In R. J. Flynn (Ed.), *Parasites of Laboratory Animals* (pp. 376–397). Iowa State Univ. Press.

Kim, K. C., & Ludwig, H. W. (1978). The family classification of the Anoplura. *Systematic Entomology, 3,* 249–284.

Kim, K. C., Pratt, H. D., & Stojanovich, C. J. (1986). *The Sucking Lice of North America: An Illustrated Manual for Identification.* Pennsylvania State Univ. Press.

Lancaster, J. J. L., & Meisch, M. V. (1986). Lice. In *Arthropods in livestock and poultry production.* (pp. 321–345). Ellis Horwood, Chichester, England.

Ledger, J. A. (1980). *The Arthropod Parasites of Vertebrates in Africa South of the Sahara. Vol. IV. Phthiraptera (Insecta).* Johannesburg: South African Institute of Medical Research.

Lloyd, J. E. (1999). Flies, lice, and grubs. In J. L. Howard, R. A. Smith (Eds.), *Current Veterinary Therapy* (pp. 706–710). W. B. Saunders Co.

Marshall, A. G. (1981). *The Ecology of Ectoparasitic Insects.* London: Academic Press.

Matthysse, J. G. (1946). Cattle lice, their biology and control. *Cornell University Experiment Station Bulletin,* 832.

Mumcuoglu, K. Y. (1996). Control of human lice (Anoplura: Pediculidae) infestations: Past and present. *American Entomologist, 42,* 175–178.

Nelson, W. A., Bell, J. F., Clifford, C. M., & Keirans, J. E. (1977). Interaction of ectoparasites and their hosts. *Journal of Medical Entomology 13,* 389–428.

Nelson, W. A. (1984). Effects of nutrition of animals on their ectoparasites. *Journal of Medical Entomology, 21,* 621–635.

Orkin, M., & Maibach, H. I. (Eds.). (1985). *Cutaneous Infestations and Insect Bites. Dermatology Services, Vol. 4.* New York: Marcel Dekker.

Pan American Health Organization/World Health Organization. (1973). *The Control of Lice and Louse-Borne Diseases.* P.A.H.O. Sci. Publ., No. 263.

Price, M. A., & Graham, O. H. (1997). *Chewing and Sucking Lice as Parasites of Mammals and Birds.* U.S. Dept. Agric., Agric. Res. Serv., Tech. Bull., No. 1849.

Price, R. D., Hellenthal, R. A., Palma, R. L., Johnson, K. P., & Clayton, D. H. (2003). *The chewing lice: World checklist and biological overview.* Illinois Natural History Survey, Special Publication No. 24.

Reed, D. L., Smith, V. S., Hammond, S. L., Rogers, A. R., & Clayton, D. H. (2004). Genetic analysis supports direct contact between modern and archaic humans. *PLoS, Biology 2*(11), e340.

Reeves, W. K., Murray, K. O., Meyer, T. E., Bull, L. M., Pascua, R. F., Holmes, K. C., et al. (2008). Serological evidence of typhus group rickettsia in a homeless population in Houston, Texas. *Journal of Vector Ecology, 33,* 206–208.

Reeves, W. K., Szumlas, D. E., Moriarity, J. R., Loftis, A. D., Abbassy, M. M., Helmy, I. M., et al. (2006). Louse-borne bacterial pathogens in lice (Phthiraptera) of rodents and cattle from Egypt. *The Journal of Parasitology, 92,* 313–318.

Reynolds, M. G., Krebs, J. W., Comer, J. A., Sumner, J. W., Rushton, T. G., Lopez, C. E., et al. (2003). Flying squirrel-associated typhus, United States. *Emerging Infectious Diseases, 9,* 1341–1343.

Roberts, F. H. S. (1952). Insects affecting livestock. Angus and Robertson, Sydney, Australia.

Rothschild, M., & Clay, T. (1952). *Fleas, Flukes and Cuckoos: A Study of Bird Parasites.* London: Collins.

Synder, J. C. (1966). Typhus fever rickettsiae. In F. L. Horsfall & I. Tamm (Eds.), *Viral and rickettsial infections of man,* (4th ed.) (pp. 105–114). J. B. Lippincott, Philadelphia.

The Merck Veterinary Manual. (2005). *Lice: Pediculosis* (C. M. Kahn Ed., 9th ed.). Merck & Company, Inc.

The Merck/Merial Manual for Pet Health. (2007). In C. M. Kahn (Ed.), (1st ed.). Merck & Company, Inc.

Townsend, L., & Scharko, P. (1999). Lice infestation in beef cattle. *Compendium on Continuing Education for the Practicing Veterinarian, 21*(Suppl.), S119–S123.

Tuff, D. W. (1977). A key to the lice of man and domestic animals. *Texas Journal of Science, 28,* 145–159.

Van der Stichele, R. H., Dezeure, E. M., & Bogaert, M. G. (1995). Systematic review of clinical efficacy of topical treatments for head lice. *British Medical Journal, 311,* 604–608.

Watson, D. W., Lloyd, J. E., & Kumar, R. (1997). Density and distribution of cattle lice (Phthiraptera: Haematopinidae, Linognathidae, Trichodectidae) on six steers. *Veterinary Parasitology, 69,* 283–296.

Werneck, F. L. (1950). *Os malofagos de mammiferos. Parte II. Ischnocera (continuacao de Trichodectidae) e Rhyncophthirina.* Rio de Janeiro: Instituto Oswaldo Cruz.

Wilkinson, F. C., De Chaneet, G. C., & Beetson, B. R. (1982). Growth of populations of lice, *Damilinia ovis,* on sheep and their effects on production and processing performance of wool. *Veterinary Parasitology, 9,* 243–252.

Zinsser, H. (1935). *Rats, Lice, and History.* New York: Bantam.

Chapter

7

True Bugs (Hemiptera)

William L. Krinsky

The order Hemiptera includes all the insects known as true bugs. Hemipterans are characterized as soft-bodied insects with piercing and sucking mouthparts and usually two pairs of wings. The order traditionally was divided into two major divisions, the Heteroptera and the Homoptera, based on wing morphology. The name Hemiptera (literally, "half-wings") is derived from the members of the Heteroptera ("different wings"), most of which have forewings called **hemelytra**. They are composed of a thickened basal portion, the **corium** and **clavus**, and a somewhat transparent or filmy distal portion, the **membrane**, hence the idea of a half wing (Figure 7.1). The hind wings are completely membranous. The difference in texture between the fore and hind wings in the heteropterans give this group its name. By comparison, the Homoptera ("same wings") have two pairs of wings that are very similar in character, both being membranous. The wings of homopterans often are held roof-like over the back of the body, whereas the wings of the heteropterans typically are held flat against the dorsum. The suborder Homoptera now is divided into two suborders, the Auchenorrhyncha, having the mouthparts clearly arising from the head, and the Sternorrhyncha, in which the mouthparts appear to arise between the front coxae, or are absent. Auchenorrhyncha are cicadas, leafhoppers, planthoppers, treehoppers, spittle bugs and their relatives. Sternorrhyncha include aphids, adelgids, mealy bugs, phylloxerans, psyllids, whiteflies, and scale insects.

The true bugs, with about 90,000 species worldwide, constitute the largest exopterygote order of insects. The North American fauna has about 16,000 species of hemipterans, about two-thirds of which are homopterans.

The piercing-sucking mouthparts of almost all true bugs enable these insects to feed on a diversity of fluids. The homopterans feed exclusively on plant juices and all of them are terrestrial. The heteropterans include phytophagous, predaceous, and hematophagous species. Common heteropterans include seed bugs, mirid plant bugs, stink bugs, assassin bugs, water striders, backswimmers, water boatmen, and giant water bugs, as well as the medically important kissing bugs and bed bugs.

Various homopterans and some predaceous and phytophagous heteropterans are known to bite humans. Predaceous or phytophagous bugs in at least 20 families of Hemiptera have been reported as occasionally biting or annoying humans by probing with their mouthparts (Table 7.1). Published reports of these bites have been reviewed by Myers (1929), Usinger (1934), Ryckman (1979), Ryckman and Bentley (1979), and Alexander (1984).

Homopteran species known to cause occasional irritation or pain are leafhoppers (Cicadellidae), treehoppers (Membracidae), spittle bugs (Cercopidae), planthoppers (Fulgoroidea), and cicadas (Cicadidae). Unlike the generally painless bites by hematophagous species, predaceous and phytophagous bug bites often cause pain or a burning sensation, presumably the result of enzymes and other substances in the saliva that normally digest insect or plant materials. Most of these bites cause only transient discomfort associated with toxic reactions to foreign proteins and the localized erythema and edema that may result.

Common terrestrial heteropterans known to probe human skin are the wheel bug (*Arilus cristatus*) and other assassin bugs (*Reduvius personatus, Sinea diadema, Melanolestes picipes*), the two-spotted corsairs (*Rasahus biguttatus* and *R. thoracicus*), and certain anthocorids (*Anthocoris musculus, Lyctocoris campestris,* and *Orius insidiosus*). There are fewer reports of nabid, lygaeid, mirid, tingid, and rhopalid bugs biting people. Humans most often are bitten by predaceous species when they enter habitats in which active predation is occurring or when the predatory species are attracted to house lights and enter dwellings. Some of the larger aquatic predaceous hemipterans can stab with their mouthparts causing pain similar to that of a wasp sting. Species that most often bite in this way are belostomatids (giant water bugs, sometimes called toe-biters) and notonectids (backswimmers). The bite of an assassin bug (*Holotrichius innesi*) found in the Sinai and Negev deserts of Israel is considered more neurotoxic and hemotoxic than the bite of venomous snakes in that region (Caras, 1974).

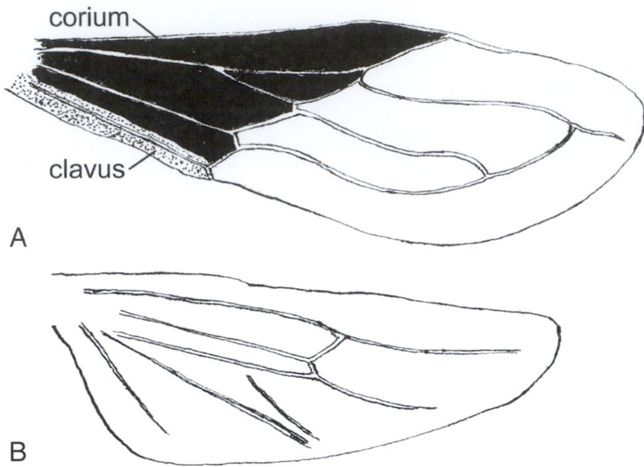

Figure 7.1 Wings of typical triatomine bug (*Triatoma rubrofasciata*). (A) Forewing (hemelytron); (B) Hindwing. (Redrawn and modified from Lent and Wygodzinsky, 1979)

The painless bites of the bloodsucking species pose the greatest threat to the health and well-being of humans and other animals because of the pathogens that often are transmitted and the blood loss associated with their feeding. The hematophagous heteropteran species of major medical and veterinary importance are the kissing bugs (triatomines) and the bed bugs (cimicids). These ectoparasitic insects are obligate blood feeders, requiring blood for growth and reproduction. The kissing bugs are vectors of the causative agent of Chagas disease, a significant medical problem in Central and South America. Bed bugs are not known to play a role in the transmission of any human disease agents. However, their bites may cause considerable discomfort, and their continued feeding may result in significant blood loss from a host. Some bed bugs feed primarily on nonhuman hosts, such as bats and swallows. The polyctenids, bat bugs that feed exclusively on bat blood, have never been associated with any medical or veterinary problems.

KISSING BUGS (REDUVIIDAE)

The kissing bugs are so named because most of them are nocturnal species that feed on humans, often biting the faces of their sleeping victims. Another common name for them is **conenoses**, referring to the shape of the anterior part of the head (Figure 7.2). Various common names in South America and where they are used locally include: *barbeiro, bicudo*, or *chupão* (Brazil), *vinchuca* (Bolivia, Uruguay, Paraguay, Chile, Argentina), *bush chinch* (Belize), *chipo* or *pito* (Colombia, Venezuela), *chinchorro* (Ecuador), *chirimacho* (Peru), *iquipito* or *chupon* (Venezuela) (Schofield et al., 1987). They are all members of the subfamily Triatominae in the family Reduviidae.

Lent and Wygodzinsky (1979) wrote an excellent monograph on kissing bugs that includes a survey of the external structures, descriptions of triatomine species, and notes on the vector importance of each species. Triatomine biosystematics, including an assessment of the evolutionary history of the subfamily, was reviewed by Schofield (1988). Biology, taxonomy, public health importance, and control were reviewed by Schofield et al. (1987), Schofield and Dolling (1993), Schofield (1994), and Yamagata and Nakagawa (2006).

Table 7.1 Nonhematophagous Hemiptera That Occasionally Bite Humans

Scientific Names	Common Names	Locations of Published Cases
Homoptera – Auchenorrhynca		
Cicadellidae	Leafhoppers	United States (California, Texas), Trinidad, England, North Africa, India, China, Japan, Philippines
Cercopidae	Spittle bugs	India
Membracidae	Treehoppers	United States (eastern)
Heteroptera		
Anthocoridae	Minute pirate bugs	North America, Panama, Brazil, England, Czechoslovakia, Sudan, South Africa
Enicocephalidae	Gnat bugs	India
Lygaeidae	Seed bugs	Hawaii, Brazil, North Africa, Kuwait, India
Miridae	Leaf or plant bugs	North America, Brazil, Europe, Sudan
Nabidae	Damsel bugs	United States, Brazil
Pyrrhocoridae	Red bugs or stainers	Brazil, North Africa
Reduviidae	Assassin bugs	North America, Brazil, North Africa, Israel, India, Philippines
Rhopalidae	Scentless plant bugs	North America
Belostomatidae	Giant water bugs	North America
Notonectidae	Backswimmers	North America

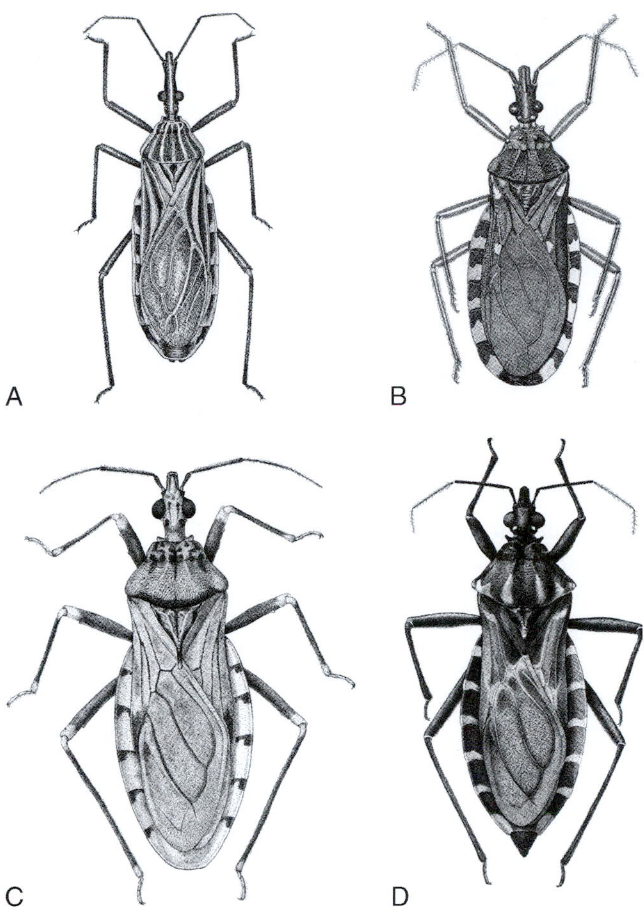

Figure 7.2 Triatomine species. (A) *Rhodnius prolixus;* (B) *Triatoma infestans;* (C) *Panstrongylus geniculatus;* (D) *Panstrongylus megistus.* (Courtesy of the American Museum of Natural History)

TAXONOMY

Members of the heteropteran family Reduviidae are commonly called assassin bugs because most species attack and feed on other insects. There are 23 subfamilies in the Reduviidae, including the Triatominae or kissing bugs. Keys for identification of triatomine species are given by Lent and Wygodzinsky (1979).

The Triatominae is divided into five tribes and 17 genera; most species are known only from the New World, six species (*Linshcosteus*) are found only in India, and seven species (*Triatoma*) are known only from Southeast Asia. The only species found in Africa is *Triatoma rubrofasciata*; it is found throughout the tropics, presumably having spread worldwide via ships.

The New World triatomine species occur from just south of the Great Lakes region of the United States to southern Argentina, with all but a few species concentrated in subtropical and tropical regions. The latter areas are considered the likely places of origin for the subfamily. All triatomines have the potential to transmit *Trypanosoma cruzi*, the etiologic agent of Chagas disease. Of the 138 described triatomine species, about half have been shown to be vectors, and fewer than 20 of these are considered vectors of major epidemiological importance (Table 7.2).

MORPHOLOGY

Triatomines range in length from 5 to 45 mm, with the majority of species falling in the range of 20 to 28 mm. Most species are black or dark brown, often with contrasting patterns of yellow, orange, or red, notably on the **connexivum** (the prominent abdominal margin at the junction of the dorsal and ventral plates) (Figure 7.2).

Table 7.2 Major Triatomine Vectors of *Trypanosoma cruzi* and Their Geographic Occurrence*

Species	Geographic Occurrence
Rhodnius prolixus	Guatemala, Honduras, Nicaragua, Colombia, Venezuela, Guyana, Surinam, French Guiana
Triatoma infestans	Ecuador, Peru, Bolivia, Brazil, Paraguay, Argentina, Uruguay, and Chile
Triatoma dimidiata	Mexico south to Ecuador and Peru
Triatoma pallidipennis	Mexico
Triatoma phyllosoma	Mexico
Rhodnius pallescens	Panama
Triatoma maculata	Colombia, Venezuela, Netherlands Antilles, Guyana, Surinam
Triatoma brasiliensis	Brazil (northeastern)
Triatoma carrioni	Ecuador (southern), Peru (northern)
Panstrongylus chinai	Ecuador, Peru
Rhodnius ecuadoriensis	Ecuador and Peru (northern)
Panstrongylus rufotuberculatus	Costa Rica, Panama, Colombia, Venezuela, Ecuador, Peru, Brazil, Bolivia
Panstrongylus herreri	Peru
Panstrongylus megistus	Brazil (especially coastal), Bolivia, Paraguay, Argentina, Uruguay
Triatoma guasayana	Bolivia, Paraguay, Argentina
Triatoma patagonica	Argentina
Triatoma sordida	Bolivia, Brazil, Paraguay, Uruguay, Argentina

*The two species with the widest geographic distribution are listed first, followed by species arranged generally by their distribution from north to south (Lent and Wygodzinsky, 1979; WHO, 2002).

The head of an adult triatomine is constricted posteriorly to form a distinct neck behind the paired ocelli. Prominent hemispherical compound eyes are situated just in front of the ocelli. The region in front of the eyes is cylindrical to conical, hence the name "cone-nosed" bugs. The antennae are filiform and four-segmented. The beak, or **rostrum**, is three-segmented and is formed by the labium, which encloses the stylet-like mouthparts. These stylets are modified portions of the maxillae and mandibles that lie within a dorsal channel of the rostrum and are grooved to form a food canal and a salivary canal. When the bug is not feeding, the straight rostrum is held under, and nearly parallel to, the head (Figure 7.3). In many nontriatomine reduviids, the rostrum is curved and strongly sclerotized.

The dorsal portions of the thorax include a collar, or neck, a somewhat triangular pronotum, and a scutellum. The undersurface of the prothorax (prosternum) has a stridulatory groove that has fine transverse sculpturing. When the tip of the rostrum is moved anteriorly to posteriorly in this groove, sound is produced, the function of which is mainly defensive.

The forewings, or **hemelytra**, have a leathery basal portion (corium and base of the clavus) and an apical membranous portion (apical clavus and membrane) typical of most heteropterans. The membrane is dusky in most species, but may be spotted, or only darkened along the wing veins. The wing veins of the membrane form two elongate closed cells. The hind wings are completely membranous (Figure 7.1). The hind wings are rarely absent, but may be greatly shortened in some species. The relatively slender legs are used for walking. In addition to paired simple claws on each tarsus that allow the bug to crawl over rough surfaces, many species have a spongy structure, the **fossula**, at the apex of the tibia on one or more pairs of legs. The fossulae have adhesive setae on their surfaces that enable the bugs to climb on smooth surfaces, such as leaves and glass.

The triatomine abdomen is 11-segmented, often pointed or lobed in the female, but smoothly rounded in the male. In many species, around the periphery of the abdomen, both dorsally and ventrally, are segmental plates (connexival plates) connected to the abdominal segments by intersegmental membranes. These membranes allow for expansion of the abdomen during engorgement. The membranes in different species are folded on themselves in various ways, allowing the plates and membranes to expand in accordion fashion during feeding.

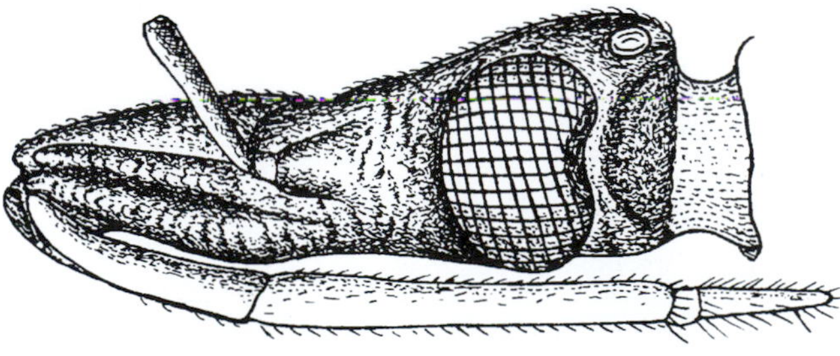

Figure 7.3 Lateral view of head of *Triatoma dimidiata*. (Lent and Wygodzinsky, 1979; courtesy of the American Museum of Natural History)

LIFE HISTORY

As in all Hemiptera, triatomines undergo hemimetabolous development. After the egg stage, development occurs through five nymphal instars. Nymphs are distinguished from adults by their smaller eyes, lack of ocelli and wings, and the presence of thoracic lobes where wings will develop. Both sexes of adults and all nymphal instars require blood for their survival and development.

Female bugs are ready to mate one to three days after the final molt. Mating involves transfer of a spermatophore from the aedeagus while the male is positioned dorso-lateral to the female with his claspers grasping the end of the female's abdomen from below. Copulation lasts from about five to 15 minutes. Although both sexes usually have had at least one blood meal before mating, unfed males also will mate with fed females.

Oviposition by females begins 10 to 30 days after copulation. Each female typically deposits only one or two eggs daily, producing a total of 10 to 30 eggs between blood meals. Depending on the species, a single female may produce up to 1000 eggs in her lifetime, but about 200 is average. Virgin, fed females may lay small numbers of infertile eggs. Each oval egg is about 2 to 2.5 × 1 mm. The eggs may be white or pink. Most species deposit eggs singly, but some females lay eggs in small clusters or masses. Different species lay eggs freely or glue them to a substrate. Gluing eggs to the substrate is seen in at least two species of *Triatoma* and many species of *Rhodnius*, *Psammolestes*, *Cavernicola*, and *Parabelminus*. In those species that glue their eggs to the substrate, the eggs may be single or in clusters. Eggs of some species turn pink or red before hatching 10 to 37 days after oviposition, depending on temperature.

The newly emerged ***nymphs*** are pink and will take a blood meal 48 to 72 hours after the eggs hatch. The nymphs must engorge fully in order to molt (Figure 7.4), often requiring more than one blood meal during all but the first instar. The entire life cycle from egg to adult may be as short as three to four months but more commonly takes one to two years. The variable developmental times within and between species are related to many factors, including environmental temperature, humidity, host availability, host species, feeding intervals, and the length of nymphal diapause.

BEHAVIOR AND ECOLOGY

The New World triatomines are found in stable, sheltered habitats that are used by reptiles, birds, and a wide variety of mammals for their nests, roosts, or burrows. The kissing bugs can be divided into three

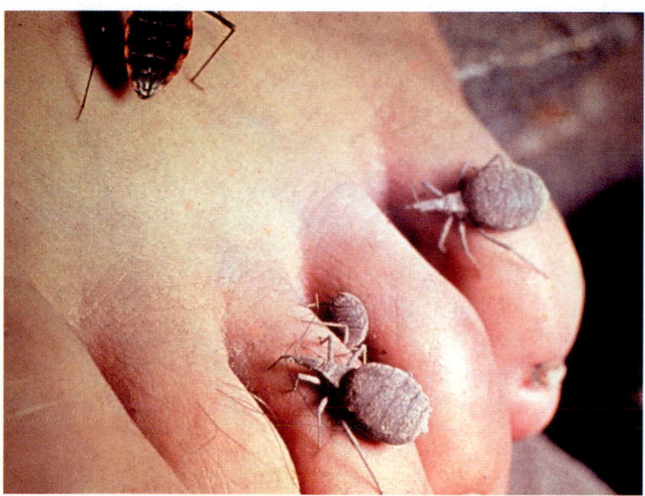

Figure 7.4 Triatomine nymphs engorging on human foot. (Courtesy of Robert B. Tesh)

general habitat groups: sylvatic, peridomestic, and domestic. **Sylvatic** forms inhabit nests and burrows, as well as a wide array of natural hiding places such as caves, rock piles, fallen logs, tree holes, hollow trees, palm fronds, bromeliads, and other epiphytes. These habitats attract amphibians, lizards (e.g., iguanas), opossums, rodents (e.g., porcupines), armadillos, sloths, bats, and other mammals, upon which the triatomines feed. The **peridomestic** species utilize domestic animals as hosts by living in chicken coops and other bird enclosures, stables, corrals, and in rabbit and guinea pig houses. Because *Triatoma infestans* infests the latter, as well as wild guinea pig habitats, this species may have entered the domestic habitat thousands of years ago when people in South America began breeding guinea pigs for use as food. The **domestic** (domiciliary) species, exemplified by *T. infestans*, have colonized human habitations where they depend on human or domestic animal blood as their source of nourishment. The domestic triatomine species are almost exclusively associated with humans and their pets and often are carried from one region to another in vehicles or concealed in household materials.

Many of the so-called peridomestic species, as well as a few domestic ones, have maintained sylvatic adaptations and may migrate from wild hosts to domestic animals and man, depending upon the availability of suitable habitats and hosts. Peridomestic species sometimes fly to the lights of houses and thereby are attracted at night to feed on sleeping humans. Passive transport of certain species to human dwellings may occur when palm fronds containing attached triatomine eggs are used as roofing material. This is commonly the case with *Rhodnius prolixus* and other avian-feeding species that cement their eggs to the leaves in and around arboreal birds' nests. The significance of birds in dispersing triatomines is not known, although eggs and young nymphs of *R. prolixus* have been found among the feathers on storks, and *Triatoma sordida* nymphs have been found in the plumage of sparrows.

In whatever habitat triatomines are found, they tend to be secretive, hiding in cracks and crevices of natural and artificial materials (e.g., debris of nests and burrows, in rock crevices and piles of vegetation); in building materials such as wood, shingles, thatch, and palm fronds; in human dwellings in cracks in the walls, behind pictures or other wall-hangings, in bedding and mattresses, in furniture, boxes, suitcases, piles of papers or clothes, and other accumulated materials that provide shelter during the day. Shaded crevices that provide extensive bodily contact with a rough, dry surface are preferred. Nymphs of many species are camouflaged by dirt and debris with which they cover themselves.

Most species of triatomines are nocturnal and actively seek blood from diurnal hosts that are resting or sleeping at night. In some cases, bugs will feed in daylight, typically on hosts that are nocturnal. Kissing bugs can survive for months without a blood meal, making them well adapted to nest habitats in which hosts may be present only intermittently with long intervals in-between. When hosts are available, bugs commonly feed every four to nine days. Individual species show definite host preferences, and may favor bats, birds, armadillos, wood rats, or humans. Those favoring the latter species in South America are most important in the epidemiology of Chagas disease.

As in other hematophagous arthropods, feeding behavior is initiated by a combination of physical and chemical factors. Heat alone stimulates *Rhodnius prolixus* to probe, the heat receptors being located on the antennae. Carbon dioxide, which induces feeding responses in various hematophagous arthropods, causes increased activity in triatomines and may alert them to the presence of a host. The possible role of aggregation pheromones in attracting bugs to a host is unclear, but a pheromone in the feces of nymphal and adult *Triatoma infestans* and in nymphal *R. prolixus* attracts unfed nymphs. These species defecate soon after feeding on or by the host, so that such a pheromone might attract other bugs to a source of blood.

The probing response begins when the rostrum is swung forward. The third segment is flexed upward so that optimal contact with a host occurs when the bug is at the side and just below a host. The serrated mandibular stylets are used to cut through the epidermis of the host, then anchor the mouthparts while the maxillary stylets probe for a blood vessel. When a vessel is penetrated, the left maxillary stylet slides posteriorly on the right stylet, disengaging the two stylets so that the left folds outward from the food canal. The purpose of this action is not known. It may allow a larger opening for ingestion of blood cells, or it may be a mechanism for holding the capillary lumen open (Lehane, 2005).

The amount of blood ingested depends on the duration of feeding. This, in turn, is governed by the presence of chemicals in the blood of the host that stimulate onset of feeding and by stretch receptors in the abdomen of the bug that stimulate cessation. Known phagostimulants of triatomines include various nucleotides and phosphate derivatives of nucleic acids.

Tritomine saliva contains an anticoagulin to help maintain blood flow during feeding, and nitric oxide that has antiplatelet and vasodilatory effects.

The time required to engorge fully varies from three to 30 minutes. During feeding, the abdomen becomes visibly distended. Adult bugs may imbibe blood equivalent to about three times their body weight, and nymphs may imbibe six to 12 times their unfed weight. Blood meals are stored in the anterior, widened portion of the midgut before the blood is passed to the narrower, posterior portion where digestion occurs. After engorging, the bug removes the rostrum from the host, and in most species defecates on or near the host before crawling away to seek shelter. The interval between feeding and defecation is a major factor in determining the effectiveness of a species as a vector of *Trypanosoma cruzi.* Schofield (1979) reviewed the behavior of triatomines, with particular attention to their role in trypanosome transmission.

PUBLIC HEALTH IMPORTANCE

Triatomine species that are efficient vectors tend to cause little or no pain when they feed. At least one substance in triatomine saliva has analgesic properties. The bugs stealthily approach their sleeping hosts and engorge without causing much, if any, awareness (Figure 7.5). However, immediate and delayed skin reactions to bites of *Triatoma infestans* and *Dipetalogaster maxima* have been observed. These reactions were not clearly correlated with previous exposure to bugs. Pruritic skin reactions following triatomine bites tend to enhance transmission of *T. cruzi* by stimulating the bitten individuals to scratch infective feces into the bite wounds.

Some individuals react to triatomine feeding with mild hypersensitivity reactions such as pruritus, edema, and erythema. These reactions occur most often in response to triatomine species that are not efficient vectors of *T. cruzi* to humans. Within the latter group of species are members of the *Triatoma protracta* complex. These triatomines fly to light and have been known to invade homes situated within natural wood rat habitats. In a small number of cases, individuals have developed severe systemic reactions, including anaphylaxis, following bug bites. Immunotherapy involving multiple injections of *T. protracta* salivary gland extract has been successful in ameliorating the effects of the bite (Marshall and Street, 1982).

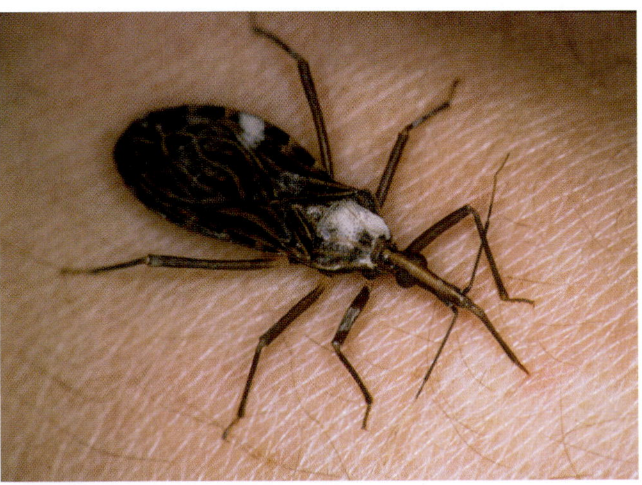

Figure 7.5 Adult triatomine feeding.

Chagas Disease (American Trypanosomiasis)

In 1907, while on an anti-malarial campaign in Minas Gerais, Brazil, Carlos Chagas was introduced to the bloodsucking triatomines (*barbeiros*). He found what is now known to be *Trypanosoma cruzi* in the hindguts of several bugs; within two years he recognized this same flagellate protozoan in domestic animals and in a sick two-year-old girl. Chagas disease and its epidemiology was thus first discovered in reverse fashion from that of most diseases. In this case the vector was found first, the nonhuman vertebrate hosts of the parasite second, and the human pathology last. The first report of the disease by Chagas was published in 1909, only 20 months after Chagas became aware of the existence of bloodsucking bugs. Chagas disease became known as American trypanosomiasis to differentiate it from African trypanosomiasis (African sleeping sickness), the disease caused by trypanosomes transmitted by tsetse flies in Africa.

Several bibliographies on the vast literature on Chagas disease and its epidemiology have been published, including Olivier et al. (1972) and Ryckman and Zackrison (1987). An excellent review of the etiologic agent and its biological associations is provided by Hoare (1972).

Triatomine species that are important vectors of *T. cruzi* are listed with their geographic occurrences in Table 7.2. *Triatoma infestans* is probably most often responsible for transmission of the trypanosome to humans because of this species' colonization of human dwellings over a wide geographic range in South America. *Rhodnius prolixus*, another important vector, is found throughout much of Central America. *Panstrongylus megistus* generally is considered a major vector in the humid coastal regions of eastern Brazil.

The basic features of the life cycle of *Trypanosoma cruzi* are shown in Figure 7.6. Broad and slender trypanosomes (**trypomastigotes**) circulating in the blood of an infected vertebrate host (Figure 7.7) are imbibed by the triatomine during feeding. In the proventriculus of the bug, the broad trypomastigotes change into **sphaeromastigotes** and slender **epimastigotes**. The latter multiply by binary fission in the midgut, and as early as the fifth or sixth day after feeding occur in tremendous numbers that carpet the walls of the rectum. As early as the seventh or eighth day after feeding, these epimastigotes become infective **metacyclic forms** (trypomastigotes) that pass out of the bug in the feces and Malpighian tubule secretions. Although alternative *T. cruzi* developmental schemes have been proposed, the life cycle generally is thought to be limited to the gut of the bug, and infective forms occur only in the hindgut and rectum.

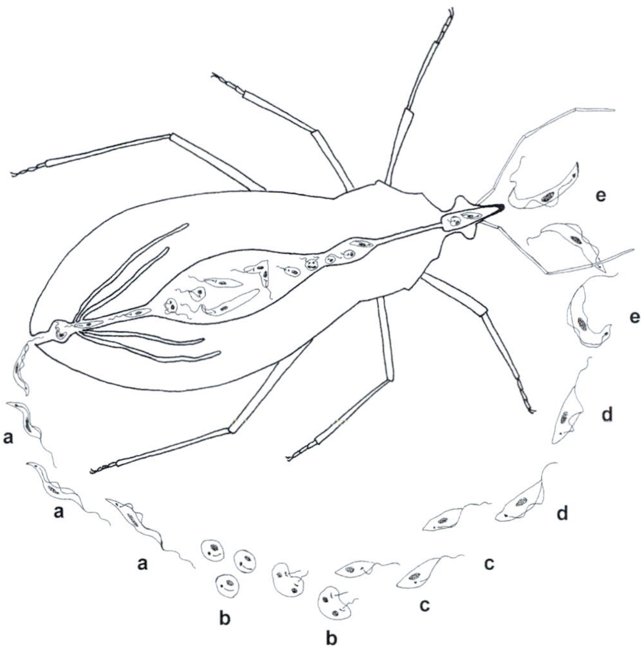

Figure 7.6 Life cycle of *Trypanosoma cruzi* in a triatomine bug and vertebrate host. (A) Metacyclic forms excreted by triatomine; (B) amastigotes (in vertebrate host); (C) epimastigotes (in vertebrate host); (D) trypomastigotes (in vertebrate host); (E) bloodstream forms ingested by triatomine. (Courtesy of W. L. Krinsky)

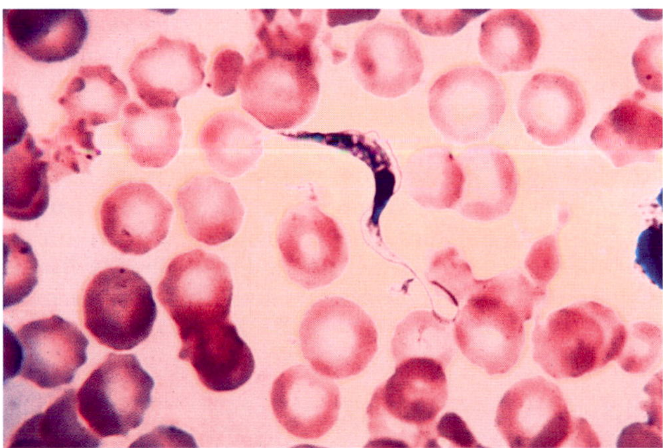

Figure 7.7 *Trypanosoma* species in blood. (U.S. Armed Forces Institute of Pathology, AFIP No. 74-5195)

Transmission to another vertebrate host occurs by this posterior-station (stercorarian) route. The trypanosomes infect the vertebrate when infective feces are rubbed into the bite site or other breaks in the skin. Transmission probably occurs most often when infective feces come in contact with the mucosal membranes of the nose or mouth, or conjunctivae of the eyes. Infected triatomines of those species that defecate while engorging or soon after feeding, while still on the host, are the most likely vectors of *T. cruzi.*

The entire development of *T. cruzi* in the lumen of the triatomine gut takes about six to 15 days or longer, depending on the ambient temperature and developmental stage of the bug (6 to 7 days in first instar-nymphs, and 10 to 15 days in older nymphs and adults). Once a nymph or adult is infected, it is infective for life. Trypanosomes do not pass via the eggs or spermatophore to the next generation. Cannibalism and coprophagy have both been observed in laboratory colonies of triatomines and have been suggested as possible modes of transmission from bug to bug. However, the minimal infection rate among uninfected bugs housed with infected ones indicates that such transmission is not of much consequence under natural conditions. Coprophagy is the means by which symbiotic bacteria essential for development of triatomines are transferred from adults to newly hatched nymphs (Beard et al., 2002).

Besides contamination of the skin or mucosal tissues, an alternative mode of transmission from bug to vertebrate is ingestion of a whole infected bug. This is probably common in the case of wild and domestic vertebrate animals (Miles, 1983) and is thought to be the most common mode of transmission to wood rats in North America. The possibility of additional trophic levels in such transmission has been demonstrated by the infection of rodents and dogs following their ingestion of house flies that previously had ingested feces of infected triatomines (Hoare, 1972). Carnivores may become infected by feeding on infected prey (Miles, 1983). Human infection has occurred after accidental contact with freshly squashed triatomine bugs. Trypanosomes in the hindguts of dead triatomines may maintain their infectivity for up to 30 days. The "fecal rain" from triatomine-infested ceilings that falls on inhabitants of some tropical houses may be another source of infection (Miles, 1983).

Although inadvertent contamination with infective triatomine feces is probably the most common route of human infection, some cultural practices involve deliberate contacts with triatomines. These include the eating of *Triatoma picturata* (*chinche de compostela*) for their supposed aphrodisiac properties in Nayarit, Mexico, and the rubbing of feces of *T. barberi* (*chinche voladora*) onto the skin to cure warts on children in Oaxaca, Mexico. Both of these activities expose individuals to a high risk of trypanosome infection (Salazar-Schettino, 1983).

Other modes of transmission from person to person include infection via blood transfusion and, less commonly, transplacental infection. Although infection of children via the breast milk of infected mothers is rare, the risk of infection during nursing is significantly increased when bleeding of the nipples occurs.

Arthropods other than triatomines have been infected with *T. cruzi* in laboratory and field studies. Both the common bed bug (*Cimex lectularius*) and the African argasid tick *Ornithodoros moubata* have been infected by feeding on infected hosts and have maintained infective metatrypanosomes in their guts following normal cyclical development of the parasite. However, these arthropods are not known to have any role in natural transmission cycles.

Many individuals who become infected with *T. cruzi* do not develop symptoms early in the course of infection. These subclinical cases may or may not develop chronic disease. In a small number of cases, especially in children, an acute clinical form of Chagas disease occurs following initial infection. Significant mortality (5–15%) occurs among those showing acute disease.

Acute Chagas Disease The acute form of Chagas disease begins with an area of erythematous and indurated skin, called a **chagoma**, at the site of parasite entry. If the infective material is rubbed into the eye, periorbital edema called **Romaña's sign** appears (Figure 7.8). This swelling, which may be accompanied by regional lymph node enlargements, may last for two to six weeks. The tissue changes result from intracellular development of amastigote trypanosomes in subcutaneous tissue and muscles. The amastigotes multiply and transform into trypomastigotes that enter the bloodstream. Other signs of acute Chagas disease include fever, general enlargement of lymph nodes, enlarged liver and spleen, and skin rashes. Complications of the acute phase that may result in death include myocarditis and meningoencephalitis. Most persons with the acute disease survive and enter the indeterminate phase of the disease, a stage in which the person appears healthy but still has the potential to develop serious chronic disease.

The indeterminate phase begins when antibodies to *T. cruzi* become detectable by serological testing, and trypanosomes, if detectable at all, may only be demonstrated by special methods, such as culturing blood or xenodiagnosis. The indeterminate phase may last indefinitely without further signs or symptoms of disease; in 10 to 30% of cases, however, chronic Chagas disease develops within the next few years, or as many as 20 years after the initial infection.

Chronic Chagas Disease This is characterized most often by cardiac symptoms including palpitations, dizziness, chest pain, and sometimes fainting. The cause of these symptoms are various forms of arrhythmias, which may lead to sudden death or persist for several years. The underlying pathology for the cardiac abnormalities is the development of amastigote trypanosomes in the cardiac muscles, accompanied by degeneration of cardiac muscle fibers (Figure 7.9), followed by fibrosis. The second most often seen type of chronic Chagas disease that occurs south of the Amazon involves enlargement of the esophagus or, less often, the colon. These conditions are known as megasyndromes. Enlargement of these organs is accompanied by gastrointestinal discomfort, including pain on eating and prolonged constipation in the case of megacolon. The often mammoth enlargement of the esophagus and colon is the result of pathologic destruction of myenteric ganglion cells, so that autonomic parasympathetic innervation (Auerbach's plexus) is greatly diminished. Individual patients may develop chronic cardiac disease, megaesophagus and megacolon, or only one or two of these syndromes. The most common outcome in patients with chronic Chagas disease is damage to heart muscle (cardiomyopathy) and conduction fibers that leads to various forms of heart block and in South America, congestive heart failure. Several years of suffering with cardiac symptoms may precede cardiac failure, which may, after as long as a few more years, result in death.

The restlessness, agitation, irritability, insomnia, and various other vague discomforts experienced by chronic Chagas disease patients have led medical historians to speculate on the basis of Charles Darwin's writings that he may have suffered from this disease. Furthermore, Darwin's palpitations and chest pains were brought on by emotional rather than physical stress, a phenomenon noted in Chagas' first patient when she was a middle-aged woman. Evidence that Darwin may have become infected with Chagas disease trypanosomes while he was in Argentina comes from

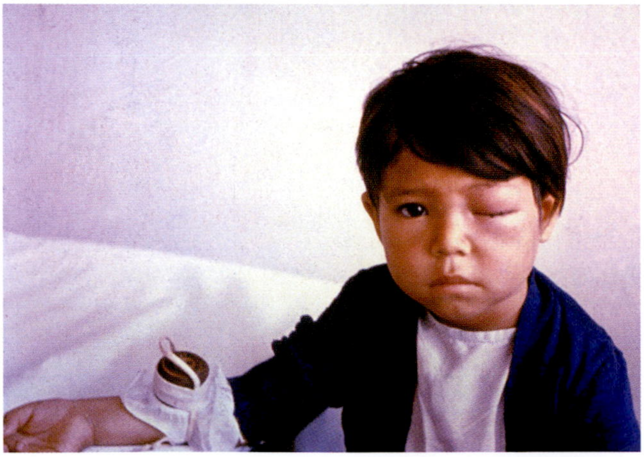

Figure 7.8 Romaña's sign in boy undergoing xenodiagnosis, the feeding of laboratory-reared triatomine bugs on a patient as means of detecting infection with trypanosomes. (Courtesy of U.S. Public Health Service)

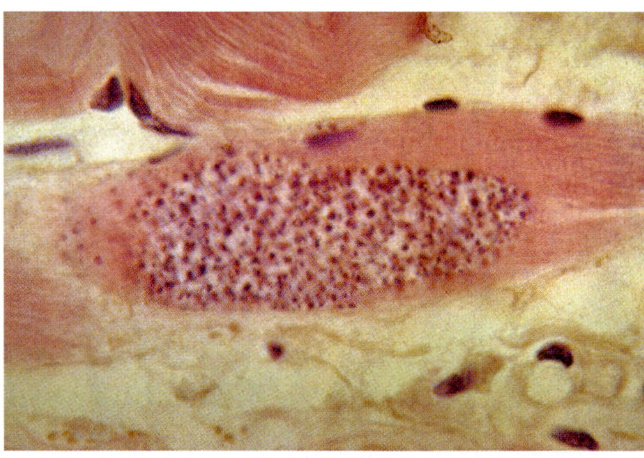

Figure 7.9 *Trypanosoma cruzi* amastigotes developing in heart muscle. (Peters and Gillies, 1981)

his own description of being attacked by "the *Benchuca*, a species of *Reduvius*, the great black bug of the Pampas" (Voyage of the H.M.S. Beagle, March 26, 1835). It is impossible to make a definitive diagnosis of Darwin's illness in the absence of pathologic material. The fact that Darwin had some of the same complaints before his voyage on the "Beagle" further complicates the speculation.

In suspected acute Chagas disease, direct examination of anticoagulated blood, buffy coat preparations, or concentrated serum may reveal living trypanosomes. Blood culture and fixed blood smears are useful for confirmation of the infection. The most sensitive procedure for recovering trypanosomes from both acute and chronic Chagas patients is **xenodiagnosis** (Figure 7.8). This involves feeding uninfected, laboratory-reared triatomines on a patient, holding the bugs in the laboratory for about 30 days, and then dissecting the hindguts of the bugs to look for trypanosomes. This procedure uses the bugs as living culture chambers. It takes advantage of the natural transmission cycle in which even very small numbers of trypanosomes ingested by a triatomine multiply in great numbers in the alimentary tract. Xenodiagnosis has been most successful when triatomine species and geographic strains from the area in which a person has been infected are used. As in all parasitic diseases, a careful history of travel or activities that may have led to infection is essential for differential diagnosis of the disease.

Diagnosis of chronic Chagas disease requires demonstration of *T. cruzi*-specific antibodies in a patient who has the characteristic cardiac dysfunction and/or megasyndromes. Positive xenodiagnosis and antibody testing alone may indicate only that an individual has been exposed to the parasite and is in the indeterminate phase.

Chagas disease is now considered the most serious parasitic disease of the Americas. An estimated 12 to 15 million people in Central and South America are infected with *T. cruzi*, with about 200,000 new cases occurring each year. More than half the populations of some rural villages are antibody positive. Historically, the nations most affected by infection and disease are Brazil, Argentina, Chile, Bolivia, Paraguay, and Venezuela (Figure 7.10). The type of disease observed varies from country to country. Cardiomyopathy and megasyndromes are common in Brazil, but cardiac disease alone is common in Venezuela. Cardiac abnormalities are present, but less prevalent, in Colombia and Panama. Cardiac disease is even less common among seropositive people in other parts of Central America and Mexico where cardiac problems, if they occur, present later in life. Although unproven, differences in clinical presentations may be related to known strain differences in *T. cruzi* isolates.

The social and economic burden caused by Chagas disease is associated primarily with morbidity rather than mortality. Chronically infected individuals often suffer for decades from weakness and fatigue that interfere with their productive enjoyment of life.

Chagas disease affects mostly the poorest people in the population. Typically the incidence of infection is directly associated with poor housing construction

Figure 7.10 Geographic distribution of human infection with *Trypanosoma cruzi*. (Courtesy of World Health Organization, Vector Biology and Control)

and proximity to domestic animal quarters or sylvatic habitats. Substandard houses such as rough-walled huts made of mud and sticks or adobe mud bricks, often roofed with thatch, provide abundant cracks and crevices in which triatomines can hide during the day and crawl out at night to feed on sleeping people and domestic animals. *Rhodnius prolixus*, naturally occurring on living palms, is especially abundant in palm roofs, whereas *P. megistus*, a species that is naturally found in hollow trees, favors the interstices of timber-framed, mud houses. Thousands of triatomines may inhabit individual houses, causing each inhabitant to be bitten by dozens of bugs each night. *Triatoma infestans*, the most important vector of *T. cruzi* in southern South America, also may occur inside houses with plastered walls and tiled roofs. The ability of triatomines to fly even has led to their intrusion into luxury high-rise buildings.

More than 100 species of mammals have been found infected with *T. cruzi.* Within domestic settings, humans, dogs, cats, and mice often are involved. Cats may become infected following ingestion of mice, whereas all domestic animals have the potential to become infected following ingestion of triatomines. In peridomestic cycles, chickens are excellent sources of blood for the triatomines, but are not susceptible to infection with *T. cruzi.* In sylvatic cycles, opossums and various rodents are often important reservoir hosts that are readily fed upon by bugs. The broad host range of the parasite and the large number of potential triatomine vectors over a vast geographic area contribute to the complexities of the natural cycles of *T. cruzi* in any given region.

In general, the greatest transmission to humans occurs in regions where domestic triatomines are abundant. These species not only are adapted to survive in close proximity to man, but most of these species have feeding patterns that cause them to defecate very soon after engorgement while still on or near the host. The lack of truly domestic species of triatomines and the relatively long delay between engorgement and defecation in triatomine species found north of Mexico have been cited as major reasons why there have been only four cases of Chagas disease acquired from triatomines in the United States.

Triatomine species repeatedly found to harbor natural infections with *Trypanosoma cruzi* in the United States are *Triatoma sanguisuga* in Pennsylvania, Ohio, Maryland, south to Florida and west into Arizona; *T. lecticularia* in Pennsylvania, Illinois, Maryland south to Florida and west into California; *T. gerstaeckeri* in Texas and New Mexico; *T. protracta* in Texas, north to Colorado and southwest to California; *T. rubida* in Texas and west to California; and *T. recurva* in Arizona. All these species except *T. recurva* commonly are associated with wood rats (*Neotoma* spp.), which are also naturally infected with *T. cruzi.* Other potential reservoirs of infection for *T. sanguisuga* and *T. lecticularia* are raccoons, armadillos, and opossums. The low incidence of human Chagas disease in North America is attributed to the relatively low percentages of infected triatomine bugs (6%) and vertebrate hosts (15%), the sylvatic behavior of the bugs, and the time delay between feeding and defecation (Wood, 1951; Lent and Wygodzinsky 1979; Neva, 2007).

In the United States, triatomine-associated Chagas disease is much less likely than transfusion-acquired infection. The more than 15 million natives of Latin America now living in the United States are estimated to have an infection rate as high as 10%. Consequently, careful screening of blood donors is essential to prevent chronically infected, asymptomatic individuals from donating blood.

Other Human Parasites Associated with Kissing Bugs

Trypanosoma rangeli, a nonpathogenic trypanosome found in Central and South America, also is transmitted by triatomines. *Rhodnius prolixus* is the chief vector. *Trypanosoma rangeli* is morphologically and serologically distinguishable from *T. cruzi* and, unlike the latter, may be transmitted via the saliva of the bug. It is also found naturally in a wide array of mammals, including monkeys, dogs, opossums, anteaters, raccoons, and humans.

As with all hematophagous arthropods, incidental infections with various blood-borne pathogens can occur when these invertebrates feed on parasitemic hosts. Although recent interest in Hepatitis B virus has led to the suggestion that triatomines might at times disseminate this virus, epidemiological evidence is lacking. Experimental feeding by fifth-instar nymphs of *T. infestans* on asymptomatic human immunodeficiency virus (HIV)-infected patients has demonstrated that HIV can survive in the bugs three to seven days after engorgement, but the bugs do not transmit the virus.

VETERINARY IMPORTANCE

Triatomines transmit *T. cruzi* to a variety of domestic and wild animals, including opossums, armadillos, rodents, carnivores, and monkeys. Depending on the strain of the trypanosome, the species and age of the host infected, and other poorly understood factors, the infection can lead to disease. Myocarditis and megaesophagus, similar to the conditions seen in humans, have been observed in dogs. **Canine trypanosomiasis** is of veterinary importance in Central and South America, and many cases have been recognized in southern Texas. Clinical indications of infection in dogs include dyspnea and ascites. There is little evidence that nonhuman wild animal hosts that serve as natural reservoirs of *T. cruzi* develop any pathology.

Trypanosoma rangeli occurs within the distribution of *Rhodnius prolixus,* its chief vector. The public health importance of *T. rangeli* in veterinary medicine lies in differentiating this common parasite from the pathogenic *T. cruzi. Trypanosoma conorhini* is a nonpathogenic parasite of rats transmitted by *Triatoma rubrofasciata* in many tropical regions of the Old and New World. It appears to have a tropicopolitan distribution identical to *T. rubrofasciata,* a domiciliary species of the tropics and subtropics. *Trypanosoma conorhini* is spread by posterior-station transmission (Hoare, 1972).

Heavy infestations of triatomines in poultry houses in Central America may cause chronic blood loss in chickens. Even though no avian pathogens are involved, the impact of constant blood-feeding may result in significant morbidity and, in the case of young birds, mortality.

PREVENTION AND CONTROL

The goal of any prevention or control program is to reduce contact between humans and kissing bugs in order to prevent discomfort from bites and the more serious problem of Chagas disease. Control of Chagas disease involves the use of insecticides, improving housing conditions, and screening blood used for transfusions. Residual insecticides sprayed on houses or applied to walls in paints are effective in controlling triatomines. However, long-term control requires careful surveillance and selective applications of insecticides. Some simple surveillance techniques, such as placing pieces of colored paper on the inside walls of houses, have been very successful in assessing the presence of triatomines. Their fecal patterns on the paper provide an indicator of triatomine activity. Houses treated with insecticides subsequently may be colonized by peridomestic or sylvatic species.

A multinational campaign to reduce Chagas disease, called the Southern Cone Initiative, was launched in 1991. The two-pronged program includes large-scale use of residual insecticides in domestic and peridomestic structures in infested villages, and universal blood screening, to eliminate transmission from infected donors. Within a decade, the incidence of Chagas disease was reduced by over 65% in the countries involved (Argentina, Brazil, Chile, Paraguay, and Uruguay). Transmission by the major vector, *Triatoma infestans*, was eliminated in Uruguay and Chile and in parts of the other countries. Similar campaigns have been initiated in Central America and in the Andean and Amazon regions (Yamagata and Nakagawa, 2006).

Defecation by *Triatoma infestans* and *Rhodnius prolixus* soon after feeding is dependent upon full engorgement by the bugs which, in turn, appears to be related to the density of the bug population in a given habitat; the greatest chance of engorgement is at low-density populations. Presumably, high bug densities with constant host resources lead to smaller blood meal sizes and a slower rate of defecation. Therefore, the chance of inhabitants becoming infected with *T. cruzi* tends to be greatest in newly colonized houses where the bug population is rising or in houses being repopulated after vector control has been instituted.

Long-term control of triatomine bugs is best achieved in houses in which rough walls have been covered with plaster, thatch roofs have been replaced with tin or tile, and mud floors have been replaced with concrete. Such changes in construction, as well as the removal of wall hangings, firewood, and accumulated debris or vegetation, which serve as hiding places for triatomines, help to reduce the size of bug populations. During the last 40 years, *Rhodnius prolixus* has almost completely disappeared from Mexico, Costa Rica, and El Salvador. This decline has been attributed to the decrease in thatched-roof houses and to widespread insecticide use by malaria eradication campaigns.

The drugs available for the treatment of acute Chagas disease, such as nifurtimox and benznidazole, are reasonably effective in preventing the development of chronic disease. However, they are associated with a high frequency of side effects, and neither is known to affect the course of chronic Chagas disease once it develops. Gentian violet effectively decontaminates donor blood, but routine screening of the blood supply is necessary to prevent transmission of *T. cruzi* via blood donors in endemic areas.

Various biological control approaches, including the use of juvenile hormone mimics, predatory arthropods, and parasitic wasps (e.g., scelionid *Telenomus fariai*), have been studied for the control of triatomines. However, no biological control method has been found for effective, widespread use in Central and South America.

An experimental approach for controlling *T. cruzi* involves genetically modifying triatomine bacterial symbionts so that triatomines infected with them produce antitrypanosomal gene products (Beard et al., 2002).

BED BUGS (CIMICIDAE)

The family Cimicidae includes species known by several common names, including bed bugs, bat bugs, and swallow bugs. All species in this family are wingless, obligate hematophagous ectoparasites. Their medical and veterinary importance relates primarily to the loss of blood and discomfort caused by their feeding on vertebrate hosts. The monograph on the Cimicidae by Usinger (1966) is still the most comprehensive and best work on the ecology, morphology, reproductive biology, systematics, and taxonomy of the group.

Over 50 common names have been given to bed bugs in different countries. Some of these are: mahogany-flat (Baltimore), heavy dragoon (Oxford), red coat (New York), wall louse (*Wandlaus*, *Wegluis*, and *Wanze*, German), *Wägglus* (Swedish), *Vaeggelus* (Danish), *Piq-seq* (Chinese), *Chinche* (Old Spanish), *Chinga* (Gallic), *Nachtkrabbler* (night crawler, German), *Tapetenflunder* (wallpaper flounder, German), *Punaise* (stinker, French), *Perceveja* (pursuer, Portuguese), *Lude* (Finnish), Plostice (flat, Czechoslovakian), *klop* (Russian), *bug* (ghost, goblin, British), *Buk* (Arabic), *Fusfus* (Syrian), *Pishpesh* (Hebrew), *Ekukulan* (Douala-Bantu), *Kunguni* (Swahili), *Uddamsa* (biter, Sanskrit), *Rep* (Vietnamese), *Nankinmusi* (Nanking bug, Japanese), and *Tokozirami* (bed louse, Japanese). These and other names were reviewed by Usinger (1966).

TAXONOMY

The family Cimicidae is divided into six subfamilies with 23 genera and 91 described species. The family is related to the predaceous family Anthocoridae, which includes species that feed on insects and mites and occasionally bite humans and other warm-blooded vertebrates. A related family, the Polyctenidae, includes species that are all ectoparasitic on bats and, like some cimicids, are also commonly called bat bugs.

The cimicids include 12 genera with species associated with bats and nine genera with species associated with birds. In addition, some species in the genus *Cimex* are found on bats and others on birds. Three species are considered ectoparasites of humans. *Leptocimex boueti*, a member of the subfamily Cacodminae, occurs on bats and people in West Africa. The other two are members of the subfamily Cimicinae, the bed bugs.

Figure 7.11 Human bed bug, *Cimex lectularius*; female, left; male, right. (Usinger, 1966)

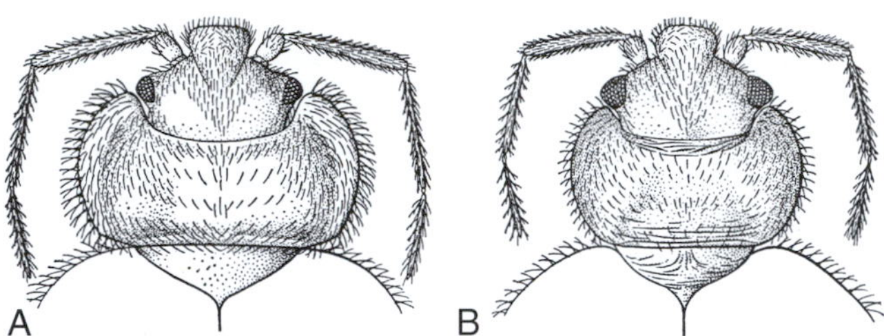

Figure 7.12 Head and prothorax of adult bed bugs. (A) *Cimex lectularius,* (B) *C. hemipterus.* (Smart, 1943, courtesy of the British Natural History Museum)

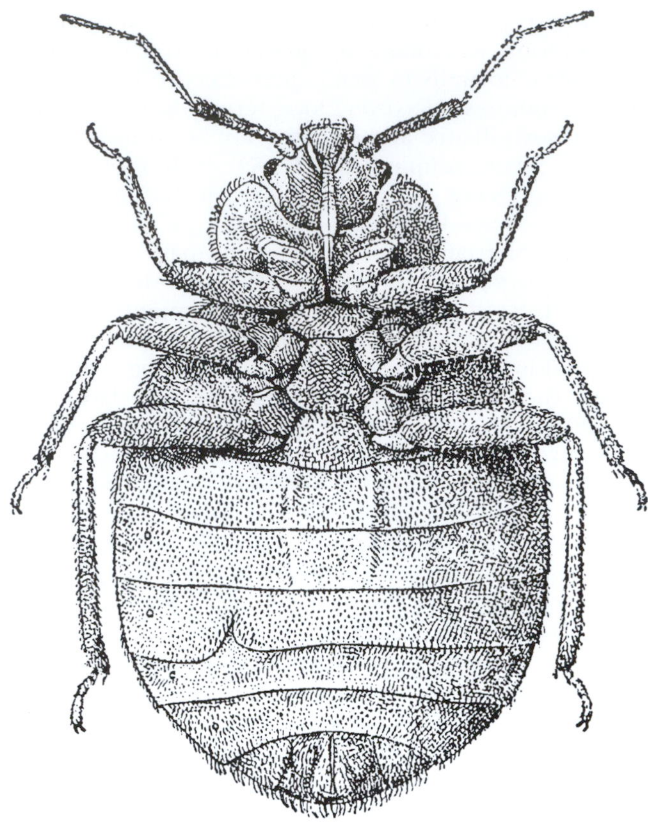

Figure 7.13 Human bed bug (*Cimex lectularius*), female, ventral view; the cleft on the hind margin of the fifth abdominal sternite, which denotes the point of entry (paragenital sinus) for the male aedeagus during insemination. (Busvine, 1966)

Cimex hemipterus (Figure 7.12) is parasitic on humans and chickens in the Old World and New World tropics; *Cimex lectularius* (Figures 7.11, 7.12, and 7.13) is a cosmopolitan species associated primarily with humans, bats, and chickens.

Both *Cimex* species that feed on humans originated in the Old World. The origin of *C. hemipterus* is uncertain; however, there is evidence that *C. lectularius* originated in the Middle East, probably being associated with bats and humans living in caves. *Cimex lectularius* apparently spread into Europe during historic times, being recorded from Greece by 400 BC, from Italy by AD 77, and from Germany for the first time in the eleventh century. The bed bug was known in France in the thirteenth century and is recorded as occurring in England in 1583. Therefore, the wide dissemination of *C. lectularius* throughout the world probably did not begin until after the sixteenth century.

MORPHOLOGY

The most striking feature of cimicids is their dorsoventral flattening. Adults of the oval, mahogany-colored *Cimex* species generally range in length from about 5.5 to 7.0 mm, with abdomens 2.5 to 3.0 mm wide. The females are larger than the males. The bat bug *Leptocimex boueti* differs from *C. lectularius* and *C. hemipterus* in having a very narrow pronotum, only slightly wider than the head, and very long legs. It is a smaller species, the total body length being 2.8 mm in males and 4.0 mm in females.

The cimicid head is small and cylindrical with two knob-like, multifaceted eyes. Ocelli are not present. The antennae are four-segmented and inserted between the eye and the clypeus. The labium is three-segmented and, as in the triatomines, dorsally encloses the maxillary and mandibular stylets; they in turn enclose a relatively large dorsal food canal and a very small ventral salivary canal. The labium has two sensory lobes at its tip. When the bug is not feeding, the rostrum, or beak, comprised of the labium and associated mouthparts, is bent below the head with the tip extending to the middle of the prosternum.

The thorax consists of a narrow canoe-shaped pronotum, a mesonotum that is covered dorso-laterally by reduced forewings called **hemelytral pads**, and a metanotum hidden below the latter. Nymphs do not have hemelytral pads. *Cimex hemipterus* can be distinguished from *C. lectularius* by the former's narrower pronotum (Figure 7.12). The hemelytral pads are oval in *Cimex* species and are reduced to small elevated ridges in *Leptocimex.* Hind wings are never present. The legs are slender, with two-segmented tarsi in the nymphs and three-segmented tarsi in the adults.

The abdomen is 11-segmented and capable of tremendous expansion during blood-feeding. In nymphs, membranous areas on the entire ventral surface and on the first, second, and part of the third abdominal terga enable expansion of the abdomen while feeding. In adults, the intersegmental membranes are wide, and the middle of the ventral side of the second to fifth segments of the abdomen are likewise membranous. Female *Cimex* adults are readily distinguished from males by the presence of an indentation on the hind margin of the fifth abdominal sternite (Figure 7.13). This narrow cleft, called the **paragenital sinus**, is surrounded by bristles, and is the point at which the male inserts his aedeagus to intra-abdominally inseminate the female. No paragenital sinus occurs in *Leptocimex boueti*.

LIFE HISTORY

Mating occurs with the male bug straddling the female's back at an oblique angle. In this position the tip of his abdomen is strongly curved against the right side of the venter of the female where the paragenital sinus is located. The male inseminates the female by injecting sperm into the sinus. This form of traumatic insemination that involves introduction of the sperm into an extragenital site occurs in many species of the superfamily Cimicoidea. Specialized structures for reception of the sperm, variously called the **spermalege, organ of Ribaga**, and **organ of Berlese**, are present in these species. Copulation usually lasts from one to several minutes but may take up to half an hour. Females that have been inseminated retain permanent scars that are visible in the integument. The sperm pass from the spermalege into the hemocoel, from which they enter paired outpouchings of the walls of the oviducts called **sperm conceptacles**. From there the sperm travel within the walls of the oviducts, via an intra-epithelial network of tubular canals, called **spermodes**, to the bases of the ovarioles.

Mated females usually feed to repletion and then begin to lay eggs three to six days later. Oviposition lasts for about six days during which six to 10 eggs are deposited. Depending on ambient temperature and relative humidity, female bugs may feed every three to four days. Eggs are laid continuously, with the mean number of eggs per week typically varying from three to eight. Some females have been observed to lay as many as 12 eggs in one day and up to 540 eggs in their lifetimes. A female is capable of producing viable eggs for five to seven weeks after feeding and mating. After that time, an increasing number of eggs are sterile.

The eggs are elongate-oval, about 1 mm long, and pearly white. They are laid singly and are coated with a transparent cement that causes them to adhere to various surfaces. The eggs usually are deposited in groups or clusters. Hatching usually takes place in four to 12 days depending on the temperature. There are five **nymphal stages**, each lasting 2.5 to 10 days. The temperature threshold for development is about 15 °C, with optimal development at 30 °C. Humidity, except at the extremes, has little or no effect on development. The total **developmental time** from egg to adult for *C. lectularius* varies from 24 days (at 30 °C) to 128 days (at 18 °C), and for *C. hemipterus* from 25 days (at 30 °C) to 265 days (at 18 °C).

The nymphs are pale straw-colored before they feed, but bear a resemblance to red berries after they have fed. Feeding generally occurs within 24 hours after hatching or molting. At low temperatures, nymphs may survive for five to six months without feeding, whereas adults can survive even longer. This makes them efficient nest parasites that are able to survive long periods when a host is absent. Nymphs feed at least once during each instar. Engorgement usually takes about three minutes for first-instar nymphs, and 10 to 15 minutes for older nymphs and adults. After fully engorging, nymphs are 2.5 to six times heavier than unfed nymphs, and the adults are 1.5 to two times heavier than unfed individuals. As in triatomines, liquid fecal matter is excreted soon after feeding. Half the weight of the entire blood meal is lost within the first five hours after feeding.

BEHAVIOR AND ECOLOGY

Cimicids are similar to triatomines in their choices of hiding places, the nature of the substrates selected, and their feeding patterns. They hide in cracks and crevices in human and animal habitations, and in nests, caves, and tree holes in natural settings. They prefer rough, dry substrates that allow maximum contact of the bugs with the surface. Their attraction to such harborages between feedings often results in large aggregations. In domestic situations, bed bugs prefer hiding in wood and paper accumulations rather than in materials made of stone, plaster, metal, or textiles. Both *Cimex lectularius* and *C. hemipterus* may infest mattresses, box springs, and upholstered furniture. Other common sites of infestations include upholstered seats in buses and public facilities such as theaters and office waiting rooms. Cimicids will crawl into the narrowest crevices, such as those formed behind loose wallpaper, pictures, or electrical switch or socket plates. The harborages and infested premises often are stained with conspicuous fecal spots that range in color from white to yellow to brown to reddish-brown to black. Areas infested by *C. lectularius* may be identified by a characteristic sweet odor.

Bugs leave their hiding places primarily to feed. They are negatively phototactic, tending to feed mostly in darkness or subdued light when the temperature is above 10 °C. Warmth and carbon dioxide, as in many other hematophagous arthropods, appear to be major factors in attracting bed bugs to a host. A temperature differential of only 1 to 2 °C is sufficient to induce probing (Lehane, 2005).

When the bug has located a host, it approaches with its antennae outstretched and its beak directed downward at a 90° angle. It grabs the host with the tarsal claws of the front legs. After contact is made, the antennae are pulled backward, and the entire bug makes rocking, pushing movements as the vertically directed stylets are embedded in the skin. As the stylets penetrate the skin, the labium becomes more and more bent at the skin surface. The mandibles, which have retrorse teeth at their tips, move in and out of the skin in alternating fashion, producing a passage for the maxillae. The bundle of feeding stylets probes actively within the skin until a blood vessel of suitable size is penetrated. Only the maxillae, and possibly only the right maxilla, actually enter the lumen of the vessel. The salivary secretion contains an anticoagulin that prevents the blood from clotting, and nitric oxide, which is an antiplatelet and vasodilatory compound. After engorging, the bug withdraws its stylets; sometimes this requires considerable effort owing to the teeth on the mandibles. Once the stylets are again encased in the straightened labial sheath, the beak is folded back under the head. The bugs are quick to retreat when disturbed at the beginning or end of feeding; however, while the stylets are fully embedded in the skin, the bug is unable to withdraw its mouthparts even if handled or rotated.

The two species of *Cimex* most commonly associated with humans have been dispersed over wide areas of the globe, with *C. lectularius* most often being found in temperate regions and *C. hemipterus* in the tropics. These species are carried concealed in luggage, furniture, and all manner of packing materials. They have been transported on land vehicles, ships, and planes.

PUBLIC HEALTH IMPORTANCE

Usinger (1966) listed 27 human pathogens, including viruses, bacteria, protozoa, and helminths that have been shown to survive for varying lengths of time in *C. lectularius* and *C. hemipterus*. However, there is little or no evidence to incriminate bed bugs as vectors of these or any other disease agents.

Recent attempts to explain transmission of Hepatitis B virus (HBV) and, to a lesser extent, human immunodeficiency virus (HIV) in otherwise unexplained situations have focused on the possibility of cimicid transmission. Hepatitis B antigens (and HBV DNA) persist for several weeks in cimicid tissues and feces under laboratory conditions after the bugs have fed on infected blood. Replication of the virus, however, does not occur. HBV DNA persists from one instar to the next (transstadial transmission), but not transovarially. Past attempts to transmit the virus from infected bugs to chimpanzees failed. These results and those from transmission studies with mosquitoes suggest that it is unlikely that Hepatitis B transmission occurs either via infective feces or interrupted feedings; however, definitive studies, attempting transmission with cimicid tissues or feces known to have infectious virus, have not been done. Hepatitis C RNA was not found in cimicid tissues after bugs fed on a viremic patient (Silverman et al., 2001).

Although transmission studies with bed bugs and HIV indicate that these insects may harbor the virus for up to eight days, replication of the virus does not occur and the virus is not present in cimicid feces. These observations, together with failed attempts to experimentally transmit HIV by interrupted feedings suggest that cimicids are neither biological nor mechanical vectors of HIV.

Despite the fact that cimicids do not play a significant role as vectors of human pathogens, bed bugs are medically important because they cause unpleasant bite reactions and significant blood loss in people living in dwellings that are chronically infested. The actual feeding by bed bugs generally does not produce any pain. If interrupted, a bug will often bite again close to the previous site, thereby creating a linear array of punctures that is characteristic of cimicid bites. People are most often bitten on the limbs, trunk, and face.

Sensitivity reactions to bed bug bites are the result of substances injected during feeding. These reactions may be localized cutaneous responses or generalized and systemic. The most common local reactions are wheals similar to uncomplicated mosquito bites or, in some individuals, large fluid-filled bullae. Erythema is not a common response but may occur as a result of multiple feedings that cause extensive hemorrhaging under the skin. Individual reactions to cimicid bites vary from no response to severe immediate or delayed sensitivity reactions, including anaphylaxis. In most cases, swelling and itching associated with the bites can be relieved by application of ice and use of an oral antihistamine. Chronic bed bug bites are sometimes misdiagnosed as allergic dermatitis or other skin disorders. Accurate clinical assessment often requires careful epidemiological evaluation of a patient's living quarters.

People living with chronic infestations of bed bugs often are subject to nightly attacks, resulting in a marked loss of blood. Children who are marginally nourished are especially vulnerable to developing anemia and other medical problems as a result of such chronic blood loss. Individuals subjected to continued feeding by bed bugs may also develop extreme irritability that results from restless nights and chronic sleep deprivation. If the source of the disturbance goes undetected, the emotional stress caused by such infestations may be misdiagnosed as a neurosis.

A significant increase in reports of bed bug infestations in major US cities (Atlanta, New York, San Francisco) and in the United Kingdom, Australia, and Brazil has occurred in recent years. Various explanations for the apparent resurgence of bed bug infestations have been suggested, including increased international travel and transport of materials globally as well as recent changes in pest control methods. Use of specific baits, instead of broad-acting insecticides for control of cockroaches, ants, and other pest species, may have allowed bed bugs to persist and increase. For several years before the current outbreaks, the bed bug was not a major pest in the developed nations. Its absence may have led to a decline in familiarity with the insect by pest-control and medical personnel, resulting in incorrect diagnoses and thereby the use of inadequate control measures. Whatever the cause, the economic impact of increased infestations on the tourist industry has been significant, resulting from reduced clientele, increased pest control costs, and litigation expenses incurred by affected patrons.

Other Cimicids that Occasionally Attack Humans

In addition to the three cimicid species directly associated with human habitations, there are several species that occasionally feed on people. These include swallow bugs of the genus *Oeciacus*; the bat bugs *Cimex pilosellus* (New World) and *C. pipistrelli* (Europe); and the bird bugs, such as the Mexican chicken bug *Haematosiphon inodorus* and *Cimexopsis nyctalis* from the nests of chimney swifts. Human bites by these species generally occur only in the vicinity of the nesting or roosting sites of their natural hosts.

The swallow bugs that occur in mud nests of swallows include two species, *Oeciacus hirundinis* in Eurasia south to Morocco and *O. vicarius* in North America south to Durango in Mexico. Both species are members of the subfamily Cimicinae and will bite people who disturb infested bird nests. Swallows may be heavily infested with the bugs, and nestlings often die as a result of blood loss. The eggs of *Oeciacus* species are attached to the outer surfaces of swallow mud nests, often being so abundant that they can be seen from a distance. There is some evidence that *Oeciacus* species are carried as nymphs by the birds from nest to nest.

Two arboviruses have been isolated from the cliff swallow bug *O. vicarius* and nestling cliff swallows and house sparrows, one originally from Colorado, and one originally from Oklahoma. They are **alphaviruses**, part of the Western Equine Encephalitis complex. The one isolated in Colorado is called Fort Morgan virus, and the one from Oklahoma is called Buggy Creek virus. They are not known to cause pathology in their avian hosts or in humans. Occurrence of these viruses in bugs and birds suggests that viruses can overwinter in swallow bugs that occupy nests left vacant by their migrating hosts.

VETERINARY IMPORTANCE

Cimicids can be significant pests in commercial poultry production. Cimicids attacking domestic poultry include *Cimex lectularius* in North America, Europe, and the former Soviet Union; *Haematosiphon inodorus* in Central America; and *Ornithocoris toledoi* in Brazil.

Raised slats and wood shavings in nest boxes in broiler breeder houses provide harborage for the bugs. Indications of cimicid infestations include fecal spots on eggs, nest boxes (Figure 7.14) and wooden supports, skin lesions on the breasts and legs of birds, reduced egg production, and increased consumption of feed. Chicken bugs are not known to transmit any avian pathogens. However, chickens and other fowl raised in poultry houses heavily infested with chicken bugs are irritable and often anemic. Morbidity in such cases may be high, and young birds may succumb from blood loss. Economic loss on infested poultry farms also may be increased by reduced productivity of workers allergic to the bugs.

Figure 7.14 Fecal spots, indicative of cimicid activity, along seams of a nesting box of laying hens in a poultry house heavily infested with *Cimex lectularius*. (Photo by Gary R. Mullen)

Two species of nonpathogenic trypanosomes that undergo development in cimicids have been isolated from bats in North America. *Trypanosoma hedricki* and *T. myoti*, both closely related to *T. cruzi*, have been found in big brown bats and little brown bats in southern Ontario, Canada. Developmental stages infective to bats form in the rectum of *Cimex brevis* and *C. lectularius*, which suggests that these trypanosomes are transmitted by the bugs via the posterior-station route. Because bats also are known hosts for *T. cruzi*, the differentiation of other bat trypanosomes and the elucidation of their transmission are important. Furthermore, because of the similarities of the life cycles and transmission of these nonpathogenic trypanosomes to that of *T. cruzi*, they could be suitable candidates for developing laboratory models of the Chagas disease pathogen.

PREVENTION AND CONTROL

Measures to prevent cimicid infestations should begin with household sanitation. Removing accumulations of paper and wood trash eliminates hiding places and harborages for the bugs. However, once an infestation occurs, eliminating cimicids requires thorough fumigation with residual insecticides that must be sprayed on surfaces over which the bugs crawl to reach their hosts. Pyrethroids have provided good control. Nonchemical treatments alone are not sufficient to control infestations, but are important adjuncts to chemical control. Nonchemical approaches include thorough inspections, vacuuming, steaming, isolating and cleaning infested garments, thermal treatments, and use of mattress covers. For temporary control, such as needed by a traveler occupying an infested room for one or a few nights, any of various insecticides applied as aerosols can be used to thoroughly spray bed frames, mattresses, and box springs.

Control of cimicids in premises in which people are bothered by the bites of bird bugs or bat bugs requires identification and removal of the source of the bugs. Such sources include bats roosting in attics or eaves, bird nests on window ledges or air conditioners, and birds roosting in chimneys. Removal of the nonhuman vertebrate hosts must be accompanied by use of an insecticide, or the hungry bugs will seek human blood more aggressively in the absence of their natural hosts.

Various arthropods are natural predators of cimicids. These include the masked bed bug hunter *Reduvius personatus*, other hemipterans, ants, pseudoscorpions, and spiders. None of these, however, has been effectively used for controlling bed bugs.

REFERENCES AND FURTHER READING

Alexander, J. O. (1984). *Arthropods and Human Skin.* Berlin: Springer-Verlag.

Asin, S. N., Catalá, S. S. (1991). Are dead *Triatoma infestans* a competent vector of *Trypanosoma cruzi*? *Memorias do Instituto Oswaldo Cruz, 86*, 301–305.

Axtell, R. C. (1999). Poultry integrated pest management: Status and future. *Integrated Pest Management Reviews, 4*, 53–73.

Beard, C. B., Cordon-Rosales, C., & Durvasula, R. V. (2002). Bacterial symbionts of the Triatominae and their potential use in control of Chagas disease transmission. *Annual Review of Entomology, 47*, 123–141.

Blow, J. A., Turell, M. J., Silverman, A. L., & Walker, E. D. (2001). Stercorarial shedding and transtadial transmission of Hepatitis B virus by the common bed bugs (Hemiptera: Cimicidae). *Journal of Medical Entomology, 38*, 694–700.

Bower, S. M., & Woo, P. T. K. (1981). Development of *Trypanosoma* (*Schizotrypanum*) *hedricki* in *Cimex brevis* (Hemiptera: Cimicidae). *Canadian Journal of Zoology, 59*, 546–554.

Brumpt, E. (1922). *Précis de Parasitologie.* Paris: Masson & Co.

Busvine, J. R. (1966). *Insects and Hygiene.* London: Methuen & Co. Ltd.

Calisher, C. H., Monath, T. P., Muth, D. J., Lazuick, J. S., Trent, D. W., Francy, D. B., et al. (1980). Characterization of Fort Morgan virus, an alphavirus of the western equine encephalitis virus complex in an unusual ecosystem. *The American Journal of Tropical Medicine and Hygiene, 29*, 1428–1440.

Caras, R. (1974). *Venomous Animals of the World* (pp. 96–99). Englewood Cliffs: Prentice-Hall, Inc.

Costa, C. H. N., Costa, M. T., Weber, J. N., Gilks, G. F., Castro, C., & Marsden, P. D. (1981). Skin reactions to bug bites as a result of xenodiagnosis. *Transactions of the Royal Society of Tropical Medicine and Hygiene, 75*, 405–408.

Doggett, S. L., Geary, M. J., & Russell, R. C. (2004). The resurgence of bed bugs in Australia: With notes on their ecology and control. *Environmental Health: A Global Access Science Source, 4*, 30–38.

Foil, L. D., & Issel, C. J. (1991). Transmission of retroviruses by arthropods. *Annual Review of Entomology, 36*, 355–381.

Hoare, C. A. (1972). *The Trypanosomes of Mammals.* Oxford: Blackwell Sci. Publ,

Hopla, C. E., Francy, D. B., Calisher, C. H., & Lazuick, J. S. (1993). Relationship of Cliff Swallows, ectoparasites, and an alphavirus in west-central Oklahoma. *Journal of Medical Entomology, 30*, 267–272.

Jupp, P. G., Purcell, R. H., Phillips, J. M., Shapiro, M., & Gerin, J. L. (1991). Attempts to transmit hepatitis B virus to chimpanzees by arthropods. *South African Medical Journal, 79*, 320–322.

Kells, S. A. (2006). Nonchemical control of bed bugs. *American Entomologist, 52*, 109–110.

Kirk, M. L., & Schofield, C. J. (1987). Density-dependent timing of defaecation by *Rhodnius prolixus*, and its implications for the transmission of *Trypanosoma cruzi*. *Transactions of the Royal Society of Tropical Medicine and Hygiene, 81*, 348–349.

Lacey, L. A., D'Alessandro, A., & Barreto, M. (1989). Evaluation of a chlorpyrifos-based paint for the control of Triatominae. *Bulletin of the Society for Vector Eccology, 14*, 81–86.

Lehane, M. J. (2005). *Biology of Blood-sucking Insects* (2nd ed.). Cambridge: Cambridge Univ. Press.

Lent, H., & Wygodzinsky, P. (1979). Revision of the Triatominae (Hemiptera, Reduviidae), and their significance as vectors of Chagas' disease. *Bulletin of the American Museum of Natural History, 163*, 123–520.

Lewinsohn, R. (1979). Carlos Chagas (1879–1934): The discovery of *Trypanosoma cruzi* and of American trypanosomiasis (footnotes to the history of Chagas's disease). *Transactions of the Royal Society of Tropical Medicine and Hygiene, 73*, 513–523.

Lowenstein, W. A., Romaña, C. A., Ben Fadel, F., Pays, J. F., Veron, M., & Rouzioux, C. (1992). Survie du virus de l'immunodéficience humaine (VIH-1) chez Triatoma infestans (Klug 1834). *Bulletin de la Societe de Pathologie Exotique (1990), 85*, 310–316.

Marshall, N. A., & Street, D. H. (1982). Allergy to *Triatoma protracta* (Heteroptera: Reduviidae). I. Etiology, antigen preparation, diagnosis and immunotherapy. *Journal of Medical Entomology, 19*, 248–252.

Maudlin, I., Holmes, P. H., & Miles, M. A. (Eds.), (2004). *The Trypanosomiases.* Cambridge, Massachusetts: CABI Publishing.

Miles, M. A. (1983). The epidemiology of South American trypanosomiasis—Biochemical and immunological approaches and their relevance to control. *Transactions of the Royal Society of Tropical Medicine and Hygiene, 77*, 5–23.

Moore, A. T., Edwards, E. A., Brown, M. B., Komar, N., & Brown, C. R. (2007). Ecological correlates of Buggy Creek virus infection in *Oeciacus vicarius*, southwestern Nebraska, 2004. *Journal of Medical Entomology, 44*, 42–49.

Myers, J. G. (1929). Facultative blood-sucking in phytophagous Hemiptera. *Parasitology, 21*, 472–480.

Neva, F. A. (2007). American trypanosomiasis (Chagas' disease). In L. Goldman & D. Ausiello (Eds.), *Cecil Medicine* (23rd ed.). Saunders Elsevier.

Olivier, M. C., Olivier, L. J., & Segal, D. B. (Eds.). (1972). *A bibliography on Chagas' disease (1909–1969).* Index-catalogue of Medical and Veterinary Zoology Special Publication No. 2 Washington: U.S. Govt. Print. Off.

Olsen, P. F., Shoemaker, J., Turner, H. F., & Hays, K. L. (1964). The incidence of *Trypanosoma cruzi* (Chagas) in wild vectors and reservoirs in east central Alabama. *The Journal of Parasitology, 50*, 599–603.

Peters, W., & Gillies, H. M. (1981). *A Colour Atlas of Tropical Medicine and Parasitology* (2nd ed.). New York: Year Book Med. Publ.

Peters, W., & Pasvol, G. (2007). *Atlas of Tropical Medicine and Parasitology* (2nd ed.). Elsevier Mosby.

Romero, A., Potter, M. F., Potter, D. A., & Haynes, K. F. (2007). Insecticide resistance in the bed bug: A factor in the pest's sudden resurgence? *Journal of Medical Entomology, 44*(2), 175–178.

Ribeiro, J. M. C., & Francischetti, I. M. B. (2003). Role of arthropod saliva in blood feeding: Sialome and post-sialome perspectives. *Annual Review of Entomology, 48*, 73–88.

Reinhardt, K., & Siva-Jothy, M. T. (2007). Biology of bed bugs (Cimicidae). *Annual Review of Entomology, 52*, 351–374.

Ryckman, R. E. (1962). Biosystematics and hosts of the *Triatoma protracta* complex in North America (Hemiptera: Reduviidae) (Rodentia: Cricetidae). *University of California Publications in Entomology, 27*, 93–240.

Ryckman, R. E. (1979). Host reactions to bug bites (Hemiptera, Homoptera): A literature review and annotated bibliography, Part I, Part II (with Bentley, D. G.). *California Vector Views, 26*, 1–49.

Ryckman, R. E. (1985). Dermatological reactions to the bites of four species of Triatominae (Hemiptera: Reduviidae) and *Cimex lectularius* L. *(Hemiptera: Cimicidae). Bulletin of the Society of Vector Ecologists, 10*, 122–125.

Ryckman, R. E., & Bentley, D. G. (1979). Host reactions to bug bites (Hemiptera, Homoptera): A literature review and annotated bibliography, Part II. *California Vector Views, 26*, 25–49.

Ryckman, R. E., Bentley, D. G., & Archbold, E. F. (1981). The Cimicidae of the Americas and Oceanic Islands, a checklist and bibliography. *Bulletin of the Society of Vector Ecologists, 6*, 93–142.

Ryckman, R. E., & Zackrison, J. L. (1987). Bibliography to Chagas' disease, the Triatominae and Triatominae-borne trypanosomes of South America (Hemiptera: Reduviidae: Triatominae). *Bulletin of the Society of Vector Ecologists, 12,* 1–464.

Salazar-Schettino, P. M. (1983). Customs which predispose to Chagas' disease and cysticercosis in Mexico. *The American Journal of Tropical Medicine and Hygiene, 32,* 1179–1180.

Schofield, C. J. (1979). The behaviour of Triatominae (Hemiptera: Reduviidae): A review. *Bulletin of Entomological Research, 69,* 363–379.

Schofield, C. J. (1988). Biosystematics of the Triatominae. In M. W. Service (Ed.), *Biosystematics of Haematophagous Insects* (pp. 285–312). Oxford: Clarendon Press.

Schofield, C. J. (1994). Triatominae*: Biology and Control.* West Sussex: Eurocommunica Publications.

Schofield, C. J., & Dias, J. C. (1999). The Southern Cone Initiative against Chagas disease. *Advances in Parasitology, 42,* 1–27.

Schofield, C. J., & Dolling, W. R. (1993). Bedbugs and kissing bugs. In R. P. Lane & R. W. Crosskey (Eds.), *Medical Insects and Arachnids* (pp. 483–516). Chapman & Hall.

Schofield, C. J., Minter, D. M., & Tonn, R. J. (1987). XIV. The triatomine bugs—biology and control. In: Vector Control Series—Triatomine bugs—Training and Information Guide. *World Health Organization Vector Biology and Control Division* 87.941.

Schofield, C. J., & White, G. B. (1984). Engineering against insect-borne diseases in the domestic environment/House design and domestic vectors of disease. *Transactions of the Royal Society of Tropical Medicine and Hygiene, 78,* 285–292.

Silverman, A. L., Qu, L. H., Blow, J., Zitron, I. M., Gordon, S. C., & Walker, E. D. (2001). Assessment of hepatitis B virus DNA and hepatitis C virus RNA in the common bedbug (*Cimex lectularius* L.) and kissing bug (*Rhodnius prolixus*). *The American Journal of Gastroenterology, 96,* 2194–2198.

Smart, J. (1943). *A Handbook for the Identification of Insects of Medical Importance.* London: British Museum (Natural History).

Solari, A., Venegas, J., Gonzalez, E., & Vasquez, C. (1991). Detection and classification of *Trypanosoma cruzi* by DNA hybridization with nonradioactive probes. *The Journal of Protozoology, 38,* 559–565.

Ter Poorten, M. C., & Prose, N. S. (2005). The return of the common bedbug. *Pediatric Dermatology, 22,* 183–187.

Theis, J. H. (1990). Latin American immigrants—Blood donation and *Trypanosoma cruzi* transmission. *American Heart Journal, 120,* 1483.

Todd, R. G. (2006). Efficacy of bed bug control products in lab bioassays: Do they make it past the starting gate? *American Entomologist, 52,* 113–116.

Trumper, E. V., & Gorla, D. E. (1991). Density-dependent timing of defecation by *Triatoma infestans. Transactions of the Royal Society of Tropical Medicine and Hygiene, 85,* 800–802.

Usinger, R. L. (1934). Blood sucking among phytophagous Hemiptera. *Canadian Entomologist, 66,* 97–100.

Usinger, R. L. (1944). *The Triatominae of North and Central America and the West Indies and their public health significance.* Public Health Bulletin 288. Washington: U.S. Govt. Print. Off.

Usinger, R. L., (1966). Monograph of Cimicidae (Hemiptera-Heteroptera). Thomas Say Foundation Vol. VII, *Entomological Society of America.*

Welch, K. A. (1990). First distributional records of *Cimexopsis nyctalis* List (Hemiptera: Cimicidae) in Connecticut. *Proceedings of the Entomological Society of Washington, 92,* 811.

WHO (World Health Organization), (2002). Control of Chagas disease. Second report of the WHO expert committee, *WHO Technical Report Series* 905., Geneva: World Health Organization.

Woo, P. T. K. (1991). Mammalian trypanosomiasis and piscine crytobiosis in Canada and the United States. *Bulletin of the Society for Vector Ecology, 16,* 25–42.

Wood, S. F. (1951). Importance of feeding and defecation times of insect vectors in transmission of Chagas' disease. *Journal of Economic Entomology, 44,* 52–54.

Yamagata, Y., & Nakagawa, J. (2006). Control of Chagas disease. *Advances in Parasitology, 61,* 129–165.

Zapata, M. T. G., Schofield, C. J., & Marsden, P. D. (1985). A simple method to detect the presence of live triatomine bugs in houses sprayed with residual insecticides. *Transactions of the Royal Society of Tropical Medicine and Hygiene, 79,* 558–559.

Zeledón, R., Alvarado, R., Jirón, L. F. (1977). Observations on the feeding and defecation patterns of three triatomine species (Hemiptera: Reduviidae). *Acta Tropica, 34,* 65–77.

Zeledón, R., Vargas, L. G. (1984). The role of dirt floors and of firewood in rural dwellings in the epidemiology of Chagas' disease in Costa Rica. *The American Journal of Tropical Medicine and Hygiene, 33,* 232–235.

Chapter

8

Beetles (Coleoptera)

William L. Krinsky

Beetles constitute the largest order of insects but are of relatively minor public health or veterinary importance. Adults and larvae of a few species occasionally bite, but more species secrete chemicals that can irritate the skin and eyes of humans and other animals. Beetles found in stored products can cause inhalational allergies, and some species found in dung and stored products act as intermediate hosts for helminths that cause pathology in domestic and wild animals. Many dung-inhabiting beetles are beneficial in interrupting the life cycles of mammalian parasitic worms and as predators or parasitoids of pestiferous flies that breed in excrement. A few beetle species are ectoparasites or mutualistic symbionts on mammals, and a few are known to temporarily invade the skin of mammals.

TAXONOMY

The order Coleoptera is divided into four suborders: Archostemata, considered the most primitive; Adephaga, named for its carnivorous members; Myxophaga, which are algae-eaters; and Polyphaga, the largest suborder, encompassing 90% of beetle families, comprised of species with diverse feeding habits. Beetles currently are grouped in about 165 families (Crowson, 1981; Arnett and Thomas, 2001). About 130 families include species that occur in North America.

More than 350,000 species of beetles have been described, representing about 40% of all known insects. More than 30,000 species of beetles occur in the United States and Canada (White, 1983; Arnett et al., 2002). Fewer than 100 species worldwide are known to be of public health or veterinary importance. Most of these are in the suborder Polyphaga. The species that have the greatest impact on the health of human and domestic animals are in the following families: Meloidae (blister beetles), Oedemeridae (false blister beetles), Staphylinidae (rove beetles), Tenebrionidae (darkling beetles), Dermestidae (larder beetles), and Scarabaeidae (scarab or dung beetles).

MORPHOLOGY

Adult beetles are distinguished from all other insects by the presence of hardened forewings called **elytra** (sing., *elytron*) that cover and protect the membranous hindwings (Figure 8.1). Coleoptera means "sheath-winged" in Greek. The size range of beetles is impressive, varying from 0.25 to 150 mm; however, most species are 2 to 20 mm long. Black and brown are the most common colors seen in the Coleoptera, but exquisite bright colors, including metallic and iridescent hues, occur especially in tiger beetles, ground beetles, plant beetles, flower beetles, metallic wood-boring beetles, long-horned beetles, and lady beetles. Beetles vary in shape from elongate, flattened, or cylindrical, to oval or round. Their bodies often are hardened, like the elytra, but some families, such as the blister and false blister beetles, have soft elytra and soft body parts that are pliable and sometimes described as leather-like.

The head of a beetle is usually conspicuous, and almost all beetles have some form of biting or chewing mouthparts. Even in specialized species adapted for piercing and sucking plants, the mandibles are retained and are functional. The antennae vary greatly in shape from filiform to pectinate to clavate or clubbed, and usually are composed of 11 visible segments. Two compound eyes are present in most species, and ocelli are rarely present.

Part of the thorax is visible dorsally as the pronotum, just posterior to the head (Figure 8.1). The divisions of the thorax are usually evident only on the ventral side. The legs vary greatly in shape from thick paddles in swimming species to slender, flexible forms in running species. The paired elytra cover the folded, membranous pair of hindwings. They usually overlay the dorsum of the abdomen and often are all that are visible in the abdominal region, when a beetle is viewed from above. In most species, the elytra are raised during flight. Some beetles have no hindwings and are flightless, and some beetles have very short elytra so that the abdominal tergites are visible dorsally (e.g., rove beetles and some blister beetles). Most beetles have

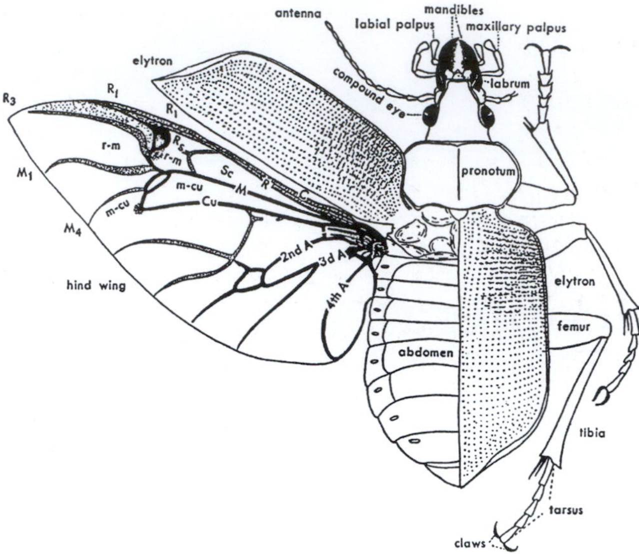

Figure 8.1 A representative adult beetle (Carabidae), dorsal view, with left elytron and wing spread. A, anal vein (2nd, 3rd, 4th); C, costa; M, media (M_1 = 1st branch, etc.); R, radius (R_1, 1st branch, etc.); R_S, radial sector; Sc, subcosta; m-cu, medio-cubital cross-vein; r-m, radio-medial cross-vein. (Modified from Essig, 1942)

eight visible abdominal tergites that can be seen when the elytra and hindwings are raised.

Defensive glands that secrete substances to repel predators are best developed in beetles in the suborder Adephaga. They are generally present as pygidial glands that open dorsally near the end of the abdomen. Secretions from these glands in the Adephaga are not known to cause notable ill effects in mammals. Within the Polyphaga, pygidial glands occur in a few families, such as the Tenebrionidae. In tenebrionids, the pygidial glands produce secretions that can deter small mammals and cause human skin irritation. The pygidial secretions of most other polyphagan species are not known to affect vertebrates.

Beetles that contain chemicals that are especially irritating to humans and other animals have toxic substances dispersed throughout their bodies rather than sequestered in specialized glands. The blister beetles, paederine rove beetles, and lady beetles fall within this group.

LIFE HISTORY

All beetles exhibit **holometabolous** development. Eggs are laid singly or in clusters on or in soil, living or dead plant matter, fabrics, water, carrion and, rarely, on living animals. The larvae of most beetles have a distinct head with simple eyes (ocelli) and chewing, mandibulate mouthparts, and the abdomen has eight to 10 segments. Beetle larvae exhibit diverse morphological types, from elongate-flattened forms (campodeiform), to cylindrical-flattened forms (elateriform), caterpillar-like forms (eruciform), and somewhat C-shaped soft forms (scarabaeiform). The larval body type usually is consistent in a particular family of beetles. In a few families, however, the larval form may vary from instar to instar in a given species, a life history progression called **hypermetamorphosis**. Certain blister beetle larvae, including scavengers in bee nests and ectoparasites or endoparasites of other insects, are hypermetamorphic. They emerge from the eggs as active campodeiform larvae, then molt into eruciform and scarabaeiform stages. Most beetle larvae molt at least three times before transforming into pupae.

Although most temperate species undergo only one generation a year, species in warmer climates are often multivoltine. Depending upon the species, any developmental stage may overwinter, but overwintering most often occurs in the pupal or adult stage. Most species exhibit diapause in one or another stage, and those that have developmental cycles exceeding one warm season usually have an obligatory diapause, initiated by changes in photoperiod and temperature. Most adult beetles live for weeks to as long as a year. However, adults of some species may live for years, spending much of their lives in diapause during periods when food is scarce.

BEHAVIOR AND ECOLOGY

Beetles live within all terrestrial and freshwater habitats. The great variation in beetle feeding behavior, whether saprophagous, herbivorous, carnivorous, or omnivorous, reflects the extremely diverse habitats in which these insects live. However, their mouthparts play a minor role in causing discomfort to humans and other animals. Beetle defense mechanisms, which involve the shedding or secretion of physically or chemically irritating materials, and beetle behavior that puts the insects in contact with developmental stages of parasitic helminths and vertebrates can lead to public health and veterinary problems.

Larder or pantry beetles (Dermestidae) are ubiquitous in human and domestic animal environments, where the larval and adult beetles eat stored food, food debris, dead insects, and other organic matter. Setae that cause human skin irritation or act as respiratory allergens are loosely affixed to the larvae of many species. The setae are elaborately barbed so that their firm adherence to many substrates, including human skin, causes them to be dislodged from the crawling, living larvae. In some species, the larvae actively raise the abdomen and make striking movements in response to touch. Other active defensive behaviors are seen in blister beetles, some chrysomelid plant beetles, long-horned beetles, and lady beetles that exude irritant chemicals from the femoro-tibial joints of the legs or from glandular openings around the mouthparts when the beetles are handled or threatened. This reflex bleeding repels predators.

One of the most dramatic defensive maneuvers is the explosion of boiling hot, acrid, quinones from the anal glands of carabid beetles called bombardier beetles. These forceful expulsions, which are aimed with extreme accuracy at potential predators, cause minimal damage to humans and other large animals but can cause physical and chemical burns in insects and small vertebrates (Evans, 1975).

Most of the beetles that serve as intermediate hosts of helminths parasitic in domestic animals and humans are grain or dung feeders. These species ingest helminth eggs present in animal feces or fecal-contaminated food. Because of the proximity of the beetles to feeding animals, whole adult beetles often are ingested incidentally by potential vertebrate hosts.

The tendency of many beetles to fly to artificial lights puts them in contact with human and domestic animal habitats and increases the chances of vertebrate contact with species that may be the sources of skin irritations, allergies, or helminthic infestations.

PUBLIC HEALTH IMPORTANCE

Human health problems caused by beetles include skin, eye, ear, and nose irritations; respiratory allergies; and minor gastrointestinal discomfort. Beetle families known to cause public health problems are listed in Table 8.1.

The greatest human discomfort associated with beetles is caused by vesicating species that secrete irritating chemicals when the insects are handled or accidentally contact human skin or sense organs. Blister beetles, false blister beetles, some rove beetles, and some darkling beetles have these irritants in their secretions, hemolymph, or body parts. Larvae of larder beetles are covered with hairs that can act as skin or respiratory allergens.

Invasion of body tissues by beetle larvae is called **canthariasis**, whereas invasion of such tissues by adult beetles is called **scarabiasis**. These forms of infestation occur most often in tropical regions. Most clinical cases involve enteric canthariasis that results from the ingestion of foodstuffs infested with beetles, or the accidental ingestion of infested materials by children. Dermestid larvae, such as *Trogoderma glabrum* and *T. ornatum*, have been associated with enteric canthariasis in infants who showed signs of extreme digestive discomfort, which in one case was the result of ulcerative colitis. It is unlikely that larvae were the cause of the latter condition, although larval hairs may have exacerbated the symptoms. Larvae were recovered from the stools of these patients and from the dry cereal they ingested. Other grain-infesting beetles, such as *Tenebrio molitor* and *T. obscurus*, have been accidentally ingested without causing noticeable symptoms.

Table 8.1 Beetle Families of Medical-Veterinary Importance, Listed in Order of Relative Importance

Family	Common Names	Clinical Importance
Meloidae	Blister beetles	Cause eye irritation and blisters on skin; can poison and kill horses and birds that ingest them
Staphylinidae	Rove beetles	Paederine species cause skin and eye lesions, and can poison livestock that ingest them; large species are known to bite humans; species attracted to dung feed on fly eggs, larvae, and pupae, and are thereby beneficial in reducing pestiferous fly populations
Scarabaeidae	Dung beetles and chafers	Spines cause irritation when adults enter ears; intermediate hosts of helminths; dung feeders are potential disseminators of pathogens; some dung feeders are beneficial in removing dung that is the source of pestiferous flies and that is infested with intermediate stages of vertebrate worm parasites
Tenebrionidae	Darkling beetles and grain beetles	Cause skin and eye irritation; larvae and adults contain inhalational allergens; grain-feeding species are intermediate hosts of helminths and potential disseminators of pathogens
Dermestidae	Larder beetles, pantry beetles, hide beetles, carpet beetles	Larval setae can cause skin, eye, ear, and nose irritation, or gastrointestinal discomfort if ingested; larvae and adults can cause inhalational allergies; grain-feeding species are intermediate hosts of helminths; carrion-feeding species are potential disseminators of pathogens
Histeridae	Hister beetles	Beneficial as predators of fly eggs and larvae developing in avian and mammalian manure
Oedemeridae	False blister beetles	Cause skin and eye irritation
Carabidae	Ground beetles	Intermediate hosts of poultry tapeworms
Silphidae	Burying beetles or carrion beetles	Potential disseminators of pathogens
Corylophidae	Minute fungus beetles	Cause eye lesions
Melyridae	Soft-winged flower beetles	Cause skin tingling, and numbing, and burning of the skin and eyes
Coccinellidae	Ladybird beetles or lady bugs	Secretions can cause skin discoloration and irritation
Cleridae	Checkered beetles	Can bite humans causing temporary distress
Cerambycidae	Long-horned beetles	Larger species can bite humans and other animals causing temporary discomfort
Merycidae	Old World cylindrical bark beetles	Can bite humans causing temporary distress
Curculionidae	Weevils	Grain-inhabiting species can cause inhalational allergies

Rarely, adult and larval beetles have been recovered from human nasal sinuses, and larvae from the urethra. Small beetles in various families have been known to fly or crawl into human eyes and ears. Some of these cause minor physical irritation, whereas others may cause extreme burning sensations, presumably due to chemicals exuded by the insects.

Painful, but temporary, eye lesions caused by tiny *Orthoperus* species (<1 mm long) in the family Corylophidae have been seen in eastern Australia, where the condition has received several names: **Canberra eye**, **Christmas eye**, and **harvester's keratitis**.

Tingling and numbing of the lips and face, and burning of the eyes and skin, are caused by contact with beetles of the family Melyridae (genus *Choresine*) in New Guinea. Melyrid beetles appear to be the source of batrachotoxins found in toxic passerine birds in New Guinea, and possibly of similar toxins found in South American poison-dart frogs. The name *nanisani* is used to describe one of the New Guinea birds with toxic feathers, as well as the facial sensations arising from contact with the bird or the melyrid beetles (Dumbacher et al., 2004).

An unusual apparent chemical defense mechanism has been described in a South American longhorned beetle. The cerambycid *Onychocerus albitarsis* has the last segment of each antenna modified into a pointed sting, which is somewhat analogous to the last segment of the tail of a scorpion in structure and in causing pain and swelling when jabbed into the skin (Berkov et al., 2008).

More than 40 species of beetles have been associated with human allergic reactions that result from inhaling beetle parts (e.g., larval setae) or excreta (Bellas, 1989). Agricultural and research workers most often are affected by inhalational allergies, because most of the beetle species involved occur in large numbers in stored products. Dermestid beetles (*Trogoderma angustum*), tenebrionids (*Tenebrio molitor* and *Tribolium* species), and grain weevils (*Sitophilus granarius*) have been incriminated in many cases of respiratory distress, such as asthma.

Beetles serve as intermediate hosts for more than 50 parasitic worms, including tapeworms (Cestoda), flukes (Trematoda), roundworms (Nematoda), and thorny-headed worms (Acanthocephala) (Hall, 1929; Cheng, 1973). These worms primarily parasitize nonhuman hosts. Only a few species, such as the rodent tapeworms *Vampirolepis nana* and *Hymenolepis diminuta*, and the *Macracanthorhynchus* species of acanthocephalan parasites occasionally infest children. The intermediate hosts of *Vampirolepis* and *Hymenolepis* species are grain beetles (Tenebrionidae), and *Macracanthorhynchus* species undergo development in dung beetles (Scarabaeidae). Children become infested because of their poor hygienic practices or by accidental ingestion of the beetles. Intentional ingestion of living tenebrionids for medicinal purposes in Malaysia is also a potential route for human infestation with rodent tapeworms (Chu et al., 1977).

Many beetles, such as scarabs, silphids, and dermestids that feed on dung and carrion have the potential to be mechanical vectors of pathogens, such as the bacteria that cause salmonellosis and anthrax. Although there is experimental evidence for maintenance and excretion of some of these microbes by beetles, given the limited sizes of the inocula and the limited contact between humans and scavenger beetles, there is no indication that these beetles play a role in direct transmission to humans.

Many families of beetles include species known to occasionally cause bites to humans. This may happen when the beetles are accidentally handled or when the beetles occur in such large numbers that many fly or crawl onto the body. Entomologists and others who pick up beetles are the persons most often bitten. Long-horned beetles (Cerambycidae), checkered beetles (Cleridae), rove beetles (Staphylinidae), and cylindrical bark beetles (Merycidae) are among those that have been reported as biting. The somewhat painful bites usually leave little or no skin marks and do not cause any long-lasting discomfort. Long-horned beetles feed on wood in their immature stages and are found as adults on flowers, dead and dying trees, and freshly cut timber. The larger species of rove beetles that can bite are predaceous on fly larvae and often are found on carrion or dung. Cylindrical bark beetles and checkered beetles are found under bark associated with wood-boring insects or fungus.

Population increases and mass migrations of checkered beetles (Cleridae), flat grain beetles (Cucujidae), and ground beetles (Carabidae) have all caused annoyance at times by their sheer numbers and, in some cases, by the strong odors of their defensive secretions.

Meloidae (Blister beetles)

Blister beetles (Figure 8.2) occur worldwide. Most, if not all, contain the terpene **cantharidin** ($C_{10}H_{12}O_4$), which can cause skin irritations. People usually develop blisters within 24 hours of contacting the secretions of these beetles or the body fluids from crushed beetles (Figure 8.3). Often this is accompanied by tingling or burning sensations. The blisters may progress to vesicular dermatitis with itching and

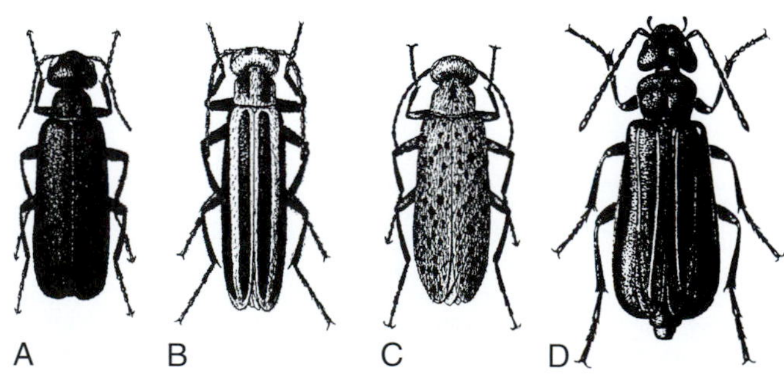

Figure 8.2 Blister beetles (Meloidae). (A) Black blister beetle (*Epicauta pennsylvanica*); (B) Striped blister beetle (*E. vittata*); (C) Spotted blister beetle (*E. maculata*); (D) European "Spanish fly" (*Lytta vesicatoria*). (A–C, modified from White, 1983; D, from Harde, 1984)

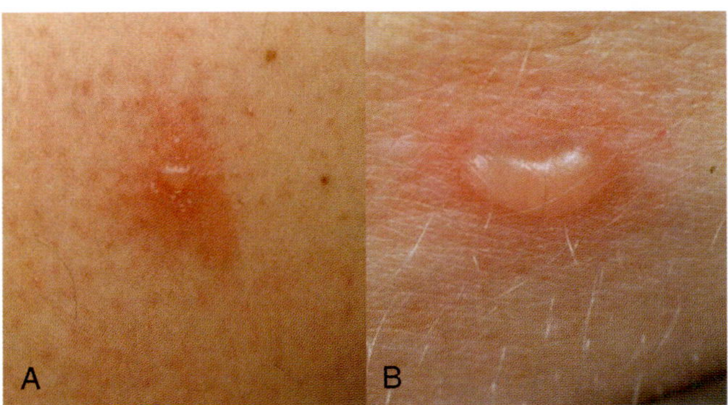

Figure 8.3 Contact dermatitis caused by blister beetles (Meloidae, *Epicauta* spp.). (Photos by Gary R. Mullen (A) and Jerry F. Butler (B))

oozing lesions. At least 20 species of meloids have been associated with dermatitis (Table 8.2). Cantharidin is present in the hemolymph and in the clear, yellow secretion that is exuded at the joints of the legs of these beetles by reflex bleeding. Reptiles and some predaceous insects are repelled by the fluid. Although cantharidin is irritating to humans, the chemical acts as a meloid courtship stimulant that is secreted by male accessory glands and passed to the female during copulation. The males, being the only source of cantharidin, generally have the highest concentrations of the chemical, with levels in female beetles varying with their mating histories. The meloid spermatophore is rich in cantharidin, and the eggs also contain the substance, presumably to deter predators.

Table 8.2 Species of Meloidae Reported to Cause Blistering of Human Skin*

Species	Geographic Occurrence of Clinical Reports
Lytta vesicatoria	Europe
L. phalerata	China
Epicauta cinerea	United States (southwestern)
E. flavicornis	Senegal
E. hirticornis	India
E. maculata	United States (western)
E. pennsylvanica	United States, Mexico
E. sapphirina	Sudan
E. tomentosa	Sudan
E. vestita	Senegal
E. vittata	United States (eastern)
Cylindrothorax bisignatus	South Africa
C. dusalti	Senegal, Mali
C. melanocephalus	Gambia, Senegal
C. picticollis	Sudan
C. ruficollis	Sudan, India
Mylabris bifasciata	Nigeria
M. cichorii	India
Psalydolytta fusca	Gambia
P. substrigata	Gambia

*Modified from Alexander (1984), with information from Selander (1988).

Blister-beetle dermatitis has been reported in Europe, Asia, Africa, North America, and Central America. The most famous blister beetle is the Spanish fly, *Lytta vesicatoria* of the Mediterranean region, an insect that has been erroneously touted as a human aphrodisiac. Cantharidin is poisonous to humans and other animals when ingested and may cause kidney damage and death. Ingestion of powder made by grinding up dried beetles or any other source of cantharidin produces extremely toxic effects on the urogenital system. The resulting inflammation causes painful urination, hematuria, and persistent penile erection (priapism), a condition mistakenly associated with increased sexual stimulation. Like many other naturally occurring toxins, cantharidin has been prescribed for centuries as a cure for various ailments, but it has never been proven to have a therapeutic effect.

Meloid species most often associated with skin lesions in the United States and Mexico are members of the genus *Epicauta*. These include the striped blister beetle (*Epicauta vittata*) in the eastern states, the black blister beetle (*E. pennsylvanica*) found throughout most of the country, and the spotted blister beetle (*E. maculata*) found in the western states (Figure 8.2). Other species of *Epicauta* cause similar problems in India and Africa (Table 8.2). The genus *Cylindrothorax* occurs over a vast area of the Old World including Africa, the Near East, India, and parts of Southeast Asia. African species that cause blistering include *C. bisignatus*, *C. dusalti*, *C. melanocephalus*, and *C. picticollis*. A *Lytta* species in China also has been associated with human dermatitis (Table 8.2).

Blister beetles are found most often on flowers or foliage where the beetles feed on pollen and other plant tissues. *Epicauta* species are usually abundant where grasshoppers flourish because the larvae of these meloids feed on grasshopper eggs. Most people who develop blister beetle lesions are agricultural workers or soldiers on maneuvers in areas where the beetles are common. Retention of cantharidin in frogs and birds that prey upon meloids may lead to human poisoning when these predators are used as human food. Nineteenth century medical reports of priapism in French legionnaires traced the cause of this clinical problem to the soldiers' ingestion of frogs that had eaten meloids. Humans have also developed signs of cantharidin poisoning following ingestion of cooked wild geese (Eisner et al., 1990).

Oedemeridae (False blister beetles)

False blister beetles in the genera *Oxycopis* (Figure 8.4), *Oxacis*, and *Alloxacis* are known to cause vesicular or bullous dermatitis in the United States, Central America, and the Caribbean region. *Sessinia kanak*, a species that is commonly attracted to lights in the Solomon Islands, and *S. lineata*, a New Zealand species, cause similar irritating lesions. Blistering has been

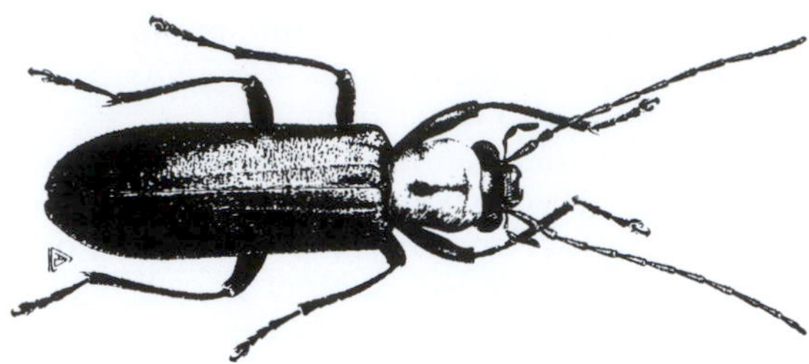

Figure 8.4 False blister beetle, *Oxycopis mcdonaldi* (Oedemeridae). (Arnett, 1984)

observed in people exposed to large numbers of swarming *Eobia apicifusca* in Australia. False blister beetles are attracted to flowers where they feed on pollen. Immediate burning of the skin following contact with *Sessinia* species swarming around coconut flowers has been reported on the Line Islands, south of Hawaii. As in meloids, cantharidin is the toxic substance in all these oedemerids.

Staphylinidae (Rove beetles)

Rove beetles in the genus *Paederus* (Figure 8.5) contain **pederin** ($C_{25}H_{45}O_9N$), a toxin more potent than that of *Latrodectus* spider venom, and the most complex nonproteinaceous insect defensive secretion known. Pederin is synthesized by endosymbiotic gram-negative bacteria (*Pseudomonas* species) occurring in female *Paederus* species. The beetles, which are mostly 7 to 13 mm long, are found in North, Central, and South America; Europe; Africa; Asia; and Australasia. Unlike most rove beetles that are dull-colored, many *Paederus* species have an orange pronotum and orange basal segments of the abdomen, which contrast sharply with the often blue or green metallic elytra and brown or black coloration of the rest of the body. This color pattern may be a form of warning (**aposematic**) coloration, but a defensive function for pederin has not been demonstrated.

At least 20 of the more than 600 described species of *Paederus* have been associated with *Paederus* dermatitis (Table 8.3). Skin reactions to the beetles, named **Ch'ing yao ch'ung**, were described in China as early as AD 739. Most cases of dermatitis have involved tropical species, including *Paederus fuscipes* (widespread from the British Isles east across Central Asia to Japan and southeast to Australia), *P. sabaeus* (Africa, where it is called Nairobi fly and champion fly), *P. cruenticollis* and *P. australis* (Australasia), *P. signaticornis* (Central America), and *P. columbinus* and *P. brasiliensis* (South America). Species in South American countries are known by various names, such as *bicho de fuego, pito, potó, podó,* and *trepa-moleque.*

Unlike blister beetles, rove beetles do not exhibit reflex bleeding as a defensive reaction. Pederin contacts human skin only when a beetle is brushed vigorously over the skin or crushed. Because of their general appearance or misunderstandings about their etiology, the resulting skin lesions have been called **dermatitis linearis, spider-lick** (India and Sri Lanka), and **whiplash dermatitis**. The dermatitis may develop on any part of the body; however, exposed areas such as the head, arms, hands, and legs are affected most often. Mirror-image lesions may form where one pederin-contaminated skin surface touches another.

Unlike meloid-induced dermatitis that develops within 18 to 24 hours after contact, the paederine-induced reaction of itching and burning, usually occurs 24 to 72 hours after contact with the beetle's body fluid. The affected skin appears reddened and vesicles form about 24 hours after the initial response

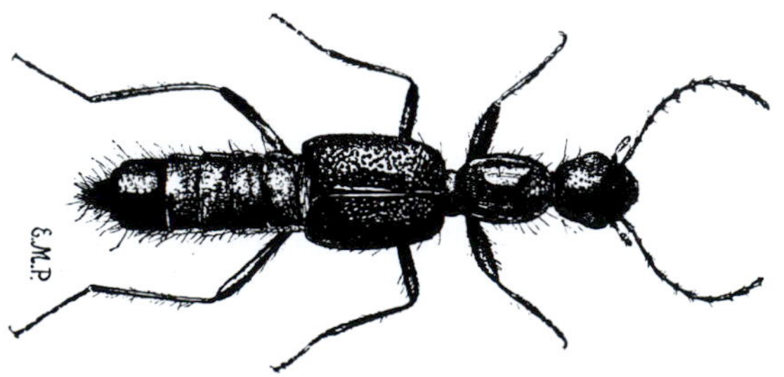

Figure 8.5 Rove beetle, *Paederus sabaeus* (Staphylinidae), West Africa. (Patton and Evans, 1929)

Table 8.3 Species of *Paederus* (Staphylinidae) Reported to Cause Skin Lesions in Humans

Species	Geographic Occurrence of Clinical Reports
Paederus alternans	India, Vietnam, Laos
P. amazonicus	Brazil
P. australis	Australia
P. brasiliensis	Brazil, Argentina
P. columbinus	Brazil, Venezuela
P. cruenticollis	Australia
P. eximius	Kenya
P. ferus	Argentina
P. fuscipes	Italy, Russia, Iran, India, China, Taiwan, Japan, Thailand, Vietnam, Laos, Indonesia
P. nr. *fuscipes*	Papua New Guinea
P. islae	Israel
P. nr. *intermediua*	Philippines
P. laetus	Guatemala
P. melampus	India
P. ornaticornis	Ecuador
P. puncticollis	Uganda
P. riparius	Russia
P. rufocyaneus	Malawi
P. sabaeus	Sierra Leone, Nigeria, Zaire, Cameroon, Namibia, Tanzania, Uganda
P. signaticornis	Guatemala, Panama
P. tamulus	China
Paederus spp.	Northern Iran, Pakistan, Malaysia, Sri Lanka

*Modified from Frank, J.H. and Kanamitsu, K. (1987). Paederus, *sensu lato* (Coleptera: Stapylinidae): Natural History and Medical Importance. *Journal of Medical Entomology 17*, 320–329.

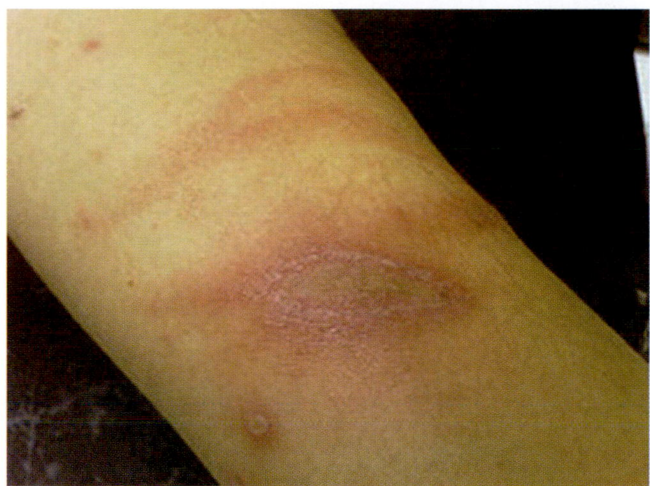

Figure 8.6 *Paederus* dermatitis on human forearm, caused by reaction of the skin on contact with toxin, called pederin, in the body fluids and tissues of rove beetles of the genus *Paederus* (Staphylinidae). (Courtesy of Chan Chee Keong, MD)

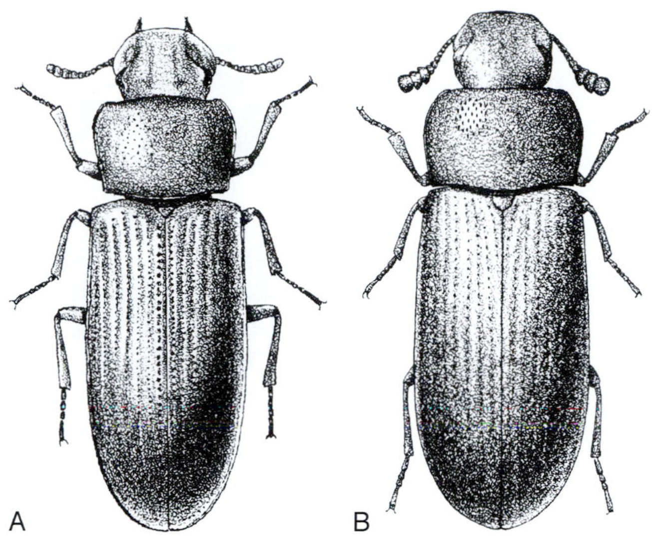

Figure 8.7 Darkling beetles (Tenebrionidae). (A) Confused flour beetle (*Tribolium confusum*), (B) Red flour beetle (*T. castaneum*). Defensive secretions containing quinones can cause skin irritation. (Gorham, 1991)

(Figure 8.6). The vesicles may coalesce into blisters and become purulent, producing a reaction that is often more severe than that seen following exposure to meloids. The itching may last for a week, after which the blisters crust over, dry, and peel off, leaving red marks or lightened skin areas that may persist for months. Rubbing the eyes with beetle fluid or contaminated hands, or beetles flying or crawling into eyes can cause pain, marked swelling of the eyelids and conjunctivae, excessive lacrimation, clouding of the cornea, and inflammation of the iris (iritis). Such ocular lesions seen in East Africa have been called **Nairobi eye**. Although eye involvement often is very irritating, permanent damage is not common.

Rove beetles live in vegetable debris and under stones and other materials, such as leaf litter. They are predaceous on insects and other arthropods, or may eat plant debris. Paederine staphylinids are most abundant in areas of moist soil, such as irrigated fields and other crop lands, where the adult beetles feed on various herbivorous insects. Consequently, agricultural workers and others working in fields and grassy areas often are affected. Because the beetles are attracted to lights, workers on brightly lit oil rigs and people occupying lighted dwellings in tropical areas are also commonly affected with what has been called night burn.

Tenebrionidae (Darkling beetles)

Darkling beetles (Figure 8.7) produce defensive secretions containing quinones. Adults of *Blaps* species found in the Middle East and Europe secrete these chemicals that cause burning, blistering, and darkening of the skin. Adult beetles of some cosmopolitan *Tribolium* species, including *T. confusum* and *T. castaneum*, have been associated with severe itching. North American desert species in the genus *Eleodes*, when threatened, take a characteristic headstand pose and exude various quinones that repel small predators and cause mild irritation to humans who handle these beetles. Darkling beetles are found in diverse habitats, including under logs and stones, in rotting wood and other vegetation, in fungi, in termite and ant nests, and among debris in and outside of homes. Most species live in dry, often desert, environments, and pest species are found in stored products, such as grain and cereals. Most tenebrionids are scavengers on decaying or dry plant material, but a few feed on living plants.

Dermestidae (Larder beetles)

Larvae of larder beetles, or pantry beetles (Figure 8.8), are covered with barbed and spear-like setae that may cause allergic reactions in the form of pruritic, papulovesicular skin lesions. Dermestid larvae often are found living in household furnishings, such as carpets, rugs, and upholstery or stored clothing of individuals suffering from these reactions. Larder beetles are named for their common occurrence as pantry pests, but they may also be found in grain storage facilities, bird and mammal nests and burrows, and on carrion. The larvae and adults are mostly scavengers on decaying or dry plant and animal matter.

Dermestid larvae and adults are known to have crawled into human ears causing itching and pain. The spear-headed setae of dermestid larvae have been observed on numerous occasions on cervical (Pap) smear slides and in sputum samples. In all these cases the setae appear to have been contaminants that were not associated with any pathological changes in the patients (Bryant and Maslan, 1994).

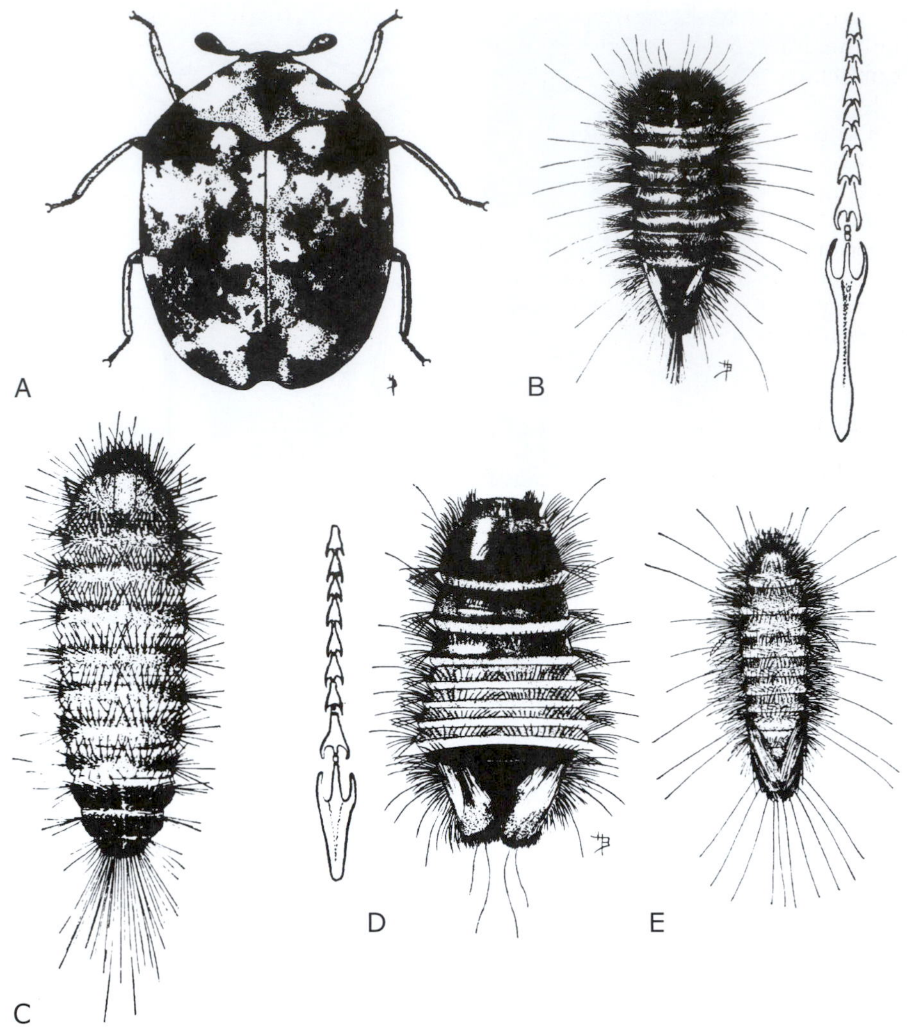

Figure 8.8 Dermestid beetles (Dermestidae). (A) Furniture carpet beetle (*Anthrenus flavipes*), adult; (B) Same, larva, with hastate seta (enlarged); (C) Cabinet beetle (*Trogoderma* sp.), larva; (D) Varied carpet beetle (*Anthrenus verbasci*), larva, with hastate seta (enlarged); (E) Common carpet beetle (*Anthrenus scrophulariae*), larva. (Sweetman, 1965)

Scarabaeidae (Scarab beetles)

In some tropical regions where human and animal excrement are abundant in the vicinity of dwellings, scarab beetles living in the dung are sometimes accidentally ingested by young children. These beetles appear in the newly passed stools of children and may disperse from the excrement in a noisy fashion that has been described in Sri Lanka as **beetle marasmus** (*kurumini mandama*). Although some of these beetles may infest the fecal matter as it is passed or after it reaches the ground, it is quite likely, as local physicians claim, that the scarab beetles (e.g., *Copris* spp.; Figure 8.9) pass through the alimentary tract and remain alive, causing little or no discomfort to the children. Evidence for such durability among the scarabs comes from cases in which frogs, horses, and cattle have ingested scarabs, which then worked their way through the stomach wall and remained alive until the hosts were killed. In Asia and Africa, humans sometimes

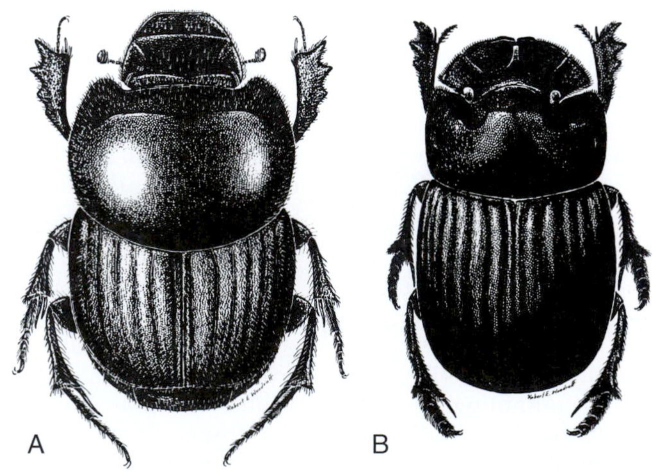

Figure 8.9 Scarab beetles (Scarabaeidae). (A) *Onthophagus polyphemi*; (B) *Copris minutus*. (Woodruff, 1973)

become infested with dung beetles (*Onthophagus* and *Caccobius* spp.) when these scarabs enter the anus and live within the rectum, causing damage to the mucosa and physical discomfort.

Large numbers of the adult scarab beetles *Cyclocephala borealis* and *Autoserica castanea* invaded the ears of 186 boy scouts sleeping on the ground at a jamboree in Pennsylvania in 1957. The beetles caused pain and some slight bleeding as a result of the tearing action of their tibial spines. After the beetles were removed, there were very few cases of secondary infection (Mattuck and Fehn, 1958).

Coccinellidae (Lady beetles)

Lady beetles, also called ladybird beetles and ladybugs, have been cited most often as causing prickling or slight stinging sensations, followed by the formation of mild erythematous lesions. These beetles may nip at the skin; however, given their small size, it is more likely that their defensive secretions cause the discomfort. The alkaloid secretions produced by reflex bleeding from the legs and around the mouthparts do stain human skin and can cause mild irritation. Lady beetles are common in gardens in warm weather, where the beetles feed on aphids and scale insects on a variety of plants. Lady beetles also occur in large aggregations as they are crawling and flying from overwintering clusters.

VETERINARY IMPORTANCE

Although not generally appreciated even by entomologists, beetles are involved in a variety of problems of a veterinary nature. These include toxicity to domestic animals on ingestion, mechanical transmission of disease agents, intermediate hosts for helminthic parasites, direct injury to animals by ectoparasitic species, and structural damage to poultry facilities. Beetles also can be beneficial by playing an important role in the recycling of animal dung and as natural control agents, especially for dung-breeding flies.

Ingestion of Toxic Beetles

Several blister beetles in the family Meloidae pose a hazard to livestock that feed on forage in which the living beetles are abundant. Horses that ingest quantities of these beetles are especially susceptible to cantharidin poisoning. Dead beetles and beetle parts retain their cantharidin content so that forage crops that are harvested for livestock feed continue to be a source of the toxin. Cantharidin is not readily degraded by heating or drying. Blister beetles in alfalfa (*Medicago sativa*) fields contain enough cantharidin to provide lethal doses to horses that feed on this material when it is used as hay. Species that pose problems in the United States include the striped blister beetle (*Epicauta vittata*), the black blister beetle (*E. pennsylvanica*) (Figure 8.2), the margined blister beetle (*E. funebris*), the three-striped blister beetle (*E. lemniscata*), as well as *E. fabricii, E. occidentalis*, and *E. temexa*. Individual beetles contain <0.1 to >11 mg of cantharidin, equivalent to <0.1 to $>12\%$ of their dry weights, with males of several species averaging over 5%. The minimum lethal dose for a horse is about 1 mg/kg, which means that depending on the size of a horse and the cantharidin content of the beetles ingested, anywhere from 25 to 375 beetles are sufficient to cause death (Capinera et al., 1985).

Horses have been poisoned by eating forage in the southern, midwestern, and western United States. Poisoning is not limited to a particular geographic area because contaminated forage is transported over long distances. Affected animals exhibit moderate to severe clinical signs, ranging from depression to shock that may be followed by death. Abdominal distress (colic), anorexia, depression, fever, dehydration, gastritis, esophagitis, and oral ulcers are commonly observed. These signs are accompanied by markedly decreased serum calcium and magnesium. Accelerated heart rate (tachycardia), increased respiratory rate (tachypnea), and increased creatinine kinase activity are indicative of severe toxicosis that is likely to lead to death. In most cases in which death has occurred, a horse has succumbed within 48 hours of the onset of clinical signs.

Horses are the most commonly affected because the kinds of forage they are fed are most often contaminated with beetles. Although the ruminant digestive tract is less susceptible to cantharidin poisoning, goats, sheep, and cattle have also died from cantharidin toxicosis. Horses and cattle have also been poisoned by pederin following ingestion of the tropical staphylinid *Paederus fuscipes*. Pederin can cause severe damage to the mucosa of the alimentary tract.

Cantharidin poisoning of emu chicks in Texas, following ingestion of the meloid *Pyrota insulata*, has been reported. Meloids feeding on nearby mesquite blossoms were attracted to lights inside a chick barn. Severely affected chicks became ataxic, vomited, and died. Chickens are the only birds, other than emus, known to have succumbed following ingestion of blister beetles.

Historically, there are reports of chickens, ducklings, goslings, and young turkeys dying as a direct result of ingesting the rose chafer (*Macrodactylus subspinosus*), a member of the Scarabaeidae (Lamson, 1922). Although this North American species is abundant in the summer months, modern enclosed poultry production facilities may have greatly reduced the incidence of such poisonings.

Transmission of Pathogens

Darkling beetles (Tenebrionidae) inhabiting farm buildings can be mechanical vectors of animal pathogens. Tenebrionid beetles infesting feed in chicken houses may become infected with *Salmonella* bacteria passed in feces from infected chickens. Both larval and adult forms of the lesser mealworm beetle (*Alphitobius*

Figure 8.10 Lesser-mealworm beetle, *Alphitobius diaperinus* (Tenebrionidae), larvae and adults; develop in litter and, as adults cause structural damage in poultry houses; can disseminate pathogenic agents via their excreta; and are intermediate hosts for the tapeworm *Choanotaenia infundibulum*. (Photo by D. Wesley Watson)

Figure 8.11 Tapeworm, *Choanotaenia infundibulum* (Cestoda: Dilepididae), a parasite in small intestines of chickens, turkeys, pheasants, and guinea fowl; utilizes tenebrionid and dermestid beetles as intermediate hosts. (Photo by Nancy C. Hinkle)

diaperinus) (Figure 8.10) have been found to maintain viable pathogens (e.g., *Salmonella typhimurium* and *S. chester*) on their external surfaces and in their digestive tracts. The bacteria survive for days after infection and are disseminated via beetle excreta. In chicken breeding facilities, grain beetles are potential disseminators of these pathogens that can infect both chicks and adult birds. These organisms can cause gastroenteritis in human consumers. The lesser mealworm also is regarded as a potential disseminator of other bacteria (*Escherichia, Bacillus, Streptococcus*), fungi (*Aspergillus*), and the viruses causing Marek's disease, Newcastle disease, fowlpox, avian influenza, enteritis, and infectious bursitis (Gumboro disease). In addition, oocysts formed by protozoans in the genus *Eimeria* are ingested by lesser mealworm beetles and when the infected beetles are ingested by birds, the latter develop avian coccidiosis, a serious disease of poultry.

Mycobacterium avium subspecies, the agents of avian tuberculosis, remain viable in larval and adult tenebrionids (*Tenebrio molitor*) after the beetles are infected experimentally. Although the bacteria have not been recovered from naturally infected beetles, there is the potential for mechanical transmission of mycobacteria to birds following ingestion of beetles.

Intermediate Hosts of Parasites

Tapeworms (cestodes), flukes (trematodes), roundworms (nematodes), and thorny-headed worms (acanthocephalans) of many species that infest domestic and wild animals use beetles as intermediate hosts. Animals become infested by ingesting parasitized beetles that contaminate feed or bedding (tenebrionids, carabids) or that are attracted to animal dung (scarabaeids), or by ingesting water in which infective beetles have disintegrated.

Two **tapeworms** that infest the small intestines of poultry are the broad-headed tapeworm (*Raillietina cesticillus*) and *Choanotaenia infundibulum* (Figure 8.11). Both parasites cause enteritis and hemorrhaging in chickens, turkeys, pheasants, and guinea fowl. A few tenebrionids and scarabaeids and more than 35 species of carabid beetles, notably in the genera *Amara* and *Pterostichus*, are intermediate hosts for *R. cesticillus* (Cheng, 1973). Some tenebrionid and dermestid species, including the lesser mealworm beetle, are intermediate hosts for *C. infundibulum*. Proglottids or tapeworm eggs ingested by beetle larvae or adults develop into cystercerci (encysted larvae) that can then infest birds that eat the beetles. Chicks are most susceptible to serious infestations and often die from worm burdens.

The beef tapeworm (*Taenia saginata*) can use dung beetles and carabids as intermediate hosts, although they are not essential for transmission. Beetles associated with infective dung or debris can ingest proglottids or eggs as in the case of poultry worms. Cattle and humans infested with the tapeworm may show mild symptoms such as weight loss, abdominal pain, and increased appetite.

The dwarf tapeworms (*Vampirolepis nana* and *Hymenolepis diminuta*) that usually infest rodents, especially rats and mice, can infest humans when the intermediate host beetles are accidentally ingested. *Tenebrio molitor* may act as an intermediate host for *V. nana*, although this worm is readily transmitted directly from one vertebrate host to another. Several species of tenebrionids (*Tenebrio* spp. and *Tribolium* spp.) are required intermediate hosts for *H. diminuta*. Larval and adult beetles infesting grain and cereals ingest worm eggs that develop into cysticercoid stages that infest rodents or humans, usually children, who ingest the beetles. Dwarf tapeworms produce minimal symptoms in rodents and people, although heavy infestations in children may cause abdominal pain, diarrhea, convulsions, and dizziness.

Beetles are known to be intermediate hosts for only a few trematodes. These are parasites of frogs that become infested by ingesting parasitized dytiscid beetles and pose no problem for other vertebrate animals.

Many **nematodes** infest livestock and wildlife, but only a few use beetles as intermediate hosts. Spirurid nematodes of various species infest livestock and, rarely, humans. *Physocephalus sexalatus* and *Ascarops strongylina* eggs develop in many species of scarabaeoid dung beetles (*Geotrupes* spp., *Onthophagus* spp., and *Scarabaeus* spp.) that then may be ingested by pigs. Both wild and domestic swine can be infested with these stomach worms that can cause digestive problems in heavily infested young animals. *Gongylonema pulchrum* is a parasite of the upper digestive tract of sheep, cattle, goats, and other ruminants as well as horses, dogs, and humans. The worms burrow in the mucosa and submucosa of the oral cavity and esophagus and may cause bleeding, irritation, numbness, and pain in the mouth and chest. Scarabaeid and tenebrionid beetles serve as intermediate hosts for the larvae. Scarabaeid dung beetles are also the intermediate hosts for *Spirocerca lupi*, the esophageal worm of dogs and wild canids. *Physaloptera caucasica*, another spirurid, often parasitizes monkeys in tropical Africa, where humans also commonly are infested. This nematode causes digestive distress by infesting the alimentary tract from the esophagus to the terminal ileum. Scarabaeid dung beetles are its intermediate hosts.

The **acanthocephalans**, aptly named for their thorny heads, include species found worldwide infesting swine, rodents, and carnivores, such as dogs. *Macracanthorhynchus hirudinaceus*, which attaches to the small intestines of swine, causes enteritis and produces intestinal nodules that lower the value of these tissues when they are sold to make sausage casings. Eggs of this parasite are ingested by scarab beetle larvae of species of various genera (*Phyllophaga, Melolontha, Lachnosterna, Cetonia, Scarabaeus,* and *Xyloryctes*), including May and June beetles, leaf chafers, dung beetles, and rhinoceros beetles. Infested beetle larvae, as well as the pupae and adults that develop from them, are infective to both pigs and humans. Humans and pigs often show no symptoms. However, in cases of heavy infestations, both human and porcine hosts may experience digestive problems, such as abdominal pain, loss of appetite, and diarrhea that can lead to emaciation.

Two other acanthocephalan worms that parasitize the small intestines of their hosts use scarab beetles or tenebrionids as intermediate hosts. They are *Macracanthorhynchus ingens* that infests raccoons and occasionally dogs and humans, and *Moniliformis moniliformis*, a parasite of rodents and dogs.

Nest Associates and Ectoparasites

In addition to those beetles that occasionally invade the alimentary tracts or sense organs of animals, other species have evolved in close association with mammals as nest dwellers or ectoparasites. These species typically have reduced eyes and wings or have lost these structures completely. The family **Leptinidae**, known as mammal nest beetles, includes *Platypsyllus* spp. that live as larvae and adults on beavers, and *Leptinus* and *Leptinillus* spp. that live as adults on various small rodents. These species feed on skin debris or glandular secretions and have been associated with skin lesions on their hosts. Two other beetle groups, the Staphylinidae (Tribe Amblyopinini) and the Languriidae (Subfamily Loberinae), have species that live on rodents and a few other mammals. These staphylinids, and possibly the languriids, appear to be mutualistically symbiotic with their mammalian hosts. The beetles infest mammalian fur to gain access to their prey, which are mammalian ectoparasites, such as fleas and mites that live in rodent nests (Ashe and Timm, 1987; Durden, 1987).

Some scarabaeid beetles are adapted to living in the fur around the anus of certain mammals. These beetles cling to the fur except when they leave to oviposit in the dung. Some of these scarabs (*Trichillium* sp.) are found on sloths and monkeys in South America and others (*Macropocopris* spp. and *Onthophagus* spp.) on marsupials in Australia.

Larvae and adults of the lesser mealworm beetle (*Alphitobius diaperinus*) have been found boring into and living in the scrotum of a rat, and feeding on sick domestic chicks and young pigeons. Similarly, the hide beetle (*Dermestes maculatus*) can feed on living poultry and has caused deep wounds in adult turkeys. In laboratory experiments, lesser mealworm beetles killed snakes and a salamander, all of which were devoured by the mealworms. The voracious and aggressive behavior of this commonly abundant tenebrionid makes it a significant pest in poultry houses.

In addition to their direct attacks on birds, the lesser mealworm and hide beetle larvae are major causes of structural damage to poultry houses. After reaching their final instar, the larvae migrate into the insulation of poultry houses to seek pupation sites. The larval tunnels and holes produced in insulation and wood framing cause enough damage to alter temperature regulation in the houses, which reduces the efficiency of poultry production (Axtell and Arends, 1990).

Dung Beetles and Biocontrol

Many beetles that are attracted to avian and mammalian excrement should be viewed as beneficial insects. Scarabaeid beetles (including coprines) remove large quantities of mammalian dung by scattering or burying the material during feeding or reproduction. The rapid removal of dung helps reduce the development of parasitic worms and pestiferous cyclorrhaphan flies that require dung for their survival and reproduction, and also opens up grazing land that would be despoiled by the rotting excrement. Staphylinid beetles and histerid beetles that are attracted to mammalian and avian dung directly reduce muscoid fly populations by feeding on the immature stages of these flies and indirectly, by introducing their phoretic mites that prey upon fly eggs in the excrement.

Dung beetle diversity is greatest in tropical regions, such as Africa, with its abundance of herbivores. More than 2,000 scarabaeid species in many genera (e.g., *Onthophagus, Euoniticellus,* and *Heliocopris*) are known to feed and reproduce in dung in Africa. Less diverse dung-feeders, such as *Aphodius, Onthophagus, Canthon,* and *Phanaeus* species, provide the same benefits in the United States. In Australia, the development of extensive cattle farming resulted in the production of millions of tons of dung that was not naturally removed, because the native coprophagous beetles were adapted to feeding only on marsupial dung. Within the last few decades, introductions of African beetles by sterile breeding programs have established several coprine species that have helped to open up grazing lands ruined by dung accumulation, and to reduce the breeding source of the pestiferous bush fly (*Musca vetustissima*).

Staphylinid beetles of several species in the genus *Philonthus* feed as both larvae and adults on fly larvae living in animal excrement. These beetles are maintained as components in biological control programs against the face fly and horn fly. Staphylinid species of *Aleochara* are also helpful in reducing dung-breeding fly populations because the parasitoid larvae of these beetles penetrate fly puparia and destroy the fly pupae. Histerid beetles, especially *Carcinops* species, are found in confined animal production facilities, such as poultry houses. The larvae and adults of these beetles feed on eggs and larvae of muscoid flies. Any beetle species observed in animal production facilities should be identified to assess if its presence is beneficial or detrimental to the maintenance of sanitary conditions and animal health.

PREVENTION AND CONTROL

Preventing public health and veterinary problems associated with beetles requires education about which species are harmful. Recognition of meloid, paederine, and oedemerid beetles allows one to immediately wash skin surfaces and eyes that come into contact with the beetles, thereby removing the chemicals that cause dermatitis or inflammation of the eyes. With the exception of the smallest species, beetles that are attracted to lights may be prevented from reaching humans or other animals by using screens and bed netting.

Control of vesicatory beetles occurring in natural and cultivated vegetation can be achieved with pesticides; however, the wide area over which these chemicals must be broadcast generally makes such control impractical. Human exposure can be prevented by combining education about the problem with personal protective measures and removal of extraneous vegetation and decaying organic matter from around agricultural fields and dwellings.

Prevention of blister beetle toxicosis of farm animals involves care in the handling of forage crops. Harvesting hay at times when meloid beetles are rare, such as in late fall in temperate climates, helps prevent contamination of dried, stored forage with dead beetles. Similarly, harvesting alfalfa before it produces the blooms that attract meloids, or raking hay more frequently after it is cut and allowing it to dry longer before it is conditioned or crimped will allow beetles to leave the hay before it is baled.

Preventing and controlling dissemination of pathogens and transmission of helminths of veterinary importance can be achieved by a combination of strict sanitary and cultural practices. Removal of dung and organic waste from animal enclosures, as well as sterilization of manure before it is used as fertilizer, help to interrupt the transmission cycle of parasites by reducing the chances of beetles ingesting worm eggs. Rotation of pastured animals also can limit contact between the definitive hosts and intermediate beetle hosts. Increased abundance of scarabaeid dung beetles that aids in the rapid removal of dung, by both ingestion and burial, has been found beneficial in reducing infestations with intestinal nematodes that do not use beetles as intermediate hosts, but that are transmitted from animal to animal by dung ingestion (Fincher, 1975).

Control of destructive poultry-house beetles requires constant monitoring for the insects and strict sanitation. Pesticides only provide temporary control and are most beneficial when applied to soil that larvae may burrow in to pupate. Careful personal hygiene, use of gowns and masks in beetle rearing facilities, and regular vacuuming of floors, floor coverings, and furniture in domestic settings help prevent exposure to dermestids and other beetles that can cause allergic responses.

REFERENCES AND FURTHER READING

Alexander, J. O. (1984). *Arthropods and Human Skin.* Berlin: Springer-Verlag.

Archibald, R. G., & King, H. H. (1919). A note on the occurrence of a coleopterous larva in the urinary tract of man in the Anglo-Egyptian Sudan. *Bulletin of Entomological Research, 9,* 255–256.

Arnett, R. H., Jr. (1984). *The false blister beetles of Florida (Coleoptera: Oedemeridae).* Fla. Dept. Agric. & Consumer Serv, Entomology Circular 259.

Arnett, R. H., Jr., & Thomas, M. C. (2001). *American Beetles, Vol. 1—Archostemata, Myxophaga, Adephaga, Polyphaga: Staphyliniformia.* Boca Raton: CRC Press.

Arnett, R. H., Jr., Thomas, M. C., Skelley, P. E., & Frank, J. H. (2002). *American Beetles, Vol. 2—Polyphaga: Scarabaeoidea through Curculionoidea.* Boca Raton: CRC Press.

Ashe, J. S., & Timm, R. M. (1987). Predation by and activity patterns of 'parasitic' beetles of the genus *Amblyopinus* (Coleoptera: Staphylinidae). *Journal of Zoology, London, 212,* 429–437.

Avancini, R. M. P., & Ueta, M. T. (1990). Manure breeding insects (Diptera and Coleoptera) responsible for cestoidosis in caged layer hens. *Journal of Applied Entomology, 110,* 307–312.

Axtell, R. C., & Arends, J. J. (1990). Ecology and management of arthropod pests of poultry. *Annual Review of Entomology, 35,* 101–126.

Bailey, W. S., Cabrera, D. J., & Diamond, D. L. (1963). Beetles of the family Scarabaeidae as intermediate hosts for *Spirocerca lupi. Journal of Parasitology, 49,* 485–488.

Barr, A. C., Wigle, W. L., Flory, W., Alldredge, B. E., & Reagor, J. C. (1998). Cantharidin poisoning of emu chicks by ingestion of *Pyrota insulata. Journal of Veterinary Diagnostic Investigation, 10,* 77–79.

Barrera, A. (1969). Notes on the behaviour of *Loberopsyllus traubi,* a cucujoid beetle associated with the volcano mouse, *Neotomodon alstoni* in Mexico. *Proceedings of the Entomological Society of Washington, 71*, 481–486.

Bellas, T. E. (1989). *Insects as a cause of inhalational allergies: a bibliography 1900–1987.* Canberra: CSIRO Div. Entomol.

Berkov, A., Rodriguez, N., & Centeno, P. (2008). Convergent evolution in the antennae of a cerambycid beetle, *Onychocerus albitarsis,* and the sting of a scorpion. *Naturwissenschaften, 95,* 257–261.

Blodgett, S. L., & Higgins, R. A. (1990). Blister beetles (Coleoptera: Meloidae) in Kansas alfalfa: Influence of plant phenology and proximity to field edge. *Journal of Economic Entomology, 83,* 1042–1048.

Blume, R. R. (1985). A checklist, distributional record, and annotated bibliography of the insects associated with bovine droppings on pastures in America north of Mexico. *Southwestern Entomologist Supplement, 9,* 1–55.

Bryant, J., & Maslan, A. M. (1994). Carpet beetle larval parts in Pap smears: Report of two cases. *Southern Medical Journal, 87,* 763–764.

Capinera, J. L., Gardner, D. R., & Stermitz, F. R. (1985). Cantharidin levels in blister beetles (Coleoptera: Meloidae) associated with alfalfa in Colorado. *Journal of Economic Entomology, 78,* 1052–1055.

Cheng, T. C. (1973). *General Parasitology.* New York: Academic Press.

Chu, G. S. T., Palmieri, J. R., & Sullivan, J. T. (1977). Beetle-eating: A Malaysian folk medical practice and its public health implications. *Tropical and Geographical Medicine, 29,* 422–427.

Clausen, C. P. (1940). Reprinted 1972. *Entomophagous Insects.* New York: Hafner Publ.

Crook, P. G., Novak, J. A., & Spilman, T. J. (1980). The lesser mealworm, *Alphitobius diaperinus,* in the scrotum of *Rattus norvegicus,* with notes on other vertebrate associations (Coleoptera, Tenebrionidae; Rodentia, Muridae). *The Coleopterists Bulletin, 34,* 393–396.

Crowson, R. A. (1981). *The Biology of the Coleoptera.* London: Academic Press.

De las Casas, E., Harein, P. K., Deshmukh, D. R., & Pomeroy, B. S. (1976). Relationship between the lesser mealworm, fowl pox, and Newcastle disease virus in poultry. *Journal of Economic Entomology, 69,* 775–779.

Dumbacher, J. P., Wako, A., Derrickson, A., Samuelson, S. R., Spande, T. F., & Daly, J. W. (2004). Melyrid beetles (*Choresine*): A putative source for the batrachotoxin alkaloids found in poison-dart frogs and toxic passerine birds. *Proceedings of the National Academy of Sciences, 101,* 15857–15860.

Durden, L. A. (1987). Predator-prey interactions between ectoparasites. *Parasitology Today, 3,* 306–308.

Eisner, T., Conner, J., Carrel, J. E., McCormick, J. P., Slagle, A. J., Gans, C. et al. (1990). Systematic retention of ingested cantharidin in frogs. *Chemoecology, 1*(2), 57–62.

Eschevarria, C. (2006). *Blister beetle poisoning: Cantharidin toxicosis in equines.* Animal Disease Diagnostic Laboratory Purdue newsletter [http://www.addl.purdue.edu/newsletters/2006/Fall/EquineCT.htm]

Essig, E. O. (1942). *College Entomology.* New York: MacMillan Company.

Evans, G. (1975). The *Life of Beetles.* New York: Hafner Press.

Fincher, G. T. (1975). Effects of dung beetle activity on the number of nematode parasites acquired by grazing cattle. *Journal of Parasitology, 61,* 759–766.

Fincher, G. T. (1994). Predation on the horn fly by three exotic species of *Philonthus. Journal of Agricultural Entomology, 11,* 45–48.

Fischer, O. A., Matiova, L., Dvorska, L., Svastova, P., Peral, D. L., Weston, R. T., et al. (2004). Beetles as possible vectors of infections caused by *Mycobacterium avium* species. *Veterinary Microbiology, 102*(3–4), 247–255.

Frank, J. H., & Kanamitsu, K. (1987). *Paederus,* sensu lato (Coleoptera: Staphylinidae): natural history and medical importance. *Journal of Medical Entomology, 24,* 155–191.

Geden, C. J., Stinner, R. F., & Axtell, R. C. (1988). Predation by predators of the house fly in poultry manure: Effects of predator density, feeding history, interspecific interference and field conditions. *Environmental Entomology, 17,* 320–329.

Gorham, J. R. (1991). *Insect and mite pests in food: an illustrated key.* US Department of Agriculture, Agriculture Handbook No. 655.

Hall, M. C. (1929). *Arthropods as intermediate hosts of helminths.* Smithsonian Miscellaneous Collections 81.

Hanski, I., & Cambefort, Y. (1991). *Dung Beetle Ecology.* Princeton: Princeton Univ. Press.

Harde, K. W. (1984). *A Field Guide in Colour to Beetles.* London: Octopus Books Ltd.

Kellner, R. L. (2002). Molecular identification of an endosymbiotic bacterium associated with pederin biosynthesis in *Paederus sabaeus* (Coleoptera: Staphylinidae). *Insect Biochemistry and Molecular Biology, 32*(4), 389–395.

Lamson, G. H., Jr. (1922). The rose chafer as a cause of death of chickens. *Storrs Agricultural Experiment Station Bulletin, 110,* 118–135.

Lawrence, J. F., & Britton, E. B. (1994). *Australian Beetles.* Carlton, Victoria: Melbourne Univ. Press.

Legner, E. F. (1995). Biological control of Diptera of medical and veterinary importance. *Journal of Vector Ecology, 20,* 59–120.

Ligett, H. (1931). Parasitic infestations of the nose. *Journal of the American Medical Association, 96,* 1571–1572.

Marshall, A. G. (1981). *The Ecology of Ectoparasitic Insects.* London: Academic Press.

Mattuck, D. R., & Fehn, C. F. (1958). Human ear invasions by adult scarabaeid beetles. *Journal of Economic Entomology, 51,* 546–547.

McAllister, J. C., Steelman, C. D., & Skeeles, J. K. (1994). Reservoir competence of the lesser mealworm (Coleoptera: Tenebrionidae) for *Salmonella typhimurium* (Eubacteriales: Enterobacteriaceae). *Journal of Medical Entomology, 31,* 369–372.

Patton, W. S., Evans, A. W. (1929). Insects, Ticks, Mites, and Venomous Animals. School of Tropical Medicine, Liverpool.

Qadir, S. N. R., Raza, N., & Rahman, S. B. (2006). Paederus dermatitis in Sierra Leone. *Dermatology Online Journal, 12*(7), 9 [http://dermatology.cdlib.org/127/case_reports/paederus/qadir.html].

Rajapakse, S. (1981). Letter from Sri Lanka: Beetle marasmus. *British Medical Journal, 283,* 1316–1317.

Samish, M., Argaman, Q., & Perlman, D. (1992). The hide beetle, *Dermestes maculatus* DeGeer (Dermestidae), feeds on live turkeys. *Poultry Science, 71,* 388–390.

Schmitz, D. G. (1989). Cantharidin toxicosis in horses. *Journal of Veterinary Internal Medicine, 3,* 208–215.

Schroeckenstein, D. C., Meier-Davis, S., & Bush, R. K. (1990). Occupational sensitivity to *Tenebrio molitor* Linnaeus (yellow mealworm). *Journal of Allergy and Clinical Immunology, 86,* 182–188.

Selander, R. B. (1988). An annotated catalog and summary of bionomics of blister beetles of the genus *Cylindrothorax* (Coleoptera: Meloidae). *Transactions of the American Entomological Society, 114,* 15–70.

Southcott, R. V. (1989). Injuries from Coleoptera. *Medical Journal of Australia, 151,* 654–659.

Sweetman, H. L. (1965). *Recognition of Structural Pests and their Damage.* Dubuque: Wm. C. Brown Co.

Théodoridès, J. (1950). The parasitological, medical and veterinary importance of Coleoptera. *Acta tropica, 7,* 48–60.

Waterhouse, D. F. (1974). The biological control of dung. *Scientific American, 230,* 100–109.

Weatherston, J., & Percy, J. E. (1978). Venoms of Coleoptera. In S. Bettini (Ed.), *Arthropod Venoms* (pp. 511–554). Berlin: Springer-Verlag.

White, R. E. (1983). *A Field Guide to the Beetles of North America.* Boston: Houghton Mifflin Co.

Whitmore, R. W., & Pruess, K. P. (1982). Response of pheasant chicks to adult lady beetles (Coleoptera: Coccinellidae). *Journal of the Kansas Entomological Society, 55,* 474–476.

Woodruff, R. E. (1973). *Scarab beetles of Florida (Coleoptera: Scarabaeidae), Part 1. Arthropods of Florida* (Vol. 8). Gainesville: Division of Plant Industry, Florida Department of Agriculture.

Chapter

9

Fleas (Siphonaptera)

Lance A. Durden . Nancy C. Hinkle

Fleas are morphologically unique ectoparasites that are unlikely to be confused with any other arthropods. They are a monophyletic group that has evolutionary ties with the mecopteroid insect orders Mecoptera (scorpion flies) and Diptera (true flies); one recent study suggests that the mecopteran family Boreidae is the sister group of the Siphonaptera. Fleas evolved from winged ancestors during the late Jurassic or early Cretaceous 125 to 150 million years ago in parallel with marsupial and insectivore hosts. Whiting et al. (2008) provide a molecular phylogeny for fleas. As a group, fleas have evolved principally as parasites of mammals on which about 94% of known species feed, representing 15 families and more than 200 genera of fleas. The remaining approximately 6%, representing five families and 25 genera, are ectoparasites of birds.

Cospeciation or coevolution has molded many host-flea associations, as reflected by their host specificity and the morphological adaptations of some fleas that conform to the morphology of the host skin, fur, or feathers. Although many flea species do not cause significant harm to their hosts in nature, most species that feed on humans and their companion animals are of medical or veterinary importance.

TAXONOMY

There are approximately 2,500 species and subspecies of fleas that currently are placed in 15 families and 220 genera (Lewis, 1993, 1998). Many of these species have been catalogued by Hopkins and Rothschild (1953–1971), Mardon (1981), Traub et al. (1983), and Smit (1987). Except for the work by Traub et al., these works are part of an eight-volume series published by the British Museum (Natural History), now The Natural History Museum, London. Another series of publications that addresses the geographical distribution, host preferences, and classification of the world flea fauna are those of Lewis (1972, 1973, 1974a, 1974b, 1974c, 1974d, 1993a) and Lewis and Lewis (1985). A publication by Ewing and Fox (1943) on North American fleas is largely outdated, and no modern text covering the fleas of this region has been published. However, Holland (1985) has produced an excellent guide to the fleas of Canada, Alaska, and Greenland that can be used for those species that also occur in the continental United States. An earlier work by Fox (1940) and a key by Benton (1983) are useful for identifying fleas from the eastern United States; Benton (1980) also has provided an atlas outlining the distribution of the fleas of this region. An older work by Hubbard (1947) addresses the fleas of western North America whereas Lewis et al. (1988) provide a guide to the fleas of the Pacific Northwest. An updated series of identification guides for North American fleas has been initiated (Lewis and Lewis, 1994; Lewis, 2000, 2002, 2003, 2008a, 2008b, 2008c; Lewis and Galloway, 2001; Lewis and Haas, 2001; Lewis and Jameson, 2002; Lewis and Wilson, 2006). Identification guides to the fleas of Britain (Whitaker, 2007), western Europe (Beaucournu and Launay, 1990) and several other regions are also available. Although flea larvae are usually difficult to assign to genus or species, Elbel (1991) provides a useful guide for identifying the larvae of some flea taxa.

Most fleas of medical or veterinary importance are members of the family Pulicidae, with other important fleas belonging to the Ceratophyllidae, Leptopsyllidae, or Vermipsyllidae. Occasionally, members of other families, notably the Hystrichopsyllidae and Rhopalopsyllidae, also feed on humans and domestic animals. Table 9.1 shows the family-level classification for flea species discussed in this chapter.

Flea classification is based almost exclusively on the chitinous morphology of cleared adult specimens. Male fleas arguably have the most complex genitalia in the animal kingdom, and the morphology of the sclerotized parts of these organs is important in most systems of flea classification. Although various classification schemes have been proposed for fleas, one that is used widely today is detailed by Lewis (1998). In this classification, the order Siphonaptera is divided into 15 families, the larger of which are the Ctenophthalmidae (744 species), Ceratophyllidae (540 species), Leptopsyllidae (346 species), Pulicidae (207 species), Pygiopsyllidae (185 species), Rhopalopsyllidae (145 species), and Ischnopsyllidae (135 species).

Table 9.1 Classification of Flea Species Mentioned in the Text

Family Ceratophyllidae:
- *Ceratophyllus gallinae* (European chicken flea – hen flea in Europe)
- *Ceratophyllus niger* (western chicken flea)
- *Nosopsyllus fasciatus* (northern rat flea)
- *Orchopeas howardi*
- *Oropsylla montana*

Family Ctenophthalmidae:
- *Stenoponia tripectinata*

Family Ischnopsyllidae:
- *Myodopsylla insignis*

Family Leptopsyllidae:
- *Leptopsylla segnis* (European mouse flea)

Family Pulicidae:
- *Cediopsylla simplex* (rabbit flea)
- *Ctenocephalides canis* (dog flea)
- *Ctenocephalides felis* (cat flea)
- *Echidnophaga gallinacea* (sticktight flea)
- *Echidnophaga larina*
- *Echidnophaga myrmecobii*
- *Euhoplopsyllus glacialis*
- *Hoplopsyllus anomalus*
- *Pulex irritans* (human flea)
- *Pulex simulans*
- *Spilopsyllus cuniculi* (European rabbit flea)
- *Tunga monositus*
- *Tunga penetrans* (chigoe)
- *Xenopsylla astia*
- *Xenopsylla bantorum*
- *Xenopsylla brasiliensis*
- *Xenopsylla cheopis* (Oriental rat flea)

Family Pygiopsyllidae:
- *Uropsylla tasmanica*

Family Vermipsyllidae:
- *Dorcadia ioffi*
- *Vermipsylla alakurt* (alakurt flea)

Modified from Lewis, R.E (1993a).

MORPHOLOGY

Adult fleas are small (1–8 mm), wingless, almost invariably bilaterally compressed, and heavily chitinized (Figures 9.1, 9.2). Many species bear one or more combs, or **ctenidia**, each appearing as a row of enlarged, sclerotized spines (Figures 9.1 to 9.3). A comb on the ventral margin of the head is called a **genal ctenidium**, whereas a comb on the posterior margin of the prothorax is called a **pronotal ctenidium**. Additional cephalic or abdominal ctenidia occur in some fleas. Smaller rows of specialized setae or bristles adorn various body regions of many fleas. The nature of the ctenidia and specialized setae often reflect the vestiture (pelage or plumage) or habits of the host, especially in host-specific fleas. They aid in preventing dislodgement of fleas from the hair or feathers of the host. It also has been suggested that ctenidia may protect flexible joints.

An important sensory feature of adult fleas is the **sensilium** (**pygidium** of older works), present on abdominal tergum 9 or 10 (Figure 9.2). This sensory organ aids fleas in detecting air movement, vibrations, and temperature gradients; in some species it also facilitates copulation. It plays an important role in host detection

Figure 9.1 Adult female cat flea. (Courtesy of Joyce Gross)

and in initiating escape responses. Just anterior to the sensilium in most fleas are the stout, paired **antesensilial setae** (= antepygidial bristles) situated on the posterior margin of tergum 7. Many adult fleas, especially those of diurnal hosts, possess well developed noncompound eyes (Figure 9.3), which are actually clusters of ocelli. Eyes are well developed in most adult fleas of medical or veterinary importance (Figures 9.1 to 9.3). Short, clubbed, three-segmented antennae are held inside protective grooves called **antennal fossae** on the sides of the head that prevent antennal damage as the flea moves through the pelage of its host.

The mouthparts of adult fleas are well adapted for piercing and sucking (Figures 2.1D and 9.3). After a suitable feeding site has been located by the sensory labial palps, three slender, elongate structures called stylets (or collectively the fascicle) are used to pierce the host skin. The three stylets consist of two lateral, blade-like maxillary laciniae and the central epipharynx (Figure 2.1D). The laciniae penetrate the host skin and the tip of the epipharynx enters a host capillary. A salivary canal is formed by the closely appressed medial surfaces of the two laciniae. A food canal is formed at the confluence of the laciniae with the epipharynx (Figure 2.1D). Anticoagulants, including the anti-platelet enzyme **apyrase**, other salivary components, and sometimes pathogens, are introduced into the bite wound via the salivary canal while host blood is imbibed through the food canal. In some sedentary fleas, such as sticktights (Pulicidae) and alakurts (Vermipsyllidae) that remain attached to the host for long periods, the mouthparts are elongate and barbed and also function as host-attachment devices.

Internally, the alimentary tract of fleas consists of an anterior pharynx that leads to the elongate esophagus and then to the proventriculus at the junction of the foregut and midgut. The proventriculus is armed with rows of spines that can be drawn together to prevent regurgitation of a blood meal from the midgut. The

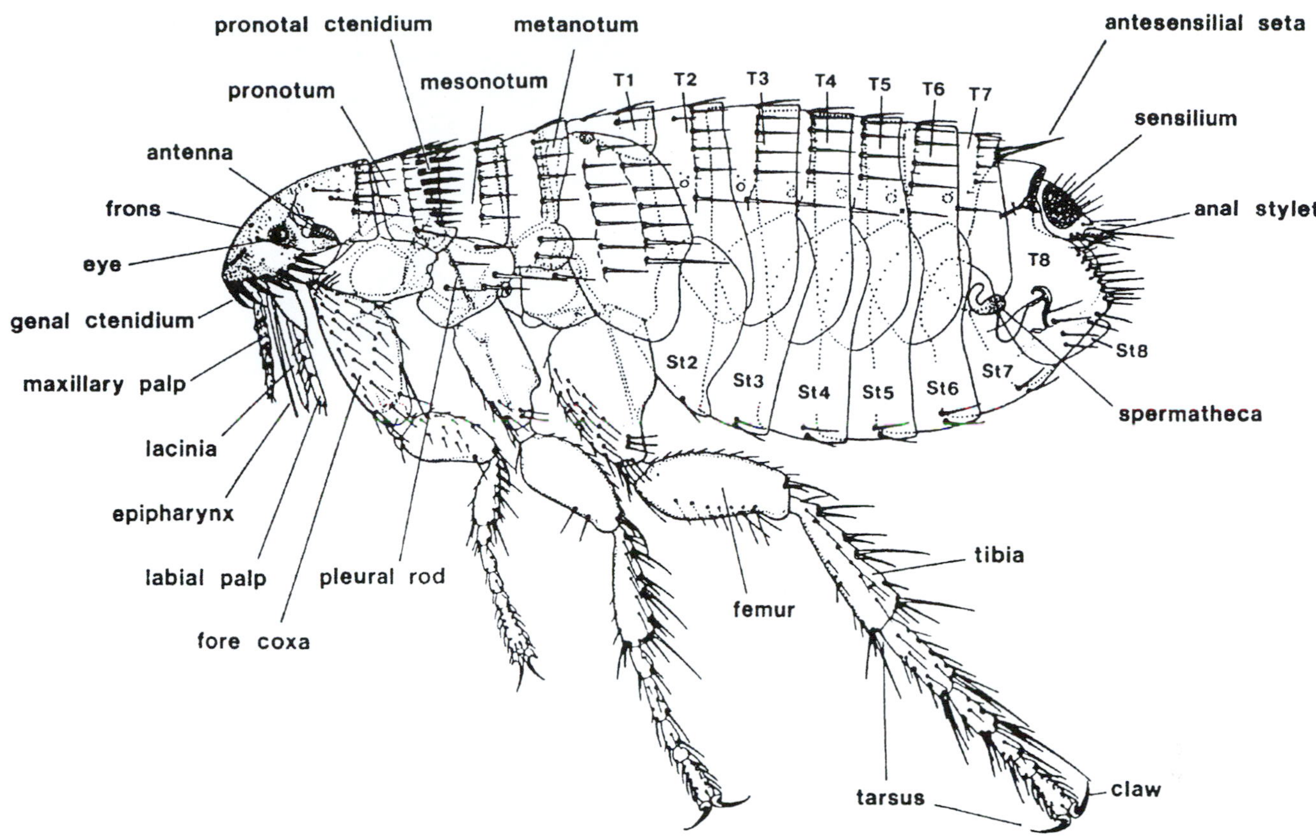

Figure 9.2 Morphology of a generalized adult female flea. (Lewis, 1993b)

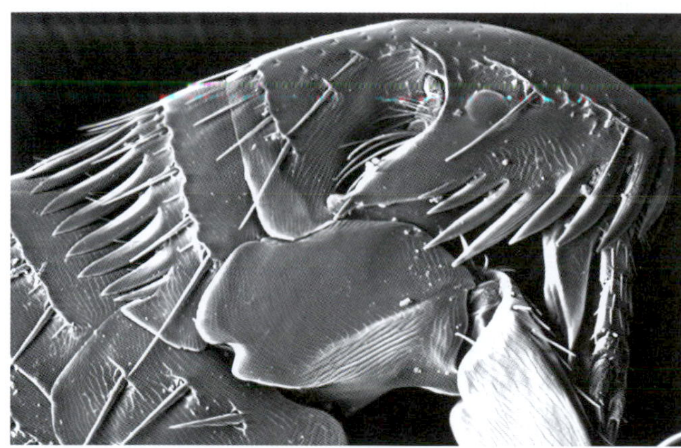

Figure 9.3 Scanning electron micrograph of the lateral aspect of the head of a cat flea (*Ctenocephalides felis*). (Courtesy of Lance A. Durden)

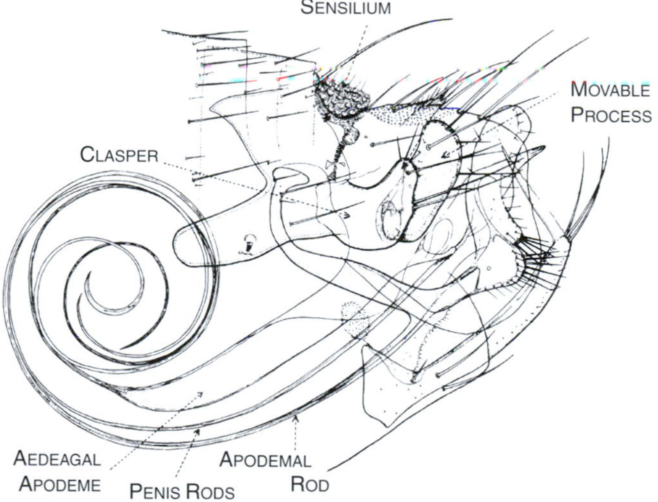

Figure 9.4 Genitalia of a male flea, the European chicken flea (hen flea in Europe) (*Ceratophyllus gallinae*). (Modified from Holland, 1985)

midgut expands to accommodate large blood meals but it lacks distendable diverticula or caeca. Many fleas imbibe larger volumes of blood than their midgut can accommodate and must void blood-rich feces during or soon after feeding. Four excretory malpighian tubules radiate from the junction of the midgut with the hindgut.

Male flea genitalia are morphologically complex (Figure 9.4). The major structures are the claspers, which are used to help secure the female during mating, the often highly specialized aedeagus, and the penis rods, which are partly inserted into the female opening during mating. The major components of the female genitalia are the vagina, the spermathecal duct, and the spermatheca; sperm are stored in the spermatheca following mating. During copulation, the male grasps the female from below using a series of sucker-like discs on his upraised antennae.

Flea eggs are small (0.1–0.5 mm), ovoid, and pearly-white (Figure 9.5). Flea larvae are elongate, legless and eyeless, with numerous stout body setae, especially on their abdominal segments. They possess a well-developed head capsule armed with chewing mandibles (Figure 9.6) and a pair of mandibular silk glands that produce silk for constructing the pupal cocoon. Most flea larvae are small, worm-like, and highly active, with voracious appetites. Although few flea larvae can be identified to species, most can be assigned to the family level based on the arrangement of head papillae (small, finger-like projections), setae, and sense organs.

Figure 9.5 Eggs (ova) and dried blood-rich feces ("flea dirt") excreted and provisioned as larval food by adults of the cat flea (*Ctenocephalides felis*). (Photo by Nathan D. Burkett-Cadena)

Figure 9.6 Larvae of the cat flea (*Ctenocephalides felis*). (Photo by Nathan D. Burkett-Cadena)

Figure 9.7 Pupae of the cat flea (*Ctenocephalides felis*). (Photo by Nathan D. Burkett-Cadena)

Flea pupae are **exarate** (i.e., have externally visible appendages; Figure 9.7) and typically are surrounded by a loose silken cocoon that is secreted by the last larval instar. Because of the sticky nature of this silk, debris from the substrate often adheres to the cocoon and helps to camouflage it (Figure 9.8). Many adult fleas possess an anterior frontal tubercle on the head, which aids them in tearing free from the cocoon during emergence; after use in this manner, the frontal tubercle breaks off in some species.

LIFE HISTORY

Fleas are holometabolous insects, with an egg, larval (typically consisting of three instars), and pupal stage (Figures 9.5 to 9.8). Gravid females of most flea species that have been studied can produce hundreds of eggs during their lifetime. Eggs typically hatch in about five days. Autogeny, or laying fertile eggs prior to ingestion of a blood meal, is not known to occur in fleas. The eggs are sticky and may adhere briefly to the host pelage; however, they usually drop into the host nest or bedding material where they hatch a few days later. Some fleas oviposit directly on leaves or debris in

Figure 9.8 Cocoons of the cat flea (*Ctenocephalides felis*), covered with debris from substrate. (Photo by Nathan D. Burkett-Cadena)

host nests such as females of *Stenoponia tripectinata*, a Palearctic rodent flea.

Most flea larvae feed on organic matter in the nest or bedding materials of their hosts. Adult cat fleas, dog fleas, European rabbit fleas, and representatives of several other species void blood-rich fecal pellets (Figure 9.5) often called **flea dirt** during feeding, which in turn provide a nutritious food source for the larvae. Larvae of the northern rat flea aggressively prod adult fleas until they excrete blood-rich feces that the larvae then ingest. Some flea larvae supplement their diet by feeding on other small arthropods in the host nest, and cannibalism among flea larvae appears to be common.

Duration of the pupal stage usually lasts one to two weeks but is influenced by ambient temperature and host availability. Eclosed adult fleas of several species can remain within the cocoon as pre-emergent adults until suitable host or environmental cues stimulate their emergence. Pre-emergent adult cat fleas can remain quiescent inside cocoons for four or five months to avoid desiccation or other environmental extremes that would kill free-living fleas.

Many fleas, including most of those of medical or veterinary importance, undergo continuous generations under favorable conditions. The cat flea is a good example. Indoors, the generation time for this flea is usually about one month, but can be as short as 20 days. Other fleas, such as alakurts associated with migrating ungulates in Asia and species parasitic on migrating birds, are more likely to pass through just one generation per year in synchrony with host availability. Some fleas, especially in temperate regions, may undergo four or five generations each summer but less or none during the winter. Host availability clearly affects the number of generations in many fleas. Longevity of fleas in the absence of available hosts is greater at low temperatures and high humidity such as during winter in temperate regions. Under optimal conditions, adult fleas of certain species may survive away from the host for more than a year.

Specialized or unusual life cycles have evolved in several fleas, including some species of medical or veterinary significance. Females of the genera *Tunga* and *Neotunga*, for example, burrow into host dermal tissue where they undergo a dramatic size increase (up to 100-fold) accompanied by extensive morphological degeneration. This type of growth, called **neosomy**, involves major integumental chitin synthesis during the adult stage. The genital opening of the female protrudes through the pore in the host skin to facilitate mating with the free-living males; fertilized eggs are likewise extruded through this opening. Because of the great size increase of neosomic females, they are able to produce many relatively large eggs. In some cases this has led to a reduction in the number of larval instars from three to two; further modifications are exhibited by *Tunga monositus*, a parasite of New World rodents, in which neither larval instar feeds. The chigoe (*Tunga penetrans*) is an important human parasite that belongs to this group of fleas.

Adult females of a few fleas oviposit randomly into the environment where the resulting larvae must search for organic matter suitable to eat. Examples are vermipsyllid fleas in the genera *Vermipsylla* and *Dorcadia*, called **alakurts**, that feed on large ungulates, remaining attached for several days. At the other end of the spectrum, females of *Uropsylla tasmanica* cement their eggs to the fur of their Australian hosts (dasyurid marsupials) and the larvae burrow into host skin where they subsist as subdermal parasites. The mature *U. tasmanica* larva drops to the ground where it spins a cocoon and pupates in a manner typical of most other fleas. Larvae of the hare-infesting *Euhoplopsyllus glacialis* are ectoparasitic, and those of some other fleas feed on host carcasses or even on the superficial tissues of moribund hosts.

BEHAVIOR AND ECOLOGY

Fleas have evolved a plethora of specialized behaviors and ecologies to locate and exploit their hosts (Krasnov, 2008). Host-finding behavior is extremely important for adult ectoparasites such as fleas in which the immature stages typically occur off the host. Important stimuli used by fleas for host location include host body warmth, air movements, substrate vibrations, sudden changes in light intensity, and odors of potential hosts or their products (e.g., carbon dioxide, urine). The sensilium, antennae, and eyes are important organs used by fleas to detect potential hosts. In cases in which adult fleas emerge from their cocoons in close association with their host, locating a food source is not difficult. However, fleas of other groups of hosts such as ungulates or migrating birds typically must employ more elaborate strategies for this purpose. These include jumping toward dark or moving objects and moving toward warmth and CO_2 sources.

Some fleas are stimulated to emerge from their pupal case and cocoon by mechanical compression and vibrational stimuli, which often indicates the presence of a potential host. This response is especially noticeable in flea-infested human premises that have been vacated temporarily for weeks or months. When humans or pets return to the premises, these stimuli are largely responsible for synchronized emergences of adult fleas from their cocoons.

Although some fleas spend much of their adult lives in the host pelage, most species visit the host principally to feed. In fact, some nidicolous (nest-associated) fleas (e.g., *Conorhinopsylla*, *Megarthroglossus*, and *Wenzella* spp.) spend very little time on the active host, living instead in crevices in the nest and feeding when the host is asleep. Nidicolous habits have evolved in several families of fleas.

Once a flea locates a host, feeding is initiated by cues such as body warmth, skin secretions, and host odors. Sensory structures of the maxillary and labial palps aid in selecting a feeding site. The labium and labial palps then guide the stylet-like mouthparts into the host skin. Most fleas are capillary feeders (solenophages); when the tip of the stylet bundle pierces a capillary, feeding is facilitated by contraction of powerful cibarial and pharyngeal muscles. Many fleas possess mutualistic microorganisms in their midgut that aid in digestion of the blood meal.

Mating behavior in most fleas that have been studied follows a distinct sequence of events. When the male and female approach one another, the male touches the female with his maxillary palps and his antennae become erect. The male then moves behind the female, lowers his head, and pushes his body beneath hers while grasping her with his antennae using sucker-like discs along the inner antennal surfaces. Next, the male raises the apex of his abdomen, partially secures the female with his claspers, and extrudes his penis rods and/or aedeagus to initiate copulation. Sperm deposited into the female are stored in her spermatheca until her eggs are ready for fertilization.

Locomotory behavior in adult fleas usually involves walking or running on the substrate or through host pelage. However, jumping is the mode of locomotion for which fleas are best known; this provides both an important means of escape and a way to reach hosts. Fleas jump using a modification of the flight mechanism of their winged ancestors. In addition to using muscles derived from subalar and basalar flight muscles, they have retained the wing-hinge ligaments that have been displaced mid-laterally due to lateral compression of the flea body. The jump is not propelled by direct muscle action, but rather by the sudden expansion of discrete pads of a highly elastic protein in the pleural arch called **resilin**. This remarkable protein can store and release energy more efficiently than any synthetic rubber and more quickly than any muscle tissue. The properties of resilin are unaffected by temperature, enabling fleas to jump even in subfreezing conditions.

Prior to jumping, the flea typically crouches, compresses its resilin pads, and keeps them compressed using one or more catch mechanisms. At "take-off" the tergo-trochanteral depressor muscle relaxes to release the catch, allowing the resilin pads to expand and rapidly transfer energy to the hind legs. This results in an acceleration of about 200 gravities, catapulting fleas of some species more than 30 cm in about 0.02 second. While airborne, the flea somersaults holding its middle or hind legs aloft to use as grappling hooks for snagging a host or the substrate. After landing, the muscles are rapidly readjusted in preparation for another jump. By repeating this action, the Oriental rat flea can make up to 600 jumps per hour for 72 hours without rest.

Nest-associated fleas typically have reduced jumping abilities because they have less resilin in the pleural arch and have undergone secondary atrophy of jumping muscles. This appears to be adaptive in ensuring that these fleas do not leap out of a nest into an unfavorable environment.

Flea populations may be naturally regulated in several ways. Hosts are often efficient groomers and are able to significantly reduce flea populations on their bodies. Cats, for example, have been shown to remove up to 18% of their fleas within 24 hours. Natural predators such as certain mesostigmatid mites, pseudoscorpions, beetles, ants, and other arthropods feed on fleas, especially the immature stages in host nests, thereby decreasing their numbers. Various parasites also contribute to flea mortality. These include the plague bacillus, *Yersinia pestis*, the protozoan *Nosema pulicis*, the nematode *Steinernema carpocapsae*, and parasitoids such as the pteromalid wasp *Bairamlia fuscipes*.

Environmental factors are often important in determining the abundance of fleas in different habitats or geographical regions. These factors typically are related to climate, weather, or soil conditions such as relative humidity, temperature, and soil moisture content. Favorable environmental conditions such as host abundance and availability, plentiful food for larvae, high relative humidity, and mild temperatures, promote high populations of many flea species. Because the immature stages typically occupy different niches from adult fleas, the ecological requirements of one or more of the immature stages, rather than of the adult, may be limiting factors that do not permit a species to become established or abundant under certain conditions.

Hormones can play an important role in synchronizing the development of fleas with that of their hosts. The life cycle of the rabbit fleas *Spilopsyllus cuniculi* and *Cediopsylla simplex*, for example, is mediated by host hormones imbibed with the host blood. These fleas can reproduce only after feeding on a pregnant doe. In this way, the emergence of adult fleas is synchronized with that of a litter of rabbits. Reproductive hormones (corticosteroids and estrogens) in the blood of the pregnant doe stimulate maturation of the ovaries and oocytes in feeding female fleas and testicular development in the males. The adult fleas are ready to mate when the rabbit litter is born. Flea mating and oviposition occurs after they have transferred onto the newborn young. The resulting flea larvae feed on organic matter in the nest debris. The next generation of adult fleas appears 15 to 45 days later, in time to infest the host littermates before they disperse from the burrow.

FLEAS OF MEDICAL-VETERINARY IMPORTANCE

Human Flea (*Pulex irritans*)

This flea (Figure 9.9C) will feed on humans and is capable of transmitting pathogens of medical importance. However it is more commonly an ectoparasite of swine and domestic dogs in most parts of the world. Although *P. irritans* is currently an infrequent parasite of humans in developed countries, this has not always been the case. *Pulex irritans* has a patchy but cosmopolitan distribution and often occurs in remote and isolated areas. Adults of this species lack both genal and pronotal ctenidia (Figure 9.9C). *Pulex simulans* is a closely related species that parasitizes large mammals, including wild canids and domestic dogs, and sometimes people, in the New World. Older records (before 1958) from this region are unreliable and could refer to either *P. irritans* or *P. simulans*.

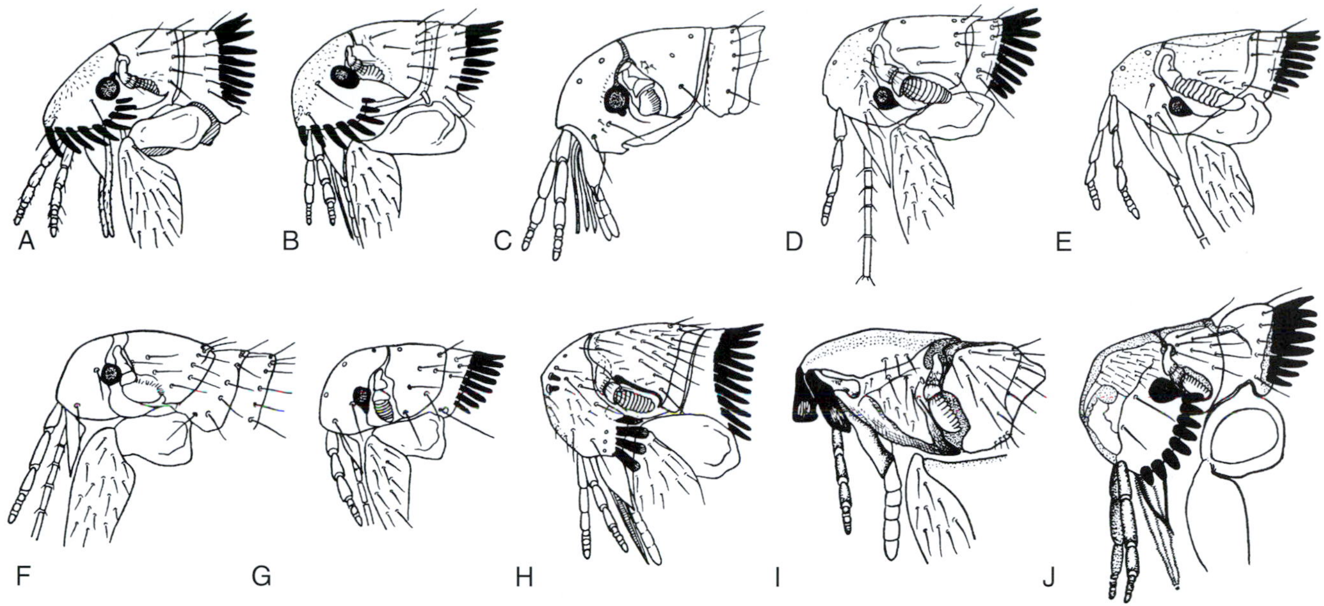

Figure 9.9 Morphology of the head and prothorax of representative adult fleas of medical and veterinary importance. (A) Cat flea (*Ctenocephalides felis*); (B) dog flea (*Ctenocephalides canis*); (C) human flea (*Pulex irritans*); (D) northern rat flea (*Nosopsyllus fasciatus*); (E) *Oropsylla montana*, a North American rodent flea; (F) Oriental rat flea (*Xenopsylla cheopis*); (G) *Hoplopsyllus anomalus*, a North American rodent flea; (H) European mouse flea (*Leptopsylla segnis*); (I) *Myodopsylla insignis*, a North American bat flea; (J) rabbit flea (*Cediopsylla simplex*). (Matheson, 1950)

Cat Flea (*Ctenocephalides felis*)

The cat flea (Figures 9.1, 9.2, 9.3, 9.9A) occurs worldwide and is currently the most important flea pest of humans and many domestic animals. It is primarily a nuisance because it feeds not only on domestic and feral cats, but also on humans, domestic dogs, and several livestock species. It also parasitizes wild mammals such as opossums and raccoons. This ectoparasite is the most common flea on dogs and cats in most parts of the world. Some strains of the cat flea appear to have adapted to ungulates such as horses or goats. Cases of severe anemia associated with huge numbers of cat flea bites have been recorded for these and other domestic animals.

Female cat fleas in most populations produce larger numbers of fertile eggs if they take their blood meals from cats rather than other host species. Under optimal conditions, a female cat flea can lay about 25 eggs per day for a month, contributing to very high densities of fleas in a relatively short time. Adult cat fleas have well-developed genal and pronotal ctenidia (Figures 9.3, 9.9A) and can be distinguished from the dog flea (*C. canis*) by the longer head and longer first spine in the genal comb in *C. felis*. For further details on the biology of the cat flea, see Dryden (1993) and Rust and Dryden (1997).

Dog Flea (*Ctenocephalides canis*)

This flea (Figure 9.9B) is much less common on domestic dogs in most parts of the world than in previous decades. Instead, the cat flea has become the most common flea on domestic dogs in most regions. No satisfactory explanation for this change has been documented; perhaps cat fleas can outcompete dog fleas under stress from modern pesticide applications. Nevertheless, dog fleas persist worldwide and remain as the predominant fleas on dogs in some countries and regions. Dog fleas also parasitize wild canids such as foxes, coyotes, and wolves, on which they can be relatively common.

Oriental Rat Flea (*Xenopsylla cheopis*)

This flea is the principal vector of the agents of plague and murine typhus in many of the tropical and subtropical parts of the world. Some other species of *Xenopsylla* also are vectors of the plague bacillus. Although it is most common on domestic rats, it also will feed on humans, dogs, cats, chickens, and other hosts especially if rats become scarce. Like the human flea, adults of *X. cheopis* lack both a pronotal and a genal ctenidium (Figure 9.9F).

European Rabbit Flea (*Spilopsyllus cuniculi*)

Originally from Europe, this flea has accompanied its host, the European rabbit, as it has been introduced throughout the world either inadvertently, for food, or as a laboratory animal. It is an example of a sedentary flea; adults attach to the host for long periods using their elongate mouthparts to anchor themselves in host skin and feed. This flea typically attaches to the ears of rabbits, where a rich peripheral blood supply provides easily accessible blood meals. Adults have a genal ctenidium with a row of five blunt spines oriented almost vertically on the head and a well-developed pronotal ctenidium.

Sticktight Flea (*Echidnophaga gallinacea*)

As indicated by its name, this is another sedentary flea. It is distributed globally wherever chickens have been introduced as domestic animals. This flea usually attaches semipermanently around the head (Figure 9.10), especially on the wattle, of chickens. Many additional hosts are also parasitized by *E. gallinacea* including other domestic birds (e.g., turkeys, quail), domestic rats, dogs, cats, and occasionally humans. Adults of this small flea are easily recognized by their sharply angled squarish head and the absence of both pronotal and genal ctenidia.

Chigoe (*Tunga penetrans*)

This flea, also called the **jigger** or **sand flea**, has major medical and veterinary significance because it burrows into tissues of humans and some domestic animals. In addition to being very small (ca. 1 mm in length), the free-living adult chigoe lacks pronotal and genal ctenidia and has a sharply angled head. It is widely distributed in tropical and subtropical regions. The life cycle of the immature stages and the male of *T. penetrans* does not deviate significantly from that of most fleas. Initially, the female is free-living but soon invades host skin. Once embedded, she begins to swell by imbibing host fluids, often expanding about 80-fold to reach the size of a pea after eight to 10 days (Figures 9.11 through 9.13). She maintains an opening to the exterior through which she respires, mates with a free-living male, and expels her eggs. The male possesses the longest intromittent organ relative to body size in the animal kingdom and mates from an inverted position. Eggs usually are expelled onto sandy soils, including coastal beaches (hence the term sand flea), frequented by potential hosts where the larvae complete their development. There are only two larval instars. Development from egg to adult usually takes four to six weeks, but sometimes only three weeks under optimal conditions.

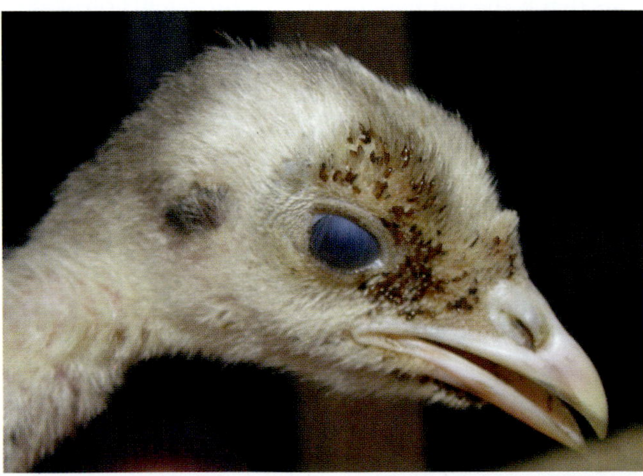

Figure 9.10 Sticktight fleas (*Echidnophaga gallinacea*) attached to the head of a young turkey. (Photo by Phillip Kaufman)

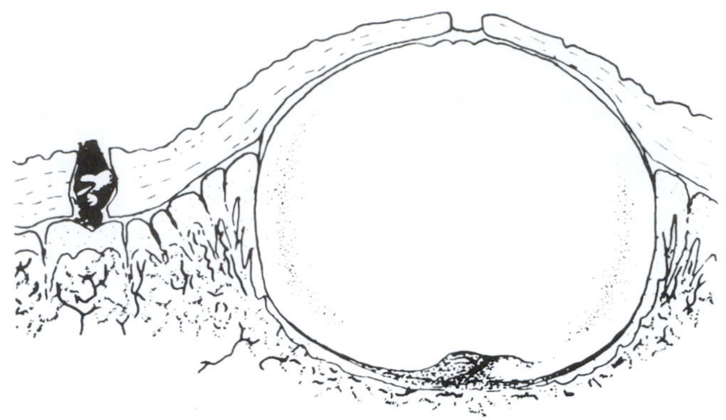

Figure 9.11 Chigoe (*Tunga penetrans*), adult females embedded in host skin; abdominal segments starting to swell in specimen at left; greatly enlarged, gravid female at right. (Harwood and James, 1979)

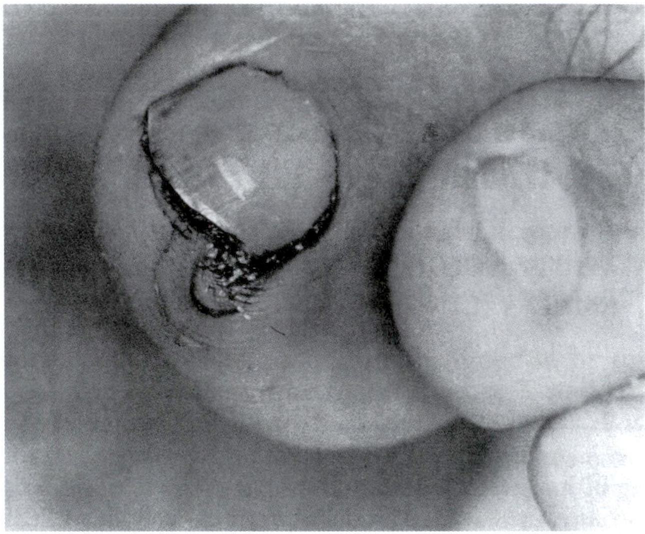

Figure 9.12 Tungiasis; female *Tunga penetrans* embedded in human toe with flea eggs around the lesion. From Ibáñez-Bernal and Velasco-Castrejón (1996).

Northern Rat Flea (*Nosopsyllus fasciatus*)

This is a common flea of domestic rats, especially in temperate and northern regions of the world. Although it will bite humans and can transmit several zoonotic pathogens such as the agents of plague and murine typhus, it is far less important in this respect than the Oriental rat flea. Occasionally, it parasitizes other rodents or domestic mammals. Adults possess a well-developed pronotal comb but lack a genal comb (Figure 9.9D).

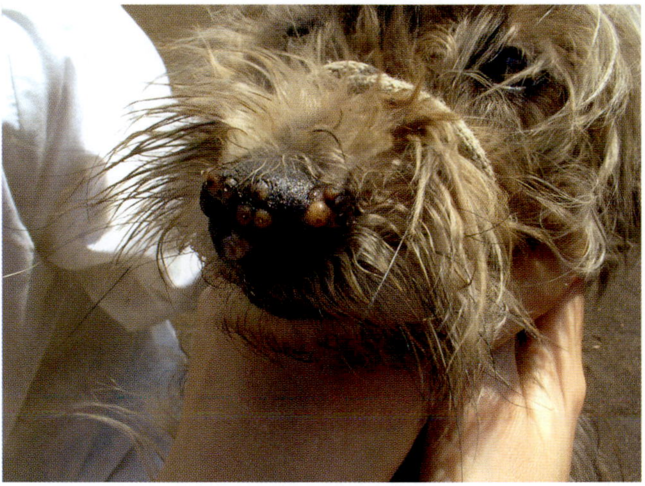

Figure 9.13 Tungiasis; several female *Tunga penetrans* embedded in a dog's snout. (Chiebao et al., 2005)

European Chicken Flea (Hen Flea in Europe) (*Ceratophyllus gallinae*)

This ectoparasite of feral and domestic birds, especially chickens, originated in Europe but has spread with poultry operations throughout the world. Because *C. gallinae* can feed on so many bird species, it is often difficult to completely eradicate. In contrast to *E. gallinacea*, this flea is highly mobile on the host and can be especially common in host nesting material. Adults have a distinct pronotal ctenidium but lack a genal ctenidium.

European Mouse Flea (*Leptopsylla segnis*)

This cosmopolitan flea typically parasitizes the house mouse (*Mus musculus*), including laboratory colonies. Rarely, large populations of *L. segnis* cause host anemia or other problems in mouse rearing facilities. Adults of this flea possess a well-developed pronotal ctenidium and a vertical genal ctenidium consisting of four bluntly rounded spines (Figure 9.9H).

PUBLIC HEALTH IMPORTANCE

Fleas can result in several threats to public health. Many species are annoying biters that can cause considerable discomfort, sometimes leading to secondary infections of bite wounds. The bites of some species can cause dermatitis or allergic reactions. Allergic responses also can result from contact with, or inhalation of, flea products (e.g., larval exuviae). Females of the chigoe actually invade human skin tissue, especially of the feet and toes, and cause painful lesions that are prone to serious secondary infection. Other fleas are intermediate hosts of tapeworms that can parasitize humans. Fleas also serve as vectors of the causative agents of several important zoonotic diseases such as murine typhus and plague (Table 9.2).

Flea bites (Figure 9.14) can cause intense irritation for several days. Bites are characterized by a tiny purplish spot, or **purpura pulicosa**, surrounded by slightly swollen skin called **roseola pulicosa**. The vast majority of flea bites experienced by humans are due to the cat flea. This flea is an unrelenting biter that generally attacks humans on the ankles (Figure 9.14), although

Table 9.2 Pathogens Transmitted by Fleas

Disease Agent	Disease	Vector(s)	Host(s)	Geographic Area
VIRUS:				
Myxoma virus	Myxomatosis	*Spilopsyllus cuniculi*	Rabbits	Europe, Australia
BACTERIA:				
Coxiella burnetii	Q fever	Several fleas	Mammals	Global
Francisella tularensis	Tularemia	Several fleas	Mammals	Global
Rickettsia typhi	Murine typhus	*Xenopsylla, Ctenocephalides*	Mammals	Global
Rickettsia prowazekii	Sylvatic epidemic typhus	*Orchopeas howardi*	Flying squirrels, humans	North America
Yersinia pestis	Plague	Mainly *Xenopsylla*	Humans, rodents, cats	Global
PROTOZOA:				
Trypanosoma lewisi	Murine trypanosomiasis	*Nosopsyllus, Xenopsylla*	Rats	Global
Trypanosoma nabiasi	Rabbit trypanosomiasis	*Spilopsyllus cuniculi*	Rabbits	Global
NEMATODA:				
Acanthocheilonema reconditum	Canine filariasis	*Ctenocephalides*	Carnivores	Global
CESTODA:				
*Dipylidium caninum**	Double-pored tapeworm	*Ctenocephalides*	Dogs, cats, humans	Global
*Hymenolepis diminuta**	Rodent tapeworm	*Nosopsyllus, Xenopsylla*	Rodents, humans	Global
*Hymenolepis nana**	Dwarf tapeworm	*Nosopsyllus, Xenopsylla*	Rodents	Global

*Fleas are not vectors for these pathogens but instead serve as intermediate hosts.

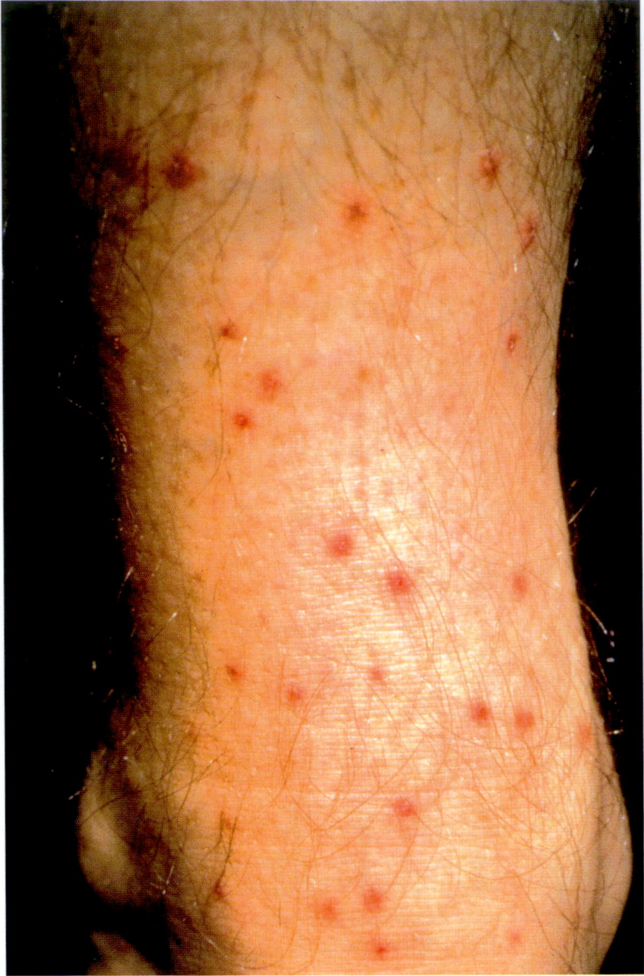

Figure 9.14 Multiple cat flea bites on a human ankle. (Photo by Elton J. Hansens)

other parts of the body may be affected. Women tend to be more commonly bitten than men, suggesting a hormonal association between this species and human females. In addition to the annoyance it causes, *C. felis* is a proven vector of the causative agent of murine typhus. The cat flea is discussed in more detail with respect to its veterinary importance.

The human flea also is an annoying biter of people in various parts of the world. The closely related *P. simulans* sometimes infests households and causes dermatitis in humans in western North America. Several other species of fleas may bite humans to the point of annoyance including the dog flea, the sticktight flea, the northern rat flea, and several species of *Xenopsylla*, including the Oriental rat flea (Table 9.1). Also, the squirrel flea, *Orchopeas howardi*, and some bird fleas belonging to the genus *Ceratophyllus*, including the European chicken flea, occasionally bite humans.

Members of households and adjacent premises harboring pets or domestic rodents can be especially prone to flea bites. Cat fleas and dog fleas readily bite humans, especially if flea populations are large or if pets are removed temporarily. Rodents inhabiting households also can be a source of fleas that can bite humans. Fleas typically abandon dead hosts; if domestic rats die in wall voids, basements, or other poorly accessible structures, their fleas may seek human hosts. A similar situation can occur involving squirrel fleas, whose hosts often nest in attics or eaves of houses. Basements, garbage, and pet food supplies and feeding bowls also can attract scavenging mammals such as opossums, raccoons, and skunks in North America. These animals may leave behind cat fleas and other fleas that will bite humans. Flea bites represent occupational hazards on many farms, in barns, rabbit hutches, and poultry operations. The fleas in such cases may come directly from livestock or indirectly from domestic rodents attracted to food supplies.

Flea-Associated Allergies

Flea-bite dermatitis usually occurs in persons who have become hypersensitive to flea saliva. In sensitized individuals, bite sites typically develop into papules, causing a form of papular urticaria, often with associated wheals, especially in children. In more serious cases, skin scaling, hardening, or discoloration can occur. In adults, it is usually the distal extremities (hands and feet) that are involved, whereas in children the entire body may be affected. With time and repeated exposure to flea bites, hyposensitization may reduce the severity of the dermatitis without medical intervention. The administration of corticosteroids or desensitizing antigens can be helpful for some hypersensitive individuals.

People can also become sensitized to flea feces and particles of exoskeletons upon contacting or inhaling them in house dust. Adult fleas have been identified as the source of some of these allergens. Airborne larval exuviae also have been implicated as causes of asthmatic symptoms. Relief from these allergic responses may be achieved by administering a course of desensitizing antigens to the patient.

Plague

Plague is caused by infection with *Yersinia pestis*, a gram-negative coccobacillus bacterium; the entire genome of *Y. pestis* was sequenced in 2001. The disease also is referred to as the black death, and in Francophone countries as *la peste*. The organism was first isolated in 1894 when Swiss bacteriologist Alexandre Yersin cultured it from sick patients at the Pasteur Institute in Hong Kong. Although plague is thought to have originated in Asian gerbils (a recent theory suggests it may have originated in Nile rats in Egypt), it is typically maintained in urban areas in peridomestic rodents, especially the black rat (*Rattus rattus*) and Norway rat (*Rattus norvegicus*). The pathogen is transmitted to these rodents by fleas, especially the Oriental rat flea and other members of the genus *Xenopsylla*. The role of *X. cheopis* as a vector of *Y. pestis* was definitively demonstrated by French physician/bacteriologist Paul-Louis Simond during his plague studies in

Pakistan (then part of India) in 1898. Plague is the most significant flea-borne disease in human and wild mammal populations.

Plague pandemics have had a significant impact on human civilization and have claimed more human lives than all wars ever fought. At least three major pandemics have been documented. The first, sometimes called **Justinian's plague** (named for the Roman emperor Justinian), originated in AD 541 in Africa and later spread throughout Mediterranean Europe killing an estimated 40 million people during the sixth and seventh centuries.

The second pandemic, usually referred to as the **black death**, originated in central Asia in the fourteenth century and spread to Europe as a consequence of developing trade routes between these two continents. In 1347, crew members of an Asian trading vessel that docked in Sicily were suffering from a mysterious disease that later was identified as plague. Over the next five years, plague spread throughout most of Europe from this point of entry, with devastating consequences; by 1352, at least 25 million people had died in Europe alone. This pandemic continued for more than 200 years, with the disease appearing or reappearing in different areas of Europe. London experienced a major epidemic in 1348 followed by another in 1665, with foci of the disease persisting in the city throughout this time period. Ancient *Y. pestis* DNA has been detected in tooth pulp from human victims in plague burial pits in Europe (Drancourt et al., 2004, 2007).

The third major pandemic of plague spread across the globe in the late 1800s after originating from a focus in China's Yunnan Province in 1855. From 1896 to 1948, this pandemic accounted for 12 million deaths in India alone. Several plague foci initiated during the peak of this pandemic still persist today. In the United States for example, plague bacilli were introduced with infected ship-borne rats in 1899 in San Francisco and from there eventually spread to at least 14 western states and two Canadian provinces. Plague continues to persist in native rodents and fleas in most of these areas of western North America. Accounts of plague include works by Duplaix (1988), Mee (1990), Poland et al. (1994), Madon et al. (1997), Dennis et al. (1999), Carniel and Hinnebusch (2004), Kelly (2004), Christakos et al. (2005), Gage and Kosoy (2005), and Eisen et al. (2007).

Nucleotide sequencing of rRNA in *Y. pestis* shows a correlation between the geographical distribution of genetic strains, or biovars, of *Y. pestis* and their spread during the three pandemics. There are three biovars of *Y. pestis*: biovar Antiqua occurs in Africa; biovar Medievalis mainly in central Asia; and biovar Orientalis in Europe, Asia, Africa, North America, and South America.

Today plague occurs as fairly discrete foci in various parts of Asia, southern and northwestern Africa, South America, and western North America. Recent outbreaks have surfaced in Algeria, Brazil, Democratic Republic of the Congo, Ecuador, India, Iran, Madagascar, Malawi, Mongolia, Peru, South Africa, Tanzania, Vietnam, and Zambia. Globally, nearly 19,000 human cases of plague were reported to the World Health Organization from a total of 20 countries during 1984 to 1994. Currently, 1,000 to 3,000 cases per year are reported, with most human plague deaths occurring in Africa. During 1970 to 1994, a total of 334 cases of indigenous plague were reported for the United States whereas during 1988 to 2002, a total of 112 cases were reported; peak years were 1983 and 1984 with 40 and 31 cases, respectively. Eighty percent of cases in the United States have occurred in Arizona, New Mexico, or Colorado.

In addition to bites from infected fleas, plague infections can result from direct contact with moribund or dead mammals infected with *Y. pestis*, or rarely, from inanimate objects harboring the pathogen. In such cases the pathogen typically enters the body through skin lesions. Inhalation infection also can occur from aerosolized *Y. pestis*.

Two ecological forms of plague are recognized: **urban plague** carried by domestic rats and their fleas in cities and towns, and **wild-rodent plague** (**sylvatic plague**, derived from the Latin *silva* meaning trees; **campestral plague, rural plague**) maintained in several species of mammals (mainly rodents) and their fleas in rural areas away from human populations. Over 200 species of rodents and other mammals (e.g., certain carnivores) may serve as reservoir hosts of wild-rodent plague. In North America, ground squirrels, rock squirrels, chipmunks, and prairie dogs are particularly important, whereas in Asia, gerbils and susliks (ground squirrels) typically fill this role. Similarly, various gerbils and the peridomestic rat *Mastomys natalensis* are important reservoirs in parts of Africa. The short-tailed field mouse, *Zygodontomys brevicauda*, is prevalent in some South American foci. Plague-infected tree squirrels have been found in some towns in the western United States.

In many plague-endemic regions, wild-rodent plague persists enzootically in discrete rodent populations. Under certain conditions, the disease can become epizootic and spread to peridomestic rats to trigger urban plague. Some populations of reservoir hosts are refractory to infection with *Y. pestis* and others are highly susceptible. This is reflected by large-scale die-offs in infected prairie dog (*Cynomys* spp.) towns in North America. Intermediate stages of susceptibility to plague exist between these two extremes in other reservoir populations. In most regions where plague persists, there are distinctly different species of enzootic and epizootic rodent reservoir hosts. Most carnivores, especially felids, are susceptible to infection with *Y. pestis*. The disease is often severe in domestic cats, which can serve as a source of infected fleas to households. Bites, scratches, or inhalation of infectious aerosols from infected cats also can disseminate *Y. pestis*. Human plague cases acquired from domestic cats have increased in recent years in the United States.

Although the Oriental rat flea and other *Xenopsylla* species are important vectors of *Y. pestis*, there are at least 125 species of fleas that are capable of transmitting the pathogen. In North America, several flea species can transmit the plague bacterium to native rodents; *Oropsylla montana* (Figure 9.9E), and perhaps *Hoplopsyllus anomalus* (Figure 9.9G), are the more important among these. In Russia and northern Asia, fleas belonging to the genera *Citellophilus*, *Neopsylla*, and *Ctenophthalmus* are significant enzootic vectors within rodent communities. In addition, the widespread fleas *Pulex irritans* and *Nosopsyllus fasciatus* are capable of transmitting plague bacilli. Because *X. cheopis* survives poorly in cool climates, historical reevaluations suggest that, contrary to former dogma, *P. irritans* rather than *X. cheopis* may have been the principal flea vector of *Y. pestis* in the great plague epidemics of northern Europe.

A susceptible flea typically becomes infected after imbibing plague bacilli in its blood meal from an infected host. The bacterium invades the flea midgut where, under suitable conditions it multiplies rapidly, often culminating in complete blockage of the gut anterior to the proventricular spines. This proventricular blockage (Figure 9.15) results from clumping of the bacteria eight to 51 days after ingestion. Although some gut blockages may clear spontaneously, persistent blockage of the flea gut is central to efficient transmission of *Y. pestis*. Fleas with blockages are incapable of ingesting a blood meal from a host. Feeding attempts by these fleas result in the drawing of blood into the flea esophagus followed by regurgitation of the blood meal into the host. This regurgitation is caused by the elastic recoil action of the esophagus when resistance from the proventricular blockage is reached. Infection results when plague bacilli are regurgitated with the blood meal into the host.

Fleas with gut blockages are hungry and make repeated, aggressive feeding attempts. This can result in the infection of several different hosts and amplification of an epidemic. Unless the gut blockage clears, the infected flea ultimately succumbs to starvation, dehydration, or toxicity from bacterial metabolites, itself becoming a victim of plague. As with most other pathogens transmitted by fleas, plague bacilli do not pass through the gut wall of infected fleas to invade the hemocoel, salivary glands, or other organs.

Key environmental parameters influence the establishment of plague foci throughout the world. For example, *X. cheopis* is principally a denizen of drier habitats, and it is in these zones that the major plague foci have persisted. Where environmental factors tend to keep the number of flea species in a given area low, plague is generally absent or rare. If the ambient temperature exceeds 28 °C, plague-infected fleas often can clear their guts of blockages and the disease does not develop to epidemic proportions.

There are three recognized clinical types of plague infection: bubonic, septicemic, and pneumonic plague. The most common of these is **bubonic plague**, which usually results from the bites of infected fleas but can result from handling infectious mammal carcasses. This type of plague is characterized by grossly enlarged, tender, peripheral lymph nodes called **buboes** (singular, bubo). They usually occur in the axillary or inguinal region (Figure 9.16), and typically are teeming with plague bacilli.

In **septicemic plague**, the pathogen initially bypasses, or overwhelms, the peripheral lymph nodes and invades deeper recesses of the body. Although internal buboes may develop, they are not easily detected. Instead, the bloodstream is invaded rapidly by bacteria and capillary walls start to leak, often turning the skin black. The absence of external buboes to aid diagnosis, coupled with swift invasion of the blood, make this form of plague especially severe, with many patients succumbing to fatal septicemia.

The most life-threatening form of the disease is **pneumonic plague** (**plague pneumonia**), in which patients have a lung infection of *Y. pestis* and can cough or sneeze viable bacteria into the air. Inhalation of *Y. pestis* by susceptible individuals results in the pulmonary form of this disease. Also, bubonic or septicemic plague can progress to pneumonic plague. Without prompt

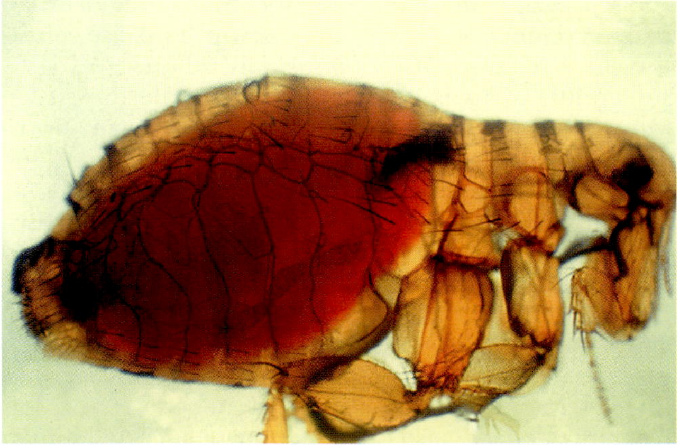

Figure 9.15 Blockage of midgut of Oriental rat flea (*Xenopsylla cheopis*) by mass of *Yersinia pestis*, the causative agent of plague (dark mass in anterior part of the blood-filled midgut). (Courtesy of U.S. Public Health Service, Public Health Image Library)

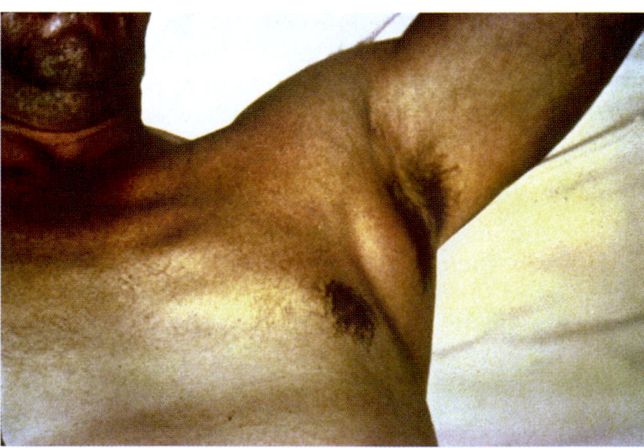

Figure 9.16 Male patient with enlarged axillary lymph node, or bubo, characteristic of bubonic plague. (Courtesy of U.S. Public Health Service, Public Health Image Library)

and aggressive medical attention, pneumonic plague is invariably fatal; some untreated patients die within a day of inhaling the pathogen. The severe pathogenicity of *Y. pestis* is caused largely by endotoxins and exotoxins released by the dividing bacilli.

Flea-transmitted infection typically results in classic bubonic plague, in which buboes develop after an incubation period of two to six days. Accompanying symptoms are severe headache, fever, and shaking chills. Without treatment, most patients deteriorate rapidly with a typical mortality rate of 50 to 60%. Septicemic plague is a particularly dangerous form of the disease because the incubation time is only two to five days and external buboes are absent. Pneumonic plague has a very short incubation period (1–3 days) and may spread rapidly from one victim to another. Overwhelming pneumonia characterized by coughing, bloody sputum, chills, and fever usually results in death within three days unless specific medication is rapidly administered. Pneumonic plague spreads directly, and usually rapidly, from person to person without the involvement of flea vectors. In some cases, humans have contracted pneumonic plague after inhaling aerosolized bacilli expelled by infected household cats.

Accurate diagnosis is important in identifying plague-infected patients. Various biochemical, serological, chromatographic, and staining tests, or lysis with a specific bacteriophage, typically are employed to detect *Y. pestis* or specific antibody directed against it in humans, animals, or fleas. DNA probes and polymerase chain reaction (PCR) techniques used to amplify specific nucleotide sequences of *Y. pestis* are becoming routine screening tools in many public health laboratories. Rapid dipstick tests have been developed for serological field screening.

The treatment of plague patients usually involves immediate hospitalization, isolation, and administration of broad-spectrum antibiotics. Formalin-inactivated plague vaccines are available. They are not, however, totally effective. None are currently protective against pneumonic plague in humans, and most must be administered in multiple doses, or at regular intervals, to ensure protection.

Efforts to control plague typically involve removal of wild-rodent reservoir hosts and their fleas. In areas of potential plague activity, samples of rodent blood and tissues often are collected in order to monitor plague in reservoir host populations. Fleas also can be collected and screened for *Y. pestis.* If samples are positive, then rodent and flea control measures should be considered.

Plague is currently an infectious disease of major concern as a potential bioweapon. Historically, plague was used as a bioweapon when conquering Mongol soldiers hurled plague cadavers into the besieged Crimean city of Caffa in 1346. Further, fleeing infected inhabitants from Caffa appear to have had a role in introducing the Black Death to Europe in 1347 (Wheelis, 2002). Additionally, "flea bombs" (canisters containing plague-infected fleas and dropped from planes) have been used as recently as the 1930s and 1940s.

Murine Typhus

Murine typhus, also known as endemic typhus, Mexican typhus, shop typhus, rat typhus, urban typhus, or flea-borne typhus, is caused by the rickettsial organism *Rickettsia typhi* (formerly *Rickettsia mooseri*). Although this zoonosis typically is maintained in peridomestic rats by flea transmission, humans occasionally are infected. Murine typhus is one of the most prevalent rickettsial diseases of humans, even though it is underdiagnosed and its importance is generally unappreciated. *Rickettsia typhi* is a small, obligate, intracellular bacterium that can cause mild febrile infection in humans. It usually is transmitted via infected flea feces. When the bite site of an infected flea is scratched, rickettsiae from flea feces gain access to the host through abraded skin. Under experimental conditions, however, some fleas also can transmit this pathogen via their bite. Reviews on the ecology and epidemiology of murine typhus have been provided by Traub et al. (1978), Azad (1990), Rawlings and Clark (1994), Azad et al. (1997), and Goddard (1998).

The geographical distribution of murine typhus is almost global. Although it occurs on all continents except Antarctica, its importance as a human pathogen has diminished in recent years. Significant foci persist, however, especially in Indonesia, the People's Republic of China, Thailand, North Africa, and Central America. In the United States, the annual number of human cases has decreased from more than 5,000 in 1945 and 1946 to 20 to 80 per year from 1958 to the present. This zoonosis was formerly widespread throughout the southern and southwestern United States. Currently in the United States it is principally recorded in southern Texas where 200 human cases were diagnosed in the five-year period from 1980 to 1984, and many children are seropositive (Purcell et al., 2007); homeless persons in Houston, Texas, also have been recorded to be seropositive (Reeves et al., 2008). Several human cases usually are reported annually from California and Hawaii as well.

Murine typhus is maintained primarily in a cycle that involves commensal rodents of the genus *Rattus* and their ectoparasites, especially fleas of the genus *Xenopsylla.* Humans typically are infected when feeding fleas void infectious feces on their skin. The black rat (*Rattus rattus*) and Norway rat (*Rattus norvegicus*) are the principal reservoirs of *R. typhi.* Infections also have been recorded in many other mammals including other peridomestic rats (*Rattus* spp.) worldwide, bandicoot rats (*Bandicota* spp.) on the Indian subcontinent, house mice (*Mus musculus*) worldwide, the oldfield mouse (*Peromyscus polionotus*) in the southern United States, the giant pouched rat (*Cricetomys gambianus*) in Africa, the house shrew (*Suncus murinus*) in the Old World, domestic cats worldwide, and the Virginia opossum (*Didelphis virginiana*) in North America. Within the last 30 years, peridomestic mammals such as opossums and feral cats and dogs have been implicated as reservoirs of murine typhus in Texas and southern California. Field infection rates in commensal rats of up to 46% have been reported in Burma

(Myanmar), Egypt, and Ethiopia, and up to 94% in some cities in Texas. New World strains of *R. typhi* are much less virulent (ca. 2% mortality rate in humans) than some Old World strains (ca. 70% mortality rate).

Peridomestic rats almost invariably are the most important reservoirs and amplifying hosts of *R. typhi.* Infection in these rats is not fatal; instead, they display a persistent transient rickettsemia. This is important in extending the period during which ectoparasites, especially fleas, can feed on infective hosts. Because seropositive Virginia opossums have been associated with human cases in some regions of the United States, it appears that opossums also can be important reservoir hosts.

At least 11 species of fleas belonging to nine different genera have been found to be infected with *R. typhi* in nature. *Xenopsylla cheopis* is the most important vector. Other vectors are *Xenopsylla astia, X. bantorum, X. brasiliensis, Ctenocephalides felis, Pulex irritans, Leptopsylla segnis,* and *Nosopsyllus fasciatus.* Except for *C. felis* and *P. irritans,* all these fleas are common ectoparasites of commensal rodents in various parts of the world. Human cases of murine typhus usually coincide with population peaks of *X. cheopis* on rats. The number of cases generally declines or the disease disappears after this flea has been controlled by chemical applications or rodent removal. Infection rates of field-collected *X. cheopis* in hyperendemic regions typically are 50 to 70%.

Infection of a flea occurs when rickettsiae are ingested while the flea is feeding on a host that has *R. typhi* circulating in its blood. The ingested rickettsiae then invade the midgut epithelial cells of the flea and start to replicate. The infection spreads rapidly until most or all of the midgut cells are infected after seven to 10 days. Ultimately, infectious rickettsiae are released from these cells and liberated into the gut lumen, from which they are excreted in the feces. *Xenopsylla cheopis* fleas are typically infective about 10 days after an infectious blood meal and infective fleas can transmit the pathogen for at least another 40 days. Infected fleas survive with a persistent *R. typhi* infection and demonstrate no obvious pathological effects. This contrasts with the related pathogen, *Rickettsia prowazekii,* which causes a fatal infection in its louse vector. Because *X. cheopis* can maintain and transmit *R. typhi* transovarially, this flea may be both a reservoir and a vector of murine typhus rickettsiae.

Although modes of *R. typhi* transmission other than via infected flea feces are known, their significance in nature remains unclear. Because *X. cheopis* has been shown to transmit *R. typhi* by bite in the laboratory, other fleas also may be capable of transmitting *R. typhi* by bite. The possibility of aerosol transmission from infective flea feces has been suggested.

Rickettsia typhi has been detected in ectoparasites other than fleas. Because most of these arthropods do not bite humans, their presumed role is in transmitting *R. typhi* enzootically among commensal rats. Ectoparasites in this category include the sucking lice *Hoplopleura pacifica* and *Polyplax spinulosa,* the mesostigmatid mites *Laelaps echidnina* and *Ornithonyssus bacoti,* and the chigger *Ascoschoengastia indica.* Although the human body louse (*Pediculus humanus humanus*) is an experimental vector of *R. typhi,* it apparently is not involved in natural transmission cycles.

Clinical symptoms of murine typhus appear after an incubation period of six to 14 days and include a rash, high fever (40–41 °C), prostration, delirium, and coma, especially in severe cases. In milder cases, patients may have low-grade fever (38–39 °C) and remain partially mobile. Although the case fatality rates are usually low (<5%) in untreated patients, severe debilitation may last two to three months.

The diagnosis of human infection usually involves the demonstration of seroconversion against *R. typhi* or isolation of the bacterium. Recent advances in *R. typhi* detection in fleas include the development of an enzyme linked immunosorbent assay (ELISA) and of a technique to demonstrate a 434 base-pair nucleotide sequence of the *R. typhi* genome using a PCR assay. These techniques are useful in patient diagnosis and in screening potential reservoir hosts.

Treatment of patients presenting with murine typhus is with antibiotics such as doxycycline and tetracycline. Surveillance and control techniques involve monitoring mammals (especially rats) and fleas for infection and then initiating reservoir or vector control measures as needed.

Other Flea-Borne Rickettsial Agents

In addition to the causative agent of murine typhus, several other rickettsial agents may be transmitted to humans by fleas. One of these, *Coxiella burnetii* the agent of **Q fever**, can be transmitted not only by fleas but also via other blood-feeding arthropods, infected mammalian tissues, infective fomites (inanimate objects), or by aerosol. **Sylvatic epidemic typhus (sporadic epidemic typhus)** is a curious but potentially serious disease that occasionally is diagnosed in humans in the United States. The agent of this disease is *Rickettsia prowazekii,* which causes classic epidemic typhus transmitted to humans by the body louse, *Pediculus humanus humanus* (see Chapter 6). However, flying squirrels (*Glaucomys volans*) rather than humans are the reservoir hosts of sylvatic epidemic typhus. Flying-squirrel fleas, especially the widespread squirrel flea *Orchopeas howardi,* and lice also harbor the causative rickettsiae. The exact mode of transmission to humans is unknown but, because ectoparasites of flying squirrels rarely feed on humans, it is hypothesized that under certain conditions infective rickettsiae in the feces of fleas and lice become aerosolized and may be inhaled by humans. Infected flying squirrels are not adversely affected. These squirrels sometimes are closely associated with humans through their predilection for constructing nests in attics or eaves of houses. Reynolds et al. (2003) provide additional discussion of this rickettsial zoonosis.

Another flea-borne rickettsial agent is *Rickettsia felis* (formerly named the ELB agent for EL Laboratories in Soquel, California). This rickettsia has been found in Virginia opossums and cat fleas (which are common ectoparasites of this opossum) in the United States and has been detected in fleas in many parts of the world (Pérez-Osorio et al., 2008). It has been shown to have caused infection in humans who are serologically positive for infection with *Rickettsia typhi*, the etiologic agent of murine typhus. Definitive demonstration of infection by either *R. felis* or *R. typhi* involves PCR amplification of specific nucleotide primers. Thus, it is likely that some human infections serologically attributed to *R. typhi* are actually caused by *R. felis*. *Rickettsia felis* is also of interest because it is transmitted transovarially in fleas, and fleas rather than mammals may be the reservoirs for the organism. Azad et al. (1997) provide further information on *R. felis* infections. In addition to these rickettsiae, some other fleas are known to harbor symbiotic rickettsiae, about which little is currently known.

Other Flea-Borne Pathogens

Table 9.2 lists other pathogens known to be transmitted by fleas. Most of these microorganisms principally occur in the flea gut rather than the salivary glands or other organs. It has been suggested that this is why fleas are ineffective vectors of viruses (Bibikova, 1977). However, murine typhus rickettsiae can sometimes escape the flea gut and multiply in other organs. Transmission of these gut-localized pathogens therefore occurs either by regurgitation (anterior station) or defecation (posterior station) during or soon after flea feeding. The apparent ease with which fleas can acquire and harbor a wide variety of infectious agents indicates why these insects play a major role in the maintenance and epidemiology of enzootic infections among rodents and other mammals. Many of these pathogens can produce disease in humans and domestic animals if these fleas, or bridge vectors, feed on these hosts.

Bacteria In addition to flea-borne rickettsial organisms, the following bacterial agents cause diseases that affect humans: *Francisella tularensis*, causing **tularemia**; *Salmonella enteriditis*, causing **salmonellosis**; and *Staphylococcus aureus*, causing **staphylococcal infection**. All these agents also can be transmitted by other means such as other ectoparasites and contact or aerosol exposure to infective fomites and mammalian tissues. All three infections are widespread, and the degree of involvement of fleas in transmission varies regionally. A relatively recent development of certain skin lesions including some flea bites is the development of Methicillin Resistant *S. aureus* (MRSA) infections that can be difficult (or impossible) to treat with conventional antibiotics.

Other bacterial agents that may be transmitted by fleas are *Bartonella* (formerly *Rochalimaea*) *henselae*, the agent of cat scratch disease, and *Bartonella elizabethae*, which can damage heart valves (endocarditis). Infections by these zoonotic agents typically cause swollen regional lymph nodes. Long-term bacteremia caused by either of these agents can occur in inapparently infected cats that appear to be important reservoir hosts. About 25,000 cases of cat scratch disease occur annually in the United States, whereas endocarditis caused by *B. elizabethae* appears to be relatively uncommon. Infection with *B. henselae* also can cause fever, hepatitis (liver inflammation), endocarditis, bacillary angiomatosis, and bacillary peliosis. The last two conditions manifest as vascular proliferations, and are most commonly seen in immunocompromised persons such as HIV-positive individuals. Cat fleas can transmit *B. henselae* by bite to cats under laboratory conditions. Although cat fleas may also be capable of transmitting this pathogen to humans, a scratch from an infected cat appears to be the usual mode of transmission. A related organism, *Bartonella clarridgeiae*, can also occur in cats and cause infection in humans, but the potential role of fleas as vectors of this agent has not been determined.

The following zoonotic bacterial agents also have been detected in fleas; however, it is assumed that fleas imbibe these pathogens in blood meals from infected hosts and are not vectors: *Borrelia burgdorferi*, the etiologic agent of Lyme disease; *Borrelia duttoni*, an agent of relapsing fever; *Listeria monocytogenes*, the agent of listeriosis; *Yersinia pseudotuberculosis*, causing pseudotuberculosis (yersiniosis); *Erysipelothrix rhusiopathiae*, causing erysipelas; and *Brucella abortis*, causing brucellosis.

Viruses Although several viral pathogens of humans have been isolated from, or detected in, fleas, the role of fleas in their transmission is either unknown or considered to be incidental. These viruses include those that cause lymphocytic choriomeningitis, tick-borne encephalitis, Russian spring-summer encephalitis, and Omsk hemorrhagic fever.

It should be emphasized that demonstration of a pathogen within an arthropod does not necessarily imply that it is a vector of the agent.

Tungiasis

Tungiasis is the pathological condition resulting from infestation by fleas belonging to the genus *Tunga*. Although there are several species of *Tunga*, only the chigoe (*Tunga penetrans*) is known to attack humans. *Tunga penetrans* occurs in many tropical and subtropical zones but is especially common in the New World tropics, the West Indies, tropical Africa, and southern India. The first record of this flea was in 1492 from crewmen of Christopher Columbus stationed in Haiti. It apparently spread from the New World to other areas of the world by shipping commerce, and was first recorded on the African continent in 1732 as a consequence of the slave trade.

Females of *T. penetrans* usually invade a site between the toes, beneath the toe nails, or on the soles of the feet. Other sites may include the hands, arms, especially around the elbow, and genital region in heavy infestations. Skin invasion by this flea can cause painful, subcutaneous lesions that often lead to more serious medical complications. The embedded chigoe (Figures 9.11 and 9.12) invariably causes intense irritation and can result in secondary infections that ooze pus. When several chigoes attack an individual host at the same time, ulcerations often develop as the resultant lesions coalesce. Some people may harbor many lesions on their feet, which can result in them being unable to walk. Multiple lesions at the fingertips can result in difficulty with gripping. Tetanus (in non-vaccinated individuals), cellulitis (inflammation of cellular or connective tissue), regional lymphadenitis (swollen lymph nodes), deformation of digits, and loss of toenails and fingernails may occur. Impaired blood flow to the site often leads to gangrene and may necessitate amputation of toes or, sometimes, an entire foot. Chigoe lesions therefore should receive prompt medical attention. Although the flea can be removed using a sterile needle or scalpel, it is important that lesions be thoroughly cleaned and dressed to avoid infection. This also applies to embedded dead fleas, which may rapidly cause affected tissues to fester and ulcerate if left untreated. Eisele et al. (2003) have identified five distinct clinical phases of *T. penetrans* infestation in humans that aid in diagnosis and treatment. The best defense against tungiasis is to avoid walking barefoot on beaches and other sandy soils in endemic regions where this flea develops.

Fleas as Intermediate Hosts of Helminths

Certain fleas are intermediate hosts for the cysticercoid stage of three species of tapeworms that occasionally infest humans. The most important of these is the ***double-pored tapeworm*** (*Dipylidium caninum*), the adults of which normally parasitize dogs. Gravid worm-like proglottids are released by *D. caninum* adults in the gut of the definitive host, then actively exit the anus, partially dry upon exposure to air, and fall to the ground where they resemble sesame seeds. The subsequently expelled eggs are ingested by flea larvae; the chewing mandibles of the larvae enable them to ingest the eggs whereas the sucking mouthparts of adult fleas do not. Fleas such as *Ctenocephalides felis*, *C. canis*, and *Pulex irritans* play a significant role as intermediate hosts for this tapeworm. The dog chewing louse (*Trichodectes canis*) occasionally ingests *D. caninum* eggs and also can serve as an intermediate host.

The tapeworm develops slowly in flea larvae, but rapidly in flea pupae. Cysticercoids can be seen in the body cavity of larvae and pupae where they remain through development of the flea to the adult stage. Some flea mortality occurs in the pupal stage due to this helminth. Infestation of the human (definitive) host occurs when a person incidentally ingests an infested flea. The cysticercoid is liberated from the flea by digestive enzymes, after which it everts and attaches to the gut of its new host. Children playing with pets are especially susceptible to infestation by this tapeworm.

Two other tapeworms that utilize fleas as intermediate hosts are the rodent tapeworm (*Hymenolepis diminuta*) and the dwarf tapeworm (*H. nana*). Both infest rodents and occasionally parasitize humans, especially children. The development and transmission of these two cestodes are similar to that for *Dipylidium caninum*. Both *H. diminuta* and *H. nana* form viable cysticercoids in several species of fleas, especially *Ctenocephalides canis, Pulex irritans, Xenopsylla cheopis*, and *Nosopsyllus fasciatus*. They also infest several other arthropods, notably coprophagous beetles.

The zoonotic nematode *Trichinella spiralis*, which causes trichinosis, has also been found in fleas although this is assumed to represent an accidental association.

VETERINARY IMPORTANCE

Several species of fleas are important ectoparasites of domestic and wild animals. Emphasis here is given to those that infest pets and livestock. Many fleas associated with domestic animals merely cause a nuisance through their biting activity; they also may cause flea-bite dermatitis, allergies, and anemia when present in large numbers. Other fleas such as sticktights and chigoes embed their mouthparts, or entire bodies, in mammalian or avian tissues causing local inflammation and other problems. Some fleas are intermediate hosts of helminths that parasitize domestic animals, whereas others transmit pathogens such as viruses and trypanosomes to their hosts.

The cat flea (*C. felis*; Figures 9.1–9.3, 9.9A) is an extremely important ectoparasite, not only of cats and dogs, but also of several other mammals including opossums, cattle, horses, sheep, goats, rabbits, and monkeys. Some populations of *C. felis* have adapted to certain hosts, such as dogs or cattle, and show a preference for feeding on these species. Occasionally cat fleas infest goats, lambs, calves, or other ungulates in large numbers and can cause anemia or even death. Individual pets, especially cats and dogs, may support hundreds or thousands of cat fleas. Because the larvae thrive on blood-rich fecal pellets voided by adult fleas on the host, it is important to vacuum or treat areas where pets rest or sleep to reduce flea numbers. The dog flea (*C. canis*; Figure 9.9B) is a relatively infrequent ectoparasite of dogs, with established populations on dogs persisting in only a few countries. Almost invariably, fleas associated with dogs are *C. felis*. Further details on the biology of fleas associated with cats and dogs are provided by Dryden (1993) and Rust and Dryden (1997).

Several species of fleas are parasites of domestic and laboratory rats and mice. These include the Oriental rat flea (*X. cheopis*; Figure 9.9F), the northern rat flea (*N. fasciatus*; Figure 9.9D), and the European mouse flea (*L. segnis*; Figure 9.9H). Flea infestations of these rodents are usually more important with respect to potential transmission of pathogens rather than their discomforting bites. The European rabbit flea (*S. cuniculi*) is a parasite of the European rabbit (*Oryctolagus cuniculus*) throughout much of the world where it has been introduced as a game or small-livestock animal. Since this is the laboratory rabbit commonly used in scientific studies, the European rabbit flea occasionally is recorded in animal research facilities.

This flea is commonly a pest in rabbit hutches and where European rabbits are raised commercially for food in many parts of the world. *Spilopsyllus cuniculi* usually attaches to the ears where it embeds its mouthparts deeply and for long periods, causing host irritability and ear scabbing.

In Central Asia, the alakurt fleas *Dorcadia ioffi* and *Vermipsylla alakurt* parasitize ungulates, especially horses, sheep, and yaks. These fleas often occur in very large numbers on these hosts and can cause anemia, hair loss, retarded growth, unthriftiness, and occasionally death, especially in newborn lambs.

Other fleas that are annoying biters of domestic mammals include *Pulex simulans*, the human flea (*P. irritans*; Figure 9.9C), and the sticktight flea (*E. gallinacea*; Figure 9.10), all of which may be recovered from cats or dogs. *Pulex simulans* and the human flea can be important ectoparasites of dogs and swine, whereas the sticktight flea can infest domestic rats and several other mammals.

Several species of fleas feed on birds. At least three of these are important pests to the poultry industry. The sticktight flea is principally a poultry pest in the subtropical and tropical regions of the New World. These small fleas typically attach to the nonfeathered areas of birds such as the head (Figure 9.10), comb, wattle, and anus. Large flea populations can cause anemia. Feeding sites can become ulcerated; when this occurs around the eyes, blindness can result and the host is unable to feed. Secondary infections may develop. The European chicken flea (hen flea, in Europe) (*Ceratophyllus gallinae*) is a nonsedentary ectoparasite of domestic fowl in several parts of the world, including Europe and eastern North America. In western North America the western chicken flea (*Ceratophyllus niger*), another nonsedentary species, is a parasite of domestic fowl and several species of wild birds. All these poultry fleas can cause host emaciation and reduced egg production when they occur in large numbers.

Flea-Bite Dermatitis

Allergic skin reactions to flea bites are a common problem of domestic animals, especially household pets. Hypersensitivity to saliva from feeding fleas is usually more apparent in pets than in humans because larger numbers of pets are bitten by fleas. A single flea bite can trigger an acute, sometimes chronic, dermatitis in hypersensitive dogs or cats. Incessant scratching and skin irritation, especially during the warmer months, often reflects this condition. In cats, flea-bite dermatitis usually manifests as purplish papules that often are covered with crusts; in dogs, crusts are typically absent. In both cats and dogs, lesions usually are concentrated on the rump and inner thighs, with accompanying fur loss from frequent scratching. Cats sometimes also have a ring of crusts around the neck. Diligent flea control is important in combating this condition. Administration of corticosteroids or a course of desensitizing antigens are other treatment options. Except in severe cases, the hypersensitivity often resolves after repeated flea bites as the host gradually becomes desensitized to antigens in flea saliva.

Tungiasis

Some domestic animals, especially hogs and dogs (Figure 9.13), are parasitized by *Tunga penetrans* causing tungiasis. Infestations in hogs primarily affect the feet but also the snout, teats, legs, and scrotum. Infestations of the teats can result in restricted milk flow in nursing sows and starvation of piglets. Swine are reservoirs of tungiasis, which can be transferred to humans in some tropical climates. Dogs are commonly infested by *T. penetrans* in some tropical regions including parts of rural Brazil. Female fleas often embed in the snout (Figure 9.13) or in the pads of the feet of dogs. There are at least eight other species of fleas belonging to the genus *Tunga* that burrow into host tissues. Females of each of these species are subdermal parasites that mostly attack New World rodents. Hopkins and Rothschild (1953) discuss other species of *Tunga*.

Myxomatosis

Myxomatosis is primarily a disease of the European rabbit caused by infection with the myxoma virus. The virus causes benign fibromas in its natural rabbit hosts in California, Central America, and South America. However, in the European rabbit, a severe and usually fatal infection with enlarging skin lesions and generalized viremia occurs. The myxoma virus was introduced to Australia in 1950 and to Europe in 1953. The aim of these introductions was to control burgeoning populations of European rabbits.

The virus is transmitted mechanically to rabbits by various blood-feeding arthropods, particularly mosquitoes. However, the European rabbit flea (*S. cuniculi*) is also a proven vector, at least in Britain where this flea occurs naturally, and in Australia where it was introduced in 1966. Although it is an inefficient vector, an Australian sticktight flea (*Echidnophaga myrmecobii*) also can transmit the myxoma virus to rabbits. Infection with this virus apparently does not adversely affect these flea vectors. As with most other flea-transmitted pathogens, myxoma virus remains confined to the gut and mouthparts of *S. cuniculi*. Survival of the virus for three to four months in infected fleas has been demonstrated.

Because strains of the virus differ in virulence whereas rabbit populations differ in their susceptibility to this pathogen, the success of this virus in controlling rabbits has been variable. When the virus was first introduced to Australia and Europe it was very effective in culling wild rabbits; today, however, many rabbit populations in both Australia and Europe have developed resistance to several strains of the virus.

Murine Trypanosomiasis

Trypanosoma lewisi is the causative agent of murine trypanosomiasis in domestic rats throughout much of the world. It is principally transmitted by the Northern rat flea (*N. fasciatus*; Figure 9.9D) and the Oriental rat flea (*X. cheopis*; Figure 9.9F). Fleas imbibe trypanosomes while feeding on infected rats; the pathogen remains in the flea midgut where development occurs. Within six hours after ingestion, the trypanosomes invade midgut epithelial cells, transform into pear-shaped forms, and begin to divide. The parasitized gut cells rupture after 18 hours to five days to release the trypanosomes; these then either invade new epithelial cells to repeat the process or move posteriorly to the rectum and anus. Trypanosomes in this "rectal phase" are voided in the flea feces. The trypanosomes enter their rat hosts when the latter lick and scratch their fur during grooming, representing a classic example of posterior-station transmission. Murine trypanosomiasis is usually a benign infection in rats. However, the *T. lewisi*-flea-rat system has been used as a laboratory model for studying more virulent trypanosome species that are pathogenic to humans and domestic animals.

At least nine species of trypanosomes other than *T. lewisi* are transmitted to rodents by fleas. Rodent trypanosomes with confirmed flea transmission cycles include *Trypanosoma musculi* (synonym: *T. duttoni*) of house mice, *T. rabinowitschi* of hamsters, *T. neotomae* of wood rats, and *T. grosi* of the European wood mouse (*Apodemus sylvaticus*). *Trypanosoma nabiasi* is one of two species of trypanosomes known to be transmitted to rabbits by fleas. Fleas are also suspected as vectors of trypanosomes associated with some birds, shrews, voles, and lagomorphs.

Other Flea-Borne Pathogens and Parasites

Many of the flea-borne pathogens listed in Table 9.2 and others cause diseases in humans, with wild or domestic animals serving as reservoirs. These include plague, tularemia, murine typhus, Q fever, and sylvatic epidemic typhus. Infections of domestic animals with most of these pathogens can be inapparent, febrile, or fatal depending on the host species, its health, and the strain of pathogen involved. Cats, for example, are typically susceptible to most strains of plague whereas dogs usually are not.

Other pathogens of veterinary importance that have been isolated from, or detected in, fleas include lymphocytic choriomeningitis virus, which affects many mammals, especially rodents; feline leukemia virus (FeLV); and the bacterial agents *Borrelia burgdorferi*, the causative agent of Lyme disease, *Listeria monocytogenes*, the agent of listeriosis mainly in ungulates, *Brucella abortis*, an agent of brucellosis mainly in bovines, *Burkholderia mallei*, the agent of glanders in equines, and *Burkholderia pseudomallei*, the agent of melioidosis in several mammals. However, the role of fleas as vectors of these pathogens is doubtful or undetermined.

Other microorganisms known to occur in fleas and that may be transmitted to vertebrates include haemogregarine sporozoans, various rickettsial organisms, and several symbionts. The protozoan *Hepatozoon erhardovae* is transmitted to European voles (*Clethrionomys* spp.) by at least five species of fleas. The parasite reproduces sexually in the hemocoel of fleas where it develops to the sporocyst stage; transmission to voles occurs when they eat infected fleas during grooming. The related *Hepatozoon pitymysi* and *H. sciuri*, which parasitize North American and Eurasian voles and North American squirrels, respectively, also have been detected in fleas and are thought to be transmitted in a similar way.

Fleas as Intermediate Hosts of Helminths

The double-pored tapeworm (*Dipylidium caninum*) normally develops as an adult parasite in the intestines of dogs, cats, and some wild carnivores. The most important intermediate flea hosts are the cat flea and dog flea, although the human flea can also serve in this capacity. In tropical Africa, a warthog flea (*Echidnophaga larina*) is sometimes responsible for *D. caninum* infestations in domestic dogs. Infestations usually are initiated when animals consume parasitized fleas while grooming.

Two species of tapeworms that typically infest rats and mice as adults are the rodent tapeworm (*Hymenolepis diminuta*) and the dwarf tapeworm (*H. nana*). Rat fleas, especially the Oriental rat flea and the northern rat flea, serve as intermediate hosts. Infestations are initiated when infested fleas are eaten by the definitive rodent hosts.

The onchocercid nematode, *Acanthocheilonema* (formerly *Dipetalonema*) *reconditum*, which causes a relatively benign form of canine filariasis in many parts of the world, has been found in several species of fleas. The cat flea and dog flea are considered to be the principal vectors. Transmission of mature larvae by these fleas occurs by bite. Dogs, jackals, and hyenas are the principal definitive hosts of *A. reconditum.*

Several other species of helminths have been isolated from wild-caught fleas, and fleas have been found to serve as suitable intermediate hosts for some of these under laboratory conditions. However, the importance of fleas in maintaining these pathogens in nature is unknown. For example, the trichina worm (*Trichinella spiralis*) has been found encysted in the Oriental rat flea in India; this helminth normally encysts in muscle tissue of rats and hogs, causing trichinosis.

PREVENTION AND CONTROL

Various methodologies are used to control fleas or to protect humans and other animals from flea bites. Frequent vacuuming in homes, especially in areas where pets rest or sleep, helps to remove immature fleas and their food; steam cleaning of carpets is even more effective. Insect growth regulator (IGR) applications (see later) to carpets following steam cleaning will prevent subsequent reinfestation. Household foggers, which produce a mist of insecticide in closed, temporarily vacated rooms are minimally effective in combating household flea infestations. Treatment of flea-infested premises or domestic animals with various insecticides generally provides good flea control. Commercially available products include botanical derivatives, carbamates, organophosphates, pyrethroids, boron compounds, and diatomaceous earth. Some nonchemical techniques are effective in reducing flea populations in homes, especially on or around pets. Sticky traps and pan traps are useful for detecting and monitoring flea infestations. Pan traps are trays of detergent water, often with an attractant light source, in which fleas drown. A flea light-trap fitted with a green-yellow filter with a transmittance spectrum centered at 515 nm is effective in attracting cat fleas. Chemicals used to kill fleas are called **pulicides**.

Several botanical derivatives such as pyrethrins have low mammalian toxicity and are useful as flea powders for dusting flea-infested pets. Products such as flea soaps that contain fatty acids and flea shampoos can be used to bathe pets. Bathing removes or drowns many fleas, while those fleas that survive often desiccate because their integumental waxes have been removed by the detergents. Pet-administered applications such as pulicidal shampoos, dips, mousses, and dusts are now declining in usage as low-volume topical applications of lipophilic formulations are replacing them. Flea combs can be used to mechanically remove fleas. Flea collars for pets are minimally effective. These collars should not be worn by humans (e.g., on the ankles), because they can cause skin irritation and allergic reactions. Some progress has been made in developing vaccines against fleas, mainly using midgut antigens of the cat flea to induce an immune response in the host. In several trials, dogs, cats, and rabbits that were experimentally challenged with cat-flea antigens had significantly more dead or reproductively compromised fleas than did nonvaccinated animals.

Insect growth regulators (IGRs), especially formulations of methoprene and pyriproxyfen, currently are popular flea-control weapons because they have low mammalian toxicity. At low concentrations, these compounds interfere with flea development and eventually (after 1–2 months) provide high levels of flea control. Although insect growth regulators and chitin synthesis inhibitors do not kill adult fleas, they can act as larvicides, preventing flea eggs from hatching and larvae from successfully molting. Lufenuron is a chitin synthesis inhibitor that is administered orally as a pill for dogs or as a liquid added to food or an injectable formulation for cats. Pyriproxyfen and methoprene are juvenile hormone analogs that can be applied topically.

Since all adult fleas must blood-feed, host-targeted control uses the animal as the lure, ensuring that all adult fleas are exposed to the pulicide when they blood-feed. Some insecticides may be given orally to the host so that the active ingredient passes into the bloodstream and is picked up as the fleas feed. Orally administered pulicides include neonicotinoids (such as nitenpyram) and spinosyns (e.g., spinosad).

Other insecticides may be topically applied so that the flea acquires the toxicant via its cuticle. Topical products generally are maintained in the dermal lipids of the host where they are available for cuticular absorption by fleas. Host-targeted topically applied ectoparasiticides include pyrethrins, pyrethroids (e.g., permethrin), neonicotinoids (e.g., imidacloprid, dinotefuran), phenylpyrazoles (e.g., fipronil), macrocyclic lactones (e.g., selamectin), and semicarbazones (e.g., metaflumizone).

Some products, such as those containing permethrin, are toxic to felines and are not labeled for use on cats or young puppies. Further, some products with the same name may include different active ingredients in the canine version than in their feline counterparts.

Two biological control agents, the parasitic nematode *Steinernema carpocapsae*, and the entomopathogenic fungus *Beaveria bassiana*, reduce cat flea numbers under laboratory conditions and show promise as future control agents.

Environmental flea control efforts should be focused on areas where pets spend most time, especially where they sleep, as these are locations most likely to support development of flea larvae. Outdoor flea control can be challenging, because feral mammals such as raccoons, opossums, dogs, and cats, may continually reinfest premises. Flea larvae cannot survive the heat and drying conditions of full sun exposure, so control efforts should be directed toward shaded areas of host activity. Typically, this includes under shrubbery, against foundations, in crawl spaces, or areas under porches/decks. Pulicides used in these sites should have sustained residual efficacy and be photostable.

Personal protectants such as those containing DEET (diethylmethylbenzamide, formerly N,N-diethyl-m-toluamide) or permethrin are often very helpful in reducing the number of flea bites. Permethrin should be applied only to clothing and not directly on the skin. Although banned for use in the United States, DDT (dichloro-diphenyl-trichloroethane) is still used to control outbreaks of plague or murine typhus in some parts of the world. As with the use of other insecticides, there is a constant risk that fleas may develop resistance to these chemicals.

Plague outbreaks usually are followed by public education and area wide programs to remove rodent hosts and flea vectors. Control programs for murine typhus typically involve eliminating the flea vectors or rodent reservoirs by insecticide applications and trapping, respectively. Rodent harborages and access of these reservoir hosts to houses should be eliminated where feasible. Dusting rodent burrows with insecticides or providing rodent bait stations spiked with either rodenticides or flea-control agents can be effective in killing the rodent hosts or fleas, respectively. Frequent surveillance of rodent and flea populations in plague-endemic regions often allows control measures to be implemented before human cases occur. Outbreaks of murine typhus may be handled in a similar manner, although there is greater emphasis on rodent control because the reservoir hosts are more likely to be commensal rats.

An ineffective approach to flea control is the use of ultrasonic repellent devices. No fleas that have been tested have shown responses to ultrasound or to devices incorporating it. Nor has the oral intake of garlic or of B-complex vitamins, including Brewer's yeast, been proven to reduce flea populations on pets, despite claims about their effectiveness. Flea control strategies are addressed by MacDonald (1995), Rust and Dryden (1997), Hinkle et al. (1997), and Dryden (1999).

REFERENCES AND FURTHER READING

Askew, R. R. (1973). *Parasitic Insects.* London: Heinemann Educational Books.

Azad, A. F. (1990). Epidemiology of murine typhus. *Annual Review of Entomology, 35,* 553–569.

Azad, A. F., Radulovic, S., Higgins, J. A., Noden, B. H., & Troyer, J. M. (1997). Flea-borne rickettsiosis: Ecologic considerations. *Emerging Infectious Diseases, 3,* 319–327.

Beaucournu, J., & Launay, H. (1990). Les puces (Siphonaptera) de France et du Bassin Méditerraneén occidental. *Faune de France, 76,* 1–548.

Benton, A. H. (1980). *An Atlas of the Fleas of the Eastern United States.* Fredonia, NY: Marginal Media.

Benton, A. H. (1983). *An Illustrated Key to the Fleas of the Eastern United States.* Fredonia, NY: Marginal Media.

Bibikova, V. A. (1977). Contemporary views on the interrelationships between fleas and the pathogens of human and animal diseases. *Annual Review of Entomology, 22,* 23–32.

Carniel, E., & Hinnebusch, B. J. (2004). *Yersinia: Molecular and Cellular Biology.* Milton Park, UK: Horizon Bioscience.

Chiebao, D. P., Rodrigues, A. R., Pinheiro, S. R., & Gennari, S. M. (2005). Ocorrência de tungíase em cães no município de São Paulo. *Clinica Veterinária, 10*(59), 50–54.

Christakos, G., Olea, R. A., Serre, M. L., Yu, H. L., & Wang, L. L. (2005). *Interdisciplinary Public Health Reasoning and Modelling: The Case of the Black Death.* Berlin: Springer.

Dennis, D. T., Gage, K. L., Gratz, N., Poland, J. D., & Tikhomirov, E. (1999). *Plague Manual: Epidemiology, Distribution, Surveillance and Control.* Geneva: World Health Organization, WHO/CDS/CSR/EDC/99.2. (Also available as electronic document on the WHO web site.)

Drancourt, M., Roux, V., Dang, L. V., Tran-Hung, L., Castex, D., Chenal-Francisque, V., et al. (2004). Genotyping, Orientalis-like *Yersinia pestis,* and plague pandemics. *Emerging Infectious Diseases, 10,* 1585–1592.

Drancourt, M., Signoli, M., Dang, L. V., Bizot, B., Roux, V., Tzortzis, S., et al. (2007). *Yersinia pestis.* Orientalis in remains of ancient plague patients. *Emerging Infectious Diseases, 13,* 332–333.

Dryden, M. W. (1993). Biology of fleas of cats and dogs. *Compendium on Continuing Education for the Practicing Veterinarian, 15,* 569–579.

Dryden, M. W. (1999). Highlights and horizons in flea control. *Compendium on Continuing Education for the Practicing Veterinarian, 21,* 296–298, 361–365.

Duplaix, N. (1988). Fleas. The lethal leapers. *National Geographic, 173,* 672–694.

Eisele, M., Heukelbach, J., van Marck, E., Mehlhorn, H., Meckes, O., Franck, S., et al. (2003). Investigations on the biology, epidemiology, pathology and control of *Tunga penetrans* in Brazil: I. Natural history of tungiasis in man. *Parasitology Research, 90,* 87–99.

Eisen, R. J., Enscore, R. E., Biggerstaff, B. J., Reynolds, P. J., Ettestad, P., Brown, T., et al. (2007). Human plague in the southwestern United States, 1957–2004: Spatial models of elevated risk of human exposure to *Yersinia pestis. Journal of Medical Entomology, 44,* 530–537.

Elbel, R. E. (1991). Order Siphonaptera. In F. W. Stehr (Ed.), *Immature Insects* (Vol. 2. pp. 674–689). Dubuque, IA: Kendall Hunt.

Ewing, H. E., & Fox, I. (1943). *The fleas of North America. Classification, identification, and geographic distribution of these injurious and disease-spreading insects.* U.S. Dept. Agric. Misc. Publ. No. 500.

Fenner, F., & Ross, J. (1994). Myxomatosis. In H. V. Thompson, & C. M. King (Eds.), *The European Rabbit. The History and Biology of a Successful Colonizer* (pp. 205–239). Oxford: Oxford University Press.

Fox, I. (1940). *Fleas of eastern United States.* Ames: Iowa State College Press.

Gage, K. L., & Kosoy, M. Y. (2005). Natural history of plague: Perspectives from more than a century of research. *Annual Review of Entomology, 50,* 505–528.

Goddard, J. (1998). Fleas and murine typhus. *Infections in Medicine, 15,* 438–440.

Harwood, R. F., & James, M. T. (1979). *Entomology in Human and Animal Health* (7th ed.). New York: Macmillan.

Hinkle, N. C. (2008). 5. Fleas. In X. Bonnefoy, H. Kampen, & K. Sweeney (Eds.), *Public Health Significance of Urban Pests* (pp. 155–173). Geneva: World Health Organization. (Also available as an electronic document on the WHO web site.)

Hinkle, N. C., Rust, M. K., & Reierson, D. A. (1997). Biorational approaches to flea (Siphonaptera, Pulicidae) suppression—present and future. *Journal of Agricultural Entomology, 14,* 309–321.

Holland, G. P. (1985). *The fleas of Canada, Alaska and Greenland (Siphonaptera). Memoirs of the Entomological Society of Canada* No. 130.

Hopkins, G. H. E., & Rothschild, M. (1953–1971). *An Illustrated Catalogue of the Rothschild Collection of Fleas (Siphonaptera) in the British Museum (Natural History)* (Vols. I–V). London: British Museum (Natural History).

Hopla, C. E., & Hopla, A. K. (1994). Tularemia. In G. W. Beran (Ed. in-chief), *Handbook of Zoonoses. Sect. A. Bacterial, Rickettsial, Chlamydial, and Mycotic* (2nd ed.) (pp. 113–123). Boca Raton: CRC Press.

Hubbard, C. A. (1947). *Fleas of Western North America.* Ames: Iowa State College Press.

Ibanez-Bernal, S., & Velasco-Castrejón, O. (1996). New records of human tungiasis in Mexico. (Siphonaptera: Tungidae). *Journal of Medical Entomology, 33,* 988–989.

Kelly, J. (2004). *The Great Mortality: An Intimate History of the Black Death, the Most Devastating Plague of all Time.* New York: HarperCollins.

Krasnov, B. R. (2008). *Functional and Evolutionary Ecology of Fleas: A Model for Ecological Parasitology.* Cambridge University Press.

Lewis, R. E. (1972). Notes on the geographic distribution and host preferences in the order Siphonaptera. Part 1. Pulicidae *Journal of Medical Entomology 9,* 511–520.

Lewis, R. E. (1973). Notes on the geographic distribution and host preferences in the order Siphonaptera. Part 2. Rhopalopsyllidae, Malacopsyllidae and Vermipsyllidae. *Journal of Medical Entomology 10,* 255–260.

Lewis, R. E. (1974a). Notes on the geographic distribution and host preferences in the order Siphonaptera. Part 3. Hystrichopsyllidae. *Journal of Medical Entomology 11,* 147–167.

Lewis, R. E. (1974b). Notes on the geographic distribution and host preferences in the order Siphonaptera. Part 4. Coptopsyllidae, Pygiopsyllidae, Stephanocercidae and Xiphiopsyllidae. *Journal of Medical Entomology 11,* 403–413.

Lewis, R. E. (1974c). Notes on the geographic distribution and host preferences in the order Siphonaptera. Part 5. Ancistropsyllidae, Chimaeropsyllidae, Ischnopsyllidae, Leptopsyllidae and Macropsyllidae. *Journal of Medical Entomology 11,* 525–540.

Lewis, R. E. (1974d). Notes on the geographic distribution and host preferences in the order Siphonaptera. Part 6. Ceratophyllidae. *Journal of Medical Entomology 11,* 658–676.

Lewis, R. E. (1993a). Notes on the geographic distribution and host preferences in the order Siphonaptera. Part 8. New taxa described between 1984 and 1990, with a current classification of the order. *Journal of Medical Entomology 30,* 239–256.

Lewis, R. E. (1993b). Fleas (Siphonaptera). In R. P. Lane, & R. W. Crosskey (Eds.), *Medical insects and arachnids* (pp. 529–575). London: Chapman and Hall.

Lewis, R. E. (1998). Résumé of the Siphonaptera (Insecta) of the world. *Journal of Medical Entomology 35,* 377–389.

Lewis, R. E. (2000). A taxonomic review of the North American genus *Orchopeas* Jordan, 1933 (Siphonaptera: Ceratophyllidae: Ceratophyllinae). *Journal of Vector Ecology 25,* 164–189.

Lewis, R. E. (2002). A review of the North American species of *Oropsylla* Wagner and Ioff, 1926 (Siphonaptera: Ceratophyllidae: Ceratophyllinae). *Journal of Vector Ecology 27,* 184–206.

Lewis, R. E. (2003). A review of the North American flea genus *Spicata* I. Fox, 1940 (Siphonaptera: Ceratophyllidae). *Proceedings of the Entomological Society of Washington 105*, 876–882.

Lewis, R. E. (2008a). On the Nearctic flea genus *Opisodasys* Jordan, 1933: its taxonomy, distribution, and host preferences (Siphonaptera: Ceratophyllidae). *Annals of Carnegie Museum 76*, 279–299.

Lewis, R. E. (2008b). *Malaraeus* Jordan, 1933: A North American genus of fleas (Siphonaptera: Ceratophyllidae). *Annals of Carnegie Museum 77*, 289–299.

Lewis, R. E. (2008c). The North American fleas of the genus *Amalaraeus* Ioff, 1936 (Siphonaptera: Ceratophyllidae). *Annals of Carnegie Museum 77*, 313–317.

Lewis, R. E., Galloway, T. D. (2001). A taxonomic review of the *Ceratophyllus* Curtis, 1832 of North America (Siphonaptera: Ceratophyllidae: Ceratophyllinae). *Journal of Vector Ecology 26*, 119–161.

Lewis, R. E., Haas, G. E. (2001). A review of the North American *Catallagia* Rothschild, 1915, with the description of a new species (Siphonaptera: Ctenophthalmidae: Neopsyllinae: Phalacropsyllini). *Journal of Vector Ecology 26*, 51–69.

Lewis, R. E., Jameson, E. W., Jr. (2002). A review of the flea genus *Eumolpianus* Smit, 1983 with a discussion of its geographic distribution and host associations (Siphonaptera: Ceratophyllidae: Ceratophyllinae). *Journal of Vector Ecology 27*, 235–249.

Lewis, R. E., Lewis, J. H. (1985). Notes on the geographic distribution and host preferences in the order Siphonaptera. Part 7. New taxa described between 1972 and 1983, with a supraspecific classification of the order. *Journal of Medical Entomology 22*, 134–152.

Lewis, R. E., Lewis, J. H. (1994). Siphonaptera of North America north of Mexico. *Journal of Medical Entomology 31*, 82–98 (Vermipsyllidae and Rhopalopsyllidae), 348–368 (Ischnopsyllidae), 795–812 (Hystrichopsyllidae, *s. str.*).

Lewis, R. E., Lewis, J. H., Maser, C. (1988). The Fleas of the Pacific Northwest. Oregon State University Press, Corvallis.

Lewis, R. E., Wilson, N. (2006). A review of the cearatophyllid subfamily Dactylopsyllinae. Part 2. *Dactylopsylla* Jordan, 1929, and *Foxella* Wagner, 1929 (Siphonaptera: Ceratophyllidae). *Annals of Carnegie Museum 75*, 203–229.

MacDonald, J. M. (1995). Flea control: An overview of treatment concepts for North America. *Veterinary Dermatology, 6*, 121–130.

Madon, M. B., Hitchcock, J. C., Davis, R. M., Myers, C. M., Smith, C. R., Fritz, C. L., et al. (1997). An overview of plague in the United States and a report of investigations of two human cases in Kern County, California. *Journal of Vector Ecology, 22*, 77–82.

Mardon, D. K. (1981). *An Illustrated Catalogue of the Rothschild Collection of Fleas (Siphonaptera) in the British Museum (Natural History). Vol. VI. Pygiopsyllidae.* London: British Museum (Natural History).

Marshall, A. G. (1981). *The Ecology of Ectoparasitic Insects.* London: Academic Press.

Matheson, R. (1950). *Medical Entomology* (2nd ed.). Ithaca: Cornell University Press.

Mee, C. L., Jr. (1990). How a mysterious disease laid low Europe's masses. *Smithsonian, 20*(11), 66–79.

Pérez-Osorio, C. E., Zavala-Velázquez, J. E., Arias León, J. J., & Zavala-Castro, J. E. (2008). *Rickettsia felis* as emergent global threat for humans. *Emerging Infectious Diseases, 14*, 1019–1023.

Perry, R. D., & Fetherston, J. D. (1997). *Yersinia pestis*—Etiologic agent of plague. *Clinical Microbiology Reviews, 10*, 35–66.

Poland, J. D., Quan, T. J., & Barnes, A. M. (1994). Plague. In G. W. Beran (Ed.) in-chief, *Handbook of Zoonoses. Sect. A: Bacterial, Rickettsial, Chlamydial and Mycotic* (2nd ed., pp. 93–112). Boca Raton: CRC Press.

Purcell, K., Fergie, J., Richman, K., & Rocha, L. (2007). Murine typhus in children, south Texas. *Emerging Infectious Diseases, 13*, 926–927.

Rawlings, J. A., & Clark, K. A. (1994). Murine typhus. In G. W. Beran (Ed.) in-chief, *Handbook of Zoonoses. Sect A: Bacterial, Rickettsial, Chlamydial and Mycotic* (2nd ed.) (pp. 457–461). Boca Raton: CRC Press.

Reeves, W. K., Murray, K. O., Meyer, T. E., Bull, L. M., Pascua, R. F., Holmes, K. C., et al. (2008). Serological evidence of typhus group rickettsia in a homeless population in Houston, Texas. *Journal of Vector Ecology, 33*, 205–207.

Reynolds, M. G., Krebs, J. W., Comer, J. A., Sumner, J. W., Rushton, T. C., Lopez, C. E., et al. (2003). Flying squirrel-associated typhus, United States. *Emerging Infectious Diseases, 9*, 1341–1343.

Rust, M. K., & Dryden, M. W. (1997). The biology, ecology, and management of the cat flea. *Annual Review of Entomology, 42*, 451–473.

Rust, M. K., Waggoner, M. M., Hinkle, N. C., Stansfield, D., & Barnett, S. (2003). Efficacy and longevity of nitenpyram against adult cat fleas (Siphonaptera: Pulicidae). *Journal of Medical Entomology, 40*, 678–681.

Smit, F. G. A. M. (1987). *An Illustrated Catalogue of the Rothschild Collection of Fleas (Siphonaptera) in the British Museum (Natural History). Vol. VII. Malacopsylloidea.* London: Oxford University Press, Oxford and British Museum (Natural History).

Traub, R. (1985). Coevolution of fleas and mammals. In K. C. Kim (Ed.), *Coevolution of Parasitic Arthropods and Mammals.* (pp. 295–437). New York: Wiley.

Traub, R., Rothschild, M., & Haddow, J. (1983). *The Rothschild Collection of Fleas. The Ceratophyllidae: Key to Genera and Host Relationships with Notes on their Evolution, Zoogeography and Medical Importance.* Cambridge: Cambridge University Press.

Traub, R., & Starcke, H. (1980). *Fleas. Proceedings of the International Conference on Fleas, Ashton Wold, England, June 1977.* Rotterdam: A. A. Balkema.

Traub, R., Wisseman, C. L., Jr., & Azad, A. F. (1978). The ecology of murine typhus—A critical review. *Trop. Dis. Bull., 75*, 237–317.

Wheelis, M. (2002). Biological warfare at the 1346 siege of Caffa. *Emerging Infectious Diseases, 8*, 971–975.

Whitaker, A. P. (2007). *Fleas (Siphonaptera). Handbooks for the Identification of British Insects. vol. 1, Part 16* (2nd ed.). St. Albans, UK: Royal Entomological Society.

Whiting, M. F., Whiting, A. S., Hastriter, M. W., & Dittmar, K. (2008). A molecular phylogeny of fleas (Insecta: Siphonaptera): Origins and host associations. *Cladistics, 24*, 1–31.

Chapter

10

Flies (Diptera)

Robert D. Hall . Reid R. Gerhardt

The Diptera, or "true flies," are one of the largest and most diverse orders of insects, both morphologically and biologically. The order name means "two-winged," and refers to the fact that the hind pair of wings is greatly modified and reduced. The number of described species worldwide is estimated to be 120,000 or more. There are perhaps 20,000 species of Diptera in the Nearctic Region, a significant proportion of which is cataloged (Stone et al., 1965). Although flies with medical or veterinary significance constitute only a small fraction of these numbers, their diversity is impressive, ranging from mosquitoes to wingless ectoparasites, larvae that parasitize various animals, and species that help to decompose carrion or feces.

No other group of insects has as much impact on human and animal health as do the Diptera (Tables 10.1 and 10.2). Mosquitoes, black flies, and biting midges annoy outdoor enthusiasts as well as livestock, pets, and other domestic or wild animals. Filth flies associated with cattle, hog, and poultry operations can annoy nearby residents and are frequently the focus of litigation. The ubiquitous house fly is an effective mechanical vector of many pathogens associated with enteric diseases. The depredation of blood-sucking and myiasis-producing flies has an adverse effect on the productivity and profitability of animal agriculture worldwide.

No other group of insects exhibits the number or diversity of vector relationships that have evolved among the Diptera (Table 10.1). The two-volume treatise *Flies and Disease* by Greenberg (1971, 1973) provides an exhaustive list of fly-pathogen associations. Mosquitoes stand as archetypical vectors, being associated with such historically notorious diseases as malaria, encephalitis, yellow fever, and human filariasis. The story of the United States Yellow Fever Commission in Cuba in 1900, and the names Carlos Finlay, Walter Reed (for whom the United States Army Medical Center in Washington, DC, is named), and L. O. Howard are familiar to most students of medicine. Such is the importance of mosquitoes that some institutions offer a separate course in **culicidology**, the study of mosquitoes. Insect-vectored tropical diseases such as malaria, filariasis, leishmaniasis, and onchocerciasis currently affect almost half a billion humans worldwide, with about 3.5 billion rated at risk.

Flies are occasionally of direct use to humans. Knowledge of the taxonomy and biology of some necrophilous species makes them useful under certain circumstances in determining how long a body has been dead. This subspecialty of medical entomology, called **medicocriminal** or **forensic entomology**, is readily accepted in judicial circles.

Additional information regarding the medical and veterinary importance of Diptera is provided by Horsfall (1962), Smith (1973), Harwood and James (1979), Williams et al. (1985), and Lancaster (1986).

TAXONOMY

The order Diptera is divided by most authorities into two suborders, the Nematocera and the Brachycera (Table 10.3). The Nematocera are typified by mosquitoes and other flies with conspicuously long antennae. The Brachycera include horse flies, deer flies, house flies, and other flies with short antennae. The Brachycera are subdivided into four infraorders: Tabanomorpha, including the horse flies and deer flies; Xylophagomorpha, which includes groups generally having no medical or veterinary importance; Stratiomyomorpha, including the soldier flies; and Muscomorpha, or "circular-seamed" flies, often called Cyclorrhapha. The Muscomorpha in turn is divided into the Aschiza and Schizophora, and the latter into two sections, the Acalyptratae and Calyptratae. This taxonomic scheme is essentially that proposed by McAlpine et al. (1981) and followed by Borror et al. (1989). A catalog of the Diptera of America north of Mexico is provided by Stone et al. (1965).

Various keys are available for identifying adult flies. Keys to the families and genera of most Nearctic Diptera are presented in McAlpine (1981b). The flies of western North America are treated by Cole (1969). The key in Borror et al. (1989) is adequate for identification

Table 10.1 Major Fly-Borne Diseases and Related Problems Affecting Human Health

Family	Diseases and Other Health-Related Problems	Geographic Occurrence
Psychodidae	Bartonellosis	Andes Mountains of Columbia, Ecuador, and Peru
	Leishmaniasis	New World tropics; Old World tropics and temperate regions
	Sand fly fever	Mediterranean area to southern China and India
Culicidae	Dengue fever	Widespread between latitudes 40°N and 40°ES
	Encephalitis	Widespread
	Filariasis	Tropics and Mediterranean area
	Malaria	Widespread in humid tropics
	Yellow fever	Widespread in humid tropics
Simuliidae	Onchocerciasis	Tropical Africa and Americas
Tabanidae	Loiasis	Tropical Africa
	Tularemia	Widespread in Northern Hemisphere
Chloropidae	Conjunctivitis	United States (southern) and Mexico; Orient
Muscidae	Enteric diseases	Worldwide
Glossinidae	Trypanosomiasis	Tropical Africa
Calliphoridae	Enteric disease	Worldwide
	Myiasis	Worldwide
Sarcophagidae	Myiasis	Worldwide
Oestridae	Myiasis	Worldwide

Table 10.2 Major Fly-Borne Diseases and Related Problems Affecting Livestock, Poultry, and Other Domestic or Wild Animals

Family	Diseases and Other Health-Related Problems	Geographic Occurrence
Psychodidae	Leishmaniasis	New World tropics; Old World tropics and temperate regions
Ceratopogonidae	Bluetongue	Widespread
Culicidae	Malaria	Widespread in tropics
	Dirofilariasis	Widespread in tropics and temperate regions
	Encephalitis	Widespread
	Fowlpox	Widespread
	Yellow fever	Widespread in humid tropics
Simuliidae	Leucocytozoonosis	Widespread, especially North America
	Feeding damage	Worldwide
Tabanidae	Anaplasmosis	Widespread
	Tularemia	Widespread in Northern Hemisphere
	Exsanguination	Worldwide
Muscidae	Annoyance	Worldwide
	Bovine pinkeye	Northern Hemisphere (widespread)
	Exsanguination	Worldwide
Glossinidae	Nagana	Tropical Africa
Calliphoridae	Myiasis	Worldwide
Sarcophagidae	Myiasis	Worldwide
Oestridae	Myiasis	Worldwide

of most North American Diptera to the family level. The larvae of many Diptera can be identified to family with the aid of Teskey (1981b) and Foote (1991); those of synanthropic species are treated by Dusek (1971). Furman and Catts (1982) present a very usable key to both adults and larvae of medically important flies, particularly in the United States, and James (1947) covers flies that cause myiasis in humans. For identification of taxa outside the Nearctic region, students should refer to Lindner's (1949) series on Palearctic Diptera, and to Zumpt (1965) for old-world myiasis-causing flies.

MORPHOLOGY

Nematoceran larvae range in length from only a few millimeters to many centimeters, depending on the species, and usually are distinguished by having a conspicuous head capsule with opposable mandibles that move in a pincer-like horizontal plane (Figure 10.1). The general body shape ranges from minute and eel-like in the Ceratopogonidae to large and fleshy in the Tipulidae. Some nematocerans have thoracic prolegs (e.g., Chironomidae and Simuliidae) and others

Table 10.3 Taxonomic Classification and Families of Diptera of Interest to Medical and Veterinary Entomologists

Higher Taxa	Family	Common Names
Suborder NEMATOCERA	Tipulidae	Crane flies
	Bibionidae	March flies
	Mycetophilidae	Fungus gnats
	Sciaridae	Darkwinged fungus gnats
	Psychodidae*	Moth flies, sand flies
	Chaoboridae	Phantom midges
	Culicidae*	Mosquitoes
	Simuliidae*	Black flies
	Ceratopogonidae*	Biting midges
	Chironomidae	Chironomid midges
Suborder BRACHYCERA		
Infraorder TABANOMORPHA	Tabanidae*	Horse flies, deer flies
	Rhagionidae	Snipe flies
	Athericidae	Athericid flies
	Stratiomyidae	Soldier flies
Infraorder XYLOPHAGOMORPHA	[none]	[none]
Infraorder STRATIOMYOMORPHA	Stratiomyidae	Soldier flies
Infraorder MUSCOMORPHA		
Division Aschiza	Phoridae	Humpbacked flies
	Syrphidae	Flower flies, hover flies
Division Schizophora		
Section Acalyptratae	Piophilidae	Skipper flies
	Drosophilidae	Small fruit flies, vinegar flies
	Chloropidae	Chloropid flies, eye gnats
Section Calyptratae	Muscidae*	House flies, stable flies, and allies
	Glossinidae*	Tsetse
	Calliphoridae*	Blow flies
	Sarcophagidae*	Flesh flies
	Oestridae* (including Cuterebrinae, Gasterophilinae, and Hypodermatinae)	Bot flies, warble flies
	Hippoboscidae*	Louse flies
	Nycteribiidae	Spider-like bat flies
	Streblidae	Bat flies

*These families are addressed in separate chapters.

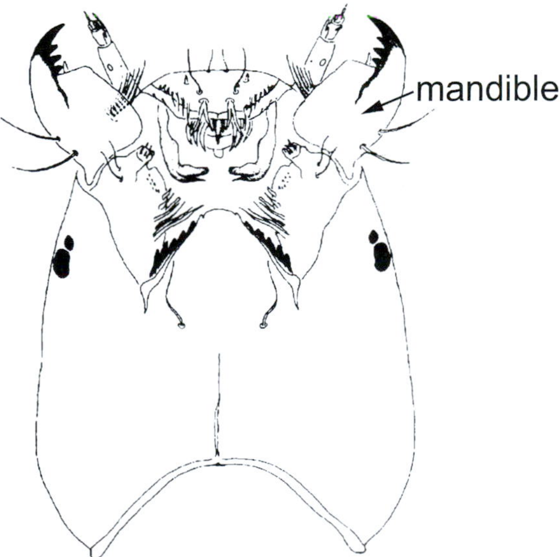

Figure 10.1 Representative nematoceran head capsule, with opposable mandibles; Chironomid midge (Chironomidae), ventral view. (redrawn from Merritt and Cummins, 1996)

have caudal structures (e.g., Simuliidae) that assist in attachment to substrates. Although the early instars of many aquatic species depend on cuticular respiration, the later instars generally respire via gills or have various adaptations that permit them to obtain atmospheric air. Mosquito larvae, for example, are highly adapted, air-breathing nematocerans that hang from the water's surface film by respiratory siphons or specialized abdominal setae.

Tabanomorpha larvae have fang-like mandibles that move in a vertical plane; the head capsule frequently is described as "incomplete posteriorly," meaning that only the anterior parts are sclerotized (Figure 10.2). The latter character is best seen in specimens that have been cleared in potassium hydroxide or lactophenol. Horse fly larvae are good examples of this group. They often have posterior respiratory tubes.

Muscomorpha larvae lack a sclerotized head capsule (Figure 10.3A) and are commonly known as maggots. At the narrow, anterior end of the 12-segmented larva is the cephalopharyngeal skeleton (Figure 10.3C) that usually bears one or two mouthhooks used for feeding and in assisting the insect in movement. The caudal end of the maggot is broader and bears the posterior

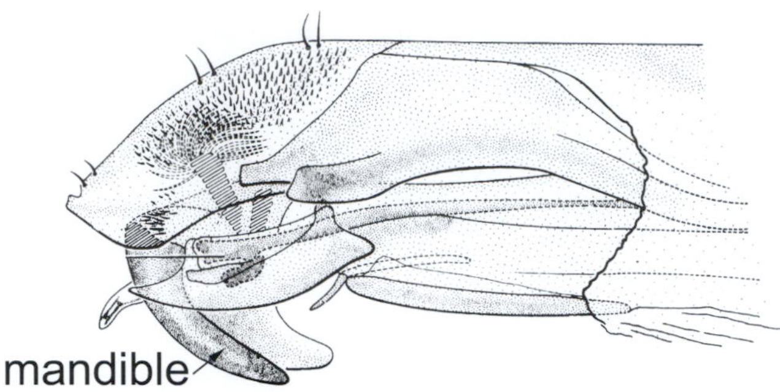

Figure 10.2 Lateral view of anterior part of *Tabanus marginalis* larva (Tabanidae), showing incomplete head capsule and vertical, fang-like mouth hook. (McAlpine et al., 1981b)

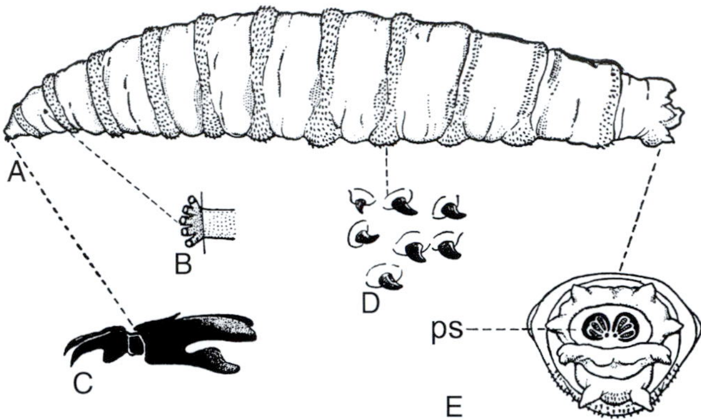

Figure 10.3 Blow-fly larva, *Chrysomya bezziana* (Calliphoridae). (A) Complete larva; (B) anterior spiracle; (C) cephalopharyngeal skeleton; (D) spines; (E) caudal end with pair of spiracular plates. (James, 1947)

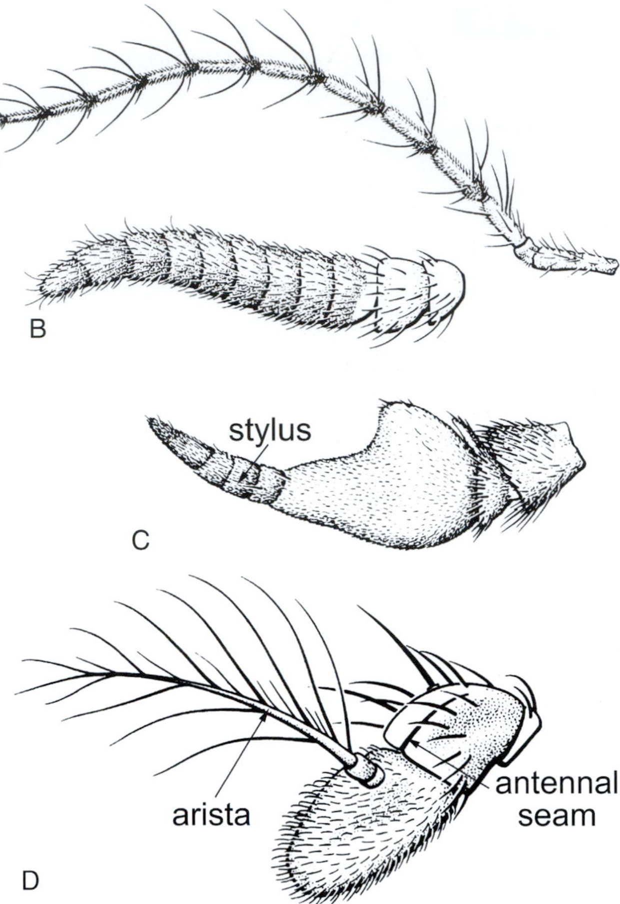

Figure 10.4 Antennae of adult flies. (A) Tipulidae (*Tipula*); (B) Simuliidae (*Cnephia*); (C) Tabanidae (*Tabanus*); (D) Drosophilidae (*Drosophila*). (McAlpine et al., 1981b)

spiracular plates (Figure 10.3E); like the cephalopharyngeal skeleton, they often are valuable for identification. The segments of the maggot typically bear spines in regular patterns (Figure 10.3D), and the larvae of some species may possess structures that vary from simple setae to large protuberances. Others, such as cattle grubs and bot flies, are rounded and robust, and their cuticle is frequently armed with stout spines. They range up to several centimeters in length.

Nematocera adults possess elongate, filamentous antennae composed of six or more segments (Figures 10.4A and 8.4B). The antennae usually are longer than the length of the head and thorax combined. A notable exception is the family Simuliidae, in which the antennae are short and compact (Figure 10.4B). In those groups that feed on blood, only the females display this behavior, doing so by means of piercing-sucking mouthparts as in mosquitoes.

Tabanomorpha adults are characterized by relatively short antennae bearing a terminal annulus, or stylus (Figure 10.4C). In general, these are large, robust flies. Like the Nematocera, only the females feed on blood. Members of the Tabanidae are good examples, being typically large, active flies whose females aggressively pursue blood meals. Their mouthparts are adapted for lacerating skin to feed on blood that pools at the wound site.

Muscamorpha adults have antennae that are aristate, bearing a large dorsal bristle (**arista**) on the apical antennal segment (Figure 10.4D). Division Aschiza, typified by the phorids and syrphids, includes those Muscomorpha lacking a frontal suture, or *lunule.* Diptera in the Shizophora have a frontal suture (Figure 10.5); this group includes a large number of species generally known as the muscoid flies. The Schizophora is perhaps the most taxonomically complex group of Diptera. Members of the Acalyptratae, the acalyptrate muscoid flies, lack a dorso-lateral seam on the second antennal segment, whereas this seam is present in the Calyptratae, the calyptrate muscoid flies (Figure 10.4D). Calyptrate muscoid flies possess posterobasal wing lobes called **calypters** (Figure 10.6) that cover the halteres. Included in the Calyptratae are the hippoboscoid flies, which are sometimes secondarily wingless.

The mouthparts of blood-feeding muscomorphan adults are of the piercing-sucking type. In contrast to other Diptera, both male and female Calyptratae suck blood in those species that exhibit this feeding style

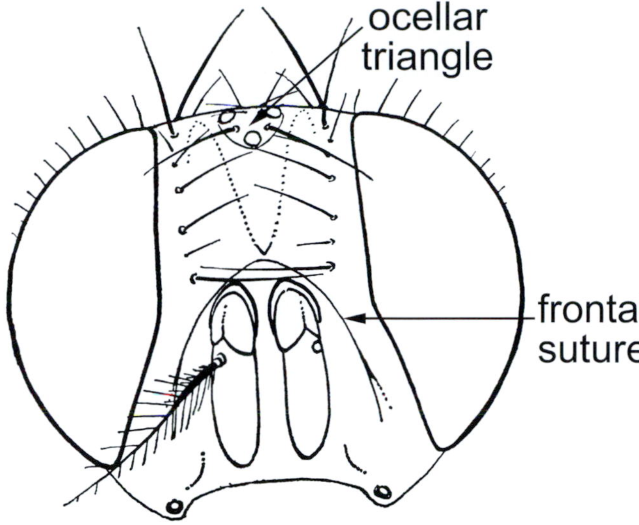

Figure 10.5 Frontal view of head of female fly, showing frontal suture and ocellar triangle at vertex. (B. Greenberg, Flies and Disease, Vol. 1, Ecology, Classification and Biotic Associations, Princeton University Press, 1971)

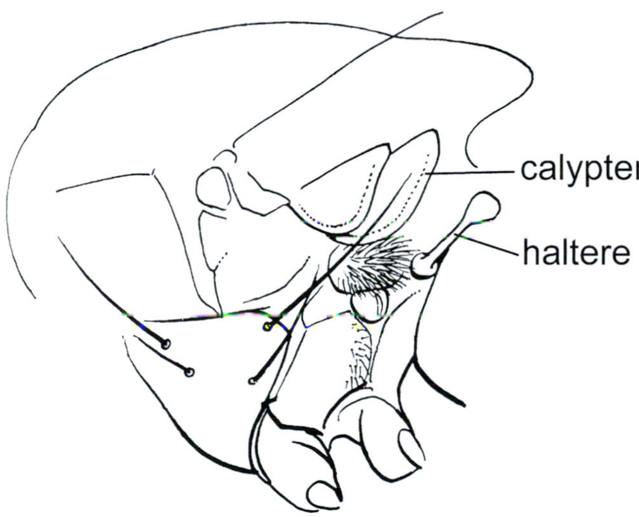

Figure 10.6 Calypterate fly, showing haltere and calypters. (B. Greenberg, Flies and Disease, Vol. 1, Ecology, Classification and Biotic Associations, Princeton University Press, 1971)

(e.g., horn flies and stable flies). Other species generally possess mouthparts that permit liquid food materials to be lapped or sponged. The latter type of mouthparts in some species have structures sclerotized enough to scarify tissue during feeding activities (e.g., face fly).

The functional pair of wings in the Diptera arise from the mesothorax. The metathoracic wings are modified to form a pair of knobbed balancing organs known as **halteres** (Figure 10.6). The wing venation is highly variable between groups and provides valuable taxonomic characters for distinguishing the families. Many dipteran adults have characteristic wing patterns, including species of biting midges, deer flies, and horse flies.

The adults of most Diptera possess distinct compound eyes; ocelli are present in a triangle on the vertex of many species (Figure 10.5). Adults are identified easily to sex, because most species exhibit some degree of sexual dimorphism. Nematoceran males often possess densely plumose antennae, and the females of blood-sucking species bear stylet-like mouthparts. The eyes of brachyceran males typically meet along the dorsal midline of the head (**holoptic**), whereas the eyes of females are more widely separated (**dioptic**). The female abdomen ends in an ovipositor (larvipositor in some species), whereas the male abdomen typically bears distinct genitalia at the terminus. In the males of some Nematocera and Brachycera, the genital segments rotate one half turn shortly after the adult fly emerges; thus, the genital capsule appears "upside down" in adults of those species. In the Schizophora, this rotation continues through a full circle, so that the genital capsule is in its normal position. A morphological approach to identification that has proven useful in the Diptera, particularly with the Muscomorpha, is the characteristic appearance of male genitalia. In many species the aedeagus, claspers, and associated structures are unique. "Pulling the tail" of male flies is a technique used by dipterists that permits detailed examination of the genital structures. Descriptions of species in some families, such as the Sarcophagidae, are based in large part on male specimens.

McAlpine (1981a) and Teskey (1981a) present comprehensive reviews of the morphology and related terminology of diptera adults and larvae, respectively.

LIFE HISTORY

The Diptera are **holometabolous**. Most dipteran females lay eggs and are thus **oviparous**. Others are **ovoviviparous**, hatching their eggs internally and thus producing motile early-instar larvae. Such flies are called **larviparous** as represented by flesh flies (Sarcophagidae). In a few dipteran groups, the developing larvae are retained within the female's body until they are ready to pupate. These flies are called **pupiparous**, and include the louse flies (Hippoboscidae) and tsetse flies (Glossinidae). The number of offspring produced per female by larviparous and pupiparous species is low compared to oviparous and ovoviviparous species.

Many dipteran species inhabit aquatic or semiaquatic environments during their immature stages. Typical examples are mosquitoes, black flies, and most horse flies and deer flies. The females of many of these hematophagous flies are capable of producing an initial batch of eggs before obtaining a blood meal, known as **autogeny**. In contrast, those species that must feed on blood prior to their producing eggs are referred to as **anautogenous**.

Although the number of larval instars varies within the Diptera, it remains generally constant for a given species. Mosquitoes and most other nematocerans have four larval instars, whereas most muscoid Diptera pass through three observable larval instars, with a fourth instar, the **prepupa**, occurring cryptically inside the pupal case. The muscoid instars usually can be distinguished morphologically: the first instar lacks anterior spiracles and generally has only one slit in each caudal spiracular plate; second and third instars bear anterior spiracles and have two and three slits, respectively, in the caudal spiracular plate. These slits are lacking in some groups; instead, the spiracular plate has many small openings (e.g., cattle grubs).

The pupae of Nematocera are **obtect**, with the appendages and other external body structures of the developing adult being discernible externally. Pupae are typically immobile; significant exceptions are the pupae of mosquitoes and a few other nematoceran families that can move by means of caudal paddles. The Brachycera have **coarctate** pupae, in which the pupa is encased within the hardened exuviae of the penultimate larval instar. The latter structure, called a **puparium**, is most frequently brown in color and is often said to resemble a pill. It retains many morphological features of the larval integument. Adult flies emerge from the pupal case by employing hydrostatic pressure from hemolymph to generate splits along predetermined lines. The head of most Schizophora has an eversible sac, the **ptilinum**, which facilitates the fly's escape from the puparium. After emergence, the ptilinum retracts through a fissure proximate to the lunule at the antennal bases.

BEHAVIOR AND ECOLOGY

An aspect of fly behavior of particular interest to medical and veterinary entomologists is the host-finding capabilities of blood-feeding species. Although various mechanisms have been described, they generally fall into two categories, olfactory and visual. A common olfactory cue used by blood-feeding insects, including many true flies, is the relative titer of carbon dioxide (CO_2) in the atmosphere surrounding, or downwind from, the host. If the goal of a mobile parasite is to locate a warm-blooded animal, exhaled CO_2 can serve as a cue for recognizing and locating potential hosts. A practical result is the widespread use of dry ice or bottled CO_2 to improve trapping success for common blood-feeding flies such as mosquitoes, black flies and no-see-ums. Other chemicals (e.g., mercaptans, octenol, and lactic acid) are used as olfactory cues by certain species. The principle means by which most insect repellents work is by inhibiting olfactory perception, thereby disrupting normal host-seeking behavior.

Visual host-finding cues are employed effectively by some flies, notably the Tabanidae. Although entomologists have not been able to prove conclusively what horse flies and deer flies actually "see," there is little question that the blood-seeking females are sensitive to black-body radiation outside the spectrum visible to humans. It has been theorized in some cases that such females sense warmth against a cool background, in the manner that thermal-vision cameras are able to scan houses for heat leaks, crops for disease-induced stress, and nocturnal battlefields for invading personnel. The shape and size of hosts also may be important to some flies as they visually recognize or orient to certain host animals. In many instances, olfactory and visual cues presumably complement each other.

Another important aspect of fly behavior is the female's ability to identify an environment suitable for development of her offspring. As with host-finding behavior, olfaction can play an important role. Necrophilous blow flies and flesh flies appear quickly after an animal dies; olfaction is almost certainly their major cue, even though the odor may not be detectable by humans. Similarly, face flies appear at cattle dung pats almost immediately after cattle defecate. Most flies, in common with many other types of insects, can perceive chemical cues at a level many orders of magnitude greater than that of humans. The females of other dipteran species similarly locate appropriate breeding sites. As examples, salt marsh mosquitoes and tree-hole mosquitoes must select aquatic habitats suitable for their eggs. The females of some blow fly and flesh fly species are highly attracted to human feces, and female screwworms are readily drawn to sores or wounds on living hosts.

Flies of medical and veterinary importance afford excellent examples of both *K*- and *r*-strategies in their life history. A few dipterans are known as **K-strategists**, typified best by tsetse, sheep keds, and other members of the Pupipara. The symbol *K* represents the carrying capacity of the environment. These flies have longer life cycles, produce fewer offspring and are particularly influenced by density-dependent mortality factors. More commonly, pest flies **r-strategists** in which large numbers of offspring are produced, each individual having a relatively small chance of survival. The symbol *r* denotes the instantaneous rate of increase for a population. These flies typically exhibit rapid growth, short life cycles, and high mortality attributable mainly to density-independent factors. House flies and other filth-breeding species, as well as mosquitoes, serve as good examples.

FAMILIES OF MINOR MEDICAL OR VETERINARY INTEREST

The major families of Diptera of medical-veterinary importance are treated in separate chapters of this book. The following discussion is provided for 14 other families of minor medical-veterinary importance, which include species that can cause problems for humans and other animals.

Tipulidae (Crane flies)

Adults are slender-bodied flies, 5 to 60 mm in length. They have long stilt-like legs, lack ocelli, and have a V-shaped mesonotal suture (Figure 10.7A). Many species are attracted to light and readily enter houses where they may be mistaken for large mosquitoes. Some are known to feed on nectar, but none bite or are able to feed on blood. Tipulid larvae have a distinct head capsule that can be retracted into the anterior thoracic segments (Figure 10.7B). They are found

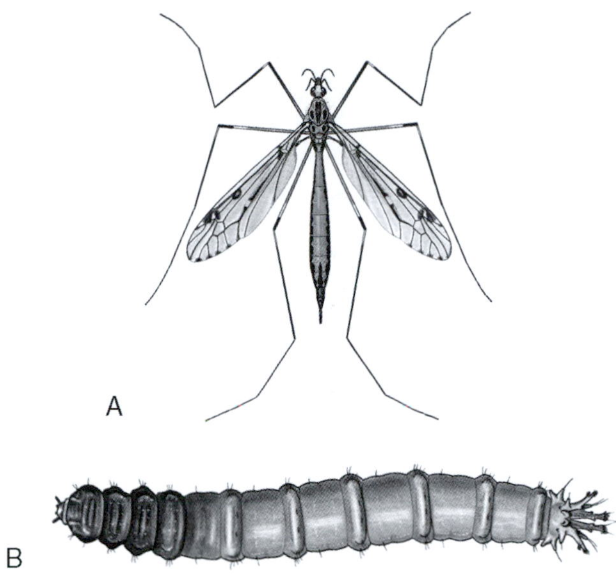

Figure 10.7 Tipulidae (*Tipula*). (A) Adult; (B) larva. (McCafferty, 1981)

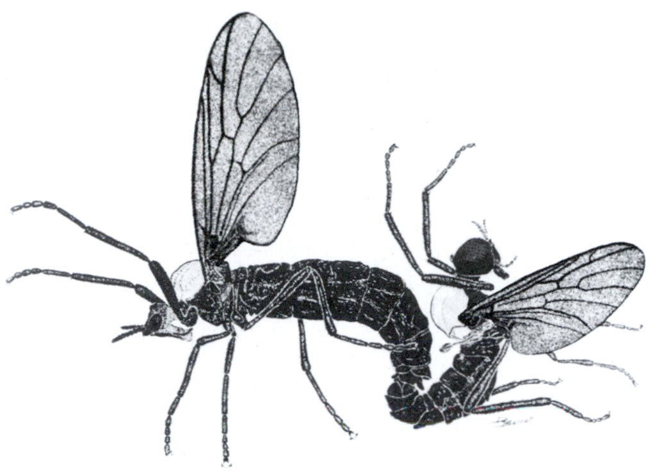

Figure 10.8 Bibionidae, love-bug (*Plecia nearctica*); pair of adults *in copula*, female at left. (Leppla et al., 1975)

in a wide range of aquatic and semiaquatic habitats and commonly are collected at the margins of streams and ponds and in moist leaf litter. A few species occur in dry soil where the larvae may be pests of grain and turf crops by feeding on the roots. Most species in temperate areas have one or two generations a year, with four larval instars and a brief pupal stage. The length of the life cycle varies from six weeks to four years, the latter being typical of some Arctic species.

The Tipulidae is a very large, cosmopolitan family of Diptera with over 60 genera and 1,500 species described in North America. Keys to both the adults and larvae of the Nearctic genera are provided by Alexander and Byers (1981) and Byers (1984). Adult and larval ecology are presented in Knizesk and Sullivan (1984) and Freeman (1967), respectively.

Bibionidae (March flies)

March flies are dark-colored flies varying in size from small to moderately large (4–10 mm). The adults (Figure 10.8) generally can be distinguished from other Nematocera by the lack of a V-shaped suture on the mesonotum, the presence of ocelli, antennae inserted below the eyes, and the presence of tibial spurs and pulvilli. Adults usually emerge in the spring and feed on flower nectar and pollen. The larvae (Figure 10.8) are scavengers and are found mostly in decaying organic materials such as forest litter, manure, and soils rich in humus (Figure 10.8). Some species cause damage to the roots of cultivated plants, especially cereal and grass crops.

Adults of the love-bug (*Plecia nearctica*) often emerge in large swarms along the Gulf and South Atlantic coasts of the United States mainly during May and September. Larvae are found in aggregations under moist, decaying materials including leaves, grass clippings, Spanish moss, and manure. They most often are seen as copulating pairs and may remain *in copula* for several days, even while feeding together on flowers. When locally abundant, flying pairs can pose a hazard to automobile travelers by obscuring vision, clogging radiators, and occasionally damaging automobile finishes. Large flights occur in the United States along the Gulf Coast of Florida westward to Louisiana and eastern Texas and southward to Central America (Denmark and Mead, 1992). *Plecia nearctica* also occurs in large numbers along the Atlantic Coast of Georgia and southern South Carolina Taxonomic keys for the larvae and adults of the six North American bibionid genera are provided in Hardy (1981). Denmark and Mead (1992) provide keys and review the biology and ecology of Nearctic *Plecia*.

Sciaridae (Darkwinged fungus gnats)

Darkwinged fungus gnats are 1 to 11 mm in length and closely resemble the Mycetophilidae except that their eyes meet above the base of the antennae. The adults (Figure 10.9A) are usually encountered in moist, shady habitats. The larvae (Figure 10.9B) feed on a wide range of materials including fungi, decaying plants, manure, and in some cases, the roots of greenhouse plants, soybeans, and clovers. *Lycoriella mali* is a major pest of commercial mushrooms, feeding on compost and all stages of mushrooms. *Bradysia* species are known to infest greenhouses where they damage plant roots and consume fungi in potting soil; they also transmit spores of plant-parasitic fungi of the genus *Pythium*. There are four larval instars. The adult-to-adult life cycle lasts 15 to 49 days in some of the economically important species. Like the myctetophilids, the sciarids may emerge inside houses from ornamental plantings and potted plants.

Sciarids pose no medical problems except for rare reports of household pets becoming ill after eating

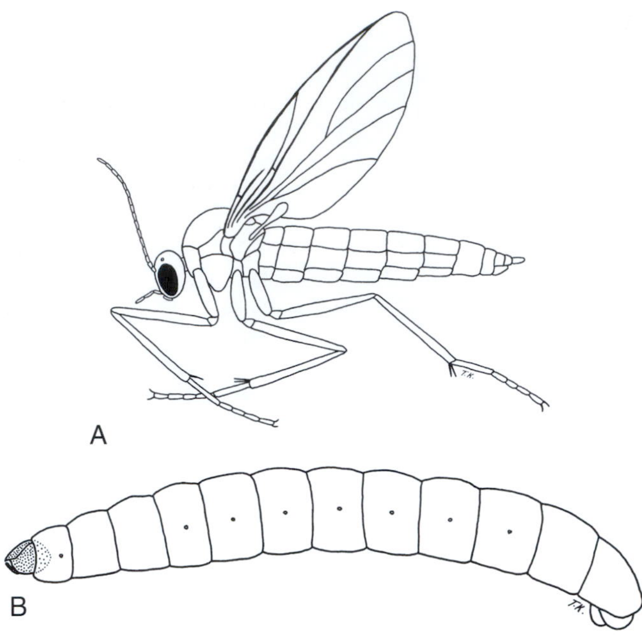

Figure 10.9 Sciaridae, darkwinged fungus gnat. (A) Adult female; (B) larva. (original by Takumasa Kondo)

adult flies. In one Florida case, a four-month-old dog died after ingesting large numbers of an unidentified sciarid species during an unusually large emergence in early May. The dog exhibited seizures and shock, and was comatose by the time it was seen by a veterinarian. The dog died a short time later after experiencing extensive internal hemorrhaging and hepatic toxicosis. Examination of the stomach contents revealed several hundred sciarid adults (G.R. Mullen, personal communication).

There are more than 100 Nearctic species, for which keys to the genera of adults and larvae are found in Steffan (1981). The ecology of some species is presented in Madwar (1937).

Chaoboridae (Phantom midges)

Adults are small (1.4–10 mm in length), mosquito-like midges without the elongate proboscis and abundant wing scales characteristic of the Culicidae (Figure 10.10A). Eggs of members of the common genus *Corethrella* are laid on the surface of water and hatch within two to four days. The larval stage averages 15 to 32 days, and the pupa is active and lasts three to six days. The transparent larvae (Figure 10.10B) are aquatic and are found commonly in lentic habitats (e.g., large lakes, small pools, bogs, small ponds). The larvae of all 19 species in the five North American genera are predators that grasp their prey with prehensile antennae. Prey include small crustaceans and aquatic insect larvae, including mosquitoes, which they sometimes eliminate from restricted habitats.

Although most adults do not feed on blood, females of the genus *Corethrella* have toothed mandibles and have been found with avian and mammalian blood

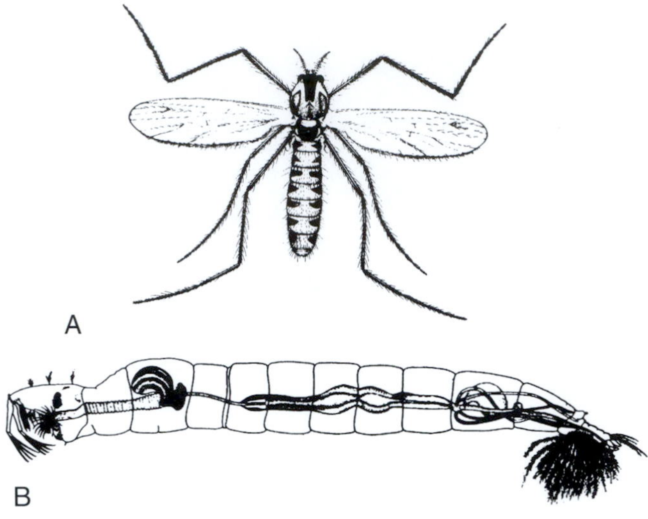

Figure 10.10 Chaoboridae, the Clear Lake gnat (*Chaoborus astictopus*). (A) Adult female; (B) larva. (Herms, 1937) (Copyright Regents of the University of California. Used with Permission)

in their digestive tracts (Williams and Edman, 1968). *Corethrella brakeleyi* and *C. wirthi* have been observed feeding on tree frogs (*Hyla* spp.) (McKeever, 1977). *Corethrella* females are attracted to the calls of male tree frogs (McKeever and French, 1991), to which *C. wirthi* can transmit a *Trypanosoma* species (Johnson et al., 1993). They do not feed on female frogs that do not call. The Clear Lake gnat (*Chaoborus astictopus*) is an inhabitant of large lakes and impoundments in the western United States. Large numbers emerge synchronously in the spring and are attracted to lights in residential and resort areas where they can cause annoyance (Herms, 1937; Linquist and Deonier, 1942).

Generic keys for chaoborid larvae and adults are found in Cook (1981). Keys to United States species of *Corethrella* are found in Stone (1968).

Chironomidae (Chironomid midges)

Adult chironomid midges (Figure 10.11A) are 1 to 10 mm long with slender legs, narrow scaleless wings, and plumose antennae in the adult males. They often are mistaken for adult mosquitoes but lack the long proboscis and are unable to feed on blood. Adults are short-lived, living only a few days to several weeks. Some imbibe honeydew and other natural sugars, but some take no food at all as adults. Most chironomid larvae are aquatic or semiaquatic and construct tubes in, or attached to, the substrate. They are often the most abundant benthic organisms and occur in all types of habitats including rivers, streams, lakes, ponds, water supplies, and sewage systems. Chironomid larvae are cylindrical and have paired prolegs on the prothoracic and last abdominal segments (Figure 10.11B). The head is heavily sclerotized and nonretractile. They have no spiracles. Many species, however, have a hemoglobin-like substance in their

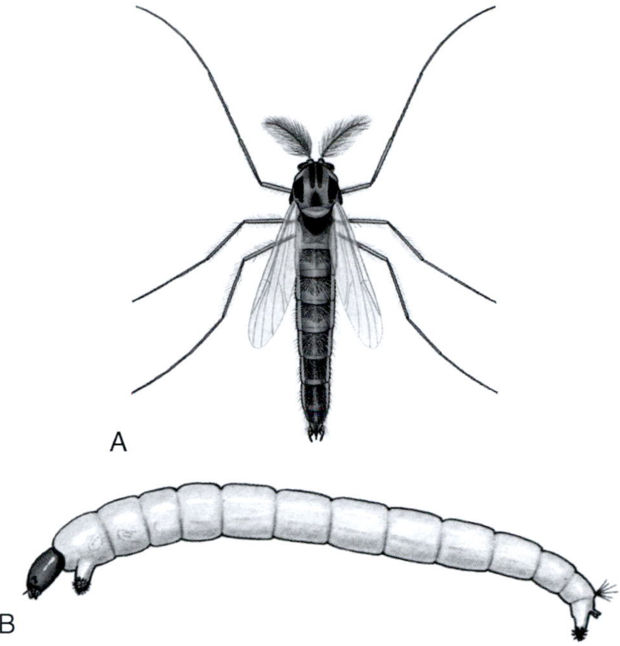

Figure 10.11 Chironomidae. (A) Adult male (*Chironomus* sp.); (B) larva (*Pseudodiamesa* sp.). (McCafferty, 1981)

hemolymph and are called bloodworms because of their pink or red color. Most species are detritus feeders that graze on aquatic substrates. Others filter drifting food particles from the water with strands of saliva or are predators on other chironomid larvae or oligochaete worms.

In addition to being mistaken for adult mosquitoes, chironomids can pose other medical and economic problems. Inhabitants of localities where large, synchronous emergences occur can develop allergies to the larval hemoglobin that is carried over from the larva to the adult and becomes airborne as the bodies of the adults decompose (Cranston, 1988). Larval hemoglobin also can induce allergies in workers who process bloodworms into fish food for aquaria. Large chironomid emergences from polluted bodies of water are common and may cause local annoyance to humans, in addition to economic damage to machinery, paint finishes, automobiles, and airplanes (Ali, 1991). Several serogroups of *Vibrio cholerae*, the bacterium responsible for cholera, have been isolated from chironomid egg masses and from the cuticle of adults. This suggests that they may be involved in the maintenance and movement of *V. cholerae* in and between bodies of water (Broza et al., 2005). Large numbers of adult midges can discourage tourism and cause contamination of materials in food processing, pharmaceutical, and manufacturing plants. Larvae that occur in water-storage and water-distribution systems can pass through taps into homes (Bay, 1993).

The Chironomidae are a large family distributed worldwide with more than 130 genera and 700 species in North America (Oliver, 1981). Armitage et al. (1995) give an overall account of the biology and ecology of chironomids.

Rhagionidae (Snipe flies)

Adult snipe flies (Figure 10.12A) are 4 to 15 mm in length with long legs, often spotted wings, and distal antennal flagellomeres forming a slender stylus (Figure 10.12A). Most prey on other insects, except that females of *Symphoromyia* in western North America and *Spaniopsis* in Australia suck blood. Larvae (Figure 10.12B) are predatory and usually are found near the surface of moist soil in meadows and steep, well-drained slopes usually associated with mosses, woodland grasses, willows, and alders.

In California, *Symphoromyia* adults are active from April through mid July. They readily attack humans, deer, cattle, and horses, usually inflicting a painful bite around the head. Most of the species studied appear to be anautogenous and univoltine. Although they may be annoying to humans, livestock, and wildlife, they have not been implicated in the transmission of any disease organisms (Hoy and Anderson, 1978). In the Yellowstone National Park (USA), biting activity starts in early July and continues until early September. Horses, mule deer, and humans are often attacked by swarms of females in localized areas along trails, with relatively fewer attacks outside these areas (Burger, 1995). Human responses to bites range from mildly annoying to very painful, with rare incidences of anaphylactic shock (Turner, 1979). For further information on the taxonomy and biology of the genus *Symphoromyia*, see Turner (1974), James and Turner (1981), and Burger (1995).

Athericidae (Athericid flies)

Adult athericids (Figure 10.13A) are 7 to 8 mm long and resemble the rhagionids. They differ from snipe flies by the presence of a strongly developed subscutellum, the R_1 cell being closed at the wing margin, and the absence of spurs on the foretibia. Larvae

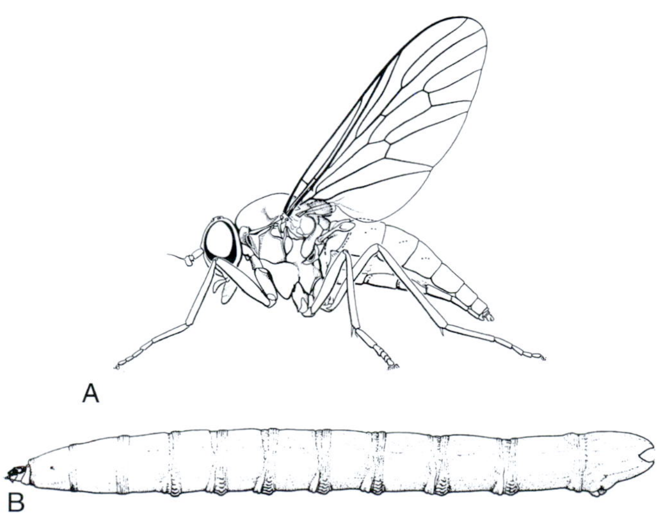

Figure 10.12 Rhagionidae. (A) Adult female (*Symphoromyia* sp.); (B) larva (*Rhagio* sp.). (McAlpine et al., 1981b)

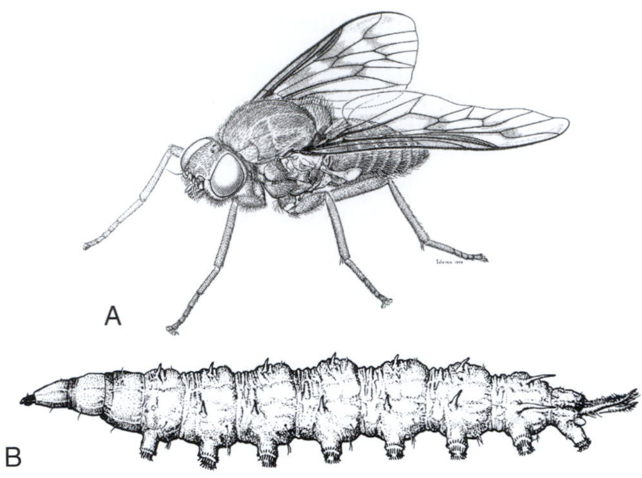

Figure 10.13 Athericidae (*Atherix*). (A) Adult female; (B) larva. (McAlpine et al., 1981b)

(Figure 10.13B) inhabit flowing water where they prey on other insect larvae. There is apparently one generation per year. Some adults prey on insects, but females of *Suragina* species are known to suck blood from humans, cattle, and some cold-blooded vertebrates (Hoy and Anderson, 1978). The family has only six Nearctic species, with all known species occurring in Texas and Mexico. Of these, three belong to the blood-feeding genus *Suragina*. Keys to the North American species are provided in Webb (1977, 1981).

Stratiomyidae (Soldier flies, Latrine flies)

Adult soldier flies (Figure 10.14A) vary from 2 to 20 mm in length. Their wings are distinctive by having all branches of the radius thickened and crowded toward the costal margin, ending before the apex of the wing. Body color may be yellow, green, blue, or black, and sometimes metallic. Many adults visit flowers, cattails, or other emergent aquatic vegetation. Larvae (Figure 10.14B) are elongate, dorso-ventrally flattened, and have a toughened or leathery integument with small, closely spaced calcareous tubercles. Many larvae are aquatic, living in a wide range of shallow, lentic habitats where they breathe at the surface through posterior spiracles. Others are terrestrial, breeding in animal wastes, decaying plants and animals, or in soil where they feed on roots of grasses.

The black soldier fly (*Hermetia illucens*) (Figure 10.14) is the stratiomyid best known to medical and veterinary entomologists and sanitary engineers. The adults are about 20 mm long, bluish-black, with yellowish-white tarsi and two lateral, translucent spots on the second abdominal segment. Mature larvae are about 20 mm long, flattened dorso-ventrally, and dull tan in color with a narrow head bearing eye spots. They develop in a broad spectrum of decaying materials including fruits, vegetables, human and animal wastes, and carrion. Eggs are laid in masses on the substrate and hatch in four days. The five larval instars last a total of about 14 days, and pupation occurs inside the last larval integument, lasting about two weeks. Black solider fly larvae can become abundant in sewer processing plants with trickle filters where they may be numerous enough to block the system. In caged-layer poultry manure, large populations of larvae can churn the manure and cause it to become liquefied and thus unsuitable for house fly larvae (Sheppard, 1983). In addition to helping to control house flies, this process reduces the total volume of manure and populations of the pathogens *Escherichia coli* O157:H7 and *Salmonella enterica* serovar *Enteritidis*. Mature larvae also may be processed into animal food (Sheppard et al., 1994; Erickson et al., 2004). Larvae occasionally are eaten by humans in overripe fruit or undercooked meat that can result in intestinal myiasis (James, 1947). James (1960, 1981) discusses the biology and provides keys to larvae and adults.

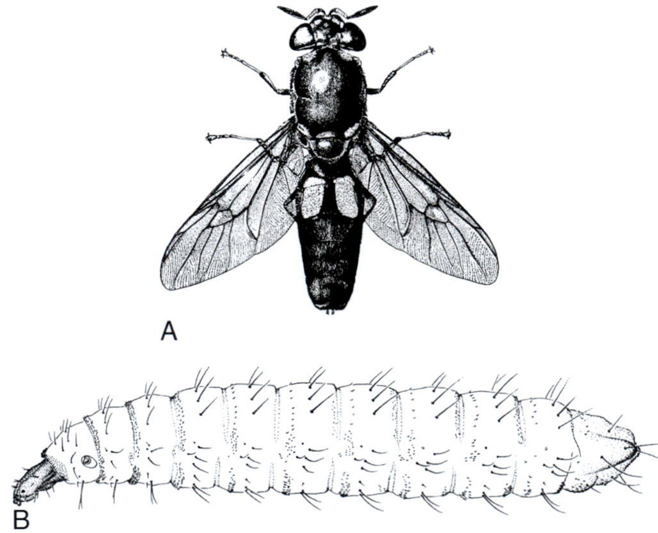

Figure 10.14 Stratiomyidae, black soldier fly (*Hermetia illucens*). (A) Adult female; (B) larva. (Gagné, 1987)

Phoridae (Humpbacked flies, Scuttle flies)

Adult phorids are 0.5 to 5.5 mm long with an enlarged thorax giving them their characteristic humpbacked appearance (Figures 10.15A and 10.15B). The hind femora are flattened, and the major bristles of the head and legs are feathered. They run in short, quick bursts and usually are found in damp places near larval habitats. Larvae (Figure 10.15C) are less than 10 mm long, lack an apparent head, and possess abdominal projections that range from being inconspicuous to large and plumose. Larval habitats are extremely varied. They include all kinds of decomposing plant and animal matter, fungi, bird nests, feces, dead insects, sewage treatment beds, and commercial mushrooms. Some larvae are internal parasitoids of other arthropods or live as commensals with social insects.

Megaselia scalaris (Figure 10.15B) is the phorid of most medical importance. The female lays eggs in fruits and vegetables, feces, and decaying plant and animal matter. Sporadic cases of facultative human

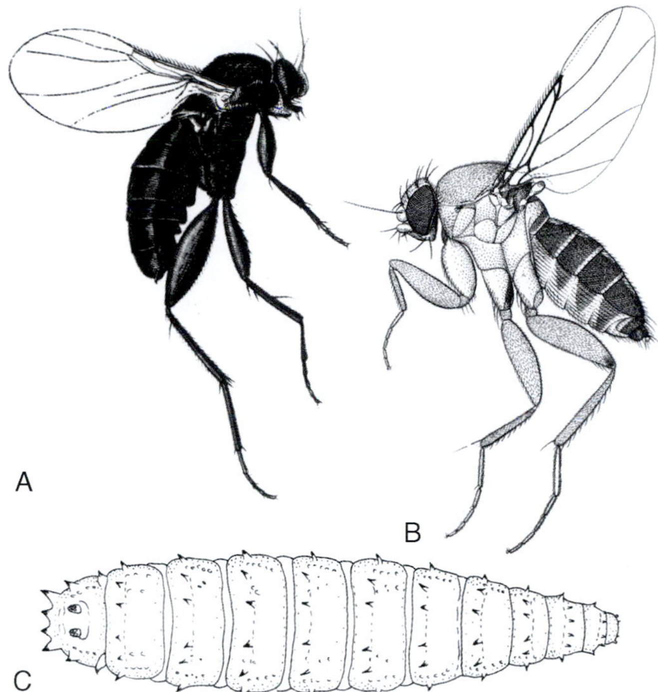

Figure 10.15 Phoridae. (A) Coffin fly (*Conicera tibialis*), adult female; (B) *Megaselia scalaris*, adult female; (C) larva (*Megaselia*). (Smith, 1986)

myiasis caused by *M. scalaris* have been documented in many areas of the world; they include cutaneous, pneumonic, nasal, gastrointestinal, urogenital, and ophthalmic myiasis (Carpenter and Chastain, 1992). Phorid larvae also commonly are associated with decomposing animal remains where they tend to be late invaders after the calliphorid flies have pupated (Smith, 1986). This fly is often a problem around mausoleums and mortuaries where the larvae develop in burial crypts, producing large numbers of adults (Katz, 1987). A small, black, European species called the coffin fly (*Conicera tibialis*) (Figure 10.15A) commonly is associated with interred human remains that have been underground for a year (Smith, 1986).

There are about 350 species and 48 genera of phorid flies in North America. Keys to adults in the Nearctic region are provided in Peterson (1987). The biology, ecology, and keys for identification of Phoridae are compiled in Disney (1994).

Syrphidae (Flower flies, Hover flies)

Adults of this family vary in length from 4 to 25 mm and are distinguished by the presence of a spurious vein between the radius and media. Many are boldly marked with black and yellow transverse bands and are effective wasp mimics. Others, including *Eristalis* and *Eristalinus*, which are called drone flies, are covered with fine yellow hairs and resemble honey bees or bumble bees (Figure 10.16A). Most adults are strong fliers and often are seen hovering near flowers where they feed on nectar. They neither bite nor are

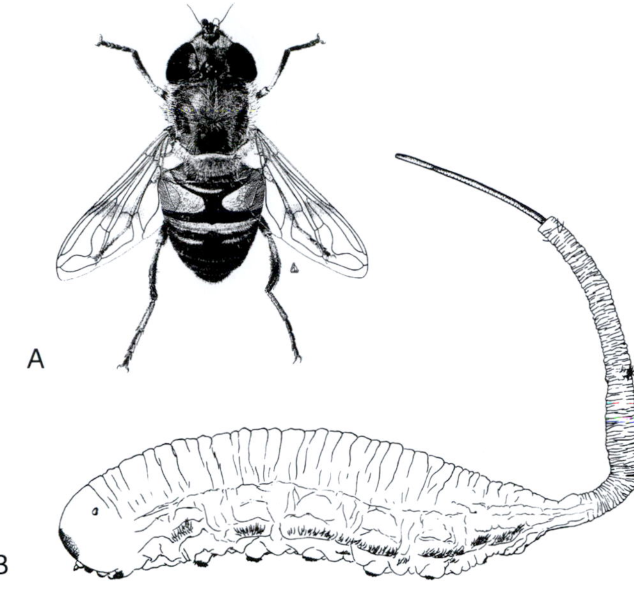

Figure 10.16 Syrphidae, drone fly (*Eristalis tenax*). (A) Adult female; (B) larva, rat-tailed maggot. (Gagné, 1987)

capable of stinging. Syrphid larvae are quite varied in form and feeding habits. Some are slug-like and live exclusively in nests of social insects. The most common larval forms are strongly flattened and are predaceous on aphids and other plant feeding insects. The larvae of *Eristalis* and *Eristalinus* species are aquatic. They are known as rat-tailed maggots because of their long, retractable caudal segment bearing the posterior spiracles, which can be extended two to three times the length of the body (Figure 10.16B). This extensible air tube allows the aquatic larvae to breathe air from the surface while inhabiting highly polluted water. Rat-tailed maggots, especially *Eristalis tenax*, often are found in manure-polluted water in and around confined livestock operations. They are common in wastewater treatment lagoons for livestock and human wastewater treatment facilities. Occasionally *E. tenax* causes gastrointestinal and urogenital myiasis in humans. There are over 900 species and more than 90 genera of syrphids in the Nearctic Region (Vockeroth and Thompson, 1987).

Piophilidae (Skipper flies)

Adult piophilids (Figure 10.17A) are small (ca. 5 mm in length), dark, acalypterate flies that are usually shiny black with strong black bristles. The vermiform larvae (Figure 10.17B) live in a variety of dead plant and animal materials, including carrion, bones, hides, fungi, and stored food products of animal origin. The species most likely to come to the notice of medical or veterinary personnel is the cosmopolitan cheese skipper, *Piophila casei* (Figure 10.17A). It is a pest of stored food, particularly cheeses and cured hams. The common name derives from the larva's ability to catapult itself into the air by assuming an O-shape by seizing its anal papillae with its mandibles and abruptly

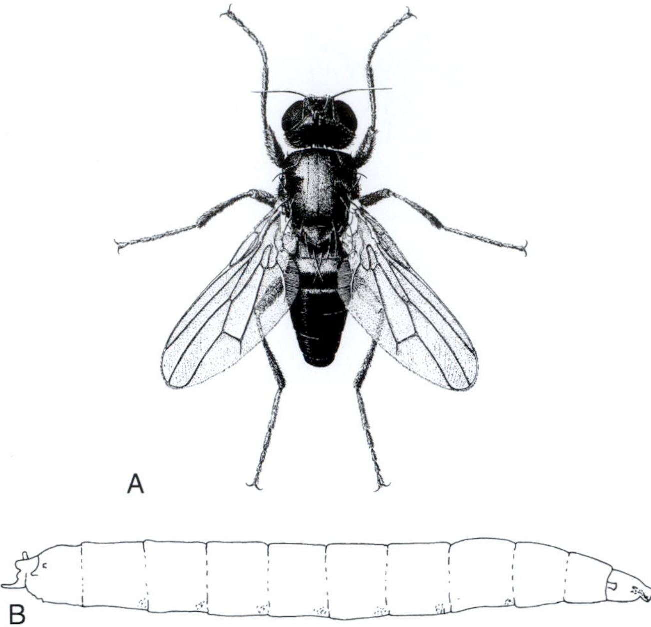

Figure 10.17 Piophilidae, cheese skipper (*Piophila casei*). (A) Adult female (Gagné, 1991); (B) larva. (Smith, 1986)

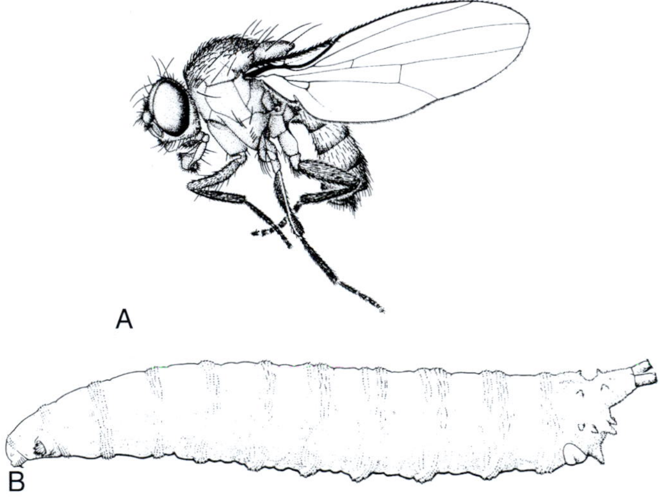

Figure 10.18 Drosophilidae, *Drosophila melanogaster*. (A) Adult female (Gagné, 1991); (B) larva. (Wheeler, 1987)

releasing its hold. Cheese skippers that are consumed by humans in contaminated food have been responsible for numerous cases of gastrointestinal myiasis (James 1947). Cheese skipper larvae (Figure 10.17B) sometimes colonize corpses in situations where the larger calliphorid and sarcophagid flies are denied access. There are 14 genera containing about 60 species in the Nearctic Region (McAlpine, 1987).

Drosophilidae (Small fruit flies)

Also commonly referred to as vinegar flies, these are generally small insects (1–6 mm) typically with red eyes. The adults (Figure 10.18A) are found around the larval habitats of decaying vegetation, plant sap, fungi, and ripe fruit. Larvae are maggot-like with stalked posterior spiracles (Figure 10.18B). Most feed on yeast and other microorganisms in the decaying substrate. Some are leaf miners, and others are parasitoids or predators of Homoptera.

Drosophilids are familiar in most households, flying around or crawling on overripe fruit. *Drosophilia melanogaster* is a common laboratory animal used extensively in genetic research. Although the flies are generally harmless, some species (especially *D. repleta*) are a potential means for mechanical transmission of pathogens when they breed in animal feces (Greenberg, 1973; Harrington and Axtell, 1994). Males of *Phortica variegata* feed on ocular secretions and have been incriminated as vectors of *Thelazia callipaeda*, an eye worm, in Europe. This is the only known instance of transmission of a vertebrate pathogen strictly by male arthropods (Otranto et al., 2006). *Drosophila* species occasionally are found in the putrid effluents from corpses. *Drosophilia funebris* has been reported to cause intestinal myiasis in humans (James, 1947). There are 17 genera and approximately 175 North American species (Wheeler, 1987).

Chloropidae (Grass flies, Eye gnats)

Adults (Figure 10.19A) are small (1.5–5 mm in length) with few large bristles and a prominent break in the costal vein of the wing just mesiad of the subcostal junction. Many adults are commonly found in grasses and other low vegetation, or visiting flowers. Larvae

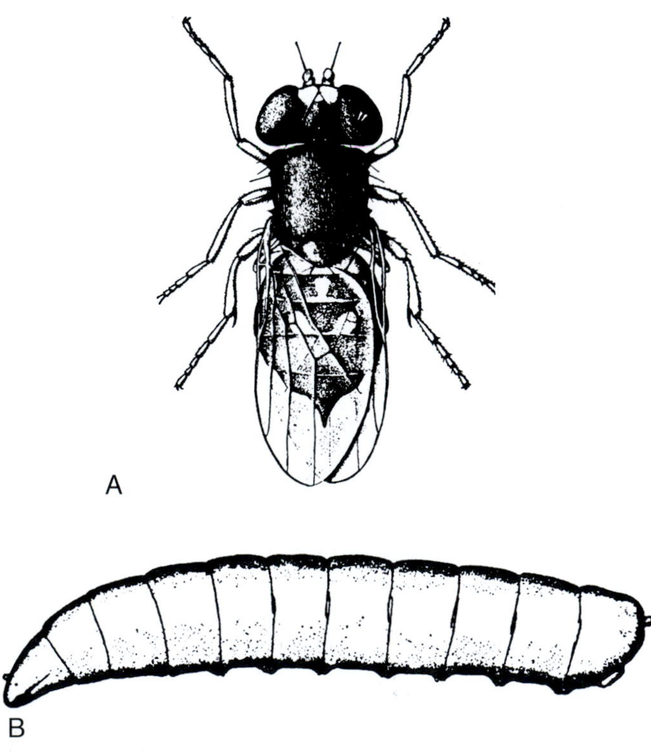

Figure 10.19 Chloropidae, eye gnat (*Liohippelates pusio*). (A) Adult female; (B) larva. (Hall, 1932)

(Figure 10.19B) lack an apparent head and have posterior spiracles and palmate anterior spiracles. Most larvae are phytophagous, feeding on stems, roots, and root hairs of grasses. The frit fly (*Oscinella frit*) and the wheat stem maggot (*Meromyza americana*) are important agronomic pests of grain crops. Other species are saprophytes, feeding mostly on decaying vegetable matter in soils, and a few are predators or gall formers.

Others are known as eye gnats, including *Liohippelates* species in North and South America and *Siphunculina* species in Asia. Eye gnats are attracted to humans and other mammals where they hover about the face, body orifices, and open wounds. The genus *Liohippelates* (formerly included in the genus *Hippelates*) occurs throughout much of North America (Sabrosky, 1980). Several species are particularly abundant in some of the sandy-soil regions of the southeastern United States (e.g., *L. pusio* and *L. pallipes*) and irrigated areas of southern California (*L. collusor*). The larvae feed on decaying organic matter in soil and can be particularly abundant in humus-enriched, cultivated soil or turf in sandy soils. The life cycle is about two weeks, and there are multiple generations each year. The adults hover around the head of humans causing annoyance, especially when they fly into eyes, nostrils, or mouths. They also are commonly found on domestic animals, especially on areas soiled with urine or manure (Greenberg, 1971).

Liohippelates species have been implicated in the mechanical transmission of several organisms that cause diseases in humans and livestock. *Treponema pertenue*, the spirochete that causes yaws, has been shown to be transmitted by *Liohippelates flavipes* in Jamaica and other Caribbean and South American locales (Kumm, 1935). Human acute conjunctivitis (pinkeye) caused by several bacterial species, is noticeably more prevalent during outbreaks of *Liohippelates* in the United States and *Siphunculina* in the Orient (Dow and Hines, 1957; Greenberg, 1973). *Liohippelates* species also have been implicated in the spread of the causative organisms of vesicular stomatitis in livestock and streptococcal infections of human skin (Taplin et al., 1967; Francy et al., 1988). Brazilian Purpuric Fever is a fulminating, highly fatal, bacterial disease of children caused by *Haemophilus influenzae* biotype *aegyptius*, which produces acute conjunctivitis in children (Harrison et al., 1989; The Brazilian Purpuric Fever Study Group, 1992). *Liohippelates puruanus* and *Hippelates neoproboscideus* have been implicated as mechanical vectors of *H. i.* biotype *aegyptius* (Tondella et al., 1994).

There are 55 genera and about 270 described species in the Nearctic Region (Sabrosky, 1987). Keys to *Liohippelates* species are provided by Sabrosky (1980). There are no effective areawide methods for controlling *Liohippelates* species. Temporary relief from their annoyance is provided by protective headnets and insect repellents containing diethyl toluamide (DEET).

PUBLIC HEALTH IMPORTANCE

On a global scale, the Diptera are the most important order of insects affecting human health (Table 10.1). Mosquitoes are the foremost group because of their role as vectors of more pathogenic organisms than any other flies. The adverse impact of malaria, mosquito-borne arboviruses (e.g., yellow fever, dengue, and encephalitis), and metazoan infections such as filariasis on humans worldwide currently exceeds that of recently publicized ailments such as Lyme disease or acquired immune deficiency syndrome (AIDS). Other human-related vector-pathogen relationships involving the Diptera are exemplified by sand flies and sand-fly fever, bartonellosis and leishmaniasis; by black flies and onchocerciasis (river blindness); and by tsetse and African trypanosomiasis (sleeping sickness). On a global scale, there are about 270 million humans infected with malaria, 90 million with lymphatic filariasis, 17 million with onchocerciasis, and 12 million with leishmaniasis. In total, almost 3.5 billion humans are currently rated at-risk from fly-borne pathogens.

The Diptera also figure prominently in public health in regard to filth flies, which are associated with materials such as dung, carrion, and garbage. House flies, stable flies, and blow flies are a few examples. Beginning shortly after Pasteur's formulation of the germ theory in the late 1870s, the link between muscoid flies and human enteric complaints gradually was understood. During the Spanish-American War (1898), this relationship became evident to United States troops in Cuba when "white-legged" muscoid flies roaming on food in mess tents were recognized as the same individuals that previously had been noted on lime-doused feces or corpses. Leland O. Howard, in what was to become one of his major entomological contributions, directed public attention to such filth-breeding flies and almost succeeded in his quest to rename the house fly the "typhoid fly" (Howard, 1911).

Many other microorganisms causing enteric disease, such as *Shigella* and *Entamoeba*, can be transmitted in a similar manner. The impact of filth flies on public health was particularly severe in the days before sanitary plumbing, effective pest management, and the ready availability of vaccines. Only the advent of the Salk vaccine for poliomyelitis in the 1950s halted long-term research on possible filth-fly involvement with the etiology of that disease.

Accounts of the enormous numbers of flies associated with corpses on battlefields are virtually as old as warfare itself; those from World War I in France and Belgium and from World War II in the Pacific region are particularly compelling. The beneficial use of certain necrophilous blow flies as surgical maggots stems almost directly from battlefield observations. Although **allantoin**, a natural antibiotic secreted by blow fly maggots, has been supplanted by more effective synthetic drugs, the use of surgical maggots remains an effective option for treating wounds, especially when they involve bone infections that are refractory to blood-borne chemotherapy.

VETERINARY IMPORTANCE

The well-being of wild and domestic animals is directly affected by the Diptera in many ways (Table 10.2). As vectors of pathogens, flies are responsible for spreading viruses such as those that cause bluetongue disease of sheep and cattle, and hemorrhagic disease of deer; rickettsial infections such as anaplasmosis; protozoans such as those causing avian malaria; and metazoan infections such as canine heartworm. *Trypanosoma*-caused nagana vectored by tsetse has eliminated most animal agriculture throughout large areas of Africa.

In many parts of the world, infestations of living tissue with fly larvae, called myiasis, can be a problem for livestock and other domestic or wild animals. At one time, myiasis caused by screwworms constituted a major impediment to cattle, hog, and sheep production in the southern United States, Mexico, and Central America, and was a major mortality factor among wildlife, especially deer. With the success of sterile-male releases and other anti-screwworm measures, this species is no longer a significant problem in North America. Other myiasis-causing flies in North America include cattle and reindeer grubs, sheep nose bots, deer and reindeer nose bots, rabbit and rodent bots, and stomach bots of horses and other equids.

In most localities, species of blow flies that cause myiasis have a direct summertime impact on most types of livestock and pets. These flies can invade wounds, sores, or body orifices such as unhealed navels of newborns, and vaginal tissues of postpartum females. The condition is perhaps best recognized by the sheep industry as fly-strike or sheep strike, and has had an extensive impact, particularly in Australia and New Zealand.

Blood-feeding flies can affect the productivity and profitability of livestock operations by causing exsanguination and, in extreme cases, anemia. Although the biting rates of diurnal species are obvious to livestock producers (e.g., horse flies), those of crepuscular or nocturnal species are largely unappreciated. Livestock on pasture, or in production systems where large numbers of animals are artificially confined, may be subject to intense biting rates from flies. Bunching, kicking, and other avoidance behavior by animals under attack by biting flies can interfere with grazing time, feed consumption, and efficiency of energy conversion.

The enormous numbers of nonbiting flies associated with livestock and livestock facilities can constitute an annoyance factor. The feeding spots and fecal spots made by flies can create sanitary and aesthetic problems. In the poultry industry, fly specks on chicken eggs are a major economic problem. Flies produced in one location may affect animals in another. For example, stable flies emanating from cattle feedlots may emigrate to proximate farmsteads where they feed on pets and companion animals. Flies such as stable flies and eye gnats breeding adjacent to dog kennels may annoy dogs and other animals even at considerable distances from the breeding sites.

PREVENTION AND CONTROL

In addition to insecticidal applications against immature or adult flies, strategies for suppression of fly-caused problems typically include personal protection. Limiting access of flies to humans or other animals by window screening, mosquito netting, or other physical barriers remains one of the cheapest and most effective forms of control. The invention and ready availability of standard 16-mesh window screening, although considered mundane by most, should actually rank as a genuine marvel of the twentieth century. Retreating behind such screening affords humans the opportunity to eat without fly-caused contamination and incessant biting by several species of blood-feeding flies. Similarly, chemical barriers to fly bites afforded by repellents are an important part of arthropod-borne pathogen management. Repellents are at the core of preventive medicine with respect to many fly-borne pathogens, especially in protecting military personnel. Typified best by mosquito repellents, these personal protectants include various formulations of DEET. The insecticide permethrin has repellent activity when used as a clothing treatment. Repellents also often are applied to livestock in an effort to reduce biting rates of flies and the annoyance of other nonbiting species.

A major emphasis of both civil and military contingency planning is preparation for the management of filth-fly populations in the event of natural or human-caused disasters. Proceeding from the axiom that, all other factors being equal, elimination of fly-breeding sites will eliminate fly populations, a major goal of vector ecologists and sanitary engineers is to reduce the amount of substrate capable of supporting nuisance species. Examples of such environmental hygiene range from the virtual replacement of outdoor human privies and pit latrines by sanitary sewage systems in urban and suburban areas to large-scale programs designed to facilitate effective manure management in areas where livestock are aggregated in large numbers.

Feces from cattle, hogs, and poultry are excellent breeding media for many species of noxious flies, most importantly the house fly. Before the advent of automobiles, dung from horses was an important source of house flies and stable flies. Dung mixed with other materials such as hay, straw, mud, and wood shavings is called **manure**. Like dung, manure is often an excellent fly-breeding medium. A biologically inescapable byproduct of livestock production is the generation of significant amounts of dung and manure. House flies in particular tend to disperse from their breeding areas, frequently invading surrounding neighborhoods where they stain and speck paint, cause annoyance, and may start local fly populations. Modern systems of livestock production therefore depend on effective removal of dung and manure; when these fail, the problem of "rural flies in the urban environment" often winds up in today's courts.

REFERENCES AND FURTHER READING

Alexander, C. P., & Byers, G. W. (1981). Tipulidae. In J. F. McAlpine et al. (Eds.), *Manual of Nearctic Diptera* (Vol. 1. pp. 153–190). Monogr. No. 27, Canada: Res. Branch, Agric.

Ali, A. (1991). Perspectives on management of pestiferous Chironomidae: (Diptera), an emerging global problem. *The Journal of the American Mosquito Control Association, 7*, 260–281.

Armitage, P. D., Cranston, P. S., & Pinder, L. C. V. (1995). *The Chironomidae: Biology and Ecology of Non-Biting Midges.* New York: Chapman & Hall.

Bay, E. C. (1993). Chironomid (Diptera: Chironomidae) larval occurrence and transport in a municipal water system. *The Journal of the American Mosquito Control Association, 9*, 275–284.

Borror, D. J., Triplehorn, C. A., & Johnson, J. F. (2005). *An Introduction to the Study of Insects.* (7th ed.) Philadelphia: Saunders.

The Brazilian Purpuric Fever Study Group. (1992). Brazilian purpuric fever identified in a new region of Brazil. *The Journal of Infectious Diseases, 165*(1), S16–S19.

Broza, M., Gancz, H., Halpen, M., & Kashi, Y. (2005). Adult non-biting midges: Possible windborne carriers of *Vibrio cholerae* non-O1 non-O139. *Environmental Microbiology, 7*, 576–585.

Burger, J. (1995). Yellowstone's snipe fly summer. *Yellowstone Science, 3*, 2–5.

Byers, G. W. (1984). Tipulidae. In R. W. Merritt, & K. W. Cummins (Eds.), *An Introduction to the Aquatic Insects of North America* (4th ed., pp. 773–800). Dubuque, IA: Kendall/ Hunt.

Carpenter, T. L., & Chastain, D. O. (1992). Faculative myiasis by *Megaselia* sp. (Diptera: Phoridae) in Texas: A case report. *Journal of Medical Entomology, 29*, 561–563.

Cole, F. R. (1969). *The Flies of Western North America.* Berkeley: Univ. of California Press.

Cook, E. F. (1981). Chaoboridae. In J. F. McAlpine (Eds.), *Manual of Nearctic Diptera* (Vol. 1., pp. 335–339). Monogr. No. 27, Canada: Res. Branch, Agric.

Cranston, P. S. (1988). Allergens of non-biting midges (Diptera: Chironomidae): A systematic survey of chironomid haemoglobins. *Medical and Veterinary Entomology, 2*, 117–127.

Denmark, H. A., & Mead, F. W. (1992). Lovebug, *Plecia nearctica* Hardy (Diptera: Bibionidae). Entomology Circular No. 350, Fla. Dept. Agric. & Consumer Serv.

Disney, R. H. L. (1994). *Scuttle Flies: The Phoridae.* London: Chapman & Hall.

Dow, R. P., & Hines, J. D. (1957). Conjunctivitis in southwest Georgia. *Public Health Reports, 72*, 441–448.

Dusek, J. (1971). Key to larvae. In B. Greenberg (Ed.), *Flies and Disease Ecology, Classification, and Biotic Associations* (Vol. 1., pp. 163–199). Princeton, NJ: Princeton Univ. Press.

Erickson, M. C., Islam, M., Sheppard, C., Liao, J., & Doyle, M. P. (2004). Reduction of *Escherichia coli* O157:H7 and *Salmonella enterica* serovar *Enteritidis* in chicken manure by larvae of the black soldier fly. *Journal of Food Protection, 67*, 685–690.

Foote, B. A. (1991). Order Diptera. In F. W. Stehr (Ed.), *Immature Insects* (Vol. 2., pp. 690–699). Dubuque, IA: Kendall/Hunt.

Francy, D. B., Moore, C. G., Smith, G. C., Jakob, W. L., Taylor, S. A., & Calisher, C. H. (1988). Epizootic vesicular stomatitis in Colorado, 1982: Isolation of virus from insects collected along the northern Colorado Rocky Mountain front range. *Journal of Medical Entomology, 25*, 343–347.

Freeman, B. E. (1967). Studies on the ecology of larval Tipulinae (Diptera, Tipulidae). *The Journal of Animal Ecology, 36*, 123–146.

Furman, D. P., & Catts, E. P. (1982). *Manual of Medical Entomology.* Cambridge: Cambridge Univ. Press.

Greenberg, B. (1971). *Flies and Disease, vol. 1. Ecology, Classification and Biotic Associations.* Princeton, NJ: Princeton Univ. Press.

Greenberg, B. (1973). *Flies and Disease, Vol. 2. Biology and Disease Transmission.* Princeton, NJ: Princeton Univ. Press.

Hall, D. G., Jr. (1932). Some studies on the breeding media, development, and stages of the eye gnat Hippelates *pusio Loew* (Diptera: Chloropidae). *American Journal of Epidemiology, 16*, 854–864.

Hardy, D. E. (1981). Bibionidae. In J. F. McAlpine, (Ed.), *Manual of Nearctic Diptera* (Vol. 1., pp. 217–222). Monogr. No. 27, Canada: Res. Branch, Agric.

Harrington, L. C., & Axtell, R. C. (1994). Comparisons of sampling methods and seasonal abundance of *Drosophila repleta* in caged-layer poultry houses. *Medical and Veterinary Entomology, 8*, 331–339.

Harrison, L. H., Da Silva, G. A., Pitmann, M., Fleming, D. W., Vranjac, A., Broome, C. V., et al. (1989). Epidemiology and clinical spectrum of Brazilian purpuric fever. *Journal of Clinical Microbiology, 27*, 599–604.

Harwood, R. F., & James, M. T. (1979). *Entomology in Human and Animal Health.* New York: Macmillan.

Herms, W. B. (1937). The clear lake gnat. *California Agricultural Experiment Station*, 607.

Horsfall, W. R. (1962). *Medical Entomology. Arthropods and Human Disease.* New York: Ronald Press.

Howard, L. O. (1911). *The House Fly, Disease Carrier. An Account of Its Dangerous Activities and the Means of Destroying It.* New York: Stokes.

Hoy, J. B., & Anderson, J. R. (1978). Behavior and reproductive physiology of the blood-sucking snipe flies (Diptera: Rhagionidae: *Symphoromyia*) attacking deer in Northern California. *Hilgardia, 46*, 113–168.

James, M. T. (1947). *The Flies that Cause Myiasis in Man.* U.S. Dept. Agric. Misc. Publ. No. 631.

James, M. T. (1960). The soldier flies or Stratiomyidae of California. *Bulletin of the California Insect Survey, 6*, 79–122.

James, M. T. (1981). Stratiomyidae. In J. F. McAlpine (Ed.), *Manual of Nearctic Diptera* (Vol. 1., pp. 497–511). Monogr. No. 27, Canada: Res. Branch, Agric.

James, M. T., & Turner, W. J. (1981). Rhagionidae. In J. F. McAlpine (Ed.), *Manual of Nearctic Diptera* (Vol. 1. pp. 483–488). Monogr. No. 27, Canada: Res. Branch, Agric.

Johnson, R. N., Young, D. G., & Butler, J. F. (1993). Trypanosome transmission by *Corethrella wirthi* (Diptera: Chaoboridae) to the green treefrog, *Hyla cinerea* (Anura: Hylidae). *Journal of Medical Entomology, 30*, 918–921.

Katz, H. (1987). Managing mausoleum pests. *Pest Control Techniques, 15*, 72–74.

Knizesk, H. M., & Sullivan, D. J. (1984). Temporal distribution of crane flies (Diptera: Tipulidae) in a southern New York woodland. *Canadian Entomologist, 116*, 1137–1144.

Kumm, H. W. (1935). The natural infection of *Hippelates pallipes* Loew with the spirochaetes of yaws. *Transactions of the Royal Society of Tropical Medicine and Hygiene, 29*, 265–272.

Lancaster, J. L., & Meisch, M. V. (1986). *Arthropods in Livestock and Poultry Production.* New York: Halsted.

Lindner, E. (1949). *Die Fliegen der Palaerktischen Region.* Stuttgart: Handbuch, E. Schweizerbartische Verlagsbuchhandlung.

Linquist, A. W., & Deonier, C. C. (1942). Flight and oviposition habits of the Clear Lake gnat. *Journal of Economic Entomology, 35*, 411–415.

Madwar, S. (1937). Biology and morphology of the immature stages of Mycetophilidae (Diptera, Nematocera). *Royal Society of London, Philosophical Transactions, Series B, 227*, 1–110.

McAlpine, J. F. (1981a). Morphology and terminology—Adults. In J. F. McAlpine et al. (Eds.), *Manual of Nearctic Diptera* (Vol. 1., pp. 9–63). Monogr. No. 27, Canada: Res. Branch, Agric.

McAlpine, J. F. (1981b). Key to families—Adults. In J. F. McAlpine et al. (Eds.), *Manual of Nearctic Diptera* (Vol. 1., pp. 89–124). Monogr. No. 27, Canada: Res. Branch, Agric.

McAlpine, J. F. (1987). Piophilidae. In J. F. McAlpine et al. (Eds.), *Manual of Nearctic Diptera* (Vol. 2. pp. 845–852). Monogr. No. 28, Canada: Res. Branch, Agric.

McAlpine, J. F. et al. (1981). *Manual of Nearctic Diptera.* (Vol. 1). Monogr. No. 27, Canada: Res. Branch, Agric.

McAlpine, J. F. et al. (1987). *Manual of Nearctic Diptera.* (Vol. 2). Monogr. No. 28, Canada: Res. Branch, Agric.

McKeever, S. (1977). Observations of *Corethrella* feeding on tree frogs (*Hyla*). *Mosquito News, 37*, 522–523.

McKeever, S., & French, F. E. (1991). *Corethrella* (Diptera: Corethrellidae) of eastern North America: Laboratory life history and field responses to anuran calls. *Annals of the Entomological Society of America, 84*, 493–497.

Oliver, D. R. (1981). Chironomidae. In J. F. McAlpine et al. (Eds.), *Manual of Nearctic Diptera* (Vol. 1. pp. 423–458). Monogr. No. 27, Canada: Res. Branch, Agric.

Otranto, D., Cantacessi, C., Testini, G., & Lia, R. P. (2006). *Phortica variegata* as an intermediate host of *Thelazia callipaeda* under natural conditions: Evidence for pathogen transmission by a male arthropod vector. *International Journal for Parasitology, 36,* 1167–1173.

Peterson, B. V. (1987). Phoridae. In J. F. McAlpine et al. (Eds.), *Manual of Nearctic Diptera* (Vol. 2., pp. 689–712). Monogr. No. 28, Canada: Res. Branch, Agric.

Sabrosky, C. W. (1980). New genera and new combinations in nearctic Chloropidae (Diptera). *Proceedings of the Entomological Society of Washington, 82,* 412–429.

Sabrosky, C. W. (1987). Chloropidae. In J. F. McAlpine et al. (Eds.), *Manual of Nearctic Diptera* (Vol. 2. pp. 1049–1067). Monogr. No. 28, Canada: Res. Branch, Agric.

Sheppard, D. C. (1983). House fly and lesser house fly control utilizing the black soldier fly in manure management systems for caged laying hens. *Environmental Entomology, 12,* 1439–1442.

Sheppard, D. C., Newton, G. L., Thompson, S. A., & Savage, S. (1994). A value added manure management system using the black soldier fly. *Bioresource Technology, 50,* 275–279.

Smith, K. G. V. (1973). *Insects and Other Arthropods of Medical Significance.* Publication 720. London: British Museum (Natural History).

Smith, K. G. V. (1986). *A Manual of Forensic Entomology.* Ithaca, NY: Cornell Univ. Press.

Steffan, W. A. (1981). Sciaridae. In J. F. McAlpine et al. (Eds.), *Manual of Nearctic Diptera* (Vol. 1., pp. 247–255). Monogr. No. 27, Canada: Res. Branch, Agric.

Stone, A. (1968). The genus *Corethrella* in the United States (Diptera: Chaoboridae). *Florida Entomologist, 51,* 183–186.

Stone, A., Sabrosky, C. W., Wirth, W. W., Foote, R. H., & Coulson, J. R. (1965). *A Catalog of the Diptera of America North of Mexico.* U.S. Dept. Agric. Handb. 276.

Taplin, D., Zaias, N., & Rebell, G. (1967). Infection by *Hippelates* flies. *Lancet, 2,* 472.

Teskey, H. J. (1981a). Morphology and terminology—Larvae. In J. F. McAlpine et al. (Eds.), Manual of Nearctic Diptera, (Vol. 1., pp. 65–88). Monogr. No. 27, Canada: Res. Branch, Agric.

Teskey, H. J. (1981b). Key to families—Larvae. In J. F. McAlpine et al. (Eds.), *Manual of Nearctic Diptera* (Vol. 1., pp. 125–147). Monogr. No. 27, Canada: Res. Branch, Agric.

Tondella, M. L. C., Paganelli, C. H., Bortolotto, I. M., Takano, O. A., Trino, K., Brandileone, M. C. C., et al. (1994). Isolamento de *Haemophilus aegyptius* associado a febre purpurica Brasileira, de cloropideos (Diptera) dos generos *Hippelates* e *Liohippelates. Revista Instit Med Trop Sao Paulo, 36,* 105–109.

Turner, W. J. (1974). A revision of the genus *Symphoromyia* Frauenfeld (Diptera: Rhagionidae). I. Introduction. Subgenera and species-groups. *Review of Biology Canadian Entomologist, 106,* 851–868.

Turner, W. J. (1979). A case of severe human allergic reaction to bites of *Symphoromyia* (Diptera: Rhagionidae). *Journal of Medical Entomology, 15,* 138–139.

Vockeroth, J. R., Thompson, F. C. (1987). Syrphidae. In J. F. McAlpine et al. (Eds.), Manual of Nearctic Diptera (Vol. 2., pp. 713–743). Monogr. No. 28, Canada: Res. Branch, Agric.

Webb, D. W. (1977). The Nearctic Athericidae (Insecta: Diptera). *The Journal of the Kansas Entomological Society, 50,* 473–495.

Webb, D. W. (1981). Athericidae. In J. F. McAlpine et al. (Eds.), *Manual of Nearctic Diptera* (Vol. 1., pp. 479–482). Monogr. No. 27, Canada: Res. Branch, Agric.

Wheeler, M. R. (1987). Drosophilidae. In J. F. McAlpine et al. (Eds.), *Manual of Nearctic Diptera* (Vol. 2., pp. 1011–1018). Monogr. No. 28, Canada: Res. Branch, Agric.

Williams, J. A., & Edman, J. D. (1968). Occurrence of blood meals in two species of *Corethrella* in Florida. *Annals of the Entomological Society of America, 61,* 1336.

Williams, R. E., Hall, R. D., Broce, A. B., & Scholl, P. J. (1985). *Livestock Entomology.* New York: John Wiley & Sons.

Zumpt, F. (1965). *Myiasis in Man and Animals in the Old World.* London: Butterworth's.

Chapter

11

Moth Flies and Sand Flies (Psychodidae)

Louis C. Rutledge ▪ Raj K. Gupta

Members of the Psychodidae are primitive Nematocera first known from the Jurassic period, with affinities to the Tanyderidae, which traditionally have been viewed as the most primitive living flies. The family is widely distributed in natural, agricultural, and urban environments of tropical, subtropical, and temperate climates. Representatives occur in all zoogeographic realms and many different terrestrial biomes, including desert, grassland, chaparral, and forest. Some species occur in mountainous regions at high altitudes.

The subfamily Psychodinae includes nonbiting species known as moth flies, or sometimes as owl flies or owl midges. Certain species of *Psychoda* and *Clogmia* (formerly *Telmatoscopus*), known as drain flies or filter flies, are common pests in buildings and in and around sewage treatment plants. Larvae of *Psychoda* and *Clogmia* have been implicated in accidental myiasis.

The subfamily Phlebotominae includes biting species known as sand flies. Many species of *Lutzomyia* and *Phlebotomus* are important biting pests and vectors of agents causing sand fly fever, Changuinola virus disease, vesicular stomatitis virus disease, Chandipura virus disease, bartonellosis, and leishmaniasis in humans, domestic animals, and wildlife. A source of confusion is that the name "sand fly" is sometimes applied to black flies (family Simuliidae) and biting midges (family Ceratopogonidae). In the vernacular the name "sand flea" is applied to all three groups and to chigoes (Siphonaptera: Pulicidae: Tunginae).

TAXONOMY

The family Psychodidae includes about 3,000 species classified into six subfamilies (Lewis et al., 1977). Subfamilies Bruchomyiinae, Trichomyiinae, Horaellinae, and Psychodinae contain only nonbiting species. Members of the Bruchomyiinae, Trichomyiinae, and Horaellinae have no known public health or veterinary relevance, but the Psychodinae includes several important nuisance species. Subfamilies Sycoracinae and Phlebotominae contain only biting species. Members of the Sycoracinae are believed to feed only on reptiles and amphibians, but many Phlebotominae are important nuisance and vector species. The Phlebotominae are known from early Cretaceous period fossils. The Psychodinae and Sycoracinae are known from late Cretaceous period fossils and are regarded as the more derived of the three subfamilies.

Sycoracinae

The subfamily Sycoracinae includes about 100 species. *Sycorax silacea* is a vector of a filarial parasite (*Icosiella neglecta*) of the frog *Rana esculenta*, which is produced commercially for human consumption in France, but the parasite is not considered to be economically important in the industry.

Psychodinae

The subfamily Psychodinae includes about 2,000 species worldwide. Quate and Vockeroth (1981) have reviewed the Nearctic Psychodinae. About 100 species classified into 18 genera occur in the United States and Canada. Two genera, *Psychoda* and *Clogmia*, include species of public health and veterinary importance. Both genera are widely distributed throughout the world. In comparison with the Phlebotominae, adult Psychodinae (Figure 11.1) have relatively short mouthparts, short antennal segments, and short legs, and hold their wings horizontal or sloping downward (roof-like) at rest. The mandibles are rudimentary or absent.

Figure 11.1 Moth fly (*Psychoda* sp.), adult female. (McCafferty, 1981)

Figure 11.2 Sand fly (*Phlebotomus papatasi*), female feeding on a human. (Public Health Image Library, Centers for Disease Control and Prevention, USA; photo by James Gathany)

Phlebotominae

The subfamily Phlebotominae includes about 700 species classified into five genera. The subfamily is distributed globally between about 50° N and 40° S, but it is absent from most oceanic islands. There are no records from Oceania, including New Zealand. About 380 species occur in the New World, including 14 in the United States and Canada. There are no records from Alaska or Hawaii. Canadian records include only British Columbia, Alberta, and Ontario. The New World genus *Lutzomyia* and the Old World genus *Phlebotomus* include species of public health and veterinary importance. Members of the New World genera *Brumptomyia* and *Warileya* and the Old World genus *Sergentomyia* are zoophilic and rarely bite humans. Adults of Phlebotominae (Figure 11.2) have relatively long mouthparts, long antennal segments, and long legs, and hold the wings sloping upward (trough-like) when at rest. The mandibles are well developed.

The Phlebotominae of North America, north of Mexico, have been reviewed by Young and Perkins (1984), and the genus *Lutzomyia* in Mexico, the West Indies, and Central and South America has been reviewed by Young and Duncan (1994). The genus *Phlebotomus* has been reviewed by Lewis (1982). Some authors treat the Phlebotominae as a separate family, the Phlebotomidae, and some treat the subgenus *Psychodopygus* of *Lutzomyia* as a separate genus. There is evidence of cryptic species in some taxa, notably in *L. longipalpis* (Bauzer et al., 2007). Many new approaches are being applied to problems of classification and identification, including electron micrography, isoenzyme electrophoresis, analysis of cuticular hydrocarbons, molecular hybridization, nucleotide sequence analysis, and oligonucleotide mapping.

MORPHOLOGY

Psychodinae

Eggs of the Psychodinae are minute, cream-colored to brown, and deposited in masses of 10 to 100 or more in or on the larval medium.

Larvae of moth flies may be platyform, or flattened dorsally, and some have spines or feathered processes along the body (Figure 11.3C). However, those of *Psychoda* and *Clogmia* are subcylindrical, elongate, up to 6 mm long, and grayish, with dark head, dorsal plates, and siphon (Figure 11.3A, B). The head is complete, exserted, and prognathous, with short antennae, lateral eyespots, and strong mandibles. The thorax and abdomen are not clearly differentiated into distinct tagmata. There are three thoracic and nine abdominal segments. The segments are secondarily divided into annuli, with two annuli comprising the thoracic and first abdominal segments and three comprising abdominal segments 2–7. The dorsal cuticle has minute spines and narrow, transverse, sclerotized plates

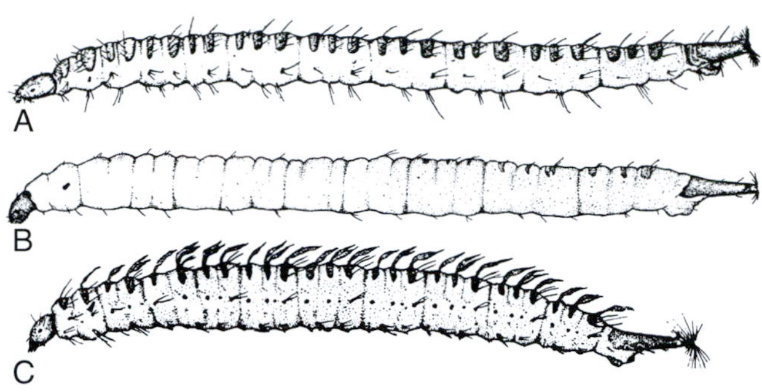

Figure 11.3 Moth fly larvae. (A) *Psychoda* sp.; (B) *Clogmia* sp.; (C) *Pericoma* sp. (Johansen, 1934).

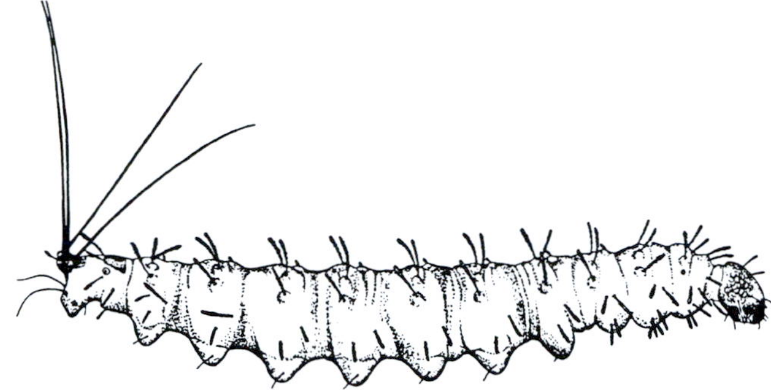

Figure 11.4 Sand-fly larva (*Phlebotomus* sp.) (Patton and Evans, 1929).

on the dorsa of the annuli. There are two pairs of spiracles, an anterior pair on the prothorax and a posterior pair at the tip of a rigid siphon terminating the abdomen. The posterior spiracles are surrounded by four lobes bearing water-repellent hairs.

Pupae of moth flies occur in or on the larval medium and may be free or attached. The pupa is generally less than 4 mm in length and obtect, with the antennae, legs, and wings visible and closely appressed to the body. The prothorax bears a pair of long, tube-like respiratory organs, and the abdomen bears numerous setae and spines.

Adults of moth flies are densely hairy, and yellowish, brownish, grayish, or black in color (Figure 11.1). They are usually 2 to 4 mm long, but vary in size from the European *Psychoda phalaenoides* with a wing span of 2 mm to the Australian *Pericoma funebris* with a wing span of 10 mm. The long, 12 to 16 segmented antennae are similar in males and females. The segments are beadlike, closely covered with short setae, with a whorl of long setae on each, those of the male longer. The palps are long, recurved and four-segmented with scattered setae. Mandibles are rudimentary or absent. Ocelli are absent. The wings are large, broadly ovate to elliptic or pointed, and densely hairy. The venation is distinctive. All the longitudinal veins separate near the base except R_2 and R_3 and M_1 and M_2. In some forms R_4 and R_5 are distinct and in others they coalesce so that the radius is only four–branched. There are no cross veins beyond the basal area. The abdomen has six to eight apparent segments.

Phlebotominae

Eggs of Phlebotominae are deposited in small, scattered groups. Eggs are about 400 μm long, elongate, dark brown, and shiny, with fine surface markings. Markings useful in classification and identification of species include polygons, ellipses, ridges, pits, mountain- or volcano-like processes, and irregular patterns (De Almeida et al., 2004).

Mature larvae of Phlebotominae are elongate, up to 5 mm in length, and whitish with a dark head and dark caudal setae (Figure 11.4). The head, thorax, and abdomen bear numerous, prominent, clavate setae that are used in classification and identification. The head is complete, exserted, and prognathous, with short antennae, lateral eyespots, clypeus, labrum, maxillae, and heavy, toothed mandibles that oppose a heavy, plate-like, serrate mentum. The thorax and abdomen are not clearly differentiated into distinct tagmata. There are three thoracic and nine abdominal segments. Unlike members of the Psychodinae, the thoracic and abdominal segments are not divided into annuli and do not have dorsal sclerites, except for a saddle-shaped plate on the eighth abdominal segment of the fourth instar. Abdominal segments 1–8 each have a medioventral proleg, or pseudopodium. There are two pairs of spiracles, an anterior pair on the prothorax and a posterior pair on the greatly reduced abdominal segment 9, which bears two (first instar) or four (second to fourth instars) long, conspicuous caudal setae adjacent to the spiracles. (Larvae of the New World genus *Brumptomyia* and the Old World species *Phlebotomus tobbi* have two caudal setae in all instars.)

Pupae of sand flies attach to substrates of the larval medium in an erect position with the exuviae of the last larval instar attached at the caudal end, and can be distinguished from those of moth flies by the clavate body setae and long caudal setae of the larval exuviae. The pupa is obtect, with antennae, legs, and wings visible and closely appressed, and pale in color, darkening prior to ecdysis. The prothorax bears a pair of short tube-like respiratory organs, and the abdomen bears numerous setae and spines.

Sand fly adults (Figure 11.2) are usually less than 5 mm long, densely hairy, and grayish, brownish, or yellowish in color. The head is small and hypognathous, with dark, conspicuous eyes, and no ocelli. The long, slender, 12–16 segmented antennae are similar in males and females. The segments are closely covered with short setae, and each segment has a whorl of long setae. The thorax is strongly humped. The wings are large, broadly ovate to elliptical or pointed, and densely hairy, with venation similar to that of members of the Psychodinae. The abdomen is 6–8 segmented. The male genitalia are conspicuous.

The mouthparts form a short proboscis with long, recurved, five-segmented palps bearing scattered setae. The mouthparts of the female include six broad, knifelike stylets (labrum, paired mandibles and maxillae, and hypopharynx) that are held within the fleshy labium when not in use. The mandibles and maxillae are toothed distally. The mandibles cut the skin with scissors-like and sawing movements while the maxillary teeth engage the sides of the wound and hold the mouthparts in place. Blood is taken from a subcutaneous pool produced by injury to the vessels. The food canal is formed by apposition of the labrum above and the hypopharynx, which contains the salivary duct, below. Components of the saliva that facilitate the blood feeding process include spreading, anti-clotting, and vasodilatory factors. Compared with those of females, the mouthparts of males are generally much reduced, with few or no teeth (Silva and Grunewald, 2000). Males do not bite but have been observed to take blood from wounds made by the females.

LIFE HISTORY

Psychodinae

Moth flies breed in aquatic and semi-aquatic habitats. Breeding sites of *Psychoda* and *Clogmia* include seashores, margins of streams and ponds, rice fields, ditches, tree holes, sewers, sewage lagoons, sewage treatment plants, outfalls, cesspits, cesspools, septic tanks, sumps, urinals, and drains. The eggs are deposited in gelatinous masses of 20 to 100 and hatch in about two days. Parthenogenesis has been observed in the Old World species *Psychoda severini*. Larvae (Figure 11.4A, B, C) develop in aquatic surface films, floating vegetation, mud, manure, and similar wet or moist organic media, where they feed on decaying organic matter, bacteria, fungi, algae, and other microorganisms. The larval period is nine to 15 days, and the pupal period is one to two days. Larvae of *Psychoda alternata* are highly tolerant of pollution, low dissolved oxygen, low pH, and high temperatures.

Phlebotominae

Sand flies breed in humid, terrestrial habitats. Breeding sites include cracks and crevices of soil, manure, rocks, masonry, rubble, forest litter, tree hollows, tree crotches, termite mounds, animal burrows, nests, poultry houses, barns, stables, homes, privies, cesspools, cellars, wells, and other dark, moist locations where organic material is present. Several important neotropical species breed in the litter of the forest floor (e.g., *L. gomezi, L. panamensis, L. pessoana*, and *L. trapidoi*). Two important Eurasian species, *P. papatasi* and *P. argentipes*, breed in organic soil in and around stables, barns, and houses. *Phlebotomus perfiliewi* breeds in farm manure in Italy, and *P. caucasicus* breeds in rodent burrows in central Asia.

Sand flies may be autogenous or anautogenous. Females of autogenous species complete the first gonotrophic cycle without taking blood but require one or more blood meals to complete each subsequent cycle. Females of anautogenous species require one or more blood meals to complete each cycle, including the first. Presently, autogeny is known in only a few species, including *Lutzomyia lichyi, Phlebotomus bergeroti, P. kazeruni*, and *P. papatasi*. In one laboratory study, 8% of female *P. papatasi* were autogenous, with a mean clutch size of 12 compared with mean clutch sizes of 60 and 70 in two groups of anautogenous females. Multiple blood meals in a single cycle have been demonstrated in several species. Females of most species complete multiple cycles. Females of *P. argentipes* may complete as many as four cycles during their normal life span.

Phlebotomus papatasi illustrates a typical pattern of egg maturation following a blood meal. Ingested blood cells begin to break down six to 18 hours after feeding. The peritrophic envelope matures at 24 hours. Digestion, absorption, and assimilation of the blood and maturation of the eggs within the ovarian follicles are usually completed in five to eight days. Approximately 30 to 60 eggs are produced in each gonotrophic cycle.

The eggs hatch in four to 20 days. Larvae feed on decaying organic matter, fungi, and associated microorganisms. There are four larval instars, and the period of larval development varies from 30 to 60 days. In climates with a cold winter or a long hot or dry season, there may be a diapause or quiescence in the egg stage or in the fourth larval instar lasting up to nearly a year. In *P. papatasi* the proportion of diapausing larvae increases from summer to fall and is independent of temperature. The pupal period is seven to eight days. The life span of the adult varies from two to six weeks. Kasap and Alten (2005) found the thermal requirement for development of *P. papatasi* to be 440 degree days above 20 °C. Belen and Alten (2006) and Kasap and Alten (2006) have reported extensive life table data for *P. papatasi*.

BEHAVIOR AND ECOLOGY

Psychodinae

Adult moth flies are common at lights and in and around breeding sites where suitable resting sites are available. Resting sites include buildings, drains, sewers, cesspools, septic tanks, sewage treatment plants, and other humid, protected sites. Adult moth flies feed on nectar and septic fluids. They walk with a characteristic hesitating motion, and fly noiselessly in short, discrete hops of a few centimeters. They have been known to invade buildings more than 90 m from the breeding site and may be carried as far as 1.5 km by prevailing winds.

Phlebotominae

Phlebotomine adults are found in and around the breeding sites where suitable resting places are available. Resting sites include forest litter, tree trunks and tree hollows, leaves of plants, caves, excavations, burrows and nests, livestock pens, buildings, cracks and crevices of rocks and masonry, and other dark, humid, protected sites. Neotropical forest species may be found in forest litter (e.g., *L. trapidoi*), on understory plants (e.g., *L. pessoana*), on the trunks of trees (e.g., *L. trinidadensis*), or in the forest canopy (e.g., *L. rorotaensis*). The Asian *P. caucasicus* is commonly found in animal burrows. Certain species are more or less peridomestic, including *L. longipalpis, L. verrucarum, P. argentipes*, and *P. papatasi*. In southeastern Asia, *P. argentipes* can be collected in the dark corners of houses and behind hanging clothing and pictures.

Sand flies walk with a characteristic hesitating motion and fly noiselessly in short, discrete hops of a few centimeters and in sustained flights of longer duration. Flight is inhibited by wind and rain. For example, *P. orientalis* does not fly at wind speeds above 15 kph. The flight ranges of neotropical forest species are usually less than 200 m, but some species migrate daily between the forest floor and canopy. Flight ranges of *P. argentipes* and *P. orientalis* may be 500 m or more, and those of *L. longipalpis* and *P. caucasicus* may be 1000 m or more. Longer records include a female *P. ariasi* recaptured 2.2 km from a release point in France, and a male *P. perniciosus* captured on the island of Jersey, 25 km from the nearest source on the mainland in France.

Adult sand flies feed on plant sap, nectar, and honey dew. Females of *P. papatasi* are known to pierce the stems and leaves of plants to obtain sap. Female sand flies feed on blood of vertebrates, in addition. Blood feeding is limited to areas of exposed skin such as the ears, eyelids, nose, lips, feet, and tail. Males of *L. longipalpis* fly to the hosts of blood-feeding females, and defend territories at blood-feeding sites on the host against other males. Males emit a terpenoid pheromone from glands on the abdomen to attract females, which then feed and mate within the males' territories. Males perform elaborate acoustic and visual displays during courtship and mating.

Most sand flies bite during twilight or darkness, but some, including several important vector species, bite habitually, or when disturbed, during daylight hours. *Lutzomyia panamensis, L. pessoana, L. sanguinaria*, and *L. trapidoi* of the tropical lowlands of Panama bite at temperatures above 20 °C, and *L. verrucarum* of the cool mountain valleys of Peru bites at temperatures as low as 10 °C. Some species are endophilic (e.g., *L. verrucarum* and *P. papatasi*) and others are exophilic (e.g., *L. trapidoi* and *P. perniciosus*).

Most sand flies have broad host ranges, but some are narrowly restrictive. For example, *Lutzomyia gomezi* is known to feed on birds and on members of five different orders of mammals, but *L. vespertilionis* feeds exclusively on bats. *Phlebotomus papatasi* is anthropophagous, feeding preferentially on man and the domestic dog throughout its range, but *Phlebotomus argentipes* is anthropophagous in some areas and zoophilic, feeding preferentially on cattle, in other areas.

Sand fly populations exhibit characteristic seasonal and biotopic patterns. In tropical areas, populations of most species increase during or shortly after the rainy season. For example, populations of *P. argentipes* and *P. papatasi* in India typically increase during the monsoon season and decrease during the dry season. In Africa, population densities of *P. duboscqi* and *P. martini* vary seasonally with those of the rodents on which they feed.

Light-trap collections in Panama have revealed the presence of a primarily anthropophilic association represented by *L. gomezi, L. panamensis*, and *L. dysponeta* and a primarily zoophilic association represented by *L. carpenteri, L. triramula*, and *L. camposi*. In some locations, the anthropophilic and zoophilic associations alternate by season. Similarly, collections in southern Turkey have revealed two overlapping altitudinal associations. One association is centered at 350 m altitude, and the other is centered at 1300 m altitude. *Sergentomyia theodori* and *P. tobbi* are the most numerous species in the lowland association, and *P. transcaucasicus* is the most numerous species in the highland association.

Sand fly populations in Panama are strongly correlated with the amount of forest cover, increasing from grassy to secondary forest biotopes and from secondary to mature forest biotopes. Similarly, in Kenya, where populations of *S. bedfordi* and *S. antennata* increase from thickets to open-canopy forests and from open-canopy forests to closed-canopy forests, and in Nigeria, where populations of *S. bedfordi* and *S. antennata* increase from the open plains to more heavily vegetated habitats. In Colombia, the sand fly association dominated by *Lutzomyia evansi* exhibits greater species richness, diversity, and abundance in natural forest areas than in nearby agricultural areas. These examples illustrate how spatial and temporal differences in local climate, seasons, soils, physiography, vegetation, and other environmental factors affecting survival and propagation collectively determine the distribution, abundance, diversity, and species composition of sand fly communities.

PUBLIC HEALTH IMPORTANCE

Psychodinae

Larvae of *Psychoda* and *Clogmia* occur in filter beds and settling tanks of sewage-treatment and water-treatment plants, where they feed on surface films of decaying organic matter, bacteria, fungi, algae, and other microorganisms. Adult moth flies are often annoying pests in the neighborhood of treatment facilities. Nuisance problems due to moth flies have also been reported in connection with turf production in Florida, greenhouse operations in California, and malt production in England. Moth flies are also frequent pests in and around homes and buildings, where they emerge from sumps, sink and floor drains, sewers, cesspools, septic tanks, aquariums, and other breeding sites. Moth flies have been identified as causes of allergic rhinitis and asthma in western and southern Africa. A number of cases of accidental myiasis due to moth flies have been reported, including enteric, urogenital, bronchial, and even ocular infestations. *Psychoda alternata* is the species most frequently involved in myiasis, but *Psychoda cinerea* and *Psychoda albipennis* have also been involved.

Psychoda alternata is widely distributed in North and South America, Europe, Asia, Africa, and Australia and is the most common species associated with sewage treatment plants worldwide. *Psychoda albipennis* and *P. severini* also are common in sewage treatment plants in Europe. *Psychoda cinerea* and *P. pacifica* occur in the eastern and western United States, respectively. *Clogmia albipunctatus* is widely distributed in the United States and Canada.

Phlebotominae

Sand flies are annoying biting pests in places where they are abundant, often probing and biting repeatedly before feeding to repletion, causing a sharp, pricking sensation each time. Biting by *L. verrucarum* reportedly makes sleep difficult in highly infested districts of Peru. In one study, the mean biting rate was estimated to be 20 to 50 bites per person per night, and one individual received an estimated 300 bites in a single night. Other highly anthropophilic species are *L. diabolica* in the United States, *L. gomezi, L. olmeca, L. panamensis, L. pessoana, L. sanguinaria, L. trapidoi,* and *L. ylephiletor* in Central America, *L. wellcomei* in Brazil, and *P. sergenti* and *P. papatasi* in the Old World.

The initial bites received by an individual during his or her lifetime typically induce sens itization, resulting in immediate or delayed skin reactions to subsequent bites. The reaction to the bite of *P. papatasi* is a pink or red papule about 2 to 3 mm in diameter and 0.5 mm high, which remains prominent for four or five days before gradually disappearing. Moderate to severe itching usually occurs. Individuals who become hypersensitive may develop hives, with pronounced swelling of the eyelids and lips where those sites are bitten. Prolonged exposure to sand fly bites results in eventual desensitization. Chronically exposed individuals living in areas with high sand fly populations may exhibit little or no reaction to the bites.

Many species of *Lutzomyia* and *Phlebotomus* are vectors of viral, bacterial, and parasitic pathogens of humans (Table 11.1). Zoophilic species, including species of *Brumptomyia, Warileya,* and *Sergentomyia,* as well as *Lutzomyia* and *Phlebotomus,* may be involved in the maintenance of zoonotic diseases.

Vesicular Stomatitis Virus Disease

Vesicular stomatitis virus is a member of the family *Rhabdoviridae,* genus *Vesiculovirus.* Vesiculoviruses have a distinctive bullet shape like that of most rhabdoviruses of animals. Vesicular stomatitis virus has been an experimental model for rabies virus (genus *Lyssavirus*), and much of what we know of the structure and replication of rabies virus has been derived from studies of the vesicular stomatitis virus. Vesicular stomatitis virus is an important pathogen of livestock and an occasional human pathogen. In humans, it produces an acute, self-limiting illness with fever, chills, and myalgia. Pharyngitis, oral mucosal vesicular lesions, and cervical adenopathy are characteristic. The Alagoas, Indiana, and New Jersey serotypes of vesicular stomatitis virus are widely distributed in temperate and tropical areas of North and South America.

Transmission of vesicular stomatitis virus in nature is complex and poorly understood. It is transmissible to humans, typically farmers, ranchers, and veterinarians, from vesicular fluids and tissues of infected animals, and there is also evidence of transmission by arthropods. Vesicular stomatitis virus has been repeatedly isolated from both nonbiting flies, including eye gnats (Chloropidae), anthomyiid flies (Anthomyiidae), and house flies (Muscidae) and biting flies, including sand flies, black flies (Simuliidae), and biting midges (Ceratopogonidae).

It has been hypothesized on the basis of circumstantial and experimental evidence that migratory grasshoppers, *Melanoplus sanguinipes* (order Orthoptera: family Acrididae), become infected by ingestion of grass contaminated with vesicular fluids of infected cattle and that uninfected cattle then become infected by ingestion of infected grasshoppers, completing the cycle of transmission. If this hypothesis eventually is confirmed, the migratory grasshopper is both an amplifying host and a reservoir host of vesicular stomatitis virus.

Peroral infection with, amplification of, and transmission of vesicular stomatitis virus have been demonstrated experimentally for both *Simulium vittatum* (Simuliidae) and *Culicoides sonorensis* (Ceratopogonidae). Direct, insect-to-insect transmission between infected and uninfected *Simulium vittatum* cofeeding on nonviremic deer mice has also been demonstrated. Outbreaks of vesicular stomatitis in cattle have been epidemiologically associated with high populations of black flies.

Two species of sand flies are proven vectors of vesicular stomatitis virus in nature (Comer and Tesh, 1991). *Lutzomyia shannoni* is a proven vector of the New Jersey serotype among feral swine on Ossabaw Island, Georgia, and *Lutzomyia trapidoi* is a proven vector of the Indiana serotype in Latin America. Vesicular stomatitis virus has been repeatedly isolated from both sandfly species and peroral infection, and subsequent transmission of the virus to experimental animals has been demonstrated for both. Transovarial transmission of *Vesiculovirus* has also been demonstrated for both species. *Lutzomyia apache* has been biogeographically associated with outbreaks of vesicular stomatitis in cattle in the southwestern United States.

The viral family *Rhabdoviridae,* members of which infect plants as well as animals, has the widest host range of any family of viruses. In a serological study conducted in Panama, vesicular stomatitis virus itself was found to infect members of seven different mammalian orders, including opossums, xenarthrans (sloths, anteaters, and armadillos), bats, primates, carnivores, rodents, and lagomorphs (rabbits). Other studies have established that odd-toed ungulates (horses) and even-toed ungulates (pigs, deer, pronghorns, cattle, and sheep) are also susceptible. Many of these alternate hosts undoubtedly function as amplifying hosts of the virus. In addition, *L. shannoni, L. trapidoi,* and other sand flies that transmit vesicular stomatitis virus transovarially may function, in effect, as insect reservoirs of the virus.

Table 11.1 Sand Fly-Borne Diseases of Humans (The reservoirs and sand fly vectors listed include both known and suspected species.)

Disease	Causative Agent	Geographic Distribution	Reservoirs	Sand Fly Vectors
Sand fly fever (New World)	Sand fly fever virus (Candiru, Chagres, Punta Toro serotypes)	Panama, Colombia	Rodents, Primates	*Lutzomyia trapidoi, L. ylephiletor*
Sand fly fever (Old World)	Sand fly fever virus (Naples, Sicilian serotypes)	Tropical and subtropical Europe, Asia, northern Africa	Rodents (Muridae)	*Phlebotomus papatasi, P. perfiliewi, P. perniciosus*
Changuinola virus disease	Changuinola fever virus	Central and South America	Sloths	*Lutzomyia umbratilis*
Vesicular stomatitis virus disease	Vesicular stomatitis virus (Alagoas, Indiana, New Jersey serotypes)	Tropical, subtropical, and temperate North and South America	Opossums, monkeys, porcupines, raccoons, bobcats, horses, swine, pronghorns, cattle, sheep	*Lutzomyia shannoni, L. trapidoi, L. ylephiletor*
Chandipura virus disease	Chandipura virus	India, West Africa	Hedgehogs	*Phlebotomus papatasi*
Bartonellosis	*Bartonella bacilliformis* (bacterium)	Colombia, Ecuador, Peru	None	*Lutzomyia verrucarum, L. peruensis, L. columbiana*
Cutaneous leishmaniasis (New World)	*Leishmania amazonensis,*[a,b] *L. braziliensis,*[b] *L. colombiensis, L. garnhami, L. guyanensis,*[a,b] *L. lainsoni, L. mexicana, L. naiffi, L. panamensis,*[b] *L. peruviana, L. pifanoi, L. shawi, L. venezuelensis*	Tropical and subtropical Central and South America, Mexico, United States (Texas)	Opossums, monkeys, sloths, armadillos, anteaters; various rodents (Sciuridae, Heteromyidae, Muridae, Dasyproctidae, Capromyidae, Echimyidae); mongooses, canines, cats, raccoons, horses	*Lutzomyia anduzei, L. anthophora, L. ayacuchensis, L. ayrozai, L. carrerai, L. christophei, L. diabolica, L. flaviscutellata, L. gomezi, L. hartmanni, L. intermedia, L. lichyi, L. llanosmartinsi, L. migonei, L. nuneztovari, L. olmeca, L. ovallesi, L. panamensis, L. paraensis, L. peruensis, L. pessoai, L. reducta, L. spinicrassa, L. squamiventris, L. townsendi, L. trapidoi, L. trinidadensis, L. ubiquitalis, L. umbratilis, L. verrucarum, L. wellcomei, L. whitmani, L. ylephiletor, L. youngi, L. yucumensis*
Cutaneous leishmaniasis (Old World)	*Leishmania aethiopica, L. killicki, L. major,*[b] *L.f. tropica*[a]	Tropical and subtropical Europe, Asia, and Africa	Monkeys, rodents (Sciuridae, Muridae), dogs, hyraxes	*Phlebotomus aculeatus, P. alexandri, P. ansarii, P. duboscqi, P. guggisbergi, P. longipes, P. papatasi, P. pedifer, P. rossi, P. salehi, P. sergenti*
Visceral leishmaniasis (New World)	*L. chagasi*[c]	Tropical and subtropical Central and South America	Opossums, canines	*Lutzomyia antunesi, L. cruzi, L. evansi, L. longipalpis*
Visceral leishmaniasis (Old World)	*Leishmania archibaldi, L. donovani,*[b,c] *L. infantum*[b,c]	Tropical and subtropical Europe, Asia, and Africa	Canines, rats (Muridae)	*Phlebotomus ariasi, P. alexandri, P. argentipes, P. caucasicus, P. celiae, P. chinensis, P. kandelakii, P. langeroni, P. longicuspis, P. longiductus, P. martini, P. neglectus, P. orientalis, P. perfiliewi, P. perniciosus, P. smirnovi, P. tobbi, P. transcaucasicus, P. vansomerenae*

[a]Also can cause visceral infections.
[b]Also can cause mucocutaneous infections.
[c]Also can cause cutaneous infections.

Chandipura Virus Disease

Chandipura virus is a *Vesiculovirus* with structure and replication similar to those of vesicular stomatitis virus. It was first isolated from the blood of two humans in the central Indian state of Maharashtra in 1965. In the following years, Chandipura virus was thought to be only an occasional pathogen of humans, but in 2003 an outbreak of disease attributed to Chandipura virus occurred in Andhra Pradesh state affecting 329 children and resulting in 183 fatalities (case fatality rate = 56%), and in 2004 a similar outbreak occurred in Gujarat state affecting 23 children and resulting in 18 fatalities (case fatality rate = 78%). A hospital-based study conducted in Andhra Pradesh in 2005 and 2006 concluded that Chandipura virus is the major cause of acute viral encephalitis in children in endemic areas during the early monsoon months (Tindale et al., 2008).

Chandipura virus disease is an emerging disease, primarily of children, in central India. Symptoms include fever, sensory disorders, convulsions, vomiting, diarrhea, and encephalitis leading to coma and death. Chandipura virus and neutralizing antibody against it also have been found in humans in Sri Lanka and in the west African countries of Senegal and Nigeria, but human disease has not been reported outside India.

Chandipura virus has been isolated from unidentified *Phlebotomus* and *Sergentomyia* in India and Senegal, and transovarial transmission has been demonstrated experimentally in *P. papatasi.* The virus also has been identified in primates (macaques) in Sri Lanka, insectivores (hedgehogs) in Nigeria, and even-toed ungulates (pigs, buffalo, cattle, goats, and sheep) in India.

Sand Fly Fever

The sand fly fever viruses are members of the viral family Bunyaviridae, genus *Phlebovirus.* All bunyaviruses except those in the genus *Hantavirus* are arboviruses. Some are transmitted transovarially with high frequency, and all have one or more vertebrate hosts. Most phleboviruses are transmitted by sand flies, but one, the Rift Valley fever virus, is transmitted by mosquitoes. Sand fly fever viruses include five viral serotypes (species) that have been isolated from humans: Chandiru, Chagres, and Punta Toro viruses of Central and South America and Naples and Sicilian viruses of southern Europe and North Africa, eastward to China.

Sand fly fever has been known since the early nineteenth century, and sand fly fever virus was one of the first arboviruses identified. In 1908, an Austrian military commission first demonstrated that the agent of sand fly fever is a filterable agent transmitted by the bite of *P. papatasi.* Albert Sabin, the developer of the oral polio vaccine, conducted extensive virological and entomological research on sand fly fever in the Middle East as a member of an American military commission in World War II. During the Cold War, the United States tested the sand fly fever virus on human volunteers as a potential biological warfare agent. A recent review lists sand fly fever virus as one of a number of viruses considered to be of bioterrorism importance.

Sand fly fever is also called phlebotomus fever and papatasi fever, from the scientific name of the first known vector, and in the older literature it often was called pappataci or papataci fever, from the Italian word for sand fly. Sand fly fever is a common, nonlethal, self-limiting illness of usually three days' duration. Sudden onset, headache, fever, malaise, nausea, limb, back and retro-orbital pain, and rapid defervescence (abatement of fever) are characteristic. Encephalitis may occur in infections with the Naples serotype. The intrinsic incubation period is usually three to four days, up to six days. The virus is present in the blood from one day before to two days after onset of fever.

Known vectors of sand fly fever viruses are *L. trapidoi* and *L. ylephiletor* in the New World and *P. papatasi, P. perfiliewi,* and *P. perniciosus* in the Old World. The extrinsic incubation period is about seven days, after which time the sand fly is infective for life, about a month. In Europe, epidemics of sand fly fever commonly occur in the summer and fall, corresponding with two generations of *P. papatasi.*

Sand fly fever virus and/or antibody have been found in antbirds (Passeriformes: Formicariidae), opossums, sloths, and rodents (New and Old World rats) in the New World. There have been relatively few positive findings in Old World vertebrates. For example, a study in France found only two positive sera, from a sheep and a deer, in a sample of 668 sera from large wild mammals. However, a study in Iran found 13 and 12 of 38 sera from gerbils positive for the Sicilian serotype and the Naples serotype (Karimabad strain), respectively. Gerbils are suspected reservoirs in Iran, but sand flies are thought to be the principal reservoirs elsewhere in the New and Old Worlds. Natural or experimental transovarial transmission has been demonstrated in a variety of species of *Lutzomyia* and *Phlebotomus.*

Changuinola Virus Disease

Changuinola virus is a member of the viral family Reoviridae, genus *Orbivirus.* Reoviruses are ubiquitous viruses that have been recovered worldwide from birds and mammals. Members of the genus *Orbivirus* are arboviruses that primarily infect mammals. Changuinola virus was first isolated in Panama in 1959 and is now known to occur widely in Panama, Colombia, and Brazil. It is associated with a single case of clinical illness in Panama, causing an acute, self-limiting, febrile illness.

Changuinola virus ordinarily is recovered from sand flies but also is recovered infrequently from mosquitoes. Numerous isolations have been made from *L. trapidoi* and *L. ylephiletor* in Panama and from *L. dasipodogeton, L. davisi, L. ubiquilatis, L. umbratilis,* and other unidentified sand flies in Brazil. Some isolations have been made from male sand flies, indicating that transovarial transmission occurs in nature. Changuinola virus also has been recovered from xenarthrans (sloths and armadillos) and rodents (rice rats). Antibodies against Changuinola virus are widespread in sloths but infrequent in other wild vertebrates, indicating that sloths are the normal vertebrate hosts.

Three related orbiviruses, Lebombo and Orungo viruses (mosquito-borne) and the Kemerovo virus (tick-borne) also infect humans, but to date fewer than 50 cases of human infection by orbiviruses have been described in the literature. However, the seroprevalence of Changuinola virus is high in parts of South America, and its true extent and frequency are currently unknown. Serologic studies suggest that most infections occur in childhood and are asymptomatic or produce a mild illness. The intrinsic incubation period of orbiviruses in humans is unknown, but has been estimated at six to nine days in animals. Little is known regarding the prognosis of orbiviral infections, but full recovery is expected in most, if not all, cases. No deaths have been reported.

Bartonellosis

Bartonellosis is a disease of humans caused by an α-proteobacterium,, *Bartonella bacilliformis* (Figure 11.5). The organism was named in honor of Alberto Barton, the Peruvian physician who first discovered it in the blood of a patient in 1905. The disease, bartonellosis, is also called Carrión's disease, in honor of Daniel Alcides Carrión, a Peruvian medical student who gave his life in 1885 in the course of research on the nature of the disease. The organisms are motile, aerobic, gram–negative bacilli that vary in size and shape from minute coccoid bodies to short rods up to 3 μ in length. Under natural conditions *B. bacilliformis* grows on the red blood cells and in the cytoplasm of the endothelial cells. A source of possible confusion is that the agents of trench fever and cat-scratch disease, formerly assigned to the genus *Rickettsia* and later to the genus *Rochalimaea*, are now assigned to the genus *Bartonella*, and the name "bartonellosis" is now applied to trench fever and cat-scratch disease also.

The intrinsic incubation period of bartonellosis is usually two to three weeks. The agent appears in the blood before onset of illness and may persist for years afterward. The disease occurs in two distinct clinical forms: an acute, febrile anemia called Oroya fever (from La Oroya, the city at the eastern terminus of the Central Railway of Peru, in the construction of which thousands of workers lost their lives to the fever in the nineteenth century), and a benign dermal eruption called verruga peruana (Spanish for "Peruvian wart"). Oroya fever is characterized by fever, headache, muscle and joint pain, enlargement of the lymph nodes, and severe anemia. The case-fatality rate of untreated Oroya fever is 10 to 90%. Verruga peruana may be preceded by Oroya fever or by an asymptomatic infection, with an interval of weeks or months between. Verruga peruana has a preeruptive stage characterized by muscle, bone, and joint pain, followed by an eruption that may be miliary, with widely disseminated small nodules, or nodular, with larger, deep-seated nodules most prominent on the limbs (Figure 11.6). Individual nodules may enlarge and ulcerate. Verruga peruana may persist for months or years but is seldom fatal.

Bartonellosis occurs at altitudes between 500 and 3000 m in the mountain valleys of Peru, Ecuador, and southwestern Colombia and at lower altitudes on the coastal plain of Ecuador (Alexander, 1995). Although bartonellosis has been known in Peru since the pre-Inca and Inca periods (Schultz, 1968), its association with sand flies was first suspected by C. H. T. Townsend in 1913. Since that time Townsend's findings have been amply confirmed on experimental, epidemiological, and biogeographical grounds, but the mode of transmission, whether mechanical or biological, and other details of the vector-parasite-vertebrate host relationship are still

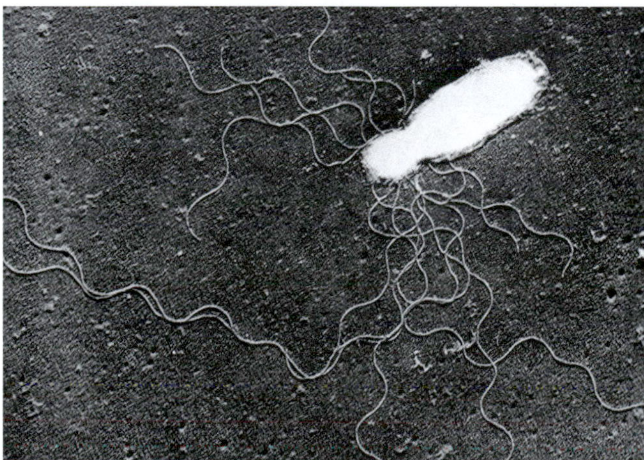

Figure 11.5 *Bartonella bacilliformis*, causative agent of Oroya fever and verruga peruana. (Courtesy of U.S. Armed Forces Institute of Pathology (AFIP) 75-8592)

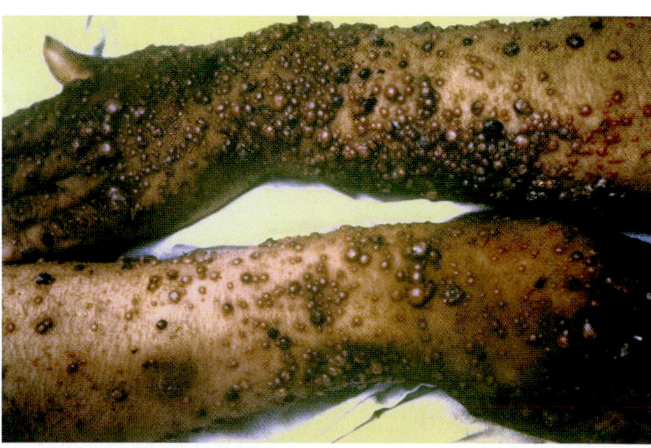

Figure 11.6 Verruga peruana, or Peruvian wart; nodular lesions on hands and forearms of Peruvian patient. (Young and Duncan, 1994)

unknown. The vectors in mountainous areas are believed to be *L. verrucarum* and *L. peruensis* in Peru and *L. columbiana* in Colombia. The vector in lowland areas of Ecuador is unknown. There are no known vertebrate reservoirs other than infected humans.

Leishmaniasis

Leishmaniasis is a complex of sand fly-borne diseases widely distributed in tropical and subtropical areas of North and South America, Europe, Asia, and Africa (Figure 11.7). The ecology of leishmaniasis varies widely in different areas. In central Asia it occurs in semiarid and arid situations. In the Mediterranean region and the Middle East it is primarily urban, and in Africa it is primarily rural. In the American tropics it is primarily a forest disease. In Peru it occurs in villages and farms of high mountain valleys. Worldwide, about 20 million people in 88 countries are at risk of leishmaniasis with an incidence of some 2 million cases per year.

Leishmaniasis is caused by numerous species of kinetoplastid protozoan parasites of the genus *Leishmania* (Table 11.1). The disease and the genus are named in honor of William B. Leishman, the British medical officer who discovered the organism early in the nineteenth century. The taxonomy of *Leishmania* is still unsettled, but advanced techniques such as electron micrography, molecular hybridization, nucleotide sequence analysis, and oligonucleotide mapping currently are being applied to the problem. The World Health Organization is promoting research to determine the genomics of leishmanial parasites. Most authors now recognize two subgenera, *Leishmania* and *Viannia*, and accept the elevation of former subspecies to species rank within these subgenera. Several taxa listed in Table 11.1 are relatively new and have not yet been accepted by all authorities. Some authorities synonymize the taxa *L. chagasi* and *L. killicki* with *L. infantum* and *L. tropica*, respectively. Exact identification of *Leishmania* species usually requires culture of the pathogen followed by immunological, biochemical, and molecular assay.

In the vertebrate host, leishmania are obligate intracellular parasites of the macrophages of the reticuloendothelial system and circulating monocytes. These forms, known as Leishman-Donovan bodies, represent the amastigote stage of development (Figure 11.8A, B). Amastigotes are round or oval, 3–7 μm in diameter, with a round nucleus, rod-like kinetoplast, and rudimentary, internal undulipodium. The parasites multiply by binary fission, producing 50 to 200 new parasites that rupture the host cell and invade, or are taken up by, other cells.

In the sand fly host the parasites develop extracellularly in the alimentary canal from ingested amastigotes. Two morphological forms have been described: the promastigote and the paramastigote. Promastigotes and paramastigotes are pleomorphic, or variable in shape, with a free undulipodium arising anteriorly. Promastigotes are elongate or pear-shaped and 5–24 μm long, with the kinetoplast situated anterior to the nucleus (Figure 11.8C, E, F, G). Paramastigotes are round or oval and 3–7 μm in diameter, with the kinetoplast situated lateral to the nucleus (Figure 11.8D, H). Promastigotes and paramastigotes may attach to the lining of the alimentary tract (haptomonad phase) or remain free-swimming (nectomonad phase). Haptomonads attach to cuticular surfaces of the foregut and hindgut by means of hemidesmosomes formed within the undulipodial tip (Figure 11.8G, H). Nectomonads may temporarily attach in the midgut by interdigitation of the undulipodium with the epithelial microvilli.

In the sandfly blood meal, amastigotes transform into short promastigotes and mature into long promastigotes within three days of ingestion. Initially, the peritrophic envelope surrounding the blood meal prevents establishment of the parasites in the gut. After natural or parasite-facilitated degradation of the envelope, the parasites develop as promastigotes and paramastigotes in the midgut (suprapylarian species, subgenus *Leishmania*) or in both hindgut and midgut (peripylarian species, subgenus *Viannia*) and subsequently migrate to the foregut.

Certain relatively short, slender, highly motile nectomonad promastigotes with very long undulapodia are regarded as infective, or metacyclic, forms

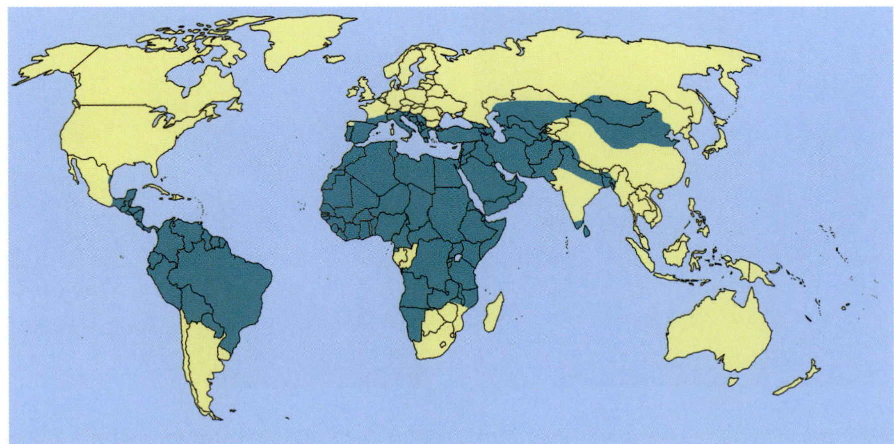

Figure 11.7 Geographic distribution of human leishmaniases.

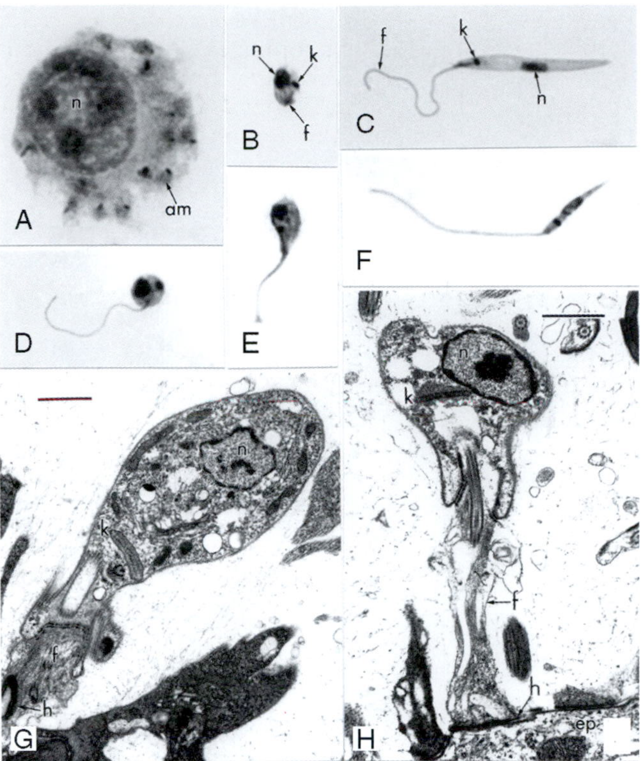

Figure 11.8 Developmental forms of *Leishmania* in sand flies. (A–F) Light micrographs. (G-H) electron micrographs, bars 1 μm. (A) Ingested macrophage containing amastigotes; (B) amastigote; (C) elongate nectomonad promastigote; (D) nectomonad paramastigote; (E) pear-shaped haptomonad promastigote dissected from foregut intima; (F) metacyclic promastigote; (G) pear-shaped haptomonad promastigote attached in hindgut; (H) haptomonad paramastigote attached in foregut. am, amastigote; ep, epithelium; f, flagellum; h, hemidesmosome; k, kinetoplast; n, nucleus. (Photos by L. L. Walters; (B) Walters et al., 1989; (C) and (G) Walters et al., 1992; (D) and (E) Walters, L.L., (1993b). Life cycle of Leishmania major (Kinetoplastida: Trypanosomatidae) in the neotropical sand fly Lutzomyia longipalpis (Diptera: Psychodidae). *Journal of Medical Entomology, 30*, 669–718. 1993; (H) Walters, 1993a)

(Figure 11.8F). Metacyclic promastigotes develop in the midgut, foregut, and mouthparts, and are transmitted to new hosts during blood feeding. Infection of the new host is facilitated by several factors present in the sand fly salivary secretion. Parasite colonization of the foregut impedes the flow of blood during feeding, resulting in repeated probing and enhanced probability of transmission. The extrinsic incubation period is species- and temperature-dependent, ranging from four to 17 days.

The relationship of parasite, vector, and vertebrate host in leishmaniasis is an ancient one. A number of fossil sand flies, *Paleomyia burmitis*, from the Cretaceous amber of Burma have been found to contain blood that is believed to be dinosaur blood. Remarkably, when the blood meals of 21 sand flies were examined microscopically, the developmental stages of a fossil leishmanial parasite, *Paleoleishmania proterus*, were identified in 10 (Poinar and Poinar, 2008). The genus *Leishmania* itself is believed to have evolved in the Palaearctic during the early Cenozoic era, from which it has subsequently dispersed to other biogeographic realms where it presently exists (Kerr, 2000).

Multiplication by binary fission occurs at many points in the life cycle. Although there is some evidence of sexual reproduction in *Leishmania*, it is believed to be very rare or absent in nature (Tibayrenc et al., 1990). Clonal reproduction by mitosis has largely replaced segregation and recombination by meiosis in the evolution of the genus. Certain successful clones are known to be stable over large geographic areas and long periods of time, and current Linnaean taxonomy does not adequately reflect the relationships of these natural clones. For example, visceral leishmaniasis is mainly caused by a single clone that is *L. chagasi* in the New World and a small component of *L. infantum* in the Old World. Research on leishmaniasis is more efficiently focused on such major clones than on Linnaean species.

Leishmaniasis is transmitted rarely in blood transfusions, in addition to the normal mode of transmission by sand flies. Recently, leishmaniasis, particularly the visceral form, has emerged as an opportunistic disease in immunocompromised people, who may acquire the disease by reactivation of latent infections or through exchange of needles and syringes among intravenous drug users. Coinfections of *Leishmania* and human immunodeficiency virus (HIV) have been reported from 35 countries, most notably Spain, Portugal, France, and Italy. Persons coinfected with *Leishmania* and HIV can infect sand flies and act as human reservoirs in zoonotic foci to shift the local epidemiology toward an anthroponotic pattern.

Leishmaniasis occurs in two principal clinical forms, known as **cutaneous leishmaniasis** and **visceral leishmaniasis**. A given species of *Leishmania* typically produces one or the other clinical form, but some, including *L. amazonensis*, *L. chagasi*, and *L. guyanensis* in the New World and *L. donovani*, *L. infantum*, and *L. tropica* in the Old World, can produce both (Table 11.1).

Cutaneous Leishmaniasis

Cutaneous leishmaniasis, also known as dermal leishmaniasis, begins with a macule at the site of inoculation by the bite of an infected sand fly. The macule develops into a papule that enlarges and typically becomes an indolent ulcer (Figure 11.9). The lesion may be single or multiple or, occasionally, nonulcerative and diffuse. The intrinsic incubation period may be a week to many months. New areas of the body become involved by extension of the primary lesions or by metastasis via the blood or lymph. Lesions of the mucous membranes of the nose, mouth, and pharynx (**mucocutaneous leishmaniasis**, also known as **mucosal** or **nasopharyngeal leishmaniasis**) may develop after the primary lesion has healed or in the absence of a recognized primary lesion (Figure 11.10). Species with known potential to produce mucocutaneous infections are indicated in Table 11.1. The term diffuse cutaneous

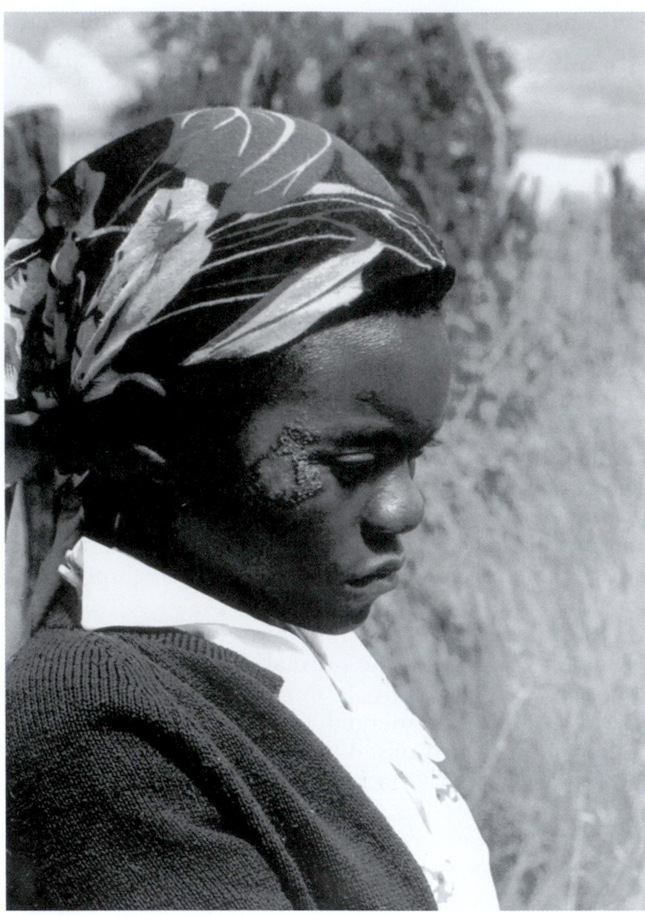

Figure 11.9 Cutaneous leishmaniasis, granulomatous and necrotic lesion on face of 16-year-old girl in Kenya. (Mehbrahtu et al., 1992)

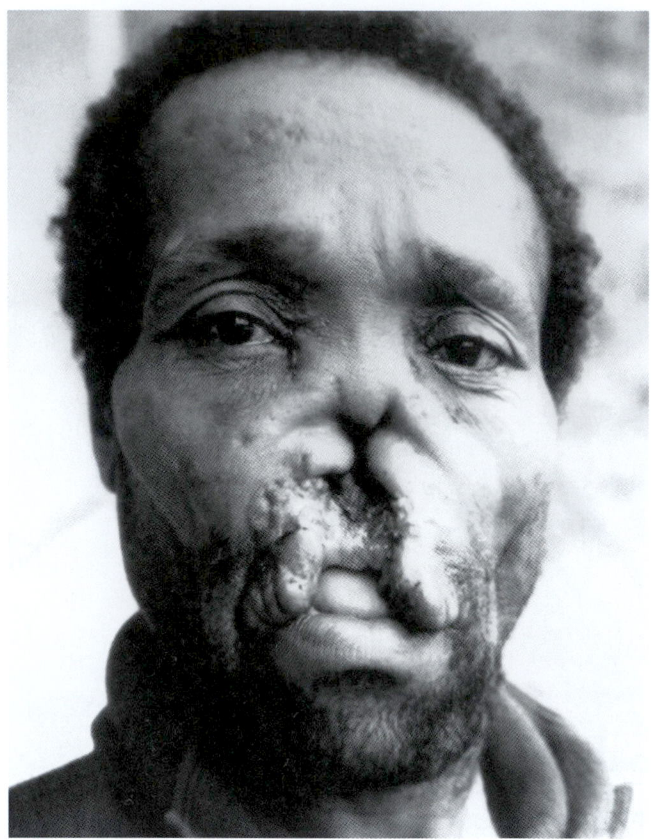

Figure 11.10 Mucocutaneous form of leishmaniasis, destruction of oral and nasal tissues of man in Bolivia. (From Walton and Valverde,1979 Used with permission of Maney Publishing.)

leishmaniasis is applied to a progressive, anergic, nonulcerative condition due to lack of delayed-type hypersensitivity in the patient.

Several clinical manifestations of cutaneous leishmaniasis have acquired specific common names. In the Old World, the condition characterized by single or multiple cutaneous ulcers due to *L. tropica* or *L. major* has been called **oriental sore**, **tropical sore**, **Aleppo boil** (or Aleppo evil), **Baghdad boil**, or **Delhi boil**. In Central America, the condition characterized by single or multiple ulcers on the face or ears due to *L. mexicana* is known as **chiclero ulcer**. In French Guiana, the condition characterized by moderate ulcers due to *L. amazonensis* or *L. guyanensis* is known as **pian bois** (French for "forest yaws"). In Peru and Ecuador, the condition is characterized by numerous, small, benign lesions due to *L. peruviana* or *L. mexicana* and is known as **uta**. In South America, mucocutaneous leishmaniasis due to *L. amazonensis* or *L. braziliensis* is known as **espundia** (Spanish for "sponge").

Cutaneous leishmaniasis may be self-limiting or chronic. Chiclero ulcer and diffuse cutaneous leishmaniasis are chronic, and diffuse cutaneous leishmaniasis is resistant to treatment. Mucocutaneous leishmaniasis persists for many years and ultimately may be fatal. Infections of *L. tropica* may recur at or near the site of the healed ulcer after apparent cure, a condition clinically known as leishmaniasis recidivans or chronic relapsing leishmaniasis. Lesions of cutaneous and mucocutaneous leishmaniasis are prone to secondary infections by bacteria and fungi and to infestation by fly larvae (myiasis). Disfiguring scars remain after healing.

Known and suspected vectors and reservoirs of cutaneous leishmaniasis are shown in Table 11.1. Transmission of *L. major* by *P. caucasicus* in semiarid regions of central Asia is a classic example of an endemic zoonosis. In this area, *P. caucasicus* breeds in burrows of gerbils (*Rhombomys opimus*) and ground squirrels (*Spermophilopsis leptodactylus*) and transmits *L. major* from animal to animal and from animals to humans. When gerbils were eradicated from the vicinity of a construction camp in Turkestan, leishmaniasis disappeared from the construction workers.

Cutaneous leishmaniasis due to *L. mexicana* occurs widely, but rarely, in south-central Texas (Bexar, Cameron, Gonzales, Uvalde, and Wells counties) and the adjoining states of Coahuila, Nuevo Leon, and Tamaulipas in Mexico. Both typical and diffuse forms of the disease have been reported. *Lutzomyia anthophora* is believed to transmit the disease among woodrats (*Neotoma micropus*), and *L. diabolica* is suspected to be a bridge vector between woodrats and humans. Seropositive coyotes and an infected cat have been found in southern Texas. An additional focus of *L. mexicana* in woodrats (*Neotoma albigula*) has been identified in Pima County, Arizona. *Lutzomyia anthophora* is considered to be the most likely vector.

Visceral Leishmaniasis

Visceral leishmaniasis, also known as **kala-azar** (from the Hindi for "black fever") and **dumdum fever** (from Dum Dum, a village and former British arsenal near Calcutta, now the site of an international airport), is a chronic systemic disease that begins with an inconspicuous cutaneous lesion at the site of inoculation by the bite of an infective sand fly. From this site the parasites are distributed through the body in the blood stream, producing chronic fever, enlargement of the lymph nodes, liver, and spleen; deficiency of red and white blood cells and blood platelets; and progressive emaciation and weakness. Early clinical manifestations are variable, and unapparent or subclinical infections occur. Unapparent and subclinical infections due to *L. tropica* have been called viscerotropic leishmaniasis. Clinically evident visceral leishmaniasis is usually fatal if not treated. The intrinsic incubation period is usually two to four months. Cutaneous lesions may appear after apparent recovery or cure and may persist for up to 20 years in the absence of treatment. Such lesions are known as post-kala-azar cutaneous leishmaniasis or post-kala-azar dermal leishmaniasis (PKDL).

Known and suspected vectors and reservoirs of visceral leishmaniasis are shown in Table 11.1. In India, Nepal, and Bangladesh, humans and sand flies are the only known hosts. Transmission of *L. donovani* by *P. argentipes* in the Ganges and Brahmaputra River basins of India is a classic example of an epidemic anthroponosis. Between 1824 and 1981 a series of nine major epidemics of visceral leishmaniasis occurred in this region. Some were so severe that entire villages were depopulated and extensive areas were abandoned. Similarly, between 1988 and 1993 a major epidemic of visceral leishmaniasis affecting 600,000 to 700,000 people and killing 40,000 occurred in the Sudan, where *L. donovani* is transmitted by *P. martini.* It is thought that the Sudanese epidemic was precipitated by conditions of malnutrition, famine, displacement of persons, and disruption of health services caused by civil war in that country.

VETERINARY IMPORTANCE

Members of the Psychodinae have no known veterinary importance.

Members of the Phlebotominae are presumably pests of livestock, pets, and wildlife in places where they are abundant, but their contribution to the overall economic loss caused by biting arthropods is not known. In addition, sand flies transmit *Leishmania* to dogs and cats and may play a role in the transmission of vesicular stomatitis virus among livestock.

Leishmaniasis

The veterinary forms of leishmaniasis are canine leishmaniasis and feline leishmaniasis. Both dogs and cats are susceptible to cutaneous leishmaniasis. The lesions usually occur on the nose and ears. Dogs are also susceptible to visceral leishmaniasis and may be important reservoirs, but cats are rarely infected and do not show signs of disease. The incubation period may be months or years. Infection in dogs is prevalent in Brazil, China, and the Mediterranean region.

Both imported and autochthonous cases of canine leishmaniasis occur in the United States. Imported cases can occur either in dogs imported into the country or in dogs returning from foreign travel. Most are due to *L. infantum* or *L. donovani.* Sporadic cases of autochthonous canine leishmaniasis have been reported from Texas, Oklahoma, Kansas, and Ohio for many years. The parasites have been variously identified as *L. chagasi, L. infantum,* and *L. mexicana.* The sand fly vectors are unknown.

Beginning in 1999 an outbreak of visceral leishmaniasis due to *L. infantum* occurred among foxhounds in a foxhunting club in New York, eventually resulting in 20 fatalities. Inquiry uncovered a prior outbreak among foxhounds in a foxhunting club in Michigan in 1989. Subsequent investigation found seropositive dogs in clubs located in 21 of the United States and in the province of Ontario, Canada. Since all known cases have been limited to foxhounds, it has been suggested that direct dog-to-dog transmission may occur during annual foxhound shows. Direct transmission of *L. infantum* has been demonstrated experimentally in mice. In addition, transplacental transmission of *L. infantum* has been demonstrated experimentally in dogs, and a case of transmission of *L. donovani* to a dog by blood transfusion has been reported. A recent study has shown that *Lutzomyia vexator* is widespread and abundant in Dutchess County, New York, where the 1999 outbreak occurred. However, *L. vexator* is believed to feed on reptiles, and there is presently no evidence other than sympatry to implicate it as a vector of the agent of canine visceral leishmaniasis.

Vesicular Stomatitis Virus Disease

In veterinary practice, vesicular stomatitis virus disease is known as vesicular stomatitis because of the prominence of oral symptoms in livestock. Vesicular stomatitis is an acute, febrile, weakening, viral disease of horses, cattle, swine, and occasionally sheep and goats. It is characterized by small, superficial, erosive blisters that form in and about the mouth and on the feet, teats, and occasionally other parts of the body. Because the symptoms closely resemble those of foot-and-mouth disease, vesicular stomatitis also is known as pseudo-foot-and-mouth disease. Susceptibility depends on the immune status of the host animal. Nonimmune cattle are 100% susceptible, and up to 90% develop clinical disease. The intrinsic incubation period is two to eight days. The disease is usually self-limiting, with recovery in about two weeks, but recrudescence or reinfection may occur. Economic losses are due to the reduced condition of infected animals, reduced meat and milk production, and secondary bacterial infections. Vesicular stomatitis virus is on the US Department of Agriculture High Consequence list and the Australian Group Core list of potential biological warfare agents.

The Indiana, New Jersey, and Alagoas serotypes of vesicular stomatitis virus are known to cause disease in domestic animals in North and South America. Vesicular stomatitis is endemic in tropical regions, but tends to be epidemic in temperate regions. In tropical regions it occurs year-round, but in the United States it occurs primarily in late summer and early fall. Opossums, monkeys, porcupines, raccoons, bobcats, and pronghorns are suspected reservoirs. Antibodies occur in domestic and wild dogs.

Lutzomyia shannoni is a proven vector of the New Jersey serotype among feral pigs on Ossabaw Island, Georgia, and *L. trapidoi* is a proven vector of the Indiana serotype in Latin America. *Lutzomyia ylephiletor* is a suspected vector. Other modes of transmission of vesicular stomatitis have been discussed earlier in connection with the public health importance of sand flies.

PREVENTION AND CONTROL

Psychodinae

Control of moth flies in buildings and homes depends on removal of gelatinous films and slimes that form in sink, floor, and bathtub drains, condensation pans of refrigeration and air-conditioning units, and other situations where oviposition and larval development occur. Mechanical cleaning is best, but infestations in drains can sometimes be eliminated by flushing with cleaning materials followed by very hot water. Recently effective drain cleaners based on bacterial cultures have been developed to remove and prevent formation of slime. An insect growth regulator, hydroprene, is also available for treatment of drains.

Larvae can be eliminated from filters of sewage treatment plants by flooding for 24 hours. Flooding does not affect the eggs and must be repeated periodically for continuous control. Larvae also can be eliminated from the filters by addition of insecticides to the flow. An approved insecticide, formulation, and dose must be used to avoid harm to the filters and the downstream environment. Destruction of large numbers of larvae in sewage treatment plants may create an odor problem in the neighborhood of the plant due to decomposition. The bacterium *Clostridium bifermentans* serovar *malaysia* has been reported to be highly toxic to larvae of *Psychoda alternata*, but its use as a biological control agent has not been demonstrated.

Adult moth flies can be controlled by application of insecticides to resting sites on structures and surrounding areas. Moth flies are susceptible to all classes of insecticides, but use of approved nonpersistent pyrethroid insecticides is recommended. Residual insecticides are rarely, if ever, required.

Phlebotominae

Methods for investigating breeding sites of phlebotomine sand flies include direct examination of soil and litter, extraction by Berlese funnel, wet sieving, flotation, and emergence trapping. Surveillance and collection methods for adult sand flies include trapping and aspiration from resting sites, humans, and bait animals. Effective trap designs include light traps, bait traps, sticky traps, and flight traps. Smoke, insect repellent spray, or a twig or stick can be used to flush sand flies from inaccessible resting sites for collection.

Insect repellents and protective clothing are effective personal protection. Sand flies cannot bite through outdoor clothing because of their relatively short mouthparts. Long sleeves, trousers and socks should be worn in areas where sand flies are active, and an approved repellent lotion or spray should be applied to exposed skin. The leading repellent for personal use is deet (diethylmethylbenzamide, formerly known as diethyltoluamide). Where sand flies are present, campsites and outdoor sleeping areas should be dry, open to the wind, and away from potential breeding sites. Sand flies can pass through or bite through untreated standard 16- and 18-mesh mosquito netting and screening. Fine-mesh nets and screens generally are not used because they impede circulation of the air. However, repellent- or insecticide-treated standard- or wide-mesh net jackets and hoods, head nets, tent openings, screens, and bed nets provide effective protection. Treated bed nets should be used when sleeping outdoors, and treated window screens, screen doors, and bed nets should be used for protection indoors. Synthetic pyrethroid insecticides, including permethrin, cypermethrin, cyfluthrin, and etofenprox, are currently recommended by the World Health Organization for treatment of bed nets. In connection with personal protection, it should be noted that the endophilic New World vector *L. verrucarum* is reported to crawl beneath untreated clothing and bedding to reach the skin.

Aerosol formulations of natural and synthetic pyrethroid insecticides provide effective control of sand flies when used indoors during the daily period of sand fly activity. Insecticidal smoke produced by "mosquito coils" is a useful and popular control measure for indoor spaces. Neither aerosols nor smokes have any appreciable residual effect. Mosquito coils are made by combining natural pyrethrum or a synthetic pyrethroid, commonly allethrin or prallethrin, and a slow-burning organic filler (punk) with a binder and forming the mixture into a flat spiral. It is believed that mosquito coils evolved as a folk remedy from the joss sticks burned as incense in Eastern temples.

Organochlorine, organophosphate, carbamate, and synthetic pyrethroid insecticides are effective for residual control of adult sand flies. Indoor treatments of houses should include the inside walls and ceilings, window and door frames, and screens. Pyrethroid–treated curtains for eaves, doors, windows, and walls have been used for sand fly control in homes in Sudan, Burkina Faso, Venezuela, and Italy. Outdoor treatments should be directed toward breeding and resting sites such as buildings, walls, caves, animal burrows, and the bark of tree trunks. Resistance to organochlorine insecticides has been reported in *P. argentipes* and *P. papatasi.*

From 1955 to 1969, the World Health Organization conducted a program for the global eradication of malaria by indoor treatment of houses with DDT. An unexpected collateral benefit of the program was a reduction in the incidence of leishmaniasis in areas where domiciliary transmission occurred. When the program was discontinued, the incidence of leishmaniasis increased, along with the incidence of malaria, in at least five countries: India, Bangladesh, Nepal, Greece, and Colombia. A major epidemic of visceral leishmaniasis occurred in India from 1973 to 1981. These events provide a dramatic, albeit unintended, demonstration of the effectiveness of residual insecticides in sand fly (and mosquito) control.

Environmental measures for sand fly control include the elimination of breeding and resting sites. Breeding and resting sites of peridomestic *P. papatasi* can be eliminated by filling cracks and crevices of walls, ceilings, and floors of houses, outbuildings, and masonry structures, and by clearing outdoor areas of accumulations of refuse, stone, and unneeded materials. In Italy, *P. perfiliewi* was eliminated from farmhouses by relocating storage piles of farm manure in which the sand flies breed to safe distances from houses. In Kenya, *P. martini* was eliminated from houses by destroying termite mounds in which the sand flies breed within 20 meters of homes. In Panama and French Guiana, breeding and resting sites of local forest species were eliminated by deforestation of areas around villages and settlements.

Biological and genetic methods for sand fly control have not been demonstrated. The bacteria *Bacillus sphaericus* and *B. thuringiensis* var. *israelensis* are potential agents for biological control of the larval stage. Rickettsial endosymbionts of the genus *Wohlbachia* have been found in both New and Old World sand flies. In the mosquito, *Culex pipiens, Wohlbachia* causes a form of male sterility known as cytoplasmic incompatibility, but it is not known if this occurs in sand flies. Several Apicomplexa (gregarines), Microspora (microsporidia), and Nematoda (roundworms) have been found in New and Old World sand flies. A cricket, *Anaxipha gracilis* (Orthoptera: Gryllidae), and a thread-legged bug, *Ploiaria domestica* (Hemiptera: Reduviidae), have been reported as active predators of adult sand flies in the New and Old Worlds, respectively.

Elimination of animal reservoirs of zoonotic leishmaniasis is feasible in some situations. Culling of infected dogs has been an effective measure in China and Brazil. It has been suggested that either mandatory treatment of all infected dogs or mandatory pyrethroid-impregnated collars or vaccinations for all dogs could be alternatives to culling. Reduction of feral dog and jackal populations by proper disposal of offal from slaughter houses and poultry farms has significantly reduced the incidence of visceral leishmaniasis in Iraq. The fat sand rat *Psammomys obesus* has been controlled in Jordan by flooding, digging, or deep plowing of the burrows. The great gerbil *Rhombomys opimus* has been controlled in Kazakhstan by deep plowing of burrows and dispersal of baits consisting of wheat treated with zinc phosphide or anticoagulant rodenticides.

Penicillin, streptomycin, chloramphenicol, and tetracyclines are effective in reducing fever and bacteremia in Oroya fever. Specific treatments for verruga peruana include streptomycin and rifampin. The first-line drugs for treatment of leishmaniasis are pentavalent antimonials, most importantly sodium stibogluconate and meglumine antimonite. An extended course of multiple treatments is always required.

A kind of folk vaccination called leishmanization has been practiced in southwest Asia since ancient times. As traditionally practiced, children are inoculated on a part of the body normally covered by clothing with living amastigotes from an active lesion of cutaneous leishmaniasis. The lesion that develops at the site of inoculation eventually heals, leaving the child immune to reinfection and protected from unsightly scarring of the face or another visible body part. In recent years, large-scale government-sponsored programs of leishmanization of children and adults have been conducted, sometimes using promastigotes from culture in lieu of amastigotes from lesions, most notably in Uzbekistan and in Iran during its 1980–1988 war with Iraq. Currently, ongoing research is directed toward development of modern, safe, effective vaccines for both cutaneous and visceral leishmaniasis.

REFERENCES AND FURTHER READING

Alexander, B. (1995). A review of bartonellosis in Ecuador and Colombia. *The American Journal of Tropical Medicine and Hygiene, 52*, 354–359.

Bauzer, L. G. H. R., Souza, N. A., Maingon, R. D. C., & Peixoto, A. A. (2007). *Lutzomyia longipalpis* in Brazil: A complex or a single species? *Memórias do Instituto Oswaldo Cruz, Rio de Janeiro, 102*, 1–12.

Belen, A., & Alten, B. (2006). Variation in life table characteristics among populations of *Phlebotomus papatasi* at different altitudes. *Journal of Vector Ecology, 31*, 35–44.

Beverly, S. M., Ismach, R. B., & McMahon-Pratt, D. (1987). Evolution of the genus *Leishmania* as revealed by comparisons of nuclear DNA restriction fragment patterns. *Proceedings of the National Academy of Sciences of the United States of America, 84*, 484–488.

Chang, K. P., & Bray, R. S. (1985). *Leishmaniasis.* London: Elsevier.

Chaniotis, B. (1978). Phlebotomine sand flies (family Psychodidae). In R. A. Bram (Ed.), *Surveillance and Collection of Arthropods of Veterinary Importance. Agriculture Handbook No. 518* (pp. 19–30). Washington, DC: US Department of Agriculture, Animal and Plant Health Inspection Service.

Charlab, R., Valenzuela, J. G., Rowton, E. D., & Rebeiro, J. M. C. (1999). Toward an understanding of the biochemical and pharmacological complexity of the saliva of a hematophagous sand fly, *Lutzomyia longipalpis. Proceedings of the National Academy of Sciences of the United States of America, 96*, 15155–15160.

Comer, J. A., & Tesh, R. B. (1991). Phlebotomine sand flies as vectors of vesiculoviruses: a review. *Parassitologia, 33*(Supplement 1), 143–150.

Cupolillo, E., Grimaldi, G., & Momen, H. (1994). A general classification of New World *Leishmania* using numerical zymotaxonomy. *The American Journal of Tropical Medicine and Hygiene, 50,* 296–311.

De Almeida, D. N., Oliveira, R. D. S., Brazil, B. G., & Soares, M. J. (2004). Patterns of exochorion ornaments on eggs of seven South American species of *Lutzomyia* sand flies (Diptera: Psychodidae). *Journal of Medical Entomology, 41,* 819–825.

Desjeux, P. (1991). *Information on the Epidemiology and Control of the Leishmaniases by Country and Territory.* Geneva: World Health Organization: Report WHO/LEISH/91.30.

Grimaldi, G., Tesh, R. B., & McMahon-Pratt, D. (1989). A review of the geographic distribution and epidemiology of leishmaniasis in the New World. *The American Journal of Tropical Medicine and Hygiene, 41,* 687–725.

Grimaldi, G., & Tesh, R. B. (1993). Leishmaniases of the New World: Current concepts and implications for future research. *Clinical Microbiology Reviews, 6,* 230–250.

Jobling, B. (1987). *Anatomical Drawings of Biting Flies.* London: British Museum (Natural History).

Kasap, O. E., & Alten, B. (2005). Laboratory estimation of degree-day developmental requirements of *Phlebotomus papatasi* (Diptera: Psychodidae). *Journal of Vector Ecology, 30,* 328–333.

Kasap, O. E., & Alten, B. (2006). Comparative demography of the sand fly *Phlebotomus papatasi* (Diptera: Psychodidae) at constant temperatures. *Journal of Vector Ecology, 31,* 378–385.

Kerr, S. F. (2000). Palaearctic origin of *Leishmania. Memórias do Instituto Oswaldo Cruz, Rio de Janeiro, 95,* 75–80.

Killick-Kendrick, R. (1987). Studies and criteria for the incrimination of vectors and reservoir hosts of the leishmaniases. In *Proceedings of the International Workshop on Control Strategies for the Leishmaniases, Ottawa, 1–4 June 1987* (pp. 272–280). Ottawa, Canada: International Development Research Centre.

Killick-Kendrick, R. (1990a). Phlebotomine vectors of the leishmaniases: A review. *Medical and Veterinary Entomology, 4,* 1–24.

Killick-Kendrick, R. (1990b). The life-cycle of *Leishmania* in the sandfly with special reference to the form infective to the vertebrate host. *Annales de Parasitologie Humaine et Comparée, 65*(Supplement 1), 37–42.

Killick-Kendrick, M., & Killick-Kendrick, R. (1991). The initial establishment of sandfly colonies. *Parassitologia, 33*(Supplemento 1), 315–320.

Kreutzer, R. D., Souraty, N., & Semko, M. E. (1987). Biochemical identities and differences among *Leishmania* species and subspecies. *The American Journal of Tropical Medicine and Hygiene, 36,* 22–32.

Leite, A. C. R., Williams, P., & dos Santos, M. C. (1991). The pupa of *Lutzomyia longipalpis* (Diptera: Psychodidae – Phlebotominae). *Parassitologia, 33,* 477–484.

Leite, A. C. R., & Williams, P. (1996). Description of the fourth instar larva of *Lutzomyia longipalpis,* under scanning electron microscopy. *Memórias do Instituto Oswaldo Cruz, Rio de Janeiro, 91,* 571–578.

Lewis, D. J. (1975). Functional morphology of the mouth parts in New World phlebotomine sandflies (Diptera: Psychodidae). *Transactions of the Royal Entomological Society London, 126,* 497–532.

Lewis, D. J. (1982). A taxonomic review of the genus *Phlebotomus* (Diptera: Psychodidae). *Bulletin of the British Museum (Natural History), 45,* 121–209.

Lewis, D. J., Young, D. G., Fairchild, G. B., & Mintter, D. M. (1977). Proposals for a stable classification of the phlebotomine sandflies (Diptera: Psychodidae). *Systematic Entomology, 2,* 319–332.

Magill, A. J. (1995). Epidemiology of leishmaniasis. *Dermatologic Clinics, 13,* 505–523.

Nieves, E., & Pimenta, P. F. P. (2000). Development of *Leishmania* (*Viannia*) *braziliensis* and *Leishmania* (*Leishmania*) *amazonensis* in the sand fly *Lutzomyia migonei* (Diptera: Psychodidae). *Journal of Medical Entomology, 37,* 134–140.

Peters, W., & Killick-Kendrick, R. (1987). *The Leishmaniases in Biology and Medicine, 2 Vols.* London: Academic Press.

Poinar, G., & Poinar, R. (2008). *What Bugged the Dinosaurs? Insects, Disease, and Death in the Cretaceous.* Princeton, NJ: Princeton University Press.

Quate, L. W., & Vockeroth, J. R. (1981). Psychodidae. In J. F. McAlpine, B. V. Peterson, G. E. Shewell, H. J. Tesky, J. H. Vockeroth, & D. M. Wood (Eds.), *Manual of Nearctic Diptera* (Vol. 1). Agriculture Canada Monograph 27 (pp. 293–300). Quebec: Canadian Government Publishing Center.

Ready, P. D., Fraiha, H., Lainson, R., & Shaw, J. J. (1980). *Psychodopygus* as a genus: Reasons for a flexible classification of the phlebotomine sandflies (Diptera: Psychodidae). *Journal of Medical Entomology, 17,* 75–88.

Rioux, J. A., Lanotte, G., Serres, E., Pratlong, F., Bastien, P., & Perieres, J. (1990). Taxonomy of *Leishmania.* Use of isoenzymes. Suggestions for a new classification. *Annales de Parasitologie Humaine et Comparée, 65,* 111–125.

Rutledge, L. C., Ellenwood, D. A., & Johnstone, L. (1975). An analysis of sand fly light trap collections in the Panama Canal Zone (Diptera: Psychodidae). *Journal of Medical Entomology, 12,* 179–183.

Rutledge, L. C., Walton, B. C., & Ellenwood, D. A. (1976). A transect study of sand fly populations in Panama (Diptera: Psychodidae). *Environmental Entomology, 5,* 1149–1154.

Sacks, D. L. (1989). Metacyclogenesis in *Leishmania* promastigotes. *Experimental Parasitology, 69,* 100–103.

Schultz, M. G. (1968). A history of bartonellosis (Carrión's disease). *The American Journal of Tropical Medicine and Hygiene, 17,* 503–515.

Silva, O. S., & Grunewald, J. (2000). Comparative study of the mouthparts of males and females of *Lutzomyia migonei* (Diptera: Psychodidae) by scanning electron microscopy. *Journal of Medical Entomology, 37,* 748–753.

Tesh, R. B. (1988). The genus *Phlebovirus* and its vectors. *Annual Review of Entomology, 33,* 169–181.

Tibayrenc, M., Kjellberg, F., & Ayala, F. J. (1990). A clonal theory of parasitic protozoa: The population structures of *Entamoeba, Giardia, Leishmania, Naegleria, Plasmodium, Trichomonas,* and *Trypanosoma* and their medical and taxonomical consequences. *Proceedings of the National Academy of Sciences of the United States of America, 87,* 2414–2418.

Tindale, B. V., Tikute, S. S., Arankalle, V. A., Sathe, P. S., Joshi, M. V., Ranadive, S. N., et al. (2008). Chandipura virus: A major cause of acute encephalitis in children in North Telangana, Andhra Pradesh, India. *Journal of Medical Virology, 80,* 118–124.

Walters, L. L. (1993). *Leishmania* differentiation in natural and unnatural sand fly hosts. *Journal of Eukaryotic Microbiology, 40,* 196–206.

Chapter

12

Biting Midges (Ceratopogonidae)

Gary R. Mullen

Biting midges are minute blood-sucking flies represented by only a few of the many genera in the family Ceratopogonidae. They are commonly known as **no-see-ums**, owing to their small size and the fact that they often go unnoticed despite the discomforting bites they can cause. Another name for this group, especially in the northeastern United States, is **punkies**. It is derived from a Dutch corruption of the Algonquin Indian root *punkwa*, which means ash-like, referring to the appearance of the fly as it is biting. The associated burning sensation is likened to that of a hot ash from a fire on contact with the skin. The early French Canadians called them **brulôt**, from *bruler* meaning "to burn." They also are called **sand flies**, particularly in the coastal areas of the southeastern United States, the West Indies, and adjacent parts of the Caribbean and Latin America. This name should not be confused with the same term applied to phlebotomine flies of the family Psychodidae. Along the Gulf Coast of Alabama and Florida, local residents refer to biting midges as **five-o's** because of their biting activity that commences late in the afternoon, about 5 o'clock. Other names for biting midges in various parts of the world include **moose flies** in Alaska, **knotts** in Norway, **jejenes** in Latin America, **maruins** in Brazil, **kuiki** in India, **makunagi** and **nukaka** in Japan, **nyung noi** in Laos, **agas** and **merutu** in Indonesia, **merotoe** in Sumatra, and **no-no's** in Polynesia.

Biting midges can be annoying pests of humans and both domestic and wild animals. In addition to the discomfort that they cause, biting midges serve as vectors of a number of viruses, protozoans, and nematodes. Among the more important viral diseases are Oropouche fever in humans, bluetongue disease and epizootic hemorrhagic disease in ruminants, and African horsesickness in equines. Blood protozoans transmitted by biting midges cause diseases in poultry, whereas certain nematodes are the cause of mansonellosis in humans and onchocerciasis in various domestic and wild animals.

TAXONOMY

The Ceratopogonidae are represented worldwide by approximately 110 genera and 6,000 described species. Ceratopogonids are divided into four subfamilies, the Leptoconopinae, Forcipomyiinae, Dasyheleinae, and Ceratopogoninae. Catalogues of the species of ceratopogonids worldwide are provided by Borkent and Wirth (1997), Borkent (2006), Yu and Liu (2006), and Borkent and Spinelli (2000, 2007). For a world list of the species and subspecies within the genus *Culicoides*, see Boorman and Hagan (1996).

With the exception of the Dasyhelinae, each subfamily includes species that feed on vertebrate blood. Only four genera are known to attack man and other animals. The most important genus in this respect is *Culicoides*. It includes most of the troublesome species throughout the world and those that serve as the principal vectors of animal disease agents. *Leptoconops* occurs primarily in the subtropics and tropics; a few species are annoying biters in the Caribbean area and along the coast of the southeastern United States. *Forcipomyia* species in the subgenus *Lasiohelea* attack vertebrates, particularly in subtropical and tropical rain forests. *Austroconops macmillani*, a blood feeder and the only known species in its genus, has been reported in Western Australia. No members of the Dasyheleinae are of medical or veterinary importance.

Because of its importance in the transmission of animal viruses in North America, the *Culicoides variipennis* complex warrants special comment. For many years, *C. variipennis* was thought to consist of five subspecies: *C. v. albertensis*, *C. v. australis*, *C. v. occidentalis*, *C. v. sonorensis*, and *C. v. variipennis*. However, based on morphological and electrophoretic analyses, this complex is now regarded as three species (*C. occidentalis*, *C. sonorensis*, and *C. variipennis*), with *C. v. albertensis* and *C. v. australis* being synonyms of *C. sonorensis* (Holbrook et al., 2000). *Culicoides sonorensis*, rather than *C. variipennis* as widely reported in the earlier literature, is the principal vector of the viruses causing bluetongue disease and epizootic hemorrhagic disease in North American ruminants.

The *Culicoides obsoletus* complex and *Culicoides pulicaris* complex have become increasingly implicated as vectors of bluetongue virus in the Palearctic region. Species-specific PCR primers are now available for identification of the members of both of these groups (Nolan et al., 2007). The *Culicoides imicola* complex includes 10 morphological species (Sebastiani et al., 2001). This complex is confined to the Old World, where it plays an important role in transmission of the viruses that cause African horsesickness and bluetongue disease. For further discussion of *Culicoides* vector complexes, see Meriswinkel et al. (2004).

Keys to the genera of ceratopogonid adults are provided by Wirth et al. (1974) and Downes and Wirth (1981). For keys to adults of North American *Culicoides* species, see Jamnback (1965), Battle and Turner (1971), and Blanton and Wirth (1979). Generic keys for larvae are provided by Glukhova (1977, 1979). For larval keys to North American species of *Culicoides*, see Jamnback (1965), Blanton and Wirth (1979), and Murphree and Mullen (1991). Major taxonomic works for identification of ceratopogonid fauna in other parts of the world include: Central America (Spinelli and Borkent, 2004), South America (Spinelli et al., 2005; Borkent and Spinelli, 2007), former Union of Soviet Socialist Republics (Glukhova, 1989), China (Yu, 2005, 2006), Southeast Asia (Wirth and Hubert, 1989), and Africa (Glick, 1990; Rawlings et al., 2003).

MORPHOLOGY

Ceratopogonid larvae (Figure 12.1B), as represented by *Culicoides* species, are typically long and slender, ranging from 2 to 5 mm in length when mature. The body is translucently whitish in contrast to the yellow to brownish head capsule (Figure 12.2). The thorax often is marked by a characteristic pattern of subcutaneous pigmentation. Thoracic and abdominal segments are similar in size, contributing to their elongate, cylindrical body shape. Although larvae of other genera may possess distinctive setae and abdominal projections, the larval chaetotaxy of *Culicoides* and related genera is generally inconspicuous, except for four pairs of setae that may be apparent at the caudal end. These setae are especially long in tree-hole species and are believed to help increase larval mobility. A pair of narrow, bifid anal papillae that function in osmoregulation can be everted through the anus; in most preserved specimens, however, they are retracted into the rectum. Larvae generally lack spiracles and are dependent on cutaneous respiration. Whereas *Culicoides* and *Leptoconops* larvae lack thoracic and abdominal appendages, *Forcipomyia* larvae possess a well-developed, ventral prothoracic proleg and associated apical hooklets or setae.

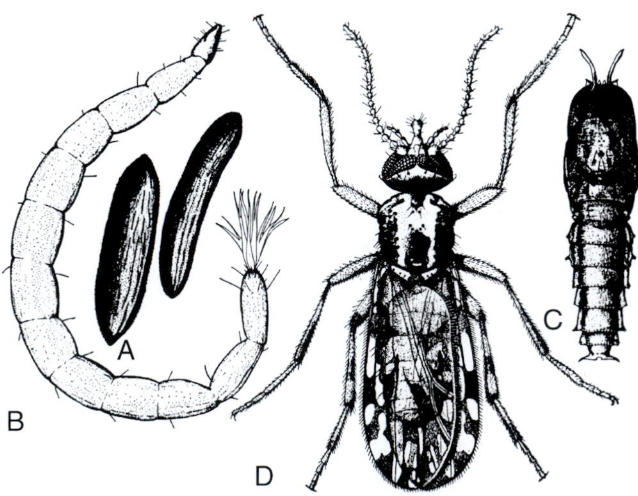

Figure 12.1 Developmental stages of the salt-marsh biting midge *Culicoides furens*. (A) Eggs; (B) larva; (C) pupa; (D) adult female. (Modified from Hall, 1932)

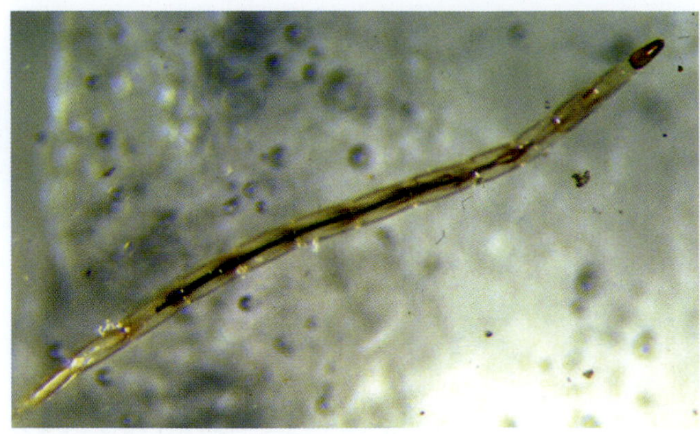

Figure 12.2 *Culicoides* larva, fourth instar. Note the slender, cylindrical body and distinct head capsule. (Photo by Richard C. Lancaster)

The mouthparts of larvae are characterized by a pair of mandibles that are not opposable; they move vertically or partially rotate while the larva feeds and are used to scrape, tear, or seize items depending on the species involved. Located within the buccal cavity is a complex, sclerotized internal structure called the epipharynx that is best observed in cleared, slide-mounted specimens. It consists of a pair of lateral arms and a median region supporting two to four combs that overlay one another. The epipharynx is rocked back and forth by muscles attached to the lateral arms and functions by helping to shred solid food and move food items posteriorly into the alimentary tract. In species that feed primarily on detritus and microorganisms, the combs apparently serve to strain material entering the mouth cavity. The number of pharyngeal combs and degree of sclerotization of the epipharynx is highly variable, reflecting the diversity of ingested food items and feeding behaviors exhibited by ceratopogonid larvae.

Pupae are typically brownish in color with a pair of relatively short but conspicuous pair of prothoracic respiratory horns arising at the anterior end (Figure 12.1C). Close inspection reveals numerous, tiny spiracular openings at the tip. The respiratory tubes repel water, enabling aquatic forms to hang at the water surface where they can obtain air during metamorphosis to the adult stage. A pocket of air beneath the developing wings provides additional buoyancy to keep the pupa at the water surface. Cuticular features in the form of tubercles, spines, and setae provide valuable taxonomic characters for identification of pupae to species.

Adult *Culicoides* midges (Figure 12.1D) are tiny, usually 1–2.5 mm in body length. Their mouthparts are adapted for biting or piercing tissues and are especially well developed in blood-sucking species (Figure 12.3). In females, the mouthparts are surrounded by a fleshy extension of the labium called a proboscis, which is relatively short, about as long as the head. It consists of an upper labrum-epipharynx, a pair of blade-like mandibles, a pair of laciniae (maxillae), and a ventral hypopharynx bearing a median, longitudinal groove along which saliva is passed as the female feeds. The mandibles bear a row of teeth along the inner edge near the tip, which is used to lacerate the skin while biting. The mouthparts of males are generally reduced and are not used in blood feeding.

Associated with the mouthparts are a pair of five-segmented maxillary palps. The third segment typically is enlarged and bears a specialized group of sensilla located in a depression, or **sensory pit**, which serves as a sensory organ. The adult antennae are 15-segmented and consist of a basal scape, an enlarged pedicel containing Johnston's organ, and 13 flagellomeres. The antennal segments bear differing numbers of small sensory pits (sensilla coeloconica), the number and pattern of which provide important taxonomic characters. The number of segments bearing sensory pits appears to be correlated with host feeding; species that feed primarily on birds generally have more sensory pits than those which feed on mammals. In males, flagellomeres 1–8 possess whorls of long setae that increase their sensitivity as mechanoreceptors and give them their plumose appearance. The wings possess a characteristic venation that distinguishes the ceratopogonids from other groups of flies. More important, however, are the distinctive wing patterns of the genus *Culicoides*, which are the basis for most species determinations in this large and important group. The darker areas of the wings are not pigmented, but represent the density of tiny setae (micro- and macrotrichia) on the wing surface.

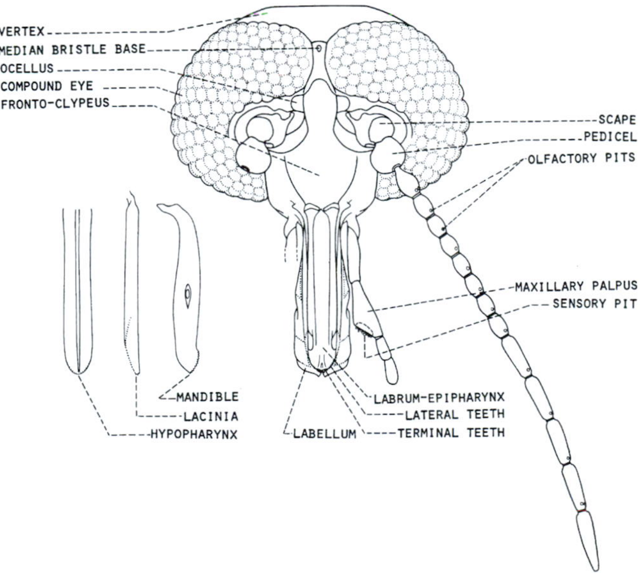

Figure 12.3 Morphology of head, mouthparts, and other associated structures of a female biting midge, *Culicoides* species. (Blanton and Wirth, 1979)

LIFE HISTORY

Adult females typically require a blood meal in order to develop their eggs (Figure 12.1A). Some, however, are autogenous and carry over enough nutrients from the larval stage to develop eggs during the first gonotrophic cycle without feeding on blood. Development of the eggs usually requires seven to 10 days but may be as short at two to three days. The eggs are deposited in batches on moist substrates. The number of eggs per female varies from 30 to 450 or more depending on the species and size of the blood meal. Autogenous females tend to produce fewer eggs. The eggs are small and elongate (250–500 μm in length), often banana-shaped, and are covered with minute projections that apparently function in plastron respiration. They are white when first deposited but gradually turn brown.

The eggs hatch in two to seven days. The larvae develop through four instars with a development time that varies from two weeks to more than a year, reflecting different species, latitudes, and times of the year. Many species overwinter as larvae and thus commonly pass seven or eight months of the year in this stage. In other cases, larvae become dormant during the hot summer months, prolonging their developmental time. Larval development of some arctic species may take as long as two years. Pupation generally occurs near the surface of the substrate where the prothoracic horns of the pupae can penetrate the water film. Pupae of *Culicoides* species that develop in water-filled tree holes may remain afloat at the water surface, loosely adhering to the sides of the tree cavity.

Overwintering larvae pupate in the spring or early summer, producing the first generation of adults. Autogenous females usually oviposit about a week following emergence; thereafter they must obtain a blood meal each time before they can develop another batch of eggs. Newly emerged, anautogenous females oviposit following their initial blood meal. The small percentage of the females that are successful in obtaining a second blood meal can produce a second batch of eggs, but they seldom do so a third time under field conditions. In the laboratory, however, *C. variipennis* is capable of completing up to seven gonotrophic cycles. Longevity of captive adults varies from two to seven weeks, with most individuals probably surviving only a few weeks at most under natural conditions. The generation time may be as short as two weeks for members of the *C. variipennis* complex, but more typically is six weeks or longer for most species.

Although some species are univoltine, most biting midges are multivoltine, producing two or more generations per year. Because of overlapping generations and multiple oviposition cycles by individual females, populations of a given species may be present throughout the warm months of the year. Usually, however, each species exhibits a general seasonal pattern with characteristic peaks of adult abundance. Some species are abundant only in the spring (e.g., *C. biguttatus*, *C. niger*, *C. travisi*), whereas others may exhibit high spring populations and be present in lower numbers throughout the summer and fall (e.g., *C. spinosus*). Others tend to be abundant throughout the spring, summer, and fall (e.g., *C. crepuscularis*, *C. furens*, *C. haematopotus*, *C. stellifer*, *C. venustus*). Still others are bivoltine, with peaks in the spring and fall (e.g., *C. hollensis*).

BEHAVIOR AND ECOLOGY

Ceratopogonid larvae develop in a wide range of aquatic and semiaquatic habitats ranging from the tropics to the arctic tundra. *Leptoconops* species occur primarily in sandy or clay-like alkaline soils of arid regions and along tidal margins or coastal marshes and beaches. *Forcipomyia* species are generally found in mosses and algae in shallow water and in more terrestrial habitats, such as rotting wood. The larval habitat of *Austroconops* is unknown. It is difficult to make generalizations about the breeding sites of *Culicoides* species except to say that they occur primarily in organically rich substrates (Figure 12.4). As a group, they utilize a broad diversity of habitats including freshwater marshes and swamps; shallow margins of ponds, streams and rivers; bogs and peat lands; tree holes and other natural cavities in rotting wood; tidal marshes and mangroves; and more specialized habitats such as rotting cacti, animal manure (Figure 12.5), and highly alkaline or saline inland pools.

Figure 12.4 Sampling organically enriched substrate of freshwater marsh for ceratopogonid larvae. (Photo by Gary R. Mullen)

Figure 12.5 Typical breeding site of *Culicoides sonorensis* in wet, manure-contaminated soil surrounding leaking water trough. (Photo by Gary R. Mullen)

The diversity of ceratopogonid larvae is similarly reflected in their feeding behavior. Many are predaceous, feeding on protozoans, rotifers, oligochetes, nematodes, immature stages of insects, and various other small aquatic or semiaquatic invertebrates. Others feed on detritus, bacteria, fungi, green algae, diatoms, and other organic materials. Based on feeding experiments and direct observations, it is apparent that many species are omnivorous and feed opportunistically on a variety of food items. Members of the *C. variipennis* complex, for example, generally are reared on a diet of microorganisms; however, it can also complete its development when fed only nematodes. For most species, the natural diet and nutritional requirements remain unknown, precluding the establishment of laboratory colonies for most of the economically important species. Among the North American species of *Culicoides* that have been colonized are *C. furens*, *C. guttipennis*, *C. melleus*, *C. mississippiensis*, *C. sonorensis*, and *C. wisconsinensis*. *C. nubeculosus* (England) and C. *oxystoma* (Japan) also have been colonized.

Despite their small size, ceratopogonid larvae often can be recognized by their serpentine locomotion consisting of side-to-side lashing movements of the body as they propel themselves through the water. *Culicoides* larvae generally are considered to be good swimmers, especially the later instars. *Culicoides circumscriptus* can lash back and forth an estimated 9 cycles/sec, and members of the *C. variipennis* complex are capable of sustained, directed swimming at speeds up to 1.7 cm/sec. Species such as *C. denningi* that burrow in the bottom of streams and rivers are excellent swimmers, enabling them to make their way to shore where they pupate.

Males typically emerge a short time before females and are ready to mate by the time the females are produced. Mature sperm are already present within 24 hours of eclosion. Unlike mosquitoes, ceratopogonid males do not undergo permanent rotation of the genitalia. Instead, the genital structures are temporarily rotated 180° to facilitate clasping the female in an end-to-end position just before mating takes place.

Mating usually involves swarming, in which large numbers of males form aerial aggregations, often near water or in open areas near potential breeding sites. Females fly into the swarm where males recognize them as being the same species by their wing-beat frequency. Sex pheromones have been shown to be involved in some species. If a female is receptive, she couples with the male and typically drops to the ground or vegetation where copulation takes place. A few species mate without forming swarms. In such cases, the male and female locate one another by crawling about on the ground or some other substrate where coupling occurs. In other cases, both sexes are attracted to a host where the male seeks out, and mates with, the female shortly after she has taken a blood meal. Most species are believed to mate only once, although members of the *C. variipennis* complex and others may mate repeatedly. Sufficient sperm from a single mating is stored by females of the *C. variipennis* complex to fertilize up to three batches of eggs.

As adults, both males and females feed on nectar of flowering plants. This serves as an energy source for flight activity and increased longevity, especially in females. Only females feed on vertebrate blood (Figure 12.6). As in other hematophagous insects, usually blood is required for egg development and subsequent oviposition. Females are pool feeders. They lacerate the skin and underlying capillaries with the serrated tips of their mandibles, causing blood to seep into the surrounding tissues. From there it is drawn into the foregut by action of the pharyngeal pump and passed back into the midgut. After feeding, the blood-laden female flies to nearby vegetation or another sheltered site where she rests for several days while her eggs develop.

Many species of biting midges feed primarily on mammals, whereas others feed preferentially on birds, reptiles, or amphibians. Those that feed on a given class of hosts often show preferences for certain groups within that class, such as small versus large mammals or certain types of birds. Some are quite host-specific, whereas others are considered generalists and may feed, for example, on both birds and mammals, depending on host availability.

Figure 12.6 *Culicoides sonorensis*, adult female feeding on human arm. (Photo by P. Kirk Visscher)

In general, *Culicoides* adults are crepuscular or night-time feeders, whereas *Leptoconops* adults tend to be more active during the daytime. The activity periods of most biting midges occur during twilight, particularly an hour before to an hour after sunrise and sunset. Some species exhibit bimodal activity at dawn and dusk (e.g., *C. barbosai, C. furens, C stellifer*) or during morning and late afternoon (e.g., *Leptoconops becquaerti*), whereas others tend to be active during only one of these periods (e.g., *C. debilipalpis* in the early dawn hours; *C. furens, C paraensis,* and *L. linleyi* in late afternoon.). Certain species will readily bite during the daytime and can be particularly annoying in late afternoon. Activity periods for a given species may vary at different times of the year reflecting seasonal changes in temperature and light intensity. The effects of daily and seasonal temperatures on flight activity are evident among salt-marsh species along the Florida coast. Whereas *C. barbosai, C. floridensis,* and *L. becquaerti* are not active below 14°C (57°F), *C. mississippiensis* remains active all year, even at winter temperatures as low as 4°C (37°F).

In addition to temperature, a number of other factors can influence flight activity by biting midges (Lillie et al., 1987; Kettle, 1972). They include light intensity, lunar cycles, relative humidity, changes in barometric pressure, and other weather conditions. Wind velocity is especially important. Because of their small size, most species do not tolerate appreciable air movement and are seldom troublesome at wind speeds above 2.5 m/sec. Such velocities interfere with normal flight activity and the ability to orient to a host. There are exceptions, however, such as *Leptoconops becquaerti*, which will continue to bite at wind speeds of 5.0 m/sec or more.

Biting midges are most abundant in proximity to productive breeding sites. From there they disperse into surrounding areas in search of mates and suitable hosts. How far they travel is highly variable depending on their success in finding a mate, availability of hosts, and prevailing weather conditions. Mark-recapture studies in which adults are released at a given location and are subsequently recovered by trapping at increasing distances from the release point, indicate that the mean distance traveled by many *Culicoides* females is about 2 km. The distance traveled by males is usually much shorter, often less than half that of females of the same species. The flight distance for any individual, however, can be much lower or higher than this mean value implies. Members of the *Culicoides variipennis* complex have been recovered up to 2.8 km from release sites within 12 hours and nearly 5 km within 36 hours. Salt-marsh species such as *C. mississippiensis* have been shown to disperse more than 3 km within 24 hours, whereas *L. kerteszi* in semi-arid regions of the southwestern United States has been reported to fly 15 km or more in this same time period.

PUBLIC HEALTH IMPORTANCE

Most complaints about biting midges relate to the annoyance caused by their persistent biting. In Scotland, the highland biting midge *Culicoides impunctatus* is a major tourism problem and an occupational concern for people working outdoors, such as in the forest industry. Other species are particularly a problem in coastal areas where salt-marsh species are notorious pests, often creating great discomfort for local residents and beach-goers, and discouraging tourism during the summer months. In the United States, *Culicoides furens*, *C. hollensis*, and *C. melleus* are the most troublesome species, attacking humans along the Atlantic Coast. *Culicoides mississippiensis* is a problem along the Gulf Coast, whereas *C. barbosai* is commonly a problem near mangroves of southern Florida. Certain *Leptoconops* species are especially pestiferous along coastal beaches; e.g., *L. linleyi* along the Gulf Coast and *L. bequaerti* in the Caribbean, whereas members of the *L. kertszi* group are annoying biters in semi-arid regions of the southwestern United States.

A common pest throughout much of the eastern United States is *C. paraensis*, a tree-hole species that readily bites humans, especially during the late afternoon and early evening. It causes considerable discomfort to hunters, campers, and hikers, and is typically the species involved in complaints by homeowners while picnicking or trying to work in their yards bordering deciduous woods.

Because of their small size, biting midges frequently are overlooked, their bites often being blamed on mosquitoes. Reactions to their bites generally consist of a localized stinging or burning sensation, producing a well-defined reddened area about the bite site without the formation of a wheal. The discomfort usually lasts from only a few minutes to a few hours. In individuals who develop hypersensitivity, the bites may continue to itch for two or three days. In the tropics, certain *Leptoconops* and *Lasiohelea* species can cause more severe reactions resulting in blisters and serous exudates at the bite sites of sensitized individuals.

Viruses and filarial nematodes are the only disease agents known to be transmitted to humans by the bite of ceratopogonid midges (Tables 12.1 and 12.2). They are primarily subtropical or tropical in distribution, with no associated diseases having been reported in North America. Although several viruses have been

Table 12.1 Arboviruses of Medical and Veterinary Interest Transmitted by Biting Midges (*Culicoides* Species)

Virus	Vertebrate Host	Geographic Area	Known or Suspected Vectors
Bunyaviridae			
Bunyawera Group			
Lokern	Lagomorphs (*Lepus*, *Sylvilagus*)	North America	*C. variipennis* complex, *C.* (*Selfia*) spp.
Main Drain	Lagomorphs (*Lepus*)	North America	*C. variipennis* complex
Simbu Group			
Aino	Cattle, sheep, buffalo	Japan	*C. brevitarsis*
Akabane	Cattle, sheep, goats, horses, buffalo, camels	Africa, Middle East, Japan, Australia	*C. brevitarsis*
Buttonwillow	Lagomorphs (*Lepus*, *Sylvilagus*)	United States	*C. variipennis* complex
Douglas	Cattle, sheep, goats, horses, buffalo, deer	Australia, New Guinea	*C. brevitarsis*
Oropouche	Human, forest primates, sloth	South America, Caribbean	*C. paraensis*
Peaton	Cattle	Australia	*C. brevitarsis*
Sabo	Cattle, goats	Nigeria	*C. imicola*
Sango	Cattle	Nigeria, Kenya	*Culicoides* spp.
Sathuperi	Cattle	Nigeria, Kenya, India	*Culicoides* spp.
Shamonda	Cattle	Nigeria	*C. imicola*
Shuni	Human, cattle	Nigeria, South Africa	*Culicoides* spp.
Thimiri	Birds	Egypt, India, Australia	*C. histrio*
Tinaroo	Cattle, sheep, goats, buffalo	Australia	*C. brevitarsis*
Utinga	Sloth	Panama, Brazil	*C. diabolicus*
Utive	Sloth	Panama	*C. diabolicus*
Other Bunyaviridae			
Crimean-Congo Hemorrhagic Fever Group	Human, cattle	Africa, Asia	*Culicoides* spp. (primarily ticks)
Dugbe (Nairobi Sheep Disease Group)	Human, cattle	Africa	*Culicoides* spp. (primarily ticks)

Table 12.1 Arboviruses of Medical and Veterinary Interest Transmitted by Biting Midges (*Culicoides* Species)—Cont'd

Virus	Vertebrate Host	Geographic Area	Known or Suspected Vectors
Rift Valley Fever	Human, cattle, buffalo, sheep, goats, antelope, camels	Africa	*Culicoides* spp. (primarily mosquitoes)
Reoviridae			
African Horsesickness	Horses, mules	Africa, Middle East, India, Europe, Asia	*C. imicola, C. bolitinos*
Bluetongue	Cattle, sheep, other domestic and wild ruminants	Africa, Middle East, Europe, Japan, Australia, North America, Central America, South America	*C. bolitinos, C. fulvus, C. gulbenkiani, C. imicola, C. insignis, C. milnei, C. obsoletus, C. sonorensis*
Epizootic Hemorrhagic Disease	Deer	North America, Africa, Asia, Australia,	*C. sonorensis*
Equine Encephalosis	Cattle	Africa, Australia	Unknown
Palyam Group			
Abadina	Unknown	Nigeria	*Culicoides* spp.
Bunyip Creek	Cattle, buffalo, sheep, deer	Australia	*C. brevitarsis, C. oxystoma*
Chuzan (=Kagoshima, Kasba)	Cattle	Japan	*C. oxystoma*
CSIRO Village	Cattle, buffalo	Australia	*C. brevitarsis*
D'Aguilar	Cattle, sheep	Australia	*C. brevitarsis*
Marrakal	Buffalo	Australia	*C. oxystoma, C. peregrinus*
Nyabira	Cattle	Zimbabwe	*Culicoides* sp.
Wallal Group			
Mudjinbarry	Marsupials	Australia	*C. marksi*
Wallal	Marsupials	Australia	*C. marksi*
Warrengo Group			
Mitchell River	Cattle, marsupials	Australia	*Culicoides* spp.
Warrengo	Cattle, marsupials	Australia	*C. dycei, C. marksi* (also mosquitoes)
Rhabdoviridae			
Bovine Ephemeral Fever	Cattle	Africa, Asia, Australia	*Culicoides* spp.
Kotonkan	Cattle, sheep, rats, hedgehogs	Africa	*Culicoides* spp.
Tibrogargan	Cattle, water buffalo	Australia	*C. brevitarsis*

Table 12.2 Filarial Nematodes Transmitted by Biting Midges to Humans and Domestic Animals (Vectors Include *Culicoides, Forcipomyia,* and *Leptoconops* Species.)

Nematode	Vertebrate Host	Geographic Area	Known or Suspected Vectors
Mansonella ozzardi	Human	South America, Caribbean Basin	*C. barbosai, C. furens, C. paraensis C. phlebotomus, Leptoconops becquaerti*
M. perstans	Human	Sub-Saharan West Africa; Central Africa to Kenya and Mozambique	*C. austeni, C. grahamii, C. inornatipennis*
		Northern coast of South America; Caribbean Islands	*Culicoides* spp.
M. streptocera	Human	West and Central Africa (rain forests)	*C. austeni, C. grahamii*
Onchocerca cervicalis	Horses	North America, Australia	*C. variipennis, C. victoriae Forcipomyia townsvillensis*
O. gibsoni	Cattle	India, Sri Lanka, Malaysia, northern Australia, South Africa	*C. pungens, Culicoides* spp.
O. gutturosa	Cattle	Australia	*Culicoides* spp.
O. reticulata	Horses, ponies	Australia	*C. nubeculosus, C. obsoletus*
O. sweetae	Water buffalo	Unknown	Unknown

isolated from *Culicoides* adults, Oropouche virus is the only significant viral agent transmitted to humans by ceratopogonid midges. Biting midges also transmit three filarial nematodes that infest humans, causing a disease known as **mansonellosis** or **mansonelliasis**. The causative agents are *Mansonella ozzardi* in the Americas, *M. perstans* in Africa and South America, and *M. streptocerca* in Africa. Linley (1983) has provided an excellent overview of the various human pathogens and parasites transmitted by this group of flies.

Oropouche Fever

Oropouche fever is caused by a virus in the Simbu group, family Bunyaviridae. Since it was first isolated from a charcoal worker in Trinidad in 1955, this virus has been documented in numerous epidemics in the Amazon region of Brazil. Outbreaks prior to 1980 were largely restricted to Pará State where over 165,000 human cases of this disease were estimated to have occurred between 1961 and 1980. Since that time, outbreaks have been reported in the Brazilian states of Amazonas, Goias, and Marahnao, and in the Amapa Territory. These epidemics have taken place primarily in urban areas where surveys indicate that up to 44% of local populations have been seropositive for antibodies to the virus.

Oropouche virus causes a nonfatal, acute febrile illness with general muscular and joint pains usually lasting two to five days. More than half of the cases involve symptoms such as headaches, dizziness, photophobia, and severe myalgia and arthralgia, which can lead to prostration in some cases. Recurrence of symptoms often prolongs the illness up to two weeks. The incubation period for this disease is believed to be four to eight days.

The epidemiology of Oropouche fever is complicated by multiple strains of the virus and uncertainty about which animals serve as reservoirs. Antibody levels in potential reservoir hosts tend to be highly variable, with some evidence to indicate that urban and sylvatic cycles are involved. During nonepidemic periods, several species of monkeys have been found to have high antibody levels, implicating them as important reservoirs. Other likely reservoirs in the sylvatic cycle are wild birds and sloths. During epidemics, high antibody levels among various carnivores and domestic birds suggest that they play a role as reservoirs in urban outbreaks of this disease.

The principal vector in urban outbreaks is *Culicoides paraensis*. This forest species breeds in tree holes and decaying cacao and calabash pods. It readily feeds on humans both inside and outside houses. Infected *C. paraensis* females can transmit the virus as early as four to six days following a blood meal. Although Oropouche virus also has been isolated from naturally infected mosquitoes such as *Culex quinquefasciatus*, *Aedes serratus*, and *Coquillettidia venezuelensis*, the role of these species as vectors remains uncertain.

Other Viral Agents

Many other arboviruses have been isolated from *Culicoides* adults. Four of them are members of the Bunyaviridae: Crimean-Congo, Rift Valley fever, Dugbe and Shuni viruses. In addition, three mosquito-borne viruses that cause eastern equine encephalitis, Japanese B encephalitis and Venezuelan equine encephalitis, have been isolated from *Culicoides* and *Lasiohelea* species. There is no evidence, however, to indicate that biting midges play a significant role in transmission of any of these viruses. For a list of other viruses and organisms transmitted by ceratopogonids, see Borkent (2005).

Mansonellosis

Three filarial nematodes in the genus *Mansonella* cause infestations in humans called mansonellosis (Table 12.2). Cases occur widely throughout the tropical and subtropical regions of both the Old World and New World, where *Culicoides*, *Forcipomyia*, and *Leptoconops* species serve as arthropod vectors. Although infestations involving these parasites are generally mild or asymptomatic, they sometimes cause serious medical problems.

Mansonella ozzardi *Mansonella ozzardi* is the only native New World ceratopogonid-borne nematode of humans (Figure 12.7). It is indigenous to the Americas, occurring in the Amazon Basin (Brazil); along the northern coast of South America (Colombia,

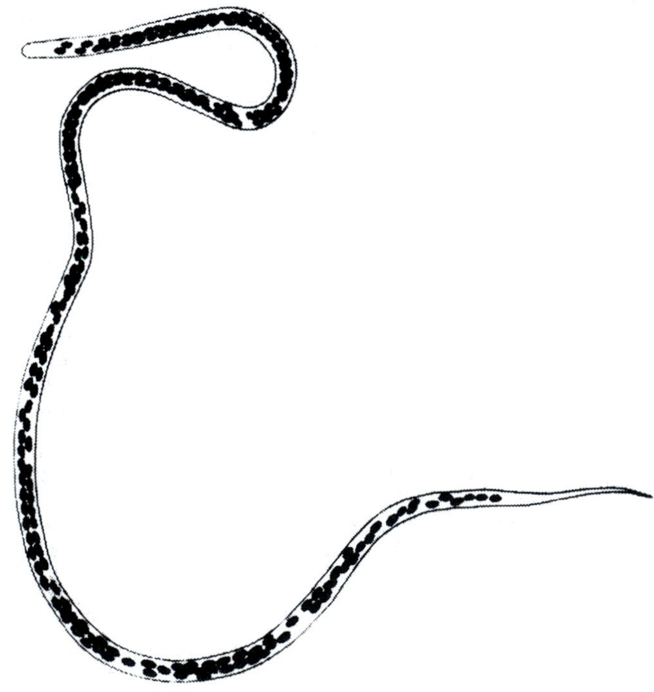

Figure 12.7 Filarial nematode (*Mansonella ozzardi*), microfilarial stage. (Courtesy of Lea & Febiger)

Venezuela, Guyana, Surinam, French Guiana); on Trinidad, Haiti, and other islands of the West Indies; Panama; and parts of Peru, Bolivia, and Argentina. It particularly affects coastal fishing communities near breeding sites of associated vectors. The infection rate among local inhabitants is highly variable, ranging from as low as 5% or less in northern Brazil and some of the Caribbean islands to over 95% among Amerindians in Colombia and Venezuela. Infection rates are generally highest among men and women in older age groups, reflecting chronic exposure to infection in endemic areas.

Infections with *M. ozzardi* usually do not result in significant pathological effects. The microfilariae typically remain in the capillaries of the skin and surrounding dermal tissues where they cause relatively little harm. Surveys usually are conducted by taking skin biopsies or blood samples and examining them for the presence of microfilariae. The adult worms are found primarily in fat tissue associated with the peritoneum and various body cavities, occasionally causing conjunctivitis and swelling of the eyes. In some cases this nematode can cause more serious problems such as severe joint pains, eosinophilia, enlargement of the liver, and blockage or inflammation of the lymphatic vessels resulting in conditions similar to bancroftian filariasis and elephantiasis. Ivermectin has been successfully used in treatment of *M. ozzardi* cases, whereas the widely used filarial nematocide diethylcarbamazine is ineffective in killing this parasite.

Vectors of *M. ozzardi* include both biting midges and black flies, with different taxa playing important roles in different areas. *Culicoides furens* and *C. phlebotomus* are the principal vectors in Haiti and Trinidad, respectively. In Argentina *C. lahillei* is believed to be the primary vector, with *C. paraensis* and the blackfly *Simulium exiguum* playing a secondary role in transmission. Other species that support development of microfilariae to infective larvae and generally are considered to play a secondary role in transmission are *C. barbosai, C. paraensis* and *Leptoconops becquaerti.* After ingestion by a biting midge as it feeds on an infected host, microfilariae are carried into the midgut where they penetrate the midgut wall and make their way to the thoracic muscles within 24 hours. There they develop to third-stage larvae during the next six to nine days before moving to the head and mouthparts. Infective third-stage larvae enter the bite wound when the midge subsequently feeds on another host. Typically only one to three larvae successfully complete development to the infective stage in a host insect, regardless of the number of microfilariae initially ingested.

The role of black flies as vectors of *M. ozzardi* remains unclear. Species in the *Simulium amazonicum* group and the *S. sanguineum* group have been found to be naturally infected with this nematode and probably play a role in transmission, particularly in the Amazon Basin. Other species incriminated as potential vectors based on field collections and experimental infection studies include *S. sanchezi* and *S. pintoi.* Despite earlier suggestions that there may be two different forms or species of nematode involved, one transmitted by biting midges and the other by black flies, it is generally accepted that they are morphologically identical and represent a single species.

Mansonella perstans This nematode (formerly placed in the genera *Acanthocheilonema, Dipetalonema,* and *Tetrapetalonema*) is the most widely distributed of the three human filarial nematodes transmitted by biting midges. It is indigenous to the Old World where it occurs in sub-Saharan Africa, extending primarily from West African countries bordering the Gulf of Guinea (Ivory Coast, Nigeria, and Equatorial Guinea) and from Gabon and Angola east through Central Africa to Kenya and Mozambique. Infection rates are commonly 50% or higher in some communities. *Mansonella perstans* was introduced to Central and South America with the slave trade and now occurs along the northern coast of South America (Colombia, Venezuela, Guyana, Surinam, and French Guiana), in the Yucatán area of Mexico, and on Trinidad and other Caribbean Islands. Prevalence of infection exceeding 50% has been reported among the Curripaco Indians of Venezuela. Although there is evidence to suggest that *M. perstans* represents a complex of species, this issue remains unresolved.

Mansonella perstans typically is regarded as nonpathogenic. The microfilariae remain primarily in the circulating blood, whereas the adult worms occur freely in the body cavities. Some infested individuals develop problems such as joint pains, fever, fatigue, transient edema, elephantoid scrota, mild urticarial skin reactions, and eosinophilia. Various ocular problems including swelling of the eyelids, excessive lacrimation, pruritis, and conjunctival granulomas or nodules have been reported. The latter is the result of adult worms coiled within the connective tissue of the conjunctiva causing a condition known as **bulge-eye** or **bung-eye**. Adult worms also have been removed from connective tissue of the pancreas, kidneys, rectum, and mesenteric lymph nodes of infested patients with little evidence of serious harm. Mebendazole has been used successfully in treating *M. perstans* cases, whereas diethycarbamazine and ivermectin are ineffective in killing either the microfilariae or adult worms.

The principal vectors of *M. perstans* remain virtually unknown in the New World. In Africa, however, several *Culicoides* species have been implicated as vectors based on natural infections and support of development to the infective stage in experimental studies. They include *C. austeni, C. grahamii,* and *C. inornatipennis* as probable vectors, and *C. fulvithorax, C. hortensis, C. krameri, C. kumbaensis, C. milnei, C. pycnostictus, C. ravus,* C. *rutshuruensis,* and *C. vitshumbiensis* as possible vectors. As in the case of *Mansonella ozzardi,* the microfilariae of *M. perstans* move from the midgut of the biting midge to the thoracic musculature where they complete their development to infective, third-stage larvae eight to 10 days after the infective blood meal.

Mansonella streptocerca This filarial nematode (formerly placed in the genera *Dipetalonema* and

Tetrapetalonema) occurs only in the rain forests of West and Central Africa, extending from the Ivory Coast and Burkina Faso to the Congo and Zaire. Little information on prevalence is available for this species, although a figure of 13 to 14% has been reported in certain villages of the Central African Republic based on peripheral blood smears. Although it is regarded as nonpathogenic to humans, *M. streptocerca* occasionally causes mild skin reactions due to activity of microfilariae in dermal tissues, usually involving the trunk and upper arms. Adult worms typically occur subcutaneously in upper parts of the body. Diethylcarbamazine is effective as a treatment. *Culicoides grahamii* is regarded as the principal vector.

VETERINARY IMPORTANCE

Biting midges serve as vectors of more than 35 arboviruses that infect domestic animals (Table 12.1). Only a few of these viruses cause significant clinical disease. Cattle, sheep, and horses usually are the most seriously affected. The majority of these viral agents are members of the Reoviridae and Bunyaviridae, including the pathogens that cause bluetongue disease, epizootic hemorrhagic disease, and African horsesickness. Two other families of viruses with which *Culicoides* species have been implicated as vectors are the Rhabdoviridae and Poxviridae. For most of these viral agents, the principal ceratopogonid species involved as vectors remain largely unknown. Other disease agents transmitted by biting midges include blood protozoans of birds such as *Haemoproteus meleagridis* in turkeys, *Leucocytozoon caulleryi* in chickens, and the nematode *Onchocerca cervicalis* in horses. Biting midges also can cause discomforting skin reactions in horses known as equine allergic dermatitis.

Bluetongue Disease

Bluetongue disease **(BT)** is caused by an orbivirus in the family Reoviridae that infects ruminants, notably sheep and cattle. It was first recognized in South Africa in the early 1930s following the introduction of European breeds. Historically it has been limited between latitudes 40°N and 35°S, including temperate and southern North America, parts of Central America and South America bordering the Caribbean, southern Europe and countries bordering the Mediterranean Sea, the Middle East, Asia, Australia, and southern Africa. In recent years, however, it has made excursions from northern Africa and the eastern Mediterranean into northern Europe where it had not been documented previously. Outbreaks occurred among cattle and sheep in the Netherlands, Belgium, France, and Germany in 2006, and in Denmark, Luxembourg, Switzerland, and the United Kingdom in 2007. This northern expansion of bluetongue, with successful overwintering of the virus in northern Europe, has been attributed in part to windborne infected *Culicoides* females from endemic areas in the Mediterranean and to global climate changes.

In the United States, bluetongue (BLU) virus was first isolated in 1952 in Texas from sheep exhibiting a condition known as **soremuzzle**. It is now known to occur throughout most of the southern and western states where prevalence of antibody to BLU virus in cattle is commonly 20 to 50%. No cases have been reported in Alaska or Hawaii. Canada remains largely bluetongue-free; detection of infected animals has been reported there only in the Okanagan Valley of British Columbia during localized outbreaks in 1975 and 1987, apparently originating from cattle imported from the United States.

Historically, the global distribution of bluetongue has been relatively well defined, with many parts of the world remaining largely bluetongue-free. This has led to restrictions on international trade and the movement of cattle and sheep from endemic areas to bluetongue-free zones. The regions where bluetongue does occur can be divided into fairly discrete epidemiological systems, or **episystems**, reflecting the geographic occurrence of major vector species. These include, for example, North America (*C. sonorensis*), South America (*C. insignis*), Africa (*C. imicola*), and Australia (*C. brevitarsis*).

Bluetongue virus represents an **antigenic complex** with 24 recognized serotypes that vary significantly in their pathogenicity. For current information on the serotypes and their global distribution, see the Pirbright, UK, web site http://www.reoviridae.org/dsRNA_virus_proteins/ under "Bluetongue virus and serotype-distribution."

Occurrence of clinical bluetongue disease in cattle tends to be sporadic, often involving only one or a few animals in a given herd. Occasionally, however, epizootics do occur. Following outbreaks in Cyprus in 1943, Turkey in 1944, and Israel in 1951, major epizootics occurred in Europe in the late 1950s where an estimated 179,000 sheep died in Spain and Portugal. The mortality rate among infected animals was 75 percent. Bluetongue virus has since been introduced to the Caribbean islands, bordering countries of South and Central America, and Australia where it was first detected in 1974.

Serosurveys for detection of antibodies to BLU virus indicate that most sheep and cattle that are exposed to the virus do not develop clinical signs. As a result, animals typically remain asymptomatic and often serve as unrecognized sources of infection for other animals. However, under circumstances that are poorly understood, some animals develop varying degrees of illness ranging from mild infections to acute, fatal disease. In more severe cases, animals develop lesions about the mouth and muzzle (Figure 12.10) with ulceration and sloughing of skin tissues, inflammation of the coronary band at the base of the hoofs, interdigital lesions (Figure 12.11), respiratory difficulties due to accumulation of fluids in the lungs, and internal hemorrhaging. The term "bluetongue" gets its name from the dusky blue appearance of the tongue and mucosal membranes lining the mouth resulting from cyanosis. Acutely infected animals often exhibit lameness and a characteristic arched back resulting from efforts to keep weight off their painful hooves (Figures 12.8 and 12.9).

Figure 12.8 Sheep with bluetongue disease; infected host with characteristic arched back, tender hooves, and hanging head. (Courtesy of U.S. Department of Agriculture, Animal and Plant Health Inspection Service)

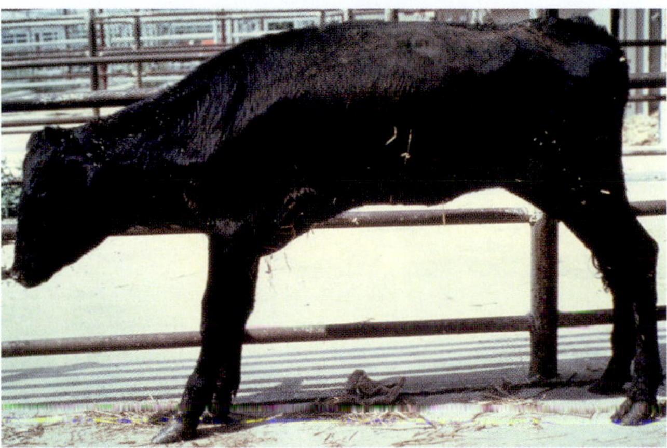

Figure 12.9 Black Angus calf in late stage of bluetongue disease, with general depression, hanging head, labored breathing, and difficulty standing. (Photo by Lloyd L. Lauerman)

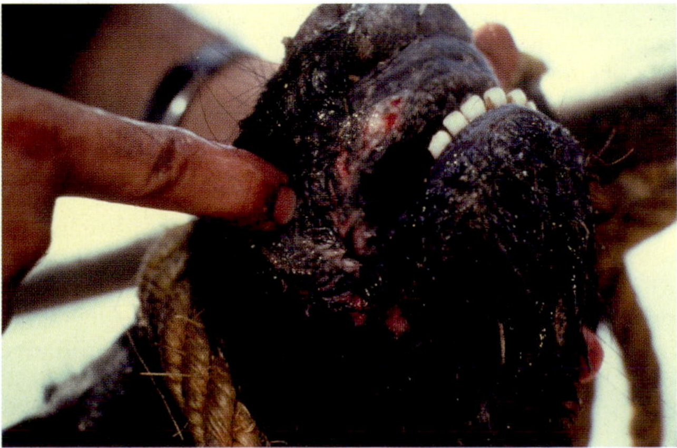

Figure 12.10 Oral and muzzle lesions in Black Angus calf suffering from bluetongue disease. (Photo by Lloyd L. Lauerman)

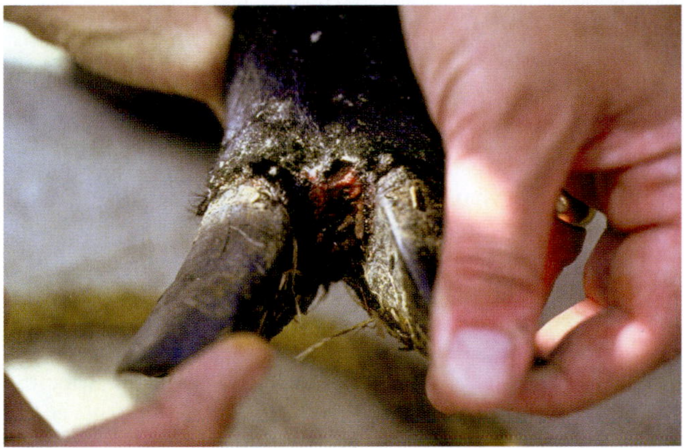

Figure 12.11 Hoof lesions in Black Angus calf in late stage of bluetongue disease. (Photo by Lloyd L. Lauerman)

Death results primarily from congestion of the lungs and massive internal hemorrhaging.

Reproduction is also affected and can result in underweight calves at birth, congenital deformities, stillbirths, and abortions. The severity of reproductive impact is due in large part to the time during gestation when the infection occurs. Infections in the early stages of fetal development can result in aborted or stillborn calves. Infections that occur later in gestation are more likely to result in congenital deformities and underweight calves at birth. The virus is also found in semen of infected bulls. Consequently restrictions are placed on the exportation of semen and live animals from endemic bluetongue areas for artificial insemination and other breeding purposes. The resulting economic impact on the livestock industry in the United States for mandatory testing of animals and losses in the foreign market are substantial, totaling millions of dollars annually.

The common occurrence of multiple serotypes of BLU virus in the same geographic area has hampered the use of immunization for protecting livestock from infection. Polyvalent vaccines have been developed but generally have not been effective. Natural immunity is acquired by sheep following their recovery from this disease, but is limited to the particular serotype with which the animal was infected. Cattle apparently do not develop significant immunity following infection.

The primary mode of transmission of BLU virus is by the bite of infected *Culicoides* midges. Based on experimental studies with *C. sonorensis*, the virus acquired while feeding on a viremic animal invades the salivary glands of the midge and there multiplies. The intrinsic incubation period is temperature-dependent (10–20 days), after which the midge can transmit the virus at subsequent feedings. Following infection, the *Culicoides* host remains infective throughout its life. Infection rates are highly variable, depending on the *Culicoides* species and the geographic populations involved. Selective breeding in laboratory colonies has produced both susceptible and highly resistant lines, indicating the complexity of factors influencing the vector competence of *Culicoides* species involved in the epizootiology of this disease.

Transmission of BLU virus also occurs venerealy via semen of infected rams and bulls. When introduced into the female genital tract, the virus potentially can infect the adult animal and, if she is pregnant, the developing embryo or fetus. No methods have been developed to destroy the virus in semen of infected animals. The virus can survive indefinitely in frozen semen samples.

Biting midges of the genus *Culicoides* are the only vectors of BLU virus. The following are known or suspected vectors. In North America the primary vector is *C. sonorensis*, with *C. insignis*, *C. debilipalpis*, *C. obsoletus*, and *C. stellifer* being possible secondary species. In the Caribbean Basin the primary vector is *C. insignis*, but also may involve *C. filarifer* and *C. pusillus*. In Europe, the Mediterranean region and Middle East the primary vector is *C. imicola* with *C. obsoletus*, *C. pulicaris*, and *C. scoticus* as probable secondary vectors. In Africa the primary vectors are *C. imicola* and *C. bolitinos*, in addition to *C. gulbenkiani*, *C. magnus*, *C. pycnostictus*, *C. zuluensis*, and members of the *C. shultzei* group. In Asia and the Indian Subcontinent, vectors include *C. actoni*, *C. brevitarsis*, *C. fulvus*, and *C. wadi*. In Australia the primary vector is *C. brevitarsis*, with *C. actoni*, *C. fulvus*, and *C. wadai* considered to be likely vectors, and *C. brevipalpus*, *C. oxystoma*, and *C. peregrinus* probably playing minor roles in transmission.

Figure 12.12 White-tailed deer fawn infected with epizootic hemorrhagic disease virus. Note hanging head and protruding tongue. (Photo by Gary R. Mullen)

Figure 12.13 White-tailed deer buck in late stage of epizootic hemorrhagic disease, with characteristic tender hooves, difficulty walking, arched back, laid-back ears, and general depression. (Photo by Gary R. Mullen)

Epizootic Hemorrhagic Disease

Epizootic hemorrhagic disease (EHD) is very similar to bluetongue in many respects, the major difference being that it occurs primarily in wild ruminants, notably deer. It is caused by an orbivirus very closely related to BLU virus, with 10 serotypes recognized worldwide. Two serotypes of the virus occur in North America, designated EHD-1 and EHD-2. EHD-1, known as the New Jersey strain, was first isolated from white-tailed deer in New Jersey during an outbreak in 1955. EHD-2, commonly referred to as the Alberta strain, was first isolated during an epizootic in that Canadian province in 1962. At least eight other EHD serotypes have been isolated in South Africa, Nigeria, Australia, and Japan.

The clinical signs in EHD cases are virtually indistinguishable from bluetongue. Isolation and identification of the etiologic agent usually is required to determine with certainty which virus is involved. Because of the similarities of these two diseases in wild ruminants, cases often are referred to simply as **hemorrhagic disease**. It also is referred to as **black tongue disease** by deer hunters in the southeastern United States.

Clinical disease in white-tailed deer and other ruminants varies from sudden death without apparent signs of illness to mild infections from which animals fully recover. Typically the disease is characterized by rapid onset of fever, loss of appetite, disorientation and weakness, a hanging head, labored breathing with the tongue often protruding (Figure 12.12), swelling of the head and neck, arched back and painful hooves (Figure 12.13). As the virus multiplies in endothelial cells lining the blood vessels, it spreads to various organ systems causing extensive internal hemorrhaging (Figure 12.14), intravascular coagulation, and thrombosis. In acute cases, death usually occurs in four to 10 days following the initial infection. In those animals that survive, recovery can be prolonged and debilitating, resulting in permanent lameness due to deformed hooves (Figure 12.15) and difficulty eating due to damage to the oral tissues.

EHD is the most important infectious disease of wild deer in the United States. It primarily affects white-tailed deer, causing sporadic die-offs. Mule deer, pronghorns, and domestic cattle also can develop fatal infections but less commonly. Other wild ruminants that have been found to be infected during EHD epizootics include elk, bison, bighorn sheep, Rocky Mountain goat, and several species of exotic animals such as yak and ibex. Wapiti and moose do not appear to be adversely affected by this virus.

Although this disease is endemic throughout the United States where white-tailed deer populations are

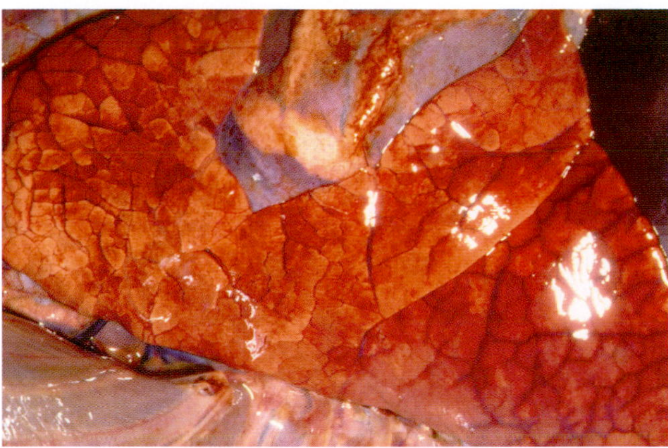

Figure 12.14 Extensive hemorrhaging and edema of lung tissue in white-tailed deer that died of epizootic hemorrhagic disease. (Courtesy of C. S. Roberts State Veterinary Diagnostic Laboratory, Auburn, AL)

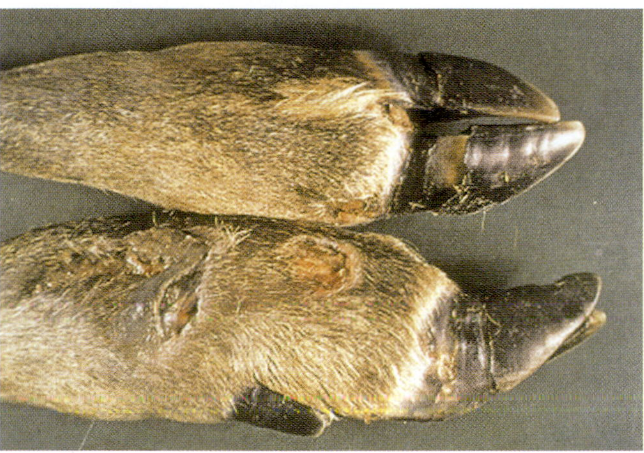

Figure 12.15 Foot lesions, swelling, and deformed hooves in white-tailed deer with epizootic hemorrhagic disease. (Courtesy of Southeastern Cooperative Wildlife Disease Study, Athens, GA)

established, it is more prevalent in the Southeast, Midwest, Northwest, and along the Pacific Coast. Epizootics, with sudden die-offs in local deer herds, tend to occur in more temperate areas, whereas asymptomatic and subclinical infections are more common in the coastal endemic areas of the southeastern states where infection rates may be as high as 70% or more. Outbreaks of EHD also have been reported in the western provinces of Canada, notably in southeastern Alberta (1962), the Okanagan Valley of British Columbia (1975), and southern Saskatchewan (1986–1987). In each case, the source of infection has been attributed to *Culicoides* from adjacent endemic areas in the United States.

Cattle are commonly exposed to EHD virus. Based on serologic surveys, infections in cattle are widespread throughout the United States. In most cases these are silent infections or involve only mild clinical disease. Occasionally, however, epizootics do occur in cattle as in central Oregon in 1969 and eastern Tennessee in 1972. Infections of cattle with EHD virus also have been reported in the South American countries of Guyana, Surinam, and Colombia, and in Taiwan, Malaysia, and Indonesia.

Biting midges of the genus *Culicoides* are the only known vectors of EHD virus. The most important species and only proven vector in North America is *C. sonorensis*. Other species have been implicated as potential vectors based on isolations of the virus from field-collected midges and limited experimental studies. The high prevalence of seropositive deer for EHD virus in areas where members of the *C. variipennis* complex is uncommon or absent supports the belief that other *Culicoides* species are involved in transmission of this virus in the United States. The vectors in other parts of the world remain unknown.

African Horsesickness

African horsesickness (AHS) is a viral disease of horses, donkeys, and mules, which can be highly fatal in susceptible animals (Figure 12.16). It is known by various names including *la pesta equine* (Spain), *pesta ecvina* (Romania), equine plague, and horsesickness fever. The disease was first recognized in South Africa in the early 1700s, with the etiologic agent being first isolated from infected horses nearly two centuries later in 1899. It occurs throughout sub-Saharan Africa and the Arabian Peninsula, extending intermittently into southwestern Asia and southern Europe where epizootics have occurred in recent years.

The etiologic agent of African horsesickness is an orbivirus in the family Reoviridae, closely related to the viruses that cause bluetongue and epizootic hemorrhagic disease. Nine AHS serotypes are recognized, all nine of which occur in eastern and southern Africa. Only serotypes 4 and 9 occur in West Africa, from which occasionally they spread to countries bordering the Mediterranean. The occurrence of multiple serotypes in a given geographic region and simultaneous

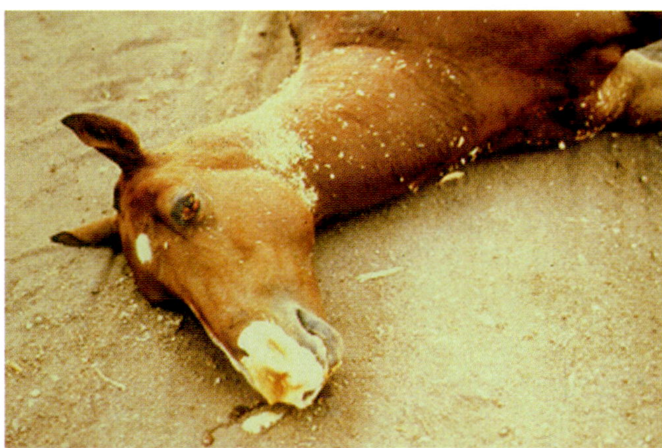

Figure 12.16 Horse that died with African horsesickness; death attributed to pulmonary edema. (Courtesy of USDA-APHIS, Foreign Animal Disease Diagnostic Laboratory, Plum Island, NY)

infections of animals with more than one serotype underscores the epizootic complexity of this disease.

Four clinical forms of African horsesickness are recognized: pulmonary (peracute), cardiac (subacute), mixed pulmonary-cardiac (acute), and **horsesickness fever**. The pulmonary form is the most fatal with mortality rates as high as 95%. Clinical signs develop within three to five days of the initial infection. The onset of symptoms is sudden, usually beginning with fever followed by congestion of the mucous membranes of the eyes, nose, and mouth. Animals sweat profusely, experience increased respiratory rates, and cough spasmodically due to the accumulation of fluids in the lungs. Froth commonly is emitted from the nostrils in the terminal stage. Death usually occurs within a few days of the onset of clinical signs.

The cardiac form is similarly characterized by initial fever and congestion of the mucous membranes following an incubation period of seven to 14 days. Animals subsequently develop extensive subcutaneous edema that is often apparent in the neck and jugular area, muscles along the back and hips, about the eyes and eyelids, and the jaws. Other signs include depression and petechial hemorrhages on the underside of the tongue. Infected animals continue to feed and drink throughout the course of the disease. Death usually occurs in four to eight days following the onset of fever, with mortality rates approaching 50%. The mixed pulmonary-cardiac form of AHS is characterized by clinical signs associated with each of the previous two syndromes. The onset of symptoms typically occurs five to seven days following infection, with death ensuing three to six days later. The mortality is approximately 80%, intermediate between the pulmonary and cardiac forms. Horsesickness fever is the mildest form of AHS. Infected animals usually recover following a low-grade fever, congested mucous membranes, loss of appetite, and mild depression over a one-week period.

The principal vertebrate hosts of AHS virus are wild and domestic equids such as zebras, horses, and mules. Donkeys are largely resistant but occasionally develop clinical disease. In endemic areas, however, native breeds seldom exhibit overt signs of infection, apparently having developed a natural or acquired immunity. Most AHS outbreaks have occurred in European breeds of equids introduced to endemic areas, or as a result of exposure of susceptible equids to infected animals imported from endemic areas in Africa and parts of the Middle East. Other animals in which the presence of antibodies indicates exposure to AHS virus are goats, sheep, domestic cattle, buffaloes, dromedaries, and elephants. None of these hosts develop more than mild clinical signs but may serve as potential reservoirs for the virus. Infected dogs, however, can develop clinical disease and are believed to be important reservoirs in urban areas. Six strains of AHS virus have been isolated from street dogs in Egypt where a number of dogs have died after consuming uncooked meat of infected horse carcasses. The progression of the disease in dogs is similar to the pulmonary form in horses.

Until the mid 1900s, epizootics of African horsesickness were confined largely to South Africa. Beginning in 1944 with an outbreak among horses in several Middle East countries, major epizootics have occurred in other parts of Africa, the Middle East, India, and Europe. The most devastating outbreak occurred in 1959 to 1960, in which over 300,000 horses died or had to be destroyed in six Middle East countries and Cyprus, Afghanistan, Pakistan, and India. Although cases of AHS were reported in Spain as early as 1965, the most severe epizootic to occur in Europe took place in Spain and Portugal in 1987 through 1990, in which more than 160 horses, mules, and donkeys died or had to be destroyed. Some of those animals were valuable thoroughbred horses participating in international equestrian competitions being held in Spain at the time. Ten zebras imported to a zoological park near Madrid, Spain, from Namibia in southern Africa are believed to have been the source of the infection. The virus subsequently was transmitted by indigenous *Culicoides* populations to Portugal and Morocco. Windborne; midges, particularly *C. imicola*, are believed to have played a role in spreading the virus. Wind dispersal of *Culicoides* vectors may help to explain the spread of AHS virus from endemic areas of Africa causing outbreaks in various parts of the Middle East, Cyprus, and Turkey in the Mediterranean region, and the Cape Verde Islands off the northwestern coast of Africa.

There presently is no cure for this disease, leaving supportive therapy as the only means of treatment. Commercially available vaccines, however, have been helpful in protecting equines from infection in areas of Africa where AHS is endemic. Annual vaccinations are effective in maintaining immunity, reflecting the natural and complete immunity acquired by animals chronically exposed to this virus over extended periods of time. Regular vaccination of susceptible equines and strict control of the movement of unvaccinated animals is currently the only practical means of containing this disease.

The major vectors of AHS virus are *Culicoides* species. *Culicoides imicola* is the principal vector in Africa and the Middle East, with *C. bolitinos* suspected of playing a secondary role. Since the first isolation from field-collected *C. imicola* during an outbreak in South Africa, other species have been implicated as vectors. A few species have been shown to support replication of AHS virus following experimental inoculation, with members of the *C. variipennis* complex, for example. The virus has been transmitted successfully 12 to 13 days after an infected blood meal. Some mosquitoes also are believed to be potential vectors even though the virus has not been isolated from them under field conditions. *Culex pipiens*, *Aedes aegypti*, and *Anopheles stephensi* have been experimentally infected with the virus, but there is no strong evidence to indicate that these particular mosquitoes are natural vectors.

AHS virus has been isolated from naturally infected camel ticks (*Hyalomma dromedarii*) in Egypt, raising a question about possible involvement of this tick as a secondary vector. The brown dog tick *Rhipicephalus sanguineus* has been shown to be capable of biologically transmitting AHS virus between horses and dogs. However, the virus has not been isolated from naturally infected ticks of this species.

Other Viral Agents

Culicoides species are suspected as potential vectors of other viruses affecting livestock (Table 12.1), although their role in most cases remains uncertain. *Culicoides sonorensis*, for example, supports replication of vesicular stomatitis (VS) virus and has been shown experimentally to be capable of transmitting the virus to cattle (Drolet et al., 2005; Leon and Tabachnick, 2006). Several other bovine arboviruses have been isolated from field-collected *Culicoides oxystoma*, for example, Akabane, Aino, Chuzan, D'Aguliar, and Ibaraki viruses in Japan (Yanase et al., 2005) and Shamonda virus in Japan and Nigeria. However, vector competence studies remain to be done before any meaningful conclusions can be drawn regarding the importance of *Culicoides* species in the natural transmission of these viruses. *Culicoides imicola* and *C. bolitinos* are suspected vectors of equine encephalosis (EE) virus among horses in southern Africa (Venter et al., 2002; Paweska and Venter, 2004).

Blood Protozoans

Biting midges are biological vectors of a number of protozoans called **haemosporidians**, which are blood parasites of reptiles, birds, and mammals. Three genera that are transmitted by biting midges are *Haemoproteus* (Figure 12.17), *Hepatocystis*, and *Leucocytozoon*. Most of the species are avian parasites that cause little or no apparent harm to their hosts. A few, however, such as *Haemoproteus meleagridis* of turkeys and *Leucocytozoon caulleryi* of chickens can cause significant problems for poultry producers. In addition, *Culicoides* species may serve as hosts for several avian trypanosomes (e.g., *Herpetomonas*, *Sergeia*).

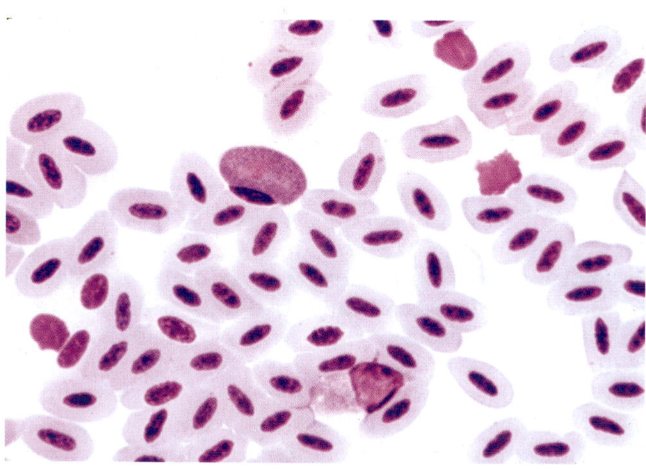

Figure 12.17 *Haemoproteus* sp., avian protozoan developing in blood cells of mourning dove, transmitted by biting midges. (Photo by Mary E. Hayes/Rogers)

Haemosporidians transmitted by biting midges are related to malarial parasites (*Plasmodium* species) with which they share a similar life cycle and developmental stages. While feeding on an infected vertebrate host, female midges ingest red blood cells containing **gametocytes**, the sexual stage of the parasite. In the midgut of the midge, the gametocytes are released where they unite to form a motile zygote, the **ookinete**. The ookinete typically penetrates the peritrophic membrane and midgut tissue to form a cyst-like structure or **oocyst** on the outer midgut wall. Within the oocyst, **sporozoites** are produced asexually, eventually rupturing from the mature oocysts into the haemocoel where they make their way to the salivary glands and accumulate there. The sporozoite is the infective stage that is transmitted via the saliva to suitable hosts when the biting midge subsequently blood feeds. Development of the parasite in *Culicoides* species usually takes about six to 10 days.

Upon entering the vertebrate host, sporozoites invade cells of fixed tissues, notably the endothelium of various organs, and myofibroblasts, precursor cells that form muscle fibers. There they undergo one or more cycles of asexual reproduction called **schizogony** to produce **merozoites**. The merozoites then invade the blood and penetrate circulating erythrocytes. There they develop into gametocytes, thereby completing the life cycle.

The species of blood protozoans that are known to be transmitted by biting midges are summarized in Table 12.3. With the exceptions of *Haemoproteus kochi*, which parasitizes Old World monkeys, and *H. brayi*, that parasitizes Malaysian squirrels, these haemosporidians are parasites of birds. The species are primarily members of the genus *Haemoproteus*, all of which are transmitted by *Culicoides* spp. Relatively few details are known about most of these arthropod-borne haemosporidians and their associated *Culicoides* vectors. What is known is based primarily on studies of *H. meleagridis* and *L. caulleryi* as parasites of poultry.

Haemoproteus meleagridis *Haemoproteus meleagridis* is primarily a parasite of wild and domestic turkeys. It also can cause at least transient infections in pheasants and chukars but apparently does not infect chickens, guinea fowl, bobwhite quail, and other gallinaceous birds.

This parasite generally is regarded as nonpathogenic. Even in cases in which large numbers of circulating red blood cells are infected with gametocytes, birds usually exhibit few signs of stress or other pathologic effects. This suggests that compensatory mechanisms are operative in which the replacement rate of erythrocytes is sufficient to maintain a stable hematocrit despite high parasitemia. In other cases, however, there is evidence to indicate that *H. meleagridis* does harm its avian hosts, especially domestic turkeys. Heavy infections can result in anemia, reduced weight gain and growth rates, inflammation of skeletal and cardiac muscles, lameness, damage to the spleen and liver, and a wasting condition associated with chronic infections. Young birds are particularly vulnerable.

Five *Culicoides* species have been identified as vectors of *H. meleagridis* based primarily on studies in

Table 12.3 Protozoans Transmitted by Biting Midges

Protozoan	Vertebrate Hosts	Geographic Area	Known or Suspected *Culicoides* Vectors
Haemoproteus			
H. danilewskyi	Crows, jays (Corvidae)	North America	*C. arboricola, C. crepuscularis, C. edeni, C. sphagnumensis, C. stilobezzioides*
H. desseri (=*H. handai*)	Parakeets (Psittacidae)	Thailand	*C. nubeculosus* (experimental)
H. fringillae	Finches, sparrows (Fringillidae)	North America	*Culicoides crepuscularis, C. stilobezzioides*
H. lophortyx California quail North America *C. bottimeri*	Grouse (Tetraonidae)	North America	*C. sphagnumensis*
H. mansoni (=*H. canachites*)			
H. meleagridis	Turkey (Meleagrididae)	North America	*C. edeni, C. arboricola, C. haematopotus, C. hinmani, C. knowltoni*
H. nettionis	Ducks, geese (Anatidae), other waterfowl	Canada	*C. downesi*
H. velans	Woodpeckers (Picidae)	North America	*C. sphagnumensis*
Hepatocystis			
H. brayi	Squirrels (Sciuridae)	Malaysia	*Culicoides* spp.
H. kochi	Monkeys (*Cercopithecus*)	Kenya	*C. adersi*
Leucocytozoan caulleryi	Chickens	Southeast Asia, Japan	*C. arakawae, C. circumscriptus, C. guttifer, C. schultzei*

Florida (USA). *Culicoides edeni* is regarded as the most important vector, with *C. hinmani*, *C. arboricola*, *C. haematopotus*, and *C. knowltoni* playing secondary roles in transmission. Other species such as *C. baueri*, *C. nanus*, and *C. paraensis* have been shown to support development of the parasite only to the oocyst stage. Transmission of *H. meleagridis* occurs throughout the year in southern Florida, whereas it is limited to the warmer months of the year throughout the rest of the United States where turkeys occur.

Other *Haemoproteus* Species *Haemoproteus danilewskyi*, an avian parasite of the blue jay *Cyanocitta cristata* in Florida (USA), is capable of sporogonic development in *Culicoides arboricola* and *C. edeni* (Garvin and Greiner, 2003). In California, *Haemoproteus lophortyx*, a parasite of the California quail (*Callipepla claifornica*), is believed to be transmitted by *C. bottimeri* (Mullens et al., 2006). In Europe, *Culicoides impunctatus* appears to be a likely vector of several *Haemoproteus* species that infect passerine birds, whereas other *Haemoproteus* species have been shown to have a detrimental effect on *C. impunctatus*, significantly decreasing its longevity (Valkiunas et al., 2002; Valkiunas and Lezhova, 2004).

Leucocytozoon caulleryi This is the only *Leucocytozoon* species known to be transmitted by biting midges. It has been recognized for many years as causing a serious poultry disease of chickens in Japan and Southeast Asia where it is known as **poultry leucocytozoonosis** and by the earlier name **Bangkok hemorrhagic disease**, where it occurred in Thailand. The principal vector of *L. caulleryi* is *Culicoides arakawae*, which commonly breeds in rice paddies.

Equine Onchocerciasis

Equine onchocerciasis is caused by the filarial nematode *Onchocerca cervicalis* (Figure 12.18), the most widely distributed nematode transmitted to domestic animals by biting midges. Horses are the only known host. Although it occurs worldwide, most of the problems associated with this nematode have been reported in the United States and Australia where it commonly causes dermatitis. Various names that refer to infestations by *O. cervicalis* include **cutaneous equine onchocerciasis, equine ventral midline dermatitis, equine nuchal disease, and fistulous withers**. Prevalence of *O. cervicalis* is high in many regions of the United States, with up to 85% or more of older horses having been reported infected with this parasite in New York, Kentucky, and the Gulf Coast states.

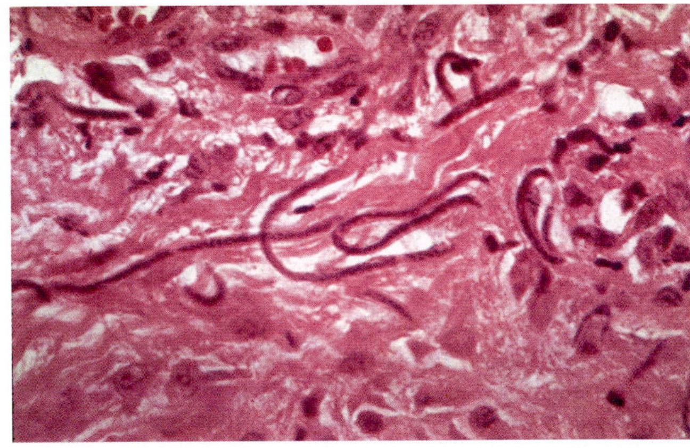

Figure 12.18 *Onchocerca cervicalis*, histological preparation showing microfilariae in skin of infested horse. (Montes and Vaughan, 1983)

Adult worms occur primarily in the nuchal ligament of the neck and between the shoulder blades, or withers. Microfilariae produced by the females move to the skin, where they are active in the dermal tissues, often eliciting a host response in the form of localized inflammation and pruritis. The highest concentrations of microfilariae tend to be along the ventral midline of the horse. High numbers of microfilariae also may occur in skin of the inner thighs, chest region, withers, and eyelids. The density of microfilariae in skin tissue varies seasonally, being highest during the spring and summer months and lowest in the winter when they move to the deeper dermal layers. This is correlated with seasonal activity of most biting midges.

Horses that become sensitized to *O. cervicalis* develop various types of skin lesions including depigmentation, pruritius, scaling, and hair loss. This usually occurs on the face, chest, withers, and ventral midline where microfilariae are most abundant. Ocular lesions also have been reported. Diagnosis is based on clinical signs and the detection of microfilariae in skin biopsies. Treatment with ivermectin has been found to be effective in killing microfilariae but not the adults. In most cases the skin lesions show significant improvement, or are completely resolved, within a few weeks following treatment.

Members of the *Culicoides variipennis* complex are the only known vectors of *O. cervicalis* in North America. Based primarily on laboratory studies, *C. victoriae*, *Forcipomyia townsvillensis*, and the black fly *Austrosimulium pestilens* also have been identified as potential vectors of *O. cervicalis* in Australia. Since these biting midges tend to ingest very few microfilariae while feeding on an infected animal, only one or two infective third-stage larvae are typically found in field-collected flies.

Other Filarial Nematodes

At least three other *Onchocerca* species that infest bovine and equine hosts are believed to be transmitted by biting midges (Table 12.2): *O. gibsoni* of cattle in Southeast Asia, Malaya, Australia, and South Africa; *O. gutturosa* of cattle; and *O. reticulata* of horses and ponies in Australia. They are considered to be nonpathogenic.

Equine Allergic Dermatitits

Horses exposed to bites of certain *Culicoides* species commonly exhibit an allergic skin reaction. This typically occurs as a seasonal dermatitis affecting the withers, mane, tail, and ears. The back, ventral midline, and other body regions also can be affected, presumably reflecting the feeding sites of different biting midges involved. Equine allergic dermatitis was first attributed to *Culicoides* bites in Australia in the early 1950s where it was known as **Queensland itch**. It is now known to occur widely throughout the world by various names such as **sweet itch, summer dermatitis, summer recurrent dermatitis, summer eczema, equine *Culicoides* sensitivity, Dhobie itch** (Philippines), **and Kasen disease** (Japan). A similar seasonal dermatitis in response to *Culicoides* bites also occurs in sheep.

The dermal response apparently is a sensitivity reaction to components of salivary fluids introduced to the bite wound while the flies are feeding (Figure 12.19).

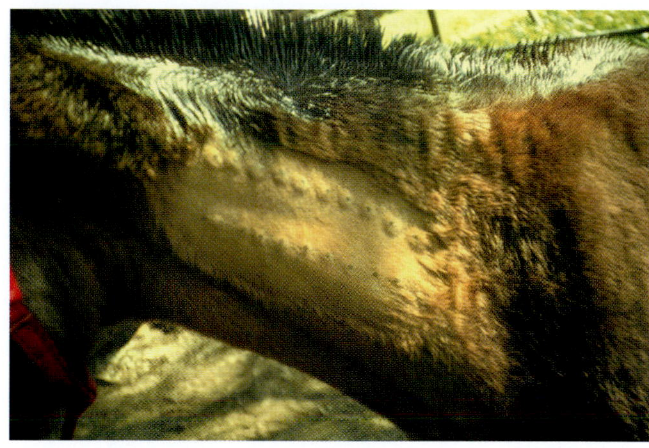

Figure 12.19 Allergic dermatitis in neck region of horse in response to injections of *Culicoides* extracts. (Courtesy of Yehuda Braverman, Kimron Veterinary Institute, Israel)

Normal, nonsensitized horses usually react to these bites by developing small welts with relatively little associated discomfort. Sensitized horses, however, react more severely by developing intense local inflammation and pruritus; this can result in irritability, rubbing and scratching of involved areas, open wounds, and secondary infections. Ponies are especially sensitive. Affected animals often are unsuitable for riding and, in the case of show horses, may decrease substantially in commercial value because of their irritable behavior, hair loss, and skin blemishes.

Once sensitized, horses experience either an immediate hypersensitivity response, peaking within four hours, or a delayed hypersensitivity response in which large welts develop after 24 hours, with inflammation persisting up to three weeks or more. There is good evidence to show that *Culicoides*-induced hypersensitivity is a polygenic hereditary trait that predisposes certain animals to this response. This sensitivity occurs primarily in older horses, usually after four to five years of age.

A number of *Culicoides* species have been implicated as the cause of equine allergic dermatitis. Most are based on correlations between seasonal occurrences of the midges and clinical signs, biting sites on horses, and positive reactions to intradermal injections of horses with extracts of the respective biting midges. The following species are suspected of being involved: *C. insignis*, *C. obsoletus*, *C. spinosus*, *C. stellifer*, and *C. venustus* in the United States; *C. pulicaris* in England; *C. nubeculosus* and *C. punctatus* in Ireland; *C. chiopterus*, *C. impunctatus*, and *C. obsoletus* in Norway; *C. imicola* in Israel; and *C. brevitarsis* in Australia.

Treatments for equine allergic dermatitis in the form of antihistamines and corticosteroids usually provide only temporary relief of symptoms. Desensitization of animals with injections of *Culicoides* extracts has not proved to be effective. Horse owners in areas where this condition is recognized as a problem should avoid breeding their animals with lineages of known sensitivity. Insecticides applied directly to horses to repel or kill biting midges affords some protection and can substantially reduce the severity if administered on a regular basis throughout the fly season. Ivermectin, however, is ineffective. Stabling horses at night or pasturing them away from the attack of biting midges can also help to alleviate the problem.

PREVENTION AND CONTROL

Larviciding generally has not been effective in reducing populations of biting midges. Often the breeding sites are difficult to locate and may be so dispersed that the application of insecticides to kill the immature stages is not practical. In some situations modifications of the habitat can help to reduce breeding sites by filling low-lying areas, diking, and regulating water levels to disrupt breeding and larval development. Eliminating seepage areas and leaking water troughs in or around livestock facilities can discourage the breeding of important species like *Culicoides sonorensis* and *C. variipennis*. Proper maintenance of farm ponds and fluctuation of the water level in dairy ponds and waste lagoons can help to reduce the numbers of adult biting midges that emerge. Disking of low-lying crop lands and the use of appropriate irrigation schedules have been effective in reducing adult populations of some pest species.

Adulticides have been used with limited success in suppressing adults. To be effective they usually are applied as mists or fogs in the evening hours when the insects are most active. In coastal areas where problems can be especially severe, aerial applications of ultra-low volume formulations of insecticides to salt marshes bordering populated areas can help to provide some relief. Ground applications using truck-mounted mist sprayers for control of biting midges are sometimes conducted in conjunction with municipal mosquito control programs in problem areas.

Individual protection of humans and other animals is often the only practical means of discouraging ceratopogonid midges from biting. Scheduling outdoor activities to avoid the peak biting periods of troublesome species is advisable. Animals such as horses can be stabled at night to protect them from species that do not readily enter buildings and shelters to feed. The mesh sizes of most window and door screens are not effective in excluding biting midges, especially in the case of species such as *C. furens* and *C. paraensis* that will enter buildings in search of hosts. Treatment of resting surfaces, such as the walls and roofs of animal shelters, with residual insecticides has been shown to reduce the numbers of certain *Culicoides* species. Insect repellents applied to exposed skin and the use of jackets and other clothing impregnated with compounds such as DEET can provide effective protection from the bites of many of the more troublesome species.

For a review of techniques to control biting midges, see Carpenter et al. (2008); and for an assessment of mermithid nematodes as biological control agents, see Mullens et al. (2008).

REFERENCES AND FURTHER READING

Akiba, K. (1960). Studies on the *Leucocytozoon* found in the chicken in Japan. II. On the transmission of *Leucocytozoon caulleryi* by *Culicoides arakawae*. *Japanese Journal of Veterinary Science, 22*, 309–317.

Akiba, K. (1970). Leucocytozoonosis of chickens. *National Institute of Animal Health Quarterly, 10*(Suppl.), 131–147.

Anderson, G. S., Belton, P., & Kleider, N. (1993). Hypersensitivity of horses in British Columbia to extracts of native and exotic species of *Culicoides* (Diptera: Ceratopogonidae). *Journal of Medical Entomology, 30*, 657–663.

Atchley, W. R., Wirth, W. W., Gaskins, C. T., & Strauss, S. L. (1981). *A bibliography and keyword index of the biting midges (Diptera: Ceratopogonidae)*. USDA, Sci. and Edu. Admin., Bibliographies and Literature of Agriculture No. 13.

Atkinson, C. T. (1988). Epizootiology of *Haemoproteus meleagridis* (Protozoa: Haemosporina) in Florida: Potential vectors and prevalence in naturally infected *Culicoides* (Diptera: Ceratopogonidae). *Journal of Medical Entomology, 25*, 39–44.

Atkinson, C. T. (1991). Vectors, epizootiology, and pathogenicity of avian species of *Haemoproteus* (Haemosporina: Haemoproteidae). *Bulletin of the Society for Vector Ecology, 16*, 109–126.

Battle, F. V., & Turner, E. C., Jr. (1971). A systematic review of the genus *Culicoides* (Diptera: Ceratopogonidae) of Virginia. The Insects of Virginia No. 3, Va. Polytechnic Instit. and State Univ. Bull, 44.

Blanton, F. S., & Wirth, W. W. (1979). The sand flies (*Culicoides*) of Florida (Diptera: Ceratopogonidae). *Arthropods of Florida and Neighboring Land Areas*, 10.

Barber, T., & Jochim, M. J. (1985). Bluetongue and Related Orbiviruses. *Progress in Clinical and Biology Research*, 178.

Boorman, J. (1993). Biting midges (Ceratopogonidae). In R. P. Lane, & R. W. Crosskey (Eds.), *Medical Insects and Arachnids* (pp. 288–301). London: Chapman & Hall.

Boorman, J., & Hagan, D. V. (1996). A name list of world *Culicoides* (Diptera: Ceratopogonidae). *International Journal of Dipterological Research, 7*, 161–192.

Borkent, A. (2005). Ceratopogonidae. In W. C. Marquart (Ed.), *Biology of Disease Vectors* (pp. 113–126). San Diego: Elsevier Inc.

Borkent, A., 2006. World species of biting midges (Diptera: Ceratopogonidae). http://www.inhs.uiuc.edu/FLYTREE/CeratopogonidaeCatalog.pdf

Borkent, A., and Spinelli, G. R., 2000. Catalog of New World biting midges south of the United States (Diptera: Ceratopogonidae). Contributions to Entomology, International 4, 1–107.

Borkent, A., & Spinelli, G. (2007). *Neotropical Ceratopogonidae, Diptera, Insecta (Aquatic Biodiversity in Latin America)*. Pensoft Publishers.

Borkent, A., & Wirth, W. W. (1997). World species of biting midges (Diptera: Ceratopogonidae). *Bulletin of the American Museum of Natural History, 233*, 1–257.

Carpenter, S., Mellor, P., & Torr, S. J. (2008). Control techniques for Culicoides biting midges and their application in the U.K. and northwestern Palaearctic. *Medical and Veterinary Entomology, 22*, 1–13.

Coetzer, J. A. W., & Tustin, R. C. (2004). *Infectious Diseases of Livestock, with Special Reference to South Africa* (2nd ed.). 3 Vols. Oxford University Press.

DeHaven, W. R., Valle Molina, J. A., & Evans, B. (2004). Bluetongue viruses and trade issues: A North American perspective. *Veterinaria Italiana, 40*, 683–687.

Downes, J. A., & Wirth, W. W. (1981). Ceratopogonidae. In J. F. McAlpine (Ed.), *Manual of Nearctic Diptera* (Vol. 1, pp. 393–421). Research Branch, Agriculture Canada, Monogr. 17.

Drolet, B. S., Campbell, C. L., Stuart, M. A., & Wilson, W. C. (2005). Vector competence of *Culicoides sonorensis* (Diptera: Ceratopogonidae) for vesicular stomatitits virus. *Journal of Medical Entomology, 42*, 409–418.

Dulac, G. C., Sterritt, W. G., Dubuc, C., Afshar, A., Myers, D. J., Taylor, E. A., et al. (1991). Incursions of orbiviruses in Canada and their serologic monitoring in the native animal populations between 1962 and 1991. In T. E. Walton, & B. I. Osburn (Eds.), *Bluetongue, African Horsesickness, and Related Orbiviruses* (pp. 120–128). Boca Raton, FL: CRC Press.

Eberhard, M. L., & Orihel, T. C. (1984). The genus *Mansonella* (syn. Tetrapetalonema). A new classification. *Annales de Parasitologie Humaine et Comparee, 59,* 483–496.

Foil, L., Stage, D., & Klei, T. R. (1984). Assessment of wild-caught *Culicoides* (Ceratopogonidae) species as natural vectors of *Onchocerca cervicalis* in Louisiana. *Mosquito News, 44,* 204–206.

Garnham, P. C. C., Desser, S. S., & Khan, R. A. (1974). On species of *Leucocytozoon, Haemoproteus* and *Hepatocystis.* In J. P. Kreier (Ed.), *Parasitic Protozoa* (Vol. 3, pp. 239–266).

Garvin, M. C., & Greiner, E. C. (2003). Ecology of *Culicoides* (Diptera: Ceratopogonidae) in Southcentral Florida and experimental *Culicoides* vectors of the avian hematozoan *Haemoproteus danilewskyi* Kruse. *Journal of Wildlife Diseases, 39,* 170–178.

Gerdes, G. H. (2004). A South African overview of the virus, vectors, surveillance and unique features of bluetongue. *Veterinaria Italiana, 40,* 39–42.

Gibbs, E. P. J., & Greiner, E. C. (1988). Bluetongue and epizootic hemorrhagic disease. In T. P. Monath (Ed.), *The Arboviruses: Epidemiology and Ecology* (Vol. 2, pp. 39–70). Boca Raton, FL: CRC Press.

Gibbs, E. P. J., & Greiner, E. C. (1994). The epidemiology of bluetongue. *Comparative Immunology, Microbiology & Infectious Diseases, 17,* 207–220.

Glick, J. I. (1990). *Culicoides* biting midges (Diptera: Ceratopogonidae) of Kenya. *Journal of Medical Entomology, 27,* 87–195.

Gloster, J., Mellor, P. S., Manning, A. J., Webster, H. N., & Hort, M. C. (2007). Assessing the risk of windborne spread of bluetongue in the 2006 outbreak of disease in northern Europe. *Veterinary Record, 160,* 54–56.

Glukhova, V. M. (1977). The subgeneric classification of the genus *Culicoides* Latreille, 1809 (Diptera: Ceratopogonidae), including morphological characters of the larvae (in Russian). *Parazitol Sbornik, 27,* 112–128.

Glukhova, V. M. (1979). *Larval Midges of the Subfamilies Palpomyiinae and Ceratopogoninae of the Fauna of the U.S.S.R. (in Russian).* Leningrad: Nauka Publishers.

Glukhova, V. M. (1989). *Blood-Sucking Midges of the Genera Culicoides and Forcipomyia (Ceratopogonidae) of the Fauna of the U.S.S.R. (in Russian).* Leningrad: Nauka Publishers.

Gomez-Tejedor, C. (2004). Brief overview of the bluetongue situation in Mediterranean Europe, 1998–2004. *Veterinaria Italiana, 40,* 57–60.

Greiner, E. C. (1995). Entomological evaluation of insect hypersensitivity in horses. *Vet Clin NA: Equine Pract, 11,* 29–41.

Greiner, E. C., Fadok, V. A., & Rabin, E. B. (1990). Equine *Culicoides* hypersensitivity in Florida: Biting midges aspirated from horses. *Medical and Veterinary Entomology, 4,* 375–381.

Greiner, E. C., Mo, C. L., Tanya, V., Thompson, L. H., & Oviedo, M. T. (1991). Vector ecology of bluetongue viruses in Central America and the Caribbean. In T. E. Walton, & B. I. Osburn (Eds.), *Bluetongue, African Horsesickness, and Related Orbiviruses* (pp. 320–324.

Halldorsdottir, S., & Larsen, H. J. (1991). An epidemiological study of summer eczema in Icelandic horses in Norway. *Equine Veterinary Journal, 23,* 296–299.

Hess, W. R. (1988). African horse sickness. In T. P. Monath (Ed.), *The Arboviruses: Epidemiology and Ecology* (Vol. 2. pp. 1–18). Boca Raton, FL: CRC Press.

Holbrook, F. R., Tabachnick, W. J., Schmidtmann, E. T., McKinnon, C. N., Bobian, R. J., & Grogran, W. L. (2000). Sympatry in the *Culicoides variipennis* complex (Diptera: Ceratopogonidae): A taxonomic reassessment. *Journal of Medical Entomology, 37,* 65–76.

Hunt, G. J. (1994). *A procedural manual for the large-scale rearing of the biting midge, Culicoides variipennis (Diptera: Ceratopogonidae).* U.S. Dept. Agric., Agric. Res. Ser. ARS-121.

Jamnback, H. A. (1965). *The Culicoides of New York State (Diptera: Ceratopogonidae).* New York State Mus. Sci. Serv. Bull. No. 399.

Jones, R. H., Luedke, A. J., Walton, T. E., & Metcalf, H. E. (1981). Bluetongue in the United States; an entomological perspective toward control. *World Animal Review, 38,* 2–8.

Kettle, D. S. (1965). Biting ceratopogonids as vectors of human and animal diseases. *Acta Tropica, 22,* 356–362.

Kettle, D. S. (1972). The biting habits of *Culicoides furens* (Poey) and *C. barbosai* Wirth and Blanton. III. Seasonal cycle, with a note on the relative importance of ten factors that might influence the biting rate. *Bulletin of Entomological Research, 61,* 565–576.

Kettle, D. S. (1984). *Medical and Veterinary Entomology.* John Wiley & Sons.

Lager, I. A. (2004). Bluetongue in South America: Overview of viruses, vectors, surveillance and unique features. *Veterinaria Italiana, 40,* 89–93.

LeDuc, J. W., Hoch, A. L., Pinheiro, F. P., & Travassos da Rosa, A. P. A. (1981). Epidemic Oropouche virus disease in northern Brazil. *Bulletin of the Pan American Health Organization, 15,* 97–193.

Leon, A., & Tabachnick, W. J. (2006). Transmission of vesicular stomatitis New Jersey virus to cattle by the biting midge *Culicoides sonorensis* (Diptera: Ceratopogonidae). *Journal of Medical Entomology, 43,* 323–329.

Lillie, T. H., Kline, D. L., & Hall, D. W. (1987). Diel and seasonal activity of *Culicoides* spp. (Diptera: Ceratopogonidae) near Yankeetown, Florida, monitored with a vehicle-mounted insect trap. *Journal of Medical Entomology, 24,* 503–511.

Linley, J. R. (1983). Biting midges (Diptera: Ceratopogonidae) and human health. *Journal of Medical Entomology, 20,* 347–364.

Linley, J. R. (1985). Biting midges (Diptera: Ceratopogonidae) as vectors of nonviral animal pathogens. *Journal of Medical Entomology, 22,* 589–599.

Linley, J. R., & Adams, M. (1972). A study of the mating behavior of *Culicoides melleus* (Diptera: Ceratopogonidae). *Transactions of the Royal Entomological Society of London, 124,* 81–121.

Lubroth, J. (1988). African horsesickness and the epizootic in Spain 1987. *Equine Practice, 10,* 26–33.

Lubroth, J. (1991). The complete epidemiologic cycle of African horse sickness: Our incomplete knowledge. In T. E. Walton, & B. I. Osburn (Eds.), *Bluetongue, African Horse Sickness, and Related Orbiviruses* (pp. 197–204). Boca Raton, FL: CRC Press.

Mehlnorn, H., Walldorf, V., Klimpel, S., Jahn, B., Jaeger, F., Eschweiler, J., et al. (2007). First occurrence of *Culicoides obsoletus*-transmitted Bluetongue virus epidemic in Central Europe. *Parasitology Research, 101,* 219–228.

Meiswinkel, R., Gomulski, L. M., Delecolle, J. C., Goffredo, M., & Gasperi, G. (2004). The taxonomy of *Culicoides* vector complexes—Unfinished business. *Veterinaria Italiana, 40,* 151–159.

Meiswinkel, R., Venter, G. J., & Neville, E. M. (2004). Vectors: *Culicoides* spp. In J. A. W. Coetzer, & R. C. Tustin (Eds.), *Infectious Diseases of Livestock, with Special Reference to South Africa* (Vol. 1, pp. 93–136). Oxford University Press.

Mellor, P. S. (1990). The replication of bluetongue virus in *Culicoides* vectors. In P. Roy, & B. M. Gorman (Eds.), *Bluetongue Viruses. Current Topics in Microbiology and Immunology 162* (pp. 143–161). Berlin: Springer-Verlag.

Mellor, P. S., & Hamblin, C. (2004). African horse sickness. *Equine Infectious Diseases (special issue), 4,* 445–466.

Miura, Y., Goto, Y., Kubo, M., & Kono, Y. (1988). Isolation of Chuzan virus, a new member of the Palyam subgroup of the genus *Orbivirus,* from cattle and *Culicoides oxystoma in Japan. American Journal of Veterinary Research, 49,* 2022–2025.

Mo, C. L., Thompson, L. H., Homan, E. J., Oviedo, M. T., Greiner, E. C., González, J., et al. (1994). Bluetongue virus isolations from vectors and ruminants in Central America and the Caribbean. *American Journal of Veterinary Research, 55,* 211–215.

Mullen, G. R., & Hribar, L. J. (1988). Biology and feeding behavior of ceratopogonid larvae (Diptera: Ceratopogonidae) in North America. *Bulletin of the Society for Vector Ecology, 13,* 60–81.

Mullens, B. A. (1991). Integrated management of *Culicoides variipennis*: A problem of applied ecology. In T. E. Walton, & I. Osburn (Eds.), *Bluetongue, African Horsesickness, and Related Orbiviruses* (pp. 896–905). Boca Raton, FL: CRC Press.

Mullens, B. A., Cardona, C. J., McClellan, L., Szijj, C. E., & Owen, J. P. (2006). *Culicoides bottimeri* as a vector of *Haemoproteus lophortyx* to quail in California, USA. *Veterinary Parasitology, 140,* 35–43.

Mullens, B. A., Sarto I Monteys, V., & Przhboro, A. A. (2008). Mermithid parasitism in the Ceratopogonidae: A literature review and critical assessment of host impact and potential for biological control. *Russian Entomological Journal, 17,* 87–113.

Murphree, C. S., & Mullen, G. R. (1991). Comparative larval morphology of the genus *Culicoides* Latreille (Diptera: Ceratopogonidae) in North America with a key to species. *Bulletin of the Society for Vector Ecology, 16,* 269–399.

Nolan, D. V., Carpenter, S., Barber, J., Mellor, P. S., Dallas, J. F., Mordue, A. J., et al. (2007). Rapid diagnostic PCR assays for members of the *Culicoides obsoletus* and *Culicoides pulicaris* species complexes, implicated vectors of bluetongue virus in Europe. *Veterinary Microbiology, 124*, 82–94.

Ottley, M. L., Dallemagne, C., & Moorhouse, D. E. (1983). Equine onchocerciasis in Queensland and the Northern Territory of Australia. *Australian Veterinary Journal, 60*, 200–203.

Paweska, J. T., & Venter, G. J. (2004). Vector competence of *Culicoides* species and the seroprevalence of homologous neutralizing antibody in horses for six serotypes of equine encephalosis virus (EEV) in South Africa. *Medical and Veterinary Entomology, 18*, 398–407.

Paweska, J. T., Venter, G. J., & Mellor, P. S. (2002). Vector competence of South African *Culicoides* species for bluetongue virus serotype 1 (BTV-1) with special reference to the effect of temperature on the rate of virus replication in *C. imicola* and *C. bolitinos*. *Medical and Veterinary Entomology, 16*, 10–21.

Pearson, J. E., Gustafason, G. A., Shafer, A. L., & Alstad, A. D. (1991). Distribution of bluetongue in the United States. In T. E. Walton, & B. I. Osburn (Eds.), *Bluetongue, African Horsesickness, and Related Orbiviruses* (pp. 128–138). Boca Raton, FL: CRC Press.

Pinheiro, K. F. P., Travassos da Rosa, A. P. A., Travassos da Rosa, J. F. S., et al. (1981). Oropouche virus I. A review of clinical, epidemiological, and ecological findings. *The American Journal of Tropical Medicine and Hygiene, 30*, 149–160.

Purse, B. V., Mellor, P. S., Rogers, D. J., Samuel, A. R., Mertens, P. P. C., & Baylis, M. (2005). Climate change and the recent emergence of bluetongue in Europe. *Nature Reviews Microbiology, 3*, 171–181.

Rawlings, P., Meiswinkel, R., Labuschange, K., Welton, N., Baylis, M., & Mellor, P. S. (2003). The distribution and species characteristics of the *Culicoides* biting midges of South Africa. *Ecological Entomology, 28*, 559–566.

Sebastiani, F., Meiswinkel, R., Gomulski, L. M., Guglielmino, C. R., Mellor, P. S., Malacrida, A. R., et al. (2001). Molecular differentiation of the Old World *Culicoides imicola* species complex (Ditpera: Ceratopogonidae), inferred using random amplified polymorphic DNA markers. *Molecular Ecology, 10*, 1773–1786.

Spinelli, G. R., & Borkent, A. (2004). New species of Central American *Culicoides* Latreille (Diptera: Ceratopogonidae) with a synopsis of species from Costa Rica. *Proceedings of the Entomological Society of Washington, 106*, 361–395.

Spinelli, G. R., Ronderos, M. M., Diaz, F., & Marino, P. I. (2005). The bloodsucking biting midges of Argentina (Diptera: Ceratopogonidae). *Memorias do Instituto Oswaldo Cruz, Rio de Janeiro, 100*, 137–150.

Tabachnick, W. J. (1992). Genetic differentiation among populations of *Culicoides variipennis* (Diptera: Ceratopogonidae), the North American vector of bluetongue virus. *Annals of the Entomological Society of America, 85*, 140–147.

Tabachnick, W. J. (1996). *Culicoides variipennis* and bluetongue-virus epidemiology in the United States. *Annual Review of Entomology, 41*, 23–43.

Tabachnick, W. J. (2004). *Culicoides* and the global epidemiology of bluetongue virus infection. *Veterinaria Italiana, 40*, 135–150.

Thomas, F. C. (1981). Hemorrhagic disease. In W. R. Davidson, F. A. Hayes, V. F. Nettles, & F. E. Kellogg (Eds.), *Diseases and Parasites of White-tailed Deer* (pp. 87–96). Miscellaneous Publications No. 7. 87–96. Tallahassee, FL: Tall Timbers Research Station.

Valikiunas, G., & Lezhova, T. A. (2004). Detrimental effects of *Haemoproteus* infections on the survival of the biting midge *Culicoides impunctatus* (Diptera: Ceratopogonidae). *The Journal of parasitology, 90*, 194–196.

Valikiunas, G., Liutkevicius, G., & Lezhova, T. A. (2002). Complete development of three species of *Haemoproteus* (Haemosporidia, Haemoproteidae) in the biting midge *Culicoides impunctatus* (Diptera, Ceratopogonidae). *The Journal of Parasitology, 88*, 864–868.

Venter, G. J., Groenewald, D., Venter, E., Hermanides, K. G., & Howell, P. G. (2002). A comparison of the vector competence of the biting midges, *Culicoides* (*Avaritia*) *bolitinos* and *C.* (*A.*) *imicola*, for the Bryanson serotype of equine encephalosis virus. *Medical and Veterinary Entomology, 16*, 372–377.

Walton, T. E. (2004). The history of bluetongue and a current global overview. *Veterinaria Italiana, 40*, 31–38.

Walton, T. E., & Osburn, B. I. (1991). *Bluetongue, African Horsesickness, and Related Orbiviruses.* Boca Raton, FL: CRC Press.

Ward, M. P. (1994). The epidemiology of bluetongue virus in Australia—A review. *Australian Veterinary Journal, 71*, 3–7.

Wirth, W. W., Dyce, A. L., & Peterson, B. V. (1985). An atlas of wing photographs, with a summary of the numerical characters of the Nearctic species of *Culicoides* (Diptera: Ceratopogonidae). *Contributions of the American Entomological Institute, 22*(4).

Wirth, W. W., Dyce, A. L., & Spinelli, G. R. (1988). An atlas of wing photographs, with a summary of the numerical characters of the Neotropical species of *Culicoides* (Diptera: Ceratopogonidae). *Contributions of the American Entomological Institute, 25*(1).

Wirth, W. W., & Hubert, A. A. (1989). *The Culicoides of Southeast Asia (Diptera: Ceratopogonidae).* Mem. Am. Entomol. Instit. No. 44.

Wirth, W. W., Ratanaworabhan, N. C., & Blanton, F. S. (1974). Synopsis of the genera of Ceratopogonidae (Diptera). *Annales de Parasitologie Humaine et Comparee, 49*, 595–613.

Yanase, T., Kato, T., Kubo, T., Yoshida, K., Ohashi, S., Yamakawa, M., et al. (2005). Isolation of bovine arboviruses from *Culicoides* biting midges (Diptera: Ceratopogonidae) in southern Japan: 1985–2002. *Journal of Medical Entomology, 42*, 63–67.

Yu, Y. (2005). *Catalog and Keys of Chinese Ceratopogonidae (Insecta, Diptera) (in Chinese).* Beijing: Military Medical Science Press.

Yu, Y. (2006). *Ceratopogonidae of China (Insecta, Diptera) (in Chinese)* 2 Vol. Beijing: Military Medical Science Press.

Yu, Y., & Liu, J. (2006). *World Species of Bloodsucking Midges (Diptera: Ceratopogonidae).* Beijing: Military Medical Science Press.

Chapter

13

Black Flies (Simuliidae)

Peter H. Adler . John W. McCreadie

As small, powerful fliers adapted for blood-feeding, black flies can be formidable pests of humans, domestic animals, and wildlife, affecting virtually all facets of outdoor life. They are distributed worldwide, with the exception of Antarctica and some oceanic islands. Their distribution is largely influenced by the availability of flowing water, which is required for development of the immature stages. Many of the worst pest species breed in large rivers, some of which can produce nearly a billion flies per kilometer of riverbed per day. Other pest species inhabit the myriad small streams of heavily wooded terrain, making management efforts difficult.

Often ranked third worldwide among arthropods in importance as vectors of disease agents, black flies also are among the few arthropods that have killed animals by exsanguination during massive attacks. Even when not biting, their persistent swarming behavior can create an intolerable nuisance as the blood-seeking females dart into facial orifices and crawl on the skin. As often is the case, the behavior of a minority defines the reputation of the group. So it is with black flies, for only about 10 to 20% of the world's species are actually pests of humans and their animals. But among these species are the vectors of the agents of human onchocerciasis and mansonellosis, bovine onchocerciasis, and avian leucocytozoonosis. The majority of species, however, go unnoticed, either because they do not feed as adults or their hosts are of little economic concern.

TAXONOMY

More than 2,000 species of black flies have been described worldwide (Adler and Crosskey, 2008). Many other species are known but unnamed, and additional species remain to be discovered. The Palearctic Region contains the most described species, about 700, followed by the Neotropical and Oriental Regions with well over 300 species each (Currie and Adler, 2008). The Nearctic Region has about 256 known species (Adler et al., 2004).

The Simuliidae comprise two subfamilies. The most primitive subfamily Parasimuliinae includes four described and one undescribed species endemic to the Pacific Northwest. The females of these species do not have biting mouthparts. The subfamily Simuliinae contains all remaining species and is divided into two tribes, the Prosimuliini and the Simuliini, the latter including the majority of pest species. The most universally accepted classification system below the tribal level is summarized by Adler and Crosskey (2008), who recognize 25 extant genera in the subfamily Simuliinae. Thirteen genera in the subfamily are found in North America. The largest genus of black flies is *Simulium*, which contains 41 subgenera and many of the species of economic importance.

The morphological uniformity of black flies creates difficulty for species identification. For this reason, a holistic approach to identification typically is used, relying on characters from larvae, pupae, males, females, and the polytene chromosomes, as well as distributional and ecological information. The need for accurate identifications, particularly in programs for pest and vector management, has driven the taxonomy of black flies. As a result, black flies are taxonomically one of the best known groups of arthropods at the species level; for example, about 98% of North American species are known as larvae and pupae.

More than 150 identification keys exist for black flies in various parts of the globe. Crosskey and Howard (1997) provide a comprehensive list of identification keys by zoogeographic region. Keys to the genera and species of adults, pupae, and larvae of the Nearctic Region are provided by Adler et al. (2004). The most comprehensive English-language treatment of the Palearctic fauna is by Rubtsov (1956), and of the Neotropical fauna by Coscarón and Coscarón Arias (2007). Keys to the supraspecific taxa of the Australasian and Afrotropical Regions are given by Crosskey (1967, 1969, respectively). Comprehensive keys for the Oriental Region are lacking, but the keys by Takaoka and Davies (1995) and Takaoka (2003) provide a helpful starting point.

The giant polytene chromosomes (usually $n = 3$), which are best developed in the larval silk glands, provide a highly useful tool for discovering and identifying species. Giant chromosomes, particularly their banding patterns, reveal that many black flies regarded as single species are actually complexes of two or more species known as **cryptic species** or **sibling species**, each of which is biologically unique. The existence of sibling species has far-reaching implications for biological studies and population management of pests and vectors. For example, *Simulium damnosum*, the black fly known for much of the twentieth century as a vector of the agent of human onchocerciasis or river blindness, is the largest species complex among all hematophagous arthropods worldwide, consisting of more than 55 cytologically distinct entities (Post et al., 2007). Many of these entities are distinct species, but not all are vectors of the pathogen. The existence of **homosequential sibling species** that are identical in both morphology and chromosomal banding patterns increases the taxonomic complexity of the family. Cytotaxonomy of black flies has been reviewed for the world fauna (Rothfels, 1979), North American species (Adler et al., 2004), and selected vector species (Procunier, 1989). Identification and species discovery is moving now toward a molecular-based approach.

MORPHOLOGY

The immature stages of black flies are adapted for aquatic life, although the nonmobile pupa also has terrestrial adaptations that are useful if the water recedes. The egg is roughly oval or triangular with rounded angles. It has a glutinous outer layer and a smooth, pigmented inner shell. A **micropyle**, consisting of a simple hole in the egg for the entry of sperm, is present in some species but not others.

The larva (Figure 13.1) hatches with the aid of an egg burster, a small tubercle on the dorsum of the head capsule. The basic larval design consists of a well-sclerotized head capsule bearing an anterior pair of labral fans, and an elongate body with one thoracic proleg and a terminal abdominal proleg. Rows of tiny hooks on the prolegs enmesh with silk pads spun from a pair of larval silk glands and applied to a substrate. These silk glands extend from the anterior of the head into the posterior portion of the abdomen where they enlarge and double back on themselves. The adhesiveness of the silk is correlated with the velocity of the flowing water to which each species is adapted.

While clinging to a pad by its posterior proleg, the larva extends its body to filter feed. The prominent labral fans, each with about 20 to 80 individual rays bearing microtrichia (minute hairs) on their inner surface, are used to filter particulate matter from the water current. Larvae of some species (e.g., *Gymnopais* spp.) that live in habitats, such as glacial meltwaters, with little suspended food have lost the labral fans over evolutionary time. These species rely on their mandibles, specialized labrum, and hypostoma to scrape food from the substrate.

Additional features of the head and body are conspicuous and taxonomically important. The antennae, which consist of three articles and a terminal cone sensillum, are elongate, slender, and variously pigmented. A pair of dark eyespots is prominent on each side of the head capsule. Pigmentation patterns of the head capsule and body and the shape of the **postgenal cleft**, an area of weakly sclerotized cuticle on the ventral side of the head capsule, are important for interpreting the taxonomy of the family. The anteroventral portion of the head capsule bears the **hypostoma**, an anteriorly toothed plate used in conjunction with the mandibles to cut strands of silk and to scrape food from the substrate. Mature larvae are recognized by the presence of a prominent, dark gill histoblast on each side of the thorax.

The pupa (Figure 13.2), which resembles an adult with its appendages held close to the body, is housed in a silk cocoon. Cocoons are shapeless sacs in the evolutionarily older species but are well-formed, slipper- or boot-shaped coverings sometimes bearing anterior processes and lateral windows in the more derived species. The pupa is held firmly in its cocoon by numerous anteriorly directed sets of hooklets. A pair of conspicuous gills arises from the thorax. The gills are among the most taxonomically useful and fascinating

Figure 13.1 Larvae of the North American black fly *Simulium venustum* attached to aquatic vegetation, filter feeding. (Photo and permission by Stephen A. Marshall)

Figure 13.2 Pupae of the North American black fly *Simulium vittatum*. (Copyright Dwight R. Kuhn)

Figure 13.3 North American black fly *Prosimulium mixtum.* (A) Female feeding on a human; (B) male; note the greatly enlarged eyes characteristic of male black flies. (Photos by Stephen A. Marshall)

structures in any life stage. They vary in arrangement from thick, clublike structures to clusters of two to more than 100 slender filaments.

Black-fly adults (Figures 13.3 and 13.4) are characterized by a small but robust body, conical or beadlike antennae with seven to nine flagellomeres, and an arched thorax bearing a pair of wings that typically span 6–10 mm and have thickened veins near the leading margin. Most species are blackish, but orange, yellow, and variously patterned species also exist. Males (Figure 13.3B) of nearly all species are holoptic, with eyes that occupy most of the head and meet at the midline. Male eyes consist of enlarged dorsal facets, in addition to the typical-sized ventral facets, an arrangement that enhances the ability of males to locate females entering a mating swarm from overhead. Females are dichoptic, with smaller eyes separated by the frons.

The mouthparts arise ventrally from the head. A conspicuous pair of long maxillary palps attaches near the base of the proboscis. The third palpal segment accommodates the sensory vesicle (Lutz's organ), which has many chemosensilla that detect odors such as carbon dioxide. The labium forms the back of the proboscis and envelops the other mouthparts, including the minutely serrated mandibles and the toothed laciniae, with a pair of large, fleshy lobes called the labella. The mouthparts of the male are similar to those of the female, except the mandibles and laciniae are not adapted for blood feeding and, therefore, do not bear teeth.

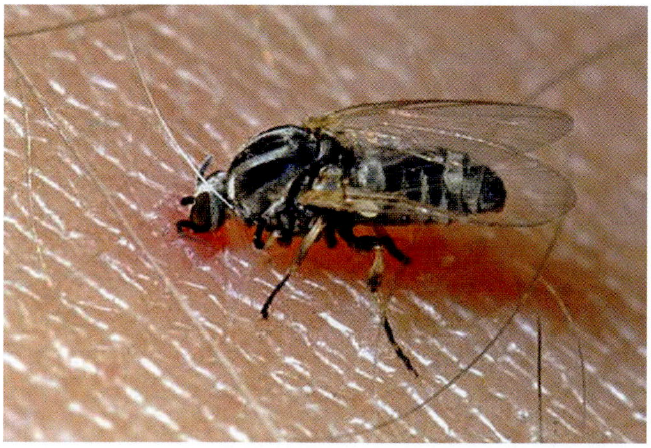

The stout thorax bears a pair of wings, either smoky or hyaline but never patterned. The venation, including the setation, is taxonomically important at the generic level. The color patterns of the legs and thoracic scutum are useful for species identification. The tarsal claws exist in one of three conditions. Species that feed on mammals have either a simple, unarmed claw or a minute tooth at the base of each claw. Bird feeders are endowed with a large thumblike lobe at the base of each claw. The abdomen is weakly sclerotized except the genitalia, which are of the utmost importance in the identification of species. To interpret the taxonomically important characters of the genitalia of both males and females, the abdomens must be treated with a clearing agent such as potassium hydroxide or hot lactic acid and examined in a depression slide with glycerin.

Figure 13.4 Female of the South American black fly *Simulium nemorale*, feeding on a human in Chile. (Photo by Stephen A. Marshall)

LIFE HISTORY

Immature black flies are found in virtually any water that flows, even if only imperceptibly and temporarily, from the smallest trickles to the largest rivers. Most species occupy specific habitats, and some higher taxa are characteristic of particular environments. For example, members of the genus *Gymnopais* occupy small, icy streams of the Far North, species of *Simulium* (subgenus *Hemicnetha*) live on the lips of waterfalls and in swift rocky flows, members of the *S. noelleri* species group are found below impounded waters, and species of *Simulium* (subgenus *Psilozia*) are found in warm, highly productive streams and rivers with open canopies.

Each species of black fly has a specific pattern of seasonal occurrence. Nearly all species in the tribe

Prosimuliini are univoltine, completing a single generation annually. The tribe Simuliini contains both univoltine and multivoltine species. Some multivoltine species can complete seven or more generations per year in areas of North America with mild climates. In certain tropical areas of the world, some species (e.g., members of the *S. damnosum* complex) might cycle through more than 20 generations each year.

Eggs typically cannot resist desiccation, although those of some species (e.g., *Austrosimulium pestilens*) can survive in moist soil of dry streambeds for several years, hatching when streams are inundated. During the summer, eggs of multivoltine species (e.g., *S. vittatum*, *S. damnosum*) can hatch in fewer than four days. In northern temperate regions, univoltine species (e.g., *Prosimulium* spp.) often spend the warm months as eggs, whereas multivoltine species spend the cold months as eggs. Accordingly, the potential for long-term survival of eggs must be considered in management programs. Eggs of some species (e.g., *S. rostratum*) remain viable in the laboratory just above freezing for up to two years.

The larval stage lasts from about a week, or even less, to nearly half a year, depending on species, stream temperature, and food availability. At one extreme, the larvae of some species in the West African *S. damnosum* complex complete development in four days. At the other extreme, larvae of many univoltine, temperate species hatch in the fall, develop during the winter, and pupate in the spring. The number of larval instars varies from six to 11, depending on species and environmental conditions, such as food supply.

Final-instar larvae typically move to slower water before pupating in a silk cocoon that is spun on a substrate. Some species (e.g., *Prosimulium magnum*) pupate in masses, but most pupate individually. The duration of the pupal stage depends largely on temperature and species, lasting from several days to a few weeks. When the adult is ready to emerge, it expels air from its respiratory system, thus splitting the pupal cuticle along the dorsal eclosion line.

The newly emerged adult, partially covered in air, rises to the surface of the water with enough force to break the water-air interface. It then seeks a resting site, often streamside, to tan and harden. Adults generally live less than a month, during which time mating, sugar-feeding, host location, blood-feeding, and oviposition must be accomplished. Crosskey (1990) provides a detailed treatment of simuliid life history and bionomics.

BEHAVIOR AND ECOLOGY

After hatching, early instars often disperse to more suitable sites for development. Larvae lead a largely sessile life attached to silk pads on substrates such as stones, trailing vegetation, sticks, aquatic plants, and leaf packs. The larvae of about 30 species, mostly in tropical Africa, are obligatorily phoretic, anchoring themselves to the bodies of larval mayflies and freshwater crabs and prawns. When disturbed, a larva repositions itself by looping over the substrate in inchworm-like fashion or by releasing itself from the silk pad and drifting downstream, often on a lifeline of silk. Downstream drift is usually greatest around dusk and during the night; its extent and timing should be considered in management programs.

The majority of larval life is spent feeding, usually by passively filtering suspended matter from the current (Figure 13.1) or actively grazing adherent material from the substrate. The larvae of some species are also predaceous, consuming small invertebrates such as chironomid midges. Larvae that filter their food lean with the current and twist their bodies longitudinally 90–180°. In this position, one labral fan receives particulate matter that is resuspended by vortices arising from the substrate, and the other fan receives material from the main flow. Larvae filter particles that are about 0.09 to 350 μm in diameter, with the majority of ingested particles less than 100 μm in diameter. Larval diet consists of detritus, bacteria, small invertebrates, larval fecal pellets, and algae, with gut contents largely reflecting particle size and composition of available material suspended in the water. Feeding efficiency (i.e., the ability to remove particles from the water column) is low, typically less than 2%. Retention of material in the gut typically varies from 20 minutes to more than two hours, depending mainly on larval age, species, and water temperature. Where larvae achieve extraordinarily high densities, as in the boreal region, their fecal pellets provide significant nutrition for freshwater organisms and might even fertilize river margins, leading ecologists to refer to the larvae as "ecosystem engineers" (Malmqvist et al., 2004a).

The distribution patterns of larvae and pupae are associated with a variety of environmental factors. Distributions in a small section of stream or on a specific substrate customarily are referred to as **microdistributions**. Factors influencing microdistributions are those that vary over a few centimeters or meters, including substrate texture, water depth, hydrodynamics, and interactions with other organisms. Microdistributions are species specific. For example, last instars of *Simulium truncatum* and *S. rostratum*, two morphologically similar, boreal species often found in the same section of stream, select different microhabitats on the basis of water velocity. The patterns of larval dispersion on a substrate are either spaced (e.g., *S. vittatum*), with a well-defined area surrounding each larva, or clumped (e.g., *S. noelleri*), with each larva occupying only enough space to attach its silk pad. Larvae with spaced patterns vigorously defend their space from other larval black flies.

Macrodistributions encompass a scale of many meters to hundreds of kilometers. The most important factors influencing macrodistributions are stream size, water velocity, temperature, water chemistry, food quality and quantity, and the presence of lake outlets. Within a stretch of stream, species distributions can be predicted by gradients of physical and chemical factors such as temperature and oxygen. Larval densities are usually greatest within a short distance downstream of impoundment outflows. For example, densities as high as 1.2 million larvae/m^2 have been recorded for *S. noelleri* at lake outlets in Europe. Some species rarely are found far from lake outlets and, therefore, species assemblages at these outflows are often distinct from those farther downstream. Distributions of species among streams can be predicted by factors such as stream width. For example, species in genera such as *Twinnia*, *Gymnopais*, *Greniera*, and *Stegopterna*, as well as many species of *Prosimulium* and *Simulium*, occur in trickles and small streams. Species such as *Metacnephia lyra*, *Simulium jenningsi*, *S. reptans*, *S. vampirum*, and members of the genus *Ectemina* occupy large streams and rivers.

The influence of stream size and impoundment outflows on the distribution of black flies is important throughout most of the world. At larger scales, biogeographic factors are useful in predicting species distributions. Streams in one ecoregion (e.g., mountains) tend to be more similar to one another than to streams in a different ecoregion (e.g., coastal plain), with respect to physical, chemical, and riparian characteristics. Simuliid faunas also show significant differences among ecoregions. The major factors associated with distributions of black flies are summarized by Adler and McCreadie (1997).

Species richness, the number of species in a specified location, varies from one to 13 for any given stream site, but is typically less than eight. At a larger scale, the mean number of species per stream reach within a given region is remarkably consistent, ranging from three to four species (McCreadie et al., 2005). Thus, even though the total number of species within regions varies across North America, the mean number of species per stream reach remains relatively invariant.

After emergence, adults of most species of black flies undertake short dispersal flights, usually less than 5 km. Males disperse to find mates and a source of sugar, whereas females of most species have the additional need to find hosts for blood and sites for oviposition. Although exceptions have been reported, black flies are diurnal fliers, generally taking to the wing when temperatures exceed 10°C. Local meteorological conditions can modify or even halt flight, but the primary factors that control daily flight patterns are wind, light, and temperature. Most species show a propensity to be on the wing at particular times of the day, these times varying with species, sex, physiological state, season, and nature of the activity.

Some species of black flies (e.g., *S. vampirum*) undertake far longer flights in search of hosts or breeding sites. These long-range movements are wind assisted, with the direction of movement being controlled by prevailing winds. Movements of hundreds of kilometers by some species (e.g., members of the *S. damnosum* complex) have implications for control strategies, requiring that breeding grounds remote from problem areas be treated. Long-distance flights typically occur in species that feed on mammals, especially those that inhabit open areas such as savannas or prairies.

The universal energy source used by males and females for flight is sugar. Adults are opportunistic in their choice of carbohydrate sources, using floral or extrafloral nectar, plant sap, and honeydew from aphids and related insects. Water markedly increases longevity, and a 10% sugar solution further increases longevity. The sugar meal is stored in the crop and passed to the gut as needed for digestion.

Mating is necessary for all but about 10 parthenogenetic species (e.g., *Prosimulium ursinum*). These parthenogenetic species lack males and are triploid and northern in distribution. In the sexual species, mating occurs shortly after emergence. Males use a variety of strategies to encounter females. The most commonly reported method is the formation of precopulatory swarms. These aerial swarms usually form 2–3 m above ground, either beside or above a marker. Swarm markers tend to be visually apparent aspects of the environment such as a tree branch, rock, waterfall, or host. Females enter the swarm, sometimes immediately after emergence, and are seized by males. Coupled pairs fly out of the swarm or fall to the ground or lower vegetation. Some species (e.g., *S. decorum*) do not form swarms, but instead couple on the ground during large, synchronous emergences. Males of some species also perch on vegetation and seize passing females. Visual cues mediate mating, but contact pheromones might also play an important role. Black flies generally are refractive to mating under laboratory conditions, which has impeded attempts to colonize most species and to elucidate details of mating behavior. The long-term successful colonization of *S. vittatum* is a notable exception (Cupp and Ramberg, 1997).

Copulation lasts from a few seconds (e.g., *S. vittatum*) to two hours (e.g., *Gymnopais* spp.). During copulation, the male passes a spermatophore (i.e., a package of spermatozoa) to the female. The tip of the spermatophore is opened enzymatically by the female and sperm move into the female's storage structure, the spermatheca. Stored sperm are released to fertilize eggs as they are being deposited.

A blood meal is required for the females of more than 90% of the world's simuliid species to mature the eggs. Males do not feed on blood. Females of those species that never take blood have feeble, untoothed mouthparts unable to cut host skin. These females are obligatorily **autogenous**; that is, they are able to produce eggs without taking blood, relying instead on energy acquired during the larval stage for all their egg production. Females of species with biting mouthparts are **anautogenous**; that is, they mature their eggs with the aid of a blood meal. Nonetheless, the females of some species with biting mouthparts can mature the first batch of eggs without a blood meal (facultative autogeny) if conditions for larval growth have been optimal. In these facultatively autogenous species, however, each subsequent batch of eggs requires a blood meal. Each ovarian or **gonotrophic cycle** (i.e., maturation of an egg batch), varies from about two days to two weeks, depending on the species and ambient temperature. Most females probably do not survive long enough to complete more than two or three ovarian cycles. Because the transmission of pathogens is usually horizontal, passing from host to host via the simuliid vector, anautogenous females have a greater potential than facultatively autogenous females to acquire and transmit disease agents.

The majority of simuliid species in the world probably feed on mammals (**mammalophily**), although those that feed on birds (**ornithophily**) also are common; no other groups of organisms serve as hosts for black flies. About two-thirds of the blood-feeding species in North America are principally mammalophilic and the other third are mainly ornithophilic (Adler et al., 2004). A number of these species (e.g., *S. johannseni, S. venustum*), however, feed on both mammals and birds. Molecular analyses of blood meals in wild-caught black flies have proved valuable in determining the hosts of different species of black flies (Malmqvist et al., 2004b). Host specificity varies from highly specific in species such as *Simulium annulus*, which feeds chiefly on loons, to those such as *S. rugglesi*, which have been recorded feeding on nearly 30 different host species. Most simuliid species attack thinly haired or sparsely feathered regions of the host body and areas that are difficult for the host to groom. Thus, mammals often are attacked along the ventral region of the body and inside the ears. Birds are attacked especially on the neck, bases of the legs, and around the eyes. Humans are bitten wherever flesh is exposed, although specific areas often are attacked, such as along the hairline (e.g., *S. venustum*), the arms and hands (e.g., *S. parnassum*), the upper torso (e.g., members of the *S. ochraceum* and *S. oyapockense* complexes), and the ankles and feet (e.g., members of the *S. damnosum* and *S. metallicum* complexes).

A number of host attractants have been identified. Carbon dioxide released from the host, as well as color, shape, and size of the host, provide some of the major cues and attractants that females use to locate an appropriate blood source (Sutcliffe, 1986). Traps used to monitor female populations often exploit these cues. For example, sticky silhouettes and carbon dioxide in gaseous or dry-ice form often are used to monitor females.

Biting and engorging require a series of appropriate cues, especially temperature and various phagostimulants, such as adenosine phosphates, in the host's blood. When the fly begins to bite, the labella are withdrawn, and small teeth and spines at the apex of the labrum and hypopharynx pull the host skin taut (Sutcliffe and McIver, 1984). The serrated mandibles cut the host flesh, allowing the labrum and hypopharynx to enter the wound, along with the laciniae, which are armed with backwardly directed teeth that anchor the mouthparts. Blood from the wound forms a small pool that is drawn up the food channel formed when the mandibles overlap the labral food canal. Because of their method of feeding from pooled blood, black flies are termed pool feeders or **telmophages**. Saliva is applied to the host flesh via a salivary groove along the anterior surface of the hypopharynx. Various salivary components promote local anesthesia, enhance vasodilation, inhibit platelet aggregation, and prevent clotting (Cupp and Cupp, 1997). Chemosensilla on the mouthparts help determine that blood will be directed to the midgut.

Female black flies are determined feeders. Once the host skin is penetrated, females typically do not leave until they are satiated. Because most black flies are not nervous, easily interrupted feeders, they make poor mechanical vectors of pathogens. Most species feed for about three to six minutes, taking approximately their own weight in host blood.

Most blood-feeding activity is restricted to outdoor settings (**exophily**), with females infrequently entering shelters to feed. This behavioral trend has implications for vector control. For example, residual house treatments effective for the control of mosquitoes are of no use for controlling black flies. Nonetheless, some cavity-nesting wild birds can experience severe assaults by ornithophilic black flies that enter through the cavity opening. Biting activity occurs within certain optimal ranges of temperature, light intensity, wind speed, and humidity, with optima differing for each species. Given the appropriate range of meteorological conditions, many species bite throughout the day. Other species show a particular pattern of biting activity, such as a single peak in the morning (e.g., members of the *S. exiguum* complex) or a bimodal pattern with peaks in the morning and early evening (e.g., those in the *S. damnosum* complex). For all black flies, feeding typically is restricted to the hours of daylight and dusk. A rapid decrease in air pressure, combined with increased cloud cover, produces a sudden flush of biting activity.

Most female black flies can produce a batch of about 100 to 600 eggs, although the number varies from 25 (some *Gymnopais* spp.) to about 800 (some *Simulium* spp.). Females of some species can produce several of these egg batches in a lifetime, depending on the number of blood meals and how long the female lives. **Oviposition** usually occurs in the late afternoon and early evening. Eggs are deposited freely into the water during flight (e.g., *S. venustum* complex) or attached in strings or masses to substrates such as rocks and trailing vegetation at the water line (e.g., *S. vittatum*). Some species, however, oviposit in moist fissures in riverbanks (e.g., *S. posticatum*) or in streamside mosses (e.g., some *Prosimulium* spp.).

PUBLIC HEALTH IMPORTANCE

The importance of black flies to humans centers largely around the pestiferous habits of the blood-seeking females and the disease agents they transmit. The human disease agents transmitted by black flies are those that cause onchocerciasis in the tropics of Africa and Central and South America, and mansonellosis in northwestern Argentina, southern Panama, and the western Amazon Region. No other human pathogens or parasites are known definitively to be transmitted by black flies, and no endemic simuliid-borne disease of humans has been reported from North America.

The biting and nuisance problems inflicted by black flies have had severe consequences for most outdoor activities including agriculture, forestry, industrial development, military exercises, mining, and tourism. Industrial and recreational development in some regions of Canada and Russia has been impeded or halted by overwhelming attacks from black flies. Yet, these negative effects gain balance through the protection that the attacking black flies afford to sensitive environmental areas that otherwise might suffer from development. Actual monetary losses due to biting and nuisance problems in different sectors of the economy, although significant and sometimes crippling, are poorly documented.

Biting and Nuisance Problems

The black flies that bite humans (i.e., anthropophilic species) constitute 10% or less of the total simuliid fauna in any zoogeographic region (Table 13.1), with some areas of the world being nearly free of biting problems. No black fly is known that feeds exclusively on humans. In North America, where the name "black fly" originated, fewer than 60 species have been recorded to bite humans. Less than one-third of these hold any real status as biting pests, but those that do bite regularly can be unrelenting in their attacks. Individual reactions to bites vary from a small red spot at the puncture site, often with initial streaks of oozing blood (Figure 13.5), to an enlarged swelling the size of a golf ball (Stokes, 1914). Swelling from bites around the eyes can impede vision, and bites on the legs can impair walking.

A general syndrome, sometimes called **black fly fever**, is common in areas such as northeastern North America where biting problems can be intense. It is presumably a response to salivary components from the fly, and is characterized by headache, nausea, fever, and swollen lymph nodes in the neck. Many people experience some itching from the bites, intensified by scratching the wounds. Severe allergic reactions, including asthmatic responses, are infrequent; however, medical treatment, including hospitalization, is sometimes necessary (Gudgel and Grauer, 1954). No human deaths from simuliid bites have been recorded since the beginning of the twentieth century, although anecdotal accounts suggest that an unclothed human can be exsanguinated in about two hours in some areas of Russia. Exposure to fierce attacks of biting and swarming black flies can affect a person's emotional state and produce short-term psychological effects that reduce individual efficiency.

Many species of black flies are attracted to humans but do not bite, or they bite infrequently in proportion to the number of flies actually attracted. These species can create enormous nuisance problems. One such species is *Simulium jenningsi*, a major pest in North America. Females of this species sometimes bite humans and occasionally cause allergic reactions, but they are more of a nuisance because of their habit of swarming about the head and entering the eyes, ears, nose, and mouth. Outdoor activities in afflicted areas, such as Pennsylvania, can become unbearable as the females ceaselessly swarm around the head. More than $5 million is spent annually in the management of *S. jenningsi*.

Occasional nuisance problems have been caused by large numbers of flies attracted to incandescent lights and by mating swarms that form over bicycle and foot paths at about the same height as a person walking or riding a bicycle. These kinds of problems usually are caused by members of the North American *S. vittatum* species complex, which breed abundantly in human-altered habitats, such as lake outlets and polluted waters.

Table 13.1 Species of Black Flies Regarded as Significant Biting and Nuisance Pests of Humans, Livestock, and Poultry

Species	Geographic Region
HUMANS	
Austrosimulium australense	New Zealand
Austrosimulium ungulatum	New Zealand
Prosimulium mixtum group	Eastern North America
Simulium arakawae	Japan
Simulium buissoni	Marquesas Islands
Simulium cholodkovskii	Russia
Simulium decimatum	Russia
Simulium jenningsi	Eastern North America
Simulium jujuyense	Argentina
Simulium meridionale	Western North America
Simulium nigrogilvum	Thailand
Simulium ochraceum	Galapagos Islands
Simulium oyapokense complex	South America (Amazonian Region)
Simulium parnassum	Eastern North America
Simulium penobscotense	Northeastern North America
Simulium pertinax	Brazil
Simulium posticatum	England
Simulium quadrivittatum	Central America
Simulium sanguineum	Northwestern South America
Simulium tescorum	Southwestern United States
Simulium venustum complex	North America
Simulium vittatum complex	North America
LIVESTOCK	
Austrosimulium pestilens	Australia (Queensland)
Cnephia pecuarum	United States (Mississippi River Valley)
Simulium cholodkovskii	Russia
Simulium chutteri	South Africa
Simulium decimatum	Russia
Simulium equinum	Europe, Russia
Simulium erythrocephalum	Europe
Simulium incrustatum	Paraguay
Simulium jenningsi group	Eastern North America
Simulium lineatum	Europe and Russia
Simulium luggeri	Western Canada
Simulium maculatum	Russia
Simulium ornatum complex	Europe, Russia
Simulium reptans	Europe, Russia
Simulium vampirum	Western Canada
Simulium vittatum complex	North America
POULTRY	
Cnephia ornithophilia	Eastern North America
Simulium meridionale	North America
Simulium rugglesi	North America
Simulium slossonae	Southeastern United States

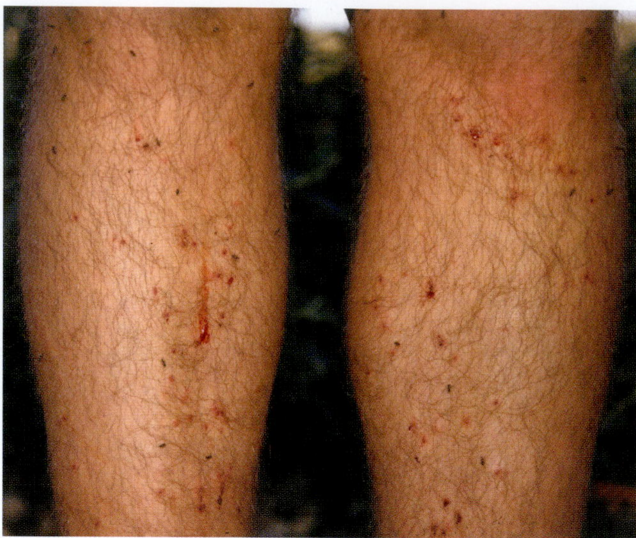

Figure 13.5 Bite wounds on legs of a river guide, caused by a North American black fly of the *Simulium venustum* complex. (Photo by P. H. Adler)

Human Onchocerciasis

The greatest public health problem associated with black flies is onchocerciasis or **river blindness**, a tropical disease caused by the filarial nematode *Onchocerca volvulus*, which is transmitted solely by black flies during blood feeding. River blindness is the second leading infectious cause of blindness in the world. In the Old World, river blindness is found in 27 countries in the central belt of Africa, with small foci in southern Yemen. In the New World, where the disease possibly was introduced during the slave trade, its distribution is patchy, with foci in northern Brazil, Colombia, Ecuador, Guatemala, Mexico, and Venezuela. The World Health Organization (1995) conservatively estimated that about 17.7 million people are infected (17.5 million in Africa and Yemen; 140,500 in tropical America), with approximately 270,000 cases of blindness and another half million individuals with severe visual impairment. Onchocerciasis occasionally is diagnosed in patients in North America who have travelled from endemic areas. Research on the disease and its vectors has generated a massive literature (Muller and Horsburgh, 1987) and numerous reviews (Shelley, 1988b; Crosskey, 1990; World Health Organization, 1995).

Onchocerca volvulus typically is found only in humans (definitive host) and adult flies of the genus *Simulium* (intermediate host). Various strains of *O. volvulus* are recognized, such as forest and savanna strains in West Africa, and these form highly compatible parasite-vector complexes with distinct clinical facies. When the female black fly ingests a blood meal from an infected human, the microfilariae (220–360 μm long) penetrate the gut of the fly and make their way to the thoracic flight muscles. Once in the thoracic muscles, the microfilariae lose their motility and transform to first-stage larvae (L1), which then molt to become second-stage larvae (L2). The final molt in the fly produces the infective third-stage larvae (L3), which migrate to the fly's head and mouthparts. Vector incrimination is based on the presence of L3 worms in the head capsules of female black flies. In West Africa, DNA tests allow animal parasites and the human parasites of savanna and forest to be distinguished. Development in the black fly, which is influenced by ambient temperature, typically requires six to 12 days, but the time between successive blood meals taken by the fly is usually three to five days. Consequently, the infective larvae will be passed to a human host no earlier than the third blood meal when the fly is about eight to 10 days old.

In humans, the infective larvae molt to the L4 stage within about a week. One more molt yields juvenile adults, which grow to mature adult worms over the next 12 to 18 months and begin reproducing. Adult worms (Figure 13.6) typically become encapsulated in fibrous nodules that vary in size from about 0.5 to 10 cm and can be subcutaneous (Figure 13.7) or deep in muscular and connective tissues; they cause no inflammatory response and no great discomfort. Mating between the small male worms (3–5 cm long) and the large females (30–80 cm) occurs in the nodules. Adult female worms can produce microfilariae for up to 14 years. These microfilariae migrate from the nodules to the skin, where they can be acquired by a vector, as well as to the eyes and various other organs (e.g., liver) of the human host. A diagnostic clinical feature of onchocerciasis is the presence of hundreds of microfilariae in skin snips.

River blindness is essentially a rural disease, afflicting those people most vulnerable to both the medical consequences and social stigmas of infection. Symptoms of the disease depend on factors such as geographical location, microfilarial transmission rates, and frequency of reinfection. Where transmission rates are low, the disease can be asymptomatic. With heavy infections, however, the classic manifestations of the disease appear—dermal changes, lymphatic reactions, nodules (Figure 13.7), and ocular disturbances. Other

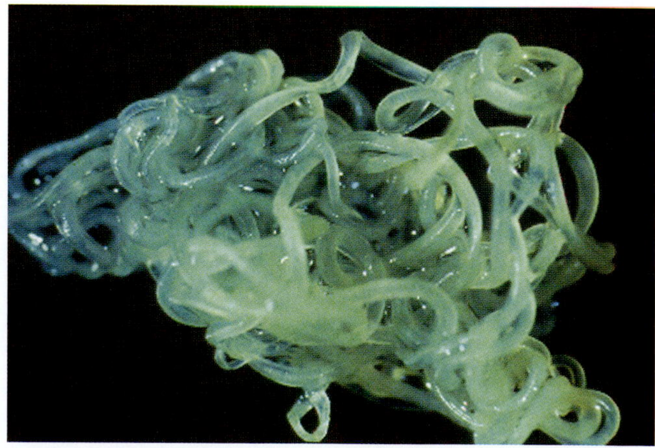

Figure 13.6 *Onchocerca volvulus*, after digesting away the nodular human tissue. (Courtesy of Armed Forces Institute of Pathology, USA)

Figure 13.7 Children at a clinic in Guatemala for surgical removal of *Onchocerca volvulus* adults from subcutaneous nodules on their heads, either waiting (scalps shaved) or returning to be checked following previous surgery (heads bandaged). (Photo by E. W. Cupp)

than the nodules in which the adults are enveloped, all symptoms are caused by the microfilariae.

Large numbers of microfilariae migrating throughout the dermis cause horrific itching that can lead to bleeding, secondary bacterial infections, inability to sleep, fever, headache, and even suicide. In addition to itching, chronic infections in Africa and Yemen can cause dermal lesions, patches of depigmentation ("leopard skin"), fibrosis, and loss of elasticity (e.g., "elephant knees"). In Yemen, the itching symptoms of the disease are known as *sowda*. In Central America, two unique, chronic skin conditions occur—a painful, reddish rash on the face (*erisipela de la costa*), and lesions associated with reddish skin on the trunk and arms (*mal morado*). The lymphatic nodes also can be affected, especially in the groin and thighs; combined with loss of skin elasticity, the result is a condition known as **hanging groin**.

Migrating microfilariae also enter the eye, resulting in a severe ocular pathology that can involve all tissues of the eye. *Wolbachia* bacteria of the filarial worms are responsible for the major inflammatory response in the eye (Pearlman, 2003). Ocular problems manifest in many forms, including cataracts, retinal hemorrhages, corneal opacities, secondary glaucoma, sclerosing keratitis, and optic neuritis. Various forms of visual impairment occur, such as night blindness and reduction in peripheral vision, but the most severe consequence is irreversible blindness with complete loss of light perception (Figure 13.8). Blindness usually takes years to occur; at age 20, for example, it is rare in infected people, but at 50 years of age, half of the infected victims can be blind (Figure 13.9). The incidence of blindness is highest in the savannas of West Africa, with some villages experiencing 15% blindness. At these high levels of disease, the village often is abandoned. Outside West Africa, ocular pathology is rare.

At least 26 species of *Simulium* are known vectors of *O. volvulus* (Table 13.2). Most of these vectors are

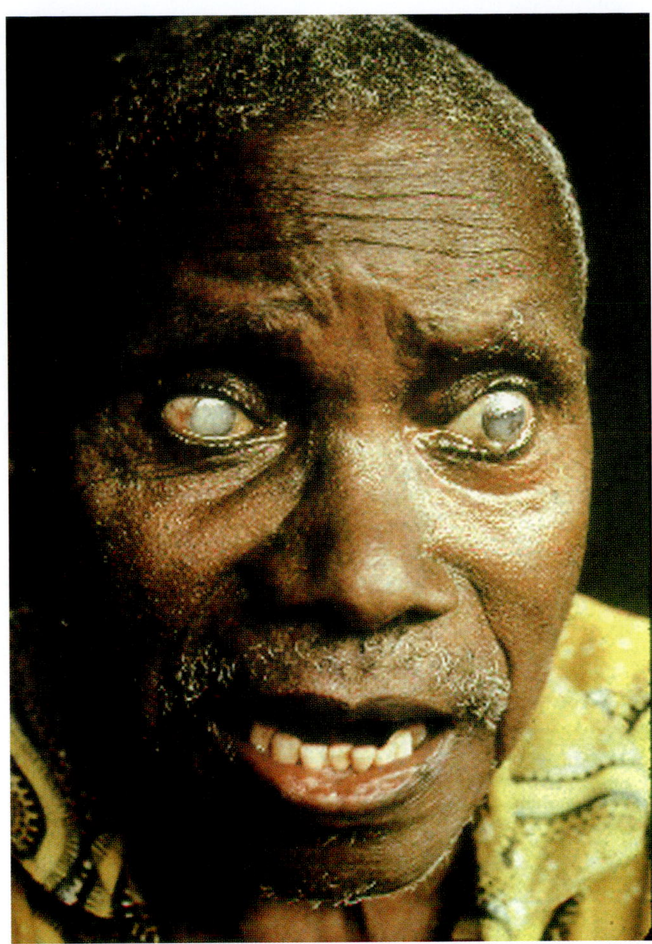

Figure 13.8 Human blindness caused by the filarial nematode *Onchocerca volvulus* and its associated *Wolbachia* bacterium transmitted by black flies of the *Simulium damnosum* complex in West Africa; note opacity of cornea due to damage by microfilariae and *Wolbachia*. (Copyright Eric Poggenphol)

Figure 13.9 Young boy leading man blinded by onchocerciasis, Burkina Faso. (Photo by E. W. Cupp)

Table 13.2 Disease Agents Transmitted by Black Flies

Disease Agent	Vectors[1]	Hosts	Geographic Areas	Select References
PROTOZOA				
Leucocytozoon cambournaci	*Helodon decemarticulatus, Cnephia ornithophilia, Simulium aureum* complex, *S. vernum* group	Sparrows	North America	Adler et al., 2004
Leucocytozoon dubreuili	*H. decemarticulatus, C. ornithophilia S. aureum* complex, *S. vernum* group	Thrushes	North America	Adler et al., 2004
Leucocytozoon icteris	*H. decemarticulatus, C. ornithophilia, S. anatinum, S. annulus, S. aureum* complex *S. venustum* complex, *S. vernum* group	Blackbirds	North America	Adler et al., 2004
Leucocytozoon lovati	*S. aureum* complex, *S. vernum* group	Grouse	North America	Adler et al., 2004
Leucocytozoon neavei	*Simulium* spp., especially *S. adersi*	Guinea fowl	Eastern Africa	Fallis et al., 1974
Leucocytozoon sakharoffi	*H. decemarticulatus, S. aureum* complex, *S. angustitarse*	Corvids	North America, England	Adler et al., 2004; Fallis et al., 1974
Leucocytozoon schoutedeni	*Simulium* spp., especially *S. adersi*	Chickens	Eastern Africa	Fallis et al., 1973
Leucocytozoon simondi	*Cnephia ornithophilia, S. anatinum, S. fallisi, S. rendalense, S. rugglesi, S. usovae, S. venustum* complex	Ducks, geese	North America, Norway	Adler et al., 2004
Leucocytozoon smithi	*S. aureum* complex, *S. congareenarum, S. jenningsi* group, *S. meridionale, S. slossonae,* possibly *S. ruficorne* group	Turkeys	North America, introduced to Africa	Adler et al., 2004
Leucocytozoon tawaki	*Austrosimulium ungulatum*	Penguins	New Zealand	Allison et al., 1978
Leucocytozoon toddi	*H. decemarticulatus, S. aureum* complex, *S. vernum* group	Hawks	North America	Adler et al., 2004
Leucocytozoon ziemanni	*H. decemarticulatus, S. aureum* complex *S. vernum* group	Owls	North America	Adler et al., 2004
Trypanosoma avium	*Metacnephia lyra, S. aureum* complex, *S. latipes, S. vernum*	Grouse, raptors	Europe	Votýpka et al., 2002; Reeves et al., 2007
Trypanosoma confusum	*H. decemarticulatus, Simulium* spp.	Birds	North America	Bennett, 1961
Trypanosoma corvi	*S. latipes*	Kestrels	England	Dirie et al., 1990
Trypanosoma numidae	*Simulium* spp., especially *S. adersi*	Chickens, guinea fowl	Eastern Africa	Fallis et al., 1973
FILARIAL NEMATODES				
Dirofilaria ursi	*S. venustum* complex	Bears	North America	Addison, 1980
Mansonella ozzardi	*S. amazonicum, S. argentiscutum,* possibly *S. exiguum* complex *S. oyapockense* complex, *S. sanguineum*	Humans	Northern South America, Northwestern Argentina, Panama	Shelley, 1988a; Shelley and Coscarón, 2001
Onchocerca cervipedis	*Prosimulium impostor, S. decorum, S. venustum* complex	Deer, moose	North America	Pledger et al., 1980
Onchocerca dukei	*S. bovis*	Cattle	Africa	Wahl and Renz, 1991
Onchocerca gutturosa	*S. erythrocephalum, S. bidentatum*	Cattle	Japan, Ukraine	Crosskey, 1990; Takaoka, 1999
Onchocerca lienalis	*S. erythrocephalum, S. jenningsi, S. ornatum* complex, *S. reptans, S. arakawae, S. daisense, S. kyushuense*	Cattle	North America, Russia, Western Europe, Japan	Lok et al., 1983; Crosskey, 1990; Takaoka, 1999
Onchocerca ochengi	*S. damnosum* complex	Cattle	West Africa	Wahl et al., 1998
Onchocerca ramachandrini	*S. damnosum* complex	Wart hogs	West Africa	Wahl, 1996
Onchocerca possibly *skrjabini*	*S. arakawae, S. bidentatum, S. daisense, S. oitanum*	Japanese deer	Japan	Takaoka, 1999
Onchocerca tarsicola	*P. tomosvaryi, S. ornatum* complex	Deer, reindeer	Western Europe	Crosskey, 1990

Table 13.2 Disease Agents Transmitted by Black Flies—Cont'd

Disease Agent	Vectors[1]	Hosts	Geographic Areas	Select References
Onchocerca volvulus	Africa: *S. albivirgulatum, S. damnosum, S. dieguerense, S. ethiopiense, S. kilibanum, S. konkourense, S. leonense, S. mengense, S. neavei, S. rasyani, S. sanctipauli, S. sirbanum, S. soubrense, S. squamosum, S. thyolense, S. woodi, S. yahense* Americas: *S. callidum, S. exiguum* complex, *S. guianense* complex, *S. incrustatum, S.limbatum, S. metallicum* complex, *S. ochraceum* complex, *S. oyapockense* complex, *S. quadrivittatum*	Humans	Africa, Central America, South America	Shelley, 1988b; Crosskey, 1990; World Health Organization, 1995; Post et al., 2008
Splendidofilaria fallisensis	*S. anatinum, S. rugglesi*	Ducks	North America	Anderson, 1968

[1]Vector species also have the potential to be pests, depending on host specificity and population size.

members of species complexes, and considerable taxonomic work still is needed to resolve all the vector species in areas such as East Africa and the Americas. In West Africa and Yemen, all vectors are members of the *S. damnosum* species complex, and include *S. damnosum sensu stricto* and *S. sirbanum*, the principal vectors associated with the savanna form of the disease and ocular pathology. The vectors in East Africa are members of the *S. damnosum* complex, the *S. neavei* group, and *S. albivirgulatum*. In the Americas, at least nine species are vectors, the most important of which are members of the *S. exiguum, S. guianense, S. metallicum, S. ochraceum,* and *S. oyapockense* complexes; their importance varies with location. Because the vectors in South America are more widespread than is onchocerciasis, the disease is predicted to spread as humans continue to push into undeveloped areas (Basáñez et al., 2000).

An understanding of the unique life history and behavior of each vector species is key to the control of onchocerciasis. Breeding sites of the vectors represent the ideal targets for control. The immature stages of the *S. damnosum* complex primarily inhabit swift sections of medium to large rivers, from dry savannas to forest highlands, depending on the species. Larvae and pupae of species in the *S. neavei* group live primarily in perennial, shaded forest streams where they have an obligatory phoretic relationship with river crabs. In the New World, members of the *S. metallicum* and *S. ochraceum* complexes breed in large and small streams, respectively, that drain forested mountain slopes, whereas members of the *S. exiguum* and *S. oyapockense* complexes breed in large rivers of the rain forest.

Although each species breeds in a specific habitat, the adults of some species can travel great distances beyond their natal waterways. The adults of *S. sirbanum* and *S. damnosum sensu stricto*, for example, can travel more than 500 km, assisted by seasonally changing winds. In the wet season, moist monsoon winds from the southwest move flies in a northeastwardly direction. In the dry season, winds from the northeast assist flies in their reverse, southwestwardly flights. Continual reinvasions by vectors, therefore, occur with each season in both the northern and southern parts of West Africa, and must be considered in control efforts.

Mansonellosis

The filarial nematode *Mansonella ozzardi* is the causal agent of mansonellosis, a questionably pathogenic disease of humans. It is transmitted by at least five species of black flies in the Neotropical rain forests of Brazil, Colombia, Guyana, Venezuela, and southern Panama, as well as in northwestern Argentina (Table 13.2). Black flies first were incriminated as vectors of *M. ozzardi* in 1959 and subsequently confirmed experimentally as vectors in 1980. Mansonellosis also is found in the Caribbean Islands where only ceratopogonid midges (*Culicoides* species) are known to transmit the causal agent (Shelley, 1988a).

Adult nematodes of *M. ozzardi* occur in the subcutaneous tissues of humans, and the microfilariae are found principally in the peripheral blood where they are acquired by blood-feeding flies. The life cycle of the nematode in black flies is similar to that of *Onchocerca volvulus*. A number of mammals and some birds and amphibians can be infected, but humans are the only significant reservoirs. In some highly endemic areas (e.g., Colombia), up to 70% of the human population can be infected. Mansonellosis generally is viewed as causing little or no pathology, but some reports have indicated that joint pains, headaches, hives, and pulmonary symptoms are associated with infections (Klion and Nutman, 1999). Treatment with ivermectin can reduce microfilaremia.

Other Diseases Related to Black Flies

Because black flies that feed on humans also feed on other hosts, the potential exists for certain disease-causing agents of domestic and wild animals to be transferred to humans. Black flies, for example, have been implicated as mechanical vectors of the bacterial agent of tularemia in the United States and Russia, suggesting that occasional cases of transmission of this pathogen to humans might occur. Similarly, eastern equine encephalitis virus in the United States and Venezuelan equine encephalitis virus in Colombia have been isolated from several species of black flies, suggesting at least the potential for transmission of these pathogens to humans. Black flies in the Marquesas Islands have been implicated in the indirect transmission of hepatitis B virus by causing numerous, itching lesions on the skin (Chanteau et al., 1993). Direct transmission of the virus by black flies also is theoretically possible.

Several additional diseases might be related to biting black flies. One such disease is **endemic pemphigus foliaceus** or **fogo selvagem**, a potentially lethal, autoimmune, blistering skin affliction. The disease is centered among poor, outdoor laborers in certain regions of Brazil (Eaton et al., 1998). Further work is needed to determine if black flies are the causal agents of the disease. Another affliction possibly associated with black flies in the New World is **thrombocytopenic purpura**, a disorder in which the platelet count is reduced. Again, more data are needed before black flies can be linked to the cause.

VETERINARY IMPORTANCE

The veterinary importance of black flies is manifested through pathogen transmission, biting, and nuisance swarming. Filarial nematodes, protozoans, and possibly several viruses are transmitted to animals. The most insidious parasites are those that cause leucocytozoonosis in domestic ducks, geese, and turkeys.

Deaths of birds and livestock have resulted from attacks by large numbers of black flies. Livestock under persistent attack sometimes stampede, trampling young animals, crashing into structures, and tumbling from precipices. Suffocation has been blamed for some deaths, with so many flies clogging the respiratory passages that breathing can become severely impaired. Deaths also have been attributed to respiratory tract infections caused by inhalation of flies. If enough blood is withdrawn, it may become too thick to transport oxygen efficiently, thereby killing the animal by exsanguination. Perhaps the most common cause of mortality can be attributed to the actual bites of the flies or, more specifically, to toxemia and acute shock caused by the various salivary components that are injected during blood feeding.

More difficult to assess in economic terms, but equally harmful to the livestock and poultry industries, are the effects of harassment through biting and swarming (Figure 13.10; Table 13.1). Biting often is aimed at weakly protected areas of the body, such as the ears, neck, and ventral midline (Figure 13.11). Persistent attacks by black flies can cause unruly host behavior, weight loss, reduced egg and milk production, malnutrition in young animals, dermatitis and epidermal necrosis, impotence in bulls, delayed pregnancies, abortions, and possibly stress-related diseases such as pneumonia. Attacks on endangered species are of particular concern when the animals are stressed, as in years of lean food supply. The endangered Attwater's prairie chicken suffers attacks from the black fly *Cnephia ornithophilia*, which also carries a *Leucocytozoon* blood parasite (Adler et al., 2007).

Figure 13.10 Cattle under attack by *Simulium vampirum* on the prairie, Alberta, Canada. (Photo by Joseph A. Shemanchuk, Department of Agriculture and Agri-Food, Government of Canada)

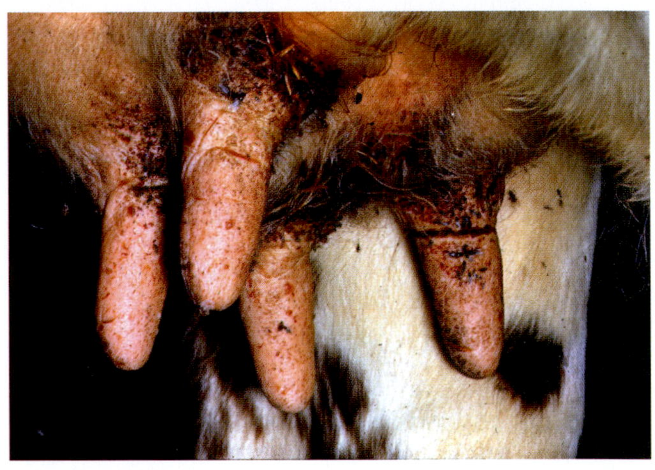

Figure 13.11 Damage to cow udder caused by black flies on the Canadian prairie. (Photo by Joseph A. Shemanchuk, Department of Agriculture and Agri-Food, Government of Canada)

Actual monetary losses from black flies are not well documented, but can be great. The beef and dairy industries of Saskatchewan, for example, lost more than $3 million in 1978 from attacks by *Simulium luggeri* (Fredeen, 1985). In spring 1993, the ostrich and emu industry lost about $1.5 million along the Trinity River in eastern Texas as a result of attacks by black flies (Sanford et al., 1993). Effects of black flies on pets have rarely been documented, although *Cnephia pecuarum* has caused hospitalizations and deaths as recently as the end of the twentieth century (Atwood, 1996).

Figure 13.12 Common loon attacked by black flies of the *Simulium annulus* group in British Columbia, Canada. (Photo by Neil K. Dawe, Canadian Wildlife Service)

The death of even a single exotic bird, such as a parrot, from blood feeding by black flies can run into thousands of dollars (Mock and Adler, 2002).

Wildlife also succumb to withering attacks from black flies (Figure 13.12). Nestling birds are particularly vulnerable, including raptors and songbirds. Nestling bluebirds and tree swallows, for example, have been killed (Figure 13.13) by *S. meridionale*, a species that routinely enters nest boxes (Adler et al., 2004).

An odd, but indirect, nuisance problem mediated occasionally by black flies involves the Neotropical human bot fly *Dermatobia hominis*. Female bot flies capture hematophagous arthropods, including black flies, to which they glue their eggs. Once the carrier has landed on a host, the larvae of the bot fly hatch and bore into the host skin, causing myiasis. At least one species of black fly (*Simulium nigrimanum*) that feeds on domestic animals has been used as a carrier.

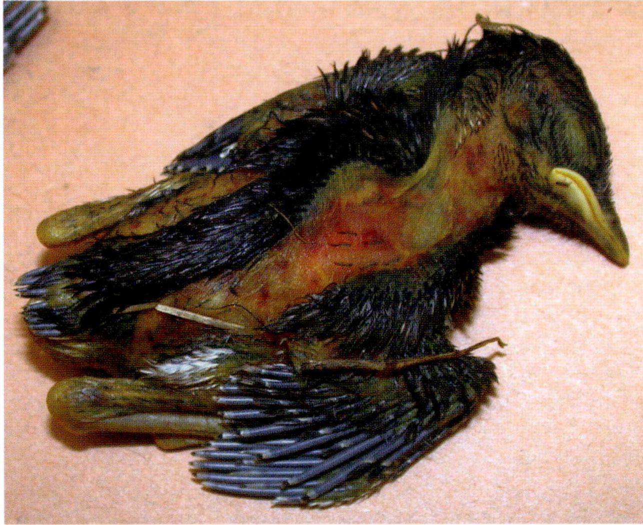

Figure 13.13 Nestling bluebird killed by *Simulium meridionale* in Wisconsin, showing multiple hematomas from bites. (Photo by Phil Pellitteri)

Bovine Onchocerciasis

Black flies transmit at least four species of filarial nematodes (genus *Onchocerca*) to cattle in the Afrotropical, Nearctic, and Palearctic Regions (Table 13.2). *Onchocerca lienalis* is the most widespread of these filarial parasites. *Simulium jenningsi* is its primary vector in the United States, whereas the *S. ornatum* complex is a principal vector in the Old World. The microfilariae of *O. lienalis* are concentrated in the umbilical region of the host. They are ingested during blood feeding and transmitted to a new host after they have developed to the infective third stage in the simuliid vector. The percentage of infected cattle is often quite high, but symptoms and general effect on the host are usually not overt. Infected animals sometimes show dermatitis and inflammation of the skin and connective ligament. *Onchocerca gutturosa*, a Palearctic species, often has been confused with *O. lienalis*. Its microfilariae occur in the skin of the neck and back of the host. It has been confirmed from Japan and perhaps the Ukraine, where *S. bidentatum* and *S. erythrocephalum*, respectively, have been implicated in its transmission. Elsewhere, ceratopogonid midges (*Culicoides*) are vectors. In West Africa, *O. ochengi* is transmitted to cattle by members of the *S. damnosum* complex, and *O. dukei* is transmitted by *S. bovis*. Both of these *Onchocerca* species can create nodules, either dermal (*O. ochengi*) or subcutaneous (*O. dukei*), in the inguinal region of the host. Economic losses resulting from bovine onchocerciasis rarely have been assessed, although a few reports have indicated that the quality of hides can be reduced.

At least four additional species of *Onchocerca* are transmitted by black flies to nonbovine hosts (Table 13.2). *Onchocerca cervipedis* in North America and *O. tarsicola* in Europe infect the subcutaneous connective tissues, mainly in the legs of deer, moose, and reindeer; consequently, they are sometimes called legworms. More than 60% of a host population can be infected. *Onchocerca ramachandrini* is a parasite of warthogs and is transmitted by members of the *S. damnosum* complex in West Africa. *Onchocerca* possibly *skrjabini* parasitizes Japanese deer and is transmitted by at least four species in Japan. The vectors of 10 or more *Onchocerca* species that infect domestic animals remain unknown but might include black flies.

Leucocytozoonosis

At least 12 described species of protozoans in the genus *Leucocytozoon* are transmitted to birds by black flies, causing a malaria-like disease termed leucocytozoonosis (Table 13.2). The disease is known colloquially as turkey malaria, duck malaria, or gnat fever. The taxonomy of the genus *Leucocytozoon* is being revised using molecular techniques, and the parasite–vector–bird associations are expected to be reworked significantly. Only two species of the parasite are of major economic concern, and both occur in North America. *Leucocytozoon simondi* is specific to ducks and geese, and its primary vectors are *S. anatinum* and *S. rugglesi*. *Leucocytozoon smithi* is specific to turkeys and is transmitted primarily by *S. meridionale* and *S. slossonae*.

Leucocytozoon species undergo a complex malaria-like life cycle. Gametocytes in the blood of an avian host are acquired by a female black fly. The parasite then undergoes both asexual and sexual development over a period of three to four days in the fly. During a subsequent blood meal, the fly transmits the parasites, as sporozoites, to another bird that serves as a host for asexual development and gametocyte production (Figure 13.14).

Leucocytozoonosis can be fatal in poultry, but its effects on wild hosts, with the exception of some populations of Canada geese (Herman et al., 1975), generally are less apparent or difficult to separate from the effects of blood-feeding (Rohner et al., 2000). Birds with chronic infections have weakened immune systems and reduced reproduction. Severe infections produce emaciation, dehydration, and convulsions that lead to death. Internally, the liver and spleen of moribund hosts are enlarged, the heart muscle is pale, and the lungs are congested.

The disease had devastating effects on the poultry industry throughout much of North America when birds were held in outdoor arenas (Noblet et al., 1975). Entire flocks were killed and production facilities shut down in areas such as Nebraska, South Carolina, and Manitoba. The United States Agricultural Research Service estimated an annual average loss of nearly three-quarters of a million dollars in the United States from 1942 to 1951 as a result of leucocytozoonosis in domestic turkeys. The last major outbreaks of the disease in domestic turkeys were in the 1970s. Turkeys now are raised primarily in poultry houses, reducing the incidence of disease because the vectors generally do not venture inside shelters.

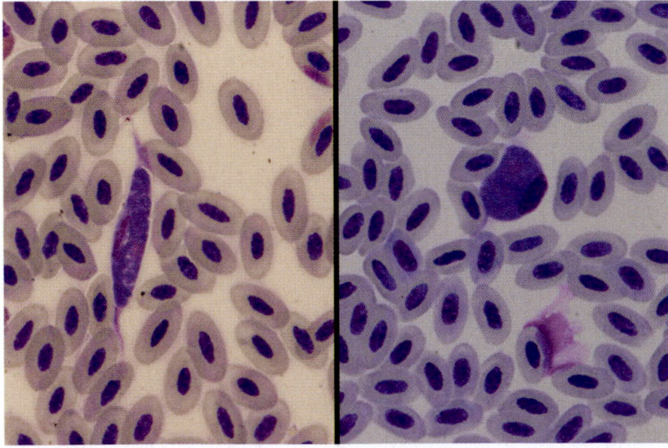

Figure 13.14 Mature macrogametocytes of *Leucocytozoon cf. toddi* from a buzzard (*Buteo buteo*), showing remarkable polymorphism. (Photo by Jan Votýpka)

Other Parasites and Pathogens of Veterinary Importance

Black flies transmit additional parasites to wild animals (Table 13.2). The protozoan *Trypanosoma confusum* is specific to birds in North America, and is transmitted when infected fecal droplets from the black fly contaminate the bite. Birds of numerous families serve as hosts. Other species of bird trypanosomes (e.g., *Trypanosoma corvi*) are believed to cause infections when the birds consume infected black flies or eat other birds that have been infected. The filarial nematodes *Splendidofilaria fallisensis* and *Dirofilaria ursi* are transmitted to ducks and black bears, respectively. The effects of these protozoan and filarial parasites on their wild hosts are poorly known.

Considerable evidence suggests that several North American species of simuliids, such as *S. notatum* and *S. vittatum*, naturally transmit **vesicular stomatitis virus** to livestock, primarily cattle, horses, and pigs (Schmidtmann et al., 1999). The virus causes lesions in various epithelial tissues, especially in the mouth. Millions of dollars can be lost during epizootics. Laboratory experiments have shown that a viremic host is not necessary for a female black fly to become infected; flies can become infected by feeding on the same host with an infected black fly (Mead et al., 2000).

Additional parasites of wildlife have been associated with black flies. The minute nematodes of the family Robertdollfusidae in the guts of African black flies might be transmitted to wildlife (Bain and Renz, 1993). Bunyaviruses, eastern equine encephalitis virus, and snowshoe hare virus have been isolated from several North American black flies. Minimal mechanical transmission has been demonstrated for Whataroa virus in laboratory mice in New Zealand (Austin, 1967) and for **myxomatosis** in rabbits in Australia (Mykytowycz, 1957). These examples suggest that much is yet to be learned about the vector potential of black flies among wildlife.

Simuliotoxicosis

Attacks by black flies have, at times, been so massive and virulent that livestock have been killed. Many of the deaths probably result from acute toxemia and anaphylactic shock caused by the toxins introduced with the saliva as black flies are feeding. The diseased condition, either temporary or terminal, that results from the bites of black flies is known as **simuliotoxicosis**, a term first used to describe the toxic effects of simuliid bites on reindeer (Wilhelm et al., 1982). Cattle, especially calves, are vulnerable to simuliotoxicosis, but goats, horses, mules, pigs, and sheep also have been affected. Susceptible animals succumb in less than two hours. Some immunity is apparent in animals living in afflicted areas. The biochemical nature of simuliotoxicosis requires more investigation.

Most of the species responsible for simuliotoxicosis breed in large rivers from which the adults emerge in astronomical numbers. They include *Austrosimulium pestilens* in Queensland (Australia), *Cnephia pecuarum* in the Mississippi River Valley (United States), *S. colombaschense* along the Danube River in central Europe, *S. vampirum* on the Canadian prairies, and *S. erythrocephalum*, the *S. ornatum* complex, and *S. reptans* in central Europe.

One of the worst attacks in recorded history killed about 22,000 animals in 1923 along Europe's Danube River in the southern Carpathian Mountains (Ciurea and Dinulescu, 1924). Prodigious attacks in this region during the 1700s prompted Empress Maria Theresa of the old Austro-Hungarian Empire to order one of the first biological studies of black flies, which eventually was published in 1795. On the Canadian prairies, thousands of livestock were killed from about 1886 into the 1970s by *S. vampirum*, a member of the *S. arcticum* complex (Fredeen, 1977). Massive mortality due to attacks by *Cnephia pecuarum* occurred in the United States during and immediately after the Civil War when the levees of the Mississippi River deteriorated, allowing the river to overflow and create extensive breeding areas for this species (Riley, 1887).

Simuliotoxicosis on a large scale is now rare, mainly because the former breeding sites of most of the responsible species have been altered by pollution, impoundment, and land development. Some of these species, however, still create nuisance problems for livestock, and occasional outbreaks cause deaths in localized areas of their ranges (Werner and Adler, 2005).

PREVENTION AND CONTROL

Management of black flies typically is aimed at the larval stage, in large part because in this life stage the pest species are concentrated in easily identifiable, specific habitats. Although adulticiding has sometimes offered temporary relief, it is typically more costly and has been used less frequently than larviciding. It usually has involved both aerial and ground fogging with DDT or permethrin products. Current efforts to manage black flies in the adult stage are restricted primarily to the application of repellents and pour-on insecticides.

The use of chemical insecticides in managing black flies dates to the dawn of the twentieth century, reaching a peak from the mid-1940s into the 1970s when DDT was the principal means of control against both larvae and adults. The development of resistance and the undesirable effects on nontarget organisms led to the abandonment of DDT and the search for surrogate compounds, the most prominent of which were methoxychlor (chlorinated hydrocarbon) and temephos (organophosphate). These compounds, as well as insect growth regulators, were not selective and, therefore, had negative effects on nontarget organisms. The use of chemical insecticides to manage black flies became infrequent toward the end of the twentieth century, although compounds such as methoxychlor and temephos continued to be used in a few areas of the world.

Black flies worldwide are managed primarily through the use of the entomopathogenic bacterium *Bacillus thuringiensis* var. *israelensis* (*Bti*, serotype H14), which is aimed at the larval stage. The actual killing agent is an endotoxin in the parasporal inclusions that disrupts the cells of the highly alkaline larval midgut. The efficacy and environmental safety of *Bti* are so superb that most other means of population suppression and management have disappeared since the commercial *Bti* product entered the scene in the early 1980s (Molloy, 1990; Gray et al., 1999). *Bti* can be applied by hand or aircraft. North America's largest suppression program for black flies is operated by the state of Pennsylvania, which treats waterways for *S. jenningsi* in about half of its counties. Because of the intensive use of *Bti* for more than 25 years, target populations should be monitored for resistance.

Natural enemies exert some control in most populations of black flies, but attempts to mass produce them have not been made since the 1970s (Laird, 1981). Commonly encountered parasites include mermithid nematodes, microsporidia, the chytrid fungus *Coelomycidium simulii*, and several viruses. The prevalence of infection with these parasites and pathogens is usually less than 10% of a population. Infections typically slow development, however, so that parasitized larvae become relatively more frequent in a population over time as healthy individuals pupate first.

Mermithid nematodes probably hold the greatest promise for biological control of black flies. However, until more can be learned about their taxonomy and host specificity and how to cultivate them economically for mass release, they are unlikely to be useful in integrated pest management programs. Preparasitic mermithid nematodes crawl on stream substrates and use a protrusible stylet to penetrate the host body. As the mermithids mature, they can be seen through the host integument, coiled within the abdomen. Mermithids either exit and kill the host larva or pass into the adult, exiting shortly thereafter. Postparasitic worms molt to adults, mate, and deposit eggs in the streambed.

Patent infections with microsporidia are recognized by the large, irregular cysts that distort the larval host abdomen. Life cycles of microsporidia that attack black flies are poorly known, although transovarial transmission has been documented. Larvae with patent infections of the fungus *Coelomycidium simulii* are packed with minute, spherical thalli throughout their bodies. Thalli produce spores that are released into the water column after death of the host. Two common viruses that infect larvae are iridescent virus, which imparts an overall blue or violet cast, and cytoplasmic polyhedrosis virus, which creates white bands around the midgut. Many predators consume black flies; most are typically opportunistic.

Physical control of the breeding habitat is occasionally effective in reducing pest populations, usually when the pest species is concentrated in a restricted area, such as directly downstream of an impoundment.

In these situations, attachment sites (e.g., trailing vegetation) can be removed, or water levels can be altered to strand larvae above the water line.

Personal protection for humans involves primarily the use of repellents, both natural and synthetic, that are applied directly to the skin or impregnated in clothing. Among the more effective repellents are those with *N, N*-diethyl-meta-toluamide (DEET) as the active ingredient. Wearing light-colored clothing and minimizing openings in the clothing, such as button holes, through which black flies can gain access to skin, is standard practice when entering areas where black flies are a problem. Fine-mesh head nets are effective in areas where pest populations are intolerable. Many additional means of protection can be found in the annals of folklore, but the utility of most remains suspect.

Various techniques have been devised to protect livestock, ranging from the use of smudges (i.e., smoldering fires that produce dense smoke) to the application of repellent substances and the use of shelters. Repellent products for livestock historically involved oils and greases, often laced with turpentine or other plant-derived products. Among the more commonly used repellents in recent times are permethrin solutions and eartags containing ivermectin. Various pour-on and spray formulations of insecticides and repellents are available commercially. White petroleum jelly can be applied inside the ears of horses to reduce biting problems. Providing shelters is an effective means of protecting livestock and poultry because many of the pest species of black flies infrequently enter enclosures. Providing the entries of shelters with self-application devices for repellents provides an added dose of protection.

Onchocerciasis Control

The largest management program in the world for black flies was the World Health Organization's Onchocerciasis Control Programme (OCP) in West Africa. Its history, as briefly summarized here, has been written by numerous authors (e.g., Davies, 1994; World Health Organization, 1995; Bump, 2004). The initial foundations for the program were laid in 1968, and in 1975 the program launched its first aerial treatments for the control of onchocerciasis. The goal of the OCP was to eliminate onchocerciasis as a major public health threat in seven West African countries: Benin, Burkina Faso, Ghana, Ivory Coast, Mali, Niger, and Togo. The program later was expanded to include the countries of Guinea, Guinea-Bissau, Senegal, and Sierra Leone, thus covering a total of 11 countries and 50,000 km of rivers. It was directed at the vectors of onchocerciasis, namely members of the *S. damnosum* species complex. The primary strategy of the OCP was a massive aerial larviciding program aimed at reducing adult vector populations, thus interrupting transmission. Maintaining vectors at a sufficiently low number for a sufficiently long time prevents new cases of transmission while worms in the human reservoir die out, breaking the disease cycle. Given the longevity of adult worms, control programs in endemic areas must be maintained for approximately 15 years to eliminate the worm from the human reservoir (Remme et al., 1990; Plaisier et al., 1991). Prior to the OCP, aerial application of DDT was the main means of control, but by 1970 resistance had begun to develop. From 1975 into the 1980s, the OCP applied primarily temephos to the rivers. The first appearance (1980) of resistance to this compound by the vectors in the OCP area (Guillet et al., 1980) eventually led to the rotation of six insecticides, including *Bti*.

Vector control was integrated with an ivermectin chemotherapy program for the human reservoir in 1988. Ivermectin, originally developed for veterinary purposes, reduces the number of microfilariae in the skin, so that ingestion of sufficient microfilariae by the vectors becomes difficult. This microfilarialcidal drug, however, does not kill the adult worms. A single oral dose of Mectizan (the formulation of ivermectin for humans) every six to 12 months is not only nontoxic at levels higher than prescribed dosages, but it also is sufficient to kill microfilariae in the skin and eyes and reverse progression of the disease. Dying microfilariae, however, can cause temporary adverse reactions in patients. Mass distribution of ivermectin has been possible through the humanitarian efforts of numerous organizations, including Merck and Co., which decided in 1987 to donate ivermectin tablets for the worldwide treatment of onchocerciasis for as long as necessary. In 1995, the African Programme for Onchocerciasis Control (APOC), which includes 19 African countries, was formed, with the goal of eliminating onchocerciasis from the continent. Its focus has been the mass distribution of ivermectin. The Onchocerciasis Elimination Program for the Americas (OEPA), which was initiated in 1993, includes Brazil, Colombia, Ecuador, Guatemala, Mexico, and Venezuela (Blanks et al., 1998). It, too, relies on mass treatment with ivermectin.

By 1995, vector control had interrupted transmission in about 90% of the original OCP area, protecting more than 30 million people from infection and sparing 100,000 from blindness at a cost of about $360 million. The combined use of ivermectin and weekly insecticide treatments of larval breeding sites was predicted to free the OCP area of onchocerciasis by 2002. Accordingly, the OCP terminated on December 31, 2002, but with approximately 46,000 new cases of onchocerciasis-related blindness in Africa each year, APOC is slated to continue until 2010. The specter of resistance to ivermectin by *Onchocerca volvulus* (Osei-Atweneboana et al., 2007), however, threatens to compromise one of the pillars of onchocerciasis control. At the same time, the discovery that *Wolbachia* bacteria symbiotic with *Onchocerca volvulus* are responsible for the ocular pathology provides the possibility to reduce and prevent ocular onchocerciasis through antibiotic treatments (Pearlman, 2003). Nonetheless, civil wars, weakened infrastructure, and other social disruptions present formidable obstacles to achieving an onchocerciasis-free continent.

REFERENCES AND FURTHER READING

Addison, E. M. (1980). Transmission of *Dirofilaria ursi* Yamaguti, 1941 (Nematoda: Onchocercidae) of black bears (*Ursus americanus*) by blackflies (Simuliidae). *Canadian Journal of Zoology, 58,* 1913–1922.

Adler, P. H., & Crosskey, R. W. (2008). World blackflies (Diptera: Simuliidae): A fully revised edition of the taxonomic and geographical inventory. http://entweb.clemson.edu/biomia/pdfs/blackflyinventory.pdf

Adler, P. H., & McCreadie, J. W. (1997). The hidden ecology of black flies: Sibling species and ecological scale. *American Entomologist, 43,* 153–161.

Adler, P. H., Currie, D. C., & Wood, D. M. (2004). *The Black Flies (Simuliidae) of North America.* Ithaca, New York: Cornell University Press.

Adler, P. H., Roach, D., Reeves, W. K., Flanagan, J. P., Morrow, M. E., & Toepfer, J. E. (2007). Attacks on the endangered Attwater's Prairie-Chicken (*Tympanuchus cupido attwateri*) by black flies (Diptera: Simuliidae) infected with an avian blood parasite. *Journal of Vector Ecology, 32,* 309–312.

Allison, F. R., Desser, S. S., & Whitten, L. K. (1978). Further observations on the life cycle and vectors of the haemosporidian *Leucocytozoon tawaki* and its transmission to the Fjordland crested penguin. *New Zealand Journal of Zoology, 5,* 663–665.

Anderson, R. C. (1968). The simuliid vectors of *Splendidofilaria fallisensis* of ducks. *Canadian Journal of Zoology, 46,* 610–611.

Atwood, D. W. (1996). Distribution, Abundance, Control and Field Observations of the Southern Buffalo Gnat, *Cnephia pecuarum* (Diptera: Simuliidae), in Arkansas and Texas. Ph. D. thesis, Fayetteville, Arkansas: University of Arkansas.

Austin, F. J. (1967). The arbovirus vector potential of a simuliid. *Annals of Tropical Medicine and Parasitology, 61,* 189–199.

Bain, O., & Renz, A. (1993). Infective larvae of a new species of Robertdollfusidae (Adenophorea, Nematoda) in the gut of *Simulium damnosum* in Cameroon. *Annales de Parasitologie Humaine et Comparée, 68,* 182–184.

Blanks, J., Richards, F., Beltrán, F., Collins, R., Álvarez, E., Zea Flores, G., et al. (1998). The Onchocerciasis Elimination Program for the Americas: A history of partnership. *Revista Panamericana de Salud Pública, 3,* 367–374.

Basáñez, M. G., Yarzábal, L., Frontado, H. L., & Villamizar, N. J. (2000). *Onchocerca-Simulium* complexes in Venezuela: Can human onchocerciasis spread outside its present endemic area? *Parasitology, 120,* 143–160.

Bennett, G. F. (1961). On the specificity and transmission of some avian trypanosomes. *Canadian Journal of Zoology, 39,* 17–33.

Bump, J. B. (2004). The Lion's Gaze: African River Blindness from Tropical Curiosity to International Development. Ph.D. thesis, Baltimore, MD: Johns Hopkins University.

Chanteau, S., Sechan, Y., Moulia-Pelat, J. P., Luquiaud, P., Spiegel, A., Boutin, J. P., et al. (1993). The blackfly *Simulium buissoni* and infection by hepatitis B virus on a holoendemic island of the Marquesas Archipelago in French Polynesia. *American Journal of Tropical Medicine and Hygiene, 48,* 763–770.

Ciurea, I., & Dinulescu, G. (1924). Ravages causés par la mouche de Goloubatz en Roumanie; des attaques contre les animaux et contre l'homme. *Annals of Tropical Medicine and Parasitology, 18,* 323–342.

Coscarón, S., & Coscarón Arias, C. L. (2007). Neotropical Simuliidae (Diptera: Insecta). In J. Adis, J. R. Arias, G. Rueda-Delgado, & K. M. Wantzen (Eds.), *Aquatic Biodiversity in Latin America* Vol. 3. Bulgaria: Pensoft, Sofia.

Crosskey, R. W. (1967). The classification of *Simulium* Latreille (Diptera: Simuliidae) from Australia, New Guinea and the Western Pacific. *Journal of Natural History, 1,* 23–51.

Crosskey, R. W. (1969). A re-classification of the Simuliidae (Diptera) of Africa and its islands. *Bulletin of the British Museum (Natural History). Entomology Supplement, 14,* 1–195.

Crosskey, R. W. (1990). *The Natural History of Blackflies.* Chichester, England: John Wiley & Sons Ltd.

Crosskey, R. W., & Howard, T. M. (1997). *A New Taxonomic and Geographical Inventory of World Blackflies (Diptera: Simuliidae).* London: The Natural History Museum.

Cupp, E. W., & Cupp, M. S. (1997). Black fly (Diptera: Simuliidae) salivary secretions: Importance in vector competence and disease. *Journal of Medical Entomology, 34,* 87–94.

Cupp, E. W., & Ramberg, F. B. (1997). Care and maintenance of blackfly colonies. In J. M. Crampton, C. B. Beard, & C. Louis (Eds.), *The Molecular Biology of Insect Disease Vectors: A Methods Manual* (pp. 31–40). London: Chapman and Hall.

Currie, D. C., & Adler, P. H. (2008). Global diversity of black flies (Diptera: Simuliidae) in freshwater. *Hydrobiologia, 595,* 469–475.

Davies, J. B. (1994). Sixty years of onchocerciasis vector control: A chronological summary with comments on eradication, reinvasion, and insecticide resistance. *Annual Review of Entomology, 39,* 23–45.

Dirie, M. F., Ashford, R. W., Mungomba, L. M., Molyneux, D. H., & Green, E. E. (1990). Avian trypanosomes in *Simulium* and sparrowhawks (*Accipiter nisus*). *Parasitology, 101,* 243–247.

Eaton, D. P., Diaz, L. A., Hans-Filho, G., dos Santos, V., Aoki, V., Friedman, H., et al. (1998). Comparison of black fly species (Diptera: Simuliidae) on an Amerindian reservation with a high prevalence of fogo selvagem to neighboring disease-free sites in the state of Mato Grosso do Sul, Brazil. *Journal of Medical Entomology, 35,* 120–131.

Fallis, A. M., Jacobson, R. L., & Raybould, J. N. (1973). Experimental transmission of *Trypanosoma numidae* Wenyon to guinea fowl and chickens in Tanzania. *Journal of Protozoology, 20,* 436–437.

Fallis, A. M., Desser, S. S., & Khan, R. A. (1974). On species of *Leucocytozoon. Advances in Parasitology, 12,* 1–67.

Fredeen, F. J. H. (1977). A review of the economic importance of black flies (Simuliidae) in Canada. *Quaestiones Entomologicae, 13,* 219–229.

Fredeen, F. J. H. (1985). Some economic effects of outbreaks of black flies (*Simulium luggeri* Nicholson & Mickel) in Saskatchewan. *Quaestiones Entomologicae, 21,* 175–208.

Gray, E. W., Adler, P. H., Coscarón-Arias, C., Coscarón, S., & Noblet, R. (1999). Development of the first black fly (Diptera: Simuliidae) management program in Argentina and comparison with other programs. *Journal of the American Mosquito Control Association, 15,* 400–406.

Gudgel, E. F., & Grauer, F. H. (1954). Acute and chronic reactions to black fly bites (*Simulium* fly). *Archives of Dermatology and Syphilology, 70,* 609–615.

Guillet, P., Escaffre, H., Ouédraogo, M., & Quillévéré, D. (1980). Mise en évidence d 'une résistance au téméphos dans le complexe *S. damnosum* (*S. sanctipauli* et *S. soubrense*) en Côte d'Ivoire (zone du programme de lutte contre l 'onchocercose dans la région du bassin de la Volta). *Cahiers ORSTOM Séries Entomologie Médicale et Parasitologie, 23,* 291–299.

Herman, C. M., Barrow, J. H., & Tarshis, I. B. (1975). Leucocytozoonosis in Canada geese at the Seney National Wildlife Refuge. *Journal of Wildlife Diseases, 11,* 404–411.

Klion, A. D., & Nutman, T. B. (1999). Loiasis and *Mansonella* infections. In R. L. Guerrant, D. H. Walker, & P. F. Weller (Eds.), *Tropical Infectious Diseases: Principles, Pathogens, & Practice* (pp. 861–872). Philadelphia: Churchill Livingstone.

Laird, M. (1981). *Blackflies: The Future for Biological Methods in Integrated Control.* New York: Academic Press.

Lok, J. B., Cupp, E. W., & Bernardo, M. J. (1983). *Simulium jenningsi* Malloch (Diptera: Simuliidae): a vector of *Onchocerca lienalis* Stiles (Nematoda: Filarioidea) in New York. *American Journal of Veterinary Research, 44,* 2355–2358.

Malmqvist, B., Adler, P. H., Kuusela, K., Merritt, R. W., & Wotton, R. S. (2004a). Black flies in the boreal biome, key organisms in both terrestrial and aquatic environments: a review. *Écoscience, 11,* 187–200.

Malmqvist, B., Strasevicius, D., Hellberg, O., Adler, P. H., & Bensch, S. (2004b). Vertebrate host specificity of wild-caught blackflies revealed by mitochondrial DNA in blood. *Proceedings of the Royal Society of London B (Supplement), Biology Letters, 271,* S152–S155.

McCreadie, J. W., Adler, P. H., & Hamada, N. (2005). Patterns of species richness for blackflies (Diptera: Simuliidae) in the Nearctic and Neotropical regions. *Ecological Entomology, 30,* 201–209.

Mead, D. G., Ramberg, F. B., Besslesen, D. G., & Máre, C. J. (2000). Transmission of vesicular stomatitis virus from infected to noninfected black flies co-feeding on nonviremic deer mice. *Science, 287,* 485–487.

Mock, D. E., & Adler, P. H. (2002). Black flies (Diptera: Simuliidae) of Kansas: Review, new records, and pest status. *Journal of the Kansas Entomological Society, 75*, 203–213.

Molloy, D. (1990). Progress in the biological control of black flies with *Bacillus thuringiensis israelensis*, with emphasis on temperate climates. In H. de Barjac, & D. J. Sutherland (Eds.), *Bacterial Control of Mosquitoes and Black Flies: Biochemistry, Genetics, and Applications of Bacillus thuringiensis israelensis and Bacillus sphaericus* (pp. 161–186). New Brunswick, New Jersey: Rutgers University Press.

Muller, R., & Horsburgh, R. C. R. (1987). *Bibliography of Onchocerciasis (1841–1985)*. C. A. B. International Institute of Parasitology.

Mykytowycz, R. (1957). The transmission of myxomatosis by *Simulium melatum* Wharton (Diptera: Simuliidae). *CSIRO Wildlife Research, 2*, 1–4.

Noblet, R., Kissam, J. B., & Adkins, T. R., Jr. (1975). *Leucocytozoon smithi*: Incidence of transmission by black flies in South Carolina (Diptera: Simuliidae). *Journal of Medical Entomology, 12*, 111–114.

O'Roke, E. C. (1934). A malaria-like disease of ducks caused by *Leucocytozoon anatis* Wickware. *University of Michigan School of Forestry Conservation Bulletin, 4*, 1–44.

Osei-Atweneboana, M. Y., Eng, J. K., Boakye, D. A., Gyapong, J. O., & Prichard, R. K. (2007). Prevalence and intensity of *Onchocerca volvulus* infection and the efficacy of ivermectin in endemic communities in Ghana: A two-phase epidemiological study. *Lancet, 369*, 2021–2029.

Pearlman, E. (2003). Immunopathogenesis of *Onchocerca volvulus* keratitis (river blindness): A novel role for endosymbiotic *Wolbachia* bacteria. *Medical Microbiology and Immunology (Berlin), 192*, 57–60.

Plaisier, A. P., Van Oortmarssen, G. J., Remme, J. H. F., & Habbema, J. D. F. (1991). The reproduction lifespan of *Onchocerca volvulus* in West Africa savanna. *Acta Tropica, 48*, 271–284.

Pledger, D. J., Samuel, W. M., & Craig, D. A. (1980). Black flies (Diptera: Simuliidae) as possible vectors of legworm (*Onchocerca cervipedis*) in moose of central Alberta. *Proceedings of the North American Moose Conference Workshop, 16*, 171–202.

Post, R. J., Mustapha, M., & Krueger, A. (2007). Taxonomy and inventory of the cytospecies and cytotypes of the *Simulium damnosum* complex (Diptera: Simuliidae) in relation to onchocerciasis. *Tropical Medicine and International Health, 12*, 1342–1353.

Procunier, W. S. (1989). Cytological approaches to simuliid biosystematics in relation to the epidemiology and control of human onchocerciasis. *Genome, 32*, 559–569.

Reeves, W. K., Adler, P. H., Rätti, O., Malmqvist, B., & Strasevicius, D. (2007). Molecular detection of *Trypanosoma* (Kinetoplastida: Trypanosomatidae) in black flies (Diptera: Simuliidae). *Comparative Parasitology, 74*, 171–175.

Remme, J. H. F., De Sole, G., & Van Oortmarssen, G. J. (1990). The predicted and observed decline in the prevalence and intensity of onchocerciasis infection during 14 years of successful vector control. *Bulletin of the World Health Organization, 68*, 331–339.

Riley, C. V. (1887). Report of the entomologist. *United States Department of Agriculture Report, 1886*, 459–592.

Rohner, C., Krebs, C. J., Hunter, D. B., & Currie, D. C. (2000). Roost site selection of great horned owls in relation to black fly activity: an anti-parasite behavior. *Condor, 102*, 950–955.

Rothfels, K. H. (1979). Cytotaxonomy of black flies (Simuliidae). *Annual Review of Entomology, 24*, 507–539.

Rubtsov, I. A. (1956). Blackflies (fam. Simuliidae). "Fauna of the USSR." New Series No. 64, Insects, Diptera 6 (6). Akademii Nauk SSSR, Moscow. [In Russian; English translation: 1990. Blackflies (Simuliidae). Second Edition. "Fauna of the USSR". Diptera, 6 (6). E. J. Brill, Leiden.].

Sanford, D., Eikenhorst, B., Lamb, T., Cates, J. E., Robinson, J., Olsen, J., et al. (1993). Black flies cause costly losses in East Texas ostriches and emus. *Texas Agricultural Extension Service Veterinary Quarterly Review, 9*(2), 1–2.

Schmidtmann, E. T., Tabachnick, W. J., Hunt, G. J., Thompson, L. H., & Hurd, H. S. (1999). 1995 epizootic of vesicular stomatitis (New Jersey serotype) in the western United States: An entomologic perspective. *Journal of Medical Entomology, 36*, 1–7.

Shelley, A. J. (1988a). Biosystematics and medical importance of the *Simulium amazonicum* group and the *S. exiguum* complex in Latin America. In M. W. Service (Ed.), *Biosystematics of Haematophagus Insects* (pp. 203–220). Oxford: Clarendon Press.

Shelley, A. J. (1988b). Vector aspects of the epidemiology of onchocerciasis in Latin America. *Annual Review of Entomology, 30*, 337–366.

Shelley, A. J., & Coscarón, S. (2001). Simuliid blackflies (Diptera: Simuliidae) and ceratopogonid midges (Diptera: Ceratopogonidae) as vectors of *Mansonella ozzardi* (Nematoda: Onchocercidae) in northern Argentina. *Memórias do Instituto Oswaldo Cruz, 96*, 451–458.

Stokes, J. H. (1914). A clinical, pathological and experimental study of the lesions produced by the bite of the "black fly" (*Simulium venustum*). *Journal of Cutaneous Diseases, 32*, 751–769, 830–856.

Sutcliffe, J. F. (1986). Black fly host location: a review. *Canadian Journal of Zoology, 64*, 1041–1053.

Sutcliffe, J. F., & McIver, S. B. (1984). Mechanics of blood-feeding in black flies (Diptera, Simuliidae). *Journal of Morphology, 180*, 125–144.

Takaoka, H. (1999). Review on zoonotic *Onchocerca* species and their insect vectors in Japan. *Medical Entomology and Zoology, 50*, 1–8. [In Japanese, with English summary]

Takaoka, H. (2003). *The Black Flies (Diptera: Simuliidae) of Sulawesi, Maluku and Irian Jaya.* Fukuoka, Japan: Kyushu University Press.

Takaoka, H., & Davies, D. M. (1995). *The Black Flies (Diptera: Simuliidae) of West Malaysia.* Japan: Kyushu University Press.

Votýpka, J., Oborník, M., Volf, P., Svobodová, M., & Lukeš, J. (2002). *Trypanosoma avium* of raptors (Falconiformes): Phylogeny and identification of vectors. *Parasitology, 125*, 253–263.

Wahl, G. (1996). Identification of a common filarial larva in *Simulium damnosum* s. l. (Type D, Duke, 1967) as *Onchocerca ramachandrini* from the wart hog. *Journal of Parasitology, 82*, 520–524.

Wahl, G., & Renz, A. (1991). Transmission of *Onchocerca dukei* by *Simulium bovis* in North-Cameroon. *Tropical Medicine and Parasitology, 42*, 368–370.

Wahl, G., Ekale, D., & Schmitz, A. (1998). *Onchocerca ochengi*: Assessment of the *Simulium* vectors in North Cameroon. *Parasitology, 116*, 327–336.

Werner, D., & Adler, P. H. (2005). A faunistic review of the black flies (Simuliidae, Diptera) of the federal state of Sachsen-Anhalt, Germany. *Abhandlungen und Berichte für Naturkunde, 27*, 205–245.

Wilhelm, A., Betke, P., & Jacob, K. (1982). Simuliotoxikose beim Ren (*Rangifer tarandus*). In R. Ippen, & H. D. Schräder (Eds.), *Erkrankungen der Zootiere* (pp. 357–360). Verhandlungbericht des XXIV Internationalen Symposium über die Erkrankungen der Zootiere. Berlin: Akademie-Verlag.

World Health Organization. (1995). Onchocerciasis and its control: Report of a WHO-Expert Committee on Onchocerciasis Control. *WHO Technical Report Series, 852*, 1–103.

Chapter

14

Mosquitoes (Culicidae)

Woodbridge A. Foster . Edward D. Walker

Since ancient times, mosquito bites or habitats have been associated with human disease, and, in 1878, mosquitoes were the first arthropods formally incriminated as intermediate hosts of vertebrate parasites. During the past century of research, it has become established that mosquitoes are the most important arthropods affecting human health. They attain their greatest impact as vectors for the organisms causing such well-known human diseases as malaria, filariasis, encephalitis, yellow fever, and dengue. These afflictions are especially severe in developing regions of the tropics. They cause early death and chronic debilitation, which strain the resources of health services and reduce human productivity, thereby perpetuating economic hardship.

Mosquito-borne diseases also persist in industrialized temperate countries. Yet, human discomfort from bites is often the chief concern. In the United States, hundreds of millions of dollars are spent annually to control them for this reason alone. Additionally, large populations of mosquitoes can cause intense irritation and extensive blood loss to livestock and wildlife, resulting in reduced productivity and even death.

Mosquitoes occur in practically every region of every continent in the world except Antarctica. They develop in an extremely broad range of biotic communities: arctic tundra, boreal forests, high mountains, plains, deserts, tropical forests, salt marshes, and ocean tidal zones. Greatest species diversity occurs in tropical forests, but extremely high densities of mosquitoes are common even in the species-poor biomes, such as the tundra. Many species have benefited from human alteration of the environment, and a few have become domesticated. Because of their immense importance, mosquitoes have been the subject of many major books. Among the more important ones that deal exclusively with mosquito biology are Christophers (1960), Clements (1992, 1999), Forattini (1962, 1965), Gillett (1971), Bock and Cardew (1996), Horsfall (1955), Lounibos et al. (1985), Mattingly (1969), Service (1990, 1993a), and Silver (2008). *Journal of the American Mosquito Control Association* is devoted mainly to studies of mosquitoes. A substantial proportion of the scientific articles in the *Journal of Medical Entomology* and *Medical and Veterinary Entomology* also report mosquito research. *Wing Beats* is a trade magazine dedicated to mosquitoes.

TAXONOMY

The family **Culicidae**, derived from *culex*, the Latin name for gnat is a member of one of the main stocks of Nematocera, the infraorder Culicomorpha. It consists of two superfamilies that include all of the piercing/sucking nematocerans, both predators and blood-feeding biters. The superfamily Chironomoidea comprises the families Chironomidae and Thaumaleidae, which have nonpiercing mouthparts, and Simuliidae and Ceratopogonidae, which pierce either vertebrates or invertebrates. The superfamily Culicoidea comprises the Dixidae, Corethrellidae, Chaoboridae, and Culicidae, the second and fourth of which feed on vertebrate blood. Several of these families are superficially similar. However, among all the culicomorphs, the long proboscis of mosquitoes is distinctive. It is considered the most specialized of biting mouthparts among Nematocera and indicates a long and close association of mosquitoes with vertebrate animals. Wood and Borkent (1989) provide an overview of nematoceran phylogeny and classification.

Culicidae consists of about 3,500 recognized species. The largest number remaining to be discovered probably inhabits tropical rainforests, where faunas are more diverse but less well surveyed than temperate regions. Species that have been studied intensively often reveal that they consist of complexes of closely related species, indicating that many reproductively isolated and niche-specific forms remain to be identified or are undergoing speciation. Current culicid classification (Table 14.1) recognizes two subfamilies: Anophelinae and Culicinae. Anophelinae is considered to be a primitive group. The former subfamily Toxorhynchitinae has been reduced to tribe status on the basis of cladistic analysis of morphological and nucleotide-sequence data (Harbach and Kitching 1998). Anopheline eggs bear characteristic floats, their

Table 14.1 Classification of Culicidae

Subfamily	Tribe	Genera
Anophelinae		*Anopheles* (*An.*), *Bironella* (*Bi.*), *Chagasia* (*Ch.*)
Culicinae	Aedeomyiini	*Aedeomyia* (*Ad.*)
	Aedini	*Aedes* (*Ae.*), *Armigeres* (*Ar.*), *Ayurakitia* (*Ay.*), *Eretmapodites* (*Er.*), *Haemagogus* (*Hg.*), *Heizmannia* (*Hz.*), *Ochlerotatus* (*Oc.*), *Opifex* (*Op.*), *Psorophora* (*Ps.*), *Tanakaius* (*Ta.*), *Udaya* (*Ud.*), *Verrallina* (*Ve.*), *Zeugnomyia* (*Ze.*)
	Culicini	*Culex* (*Cx.*), *Deinocerites* (*De.*), *Galindomyia* (*Ga.*), *Lutzia* (*Lu.*)
	Culisetini	*Culiseta* (*Cs.*)
	Ficalbiini	*Ficalbia* (*Fi.*), *Mimomyia* (*Mi.*)
	Hodgesiini	*Hodgesia* (*Ho.*)
	Mansoniini	*Coquillettidia* (*Cq.*), *Mansonia* (*Ma.*)
	Orthopodomyiini	*Orthopodomyia* (*Or.*)
	Sabethini	*Isostomyia* (*Is.*), *Johnbelkinia* (*Jb.*), *Limatus* (*Li.*), *Malaya* (*Ml.*), *Maorigoeldia* (*Mg.*), *Onirion* (*On.*), *Runchomyia* (*Ru.*), *Sabethes* (*Sa.*), *Shannoniana* (*Sh.*), *Topomyia* (*To.*), *Trichoprosopon* (*Tr.*), *Tripteroides* (*Tp.*), *Wyeomyia* (*Wy.*).
	Toxorhynchitini	*Toxorhynchites* (*Tx.*)
	Uranotaeniini	*Uranotaenia* (*Ur.*)

The classification of all mosquitoes into two subfamilies, 10 tribes of Culicinae, and 43 genera is based on Knight, K.L. & Stone, A. (1977). A Catalog of the Mosquitoes of the world (Diptera: Calicidae) (2nd ed.). Thomas Say Foundation. Vol. 6. Entomological Society of America and modified according to the 2001 Systematic Catalog of the Culicidae, Walter Reed Biosystematics Unit, and other sources. In parentheses are the two-letter generic abbreviations recognized by the American Mosquito Control Association and used in several journals and books.

larvae lack air tubes, and adults have elongate palps in both sexes. Typical culicine larvae have air tubes, and adult females have short palps.

There are 43 genera of mosquitoes, 40 of which are in the subfamily Culicinae. Culicines are organized into 11 tribes, the most diverse of which are Aedini and Sabethini in terms of numbers of genera and species worldwide. The 15 genera in North America north of Mexico, and the number of species in each, are *Anopheles* (21), *Aedes* (4), *Ochlerotatus* (77), *Psorophora* (13), *Haemagogus* (1), *Tanakaius* (1), *Culex* (29), *Deinocerites* (3), *Culiseta* (8), *Coquillettidia* (1), *Mansonia* (2), *Orthopodomyia* (3), *Wyeomyia* (3), *Uranotaenia* (4), and *Toxorhynchites* (2) (Darsie and Ward, 1981, 2004; online-updated classification of Knight and Stone, 1977).

Two recent proposals for raising subgenera of the tribe *Aedini* to the rank of genus are controversial. The first of these (Reinert, 2000) was the elevation of *Aedes (Ochlerotatus)* from subgenus to the rank of genus *Ochlerotatus*, now frequently used in mosquito literature, though not without resistance (Savage and Strickman, 2004). The second proposal (Reinert et al., 2004), based on extensive cladistic analysis, recognizes another 34 genera formerly within *Aedes* and *Ochlerotatus*. This has been met with greater objection, in view of the extensive nomenclatural confusion that is expected to result (Journal of Medical Entomology Editorial Board, 2005). While issues are being resolved, this chapter recognizes the genus *Ochlerotatus* and the new genus *Tanakaius* while using the name *Aedes* for other species previously classified as members of that genus.

Three important species groups of mosquitoes worldwide are the *Anopheles gambiae* and *Culex pipiens* complexes and the *Aedes* subgenus *Stegomyia*. The ***Anopheles gambiae* complex** of Africa consists of at least six species. Two of these, *An. gambiae* (Figure 14.33) and *An. arabiensis*, are important vectors of malaria and lymphatic filariasis. *Anopheles arabiensis* tends to occur in somewhat drier regions than does *An. gambiae*. Both prefer to bite humans, but *An. gambiae* is more anthropophilic, endophilic, and endophagic, and therefore it is the more important vector. The ***Culex pipiens* complex** is a ubiquitous group of closely related domestic and peridomestic species. The medically most important taxa worldwide are the temperate species *Cx. pipiens*, the northern house mosquito, and the tropical and subtropical *Cx. quinquefasciatus* (= *fatigans*), the southern house mosquito (Figure 14.29). Their ranges are overlapping in the central latitudes of the United States, where they commonly hybridize. They are vectors of several human pathogens, such as St. Louis encephalitis virus, West Nile virus, and lymphatic filariasis. *Culex molestus* is a name sometimes applied to a variant of *Cx. pipiens*, which is facultatively autogenous and often breeds in subterranean habitats. *Culex pallens*, apparently a stable hybrid of *Cx. pipiens* and *Cx. quinquefasciatus*, occurs in temperate China and Japan, whereas *Cx. globocoxitus* and *Cx. australicus* inhabit Australia.

Several brightly marked *Aedes* species in the large **subgenus *Stegomyia*** are medically important, including *Ae. aegypti* and *Ae. albopictus*. *Aedes aegypti*, the yellow fever mosquito (Figure 14.23), has a worldwide distribution in the tropics and subtropics. It is the primary vector of both dengue and urban yellow fever viruses. It exists in at least two forms, *aegypti* and *formosus*, considered to be either subspecies or separate species. *Aedes aegypti formosus* is the original feral form and is found in large parts of interior Africa. It has a black body, develops in tree holes, feeds on a wide variety of animals, and rarely enters houses. It has adapted to some domestic situations in Africa, where it develops in rain-filled containers. *Aedes aegypti aegypti* is a paler, brownish-black domestic form. It occurs mainly in coastal regions of Africa and is distributed throughout much of southern Asia and most warmer parts of the New World, including the southern United States.

In Africa it has become independent of rain, developing in hand-filled water jars without regard to season. On other continents, where it does not compete with *Ae. a. formosus*, it utilizes both rain-filled and hand-filled containers. Some authorities recognize a still paler and more domestic type of *Ae. a. aegypti* as the subspecies *Ae. a. queenslandensis*, but this is probably only a localized variant.

Aedes albopictus, the Asian tiger mosquito (Figure 14.26), is similar to *Ae. aegypti*, occupies the same kinds of containers, and also transmits dengue virus. It was largely confined to Asia, where it occurs in tropical and subtropical rural settings. It readily oviposits in treeholes. A cold-hardy, egg-diapausing strain of this mosquito has been carried from northern Japan to other parts of the world by the trade in used automobile and truck tires. The first established population was detected in Texas in 1985. It has since spread through much of the southern, central, and eastern United States, including foci in the upper Midwest, much farther north than the nondiapausing *Ae. aegypti*. It also has gained a foothold in several other parts of the world. In most of its range in the southern United States, *Ae. albopictus* has replaced *Ae. aegypti* as the predominant mosquito in artificial containers in suburban and rural environments.

Other important members of the subgenus *Stegomyia* include *Ae. africanus, Ae. bromeliae*, and *Ae. luteocephalus*, which transmit yellow fever and dengue viruses in parts of Africa, and *Ae. polynesiensis* and *Ae. pseudoscutellaris*, which transmit lymphatic filariasis in South Pacific islands.

Keys to the mosquito genera worldwide were provided by Mattingly (1971). Keys for the identification of species of restricted geographical regions are available for most states, provinces, and many countries throughout the world. These include many fine handbooks that also present biological and medical information on individual species. Good examples of statewide handbooks are written for New Jersey (Headlee, 1945), California (Bohart and Washino, 1978), Indiana (Siverly, 1972), Minnesota (Barr, 1958), New York (Means, 1979), Florida (Darsie and Morris, 1998), and Alaska (Gjullin et al., 1961). United States regional handbooks include the southeastern United States (King et al., 1960) and the northwestern states (Stage et al., 1952). Some handbooks include keys to pupae, and the handbook for Illinois (Ross and Horsfall, 1965), is noteworthy in particular for its egg keys.

The most recent comprehensive treatments of North American species are Wood et al. (1979), which contains keys to larvae and adults of Canada, plates of taxonomic structures for each species, distribution maps, and biological information; and Darsie and Ward (1981), updated by Darsie et al. (2004), which covers all of North America north of Mexico and has illustrated keys and distribution maps. These works were preceded by Carpenter and LaCasse (1955), which contains formal descriptions, biology, and meticulously crafted full-page plates of adults of each species. A thorough treatment of North American genera was presented by Stone (1981). Other parts of the world covered by notable works include the South Pacific (Belkin, 1962), United Kingdom (Marshall, 1938), the Neotropical Region (Lane, 1953), and Japan and Korea (LaCasse and Yamaguti, 1950). For details on morphological terminology and anatomical features of mosquitoes, Harbach and Knight (1980) is recommended. Members of species complexes are often indistinguishable morphologically. Specialists have overcome some of these problems by using chromosome banding patterns, isozyme profiles, or DNA probes and DNA restriction fragment patterns to distinguish these species from one another. These methods for identification are not yet sufficiently simple or widely available to be used routinely in field work.

All known mosquito species in the world are listed in *A Catalog of the Mosquitoes of the World* (Knight and Stone, 1977), plus its four supplements (Knight, 1978; Ward, 1984, 1992; Gaffigan and Ward 1985). Updates to the catalog are presented online by the Walter Reed Biosystematics Unit. This work provides the taxonomic history and current standing of all recognized species, their distributions by country, and references to the general literature and to taxonomic works for all regions of the world. The most recent large treatment of a major geographic area is the 12-volume catalog on the Australasian Region, edited by Debenham, Hicks, Lee, and others (1980–1989). Original systematic studies of mosquitoes are published in several scientific journals, but the one devoted solely to this subject, *Mosquito Systematics*, was subsumed under the *Journal of the American Mosquito Control Association* in 1995. Two long series of valuable papers of international scope on mosquito taxonomy and distribution have been published in the journal *Contributions of the American Entomological Institute: Contributions to the Mosquito Fauna of Southeast Asia* and *Mosquito Studies*. Preferred common names of mosquito species are listed by the Entomological Society of America (Bosik, 1997) and by Pittaway (1992). An internationally accepted set of two-letter abbreviations for all mosquito genera is shown in Table 14.1. These abbreviations appear in most mosquito publications and are used in this chapter.

Morphology

The eggs of most mosquitoes are elongate, ovoid, or spindle-shaped; others are spherical or rhomboid. The outermost layer of the egg shell, or **chorion** (Figure 14.1), often has intricate surface structures and patterns diagnostic of the particular species. The chorions of *Anopheles* species have unique, transparent, air-filled compartments flanking the egg that serve as **floats** (Figure 14.1A). Eggs of *Anopheles, Toxorhynchites, Wyeomyia, Aedes, Ochlerotatus, Psorophora*, and *Haemagogus* species are laid individually, whereas in *Culex, Culiseta, Coquillettidia*, and *Mansonia* species, they are attached together in a single clump, forming a floating *egg raft* (Figure 14.2A) or submerged cluster (Figure 14.2B). *Culex* eggs have a cup-shaped **corolla** at one end (Figure 14.1B), allowing them to sit vertically on the water surface in a raft (Figure 14.2A); the upper ends have apical droplets with a chemical thought to maintain the raft upright.

Mosquito larvae, commonly known as **wigglers** or **wrigglers**, pass through four instars, which closely

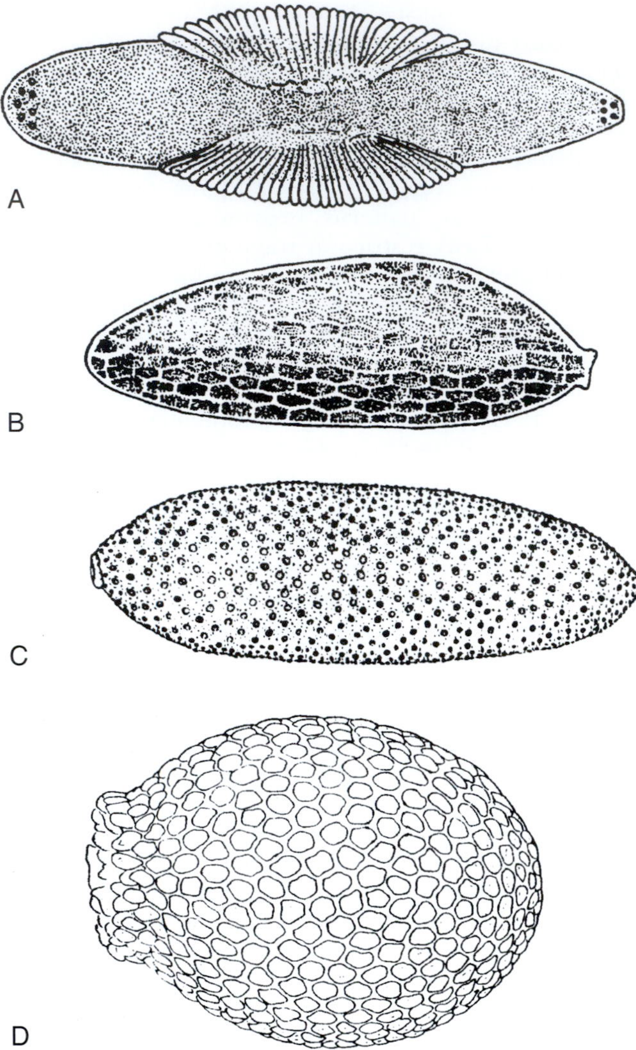

Figure 14.1 Eggs of mosquitoes, showing variations in shape and chorionic sculpturing. (A) *Anopheles*; (B) *Culex*; (C) *Aedes aegypti*; (D) *Toxorhynchites brevipalpis*. (A and B from Ross, 1947; C and D from Harbach and Knight, 1980)

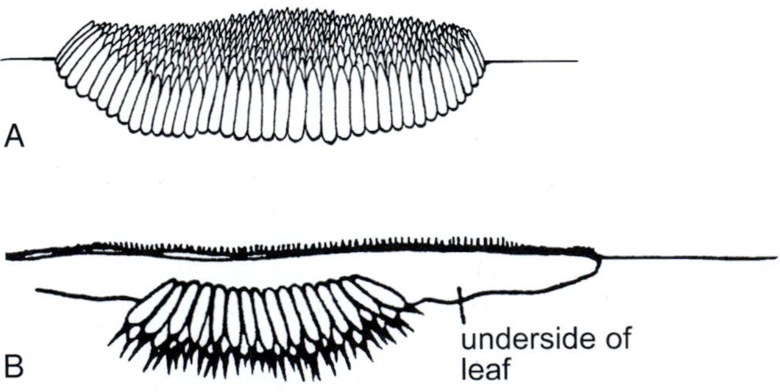

Figure 14.2 Mosquito egg rafts and clusters. (A) Floating egg raft of *Culex restuans* (Ross, 1947); (B) submerged egg cluster of *Mansonia*, attached to underside of floating leaf. (Gordon and Lavoipierre, 1962)

resemble one another except for their size. Larvae are rich in taxonomic characters that are easy to see on slide-mounted specimens (Figure 14.3). The head is defined by a distinct capsule bearing a pair of "eyes" comprised of clusters of lateral ocelli, a pair of antennae of variable shape and length, and chewing mouthparts bearing a variety of brushes, combs, and sweepers used in feeding (Figure 14.4). The lateral **palatal brushes** on the labrum create water currents that draw floating or suspended particles toward the mouth. Sweepers and brushes on the mandibles, and brushes on the maxillae, are thought to collect and pack particles to create a bolus of food in the pharynx. In predatory larvae, the mandibles and maxillae are heavy and sharply toothed for seizing or holding prey. The thorax is wide, with three indistinct, legless segments.

The larval abdomen is narrower than the thorax, cylindrical, and composed of eight apparent segments, the second-to-last being a composite of segments 8 and 9. A pair of spiracles opens on the dorsal side of this segment. In culicines the spiracles open at the end of the **respiratory siphon**, an elongate air tube extending dorsally. The siphon of *Coquillettidia* and *Mansonia* is short, ending in a heavily sclerotized point with a dorsal saw-like edge used to pierce and remain lodged in plant tissue. In anophelines the siphon is lacking, and the spiracles are borne on a short spiracular plate. Segment 10, the anal segment, extends ventrally at an angle from the rest of the abdomen. It typically bears four anal papillae used primarily in osmoregulation. The terminal region of the larva bears several structures useful in identification (Figure 14.5). They include comb scales on segment 8, pecten spines on the siphon, a saddle sclerite encircling the anal segment, and various tufts and brushes of setae. Some of these terminal structures apparently are used to groom the mouthparts when the larva bends its body around to form a loop.

Larval internal anatomy conforms to the general insect plan. The alimentary tract is almost straight, the only notable features being eight large gastric caeca at the junction of the foregut and midgut in the thorax and five Malpighian tubules at the midgut-hindgut junction. Because most of the cuticle is semitransparent, two large tracheal trunks are obvious, extending forward from the spiracles to the thorax.

Mosquito pupae, commonly known as **tumblers**, are comma-shaped, with the head and thorax fused to form a cephalothorax and the abdomen curled beneath it (Figure 14.6). Projecting from the dorsal mesothorax is a pair of respiratory tubes, or **air trumpets**, through which the pupa obtains oxygen at the water surface. Within the cephalothorax the developing appendages of the adult head and thorax usually can be seen coiled ventrally; they envelop an air pocket, the ventral air space, that provides buoyancy to help maintain the pupa at the water surface when resting. At the end of the abdomen two broad paddles are attached to the eighth segment. The pupa can flex its abdominal segments, causing the paddles to flap downward, propelling it through the water when it is disturbed.

Adult mosquitoes are slender, with thin legs and narrow, elongate wings (Figure 14.7). The body surface is

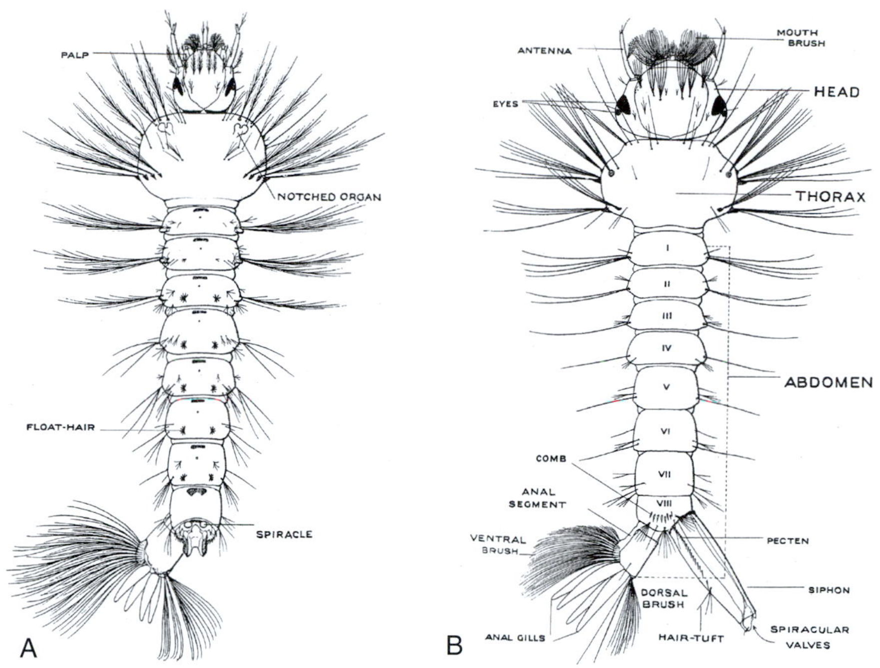

Figure 14.3 External anatomy of mosquito larvae, dorsal view, with anal segment and siphon at posterior end rotated to provide better view. (A) Anopheline form (*Anopheles maculipennis*). (B) Culicine form (*Aedes cinereus*). (Marshall, 1938)

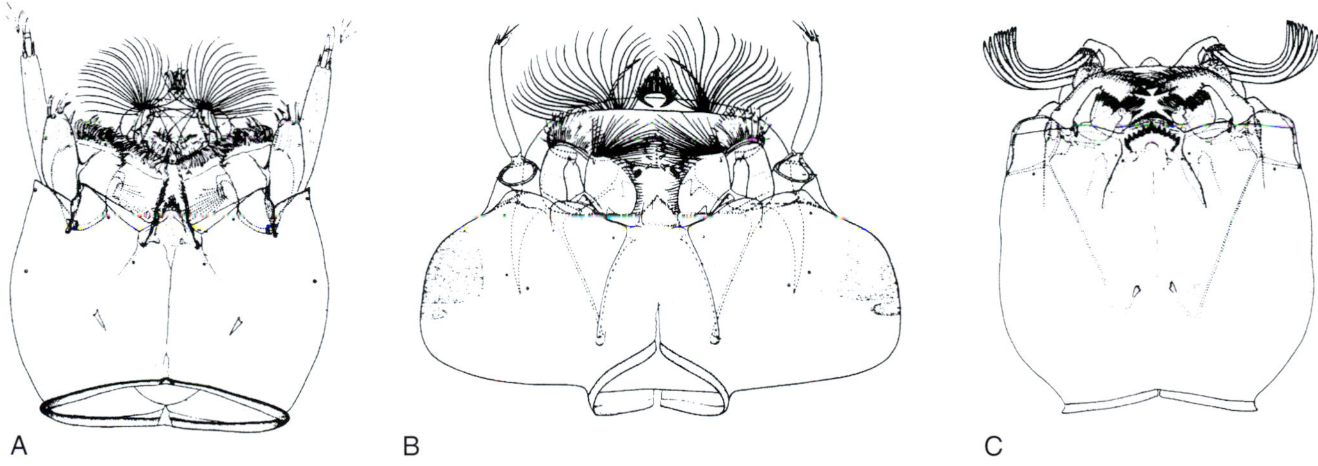

Figure 14.4 Heads of mosquito larvae, postero-ventral view. (A) Anopheline form (*Anopheles quadrimaculatus*); (B) typical culicine form (*Aedes fulvus pallens*); (C) toxorhynchitine form (*Toxorhynchites brevipalpus*). The lateral palatal brushes of most larvae are used to generate water currents for filter feeding; in *Toxorhynchites* they are modified for seizing prey. (Harbach and Knight, 1980)

covered with scales, setae, and fine pile, creating the characteristic markings and colors of each species. The two compound eyes, each represented by 350–900 ommatidial lenses, wrap around the front and sides of the head. The antennae arise between the eyes, are long and filamentous, and are usually sexually dimorphic. In species in which sound is used to locate females in flight, the flagellum of the male antenna has whorls of much longer fibrillae, giving it a plumose appearance (Figure 14.8). The pedicel at the base of the antenna is a large globular structure that contains **Johnston's organ**, a mass of radially arranged mechanoreceptors that respond to vibrations of the flagellum induced by sound. In addition to the long fibrillae, the antenna has a variety of sensory structures, including those for detecting host odors.

The mosquito proboscis is prominent, projecting anteriorly at least two-thirds the length of the abdomen. It consists of the basic complement of insect mouthparts: the labrum, paired mandibles, hypopharynx, paired maxillae, and labium. The first four structures have evolved into fine stylets, forming a tightly fitting fascicle that in females is used to penetrate host skin (Figure 14.9). The fascicle is cradled within the groove

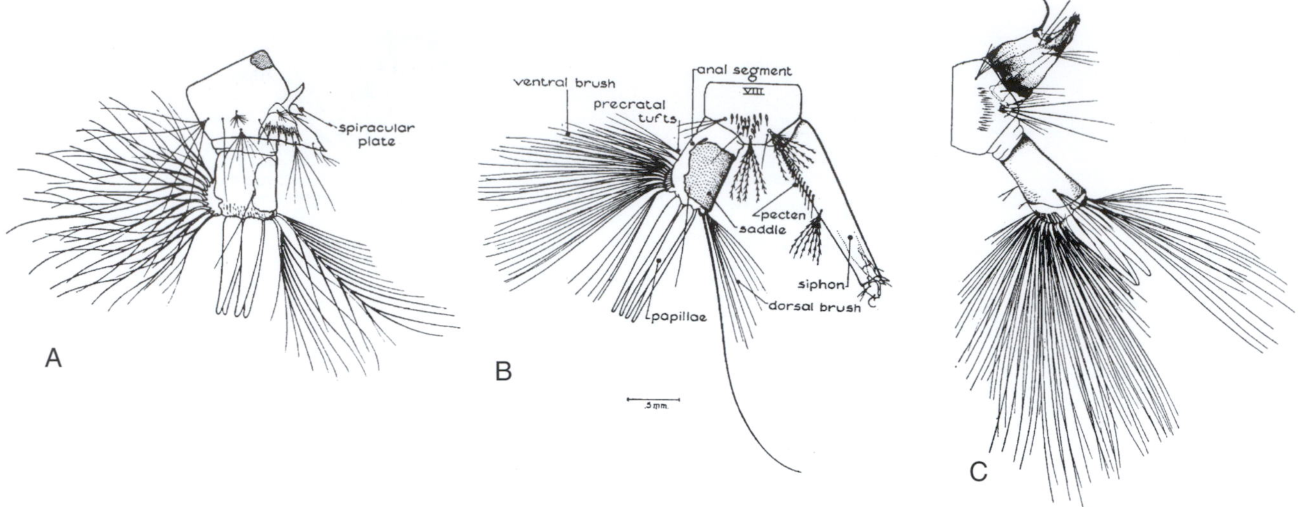

Figure 14.5 Terminal segments of mosquito larvae. (A) Anopheline form (*Anopheles earlei*), lacking a siphon; (B) typical culicine form (*Aedes fitchii*), showing elongate siphon; (C) specialized culicine form (*Coquillettidia perturbans*), showing short, stout siphon suited for piercing and clinging to plants. (Barr, 1958)

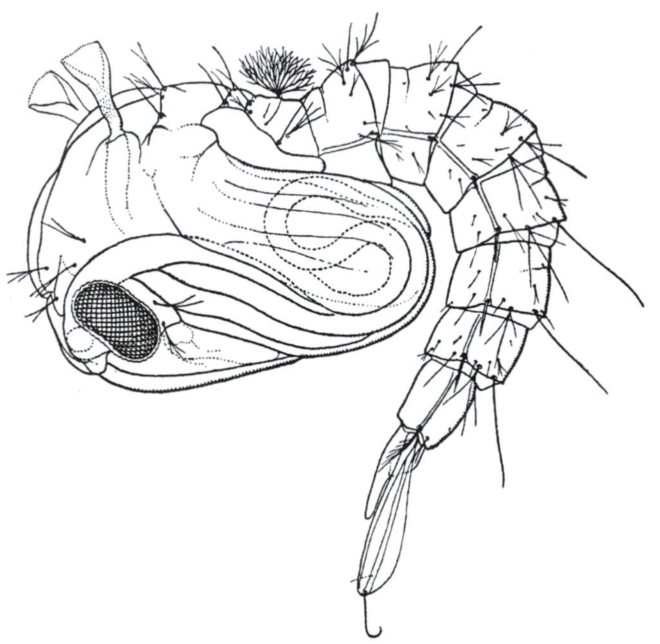

Figure 14.6 Mosquito pupa. Lateral view of *Anopheles gambiae* in resting position at water surface; presence of adult structures visible within pupal cuticle. (Smart, 1948)

of the large and conspicuous labium (Figure 14.10), which comprises the bulk of the proboscis. The tip of the labium bears two small taste-sensitive labellar lobes and a short, pointed ligula (function unknown) between them. Of the fascicle of stylets, the hypopharynx and mandibles are narrowly pointed at their tips, whereas the maxillae end in serrated blades. Both mandibles and maxillae puncture the skin and advance the fascicle into the host's tissue. A salivary channel runs

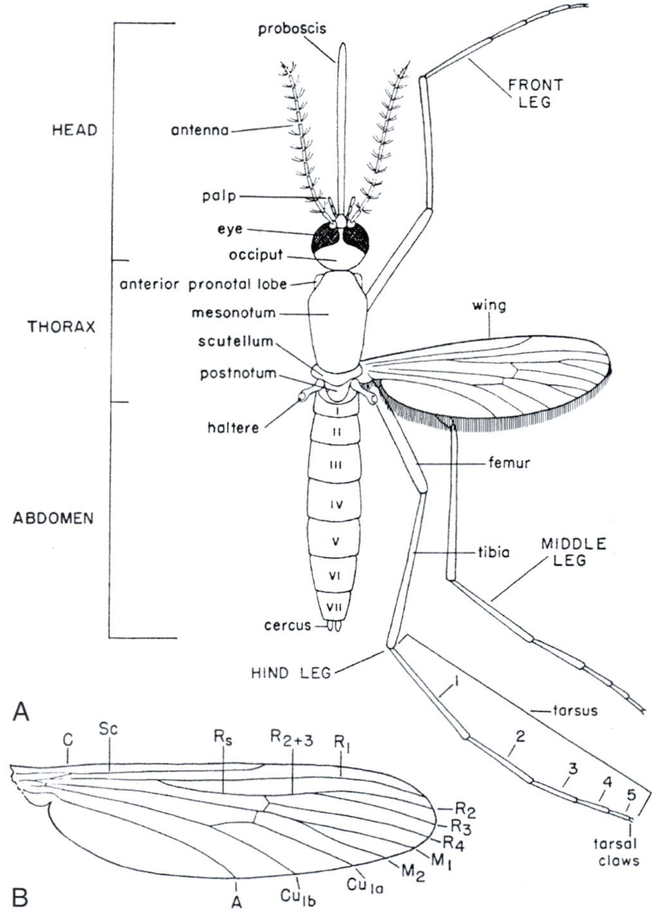

Figure 14.7 External anatomy of adult mosquito. (A) Generalized adult, dorsal view; (B) wing, showing typical venation and vein nomenclature). (Ross and Horsfall, 1965)

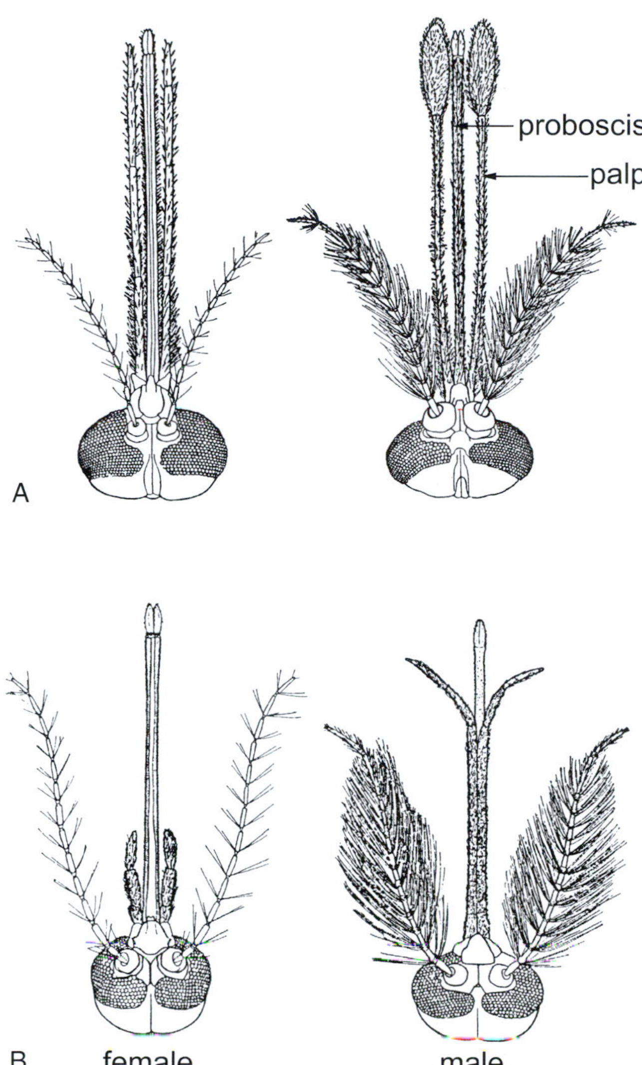

Figure 14.8 Heads of anopheline and culicine mosquitoes, females (left), males (right); males typically with plumose antennae. (A) *Anopheles* (anopheline), both males and females with palps about as long as the proboscis; male with plumose antennae and tips of palps broadened; (B) *Culex* (culicine), females typically with short palps and males with long, curved or brushlike palps. (Gordon and Lavoipierre, 1962)

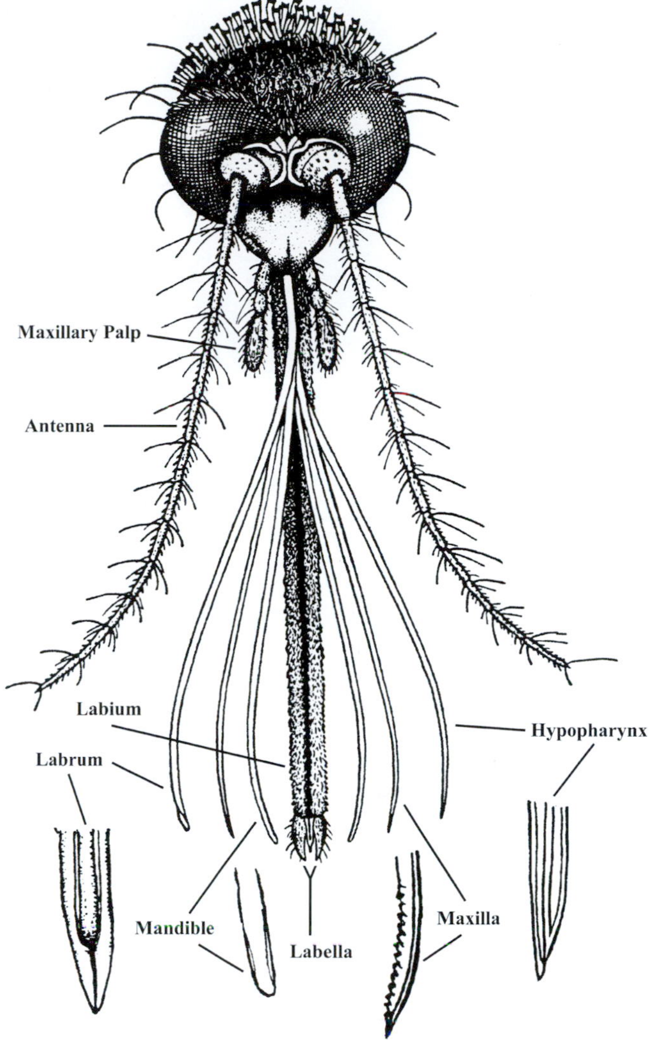

Figure 14.9 Mouthparts of adult female mosquito, showing labium, splayed stylets, and variations in structure of their tips. (Matheson, 1944)

the length of the hypopharynx, delivering saliva to the tissue during probing. The labrum is curled laterally to form a food canal for drawing the host's blood or a sugar solution up the proboscis. In males, and in females of nonblood-feeding species, the mandibles and maxillae have atrophied, so they cannot pierce skin. In both sexes of *Toxorhynchites*, the nonpiercing proboscis is curved downward (Figure 14.11). Maxillary palps arise at the base of the proboscis and bear several kinds of sensilla. Though there are many exceptions, palps usually are short in female culicines, but longer than the proboscis in most male culicines and also in both sexes of anophelines (Figure 14.8).

The mosquito thorax forms a single, relatively rigid muscle-filled locomotor unit with obscured segmentation. The meso- and metathorax each has a pair of

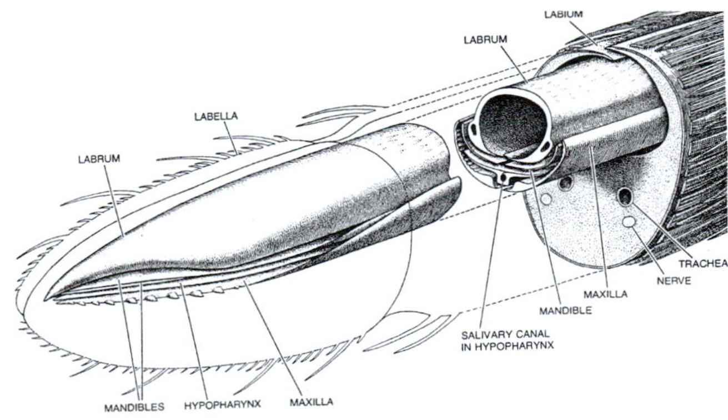

Figure 14.10 Fascicle of stylets of adult female mosquito. Mouthparts near tip of proboscis, showing natural arrangement of stylets in a single bundle, or fascicle, within a groove in the labium, which forms a sheath. (Jones, 1978; illustration by Tom Prentiss)

Figure 14.11 *Toxorhynchites amboinensis*, adult male. The form of the palps and antennae in females and males is similar to that of other culicines; however, the proboscis of both sexes is bent downward at an angle of 90° or more. (Photo by W. A. Foster)

lateral spiracles. The slender legs are attached close together on the underside of the thorax by elongate, downward-projecting coxae; the tarsi are tipped with two claws and a central pad, the **empodium**. The wings are narrow, have a distinctive pattern of veins, and bear scales along the veins and the hind margin, the latter forming a fringe. The **halteres**, tiny modified hindwings used in flight control, are located right behind the insertions of the wings.

The abdomen is clearly segmented and capable of extensive expansion and some movement, owing to the membranous areas between each set of tergites and sternites. This allows for expansion of the abdominal wall to accommodate large blood and sugar meals and developing clutches of eggs. Abdominal segments 5 through 8 are progressively smaller so that the abdomen tapers toward the posterior end. Segment 9 is quite small and bears the **cerci**, the postgenital lobe of the female, and the claspers and other genitalic structures, or **terminalia**, of the male (Figure 14.12). At emergence, the male genitalia are inverted. During the first hours of adulthood, segments 8 and 9 of males together rotate 180° to reach the mature position. The complex and varied male genitalia are a useful source of characters for species identification.

Located within the thorax are a pair of three-lobed salivary glands, whose ducts join anteriorly to form a common salivary duct that enters the hypopharynx (Figure 14.13). In males, these glands produce saliva used only in sugar-feeding; in females, some portions are devoted to sugar-feeding and others to blood-feeding. The foregut, which begins in the head with the muscular cibarium and pharynx, pumps food up the labral food canal. The tubular esophagus extends through the cervix, or neck, into the thorax. There it is modified to form three diverticula, including two small dorsal outpocketings and a large ventral crop; the crop extends through the thorax and expands to form a large sac within the abdomen. Imbibed sugar solutions are stored in these diverticula and pass, a little at a time, through the proventricular valve into the midgut. A blood meal, on the other hand, passes directly into the widened posterior midgut, or stomach (Figure 14.13). There it becomes surrounded by a semipermeable, sac-like **peritrophic membrane** secreted by the midgut epithelium. The resulting blood bolus then is digested and absorbed.

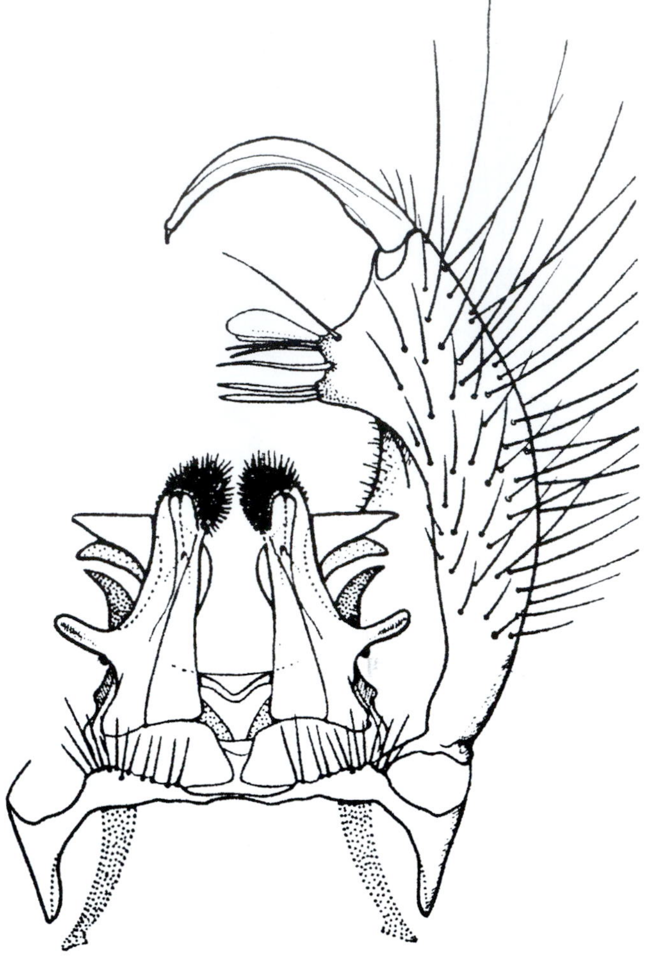

Figure 14.12 Male genitalia of *Culex quinquefasciatus*, showing principal copulatory structures used in male taxonomic identification. Gonocoxite and gonostylus on left side has been omitted. (Ross and Roberts, 1943)

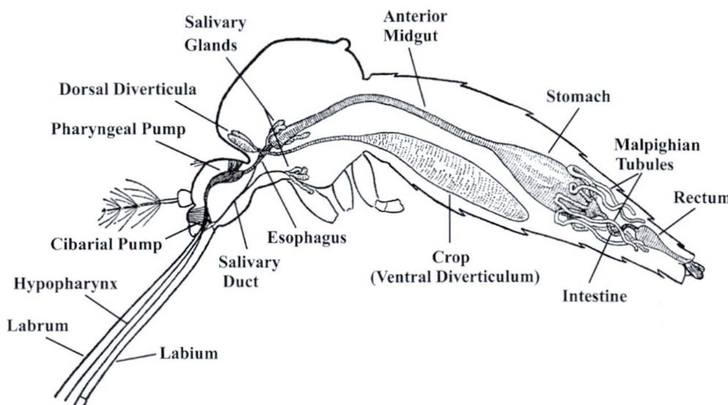

Figure 14.13 Digestive system of the adult mosquito. Semidiagrammatic view of major structures, including the salivary glands, foregut-midgut junction, and rectum. (Snodgrass, 1959)

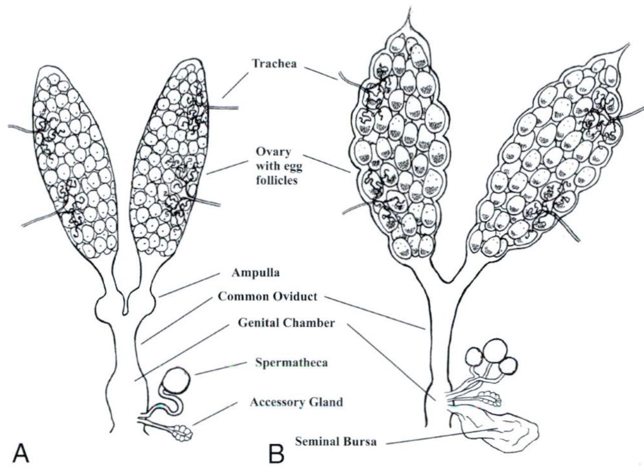

Figure 14.14 Female reproductive system of the mosquito. (A) Anopheline form, based on *Anopheles gambiae*. (B) Culicine form, based on *Aedes aegypti*. (Original by W. A. Foster)

The pyloric valve separates the midgut from hindgut; five Malpighian tubules empty into the hindgut just beyond the valve in the pyloric chamber. The anterior portion of the hindgut is tubular and loosely coiled; the posterior part is enlarged to form a bulbous rectum with large papillae projecting into it. The papillae probably are involved in resorption of salt ions.

Paired gonads are located in the posterior one-third of the abdomen. The testes of males contain packets of sperm in various stages of maturation. A duct extending posteriorly from each testis widens to form a seminal vesicle, which stores mature sperm. The two seminal vesicles lie together and unite posteriorly to form the ejaculatory duct. Two large accessory glands open into this duct, which leads to the aedeagus. In females, the reproductive system consists of a pair of ovaries and accessory structures (Figure 14.14). Each ovary includes a few dozen to over 200 polytrophic ovarioles, the egg-forming units. A duct from each ovary extends posteriorly, and the two unite to form a common oviduct, which connects to the gonopore via a genital chamber. Opening into the genital chamber by tiny ducts are one to three sperm-storing spermathecae, a small accessory gland, and the seminal bursa (lacking in most *Anopheles*), which receives semen from the male during mating.

LIFE HISTORY

The holometabolous life cycle of mosquitoes is completed in two different environments, one aquatic, the other terrestrial. The larvae and pupae develop in a wide range of aquatic habitats. These include temporary surface water (e.g., tidal pools in salt marshes, rain pools, and flood water), permanent surface water (e.g., pools, streams, swamps, and lakes), and diverse natural and artificial water-holding containers (e.g., tree holes, leaf axils, fruit husks, mollusk shells, drinking water pots, and discarded tires). An extensive analysis and classification of larval habitats has been presented by Bates (1949), Mattingly (1969), Laird (1988), and Service (1993). The only absolute requisite of all development sites is that they maintain at least a film of water for the duration of the larval and pupal periods. However, individual species tend to oviposit, and therefore develop, in sites with characteristic physical and chemical properties. Adults are much more mobile than the immatures, but they also tend to occupy characteristic resting, foraging, and overwintering habitats.

Mosquito eggs are laid either on or in water, or on solid substrates that are likely to become inundated. Females in the subfamilies Anophelinae and most of the Culicinae in the tribe Sabethini scatter their eggs individually on the water surface, whereas those in the tribe Aedini (e.g., *Aedes, Ochlerotatus, Psorophora*, and *Haemagogus*) attach their eggs individually on a substrate that later will become inundated with water. Clumped eggs are laid in boat-like rafts on the water surface by several genera of Culicinae (e.g., *Culex, Culiseta, Coquillettidia, Uranotaenia, Armigeres*, and *Trichoprosopon*) or in a radial cluster attached underwater to vegetation in the case of *Mansonia*.

At first the eggs are white, but most turn dark within hours as the chorion tans. In lowland tropics, subtropics, and summers of some temperate regions, eggs usually complete embryonic development within two to three days after being laid, but may take up to a week or more in cool climates. Larvae hatch soon after embryonation in species that lay their eggs directly in water, including all Anophelinae and most tribes of Culicinae. The best known exceptions are members of the genera *Aedes, Ochlerotatus*, and other Aedini that typically develop in temporary water. Their eggs are laid on solid substrates out of water, and the larvae within them remain quiescent until inundated. The eggs can tolerate periods of cold and desiccation and may remain viable for years. Hatching usually occurs at warm temperatures after the eggs have been submerged and microbial activity has caused the oxygen level in the water to drop.

Depending on the species and particular conditions of the water, most mosquito larvae spend most of their time either at the water surface or at the bottom of the water column, coming to the surface for air only occasionally, or not at all. At ideal conditions of food and temperature (26–28°C), the entire larval phase of *Aedes aegypti*, a tropical and subtropical mosquito, may last as few as five to six days. The first three instars are completed in about a day each and the fourth lasts about three days. In males these periods are slightly shorter, so the males pupate about one day earlier than females. Larvae of many species grow even faster, as when the water is heated by direct sunlight, whereas others develop slowly. *Toxorhynchites* and *Wyeomyia* species usually take two to three weeks even under ideal conditions. At cooler temperatures, or when food is scarce, growth becomes slower and can practically cease, with larvae remaining alive for months. Larvae of some species that inhabit high latitudes or high altitudes, or that develop in the early spring in temperate regions, have growth thresholds close to freezing and can tolerate even temporary entrapment in solid ice. This is typical of the snowpool *Ochlerotatus* species and of mosquitoes that overwinter as larvae, such as

Wyeomyia smithii in pitcher plants and *Orthopodomyia alba* in tree holes.

The pupa spends nearly all its time at the water surface. By the time it has molted to form a pharate adult within the pupal cuticle, it is very dark. In warm water the entire pupal stage typically lasts about two days in both sexes. In some mosquitoes, such as *Toxorhynchites* and *Wyeomyia* species, the shortest pupal periods may be five to six days. In all species the pupal period lasts longer at lower temperatures.

Adult males tend to emerge earlier than females, because of their shorter larval growth periods. As adult emergence approaches, the pupa remains stationary at the water surface, and the abdomen gradually straightens over 10 to 15 min. The adult emerges from the pupal cuticle by ingesting air, causing the cephalothorax to split and the adult to rise up out of the cuticle and stand on the water surface. The entire process takes only a few minutes. The newly emerged adult is capable of short flights a few minutes later but cannot sustain long flights for many hours, until after the cuticle becomes fully sclerotized. Lipids and glycogen, carried over from larval reserves, provide sufficient energy for a few days of flight and survival.

It is typically during the first three to five days of adult life that both sexes obtain sugar from plant nectar or honeydew, become sexually mature, and then mate. In some species (e.g., *Culiseta inornata, Wyeomyia smithii*, and *Deinocerites cancer*) sexual maturation is complete at the time of emergence or only a few hours later, and mating occurs almost immediately. Mosquitoes typically first feed on sugar to obtain enough energy for sexual maturation and for the flight necessary for mating, dispersal, and finding vertebrate blood. Natural sugar is taken repeatedly throughout adult life by both sexes of most species. Females typically mate only once. Males can inseminate several females before their supplies of mature sperm and accessory gland secretion become depleted. The semen supply is replenished in a few days.

Amorphous masses of fat body line the inner walls of the abdomen. The fat body synthesizes and stores both glycogen for flight and lipids for maintenance, using the digestive products of sugar and blood meals. Glycogen is stored also in the fibrillar flight muscles of the thorax, serving as a source of energy for immediate flight if the sugars in the crop and hemolymph have been exhausted.

Only females feed on vertebrate blood. In most mosquitoes, ingestion and digestion of a blood meal initiates egg development by stimulating a cascade of hormones from the brain and ovaries. The large amount of protein contained in hemoglobin and the blood serum provides the amino acids for synthesizing **vitellogenin**, the proteinaceous precursor of egg yolk. The protein also serves as the substrate for building lipid and glycogen, which contribute both to egg yolk and to the maternal energy reserves used for survival and flight. A blood meal will stimulate egg development only if it is sufficiently large and if the female's ovarian follicles have reached the resting stage, at which point they are considered to be **gonoactive**. If a female has had poor larval nutrition, the follicles may not have reached the resting stage, and she will be unable to develop any eggs until having ingested sugar or a preliminary blood meal. Such a **gonoinactive** female, needing food to bring the ovarian follicles to the resting stage, is sometimes said to be "pre-gravid." Details of the hormonal control of these processes are discussed by Brown and Lea (1990), Hagedorn (1994, 1996), and Klowden (1996).

In most species, females are **anautogenous**, the egg follicles remaining in the resting stage until a blood meal is taken. Following each blood meal, the female develops one mature clutch of eggs, exhibiting what is known as **gonotrophic concordance**. However, females of **autogenous** species or populations can develop eggs without a blood meal; among these there are obligate and facultative types. A facultatively autogenous female typically develops only the first clutch of eggs without blood; she does so only if she emerges with sufficient reserves and cannot readily find blood. Thereafter, a blood meal is required for each gonotrophic cycle. Species that are obligately autogenous have atrophied feeding stylets, never take blood, and subsist entirely on their larval reserves and plant sugar. Autogeny has been reviewed by O'Meara (1985). At the other extreme, there are some anautogenous species that take blood not only at the beginning of each gonotrophic cycle but also once or twice during egg development. These supplementary blood meals can provide extra energy, acting as a substitute for sugar (e.g., domestic *Aedes aegypti, Anopheles gambiae*).

Ordinarily, all eggs develop synchronously and become mature in two to five days after blood-feeding at favorable temperatures. During this time, the most advanced follicle within each ovariole passes through a series of five easily observed physiological stages (Figure 14.15), originally described by Christophers (1911) and concisely summarized by Clements (1992).

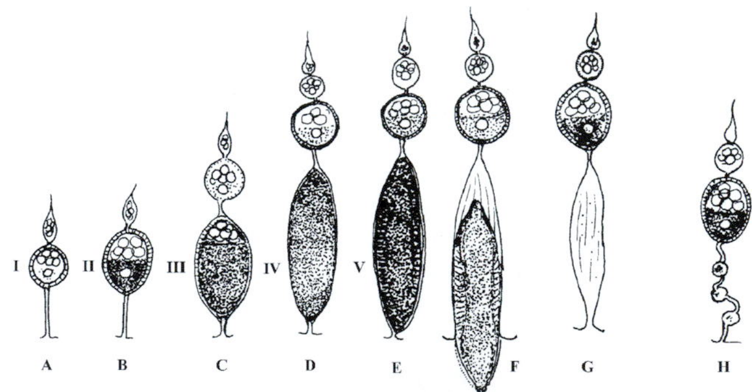

Figure 14.15 Egg follicle development in mosquitoes, showing stages in the development of an egg follicle within a single ovariole of the ovary. (A)–(E), Stages I–V of Christophers in an *Anopheles* female. After oviposition (F), the ovariole stalk remains swollen for a while, called the "sac stage" (G) Note: Formerly, the sac was thought to shrink to a dilatation. Now it is believed that the dilatations (three of them shown in H) generally form from follicles that fail to develop into eggs after a blood meal and are then resorbed. (World Health Organization, 1975)

These stages comprise four physiological phases. The follicles develop synchronously, beginning with the **previtellogenic phase** (stages G through II). The follicle typically stops growing at stage IIa or IIb, the **resting stage**, until the female takes a blood meal. Within hours of blood-feeding, the follicle enters the **initiation phase** (stage IIIa), when yolk (vitellogenin) synthesis by the fat body begins, followed by the **trophic phase** (stages IIIb and IV), the main period of yolk incorporation and follicle growth. In the **post-trophic phase**, the egg has reached its mature shape, and the **chorion**, or egg shell, is formed. The eggs are then ready to be oviposited. By this time, the next follicle in each ovariole either has progressed to stage I or already has reached the resting stage (stage II). In the first case, it awaits oviposition before developing to the resting stage. Once oviposition has occurred and these new follicles are in the resting stage, the next blood meal is necessary for subsequent follicle development. The entire cycle of egg production, from blood meal through oviposition, is the **gonotrophic cycle**.

The number of ovarioles that produce mature eggs depends on the sizes of the female body, energy reserves, and blood meal. The follicles that do not pass beyond stage II degenerate. When most or all of the ovarioles contain mature eggs, the ovaries may occupy nearly the entire volume of the distended abdomen. The gravid female fertilizes one egg at a time, as each passes down the oviduct to be oviposited on the water or a damp substrate.

When all eggs have been fertilized and expelled during oviposition, the ovaries return to their pretrophic size, but the tracheae on the ovaries, which had been tightly coiled as **tracheal skeins** before the eggs developed, become stretched and straightened. In *Anopheles* species, a swelling at the base of each lateral oviduct, the **ampulla**, becomes permanently stretched during the first oviposition. These signs serve to distinguish **parous** females, those that have completed at least one gonotrophic cycle, from **nulliparous** females, those that have not.

The number of completed gonotrophic cycles also can be determined. According to current interpretations, each ovariole ovulating a mature egg is left with an egg sac, which becomes reduced to a zone of granules in the calyx, the ovariole's connection to the lateral oviduct. Furthermore, a dilatation is formed in the stalk of each ovariole where a follicle has degenerated after a blood meal, instead of developing into an egg (Figure 14.15). Thus, a count of the maximum numbers, per ovariole, of dilatations in the stalk and zones of granules in the calyx yields an estimate of the number of gonotrophic cycles completed. This **physiological age grading** can provide the medical entomologist with valuable information on the age of individuals and the age structure of a mosquito population. Details of these processes and their interpretation and application are given by Detinova (1962), Sokolova (1994), Fox and Brust (1994), and Hoc (1996).

Univoltine mosquito species complete only one generation per year. This occurs either if the developmental time is slow in relation to the season favorable for development or if the life cycle includes an obligate form of diapause, a compulsory phase of arrested development. **Bivoltine** and **multivoltine** species can complete two or more generations, respectively, during each breeding season, but the number actually completed may depend on temperature, available larval habitats, or available hosts. Mosquitoes pass through the winter or dry season as eggs, larvae, or adults, depending on the species and the climate. In cold climates, overwintering takes place in a state of diapause.

BEHAVIOR AND ECOLOGY

Eggs that are laid on or in water generally are not resistant to desiccation and hatch shortly after embryonation, provided that they are wet and not too cold. This is typical of *Anopheles, Culex, Culiseta,* and *Toxorhynchites* species. *Aedes, Ochlerotatus, Psorophora,* and *Haemagogus* eggs, on the other hand, typically are laid on damp substrates, display great resistance to desiccation, and remain quiescent for months or years after embryogenesis until they receive a hatching stimulus. Sometimes moisture by itself is sufficient to induce hatching. Usually, however, the requisite stimulus is a reduction of dissolved oxygen in the water caused by microbial activity and decomposition of organic matter. Among quiescent eggs that are eventually submerged, only a portion of a single egg clutch may hatch during any one inundation, resulting in **installment hatching**. This apparently is the combined result of intrinsic variations among eggs in their hatching-stimulus thresholds and of local variations in microbial activity, causing differences in oxygen tension around the eggs. Even during a single inundation, hatching may not occur all at once but over a period of many days.

When an egg is ready to hatch, the first-instar larva uses a dorsal hatching spine on its head, the egg breaker or egg burster, to apply pressure to a preformed weakness in the chorion. This causes the chorion to pop open at one end, and the larva wriggles free. Because the eggs of *Culex, Culiseta,* and *Coquillettidia* usually stand vertically on the water surface in rafts, the larvae develop inside them with their anterior end oriented downward and hatch directly into the water.

Mosquito larvae are not buoyant and must, at rest, be suspended at the surface by special hairs and spiracular structures that cling to the surface tension while obtaining oxygen directly from the air. Culicinae typically migrate up and down in the water column, so they occur both at the surface and at the bottom of a body of water, depending on the availability of food. At the surface the tip of the siphon opens above the surface film, as the larvae hang diagonally downward most of the time (Figure 14.16B). *Mansonia, Coquillettidia,* and some *Mimomyia* species are unusual in remaining submerged throughout larval and pupal development, with their siphons embedded in the tissues of aquatic plants from which they derive some oxygen (Figure 14.16C). Mosquitoes that live in water-filled leaf axils (e.g., *Wyeomyia* spp.) are adept at flattening themselves against vertical surfaces and maneuvering in narrow spaces. Anopheline larvae spend most of their time at the water

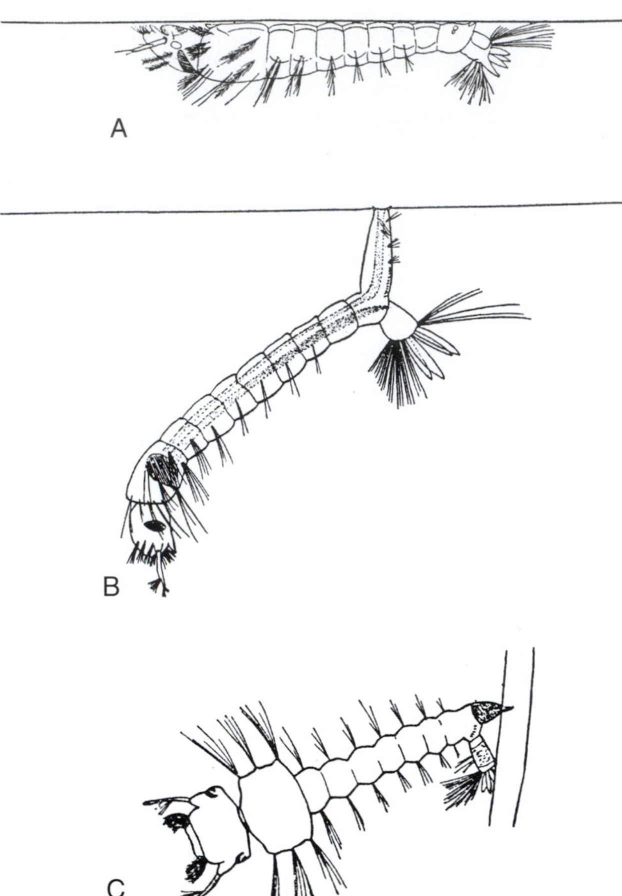

Figure 14.16 Resting and feeding positions of mosquito larvae. (A) *Anopheles*, showing horizontal position at the water surface, dorsal-side up; larva has rotated its head 180° so that the ventral side of the head is uppermost and the mouthparts are applied to the water's surface for filter feeding on floating detritus; (B) *Culex*, showing typical diagonal position while suspended from water surface; (C) *Mansonia*, attached to submerged part of aquatic plant (stem or root) by its siphon. (A and B from Ross, 1947; C from Gordon and Lavoipierre, 1962)

surface, often close to vegetation or floating material. They are able to remain suspended horizontally at the surface (Figure 14.16A) due to pairs of dorsal palmate setae (float hairs) on several abdominal segments (Figure 14.3A).

Larvae propel themselves by a back-and-forth lashing movement of the abdomen. Anopheline larvae usually swim horizontally at the surface film. When larvae of typical culicine mosquitoes are feeding below the surface, they periodically swim actively back to the surface to obtain oxygen. However, in many microenvironments dissolved oxygen also is absorbed from the water through the cuticle, requiring infrequent trips to the surface by some species.

Mosquito larvae feed on a variety of organic detritus, suspended material, and small organisms in their aquatic habitats. The organisms include bacteria, protists, fungi, algae, microinvertebrates, and small macroinvertebrates; the organic detritus usually consists of dead plant material and dead macroinvertebrates. They collect these food items in five basic ways: filtering, gathering, scraping, shredding, and preying. Filterers generate water currents with their lateral palatal brushes on the labrum, drawing suspended particles though fine combs where they are collected and directed to the mouth. Gatherers use their mouthparts in a similar manner, but only after stirring up the particles from solid surfaces. Scrapers obtain food by scraping it off solid surfaces, whereas shredders gnaw, chew, and bite off pieces of organic matter. Predators grasp insects and other small, mobile prey in their large and sharp mandibles or maxillae (e.g., some *Psorophora* spp.) or with long, curved palatal brushes (e.g., *Toxorhynchites*) (Figures 14.4C, 14.17). Most species use more than one of these techniques. *Anopheles* primarily filter-feed at the water surface by rotating their heads 180° so that the oral opening becomes dorsal. Many *Aedes*, *Ochlerotatus*, and *Culex*, on the other hand, filter-feed near the surface but also gather, scrape, or shred organic matter at the bottom, depending on food availability. *Coquillettidia* and *Mansonia*, which are anchored to submerged vegetation, employ a combination of filter-feeding, gathering, and scraping techniques within their immediate surroundings. Larval feeding has been reviewed by Merritt et al. (1992).

Figure 14.17 *Toxorhynchites amboinensis* larva feeding on larva of *Culex pipiens*. This predaceous species has been used in biological control trials and naturally exerts a damping effect on populations of pest and vector mosquitoes. (Photo by W. A. Foster)

Mosquito pupae normally remain motionless at the water surface with the tips of their thoracic air trumpets in contact with the air. Like larvae, they dive when disturbed, propelling themselves with their caudal paddles by extending the abdomen, then snapping it back inward toward the cephalothorax. Pupae of most species are buoyant, due to the ventral air space beneath the cephalothorax, and rise to the surface without swimming. They remain submerged by repeatedly swimming downward or by wedging or lodging themselves under debris. After sufficient submergence time, they lose their buoyancy as their air supply dwindles, and they must swim actively to the surface. Pupae of a few mosquitoes (e.g., *Limatus* spp.) are never buoyant and can keep from sinking only by clinging to the surface film, much as most mosquito larvae do. The plant-piercing pupae of *Mansonia* and *Coquillettidia* species do not rise to the water surface until they release their attachment to plants when ready for adult emergence.

Upon emergence from the pupal stage, adults typically seek shelter in vegetation, cavities, and other resting sites, where they remain except during periods of activity. When resting, they typically are positioned head-up on vertical surfaces, with forelegs and midlegs on the substrate and hindlegs raised (Figure 14.18). Culicines hold the abdomen in various positions, but the proboscis is always at an angle to it (Figure 14.18A). Anophelines, on the other hand, hold the proboscis and abdomen in line, oblique to the substrate (Figure 14.18B). This distinctive position also is apparent during feeding. While at rest, adults perform various stereotyped grooming movements and frequently wave their hindlegs.

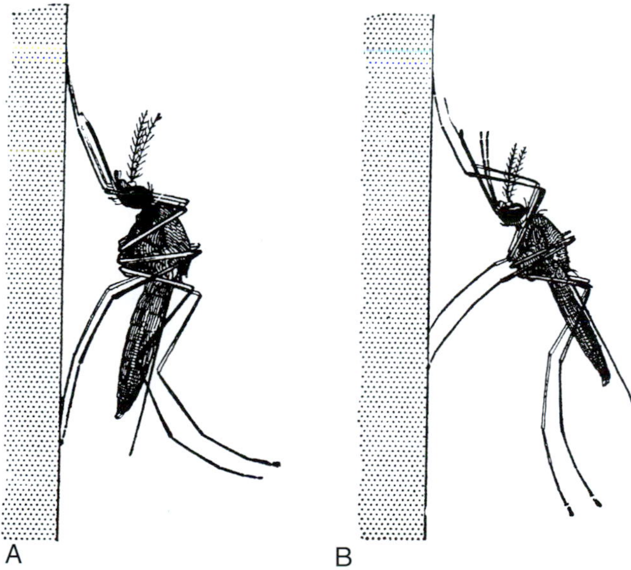

Figure 14.18 Comparison of typical positions of mosquito adults resting on a vertical surface. (A) Culicine; (B) anopheline. Culicines typically hold the abdomen parallel to the substrate or pointed toward it, and the proboscis and abdomen form an angle. Anophelines characteristically tilt the body at a sharp angle to the substrate, with the abdomen pointing away from it, and with the proboscis and abdomen in line. (Marshall, 1938)

Each mosquito species has a characteristic pattern of diel activity, under the control of an endogenous circadian rhythm that is entrained by the daily light-dark cycle. Generally one or two flight periods occur each 24 hours, characterized as being diurnal, nocturnal, or crepuscular (dawn and dusk). During these periods, both sexes will take flight without external cues. Mosquitoes likely engage in a generalized search pattern during foraging flights, then respond to specific stimuli associated with either mating sites, sugar sources, hosts, or oviposition sites as they encounter them, depending on their needs. Mosquito species vary in the habitats where they forage for mates and food. Some fly over varied terrain; others tend to be active in either wooded or open areas; still others perform all activities close to larval and resting sites. There is some evidence that adult females become familiar with local habitats that provide both food and oviposition sites and tend to remain there.

Dispersal of some mosquitoes is only a few dozen meters from their larval habitats. Most species fly less than 2 km in a lifetime. Such ranges are typical of domestic species and result from random elements in their repeated foraging flights for hosts, sugar sources, resting sites, and oviposition sites. Other species enter a specific dispersal mode that is wind-assisted or light-directed and carries them dozens or even hundreds of kilometers from their origins. These one-way movements are most obvious following massive adult emergences of species with quiescent eggs, such as the salt-marsh mosquito *Ochlerotatus taeniorhynchus* after high tides, and floodwater mosquitoes such as *Aedes vexans* in bottomlands after heavy rain. The average dispersal distance of such broods is difficult to determine accurately, because efforts to recapture marked specimens must be made over vast areas, and relatively few are caught. Some salt-marsh species make extended, round-trip migrations to complete their gonotrophic cycles when development sites and blood-feeding sites are many kilometers apart. Average flight speed also is difficult to determine under natural conditions. *Aedes aegypti* can fly upwind at air speeds up to 5.4 km/hr, whereas other species in a dispersing mode are estimated to fly much faster. However, their ground speed drops nearly to zero as headwind velocity approaches their maximum air speed. Thus mosquitoes tend to avoid flight under windy conditions, except when a tailwind assists their flight. Mosquito dispersal in its various forms has been reviewed by Service (1997).

Mating usually takes place a few days after adult emergence. The males typically form flight swarms at particular times at swarm markers (prominent objects or other contrasting features of the environment). Each male follows a looping flight path over the marker. In species such as *Aedes aegypti* and *Ae. albopictus*, the females' preferred host serves as the swarm marker. When a female enters a swarm, males detect the characteristic frequency of her wing beat and her position with their plumose antennae and Johnston's organs. Her tone varies from about 150 to 600 Hz, depending on the temperature and her size and species. It is about 100–250 Hz lower than the males' flight sound. The male turns toward the female, pursues her, and couples with her. Swarms are usually species-specific, but mixed swarms occur. If the female is of another species, males either do not respond to her flight tone or release her upon detecting that she lacks the appropriate species-specific contact pheromones. Otherwise, he may attempt copulation. Successful copulation usually occurs only with the first conspecific male to orient to the female, venter-to-venter, clasping her genitalia in his. The couple drifts or flies from the swarm, often shifting to an end-to-end position, with the female flying forward and the male facing backward, clinging to her only by his genitalia. Mating may be completed in the air or on vegetation. Copulation lasts from 12 seconds to several minutes.

There are many exceptions to this standard method of mating. Males of *Deinocerites* and *Opifex* guard pupae at the water surface and mate with the females as they emerge. Males of *Culiseta inornata* remain at the emergence site for long periods and locate newly emerged females at random while crawling about, recognizing them by a specific contact pheromone on their legs. A male *Eretmapodites chrysogaster* will follow a female in tandem flight to a host, wait beside her while she takes a blood meal, then copulate when she is finished. Males of the sabethine genera *Sabethes, Limatus, Wyeomyia,* and *Topomyia* locate females at rest on vines, sticks, and tree trunks and perform a variety of leg-waving and genitalic courtship rituals before insemination. Several aspects of mosquito mating behavior have been reviewed by Downes (1969).

During copulation the male deposits a mixture of sperm and accessory gland secretion in the female's seminal bursa or genital chamber. The semen often produces a distinct seminal mass in the bursa or genital chamber, sometimes called a **mating plug**. It disappears in one to two days and therefore does not, by itself, prevent subsequent insemination. Within an hour the sperm move into the spermathecae where they are stored. At least in culicine mosquitoes, the swollen bursa itself and, later, a substance in the accessory fluid called **matrone**, cause the female to become unreceptive to males. In anophelines, a spermatheca filled with sperm is responsible for a female's unreceptive behavior. Substances in the accessory fluid also can affect feeding behavior and promote egg development and oviposition. A single insemination usually is sufficient for the life of a female. Most evidence indicates that females only rarely receive sperm from more than one male.

Adult mosquitoes of both sexes of most species regularly feed on plant sugar throughout life, but only females feed on vertebrate blood. Water presumably is taken from the surface of moist substrates as well as in sugar meals and blood meals. Field studies and experiments indicate that some species are guided in their foraging flights by specific visual features along the horizon, or they fly along the edges of tree stands bordering open terrain. Others apparently simply fly crosswind or downwind, depending on wind speed. When they can see likely sources of sugar or blood, they alter their flight paths to move directly toward the object. If they detect odors of flowers or hosts in the absence of visual information about a food source, they turn to fly upwind within the downwind drifting odor plume, eventually arriving at the odor source.

Sugar feeding starts soon after emergence, usually before females begin responding to host stimuli. Sugar is taken frequently by both sexes throughout adult life, both between and during gonotrophic cycles. Females of some domestic species may take sugar infrequently or never (e.g., *Aedes aegypti* and *Anopheles gambiae* in some localities); they typically live in close association with their hosts and utilize blood for both energy and reproduction. Sugar sources include floral and extrafloral nectaries, homopteran honeydew, spoiled or damaged fruit, tree sap, and damaged or even undamaged plant leaves and stems. *Malaya* species solicit regurgitated nectar and honeydew from ants. Nectar and honeydew are the most important sugar sources. They contain not only sugar but also amino acids; these are insufficient to initiate egg development but probably promote longevity. Most mosquitoes obtain nectar from a variety of plant species; others seem to be fairly specific in their choices, and a few are important pollinators of particular plant species whose flowers they visit. Mosquitoes generally feed from fragrant, light-colored, clustered flowers with short corollas that allow easy access to the nectar. Details of sugar feeding have been reviewed by Foster (1995).

Female mosquitoes rarely begin responding to vertebrate hosts and taking blood meals until at least one to three days after adult emergence and often not until after mating and sugar feeding. Their hosts include all classes of vertebrates: mammals, birds, reptiles, amphibians, and even amphibious fish. They have been reported to take hemolymph from other insects, but perhaps this occurs only when the insects have been contaminated with vertebrate odors. Host specificity and host preference vary widely. Some species feed almost entirely on members of one genus of animal, others opportunistically attack members of two or three vertebrate classes. Blood meal identification methods, used to determine mosquito-host specificity, have been reviewed by Washino and Tempelis (1983) and Clements (1999). Host specificity is a function of both the mosquito's innate host preference and the hosts available to the mosquito when and where it is active. Some species forage over a broad range of habitats. Others are active principally in either wooded or open areas or remain close to sites of larval development or adult resting. Still other species attack their hosts in rather narrowly defined zones within a habitat. For example, within tropical forests, different species feed in different strata: ground level, intermediate levels, and just below the leaf canopy.

Host-finding behavior in mosquitoes involves the use of volatile chemicals to locate vertebrate hosts. Carbon dioxide, lactic acid, and octenol are among the best-documented host attractants. Other yet-to-be-identified skin emanations also are known to be important, because odors from live hosts are always more attractive than any combination of these chemicals in a warm, humid airstream. Fatty acids produced by the normal bacterial flora of the skin are particularly effective in attracting *Anopheles gambiae* to human feet. Mixtures of these fatty acids probably play a major role in attracting most mosquitoes. Subtle differences in these odors of different host species and even different individuals undoubtedly play a role in host preference. These odors commonly have a combined effective range of 7 to 30 m, but up to 60 m for some species. Vision also is important in orienting to hosts, particularly for diurnal species and especially in an open environment and at intermediate or close ranges. Dark, contrasting, and moving objects are particularly attractive. As the female approaches to within 1 to 2 m of a potential host, chemical and visual cues are still important, but convective heat and humidity surrounding the body also come into play. Odor, carbon dioxide, heat, and humidity all are detected by sensilla on the antennae and palps. Host-finding behavior in mosquitoes was reviewed by Bowen (1991). Specific behavioral and physiological aspects of attraction have been reviewed by a series of authors in the proceedings of a symposium (Anon., 1994).

If the suite of host stimuli is acceptable, the female attempts to land on the host animal, often preferring certain body parts, such as the head or legs. Upon landing, she proceeds through four phases of feeding behavior: exploration, penetration and vessel-seeking, imbibing, and withdrawal. She typically remains motionless for a few seconds, then begins exploratory movements, including contacting the skin surface with her proboscis in probing motions. If the host is not suitable, she may wander for a considerable time and leave without feeding. Even on a suitable animal she usually explores at least briefly before selecting a spot that is likely to be well vascularized. Probing activity is stimulated by heat and moisture, and probably also by chemicals on the surface of the skin. As in the case of airborne attraction, these stimuli are detected by antennal and palpal receptors, but receptors on the proboscis, tarsi, and elsewhere on the legs apparently also are important.

Mosquitoes can feed from a variety of skin surfaces, including the moist skin of frogs and the scaly legs of reptiles and birds. They can penetrate mucus, matted hair, light layers of feathers, and heavy cloth such as denim, provided it is not thicker than the length of the proboscis. Once a feeding site is selected, the fascicle of stylets pierces the skin and the labium serves as its guide and is bent backward without penetrating (Figure 14.19). The maxillae and mandibles on each side of the fascicle alternately slide by each other in quick stabbing/puncturing movements. While doing so the tissue is gripped with the backward-directed maxillary teeth as the stylets penetrate epidermal and subepidermal tissue.

Saliva flows from the tip of the hypopharynx as the flexible end of the fascicle bends at sharp angles, probing in various directions within the subepidermal tissue in search of a small arteriole or venule. The saliva contains an anti-hemostatic enzyme, apyrase, which inhibits platelet aggregation and causes randomly punctured vessels to bleed freely into the surrounding tissue spaces. This makes it easier for the mosquito to locate a vessel and shortens the total time on the host. The saliva also contains anticoagulants, which facilitate vessel location and blood ingestion by preventing the blood from clotting. Sensilla on the labrum and in the cibarium apparently detect plasma and cellular factors, including adenyl nucleotides such as ATP, which help the mosquito to locate a blood vessel and stimulate ingestion. Upon finding a vessel, the female slips the tip of her fascicle into the lumen and draws blood up through the food canal by pumps in the cibarium and pharynx. The blood accumulates in the midgut, allowing the mosquito to engorge fully in one to four minutes. During this time, the female begins to extract water from the blood meal and may deposit small droplets of urine on the host's skin. In *Anopheles* species, this urine is copious, includes some blood cells from the accumulating meal, and appears red. When

Figure 14.19 Female mosquito (*Haemagogus lucifer*, a vector of jungle yellow fever), showing position of mouthparts during blood-feeding. The fascicle is exsheathed from the labium along part of its length while the labium buckles backward, allowing the fascicle to enter the skin in search of a blood vessel. (Photo by W. A. Foster)

abdominal stretch receptors signal the presence of sufficient blood in the midgut, the female pushes with her forelegs to withdraw her stylets and flies away. Usually she is too heavy to fly far until a substantial amount of water and salt in the blood meal has been excreted, after one to two hours.

While digesting the meal and developing eggs, females locate species-characteristic resting sites and may remain there until the eggs are mature. However, females of many species are known to leave their resting sites during each daily activity period throughout the gonotrophic cycle. These flights allow them to obtain sugar meals or supplementary blood meals, to relocate closer to an oviposition site, or perhaps simply to find a more suitable resting site. In at least some species, a hormone from the ovaries in the trophic phase causes the head to release a neuropeptide that inhibits the mosquito's responsiveness to host attractants, by blocking host-odor receptors on the antennae, provided she has substantial energy reserves (see Klowden, 1996, for a concise review).

Oviposition generally occurs during the same part of the day as mating and feeding. Gravid females locate and evaluate suitable sites by using chemical and visual cues, including organic chemicals, salts, high humidity, dark cavities, and reflective surfaces. The organic chemical cues are derived from decaying organic matter, microorganisms, the chemical byproducts of larvae or pupae that have previously developed there, and the presence of mosquito eggs that have been deposited by other females. The apical droplets on the eggs of *Culex quinquefasciatus* contain an oviposition-attractant pheromone.

Within each genus of mosquito, there is considerable variation among species in their oviposition-site preferences, and therefore their larval habitats. In general, *Anopheles* species occur in permanent or semipermanent water, such as the edges of lakes, ponds, streams, and pools; others develop in temporary rain puddles, leaf axils, and tree holes. *Culex* typically lay eggs in permanent or semipermanent pools, ponds, and water containers. Several medically important species of both *Anopheles* and *Culex* develop in large bodies of surface water and take advantage of irrigated fields and of reservoirs created by dams. *Culiseta* are found in several kinds of permanent surface pools; some species have very narrow requirements. *Coquillettidia* and *Mansonia* oviposit in permanent bodies of water that contain floating or emergent aquatic plants to which the submerged immatures can attach. *Aedes, Ochlerotatus, Psorophora,* and *Haemagogus* species lay their eggs on damp surfaces where they will be inundated by temporary water or a rising water level. *Aedes* species and ecologically similar mosquitoes form two general categories, according to typical habitat: (1) floodwater mosquitoes, which include floodplain species, saltmarsh species, and snowpool and spring species (some floodplain species have become prolific in rice fields and other forms of irrigation); (2) container mosquitoes, including leaf-axil species, tree-hole species, and artificial-container species. Several medically important species utilize both tree holes and artificial containers. Among genera of minor importance, *Toxorhynchites* lay their eggs only in natural and artificial containers in wooded areas; *Wyeomyia* oviposits primarily in leaf axils; and *Deinocerites* oviposits exclusively in crab holes in tidal mudflats and mangrove swamps.

The distribution of mosquito eggs reflects the availability, size, and stability of the larval habitats used by a species. Though mosquito life histories are highly variable, the oviposition behavior of mosquitoes tends to follow along taxonomic lines. Most mosquitoes fall into one of three behavioral categories:

(1) **Eggs laid out of water**. Species in this group may distribute eggs of a single clutch individually among several widely scattered potential development sites, particularly if those sites are common but small. Container species such as *Aedes aegypti, Ae. albopictus,* and *Haemagogus* deposit the eggs at varying distances above the water line, and at least *Ae. aegypti* lays only a portion of the clutch in each water container, sometimes called "skip oviposition." Floodwater species such as *Ae. vexans, Ochlerotatus dorsalis,* and the saltmarsh-inhabiting *Oc. taeniorhychus* generally scatter their eggs widely over areas where water will accumulate, inserting them into crevices of drying mud or plant debris in low ground.

(2) **Eggs placed on or in water**. Mosquitoes in this category lay the entire clutch in a clump at one site while standing on the water surface or on floating vegetation. The egg rafts of *Culex, Culiseta, Coquillettidia,* and *Uranotaenia* species are formed between the female's hindlegs as she deposits each egg on end in the water, one against the next. Some *Armigeres* species suspend the egg raft above the water with their hindlegs while forming it and then carry it with them before placing it on the water. *Trichoprosopon digitatum* females stand guard over the raft until the eggs hatch. *Mansonia* species prepare their egg clusters underwater, attached to a plant, while standing on the floating leaves of aquatic plants. Exceptional species in various genera deposit egg rafts on top of floating vegetation, lay their eggs singly underwater on the sides of rock pools, or enter beetle holes in bamboo and extrude the eggs in ribbons.

(3) **Eggs dropped onto water**. Species in this group oviposit aerially while hovering. *Anopheles* drop all of them at one site or distribute them among several smaller sites. *Toxorhynchites* and most culicine species in the tribe Sabethini (e.g., *Wyeomyia* and *Sabethes*) propel a few eggs into each of many container habitats with a flick of the abdomen, often through small openings in tree limbs or bamboo.

If a mosquito cannot find suitable oviposition sites when the eggs are mature, it may lay them in suboptimal situations or retain them until a suitable site is found. Retained eggs gradually lose their viability over several weeks or months. An extensive review of oviposition behavior is given by Bentley and Day (1989).

Mosquito dormancy occurs in all but those tropical and subtropical habitats that provide conditions for year-round larval development. The life stage that becomes dormant depends on the severity of a region's winter or dry season and also on the species. Species of *Aedes, Ochlerotatus, Psorophora,* and *Haemagogus,* which all have quiescent eggs, typically overwinter (hibernate) or oversummer (aestivate) as eggs. Larvae serve as the dormant stage in mosquitoes whose adult activity is precluded seasonally but whose breeding sites are protected from severe cold or complete drying. When adults overwinter as the dormant stage, typically in *Anopheles* and *Culex,* they seclude themselves in well-protected harborages, or **hibernacula**. Prior to dormancy, mosquitoes often enter **diapause**, a physiological state of arrested development that is induced or broken only by specific environmental cues. **Facultative egg diapause**, a feature of multivoltine species, is induced by exposure of the pupae or adult females to lowered temperatures and short photoperiods. They lay diapausing eggs, which will not hatch until the daylength is appropriate, even during unseasonably warm periods in autumn, winter, or early spring, when the resulting larvae and adults might not survive. **Obligate egg diapause** occurs in univoltine species, regardless of preceding conditions, and is maintained despite warm, long-day conditions. This is typical of snowpool and spring *Ochlerotatus* species in cold and temperate climates. Diapause is broken after the eggs have been subjected to winter conditions (in the case of obligate diapause) and when favorable temperatures and long days resume. **Larval diapause** is similar to facultative egg diapause in its induction and termination. Diapausing larvae feed and grow little or not at all, and they do not molt.

Temperate species destined to overwinter as adult females emerge in a state of **reproductive diapause** induced by larval and pupal exposure to shortening photoperiod and cool temperatures. Although these females mate, their egg follicles do not reach the resting stage, despite frequent sugar feeding and accumulation of extensive fat reserves. Fattened female *Culex* species that hibernate through hard winters forego all further feeding until the onset of spring, whereas in milder climates they periodically leave their overwintering sites to take sugar meals. Although diapausing *Culex* adults rarely feed on blood, some *Anopheles* species may take blood meals fairly regularly from hosts near these sites. They develop no eggs, however, exhibiting gonotrophic dissociation. Other overwintering *Anopheles* continue to feed and develop eggs, but these are not laid. Similarly, some tropical *Anopheles* take blood repeatedly during the dry season and remain continually gravid because there is nowhere to oviposit. These phenomena are sometimes referred to as **gonotrophic discordance**, a term that also applies to the taking of nonvitellogenic or otherwise supplementary blood meals, mentioned previously.

PUBLIC HEALTH IMPORTANCE

Mosquitoes are of public health significance because they feed on human blood. Blood feeding compromises skin, presenting the possibility of secondary infection with bacteria. It introduces foreign proteins with saliva that stimulate histamine reactions, causing localized irritation, and that may be antigenic, leading to hypersensitivity; and it allows for acquisition and transmission of microorganisms that cause infection and disease in humans, domestic animals, and wild animals. Mosquito-borne diseases are caused by three groups of pathogens: viruses, malaria protozoans, and filarial nematodes. Mosquitoes are not known to transmit pathogenic bacteria to humans, with the exception of mechanical transmission of the causative agents of tularemia (*Francisella tularensis*) and anthrax (*Bacillus anthracis*).

Mosquito Bites

In addition to the tremendous impact of mosquitoes on human health as vectors of disease pathogens, the bites themselves are important. Aside from the annoying flight and buzzing sound, a single bite can be irritating and a distracting nuisance. In Rangoon, Myanmar, *Culex quinquefasciatus* has been estimated to have densities of 15 million per square kilometer, and residents in poor districts receive 80,000 bites by this species per year. In Burkina Faso, in West Africa, residents of cities are estimated to experience 25,000 bites by this species per year. In northern Canada, the spring melt of snow brings with it hordes of snowpool *Ochlerotatus*; counts on an exposed human forearm can be as high as 280 to 300 bites per minute. It has been estimated that this rate of biting could reduce the total blood volume in a human body by half in 90 minutes, unless protective measures are taken.

As with the other blood-feeding arthropods, the wound created at the bite site may allow secondary infection by bacteria, which can be exacerbated by scratching. In the absence of prior exposure to mosquitoes, a bite rarely produces more than a temporary tingling or burning sensation and sometimes a tiny spot of blood on or just beneath the surface of the skin. After one or more previous exposures to mosquito bites, the proteins in mosquito saliva, which are injected both before and during feeding, normally stimulate development of immunity so that subsequent bites give rise to one or both of two general kinds of allergic response: immediate reactions and delayed reactions. The immediate reaction, called Type I Hypersensitivity, is an inflammation of the skin known as **wheal-and-flare**. It usually starts within minutes of the bite and lasts a few hours at most. The typical delayed reaction, designated Type IV Hypersensitivity, involves a cellular immune response caused by lymphokines that are secreted by antigen-sensitized T cells. Both delayed and immediate reactions result in itching, redness, and swelling. The typical delayed reaction takes about one day to develop, may last for up to a week, and tends to result in a larger wheal with a deeper discoloration.

Mosquito-Borne Viruses

Among the more than 520 viruses associated with arthropods and registered in the International Catalogue of Arboviruses (Karabatsos, 1985), somewhat less than half have biologic relationships with mosquitoes, and about 100 infect humans. The term **arbovirus** is a contraction of "arthropod-borne virus" and has no formal taxonomic meaning. The most significant mosquito-borne viruses causing human illness belong to four genera in three families (Table 14.2): the **Togaviridae**, genus *Alphavirus*; the **Flaviviridae**, genus *Flavivirus*; and the **Bunyaviridae**, genera *Orthobunyavirus* and *Phlebovirus*. These taxa have replaced categories in the older arbovirus literature, that is, group A viruses for the alphaviruses, group B for the flaviviruses, and Bunyamwera Supergroup for the orthobunyaviruses. Some of the arboviruses infect both humans and domestic animals and may cause disease in both. A representative arbovirus, Rift Valley fever virus, is shown in Figure 14.20.

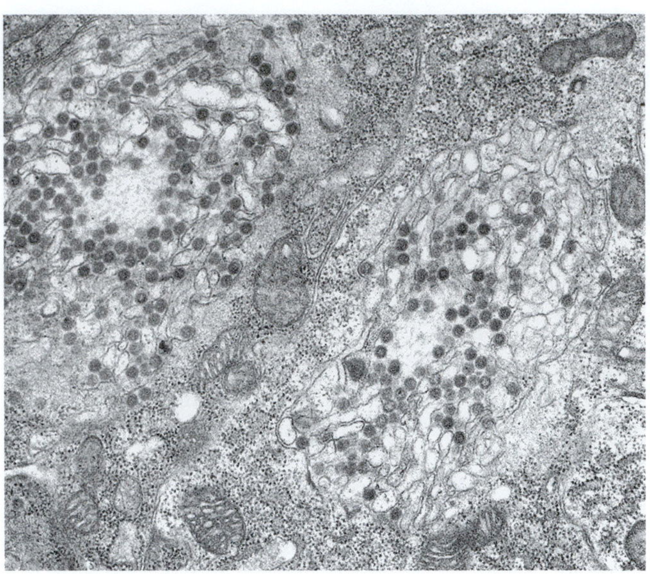

Figure 14.20 Rift Valley Fever virus virions in salivary gland tissues of *Culex pipiens* female, 42,000 ×. (Courtesy of K. Lerthudsnee and W. S. Romoser)

Aside from their genomic organization and morphology, mosquito-borne viruses may be viewed in terms of the kind of disease symptoms they cause. In humans, generally speaking, the mosquito-borne viruses cause infection with either no apparent symptoms, or acute disease of either systemic febrile illness (fever), encephalomyelitis (inflammation of the brain and spinal cord), hemorrhagic fever (bleeding and

Table 14.2 Selected Mosquito-Borne Viruses of Importance to Humans or Domestic Animals

Family (Genus)	Virus Species & Serotypes	Distribution
Togaviridae (*Alphavirus*)	Eastern equine encephalomyelitis	Americas
	Venezuelan equine encephalomyelitis	South and Central America, Mexico, United States (Florida)
	Western equine encephalomyelitis	North America, Mexico, South America (eastern)
	Chikungunya	Africa, Asia, including the Philippines
	O'nyong nyong	Africa
	Ross River	Australia, New Guinea, Fiji, American Samoa
	Semliki Forest	Africa, Asia, including Philippines
	Mayaro	South America (northern), Trinidad
Flaviviridae (*Flavivirus*)	Dengue (4 serotypes)	Tropics, especially southern Asia and Caribbean
	Yellow fever	Africa, Central and South America
	St. Louis encephalitis	Americas
	Murray Valley encephalitis	Australia, New Guinea
	Japanese encephalitis	Asia (eastern), including the Philippines,
	West Nile	Africa, Europe, Israel, Asia
	Ilheus	Central and South America
	Rocio	Brazil
	Wesselsbron	Africa, Asia (southern)
Bunyaviridae (*Orthobunyavirus*)	Bunyamwera	Africa
	Germiston	Africa
	Ilesha	Africa
	Wyeomyia	Central America
	Itaqui	South America
	Marituba	South America
	Murutucu	South America
	Oriboca	South America
	Madrid	Central America
	Nepuyo	Central and South America
	California encephalitis	United States (western)
	Jamestown Canyon	North America
	La Crosse encephalitis	United States (eastern)
	Inkoo	Finland
	Tahyna	Europe
	Guaroa	South America
Bunyaviridae (*Phlebovirus*)	Rift Valley fever	Africa (northern, eastern)

fever), or febrile myalgia and arthralgia (fever with muscle and joint pain, or arthritis). The case-fatality rate tends to be low for fevers, although morbidity (illness) may be high. For the encephalitis and hemorrhagic fevers, morbidity and mortality may range from low to high, depending upon the virus and factors such as age. After the acute phase, humans either recover fully or show various sequelae, such as neurological problems after acute encephalitis. Long-term, chronic infection with the mosquito-borne viruses does not occur in humans, although the consequences of infection may be long lasting.

Table 14.2 is a list of the more important mosquito-borne arboviruses and their geographic distribution. The viruses may be classified hierarchically as follows: family, genus, serogroup, complex, species, serotype, and strain. The word **strain** is used to refer to different viruses of the same serotype that were isolated from different locations or different biologic sources, or that show minor differences in antigenicity or genotype but not enough to justify elevating them to the varietal level. For example, Altamont virus is an eastern New York State strain of La Crosse virus, which along with Snowshoe hare virus is a serotype of the species California encephalitis virus in the California encephalitis virus complex of the California serogroup in the genus *Orthobunyavirus*, family Bunyaviridae. The relationships among the viruses are determined on the basis of similarities and differences in antigenic reactions to antibodies in immunological tests, and on genetic relationships determined by molecular analyses (i.e., nucleotide sequences or oligonucleotide fingerprint patterns). Arbovirus classification is presented by Fauquet et al. (2005). Karabatsos (1985) lists their histories and basic antigenic properties.

Mosquito-borne viruses multiply in both invertebrate and vertebrate cells. Many arboviruses cause cytopathic effects and cell destruction in vertebrate cells; in invertebrate cells the same viruses typically cause a chronic cellular infection without cytopathology. Competent mosquitoes become infected when they feed on blood of a viremic vertebrate host in which there is sufficient circulating virus in its blood to provide an infectious dose to the mosquito. After blood has entered the midgut, the virions bind to and then pass through the microvillar membrane and into midgut epithelial cells. Within these cells, viruses replicate and virions bud off from the cells, pass through the basal lamina, and enter the hemolymph. Virions disseminate throughout the body of the mosquito and may infect and replicate in a variety of tissues, including salivary glands, fat body, ovaries, and nerves. A mosquito with a salivary-gland infection (Figure 14.20) may transmit infectious virions during salivation as it probes the tissues of another vertebrate host. In some mosquitoes and for some viruses, transovarial transmission of virions occurs from the female mosquito to her progeny, and females of the next generation can transmit the virus orally without having become infected by a prior blood meal. Also, venereal transmission of some arboviruses from male to female mosquito has been documented experimentally. The rate of virus infection and dissemination in mosquitoes is temperature-dependent; higher temperature results in shorter extrinsic incubation. Virus infection may harm mosquitoes in some cases, and they become infected for life.

Because the flaviviruses, alphaviruses, and bunyaviruses have RNA genomes and can replicate in both invertebrate and vertebrate hosts, these viruses have a high capacity for rapid evolution into antigenically-variable strains of varying virulence. This capacity for change is important, because it may result in emergence of highly virulent, epidemic strains; indeed there is evidence for this process in recent history (see sections on O'nyong nyong virus and Rocio encephalitis virus, later). Whether rapid evolution of arboviruses leads inevitably to coevolution of viruses and mosquito hosts is debatable, because rapid evolution implies capacity to shift to new hosts, whereas coevolution implies adaptation to specific hosts through reciprocal selection and perhaps cospeciation.

The literature on the mosquito-borne viruses is voluminous. Extensive reviews are presented in the multivolume series edited by Monath (1988), in Strickland (1991), in the sections on Togaviridae, Flaviviridae, and Bunyaviridae by Fields et al. (1996), and in Reeves (1990), as well as in references mentioned under specific groups as follows, and at the end of this chapter.

Togaviridae (*Alphavirus*)

The Togaviridae, genus *Alphavirus*, contains seven antigenic complexes of viruses involving 37 types, subtypes, and varieties distributed worldwide, 35 of which have been isolated from mosquitoes. Many of these are medically important. Table 14.3 summarizes relationships between primary mosquito vectors and important alphaviruses, and Figure 14.21 shows their distribution in the Western Hemisphere.

Eastern Equine Encephalomyelitis (EEE) Virus This virus is distributed in South America, Central America, the Caribbean basin, and eastern North America. Analyses of geographic strains of EEE virus have revealed two varieties, South American and North American/Caribbean. The virus has been isolated from many different states in the United States, but most cases of human or equine disease are in coastal states from Massachusetts to Louisiana, an area of upstate New York near Syracuse, a swamp focus in east-central Ohio, and southern Michigan and part of northern Indiana.

Eastern equine encephalomyelitis virus is one of the most pathogenic among all the mosquito-borne encephalitis viruses. In humans, disease caused by EEE virus infection results in high morbidity and mortality. The type and severity of illness in humans depends upon the age and health status of the individuals. Children, the elderly, immunocompromised individuals, and sometimes apparently healthy adults

Table 14.3 Relationships of Selected Alphaviruses to Mosquito Vectors and Vertebrate Reservoir Hosts

Virus Species	Vector(s)	Vertebrate Reservoirs
Eastern equine encephalomyelitis	*Culiseta melanura, Ochlerotatus sollicitans, Coquillettidia perturbans, Culex nigripalpus, Culex (Melaniconion)* spp.	Birds
Western equine encephalomyelitis	*Culex tarsalis, Culex (Melaniconion)* spp. *Aedes albifasciatus, Ochlerotatus melanimon, Ochlerotatus dorsalis*	Birds, lagomorphs
Venezuelan equine encephalomyelitis[a]	*Culex (Melanoconion)* spp., *Aedes, Ochlerotatus Psorophora, Anopheles, Mansonia* spp.	Rodents, equids
Chikungunya	*Aedes aegypti, Aedes africanus Aedes albopictus*	Primates, including humans
O'nyong nyong	*Anopheles* spp.	Humans
Ross River	*Culex annulirostris, Ochlerotatus vigilax, Aedes polynesiensis*	Humans, rodents

[a]See Table 14.4 for elaboration of this virus complex.

develop acute encephalitis with high fever, drowsiness, lethargy, vomiting, convulsions, and coma. Mortality rates among clinical cases exceed 50%. Individuals who survive infection often show neurologic sequelae, although some survivors recover completely, sometimes showing rapid and dramatic improvement from coma.

In the eastern United States, EEE virus occurs in a bird-mosquito enzootic cycle in swamps that support the biology of the enzootic vector, *Culiseta melanura*. The swamps comprising EEE foci are characterized in the northern distribution of its range by northern white cedar, black spruce, tamarack, or red maple trees, typical of swamp or bog ecosystems. Larvae of *Cs. melanura* occur in water-filled cavities underneath raised tree hummocks, water-filled depressions formed by uprooted trees, and in holes in bog mats. Adults feed on birds in the swamp and may leave swamps for open areas to locate hosts, returning later to oviposit. *Culiseta melanura* females are highly efficient vectors of EEE virus and transmit it primarily to swamp-dwelling passeriform birds. More than 48 species of wild, native birds have shown evidence of infection. The mechanism of overwintering of EEE virus is unknown. Recent studies in New Jersey indicate that the virus may recrudesce in resident birds in the spring, whereas other studies incriminate reptiles as overwintering hosts. Still other scenarios suggest that this virus is introduced by migrating birds from southern regions where viral transmission may occur year-round.

In some summers, for reasons possibly related to weather patterns and density of bird and mosquito hosts, EEE virus in its enzootic swamp setting becomes amplified to high levels so that epizootics and epidemics develop. Certain mosquito species function as bridge vectors, especially *Ochlerotatus sollicitans* in coastal areas and *Coquillettidia perturbans* in inland areas. They acquire virus infection by feeding on viremic birds, later blood-feeding on mammals and transmitting the virus to them. In Central and South America, EEE virus apparently circulates among

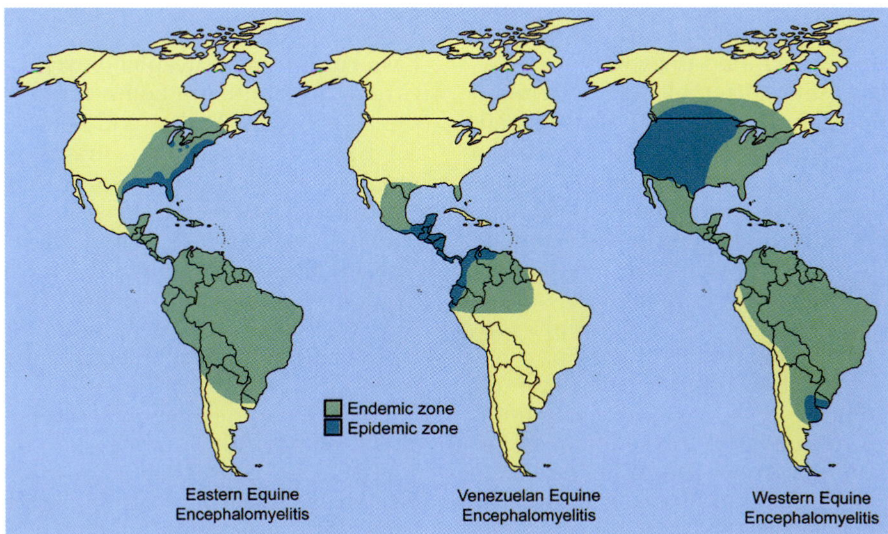

Figure 14.21 Endemic distribution and epidemic zones of mosquito-borne alphaviruses that cause encephalitis in the New World. (Reconstructed from Mitchell, 1977, and other sources)

rodents and birds through mosquitoes in the *Culex* (*Melanoconion*) group; however, the relationships among vectors and vertebrate hosts involved in enzootic, epizootic, and epidemic cycles of EEE virus in these regions are poorly known compared to North America.

Epidemics and concurrent epizootics have been documented in the eastern United States since the 1930s, generally involving cases in horses, pheasants, and humans. The virus was first isolated from brains of horses that died during a 1933 epizootic along the eastern coast of the United States, in Virginia, Delaware, and Maryland. Disease in humans was first recognized in 1938 in Massachusetts. Generally, outbreaks involve many horse cases and very few human cases. Nearly all outbreaks involving human cases have occurred along the eastern coast of the United States. From 1964 to 1995, a total of 151 human cases of EEE infection were reported in the United States, with an average of about five cases (range, usually 0 to 14) per year. The greatest number of human cases occurred in 1959, when 36 were documented, mainly in New Jersey. Morris (1988) reviewed the ecology, epidemiology, and vector relationships of EEE.

Western Equine Encephalomyelitis (WEE) Virus This virus is a complex of six types and six subtypes. All are mosquito-borne, with the exception of **Fort Morgan virus**, which is associated with cliff swallows and their parasitic cimicid bugs (*Oeciacus vicarius*) in western North America. The other viruses are widely distributed across the high plains of the United States and Canada, in California, and through Central and South America. The WEE virus type that occurs in North America has been responsible for acute encephalitis in horses and humans. An apparently less pathogenic virus, **Highlands J virus**, is closely related to Fort Morgan virus, yet it exists in the same basic North American enzootic cycle as EEE virus.

Disease caused by WEE virus in humans is less severe than that caused by EEE virus, and some infections are unapparent. Symptoms are generally similar, with acute onset of meningitis or encephalitis with headache, fever, drowsiness, and coma and death in severe cases. Fatality rates are in the range of 5% or less. Most morbidity and mortality occurs in infants rather than teens, adults, or the elderly. Neurologic sequelae are often evident in survivors.

In North America, WEE virus apparently exists enzootically in at least two different cycles. The primary cycle involves passerine and columbiform birds (especially house sparrows, house finches, blackbirds, orioles, and mourning doves), with *Culex tarsalis* functioning as the enzootic, epizootic, and epidemic vector. Nestling birds, which are more exposed to mosquito bites, may be more important as virus amplifier hosts than adult birds. The other cycle involves jackrabbits as vertebrate hosts and *Ochlerotatus melanimon* or *Oc. dorsalis* as vectors in California, Utah, and Colorado. Ground and tree squirrels also may function as vertebrate hosts in some areas, and in the central United States *Oc. trivittatus* may be a secondary vector. In Central and South America, WEE virus apparently circulates in nature among *Culex* (*Melanoconion*) mosquitoes, although a cycle involving *Oc. albifasciatus* and lagomorphs has been elucidated in Argentina.

Epizootics and epidemics of WEE in North America appear to be related to cumulative summertime precipitation and wintertime snowpack development, both of which can increase populations of *Culex tarsalis* and favor virus transmission. The larvae of this mosquito often are associated with agricultural irrigation and vernal flooding. Areas that normally are dry can produce large populations of *Cx. tarsalis* if irrigation activities result in the accumulation of pools of still water long enough to support larval development. For example, *Cx. tarsalis* populations burgeon after winter rain and vernal snow melt result in inundation of saltwater marshes along the north shore of Salton Sea in southern California. Western equine encephalitis virus is detected in *Cx. tarsalis* populations about two months after these populations increase, during March through June. The virus spreads to upland sites to the northwest of the sea, where mosquitoes are produced primarily from poorly maintained irrigation systems. The movement of WEE virus along this corridor is probably due to dispersal of infected *Cx. tarsalis* rather than to movement of birds. The mechanism by which WEE virus overwinters at this site, or is introduced there, is not known.

WEE virus was first isolated in 1930 from the brain of a dead horse in the San Joaquin Valley of California. It later was identified as the causative agent of human disease in a child who died of encephalitis in the San Joaquin Valley of California in 1938. Since that time, WEE virus has been implicated in epizootics in horses and concurrent epidemics in humans, with cases numbering from hundreds to thousands in some instances. Large outbreaks occurred in 1941 and 1975 in the Red River Valley in Minnesota and North Dakota, USA, and Manitoba, Canada; in 1952 in the Central Valley of California, USA; and in 1965 in Hale County in western Texas, USA. There are horse cases almost every summer within the range of the virus, but epizootics do not occur every summer. From 1964 to 1995, a total of 639 human cases of WEE infection were reported in the United States, with an average of about 20 cases (range, 0–172) per year. Reisen and Monath (1988) provided a review of WEE.

Venezuelan Equine Encephalomyelitis (VEE) Virus This is a complex of 12 viruses that cause disease in humans and equids (horses, burros, and mules) and occurs in northern South America, Central America, and Mexico, occasionally extending into Texas (Walton and Grayson, 1988). These viruses exist as either enzootic or epizootic varieties and strains, with overlapping or disjunct geographic distributions and with variable vector and vertebrate-host relationships. Table 14.4 shows the classification and vector associations of these viruses.

Many VEE "enzootic" virus subtypes and varieties exist in cycles involving rodents and mosquitoes of the *Culex* (*Melanoconion*) group, such as *Culex ocossa*, *Cx. panocossa*, and *Cx. taeniopus* in Central and South

Table 14.4 Venezuelan Equine Encephalomyelitis Virus Classification, Mosquito Associations, and Geographic Distribution of Subtypes and Varieties of this Virus

Subtype	Variety	Name	Vector Assocations	Geographic Distribution
I	A-B[a]	—	*Aedes, Ochlerotatus, Psorophora, Mansonia, Anopheles, Deinocerites pseudes*	Central America, South America (northern)
	C[a]	—	Same as IA-B	Central America, South America (northern)
	D	—	*Culex* (*Mel.*) *ocossa* *Culex* (*Mel.*) *panocossa*	Central America, South America (northern)
	E	—	*Culex* (*Mel.*) *taeniopus*	Central America
	F	—	unknown	Brazil
II		Everglades	*Culex* (*Mel.*) *cedecei*	United States (southern Florida)
III	A	Mucambo	*Culex* (*Mel.*) *portesi*	South America (northern)
	B[b]	Tonate	unknown	South America (northern)
	C	—	unknown	Peru
IV		Pixuna	unknown	Brazil
V		Cabassou	unknown	French Guiana
VI		—	unknown	Argentina

[a]Virulent to equids and humans and involved in epizootics. The other subtypes and varieties are enzootic.
[b]A related variety IIIB (Bijou Bridge virus) is not listed here. It is associated with cliff swallow bugs (*Oeciacus vicarius;* Hemiptera: Cimicidae) in western North America.

America. Rodents in the genera *Sigmodon, Oryzomys, Zygodontomys, Heteromys, Peromyscus,* and *Proechimys* are important vertebrate hosts; birds, opossums, and bats also may be reservoir hosts. The ecology of the epizootic viruses is quite different. A large number of species of mosquitoes in several different genera (see Table 14.4) have been implicated as vectors of the epizootic/epidemic virus strains. Equids attain sufficient viremia to infect these mosquitoes. VEE epidemics can be maintained by mosquito-equid-mosquito transmission, unlike WEE and EEE epidemics, in which equids are for the most part dead-end hosts. Wading birds, particularly green herons, have been incriminated as vertebrate hosts of epizootic strain IA-B in Panama, with the crab hole mosquito *Deinocerites pseudes* functioning as vector. Persistence of epizootic strains of VEE in interepidemic periods is not well understood, thus their emergence in epidemics among equids and humans is difficult to predict.

The single representative of the VEE viruses in the United States, other than during epizootics in Texas, occurs in the Everglades region of southern Florida. Called **Everglades virus**, it is associated with *Culex* (*Melanoconion*) *cedecei* (formerly, *Cx. opisthopus*) as vector, and cotton rats (*Sigmodon hispidus*) and cotton mice (*Peromyscus gossypinus*) as vertebrate hosts. The zoonotic setting is the hardwood hammocks of the Everglades, where mosquito and rodent habitats overlap. Serosurveys of Seminole and Miccosukee Indians in these regions have shown that many Indians have antibodies to VEE virus, but there have been very few cases of human disease attributable to this virus.

Humans infected with an epizootic or certain enzootic strains of VEE virus may show no symptoms, only mild flu-like symptoms, or severe encephalitis with acute onset of vomiting, headache, seizures, and fever. Symptoms tend to be most severe in children. During epidemics, the mortality rate is typically less than 1%, although in some epidemics the mortality rate has been considerably higher.

The VEE viruses in Central America and northern South America have been intensively studied because of the history of epidemics among equids and humans in these regions. Outbreaks of VEE have occurred periodically in South America, Central America, and Mexico since the 1930s. The first VEE virus was isolated in 1938 from a dead horse in Venezuela. In 1969, a large outbreak of VEE involving both equids and humans in Central America spread northward through Mexico in the next two years, moving into Texas in 1971. Cases continued in Mexico through 1972. There were thousands of both horse and human cases throughout this region during that time, but epizootic virus activity did not occur again there until an outbreak in Venezuela in 1992 to 1993 and in Chiapas, Mexico, in 1993. More recently, an outbreak of VEE occurred in northern Colombia in 1995. The rapidity of spread of these outbreaks over large geographic areas undoubtedly is due to the role of horses as competent reservoir hosts.

Chikungunya (CHIK) Virus This virus, which has caused outbreaks for many years in Africa and parts of India and southeastern Asia (Jupp and McIntosh, 1988), has recently surged in human populations. In 2004, almost half a million human cases were reported in Africa, primarily coastal Kenya. An outbreak starting in 2005 spread across the islands of the Indian Ocean, resulting by early 2006 in 244,000 cases on Réunion, with an attack rate of 35% and more than 200 deaths. From there it spread to India, causing over a million cases. Travelers returning to Italy and France from the islands as Réunion, Mauritius, the Comoros, and

Seychelles brought the virus with them. This resulted in cases diagnosed in Europe and a focal epidemic involving local transmission.

CHIK virus generally does not cause the encephalitis-type symptoms characteristic of EEE, WEE, and VEE viral infections, but rather a dengue-like arthralgic illness with fever, rash, and severe joint pain. Neurologic and other severe complications occurred in the recent Indian Ocean epidemic, with deaths having been reported for the first time.

In Africa, CHIK virus infects nonhuman primates such as the vervet monkey (*Cercopithecus aethiops*) and baboon (*Papio ursinus*). The enzootic vectors include *Aedes africanus, Ae. luteocephalus, Ae. opok, Ae. furcifer, Ae. taylori,* and *Ae. cordellieri* in the savanna and forest cycles. Reservoirs are rodents and cattle and mosquito species that feed on them. The virus appears to lack a zoonotic reservoir in Asia and the Indian Ocean region. Humans infected with CHIK virus often develop viremia sufficient to infect *Ae. aegypti* and *Ae. albopictus*, important vectors in urban and suburban areas.

Aedes aegypti is the primary vector of CHIK virus in urban areas of India and Asia, where it causes epidemics of arthralgic disease during rainy seasons. A 1994 epidemic in Vellore, southern India, was characterized by human cases increasing during August and September, and reaching a peak in October as the human-biting frequency and viral infection rate of *Ae. aegypti* increased. The epidemic lasted about five months and affected 44% of the city's population. This epidemic and others, along with experimental studies, indicate that interrupted feeding by mosquitoes, resulting in partial blood meals, may facilitate both mechanical and biological transmission of CHIK virus, thus rapidly amplifying the virus in human populations.

The recent increases in CHIK virus activity that started along the East African coast is thought to have been caused not by abundant rainfall but rather by several years of drought. Drought promotes the use of water-storage containers around human habitations, producing large numbers of *Ae. aegypti* as vectors.

The epidemics on the Indian Ocean islands and in Italy have involved *Ae. albopictus* as the primary vector, rather than *Ae. aegypti*. A small genetic change in the virus apparently has significantly increased the vector competence of this typically more rural and less anthropophilic mosquito. The establishment of *Ae. albopictus* in many temperate areas of the world in recent years increases the opportunities for CHIK virus to spread in other human populations.

Other Alphaviruses

There are other important alphaviruses that occur endemically and epidemically and cause fever, arthralgia, and other symptoms in humans.

O'nyong-nyong (ONN) Virus This is an antigenic subtype of chikungunya virus that is transmitted among humans by *Anopheles* species in widespread parts of Africa. The vectors are the same ones that transmit human malaria parasites (i.e., *An. gambiae* and *An. funestus*). A large epidemic occurred from 1959 through the 1960s, infecting about 2 million people in Uganda, Kenya, Tanzania, Malawi, Zambia, and Mozambique. The virus, whose name comes from an Acholi African word meaning "weakening of the joints," was first isolated from humans during this time and was later isolated from *An. funestus* in 1974. Only isolated cases had occurred since then, until an epidemic in Uganda in 1997. Neither a vertebrate animal host nor the mechanism of persistence of ONN virus between these epidemics is known. The reservoir host is probably humans. A closely related virus, called **Igbo Ora**, also is transmitted by *Anopheles* mosquitoes and occurs in parts of West Africa, where it was associated with an outbreak of chikungunya-like disease in the Ivory Coast in 1984. ONN and Igbo Ora are the only arboviruses causing human disease that have *Anopheles* mosquitoes as the primary vectors.

Sindbis (SIN) Virus This virus is distributed widely in Eastern Europe, Scandinavia, the former Soviet Union, Asia, Africa, the Middle East, and Australia. The virus is a member of the WEE virus complex. It originally was isolated from *Culex univittatus* in Egypt in 1952, and has been associated with a human disease of rash, fever, and muscle and joint pain in Uganda, South Africa, and Australia. Birds are vertebrate hosts. A subtype of SIN virus is **Ockelbo**, which is distributed in Sweden, Finland, and northern Russia. It is transmitted by *Culiseta ochroptera*, and possibly *Culex* and *Aedes* species, and has been associated with human disease similar to that caused by Sindbis virus.

Ross River (RR) Virus This virus occurs in Australia, Fiji, and the Cook Islands, where it causes an illness known as **epidemic polyarthritis**, consisting of fever, rash, and arthralgia (Kay et al., 1988). In Australia, RR virus is transmitted by *Ochlerotatus camptorhynchus, Oc. vigilax* and *Culex annulirostris*, whereas in the islands the vectors are *Ae. aegypti* and *Ae. polynesiensis*. The virus was first isolated from *Oc. vigilax* in 1963. The vertebrate reservoir hosts are unknown, but in Australia they may be marsupials. **Barmah Forest virus** is another mosquito-borne alphavirus in Australia that causes symptoms similar to RR virus.

Mayaro (MAY) Virus This virus occurs in the Caribbean and parts of South America. In humans the illness is similar to chikungunya. The virus was first isolated from febrile patients in Trinidad in 1954. Marmosets are the reservoir hosts of the virus, and *Haemagogus* mosquitoes are vectors.

Flaviviridae (*Flavivirus*)

The Flaviviridae, genus *Flavivirus*, contains eight antigenic complexes plus many unassigned viruses involving 70 types, subtypes, and varieties distributed worldwide. Some of these have mosquitoes as vectors whereas others are associated with ticks or with rodents or bats. The flavivirus diseases include some of the most dangerous and historically significant infections of humans. Table 14.5 shows the mosquito-vector and vertebrate-host associations of some of the more

Table 14.5 Relationships of Selected Flaviviruses to Mosquito Vectors and Vertebrate Reservoirs Hosts

Virus Species	Vector(s)	Vertebrate Reservoirs
Yellow fever	*Aedes aegypti*	Humans in urban environments
	Aedes africanus	Monkeys
	Aedes bromeliae	Monkeys
	Aedes furcifer	Monkeys
	Aedes luteocephalus	Monkeys
	Aedes metallicus	Monkeys
	Aedes taylori	Monkeys
	Aedes vittatus	Monkeys
	Haemagogus spp.	Monkeys
	Sabethes spp.	Monkeys
Dengue	*Aedes aegypti*	Humans
	Aedes albopictus	Humans (Monkeys?)
	Ochlerotatus niveus group	Monkeys
	Aedes africanus	Monkeys
	Aedes furcifer	Monkeys
	Aedes taylori	Monkeys
	Aedes luteocephalus	Monkeys
	Aedes opok	Monkeys
	Aedes scutellaris	Humans
	Aedes polynesiensis	Humans
	Aedes pseudoscutellaris	Humans
	Aedes rotumae	Humans
Japanese encephalitis	*Culex tritaeniorhynchus*	Birds, Pigs
	Culex gelidus	Birds
	Culex vishnui complex	Birds
St. Louis encephalitis	*Culex pipiens*	Birds
	Culex quinquefasciatus	Birds
	Culex tarsalis	Birds
	Culex nigripalpus	Birds
Murray Valley encephalitis	*Culex annulirostris*	Birds
West Nile	*Culex univittatus*	Birds
	Culex pipiens	Birds
	Culex quiquefasciatus	Birds
	Culex spp.	Birds

important mosquito-borne flaviviruses. Among them are yellow fever, dengue, Japanese encephalitis, St. Louis encephalitis, Murray Valley encephalitis, and West Nile viruses.

Yellow Fever

This disease is caused by Yellow fever (YF) virus and occurs over broad portions of lowland equatorial Africa and South and Central America (Figure 14.22), either as isolated cases or epidemics (Monath 1988). Yellow fever virus exists principally in two epidemiological forms: an enzootic form, maintained in monkey populations by forest mosquitoes in a sylvan cycle and responsible for most of the isolated human cases (jungle yellow fever); and an epidemic form, spreading rapidly through human populations by the domestic form of *Aedes aegypti* (Figure 14.23) in an urban cycle.

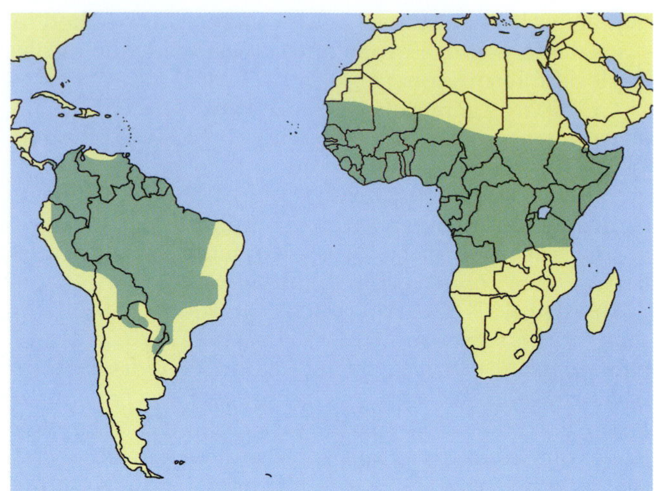

Figure 14.22 Geographic distribution of human yellow fever cases, 1982 to 2007. (Reconstructed from the Centers for Disease Control and Prevention, http://wwwn.cdc.gov/travel/yellowBookCh4-YellowFever.aspx#669 and WHO, http://www.afro.who.int/yellowfever/surveillance/index.html)

The urban vector is particularly efficient because it readily enters and typically rests in houses, feeds almost exclusively on humans, and oviposits in artificial containers. It frequently takes two to three blood meals per gonotrophic cycle to supplement its energy reserves in lieu of sugar-feeding.

In previous centuries, urban yellow fever affected subtropical and temperate regions of North America with devastating effects. It remains a serious cause of mortality, particularly in village settings in tropical Africa, and is a constant threat in South America. There are sporadic, annual cases and occasional epidemics in Africa and 50 to 300 cases of jungle YF annually in South America. Jungle YF occurred in Brazil (1932), Panama and northward into Central America (1948–1955), and

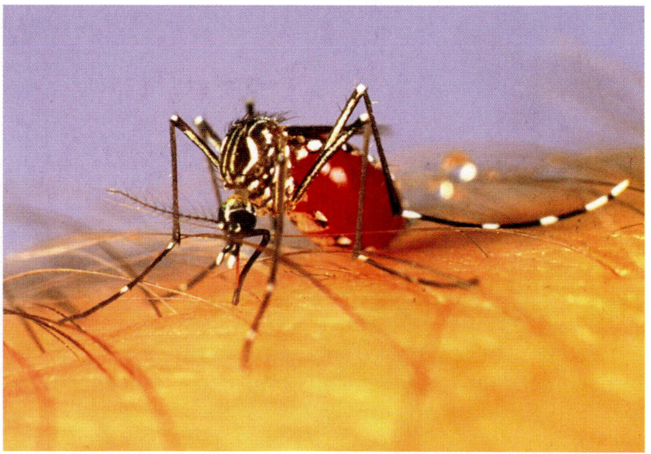

Figure 14.23 *Aedes aegypti* female feeding on blood. This is the primary vector of dengue and urban yellow fever. The "lyre" pattern on the thorax is distinctive. (Photo by Robert G. Hancock)

several episodes (including some urban transmission) in Trinidad (1954, 1959, 1978–1979). Epidemics in Africa have involved diverse areas, including Ethiopia (1960–1962), the Gambia (1978), Ghana (1979–1980, 1983, 1993, 1996), Benin (1996), Burkina Faso (1983), Nigeria (1986, 1994), and Kenya (1992–1993).

Yellow fever is a hemorrhagic disease. Mortality in infected humans ranges from 5 to 75% or more. After infection by mosquito bite, there is an incubation period of three to six days, followed by sudden high fever (>104° F or 40° C), headache, nausea, and pain. Humans are viremic during this initial acute phase only for about three days. The virus is viscerotropic and causes parenchymal cell necrosis in the liver, resulting in elevated blood bilirubin levels and jaundice, thus the name yellow fever. Jaundice leads to systemic toxemia. Hemorrhage manifests as bleeding gums, easy bruising, and sloughing of the stomach lining, thus causing a characteristic black vomit. Delirium and coma often precede death.

The causative agent of YF is the prototype virus of the Flaviviridae and the first arbovirus to be associated with human disease. Although the disease was first recognized in the New World in the seventeenth century, the virus was isolated first in 1927 in Ghana, Africa, when the blood of a man with YF was inoculated into rhesus monkeys. The virus apparently originated in Central Africa among monkeys (e.g., *Cercopithecus* spp., family Cercopithecidae) and mosquitoes dwelling in the forest canopy. In this environment, the vector is the monkey-feeding *Aedes africanus*, a relative of *Ae. aegypti*. There also is evidence for infection with chimpanzees (*Pan troglodytes*, family Pongidae), baboons (*Papio*, family Cercopithecidae), and bushbabies (*Galago*, family Lorisidae). Generally, Old World monkeys do not suffer mortality from infection. In western Africa, where YF virus also is enzootic, the likely sylvatic vectors are *Aedes vittatus*, *Ae. metallicus*, *Ae. furcifer*, *Ae. taylori*, and *Ae. luteocephalus*, which develop in treeholes and sometimes other natural containers and rock pools.

Human disease occurs when humans enter the habitat where the mosquito–monkey cycle is ongoing and are bitten by infected mosquitoes. Alternatively, infected monkeys or baboons sometimes enter human habitat on raids in gardens or banana plantations, bringing YF virus to mosquitoes at the interface between forest and human habitation. The African mosquito *Aedes bromeliae* (formerly *Ae. simpsoni*), the larvae of which develop in water-filled leaf axils of plants such as banana, occupies this interface and frequently has become involved in forest-edge and rural transmission of YF virus to humans. *Aedes bromeliae* sometimes has served as the primary vector in massive rural epidemics. Thus, it functions both as a bridge vector between sylvatic and rural cycles and as an interhuman vector in that peridomestic environment. It was the principal vector during the 1960 to 1962 epidemic in Ethiopia, in which about 100,000 people became infected and 30,000 died.

The spread of both the domestic form of *Aedes aegypti* and YF virus from Africa to other parts of the world apparently occurred within the last 400 years. Trading and slaving ships, with their potentially virus-infected cargoes of slaves, and with water barrels as a mosquito development site, greatly facilitated spread and establishment of both vector and virus. The need for slave labor in the newly established sugar cane-molasses-rum economy of the Caribbean region undoubtedly promoted movement of YF virus about the New World. Oftentimes, crews became ill with YF while their ships were in transit. Epidemics regularly occurred in port cities in both West Africa and in coastal South America, North America, and the Caribbean. *Aedes aegypti* also moved into the Arabian peninsula and Indian subcontinent, and then to Asia and the Pacific region, probably via dhow traffic along sea trade routes. Yellow fever has not become established in Asia, even though vector-competent mosquitoes and humans occur there.

Yellow fever epidemics have occurred in the New World over a period of three centuries, beginning in the mid 1600s. Epidemics occurred in such places as Barbados and Trinidad (Caribbean), Havana (Cuba), Yucatan (Mexico), Guadeloupe (Caribbean), Guayaquil (Ecuador); and in Charleston (South Carolina), Mobile (Alabama), Pensacola (Florida), and other areas of the United States as far north as Boston and New York. In the 1700s and 1800s, epidemics continued in ports of tropical and temperate America, including an outbreak in Philadelphia (Pennsylvania) in 1793, where some 4,000 deaths occurred in a population of 55,000. A large outbreak in Haiti in 1802 decimated the French military force there, causing Napoleon to abandon his New World ambitions, contributing to the Louisiana Purchase by the United States government. New Orleans (Louisiana) had regular epidemics of YF from 1796 through 1905. The shipping blockade enforced by the navy on the federalist side of the United States' Civil War (1861–1865) prevented yellow fever in New Orleans during that period. Probably, the blockade stopped importation of infected ship crews. An epidemic in the Mississippi River valley in 1878, extending north as far as Gallipolis, Ohio, caused over 13,000 deaths. The last epidemic in the United States was in New Orleans in 1905, involving 3,402 cases and 452 deaths.

As YF virus invaded the New World, it became established in a mosquito–monkey cycle in forested parts of Central and South America. In the sylvatic cycle there, the vectors are forest canopy-dwelling mosquitoes, particularly *Haemagogus* species and *Sabethes chloropterus*, whose larvae develop in treeholes. New World Monkeys in the family Cebidae are highly susceptible to infection. During epizootics, howler monkeys (*Alouatta* spp.), squirrel monkeys (*Saimiri* spp.), spider monkeys (*Ateles* spp.), and owl monkeys (*Aotus* spp.) may show considerable mortality in their populations. However, capuchin monkeys (*Cebus* spp.) and woolly monkeys (*Lagothrix* spp.) circulate mosquito-infective viremias, but they do not die from infection. Therefore, in this cycle noticeable die-offs of monkeys may or may not precede sylvatic transmission of YF virus to humans. Sylvan YF in the Americas can lead to urban outbreaks when humans enter the tropical forests where transmission is ongoing, become infected by mosquito bite, then return to villages or cities. If

Aedes aegypti is present in these settlements, it can initiate an epidemic of urban yellow fever.

The history of the discovery that YF virus is transmitted by mosquitoes is intriguing. The discovery is particularly important because YF was the first arbovirus to be recognized as a mosquito-borne agent, and mosquito control measures imposed quickly afterward caused a dramatic reduction in this devastating disease. Although suspicion that the agent causing yellow fever might be transmitted by mosquitoes can be traced to several independent sources in the 1800s, it was the intuition of the Cuban physician Carlos Finlay, followed by the research activities of the United States Yellow Fever Commission in Havana, Cuba, in 1900, that resulted in experimental evidence that *Aedes aegypti* was the vector. This commission was comprised of the U.S. Army officers Walter Reed, James Carroll, Jesse Lazear, Aristides Agramonte, and others. After consulting with Finlay, this team carried out a series of experiments that demonstrated the transmissibility of the agent from infected to uninfected humans by mosquito bite, after a suitable incubation period in mosquitoes. During this work, James Carroll allowed himself to be bitten by an infected mosquito and later developed YF, but he recovered. Jesse Lazear was accidentally bitten, and he died of the disease.

The team's findings stimulated William Gorgas, a physician in the United States Army, to impose a control program against *Ae. aegypti*, resulting in the elimination of urban YF in Havana and, soon after, in the Panama Canal Zone. Later, the Rockefeller Foundation sponsored teams of biomedical scientists and public health practitioners to begin intensive studies of YF by establishing a research center in Guayaquil, Ecuador, in 1918. This foundation eventually established research institutes in Brazil, Nigeria, Uganda, the United States, and elsewhere, and supported research and disease control programs in many sites. For a time, the causative agent of YF was thought to be a spirochete. Within 10 years of isolation of the virus in 1927, an attenuated, live vaccine was produced that was shown to provide excellent protection against infection. Max Theiler was awarded the Nobel Prize in Medicine for his efforts in this regard. In many areas of South America, antimosquito programs resulted in virtual elimination of urban outbreaks, although sylvan transmission continued. Despite these early advances, cases of jungle and rural YF, and epidemics in Africa continue to occur. The reintroduction and resurgence of *Ae. aegypti* populations throughout Central and South America and the Caribbean islands in recent decades, and the immense growth of cities that provide a habitat for it, have created an increased potential for urban epidemics.

Dengue

This disease is caused by **Dengue (DEN) virus**, represented by four closely related serotypes called Dengue 1, 2, 3, and 4 (Gubler, 1988; Gubler and Kuno, 1997). The disease in humans is either classic dengue fever or the more severe dengue hemorrhagic fever or dengue shock syndrome. The DEN viruses are transmitted by mosquitoes, principally *Aedes aegypti*. The current distribution of DEN viruses includes Southeast Asia, the south Pacific, the Caribbean basin, Mexico, Central America, and South America. However, epidemics of DEN have occurred elsewhere in the past, including the United States, Japan, Australia, Greece, and both eastern and western Africa. Figure 14.24 shows its current distribution and the regions with projected risk for occurrence, given the appropriate climate and availability of vectors. Dengue is commonly reported as an introduced disease in the United States, but indigenous transmission apparently has occurred in Texas in 1980, 1986, and 1995. All four serotypes now occur in the Western Hemisphere. In addition to large-scale epidemics in the Americas, there have been recent, large outbreaks involving dengue hemorrhagic fever in Africa, China, Taiwan, India, Maldives, and Sri Lanka. In the hyperendemic areas of southeastern Asia, such as Thailand and the

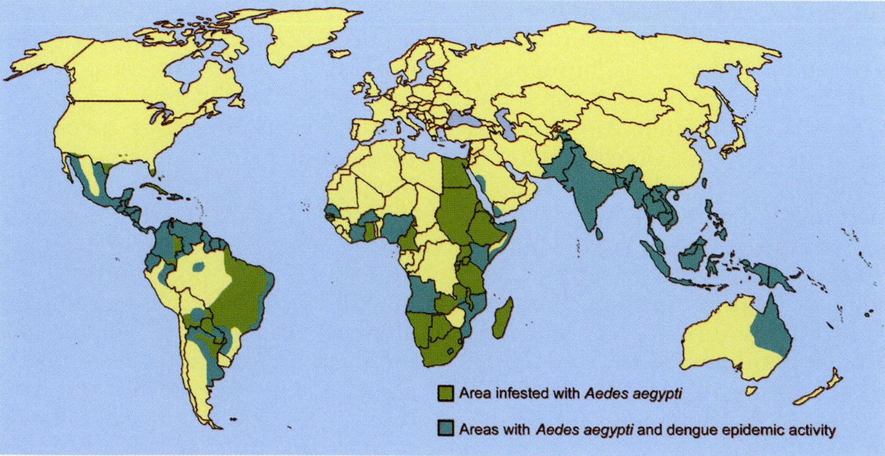

Figure 14.24 Geographic distribution of countries reporting human dengue fever cases and infestations with *Aedes aegypti* mosquitoes. (Reconstructed from the Centers for Disease Control and Prevention, http://www.cdc.gov/ncidod/dvbid/dengue/map-distribution-2005.htm)

Philippines, the severe forms of disease have become more common and appear in epidemics at three- to five-year intervals.

Dengue is characterized by fever, rash, severe headache, and excruciating pain in muscles and joints, earning it the name **breakbone fever**. Clinical disease develops five to eight days after bite of an infected mosquito. Often the disease runs a mild course of about a week, leading to complete recovery. However, DEN can become severe in cases of **dengue hemorrhagic fever** (DHF) and **dengue shock syndrome** (DSS), both of which generally occur in children and may be fatal. These manifestations first were observed as complications of dengue fever in 1954 and are characterized by blotchy rash, bleeding from the nose and gums (Figure 14.25), and shock.

The increased frequency of DHF and DSS in Southeast Asia and parts of the Caribbean and Latin America have stimulated a debate in the biomedical community regarding the mechanisms by which DHF emerges during epidemics of classic dengue fever. One hypothesis is that there are variable forms of the different serotypes of DEN viruses, some of which are more pathogenic than others. Another idea is that people of different races differ in propensity to develop severe symptoms and that the virus has been spreading into more susceptible populations. Still another hypothesis is that, as human populations in tropical cities in Asia and the Americas increase and as DEN epidemics in general increase in frequency, there are simply more cases with the noticeable manifestations of DHF and DSS. The fourth hypothesis, and the one currently viewed as most likely, is that prior exposure to one serotype, followed by exposure to another serotype within a critical period of 5 years, leads to the development of hemorrhagic fever and, in some cases, shock syndrome. The more recent epidemics taking place in various parts of the world, especially in the Americas, indicate that DEN has replaced yellow fever as the major urban, epidemic flavivirus of importance worldwide.

The main epidemic vector of DEN viruses, *Aedes aegypti* (Figure 14.23), is ideal, because it commonly rests indoors, feeds preferentially on humans, has a tendency to take supplementary blood meals, and often moves from one residence to another as it oviposits. The larvae develop in such vessels as water barrels and jars, potted-plant containers, cemetery urns, and discarded tires. The close proximity of these larval habitats to human dwellings further facilitates *Ae. aegypti*-human contact and allows large mosquito populations to develop. To the extent that larval sites are hand-filled, DEN transmission is independent of rainfall patterns. Indeed, in some places where water stores are replenished independently of rainfall, epidemics may occur in the hot, dry season when temperatures are higher and the extrinsic incubation period of the viruses in the mosquitoes is shortened. However, in areas where breeding containers depend primarily on rainwater, DEN epidemics occur during rainy seasons.

Other vectors of DEN viruses are *Aedes albopictus* (Figure 14.26) in rural areas of Southeast Asia and *Ae. polynesiensis, Ae. scutellaris, Ae. pseudoscutellaris*, and *Ae. rotumae* in the Pacific region. The role of the newly introduced *Ae. albopictus* in the New World as a vector of DEN viruses remains to be determined. In peninsular Malaysia, a series of studies showed that all four serotypes of dengue virus circulated between monkeys and mosquitoes of the *Ochlerotatus niveus* group, suggesting the possibility of an enzootic sylvatic cycle. Similarly, studies in West Africa provide equally strong evidence that monkey populations there maintain a sylvatic form of DEN type 2 virus. Where monkeys and humans are loosely associated in forested areas and human habitations are intermixed with patches of forest, this strain of the virus has been isolated from humans, monkeys, and from arboreal mosquitoes such as *Ae. africanus, Ae. opok, Ae. luteocephalus, Ae. furcifer*, and *Ae. taylori*, which appear to transmit the virus indiscriminately among monkeys and humans. Thus, DEN may have a zoonotic reservoir and system of

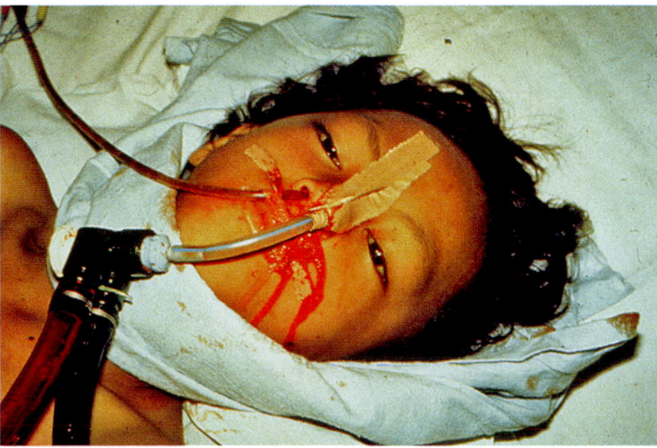

Figure 14.25 Dengue hemorrhagic fever (DHF), showing external bleeding from mucous membranes. (Photo by D. J. Gubler)

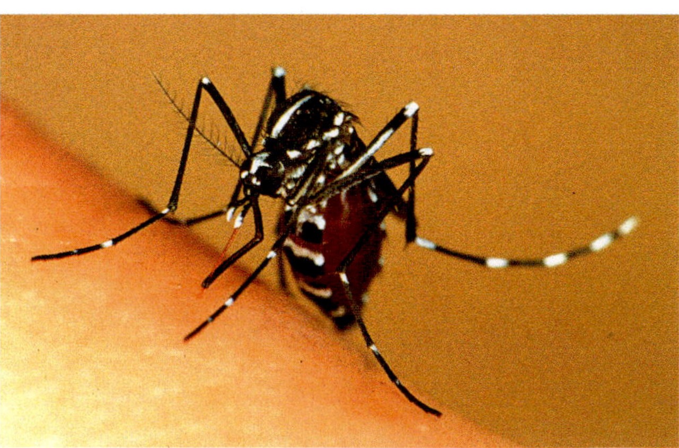

Figure 14.26 Asian tiger mosquito (*Aedes albopictus*), female feeding on human. The median longitudinal silver stripe on the thorax is its most prominent identifying feature. (Photo by W. A. Foster)

transmission in West Africa similar to that of yellow fever and chikungunya viruses in the same region. But in this case the enzootic and endemic systems probably remain distinct and separate from one another.

Humans generally are considered to be the only vertebrate host of the endemic/epidemic form of the virus in situations where monkeys do not occur, such as congested urban slums in huge tropical cities including Bangkok, Manila, Jakarta, Caracas, and Guayaquil. Thus, a mosquito-human-mosquito cycle of DEN transmission by *Ae. aegypti* is the usual means of virus maintenance and epidemic spread. Transovarial transmission of some of the DEN viruses has been demonstrated in *Ae. aegypti* and *Ae. albopictus*, and therefore mosquitoes may be reservoirs, particularly in periods of low-level transmission among humans.

Japanese Encephalitis Virus Complex

Another important group of mosquito-borne flaviviruses is the Japanese encephalitis virus antigenic complex, including West Nile, Japanese encephalitis, St. Louis encephalitis, and Murray Valley encephalitis viruses. Some authorities refer to this complex as the West Nile antigenic complex. These viruses occur in widely separated geographic regions but show similarities in the nature of their enzootic cycles. Each has *Culex* vectors (Table 14.5) and birds as vertebrate reservoirs. In the case of Japanese encephalitis virus, pigs often serve as amplifying hosts. Figure 14.27 shows the worldwide distribution of this complex of viruses. The most important of these in terms of human morbidity and mortality is Japanese encephalitis virus. However, all cause human illness of varying severity, depending upon the virus and age and health of the person.

West Nile (WN) Virus This virus was first isolated from the blood of a febrile man in Uganda in 1937. The primary mosquito vectors are *Culex* species, particularly *Culex pipiens* and *Cx. univittatus*. Birds are the vertebrate reservoirs and amplifying hosts during epidemics. WN virus is widely distributed in Africa, the Middle East, Europe, parts of the former Soviet Union, India, Indonesia, and North and South America (Figure 14.28) (Hayes, 1988). It has been the cause of endemic and epidemic fever, myalgia, and rash, especially in children in the Middle East, and has caused encephalitis in some instances. Seroprevalence of antibody to WN virus in Egypt can exceed 60% in adults. Recent epidemics have occurred in Israel (1950s), southern France (1962), South Africa (1974, 1983–1984), Romania (1996), and other eastern European countries, and most recently in North America.

In 1999, West Nile virus was introduced through unknown means into New York City, and was linked to human encephalitis in 61 confirmed cases (mostly in the city borough of Queens); there were seven fatalities. Initially, the outbreak was thought to be due to St. Louis encephalitis. However, an unusual feature of the outbreak was the large numbers of birds that died, especially crows (*Corvus brachyrhynchos*). The vectors are believed to have been *Cx. pipiens* and other *Culex* species, although WN virus later was isolated from other mosquito genera as well. A serosurvey conducted by the US Centers for Disease Control and Prevention (CDC) in the Queens area indicated that as many as 1,250 people had been infected during the course of the outbreak, and that approximately 240 of these people (19%) may have experienced clinical illness due to infection.

The virus successfully overwintered in the New York area, despite efforts to control *Culex* mosquitoes, and spread to adjoining northeastern states in 2000. By 2001 it had spread into the South and Midwest regions of the United States and Canada. In the Midwest human cases and high bird mortality peaked in 2002. By that time the first human case had already occurred on the West Coast and virus was isolated from birds and mosquitoes. The following year, infections peaked

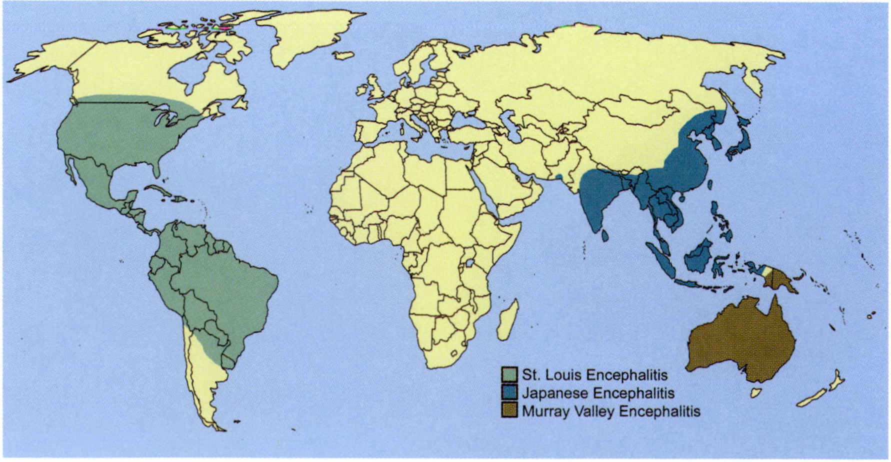

Figure 14.27 Geographic distribution of mosquito-borne flaviviruses causing encephalitis. (Reconstructed from Mitchell, 1977, the Centers for Disease Control and Prevention, http://www.cdc.gov/ncidod/dvbid/jencephalitis/map.htm, and other sources)

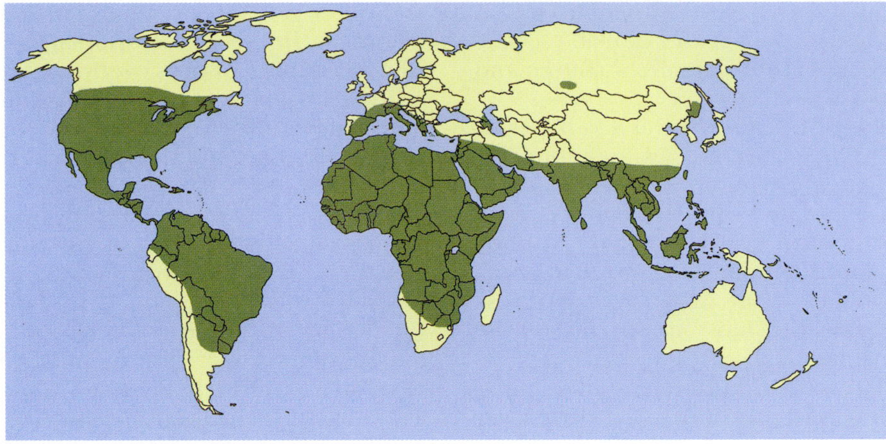

Figure 14.28 Geographic distribution of West Nile virus, a mosquito-borne flavivirus.

in the South, and highest numbers of human cases occurred in the prairie and Rocky Mountain States, followed by a high incidence of cases in California during 2004 and 2005. The highest virus activity in the United States was recorded in 2006, with more than 4,200 human cases of WN, including 177 deaths. The numbers were moderately lower the following year. During the period 2002 to 2007, the virus also spread steadily southward into Mexico, the Caribbean islands, Central America, and finally South America as far as Argentina. In the United States, epidemiological studies indicate that about 80% of human infections are asymptomatic. The great majority of clinical cases are classified by CDC as West Nile fever, with fewer than 5% presenting West Nile neuroinvasive disease (including these categories: encephalitis, meningitis, and poliomyelitis).

In the North American outbreaks over 200 species of birds suffered mortality from WN virus infections, with corvids (crows and jays), raptors, and exotic species being particularly vulnerable. About 40% of infected equines (horses, donkeys) died. Bird species vary widely in their tolerance of infections and in their ability to generate high and prolonged viremias. Jays, grackles, crows, house finches, and house sparrows are all highly competent in the latter regard, making them good reservoir hosts.

Many mosquito species are capable of supporting and transmitting WN virus. In the United States, *Culex* species are the principal vectors that maintain it in bird populations: particularly *Cx. pipiens* in the Northeast and upper Midwest, *Cx. quinquefasciatus* (Figure 14.29) in the South, and *Cx. tarsalis* in several western regions. Mosquitoes that feed on both birds and mammals, such as *Coquillettidia perturbans* and *Aedes vexans*, may act as bridge vectors, transmitting the virus from birds to equines and humans. However, many *Cx. pipiens* populations feed readily on both birds and mammals and can likewise serve as bridge vectors.

Studies in northeastern North America have shown that a small proportion of infected *Cx. pipiens* adults can pass WN virus to their offspring by transovarial transmission. Diapausing adults emerging in the fall can carry the virus through the winter in their hibernacula and transmit it to birds the following spring.

Japanese Encephalitis (JE) Virus Japanese encephalitis is a severe disease of acute encephalitis, with children and the elderly primarily affected and with mortality rates reaching over 25% of those with overt disease. Many infections are asymptomatic or mild. Survivors often show neurologic sequelae. The disease is distributed throughout the rice-growing areas of Asia, from Japan south to Papua New Guinea and west to India and Nepal (Figure 14.27). In Japan, JE epidemics have occurred in August and September in many different years since its discovery there. The virus first was isolated from the brain of a human who died of encephalitis in 1935, then from brain tissue of a horse in 1937, and from *Culex tritaeniorhynchus*

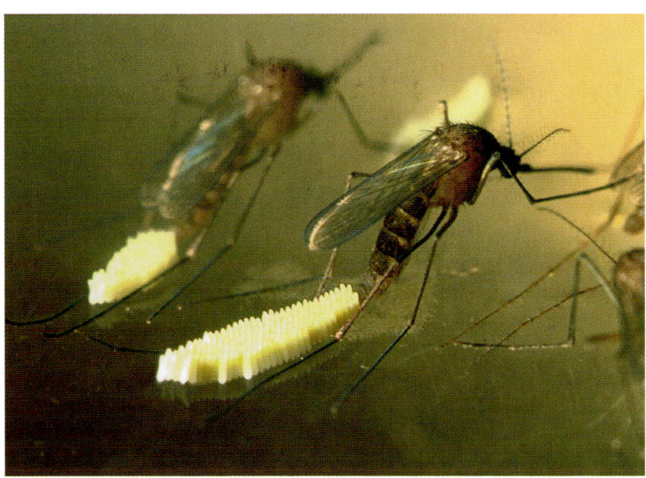

Figure 14.29 *Culex quinquefasciatus* females laying eggs in the form of floating rafts; this species is a vector of filariasis, West Nile virus, and St. Louis encephalitis, among other diseases. (Photo by W. A. Foster)

in 1938. Numbers of cases have declined in Japan, Korea, and Taiwan, because of vector control, vaccination, and changes in agricultural practices (Burke and Leake, 1988).

The enzootic cycle of JE virus involves *Cx. tritaeniorhynchus* and several other *Culex* species including *Cx. vishnui, Cx. pseudovishnui, Cx. fuscocephala,* and *Cx. gelidus,* all of which are associated with rice culture. Wading birds such as herons are important enzootic reservoirs. Pigs are important amplifier hosts in rural, rice-growing areas where swine are kept. Japanese encephalitis is probably the most important of the mosquito-borne encephalitis diseases, owing to its epidemic nature, widespread distribution, and large number of humans who acquire infection, die, or recover yet suffer neurologic sequelae.

St. Louis Encephalitis (SLE) Virus This virus was identified as the causative agent of disease during an outbreak of encephalitis-like illness in Paris, Illinois, in 1932 and St. Louis, Missouri, in 1933. It was isolated from a patient with encephalitis in the Yakima Valley of Washington in the early 1940s and found to be a frequent cause of human illness in the Central Valley of California in the 1930s and 1940s. This virus is distributed widely in North America and also occurs in parts of Mexico, Central America, the Caribbean, and South America to Argentina (Figure 14.27). The encephalitic illness caused by SLE virus shows a bimodal age distribution, with children and elderly people most frequently affected. Attack rates during epidemics range from 5 to 800 per 100,000 population, depending upon location, year, strain virulence, and population immunity due to earlier epidemics. In the eastern United States, mortality rates have ranged from about 3 to 20% of laboratory-diagnosed cases, but in the western United States mortality rates are lower.

In North America, three enzootic cycles of SLE virus have been described. In the eastern United States, north of Florida but including Texas, the primary vectors are *Culex pipiens* and *Cx. quinquefasciatus* (Figure 14.28). The former mosquito occurs in a more northerly distribution, whereas the latter is more southerly, with a hybrid zone at about the latitude of Memphis, Tennessee. Females of both species feed on birds. In addition, *Cx. quinquefasciatus* females frequently feed on mammals as the summer progresses. Whether these mosquitoes alone or other vectors function in transmission to humans depends upon the abundance of these two species and other competent vectors. House sparrows are important vertebrate amplifier hosts in peridomestic settings. In the western United States, a mosquito-bird-mosquito cycle similar to that of western equine encephalomyelitis virus, involving *Cx. tarsalis,* has been elucidated. In addition, in California both *Cx. pipiens* and *Cx. quinquefasciatus,* and possibly *Cx. stigmatosoma,* function secondarily as either enzootic or epidemic vectors. In Florida, *Cx. nigripalpus* apparently is the enzootic, epizootic, and epidemic vector of SLE virus. In Latin America and in the Caribbean basin, SLE virus has been isolated from many different species of *Culex, Sabethes, Mansonia, Wyeomyia,* and other genera, and from a wide variety of birds and mammals. In these areas, human SLE generally is rare. The mechanism of virus overwintering in North America is not well known. There is some evidence of virus persistence in overwintering, diapausing female mosquitoes.

The history of SLE in North America has been that of epidemics, either local or widespread, with intervening years when there was apparently no virus activity or epidemics, and either no or a few isolated human cases. The first epidemic, in the early 1930s in St. Louis, Missouri, was accompanied by hot and dry weather, which favored the development of populations of mosquitoes of the *Culex pipiens* complex, the larvae of which develop in water rich in sewage or other organic matter. The epidemic involved about 1,100 human cases and 200 deaths. Since that time, there have been some 50 outbreaks of SLE in the United States. Cases during three recent decades are shown by state in Figure 14.30. Human cases also have occurred in Manitoba and Ontario, Canada. These epidemics have been both rural and urban. A very large outbreak occurred in 1975, involving 30 states and the District of Columbia, with over 1,800 cases reported. More recently, epidemics have occurred in such disparate locations of the United States as Pine Bluff (Arkansas), Florida, Los Angeles (California), Houston (Texas), New Orleans (Louisiana), and Grand Junction (Colorado). A total of 4,437 human cases of SLE infection were reported to the US Centers for Disease Control and Prevention from 1964 to 1995, with an average of 139 cases per year (range, 4–1,967). Monath (1980) and Tsai and Mitchell (1988) have reviewed the ecology and public health significance of SLE.

Murray Valley Encephalitis (MVE) Virus This virus has been associated with encephalitis-type illness in humans in eastern and western parts of Australia and in New Guinea (Marshall 1988) (Figure 14.27). Mortality rates vary greatly during outbreaks, ranging from

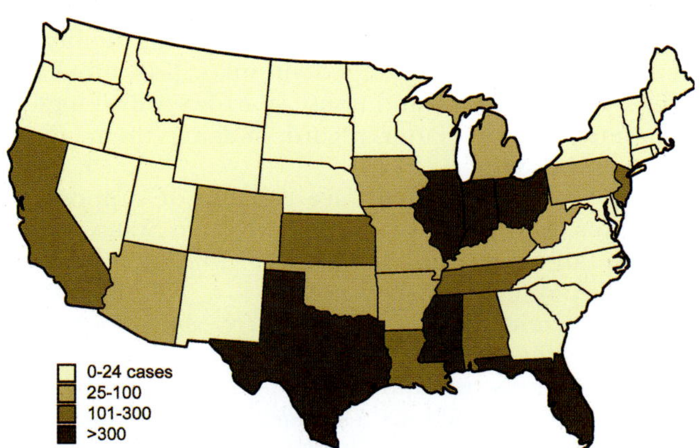

Figure 14.30 Historical distribution of human cases of St. Louis encephalitis virus infection in the United States, 1964 to 2007. (Reconstructed from the Centers for Disease Control and Prevention, http://www.cdc.gov/ncidod/dvbid/sle/MapsActivity/Sle_SurvControl.html)

18 to 80%. Epidemics of encephalitis in Australia in 1917, 1918, 1922, and 1925 were probably caused by MVE virus. It was isolated from the brain of a human in 1951 and from a pool of *Culex annulirostris* in 1960. Later epidemics occurred in Australia in 1956, 1971, 1974, 1978, 1981, and 1984. The enzootic cycle involves *Cx. annulirostris* as vector and birds as vertebrate reservoir and amplifier hosts. Epidemics appear to be associated with excessive rainfall, which allows increase of mosquito populations to high densities and the immigration of wading birds. **Kunjin virus** is closely related to MVE virus. It is also mosquito-borne, but it is associated rarely with human disease in Australia.

Other Flaviviruses

There are many other mosquito-borne flaviviruses of importance to human health.

Rocio (ROC) Virus This virus occurs in Brazil (Iversson, 1988). It appeared suddenly in São Paulo State in 1975 and caused an encephalitis-type illness in about 1,000 cases, with some fatalities. Most cases occurred among young, male agricultural workers. Only a single case has been reported since. The vector and vertebrate host relationships for this virus are poorly known, although *Psorophora ferox* and *Ochlerotatus scapularis* have been incriminated as vectors, based on virus isolation in the field and laboratory transmission experiments. A closely related pathogen is **Ilheus (ILH) virus**, which occurs in Trinidad, Panama, and parts of South America. It has been associated with about 10 documented human cases of illness and has been isolated from mosquitoes in eight different genera, but mostly from *Psorophora* species.

Bunyaviridae (*Orthobunyavirus* and *Phlebovirus*)

The Bunyaviridae includes the genera *Orthobunyavirus* and *Phlebovirus*. The viruses in the genus *Orthobunyavirus* comprise a complex and diverse group of more than 50 virus species distributed worldwide. Mosquitoes or biting midges serve as vectors, and small mammals, ungulates, or birds are the vertebrate hosts. *Phlebovirus* contains 37 viruses, most of which have phlebotomine sand fly vectors. It also includes the important mosquito-borne Rift Valley Fever virus.

Among the 30 virus species associated with mosquitoes and with human or animal diseases in the genus *Orthobunyavirus* are 14 serotypes in the species *California encephalitis virus* (Grimstad, 1988; Eldridge, 1990; Fauquet et al., 2005). Table 14.6 lists these virus serotypes and indicates their geographic distribution, mosquito vectors, and vertebrate host relationships. Characteristically they are transmitted by *Ochlerotatus* species, but other genera such as *Culiseta* and *Anopheles* may be involved. Nine occur only in North America. The others are distributed in various places worldwide. The viruses

Table 14.6 California Encephalitis Virus (*Orthobunyavirus*, Bunyaviridae) Mosquito Associations, Vertebrate Hosts, and Geographic Distribution of the Serotypes of this Virus Species

Serotype	Vector Associations	Vertebrate Assocations	Geographic Distribution
California encephalitis	*Ochlerotatus melanimon, Oc. dorsalis*	Lagomorphs	United States (western, southwestern)
Inkoo	*Ochlerotatus* and *Aedes* spp.	Lagomorphs?	Finland
La Crosse	*Oc. triseriatus*	Sciurid Rodents, Foxes	United States (eastern)
Snowshoe Hare	*Oc. stimulans* group *Oc. canadensis* *Culiseta inornata*	Lagomorphs	United States (northern), Canada
San Angelo	*Aedes, Anopheles, Psorophora?*	Unknown	United States (southwestern)
Tahyna	*Ae. vexans, Cs. annulata*	Lagomorphs	Europe, Tajikistan, Azerbaijan
Lumbo	*Ae. pembaensis?*	Unknown	East Africa
Melao	*Oc. scapularis?*	Unknown	Trinidad, Brazil, Panama
Jamestown Canyon[a]	*Cs. inornata* *Oc. communis group* *Oc. provocans* *Oc. abserratus* *Oc. intrudens* *Oc. stimulans group* *Anopheles* spp.	Deer	United States, Canada
South River	Unknown	Deer?	United States (northeastern)
Keystone	*Oc. atlanticus* *Oc. tormentor*	Lagomorphs, Cotton Rats	United States (coastal, eastern)
Serra do Navio	*Oc. fulvus*	Unknown	Brazil (Amapa state)
Trivittatus	*Oc. trivittatus*	Lagomorphs	United States
Guaroa	*Anopheles* spp.	Unknown	Panama, Colombia, Brazil

[a]Jerry Slough virus is a strain of Jamestown Canyon virus in the western United States.

in this species were isolated and described during the period from 1943 through the 1970s, culminating in a monograph by Calisher and Thompson (1983). Some of them are described as follows.

California Encephalitis (CE) Virus This is the prototype virus for the complex of the same name and was the first of the California serogroup viruses to be associated with human disease, involving three cases in California in 1943. It was isolated from *Ochlerotatus melanimon* and *Culex tarsalis* at that place and time. Extensive studies of the mosquito and vertebrate-host relationships of this virus in California have shown that *Oc. melanimon* and *Oc. dorsalis* are the principal vectors and that the virus is transmitted transovarially by these species. Serologic surveys have implicated jackrabbits, cottontail rabbits, California ground squirrels, and kangaroo rats as vertebrate hosts. The virus also has been isolated in New Mexico, Utah, and Texas, and in Canada. However, since the time of original discovery, CE virus only rarely has been associated with human disease.

La Crosse Encephalitis (LAC) Virus This is the most important human pathogen in the California serogroup, causing an acute, febrile illness in children. Most cases are subclinical or mild, but some progress to severe encephalitis and, rarely, death. In 1964, LAC virus was isolated from preserved brain tissue of a child who had died of encephalitis in 1960 in the vicinity of La Crosse, Wisconsin. It currently is distributed in the eastern United States, including the midwestern states bordering the Great Lakes, east to New York and Pennsylvania, south to West Virginia and North Carolina, and west to Texas. However, most human cases occur in West Virginia, Wisconsin, Illinois, Indiana, and Ohio.

The disease tends to be highly focal within its known range, such that particular regions or towns are known to be endemic. Prevalence varies regionally. In the United States, there were 2,245 cases of LAC reported to the Centers for Disease Control and Prevention (CDC) from 1964 to 1995, with an average of 70 per year (range, 29–160). In Ohio, where cases are particularly well documented, there was an average of 26 cases per year between 1963 and 1995. La Crosse encephalitis probably is underreported to public health agencies.

The principal vector of LAC virus is the eastern tree hole mosquito, *Ochlerotatus triseriatus*. The virus is transmitted both horizontally to sciurid rodents, particularly chipmunks and squirrels, and vertically from female mosquitoes to their progeny. The discovery of transovarial transmission of LAC virus was one of the first documentations of this phenomenon in mosquitoes and revealed an overwintering mechanism for LAC virus. It also demonstrated that vertebrate reservoirs were not always essential to the persistence of mosquito-borne viruses in nature, and that the mosquito itself could be a reservoir host. Thus an infected female is able to transmit the virus at its first blood feeding without previously having taken an infectious blood meal. Another important new finding was that *Oc. triseriatus* males, infected transovarially, transferred LAC virus to females via mating (i.e., venereal transmission).

Epidemiologic investigations of cases of encephalitis or aseptic meningitis of unknown origin often reveal LAC encephalitis in areas where previously it was unknown. Such investigations almost always reveal populations of *Ochlerotatus triseriatus* in the immediate vicinity where infection was thought to occur, such as backyards or wooded areas where children play. Water-filled artificial containers, particularly discarded tires, have become important habitats for *Oc. triseriatus* larvae and provide a link between the sylvan La Crosse cycle and humans. In Ohio and New York State, LAC virus also has been isolated repeatedly from *Ochlerotatus canadensis*; however, the role of this mosquito as a vector to humans and its role in an enzootic cycle are not well understood.

Snowshoe Hare (SSH) Virus This pathogen is closely related to La Crosse virus, but its ecology is very different. It originally was isolated from the blood of a snowshoe hare in Montana in 1958. Lagomorphs (hares and rabbits) are the enzootic vertebrate hosts. Snowshoe hare virus is distributed in the northern parts of the United States and in Canada, where it has been isolated from a variety of *Ochlerotatus* species and from *Culiseta inornata*. Even though SSH virus is very similar antigenically to La Crosse virus, human disease rarely has been documented except in Ontario, Quebec, and Nova Scotia (Canada), where 10 cases of an encephalitis-like illness have been attributed to SSH virus.

Keystone (KEY) Virus This was first isolated in 1964 from a collection of blood-fed *Ochlerotatus atlanticus* and *Oc. tormentor* in Florida. It is not considered to be a human pathogen. It occurs along the eastern seaboard of the United States where it has been isolated from *Oc. atlanticus, Oc. tormentor, Oc. infirmatus*, and other mosquitoes. Transovarial transmission of the virus has been demonstrated for *Oc. atlanticus* in the field. Gray squirrels and cottontail rabbits in northern coastal areas, and cottontail rabbits and cotton rats in Florida and Texas, have been identified as vertebrate hosts.

Trivittatus (TVT) Virus This was first isolated from *Ochlerotatus trivittatus* in North Dakota in 1948. It also has been isolated from other mosquitoes, including *Oc. infirmatus* in the southeastern United States, where *Oc. trivittatus* is absent. Transovarial transmission has been demonstrated in the latter species. Trivittatus virus shows a widespread distribution in the eastern half of the United States. Cottontail rabbits are vertebrate hosts.

Jamestown Canyon (JC) Virus This originally was isolated from *Culiseta inornata* in Colorado in 1961. Since that time it has been isolated in both Canada and the United States from *Ochlerotatus, Culiseta*, and *Anopheles* species in regions from Alaska east to Ontario and New England, south to Maryland, and in western and southwestern states, including California. The principal vectors are *Ochlerotatus* species with univoltine life cycles (i.e., snowpool and spring species). An antigenic strain known only from California is **Jerry Slough virus**, which is transmitted by *Culiseta inornata*. In the eastern United States, a JC-related serotype is **South River virus**. Transovarial transmission of JC virus has been demonstrated in some mosquito species. Its vertebrate hosts are large wild ungulates, especially deer. JC virus has been associated with encephalitis-type illness in humans in Ontario, New York, and Michigan.

Tahyna (TAH) Virus This is distributed widely in Europe and parts of western Asia. It has been associated with human febrile and central nervous system illnesses in France, the former Czechoslovakia, and Tajikistan. Foci are now known from Finland south to Tajikistan. Although the prevalence of infection in humans is poorly known, serosurveys in the Rhine River valley of Germany documented antibody to TAH virus in up to 23% of humans living in the area. In the former Czechoslovakia, TAH virus was implicated in 1% of febrile illnesses of children in an endemic area, and 20% of central nervous system illnesses. This virus was first isolated from *Aedes vexans* and *Ochlerotatus caspius* in Slovakia in 1958. The mosquito vectors are *Ae. vexans, Oc. caspius*, and *Culiseta annulata*. Hares and pigs are vertebrate reservoir hosts. **Lumbo virus** is a variety of TAH and occurs in parts of Africa.

Rift Valley Fever (RVF) Virus This pathogen (Figure 14.20) is classified with viruses in the genus *Phlebovirus* in the Bunyaviridae (Meegan and Bailey, 1988). It is distributed in eastern Africa north to Egypt, and in parts of West Africa, where it has been associated with large outbreaks of acute illness in livestock (see "Veterinary Importance," later). Humans may become infected by mosquito bites, or more commonly by contact with virus-contaminated blood or through inhalation of virus in aerosols during slaughter of livestock. Humans rarely die of infection but develop an illness including fever, headache, myalgia, retinitis, and in rare cases liver involvement. Large epizootics and epidemics have occurred in South Africa (1950–1951, 1953), Zimbabwe (1968–1969), Egypt (1977–1978, 1993), Mauritania (1987), and Kenya and Somalia (1997–1998). These outbreaks generally involved thousands to hundreds of thousands of cases in livestock. The epidemic in Egypt from 1977 to 1978 probably involved some 200,000 human cases and 600 deaths. The epidemic in Mauritania involved about 1,000 human cases and 50 deaths. In late 2006, an epidemic of RVF virus began in Kenya and spread to Somalia and Tanzania, causing 1,062 confirmed human illnesses and 315 deaths, a higher case fatality rate than in previous outbreaks.

A variety of mosquito vectors have been associated with RVF virus, including *Culex pipiens* in the Egyptian outbreak. In parts of the eastern African savanna, an enzootic cycle of RVF virus has been identified in and around *dambos*, low-lying temporary wetlands. Prolonged rainfall floods these areas and allows development of large populations of *Aedes mcintoshi* and other *Aedes* species. These mosquitoes maintain RVF virus through transovarial transmission and transmit it to domestic and wild ungulates that come to the dambos for water, functioning as enzootic vectors. As amplification ensues, epizootic vectors such as *Culex theileri* become important in transmission to domestic livestock.

Malaria

Malaria is one of the most widespread and prevalent of infectious human diseases. It is caused by sporozoan protists that infect blood tissues and other organs of the body, primarily the liver. The organisms are transmitted by *Anopheles* mosquitoes. The word "malaria" derives from the Italian *mala aria* or "bad air." Another term for malaria is *paludism*, from the French *paludisme* and the Spanish *paludismo*. Both of these words are derived from the Latin *palus*, meaning "swamp." The connotation is that malaria was contracted through association with swamps and inhaling the bad air emanating from them. In English-speaking countries, malaria was called **ague** from the Old French *agu* ("sharp"), from the context of the Latin *febris acuta*, or acute fever. Ague referred to the cyclic fevers and chills, or **paroxysms**, which are characteristic of malaria.

The organisms that cause human malaria are protozoans of the genus *Plasmodium*, family Plasmodiidae (Order Haemosporidida, Class Haemosporidea, Phylum Sporozoa). They are obligate intracellular parasites. There are four species of human malaria: *P. falciparum*, causing malignant tertian malaria; *P. vivax*, causing benign tertian malaria; *P. malariae*, causing quartan malaria; and *P. ovale*. The first two species are widespread in the tropics, whereas *P. vivax* also occurs in some temperate areas. *Plasmodium malariae* also is distributed widely but less commonly, and *P. ovale* is rare, occurring mainly in Africa. Recently, a fifth species of *Plasmodium* that causes quotidian malaria, *P. knowlesi*, has been discovered to be relatively common in the human population of some countries of Southeast Asia (28–84% of malaria cases in Malaysian Borneo). Heretofore this species had been known to occur almost exclusively in monkeys, with only occasional human infections.

Currently, 1.6 billion people are at direct risk of malaria infection via mosquito bite. Globally, an estimated 300 to 500 million cases occur annually, including perhaps 100 million cases and over 1 million deaths per year in sub-Saharan Africa, where 70-85% of all malignant tertian malaria cases occur. Travelers may become infected during visits to endemic areas. Malaria currently occurs in sub-Saharan Africa and parts of northern Africa; the Middle East to Iran, Afghanistan, and Pakistan; India and Sri Lanka; parts of China, and Southeast Asia, including Indonesia, the Philippines, Irian Jaya, New Guinea; and Latin America from Mexico through Central America to most of the northern half of South America (Figure 14.31).

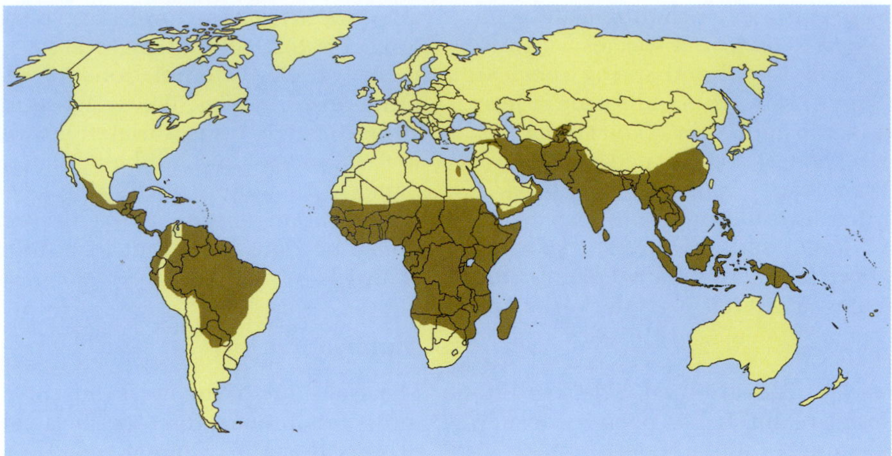

Figure 14.31 Geographic distribution of human malaria. (Reconstructed from the Centers for Disease Control and Prevention, http://www.cdc.gov/malaria/distribution_epi/distribution.htm and http://wwwn.cdc.gov/travel/yellowBookCh4-Malaria.aspx)

The degree of endemicity, or disease prevalence, depends upon a variety of factors, including the species of malaria present, environmental and social factors, and species of vectors present. Human malaria is thought not to have occurred in the Western Hemisphere prior to the period of European exploration, colonization, importation of African slaves, and establishment of intercontinental trade.

The literature on human malaria and its vectors is voluminous. A comprehensive source is Gilles and Warrell (1994). Other sources include Boyd (1949), Macdonald (1957), Molineaux and Gramiccia (1980), Wernsdorfer and McGregor (1988), and Strickland (1991).

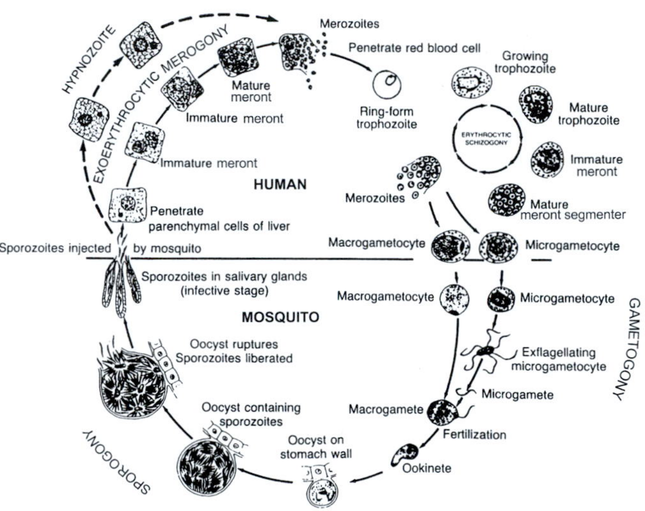

Figure 14.32 Life cycle of *Plasmodium vivax* in the human and *Anopheles* hosts. (Modified from Strickland, 1991)

Plasmodium Life Cycle The complex life cycle of *Plasmodium* species involves both sexual and asexual reproduction. The sexual phase (**gametogony**), begins in the blood of the human host and is completed within the lumen of the midgut of the mosquito. The first phase of asexual reproduction (**sporogony**), occurs on the outer wall of the midgut. The second phase of asexual reproduction occurs first in the liver and later in the blood of the human host. The process in both sites is termed **schizogony** or **merogony**. Sporogony often is referred to as the exogenous phase of malaria parasite development, because it occurs outside of the human host. Conversely, merogony is often referred to as the endogenous phase of development within the human host. These two developmental phases are depicted in Figure 14.32 and discussed in more detail later.

An *Anopheles* female becomes infected with malarial parasites when she ingests blood containing red blood cells that are infected with **gametocytes**, specifically the sexual microgametocyte and macrogametocyte stages of the parasite. A microgametocyte bursts from its host red blood cell within the blood meal in the midgut lumen of the mosquito, where it extends four to eight flagella-like forms called microgametes in a process termed **exflagellation**. A macrogametocyte sheds the erythrocytic membrane and transforms into a single mature macrogamete. One microgamete locates and fertilizes a macrogamete, forming a diploid zygote. The zygote transforms into a motile **ookinete**. The ookinete passes through the peritrophic membrane, then through the midgut epithelial cell membrane, and forms an **oocyst** between the midgut epithelial cells and the basement membrane of the epithelium. A single, malaria-infected *Anopheles* may have few to hundreds of oocysts, depending on the original number of gametocyte-infected red blood cells in the blood meal and on the number of macrogametocytes that become fertilized. Prior to sporogony, the encysted parasite becomes haploid again then undergoes multiple mitoses until the oocyst contains thousands of motile **sporozoites**. The oocyst bursts, releasing sporozoites, which make their way to the salivary glands and penetrate the secretory cells. The sporozoites accumulate within these cells, with some also passing into the salivary ducts. The mosquito then is infective. The sporozoites enter a human host when the mosquito probes and the salivary gland cells release saliva into the skin prior to ingesting blood. The period of time between ingestion of gametocytes and infection of the salivary glands with sporozoites is the extrinsic incubation period.

In the human host, sporozoites migrate in the blood to the liver within minutes of entering subdermal capillaries. They invade liver parenchymal cells where they typically form a **primary tissue meront**. In the case of *P. vivax* and *P. ovale* they may form a **hypnozoite**, a quiescent or resting stage of the parasite. Inside each meront, **merozoites** develop through the process of exoerythrocytic merogony. The meront then bursts, releasing merozoites into the blood stream, where they circulate and invade red blood cells. The merozoites form meronts in the red blood cells, where they produce more merozoites through a process called erythrocytic merogony. An invasive merozoite, once inside a red blood cell, transforms into a **trophozoite**, which utilizes hemoglobin for nutrients, then into a segmenter form distinguished by dark dots of heme in the red blood cell, and finally into a mature meront, which produces still more merozoites. Merozoites released from infected red blood cells invade new red blood cells where the process of erythrocytic merogony begins anew. Other invasive merozoites form microgametocytes or macrogametocytes within the blood cells, which will mature into microgametes and macrogametes if they are ingested by *Anopheles* mosquitoes during blood-feeding.

Clinical Disease Malaria is characterized by sudden paroxysms of fever and chills, which recur at highly predictable intervals, often in the afternoon. Other acute symptoms include headache, lethargy, fatigue, and profuse sweating after each bout of fever. After infection of erythrocytes by merozoites, erythrocytic merogony leads to the synchronous rupture of the erythrocytes and release of new merozoites, toxins, and heme digestion products. This event in the circulatory system prompts each episode of chills and fever. The next episode occurs in 24, 48, or 72 hours, depending upon the species of *Plasmodium*.

Malarial infections in humans can result in severe illness and sometimes death. However, the particular symptoms, including timing and severity, vary with the species of *Plasmodium*. The most severe form of malaria is caused by *P. falciparum*, whose merozoites invade both young and old red blood cells. Over time, repeated reinvasions and mass destruction of red blood cells may lead to high parasitemia, severe anemia, and anoxia of tissues. In some cases, hemolysis results in a condition of hemoglobinuria, blackwater fever, when the urine contains hemoglobin and turns reddish-brown. Toxins from dead red blood cells stimulate macrophages to produce chemicals such as Tumor Necrosis Factor and other cytokines, which cause characteristic malaria symptoms such as fever. In falciparum malaria, infected red blood cells stick to the vascular epithelium of capillaries in organs including the brain, impeding blood flow and causing a serious and sometimes lethal condition called **cerebral malaria**. Because of this affinity for internal organs, only very young trophozoites and gametocytes are common in peripheral blood. Malaria caused by *P. falciparum* is called **malignant tertian malaria** because of the severity of symptoms and because of the typical 48-hour interval between paroxysms. The term "tertian" for a two-day cycle originated from counting the day when the paroxysm occurs as the first day, so that the next paroxysm occurs on the third. Left untreated, nonfatal infections with *P. falciparum* last five months or more, depending on the immune status of the individual.

Plasmodium vivax malaria is called benign tertian malaria because symptoms are less severe than *P. falciparum* malaria, and death rarely occurs. Paroxysms occur on a 48-hour cycle. In this type of malaria, the merozoites invade only immature red blood cells, called **reticulocytes**, which typically comprise less than 6% of the total red blood cell count in circulation. Thus vivax malaria, compared to falciparum malaria, has less severe symptoms of anemia and toxemia, making death unlikely. The infected red blood cells do not stick to the epithelial lining of capillaries as they do in falciparum malaria. Vivax malaria can evolve into chronic infection with development of an enlarged spleen, or **splenomegaly**. However, persons infected with other malarias also may have enlarged spleens, as these organs work to replace red blood cells lost to infection. The hypnozoite stage of *P. vivax* provides a mechanism for the parasite to overwinter in humans in temperate areas with short transmission seasons. The period between infection and onset of symptoms can last up to nine months, and untreated infection persists in the body for many months to many years, with relapses recurring at irregular intervals after initial infection and acute onset of disease.

Plasmodium ovale is an uncommon tertian malaria with milder symptoms. Its course of infection is similar to that of *P. vivax*.

P. knowlesi, a species previously known only as one of the monkey malarias, morphologically similar to *P. malariae*, has a 24-hour or quotidian cycle of paroxysms. The high frequency of asexual reproduction in red blood cells results in hyperparasitemia, making it especially virulent if left untreated.

Plasmodium malariae, which causes quartan malaria, differs from *P. vivax* and *P. falciparum* in that the parasites invade only mature erythrocytes. Therefore, symptoms can be more severe than in vivax malaria in the acute phase. However, infections tend to develop more slowly and become chronic. Malaria caused by *P. malariae* has a 72-hour erythrocytic cycle. The term *quartan* refers to the four days included within one cycle, with a three-day interval from the beginning of the first paroxysm to the beginning of the next. Recrudescences of *P. malariae* may occur in individuals up to 50 years after initial infection, owing to low levels of parasitemia that increase under periods of immunosuppression.

After a person is bitten and inoculated with sporozoites, and exoerythrocytic merogony commences in the liver, symptoms do not appear until days to weeks later (up to a month in *P. malariae*), when erythrocytic merogony begins in the blood. In *P. vivax* and *P. ovale*, if the sporozoites develop into hypnozoites in the liver cells, relapses are possible long after inoculation and initial onset of symptoms, with an intervening period of no apparent symptoms of infection. For *P. falciparum* and *P. malariae*, there are no persistent exoerythrocytic stages of the parasites and relapses do not occur. However, infection with *P. malariae* may recrudesce years after initial infection owing to persistent erythrocytic infections. Therefore, in human malaria there is a clear distinction between relapse and recrudescence of infection. The course of infection of malaria in humans varies with many factors, including history of past exposure and presence of antibodies; age, health, and nutritional status; and genetic resistance factors such as the sickle-cell anemia trait, Duffy-negative blood type, certain hemoglobin types such as hemoglobin S and fetal hemoglobin; and deficiency of the erythrocytic enzyme glucose-6-phospate dehydrogenase.

Mosquito Vectors and Epidemiology Many different species of *Anopheles* mosquitoes are competent vectors of malaria organisms (Table 14.7). However, most *Anopheles* species are not, because of variation in host-selection patterns, longevity, abundance, and vector competence. In North America, *Anopheles quadrimaculatus*, which forms a complex with four more localized but nearly identical species (Reinert et al., 1997), is the principal vector of malaria in the eastern two-thirds of the continent. It develops along the edges of permanent pools, lakes, and swamps that provide relatively clean, still, sunlit water, with lush emergent vegetation, marginal brush, or floating debris to provide partial shade and protection from wave action. In western North America, *An. freeborni* is the main vector, an inhabitant of clear water in open, shallow, sunlit pools, ponds, ditches, and seepage areas that are partially shaded by vegetation. *Anopheles hermsi* also is a vector in California.

Other important vectors include *An. albimanus* in Central America, *An. darlingi* in South America, *An. gambiae* (Figure 14.33) and *An. funestus* in Africa, *An. culicifacies* in Asia, and *An. dirus* in Southeast Asia. *Anopheles gambiae* is considered the most important of all, because of its involvement in such large numbers of malaria cases and deaths, mainly in Africa. This species lives in close association with humans, on which it primarily feeds, and can complete a gonotrophic cycle in only two days. Larvae develop in a wide variety of sunlit surface pools during the rainy seasons, many of which are associated with human activity. These include borrow pits, roadside ditches, wheel ruts, and the hoof prints of domestic animals. Larval development normally takes only about one week.

Malaria has been viewed in the context of stable or unstable transmission, reflecting in part attributes of an *Anopheles* population that affect its vectorial capacity. These include density, longevity, tendency to feed on humans, and duration of the extrinsic incubation period of the parasite in the vector. Stable malaria most often is associated with *P. falciparum* infection in highly endemic settings. It is characterized by low fluctuations in parasite incidence in human and vector populations,

Table 14.7 *Anopheles* Vectors of Human Malaria Parasites in 12 Epidemiologic Zones: Subgenera, Species, and Geographic Distributions are Given (Macdonald 1957)

Malaria Epidemiologic Zone	*Anopheles* Vectors
North American	Subgenus *Anopheles: freeborni, punctipennis, quadrimaculatus*
	Subgenus *Nyssorhynchus: albimanus*
Central American	Subgenus *Anopheles: aztecus, pseudopunctipennis, punctimacula,*
	Subgenus *Nyssorhynchus: albimanus, albitarsis, allopha, aquasalis, argyritarsis, darlingi*
South American	Subgenus *Anopheles: pseudopunctipennis, punctimacula*
	Subgenus *Nyssorhynchus: albimanus, albitarsis, aquasalis, argyritarsis, braziliensis, darlingi, nuneztovari*
	Subgenus *Kerteszia: bellator, cruzii*
North Eurasian	Subgenus *Anopheles: atroparvus, messeae, sacharovi, sinensis*
	Subgenus *Cellia: pattoni*
Mediterranean	Subgenus *Anopheles: atroparvus, claviger, labranchiae, messeae, sacharovi*
	Subgenus *Cellia: hispaniola, pattoni*
Africo-Arabian	Subgenus *Cellia: hispaniola, multicolor, pharoensis, sergentii*
Africo-Tropical	Subgenus *Cellia: arabiensis, christyi, funestus, gambiae, melas, merus, moucheti, nili, pharoensis,*
Indo-Iranian	Subgenus *Anopheles: sacharovi*
	Subgenus *Cellia: annularis, culicifacies, fluviatilis, pulcherrimus, stephensi, superpictus, tesselatus*
Indo-Chinese Hills	Subgenus *Anopheles: nigerrimus*
	Subgenus *Cellia: annularis, culicifacies, dirus, fluviatilis, maculatus, minimus*
Malaysian	Subgenus *Anopheles: campestris, donaldi, letifer, nigerrimus, whartoni*
	Subgenus *Cellia: aconitus, balabacensis, dirus, flavirostris, leucosphyrus, ludlowae, maculatus, minimus, philippinensis, subpictus, sundaicus*
Chinese	Subgenus *Anopheles: anthropophagus, sinensis*
	Subgenus *Cellia: pattoni*
Australasian	Subgenus *Anopheles: bancrofti*
	Subgenus *Cellia: annulipes, farauti, karwari, koliensis, punctulatus, subpictus*

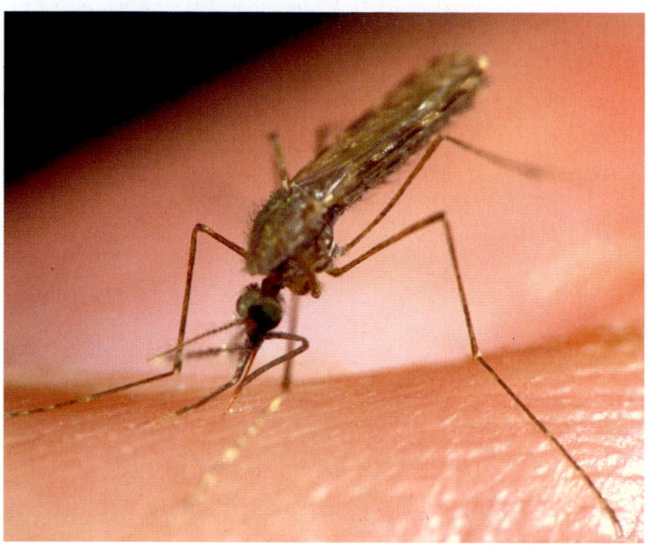

Figure 14.33 *Anopheles gambiae* female feeding on blood. This is the major vector of malaria in Africa, where most cases occur. (Photo by W. A. Foster)

high prevalence, and high seroprevalence for antibodies. Epidemics are unlikely under these conditions, even though transmission continues at high rates. In such settings, vectors tend to be highly anthropophagic, exhibit greater longevity, and have relatively low, stable densities but still exhibit considerable seasonal variation. Unstable malaria tends to be associated with *P. vivax* infections in endemic settings of high fluctuation in disease incidence. Vectors tend to be zoophagic, have seasonally profound variation in population densities, have low or nondetectable field infection rates, and may have shorter longevity than do those in stable malaria settings. Epidemics can occur in conditions of unstable malaria if environmental changes favor increased vector-human contact; for example, during civil strife, following water projects such as dams or irrigation schemes, or when a new vector species is introduced into an area.

The epidemiological implications of infection of humans in Southeast Asia with *P. knowlesi* have not yet been fully investigated. Because the same species is maintained in monkeys, possibly this form of malaria can be considered an anthropozoonosis. If so, even complete interruption of transmission between humans will fail to eradicate the parasite, as in the case of yellow fever.

Historical Perspective After the development of the germ theory of disease by Louis Pasteur, the French-Algerian physician Charles Louis Alphonse Laveran examined and described malarial organisms in the red blood cells of his patients in 1870. This finding, along with the work of Patrick Manson on filarial nematodes and mosquitoes in China, inspired Ronald Ross, then a physician in British colonial India, to examine the hypothesis of mosquito transmission of malaria parasites in the 1890s. His persistent and careful experimentation and observation with both human and bird malarias, using *Anopheles* and *Culex* mosquitoes, respectively, provided conclusive proof that mosquitoes transmit *Plasmodium* species by bite. In concurrent research, Giovanni Batista Grassi and colleagues demonstrated transmission of *Plasmodium falciparum* by *Anopheles maculipennis*-complex mosquitoes in the environs of Rome, Italy. Ross was awarded the Nobel Prize for Medicine in 1902.

Malaria was formerly endemic in many temperate areas of the United States, particularly in the South and Southeast. Malaria became epidemic after the Civil War, as malaria-infected soldiers returned to their homes and brought the infection with them to their local communities. Malaria was an important rural disease in the eastern and southern states, California, and other areas of the United States through the 1930s, but gradually disappeared by the 1940s. This was due to a combination of antimosquito measures, improved medical care, a higher standard of living, and transformation of marshes and swamps to agricultural land largely through organized ditching efforts. Changes in life style because of technological advances such as window screens and the invention of the radio, television, and air conditioning also contributed to the decline in malaria. Boyd (1941) reviewed the history of malaria in the United States, and to a brief degree elsewhere in the New World.

Roughly 1,000 cases of malaria are introduced into the United States each year. In addition, cases involving local or indigenous transmission occur sporadically, including recent outbreaks in California (1988, 1989, 1990), Florida (1990, 1996), Michigan (1995), New Jersey (1993), New York (1993), and Texas (1995). These incidents were due to introductions of infected humans into areas with competent *Anopheles* vectors. However, airport malaria has occurred near major international airports (e.g., London-Heathrow and Paris-DeGaulle) where infected mosquitoes have been imported on aircraft from endemic regions.

Filariasis

Filariasis is the infection of vertebrate tissues by filarial nematodes or roundworms (Phylum Nematoda, Order Spirurida, Superfamily Filarioidea, Family Onchocercidae). Mosquito-borne filarial nematodes are associated with acute and chronic human disease, termed **lymphatic filariasis**, which is widespread in tropical and subtropical regions (Grove, 1990). The three causative agents of lymphatic filariasis are *Wuchereria bancrofti*, *Brugia malayi*, and *Brugia timori*.

The areas of the world endemic for lymphatic filariasis (Figures 14.34 and 14.35) include parts of western, central, and southern Africa; parts of northeastern South America (principally Brazil, Surinam, and French Guyana), the Dominican Republic, and Haiti; southern and eastern India, southeastern Asia, eastern China, and southern Japan; the Malay archipelago, Indonesia, the Philippines, Irian Jaya and Papua New Guinea; and many island groups of the south Pacific Ocean, including Melanesia, Micronesia, and Polynesia. Within the United States, filariasis was locally endemic in Charleston, South

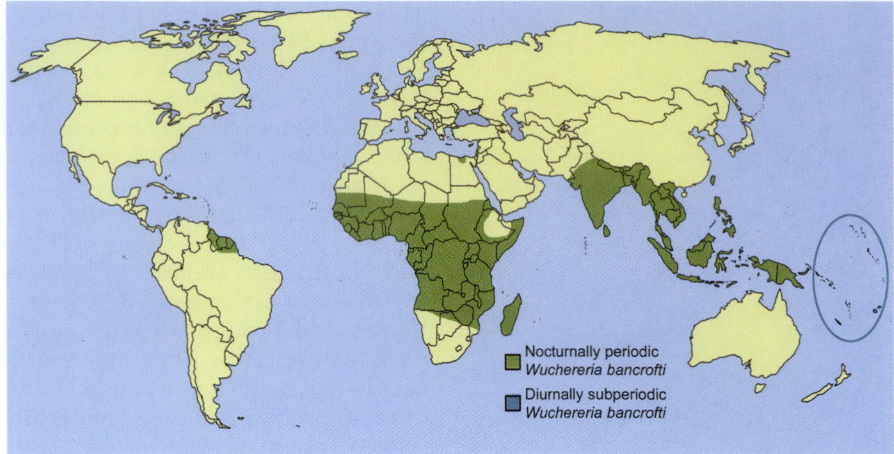

Figure 14.34 Geographic distribution of human lymphatic filariasis caused by *Wuchereria bancrofti*, showing both nocturnal and diurnal periodicity. (Reconstructed from Strickland, 1991, and other sources)

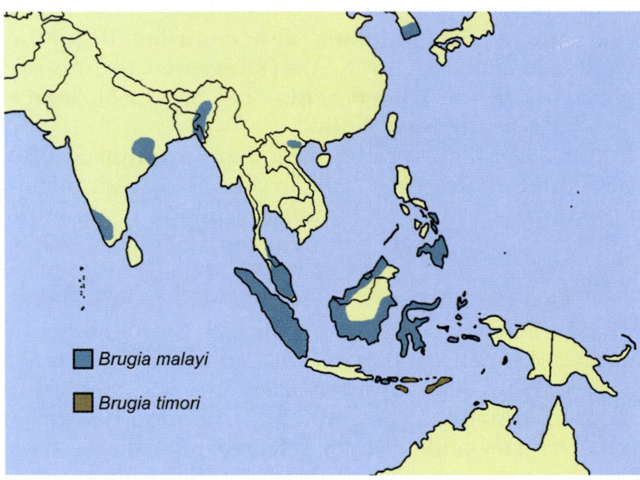

Figure 14.35 Geographic distribution of human lymphatic filariasis caused by *Brugia malayi* and *B. timori*. (Reconstructed from Strickland, 1991, and other sources)

Carolina, but the disease disappeared there in the late 1930s. It disappeared at about the same time from northern Australia. It no longer occurs in regions of the Mediterranean basin and on the Arabian Peninsula. Recently, however, incidence of lymphatic filariasis has increased in the Nile River Delta of Egypt.

Within the area of current distribution, there are an estimated 905 million people at risk of contracting lymphatic filariasis, and some 128 million active infections. Of these, about 115 million are caused by *W. bancrofti*, the causative agent of Bancroftian filariasis, which is widespread in both the Old World and New World tropics (Figure 14.34). Another 13 million cases are caused by *B. malayi*, the causative agent of Brugian filariasis or Malayan filariasis, which is restricted to southeastern Asia (Figure 14.35). About 43 million people have chronic symptoms of elephantiasis, hydrocele, or lymphedema (see later). Another Brugian filariasis, Timorian filariasis, is caused by infection with *B. timori* and occurs in localized foci among southern islands of Indonesia.

Filarial Life Cycle Infection with filarial nematodes in humans begins when infective third-stage larvae enter the skin at the site of the mosquito bite, molt twice, and migrate to the lymphatic vessels and lymph nodes, particularly of the lower abdomen. There, the nematodes develop to the adult stage. Female worms (80–100 mm long at maturity for *W. bancrofti* and about half of that length for *B. malayi*) release active, immature worms called **microfilariae** into the peripheral circulatory system. A female may release 50,000 or more microfilariae each day. Microfilariae of *W. bancrofti* are 250 to 300 μm long and about 7 to 9 μm wide; those of *B. malayi* are somewhat shorter and thinner. Presence of microfilariae in the blood is called **microfilaremia** and first appears about six months to a year after adult worms become established in the lymphatic system. An infected human may be microfilaremic for more than 10 years. The density of microfilariae in peripheral blood is highly variable, but it can range from 1 to over 500 microfilariae per 20 mm^3.

The appearance of microfilariae in the peripheral blood has a 24-hour cycle; that is, they exhibit diel periodicity. If microfilariae completely disappear from the peripheral circulation at some time during the day, they are said to be periodic. If the microfilariae fluctuate in density during a 24-hour period but are detectable at all times, they are said to be subperiodic. In most areas, the microfilariae appear only at night and are transmitted by mosquitoes that have night-biting habits. These are the nocturnally periodic forms of *W. bancrofti* and *B. malayi*. Both *W. bancrofti* and *B. malayi* also have nocturnally subperiodic and diurnally subperiodic forms, though nocturnally subperiodic Bancroftian and diurnally subperiodic Brugian forms have very restricted distributions. In all the subperiodic forms, the microfilariae appear in the peripheral blood of the human host mainly in the evening and night (nocturnally subperiodic) or mainly during the daytime (diurnally subperiodic). These nematodes are associated with two or more species of mosquito vectors, which differ in their typical biting times and whose combined diel patterns of man-biting density are matched by the periodicity of microfilariae. Both nocturnally periodic *W. bancrofti* and nocturnally periodic *B. malayi* are now considered to be strictly human pathogens. However, *B. malayi* in its subperiodic form is a zoonosis, with both leaf monkeys and humans as reservoirs, and with domestic cats and other carnivores also implicated as hosts. *Brugia timori* is nocturnally periodic and has no animal reservoir.

Development of *Wuchereria bancrofti* and *Brugia malayi* in mosquitoes is similar. Microfilariae ingested with the mosquito blood meal usually shed their outer, sheath-like membrane as they penetrate through the midgut epithelium. Some microfilariae retain the sheath during penetration. The microfilariae move to the indirect flight muscles of the thorax, penetrate individual cells, and transform to a short sausage stage, the L_1 or first-stage larva. They molt to more slender L_2 or second-stage larvae and then again to the elongate, filariform L_3 or third-stage infective larvae (about 1.5 mm long). These larvae leave the thoracic flight muscles, traverse the hemocoel of the mosquito, enter the lumen of the mouthparts, and eventually arrive at the apex of the proboscis. When the mosquito blood-feeds, the L_3 larvae exit through the cuticle of the labium, crawl onto the skin, and enter through the hole made by the mosquito during feeding. Therefore, technically the transmission method may be said to be contaminative, rather than inoculative. Heavy infections of larval nematodes can be fatal to the mosquito.

Many species of mosquitoes are refractory to filarial development, owing to genetic factors and to their ability to mount adequate immune responses, whereas other species are susceptible to parasite infection and support the development described earlier. The relationships between *W. bancrofti* and certain mosquito species exhibit evidence of local adaptation. For example, in West Africa, *Culex quinquefasciatus* is not a competent vector of *W. bancrofti*, whereas *Anopheles gambiae* is competent there. By contrast, in India, *Cx. quinquefasciatus* is a competent vector and most *Anopheles* species are not. Thus, there is geographic variation in susceptibility of mosquitoes to filarial nematodes.

Clinical Disease Generally, a case of human infection does not occur after the bite of a single, infective mosquito. Rather it results after accumulation of hundreds to thousands of such infective bites, under conditions in which there is a high probability of parasite maturation and mating. Humans may show disease symptoms without having microfilaremia, or may have microfilaremia without showing signs of disease. Lymphatic filariasis has both acute and chronic manifestations in the human host. The disease may cause chronic debilitation in untreated cases. Acute lymphatic filariasis is characterized by episodes of fever, swelling, pain, and inflammation in the affected lymph nodes and lymph vessels, a condition called **adenolymphangitis**. The episodes may last for several days and incapacitate the affected individual because of the local and systemic effects. Over time, deep abscesses may develop at the sites of inflammation. Dermal ulcers may form through the skin over these sites, and secondary bacterial infection may ensue.

In the chronic phase, which often occurs years after onset of acute symptoms, the pathology may involve accumulation of lymphatic fluid (lymphedema) in the limbs, breasts, vulva, and scrotum, resulting in swelling and enlargement. In the scrotum, this condition is termed **hydrocele**. The grotesque distentions and thickening, folding, and nodulation of the skin, notably the lower limbs, is a condition called **elephantiasis** (Figures 14.36 and 14.37). Bacterial and fungal infections at affected sites exacerbate these conditions. Appearance of lymph fluid in urine may occur as a consequence of the disruption of the abdominal lymphatic vessels and leakage of lymph fluid into the urinary tract, causing urine to appear whitish, a condition called **chyluria**. In contrast with *W. bancrofti*, infection with *B. malayi* is not associated with scrotal distension, but rather involves only the limbs. Hypersensitivity to parasite-associated antigens also may be part of the syndrome of lymphatic filariasis. It is mediated through elevated IgE and IgG4 antibodies and is characterized by increased production of eosinophils, coughing, and shortness of breath. This type of filarial disease is called tropical eosinophilia.

Lymphatic filariasis has important social implications in communities where it is endemic. Acutely affected individuals are often feverish and in pain, and they may have difficulty working and thus suffer economic loss. Hard work may bring on attacks of

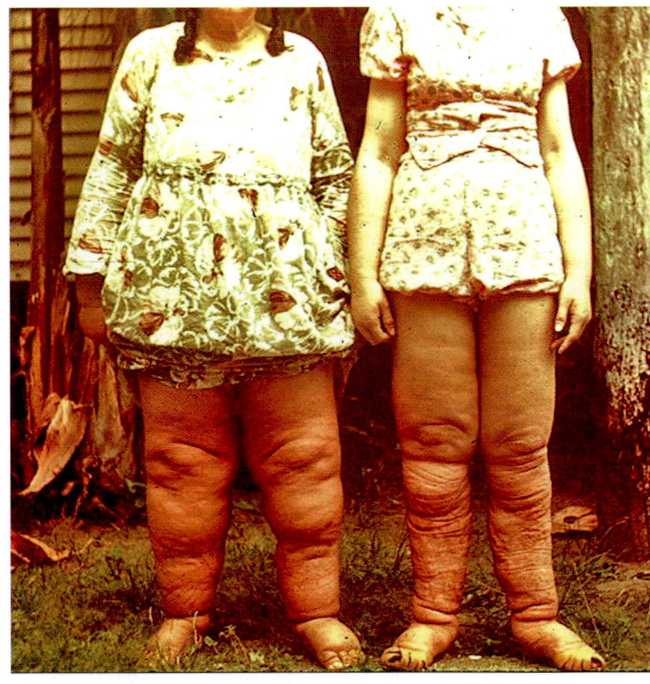

Figure 14.36 Human lymphatic filariasis in two Tahitian women, 60-year-old (left) and 30-year-old (right), in the early 1940s; both individuals after enlargement of legs reduced 30–50% of their former size by tight bandaging for 6 months, enabling them to walk again. (Photo by A. Edgar)

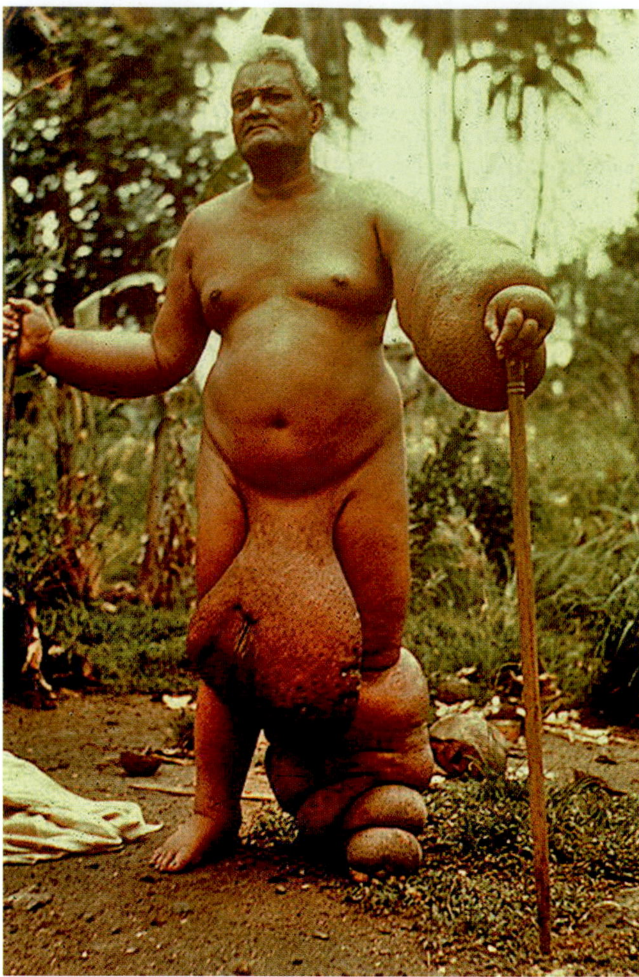

Figure 14.37 Human lymphatic filariasis in Tahitian man, with extreme case of elephantiasis involving notably the left arm, left leg, and scrotum; photograph taken during World War II. (Photo by A. Edgar)

filarial fever, which require rest for recovery. In a study conducted in Ghana, West Africa, the episodes of acute adenolymphangitis lasted about five days, with three days of incapacitation, and occurred during those months of the year when peak agricultural work was required and when mosquito transmission of infective stage larvae was highest (Gyapong et al., 1996). Likely, chronically-infected individuals are immunologically sensitized to their worm infections, and exposure to new L_3 larvae results in hypersensitive reactions such as adenitis. Chronically affected individuals with symptoms of scrotal hydrocele and elephantiasis may have difficulty in their social and personal lives and may suffer incontinence and impotence. Although hydrocele can be treated through fluid aspiration from the scrotum, the gross distention of elephantiasis is more difficult to remedy even with surgery.

Mosquito Vectors and Epidemiology The filarial nematodes that cause lymphatic filariasis have evolved associations with mosquitoes in the genera *Culex, Mansonia, Aedes, Ochlerotatus,* and *Anopheles* (Table 14.8). This probably has occurred through a process of adaptive radiation from the original *Anopheles* vectors in Southeast Asia. A very likely scenario for *W. bancrofti* is that it arose as a human pathogen in forested regions of Indonesia and perhaps other parts of Southeast Asia, where the *Anopheles umbrosus* group of mosquitoes serve as vectors of *Wuchereria kalimantani.* This filarioid nematode is a parasite of silvered leaf monkeys (also called brow-ridged langurs), *Presbytis cristata.* Similarly, *B. malayi* infection in humans probably evolved from subperiodic *B. malayi* infections in leaf monkeys (*Presbytis* spp.), with *An. hyrcanus* as the vector. Human infection and new disease foci probably arose in forests and along forest ecotones with these same *Anopheles* vectors. Through time, as people developed agricultural systems and migrated to other regions, both *W. bancrofti* and *B. malayi* adapted to these new settings and the competent mosquito vectors there. In some parts of Southeast Asia, members of the *Ochlerotatus niveus* group are important vectors of the subperiodic form of *W. bancrofti,* including *Oc. niveus* in Thailand; this system may have been the origin of the subperiodic strains that radiated into the Pacific regions.

Transmission of *W. bancrofti* occurs in both urban and rural areas. The primary mosquito vector in urban areas is *Culex quinquefasciatus.* It is the most important vector of nocturnally periodic *W. bancrofti* in the Americas and parts of Africa and Asia, particularly India. It feeds opportunistically at night on both mammals and birds. This mosquito occurs abundantly in areas with poor sanitation, open sewers, untreated waste water, and pit latrines, which provide the high organic content and low oxygen characteristic of the larval habitat. For this reason, *Cx. quinquefasciatus* is often more abundant during parts of the year when water stagnates from lack of rain. In the Nile River Delta of Egypt, *Culex molestus,* a name applied to the autogenous variant of *Cx. pipiens,* is the primary vector. In rural settings, nocturnally periodic Bancroftian and Brugian filariases are transmitted by *Anopheles* mosquitoes, which are also nocturnally active. Often, the same *Anopheles* species that transmit *W. bancrofti* or *B. malayi* in an area also are responsible for local malaria transmission (e.g., *An. darlingi* in South America and *An. gambiae* and *An. funestus* in parts of West and East Africa, respectively, for Bancroftian filariasis; and *An. sinensis* in rice-growing areas of China, for Brugian filariasis).

Nocturnally periodic Brugian filariasis occurs in rural parts of southern India, Malaysia, the Philippines, and Indonesia; there the nocturnally active *Mansonia annulifera* and *Ma. uniformis,* and also *Anopheles* species, are vectors near rice fields and open swamps. The nocturnally subperiodic form occurs in swamp forest areas of Southeast Asia and Indonesia, involving *Ma. bonneae* and *Ma. dives,* which are nocturnally active but also feed during the day within the swamp forests. All these *Mansonia* species are associated with particular kinds of plants, where the larvae and pupae attach to their submerged roots and stems. In the Pacific region, where many island groups are endemic for *W. bancrofti,* the primary vectors are day-biting *Aedes* and *Ochlerotatus* species, but nocturnal

Table 14.8 Mosquito Vectors of Filarioid Nematodes of Humans: Geographic Distribution and Associations with Periodicity of Microfilaremia

Geographic Region	Filarioid Species	Periodicity	Mosquito Vectors
Neotropical	*Wuchereria bancrofti*	Nocturnally Periodic	*Anopheles aquasalis, An. bellator, An. darlingi, Ochlerotatus scapularis, Culex quinquefasciatus, Mansonia titillans.*
Afrotropical	*Wuchereria bancrofti*	Nocturnally Periodic	*Anopheles funestus, An. gambiae, An. arabiensis, An. bwambae, An. melas, An. merus, An. nili, An. pauliani, Culex quinquefasciatus.*
Middle Eastern	*Wuchereria bancrofti*	Nocturnally Periodic	*Culex molestus.*
Oriental	*Wuchereria bancrofti*	Nocturnally Periodic	*Anopheles anthropophagus, An. kweiyangensis, An. nigerrimus, An. letifer, An. whartoni, An. aconitus, An. flavirostris, An. minimus, An. candidiensis, An. balabacensis, An. leucosphyrus, An. maculatus, An. philippinensis, An. subpictus, An. vagus, Ochlerotatus niveus, Oc. togoi, Oc. poicilius, Culex bitaeniorhynchus, Cx. sitiens* complex, *Cx. pallens, Cx. quinquefasciatus, Mansonia uniformis.*
	Wuchereria bancrofti	Nocturnally Subperiodic	*Anopheles sinensis* complex, *Ochlerotatus harinasutai.*
	Brugia malayi	Nocturnally Periodic	*Anopheles barbirostris, An. campestris, An. donaldi, An. anthropophagus, An. kweiyangensis, An. nigerrimus, Ochlerotatus togoi, Mansonia uniformis, Ma. bonneae, Ma. dives.*
	Brugia malayi	Nocturnally Subperiodic	*Anopheles sinensis* complex, *Mansonia uniformis, Ma. bonneae, Ma. annulata, Ma. indiana, Ma. dives.*
	Brugia timori	Nocturnally Periodic	*Anopheles barbirostris.*
Western Pacific	*Wuchereria bancrofti*	Nocturnally Periodic	*Culex pallens, Ochlerotatus poicilius, Oc. togoi.*
	Brugia malayi	Nocturnally Subperiodic	*Oc. togoi.*
Papuan	*Wuchereria bancrofti*	Nocturnally Periodic	*Anopheles bancrofti, An. punctulatus, An. farauti, An. koliensis, Culex annulirostris, Cx. bitaeniorhynchus, Mansonia uniformis.*
South Pacific	*Wuchereria bancrofti*	Nocturnally Subperiodic	*Ochlerotatus samoanus.*
		Diurnally Subperiodic	*Ochlerotatus fijiensis, Oc. oceanicus, Oc. vigilax* group, *Aedes futunae, Ae. polynesiensis, Ae. pseudoscutellaris, Ae. tabu, Ae. tongae, Ae. upolensis.*
	Brugia malayi	Nocturnally Periodic	*Ochlerotatus oceanicus.*
		Nocturnally Subperiodic	*Ochlerotatus oceanicus.*

biters also are involved, and the form of the parasite is diurnally subperiodic. *Anopheles barbirostris* is the vector of *Brugia timori.*

The endemicity of mosquito-borne filariasis depends on a high and steady rate of transmission of infective-stage larvae in the human population. In endemic areas, the inoculation rate (parasite transfer rate) can range as high as hundreds of infective bites and thousands of larval inoculations per person per year. The estimated number of L_3 larvae transmitted per person per year is called the annual transmission potential. With each infective bite, only a few L_3 larvae actually enter the skin. These larvae must then develop further and migrate to a person's lymphatic system, where mature male and female nematodes mate and initiate microfilarial production. Accumulation of thousands of infective bites over months or years eventually results in an infection of mature worms in a human, who normally then will have a microfilaremia, and possibly, chronic disease. Most microfilariae entering the circulatory system are never ingested by a mosquito, but those that are ingested become infective-stage larvae only in a competent vector and only if the individual mosquitoes survive the extrinsic incubation period. Furthermore, many infective-stage larvae fail to reach a new human host or fail to mature if they do. The inefficiency of transmission and parasite perpetuation is compensated by the prodigious production of microfilariae and the long life of adult worms.

Historical Perspective Association of infection with filarial nematodes and lymphatic filariasis was first established in the late 1800s. Our understanding of the natural history of lymphatic filariasis is related intimately to the initial discovery of a link between human pathogens and insect vectors. During 1877 to 1878, Patrick Manson, working in China as a medical officer for the Chinese Imperial Customs Service, conducted experiments on the development of filarial nematodes. Manson had already discovered that the microfilariae occurred in the peripheral blood only at night. He speculated that this was timed to coincide with the night-time biting activity of mosquitoes. After feeding mosquitoes (*Culex quinquefasciatus*) on his gardener, who had a microfilaremia, and then dissecting the mosquitoes on successive days, he found that the worms developed within the mosquitoes into longer, different forms. He speculated that the mosquito functioned as a kind of "nurse" for the filarial worms, so that when a mosquito died on the water after laying an egg raft, the worms entered the water and later infected a person drinking it. At that time, it was not known that mosquitoes could bite more than once during their lives, so the principle of transmission by bite was not established. Yet, the idea that mosquitoes can function as intermediate hosts for a human pathogen is founded on Manson's experiments.

VETERINARY IMPORTANCE

Aside from their importance as vectors of disease agents of animals, mosquitoes are a cause of irritation, blood loss, and allergic reactions. They are not only annoying but also disrupt normal behavior of livestock and companion animals. Large swarms may cause livestock to discontinue feeding and to seek relief. Increased scratching behavior may result in skin abrasions, hair loss, and secondary infection with bacteria at the bite and scratch sites. For cattle, mosquito bites can result in decreased weight gains and milk production and prompt producers to alter pasturing practices. Deaths of cattle due to anemia and stress have been reported.

Mosquito-Borne Viruses of Animals

Mosquito-borne viruses affecting domesticated animals include the groups of alphaviruses that are associated with the equine encephalitides (eastern equine encephalomyelitis, western equine encephalomyelitis, and Venezuelan equine encephalomyelitis), all of which cause an acute encephalitis with high fever in equids (horses, donkeys, mules). The history, distribution, vector relationships, and vertebrate reservoir hosts of these viruses were discussed earlier in the section "Public Health Importance." Other mosquito-borne viruses of veterinary significance include Japanese encephalitis virus, Rift Valley fever virus, Wesselsbron virus, fowl pox virus, and myxomatosis virus. Equine infectious anemia (EIA) virus, a lentivirus in the family Retroviridae, may be mechanically transmitted by mosquitoes, but its more important mechanical vectors are larger biting flies (deer flies, horse flies, and stable flies).

Eastern Equine Encephalomyelitis (EEE) Virus This virus is an important cause of mortality of horses and other equids, caged pheasants, whooping cranes, and emus. It occurs in endemic areas of the United States in Texas, along the Gulf coast and Atlantic seaboard to Massachusetts, and at inland sites in upstate New York, Ohio, Michigan, Indiana, Georgia, and Alabama. Horses rapidly succumb to infection after a short incubation period of two to five days. They exhibit abnormal behavior and high fever, then drop to the ground and lapse into coma before death (Figure 14.38). Few horses survive infection involving these acute symptoms. Viral infection in the brain shows characteristic lesions in nerve tissue, accompanied by perivascular cuffing with macrophages; viral antigen is detectable in neurons (Figure 14.39). In pheasant flocks, often a single infected, sick bird will be pecked by other birds, thus transmitting the virus directly to healthy birds without mosquito bite. During such occurrences, called **epiornitics**, thousands of pheasants in a single outdoor pen may die, yet none of the pheasants in adjacent pens become infected. Aside from equids and exotic birds, EEE viral infection has been reported in young dogs and pigs.

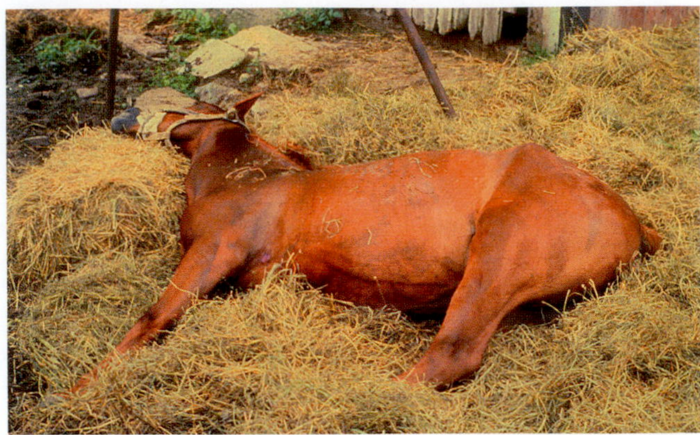

Figure 14.38 Horse dying from infection with eastern equine encephalomyelitis virus in Michigan outbreak in 1980. (Photo by H. D. Newson)

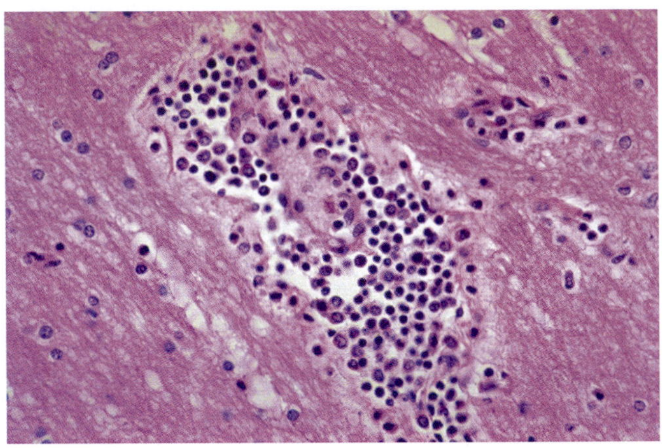

Figure 14.39 Section of horse brain (cerebrum) infected with eastern equine encephalomyelitis virus, showing neutrophil invasion around capillary. (Photo by J. D. Patterson)

Cases of EEE in horses in cool temperate climates tend to occur in mid to late summer and early fall, whereas in milder climates horse cases begin to occur earlier in the spring and summer. In the tropics and subtropics, cases may occur year-round. Horse deaths due to EEE viral infection are an important indicator of virus activity promoted by bridge vectors in endemic areas. Rapid, differential diagnosis of horse cases is crucial if these animals are to be used as sentinels for potential transmission of EEE virus to humans.

Western Equine Encephalomyelitis (WEE) Virus As with EEE virus, the primary epidemic host of significance for WEE virus is the horse. Human cases are rare. Since the first isolation of WEE virus from the brain of a dead horse in 1930 in the San Joaquin Valley of California, this virus has been implicated in epizootics in horses, with cases numbering from hundreds to thousands in some instances. Large outbreaks occurred in 1941 and 1975 in the Red River Valley in Minnesota, North Dakota, and Manitoba, in 1952 in the Central Valley of California, and in 1965 in Hale County in western Texas. There are horse cases almost every summer within the range of the virus, but epizootics do not always occur. For both EEE and WEE viruses, immunization has reduced the frequency of horse cases.

Venezuelan Equine Encephalomyelitis (VEE) Complex Encephalomyelitis in equids caused by viruses of the VEE complex occurs in northern South America, Central America, and Mexico. The epizootic viruses are transmitted by many species of mosquitoes (see Table 14.4) among horses, burros, and mules. These animals develop a viremia sufficient to infect the mosquitoes. Consequently, VEE epidemics can be maintained by transmission between mosquito and horses, differing in this regard from WEE and EEE, for which horses are largely dead-end hosts. An epidemic strain of VEE virus was first isolated in 1938 from a horse in Venezuela. In 1969, a large outbreak of VEE involving both equids and humans in Central America spread northward in the succeeding two years through Mexico, and in 1971 across the border into Texas (PAHO, 1972). Cases continued in Mexico through 1972. There were thousands of horse cases throughout this region during that time. Epizootic virus activity did not occur again in the region until an outbreak in Venezuela in 1992 and 1993 and in Chiapas, Mexico, in 1993. Another outbreak of VEE occurred in northern Colombia and Venezuela in 1995. The rapidity of spread of these outbreaks over large geographic areas is undoubtedly due to the role of both horses and birds as competent reservoir hosts. It is expedited by the evacuation of horses, already infected but not yet ill, away from an epizootic area.

Japanese Encephalitis (JE) Virus Encephalitis caused by JE virus occurs in widespread parts of Asia, including Malaysia and Indonesia. In Japan, epizootics and epidemics have occurred in August and September in many years since the discovery of this disease in 1935 in that country. The virus was isolated from brain tissue of a horse in 1937. Japanese encephalitis virus causes acute infection in horses and swine. It is particularly an economic problem because of the importance of swine as a food source and market commodity in rural Asia. Pigs develop viremia sufficient for mosquito transmission, therefore serving as important amplifying hosts, and may develop encephalitic symptoms. Transplacental infection causes stillbirth and abortion. Infected boars may become sterile.

Rift Valley Fever (RVF) Virus This pathogen has caused epizootics of acute illness, elevated rates of abortion, and death in cattle, goats, and sheep in Egypt and parts of sub-Saharan Africa. Outbreaks in Egypt and Mauritania were particularly noteworthy. The virus is both viscerotropic and neurotropic in these animals. Their viremias are of sufficient titer to infect mosquitoes. Disease outbreaks generally have involved thousands to hundreds of thousands of livestock cases, causing substantial economic losses, but these losses are rarely counted accurately.

Wesselsbron (WSL) Virus This is a flavivirus with distribution in parts of sub-Saharan Africa, Madagascar, and Thailand. It causes a disease similar to that of RVF in sheep and goats and also causes a mild illness in cattle. Infected ewes may abort their fetuses, and lambs suffer high mortality. Humans infected with Wesselsbron virus may develop a febrile illness with rash, fever, and myalgia. The virus is transmitted by *Aedes* species, including *Ae. mcintoshi* and *Ae. circumluteolus* in South Africa.

Fowlpox Virus This virus belongs to a group of poxviruses that infect vertebrates and invertebrates and are classified within the family Poxviridae. Among the poxviruses are those in the bird-infecting genus *Avipoxvirus*, such as fowlpox, canarypox, and pigeonpox viruses. Mosquitoes may mechanically transmit the avipox viruses by contamination of mouthparts and subsequent transfer of infectious virions to noninfected birds. Fowlpox is an important disease of domestic fowl, particularly chickens. It causes development of papules along the comb and beak. While probing these papules, mosquitoes may contaminate their mouthparts with virions. If disturbed during feeding, they may move to another animal to feed, thus transferring the virus to a new host. Another form of fowlpox virus is transmitted directly by droplets of pus containing the virus.

Myxoma (MYX) Virus This is a leporivirus and the causative agent of **myxomatosis**, an enzootic disease of lagomorphs in parts of South America and the western United States. It is transmitted mechanically by the bite of arthropods, principally by mosquitoes and fleas, depending on the region. These viruses produce dermal, vascularized tumors. When vectors probe these tumors, the mouthparts become contaminated with virus particles. Later, if the mosquitoes probe another uninfected lagomorph, that animal may become infected. Natural infections of myxoma virus occur without acute disease in rabbits of the genus *Sylvilagus* in South America and California. However, Old World rabbits (*Oryctolagus cuniculus*) are highly susceptible to infection and generally die. Outbreaks of acute disease among domesticated Old World rabbits have been documented in South America and California.

Myxoma virus was introduced into Australia in the 1950s as a means of controlling introduced European rabbits, a pest in that country. The virus spread rapidly through the rabbit populations via mechanical transmission by mosquitoes, fleas, and other means, and greatly reduced rabbit populations there. The mosquito vectors in Australia are *Culex annulirostris, Anopheles annulipes*, and *Aedes* species.

Nonhuman Malarias

Many *Plasmodium* species infect animals other than humans, including reptiles, birds, rodents, and nonhuman primates.

Reptilian Malarias The malarias of reptiles, formerly called saurian malarias, are caused by a group of 29 *Plasmodium* species. They infect a wide range of lizards and some snakes in 15 families (Telford, 1994). Vectors are biting midges, phlebotomine sand flies, and *Culex* mosquitoes. Haemoproteid and leucocytozooid malarias also occur in reptiles, but their vectors have not been established.

Avian Malarias Malarial infection of birds is widespread geographically (Van Riper et al., 1994). Parasites in three common genera of hemosporine blood parasites of birds (*Hepatocystis, Haemoproteus,* and *Leucocytozoon*) are transmitted by biting midge, louse fly, and black fly vectors, respectively. The avian malarias in the genus *Plasmodium* are all mosquito-borne. *Plasmodium* species that infect birds have been important research models for studying malaria. Indeed, the original observations by Ronald Ross on the role of mosquitoes as malaria vectors were made with bird malaria.

Currently, about 30 species of avian *Plasmodium* are recognized. However, the taxonomic status of some species is uncertain and others remain to be described. Among the important species that cause disease in domestic fowl or wild birds are *P. gallinaceum* (sometimes called chicken malaria), *P. hermansi* (a parasite of wild and domestic turkeys in the United States), *P. relictum, P. lophurae, P. cathemerium, P. circumflexum,* and *P. elongatum.* As with human malarias, there is variation in life cycles and pathogenesis of the avian malarias. This variation is related to intrinsic qualities of the species and to variation in susceptibility among host species, age, and general health status.

A bird becomes infected after inoculation of sporozoites from an infective mosquito. Merogony occurs in bone marrow, endothelial cells, and in the erythrocytes. In acute infections, these parasites may cause severe anemia, damage to bone marrow tissues, and other pathology that may result in death. Younger birds tend to be more susceptible to overt illness than older birds.

Although *Anopheles* mosquitoes can be competent laboratory vectors for some bird malarias, field and laboratory data show that culicines in the genera *Culex, Culiseta, Aedes,* and *Ochlerotatus* are the natural vectors. In Africa, *Aedes aegypti* is an important local vector of *P. gallinaceum* to chickens. The impact of bird malaria on natural bird populations is poorly known. It was introduced into Hawaii along with exotic birds and *Culex* mosquitoes, and is thought to be responsible for the reduction and extinction of native bird populations there. Bird malaria occasionally has been documented as the cause of morbidity and mortality among penguins in zoos.

Rodent Malarias The 12 *Plasmodium* species infecting rodents, called rodent or murine malarias, all occur in Africa and Asia. The vectors are assumed to be *Anopheles* mosquitoes, but in most cases the vector species is unknown. *Plasmodium berghei, P. vinckei, P. yoelli, P. chabaudi,* and *P. aegyptensis* parasitize African murine rodents. The first two are transmitted by *Anopheles dureni* in Zaire, and *P. vinckei* is transmitted by *An. cinctus* in Nigeria. *Plasmodium atheruri* infects the African brush-tailed porcupine (*Atherurus africanus*) and is transmitted by *An. smithii. Plasmodium anomaluri, P. landauae,* and *P. pulmophilum* occur in African flying squirrels (*Anomalurus* spp.); *An. marchadyi* is the probable vector of the *P. atheruri.* The three species of *Plasmodium* found in Asian flying squirrels are *P. booliati, P. watteni,* and *P. incertae.* The significance of rodent malarias to the health and population dynamics of their natural hosts is largely unknown, although the prevalences of infection can be high. They have become important laboratory models for human malaria, particularly in host immunological responses, drug screening studies, and vaccine development. Cox (1993) provides a succinct review of the rodent malarias.

Primate Malarias The nonhuman primate malarias are caused by a group of 25 *Plasmodium* species, many of which are closely related to the human malarias (Collins and Aikawa, 1993). Seven of them infect lemurs in Madagascar and are poorly known. All 18 others have life cycles similar to those of the human malarias. Most have a tertian periodicity, but two species (*P. brasilianum* and *P. inui*) are quartan and one (*P. knowlesi*) is quotidian (i.e., has a periodicity of 1 day). *Plasmodium knowlesi* appears to have become a frequent parasite of humans, as well. Probably all are transmitted by *Anopheles* mosquitoes, but for 10 of them the vector species are unknown.

Of the 18 well-known primate malaria species, 13 occur in southern or southeastern Asia, where macaques, langurs (leaf monkeys), gibbons, and orangutans are the vertebrate hosts. These plasmodia include *P. pitheci* in the orangutan, the first nonhuman primate malaria to be described; *P. knowlesi,* a macaque parasite that has become an important laboratory model for development of human vaccines; and *P. cynomolgi,* a parasite of macaques and langurs that serves as an important model for human *P. vivax* malaria. The vectors of these Asian primate malarias include *Anopheles hackeri, An. dirus, An. balabacensis, An. elegans,* and *An. introlatus.*

Three primate malarias occur in Africa, where *Plasmodium gonderi* infects mangabeys and mandrills, and *P. reichenowi* and *P. schwetzi* infect chimpanzees and gorillas. Their natural vectors are unknown. Two *Plasmodium* species infect nonhuman primates in South America. *Plasmodium simium* infects howler monkeys and woolly spider monkeys in Brazil. It is similar to the human parasite *P. malariae. Plasmodium brasilianum* infects a wide range of New World monkeys in the family Cebidae, including howler monkeys, spider monkeys, woolly spider monkeys, titis, capuchins, woolly monkeys, bearded sakis, and squirrel monkeys. It is similar to the human parasite *P. vivax.* Both South American species are transmitted by *Anopheles cruzii,* which also is an important vector of human malaria in parts of South America. The larvae inhabit water-filled leaf axils of bromeliad plants at heights of 5 m or more, where the adults are likely to encounter arboreal primates.

Many species of *Anopheles* are competent laboratory vectors of primate malarias, including *An. stephensi, An. maculatus, An. gambiae,* and *An. dirus,* which serve as vectors of human malaria. *Plasmodium knowlesi* can infect humans both experimentally and naturally, and can be transmitted by the bite of *An. dirus* to other humans. *Plasmodium cynomolgi* has infected laboratory workers, and experimental studies showed that mosquito transmission from monkeys to humans, and from humans to humans, can occur. *Plasmodium brasilianum* also infects humans. Human malaria due to infection with *P. brasilianum* and *P. simium* possibly occurred as a zoonosis in the New World prior to the arrival of Europeans, with *An. cruzii* acting as the vector. Alternatively, these two simian parasites might be derived from human *Plasmodium* species to which they are closely related. Coatney et al. (1971) reviewed the infectivity of nonhuman primate malarias to humans, and Collins and Aikawa (1993) reviewed the primate malarias.

Dog Heartworm

Dog heartworm is caused by the mosquito-borne filarial nematode *Dirofilaria immitis,* a member of the family Onchocercidae (Boreham and Atwell, 1988). Adult *D. immitis* occupy the right ventricle of the canine heart and the pulmonary arteries (Figure 14.40). The worms are 12 to 31 cm long and form aggregations of up to 50 or more individuals. In large aggregations, infection may extend to the right atrium. Contrary to popular belief, heartworm disease in dogs is not simply a consequence of a heavy worm burden in the ventricle resulting in impedance of blood flow. Rather, it is the result of deleterious changes in the endothelium and integrity of the walls of the pulmonary arteries, leading to pulmonary hypertension and right ventricular hypertrophy. These pathologic changes cause decreased cardiac output to the lungs, weakness, lethargy, chronic coughing, and ultimately congestive heart failure. Dogs may die if left untreated.

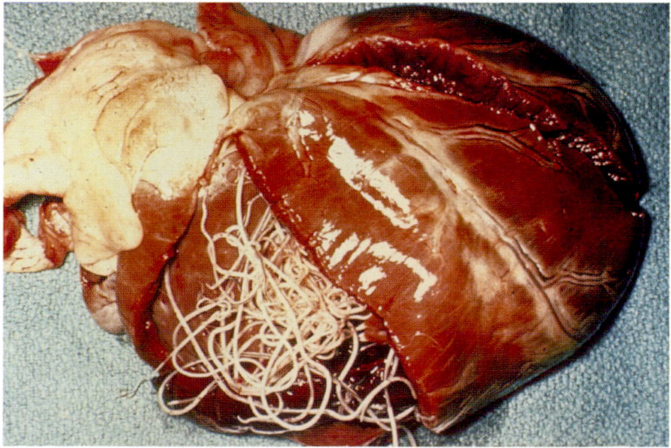

Figure 14.40 *Dirofilaria immitis* adults in right ventricle of dog heart. (Photo from H. D. Newson)

The life cycle of *Dirofilaria immitis* involves canids and mosquitoes. Dogs become infected by the bite of a mosquito whose labium carries third-stage larvae. These larvae break out of the labium while it is bent during feeding and are deposited onto the dog's skin, along with a small droplet of mosquito hemolymph from the ruptured labium. Only about 10% of the larvae successfully enter the skin, generally through the hole made by the mosquito's fascicle. They remain *in situ* subcutaneously, where they molt to fourth-stage larvae. The larvae then migrate to other subcutaneous, adipose, or muscle tissues and molt again to a fifth-stage larva. These worms, now approximately 18 mm long, enter the venous circulation and become established in the heart and pulmonary arteries. Generally, the fifth-stage larvae reach the heart at about 70 to 90 days after infection.

In the heart and pulmonary arteries, the fifth-stage larvae develop into sexually mature adults. After mating, at six to seven months, the females begin to release into circulation the microfilariae, active embryonic life stages about 300 μm long and 7 μm wide. The microfilaremia varies considerably, from 1,000 to 100,000 microfilariae per ml. It is nocturnally subperiodic, with peak concentrations occurring in the peripheral blood in the evening. Some dogs never develop microfilaremia, even though they support *D. immitis* adults and may have patent disease. These dogs are said to have occult infections.

Mosquitoes become infected with *Dirofilaria immitis* when they imbibe blood from a microfilaremic dog. In an average blood meal of 5 μl, a mosquito may ingest between five and 500 microfilariae. Within 48 hours of ingestion, microfilariae migrate posteriorly in the midgut lumen to the Malpighian tubules and then into the distal cells of these tubules, where they develop intracellularly to "sausage forms" or first-stage larvae, taking about four days at 26°C. Some remain trapped in the midgut. If more than a few begin to develop in the tubules, the mosquito is likely to be killed. The first-stage larvae molt to the second-stage at about eight to 10 days after ingestion. As they continue to grow they cause swelling and distention of the Malpighian tubules. At 12 to 14 days after ingestion, they molt to the third stage (Figure 14.41). These forms break out of the Malpighian tubules and migrate through the hemolymph to the head and base of the mouthparts, then into the interior of the labium. The mosquito is then infective. The rate of these developmental processes is temperature-dependent and varies with factors affecting competence for parasite development.

Vectors of *Dirofilaria immitis* differ with geographic region; many mosquito species in several genera are competent to transmit it. Grieve et al. (1983) listed 20 species field-caught in the United States, in the genera *Aedes, Ochlerotatus, Psorophora, Anopheles,* and *Culex,* in which infective-stage larvae of *D. immitis* have been detected.

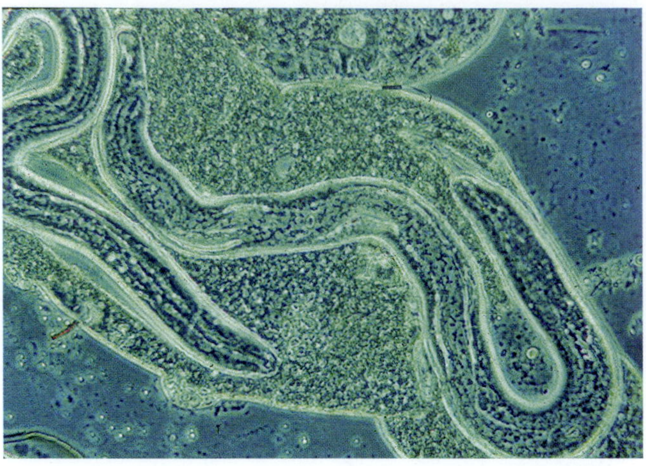

Figure 14.41 Filariform larva of *Dirofilaria immitis* within a Malpighian tubule of *Ae. aegypti*, prior to rupturing the tubule and migrating to the labium, where transfer to dog occurs during blood-feeding. (Photo by Bonnie Buxton)

Other Filarial Nematodes of Animals

Other species of *Dirofilaria* infect mammals. These include *Dirofilaria ursi*, a bear parasite transmitted by the black fly *Simulium venustum*, and *D. roemeri*, a wallaroo (a type of small kangaroo) parasite transmitted by the horse fly *Dasybasis hebes*; and the following mosquito-transmitted *Dirofilaria* species: *D. repens* in canids; *D. carynodes* and *D. magnilarvatum* in monkeys; *D. scapiceps* in rabbits; *D. tenuis* in raccoons; and *D. subdermata* in porcupines.

In addition to *Dirofilaria* species, a large number of filarial nematodes in other genera of the Onchocercidae infect wild and domestic animals. Vectors include mosquitoes and a wide range of other blood-feeding Diptera, lice, fleas, mites, and ticks. The mosquito-borne onchocercid nematodes include species in the following genera; *Aproctella, Breinlia, Brugia, Cardiofilaria, Conispiculum, Dirofilaria, Deraiophoronema, Folyella, Loiana, Molinema, Pelecitus, Oswaldofilaria, Saurositus, Skrjabinofilaria, Waltonella,* and *Wuchereria* (Lavoipierre, 1958; Hawking and Worms, 1961; Bain and Chabaud, 1986; Anderson, 1992). *Brugia pahangi* of birds (*Meriones*) is an important laboratory organism for studies on filariasis. *Brugia malayi* develops in the peritoneal cavity of gerbils, providing a laboratory infection model.

PREVENTION AND CONTROL

The four overlapping aims of mosquito control are to prevent mosquito bites, keep mosquito populations at acceptable densities, minimize mosquito-vertebrate contact, and reduce the longevity of female mosquitoes. All these actions minimize the annoying and harmful effects of bites and blood loss, and interrupt pathogen transmission. The eradication of either mosquito species or their associated diseases is no longer viewed as a viable objective, except in small, isolated regions or in the case of recent invasions. Two exemplary failures of the eradication approach, on a grand scale, were the World Health Organization's global malaria eradication program and the Pan American Health Organization's attempt to eradicate *Aedes aegypti* from the Western Hemisphere. A notable exception was the successful elimination of the African immigrant *Anopheles gambiae* from Brazil. The more realistic objective of modern mosquito control programs is integrated pest management to reduce mosquito abundance and disease prevalence, using prudent combinations of methods.

Personal protection is the most direct and simple approach to prevention. Outdoor exposure can be avoided during peak mosquito activity, and window screens can prevent mosquito entry into houses and animal shelters. Head nets reduce annoyance and prevent bites about the face and neck. Bed nets, impregnated with synthetic pyrethroid and strung over beds at night, repel mosquitoes and kill those that land on the nets. Impregnated mesh suits with hoods work similarly and can be worn over clothing. Other insecticidal devices create a repellent smoke or vapor that reduces mosquito attack in the immediate vicinity. Chemical repellents applied to skin or clothing prevent mosquitoes from landing or cause them to leave before probing. The most common one is N,N diethyl-meta-toluamide or DEET.

Organized control provides efficient, area-wide mosquito management at local, regional, or national levels. In the United States, mosquito programs typically are county-level abatement districts. These focus on the control of nuisance and vector species, but often also participate in surveillance for mosquito-borne disease pathogens. National organizations are usually parts of ministries of health and coordinate their disease and vector control efforts at that level. Especially in developing countries, there is now increasing emphasis on community cooperation, low technology, sustainability, and the integrated use of a variety of control tools that are adapted to local customs, conditions, and resources. Bed nets treated with synthetic pyrethroid insecticides have emerged as important community protection devices for malaria control and are now available in long-lasting, wash-durable formulations. They have been distributed in mass campaigns in countries of sub-Saharan Africa, resulting in reductions in malaria morbidity and mortality in large sections of some countries.

Habitat modification is a traditional and reliable tool in mosquito management. Adult resting places can be rendered unsuitable by harborage alteration. Changes in larval habitat that prevent oviposition, hatching, or larval development are called source reduction. Water is altered or eliminated in a variety of ways. This includes plastic foam beads that provide a floating barrier over latrine water, underground sewage lines, land drainage through ditches or underground tile pipes, waste tire shredding, trash-container disposal and natural container elimination, lids for water-storage barrels, vegetational changes in ponds, altered flow of tidal water through salt marshes, and water-level manipulation in reservoirs and rice fields. Each method is designed to interfere with specific features of a mosquito's natural history. Through appropriate application of ecological principles and an intimate knowledge of mosquito behavior and life cycles, desirable natural wetlands and newly created ones can be modified to minimize mosquito production while benefiting other wildlife.

Biological control of mosquitoes by predators or parasites has been studied extensively and has been reviewed by Chapman (1985), Beaty and Marquardt (1996), Hemingway (2005), and Floore (2007). Aerial predators, such as dragonflies, birds, and bats, receive much attention but do not specialize in adult mosquitoes and have little if any effect on their densities. Most efforts have been directed at the larval stage. Aquatic predators, both naturally occurring and introduced, include the mosquito fish (*Gambusia affinis*) and killifish (*Fundulus* spp.). Other fish, such as grass carp (e.g., *Tilapia* and *Cyprinus*), remove aquatic vegetation that provides harborage for larvae. Invertebrate predators include the predatory mosquito *Toxorhynchites*, several families of aquatic bugs and beetles, predatory copepods, hydras, and turbellarian flatworms; however, only copepods have been implemented with substantial success. There have been attempts to develop the use of parasites and pathogens of mosquito larvae as control agents, including the nematode *Romanomermis culicivorax*; protozoans such as the ciliates *Lambornella* and *Tetrahymena*; the gregarine sporozoan *Ascogregarina*; and the microsporidian *Nosema*. Fungal pathogens include *Coelomomyces*, *Lagenidium*, *Culicinomyces*, and *Metarrhizium*. Viruses pathogenic to larvae include the iridescent viruses, densonucleosis viruses, polyhedrosis viruses such as the baculoviruses, and entomopox viruses. Generally, the previously mentioned parasites or pathogens of mosquito larvae are still in experimental stages of development, or they have limited effectiveness and have not been used routinely in operational programs.

An exception is the bacterium *Bacillus thuringiensis israelensis*, or *Bti*, which has been developed into commercial formulations since its original discovery in 1975. It is used extensively in mosquito control programs. Larvae die when they ingest crystalline, proteinacious toxins produced by the bacterial cells during sporulation. The bacterium *Bacillus sphaericus* has a similar mode of action but is more specific. It is particularly effective against *Culex* larvae, and it is more persistent in water, and more tolerant of water with a high organic content, than is *Bti*.

Genetic control, a category of biological control using a variety of genetic methods, has been successful against some pests; however, its use against mosquito vectors of disease remains experimental. There are two hypothetical approaches: release of sterilized males or incompatible strains, resulting in attrition of the natural population, and replacement of natural vector populations with species or strains that are poor vectors or are not susceptible to infectious agents. These methods have been reviewed by Rai (1996) and Wood (2005). The most prominent innovation in the sterile male approach is to release males carrying a dominant lethal gene or a genetic system that allows the altered males to sire only male offspring, which then continue spreading the gene in the dwindling natural population of females. For vector replacement, progress has been most rapid in developing transgenic *Anopheles* spp. that are refractory to infection by malaria parasites and *Aedes aegypti* that are incompetent to transmit dengue virus.

Chemical control is achieved with insecticides against either larvae or adults. Larvicides are placed in water where larvae develop, or where water will accumulate and provide habitat for larvae. Formerly used larvicides included inorganic compounds such copper arsenate, fuel oil, and organochlorine chemicals such as DDT and dieldrin. Currently, categories of registered larvicides are light mineral oils, organophosphates, and insect-growth regulators. Rapidly degradable oils spread over the water surface, penetrating the tracheal systems of larvae and pupae and suffocating them. Organophosphates, such as temephos, malathion, and chlorpyrifos, function as nerve poisons. The insect-growth regulator methoprene is a mimic of juvenile hormone and interferes with metamorphosis and emergence. The specific kind and formulation (dust, powder, water-soluble liquid, emulsion, oil-soluble liquid, granule, pellet, briquet) of the larvicide recommended depends on the biology of the target mosquito, the kind and size of habitat, the method of application, the chemical composition of the water, and the presence of nontarget organisms that might be adversely affected. Some can be formulated for slow release from a carrier. These may be applied to dry ground, releasing the active ingredient when inundated.

Adulticides are applied to surfaces where adults will rest or in the air where they fly. Residual insecticides applied to resting surfaces may retain their toxicity for days to months. They were central to the global malaria eradication program, in which DDT spraying of the inner walls of human dwellings at six-month intervals killed all mosquitoes landing on these walls before or after taking blood. In areas where the vectors bit humans primarily indoors, this effectively interrupted most malaria transmission until mosquito populations developed resistance to the insecticide or when programs were abandoned. This approach is still used widely in some areas. Residual adulticides also can be used outdoors on vegetation or structures that serve as harborages. They tend to have short-term effects, because sunlight, wind, and rain cause the insecticide to degrade.

Adulticides intended for direct contact between airborne droplets and the mosquito are of two types: thermal fogs and low or ultra-low volume (ULV) sprays. Both can be applied from hand-carried equipment, motor vehicles, or aircraft. Thermal fogging involves mixing an insecticide with a combustible liquid such as kerosene. The mixture is heated, creating a fog of insecticide that drifts through the area to be treated. The ULV approach involves special nozzles and pumps that dispense fine droplets of insecticide, forming a mist that passes through the target area. Currently, insecticides registered for use in fogs and low-volume sprays are organophosphates, carbamates, pyrethrins, and synthetic pyrethroids. Resistance to insecticides is an important consequence of their use and has developed in many mosquito populations. The mechanisms of physiological resistance have been well characterized biochemically and genetically. Behavioral resistance also can develop. This is typically a change in adult feeding or resting behavior, so that mosquitoes no longer contact insecticide residues.

Surveillance, which is at the core of effective mosquito control programs, determines mosquito distribution and abundance and degree of pathogen activity. The goal is to provide data so that control agencies can take action to prevent mosquito-related problems from occurring. Unfortunately, there have been few control programs establishing action thresholds for mosquito density or infection rate, the levels of threat at which controls should be initiated. More often, action is based on human perception of a pest problem, conditions similar to past experience with disease outbreaks, or first detection of pathogen activity. Surveillance strategies and techniques for mosquito-borne encephalitis viruses have been presented by Moore et al. (1993) and Moore and Gage (1996). Bruce-Chwatt (1980) and Sasa (1976) reviewed traditional techniques for detecting malaria and filarial parasites, respectively, and several new ones are in use.

Control of Pathogen Transmission

Although all methods of reducing vector populations can lower the incidence of mosquito-borne disease, quantitative models of the dynamics of disease transmission have become important tools for setting realistic control objectives. They allow programs to focus efforts on parts of the pathogen-transmission system most vulnerable to attack. Useful references on this subject are Ross (1911), Macdonald (1957), Molineaux and Gramiccia (1980), Fine (1981), Koella (1991), and Dye (1992). The Ross-Macdonald equation describes the case reproduction rate, the total number of new cases of a disease arising from a single infective case in a totally susceptible population. The vectorial capacity equation expresses that function on a daily basis using entomological parameters. Although the vectorial capacity measure is not epidemiologically comprehensive, it allows a comparison of the relative importance of different vectors and provides estimates of critical vector density, the adult mosquito density below which the case reproduction is less than 1 and the disease should die out. It also illustrates, mathematically, that even small changes in the interval between bites on susceptible hosts (which is squared) or in the longevity of vectors (which changes exponentially) cause large changes in transmission rates. The latter relationship has been critical in mounting effective disease-control operations, which target older females, rather than just female density in general. More complex models sometimes show good agreement between predicted and observed results in extensive field studies of malaria (Molineaux and Gramiccia, 1980; Koella, 1991) and may become useful in establishing action thresholds. A simple and direct measure of transmission is the entomological inoculation rate (Onori and Grab, 1980), which is the product of the vector's human-biting rate and proportion of vectors that are infective.

Vaccines and drugs are important tools in protecting or treating humans and other animals susceptible to mosquito-borne disease. They serve not only to protect the individual but also to reduce transmission to others. Vaccines are available for several arboviral diseases, including yellow fever and Japanese encephalitis for humans, and eastern, western, and Venezuelan encephalitis for equids. These vary in the duration of protection they provide. An experimental human vaccine against eastern equine encephalitis has been produced. None currently exists for the dengue viruses. Human malaria vaccines are under development, and some field trials have achieved limited success, but their wide-scale efficacy remains uncertain. The three kinds of malaria vaccines being considered use antigens from sporozoites, blood stages, or gametes; the last kind is called a transmission-blocking vaccine because the human antibodies take effect against stages that form within the midgut of the mosquito. Among drugs, there exists a wide spectrum of antimalarials used for prophylaxis, therapy, or both. The most commonly used chemoprophylactic is chloroquine, against which there is now widespread resistance in *Plasmodium falciparum* and some *P. vivax* populations. Mefloquine is prescribed as a prophylactic for areas with resistant populations. Artemesinin, a drug derived from a plant long used by Chinese herbalists, is now widely used in derivative form in combination with other antimalarial compounds to treat uncomplicated malaria cases where resistance is widespread.

Strickland (1991) presented a detailed review of malaria chemotherapy and chemoprophylaxis. For lymphatic filariasis, diethylcarbamazine (DEC) is the standard chemotherapy, which reduces microfilaremia but does not kill adult worms. Advanced disease manifestations (e.g., elephantiasis) cannot be completely reversed, except by surgery, but sustained mass treatment of human populations can drive transmission to zero. This was achieved in parts of China in one year by the use of DEC-fortified cooking salt. Owing to the longevity of adult worms, mass treatment for five to 10 years is necessary to completely break the infection cycle in a community. Ivermectin and albendazole are two other drugs showing efficacy in lymphatic filariasis cases. Both DEC and ivermectin are used as chemoprophylaxis against dog heartworm infections.

REFERENCES AND FURTHER READING

Allan, S. A., Day, J. F., & Edman, J. D. (1987). Visual ecology of biting flies. *Annual Review of Entomology, 32*, 297–316.

American Mosquito Control Association. (1979). *Mosquitoes and Their Control in the United States.* Fresno, CA: American Mosquito Control Association.

Anderson, R. C. (1992). *Nematode Parasites of Vertebrates: Their Development and Transmission.* Wallingford, Oxon, UK: C.A.B. International.

Anonymous. (1994). Attractants for mosquito surveillance and control: A symposium. *Journal of the American Mosquito Control Association, 10,* 253–338.

Asman, S. M., McDonald, P. T., & Prout, T. (1981). Field studies of genetic control systems for mosquitoes. *Annual Review of Entomology, 26,* 289–318.

Bain, O., & Chabaud, A. G. (1986). Atlas des larves infestantes de filaires. *Tropical Medicine and Parasitology, 37,* 301–340.

Barr, A. R. (1958). *The Mosquitoes of Minnesota (Diptera: Culicidae: Culicinae).* Agric. Expt. Sta., Tech. Bull. 228, Univ. Minnesota.

Bates, M. (1949). The *Natural History of Mosquitoes.* New York: Macmillan. (1965 edition: Harper & Row, New York).

Beaty, B. J., & Marquardt, W. C. (Eds.). (1996). *The Biology of Disease Vectors.* Niwot, CO: University Press of Colorado.

Beaty, B., Miller, B. R., Shope, R. E., Rozhon, E. J., & Bishop, D. H. (1982). Molecular basis of bunyavirus per os infection of mosquitoes: Role of the middle-sized RNA segment. *Proceedings of the National Academy of Sciences of the United States of America, 79,* 1295–1297.

Belkin, J. N. (1962). *The Mosquitoes of the South Pacific (Diptera: Culicidae).* Los Angeles: Univ. California Press.

Belkin, J. N., Schick, R. X., Galindo, P., & Aitken, T. H. (1965). Mosquito studies (Diptera: Culicidae) I. A project for a systematic study of the mosquitoes of Middle America. *Contributions of the American Entomological Institute, 2,* 1–17.

Bentley, M. D., & Day, J. F. (1989). Chemical ecology and behavioral aspects of mosquito oviposition. *Annual Review of Entomology, 34,* 401–421.

Besansky, N. J., Finnerty, V., & Collins, F. H. (1992). A molecular genetic perspective on mosquitoes. *Advances in Genetics, 30,* 123–184.

Bock, G. R., & Cardew, G. (Eds.). (1996). *Olfaction in Mosquito-host Interactions.* Chichester, UK: Wiley.

Boddy, D. W. (1948). An annotated list of the Culicidae of Washington. *Pan-Pacific Entomologist, 24,* 85–94.

Bohart, R. M., & Washino, R. K. (1978). *Mosquitoes of California.* Berkeley: Div. Agric. Sci, Univ. Calif.

Boreham, P. F. L., & Atwell, R. B. (Eds.). (1988). *Dirofilariasis.* Boca Raton, FL: CRC Press.

Bosik, J. J. (1997). Common names of insects and related organisms. *Entomological Society of America.*

Bowen, M. F. (1991). The sensory physiology of host-seeking behavior in mosquitoes. *Annual Review of Entomology, 36,* 139–158.

Bowen, G. S., & Francy, D. B. (1980). Surveillance. In T. P. Monath (Ed.), *St. Louis encephalitis* (pp. 473–499). Washington, DC: Am. Pub. Hlth. Assoc.

Boyd, M. F. (1941). An historical sketch of the prevalence of malaria in North America. *American Journal of Tropical Medicine and Hygiene, 21,* 223–244.

Boyd, M. F. Ed. (1949). *Malariology: A Comprehensive Survey of All Aspects of this Group of Diseases from a Global Standpoint.* Philadelphia: Saunders.

Bradley, T. J. (1987). Physiology of osmoregulation in mosquitoes. *Annual Review of Entomology, 32,* 439–462.

Brown, M. R., & Lea, A. O. (1990). Neuroendocrine and midgut endocrine systems in the adult mosquito. *Advances in Disease Vector Research, 6,* 29–58.

Bruce-Chwatt, L. J. (1980). *Essential Malariology.* London: William Heinemann Medical Books Ltd.

Burke, D. S., & Leake, C. J. Japanese encephalitis. In T. P. Monath (Ed.), (1988). *The Arboviruses: Epidemiology and Ecology* (pp. 63–92). Vol. 3. Boca Raton, FL: CRC Press.

Burkot, T. R., & Graves, P. M. (1994). Human malaria transmission: Reconciling field and laboratory data. *Advances in Disease Vector Research, 10,* 149–182.

Calisher, C. H., & Karabatsos, N. (1988). Arbovirus serogroups: Definition and geographic distribution. In T. P. Monath (Ed.), *The Arboviruses: Epidemiology and Ecology* (Vol. 1, pp. 19–57). Boca Raton, FL: CRC Press.

Calisher, C. H., & Thompson, W. H. (Eds.). (1983). *California Serogroup Viruses: Proceedings of an International Symposium, held in Cleveland, Ohio, November 12 and 13, 1982.* New York: A. R. Liss.

Carlson, J., Olson, K., Higgs, S., & Beaty, B. (1995). Molecular genetic manipulation of mosquito vectors. *Annual Review of Entomology, 40,* 359–388.

Carpenter, S. J. (1941). *The mosquitoes of Arkansas.* Little Rock, AR: Arkansas State Board of Health.

Carpenter, S. J., & LaCasse, W. J. (1955). *Mosquitoes of North America (North of Mexico).* Berkeley, California: Univ. California Press.

Centers for Disease Control. (1977). *Mosquitoes of Public Health Importance and Their Control.* HHS Publ. No. (Centers for Disease Control) 82–8140. Atlanta, GA: U.S. Dept. of Health & Human Services.

Chapman, H. C. (1966). *The Mosquitoes of Nevada.* Carson City, NV: U. S. Dept. Agric. & Univ. Nevada.

Chapman, H. C. (Ed.), (1985). *Biological Control of Mosquitoes.* Bull. No. 6, Fresno, CA: American Mosquito Control Association.

Christophers, S. R. (1911). Development of the egg follicle in Anophelines. *Paludism, 2,* 73–88.

Christophers, S. R. (1960). *Aedes Aegypti (L.). The Yellow Fever Mosquito. Its Life History, Bionomics and Structure.* London: Cambridge Univ. Press.

Clements, A. N. (1963). *The Physiology of Mosquitoes.* New York: Pergamon.

Clements, A. N. (1992). *The Biology of Mosquitoes. Vol. 1.* Development, Nutrition and Reproduction. New York: Chapman & Hall.

Clements, A. N. (1999). *The Biology of Mosquitoes. Vol. 2.* Sensory Reception and Behaviour. Wallingford, U.K: CABI Publ.

Coatney, G. R., Collins, W. E., Warren, M. C. W., & Contacos, P. G. (1971). *The Primate Malarias.* Washington, DC: US Govt. Printing Office.

Collins, W. E., & Aikawam, M. (1993). Plasmodia of nonhuman primates. In J. P. Kreier (Ed.), *Parasitic Protozoa* (2nd ed., Vol. 5, pp. 105–134). New York: Academic Press.

Corbet, P. S., Williams, M. C., & Gillett, J. D. (1961). O'nyong nyong fever: An epidemic virus disease in East Africa: IV. Vector studies at epidemic sites. *Transactions of the Royal Society of Tropical Medicine and Hygiene, 55,* 463.

Cox, F. E. G. (1993). Plasmodia of rodents. In J. P. Kreier (Ed.), *Parasitic Protozoa* (2nd ed., Vol. 5, pp. 49–104). New York: Academic Press.

Cox, G. W. (1944). *The Mosquitoes of Texas.* Austin, TX: Texas State Health Dept.

Curtis, C. F. (Ed.). (1990). *Appropriate Technology in Vector Control.* Boca Raton, FL: CRC Press.

Dale, P. E. R., & Hulsman, K. (1990). A critical review of salt marsh management methods for mosquito control. *Aquatic Sciences, 3,* 281–311.

Darsie, R. F., Jr. (1951). Pupae of the culicine mosquitoes of the northeastern United States (Diptera, Culicidae, Culicini). *Memoirs of the Cornell University Agricultural Experiment Station, 304,* 1–67.

Darsie, R. F., Jr. (1989). Keys to the genera, and to the species of five minor genera, of mosquito pupae occurring in the Nearctic region (Diptera: Culicidae). *Mosquitoes Systematics, 21,* 1–10.

Darsie, R. F., Jr., & Morris, C. D. (1998). *Keys to the Adult Females and Fourth Instar Larvae of the Mosquitoes of Florida (Diptera, Culicidae), Technical Bulletin of the Florida Mosquito Control Association, No. 1,* Vero Beach: Florida Med. Entomol. Lab, IFAS, Univ. Florida.

Darsie, R. F., Jr., & Ward, R. A. (1981). Identification and geographical distribution of the mosquitoes of North America, north of Mexico. *Mosquitoes Systematics, Supplement, 1,* 1–313.

Darsie, R. F., Jr., & Ward, R. A. (2004). *Identification and geographical distribution of the mosquitoes of North America, north of Mexico.* Gainesville: Univ. Press of Florida.

Davidson, E. W., & Becker, N. (1996). Microbial control of vectors. In B. J. Beaty & W. C. Marquardt (Eds.), *The Biology of Disease Vectors* (pp. 549–663). Niwot: Univ. Press of Colorado.

de Barjac, H., & Sutherland, D. J. (Eds.). (1989). *Bacterial Control of Mosquitoes and Black Flies.* New Brunswick: Rutgers Univ. Press.

Debenham, M. L. (Ed.). (1987a). *Culicidae of the Australasian Region. Vol. 4. Nomenclature, Synonymy, Literature, Distribution, Biology and Relation to Disease: Genus Aedes, Subgenera Scutomyia, Stegomyia, Verrallina.* Entomology Monograph No. 2 (in part), Canberra: Australian Gov. Publ. Service.

Debenham, M. L. (Ed.). (1987). *Culicidae of the Australasian Region.* Vol. 5. *Nomenclature, Synonymy, Literature, Distribution, Biology and Relation to Disease: Genus Anopheles, Subgenera Anopheles, Cellia.* Entomology Monograph No. 2 (in part). Canberra: Australian Gov. Publ. Service.

Debenham, M. L. (Ed.). (1988a). *Culicidae of the Australasian Region.* Vol. 6. *Nomenclature, Synonymy, Literature, Distribution, Biology and Relation to Disease: Genera Armigeres, Bironella and Coquillettidia.* Entomology Monograph No. 2 (in part). Canberra: Australian Gov. Publ. Service.

Debenham, M. L. (Ed.). (1988b). *Culicidae of the Australasian region. Vol. 9. Nomenclature, Synonymy, Literature, Distribution, Biology and Relation to Disease: Genus Culex (Subgenera Lutzia, Neoculex, Subgenus Undecided), Genera Culiseta, Ficalbia, Heizmannia, Hodgesia, Malaya, Mansonia.* Entomology Monograph No. 2 (in part). Canberra: Australian Gov. Publ. Service.

Debenham, M. L. (Ed.). (1988c). *Culicidae of the Australasian region. Vol. 10. Nomenclature, Synonymy, Literature, Distribution, Biology and Relation to Disease: Genera Maorigoeldia, Mimomyia, Opifex, Orthopodomyia, Topomyia, Toxorhynchites.* Entomology Monograph No. 2 (in part). Canberra: Australian Gov. Publ. Service.

Debenham, M. L. (Ed.). (1989a). *Culicidae of the Australasian Region. Vol. 7. Nomenclature, Synonymy, Literature, Distribution, Biology and Relation to Disease: Genus Culex, Subgenera Acallyntrum, Culex.* Entomology Monograph No. 2 (in part). Canberra: Australian Gov. Publ. Service.

Debenham, M. L. (Ed.). (1989b). *Culicidae of the Australasian Region. Vol. 8. Nomenclature, Synonymy, Literature, Distribution, Biology and Relation to Disease: Genus Culex, Subgenera Culiciomyia, Eumelanomyia, Lophoceraomyia.* Entomology Monograph No. 2 (in part). Canberra: Australian Gov. Publ. Service.

Debenham, M. L., Hicks, M. M., & Griffiths, M. (Eds.). (1989). *Culicidae of the Australasian Region. Vol. 12. Summary of Taxonomic Changes, Revised Alphabetical List of Species, Supplementary Bibliography, Errata and Addenda, Geographic Guide to Species, Synopsis of Disease Relationships, Indexes.* Entomology Monograph No. 2 (in part). Canberra: Australian Gov. Publ. Service.

DeFoliart, G. R., Grimstad, P. R., & Watts, D. M. (1987). Advances in mosquito-borne arbovirus/vector research. *Annual Review of Entomology, 32,* 479–505.

Detinova, T. S. (1962). Age-grading methods in Diptera of medical importance. *World Health Organization Monograph Series, 47,* 1–216.

de Zulueta, J. (1994). Malaria and ecosystems: From prehistory to posteradication. *Parasitologia, 36,* 7–15.

Dickinson, W. E. (1944). The mosquitoes of Wisconsin. *Bulletin of the Public Museum of the City of Milwaukee, 8,* 269–365.

Dixon, R. D., & Brust, R. A. (1972). Mosquitoes of Manitoba. III. Ecology of larvae in the Winnipeg area. *Canad Enomol, 104,* 961–968.

Downes, J. A. (1969). The swarming and mating flight of Diptera. *Annual Review of Entomology, 14,* 271–298.

Dye, C. (1992). The analysis of parasite transmission by bloodsucking insects. *Annual Review of Entomology, 37,* 1–19.

Evenhuis, N. L., & Gon, S. M. III. (1989). Family Culicidae. In N. L. Evenhuis (Ed.), *Catalog of the Diptera of the Australasian and Oceanian regions* (pp. 191–218). Honolulu: Bishop Museum Press.

Fauquet, C. M., Mayo, M. A., Maniloff, J., Desselberger, U., & Ball, L. A. (Eds.). (2005). *Virus Taxonomy, 8th Reports of the International Committee on Taxonomy of Viruses.* New York: Academic Press.

Fine, P. E. M. (1981). Epidemiological principles of vector–mediated transmission. In J. J. McKelvey, B. F. Eldridge, & K. Maramorosch (Eds.), *Vectors of Disease Agents* (pp. 77–91). New York: Praeger Publishers.

Floore, T. G. (Ed.). (2007). Biorational control of mosquitoes. *Journal of the American Mosquito Control Association, 23* (2) (Suppl), 1–328.

Foote, R. H., & Cook, D. R. (1959). Mosquitoes of medical importance. *Agricultural Handbook, US Dept of Agriculture, 152,* 1–158.

Forattini, O. P. (1962, 1965). *Entomologia Medica. Vol. 1: Parte Geral., Diptera, Anophelini, 662 pp.; Vol. 2: Culicini: Culex, Aedes e Psorophora. Vol. 3: Culicini: Haemagogus, Mansonia, Culiseta, Sabethini. Toxorhychitini. Arboviruses. Filariose bancroftiana. Genetica.* Universidade de Sao Paulo, Faculdade de Higiene e Saude Publica.

Foster, W. A. (1995). Mosquito sugar feeding and reproductive energetics. *Annual Review of Entomology, 40,* 443–474.

Fox, A. S., & Brust, R. A. (1994). How do dilatations form in mosquito ovarioles? *Parasitology Today, 10,* 19–23.

Gaffigan, T. V., & Ward, R. A. (1985). Index to the second supplement to "A catalog of the mosquitoes of the world," with corrections and additions. *Mosquito Systematics, 17,* 52–63.

Gartrell, F. E., Cooney, J. C., Chambers, G. P., & Brooks, R. H. (1981). TVA mosquito control 1934–1980—Experience and current program tends and developments. *Mosquito News, 41,* 302–322.

Gerhardt, R. W. (1966). *South Dakota Mosquitoes and Their Control, Agric. Expt. Sta. Bull. 531.* Brookings: South Dakota State Univ.

Gilles, H. M., & Warrell, D. A. (1994). *Bruce-Chwatt's Essential Malariology* (3rd ed.). Boston: Little, Brown, & Co.

Gillett, J. D. (1971). *The Mosquito: Its Life, Activities and Impact on Human Affairs.* New York: Doubleday.

Gillies, M. T. (1988). Anopheline mosquitos: Vector behaviour and bionomics. In W. H. Wernsdorfer & I. McGregor (Eds.), *Malaria Principles and Practice of Malariology* (Vol. 1, pp. 453–485). Edinburgh: Churchill Livingstone.

Gillies, M. T., & Coetzee, M. (1987). A supplement to the Anophelinae of Africa south of the Sahara (Afrotropical region). *Publications of the South African Institute for Medical Research, 55,* 1–143.

Gillies, M. T., & de Meillon, B. (1968). The Anophelinae of Africa south of the Sahara (Ethiopian geographical region). *Publications of the South African Institute for Medical Research, 54,* 1–343.

Gjullin, C. M., Sailer, R. I., Stone, A., & Travis, B. V. (1961). The mosquitoes of Alaska. *Agricultural Handbook, US Dept of Agriculture, 182,* 1–98.

Gordon, R. M., & Lavoipierre, M. M. J. (1962). *Entomology for Students of Medicine.* Oxford: Blackwell Science.

Grieve, R. B., Lok, J. B., & Glickman, L. T. (1983). Epidemiology of canine heartworm infection. *Epidemiology Reviews, 5,* 220–246.

Grimstad, P. R. (1988). California group virus disease. In T. P. Monath (Ed.), *The Arboviruses: Epidemiology and Ecology* (Vol. 2, pp. 99–136). Boca Raton, FL: CRC Press, Inc.

Grove, D. (1990). *A History of Human Helminthology.* Wallingford, UK: C.A.B. International.

Gubler, D. J. (1988). Dengue. In T. P. Monath (Ed.), *The Arboviruses: Epidemiology and Ecology* (Vol. 2, pp. 223–260). Boca Raton, FL: CRC Press.

Gubler, D. J., & Bhattacharya, N. C. (1974). A quantitative approach to the study of bancroftian filariasis. *The American Journal of Tropical Medicine and Hygiene, 23,* 1027–1036.

Gubler, D. J., & Clark, G. G. (1994). Community-based integrated control of *Aedes aegypti*: A brief overview of current programs. *The American Journal of Tropical Medicine and Hygiene, 50,* 50–60.

Gubler, D. J., & Kuno, G. (Eds.). (1997). *Dengue and Dengue Hemorrhagic Fever.* Wallingford, UK: CAB International.

Gutsevich, A. V., Monchadskii, A. S., & Shtakel'berg, A. A. (1971). Mosquitoes, family culicidae. Fauna of the USSR: Diptera. *Academy of Sciences, Zoological Institute, Leningrad, 3*(4), 1–408. (English translation 1974, Israel Program for Scientific Translations.)

Gwadz, R., & Collins, F. H. (1996). Anopheline mosquitoes and the agents they transmit. In B. J. Beaty & W. C. Marquardt (Eds.), *The Biology of Disease Vectors* (pp. 73–84). Niwot: Univ. Press of Colorado.

Gyapong, J. O., Gyapong, M., & Adjei, S. (1996). The epidemiology of acute adenolymphangitis due to lymphatic filariasis in northern Ghana. *The American Journal of Tropical Medicine and Hygiene, 54,* 591–595.

Hagedorn, H. H. (1994). The endocrinology of the adult female mosquito. *Adv Dis Vector Res, 10,* 109–148.

Hagedorn, H. H. (1996). Physiology of mosquitoes. In B. J. Beaty & W. C. Marquardt (Eds.), *The Biology of Disease Vectors* (pp. 273–297). Niwot: Univ. Press of Colorado.

Harbach, R. E., & Kitching, I. J. (1998). Phylogeny and classification of the Culicidae (Diptera). *Systematic Entomology, 23,* 327–370.

Harbach, R. E., & Knight, K. L. (1980). *Taxonomist's Glossary of Mosquito Anatomy.* Marlton, NJ: Plexus Publishing.

Harbach, R. E., & Knight, K. L. (1981). Corrections and additions to Taxonomists' Glossary of Mosquito Anatomy. *Mosquito Systematics, 13,* 201–217.

Harris, K. F. (Ed.). (1985). *Advances in Disease Vector Research (Formerly Current Topics in Disease Vector Research)* (Vol. 1). New York: Springer-Verlag.

Hardy, J. L. (1988). Susceptibility and resistance of vector mosquitoes. In T. P. Monath (Ed.), *The Arboviruses: Epidemiology and Ecology* (Vol. 1, pp. 87–126). Boca Raton, FL: CRC Press.

Harrison, G. (1978). *Mosquitoes, Malaria and Man: A History of the Hostilities Since 1880.* London: John Murray.

Hawking, F., & Worms, M. (1961). Transmission of filarioid nematodes. *Annual Review of Entomology, 6,* 413–432.

Hawley, W. A. (1988). The biology of *Aedes albopictus. Journal of the American Mosquito Control Association, 4*(Suppl. 1), 1–40.

Hawley, W. A., Reiter, P., Copeland, R. S., Pumpuni, C. B., & Craig, G. B. Jr. (1987). *Aedes albopictus* in North America: Probable introduction in tires from northern Asia. *Science, 236,* 1114–1116.

Hayes, C. G. (1988). West Nile fever. In T. P. Monath (Ed.), *The Arboviruses: Epidemiology and Ecology* (Vol. 5. pp. 59–88). Boca Raton, FL: CRC Press.

Headlee, T. J. (1945). *The Mosquitoes of New Jersey and Their Control.* New Brunswick, NJ: Rutgers Univ. Press.

Hearle, E. (1926). *The Mosquitoes of the Lower Fraser Valley, British Columbia, and Their Control.* National Res. Council Rpt. No. 17, Ottawa.

Hemingway, J. (2005). Biological control of mosquitoes. In W. C. Marquardt (Ed.), *Biology of Disease Vectors* (2nd ed., pp. 649–660). New York: Elsevier Academic Press.

Hicks, M. M. (Ed.), (1989). *The Culicidae of the Australasian region: Vol. 11*; Nomenclature, Synonymy, Literature, Distribution, Biology and Relation to Disease: Genera Tripteroides, Uranotaenia, Wyeomyia, Zeugnomyia. Entomology Monogr. No. 2. Canberra: Austalian Govt. Publ. Service.

Higgs, S., & Beaty, B. J. (1996). Rearing and containment of mosquito vectors. In B. J. Beaty & W. C. Marquardt (Eds.), *The Biology of Disease Vectors* (pp. 595–605). Niwot: Univ. Press of Colorado.

Hoc, T. Q. (1996). Application of the ovarian oil injection and ovariolar separation techniques for age grading hematophagous Diptera. *Journal of Medical Entomology, 33*, 290–296.

Hopkins, C. C., Hollinger, F. B., Johnson, R. F., Dewlett, H. J., Newhouse, V. F., & Chamberlain, R. W. (1975). The epidemiology of St. Louis encephalitis in Dallas, Texas, 1966. *American Journal of Epidemiology, 102*, 1–15.

Hopkins, G. H. E. (1952). *Mosquitoes of the Ethiopian Region I. Larval Bionomics of Mosquitoes and Taxonomy of Culicine Larvae* (2nd ed.). London: British Museum (Nat. Hist.).

Horsfall, W. R. (1955). *Mosquitoes. Their Bionomics and Relation to Disease.* New York: Ronald Press.

Horsfall, W. R., Fowler, H. W. Jr., Moretti, L. J., & Larsen, J. R. (1973). *Bionomics and Embryology of the Inland Floodwater Mosquito Aedes Vexans.* Urbana, Illinois: Univ. Illinois Press.

Iversson, L. B. (1988). Rocio encephalitis. In T. P. Monath (Ed.), *The Arboviruses: Epidemiology and Ecology* (Vol. 4, pp. 77–92). Boca Raton, FL: CRC Press.

Johnston, R. E., & Peters, C. J. (1996). Alphaviruses. In B. N. Fields, D. M. Knipe, & P. M. Howley, et al. (Eds.), *Fields Virology* (3rd ed., pp. 843–898). Philadelphia: Lippincott–Raven Publishers.

Jones, J. C. (1978). The feeding behavior of mosquitoes. *Scient Am, 238*, 138–148.

Journal of Medical Entomology Editorial Board. (2005). Journal policy on names of aedine mosquito genera and subgenera. *Journal of Medical Entomology, 42*, 511.

Jupp, P. G., & McIntosh, B. M. (1988). Chikungunya virus disease. In Monath, T. P. (Ed.), *The Arboviruses: Epidemiology and Ecology* (Vol. 2, pp. 137–158). Boca Raton, FL: CRC Press.

Karabatsos, N. (Ed.). (1985). *International Catalogue of Arthropod–borne Viruses* (3rd ed.). San Antonio, TX: American Society for Tropical Medicine and Hygiene.

Kay, B. H., & Aaskov, J. G. (1988). Ross River virus (epidemic polyarthritis). In T. P. Monath (Ed.), *The Arboviruses: Epidemiology and Ecology* (Vol. 4, pp. 93–112). Boca Raton, FL: CRC Press.

Kettle, D. S. (1995). *Medical and Veterinary Entomology* (2nd ed.). Wallingford, UK: CAB International.

King, W. V., Bradley, G. H., Smith, C. N., & McDuffie, W. C. (1960). A handbook of the mosquitoes of the southeastern United States. *USDA, Agricultural Handbook No. 173*, 1–188.

Klowden, M. J. (1996). Vector behavior. In B. J. Beaty & W. C. Marquardt (Eds.), *The Biology of Disease Vectors* (pp. 34–50). Niwot: Univ. Press of Colorado.

Knight, K. L. (1978). *Supplement to a Catalog of the Mosquitoes of the World (Diptera: Culicidae).* Suppl. to Vol. VI. Entomol. Soc. Am. College Park, MD: Thomas Say Found.

Knight, K. L., & Stone, A. (1977). *A Catalog of the Mosquitoes of the World (Diptera: Culicidae)* (2nd ed.). Vol. VI. Entomol. Soc. Am., College Park, MD: Thomas Say Found. Updates are presented online by the Walter Reed Biosytematics Unit: http://www.mosquitocatalog.org/main.asp)

Knight, K. L., & Wonio, M. (1969). *Mosquitoes of Iowa (Diptera: Culicidae), Agric. & Home Econ. Expt. Sta., Special Rpt. No. 61.* Ames: Iowa State Univ. Sci. Tech.

Koella, J. C. (1991). On the use of mathematical models of malaria transmission. *Acta Tropica, 49*, 1–25.

LaCasse, W. J., & Yamaguti, S. (1950). *Mosquito Fauna of Japan and Korea.* Kyoto: Off. Surgeon, 8th U. S. Army.

Lacey, L. A., & Undeen, A. H. (1986). Microbial control of black flies and mosquitoes. *Annual Review of Entomology, 31*, 265–296.

Laird, M. (1988). *The Natural History of Larval Mosquito Habitats.* New York: Academic Press.

Laird, M., & Miles, J. W. (Eds.). (1983). *Integrated Mosquito Control Methodologies. Vol. 1. Experience and Components from Conventional Chemical Control.* New York: Academic Press.

Laird, M., & Miles, J. W. (Eds.). (1985). *Integrated Mosquito Control Methodologies. Vol. 2. Biocontrol and Other Innovative Components, and Future Directions.* New York: Academic Press.

Lane, J. (1953). *Neotropical Culicidae* (3 vols.). Sao Paulo: University of Sao Paulo.

Laven, H. (1967). Eradication of *Culex pipiens fatigans* through cytoplasmic incompatibility. *Nature, 216*, 383–384.

Lavoipierre, M. M. (1958). Studies on the host-parasite relations of filarial nematodes and their arthropod hosts. II. The arthropod as a host to the nematode; a brief appraisal of our present knowledge, based on a study of the more important literature from 1878 to 1957. *Annals of Tropical Medicine and Parasitology, 52*, 326–345.

Lee, D. J., Hicks, M. M., Griffiths, M., Russell, R. C., & Marks, E. N. (1980). *The Culicidae of the Australasian Region. Vol. 1. Entomol. Monogr. No. 2 (in part).* Canberra: Austal. Gov. Publ. Service.

Lee, D. J., Hicks, M. M., Griffiths, M., Russell, R. C., & Marks, E. N. (1982). *The Culicidae of the Australasian Region. Vol. 2. Nomenclature, Synonymy, Literature, Distribution, Biology and Relation to disease: Genus Aedeomyia, Genus Aedes (Subgenera Aedes, Aedimorphus, Chaetocruiomyia, Christophersiomyia, Edwardsaedes and Finlaya).* Entomol. Monogr. No. 2 (in part). Canberra: Austral. Gov. Publ. Service.

Lee, D. J., Hicks, M. M., Griffiths, M., Russell, R. C., & Marks, E. N. (1984). *The Culicidae of the Australasian Region. Vol. 3. Nomenclature, Synonymy, Literature, Distribution, Biology and Relation to Disease: Genus Aedes, Subgenera Geokusea, Halaedes, Huaedes, Leptosomatomyia, Levua, Lorrainea, Macleaya, Mucidus, Neomelanoconion, Nothoskusea, Ochlerotatus, Paraedes, Pseudoskusea, Rhinoskusea.* Entomol. Monogr. No. 2 (in part). Canberra: Austral. Gov. Publ. Service.

Lindsay, S. W., & Gibson, M. E. (1988). Bednets revisited: Old idea, new angle. *Parasitology Today, 4*, 270–272.

Lounibos, L. P., Rey, J. R., & Frank, J. H. (Eds.). (1985). *Ecology of Mosquitoes: Proceedings of a Workshop.* Vero Beach: Florida Med. Entomol. Lab.

Lu, B. L., & Su, L. (1987). *A Handbook for the Identification of Chinese Aedine Mosquitoes.* Beijing: Science Press [in Chinese].

Lu, B. L., Chen, B. H., Xu, R., & Ji, S. (1988). *A Checklist of Chinese Mosquitoes (Diptera: Culicidae).* Beijing: Guizhu People's Publ. House. [In Chinese, English introduction]

Macdonald, G. (1957). *The Epidemiology and Control of Malaria.* London: Oxford University Press.

Mail, G. A. (1934). *The Mosquitoes of Montana.* Bozeman, Montana: Montana State College, Agric. Expt. Sta. Bull. No. 288.

Marquardt, W. C. (Ed.). (2005). *Biology of Disease Vectors* (2nd ed.). New York: Elsevier Academic Press.

Marshall, I. D. (1988). Murray Valley and Kunjin encephalitis. In T. P. Monath (Ed.), *The Arboviruses: Epidemiology and Ecology* (Vol. 3, pp. 151–190). Boca Raton, FL: CRC Press.

Marshall, J. F. (1938). *The British Mosquitoes.* London: British Museum (Natural History).

Matheson, R. (1944). *Handbook of the Mosquitoes of North America* (2nd ed.). Ithaca, NY: Comstock Publishing Co.

Mattingly, P. F. (1969). *The Biology of Mosquito-borne Disease.* London: Allen & Unwin.

Mattingly, P. F. (1971). Contributions to the mosquito fauna of Southeast Asia. XII. Illustrated keys to the genera of mosquitoes (Diptera: Culicidae). *Contributions of the American Entomological Institute, 7*, 1–84.

Mattingly, P. F. (1973). Culicidae (Mosquitoes). In K. G. V. Smith (Ed.), *Insects and Other Arthropods of Medical Importance* (pp. 37–107). London: British Museum (Natural History).

McDonald, J. L., Sluss, T. P., Lang, J. D., & Roan, C. C. (1973). *Mosquitoes of Arizona, Agricultural Experiment Station Technical Bulletin No. 205.* Tucson: Univ. Arizona.

McIver, S. B. (1982). Sensilla of mosquitoes (Diptera: Culicidae). *Journal of Medical Entomology, 19*, 489–535.

McKelvey, J. J., Eldridge, B. F., & Maramorosch, K. (Eds.). (1981). *Vectors of Disease Agents. Interactions with Plants, Animals, and Man.* New York: Praeger.

McKiel, J. A., Hall, R. R., & Newhouse, V. F. (1966). Viruses of the California encephalitis complex in indicator rabbits. *The American Journal of Tropical and Medical Hygiene, 15*, 98–102.

Means, R. G. (1979). *Mosquitoes of New York. Part I. The Genus Aedes Meigen with Identification Keys to Genera of Culicidae. Part II. Genera of Culicidae Other than Aedes Occurring in New York.* Albany: Univ. State of New York, State Educ. Dept., State Sci. Serv., New York State Museum.

Meegan, J. M., & Bailey, C. L. (1988). Rift valley fever. In T. P. Monath (Ed.), *The Arboviruses: Epidemiology and Ecology* (Vol. 4, pp. 51–76). Boca Raton, FL: CRC Press.

Meola, R., & Readio, J. (1988). Juvenile hormone regulation of biting behavior and egg development in mosquitoes. *Advances in Disease Vector Research, 5*, 1–24.

Merritt, R. W., Dadd, R. H., & Walker, E. D. (1992). Feeding behavior, natural food, and nutritional relationships of larval mosquitoes. *Annual Review of Entomology, 37*, 349–376.

Minar, J. (1991). Family Culicidae. In A. Soos & L. Papp (Eds.), *Catalogue of Palearctic Diptera: Psychodidae—Chironomidae* (Vol. 2, pp. 73–113). Amsterdam: Elsevier.

Mitchell, C. J. (1977). Arthropod-borne encephalitis viruses and water resource developments. *Cahiers ORSTOM Ser Ent Med et Parasitol, 15*, 241–250.

Mitchell, C. J. (1983). Mosquito vector competence and arboviruses. *Current Topics in Vector Research, 1*, 63–92.

Molineaux, L., & Gramiccia, G. (1980). *The Garki Project.* Geneva: World Health Organization.

Monath, T. P. (Ed.). (1980). *St. Louis Encephalitis.* Washington, DC: American Public Health Association.

Monath, T. P. (1988). Yellow fever. In T. P. Monath (Ed.), *The Arboviruses: Epidemiology and Ecology* (Vol. 3, pp. 139–231). Boca Raton, FL: CRC Press.

Monath, T. P., & Heinz, F. X. (1996). Flaviviruses. In B. N. Fields, D. M. Knipe, & P. M. Howley et al. (Eds.), *Fields Virology* (3rd ed., pp. 961–1034). Philadelphia: Lippincott–Raven Publishers.

Moore, C. G., & Gage, K. L. (1996). Collection methods for vector surveillance. In B. J. Beaty & W. C. Marquardt (Eds.), *Biology of Disease Vectors* (pp. 471–491). Niwot: University Press of Colorado.

Moore, C. G., McLean, R. G., Mitchell, C. J., Nasci, R. S., Tsai, T. F., & Calisher, C. H. et al. (1993). *Guidelines for Arbovirus Surveillance in the United States.* Fort Collins, CO: Centers for Disease Control and Prevention, U. S. Department of Health and Human Services.

Morris, C. D. (1988). Eastern equine encephalomyelitis. In T. P. Monath (Ed.), *The Arboviruses: Epidemiology and Ecology* (Vol. III, pp. 1–20). Boca Raton, FL: CRC Press, Inc.

Morris, C. D., Baker, R. H., & Opp, W. R. (Eds.). (1992). *H. T. Evans' Florida Mosquito Control Handbook.* Florida Mosq. Control Assoc.

Muirhead-Thomson, R. C. (1951). *Mosquito Behaviour in Relation to Mosquito Transmission and Control in the Tropics.* London: Arnold.

Muirhead-Thomson, R. C. (1968). *Ecology of Insect Vector Populations.* New York: Academic Press.

Muirhead-Thomson, R. C. (1982). *Behaviour Patterns of Blood-Sucking Flies.* Oxford: Pergamon Press.

Nasci, R. S., & Miller, B. R. (1996). Culicine mosquitoes and the agents they transmit. In B. J. Beaty & W. C. Marquardt (Eds.), *The Biology of Disease Vectors* (pp. 85–97). Niwot: Univ. Press of Colorado.

Nayar, J. K. (1982). *Bionomics and Physiology of Culex Nigripalpus (Diptera: Culicidae) of Florida: An Important Vector of Diseases. Florida Agricultural Experiment Station Bulletin No. 827.*

Nayar, J. K. (Ed.). (1985). *Bionomics and Physiology of Aedes Taeniorhynchus and Aedes Sollicitans, the Salt Marsh Mosquitoes of Florida. Florida Agricultural Experiment Station Bulletin No. 852.*

Nedelman, J. (1990). Gametocytemia and infectiousness in falciparum malaria: Observations and models. *Advances in Disease Vector Research, 6*, 59–89.

O'Meara, G. F. (1985). Ecology of autogeny in mosquitoes. In E. P. Lounibos, J. R. Rey, & J. H. Frank (Eds.), *Ecology of Mosquitoes: Proceedings of a Workshop* (pp. 459–471). Vero Beach: Florida Medical Entomology Laboratory.

Onori, E., & Grab, B. (1980). Indicators for the forecasting of malaria epidemics. *Bulletin of the World Health Organization, 58*, 91–98.

Ottesen, E. A., & Ramachandran, C. P. (1995). Lymphatic filariasis infection and disease: Control strategies. *Parasitology Today, 11*, 129–131.

Owen, W. B., & Gerhardt, R. W. (1957). The mosquitoes of Wyoming. *University of Wyoming Publication, 21*(3), 71–141.

Pampana, E. (1963). *A Textbook of Malaria Eradication.* Oxford, UK: Oxford Univ. Press.

Pan American Health Organization. (1972). *Venezuelan Encephalitis.* PAHO Scientific Publication No. 243, Washington, DC.

Peters, W. (1985). The problem of drug resistance in malaria. *Parasitology, 90*, 705–715.

Pittaway, A. R. (1992). *Arthropods of Medical and Veterinary Importance: A Checklist of Preferred Names and Allied Terms.* CAB International.

Pratt, H. D., Barnes, R. C., & Littig, K. S. (1963). *Mosquitoes of Public Health Importance and Their Control, Public Health Services Publication No. 772.* Washington, DC.

Rai, K. S. (1991). *Aedes albopictus* in the Americas. *Annual Review of Entomology, 36*, 459–484.

Rai, K. S. (1996). Genetic control of vectors. In B. J. Beaty & W. C. Marquardt (Eds.), *Biology of Disease Vectors* (pp. 564–574). Niwot: University Press of Colorado.

Raikhel, A. S. (1992). Vitellogenesis in mosquitoes. *Advances in Disease Vector Research, 9*, 1–39.

Reeves, W. C. (Ed.). (1990). *Epidemiology and Control of Mosquito–borne Arboviruses in California, 1943–1987.* Sacramento: California Mosquito and Vector Control Association.

Reinert, J. F. (2000). A new classification for the composite genus Aedes (Diptera: Culicidae: Aedini), elevation of subgenus Ochlerotatus to generic rank, reclassification of the other subgenera, and notes on certain subgenera and species. *Journal of the American Mosquito Control Association. 16*, 175–188.

Reinert, J. F., Harbach, R. E., & Kitching, I. J. (2004). Phylogeny and classification of Aedini (Diptera: Culicidae), based on morphological characters of all life stages. *Zoological Journal of the Linnean Society, 142*, 289–368.

Reinert, J. F., Kaiser, P. E., & Seawright, J. A. (1997). Analysis of the *Anopheles* (*Anopheles*) *quadrimaculatus* complex of sibling species (Diptera: Culicidae) using morphological, cytological, molecular, genetic, biochemical, and ecological techniques in an integrated approach. *Journal of the American Mosquito Control Association, 13*(Suppl.), 1–102.

Reisen, W. K., & Monath, T. P. (1988). Western equine encephalomyelitis. In T. P. Monath (Ed.), *The Arboviruses: Epidemiology and Ecology* (Vol. 4, pp. 89–137). Boca Raton, FL: CRC Press, Inc.

Rempel, J. G. (1950). A guide to the mosquito larvae of western Canada. *Canadian Journal of Zoology, 28*, 207–248.

Rempel, J. G. (1953). The mosquitoes of Saskatchewan. *Canadian Journal of Zoology, 31*, 433–509.

Restifo, R. A. (1982). *Illustrated Key to the Mosquitoes of Ohio [Adapted from Stojanovich (1960, 1961)], Ohio Biological Survey, Biological Notes No. 17.* Ohio: Columbus.

Ribeiro, J. M. C. (1987). Role of saliva in blood-feeding by arthropods. *Annual Review of Entomology, 32*, 463–478.

Ross, E. S., & Roberts, H. R. (1943). *Mosquito Atlas. Part I: The Nearctic Anopheles, Important Malaria Vectors of the Americas and Aedes aegypti, Culex Quinquefasciatus. Part II: Eighteen Old World Anophelines Important to Malaria.* Philadelphia: *American Entomological Society, Academy of Natural Science.*

Ross, H. H. (1947). The mosquitoes of Illinois (Diptera, Culicidae). *Bulletin of the Illinois Natural History Survey, 24*, 1–96.

Ross, H. H., & Horsfall, W. R. (1965). *A Synopsis of the Mosquitoes of Illinois (Diptera: Culicidae).* Ill. *Natural History Survey, Biological Notes No. 52.*

Ross, R. (1911). *The Prevention of Malaria.* London: Murray.

Rozeboom, L. E. (1942). *The Mosquitoes of Oklahoma. Technical Bulletin No, T-16, Oklahoma Agricultural Experiment Station.*

Russell, P. F. (1955). *Man's Mastery of Malaria.* Oxford: Oxford Univ. Press.

Russell, P. F., Rozeboom, L. E., & Stone, A. (1943). *Keys to the Anopheline Mosquitoes of the World.* Distribution, Biology, and Relation to Malaria, *American Entomological Society, Academy of Natural Science,* Philadelphia: with Notes on Their Identification.

Rutschky, C. W., Mooney, T. C. Jr., & Vanderberg, J. P. (1958). *Mosquitoes of Pennsylvania. An Illustrated Key to Species with Accompanying Notes on Biology and Control.* University Park, PA: *Pennsylvania State University Agricultural Experiment Station Bulletin No. 630.*

Sasa, M. (1976). *Human Filariasis: A Global Survey of Epidemiology and Control.* Tokyo, Japan: Tokyo University Press.

Savage, H. M., Strickman, D. (2004). The genus and subgenus categories within Culicidae and placement of Ochlerotatus as a subgenus of Aedes. *Journal of the American Mosquito Control Association. 20,* 208–214.

Service, M. W. (Ed.). (1988). *Biosystematics of Haematophagous Insects.* Oxford: Clarendon Press.

Service, M. W. (1989). *Demography and Vector-borne Diseases.* Boca Raton, FL: CRC Press.

Service, M. W. (1990). *Handbook of the Afrotropical Toxorhynchitine and Culicine Mosquitoes, Excepting Aedes and Culex.* London: British Museum (Natural History).

Service, M. W. (1993). *Mosquito Ecology: Field Sampling Methods* (2nd ed.). New York: Elsevier Applied Science.

Service, M. W. (1993). Mosquitoes (Culicidae). In Lane, R. P., Crosskey, R. W. (Eds.), *Medical Insects and Arachnids* (pp. 120–240). New York: Chapman & Hall.

Service, M. W. (1997). Mosquito (Diptera: Culicidae) dispersal—The long and short of it. *Journal of Medical Entomology, 34,* 579–588.

Silver, J. B. (2008). *Mosquito Ecology: Field Sampling Methods* (3rd ed.). Springer.

Siverly, R. E. (1972). *Mosquitoes of Indiana.* Indianapolis: Indiana State Board Health.

Smart, J. (1948). *Insects of Medical Importance.* London: British Museum (Natural History).

Snodgrass, R. E. (1959). *The Anatomical Life of the Mosquito, Smithsonian Miscellaneous Publications, Vol. 139, No. 8. Baltimore Press.* Maryland: Baltimore.

Snow, K. R. (1990). *Mosquitoes, Naturalists' Handbooks 14.* Slough, United Kingdom: Richmond Publ.

Sokolova, M. I. (1994). A redescription of the morphology of mosquito (Diptera: Culicidae) ovarioles during vitellogenesis. *Bulletin of the Society for Vector Ecology, 19,* 53–68.

Soper, F. L. (1963). The elimination of urban yellow fever in the Americas through the eradication of *Aedes aegypti. American Journal of Public Health, 53,* 7–16.

Soper, F. L., & Wilson, D. B. (1943). *Anopheles Gambiae in Brazil 1933–1940.* New York: Rockefeller Foundation.

Stage, H. H., Gjullin, C. M., & Yates, W. W. (1952). *Mosquitoes of the Northwestern States, U. S. Dept. Agric. Handbook 46.* Washington, DC.

Steffan, W. A., & Evenhuis, N. L. (1981). Biology of *Toxorhynchites. Annual Review of Entomology, 26,* 159–181.

Stojanovich, C. J. (1960). *Illustrated Key to Common Mosquitoes of Southeastern United States.* Atlanta, GA: Cullom & Ghertner Co.

Stojanovich, C. J. (1961). *Illustrated Key to Common Mosquitoes of Northeastern North America.* Atlanta, GA: Emory University Branch, Cullom & Ghertner Co.

Stone, A. (1981). Culicidae. In J. F. McAlpine et al. (Eds.), *Manual of Nearctic Diptera* (Vol. 1, Chapt. 25, pp. 341–350.

Stone, A., & Delfinado, M. D. (1973). Family Culicidae. In M. D. Delfinado & D. E. Hardy (Eds.), *A Catalog of the Diptera of the Oriental Region. Suborder Nematocera* (Vol. 1, pp. 266–343). Honolulu: Univ. Press of Hawaii.

Strickland, G. T. (Ed.). (1991). *Hunter's Tropical Medicine* (7th ed.). Philadelphia: W. B. Saunders Co.

Tabachnick, W. J. (1994). Genetics of insect vector competence for arboviruses. *Advances in Disease Vector Research, 10,* 93–108.

Tate, H. D., & Gates, D. B. (1944). *The Mosquitoes of Nebraska. Agricultural Experiment Station Research Bulletin No. 133,* Univ. Nebraska.

Telford, S. R., Jr. (1994). Plasmodia of reptiles. In J. P. Kreier (Ed.), *Parasitic Protozoa* (2nd ed., Vol. 7, pp. 1–72). New York: Academic Press.

Tempelis, C. H. (1975). Host-feeding patterns of mosquitoes, with a review of advances in analysis of blood meals by serology. *Journal of Medical Entomology, 11,* 635–653.

Tsai, T. F., & Mitchell, C. J. (1988). St. Louis encephalitis. In T. P. Monath (Ed.), *The Arboviruses: Epidemiology and Ecology* (Vol. 4, pp. 113–143). Boca Raton, FL: CRC Press.

Tulloch, G. S. (1939). A key to the mosquitoes of Massachusetts. *Psyche, 46,* 113–136.

Turell, M. J. (1988). Horizontal and vertical transmission of viruses by insect and tick vectors. In T. P. Monath (Ed.), *The Arboviruses: Epidemiology and Ecology* (Vol. 1, pp. 127–152). Boca Raton, FL: CRC Press.

United States Department of Agriculture, Animal and Plant Health Inspection Service. (1973). *The Origin and Spread of Venezuelan Equine Encephalomyelitis.* APHIS 91–10.

United States Public Health Service and Tennessee Valley Authority. (1947). *Malaria Control on Impounded Water.* Washington, DC: U.S. Government Printing Office.

Van Dine, D. L. (1922). *Impounding Water in a Bayou to Control Breeding of Malaria Mosquitoes, Bull. No. 1098.* Washington, DC: USDA, U.S. Govt. Printing Office.

Van Riper III, C., et al. (1994). Plasmodia of birds. In J. P. Kreier (Ed.), *Parasitic Protozoa* (2nd ed., Vol. 7, pp. 73–140). New York: Academic Press.

Wallis, R. C. (1960). *Mosquitoes in Connecticut.* New Haven, Connecticut: Conn. Agric. Expt. Sta, Bull. No. 632.

Walton, T. E., & Grayson, M. A. (1988). Venezuelan equine encephalomyelitis. In T. P. Monath (Ed.), *The Arboviruses: Epidemiology and Ecology* (Vol. 4, pp. 203–231). Boca Raton, FL: CRC Press.

Ward, R. A. (1984). Second supplement to "A catalog of the mosquitoes of the world (Diptera: Culicidae)." *Mosquito Systematics, 16,* 227–270.

Ward, R. A. (1992). Third supplement to "A catalog of the mosquitoes of the world (Diptera: Culicidae)." *Mosquito Systematics, 24,* 177–230.

Ward, R. A., & Darsie, R. F. (1982). Corrections and additions to the publication, "Identification and Geographical Distribution of the Mosquitoes of North America, North of Mexico." *Mosquito Systematics, 14,* 209–219.

Washino, R. K., & Tempelis, C. H. (1983). Mosquito host bloodmeal identification: Methodology and data analysis. *Annual Review of Entomology, 28,* 179–201.

Watts, D. M., Pantuwatana, S., DeFoliart, G. R., Yuill, T. M., & Thompson, W. H. (1973). Transovarial transmission of LaCrosse virus (California encephalitis group) in the mosquito, *Aedes triseriatus. Science, 182,* 1140–1141.

Wernsdorfer, W. H., & McGregor, I. (Eds.). (1988). *Malaria: Principles and Practice of Malariology* (Vol. 2). New York: Churchill Livingstone.

White, G. B. (1980). Family Culicidae. In R. W. Crosskey (Ed.), *Catalogue of the Diptera of the Afrotropical Region* (pp. 114–148). London: British Museum (Natural History).

Wilkerson, R. C., & Strickman, D. (1990). Illustrated key of the female anopheline mosquitoes of Central America and Mexico. *Journal of the American Mosquito Control Association, 6,* 7–34.

Wood, R. J. (2005). Genetic control of vectors. In W. C. Marquardt (Ed.), *Biology of Disease Vectors* (2n ed., pp. 661–669). New York: Elsevier Academic Press.

Wood, D. M., & Borkent, A. (1989). Phylogeny and classification of the Nematocera. In J. F. McAlpine (Ed.), *Manual of Nearctic Diptera* (Vol. 3, pp. 1333–1370). Research Branch, Agriculture Canada, Monograph No. 32.

Wood, D. M., Dang, P. T., & Ellis, R. A. (1979). *The Insects and Arachnids of Canada. Part 6. The Mosquitoes of Canada. Diptera: Culicidae.* Hull, Quebec: Canad. Gov. Publ. Centre.

Woodring, J. L., & Davidson, E. W. (1996). Biological control of mosquitoes. In B. J. Beaty & W. C. Marquardt (Eds.), *The Biology of Disease Vectors* (pp. 530–548). Niwot: Univ. Press of Colorado.

World Health Organization. (1973). *Manual on Larval Control Operations in Malaria Programmes.* Offset Publication No. 1. Geneva, Switzerland: World Health Organization.

World Health Organization. (1975). *Manual on Practical Entomology in Malaria, Parts I and II.* Offset Publication No. 13. Geneva, Switzerland: World Health Organization.

World Health Organization. (1982). *Manual on Environmental Management for Mosquito Control.* Offset Publication No. 66. Geneva, Switzerland: World Health Organization.

World Health Organization, Vector Biology and Control Division. (1989). *Geographical Distribution of Arthropod–borne Diseases and Their Principal Vectors.* World Health Organization/VBC Publication 89.967.

World Health Organization. (1991). *Prospects for Malaria Control by Genetic Manipulation of Its Vectors.* Unpublished document TDR/BCV/MAL-ENT/91.3. Geneva, Switzerland: World Health Organization.

Wright, J. W., & Pal, R. (Eds.). (1967). *Genetics of Insect Vectors of Disease.* New York: Elsevier.

Yamaguti, S., & LaCasse, W. J. (1951). *Mosquito Fauna of North America, Parts I–V.* Off. Surg., Hq., Japan Logistical Command, APO 343.

Chapter

15

Horse Flies and Deer Flies (Tabanidae)

Bradley A. Mullens

Because of their fairly large size, striking appearance, and diurnal biting habits, horse flies and deer flies (Figure 15.1) are familiar to most people who have livestock or engage in outdoor activities. Diversity within the family is greatest in the tropics, but moist temperate regions typically have a rich fauna as well. Tabanids are present on every continent except Antarctica and have managed to colonize remote islands such as the Galápagos and the Melanesian Archipelago. Large seasonal populations of some species occur as far as 60° N latitude, but they disappear above treeline.

The eyes of many species, when alive, are brilliantly patterned with shades of green, yellow, orange, and violet. Some species with strikingly green eyes are commonly called greenheads, and others are called yellow flies, due to their yellow bodies (Figure 15.2). The blue-tail fly of American folk music probably was *Tabanus atratus,* a large black species with a blue cast to the abdomen. Some common names reflect times or places where these biting flies are found (e.g., March fly, May fly, and mango fly), whereas the sources of some colloquial names are obscure (e.g., Cleg and whamefly). Other common names include breezefly, bulldog, and gadfly, the latter two reflecting the persistent annoyance of tabanids in seeking a blood meal.

The term **horse fly** is applied to relatively large species of tabanids, typically 10 to 30 mm in length. They can be a serious nuisance to livestock and can mechanically transmit several significant animal pathogens, including those that cause surra, anaplasmosis, and equine infectious anemia. Even moderate numbers of flies feeding on livestock can result in significant production losses. A few horse flies readily bite people, examples being the infamous greenheads (*Tabanus nigrovittatus* and *T. simulans*) in the coastal regions of the eastern United States.

The smaller tabanid species, called deer flies, typically are 6 to 10 mm long. In contrast to horse flies, they frequently attack humans. Fortunately there are just a few human diseases known to be associated with deer flies. The most important tabanid-transmitted human diseases are loiasis and tularemia. Outdoor activity and tourism suffer in areas where tabanid populations are high, although such losses are hard to quantify.

TAXONOMY

The family Tabanidae includes approximately 4,300 species and subspecies in 133 genera worldwide. Of these, 335 species in 25 genera are found in the Nearctic Region (Burger, 1995). Although the temperate fauna is well known, the tropical fauna has been less studied; this is particularly true for the immature stages. Most pest species in North America are members of the genera *Chrysops, Hybomitra,* and *Tabanus.* The Tabanomorpha, which includes the Tabanidae and a few related families, diverged around 200 million years ago and constitute a natural grouping supported by both morphology and molecular evidence.

The family Tabanidae is divided into three subfamilies (Mackerras, 1954; Fairchild, 1969) (Table 15.1). The subfamily Pangoniinae is regarded as ancestral, containing fascinating but poorly known genera such as *Stonemyia* and *Goniops,* many of which are not known to feed on blood. Most of the economically important tabanids are members of the two other subfamilies. Tabanids in the subfamily Chrysopsinae are called deer flies; nearly all are members of the genus *Chrysops,* which includes more than 80 Nearctic species. The term "deer fly" also is applied to members of the genus *Silvius,* a few species of which can be quite pestiferous on man and animals in the western United States.

Members of the Tabaninae are the most evolutionarily derived. This subfamily includes the horse flies, represented by *Tabanus,* which has 107 Nearctic species, and *Hybomitra,* with 55 Nearctic species. Species of *Haematopota,* together with *Tabanus* and *Hybomitra,*

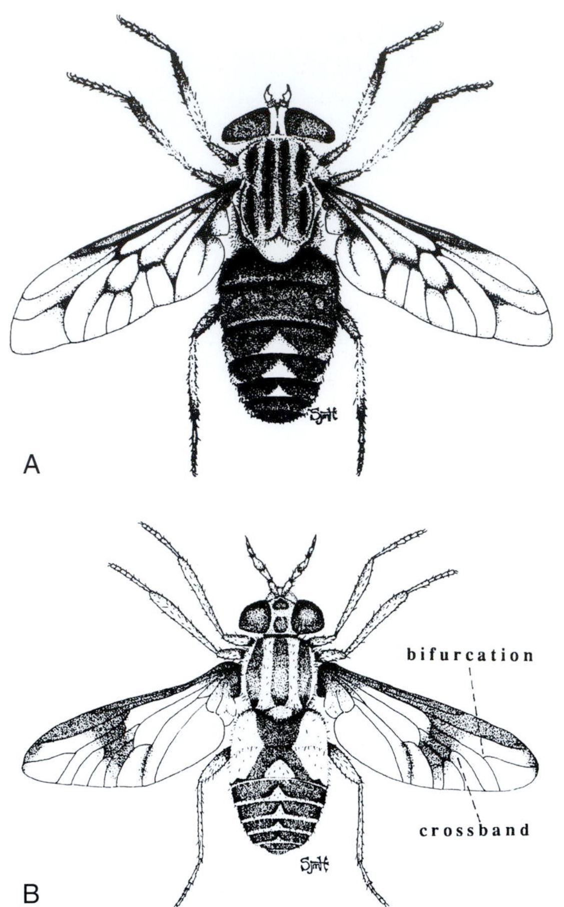

Figure 15.1 Adult tabanid flies, females, dorsal view. (A) Horse fly, *Tabanus trimaculatus*; (B) Deer fly, *Chrysops callidus*. Note distinctive wing venation, including bifurcation of vein near wing tip, and the darkened crossband in deer flies. (Original by S. J. M. Hope)

Figure 15.2 *Diachlorus ferrugatus* feeding on a human hand. Note the yellow body and especially the brightly-colored eye bands that characterize many tabanid species. (Photo by B. A. Mullens)

Table 15.1 Subfamilies, Tribes and Selected Genera of Tabanidae in North America

Taxon	No. Species
Subfamily Pangoniinae	
Tribe Pangoniini	
Genus *Apatolestes*	13
Genus *Stonemyia*	6
Tribe Scionini	
Genus *Goniops*	1
Subfamily Chrysopsinae	
Tribe Bouvieromyiini	
Genus *Mercomyia*	2
Tribe Chrysopsini	
Genus *Silvius*	12
Genus *Chrysops*	83
Subfamily Tabaninae	
Tribe Diachlorini	
Genus *Diachlorus*	1
Genus *Chlorotabanus*	1
Genus *Leucotabanus*	2
Tribe Haematopotini	
Genus *Haematopota*	5
Genus *Tabanus*	107
Genus *Atylotus*	14
Genus *Hybomitra*	55

The largest genera in terms of numbers of species are *Chrysops, Tabanus,* and *Hybomitra.*

are important pests in the Old World. Only five species of *Haematopota* occur in the Nearctic region, where *H. americana* is the only species known to be a pest of mammals.

Burger (1995) compiled a complete catalog of species of Nearctic Tabanidae. Pechuman and Teskey (1981) provide generic keys to larvae, pupae, and adults of Nearctic tabanids. There are several regional keys for identification of adults of North American tabanids at the species level: Florida (Jones and Anthony, 1964), California (Middlekauf and Lane, 1980), New York (Pechuman, 1981), Illinois (Pechuman et al., 1983), Tennessee (Goodwin et al., 1985), Texas (Goodwin and Drees, 1996), and Canada and Alaska (Teskey, 1990). Most of these works include valuable information on biology and ecology. Immatures of Nearctic tabanids are more difficult to identify than adults. Many North American species remain undescribed, and the immature stages are less likely to be encountered by the casual collector. Immatures of some species can be identified using keys or references found in Burger (1977), Pechuman et al. (1983), Goodwin et al. (1985), and Teskey (1990).

Taxonomic references for other regions include Europe (Chvala et al., 1972), Neotropics (Fairchild, 1986; Coscaron and Papavero, 1993; Fairchild and Burger, 1994), Australia (Mackerras, 1954), Australasia (Burger and Chainey, 2000), Mali (Goodwin, 1982),

Ethiopian region (Oldroyd, 1954–1957), the former Soviet Union (Olsufiev, 1977), and Japan (Takahashi, 1962; Hayakawa, 1985). Immature stages of Palearctic Tabanidae are treated by Andreeva (1990).

MORPHOLOGY

Tabanid larvae (Figure 15.3) are spindle-shaped (fusiform) and generally whitish in color, although some are shades of brown or green. Mature larvae of common species typically measure 15 to 30 mm in length, but some larger tabanid larvae may be as long as 60 mm. The head capsule is incomplete and partially sclerotized. The mandibles are strong, parallel, and ventrally curved and are used to capture and subdue prey. The larval cuticle has distinctive longitudinal striations and often exhibits species-specific pubescence patterns that give some tabanid larvae a mottled appearance. Abdominal segments have lateral and ventral pseudopods for locomotion (3 pairs in *Chrysops* spp., 4 pairs in Tabaninae). Larvae of the more terrestrial species tend to be relatively stocky with short pseudopods. Species adapted to a fully aquatic existence in streams (e.g., *Tabanus fairchildi*) have elongated pseudopods armed with cuticular, recurved, distal hooks. Semiaquatic larvae, represented by the majority of tabanid species, have intermediate characters.

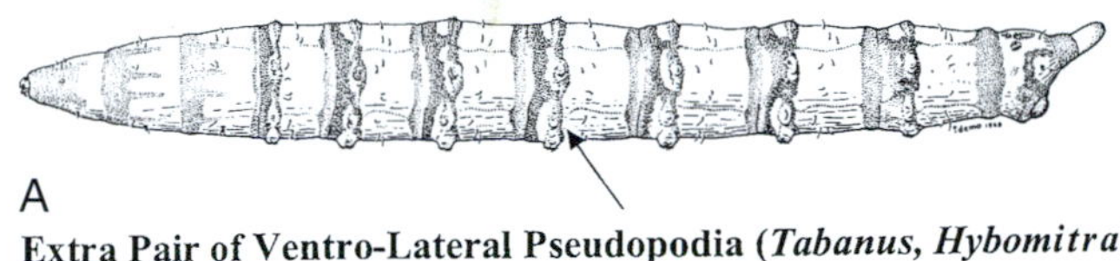

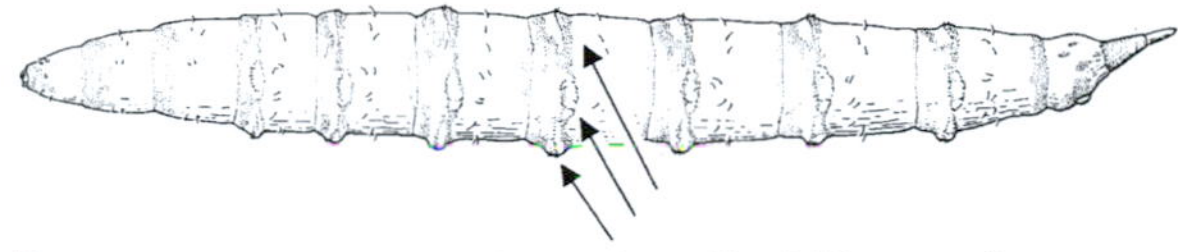
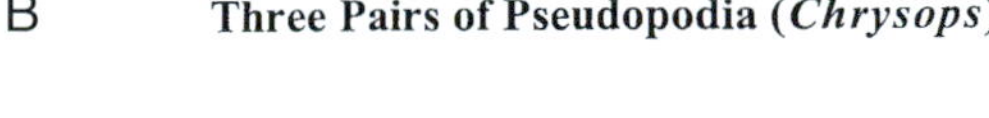

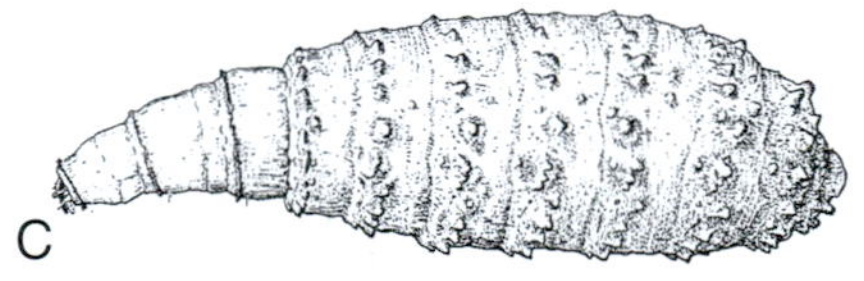

Figure 15.3 Larvae of Tabanidae. Typical semiaquatic larval forms are shown in A and B. Note the extra pair of pseudopodia useful in differentiating the common horse-fly genera *Tabanus* and *Hybomitra* from the common deer-fly genus *Chrysops*. Larva of the unusual terrestrial genus *Goniops* is shown in C. (From the Manual of Nearctic Diptera. Reproduced with permission of the Minister of Public Works, and Government Services Canada)

Located in the antero-dorsal portion of the anal segment of larvae is a pear-shaped vesicle called **Graber's organ**. It is seen readily through the cuticle only in tabanid larvae; the number of black bodies within it increases for each larval instar. Its function is unknown, although it also exists in some adult tabanids. Two main tracheal trunks run the length of the body, terminating in a dorsally directed respiratory siphon. A terminal spine is present on the siphon of some species.

Tabanid pupae are usually tan or brown, with the eyes, legs, and wing pads visible externally. A fringe of spines on the posterior margin of many abdominal segments and a star-like series of three or four pairs of caudal projections called pupal asters are useful in identification.

Tabanid adults are stout-bodied flies. They generally can be distinguished as horse flies or deer flies based on several morphological characters (Table 15.2). The antennae are prominent and extend anteriorly. The flagellum, with four to eight flagellomeres, usually is enlarged at the base in Tabaninae but only slightly enlarged in *Chrysops* species (Figure 15.4). The eyes often consist of large ommatidial facets dorsally and smaller facets ventrally. This arrangement is believed to enhance visual acuity in locating potential mates. The male has holoptic eyes, which occupy most of the head, touching each other medially. The female has dichoptic eyes that are smaller than those of the male and are separated by the frons. The frons of most species is covered by very fine pubescence. Slightly raised, sometimes bare areas of cuticle, called the median callus and basal callus, aid identification.

The remarkable color patterns of the adult eyes are very distinctive and beautiful in many species (Figure 15.2, 15.4) and sometimes are useful taxonomically. Eye patterns and colors unfortunately disappear when a specimen dries, however the basic pattern often can be restored by rehydration of the specimen. The Pangoniinae and Chrysopsinae possess well developed ocelli at the vertex of the frons, whereas the ocelli of Tabaninae are vestigial or absent. *Hybomitra* species usually exhibit a raised, denuded ocellar tubercle, which is lacking in *Tabanus* species.

Table 15.2 Morphological Characters Used to Differentiate Adult Horse Flies and Deer Flies

Character	Horse Flies (e.g., *Tabanus*)	Deer Flies (e.g., *Chrysops*)
Body length	10–30 mm	6–11 mm
Antennae	Short, base of flagellum greatly enlarged	Long, base of flagellum not greatly enlarged
Ocelli	Vestigial or lacking	Present
Wings	Clear, uniformly cloudy, or spotted	Distinctly banded
Apical spurs on hind tibiae	Lacking	Present

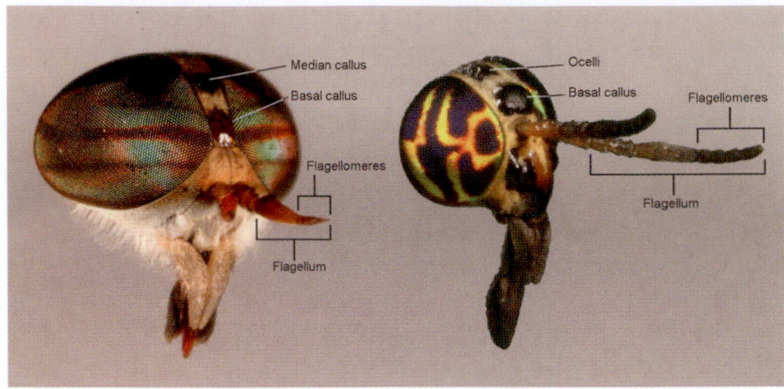

Figure 15.4 Morphology of head and antennae of tabanid flies, with important taxonomic characters. (left) Horse fly, *Tabanus* sp.; (right) Deer fly, *Chrysops* sp.; note the elongate flagellum (terminal segment of the antenna) of *Chrysops*, with multiple flagellomeres, or pseudosegments. (Photo by Nathan D. Burkett-Cadena)

The maxillary palps are two-segmented and enlarged at the base of the apical palpomere. The proboscis is stout and includes toothed, bladelike mandibles and maxillary laciniae used to lacerate the skin and capillary beds during blood-feeding. The female hypopharynx is rigid, with the salivary duct opening at the tip to introduce saliva into the feeding wound. Blood is drawn up between the labellar lobes into the food canal between the labrum and hypopharynx. This feeding method is known as pool feeding, or telmophagy. Males, which do not feed on blood, lack mandibles, recurved teeth on the laciniae, and a rigid hypopharynx.

The thorax is stout with a prominent notopleural lobe and strong flight muscles. Legs are also stout with fairly prominent tibial spurs. Apical spurs are present on the hind tibia of the Pangoniinae and Chrysopsinae, but are lacking in the Tabaninae. The wing venation is quite consistent within the Tabanidae; a key feature is the widely divergent R4 and R5 veins, which fork and enclose the apex of the wing. Wing membranes are clear in some species and variously darkened in others, providing useful taxonomic characters, particularly for *Chrysops* species. Wings of *Haematopota* species (clegs) are mottled, while those of most other Tabaninae are not. The abdomen of tabanids is as wide as the thorax, slightly compressed dorso-ventrally, and often distinctly colored or patterned.

Internally, tabanids have a large crop for water and sugar storage. Blood is directed to the midgut. As with many other blood-feeders, tabanids can rapidly eliminate excess fluids from the blood meal via the anus. Genitalia are fairly simple structurally in both sexes and are of little use in routine species identifications.

LIFE HISTORY

Tabanid eggs are 1 to 3 mm long and are deposited in masses (Figure 15.5). The female usually lays 100 to 800 eggs in a single mass, the numbers varying substantially with the species and size of the blood meal.

Figure 15.5 Female horse fly, *Tabanus imitans*, ovipositing on plant stem. (Photo by Sturgis McKeever)

Some species lay several smaller batches of eggs, particularly in captivity; *Atylotus thoracicus* has been observed to deposit eggs singly on sphagnum moss in the laboratory. Eggs are white when laid but darken to grey, brown, or black within several hours. Egg masses most often are found on leaves or stems of emergent vegetation at the edges of ponds (lentic habitats) or streams (lotic habitats) or on leaves or bark of trees overhanging the water. The larvae of Nearctic species of *Chrysops* have been categorized as 65% lentic and 18% lotic, with 13% of the species being found in both lentic and lotic habitats. Stream-dwelling species often deposit their eggs above waterline on stones in the stream where water flow is moderate. Some species can be terrestrial and may be found in fairly dry soil (e.g., *Tabanus abactor, T. sulcifrons, T. subsimilis*). Terrestrial species usually lay their eggs on vegetation or in leaf litter. *Apatolestes actites* is known to oviposit in crustacean burrows in beach habitats.

Many *Chrysops* species deposit eggs in a single layer, for example, *C. callidus* (Figure 15.6), whereas others deposit eggs in tiers, for example, *C. cincticornis* (Figure 15.7). Species of *Tabanus* and *Hybomitra* lay their eggs in tiers, commonly three or four tiers high. Such masses taper in pyramid fashion from base to apex. The exact shape of the egg mass often reflects the oviposition substrate; for example, the same species may deposit an elongate egg mass on a grass stem or a broader one on a deciduous leaf. Although the

Figure 15.6 Two egg masses of *Chrysops callidus*, deposited on vegetation. Note single layer of eggs typical of most deer flies. (Photo by L. L. Pechuman)

Figure 15.7 Egg mass of *Chrysops cincticornis*. Note the multiple layers of eggs typical of certain deer flies and most horse flies. (Photo by Elton J. Hansens)

eggs often are easily seen, it is uncommon to observe females in the act of oviposition. Recent DNA fingerprinting efforts have allowed more species-specific descriptions of egg mass morphology and location. Females of an unusual and primitive tabanid, *Goniops chrysocoma*, lay eggs on the underside of a leaf and secure themselves above the mass using the tarsal claws. They remain with the mass until it hatches, buzzing noisily if disturbed. The female dies soon after the eggs hatch.

Embryogenesis typically requires five to 12 days at temperatures of 21 to 24 °C, and is both temperature- and species-dependent. Egg hatch can occur within two to three days at temperatures of 30 to 35 °C. First-instar larvae are equipped with an egg burster, a projection on the head capsule with which they split the chorion, and then drop to the water or moist substrate below. They molt once, apparently without feeding, before beginning to move about in the substrate.

Tabanid larvae are found in a wide variety of aquatic and semiaquatic habitats. These include mud or saturated vegetation in marshes or near pond or creek margins, under stones in and along streams, and in terrestrial habitats such as under forest litter. Some species are common in a variety of semiaquatic habitats (e.g., *Tabanus punctifer*). Others are quite specific; larvae of *Leucotabanus annulatus*, for example, are found only in rotten tree stumps. Tabanid larvae are general predators that feed on a variety of invertebrates such as larvae of chironomid midges and crane flies, or annelids. Horse-fly larvae also are cannibalistic, a factor that influences their densities and distribution and complicates efforts to rear them in the laboratory. It is unusual to find high densities of tabanine larvae in nature. Nonetheless, densities of larvae in flooded hardwood forest in Louisiana have been estimated at about 10 per m^2, leading to high adult populations. In contrast to tabanine larvae, *Chrysops* larvae apparently are not as cannibalistic and may be found at high field densities. *Chrysops* larvae probably are predaceous and feed primarily on invertebrates, although some authors have speculated that they feed on detritus.

Tabanids undergo six to 13 larval molts and overwinter as larvae. Most temperate species have one generation per year (univoltine), whereas others may produce two or more generations (multivoltine). Particularly large species, such as *Tabanus atratus* and *T. calens*, may spend two or three years as larvae. Development may be prolonged for three years or more in very cold, seasonally dry, or otherwise unfavorable conditions. In the spring, the larvae leave the water-saturated soil to pupate above waterline. Pupal periods vary with species and temperature but typically last four to 21 days.

Most temperate species have very distinctive seasonal flight periods that vary little from year to year. In the eastern United States and Canada, for example, *Hybomitra lasiophthalma* is one of the first pestiferous horse flies to emerge. It begins activity as early as March in the southern United States or as late as June in southern Canada, is abundant for three to six weeks, and then adults disappear until the following spring. In contrast, *Tabanus subsimilis* has been collected in other parts of the United States from mid May through early October. Some tabanids can develop from egg to adult in as little as six weeks, and a few species have multiple broods within a season. A prolonged period of adult emergence, or the presence of unrecognized sibling species, can contribute to what appears to be a long adult flight period.

Many tabanids are anautogenous and require a single large blood meal in order to develop a batch of eggs. Blood meal size varies from 20 to 25 mg for many *Chrysops* species to almost 700 mg for *Tabanus atratus*. Other species, such as *T. nigrovittatus*, are obligately autogenous in the first gonotrophic cycle and must seek a blood meal for each subsequent cycle. Still others seem to be facultatively autogenous, probably reflecting genetic variability and the carry over of available nutrients from the larval stage. Following a blood meal, egg development is believed to be typically three to four days. Time to oviposition may be several days longer for some species, especially under laboratory conditions where the full range of oviposition cues may be lacking. One California species, *Apatolestes actites*, apparently undergoes two autogenous gonotrophic cycles, a rarity among families of haematophagous Diptera.

BEHAVIOR AND ECOLOGY

The biology and behavior of tabanid larvae are generally poorly known. They are laborious to rear in the laboratory due to their long developmental times and predaceous and cannibalistic habits. No tabanid species has been colonized successfully. As soil dwellers they are difficult to observe and sample. Tabanid larvae are rarely free-swimming in nature, but some, such as *Tabanus punctifer*, are buoyant and can swim effectively by repeatedly flipping the rear half of the body and propelling themselves in short, gliding spurts.

On contacting a prey item, a tabanid larva will strike, often with an audible click, seizing the prey with its mandibles. Tabanid larvae are capable of capturing prey larger than themselves, and prey struggles usually cease very quickly. A toxin likely is involved in prey capture, but this has not been demonstrated conclusively.

Prior to pupation, larvae generally seek out drier soil, such as above waterline at the edge of ponds or streams. Pupation occurs near the soil surface with the head end oriented upward. Larvae of a few species, including *T. atratus* in the New World and several other *Tabanus* spp. in the Old World, construct a mud cylinder above waterline. They spiral downward 5 to 13 cm to delineate the perimeter of the 3 to 9 cm diameter cylinder and then burrow into the center to form a pupation tunnel. This unusual behavior may preserve the structural integrity of the pupation tunnel and facilitate future emergence, or yield a drier pupation site in periodically flooded habitats. Mortality during tabanid egg and larval stages is high. A production ratio of only three pupae/egg mass has been calculated for *Hybomitra bimaculata* in Swiss bog-meadow habitats.

The biology and behavior of tabanid adults are better understood than the immatures. The sex ratio at emergence is approximately 1:1, and emergence of males precedes that of females by one to a few days. An important activity for both sexes is carbohydrate feeding, which provides energy for general body maintenance, flight, and mating. Sugars are obtained at floral or extrafloral nectaries or other natural plant sugar deposits. *Tabanus nigrovittatus* and some *Chrysops* spp. adults are known to feed on honeydew from plant-sucking insects, such as aphids and scales. Males, in particular, engage in "dipping" behavior, touching the surface of pools of water with the mouthparts while in flight. This may serve to fill the crop with enough water to allow flies to regurgitate on honeydew deposits. The ingested sugars replenish energy reserves expended during daily flight and mating activities.

Tabanid mating occurs in flight, especially in the morning, but has never been induced in the laboratory. This is a key barrier to colonization. Most observations are of individuals or small groups of males (Tabaninae) hovering within 3 m of the ground along forest roads, ecotone areas, or above natural features (e.g., plant clumps) that serve as markers. Some species apparently hover above treetops or forest canopies as high as 90 to 100 m above ground level. Males of other species, such as *Chrysops fuliginosus* and *Hybomitra illota*, perch on vegetation and other objects. In both case, males detect and pursue passing females. They also chase conspecific males and other passing insects. Males of *Hybomitra hinei* have been observed to pursue 8 mm beads shot past them at speeds of 27 to 30 m/sec.

Males of some species are thought to exhibit territoriality. However, what may appear to be agonistic interactions between males may actually be normal pursuit behaviors directed toward any appropriate-sized object moving through their response zone. Individual males do not necessarily use a particular site continuously over time. The occurrence of male aggregations at the tops of hills, known as "hilltopping" behavior, is common for some species. The larger eyes of males reflect a mating strategy dominated by visual cues. The larger, dorsal ommatidia may be more sensitive to ultraviolet light, allowing the male to detect a fast-moving female against the sky and the smaller, ventral ommatidia may be used to resolve visual details. Although pheromones are suspected for a few tabanid species, this has yet to be proved.

Adult feeding activity is typically diurnal, but occasionally crepuscular or nocturnal. It is affected by changes in environmental conditions, particularly temperature and barometric pressure. Species that feed diurnally generally attack hosts throughout the daytime, with discernible periods of higher activity. Females of *Tabanus wilsoni* and *T. pallidescens* tend to feed near midday, whereas *T. abactor* feeds more frequently in late afternoon or early evening. Species that feed during crepuscular periods, particularly dusk, include *Chlorotabanus crepuscularis, Leucotabanus annulatus, Tabanus moderator*, and *T. equalis*. Some crepuscular species feed into the early night.

Tabanid females usually mate before they seek a vertebrate host. Males do not feed on blood. Most species, particularly the Tabaninae, feed on large mammals such as cattle, horses, and deer (Figure 15.8). Deer flies often attack large mammals, including humans, but there also are records of *Chrysops* species feeding on ravens, crows, ducks, and robins. Reptiles such as turtles also may be attacked. Occasionally tabanids may even be rather host-specific, as appears to be the case with *Phorcotabanus cinereus* in Brazil, which feeds preferentially on ducks, attacking exposed, fleshy skin at the base of the bill.

Tabanids are strong fliers and readily disperse several kilometers in short-term flights. Adult dispersal

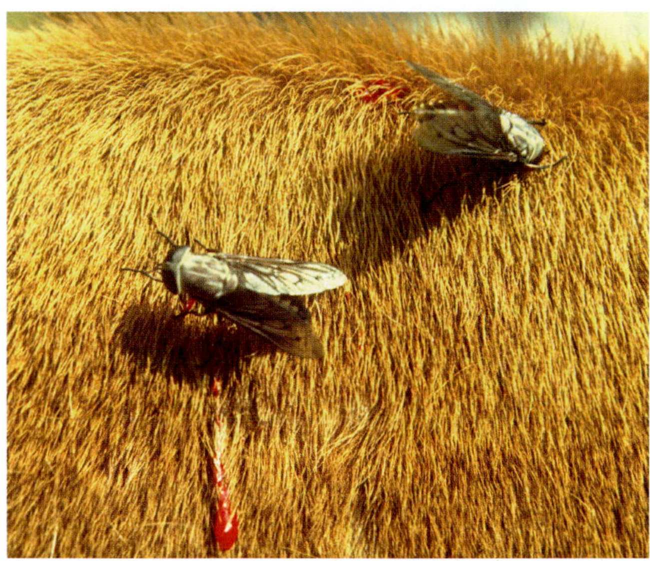

Figure 15.8 *Tabanus sulcifrons* feeding on cow. Note the blood droplet from a prior bite wound. (Photo by B. A. Mullens)

probably is influenced by host availability. Mark-release-recapture studies with tabanids may yield 3 to 6% recovery, which is very high for insects. This suggests that local populations tend to remain in a given area, with dispersal occurring in a series of short flights. Marked *Tabanus abactor* females have been shown to return to a host at the same site where they had obtained a blood meal three or four days earlier.

The attack rates by *Chrysops* species, and to a lesser extent by *Tabanus* and *Hybomitra*, vary substantially in different habitats. Many *Chrysops* species, for example, tend to frequent forest edges or ecotones, attacking in large numbers a host entering these specific areas from adjacent open fields. Dark-colored hosts, or even dark areas on a black and white animal such as a Holstein cow, often are favored for attack.

Many tabanids are selective in attacking specific body regions of their hosts (Figure 15.9), regardless of color. Species-specific attack behaviors probably contribute to the disproportionate collections of certain tabanid species in traps. Deer flies usually feed high on the body, especially on the head or shoulders, and are poorly represented in canopy or box traps that require entry from below. Horse flies feed in various regions depending on the species. Legs are favored feeding sites for many horse fly species that attack livestock. Feeding-site selection appears to reflect resource partitioning in both time and space, thereby reducing competition for hosts among tabanid species. Competition both within and among species is mediated by intensified host defensive behaviors such as kicking. In fact, daily animal movement patterns and herd structure may be substantially influenced by fly attack. Animals in larger groups, especially individuals distant from the herd perimeter, tend to suffer fewer bites.

Field studies have documented the importance of persistent feeding behavior in mechanical transmission of disease agents. It is difficult to imagine a better group of potential mechanical vectors than horse flies. They are large, painful biters that are frequently interrupted in the act of feeding. If disturbed or dislodged, they will return to the same host or one nearby within seconds. Their success in initial feeding attempts often is poor. It has been estimated that only 10% of horse flies successfully feed to repletion during the initial attempt on cattle, though this varies among species.

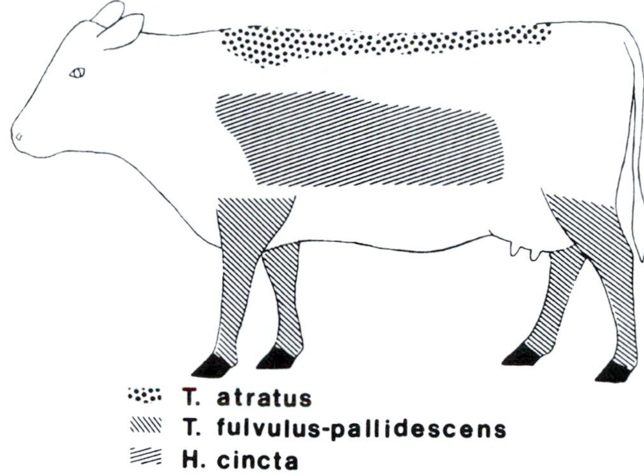

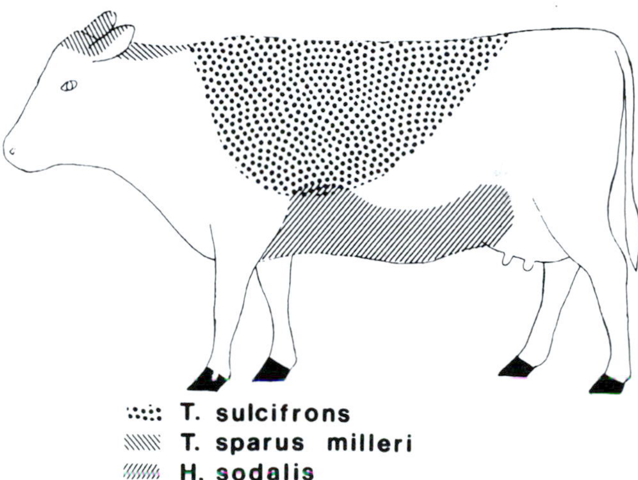

Figure 15.9 Feeding sites on cattle characteristic of individual tabanid species. H, *Hybomitra*; T, *Tabanus*. (From Mullens, B.A. and Gerhardt, R.R., 1979, Feeding behavior of some Tennessee Tabanidae. *Environmental Entomology, 8,* 1047–1051, Figure 13.8)

The propensity of horse flies to return to the same host animal or transfer to another depends on the fly species and the distance between animals. Larger tabanids are more likely to transfer. It has been estimated that almost 90% of certain horse flies that attack horses will return to the original horse if other hosts are more than 35 to 50 meters away. The moderate-sized tabanid *T. fuscicostatus* has an average of 10 nl of blood residue on its mouthparts following feeding. Biting rates and interrupted feeding are important factors that influence the transmission of disease agents.

Tabanids respond to volatile compounds that serve as chemical host cues. The best known and most effective attractant is carbon dioxide. Species vary in their response to CO_2, but even a low release rate of 100 ml/min can result in a two- to fourfold increase in collections of tabanid females in traps. Compounds such as 1-octen-3-ol, ammonia, and phenols, or complex mixtures in animal urine, have been shown to be attractive to certain tabanids and may act synergistically with CO_2.

In addition, visual cues such as shape, color, and movement of potential hosts are very important. Shades of blue, black, or red are particularly attractive to tabanids, and contrast with the background and reflectance also play a role in orientation of tabanids to their hosts. Many imaginative traps have been designed to collect tabanids based on their visual response and orientation behavior. Among the more widely used devices are box traps (Figure 15.10) and canopy traps (Figure 15.11), both of which collect primarily host-seeking females. The effectiveness of canopy traps is enhanced by incorporating movement in the form of a suspended, reflective, black sphere that responds to air movements. The distinctive eye patterns of many species are due to corneal structures acting as interference filters. Older, host-seeking females of some species have been shown to be relatively more sensitive to green wavelengths. This presumably aids in discriminating hosts against a background of green vegetation.

Figure 15.11 Canopy trap, used for collecting tabanid adults and possibly suppressing local populations. Note the black-ball target to enhance attraction of tabanids. (Photo by B. A. Mullens)

Figure 15.10 Box trap for collecting tabanid adults along salt marsh of Atlantic Coast, New Jersey. (Photo by Elton J. Hansens)

Once tabanid females have located a suitable host, they alight and begin probing. Initial blood-feeding attempts are painful and often result in vigorous host response and attempts to dislodge the fly. Flies frequently persist and attack repeatedly. Once blood flow begins, tabanids resist being dislodged, even to the point of sustaining direct strikes by an animal's tail, feet, or head. Chemicals in the saliva maintain blood flow, sometimes for several minutes after feeding ceases. The chemical **chrysoptin**, for example, blocks platelet aggregation, and **vasotab** serves as a vasodilator. It is not unusual for other flies, such as the house fly and face fly, to gather around tabanid feeding wounds to imbibe blood flowing from the wound site.

PUBLIC HEALTH IMPORTANCE

In most temperate areas, adult tabanids are primarily nuisance pests of humans. In this regard they can pose economically significant problems for local tourism. The painful bites, sometimes exceeding 10 per minute, can entirely prevent recreational outdoor activity. Horse-fly larvae can be local pests by inflicting painful bites to the feet of people working in rice paddies. If handled carelessly the larvae will bite defensively, but rarely can penetrate the skin of human fingers.

Tabanids transmit some pathogens and parasites biologically, in which cases the disease agent replicates and/or develops within the fly for a period of time prior to transmission (e.g., the filarial nematode *Loa loa*). More commonly, however, tabanids transmit pathogens mechanically via contaminated blood on their mouthparts. Although many pathogenic viruses, bacteria, protozoa, and filarial nematodes have been recovered from tabanids, documentation of transmission is relatively uncommon, in part because of the difficulties in working with tabanids in the laboratory. Many of the disease associations, particularly those

Table 15.3 Disease Agents Transmitted by Tabanids

Disease Agent	Vectors	Geographic Occurrence	Transmission
Viruses			
Equine infectious anemia	*Tabanus, Hybomitra, Chrysops* spp.	Worldwide	Mechanical
Bovine leukemia	*Tabanus* spp.	Worldwide	Mechanical
Hog cholera	*Tabanus* spp.	Worldwide; eradicated from North America, Australia, New Zealand, South Africa	Mechanical
Bacteria/Rickettsia			
Anaplasma marginale	*Tabanus* spp.	Worldwide (Tropics, Subtropics)	Mechanical
Francisella tularensis	*Chrysops* spp.	North America, Russia, Japan	Mechanical
Bacillus anthracis	*Tabanus, Haematopota, Chrysops* spp.	Worldwide	Mechanical
Protozoa			
Besnoitia besnoiti	*Tabanus, Atylotus* spp.	South America, southern Europe, Africa, Asia,	Mechanical
Trypanosoma evansi	*Tabanus, Haematopota, Chrysops* spp.	South America, North Africa, Asia, India	Mechanical
Trypanosoma vivax	*Tabanus* spp.	South America, Africa	Mechanical
Filarial nematodes			
Loa loa	*Chrysops* spp., esp. *C. dimidiatus, C. silaceus*	Central Africa	Biological
Elaeophora schneideri	*Hybomitra, Tabanus* spp.	North America, southern Europe	Biological

involving viruses and bacteria recovered from tabanids, need to be viewed cautiously, as they may be relatively insignificant epidemiologically.

Fortunately, there are relatively few human pathogens transmitted regularly by Tabanidae, as reviewed by Krinsky (1976) and Foil (1989). The more significant tabanid-associated diseases are shown in Table 15.3.

Loiasis

The most important tabanid-transmitted disease agent of humans is the African eyeworm, *Loa loa*, which causes human loiasis. This filarial nematode is biologically transmitted by *Chrysops* species in equatorial rain forests of western and central Africa. Simian loiasis is caused by a closely related form (*Loa loa papionis*) and also involves *Chrysops* species as vectors. As with other filarial nematodes, transmission of *L. loa* is cyclo-developmental, requiring the fly as an intermediate host. Interestingly, *Chrysops atlanticus* and several other common deer flies in the southeastern United States have been shown to support development of *L. loa*.

Adult nematodes live in subcutaneous tissues of the vertebrate host, particularly the thorax, scalp, axillary regions, or eyes. They are 2 to 7 cm long and produce inflammatory responses as they move through these tissues. If they remain in one area for a time, localized enlargements known as **Calabar swellings** occur; these swellings disappear when the nematode leaves. The common name "eyeworm" is due to adult nematodes sometimes moving across the conjunctiva of the eye (Figure 15.12). There, or just beneath the skin surface, they often are clearly visible and sometimes can be surgically removed. Migrating nematodes can cause considerable pain, in addition to discoloration and

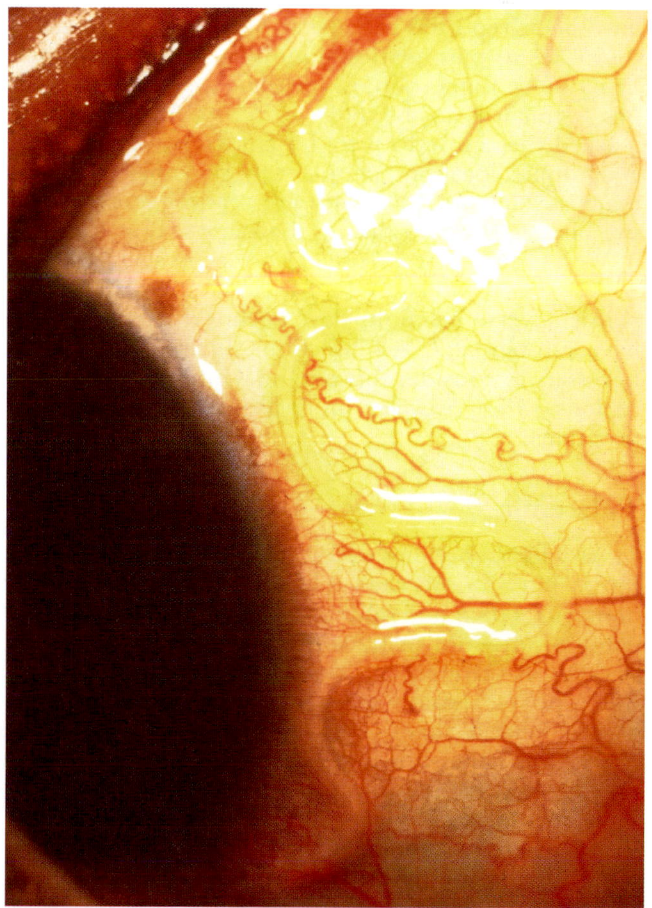

Figure 15.12 African eyeworm, *Loa loa*. Adult filarial nematode just beneath conjunctiva of human eye, Cameroon. (Courtesy of U.S. Armed Forces Institute of Pathology (AFIP) 73-6654)

bruising of the affected tissues, particularly evident in the eye.

Mature *L. loa* adults mate and produce microfilariae. Females of *Chrysops* species ingest the microfilariae with blood when feeding on an infected person. The microfilariae penetrate the midgut and develop in the abdominal fat bodies, or sometimes the thorax. There they molt to second-stage larvae (L_2) and eventually move to the head and mouthparts as infective third-stage larvae (L_3). In the laboratory, infected deer flies may produce 100 or more *L. loa* infective-stage larvae per fly. This process is temperature-dependent and requires at least seven to 10 days. On a subsequent feeding, the infective larvae escape from the fly mouthparts and enter a new host through the bite wound during blood-feeding.

The primary vectors of *L. loa* were incriminated in a series of studies in the 1950s in the Congo region of equatorial Africa, which includes parts of Zaire, Congo, Gabon, Cameroon, and southern Nigeria (see Krinsky, 1976). *Chrysops silaceus* and *C. dimidiatus* are particularly attracted to people near fire. From 80 to 90% of their blood meals are obtained from humans, and in some hyperendemic areas 90% of the people harbor microfilariae or exhibit loiasis symptoms. Infection rates of 0.5 to 1.0% of these two *Chrysops* species have been reported in central Africa, although infection rates in parous flies alone may be much higher. In southern Cameroon, *C. dimidiata* is present at densities estimated to be 800 to 3,700 flies/km^2, and marked flies moved up to 4.5 km. Depending on the geographic location, people can be subjected to as many as 2,000 to 3,000 bites per year, or several infective bites per month during the rainy season.

With the widespread use of ivermectin against *Onchocerca volvulus* infection in Africa, serious encephalopathic adverse reactions have occurred in people who have substantial *L. loa* infections. It thus has become necessary to consider this potential interaction in some regions where ivermectin is used in onchocerciasis programs.

Tularemia

Tularemia, sometimes called "rabbit fever" or "deer fly fever," is a zoonosis caused by the bacterium *Francisella tularensis*. The name rabbit fever reflects the fact that a common method of transmission, particularly in past decades, is through cuts on the hands of hunters and other people handling infected wild rabbits. In the central United States where most American tularemia cases occur, transmission is usually via ticks or direct animal contact. Transmission occurs less commonly by ingestion or aerosol exposure to the bacteria. The epidemiological role of tabanids as vectors in the central United States is unknown. However, transmission by tabanids has been well documented in parts of western North America and is suspected in parts of Russia. Periodic outbreaks of tularemia in Utah since the early 1900s have been linked convincingly to deer fly bites, particularly those of *Chrysops discalis*, hence the name deer fly fever.

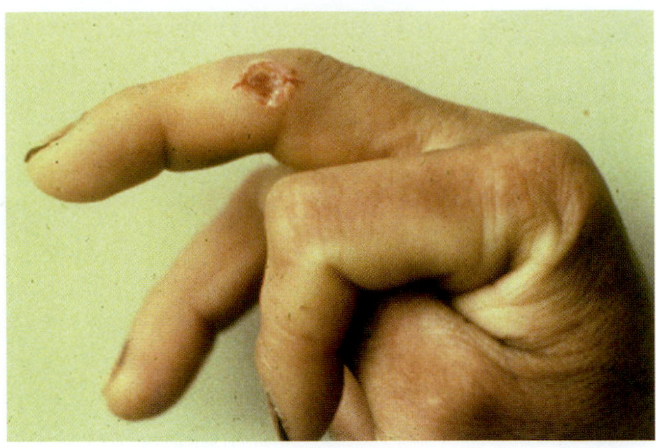

Figure 15.13 Tularemia lesion on middle finger of human hand, caused by bacterium *Francisella tularensis*. (Courtesy of Y. Ohara)

At the site of bacterial introduction, a distinctive lesion develops with an ulcerated, pinkish pit in the center and a raised, ridge-like wheal around the perimeter (Figure 15.13). Bacterial septicemia and resultant fever cause severe illness and occasionally death if the person is not adequately treated with antibiotics. Often the initial lesions are found on the head or upper torso where deer flies commonly bite people. The ability of *C. discalis* to acquire and later transmit the bacterium to humans is dependent on the propensity of this deer fly to feed on other animals, particularly rabbits, which serve as pathogen reservoirs. Occasionally other biting flies, including the horse fly *Tabanus punctifer*, also can transmit *F. tularensis*. Such transmission is likely to be mechanical, with the bacterium being introduced into a bite wound via contaminated mouthparts. Feces of deer flies that are fed the bacterium experimentally have been shown to be infective when the feces is rubbed into abraded skin.

Other Tabanid-Transmitted Human Pathogens

Anthrax, a potentially dangerous disease of humans, still occurs worldwide, including localized areas in the United States. Although tabanids are capable of mechanically transmitting the causal bacterium, *Bacillus anthracis*, this mode of transmission is minor in the epidemiology of the disease. Tabanids also were implicated in the 1980s as possible vectors of *Borrelia burgdorferi*, the spirochete that causes Lyme disease. The vast majority of transmission is accomplished by ixodid ticks. Tabanids are suspected of contributing to transmission of *B. burgdorferi* in some parts of Europe and North America, but this has not been well documented and requires further study.

VETERINARY IMPORTANCE

Owing to their painful, persistent biting behavior, tabanids are significant pests of livestock, particularly cattle and horses, and are extremely bothersome to

many wildlife species. Heavy attack by tabanids can cause direct reductions in weight gains of beef cattle, reduced milk yield, reduced feed-utilization efficiencies, and hide damage from the feeding punctures. Cattle protected from tabanid attack in screened enclosures have been shown to gain up to 0.1 kg/day more than control animals exposed to tabanids. Such direct losses can be increased by a concomitant reduction in feed-utilization efficiency of up to 17%. A daily loss of 200 ml of blood per animal may be common during periods of intense tabanid attack.

Tabanids serve as vectors of a number of disease agents of animals, including viruses, bacteria, protozoans, and nematodes (Table 15.3). For an overview of these pathogens, see Krinsky (1976) and Foil (1989).

Surra and Related Trypanosomiases

One of the more serious disease agents of livestock transmitted by tabanids is *Trypanosoma evansi*, the causal agent of surra. It is morphologically indistinguishable from *T. brucei*. Unlike tsetse fly (*Glossina* species) transmission of several other *Trypanosoma* species, including *T. brucei*, tabanid transmission of *T. evansi* is mechanical. Surra also is transmitted mechanically by vampire bats in parts of South America, where the disease is known as **murrina**. Surra affects a variety of wild and domestic mammals in northern Africa, southern Asia, the Philippines and Indonesia, and parts of Central and South America. It apparently was introduced into the Western Hemisphere by Spaniards in the sixteenth century via infected horses. Untreated infections are usually 100% fatal in horses, elephants, and dogs. The disease can be serious and also chronic in camels, which are thought to be the original hosts. Cattle and buffalo, in contrast, are not severely affected and may remain asymptomatic for months. Relatively resistant animals may serve as reservoirs of infection. Symptoms of infection are similar to other trypanosomiases. An old Arab name for surra, dating from the early 1900s or before, is *mard el debab*, which means "sickness of the gadflies."

Particularly in northern South America, mechanical transmission by tabanids of a related pathogen, *T. vivax*, is a serious problem for sheep producers and, to a lesser extent, for cattlemen. *Trypanosoma equinum* infects horses in South America where it causes a disease called *mal de caderas*, with symptoms similar to surra. In parts of Africa, mechanical transmission by tabanids of species such as *T. brucei*, normally transmitted by tsetse flies, may be significant. Recent research in field cages (to exclude tsetse) demonstrated mechanical transmission of *T. congolense* and *T. vivax* among cattle in Africa by the tabanid *Atylotus agrestis*.

Trypanosoma theileri causes widespread, generally nonpathogenic infections of cattle and wild hosts such as deer. This trypanosome is found commonly in the hindguts of tabanids but is absent from the salivary glands. Transmission therefore occurs primarily through feces entering the bite wound or perhaps by crushing or ingestion of infected tabanids by the animal. In the latter case, infection occurs through abrasions and breaks in the skin or through oral mucosa. Research on the epizootiology of this and other trypanosome diseases is complicated by the presence of *Blastocrithidia* species, nonpathogenic trypanosomes that occur naturally in tabanids and are easily confused with pathogenic trypanosomes.

Equine Infectious Anemia

Equine infectious anemia (EIA), commonly known as swamp fever, is a serious viral disease of horses and other equids (Figure 15.14). It is a febrile illness that causes lethargy, weight loss, and sometimes death in affected animals. Different strains of EIA virus differ in pathogenicity, and infected animals differ significantly in how they are affected. Acutely infected animals nearly always die fairly quickly. Chronically infected animals eventually succumb to complications, and unapparent carriers may live a number of years with few obvious health problems. This disease is found in many areas of the world and is common in the southeastern United States. A series of studies conducted mostly in Louisiana provides a good model for understanding mechanical transmission of pathogens by tabanids (Foil, 1989), especially EIA and other retroviruses such as bovine or feline leukemia viruses.

Because virus infectivity declines rapidly on the insect mouthparts, rapid and frequent transfer of vectors between hosts is essential for significant transmission. Even though the amount of blood transferred by an individual fly is low, the potential for transmission is high when multiplied by a large number of persistent flies feeding at a given time.

Anaplasmosis

Ticks and tabanids are regarded as the primary vectors of *Anaplasma marginale*, a rickettsia that causes anaplasmosis in cattle. This disease is most prevalent in the

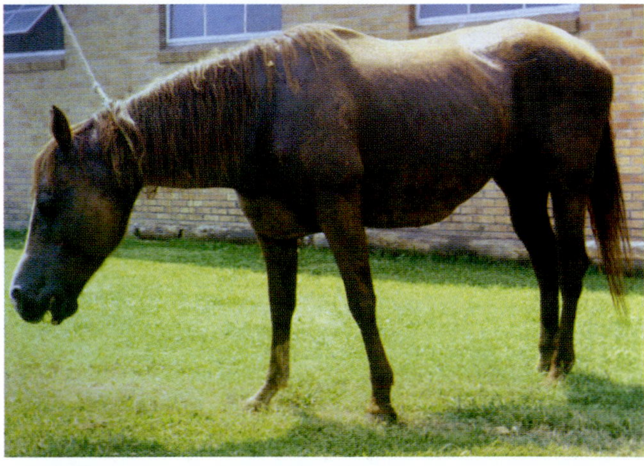

Figure 15.14 Horse suffering from equine infectious anemia. (Photo by W. V. Adams, Jr.)

tropics and subtropics, including Africa, Australia, and the Americas. In the United States, the incidence of disease is highest in the southeastern states. Whereas calves seldom are affected severely, adult cattle show marked anemia, fever, and weight loss. Mortality may be as high as 50%.

Ticks transmit the rickettsia biologically and are probably more important vectors than biting flies. Recent work (Scoles et al., 2008) showed that tick transmission was well over two orders of magnitude more efficient, compared with failed attempts to transmit *A. marginale* from an acutely infected calf to recipient cattle via interrupted feedings by tabanids. Still, mechanical transmission by tabanids up to two hours postfeeding has been documented in the literature. Particularly when biting rates are high, horse flies may be significant in transmitting *A. marginale* among susceptible animals during anaplasomosis outbreaks.

Elaeophorosis

A filarial nematode of domestic animals and wild ruminants, *Elaeophora schneideri* (Figure 15.15) is widely distributed in North America (Pence, 1991). In the Rocky Mountain states of the United States, infection rates in mule deer (*Oedocoileus hemionus*) often exceed 50%. Mule deer apparently are the normal reservoir host in western North America and rarely show pathological effects. Compared with mule deer, white-tailed deer (*O. virginianus*) are infected at a lower level (2–10%) in many regions of the southern United States and sometimes show clinical signs of infection. Domestic sheep and goats may harbor the nematodes, but generally do not develop notable pathology. Some wild ruminants, particularly elk and moose, can be quite severely affected; up to 90% or more of the elk in parts of Arizona and New Mexico have been shown to harbor the nematode. In susceptible hosts, *E. schneideri* causes obstruction of arterial flow, hence the common name arterial worm. Reduced blood flow results in dermatosis, sloughing of distal tissues such as ear or antler tips, blindness, neurological disease, and sometimes death.

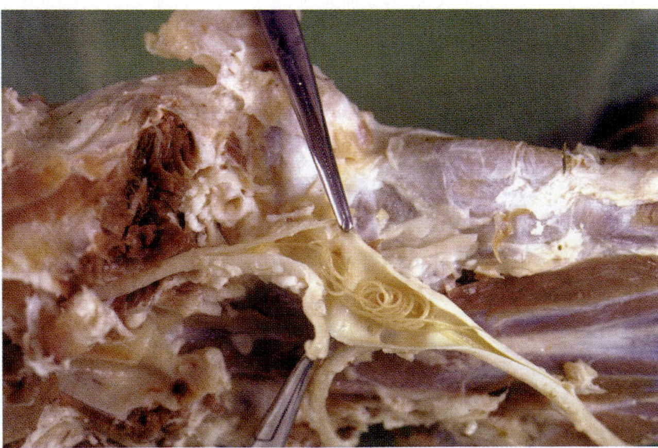

Figure 15.15 Arterial worm, *Elaeophora schneideri*. Adult filarial nematode in carotid artery of white-tailed deer. (Courtesy of Southeastern Cooperative Wildlife Disease Study, University of Georgia)

At least 16 species of tabanids in the genera *Tabanus, Hybomitra,* and *Silvius* have been implicated in transmission of *E. schneideri*. In the United States, prevalence of nematodes in horse flies captured from hyperendemic areas of New Mexico has been as high as 10 to 20%. Prevalence in other states, such as parts of southern Montana and South Carolina, is less than 1% in field-collected flies. In a given region, one or two species, such as *T. laticornis* in New Mexico, seem to be responsible for most of the transmission. Following ingestion of microfilariae by a tabanid, nematode development proceeds in the fat body and hemocoel. Infective larvae (L_3) are found in the mouthparts after about 14 days.

Other Pathogens of Veterinary Importance

Several other pathogens are transmitted mechanically by tabanids to livestock and wild animals. Most of them, however, also may be transmitted in other ways. Notable among these are the viruses that cause bovine leukemia and hog cholera, and the sporozoan genus Besnoitia that causes besnoitosis. The role of tabanids in transmitting these pathogenic agents is usually very secondary.

There are several other interesting, but economically minor, disease agents transmitted to wild animals by tabanids. A protozoan parasite of turtles, *Haemaproteus metchnikovi*, is transmitted by deer flies (*Chrysops* spp.), and tabanids biologically transmit the filarial nematode *Dirofilaria roemeri*, a pathogen of kangaroos and wallabies.

PREVENTION AND CONTROL

Tabanid control is difficult to achieve. A given area usually has multiple species with different seasonal occurrences and biological characteristics. Typical host contact is only about four minutes per fly during blood feeding, which may occur only once every three to four days. Short-term control on livestock for several days may be achieved through use of insecticides, but insecticide sprays often are not particularly effective. Aerial applications of pyrethroid insecticides can suppress local populations of certain tabanids over a period of several hours to perhaps a few days, but are considered far less effective than similar applications made for mosquito control. Use of insecticides for control of larvae or pupae, which are typically inaccessible in soil, is generally ineffective and can result in environmental damage. Insect repellents applied to human skin or impregnated into clothing may offer temporary relief for humans, but some individuals report poor results when using them.

Providing animals with structures for shelter, or pasturing them away from pasture-forest ecotones, can help to reduce tabanid biting intensity. Because tabanids prefer to fly around rather than over vegetation or screen barriers over two meters high, such barriers can help to reduce tabanid access to livestock.

Another approach to tabanid control is water management in areas where drainage or manipulation of water levels is feasible. This must be done carefully, however, since such practices can actually enhance tabanid populations. Larvae of the salt marsh greenhead *Tabanus nigrovittatus*, for example, are more abundant in better drained sites, and ditching salt marshes can increase its populations. Even in restricted habitats, such as livestock ponds in pastures, water management is not always feasible. Heavy rains and flooding during periods when tabanids are pupating can kill the pupae, resulting in fewer adults a few weeks later. In situations where oviposition sites are limited, properly timed removal of emergent vegetation on which tabanid eggs are deposited can result in a significant reduction in the number of eggs and resultant larvae.

Traps commonly are used to monitor tabanid adults, but they also provide another potential tool for control. Some new designs of large, curtain-type traps for tabanid control in pastures have been developed and marketed in the United States. There has been recent, promising research on using modifications of traps initially developed for tsetse in Africa, since they also collect other biting flies, including tabanids. In some coastal areas of the eastern United States, box traps are widely used for suppression of the salt marsh greenheads *Tabanus nigrovittatus* and *T. conterminus* (Figure 15.10). These flies are obligately autogenous, and responding females already have laid eggs. Under certain conditions, the traps may distract enough host-seeking adults from humans to result in temporary relief. Similar efforts have been made on a smaller scale using canopy traps, with limited success. Where large, fairly mobile populations of horse flies occur, even large collections of adults may result in questionable suppression. In a study in the southeastern United States, 95,000 tabanids were captured with 20 CO_2-baited sticky traps from a pasture area over a period of several days, but resulted only in very temporary reduction in numbers of tabanids attacking cattle. The development of better attractants or attractant combinations (chemicals, colors, fabrics, decoys, etc.) for adult tabanids is enhancing the control potential of traps.

The use of biological control agents also offers some potential for reducing tabanid populations. All stages of tabanids are subject to mortality by predators, including ladybird beetle larvae preying on eggs, wading birds feeding on larvae, and dragonflies and certain solitary wasps attacking adults. A few species of bembicine wasps (Sphecidae, Bembicinae) called horse guards specialize in capturing adult tabanids, on which they rear their young. They tend to fly around pastured animals where they seize tabanids as they attempt to feed. Two species of horse guards in the southeastern United States are *Strictia carolina* and *Bembix texana*. Cannibalism among horse-fly larvae also may be important in biological control.

Tabanid eggs are parasitized by wasps in the families Trichogrammatidae and Scelionidae, which can sometimes cause high egg mortality (>50%). Tabanid larvae are parasitized by flies in the Tachinidae and Bombyliidae, and pupae are parasitized by wasps in the Diapriidae and Pteromalidae. Tabanid larvae also are subject to mortality from nematode parasites in the family Mermithidae. At least one mermithid species, *Pheromermis myopis*, parasitizes other invertebrates that are likely to be eaten by tabanid larvae, thereafter killing the tabanid larvae while completing its development. A number of fungal, bacterial, and protozoan pathogens also are known from tabanids.

REFERENCES AND FURTHER READING

Allan, S. A., Day, J. F., & Edman, J. D. (1987). Visual ecology of biting flies. *Annual Review of Entomology, 32*, 297–316.

Anderson, J. F. (1985). The control of horse flies and deer flies (Diptera: Tabanidae). *Myia, 3*, 547–598.

Andreeva, R. V. (1984). *The Ecology of Horse Fly Larvae and Their Parasitoses.* Kiev: Naukova Dumka.

Andreeva, R. V. (1989). The morphological adaptations of horse fly larvae (Diptera: Tabanidae) to developmental sites in the Palearctic Region and their relationship to the evolution and distribution of the family. *Canadian Journal of Zoology, 67*, 2286–2293.

Andreeva, R. V. (1990). *Identification of the Larvae of Horse Flies of the European Part of the USSR, Caucasus and Central Asia.* Kiev: Naukova Dumka.

Auroi, C. (1983). [The life cycle of *Hybomitra bimaculata* (Marqu.) (Diptera: Tabanidae). III. Pupation, emergence, blood meal and oogenesis]. *Mitteilungen der Schweizerischen Entomologischen Gesellschaft, 56*, 343–359.

Barros, A. T. M., & Foil, L. D. (2007). The influence of distance on movement of tabanids (Diptera: Tabanidae) between horses. *Veterinary Parasitology, 144*, 380–384.

Burger, J. F. (1977). The biosystematics of immature Arizona Tabanidae (Diptera). *Transactions of the American Entomological Society, 103*, 145–258.

Burger, J. F. (1995). Catalog of Tabanidae (Diptera) of North America North of Mexico. *Contributions on Entomology, International, 1*, 1–100.

Burger, J. F., & Chainey, J. E. (2000). Revision of the Oriental and Australasian species of *Chrysops* (Diptera: Tabanidae). *Invertebrate Taxonomy, 14*, 607–654.

Burger, J. F., Lake, D. J., & McKay, M. L. (1981). The larval habitats and rearing of some common *Chrysops* species (Diptera: Tabanidae) in New Hampshire. *Proceedings of the Entomological Society of Washington, 83*, 373–389.

Chippaux, J. P., Bouchité, B., Demanov, M., Morlais, I., & LeGoff, G. (2000). Density and dispersal of the loiasis vector *Chrysops dimidiata* in southern Cameroon. *Medical and Veterinary Entomology, 14*, 339–344.

Chvala, M., Lyneborg, L., & Moucha, J. (1972). *The Horse Flies of Europe (Diptera, Tabanidae).* Denmark: Entomological Society of Copenhagen.

Cooksey, L. M., & Wright, R. E. (1989). Population estimation of the horse fly, *Tabanus abactor* (Diptera: Tabanidae) in north central Oklahoma. *Environmental Entomology, 16*, 211–217.

Coscaron, S., & Papavero, N. (1993). *An Illustrated Manual for the Identification of the Neotropical Genera and Subgenera of Tabanidae (Diptera).* Brazil: Museu Paraense Emilio Goeldi, Belem.

Diggle, P. J., Thompson, M. C., Christiansen, O. F., Rowlingson, B., Obsomer, V., Gardon, J., et al. (2007). Spatial modeling and the prediction of *Loa loa* risk: Decision making under uncertainty. *Annals of Tropical Medicine and Parasitology, 101*, 499–509.

Dukes, J. C., Edwards, T. D., & Axtell, R. C. (1974). Distribution of larval Tabanidae (Diptera) in a *Spartina alterniflora* salt marsh. *Journal of Medical Entomology, 11*, 79–83.

Duquesnes, M., & Dia, M. L. (2003). Mechanical transmission of *Trypanosoma congolense* in cattle by the African tabanid *Atylotus agrestis. Experimental Parasitology, 105*, 226–231.

Fairchild, G. B. (1969). Climate and the phylogeny and distribution of Tabanidae. *Bulletin of the Entomological Society of America, 15,* 7–11.

Fairchild, G. B. (1986). Tabanidae of Panama. *Contributions of the American Entomological Institute, 22,* 1–139.

Fairchild, G. B., & Burger, J. F. (1994). *A catalog of the Tabanidae (Diptera) of the Americas south of the United States.* Memoirs of the American Entomological Institute No. 55.

Foil, L. D. (1989). Tabanids as vectors of disease agents. *Parasitology Today, 5,* 88–96.

Goodwin, J. T. (1982). The Tabanidae (Diptera) of Mali. *Miscellaneous Publications of the Entomological Society of America, 13,* 1–141.

Goodwin, J. T., & Drees, B. M. (1996). The horse flies and deer flies (Diptera: Tabanidae) of Texas. *Southwestern Entomologist, 20*(Suppl.).

Goodwin, J. T., Mullens, B. A., & Gerhardt, R. R. (1985). The Tabanidae of Tennessee. *Tennessee Agricultural Experiment Station Bulletin,* 642.

Hayakawa, H. (1985). A key to the females of Japanese tabanid flies with a checklist of all species and subspecies (Diptera, Tabanidae). *Japanese Journal of Sanitary Zoology, 36,* 15–23.

Hayes, R. O., Doane, O. W. Jr., Sakolsky, K., & Berrick, S. (1993). Evaluation of attractants in traps for greenhead fly (Diptera: Tabanidae) collections on a Cape Cod, Massachusetts, salt marsh. *Journal of the American Mosquito Control Association, 9,* 436–440.

Hollander, A. L., & Wright, R. E. (1980). Impact of tabanids on cattle: Blood meal size and preferred feeding sites. *Journal of Economic Entomology, 73,* 431–433.

Iranpour, M., Shurko, A. M., Klassen, G. R., & Galloway, T. D. (2004). DNA fingerprinting of tabanids (Diptera: Tabanidae) and their respective egg masses using PCR-restriction fragment profiling. *Canadian Entomologist, 136,* 605–619.

Jones, C. M., & Anthony, D. W. (1964). The Tabanidae of Florida. *USDA Technical Bulletin,* 1295.

Krinsky, W. L. (1976). Animal disease agents transmitted by horse flies and deer flies (Diptera: Tabanidae). *Journal of Medical Entomology, 13,* 225–275.

Krcmar, S. A., Mikuska, A., & Merdic, E. (2006). Response of Tabanidae (Diptera) to different natural attractants. *Journal of Vector Ecology, 31,* 262–265.

Lane, R. S., Anderson, J. R., & Philip, C. B. (1983). Biology of autogenous horse flies native to coastal California: *Apatolestes actites* (Diptera: Tabanidae). *Annals of the Entomological Society of America, 76,* 559–571.

LePrince, D. J., & Lewis, D. J. (1986). Sperm presence and sugar feeding patterns in nulliparous and parous *Tabanus quinquevittatus* Wiedemann (Diptera: Tabanidae) in southwestern Quebec. *Annals of the Entomological Society of America, 79,* 912–917.

Mackerras, I. M. (1954). The classification and distribution of Tabanidae (Diptera). I. General review. *Australian Journal of Zoology, 2,* 431–454.

McElligott, P. E. K., & Lewis, D. J. (1996). Distribution and abundance of immature Tabanidae (Diptera) in a subarctic Labrador peatland. *Canadian Journal of Zoology, 74,* 1364–1369.

McKeever, S., & French, F. E. (1997). Fascinating, beautiful blood feeders—Deer flies and horse flies, the Tabanidae. *American Entomologist, 43,* 217–226.

McMahon, M. J., & Gaugler, R. (1993). Effect of salt marsh drainage on the distribution of *Tabanus nigrovittatus* (Diptera: Tabanidae). *Journal of Medical Entomology, 30,* 474–476.

Middlekauf, W. W., & Lane, R. S. (1980). *Adult and Immature Tabanidae (Diptera) of California. Bulletin of the California Insect Survey 22.* Berkeley: University of California.

Mihoc, S., & Carlson, D. A. (2007). Performance of plywood and cloth Nzi traps relative to Manitoba and greenhead traps for tabanids and stable flies. *Journal of Economic Entomology, 100,* 613–618.

Mullens, B. A., & Gerhardt, R. R. (1979). Feeding behavior of some Tennessee Tabanidae. *Environmental Entomology, 8,* 1047–1051.

Oldroyd, H. (1954–1957). *Horseflies of the Ethiopian Region, I-III.* London: British Museum of Natural History.

Olsufiev, N. G. (1977). *Horse flies. Family Tabanidae.* Fauna of the USSR. New Series. No. 113, Insects, Diptera 7(2). Izdatelistvo Nauka, Leningrad.

Pechuman, L. L. (1981). *The Horse Flies and Deer Flies of New York (Diptera: Tabanidae). Search Agriculture, Cornell University Agricultural Experiment Station No. 18.* Ithaca, NY: Cornell University.

Pechuman, L. L., & Teskey, H. J. (1981). Tabanidae. In J. F. McAlpine (Ed.), *Manual of Nearctic Diptera* (Vol. 1. pp. 463–478). Agriculture Canada Research Branch, Monograph 27, Ottawa: Agriculture Canada Research Branch.

Pechuman, L. L., Webb, D. W., & Teskey, H. J. (1983). The Diptera, or true flies, of Illinois. I. Tabanidae. *Illinois Natural History Survey Bulletin,* 33.

Pence, D. B. (1991). Elaeophorosis in wild ruminants. *Bulletin of the Society for Vector Ecology, 16,* 149–160.

Perich, M. J., Wright, R. E., & Lusby, K. S. (1986). Impact of horse flies (Diptera: Tabanidae) on beef cattle. *Journal of Economic Entomology, 79,* 128–131.

Poinar, G. O., Jr. (1985). Nematode parasites and infectious diseases of Tabanidae (Diptera). *Myia, 3,* 599–616.

Rajska, P., Pechanova, O., Takac, P., Kazimirova, M., Roller, L., Vidlicka, L., et al. (2003). Vasodilatory activity in horsefly and deerfly salivary glands. *Medical and Veterinary Entomology, 17,* 395–402.

Ralley, W. E., Galloway, T. D., & Crow, G. H. (1992). Individual and group behavior of pastured cattle in response to attack by biting flies. *Canadian Journal of Zoology, 71,* 725–734.

Schutz, S. J., & Gaugler, R. (1989). Honeydew-feeding behavior of salt marsh horse flies (Diptera: Tabanidae). *Journal of Medical Entomology, 26,* 471–473.

Scoles, G. A., Miller, J. A., & Foil, L. D. (2008). Comparison of the efficiency of biological transmission of *Anaplasma marginale* (Rickettsiales: Anaplasmataceae) by *Dermacentor andersoni* Stiles (Acari: Ixodidae) with mechanical transmission by the horse fly, *Tabanus fuscicostatus* Hine (Diptera: Tabanidae). *Journal of Medical Entomology, 45,* 109–114.

Smith, S. M., Turnbull, D. A., & Taylor, P. D. (1994). Assembly, mating, and energetics of *Hybomitra arpadi* (Diptera: Tabanidae) at Churchill, Manitoba. *Journal of Insect Behavior, 7,* 355–383.

Stoffolano, J. G. Jr., & Lin, L. R. S. (1983). Comparative study of the mouthparts and associated sensillae of adult male and female *Tabanus nigrovittatus* (Diptera: Tabanidae). *Journal of Medical Entomology, 20,* 11–32.

Sutton, B. D., & Carlson, D. A. (1997). Cuticular hydrocarbon variation in the Tabanidae (Diptera): *Tabanus nigrovittatus* complex of the North American Atlantic coast. *Annals of the Entomological Society of America, 90,* 542–549.

Takahasi, H. (1962). *Fauna Japonica. Tabanidae (Insecta: Diptera), Biogeographical Society of Japan.* Tokyo: National Science Museum.

Teskey, H. J. (1990). *The Insects and Arachnids of Canada, Part 16. The Horse Flies and Deer Flies of Canada and Alaska (Diptera: Tabanidae). Agriculture Canada Publication 1838.* Ottawa: Minister of Supply and Services Canada.

Waage, J. K., & Davies, C. R. (1986). Host-mediated competition in a bloodsucking insect community. *The Journal of Animal Ecology, 55,* 171–180.

Wiegmann, B. M., Yeates, D. K., Thorne, J. L., & Kishino, H. (2003). Time flies—A new molecular time scale for brachyceran fly evolution without a clock. *Systematic Biology, 52,* 745–756.

Wilkerson, R. C., Butler, J. F., & Pechuman, L. L. (1985). Swarming, hovering, and mating behavior of male horse flies and deer flies (Diptera: Tabanidae). *Myia, 3,* 515–546.

Wilson, B. H. (1968). Reduction of tabanid populations on cattle with sticky traps baited with dry ice. *Journal of Economic Entomology, 61,* 827–829.

Chapter

16

Muscid Flies (Muscidae)

Roger D. Moon

The family Muscidae includes significant blood-feeding parasites, vectors of disease agents, and species that annoy humans and domesticated animals. These flies and others in related families are often called **synanthropic flies**, species that exploit foods and habitats created by agriculture and other human activities. Muscid flies and their relatives can be grouped according to their habitat affinities. There are **filth flies**, such as the house fly, whose adults and immatures occur in a variety of filthy organic substrates, including latrines, household garbage, manure, and manure-soiled animal bedding. A subset of the filth flies are **dung flies**, such as the horn fly, whose immatures occur exclusively in cattle droppings. Another group is the **sweat flies**, whose adults feed persistently on perspiration.

Muscid flies also can be grouped by the nature of their mouthparts. The nonbiting muscid flies have sponging mouthparts used to ingest liquids from inanimate substrates and animal tissues. These mouthparts are soft, fleshy, and incapable of penetrating skin. In contrast, biting muscid flies have piercing-sucking mouthparts that pierce the skin to obtain blood.

Useful reviews of the literature on muscid flies include a two-volume treatise on the biology and disease associations of synanthropic flies (Greenberg, 1971, 1973) and a monograph on the identification and biology of immature muscid flies (Skidmore, 1985). A comprehensive review of the veterinary effects and control of muscid flies and other arthropods on livestock is provided by Drummond et al. (1988). Additional reviews and bibliographies concentrate on selected species and their close relatives: the house fly (Thomas and Skoda, 1993; West, 1951; West and Peters, 1973); the stable fly (Morgan et al., 1983a; Petersen and Greene, 1989; Thomas and Skoda, 1993; Zumpt, 1973); the horn fly (Bruce, 1964; Morgan and Thomas, 1974, 1977); and the face fly (Morgan et al., 1983b; Pickens and Miller, 1980; Krafsur and Moon, 1997).

TAXONOMY

The Muscidae include approximately 4,200 species in 190 genera. The North American fauna contains about 700 species in 46 genera. Fortunately, only a few of these genera contain important medical or veterinary pests. The important North American taxa are listed in Table 16.1. Five of them have been introduced from the Old World through human commerce. Outside North America, the same species, or close relatives with similar life cycles, habits, and ecology may be encountered.

Important muscid flies (Table 16.1) occur in two subfamilies, the Muscinae and the Fanniinae. Important nonbiting Muscinae are the house fly, the garbage flies, the false stable fly and relatives, the face fly, and the sweat flies. The important biting Muscinae are the stable fly and horn fly. A third biting species in North America, the moose fly (*Haematobosca alcis*), occurs exclusively on the moose (*Alces alces*). The second subfamily, the Fanniinae, are represented by the nonbiting little house fly and its relatives (*Fannia* spp.). Although some authors consider the Fanniinae to be a separate sister family (Fanniidae), it is treated here as a subfamily of Muscidae.

Adults and larvae of the North American Diptera can be identified to family using keys in McAlpine et al. (1981), and adult Muscidae can be keyed to genus using McAlpine et al. (1987) and references therein. The Muscidae of North America are cataloged in Stone et al. (1965). Other aids for identification are Skidmore's (1985) keys and descriptions of larvae and pupal cases (puparia), and James' (1947) classic treatment of adults and larvae of flies in Muscidae and other families that cause myiasis.

MORPHOLOGY

The life stages of a typical muscid fly consist of egg, larva, pupa, and adult (Figure 16.1). Eggs are similar to those of closely related families. They may occur

Table 16.1 Important Muscid Pests of Humans and Domestic Animals in North America

Ecological Group	Common Name	Scientific Name	Hosts for Adults
Filth flies	House fly[a]	*Musca domestica*	None required
	Stable fly[a]	*Stomoxys calcitrans*	Cow, horse, dog, humans, and others
	Garbage flies	*Hydrotaea (Ophyra[b])* *Hdrotaea aenescens,* *Hydrotaea ignava[a] (Ophyra leucostoma)*	None required
	False stable fly and relative	*Muscina stabulans* *Muscina levida (assimilis[c])*	None required
	Little house fly	*Fannia canicularis*	None required
	Latrine fly and relative	*Fannia scalaris* *Fannia femoralis*	
Dung flies	Horn fly[a]	*Haematobia irritans irritans*	Cow, bison, horse
	Face fly[a]	*Musca autumnalis*	Cow, bison, horse
Sweat flies	Sweat flies	*Hydrotaea meteorica,* *Hydrotaea scambus,* and others	Large mammals

[a]Species introduced to North America from Old World; others are cosmopolitan or native to New World.
[b]According to Huckett and Vockeroth (1987).
[c]According to Skidmore (1985).

singly or in groups. They are generally creamy in color, 0.8 to 2.0 mm long, elongate-ovate in shape, and concave dorsally where two ribs form hatching pleats (Figure 16.2).

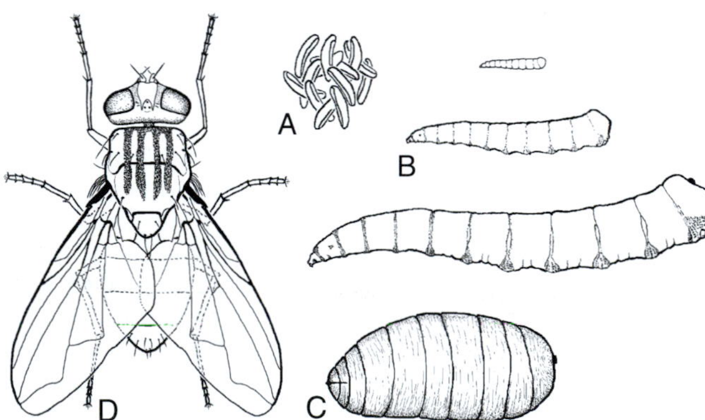

Figure 16.1 Life cycle of house fly (*Musca domestica*). (A) Eggs; (B) larvae, three instars; (C) pupa; (D) adult. (Traced from drawing by F. Gregor in Greenberg, 1971, Used with permission)

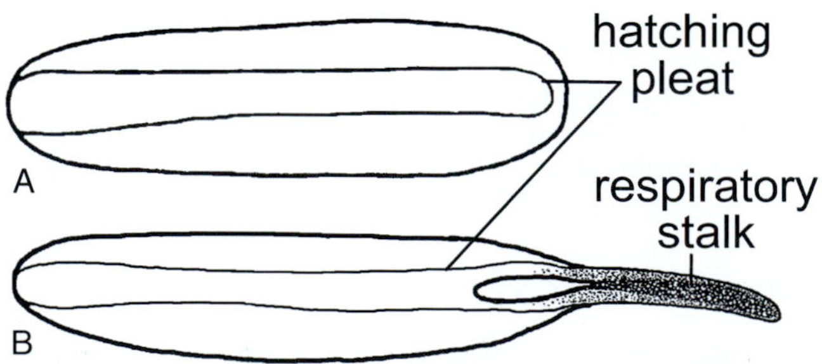

Figure 16.2 Eggs of muscid flies, showing hatching pleat. (A) House fly (*Musca domestica*); (B) face fly (*Musca autumnalis*), with distinctive respiratory stalk.

Larvae of muscid flies and related families are known as maggots, and there are three instars in all species. The body is tapered, with the head and mouth hooks at the pointed end and the anus and spiracles at the blunt end (Figure 16.1). The head is greatly reduced; it lacks eyes and has minute antennae that resemble papillae. The thorax is legless and has a pair of lateral prothoracic spiracles. There are eight segments of the abdomen, and each is marked ventrally with transverse rows of spines forming creeping welts.

Although the head lacks a sclerotized capsule, it is supported internally by a sclerotized cephalopharyngeal skeleton (Figure 16.3). This complex structure is partially visible through the integument of live larvae and is best visualized in cleared, slide-mounted specimens. The size, shape, and arrangement of elements of the cephalopharyngeal skeleton are useful in the

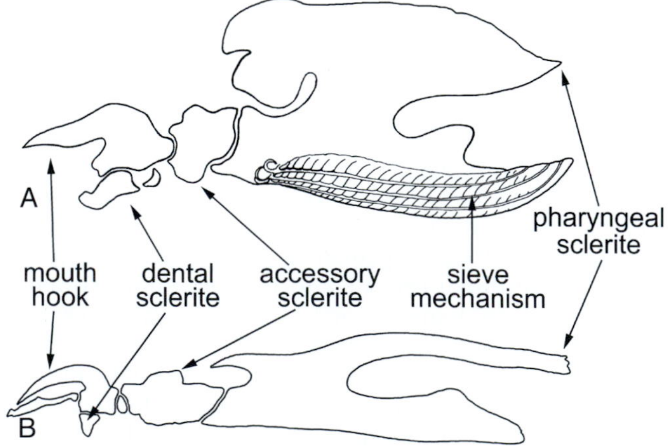

Figure 16.3 Cephalopharyngeal skeletons of muscid fly larvae. (A) Filter-feeding larva, characterized by sieve-like structures; (B) predatory larva, with pair of strong mouth hooks. (Redrawn from Skidmore, 1985)

identification of larvae. Paired mouth hooks, reduced from ancestral mandibles, can be extended and retracted from the oral cavity. They help in crawling, burrowing, and tearing into food and other substrates. Internal dental sclerites, accessory sclerites, and pharyngeal sclerites make up the rest of the cephalopharyngeal skeleton and are variously modified for muscle attachment and feeding.

Most muscid larvae occur in wet substrates where they filter particles of food from the substrate. These filter feeders possess a porous, ventral sieve mechanism between their pharyngeal sclerites (Figure 16.3A). Exceptions are *Muscina* species and *Hydrotaea* species, whose third instars are facultative or obligate predators. The sieving mechanism is absent in these predatory forms (Figure 16.3B).

Characters of the spiracles (Figures 16.4, 16.5) are useful for determining the species and instar of muscid larvae. Paired spiracles occur on the prothorax and at the end of the abdomen. Prothoracic spiracles are absent on first instars, whereas they are present on second and third instars. The shape and number of spiracular tubercles (spiracular lobes) vary considerably in different species.

Structures associated with the paired caudal spiracles on the abdomen are of greatest taxonomic value. Each spiracular plate (Figure 16.4) consists of a peritreme forming the plate's perimeter, two or three slits that are the openings for gas exchange, and a scar that is a remnant from a previous molt. First and second instars have two slits, whereas third instars have three. An exception is the horn fly, in which both the second and third instars have three slits. Peritreme shape, position of the scar, and shape and orientation of the slits are all useful in identifying muscid larvae (Figure 16.5).

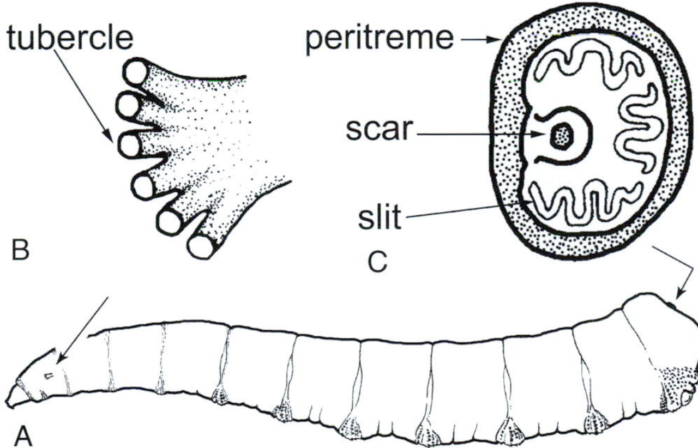

Figure 16.4 Respiratory spiracles of house fly (*Musca domestica*), third-instar larva. (A) Larva, showing location of anterior pair of spiracles on first thoracic segment and posterior pair of spiracular plates on caudal segment (arrows); (B) anterior spiracle, enlargement showing multiple openings to trachea on individual tubercles; (C) posterior spiracular plate, showing sclerotized ring (peritreme) surrounding spiracular area, three sinuous spiracular slits, and scar left by previous molt. (Redrawn from Skidmore, 1985)

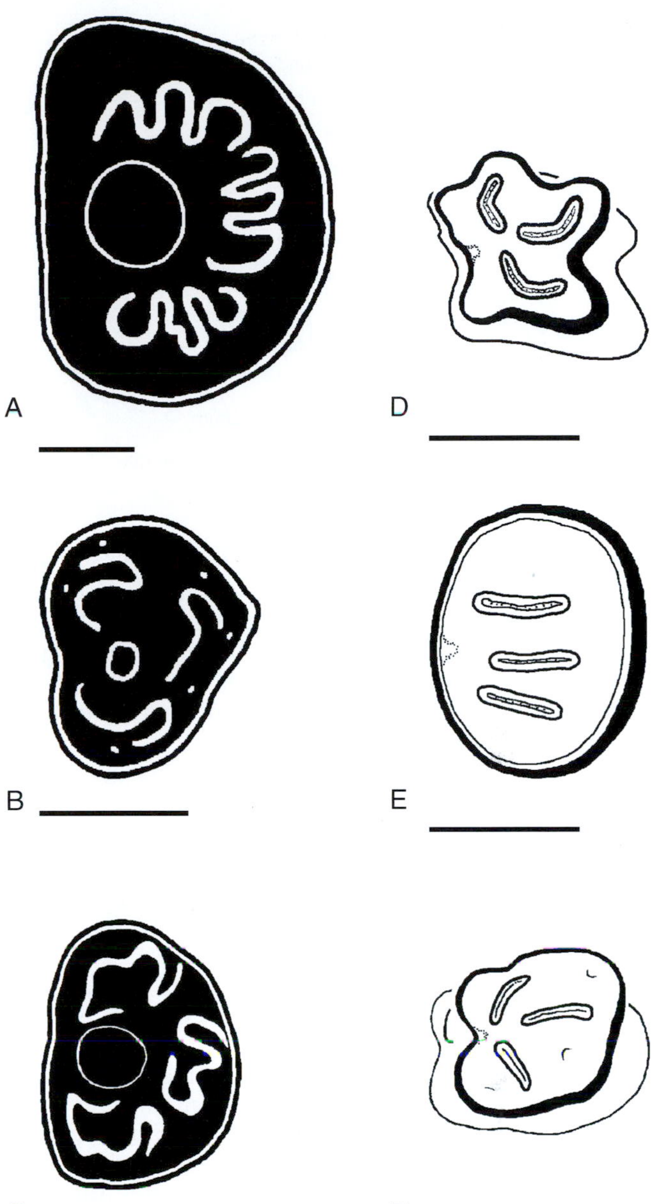

Figure 16.5 Caudal spiracular plates of third-instar larvae of six important muscid flies; only right plate of each pair is shown. (A) Face fly (*Musca autumnalis*); (B) stable fly (*Stomoxys calcitrans*); (C) horn fly (*Haematobia irritans*); (D) false stable fly (*Muscina stabulans*); (E) black garbage fly (*Hydrotaea ignava*); (F) *Hydrotaea* sp. Scale bar, 0.2 mm. (Redrawn from Skidmore, 1985)

The muscid pupa (Figure 16.1C) occurs in a case, the **puparium,** that forms during a process called **pupariation**, which involves contraction and hardening of the third instar's integument (Fraenkel and Bhaskaran, 1973). A space forms around the pupa during subsequent apolysis, or separation of the new pupal integument inside the puparium. The larval cephalopharyngeal skeleton remains attached inside the puparium's cephalic cap, and prothoracic and posterior spiracles remain embedded in the wall of the

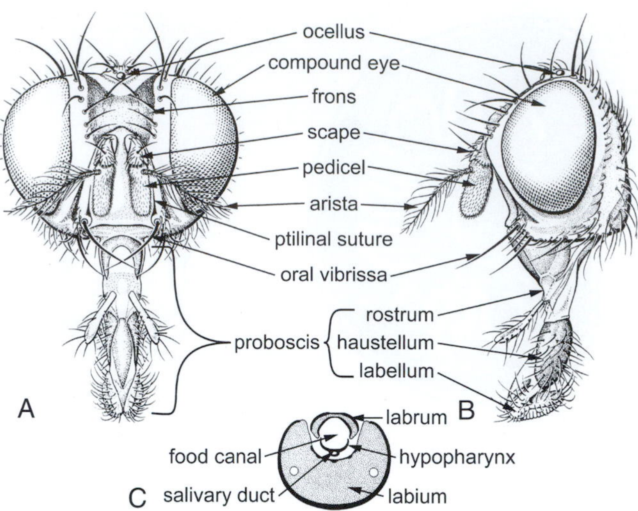

Figure 16.6 Head and mouthparts of a nonbiting muscid fly, adult. (A) Anterior view; (B) lateral view; (C) cross-section of haustellum, showing relationship of the individual mouthparts, food canal, and salivary duct. (Adapted from original drawings by R. Idema, McAlpine et al., 1981. Reproduced with permission of Minister of Public Works and Government Services, Canada)

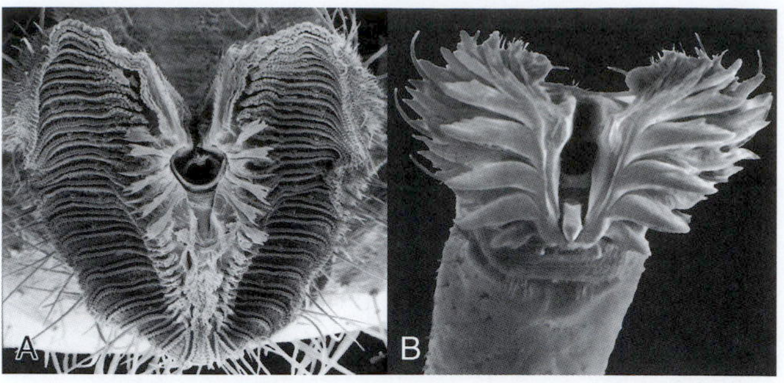

Figure 16.7 Scanning electron micrographs of everted pair of labellae at tip of proboscis of two muscid flies, adults. (A) Nonbiting fly (face fly, *Musca autumnalis*), showing parallel arrangement of feeding channels (pseudotracheae) and relatively small prestomal teeth surrounding opening to food canal (prestomum); (B) biting fly (stable fly, *Stomoxys calcitrans*), lacking pseudotracheae and with enlarged prestomal teeth used to cut through skin of hosts to feed. (Courtesy of Alberto B. Broce)

puparium. Thus, when an adult specimen has been associated with its puparium, distinguishing traits of the adult, the puparium, and the mature larva can be defined and used to identify the species from a specimen in any of the three stages.

Adult muscid flies (Figure 16.1) are 4 to 12 mm long, with wings longer than the abdomen. Integument colors vary from brownish gray to black, often with dark longitudinal stripes on the thorax, called **vittae,** and dark spots or blotches on the abdomen. The head (Figure 16.6) has three ocelli and a prominent pair of compound eyes, which in males are holoptic, nearly meeting at the dorsal midline; in females they are dichoptic, or more widely separated. Each antenna consists of a scape, a pedicel, and an arista. The arista arises from the pedicel as an undivided flagellum homologous to the flagellum of nematocerous flies. In most muscid flies, the arista is a single hair with fine setae along its shaft. A ptilinal suture encircles the bases of the antennae. The ptilinal suture is a remnant of the ptilinum, an eversible sac used by the emerging adult to break open and exit through the cephalic cap of the hardened puparium. In most muscid flies, a pair of strong bristles called oral vibrissae project ventrally from the lower edge of the face toward the mouthparts.

Mouthparts vary considerably among species. Generally they consist of a proboscis with a basal rostrum, a slender haustellum, and a terminal labellum (Figure 16.6). The maxillary palps arise from the rostrum and appear to be one-segmented. The haustellum is formed by three structures held in union (Figure 16.6C): an anterior labrum, a slender hypopharynx, and a posterior labium. The labium encloses both the labrum and hypopharynx and terminates in a two-lobed labellum. Saliva flows from the salivary glands through the salivary canal in the hypopharynx that terminates in the prestomum, or preoral cavity, at the center of the labellum. A cibarial pump inside the head creates suction that draws liquids through the prestomum and up through the food canal between the labrum and labium.

Important structural differences distinguish the labellae of nonbiting and biting muscid flies (Elzinga and Broce, 1986). Among nonbiting species, the labellum is an enlarged fleshy, two-lobed structure (Figures 16.6, 16.7A). On the mesal surface of each lobe are **pseudotracheae**, rows of fine setae used to scrape food and direct fluid toward the prestomum. The labellum also houses mechanoreceptors, chemoreceptors, and prestomal teeth. These teeth may be short, or elongate and dentate in some *Musca* species and *Hydrotaea* species. In contrast, mouthparts of the biting species (Figures 16.7B, 16.8A, B) are adapted for piercing or tearing host skin to obtain blood. Their labellar lobes are comparatively small, but their prestomal teeth are sharp, sclerotized, and greatly enlarged.

Other morphological characters of the thorax and wings are typical of muscid flies. On the thorax (Figure 16.9), the hypopleuron is bare, entirely lacking bristles and setae. The first anal wing vein (vein A1; Figure 16.10), including a faint trailing fold if present, vanishes before it reaches the wing margin. The combination of these characters, along with aristate antennae, a ptilinal suture, and usually robust oral vibrissae distinguish members of the Muscidae from all other flies.

The abdomen is reduced to five visible segments in both sexes, with succeeding segments 6 through 12 modified into eversible reproductive terminalia. The male possesses an aedeagus, or intromittent organ, which when at rest is rotated 180° and is partially enclosed in a genital pouch. In the female, the terminalia are modified into a tubular, telescoping

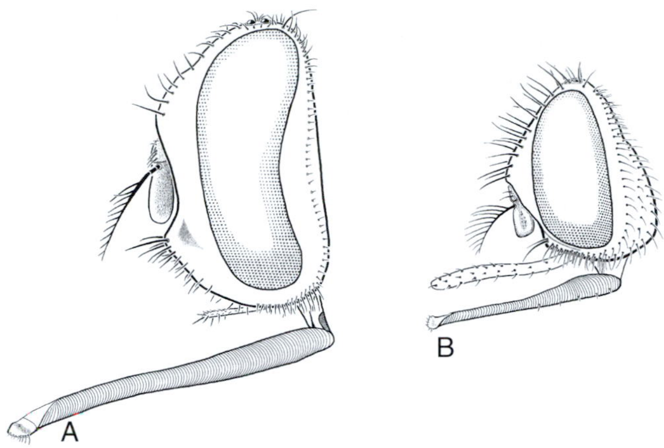

Figure 16.8 Heads of two muscid flies with biting mouthparts; in both cases the mouthparts are held in a horizontal position beneath the head when not feeding. (A) Stable fly (*Stomoxys calcitrans*); (B), horn fly (*Haematobia irritans*). (Redrawn from Edwards et al., 1939. © The Natural History Museum, London)

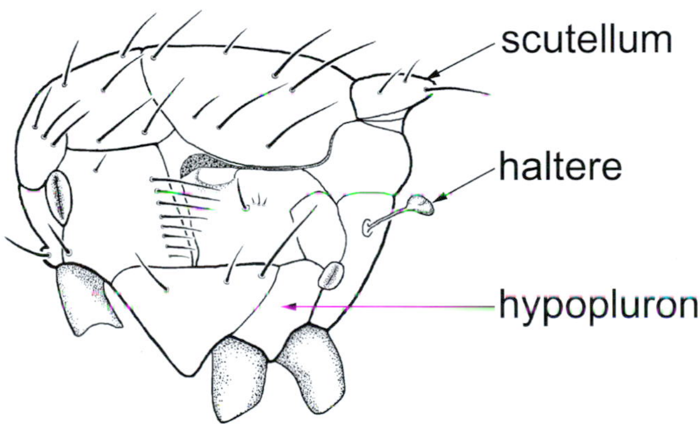

Figure 16.9 Lateral view of thorax of adult muscid fly, with wings removed; anterior to left; showing typical sclerites, scutellum, hindwing reduced to form a haltere, and a taxonomically important sclerite, the hypopleuron. (Redrawn from Borror et al., 1989, with permission of Wadsworth, an imprint of the Wadsworth Group, a division of Thomson Learning)

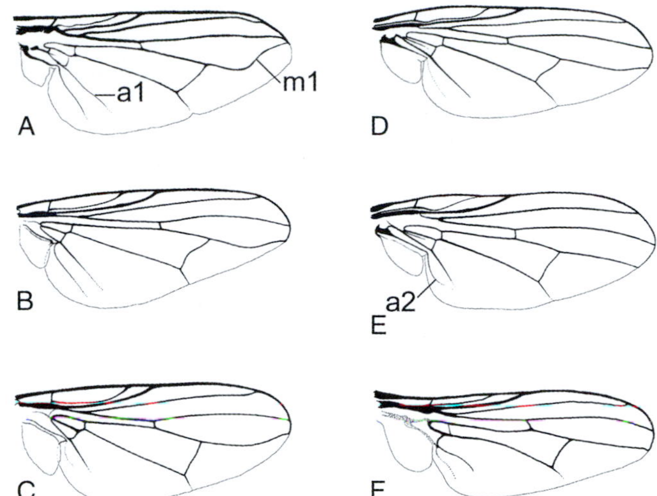

Figure 16.10 Wings and wing venation of important muscid flies; right wing shown in each case. (A) House fly (*Musca domestica*); (B) stable fly (*Stomoxys calcitrans*); (C) horn fly (*Haematobia irritans*); (D) false stable fly (*Muscina stabulans*); (E) little house fly (*Fannia canicularis*); (F) black garbage fly (*Hydrotaea ignava*). a1 and a2, anal wing veins 1 and 2; m1, medial wing vein 1. (A), (B), and (F) Redrawn from Axtell, 1986, all rights reserved, copyright © Novartis. (C) Redrawn from Lane and Crosskey, 1993 © The Natural History Museum, London. (D) and (E) Redrawn from McAlpine et al., 1987, with permission of Minister of Public Works and Government Services, Canada.

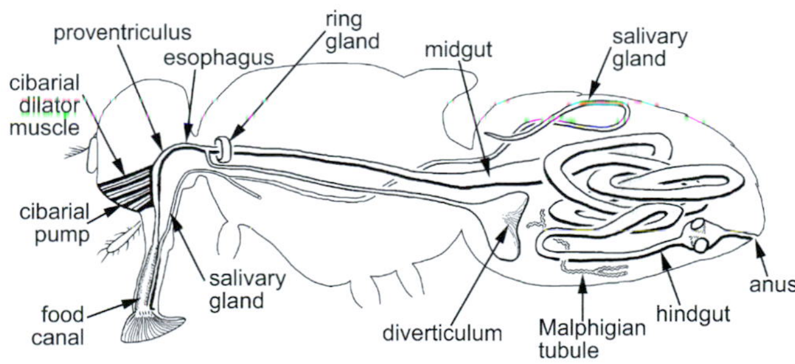

Figure 16.11 Digestive system of adult muscid fly; left salivary gland and distal ends of Malphigian tubules omitted for clarity. (Redrawn from Patton, 1929)

ovipositor that is extended to lay eggs. Chemoreceptors and mechanoreceptors occur on the terminalia, and also on the antennae, labellum, and tarsi.

The digestive system of adult muscid flies (Figure 16.11) is much like that of other Diptera. Saliva is produced in the salivary glands, which extend posteriorly from the head into the abdomen. Ingested foods that are nutrient-rich, such as blood and dung fluids, are routed via the proventriculus of the foregut to the midgut where digestion occurs. Fluids that are dilute, such as plant nectar and milk, are shunted for temporary storage to the diverticulum, typically a ventral, sac-like extension of the esophagus. Contents of the diverticulum are regurgitated through the proboscis onto the fly's substrate and then reingested to aid in evaporation. Waste products and undigested food ultimately pass through the hindgut and out the anus as drops of liquid fly feces. The brown fly specks that collect on feeding and resting substrates consist of droplets of two substances—fly vomit and feces.

LIFE HISTORY

All the important North American muscid species are oviparous, meaning that females deposit fertilized eggs in the environment before they hatch. Some members of the Muscidae in the Old World are larviparous.

Oviposition substrates vary among the different species. Filth flies oviposit in organic debris that is wet enough to support aerobic microbial fermentation. Common substrates are human feces and garbage, and decomposing organic matter such as rotting algal mats, piles of lawn clippings, and food-processing wastes. Where livestock and poultry are confined, attractive substrates include manure (a mixture of aging feces and urine); soiled bedding (bedding + manure + feed); and wet, rotting feeds such as hay, silage, and grain. Filth flies commonly exploit the kinds of wastes that accumulate around human habitations and animal-confinement facilities.

In contrast to the filth flies, the important muscid dung flies and sweat flies lay their eggs in a much narrower range of substrates. Dung flies oviposit on or into cattle dung pats within minutes to a day or so after the animal defecates. The sweat flies oviposit in plant litter and decomposing dung in grasslands and forests.

Larvae of all the important muscid flies burrow, feed, and develop in their respective ovipositional substrates. However, larval feeding differs among species and even among instars of the same species (Skidmore, 1985). Most muscid flies are saprophages that feed by filtering bacteria, yeasts, and other small organic particles suspended in their semiliquid habitat. All three instars of the house fly, the stable fly, *Fannia* species, the horn fly, and the face fly are saprophages. In contrast, the third instars of the garbage flies, *Muscina* species and the sweat flies are facultative predators. These larvae can mature as saprophages, but they will switch to predation and consume other soft-bodied insects if available. These facultative predators contribute to the natural biological control of other flies that occur in the same habitats.

Once mature, third-instar muscid larvae cease feeding, empty their alimentary tract by defecating, and enter a wandering phase before they pupate. Depending on the species and substrate moisture, they may disperse to adjacent drier locations or pupate directly in their larval medium.

Developmental times from egg to adult of the different muscid flies range from one to six weeks during the summer (Table 16.2). Most of the species develop fastest at temperatures of 27 to 32°C and all virtually cease activity at temperatures below 10°C. Sustained temperatures beyond these limits are usually lethal, with the exceptions of cold-tolerant species that overwinter as larvae or pupae. Developmental times also can be affected by the supply of food. For example, when crowded, or otherwise starved, filth flies will delay pupation to achieve a minimum size. The dung flies have a different strategy; their larvae sacrifice size and will pupate earlier, resulting in adults that are smaller and less fecund than their better nourished counterparts. Regardless of larval conditions, emerging adults typically occur in an equal sex ratio (1:1).

The reproductive capacity of a muscid fly population is determined in part by fecundity, which is the number of eggs that the average female can produce per batch. The house fly and other muscid filth flies are the most fecund, whereas the dung flies produce fewer eggs per batch (Table 16.2). Reproductive capacity also is determined by the length of the gonotrophic cycle (i.e., how many days a female requires to develop and deposit a batch of eggs) and by longevity, which governs how long she will live and how many cycles she can complete. Under summer conditions, muscid flies can develop a new batch of eggs every two to five days. The average female probably lives long enough to produce one or two batches, although longevity is difficult to measure under field conditions.

BEHAVIOR AND ECOLOGY

Because adult muscid flies emerge with little stored energy and nutrients, they must find water, salts, carbohydrates, and protein if they are to survive and reproduce. The nonbiting species obtain sugars from off-host sources such as nectar from plants and honeydew from sap-sucking aphids and scale insects. The species that feed on vertebrates obtain the bulk of their protein from blood, serum, saliva, mucus, and lachrymal secretions. Both sexes of the biting horn fly and stable fly obtain nearly all their nutrients from blood.

The feeding behavior of a nonbiting fly differs from that of biting species. A nonbiting fly opens its labellum and presses it against the substrate. If the food is

Table 16.2 Life History Attributes of Important North American Muscid Flies, Compiled from Skidmore (1985) and Other Sources

Name	Days from Egg to Adult: Minimum	Days from Egg to Adult: Normal	Eggs per Cycle	Generations per Year	Overwintering Stages(s)
House fly	7	10–21	120–150	Multiple	All?
Stable fly	12	15–30	80–100	Multiple	All?
Garbage flies	9	14–25	70–190	Multiple	Larva
False stable fly and relatives	10	14–40	98–150	Multiple	Adult in diapause
Little house fly and relatives	22	25–50	58–72	Multiple	Larva?
Horn fly	9	10–20	20–32	Multiple	Pupa in diapause
Face fly	7	12–28	16–42	Multiple	Adult in diapause
Sweat flies	25	35–300	17–150	1–2	Larva

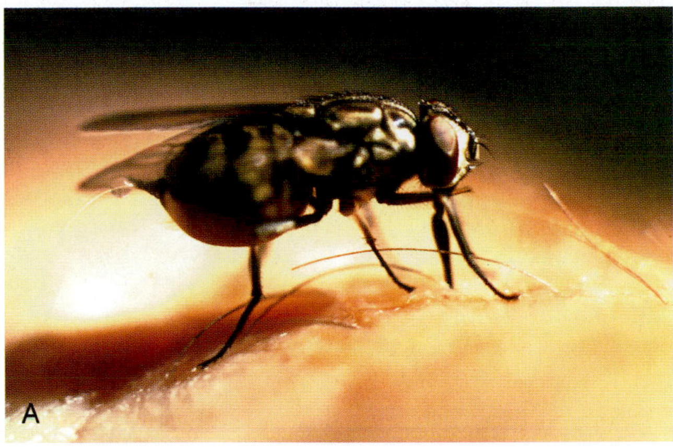

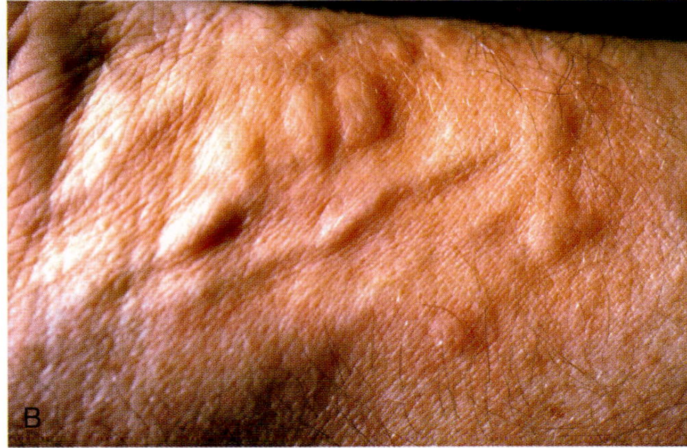

Figure 16.12 Stable fly (*Stomoxys calcitrans*) feeding on human. (A) Fully fed female, abdomen distended with blood; (B) resultant bite reaction in form of welts. (Courtesy of Elton J. Hansens)

not liquid, the fly releases enzyme-laden saliva and repetitively opens and closes the labellum to scarify solid food into the saliva. The suspension is then drawn along the pseudotracheae into the food canal. Feeding by nonbiting flies can be characterized as a process of salivating, scrubbing, and sucking; they cannot physically penetrate skin. When a biting species feeds (Figure 16.12), it presses its proboscis against skin and rapidly opens and closes the labellum, directing the prestomal teeth in a downward and outward rasping motion. Once the skin is penetrated, the teeth anchor the proboscis while blood flows into the subsurface lesion and up the food canal.

All the important muscid flies are **anautogenous**, meaning that females require protein to complete their first gonotrophic cycle. Protein is the substrate of yolk synthesis and egg maturation. In the nonbiting species, eggs in the ovarioles mature in synchrony two to five days after protein is first obtained. Development of a subsequent batch is arrested hormonally until the preceding batch has been laid. There are corresponding cycles of attraction to different substrates, first to sources of protein and then to oviposition sites as eggs mature. In the horn fly and stable fly, eggs develop asynchronously, so feeding and oviposition are distributed more evenly in time.

Behavior associated with mating differs among species. Males of *Fannia* species and some of the sweat flies and garbage flies hover in swarms, usually in locations shaded by trees or the roofs and eves of buildings. These males are attracted to females that fly into the swarm. Once coupled, a mating pair will fall to the ground and complete copulation. Males of the other muscid flies do not form swarms. Instead, they generally perch or rest in sunny locations on substrates such as tree trunks, fence posts, and rocks, and the males intercept passing females. Females of all the important muscid flies typically store enough sperm from a single mating to fertilize all the eggs they can produce during the remainder of their lives.

Activities and locations of adult muscid flies vary markedly with time of day. All the important species are active during daylight hours, and almost all are inactive at night. Activities include flying, host location, feeding, mating, and ovipositing. Sight and olfaction are used to locate hosts and oviposition substrates. Most muscid flies are exophilic, being reluctant to enter buildings. A few species are more endophilic and will enter buildings. Species that feed on animals may be on a host as briefly as a few minutes, just long enough to obtain available foods. The flies leave their hosts when replete, and rest in the surrounding environment while digestion proceeds. Because feeding times are much shorter than digestion times, the adults on a host at any instant are likely to be only a small fraction of all the adults present in the host's environment.

Muscid flies apparently choose daytime resting sites, in part, according to their needs for thermoregulation. They rest in sunny sites when the air temperature is below about 20°C and in shady sites when temperatures exceed about 30°C.

An exception to the generalized pattern of daytime activity and host-visiting behavior occurs in the horn fly. Once a host is located, the adult remains on its host almost continuously, except when disturbed or laying eggs. Horn flies feed and oviposit at all hours of day and night.

The flight range of muscid flies is extensive. Detectable numbers of all the important North American species have been collected more than 5 km from known or presumed points of origin. Large numbers of stable flies can appear on beaches 10 or more miles downwind from the nearest likely breeding sites.

The seasonal patterns in abundance and age structure of adult subpopulations vary among species, years, and locations. In localities with cold winters, population growth outside buildings is restricted to a distinct breeding season—the warmer, wetter months of spring, summer, and autumn. In these cases, populations grow to a single peak of abundance, normally in early autumn. A notable exception is the house fly, which can breed continuously in heated buildings. In warmer climates, the breeding season for most species is longer, and may be continuous year-round. For example, adults of the house fly, *Fannia* species, and horn fly occur throughout the year in the southeastern United States and southern California. Densities of adults have two seasonal peaks, with growth phases in spring and autumn separated by periods of decline during summer and winter.

Most muscid flies of medical-veterinary importance are multivoltine, developing through two or more generations per breeding season (Table 16.2). These generations usually overlap, so recruitment of new adults is continuous; eggs, larvae, pupae, and adults of all ages are present simultaneously throughout most of the breeding season. Population growth within the breeding season is influenced by availability of breeding media, by weather and its effects on survival of immature stages, and the fly reproductive rate. Survival of larvae is enhanced if their breeding habitat remains wet enough to support filter feeding, yet dry enough to allow aerobic respiration. Suitable moisture levels are about 30 to 75%. Substrate moisture is critical because the saprophagous larvae feed by filtering particles suspended in their medium.

Muscid flies overwinter in different ways (Table 16.2). The house fly and the stable fly breed continuously in frost-free southern regions of North America. Breeding by these flies is restricted to the warmer months in more northern latitudes, because they lack a stage that can endure temperatures below freezing for much more than a day. It was once thought that these two filth flies, lacking a freeze-tolerant life stage, died out each winter in temperate latitudes and were repopulated each spring from milder regions. However, it is now known that local populations can also persist through winter in protected, semiheated substrates associated with humans and livestock. Regional repopulation does occur, however, with the bushfly in Australia, where immigrants disperse southward from more northerly latitudes that remain warm during winter (Hughes, 1977).

Some other muscid flies of medical-veterinary importance overwinter in **diapause**, a state of developmental arrest typically associated with a tolerance for freezing. The face fly and *Muscina* species overwinter as adults. In autumn, these flies enter hibernacula, such as occur under bark of dead trees and siding on buildings, and emerge the next spring to begin reproduction. The horn fly, in contrast, overwinters in temperate regions as a diapausing pupa. Garbage flies, *Fannia* species, and sweat flies are thought to overwinter as larvae, but further study is needed to determine if they exhibit a true diapause, or are in a simpler state of cold-tolerant quiescence.

SPECIES OF MEDICAL-VETERINARY IMPORTANCE

Adults and larvae of some important muscid flies can be identified tentatively from external characters, and from features of their behavior, habitat, and geographic location.

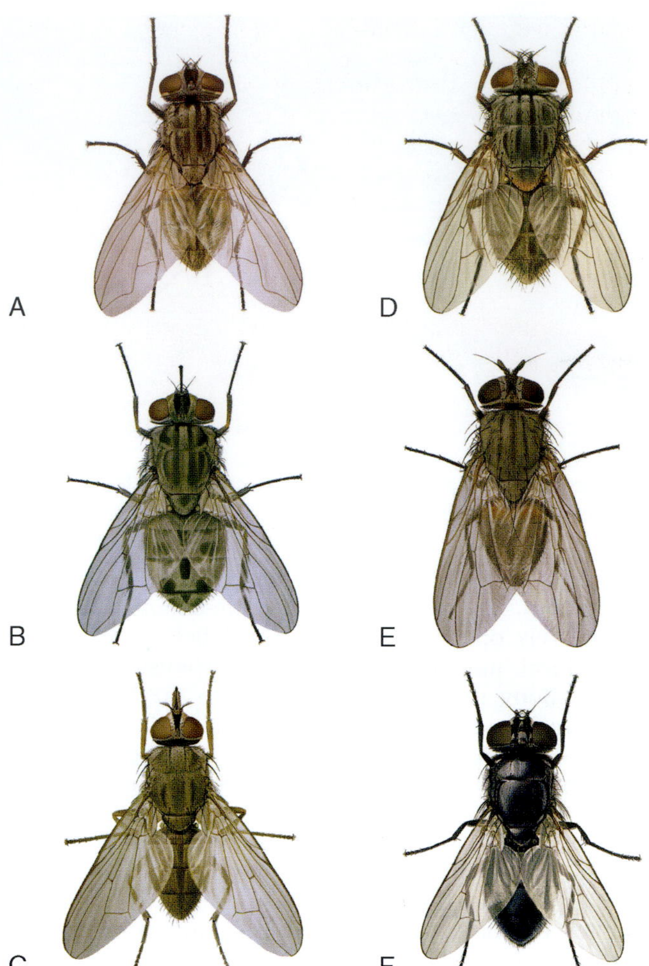

Figure 16.13 Muscid flies, adult females. (A), House fly (*Musca domestica*); (B) false stable fly (*Muscina stabulans*); (C) stable fly (*Stomoxys calcitrans*); (D) little house fly (*Fannia canicularis*); (E) horn fly (*Haematobia irritans*); (F) black garbage fly (*Hydrotaea ignava*). (Original drawings by F. Gregor, and published in B. Greenburg, 1971; Flies and Disease, Vol. 1, Ecology. Classification and Biotic Associations, Princeton University Press)

House Fly (*Musca domestica*)

This nonbiting filth fly occurs on all continents except Antarctica. It is native to the Afrotropical and Oriental regions, and was probably introduced into the Americas by Europeans during colonial times. Adults are gray and black flies, 6 to 9 mm long, with four black vittae on an otherwise gray thorax (Figures 16.1, 16.13A). The wing has a sharp forward bend in vein M1 (Figure 16.10A). The abdomens of typical females are checkered gray and black at the dorsal midline, and creamy yellow on the sides, which in North America is sufficient to distinguish this species from the face fly. Larvae have large caudal spiracles that resemble back-to-back "D"s, and the slits are sinuous (Figure 16.4B).

Immatures can be found in a wide variety of decaying organic substrates. Major breeding sites include human garbage dumps, open privies, livestock manure, soiled bedding, poultry litter, and wastes around fruit and vegetable processing plants. Breeding continues year-round in tropical and subtropical regions, but is interrupted by winter in temperate regions. From a public health standpoint, the house fly is probably most significant as a nuisance and potential vector of enteric pathogens. Although the house fly can become quite abundant where livestock, poultry, and companion animals are housed, its direct effects on animal health are comparatively unimportant.

Bazaar Fly (*Musca sorbens*)

This nonbiting filth fly is the most abundant synanthropic muscid fly in many parts of the Afrotropical, Oriental, and Pacific regions. It was introduced through commerce into Hawaii and probably would flourish elsewhere in tropical latitudes of the Americas. Greenberg's (1971) key provides characters to distinguish the bazaar fly from other *Musca* species in the Afrotropical and Oriental regions.

The species recognized as *M. sorbens* before 1970 apparently consists of a complex of at least three species that are partially distinguishable by the ratio of the width of the male's frons (area between compound eyes) to the width of the head (including eyes). The "broad-frons" form is known correctly as the bazaar fly (*M. sorbens*), and this species occurs from Africa east through the Orient and on many Pacific islands. A "narrow-frons" form occurs in Australia and Papua New Guinea, and is considered a distinct species, the bushfly (*M. vetustissima*). A second "narrow-frons" form (*M. biseta*) coexists with the bazaar fly in Africa and eastward, but further study is needed to resolve distinctions between *M. biseta* and the bushfly in southern Asia and the Orient (Pont, 1991).

Adult bazaar flies are strongly exophilic, being far less inclined than the house fly to enter buildings. Larvae have been recorded in unburied human stools and dog feces, and less commonly in feces of other animals, in carrion, and in garbage. The bazaar fly is important to public health, but it is probably unimportant to the health of domestic animals.

Bush Fly (*Musca vetustissima*)

This nonbiting dung fly occurs in Australia, where it is a major nuisance to humans and livestock. It is related closely to the bazaar fly, and keys out as *M. sorbens* in Greenberg (1971). Adults are attracted to large mammals as sources of fluids for nourishment and feces for oviposition. Several authors have speculated that the bush fly originally was associated with aboriginal encampments, and that its abundance increased when domestic cattle were imported. Larvae have been recorded from the feces of a wide variety of large mammals, but in nature cattle dung pats are overwhelmingly the most productive. Breeding is continuous in subtropical Australia, and southward migrations serve to repopulate temperate Australia and Tasmania each spring (Hughes, 1977).

Face Fly (*Musca autumnalis*)

This nonbiting dung fly is native to Europe and central Asia, and was introduced into North America before 1952 (Krafsur and Moon, 1997). It occurs in all southern Canadian provinces and in the United States north of Arizona–Georgia (35 N). The adult resembles the house fly (Figures 16.1, 16.13A), is 6 to 10 mm long, has four black vittae on an otherwise gray thorax, and a sharp forward bend in wing vein M1 (Figure 16.10A). The male's abdomen has a distinct, black longitudinal band along the midline and bright yellow sides. The female has a characteristic yellow patch on the ventrolateral aspect of the first visible abdominal segment; the remaining segments are gray-black to the ventral midline. The egg has a distinct brown-black respiratory stalk (Figure 16.2B). Mature larvae are bright yellow with black, D-shaped spiracular plates (Figure 16.5A), and puparia are white due to calcification.

During the fly breeding season, adult face flies occur around grazing cattle and horses. Their larvae develop exclusively in fresh cattle dung pats. In autumn, newly emerged adults enter diapause, aggregate on buildings, and eventually accumulate behind siding, in wall voids, in attics, and occasionally in interior rooms. Thousands of flies can occur in such aggregations. The face fly was first recognized as a nuisance in European households due to its overwintering habits. It became recognized as a pest of cattle and horses after it was introduced into North America.

Cluster Fly (*Pollenia rudis*)

The cluster fly is discussed here because it often occurs along with the face fly in household infestations. This calliphorid fly is native to Europe and North Africa, and was introduced into North America where it now occurs in all southern provinces of Canada and throughout the United States. The adult is 7 to 9 mm long, the abdomen is completely black with silvered checking, and crinkly golden hairs occur on the head and thorax. Cluster fly larvae are internal parasites of earthworms (*Allolobophora* spp., Lumbricidae), and produce two to four generations per year. Adults can be a nuisance in households, but they do not affect domestic animals.

Stable Fly (*Stomoxys calcitrans*)

This biting filth fly (Figures 16.12, 16.13C) is native to Africa, Europe, Asia, and the Orient, and was probably introduced into the Americas and Australia during colonial times. At least three common names are used regionally for the stable fly. It is known as the beach fly because of outbreaks on recreational beaches, the dog fly because it pesters dogs, and the lawn-mower fly because larvae have been found in damp, matted grass on the undersides of lawn mowers. The stable fly is also misleadingly called the biting house fly because of its superficial resemblance to the house fly.

The adult is 5 to 7 mm long, has seven circular black spots on an otherwise gray abdomen (Figure 16.13B), and a piercing-sucking proboscis with short maxillary palps (Figure 16.8A). Larvae and pupae have uniquely shaped, subtriangular posterior spiracles that are far apart; the horizontal space between them is greater than twice a plate's width (Figure 16.5B). Larvae occur in decaying fibrous substrates such as straw bedding, wet hay, algal mats, and wet grass clippings. Other larval habitats include accumulations of manure from dairy and beef cattle, mixtures of soil and partially composted bedding and animal manure, and byproducts of crop processing such as peanut hulls, beet pulp, and sugarcane bagasse. Breeding is continuous in tropical and subtropical climates, and the species is thought to overwinter as immatures wherever larval substrates do not freeze. Stable flies are important to public health because they will attack and annoy people, but they are much more important from a veterinary perspective.

Horn Fly (*Haematobia irritans irritans*) and Buffalo Fly (*Haematobia irritans exigua*)

These biting dung flies were once recognized as two separate species, the horn fly (*H. irritans*) and the buffalo fly (*H. exigua*). However, Zumpt (1973) concluded that they are subspecies of *H. irritans*, based mainly on subtle morphological differences and allopatric distributions. The horn fly is native to northern Africa, Europe, and central Asia, and was introduced into North America from Europe in the middle 1880s. It has since spread to all cattle-producing regions in the Americas, including Hawaii. The buffalo fly is native to southern Asia, the Orient, Indonesia, and several Pacific islands. It spread through commerce into New Guinea and Australia before 1840. It is possible that the two subspecies intergrade in parts of Asia.

The adult horn fly (Figure 16.14) is 3 to 5 mm long and has a piercing-sucking proboscis (Figure 16.8B). The maxillary palps are held appressed to the haustellum and are almost as long as the haustellum. Wing vein M1 is gently curved (Figure 16.10C). Adults of both subspecies are specific to cattle, bison, and water buffalo; aberrant hosts include horses and other large mammals. The flies occur mainly on the withers, back and sides, but will move to the belly when weather is hot. Females of both flies lay their eggs exclusively under edges of dung pats, usually within minutes of host defecation. For unknown reasons, dung from horses, sheep, and other large mammals is unsuitable. Horn flies occur in far greater numbers on grazing cattle than on animals confined in drylots or indoors. Reproduction is continuous and populations are multivoltine. The horn fly overwinters as a diapausing pupa in temperate latitudes. Neither subspecies poses any threat to human health, but both species are serious economic pests of grazing beef and dairy cattle.

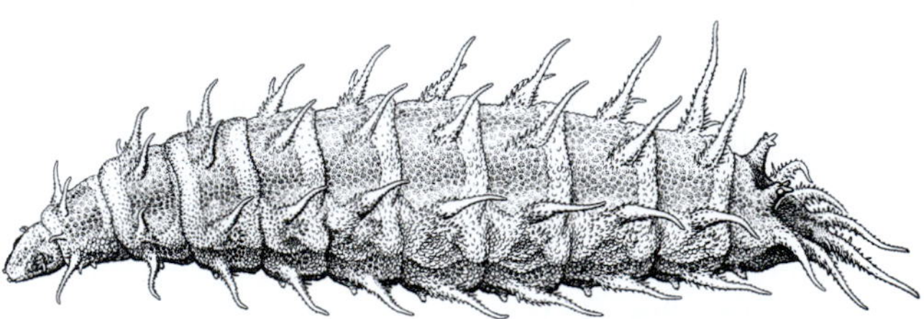

Figure 16.14 Little house fly (*Fannia canicularis*), third-instar larva. (From McAlpine et al., 1981; reproduced with permission of Minister of Public Works and Government Services, Canada)

False Stable Fly (*Muscina stabulans*) and Its Relatives

These nonbiting filth flies include the false stable fly, which has been spread worldwide through commerce, and another 10 species that occur mainly in the Holarctic region. The name *Muscina assimilis*, used widely in older literature for a relative of the false stable fly, has been relegated to a synonym of *M. levida* (Skidmore, 1985). The false stable fly (Figure 16.13B) and its relatives are stout flies, 8 to 12 mm long, with brown-black bodies and a rounded bend in wing vein M1 (Figure 16.10D). The tip of the scutellum of the false stable fly is red-orange. The posterior spiracular plates are roughly circular; they are separated by one plate's width, and the slits are bowed and arranged radially (Figure 16.5D). Third-instar larvae are facultatively predatory, and adults overwinter in a prereproductive diapause. These species can affect public health, but they are not thought to affect the health of domestic animals.

Little House Fly (*Fannia canicularis*) and Its Relatives

There are about 100 species of these nonbiting filth flies in North America (Chillcott, 1961), and additional ones in Latin America, Africa, Europe, and Asia. The little house fly (*F. canicularis*) and the latrine fly (*F. scalaris*) have spread by commerce throughout the world. *Fannia* species are 5 to 8 mm long, with dark thoraces and abdomens variously marked with yellow (Figure 16.13D). The arista lacks setae and the second anal vein (A2) curves toward the first anal vein (A1) (Figure 16.10E). Larvae and puparia have characteristic lateral and dorsal processes (Figure 16.14) whose function is unknown. Males form mating swarms in shady locations, and it is this swarming behavior that most often brings them into contact with people. The little house fly is probably the most endophilic and commonly encountered species of this genus in North America. The latrine fly is more exophilic. Although these flies are most noticeable where domestic and zoo animals are confined, they are not otherwise important to public and veterinary health.

Garbage Flies (*Hydrotaea* spp.)

There are seven known species of garbage flies, and at least one species occurs in every biogeographic region. This group of nonbiting filth flies, once placed in the genus *Ophyra*, has been merged into *Hydrotaea* (Huckett and Vockeroth, 1987). Accordingly, the scientific names of the common species have changed. The black garbage fly, known in older literature as *Ophyra leucostoma*, is now named *Hydrotaea ignava*. It is native to the Old World and has been introduced into North America (Skidmore, 1985). The black dump fly, formerly *Ophyra aenescens*, is now *Hydrotaea aenescens*. It is native to the New World, occurs in the eastern Pacific Islands including Hawaii, and has been introduced into Europe.

The garbage flies are 4 to 7 mm long with shiny black thoraces and abdomens (Figure 16.13F). Wing vein M1 is virtually straight (Figure 16.10F). Posterior spiracles of mature larvae and puparia are roughly circular; they are separated by less than one plate's width, and have slightly curved slits that barely diverge from a faint scar (Figure 16.5E). Adults are strongly exophilic. Larvae have been recorded in a great variety of filthy substrates, including carrion. Third instars are facultative predators, and will consume larvae of other flies that cohabit their breeding medium. These filth flies pose a modest threat to public health, but they are not known to harm domestic animals.

Sweat Flies (*Hydrotaea* spp.)

About 50 species of sweat flies occur in the Palearctic region, and fewer species occur in the remaining biogeographic regions. Sweat flies are gray to black, 3 to 8 mm long, with an arista that lacks setae. Although the female has no simple distinguishing characters, the male has a ventral notch or depression at the distal end of the fore femur. Third-instar larvae are facultative predators. Their spiracular plates are stalked, with radially arranged slits (Figure 16.5F).

Females of six of the 24 North American species, including *Hydrotaea meteorica* and *H. scambus*, are persistent in their attempts to imbibe perspiration and secretions from the eyes, nostrils, lips, and other parts of mammalian hosts. The remaining North American species are apparently not attracted to animals (Huckett, 1954). In Europe, the sheep headfly (*H. irritans*) is a primary pest of sheep, cattle, and deer.

PUBLIC HEALTH IMPORTANCE

Muscid flies affect people most frequently as nuisances, occasionally as vectors of pathogenic organisms, and rarely as agents of human myiasis. The cosmopolitan house fly and stable fly are of greatest medical significance. Other notable examples are the bazaar fly in Africa, Asia, and Pacific islands including Hawaii; the bushfly in Australia; and *Stomoxys nigra* and *S. sitiens* in Africa and Asia.

Filth flies pose particular risks as mechanical vectors of pathogens that cause enteric disease in humans. Among the 1.6 million cases of notifiable infectious diseases reported in 2005 to the United States Centers for Disease Control and Prevention, approximately 96,000 (6%) were enteric infections causing diarrhea or dysentery. These diseases arise from direct or indirect fecal contamination of food and water, either at points of consumption and preparation, or earlier at points of production and distribution. Globally, the World Health Organization reports that diarrhea and dysentery account for more childhood deaths and morbidity than any other infectious diseases.

Enteric diseases are caused by certain bacteria, viruses, and protozoa. The bacteria include *Escherichia coli*, *Salmonella* species, and *Shigella* species; the viruses include Cocksackie, Enterovirus 72 (Hepatitis A), and enteric cytopathogenic human orphan (ECHO) viruses; and the protozoa include *Chilomastix*, *Cryptosporidium*, *Entamoebae*, and *Giardia* species. Infections range in severity from benign to fatal, being most severe among children, the elderly, and others who are infirm. Common sources of enteric pathogens are food and water contaminated with feces from infected people or animals, or indirectly via hands, utensils, and flies.

Greenberg (1971, 1973) summarized the extensive literature on pathogens associated with muscid flies. Evidence is strong that filth flies in particular are mechanical vectors. Mouthparts, tarsi, and gastrointestinal tracts become contaminated when the flies feed on contaminated substrates. Upon dispersal, the flies can inoculate new substrates with contaminated tarsi, mouthparts, fly vomit, and feces.

The medical significance of filth flies at a given time and place depends on which flies and people are involved, and on circumstances in which flies and people come into contact. A substantial majority of people in the United States and Canada now live and work in urban and suburban settings where indoor and outdoor environments are essentially free of filth flies. Exceptions are rural settings lacking adequate sanitation systems or neighboring mismanaged livestock and poultry operations. Intolerance for flies is, in part, the basis for municipal health codes used to enforce proper management of organic wastes on the affected premises. Sanitary standards established by the mid-1900s have dramatically reduced the epidemiological importance of filth flies in many parts of the developed world. Too often, however, basic sanitation and filth-fly management are unsatisfactory due to poverty, famine, or war. Under these circumstances, filth flies can reach tremendous densities, breeding in and around accumulated human waste and carrion.

The following muscid flies warrant attention with regard to public health.

House Fly (*Musca domestica*)

The house fly is the most common cause of fly annoyance in North America. Adults aggregate around garbage, compost piles, and other food sources, and they readily enter buildings. House flies are conspicuous when alighting directly on people; crawling on human food; or resting on walls, windows, and ceilings. These substrates become soiled with fly specks, dried droplets of fly vomit, and feces.

In a classic pair of experiments, Watt and Lindsay (1948) and Lindsay et al. (1953) provided strong evidence that the house fly is a significant vector of enteric pathogens. They controlled filth flies with residual insecticides in selected towns in southern Texas and southern Georgia, and left neighboring towns untreated as controls. Fly surveillance in the treated and untreated towns showed that treatments greatly reduced the densities of house flies and other species. Surveillance of the residents in the treated towns showed concurrent declines in the incidence of diarrhea in people of all ages, and in isolates of *Shigella* from children under 10 years of age.

Other studies have confirmed the importance of the house fly in the epidemiology of enteric diseases. Intensive trapping to remove flies at two military field bases in Israel caused declines in fly populations at mess tents, concurrent declines in frequencies of diarrhea and shigellosis among base recruits, and declines in rates of seroconversion for antibodies to *Shigella* and enterotoxigenic *E. coli.* (Cohen et al., 1991). Elsewhere, village-wide spraying of six Pakistani villages during two consecutive fly seasons reduced house fly populations by 95% and lowered the incidence of childhood diarrhea by 23% (Chavasse et al., 1999; also see West et al., 2006). Finally, simple screening of caged broiler houses in Denmark caused a substantial decrease in incidence of chicken-borne campylobacteriosis (Hald et al., 2007).

These studies provide strong evidence that house flies can be important routes for spread of fecal-borne pathogens. Prudence dictates that the house fly and other filth flies should be controlled through sanitation in the synanthropic environment, and that they should be prevented from contacting human food at all points of production, distribution, preparation, and consumption.

Bazaar Fly (*Musca sorbens*)

This nonbiting, synanthropic fly is common in Africa, Asia, and many Pacific islands. Adults feed persistently at the eyes, noses, and mouths of people (Figure 16.15) and other large mammals. The flies are also conspicuous wherever human food is exposed outdoors. Fortunately, the species is strongly exophilic. Greenberg (1971, 1973) summarized the extensive literature that associates the bazaar fly and its close relatives with human pathogens. Most notably, these flies are strongly suspected of mechanically transmitting enteric pathogens and the causal agents of acute bacterial conjunctivitis and trachoma. A recent study involving paired villages in The Gambia (Emerson et al., 1999) showed that community spraying, which reduced bazaar fly populations by around 75%, lowered incidence of trachoma eye disease (caused by *Chlamydia trachomatis*) by 75%, and incidence of childhood diarrhea by 22%. However, a study in Tanzania failed to show reduced prevalence of trachoma with fly spraying following administration of antibiotics in treated communities (West et al., 2006).

Figure 16.15 Aggregating bazaar flies (*Musca sorbens*) on human hosts. (Photo by R. Lewis and D. Dawnway, with permission. © The Natural History Museum, London)

Bush Fly (*Musca vetustissima*)

The earliest European travelers in Australia recorded the annoying presence of the bush fly. This nonbiting dung fly, like the closely related bazaar fly, is strongly exophilic and is a probable irritant to humans almost anywhere in Australia. Flies that are attracted to people swarm around the head, feed at eyes and nostrils, and settle on the head, back, and shoulders. Once on hosts, the flies are peculiarly sedentary. More than a casual brush of the face with the hand is required to dislodge them, leading to a hand gesture that is humorously called an "Aussie salute." Larvae are known to occur in human and animal feces, so adults are a potential mechanical vector of enteric pathogens. Furthermore, the propensity of the adults to feed at a host's eyes makes the bush fly a prime suspect in transmission of eye pathogens (Greenberg, 1971).

Face Fly *(Musca autumnalis)* and Cluster Fly (*Pollenia rudis*)

The face fly and the cluster fly are two of several species of flies that can be a nuisance in households during winter and early spring. Other species include various blow flies (Calliphoridae), *Muscina* species, and *Ceroxys latiusculus* (Otitidae). Overwintering flies in buildings can be activated by heaters or warm weather and become attracted by light to inhabited rooms. Often people first take notice when live and dead specimens occur at sunny windows. Dead flies that have accumulated over years in an infested building can attract dermestid beetles. Although the flies and beetles can be a source of allergens, these insects do not pose any other known medical threat.

Stable Fly (*Stomoxys calcitrans*)

The stable fly is an important nuisance in outdoor environments throughout North America. This fly will readily attack people, usually on the lower part of the legs, causing a searing pain with each probe of its bayonet-like proboscis (Figure 14.12A). It does not take many stable flies to disrupt activities of sunbathers, anglers, and others seeking outdoor leisure. Outbreaks have been recorded in the United States at tourist spots in the Great Lakes area, the Atlantic seaboard, and the Gulf Coast. Annoyance by stable flies is not confined to resorts and beaches; the flies can occur wherever people, fly breeding sites, and favorable weather coincide.

Adoption of the United States Declaration of Independence on July 4, 1776, by delegates to the Continental Congress may have been hastened by stable flies. According to Fuller (1913), debate on the Declaration drafted by Thomas Jefferson and his committee might have lasted much longer were it not for torment from stable flies. Jefferson noted the weather was oppressively warm that day in Philadelphia, and the meeting room was next to "... a stable, whence the hungry flies swarmed thick and fierce, alighting on their legs and biting hard through their thin silk stockings. Treason was preferable to discomfort." Clearly Jefferson had a wit, but he also knew enough entomology to infer that the nearby stable was the source of the flies.

As with any blood-feeding arthropod, stable flies provide an opportunity for transmission of bloodborne human pathogens. Experimental evidence suggests that the fly can acquire animal pathogens as mouthpart contaminants. Ingested particles can remain viable in the lumen of a fly's gut and diverticulum for hours to several days. However, none of these pathogens infect the fly, so only mechanical transmission would be possible. Experimental evidence using animal disease models in realistic settings suggest that the stable fly is not a vector of any consequence to human health.

False Stable Fly (*Muscina stabulans*) and Its Relatives

The false stable fly and its relatives are common around filthy habitats, including latrines, household wastes, and accumulations of animal manure. The adults have feeding habits similar to those of the house fly and present similar risks for mechanical transport of food-borne pathogens. These flies remain outdoors and rarely feed on human food. However, they do feed and defecate on fruit, and serve as potential vectors wherever breeding sites are near open-air markets and roadside fruit stands. Larvae of the false stable fly and of *M. levida* have been involved in rare cases of intestinal and urinary myiasis.

Little House Fly (*Fannia canicularis*) and Its Relatives

These nonbiting filth flies can become nuisances when swarms occur inside inhabited buildings. Hovering *Fannia* species often occur at head height indoors where they can be particularly distracting and bothersome.

Adults of both sexes can be contaminated with pathogenic microbes from filthy larval breeding sites such as latrines, rotting garbage, and poultry litter. It is important, therefore, to exclude *Fannia* species from areas where human food is prepared or consumed. Nonetheless, *Fannia* species generally pose less of a health hazard than house flies because *Fannia* species rarely land and feed on human food.

In the western United States, *Fannia* species in the *benjamini* group commonly are attracted to human sweat and mucus. One species in this group, *F. thelaziae*, is a developmental vector for the mammalian eyeworm *Thelazia californiensis*. Definitive hosts of this nematode include deer, canids, horse, rabbit, sheep, and black bear; people are rare, accidental hosts. Females of *T. californiensis* live in their host's lachrymal ducts where they cause mild irritation and ophthalmia (Soulsby, 1965). Eggs are shed and hatch in eye fluids. First-stage larvae are ingested by eye-feeding flies, penetrate the midgut, and develop further in the fly's haemocoel. After two to four weeks extrinsic incubation, infectious third-stage nematodes exit the fly's mouthparts when the vector feeds on another host.

The little house fly and the latrine fly have been involved in cases of intestinal, aural, and urinary myiasis of people. Most of the cases are thought to have arisen from eggs laid on clothing or bedding soiled with human feces.

Garbage Flies (*Hydrotaea* spp.)

Garbage flies and their larvae are common around municipal garbage dumps, compost sites, poultry houses, and dairies. As occurs with other flies from these kinds of environments, garbage flies can be contaminated with microbial pathogens. However, garbage flies are more sedentary than house flies, and are far less inclined to enter buildings and contaminate human food.

Sweat Flies (*Hydrotaea* spp.)

Very little is known about the medical importance of sweat flies in North America. Females of six North American species, including *Hydrotaea meteorica* and *H. scambus*, are persistent in their attempts to obtain perspiration and secretions from the eyes, nostrils, lips, and other parts of their hosts. Because sweat flies are exophilic and occur most frequently in wooded areas, they are encountered by people in wooded parks, golf courses, and similar outdoor habitats. Except for their annoyance, sweat flies are not regarded as medically significant.

VETERINARY IMPORTANCE

Muscid flies affect the health and comfort of domestic and wild animals. Domestic hosts include cattle, sheep, goats, horses, dogs, pigs, and poultry. Wildlife hosts include bison, elk, deer, moose, and rabbits in North America, and other mammals elsewhere in the world. Muscid flies cause discomfort, injure skin, affect growth and thriftiness, and transmit pathogenic viruses, bacteria, helminths, and cestodes. Repeated feedings in localized areas of skin can lead to secondary infections and scabs on the ears, legs, back, and other body regions of affected animals. Feeding by nonbiting muscid flies can retard the healing of wounds caused by biting arthropods and other agents. Muscid fly larvae also can be involved in cases of secondary myiasis.

Animals display a variety of aversive responses to attack by biting and nonbiting flies. As examples, horses will stamp and switch their tails, dogs will cower under cover, and cattle in a herd will mill together, bunched in a rosette formation with tails outward (Figure 16.16). Frequencies of these and other aversive behaviors increase with fly density (Mullens et al. 2006). Aversive behaviors can be disruptive and interfere with the handling, working, and showing of animals.

Biting muscid flies cause host vital signs to elevate (Schwinghammer et al., 1986a, 1986b). These changes, if prolonged, may be accompanied by changes in water and nitrogen balance. The net effect can be to reduce the amount of metabolic energy available for growth and lactation, and to reduce the efficiency with which animals convert their feed into animal products. Livestock owners recognize a condition called **fly worry**, where animals appear irritated and generally unthrifty.

Economic effects of muscid flies on livestock and poultry industries of the United States are substantial. Estimates indicate that the stable fly and the horn fly alone cause annual losses of $1.3 billion in reduced yields and increased production costs for beef and dairy industries (Drummond, 1987). Losses attributable to the face fly are $123 million annually, resulting from the role of this fly in the epizootiology of bovine **pinkeye**. Costs incurred to manage the house fly and other filth flies around livestock and poultry operations have not been estimated. They are no doubt substantial and will probably increase as suburban development continues to expand into traditionally agricultural lands (Thomas and Skoda, 1993).

House Fly (*Musca domestica*)

This cosmopolitan fly is often the most abundant insect where livestock, poultry, or companion animals are housed. Adults occur on virtually all substrates surrounding the animals, including feed, feces, vegetation, and the walls and ceilings of buildings. Adults also occur directly on animals, where they feed on available blood, sweat, tears, saliva, and other body fluids. In response to the fly annoyance, animals flap their ears, shake their heads, and avoid pen locations where flies are particularly abundant. Beyond these behavioral symptoms, however, house flies appear to cause no measurable harm. Even when present in large numbers, house flies cause little or no adverse effects on animal growth or feed conversion in cattle, pigs, and other animals. Thus houseflies have much less impact on these animals than on the health and comfort of people living in the vicinity.

House flies can be significant mechanical vectors of microbial pathogens. The adults feed on feces and manure and foul their environment with fly specks. These habits degrade the appearance of facilities and contribute to microbial contamination of eggs (Figure 16.17) and milk at points of production. House flies may also be important in the spread of

Figure 16.16 Holstein cattle bunching in response to attack by flies. (Photo by Roger D. Moon)

Figure 16.17 Chicken egg speckled with vomit and feces from house fly (*Musca domestica*). (Photo by Ralph Williams)

animal pathogens and genes for antibiotic resistance among animal production facilities (Ahmad et al., 2007; Hald et al., 2007; Macovei and Zurek, 2006; Rochon et al., 2005; Schurrer et al., 2004).

House flies are also developmental hosts for *Habronema muscae* and *Draschia megastoma*, two spirurid nematodes that cause gastric and cutaneous forms of habronemiasis in horses. In gastric infections, female worms invade the mucosa of the horse stomach and lay eggs that eventually pass out in the feces. Fly larvae become infested by ingesting these eggs. First-stage nematode larvae pass through the maggot's midgut into the haemocoel, and subsequently metamorphose into infectious third instars while the maggot metamorphoses into an adult. After the fly emerges, the infectious third-stage larvae migrate through the thorax and eventually reach the mouthparts. A new gastric infestation in a horse can arise as nematodes exit the mouthparts of flies that feed around the hosts' mouth, or if the horse ingests an infected fly in its feed. The nematodes eventually mature and become established in the mucosa. A new cutaneous infestation occurs if an infested fly feeds on the host's skin. Cutaneous infestations are a dead end for these nematodes because larvae in skin do not develop to maturity. Bush flies and bazaar flies are also hosts for *H. muscae* and *D. megastoma*. Further details of the worms' life cycles and the development of habronemiasis are provided in Soulsby (1965).

The house fly is also a developmental host for a chicken tapeworm, *Choanotaenia infundibulum*. The prevalence of this tapeworm is greatly reduced where chickens are housed in elevated cages, which prevents the birds from eating infected fly larvae and pupae.

House fly larvae have been recorded in cases of secondary wound myiasis. Females attracted to purulent wounds can feed and oviposit, and subsequent larvae feed on wound discharges and retard healing. Cases have been reported from nearly all species of domestic animals.

Bush Fly (*Musca vetustissima*)

As adults, the bush fly aggregates around large mammals, including cattle and horses. These nonbiting dung flies feed on facial and urogenital secretions and on serum and blood from wounds. Irritation by feeding flies can lead to skin lesions around the eyes and vulva, particularly of horses, and can retard wound healing. Annoyance by bush flies can induce animals to bunch and mill about, but the economic consequences are not documented. Studies in Australia in the first half of the 1900s suggested that the bush fly is a developmental host for the equine parasites *Habronema muscae* and *D. megastoma*. Their extrinsic life cycle in the bush fly is the same as in the house fly.

Face Fly (*Musca autumnalis*)

Adult face flies are conspicuous on cattle, bison, and horses (Figure 16.18), swarming around their heads and feeding at eyes, faces, and wounds. Hosts respond by blinking their eyes, flapping their ears, shaking their heads and switching their tails. However, modest numbers of face flies do not appear to affect the thriftiness of grazing dairy and beef cattle. Experimental dairy herds protected with repellents grazed as much, grew as fast, and produced as much milk as unprotected herds. In other experiments with beef cattle, steers in screen cages with populations of face flies consumed as much feed and grew as fast as steers in cages without flies.

Figure 16.18 Face flies (*Musca autumnalis*) on the head of a horse. (Photo by Gary R. Mullen)

The face fly is much more important as a vector of bovine and equine pathogens. Of great concern to North American cattle producers is pinkeye (Figure 16.19). This eye disease, also known as infectious bovine keratoconjunctivitis (IBK), is caused by the bacterium *Moraxella bovis*. Symptoms include reddened conjunctiva, excessive tearing, photophobia, opacity, and ulceration of the cornea. Pinkeye may involve one or both eyes, and is most frequent among calves of white-faced breeds. Expenses associated with pinkeye involve surveillance and treatment of affected animals, and retarded growth and blindness in cases that go undetected.

Face flies are mechanical vectors of *M. bovis*, as evidenced by isolation of the pathogen from face flies collected near infected cattle. In laboratory studies, viable bacteria have been recovered from the tarsi, mouthparts, diverticula, and regurgitant of flies exposed to bacterial cultures several hours earlier. Thus, *M. bovis* can acquire the bacterium from cattle, and the bacterium can remain viable for several hours on and in contaminated flies. Face flies also may create avenues of infection when they scarify the host conjunctiva while feeding. Furthermore, by stimulating host bunching, face flies may contribute to direct eye-to-eye spread

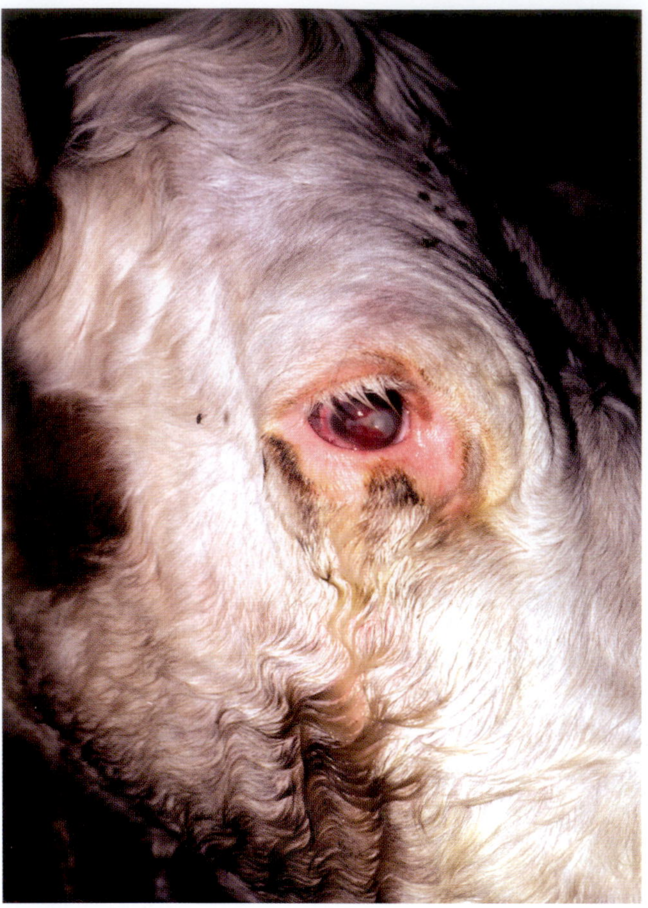

Figure 16.19 Pinkeye in a Hereford cow. Note opacity of cornea, reddened conjunctiva, and tearing below eye. (Photo by R. Moon)

of the bacterium among herdmates. Although the face fly is potentially important in the spread of *Moraxella*, it is not necessary that the face fly be present to have outbreaks of the disease. Pinkeye was known to occur in the United States at least 50 years before face fly was introduced.

The face fly is also a developmental host for several spirurid nematodes. Eyeworms in the genus *Thelazia* live in the lachrymal ducts of horses, cattle, and other mammals. In North America, the face fly transmits *T. lacrymalis* among horses, and *T. gulosa* and *T. skrjabini* among cattle. Fortunately, infections are benign in both hosts. The life cycles of these nematodes are similar to that of *T. californiensis*, which is transmitted to humans and other animals by *Fannia thelaziae*. The filarial nematode *Parafilaria multipapillosa* infests domestic and wild equids in southern Europe, northern Africa, the Middle East, central and southern Asia, and South America (Soulsby, 1965). It is transmitted in Russia by *Haematobia atripalpis*, but vectors elsewhere have not been identified.

In Sweden and South Africa, the face fly and closely related species are intermediate hosts of another filarial nematode, *Parafilaria bovicola*. This worm causes bovine parafilariasis, also known as green muscle disease, named after the appearance of subcutaneous green carcass lesions trimmed at slaughter. Following mating, female nematodes become established under the skin of the back or sides of infected hosts. There they create bleeding points, holes in the skin that exude blood and serum that attract nonbiting flies. Appearance of bleeding points coincides with the presence of vectors in the spring. Microfilariae in the exudates are ingested by the flies, penetrate the midgut and undergo development for two to three weeks in the fly haemocoel. During this extrinsic incubation period, the nematode metamorphoses to the infectious third stage and then migrates through the thorax to the fly's proboscis. New hosts become infected when the fly feeds on another animal. In a new host, *P. bovicola* takes about 10 months to reach the host's back where it matures to the adult. For further details on the biology of this parasite, see Bech-Nielsen et al. (1982).

In South Africa where *P. bovicola* is endemic, the worms are transmitted by three endemic *Musca* species, all in the subgenus *Eumusca*. Four other *Musca* species in other subgenera are not competent vectors for reasons that are unclear. *Parafilaria bovicola* was recently introduced into Sweden and France where the South African vectors are absent. In the worm's new range, it is transmitted exclusively by the face fly, itself a member of *Eumusca*. If *Parafilaria* were to be imported into North America, its establishment and spread would be almost certain because the face fly is already present.

Stable Fly (*Stomoxys calcitrans*)

This cosmopolitan species attacks most large mammals, including domestic cattle, horses, donkeys, dogs, swine, sheep, goats, and camels. Wild hosts include Bovidae, Cervidae, Equidae, Canidae, and Felidae, both in their native ranges and in zoos. The stable fly often is abundant where mammals are confined indoors and outdoors. Confined animals attract and sustain immigrating adults, and organic debris associated with the animals promotes local breeding.

Stable flies attack large animals on the legs (Figure 16.20), sides, back, and belly, whereas small ruminants and dogs are attacked most frequently on the legs, head, and ears. Individual stable flies typically feed once per day, and remain on their host for two to five minutes, just long enough to obtain a blood meal. Engorged flies can be found on nearby vegetation, fences, and walls. Stable fly bites are painful, and localized wounds can coalesce to form scabs that are slow to heal, especially when aggravated by host scratching and rubbing. Such lesions commonly occur on the tips of dogs' ears and elsewhere on other hosts where hair is short or naturally parted.

Behavioral responses to fly attack include a variety of aversive behaviors and physiological responses. Attacked animals are likely to stamp and kick their legs, which makes dairy cows difficult to milk and horses difficult to groom and show. Unrestrained

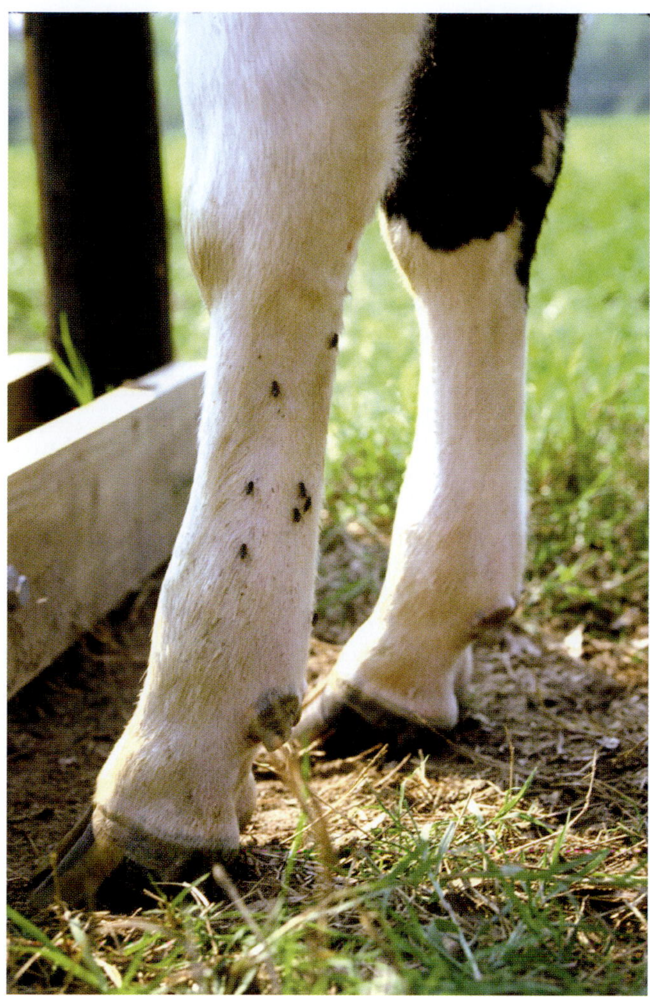

Figure 16.20 Stable flies (*Stomoxys calcitrans*) on the foreleg of a Holstein calf. (Photo by Gary R. Mullen)

cattle, horses, and small ruminants commonly will bunch when attacked. This behavior, combined with leg stamping and tail switching, is a clear indicator of stable fly activity (Mullens et al., 2006).

Experiments with penned beef cattle have shown that irritation by stable flies causes cattle to consume less feed, to grow more slowly, and to convert feed into body mass less efficiently. These effects are greater when weather is hot and humid, presumably because bunching in response to fly attack interferes with the ability of the animals to dissipate excess heat. From an economic perspective, stable flies increase beef production costs because affected cattle require more time and more feed to reach slaughter weight. It is likely that stable flies similarly affect growing beef calves and lactating dairy cows. As a general guideline, economic losses in feedlots are likely to occur whenever the average number of stable flies per foreleg is three or more (Catangui et al., 1997). Stable fly control usually is warranted when bunching, stamping, and tail switching are excessive.

The stable fly is not an important vector of animal pathogens. Experimental evidence has shown that it is possible for the stable fly to mechanically transmit retroviruses that cause equine infectious anemia in horses and bovine leukosis in cattle. In nature, however, the role of stable flies as vectors of these agents is negligible. Tabanid flies are far more important than stable flies in the spread of equine infectious anemia virus, and with bovine leukosis virus, transplacental transmission; and transmission during vaccination, tattooing, and rectal palpation are much more important in establishing infection than biting flies.

The stable fly also is a developmental vector for *Habronema microstoma*, a spirurid nematode that causes gastric and cutaneous forms of habronemiasis in horses throughout the world. The gastric form of infestation is benign, whereas the cutaneous form presents conspicuous granular lesions known as summer sores. Onset of summer sores coincides with the stable fly breeding season. These diseases, the parasite that causes them, and the roles played by the stable fly in their transmission are very similar to the situations with *D. megastoma*, *H. muscae*, and the house fly (Soulsby, 1965).

Horn Fly (*Haematobia irritans irritans*) and Buffalo Fly (*Haematobia irritans exigua*)

Of all parasitic arthropods, these two biting dung flies (Figure 16.21) may have the greatest effects on the health and productivity of cattle. Both sexes of the flies feed frequently each day, consuming an average of 10 μl of blood per fly per day (Kuramochi, 1985). At this rate, a cow with an exceptionally large population of 3,000 flies would lose about 30 ml of blood each day, a small amount given that the blood volume of an adult cow is about 25 liters. Nonetheless, the bites are painful and irritating, and feeding lesions become cosmetic defects in tanned and dyed leather. Feeding by the buffalo fly on zebu cattle in Australia can lead to scabs on their hosts' withers and faces. Infested hosts

Figure 16.21 Heavy infestation of horn flies (*Haematobia irritans*) on pastured cow. (Courtesy of the University of Georgia)

react to horn flies and buffalo flies by licking their backs, twitching their flanks, switching their tails, and kicking at their bellies with their hind legs. These defensive behaviors usually suggest an animal is being attacked by the horn fly or another biting fly.

Studies in the United States and Canada demonstrate that control of horn flies on mother cows can lead to a substantial increase in the average daily growth rate of nursing calves (Drummond, 1987). Similarly, growth rates of yearling stocker cattle and lactation rates of dairy cows may increase following control (Jonsson and Mayer, 1999). However, the size of the benefit of horn fly control has varied among studies, perhaps due to differences in densities of flies, degree of control, and presence of other biting flies and internal parasites. Benefits also may vary with weather, availability of forage, and growth potential determined by cattle genotype, age, and condition when flies are present.

Increases in animal performance following horn fly control make sense in light of animal metabolism, behavior, and energy budgets. When attacked by horn flies, stanchioned steers have elevated heart beats, respiratory rates, rectal temperatures, urine production, and urine nitrogen concentration (Schwinghammer et al., 1986b). Pastured steers switch tails more frequently, spend less time grazing, and spend more time walking and resting during the day. These metabolic and behavioral responses suggest that horn flies increase the amount of energy spent by cattle in defending themselves against flies, leaving less energy available for growth. With nursing calves, the response to horn fly control is likely to be reflected in increased milk production by their dams; horn flies occur mainly on the cows and only incidentally on their calves.

Progressive management programs for beef cattle usually rely on static thresholds to judge if control of the horn fly will be economically justified. Measured as an average number of horn flies per animal side, recommended thresholds range from 25 flies per side in Alberta, Canada, to 100 in Nebraska and 200 in Texas. When densities exceed these thresholds, it is likely that increases in calf or steer growth rates in response to fly control during the fly season will more than pay for the cost of treating the cattle, whatever method is used.

The horn fly is a developmental vector for *Stephanofilaria stilesi*, a spirurid nematode that causes stephanofilariasis in cattle. This is a form of granular dermatitis that occurs mainly on the belly, scrotum, prepuce, and udder. This nematode is most prevalent in the western United States and Canada, but is also recorded from cattle in the Old World. Mature *S. stilesi* occur in the skin. First-stage larvae are acquired by feeding horn flies, and the nematodes metamorphose to the third stage in the fly's haemocoel before being introduced into the definitive host when the fly feeds at a later time. Extrinsic incubation is about three weeks. Other species of *Stephanofilaria* occur in the Old World where they are thought to be transmitted by other species of muscid flies (Soulsby, 1965).

Sweat Flies (*Hydrotaea* spp.)

A few species of these flies feed on large mammals. In Europe, the sheep head fly (*H. irritans*) swarms and feeds at the faces of sheep, cattle, and deer. Affected sheep develop a condition known as head fly disease. Irritated sheep scratch, rub, and open wounds that are further aggravated by the flies. The disease is worst among horned animals of open-faced breeds lacking wool on their faces and among flocks grazing in wooded pastures. Larvae of sheep head fly are sparsely dispersed in soil and decomposing plant litter in grasslands and forests. Larvae of other species have been recorded from undisturbed cattle dung pats (Robinson and Luff, 1979).

The sheep head fly and a complex of other European sweat flies also feed at cow teats. Circumstantial evidence suggests that teat-feeding sweat flies may be mechanical vectors of *Actinomyces* (formerly *Corynebacterium*) *pyogenes*, the putative cause of summer mastitis in pastured dairy cattle. It is likely that *Hydrotaea* species are secondary vectors of this pathogen. The importance of sweat flies as vectors of these pathogens in North America is not well documented.

PREVENTION AND CONTROL

Three general approaches are used to avoid or reduce problems caused by muscid flies: (1) prevention of breeding; (2) killing adults before they cause harm or produce offspring; and (3) exclusion of adults with screens and other barriers. Prevention of breeding can be either indirect, by making candidate media unavailable or unsuitable for survival of preadult stages, or direct by killing immatures before they can develop to adults. A variety of methods can be used to accomplish these objectives (Drummond et al., 1988).

The best approach is to use several methods simultaneously in an integrated pest management program to achieve desired levels of control in poultry houses, stables, and dairies (Axtell, 1986). For example, sanitation and surveillance of adult abundance commonly are used in combination. When densities exceed the tolerance threshold, sanitation can be increased and adulticides can be used to keep flies below intolerable densities. Choices among alternative practices are determined by effectiveness against the target insect, practicality in a given situation, costs of the practices in materials and labor, and environmental acceptability.

Emphasis should be placed on source reduction wherever possible. Housing for people or animals should be designed to limit accumulation of fly breeding media. Particular attention should be given to locations where human and other animal feces, domestic garbage, and rotting animal feed accumulates. The crucial first step to preventing enteric diseases is to prevent filth flies from breeding near human communities. The best defense is a closed sewage system or privy that will exclude ovipositing flies from reaching human excrement. Curtis (1989) presents designs of privies that do not require running water.

Facilities should be designed to minimize the labor required to maintain adequate sanitation. In livestock and poultry housing, lanes, alleys and pens where manure can collect should be made easy to scrape. Feed and water should be provided in separate areas, if possible. Straw bedding for animals is particularly difficult to handle and is a notorious source of filth flies, so alternatives such as sawdust, sand, or washable mats should be considered.

In practice, even well-designed facilities have residual places in corners, around feeders, or along fence lines where organic debris can accumulate and fly breeding can occur. These places should be inspected regularly. Waste disposal should involve proper burial, spreading in a thin layer (<3 cm) on open fields, submersion in water, or aerobic composting. Compost piles must be turned frequently to keep the material hot and in a state of active fermentation. Special attention should be given to seepage areas that can form at the margins of compost piles if the material is not contained in bunkers with vertical sides.

Many beneficial organisms such as predators, parasites, and natural competitors occur in the breeding media of muscid flies. These natural biological control organisms kill developing fly eggs, larvae, and pupae. The faunas in poultry litter and feedlot manure are best known (Rueda and Axtell, 1985; Axtell, 1986). Important groups include nymphs and adults of predatory mites, larvae, and adults of predatory beetles, predatory third-instar larvae of *Hydrotaea* spp. and *Muscina* spp., and adults and larvae of parasitic wasps. The latter group, called **parasitoids**, can be particularly effective. Once a female wasp finds a host, she drills into the host and deposits one or more eggs (Figure 16.22). The offspring eventually consume the host and emerge as adults. Pupal parasitoids are species that attack and emerge from host pupae, whereas larval-pupal parasitoids attack larvae and emerge from pupae. The parasitoids most frequently encountered in poultry and cattle manure are pupal parasitoids in the genera *Muscidifurax* and *Spalangia* (Pteromalidae). Other genera and families of parasitic wasps and beetles are prominent in dung pats, and these are mainly larval-pupal parasitoids.

Figure 16.22 Parasitoid wasp (*Muscidifurax zaraptor*), female, ovipositing into puparium of house fly. (Photo by V. Cervenka, University of Minnesota)

Populations of beneficial insects and mites can be favored by keeping potential fly breeding media as dry as possible. Soil should be sloped to drain water away from possible breeding areas, and waterers should be kept in good repair. Certain species of parasitoids are available commercially, and these can be released to augment natural populations. However, the cost effectiveness of releasing parasitoids in commercial poultry and livestock facilities has been questioned (Thomas and Skoda, 1993).

In emergencies, larvicides can be sprayed directly into infested breeding media to kill fly larvae before adults emerge. Alternatively, larvicides can be administered to animals as feed additives or boluses. The active ingredients in these formulations pass through the animal's digestive system to create an insecticidal residue in feces or soiled bedding. Limitations of feed-through larvicides are that they are effective only against flies breeding in feces and bedding but not in other substrates, and that some larvicides can disrupt natural biological control. Whatever application method is used, larvicidal residues need to be considered when disposing of treated media.

Management of adult flies is accomplished mainly with traps and adulticides (Drummond et al., 1988). Inside closed buildings, house flies and *Fannia* species can be killed with sticky traps, light traps, sugar- and pheromone-based insecticidal baits, and adulticides formulated as knockdown or residual sprays. The same methods can be effective against adult stable flies, except for baits, which do not attract this blood-feeding species. The use of space sprays of ingredients with short half-lives, such as synergized pyrethrins, can be effective when applied as mists or fogs in closed spaces. These materials have a rapid, knock-down effect on flies contacted directly with the mist droplets. In contrast, residual sprays of more persistent insecticides such as the pyrethroids and some organophosphates can be applied as coarse sprays to structural surfaces. These formulations provide a more prolonged effect because the residues remain toxic to flies that later walk or rest on the treated surfaces. In outdoor situations, residual sprays should be directed at fly resting sites such as building walls, fence lines, and vegetation where flies seek shelter during hot weather. To limit costs and to retard development of insecticide resistance, residual sprays should be used sparingly and only when necessary.

Traps are generally effective in closed environments, but they can be overwhelmed by immigration from outside sources. Options in outdoor environments are more limited. Walk-through traps (Figure 16.23) can be used to collect and kill muscid flies from

Figure 16.23 Cow emerging from walk-through trap used to control flies on cattle. Note baffles in side and fabric draped over exiting cow. (Photo by H. J. Meyer)

pastured cattle. These traps are most effective against the horn fly in situations where host animals are forced to pass through the traps on a daily basis. This is accomplished by placing traps in an entryway of a fenced enclosure surrounding water or feed supplement, or in the doorway of a milking parlor.

Materials formulated as topical insecticides can be applied directly to animals. A variety of compounds can be applied as sprays, pour-ons, dusts, or wipe-ons. Some of the compounds are formulated into plastic, slow-release ear tags, whereas others can be dispensed in self-applicators such as oilers, back rubbers, and dust bags (Figure 16.24). To be most effective, self-applicators should be maintained in areas where animals are forced to use them on a daily basis.

Topical insecticides usually are ineffective against house flies, stable flies, and face flies because these species spend so little time directly on animals. In contrast, topical insecticides are more effective against adult horn flies because they remain continuously on their hosts. Insecticidal ear tags were adopted widely in the cattle industry in the 1980s because they were inexpensive, required little labor to apply at turnout for spring grazing, and provided several months of effective horn fly control. However, their success was a mixed blessing. Pyrethroid insecticides were the active ingredients in the first widely used tags. Unfortunately, horn flies developed resistance to those compounds in the first three to four years of tag use. In response, manufacturers substituted organophosphate insecticides into new tags. Some reports indicate that horn flies in selected regions of North and South America are beginning to evolve resistance to the organophosphate compounds as well.

Figure 16.24 Dust bag positioned in fence line, used for self-application of topical insecticides for control of flies on cattle. (Photo by H. J. Meyer)

Chemical repellents can be applied by hand directly to individuals to provide temporary relief from muscid flies and other pests. Repellents function mainly by interfering with host-finding behaviors, and less so as toxicants. A variety of formulations are marketed mainly for use on horses and companion animals. Effectiveness varies with weather and level of animal activity. Few repellents are effective against muscid flies for more than a few hours.

The most effective way to prevent the house fly and the other filth flies from entering buildings is adult exclusion with door and window screens. Double doorways or positive-pressure air doors can further reduce fly entry into closed structures. These approaches are appropriate at entrances to restaurants, hospitals, and other institutions where flies cannot be tolerated. To prevent household infestations of overwintering adult face flies, cracks and crevices around doors, windows, and eaves should be sealed tightly with caulk. Residual insecticides can be sprayed on the sunny sides of buildings to intercept the flies as they arrive in autumn.

REFERENCES AND FURTHER READING

Ahmad, A., Nagaraja, T. G., & Zurek, L. (2007). Transmission of *Escherichia coli* O157: H7 to cattle by house flies. *Preventive Veterinary Medicine, 80*, 74–81.

Axtell, R. C. (1986). *Fly Control in Confined Livestock and Poultry Production.* Ciba-Geigy, Tech. Monogr.

Bech-Nielsen, S., Bornstein, S., Christensson, D., Wallgren, T. B., Zakrisson, G., & Chirico, J. (1982). *Parafilaria bovicola* (Tubangui 1934) in cattle: Epizootiology-vector studies and experimental transmission of *Parafilaria bovicola* to cattle. *American Journal of Veterinary Research, 43*, 948–954.

Bruce, W. G. (1964). The history and biology of the horn fly, *Haematobia irritans* (Linnaeus); with comments on control. North Carolina Agricultural Experiment Station Technical Bulletin No. 157.

Burger, J. F., & Anderson, J. R. (1974). Taxonomy and life history of the moosefly, *Haematobosca alcis*, and its association with the moose, *Alces alces shirasi*, in Yellowstone National Park. *Annals of the Entomological Society of America, 67*, 204–214.

Catangui, M. A., Campbell, J. B., Thomas, G. D., & Boxler, D. J. (1997). Calculating economic injury levels for stable flies (Diptera: Muscidae) on feeder heifers. *Journal of Economic Entomology, 90*, 6–10.

Chavasse, D. C., Shler, R. P., Murphy, O. A., Huttly, S. R. A., Cousens, S. N., & Akhtar, T. (1999). Impact of fly control on childhood diarrhea in Pakistan: Community-randomised trial. *Lancet, 353*, 22–25.

Chillcott, J. G. (1961). A revision of the Nearctic species of Fanniinae (Diptera: Muscidae) from North America. *Canadian Entomologist*, (Suppl. 14), 1–295.

Cohen, D., Green, M., Block, C., Slepon, R., Ambar, R., Wasserman, S. S., et al. (1991). Reduction of transmission of shigellosis by control of houseflies (*Musca domestica*). *Lancet, 337*, 993–997.

Curtis, C. F. (1989). *Appropriate Technology in Vector Control.* Boca Raton, FL: CRC Press.

Drummond, R. O. (1987). Economic aspects of ectoparasites of cattle in North America. In W. H. D. Leaning & J. Guerrero (Eds.), *The Economic Impact of Parasitism in Cattle.* Proceedings of a Symposium, XXIII World Veterinary Congress, Montreal.

Drummond, R. O., George, J. E., & Kunz, S. E. (1988). *Control of Arthropod Pests of Livestock: A Review of Technology.* Boca Raton, FL: CRC Press.

Edwards, F. W., Oldroyd, H., & Smart, J. (1939). *British Blood-Sucking Flies.* British Museum (Natural History).

Elzinga, R. J., & Broce, A. B. (1986). Labellar modifications of Muscomorpha flies (Diptera). *Annals of the Entomological Society of America, 79*, 150–209.

Emerson, P. M., Lindsay, S. W., Walraven, G. E. L., Faal, H., Bogh, C., Lowe, K., et al. (1999). Effect of fly control on trachoma and diarrhea. *Lancet, 353*, 1401–1403.

Frankel, G., & Bhaskaran, G. (1973). Pupariation and pupation in cyclorrhaphous flies (Diptera): Terminology and interpretation. *Annals of the Entomological Society of America, 66*, 418–422.

Fuller, H. B. (1913). Myths of American history. *Munsey's Magazine, May*, 278–284.

Greenberg, B. (1971). *Flies and Disease, Vol. I. Ecology, Classification and Biotic Associations.* Princeton, NJ: Princeton University Press.

Greenberg, B. (1973). *Flies and Disease, Vol. II. Biology and Disease Transmission.* Princeton, NJ: Princeton University Press.

Hald, B., Sommer, H. M., & Skovgard, H. (2007). Use of fly screens to reduce *Campylobacter* spp. introduction in broiler houses. *Emerging Infectious Diseases, 13*, 1951–1953.

Hall, R. D. (1984). Relationship of the face fly (Diptera: Muscidae) to pinkeye in cattle: A review and synthesis of the relevant literature. *Journal of Medical Entomology, 21*, 361–365.

Huckett, H. C. (1954). A review of the North American species belonging to the genus *Hydrotaea* Robineay-Desvoidy. *Annals of the Entomological Society of America, 47*, 316–342.

Huckett, H. C., & Vockeroth, J. R. (1987). Muscidae. In J. F. McAlpine (Ed.), *Manual of Nearctic Diptera* (Vol. 2, pp. 1115–1131). Chapter 105, Monograph 28. Ottawa: Agriculture Canada.

Hughes, R. D. (1977). The population dynamics of the bushfly: The elucidation of population events in the field. *Australian Journal of Ecology, 2*, 43–54.

James, M. T. (1947). *The Flies that Cause Myiasis in Man.* U.S. Department of Agriculture, Miscellaneous Publication 631.

Jonsson, N. N., & Mayer, D. G. (1999). Estimation of the effects of buffalo fly (*Haematobia irritans exigua*) on the milk production of dairy cattle based on a meta-analysis of literature data. *Medical and Veterinary Entomology, 13*, 372–376.

Krasfur, E. S., & Moon, R. D. (1997). Bionomics of the face fly, *Musca autumnalis. Annual Review of Entomology, 42*, 503–523.

Kuramochi, K. (1985). Studies on the reproductive biology of the horn fly, *Haematobia irritans* (L.) (Diptera: Muscidae). II. Effect of temperature on follicle development and blood meal volume of laboratory-reared flies. *Applied Entomology and Zoology, 20*, 264–270.

Lindsay, D. R., Stewart, W. H., & Watt, J. (1953). Effect of fly control on diarrheal disease in an area of moderate morbidity. *Public Health Reports, 68*, 361–367.

Macovei, L., & Zurek, L. (2006). Ecology of antibiotic resistance genes: Characterization of enterococci from house flies collected in food settings. *Applied Environmental Microbiology, 72*, 4028–4035.

McAlpine, J. F., Peterson, B. V., Shewell, G. E., Teskey, H. J., Vockeroth, H. J., & Wood, D. M. (1981). *Manual of Nearctic Diptera, Vol. 1. Monograph 27.* Agriculture Canada, Ottawa.

McAlpine, J. F., Peterson, B. V., Shewell, G. E., Teskey, H. J., Vockeroth, H. J., & Wood, D. M. (1987). *Manual of Nearctic Diptera, Vol. 2. Monograph 28.* Agriculture Canada, Ottawa.

Morgan, C. E., & Thomas, G. D. (1974). *Annotated Bibliography of the Horn Fly, Haematobia irritans (L.), Including References on the Buffalo Fly, H. exigua (de Meijere), and Other Species Belonging to the Genus Haematobia.* Agricultural Research Service, Miscellaneous Publication No. 1278.

Morgan, C. E., & Thomas, G. D. (1977). *Annotated Bibliography of the Horn Fly, Haematobia irritans (L.), Including References on the Buffalo fly, H. exigua (de Meijere), and Other Species Belonging to the Genus Haematobia.* Suppl. I. Agricultural Research Service, Miscellaneous Publication No. 1278.

Morgan, C. E., Thomas, G. D., & Hall, R. D. (1983a). Annotated bibliography of the stable fly, *Stomoxys calcitrans* (L.), including references on other species belonging to the genus *Stomoxys. Missouri Agricultural Experiment Station Bulletin No., 1049*, 1–190.

Morgan, C. E., Thomas, G. D., & Hall, R. D. (1983b). Annotated bibliography of the face fly, *Musca autumnalis* (Diptera: Muscidae). *Journal of Medical Entomology*, (Suppl. 4), 1–25.

Mullens, B. A., Lii, K. S., Mao, Y., Meyer, J. A., & Peterson, N. G. (2006). Behavioral responses of dairy cattle to the stable fly, *Stomoxys calcitrans*, in an open field environment. *Medical Veterinary Entomology, 20*, 122–137.

Paterson, H. E., & Norris, K. R. (1970). The *Musca sorbens* complex: The relative status of the Australian and two African populations. *Australian Journal of Zoology, 18*, 231–245.

Petersen, J. J., & Greene, G. L. (1989). Current status of stable fly (Diptera: Muscidae) research. *Entomological Society of America, Miscellaneous Publications No., 74*, 1–54.

Pickens, L. G., & Miller, R. W. (1980). Biology and control of the face fly, *Musca autumnalis* (Diptera: Muscidae). *Journal of Medical Entomology, 17*, 195–210.

Pont, A. C. (1991). A review of the Fanniidae and Muscidae (Diptera) of the Arabian Peninsula. In W. Buttiker & F. Krupp (Eds.), *Fauna of Saudi Arabia* (Vol. 12, pp. 312–365). Basle, Switzerland: Pro Entomologia c/o Natural History Museum.

Robinson, J., & Luff, M. L. (1979). Population estimates and dispersal of *Hydrotaea irritans* Fallen. *Ecological Entomology, 4*, 289–296.

Rochon, K., Lysyk, T. J., & Selinger, L. B. (2005). Retention of *Escherichia coli* by house fly and stable fly (Diptera: Muscidae) during pupal metamorphosis and eclosion. *Journal of Medical Entomology, 42*, 397–403.

Rueda, L. M., & Axtell, R. C. (1985). *Guide to Common Species of Pupal Parasites (Hymenoptera: Pteromalidae) of the House Fly and Other Muscoid Flies Associated with Poultry and Livestock Manure.* North Carolina Agricultural Research Service, Technical Bulletin No. 278.

Schurrer, J. A., Dee, S. A., Moon, R. D., Rossow, K. D., Mahlum, C., Mondaca, E., et al. (2004). Spatial dispersal of porcine reproductive and respiratory syndrome virus-contaminated flies after contact with experimentally infected pigs. *American Journal of Veterinary Research, 65*, 1284–1292.

Schwinghammer, K. A., Knapp, F. W., Boling, J. A., & Schillo, K. K. (1986a). Physiological and nutritional response of beef steers to infestations of the stable fly (Diptera: Muscidae). *Journal of Economic Entomology, 79*, 1294–1298.

Schwinghammer, K. A., Knapp, F. W., Boling, J. A., & Schillo, K. K. (1986b). Physiological and nutritional response of beef steers to infestations of the horn fly (Diptera: Muscidae). *Journal of Economic Entomology, 79*, 1010–1015.

Skidmore, P. (1985). The biology of the Muscidae of the world. *Dr. W. Junk Series Entomologica, 29*, 1–550.

Soulsby, E. J. L. (1965). *Textbook of Veterinary Clinical Parasitology, Vol. 1. Helminths.* Philadelphia, PA: F. A. Davis, Co.

Stone, A., Sabrosky, C. W., Wirth, W. W., Foote, R. H., & Coulson, J. R. (1965). *A Catalog of the Diptera of America North of Mexico, Agriculture Handbook No. 276.* Washington, DC: U.S.D.A.

Thomas, G. D., & Skoda, S. R. (Eds.). (1993). *Rural Flies in the Urban Environment?.* North Central Regional Research Publication No. 335, Lincoln: Institute of Agriculture and Natural Resources, University of Nebraska.

United States Centers for Disease Control and Prevention. *Morbidity and Mortality Weekly Report: Summary of Notifiable Diseases*, Updated annually (http://www.cdc.gov/mmwr/summary.html).

Watt, J., & Lindsay, D. R. (1948). Diarrheal disease control studies. I. Effect of fly control in a high morbidity area. *Public Health Reports, 63*, 1319–1334.

West, L. S. (1951). *The House Fly, Its Natural History, Medical Importance, and Control.* Ithaca, NY: Comstock.

West, L. S., & Peters, O. B. (1973). *An Annotated bibliography of Musca Domestica Linnaeus.* Folkstone, UK: Dawsons.

West, S. K., Emerson, P. M., Mkocha, H., Mchiwa, W., Munoz, B., Bailey, R., et al. (2006). Intensive insecticide spraying for fly control after mass antibiotic treatment for trachoma in a hyperendemic setting: a randomized trial. *Lancet, 368*, 596–600.

Zumpt, F. (1973). *The Stomoxyine Biting Flies of the World.* Stuttgart: G. Fischer Verlag.

Chapter

17

Tsetse Flies (Glossinidae)

William L. Krinsky

Tsetse flies (Figure 17.1) are obligate blood-sucking flies of medical and veterinary importance because they transmit trypanosomes that cause African sleeping sickness in humans and nagana in livestock. Fossil tsetse flies in the Florissant shale of Colorado in the western United States indicate that this family was present in the Western Hemisphere as recently as 26 million years ago. Tsetse flies now occur over 10 million square kilometers in the tropical and subtropical regions of sub-Sahelian Africa (ca. 15°N to 26°S). Recently, isolated populations of two species of tsetse flies were observed in southwestern Saudi Arabia (Elsen et al., 1990).

Tsetse (pronounced *tsé-tsee*) commonly is used as both a singular and plural term to denote one or more individuals or species of these flies. Although the origin of the name is obscure, it was used as early as the nineteenth century by the Tswana people living along the edge of the Kalahari Desert. *Tsénse*, the Mozambique word for "fly," as well as other similar sounding African names meaning "fly," are apparently onomatopoetic terms derived from imitations of the unique buzzing sound made by the adult flies (Austen, 1903).

Tsetse generally are considered one of the greatest factors affecting the course of economic and social development in Africa. The morbidity and mortality caused by African sleeping sickness continues to be significant. Nagana, which has stifled agricultural productivity for decades, still stands as a major deterrent to the development of animal agriculture on that continent.

There is an extensive literature on tsetse, but a few monographic works provide particularly useful introductions to the field. The classic work by Buxton (1955) reviews the natural history of tsetse and provides a detailed historical account of the diseases associated with it. Mulligan (1970) includes an historical perspective in addition to an overview of the biology of tsetse and its parasites, pathology of these parasites in humans and domestic animals, treatment, and control. The epidemiology, pathology, and treatment of the trypanosomiases are reviewed by Maudlin et al. (2004). The historical, social, and economic effects of tsetse in five different African regions are extensively reviewed by Ford (1971), and the impact of tsetse on African rural development is discussed by Jordan (1986). A comprehensive monograph was written by Leak (1999).

TAXONOMY

Tsetse were formerly included in their own subfamily, Glossininae, or the Stomoxyini of the Muscidae because of the resemblance of tsetse to the stable fly and other biting muscids. However, because of their unique antennal structure, tsetse are now placed in their own family, Glossinidae. The reproductive and morphological similarities of tsetse to the keds and other hippoboscid flies has led to placement of Glossinidae within the Hippoboscoidea (McAlpine, 1989). Glossinidae includes the single genus *Glossina* with 31 species and subspecies (23 species, six of which are further divided into 14 subspecies) (Jordan, 1993). *Glossina* means "tongue fly," in reference to its prominent proboscis. Keys to species and subspecies are included in Jordan (1993). *Glossina* species are arranged in three subgenera (*Austenina, Nemorhina*, and *Glossina*) that correspond roughly with groups of species found in different ecological settings. The subgenera often are cited by their group names, each designated by one of the better known species in each subgenus; that is, the *fusca* group (*Austenina*), the *palpalis* group (*Nemorhina*), and the *morsitans* group (*Glossina*). Species in the *fusca* group are found most often in forested habitats, such as rain, swamp, and mangrove forests. Species in the *palpalis* group occur among vegetation around lakes and along rivers and streams. The *morsitans* group, with the exception of the forest-dwelling *G. austeni*, occurs in open country and is found most often in dry thickets, scrub vegetation, and areas of savanna woodland (commonly composed of *Berlinia, Isoberlinia*, and *Brachystegia* species).

The geographical distributions of the three taxonomic groups are shown in Figure 17.2. The *palpalis* group, which includes *G. palpalis, G. tachinoides, G. fuscipes,*

Figure 17.1 Adult tsetse fly (*Glossina* sp.) on rabbit. (Courtesy of The Rockefeller Foundation)

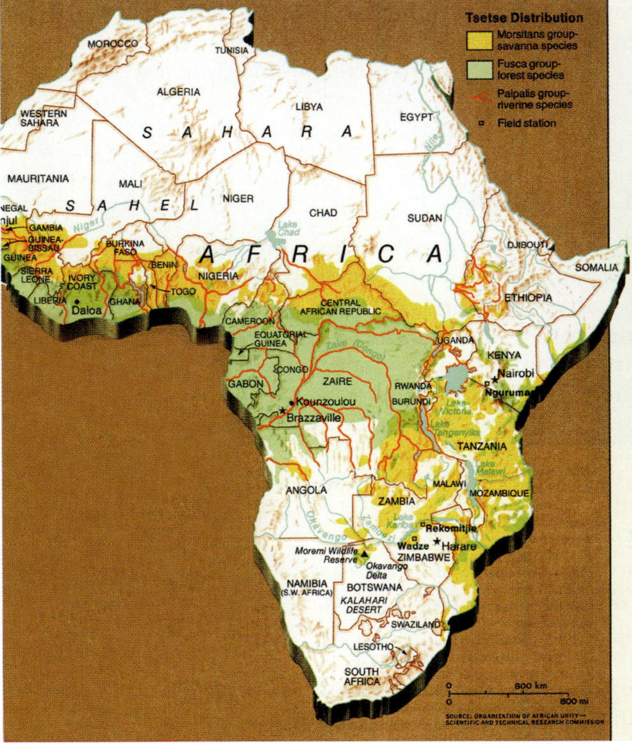

Figure 17.2 Distribution of the following tsetse species groups in Africa: *morsitans* group (savanna, shown in yellow); *fusca* group (forest, shown in green); *palpalis* group (riverways, shown in red). (Used with permission of the National Geographic Society)

and two less well-known species, occurs primarily along watercourses in western and central Africa. The *morsitans* group of savanna species, which includes *G. morsitans*, *G. pallidipes*, *G. longipalpis*, *G. swynnertoni*, and *G. austeni*, is primarily central and southeastern in distribution. The *fusca* group, which includes *G. fusca*, *G. tabaniformis*, *G. medicorum*, *G. longipennis*, *G. brevipalpis*, and eight other species, is found in forested areas that overlay most of the western and central African distribution of the *palpalis* group.

MORPHOLOGY

Glossina species are tan or brown flies, which range in length from 6 to 14 mm, excluding the proboscis. Members of the *fusca* group, which is considered phylogenetically primitive, are the largest, being 9.5 to 14 mm long. The *palpalis* and *morsitans* group species are small to medium in size, about 6.5 to 11 mm long. Species in the *palpalis* group generally have a uniformly dark brown abdominal tergum, and the dorsal aspect of each hind tarsal segment is dark brown or black. Species in the *morsitans* group usually have dark segmental bands on the abdomen, and only the distal segments of the hind tarsi are darkened dorsally.

Tsetse adults are characterized by several distinctive morphological features. These include the shape of the proboscis, the position and branching of the fringe on the arista of their antenna, and the wing venation and folding pattern. The swollen, bulbous base of the proboscis that lies under the head is very different from the angled and thinner bases of the proboscises of the Stomoxyini. When the fly is not feeding, the proboscis extends directly forward between the palps in front of the head (Figure 17.1). The proboscis (Figure 17.3) is composed of two elongate, stylet-like mouthparts: the labrum and hypopharynx. The stylets are protected ventrally by the labium. The labellum at the tip of the labium is armed with teeth for cutting into host skin. The labrum, bounded by the hypopharynx and the labium, forms the food canal through which blood is drawn as the fly feeds (Figure 17.3). The hypopharynx has a hollow central portion that forms the salivary canal through which saliva is secreted into the feeding site.

The three-segmented antennae arise on the frons, just below the ptilinal suture, as in muscoid flies. The first segment is very small; the second is at least two to four times larger than the first and generally about as long as wide; the third is very elongately oval to pea pod-shaped and bears the distinctive arista. The arista has a conspicuous fringe of hairs along its dorsal

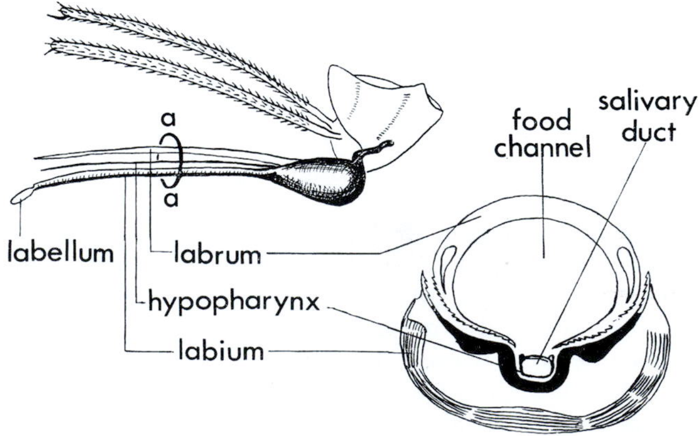

Figure 17.3 Details of proboscis and palps of *Glossina* species, with palps separated from the proboscis (left) and cross-section about midway along length (at a) of proboscis (right). (From Potts, 1973; after Newstead et al., 1924)

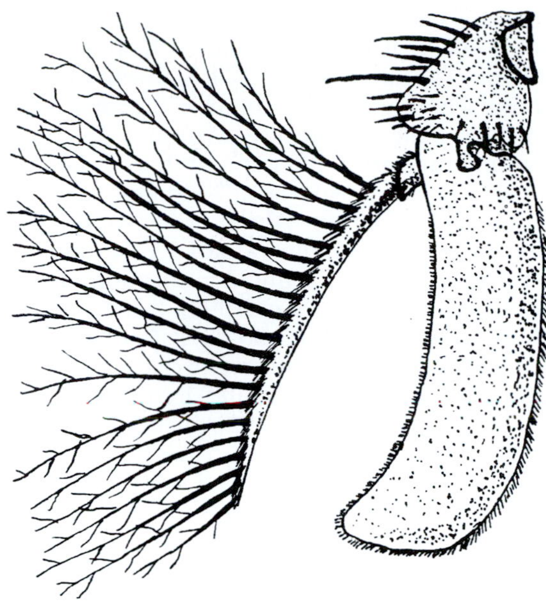

Figure 17.4 Antenna of *Glossina fuscipleuris*, showing plumose setae on arista, characteristic of tsetse fly adults. (After Zumpt, 1936)

surface, and these hairs have small branch hairs, which are not found on any other aristate fly (Figure 17.4). The large brown or reddish eyes are separated in both sexes and comprise most of the posterior portion of the head.

The base of the thorax is only slightly wider than the width of the head across the eyes. The thorax tapers to a waist-like constriction at the level of the scutellum. The wings vary from hyaline to dusky depending on the species. They are folded scissors-like over the back, with the tips extending slightly beyond the end of the abdomen. The tsetse wing has a distinctive, hatchet-shaped discal cell. This is formed by the fourth (medial) vein that curves anteriorly to produce a wing cell (discal cell) resembling the elongate handle of a hatchet attached to a thickened blade (Figure 17.5).

The base of the abdomen is about equal in width to the head and thorax. Male tsetse can be readily distinguished from females by the presence of a prominent button-like hypopygium on the ventral surface of the posterior of the abdomen (Figure 17.6). The morphological details of both male and female genitalia

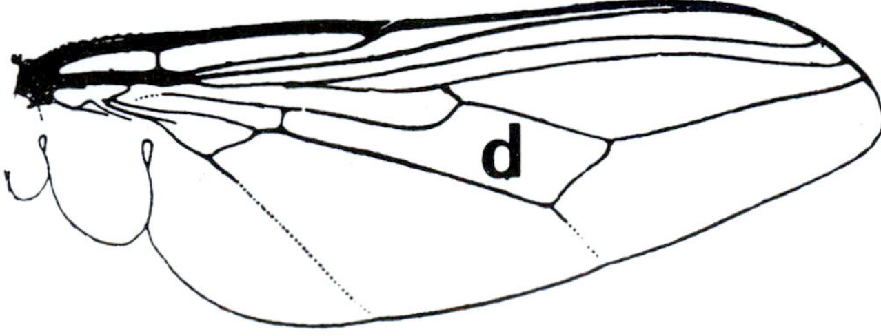

Figure 17.5 Wing structure and venation of adult tsetse flies (*Glossina*), showing characteristic hatchet-shaped discal cell (d). (Modified from Potts, 1973)

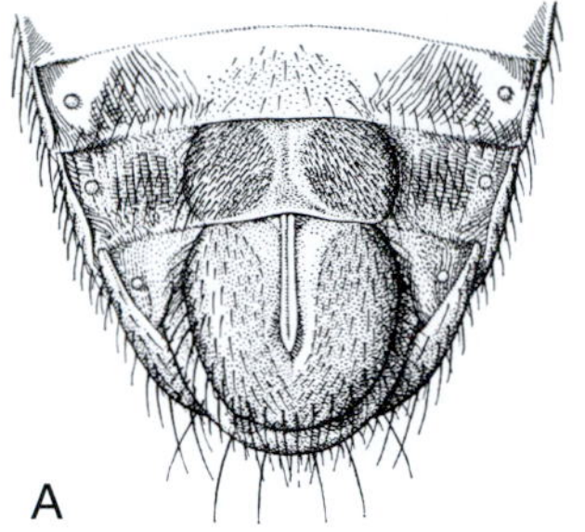

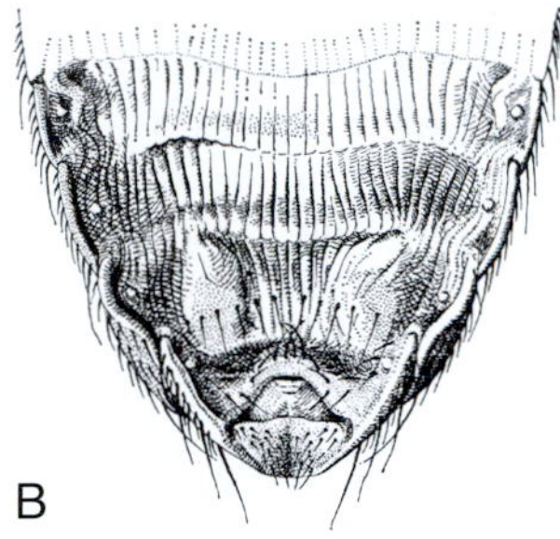

Figure 17.6 Posterior ends of abdomens of *Glossina* adults, ventral view, showing sexual differences. (A) Male, with knob-like appearance of hypopygium drawn up into the abdomen; (B) Female, lacking knob-like hypopygium. (Potts, 1973)

provide taxonomic characters that are used for distinguishing tsetse species.

The alimentary tract is adapted for hematophagy. The strong musculature of the pharynx forms a cibarial pump used for imbibing blood. The proventriculus secretes a peritrophic membrane that lines and protects the midgut. The midgut contains symbionts (Enterobacteriaceae) that provide compounds associated with Vitamin B metabolism. Females devoid of these symbionts are unable to reproduce. The reproductive tract of the female fly is unusual compared to that of most oviparous dipterans; it is very similar, however, to the reproductive system seen in the other hippoboscoid families (Hippoboscidae, Streblidae, Nycteribiidae). Each of the two ovaries has only two ovarioles. The ovarian ducts form a common duct that expands to form a uterus in which one embryo at a time is retained during development. Associated with the uterus are a pair of specialized branched glands that produce nutrients for the developing tsetse larva. Because of this function, they are commonly called milk glands.

LIFE HISTORY

Tsetse adults of both sexes bite vertebrates and imbibe blood, the fly's only food. Unfed females are sexually receptive about one day after emergence from the puparium, whereas male tsetse require several blood meals before they are fully fertile. At close range, the male visually locates a female and, once contact is made, a pheromone in the cuticle of the female stimulates mating. The female endocrine system will induce ovulation only if mating lasts longer than an hour. Sperm are transferred in a spermatophore and are stored in the female's spermathecae. Once inseminated, the female remains fertile for life, but females will mate more than once.

About nine days after copulation, the first ovulation of a single egg occurs, and sperm are released through the spermathecal duct by dilation of a sphincter. The egg is positioned with the micropyle against the spermathecal duct opening, allowing for fertilization. The fertilized egg moves posteriorly into the uterus, where hatching occurs about four days later. The first-instar larva uses an "egg tooth" on its anterior end to rupture the chorion of the egg. The larva is retained in the uterus where it is held against the uterine wall by a supporting structure called the **choriothete**. Secretions from the milk glands pool around the larval mouth and are easily ingested. The developing larva molts twice within the uterus, becoming a second-instar larva one day after hatching, and a third instar about 1.5 days later. The third-instar larva is fully developed about 2.5 days after the second molt, at which time it occupies most of the female's abdomen and is about equal in weight to the rest of the female's body. The female continues to ingest blood, albeit in progressively smaller amounts, as the larva grows.

About nine days after ovulation, the fully developed, third-instar larva is deposited on the ground by the female. Shortly thereafter, the female ovulates again (within as little as one hour after larviposition). A well-nourished female, after this first larviposition, will deposit a third-instar larva about every seven to 11 days, depending on the ambient temperature. The average interval for all tsetse species is nine to 12 days. The ovaries, and the ovarioles in each, alternate in releasing a single egg at each ovulation, starting with the right ovary. Follicular relicts seen in dissected flies reflect the ovulation history of individual females and can help in estimating the longevities of wild-caught female flies.

Tsetse females generally live for about 20 to 40 days, but may have a maximum life span of three to four months. The males typically mate only once or twice during their lives, and apparently survive in the wild for two to three weeks (Glasgow, 1963; Potts, 1973). More accurate estimates of longevity of some species have become possible with newly developed fluorescence techniques that measure accumulated pteridines in tsetse tissues (Leak, 1999).

The full-size third-instar larva is cream-colored and oval-shaped. It measures 3 to 8.5 mm in length, depending upon the species, and has two prominent black knobs at the posterior end (Figure 17.7). These conspicuous knobs are respiratory lobes that function only during, and for a short time following, intrauterine life. The active larva is deposited on the ground, usually in loose soil shaded by trees or other vegetation. The larva, which is negatively phototactic and positively thigmotactic, quickly burrows to 1.5 to 2.5 cm below the soil surface. Within a few hours of deposition, the larval integument hardens and darkens, and the third-instar larva becomes an immobile brown to black puparium. About two to four days later, molting occurs within the puparial case and a true pupa is formed. A key for the identification of puparia to species is given by Jordan (1993).

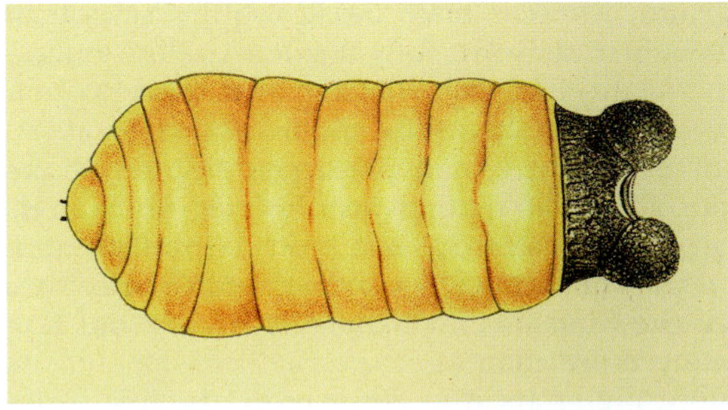

Figure 17.7 Larva of tsetse fly (*Glossina morsitans*), with distinctive knob-like respiratory lobes at posterior end. (Newstead et al., 1924)

Adult flies emerge about 30 days after formation of the puparium. As in all other cyclorrhaphan flies, eclosion involves the breaking of the circular puparial cap by a ptilinum. The teneral adult pushes its way to the surface of the substrate where it rests for a short time, usually less than an hour, before it can fly. The teneral fly does not fully harden and the thoracic flight muscles do not completely develop until about nine days later, after the fly has had at least a few blood meals (Glasgow, 1963; Potts, 1973; Lehane, 2005).

The low reproductive rate in tsetse is compensated by the extreme protection given to each larva by the female, by virtue of the viviparous mode of development. However, the low reproductive rate makes the impact of any loss of female flies greater than in species that mass produce eggs.

BEHAVIOR AND ECOLOGY

Although tsetse are found over an area estimated to be at least 10 million square kilometers, the distribution of the flies is discontinuous. The areas they inhabit may extend to several hundred kilometers and form what traditionally have been called fly belts. Within these belts are patches of forest and bush where environmental conditions, such as shade and high humidity, are suitable for tsetse survival and reproduction. Local residents living in their vicinity often are aware of these areas of high tsetse concentrations. One or more species of tsetse usually are found where woody vegetation is at least 4.5 m high. In many cases, Africans can predict the presence of particular species of tsetse by observing the types of shrubs and trees that occur in a given habitat. Rather than representing direct associations of tsetse species with specific plants, the plant communities observed probably reflect differences in a variety of microhabitat factors that directly affect the survival of tsetse, such as the water content of the soil, the availability of mammalian hosts, and the occurrence of natural predators. Remotely sensed satellite data that provide identification of different types of vegetation over large geographic areas have been used to estimate distributions of different species of tsetse (Rogers et al., 1994).

Tsetse flies are restricted in northern Africa by desert conditions, and in southern Africa by the deserts of Namibia and Botswana and their lower ambient temperatures. Tsetse live in areas where the annual rainfall is at least 0.5 m per year. They require temperatures between 16° and 40°C, with optimal development occurring from 22 to 24°C; for this reason, the flies are not found at elevations above approximately 1500 m.

The potential difficulty of males and females finding each other in low-density populations is apparently overcome in some species by the attraction of both males and females to large moving animals. Mating usually occurs on or in the vicinity of a host. Once they have mated, however, females and males tend to be more attracted to stationary animals. Tsetse feed on an array of hosts including reptiles and mammals, but rarely birds. Individual species and species groups have definite host preferences. These preferences are of considerable epidemiological significance in relation to the reservoir hosts of the pathogenic trypanosomes transmitted by the flies to humans and domestic animals.

Host preferences vary among tsetse species. Members of the *palpalis* group feed mostly on reptiles (e.g., crocodiles and monitor lizards) in their riverine and lacustrine habitats and on bushbuck, oxen, and occasionally smaller mammals and humans that visit these watering spots. Species of the *morsitans* group, living in scattered patches of vegetation in open country, feed mostly on the mammals of the savanna. In addition to showing a strong preference for warthogs, the savanna-dwelling tsetse feed on a diversity of mammalian species including bushbuck, buffalo, giraffe, kudu, rhinoceros, duiker, bushpig, and oxen. The one forest-dwelling species in the savanna group, *G. austeni*, feeds almost exclusively on suids such as bushpigs and forest hogs. The *fusca* group feeds on a variety of host species including bushbuck, buffalo, and other cattle, giraffe, rhinoceros, elephant, hippopotamus, bushpig, river hog, porcupine, aardvark, and even the ostrich (Lehane, 2005). Humans are not the preferred hosts of any of these fly species. In some cases, people in the vicinity of other mammals will actually repel tsetse, whereas hungry flies will suck blood from humans who enter tsetse habitat.

Host attraction and host recognition are mediated by visual and olfactory cues. Their vision enables tsetse to react to a herd of moving cattle as far away as 180 meters. The attraction of both male and female flies to large moving objects accounts for the common occurrence of tsetse attacking occupants of trucks and tourists in jeeps on safaris. Tsetse species in the *morsitans* group, living in open spaces, have shown the greatest attraction to host odors. Certain tsetse species are attracted to components of ox breath, such as carbon dioxide, acetone and octenol, and phenols found in mammalian urine (Willemse and Takken, 1994). Host odors have been shown to be attractive to tsetse from distances up to 100 meters away.

Although tsetse feed mostly in the daylight, adult feeding does occur at night, as in the case of *G. medicorum* that feeds on the nocturnal aardvark. Tsetse rarely fly for more than 30 minutes a day and are known to disperse up to about 1 km/day. They spend most of their time resting on vegetation. Most species are found below 3 meters where they rest on wood surfaces of trees during the day and on leaves at night. Recently engorged flies mostly rest with their heads directed upward, allowing excess water to be excreted away from their bodies. Hungry flies often rest horizontally, with the dorsal side down. When seeking a host, *Glossina* species can fly very rapidly, reaching speeds above 6.5 m/sec (ca. 25 km/hr).

Host behavioral differences may account in part for the feeding preferences shown by tsetse species. Mammals that are heavily fed upon and irritated by other kinds of biting flies sometimes react with strong defensive behaviors, such as muscle twitching and rapid tail movements, that repel tsetse. Tsetse are more prone to start feeding on calm animals and often seem to prefer to feed on a host that is in the shade. The latter may be an adaptation to avoid reaching lethal body temperatures and may serve as a means of avoiding predation during feeding, or just after, when the fly takes off and alights a short distance from its host (Glasgow, 1963).

Upon landing on a host, a tsetse fly grips the skin with its claws and applies pressure to the skin surface with its proboscis. The teeth and rasps on the labellum aid the labium in penetrating the skin. Strong back-and-forth movements of the fly's head cause the labium to rupture one or more capillaries in the skin, resulting in a hemorrhage within the bite site. The blood is rapidly sucked into the food canal of the labrum by the negative pressure produced by the cibarial pump in the fly's head. Saliva is pumped intermittently through the salivary canal of the hypopharynx into the wound. The saliva contains anticoagulant substances, including an antithrombin and an apyrase that inhibits platelet aggregation. As in other haematophagous insects that have anticoagulins in their saliva, tsetse flies presumably benefit from these substances by their role in increasing blood flow at the feeding site, thereby reducing feeding time and the vulnerability of the fly to host defenses. If a tsetse fly is disturbed while penetrating the skin, it will rapidly withdraw the proboscis and fly away; however, once feeding begins, a tsetse is less likely to react to movement and physical stimuli that would normally cause it to escape (Glasgow, 1963; Lehane, 2005).

Tsetse engorge fully within about one to 10 minutes, the length of time depending in large part on how quickly the labium is able to rupture a capillary. The actual penetration of host skin occurs quite rapidly, whether it involves the thick hide of a rhino or a thin artificial feeding membrane. During feeding, a clear fluid is excreted from the anus. A tsetse fly imbibes about 0.03 ml of blood and when fully engorged weighs about two to three times its unfed body weight (Figure 17.8). The ungainly fully-fed insect slowly flies from the host (ca. 1.6 m/sec) and lands on a nearby tree or other substrate. There the fly continues to excrete anal fluid as a means of ridding itself of excess water, while concentrating its blood meal. About 40% of the blood meal weight is lost in the first 30 minutes after feeding. The rapid loss of excess fluid that begins during blood-feeding helps

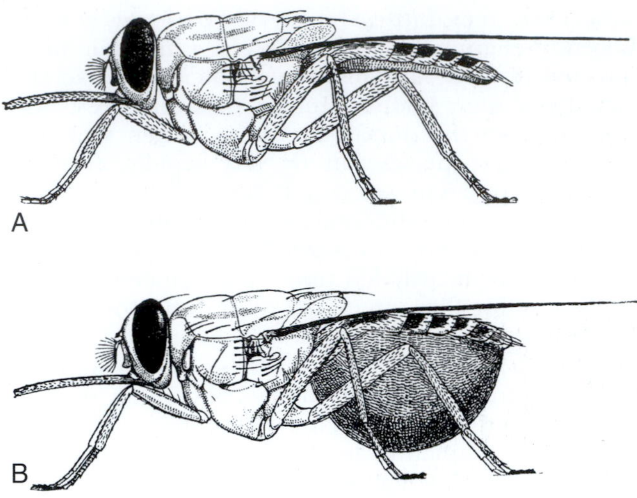

Figure 17.8 Tsetse fly (*Glossina morsitans*), female. (A) Before feeding; (B) after feeding (Austen, 1903).

the fed fly regain flight agility as quickly as possible. This helps it evade the defensive movements of the host and destruction by predatory flies, other arthropods, and vertebrate predators. A larva being carried by a female is especially vulnerable to loss just after the female has taken on the extra burden of a blood meal and has lost much of her maneuverability. Complete digestion of the blood meal occurs by about 48 hours. The interval between blood meals varies, with a mean of three to five days.

Because tsetse feed exclusively on blood, their main source of energy is derived from protein. They depend on the amino acid proline as the major energy source for flight. The energy is produced by the partial oxidation of proline to alanine in flight muscle (Glasgow, 1963; Lehane, 2005). The unique metabolism of tsetse flies enables them to live in dry habitats in which blood is their sole source of nutrition and water, and to develop massive thoracic flight muscles that allow them to fly with heavy loads of blood and/or an internally developing larva.

PUBLIC HEALTH IMPORTANCE

For centuries, tsetse flies have had a great impact on human health in Africa, both as efficient vectors of trypanosomes that cause extreme human suffering in the form of African sleeping sickness and as vectors of trypanosomes that kill nonnative animals, preventing the development of animal domestication. The exclusion of cattle from most of the African continent has prevented their use as draught animals, as sources of human dietary protein from meat and milk, and as sources of manure and transport. Human disease cases have increased since 1994 to levels not seen since before 1950, and domestic animal disease continues to inhibit agricultural productivity and economic development.

African Sleeping Sickness

Trypanosomes were first associated with African sleeping sickness in 1903, after the British Tsetse Fly Commission, composed of David Bruce, David Nabarro, Aldo Castellani, and others, was sent to Africa to investigate outbreaks among British colonists. The clinical presentation of the disease was different in west African and east African countries, with the East African form being much more severe. The trypanosomes isolated from the two forms of African sleeping sickness were named to reflect their geographic distributions (Figure 17.9). The West African form was named *Trypanosoma gambiense* for Gambia, where it was originally seen. It now occurs from west Africa through central Africa to northern Uganda, and south to Zaire and northern Angola. The East African form was named *Trypanosoma rhodesiense* for Rhodesia, now Zimbabwe. This trypanosome now occurs in Sudan, Uganda, Kenya, Tanzania, southern Congo, Zambia, Malawi, Mozambique, as well as Zimbabwe and Botswana.

The two species of trypanosomes pathogenic to humans are morphologically identical and microscopically indistinguishable from *Trypanosoma brucei*, the species that causes some of the trypanosomal disease seen in domestic animals in Africa. For taxonomic purposes, *T. gambiense* and *T. rhodesiense* are considered subspecies of *T. brucei*, namely *T. brucei gambiense* and *T. brucei rhodesiense* (Hoare, 1972). However, for convenience in the medical and parasitological literature, the subspecies names often are retained as species designations, as in the original *T. gambiense* and *T. rhodesiense*.

The term **African sleeping sickness** refers to the drowsy to comatose condition of acutely ill patients. This disease occurs in two clinical forms that result from differences in the pathogenicity of the trypanosomes transmitted by tsetse flies in West and East Africa. West African trypanosomiasis is a chronic illness involving mental deterioration and progressive

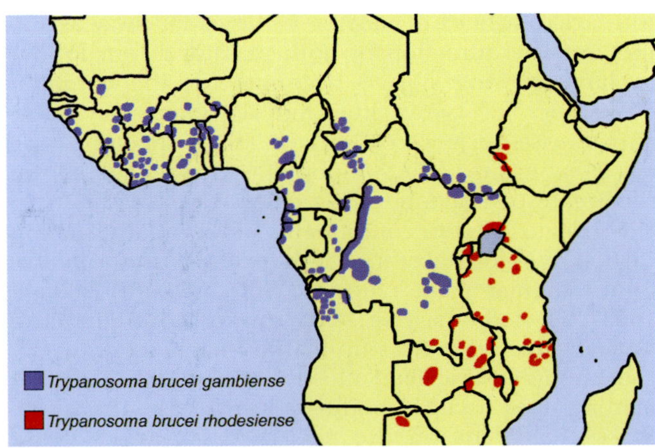

Figure 17.9 Geographic distribution of foci of African sleeping sickness. (Data from World Health Organization, 1996; modified from Peters and Pasvol, 2007)

weakening, and East African trypanosomiasis is an acute, rapidly fatal, febrile illness characterized by myocarditis and meningoencephalitis. The West African disease has been known since the fourteenth century when it was described by the Arab writer al-Qalqashandi. The first medical description of the disease was written in 1734 by John Atkins, who had served as a British Navy surgeon on slave ships traveling from West Africa to the West Indies. His description of the clinical signs of the disease were accurate, but his ideas concerning its cause created an extremely prejudiced view of those afflicted with the "Negro lethargy." Atkins attributed the disease to the "natural weakness of the brain . . . brought about by lack of use" (McKelvey, 1973).

The first recorded large-scale epidemics of sleeping sickness occurred in the middle to late nineteenth century during the period of active European exploration and colonization of the African continent. Controversy exists over the role of British and other imperialistic activities in either stimulating or just recognizing and alleviating these outbreaks. If nothing else, expanded navigation of waterways such as the Congo River, which led to increased trade and development by late-nineteenth century colonists, facilitated the spread of tsetse and sleeping sickness from western and central Africa to eastern regions of the continent. Although there is strong historical evidence for this pattern of dispersal, the extreme differences in the clinical presentations and pathogeneses of West and East African trypanosomiases lead to the more probable explanation that the geographical forms of the parasite evolved independently as humans were exposed under different ecological conditions to parasites harbored by wild ungulate mammals.

More than three quarters of a million people died from sleeping sickness in Africa between 1896 and 1906. Currently, the number of Africans succumbing annually to trypanosomiasis is about 100,000. An estimated 50 million Africans in 38 countries are at risk of infection by exposure to tsetse flies, with more than 30,000 individuals being infected annually. Imported cases in the United States, mostly involving tourists to East Africa, have numbered only about 15 in the last 25 years, but the risk of serious illness or death is great because of the American medical community's unfamiliarity with the disease.

In both West and East African trypanosomiasis, trypanosomes injected into the skin by a tsetse fly reproduce locally in connective tissue to form a nodule or chancre, called a **trypanoma**.

West African Trypanosomiasis

In the Western form, the skin lesion and associated erythematous swelling occur within a week or so after the bite. The trypanosomes then spread to the lymphatics and the resulting lymphadenopathy in the back of the patient's neck is called **Winterbottom's sign**. This distinctive sign is diagnostic for the disease in a person who has been exposed to tsetse. Urticaria and rashes are common in this chronic form of the disease. The illness progresses as the trypanosomes continue to multiply in the lymphatic and circulatory systems. After months or years, the parasites enter the central nervous system and produce symptoms such as behavioral and personality changes, hallucinations and delusions, drowsiness by day (i.e., sleeping sickness), tremors, and stupor. In untreated patients, stupor often is followed by convulsions and, inevitably, by death.

East African Trypanosomiasis

In the East African form, there is acute onset of fever, headache, and dizziness within a few days after the fly bite. There is usually little or no lymph node involvement. Instead, early circulatory system disease that includes myocardial and pericardial inflammation becomes clinically apparent as tachycardia and arrhythmias. Immune complexes composed of trypanosomal variant antigens and complement-fixing host antibodies stimulate release of proteolytic enzymes. Either the latter, or toxin production, in combination with host autoantibodies, cause damage to red blood cells, brain and heart tissue, and other organs. Anemia, thrombocytopenia, disseminated intravascular coagulation, followed by renal disease, may precede localization of the trypanosomes in the blood vessels of the central nervous system. Hemorrhages, edema, and thrombosis leading to neuronal degeneration are common following inflammation of the cerebral arteries. Damage to brain tissue results in the convulsions and other signs seen in the West African form, but much sooner, with death often occurring in weeks to months. Definite diagnosis of either form of the disease requires observation or isolation of trypanosomes from blood, cerebrospinal fluid, scrapings from a trypanoma, lymph node aspirates, or bone marrow.

Life Cycle of Trypanosomes

The developmental cycles of *T. gambiense* and *T. rhodesiense* in the flies appear to be identical, even though the tsetse vector species often are different. Trypanosomes ingested by a tsetse fly in a blood meal pass into the midgut where their life cycle continues. The relatively short trypanosomes that enter the fly transform into thin procyclic forms, which then multiply and become trypomastigotes in about three to four days. These forms then multiply for about 10 days before they move to the posterior part of the midgut. There, they either pass through the peritrophic membrane, or around its posterior open end, into the ectoperitrophic space. The trypomastigotes move anteriorly in the ectoperitrophic space of the midgut and reach the junction with the proventriculus about five days later. The parasites elongate and penetrate the soft basal ring of the peritrophic membrane to return to the endoperitrophic space. The trypomastigotes, having returned to the endoperitrophic space of the proventriculus, migrate anteriorly through the esophagus and pharynx and enter the proboscis. They then enter the open end of the hypopharynx and migrate posteriorly through the salivary ducts into the salivary glands. In the lumen of the glands and attached to the epithelium, the trypomastigotes transform into epimastigotes, which multiply and form short metatrypanosomes that are infective to vertebrates. The metatrypanosomes are injected by salivarian transmission when the infected tsetse feeds on another vertebrate host.

An alternative trypanosome life cycle has been observed in some flies. In these flies, migration of trypomastigotes from the midgut to the salivary glands involves a different anatomical route. The trypomastigotes that have moved anteriorly in the ectoperitrophic space of the midgut enter the midgut cells, penetrate the wall of the midgut, and enter the hemocoel. From there, they migrate anteriorly and penetrate the haemocoel side of the salivary glands to reach the lumen. Whether the parasites move anteriorly via the gut lumen or the haemocoel, the complete trypanosome cycle in the fly usually takes 11 to 38 days, but may be as long as 80 days (Hoare, 1972; Aksoy et al., 2003).

In nature, very small numbers of tsetse may be infective (e.g., with *T. brucei*, <0.4%) and still maintain high rates of trypanosome transmission. The physiological and ecological factors that attract tsetse flies to hosts enable continued cycling of the trypanosomes between insect and vertebrate hosts. In epidemic situations, tsetse and other bloodsucking flies, such as tabanids and stomoxyines, may transmit trypanosomes from person to person by mechanical transmission.

Epidemiologically, the West African and East African forms of the disease are different. The species of tsetse flies that act as vectors are different. The vertebrate host species and the degree to which humans are part of the natural life cycles of the trypanosomes are different (Table 17.1).

In West Africa, the major vectors of sleeping sickness are tsetse species in the *palpalis* group. The medically most important species in this group include *Glossina palpalis, G. tachinoides*, and *G. fuscipes.* These species are found in shaded forested areas close to rivers, streams, and lakes. In West African trypanosomiasis, transmission to humans usually involves a solely human–tsetse cycle without any nonhuman reservoir hosts, although some domestic and wild animals are known to be infected with *T. gambiense.* Therefore, West African trypanosomiasis is considered an anthroponosis. Humans are infected most often by exposure within a peridomestic environment that encompasses forested waterways near their dwellings. Activities, such as washing, bathing, gathering water for cooking, and wood-gathering for fires and construction purposes, place humans in the riverine habitats of the *palpalis* group of tsetse flies.

In East Africa, the major vectors of sleeping sickness are tsetse species in the *morsitans* group. The medically most important species in this group include *Glossina morsitans, G. pallidipes*, and *G. swynnertoni.* These species

Table 17.1 Tsetse Vectors and Vertebrate Hosts of Human and Animal Trypanosomiases (Hoare, 1972)

Disease	Disease Agent	Vectors	Hosts
West African sleeping sickness	*Trypanosoma gambiense*	*Glossina fuscipes, G.palpalis, G. tachinoides*	Humans
East African sleeping sickness	*Trypanosoma rhodesiense*	*Glossina morsitans, G. pallidipes, G. swynnertoni*	Humans, antelopes (bushbuck, hartebeest), cattle
Nagana	*Trypanosoma brucei*	*Glossina fuscipes, G. longipalpis, G. morsitans, G. palpalis, G. pallidipes, G. tachinoides*	All domestic mammals, antelopes (e.g., impala, hartebeest, wildebeest), warthog, hyena, lion
	Trypanosoma suis	*Glossina brevipalpis, G. vanhoofi*	Suids (domestic pigs, warthogs)
	Trypanosoma congolense	*Glossina morsitans* group; *G. brevipalpis, G. fuscipes, G. palpalis, G. tachinoides, G. vanhoofi*	All domestic mammals, elephant, zebra, antelopes (e.g., impala, hartebeest, duiker, gnu), giraffe, bushpig, hyena, lion
	Trypanosoma simiae	*Glossina austeni, G. brevipalpis, G. fusca, G. fuscipleuris, G. longipalpis, G. morsitans, G. pallidipes, G. palpalis, G. tabaniformis, G. tachinoides, G. vanhoofi*	Domestic pig, warthog, camel, horse, cattle
	Trypanosoma uniforme	*Glossina fuscipes, G. palpalis*	Cattle, goats, sheep, antelopes (e.g., bushbuck, situtunga, waterbuck), buffalo, giraffe
Nagana or Souma	*Trypanosoma vivax*	*Glossina morsitans* group; *G. fuscipes, G. palpalis, G. tachinoides, G. vanhoofi*	Domestic mammals (esp. cattle, horses, mules), wild bovids, zebra, antelopes (e.g., impala, hartebeest, gnu), giraffe, warthog, lion

are found in vegetation such as tall grasses, thickets, and small groups of trees that occur in open savanna. An exception to the vector distinction between East and West Africa is the occurrence of *G. fuscipes* in Uganda and Kenya along the northern shore of Lake Victoria, where outbreaks of East African trypanosomiasis are associated with this *palpalis*-group species. In East African trypanosomiasis, transmission involves a wild animal–tsetse–human cycle. Antelopes, the bushbuck (*Tragelaphus scriptus*), and the hartebeest (*Alcelaphus buselaphus*) are natural reservoirs for the trypanosome. Humans are incidental hosts. Therefore, East African trypanosomiasis is a zoonotic disease that can be designated an anthropozoonosis (Baker, 1974). In East Africa, African men and tourists generally have the highest risk of infection because it is the men, who are the hunters and honey-gatherers, and the tourists on safaris who most often enter the savanna areas where game animals and the *morsitans*-group species thrive. Identification of a gene difference between *T. brucei* and *T. rhodesiense* has allowed rapid differentiation between these parasites found in nonhuman hosts. Recognition that cattle often are infected with *T. rhodesiense* has led epidemiologists to consider domestic livestock also as important reservoirs of human infection (Maudlin, 2004).

The distribution of human trypanosomiasis in Africa is not as widespread as the distribution of tsetse species. The reason is that many tsetse species do not readily feed on humans and that many potential vector species inhabit areas where there is little or no contact between the flies and humans.

Besides transmitting trypanosomes, tsetse flies have other direct, but minor, effects on public health. Some people, who are particularly sensitive to arthropod antigens, develop large skin rashes when bitten, and anaphylactic reactions to tsetse bites are known. However, as with most haematophagous species that are efficient vectors of human pathogens, people who are bitten usually feel little pain or only slight irritation.

VETERINARY IMPORTANCE

Tsetse-borne trypanosomiasis in both domestic and wild nonhuman animals is caused by a number of trypanosome species. Most wild hosts in Africa are immune to these parasites or have inapparent or mild infections. Infections that cause disease in wildlife are rare because wild animals that do develop pathological changes following infection are species that rarely are fed upon by tsetse (Ford, 1971). However, more than 30 species of wild animals native to Africa harbor trypanosomes that are pathogenic when transmitted to domestic animals. The disease associated with any of these infections is called **nagana**.

Nagana

Nagana, which killed camels and horses used by nineteenth-century missionaries, is now considered to have been a major factor in halting the spread of Islam through sub-Saharan Africa. The disease was known to nineteenth-century European explorers in Africa, who similarly lost large numbers of pack animals, such as horses, mules, and oxen. In 1895, David Bruce recognized an association between the disease and tsetse flies. He named the new disease *nagana*, which is a Zulu word for "low or depressed in spirits" (McKelvey, 1973). Bruce also identified the etiologic agent that was later named *T. brucei* in his honor. In 1909, Kleine demonstrated the biological transmission of trypanosomes by tsetse, which led to the elucidation of the life cycle of the trypanosome in the fly by Robertson in 1913. Bruce's earlier observations had involved only short-term mechanical transmission of the parasite.

Nagana continues to have a major impact in preventing the development of commercial domestic animal production over about one-third of the African continent. The scarcity of domestic animals results in a severe lack of animal protein for use as human food, a lack of draught animals for use in crop production, and the absence of manure suitable for use as fertilizer. At present, between 40 and 60 million cattle and millions of sheep, goats, horses, mules, pigs, and camels are at risk of infection in Africa. Unlike African sleeping sickness in which human disease does not occur over the entire distribution of tsetse vectors, nagana occurs wherever tsetse are found, in addition to other areas where infection can be maintained by mechanical transmission by biting flies other than tsetse.

Chronic disease involving anemia and weakness is common (Figure 17.10). Affected animals have reduced muscle mass, pendulous fluid-filled abdomens, scurfy skin, rough coats, and enlarged lymph nodes. Chronic fever and watery diarrhea are common. Most organ systems are infected, and enlargement of the heart, lungs, liver, and spleen often is seen at necropsy. Chronically infected animals are unsuitable for

Figure 17.10 A naturally infected yearling cow sick with nagana (bovine trypanosomiasis), showing stunted growth and characteristic "nagana" pose. (From Murray et al., 1979, with permission of the Food and Agriculture Organization of the United Nations)

use as food or as suppliers of manure for agricultural use. Early death may result from secondary infections. For further information on the clinical pathology of nagana, see Losos and Ikede (1972) and Jubb et al. (1985).

The six species of trypanosomes that cause nagana are listed with some of their wild and domestic animal hosts in Table 17.1. The trypanosomes are identifiable by differences in morphology and antigenicity, and the trypanosome species differ to some extent in the anatomical sites where they develop in the tsetse vector. Hoare (1972) reviewed the morphology and life cycles of the tsetse-borne trypanosomes of Africa. *Trypanosoma brucei*, a parasite of diverse domestic and wild mammals, has a developmental cycle in its tsetse vector identical to that of its human forms, *T. gambiense* and *T. rhodesiense*. It undergoes development during migration from the proboscis to the midgut and subsequently to the salivary glands. The vector cycle of *T. suis*, a parasite of pigs, is almost identical to the latter forms and occupies the same anatomical sites. The vector cycles of *T. congolense*, a parasite of all domestic animals, antelopes and other wildlife, and *T. simiae*, found in pigs, cattle, horses, camels, and warthogs, are almost identical to each other. These trypanosomes are never found in the salivary glands of the fly. Development is restricted to the midgut, proventriculus, esophagus, and the food canal of the labrum from which infective forms pass into a vertebrate host. In tsetse infections with *T. vivax* and *T. uniforme*, parasites of diverse vertebrate hosts (Table 17.1), development of the parasite is even more restricted, being limited to the proboscis where the trypanosomes are found in the labrum and hypopharynx. The only tsetse-borne trypanosome found outside of Africa is *T. vivax*, which also occurs in Central and South America. As in Africa, this trypanosome can be transmitted mechanically among cattle by tabanids and other haematophagous flies in Latin America.

The only nonmammalian trypanosome transmitted by tsetse is *T. grayi*, a parasite of crocodiles. Its development in the fly is restricted to the posterior midgut, hindgut, and rectum. Infective metatrypanosomes are transmitted to crocodiles when they ingest infected flies.

The tsetse vectors of the trypanosomes causing nagana include species of the *palpalis*, *morsitans*, and *fusca* groups (Table 17.1). *Glossina* species of all three groups transmit *T. congolense*, *T. simiae*, and *T. vivax*. Species in both *palpalis* and *morsitans* groups transmit *T. brucei*. Only tsetse species of the *palpalis* group transmit *T. uniforme*, and only *Glossina* species of the *fusca* group transmit *T. suis*. Transmission of the latter two trypanosomes by single vector group species restricts the distribution of *T. uniforme* to areas near waterways, and of *T. suis* to dense forested habitats, where their respective vectors live.

The wild animal hosts listed in Table 17.1 are reservoir hosts for the trypanosomes that cause disease in the domestic animals listed. Recognition of the geographic and ecological distribution of reservoir hosts and specific tsetse vector species can help determine where introduction of domestic animals is most likely to succeed; however, after the wide areas inhabited by the diverse reservoir and vector species are excluded, little habitat remains for the maintenance of healthy domestic animals.

PREVENTION AND CONTROL

Several approaches have been taken to prevent African sleeping sickness and nagana and to control tsetse vectors. These include intensive treatment and isolation of infected human and domestic animal hosts to try to break transmission cycles, use of trypanotolerant animals for agricultural purposes, laboratory research on development of vaccines for human and nonhuman hosts, as well as chemical and ecological attacks on the tsetse flies themselves.

The drugs used to treat human patients have not changed very much since the early 1900s. They include compounds, such as arsenicals, that kill the trypanosomes, but cause severe side effects by interfering with the normal metabolism of the patients. Pentamidine, a synthetic aromatic diamidine, is the first choice drug for treatment of first stage infection with *T. gambiense*. Suramin, a polysulfonated naphthyl urea, is now the drug of choice for the hemolymphatic stage of *T. rhodesiense* human trypanosomiasis. The arsenical melarsoprol, known as Mel B, is the drug of choice for destroying trypanosomes in the central nervous system stage of both forms of the disease. Samorin (isometamidium chloride) is given to cattle as a preventative, and Berenil (diaminazene aceturate) is effective in treating infected domestic animals. Although marketed exclusively for veterinary purposes, the latter compound has been administered to large numbers of patients in Africa and has been effective in the treatment of early stages of both forms of human trypanosomiasis.

Maintenance of noninfected human or domestic animal populations requires regular surveillance for trypanosomes. Examining blood for trypanosomes is still very common. However, use of the recently developed Card Agglutination Trypanosomiasis Test (CATT) for antibodies is proving more efficient for surveillance under field conditions.

The use of trypanotolerant breeds of domestic animals is being studied, and new agricultural practices are being developed that enable Africans to breed native, normally wild, trypanotolerant animals for food and other domestic purposes. Indigenous cattle breeds, such as N'dama, which tolerate nagana trypanosomes without developing disease, are the focus of investigations to determine mechanisms of tolerance, with the hope of selective breeding or genetically engineering new tolerant breeds. Trypanotolerant game animals such as antelopes are being assessed for use in meat production. Trypanotolerant animals require careful handling because stress can trigger disease or death from their infections.

The development of an effective vaccine for domestic animals or humans has been thwarted by the presence of variant surface glycoproteins (VSG) on the trypanosomes that enable the parasites to change their antigenicity in response to each wave of antibodies during the course of infection.

Tsetse populations have been directly attacked with insecticides and, historically, indirectly attacked by destroying tsetse resting and oviposition sites and wild hosts. The direct approach, used in attempts to eradicate the fly, involves aerial spraying and ground spraying from backpacks and trucks. The indirect approach, which generally would not be tolerated today, involved cutting or burning vast areas of vegetation to destroy adult resting sites and puparia, and game hunting to rid large areas of sources of blood for tsetse. Recent development involving building and paving, with its concomitant destruction of vegetation, has had similar effects to those from purposeful clearing of tsetse habitats. In order to be effective, most direct attacks on tsetse flies require frequent repetition of treatments, which are either costly or harmful to the environment. One direct approach that has been considered for general use is to treat cattle with an insecticide when they are run through an acaricide dip to kill ticks. In that way, cattle that are attacked by tsetse can also help to remove flies from the environment.

Another approach to tsetse control is the mass rearing and release of sterile male flies. In experimental tests, eradication has been achieved in fenced grazing areas, as large as several hundred square miles, surrounded by tsetse-free buffer zones. The buffer zones are continually monitored for tsetse, and tsetse eradication is maintained with aerial or ground insecticides or insecticide-impregnated attractant baits to keep fertile flies from migrating into the eradication area. A new African initiative (Pan African Tsetse and Trypanosomiasis Eradication Campaign) has been proposed with the sterile male technique as the key to its success. The realization of such a program has been questioned for various practical and economic reasons. The extreme effort and expense required to maintain tsetse-free zones across the continent of Africa make the program impractical and the risks associated with releasing laboratory-bred flies that are capable vectors are too high for general use in Africa. Rational, cost-effective control measures that have succeeded in the past would appear preferable to any attempt at continent-wide eradication (Rogers and Randolph, 2002).

Any effective means of controlling tsetse flies requires constant monitoring of fly populations. Surveillance for tsetse originally involved African youths ("fly boys") walking around a designated route of about 1.6 km called a fly round and catching flies with hand nets. This method evolved into walking with an ox as bait (ox round), bicycling over longer fly rounds, or riding on the back of trucks and catching tsetse attracted to the vehicles. Because of the low density of tsetse, fly-round surveillance over small areas, sometimes combining hand-netting with sprayer backpacks, became a useful control technique.

Surveillance by adult trapping has a long history involving different trap designs and identification of tsetse attractants. As in many other hematophagous flies, the attraction of tsetse to dark objects and to carbon dioxide, as well as other products of mammalian metabolism, has been put to practical use in designing tsetse traps. Tsetse behavior in relation to trapping methods and host odors has been reviewed by Colvin and Gibson (1992), Willemse and Takken (1994), and Green (1994).

Studies of traps and baits for surveillance have led to improvements that now make control with these devices the method of choice. Much of the work on attractants and trapping over the last 25 years has been done by Glyn Vale and his colleagues in Zimbabwe. His studies and those of others have led to the development of traps of various kinds, such as those resembling hosts (e.g., Morris trap), biconical designs (e.g., Challier and Laveissiére trap), and those with square or rectangular cloth targets. One of the most effective targets is black and blue and baited with attractant components of ox breath or urine. Attractants include acetone, 1-octen-3-ol, and phenols (4-methyl- and 3-n-propyl). The target is designed so it can be used either with an electrocution device or an insecticide. An unattended trap charged with a residual insecticide can be used to remove flies from the environment for 12 to 18 months, long enough to eradicate local populations of tsetse. Other methods being studied involve targets impregnated with insect growth regulators or chemosterilants, and combinations of these substances with tsetse sex pheromone. Former large-scale use of insecticides and animal baits has given way to less costly, nonpolluting, and more efficient selective control with traps or targets. The possibility that traps can be maintained at little cost by local landowners, such as pastoral farmers, removes the need for extensive interference by foreign agencies and places the responsibility for maintenance on those whose economic well-being is directly related to the success of the trapping program.

As in experimental studies of triatomine symbionts, tsetse symbionts are being studied with the goal of engineering their genes to produce antitrypanosomal substances or to eliminate the vector capability of the host flies (Aksoy, 2003).

Natural enemies of tsetse include puparial parasites, such as ants and beetles, over 20 species of wasps, and at least 10 species of bombyliids (*Thyridanthrax* spp.). Predators of adults include spiders, odonates, asilids, sphecid, and vespid wasps. Field studies of predation of *G. pallidipes* puparia in Kenya suggest that more than 20% of all puparia may be killed by predators during their buried, 30-day developmental period. The parasites and predators of tsetse are reviewed by Mulligan (1970), Laird (1977), and Leak (1999).

The presence of tsetse has helped to preserve the wildlife of Africa by preventing domestic animal production and consequent overgrazing that has occurred in some tsetse-free areas. A debate continues between those who view tsetse as an environmental benefit and those who see tsetse flies as a barrier to agricultural and other forms of development considered essential for the future economic, social, and political success of the African continent.

REFERENCES AND FURTHER READING

Aksoy, S. (2003). Control of tsetse flies and trypanosomes using molecular genetics. *Veterinary Parasitology, 115*, 125–145.

Aksoy, S., & Rio, R. V. M. (2005). Interactions among multiple genomes: Tsetse, its symbionts and trypanosomes. *Insect Biochemistry and Molecular Biology., 35*, 691–698.

Aksoy, S., Gibson, W. C., & Lehane, M. J. (2003). Perspectives on the interactions between tsetse and trypanosomes with implications for the control of trypanosomiasis. *Advances in Parasitology, 53*, 1–84.

Austen, E. E. (1903). *A Monograph of the Tsetse-Flies [Genus Glossina, Westwood]*. Longmans & Co. (reprinted 1966 by Johnson Reprint Corp.).

Baker, J. R. (1974). *Epidemiology of African sleeping sickness. In: Trypanosomiasis and Leishmaniasis with Special Reference to Chagas' Disease. Ciba Foundation Symposium, New Series 20.* Amsterdam: Elsevier.

Beard, C. B., O'Neill, S. L., Mason, P., Mandelco, L., Woese, C. R., Tesh, R. B., et al. (1993). Genetic transformation and phylogeny of bacterial symbionts from tsetse. *Insect Molecular Biology, 1*, 123–131.

Buxton, P. A. (1955). *The Natural History of Tsetse Flies.* London: H. K. Lewis & Co., Ltd.

Colvin, J., & Gibson, G. (1992). Host-seeking behavior and management of tsetse. *Annual Review of Entomology, 37*, 21–40.

Elsen, P., Amoudi, M. A., & Leclercq, M. (1990). First record of *Glossina fuscipes fuscipes* Newstead, 1910 and *Glossina morsitans submorsitans* Newstead, 1910 in southwestern Saudi Arabia. *Annales de la Societé Belge de Médecine Tropicale, 70*, 281–287.

Ford, J. (1971). *The Role of the Trypanosomiases in African Ecology—A Study of the Tsetse Fly Problem.* Oxford: Clarendon Pr.

Glasgow, J. P. (1963). *The Distribution and Abundance of Tsetse.* New York: Macmillan Co.

Green, C. H. (1994). Bait methods for tsetse fly control. *Advances in Parasitology, 34*, 229–291.

Hoare, C. A. (1972). *The Trypanosomes of Mammals.* Oxford: Blackwell Sci. Publ,

Jordan, A. M. (1986). *Trypanosomiasis Control and African Rural Development.* London: Longman.

Jordan, A. M. (1993). Tsetse-Flies (Glossinidae). In R. P. Lane & R. W. Crosskey (Eds.), *Medical Insects and Arachnids* (pp. 333–338). London: Chapman & Hall.

Jordan, A. M. (1995). Control of tsetse flies (Diptera: Glossinidae) with the aid of attractants. *Journal of the American Mosquito Control Association, 11*, 249–255.

Jubb, K. V. F., Kennedy, P., & Palmer, N. (1985). Pathology of Domestic Animals. (*Vol. 3*). Academic Press.

Laird, M. (1977). *Tsetse: The Future for Biological Methods in Integrated Control.* Ottawa: International Development Research Centre.

Leak, S. G. A. (1999). *Tsetse Biology and Ecology: Their Role in the Epidemiology and Control of Trypanosomiasis.* CABI Publishing.

Lehane, M. J. (2005). *Biology of Blood-Sucking Insects.* Cambridge University Press.

Losos, G. J., & Ikede, B. O. (1972). Review of pathology of diseases in domestic and laboratory animals caused by *Trypanosoma congolense, T. vivax, T. brucei, T. rhodesiense* and *T. gambiense. Veterinary Pathology, 9*(Suppl.).

Maudlin, I., Holmes, P. H., & Miles, M. A. (2004). *The Trypanosomiases.* CABI Publ.

McAlpine, J. F. (1989). Phylogeny and classification of the Muscomorpha. In J. F. McAlpine & D. M. Wood (Eds.), *Manual of Nearctic Diptera* (Vol. 3, pp. 1397–1518). Monograph No. 32, Agriculture, Ottawa.

McKelvey, J. J., Jr. (1973). *Man Against Tsetse: Struggle for Africa.* Ithaca: Cornell University Press.

Mulligan, H. W. (1970). *The African Trypanosomiases.* New York: Wiley-Interscience.

Murray, M., Morrison, W. I., Murray, P. K., Clifford, D. J., & Trail, J. C. M. (1979). Trypanotolerance—A review. *World Animal Review (FAO), 31*, 2–12.

Newstead, R. (1924). *Guide to the study of tsetse-flies.* Liverpool School of Tropical Medicine Memoir (New Series) No. 1.

Pépin, J., & Milord, F. (1994). The treatment of human African trypanosomiasis. *Advances in Parasitology, 33*, 1–47.

Peters, W., & Pasvol, G. (2007). *Atlas of Tropical Medicine and Parasitology* (2nd ed.). New York: Year Book Publ.

Potts, W. H. (1973). Glossinidae (tsetse-flies). In K. G. V. Smith (Ed.), *Insects and Other Arthropods of Medical Importance* (pp. 209–249). London: British Museum (Natural History).

Quinn, T. C. (1996). African trypanosomiasis (sleeping sickness). In J. C. Bennett & F. Plum (Eds.), *Cecil Textbook of Medicine* (20th ed., pp. 1896–1899). Philadelphia: W. B. Saunders Co.

Rogers, D. J., Hendricks, G., & Slingenbergh, J. H. W. (1994). Tsetse flies and their control. *Rev Sci Tech Off Int épizooties, 13*, 1075–1124.

Rogers, D. J., & Randolph, S. E. (1985). Population ecology of tsetse. *Annual Review of Entomology, 30*, 197–216.

Rogers, D. J., & Randolph, S. E. (2002). A response to the aim of eradicating tsetse from Africa. *Trends in Parasitology, 18*, 534–536.

Vale, G. A. (1993). Development of baits for tsetse flies (Diptera: Glossinidae) in Zimbabwe. *Journal of Medical Entomology, 30*, 831–842.

Willemse, L. P. M., & Takken, W. (1994). Odor-induced host location in tsetse flies (Diptera: Glossinidae). *Journal of Medical Entomology, 31*, 775–794.

World Health Organization. (1996). Geographical Distribution of Foci of Gambiense and Rhodesiense Sleeping Sickness, *WHO* 96.140. Vector Biology and Control Division, Geneva.

Zumpt, F. (1936). *Die Tsetsefliegen.* Jena: Fischer.

Zumpt, F. (1973). *The Stomoxyine Biting Flies of the World.* Stuttgart: G. Fischer Verlag.

Chapter

18

Myiasis (Muscoidea, Oestroidea)

Philip J. Scholl ▪ E. Paul Catts ▪ Gary R. Mullen

Myiasis is the invasion of a living vertebrate animal by fly larvae. This invasion may or may not be associated with feeding on the tissues of the host. Myiasis-causing flies are represented by a diversity of species. Some are rarely involved in myiasis, whereas for others it is the only way of life. Many of these same fly species also feed on carrion. Among flies, dietary proteins are required both for growth, egg production, and development. Proteins may be obtained by adult flies, by their larvae, or by both. In the case of larval diets, proteins are assimilated, stored, and carried through the pupal stage for subsequent use by the reproducing adult fly. A larval diet rich in proteins dictates that there is less need for adults to seek proteins. Thus myiasis is a means of exploiting a rich protein source by the larva for its own growth and, in some cases, for reproduction by the adult.

Myiasis is classified based on the degree to which a fly species is dependent on a host. Three types of myiasis generally are recognized: accidental, facultative, and obligatory myiasis.

In **accidental myiasis**, also called **pseudomyiasis**, the fly larvae involved normally are not parasitic but under certain rare conditions can become so. This type of myiasis can occur, for example, when fly eggs or larvae contaminate foods that are subsequently ingested by an animal, as in the case of pomace flies and fruit flies (*Drosophila* spp.). The 50 or so fly species involved include those that typically are free-living in all stages and rarely are parasitic. In most cases these flies pass unharmed through the host's alimentary tract, but can cause discomfort, nausea, diarrhea, and a plethora of related problems on their way through. In some cases, symptoms can be severe. Invasion of the alimentary tract can occur in two ways: either through ingestion of contaminated food or by retro-invasion through the host's anus. There is some doubt as to whether or not these cases are true myiasis because there is scant evidence that any fly development takes place after the ingested eggs hatch.

Facultative myiasis involves larvae that can be either free-living saprophages or parasites. These flies are opportunistic, having the ability to exploit living tissue. An example of facultative myiasis is the invasion of open sores on livestock by maggots of blow flies that normally frequent carrion.

Kettle (1995) recognizes three types of facultative myiasis: **primary myiasis**, involving those species that can initiate myiasis; **secondary myiasis**, involving species that continue myiasis but only after it is started by primary species; and **tertiary myiasis**, involving species that join the primary and secondary species just prior to host death. Facultative-myiasis species are the evolutionary bridge linking saprophagous feeders to those restricted to feeding on living tissues.

Many or even most of these facultative-myiasis flies feed on dead, decaying tissues rather than invading healthy tissue (e.g., surgical maggots) but are able to shift from dead to living tissue and back again with alacrity (e.g., *Cochliomyia macellaria*, *Wohlfahrtia nuba*). In a sense, these are borderline parasites that are capable of invading a sick or injured host and continuing their larval development after the death of that host. The adult flies are attracted to open wounds or chronic surface sores with purulent exudates.

In **obligatory myiasis**, the maggots of the fly species involved are always parasitic; they require a living host for their development. Examples are primary screwworms and bot flies. Included here are those species that cause **temporary myiasis**, which involves the intermittent contact between a fly larva and its host, such as nestling maggots and floor maggots. In this type of myiasis, the maggots do not keep continual contact with their host. Occasional parasitism of atypical hosts by obligate myiasis-producing flies is called **incidental myiasis**.

Myiasis also can be categorized in relation to the site of larval invasion, or subsequent development in the host. Thus, the descriptives *gastrointestinal*, *urogenital*, *ocular*, *nasopharyngeal*, *auricular*, and *cutaneous* are antecedent to the word myiasis, indicating the general site of maggot infestation.

Gastrointestinal myiasis refers to fly larvae in the alimentary tract of a host. This can be accidental myiasis, such as the ingestion of false stable fly eggs or larvae in uncooked fruits, or obligatory myiasis, such as the development of horse stomach bots. Enteric myiasis refers specifically to the intestinal tract. Urogenital myiasis is the invasion of the urethra and/or genitalia by fly larvae. This can occur when a host is debilitated and the urogenital openings exposed, or when the host has a urogenital infection producing exudates that attract flies. Cases of urogenital myiasis usually involve blow flies and flesh flies. Ocular myiasis is the invasion of eye tissues by fly larvae; most cases are caused by the sheep nose bot fly (*Oestrus ovis*) and infrequently by rodent bot flies (*Cuterebra* spp.). Nasopharyngeal myiasis is the invasion of nasal and deep oral cavities and recesses by fly larvae. As with urogenital myiasis, this often is associated with a microbial infection, but also can be caused by nose bot flies in healthy hosts. Auricular myiasis is the invasion of ears by fly larvae, usually caused by blow flies or flesh flies. Cutaneous myiasis involves invasion of the skin, usually by blow flies, flesh flies, screwworms, or certain bot flies. When cutaneous myiasis is associated with a break, laceration, or open sore in the host's skin, it is called traumatic myiasis.

Myiasis apparently has evolved along different lines in different groups of flies. Gastrointestinal myiasis and urogenital myiasis, for example, appear to represent a transition from species contaminating host foods to those associated with host excretions. They usually involve free-living species. Obligatory cutaneous myiasis appears to have evolved from carrion-breeding flies and from predaceous flies that prey on them. The origin of obligatory nasopharyngeal myiasis is less clear, but probably is linked with host secretions associated with upper respiratory infections. Ocular myiasis and auricular myiasis are characteristically either accidental or incidental in nature, resulting in damage to tissues at those respective sites. Temporary myiasis appears to have evolved from nest associates or lair-frequenting scavenger species that fed on organic morsels.

TAXONOMY

The vast majority of species involved in myiasis are members of two superfamilies and six families of calypterate flies: Muscoidea (Anthomyiidae, Fanniidae, and Muscidae) and Oestroidea (Calliphoridae, Sarcophagidae, and Oestridae). A dozen other families in eight superfamilies include species reported to cause myiasis; however, with the exception of the nest skipper fly (Neottiphilidae) and 10 species of Australian frog flies (Chloropidae), these cause accidental myiasis only. In contrast, all species of bot flies (Oestridae) cause obligatory myiasis. Myiasis-causing species among the muscids, calliphorids, and sarcophagids are typically facultative or obligatory myiasis producers. Table 18.1 shows the taxonomic relationships and associated types of myiasis for those flies known to invade living hosts.

The Anthomyiidae are a large family with more than 100 genera worldwide. Although members are called root maggot flies, the larvae occur in a wide range of habitats other than roots. Cladistically, the fanniids are a sister group to the muscids and often are included as a subfamily of the Muscidae. Most muscids are house fly-like in appearance, having a rather drab coloration with dorsal, longitudinal stripes on the thorax. Muscidae is a very large family with worldwide distribution that includes species typically associated with excrement and decaying plant matter.

The superfamily Oestroidea is composed of six families: Calliphoridae, Sarcophagidae, Oestridae, Tachinidae, Rhinophoridae, and Mystacinobiidae. The first five of these include parasitic species. Tachinids and rhinophorids parasitize invertebrate hosts, mostly other insects, and are not involved in myiasis. Adult mystacinobiids are phoretic, not parasitic, on bats.

Members of the Calliphoridae are called blow flies. This large family includes over 1,000 species worldwide. Most have a polished or metallic blue, green, or bronze appearance and sometimes also are called bottle flies. There is confusion and continuing speculation concerning the phylogenetic relationships within this family. The major subfamilies include Calliphorinae, Chrysomyinae, Mesembrinellinae, Polleniinae, and Rhiniinae. Other subfamilies contain specialized genera or genera with limited distributions. The largest subfamily, Rhiniinae, includes about 40 genera of Old World blow flies. The Mesembrinellinae are large, showy tropical blow flies, whereas the Calliphorinae are the more temperate-climate bottle flies. The subfamily Chrysomyinae includes the economically important screwworms and the bird-nest blow flies. Members of the subfamily Rhiniinae are of no significance regarding myiasis.

The Sarcophagidae are the flesh flies, with some 2,000 species distributed worldwide. Many of these are parasitic on hymenopterous hosts or are predaceous on other insects, but a few cause myiasis. Members of two of the four subfamilies include myiasis-causing species, the Sarcophaginae and Miltogramminae.

The Oestridae are the bot flies, with fewer than 150 species worldwide. Adult flies are robust and hairy with a general appearance of honey bees or bumble bees. Larvae of all these cause obligatory myiasis. They are placed in four subfamilies: Cuterebrinae, Hypodermatinae, Oestrinae, and Gasterophilinae (Wood, 1987; Pape, 2001). Phylogenetic relationships among these subfamilies have been recently described by Pape (2006). The rationale for combining the previously grouped three (e.g., Zumpt, 1965) or four (e.g., Guimarães and Papavero, 1999) families into a single family have been discussed by Colwell et al. (2006).

For general taxonomic information and keys for identifying adults and larvae of muscoid and oestroid flies, see James (1947), Curran (1965), Zumpt (1965), Greenberg (1971), McAlpine et al. (1981, 1987, 1989), Furman and Catts (1982), Smith (1986), Lane and Crosskey (1993), Wall and Shearer (1997), and Guimarães and Papavero (1999). Additional sources for

Table 18.1 Taxonomic Relationships of Flies Known to Cause Myiasis

Superfamilies and Families	Common Names	Genera and Species	Type of Myiasis
Tipuloidea			
Tipulidae	Crane flies	(unspecified)	Gastrointestinal (A)
Psychodoidea			
Psychodidae	Moth flies	*Psychoda* (3 spp.)	Gastrointestinal (A)
Stratiomyoidea			
Stratiomyidae	Black soldier fly	*Hermetia illucens*	Gastrointestinal (A)
Asiloidea			
Therevidae	Stiletto flies	*Thereva* sp.	Gastrointestinal (A)
Platypezoidea			
Phoridae	Humpback flies, scuttle flies	*Megaselia* (3 spp.)	Gastrointestinal (A), traumatic (F)
Syrphoidea			
Syrphidae	Flower flies	*Eristalis* (2 spp.)	Gastrointestinal (A)
Tephritoidea			
Piophilidae	Cheese skipper fly	*Piophila casei*	Gastrointestinal (A)
Neottiophilidae	Nest skipper fly	*Neottiophilum praeustrum*	Cutaneous (O)
Ephydroidea			
Drosophilidae	Pomace flies, fruit flies	*Drosophila melanogaster*	Gastrointestinal (A)
Carnoidea			
Chloropidae	Australian frog flies	*Batrachomyia* (10 spp.)	Cutaneous (O)
Muscoidea			
Anthomyiidae	Anthomyiid flies	*Hylemya* (2 spp.)	Gastrointestinal (A)
	Lesser house fly	*Fannia canicularis*	Gastrointestinal, urogenital, traumatic (A)
Fanniidae	Latrine fly	*Fannia scalaris*	Gastrointestinal, urogenital, traumatic (A)
Muscidae	House fly	*Musca domestica*	Gastrointestinal, traumatic (A)
	False stable fly	*Muscina stabulans*	Gastrointestinal, traumatic (A)
	—	*Muscina* (2 spp.)	Gastrointestinal, traumatic (A)
	—	*Hydrotaea rostrata*	Gastrointestinal, traumatic (A)
	Tropical nest flies	*Passeromyia* (3 spp.)	Cutaneous (O)
	Tropical nest flies	*Mydaea* (25 spp.)	Cutaneous (O)
	Neotropical nest flies	*Neomusca* (*Philornis*) (35 spp.)	Cutaneous (O)
Oestroidea			
Calliphoridae	Blow flies		
	Toad blow flies	*Bufolucilia* spp.	Traumatic, cutaneous, gastrointestinal, nasopharyngeal, auricular (O)
	Green bottle flies	*Lucilia (Phaenicia)* spp.	Traumatic, cutaneous, gastrointestinal, nasopharyngeal, urogenital, auricular (F)
	Blue bottle flies	*Calliphora* spp.	Traumatic, cutaneous, gastrointestinal, nasopharyngeal, urogenital, auricular (F)
		Eucalliphora spp.	
		Paralucilia sp.	
		Protophormia sp.	
		Cynomya sp.	
	Black blow fly	*Phormia regina*	
	Primary screwworm	*Cochliomyia hominivorax*	Traumatic (O)
	Secondary screwworm	*Cochliomyia macellaria*	Traumatic (F)
	Old World screwworm	*Chrysomya bezziana*	Traumatic (O)
	—	*Chrysomya* (9 spp.)	Traumatic (F)
	Nest blow flies	*Protocalliphora* (90 spp.)	Traumatic (O)
	Tumbu fly	*Cordylobia anthropophaga*	Traumatic (O)
	African mouse/Lund's fly	*Cordylobia* (2 spp.)	Cutaneous (O)
	Congo floor maggot	*Auchmeromyia senegalensis*	Traumatic (O)
	African suid maggots	*Auchmeromyia* (4 spp.)	Traumatic (O)
	Asian deer/water buffalo skin maggots	*Pachychoeromyia praegrandis*	Traumatic (O)
		Booponus (4 spp.)	Cutaneous (O)
	Indian elephant skin maggot	*Elephantoloemus indicus*	Cutaneous (O)
Sarcophagidae	Flesh flies	*Sarcophaga* spp.	Traumatic (F), gastrointestinal (A)
		Wohlfahrtia (4 spp.)	Traumatic (O)
	Lizard flesh fly	*Anolisomyia* sp.	Cutaneous (O)

Continued

Superfamilies and Families	Common Names	Genera and Species	Type of Myiasis
	Turtle flesh fly	*Cistudinomyia cistudinis*	Cutaneous (O)
	Toad flesh fly	*Notochaeta bufonovoria*	Cutaneous (O)
	Lizard egg fly	*Eumacronychia nigricornis*	Cutaneous (O)
	Sea turtle egg fly	*Eumacronychia sternalis*	Cutaneous (O)
	Terrapin egg fly	*Metoposarcophaga importuna*	Cutaneous (O)
OESTRIDAE	Bot and grub flies	(25 genera, 140 spp.)	Cutaneous, gastrointestinal, nasopharyngeal (O)
Cuterebrinae	New World skin bot flies	(2 genera, 58 spp.)	Cutaneous (O)
	Rodent and rabbit bot flies	*Cuterebra* (57 spp.)	Cutaneous (O)
	Howler monkey bot	*C.* (*Allouattamyia*) *baeri*	Cutaneous (O)
	—	*C.* (*Pseudogametes*) (2 spp.)	Cutaneous (O)
	—	*C.* (*Metacuterebra*) (13 spp.)	Cutaneous (O)
	—	*C.* (*Rogenhofera*) (6 spp.)	Cutaneous (O)
	Tórsalo, human bot fly	*Dermatobia hominis*	Cutaneous (O)
Hypodermatinae	Old World skin bot flies	(9 genera, 31 spp.)	Cutaneous (O)
	Common cattle grub	*Hypoderma lineatum*	Cutaneous (O)
	Northern cattle grub	*Hypoderma bovis*	Cutaneous (O)
	Reindeer grub	*Hypoderma tarandi*	Cutaneous (O)
	Deer and yak warble flies	*Hypoderma* (3 spp.)	Cutaneous (O)
	—	*Ochotonia lindneri*	Cutaneous (O)
	Old World rodent grubs	*Oestromyia* (5 spp.)	Cutaneous (O)
	—	*Oestroderma potanini*	Cutaneous (O)
	Saiga warble fly	*Pallisiomyia antilopum*	Cutaneous (O)
	Pavlovsky's gazelle warble fly	*Pavlovskiata subgutturosae*	Cutaneous (O)
	—	*Portschinskia* (7 spp.)	Cutaneous (O)
	—	*Przhevalskiana* (6 spp.)	Cutaneous (O)
	Antelope warble flies	*Strobiloestrus* (3 spp.)	Cutaneous (O)
Oestrinae	Nose bot flies	(9 genera, 34 spp.)	Nasopharyngeal (O)
	Deer nose bots	*Cephenemyia* (8 spp.)	Nasopharyngeal (O)
	Camel nose bot	*Cephalopina titillator*	Nasopharyngeal (O)
	Tuberculous nasal bots	*Gedoelstia* (2 spp.)	Nasopharyngeal (O)
	Hairy nasal bots	*Kirkioestrus* (2 spp.)	Nasopharyngeal (O)
	Sheep nose bot	*Oestrus ovis*	Nasopharyngeal, ocular (O)
	—	*Oestrus* (5 spp.)	Nasopharyngeal (O)
	African elephant throat bot	*Pharyngobolus africanus*	Nasopharyngeal (O)
	Deer and gazelle throat bots	*Pharyngomyia* (2 spp.)	Nasopharyngeal (O)
	Horse nasal bot fly	*Rhinoestrus purpureus*	Nasopharyngeal (O)
	Nasal bot flies	*Rhinoestrus* (10 spp.)	Nasopharyngeal (O)
	Kangaroo throat bot fly	*Tracheomyia macropi*	Tracheal (O)
Gasterophilinae	Stomach bot flies	(5 genera, 17 spp.)	Gastrointestinal (O)
	Horse stomach bots	*Gasterophilus* (9 spp.)	Gastrointestinal (O)
	Common horse bot	*G. intestinalis*	Gastrointestinal (O)
	Throat horse bot	*G. haemorrhoidalis*	Gastrointestinal (O)
	Nose horse bot	*G. nasalis*	Gastrointestinal (O)
	Broad-bellied horse bot	*G. nigricornus*	Gastrointestinal (O)
	Dark-winged horse bot	*G. pecorum*	Gastrointestinal (O)
	Black elephant stomach bot	*Cobboldia elephantis*	Gastrointestinal (O)
	Blue elephant stomach bot	*C.* (*Platycobboldia*) *loxodontis*	Gastrointestinal (O)
	Green elephant stomach bot	*C.* (*Rhodhainomyia*) *roveri*	Gastrointestinal (O)
	Rhinoceros stomach bots	*Gyrostigma* (3 spp.)	Gastrointestinal (O)
	African elephant skin bot	*Neocuterebra squamosa*	Cutaneous (O)
	African elephant foot bot	*Ruttenia loxodontis*	Cutaneous (O)

The types of myiasis in each case are indicated in parentheses as accidental (A), facultative (F), or obligatory (O).

identifying the immature stages are Liu and Greenberg (1989) and Stehr (1991). Calliphorid eggs can be identified with Greenberg and Singh (1995). The Calliphoridae are treated by Hall (1948), Dear (1985), Ribeiro and Carvalho (1998), Carvalho and Ribeiro (2000), and Whitworth (2006). The Oestridae are treated by Papavero (1977), Wood (1987), Pape (2001), and Colwell et al. (2006).

MORPHOLOGY

Adult members of the superfamilies Muscoidea and Oestroidea are distinguished from other calypterate flies by possessing both a ptilinal suture and facial lunule (Figure 18.1B, C); the antennal pedicel with a complete dorsal seam and flagellum composed of a single compound segment bearing an arista or stylus; and a bulbous, greater ampulla below the wing base (Figure 18.1A). All oestroid flies have a vertical row of bristles on the thoracic meron (Figure 18.1A). The Muscoidea are distinguished from the Oestroidea by lacking bristles on the thoracic meron even though the rest of the body is bristled.

Larvae of the muscoid and oestroid families are called maggots. The larvae of muscoid flies usually lack armature, or body spines, whereas oestroid larvae possess armature ranging from sparse segmental belts of spines to a rather complete spiny vestiture. The three larval instars of both superfamilies can be distinguished by the form and number of slits in the posterior spiracular plates and by the size and shape of the internal cephalopharyngeal skeleton (Figure 18.2). The body form of muscoid maggots is peg-like and gradually tapers from a blunt posterior end to a pointed anterior end. Larvae of *Chrysomya rufifacies* and in the genus *Fannia* are atypical, bearing segmental protuberances. Larvae of blow flies and flesh flies are more cylindrical than those of muscoid flies; they are tapered more abruptly at each end. Larvae of these two families can be distinguished by the slant of the inner posterior spiracular slits. The slits of blow flies slant toward the midline, whereas those of flesh flies slant away from the midline.

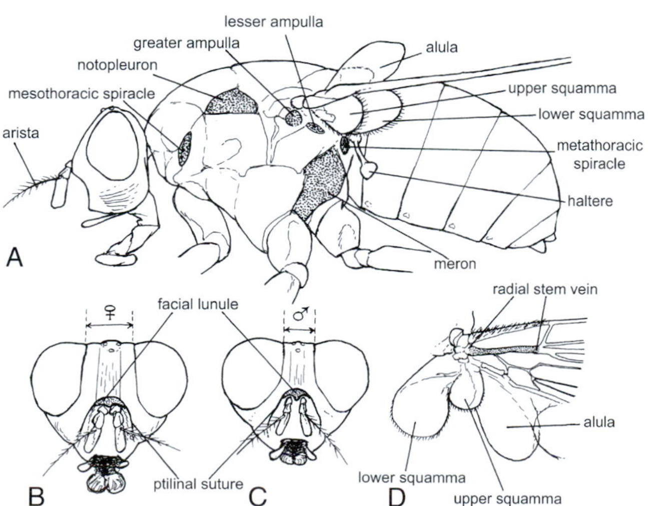

Figure 18.1 External morphology of representative adult muscoid fly. (A) Lateral view; (B) Frontal view, female; (C) Frontal view, male; (D) Dorsal view of wing base. (Original by E. P. Catts)

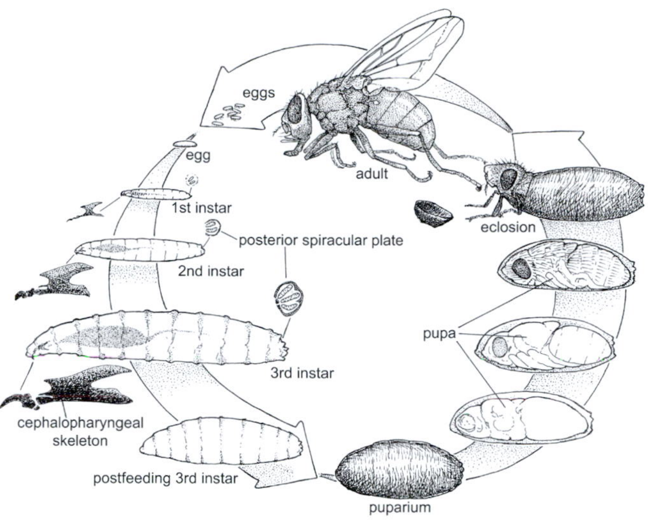

Figure 18.2 Typical life cycle of muscoid and oestroid flies. (Original by E. P. Catts)

Mature bot fly maggots, some of which are called grubs, are much larger and thicker in cylindrical form than those of other oestroids. Except for the larvae of Gasterophilinae and Cuterebrinae, which possess elbowed or highly sinuous posterior spiracular slits, respectively, oestrid larvae have porous posterior spiracular plates that mask the more typical arrangement of slits.

The puparium, which contains the developing pupa, is formed from the integument of the last larval instar. All puparia except those of oestroids can be separated at the family level by examining remnant larval features (mouthparts and posterior spiracles). In contrast, oestrid puparia exhibit overall morphological forms that characterize each subfamily.

Among the adult muscoids, the wing venation of anthomyiids differs from muscids by the length of the anal vein. Some muscid genera (e.g., *Morellia*, *Neomyia*, and *Hydrotaea*) have a metallic or polished abdomen, superficially resembling blow flies. Most adult muscoid flies have sucking, rasping, or sponging mouthparts. A few (e.g., *Stomoxys*, *Haematobia*) possess a hardened labium as a piercing structure.

Among the oestroid flies, calliphorid adults have well-developed sponging-sucking mouthparts, the thoracic meron has a vertical row of bristles, and the postscutellum is small. Typically these flies are metallic blue, green, or bronze in body coloration. Members of the major subfamilies Chrysomyinae and Calliphorinae differ in the presence or absence of hairs on the radial stem vein of the wing (Figure 18.1D).

Adult sarcophagids also have sponging mouthparts, the thoracic meron is bristled, the postscutellum is small or absent, and three or four notopleural setae are usually present. Body coloration is typically gray and black, giving them the general appearance of large houseflies. Their color patterns, however, are more distinct than those of muscoid flies. The abdomen typically displays a tessellated, or checkerboard, pattern of gray and black.

Unlike other oestroid flies, adult oestrids lack head and body bristles, except for the bristles on the meron. They are medium to large hairy flies that resemble bees. They possess small antennae and relatively small sponging-sucking mouthparts, which in most instances are atrophied and nonfunctional. All except the members of the Gasterophilinae have large wing **squammae** (Figure 18.1D).

LIFE HISTORY

The life history of muscoid and oestroid flies follows the typical holometabolous and cyclorrhaphous pattern of four stages: egg, larva (3 instars), pupa (in a puparium), and adult (Figure 18.2). Male flies are usually smaller and emerge earlier than females. All these flies form the puparium free from the larval substrate. Generally, given an adequate larval diet, more time in the developmental life cycle of an individual is spent in the pupal stage than in either that of the egg or larva. Among the oestrid flies, the cuterebrines have a long pupal stage that allows overwintering free from the host. Other bots have a prolonged larval period as an adaptation for overwintering within the protection of their host. Pupation typically is preceded by a wandering, post-feeding period by the last larval instar.

Adult longevity among these flies varies depending on environmental conditions. Longevity ranges from as little as one to several days among stomach bot and cattle grub flies, to one or two weeks for other bot flies, to one month or more among muscoids and other oestroids. In general, bot fly adults have a shorter life because they do not feed. All myiasis-causing flies either oviposit or larviposit, but the degree of embryo development within the egg prior to deposition differs. Usually eggs are fertilized at the time of oviposition. Consequently some, such as muscids and most oestroids, lay eggs that require an additional period of embryonic development before hatching (i.e., oviposition). In contrast, the nose bot flies and flesh flies retain fertilized eggs in an expanded pouch, or uterus. They deposit active larvae within membranous "shells" that hatch immediately (i.e., larviposition).

In most cases, eggs or larvae are laid in contact with the larval nutrient substrate. However, in a few cases (e.g., cuterebrines and nest maggots), the first-instar larvae must locate and invade their nutrient resources.

After hatching, the larvae begin to feed by scratching the nutrient substrate with their hook-like mandibles. This loosens fragments of substrate that are bathed in the maggot's salivary enzymes and are sucked up as a nutrient soup. Only the cephalopharyngeal skeleton is fixed in size at the time of molting relative to the total length and girth of the feeding larva. The ratio between skeleton and body sizes indicates the larval age within each instar. The body integument remains pliable, facilitating considerable growth and expansion during each stadium. While inside a living host, larvae are relatively protected from exploitation by predators or parasites, but they must cope with their host's immune defenses. This is an advantage over free-living related species or siblings developing in carrion where competition and exploitation are major risks.

The next period in development is that of the wandering, mature, postfeeding larva. Larvae may continue to feed on available nutrients even after reaching and passing a critical size necessary for wandering and subsequent pupation. This critical size for some species is reached early in the third stadium so that, if the nutrient source becomes exhausted, the result can be very small adult flies. The duration of wandering is an innate character for each species and is influenced by environmental factors such as heat, moisture, soil condition, and light. It may be shortened considerably if a suitable pupation site is obtained early on. Once a site is found the larva becomes immobile and contracts in size. The resulting puparium is formed by the synchronized, concerted action of muscular and cuticular structures resulting in a rigid, hardened protective shell formed from the cuticle of the last larval instar.

Adult flies typically emerge from the puparium in the morning hours. The teneral adult exits the puparium through a predetermined portal formed by the opening of an anterior operculum or by detachment of the anterior end of the puparium. After a short period needed to harden their wings, adults are ready to take to flight in search of a mate and, in the case of females, a suitable adult and larval nutrient source. Mating usually occurs after adult flies have matured for a period a few days to about one week. Many adult females (e.g., *Cochliomyia hominivorax* and the nonfeeding bot flies) do not require protein for the first ovarian cycle and are referred to as being **autogenous**, whereas other species such as *C. macellaria* do require protein in order to develop eggs and are referred to as **anautogenous**. Adult females in this group require additional nutrients and a period of ovarian development of several more days.

Flies that lay eggs that require time to complete their embryonic development outside of the female are oviparous. In bot flies, egg development is initiated during the larval stage and continues through pupation, so that eggs can be deposited with less delay. The more primitive cuterebrine bot flies require about five days between eclosion and oviposition. In contrast, cattle grub and stomach bot flies have fully developed eggs at eclosion and their females are ready to oviposit as soon as mating is completed. As already noted, nose bot flies and flesh flies retain eggs in utero so that their eggs hatch immediately following fertilization. This delayed oviposition is correctly termed **ovoviviparity,** but usually is referred to as **larviposition**. Unlike other oestroids, the bot flies have only one egg-producing (gonotrophic) cycle. Although subsequent cycles do not occur, among the cattle grubs each ovarian follicle may develop two eggs simultaneously, thus doubling the egg production of the one gonotrophic cycle. This limitation to a single cycle among bot flies reflects their shorter, nonfeeding adult life span. In contrast, muscoid and other oestroid flies (calliphorids and sarcophagids) can hold their unfertilized eggs for several weeks until a suitable larval breeding medium is found. Following oviposition or larviposition they are able to initiate additional gonotrophic cycles.

The primary activities of male muscoid and oestroid flies are to locate and copulate with receptive females. Males are capable of multiple matings during their life, but play no further role in reproduction or dispersal of progeny. Typically females mate only once. Dispersal of progeny is accomplished by the female fly and by the mobile host. This is another peculiar aspect of obligatory myiasis because the female fly usually attacks more than one host, thus distributing her progeny. The larvae, when mature, drop from their ambulatory hosts and thus can be scattered over an extensive area.

A number of vernacular terms are associated with myiasis-causing flies. Some refer to the adult flies, or more often their larvae, and others to the pathology that they cause. The term **maggot** is of Scandinavian origin and used for the larval stage of muscoid, calliphorid, and sarcophagid flies. The term **bot** is applied only to oestrid larvae and is derived either from a Gaelic word *botus* meaning belly worm, or from Italian *botta*, which refers to the cutaneous ulcer or furuncle caused by these flies. This ulcer, or open cyst, is termed a **warble**, which is derived from the Scandinavian *varbulde*, meaning boil. This term is misused at times to refer to the larva instead of its furuncle. In the southeastern United States warbles have been called **wolves** since colonial times in reference to the boil-like cuterebrine infections in rodents and rabbits. **Grub** is a term applied to cattle bots (as well as to scarab beetle larvae). This term probably is derived from the Indo-European word *ghrebl* meaning to scratch, scrape, or bore into. The common name of **screwworm** probably describes the threaded appearance of the maggot stage of these blow flies and the belief that they twist and bore their way into host flesh.

When flies oviposit on a host or on carrion the action is referred to as **striking** or **fly strike**. This is an English term originating from the Latin *stringere* meaning to touch lightly or brush against. Its use probably was more in reference to the ovipositing of hovering nose bot flies or stomach bot flies.

A host or other substrate on which fly eggs have been laid is said to be **fly blown**. The use of "blow" in this context comes from the Old English term *blawan*, and probably refers to the production of gas from bloated carrion containing maggots. This is also the source of the name **blow fly**, for those flies that are most conspicuous at carrion, the Calliphoridae.

ECOLOGY AND BEHAVIOR

The environmental constraints on myiasis-causing flies during the free-living periods of their life cycle are the same as for other Diptera, primarily moisture and heat. These constraints are of little importance during the parasitic period of their life cycle because both moisture and heat are provided by the host at levels well within the tolerance limits of eggs and larvae. Some myiasis-causing species lay their eggs free from the host and thus are adapted to a wider range of environmental conditions (e.g., cuterebrines and nest maggots). Muscoid and oestroid larvae are adapted in form and behavior to life in a moist, organic substrate ranging from wet feces to living tissue. Constraints on larval development, unlike on adults, stem more from host resistance and inter- or intraspecific competition.

Facultative myiasis-causing flies generally develop in carrion and feces as massed aggregations of maggots. In such situations the increased temperature that allows for rapid larval development mostly comes from metabolic heat produced by the clustered maggots themselves. For these species, a living, protein-rich, moist, warm host substrate merely substitutes for the nonliving substrate. Facultative myiasis often results in the death of the host either by direct effects of maggots or by the indirect effect of stress, which predisposes the host to predation or disease. Sterile, laboratory-reared maggots, however, can have therapeutic benefits and have been used in treating deep wounds and sores because they feed only on dead tissue.

For flies that cause facultative myiasis, finding a receptive mate generally is not difficult. The synchronous emergence of sibling adults developed from the same mass of maggots puts newly emerged males and females in close proximity. These species also find their oviposition sites by responding to the same stimuli that attracts them to any nonliving resource (e.g., fetid or putrid odors, purulent discharges, and/or accumulations of animal excretions). They show very little discrimination regarding host species. Host size and habitat are the principal limiting factors. Even though their resources have a patchy distribution, these flies are capable of responding quickly to chemical stimuli at very low concentrations in order to locate widely scattered resources.

The greater longevity of muscoid flies, blow flies, and flesh flies also requires that they imbibe fluids to maintain an internal water balance and that they obtain an energy source beyond that acquired during larval development. The energy source is often honeydew and plant juices, especially nectar. This is why adults of these flies also occur on flowering plants and are important pollinators for many plants, including certain crops. Most female flies also require a protein supplement to complete oögenesis. This is supplied by a wide array of sources (e.g., blood secretions, excretions, wounds, and carrion).

The obligatory myiasis-causing screwworms (e.g., primary screwworm, Old World screwworm, and wound flesh fly) are closely related to the facultative myiasis-causing species. The obligatory primary species invade healthy tissue and enlarge wounds, and often are found deep in these wounds. In contrast, the secondary species do not feed on healthy tissue, almost always invade necrotic wounds, and feed on dead tissue. These secondary invaders usually are found on the surface or outer part of the wound. Because their progeny often are widely scattered, the adults aggregate at certain flowering shrubs where they mate and feed. Their egg development, mating, and indiscriminate oviposition on any suitable host are similar to that of facultative myiasis-causing flies normally occurring in carrion. For the obligatory myiasis-causing oestrid flies, mating, host acquisition, and host-parasite interactions are more complicated. Because adults usually emerge from scattered sites, contact between males and females requires aggregation behavior. Species in each of the oestrid subfamilies aggregate at specific topographic sites where mating takes place. This probably is the case for all oestrids. Some bot flies also mate near potential hosts. Where hosts are plentiful, aggregation behavior can cause crowding of adult bot flies, thus exposing the flies to predation and to less-than-ideal environmental conditions for survival. At these sites bot flies exhibit a male spacing behavior that counteracts crowding, interferes with intraspecific pairing, minimizes predation, and assures that all available aggregation sites are occupied. Adult oestrid flies do not feed and have relatively short life spans, living on fat reserves accumulated as larvae.

Most bot flies display a high degree of host specificity, with each species parasitizing only one or, rarely, a few host species. *Dermatobia hominis* is the very important exception with a very large host range. Host species other than the native hosts are either intolerant or refractory to parasitism by a given species of bot fly. Although other host species are susceptible to bot fly invasion, this does not mean that they are suitable hosts or that mature larvae and adults are produced. Susceptible, but unsuitable, hosts tend to be species that are unrelated phylogenetically and ecologically to the native host species. Thus dogs and cats are susceptible, unsuitable hosts for cuterebrine bot infections, whereas a given mouse species can be completely refractory to infection by the bot fly of a related mouse species. This high degree of host specificity reflects a long coevolutionary relationship between bot flies and their respective hosts. This helps to explain why no larvae in any of the four subfamilies of Oestridae have ever been reared *in vitro* from newly hatched first instars to pupae.

Each bot fly species typically develops only at a specific site in its native hosts (e.g., nose bots, stomach bots, and foot bots). Even among the cutaneous bot flies (excluding *Dermatobia)* the location of warbles on the host occurs at specific anatomical sites. In nonnative and atypical hosts, however, this site occurrence can be erratic, often with grave consequences for the host and its parasites.

Host–parasite interaction involving bot flies is a delicate balance of tolerance by the host and limited pathology by the parasite. It also involves immunosuppression induced by some larvae. In bot fly infestations, excessive parasite burden usually is avoided by limiting the number of larvae to which a host is exposed. Bot fly females rarely dump their total reproductive complement on a single host; if this occurs, the host is likely to succumb to predators or other parasites due to the stress imposed by the developing bot larvae.

The number of bot fly larvae infesting an individual host varies from one to perhaps several hundred depending on the fly species involved. This is far less than the thousands of maggots that can make up the maggot mass in other myiasis-causing flies. Although much larger than other myiasis-causing flies, oestrid larvae appear to cause much less stress to their host. Although the reasons are unclear, this has the advantage of reducing the likelihood that the host will die during their development.

First-instar bot larvae move from their point of host invasion to the site of development in their host. Studies have shown that this movement within host tissues is typically along predetermined pathways within connective tissue or along the fibrous membranes covering and separating muscles (i.e., fascial planes). This keeps the larvae from prolonged contact with the host's hemopoetic defenses and helps to explain the seeming lack of immune resistance. Immune resistance by hosts to bot fly-caused myiasis apparently lasts only a short time following an infestation. The wall of the warble is formed by the host's response to foreign body invasion. The warble wall more or less confines the developing larvae in a multilayered pocket of fibrous connective tissue. The wall is characterized by the presence of giant cells typical of a chronic inflammatory response. In addition, the developing larva secretes a bacteriostatic substance that prevents secondary microbial infection of the warble. In a native host, once the mature larva has dropped free, the empty warble collapses and heals rapidly.

Some hosts show exaggerated, usually futile, behavioral responses to certain bot flies in the act of ovipositing. For example, cattle panic when under attack by adults of the northern cattle grub (*Hypoderma bovis*). They respond by headlong flight with tail erect, called flagging, apparently attempting to outdistance the hovering fly. This response more commonly is called **gadding**, a term derived from the Old English *gad,* a sharply pointed stick used to goad or prod livestock. Reindeer and caribou show a similar response to attacks by *H. tarandi.* In contrast, cattle display far less concern for the attacks by adults of the common cattle grub (*H. lineatum*), which often oviposit while animals are ruminating in a recumbent position. Gadding behavior also is shown by horses when under attack by the stomach bot fly (*Gasterophilus nasalis*), whereas horses usually pay little heed to oviposition by the common horse bot fly (*G. intestinalis*).

Another behavioral avoidance response is shown by deer and sheep when under attack by nose bot flies. A threatened host will lower its head and press its muzzle into a clump of grass or against the ground to deter strikes by the hovering female flies (e.g., *Oestrus ovis, Cephenemyia* spp.).

Host–parasite interaction involving species that cause facultative myiasis show little adaptation to neutralizing host defenses. When present in large numbers, however, rapidly developing larvae overcome the nonspecific host response to tissue invasion characterized by self-grooming and inflammation. Unlike oestrid larvae, most other myiasis-causing larvae can complete their development in the dead host.

Myths

A number of misconceptions have developed concerning bot flies and myiasis. One is the notion that adults of deer nose bots (*Cephenemyia* spp.) are capable of flying at speeds in excess of the speed of sound. The belief had its origin from a single questionable field observation of the deer nose bot fly coursing by a man standing on a hilltop. This tale gained some ill-founded credibility from entomologists and was published as fact for some time. The reasoning was that the bot fly needed to fly fast in order to overtake a swift running host. This reasoning ignored the fact that deer are incapable of running speeds in excess of 80 km/h (50 mph) and that the fly usually oviposits while the host is standing still. Although there have been no definitive studies of this subject, field observers conclude that *Cephenemyia* species probably cannot exceed a flight speed of 48 km/h (30 mph).

Another myth is that the larva of the rodent bot fly *Cuterebra emasculator* emasculates its squirrel and chipmunk hosts by consuming the host's testes as it develops. This tale arose because the specific site for larval development by this species is the inguinal area. Studies have shown, however, that the enlarging bot larva and warble in the scrotal sack of male hosts merely prevents the seasonal descent of gonads into the scrotum from the host's body cavity and does not cause sterility.

Because of the conspicuous, and sometimes self-destructive gadding response of cattle and horses under certain bot fly attacks, the belief arose that the bot flies involved were stinging their hosts. The extensible terminal segments of the abdomen of *Gasterophilus* females suggest a formidable sting. However, these flies only contact individual host hairs during oviposition. They do not injure or even touch the skin surface. Hosts are frightened by the buzzing sound made by these flies as they hover, and when one individual in a herd bolts, the herd instinct prompts the same reaction from others.

There is a common belief in the southeastern United States that the flesh of squirrels and rabbits infected with "wolves" (*Cuterebra* spp.) is unfit to eat. Because of this, the entire squirrel carcass often is discarded. This too is a myth. In fact, skin-bot larvae are eaten as a delicacy by sub-Arctic human cultures that traditionally herd reindeer in which bots commonly occur. Although the host flesh surrounding the warble may be discolored, there is no health hazard in eating either it or the bot larva. Squirrels and rabbits have long been a staple in the fall and winter diet of rural people in the United States. In some states the legal hunting season even has been postponed to begin after the bot season has ended to avoid the wasteful discarding of bot-infested game.

FLIES INVOLVED IN MYIASIS

Flies that typically develop in dung or decaying plant matter generally are involved in gastrointestinal myiasis. A few muscids also are adapted for blood-feeding on nestling birds. The carrion-breeding flies and their more fastidious relatives, screwworms and flesh flies, are adapted for both facultative and obligatory myiasis. The bot flies (Oestridae) are all obligatory myiasis producers.

The families and groups of flies that follow include species that cause myiasis in humans and other animals. A more extensive listing of families, species, and the types of myiasis that they cause is presented in Table 18.1. For reviews of this subject see James (1947), Zumpt (1965), Leclercq (1969), Hall and Wall (1995), Wall and Shearer (1997), Guimarães and Papavero (1999), and Colwell et al. (2006).

Stratiomyidae (Soldier Flies)

The only species in this family that reportedly causes myiasis is the black soldier fly (*Hermetia illucens*). Although originally a New World species, it is widely distributed in warmer temperate and tropical areas of the world. The larvae (Figure 18.3) develop in decaying fruits, vegetables, and other plant material, decomposing animal carcasses, and excrement. Only a few human cases of intestinal myiasis involving *H. illucens* have been documented. At least one case resulted in rather severe enteric disturbances, apparently due to the large size and vigorous activity of the larvae (James, 1947).

Syrphidae (Flower Flies, Hover Flies, Rat-tailed Maggots)

This is a large family (180 genera, 6,000 species) that includes only a few taxa that cause gastrointestinal myiasis. Adults also are called drone flies because of their bee-like appearance and resemblance to honeybee drones. The terms flower flies and hover flies refer to their common habit of visiting flowers for nectar and pollen, and their ability to hover motionless in

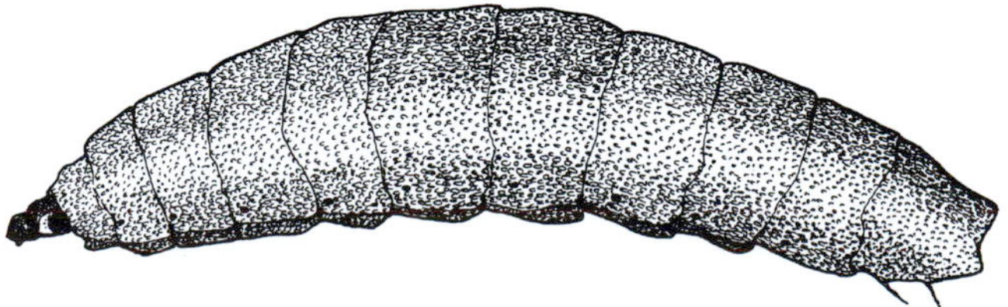

Figure 18.3 Black soldier fly (*Hermetia illucens*), larva (Stratiomyidae). (Original by E. P. Catts)

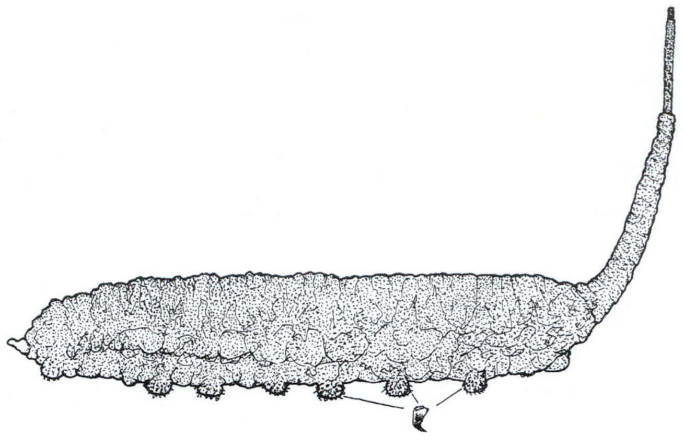

Figure 18.4 Rat-tailed maggot, *Eristalis tenax* (Syrphidae). (Original by E. P. Catts)

flight. The larvae of *Eristalis* (Figure 18.4) and other aquatic genera are called rat-tailed maggots, referring to the long, telescopic, three-segmented respiratory tube at their posterior end by which they breathe at the water surface.

The syrphid species most frequently involved in myiasis is *E. tenax*. Its larvae develop in sewage, liquefied excrement, decaying animal carcasses, and other decomposing plant and animal material of a liquid consistency. A number of human cases of gastrointestinal myiasis have been reported, with live larvae being passed in stools. Two other *Eristalis* species have been identified in human myiasis: *E. arbustorum* in Europe and *E. dimidiata* in the United States (James, 1947). Leclercq (1969) described several human cases involving rat-tailed maggots, including a man in Germany who passed more than 40 *Eristalis* larvae in one day and another 10 to 30 larvae on each of the six following days. *Eristalis* larvae also have been the cause of vaginal myiasis in cattle.

Piophilidae (Skipper Flies)

This is a small family of about 70 species in 35 genera worldwide. Females of the common cheese skipper (*Piophila casei*), oviposit on putrid, dried, cured, or smoked meats and cheeses, typically depositing 400 to 500 eggs per female. Adult cheese skippers are small (3–5 mm), slender, glossy black flies with yellow on the lower face and part of the legs. Larvae are slender, cylindrical, white, and truncated caudally with three pairs of short caudal protuberances, the ventral pair being the largest (Figure 18.5). Larvae require about five days to develop under warm conditions. In temperate regions the mature larva overwinters. The name "skipper" originates from the ability of the larva to flex head-to-tail in a circle and, following total-body muscular contraction and release, the larva propels itself off the substrate for a considerable distance (up to 24 cm). This behavior is used as a means of escape when disturbed or when dispersing to suitable pupation sites. The pupal stage lasts about five days. The life cycle, egg-to-egg, can be completed in as little as two weeks.

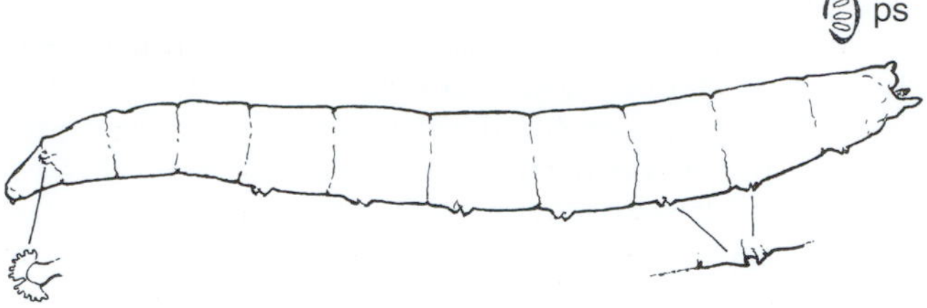

Figure 18.5 Cheese skipper, *Piophila* sp., larva (Piophilidae); ps, posterior spiracular plate. (Original by E. P. Catts)

Piophila casei is a widely distributed species that commonly infests cured meats, cheeses, and dried fish. It probably is the species most commonly involved in gastrointestinal myiasis of humans. The tendency of people to leave cured meats and cheeses unrefrigerated makes these foods available to gravid females for oviposition. These flies can survive the rigors of alimentary tract passage and can even pupate and emerge as adults prior to leaving the host. The related *P. vulgaris* and *Stearibia* species are common inhabitants of dried carrion.

Neottiophilidae (Nest Skipper Flies)

This small Palaearctic family includes two genera, *Neottiophilum* and *Actinoptera*. The larvae cause cutaneous myiasis by sucking the blood of nesting birds. Larvae attack with their mandibles and can penetrate deeply enough into their host to cause septicemia and death. Larvae pupate in the host's nest in the fall and emerge as adults in the spring. The adult is a yellow-brown fly, about 7 to 8 mm in length, with wings "pictured" with a few brown spots. Hosts typically are passeriform birds. Local strains of *N. praeustum* show narrow host specificity to different avian species.

Drosophilidae (Pomace Flies, Vinegar Flies, Fruit Flies, and Wine Flies)

This is a large family (3,000 spp. in 60 genera) of small red-eyed flies (1.6 mm) whose adults favor the odors of overripe or fermenting plant products, usually fruits. Larvae feed on microorganisms found in such substrates. The genus *Drosophila* is the largest and includes more than half of the species in this family. Larvae have posterior spiracles on paired caudal protuberances (Figure 18.6), which also are evident in the puparia. The best known species is the highly

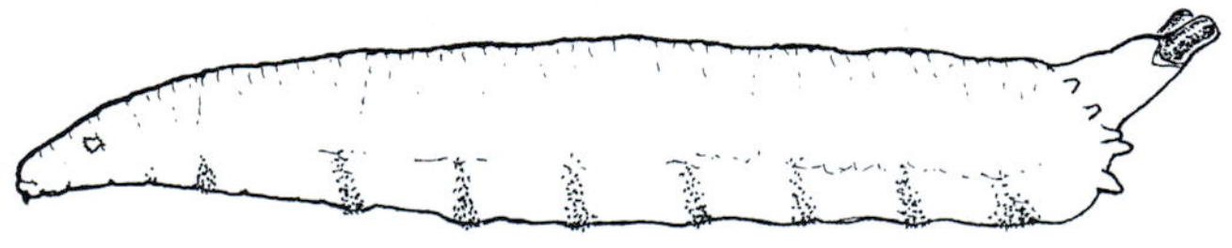

Figure 18.6 Pomace fly or fruit fly, *Drosophila* sp., larva (Drosophilidae). (Original by E. P. Catts)

domesticated kitchen gnat *D. melanogaster*. An additional seven species also are locally common domestic pests (e.g., *D. busckii, D. funebris, D. hydei, D. immigrans, D. repleta, D. simulans*, and *D. virilis*). The life cycle for these species is typically 12 to 14 days, making these small flies useful as biological models in studies of genetics, physiology, cytology, and population dynamics. Because of their attraction to fruits and vegetables these species can cause accidental gastrointestinal myiasis.

Chloropidae (Grass Flies and Australian Frog Flies)

The genus *Batrachomyia* includes 10 species whose larvae occur individually in swollen, subcutaneous pockets on the body (not the legs), of Australian frogs. The adult flies are yellow-brown in color and possess hairy eyes. Adults feed on plant juices. Their eggs require high humidity and are laid near, but not on, the host. After moving to a frog host, the larvae attack and appear to feed on blood, reaching a length of 10 mm when fully mature. The mature larvae are peculiar in appearance, having paired anterior and posterior "tentacles," each bearing a spiracle (Figure 18.7). Seasonal prevalence in frog populations can be as high as 25 percent with a parasite load of one to four maggots per host. Death results in about 10 percent of the frogs at the time of larval drop.

Anthomyiidae (Root Maggots)

The Anthomyiidae and the following two closely related families, Fanniidae and Muscidae, are quite similar in their development, behavior, and occasional association with cases of gastrointestinal myiasis. Both the immatures and adults are similar in appearance to muscids. As the name "root maggot" implies, most anthomyiid larvae feed on plants. The anthomyiid genus *Hylemya*, however, is very large with 180 species occurring in a diversity of habitats in the Nearctic region. The gastrointestinal myiasis they can cause generally results from the ingestion of larvae infesting vegetables or fruits.

Fanniidae (Faniid Flies)

Four species of *Fannia* have been reported to cause myiasis: little housefly (*F. canicularis*), latrine fly (*F. scalaris*), *F. incisurata*, and *F. manicata*. Adults of these flies look like small, slender house flies. They are drab gray in color and lack black stripes on the thorax. The larvae commonly occur in feces, rotting fruits or bulbs, and in bird nests. Larvae occasionally occur in older, somewhat dried carrion. The larvae have a characteristic, fringed appearance (Figure 18.8) that easily distinguishes them from other muscoid maggots. The fringes apparently allow these maggots to float in a near-liquid medium. The life cycle requires about one month. Although cases of enteric and urethral myiasis in humans have been documented (James, 1947; Leclercq, 1969), the involvement of *Fannia* species in myiasis is rare, except as important mechanical vectors of *Dermatobia hominis* in its southern range.

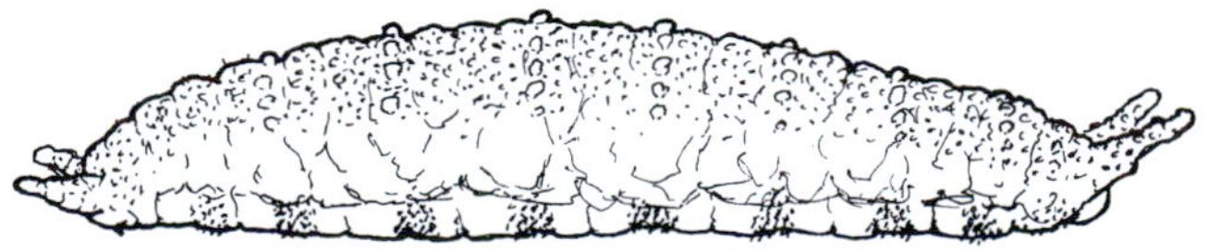

Figure 18.7 Australian frog fly, *Bratrachomyia* sp., larva (Chloropidae). (Original by E. P. Catts)

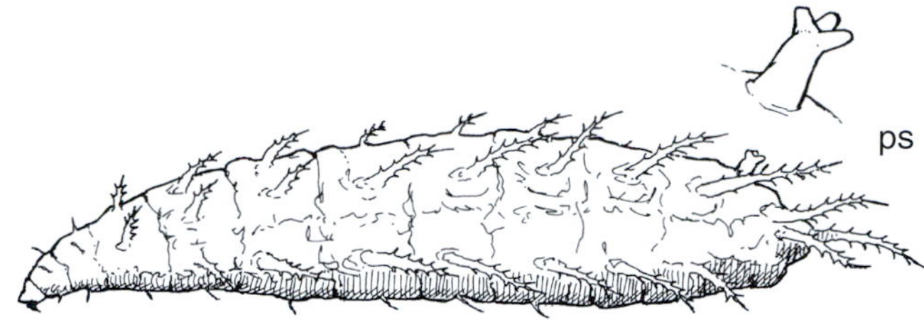

Figure 18.8 *Fannia* sp., larva (Fanniidae); ps, posterior spiracle. (Original by E. P. Catts)

Muscidae (Dung Flies)

The large family Muscidae includes at least seven genera in which species cause myiasis (Table 18.1). Muscid larvae (Figure 18.9) develop in a wide diversity of decaying organic matter, usually of plant or fecal origin. Occasionally they develop in old or buried carrion that is unsuitable for blow-fly exploitation. All stages of myiasis-causing muscids are typically house fly-like in appearance. Exceptions are *Neomyia* and *Hydrotaea* species, in which some adults have metallic coloration similar to blow flies, and the nest flies which are larger and yellow to yellow-brown in color. Gastrointestinal myiasis caused by muscids usually results from oviposition on wet foods. It also may result from retroinfection through the host's anus following fly attraction to foul odors or soiling by feces. Urogenital myiasis may occur in association with purulent discharges, urine-soaked clothing, and secondary microbial infections.

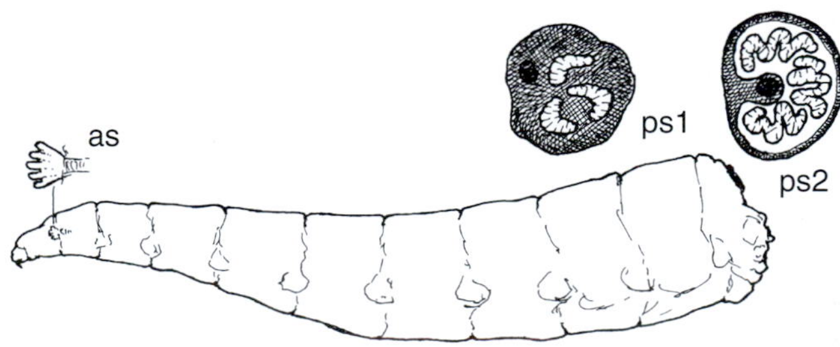

Figure 18.9 Typical muscid larva (Muscidae), third instar. as, anterior spiracle; ps1, posterior spiracular plate, house fly (*Musca domestica*); ps2, posterior spiracular plate, false stable fly (*Muscina stabulans*). (Original by E. P. Catts)

The genus *Musca* includes about 60 species that are confined mostly to the Old World. The two most important species that have invaded the temperate regions of the New World are the house fly or typhoid fly (*Musca domestica*) (Figure 18.9) and the face fly (*Musca autumnalis*). In the Old World tropics the prevalent species is the bazaar fly (*Musca sorbens*). Both the house fly and bazaar fly oviposit in a wide range of wet, decaying organic matter. Greenberg (1971) suggests that the house fly and bazaar fly originally were adapted, as larvae, to develop in wet ungulate feces. They do show preference for accumulations of animal excrement, especially that from horse, cow, human, pig, and poultry. Their egg complement ranges from 1,000 to 3,000 per female and is laid over the adult life span in clusters of 120 to 150 eggs. The quality of larval and adult diets largely determines the number of eggs produced by any single female. Maggots develop rapidly in three to five days under wet, warm conditions but are intolerant of desiccation. In the tropics, their life cycle can be as brief as 10 to 12 days, whereas three weeks is more typical in other regions. All stages can overwinter, but in colder areas there is a dramatic winter die-off. Larvae of *Musca* species that invade wounds feed primarily on necrotic tissues. *Musca* species with rasping-sucking mouthparts that feed on blood or other secretions (e.g., *M. autumnalis*) are not known to cause myiasis.

The genus *Muscina* includes eight species, three of which are implicated in accidental myiasis (*M. stabulans*, *M. assimilis*, and *M. pabulorum*). The false stable fly (*M. stabulans*) (Figure 18.9) is the most important and is involved primarily in gastrointestinal myiasis. Occasionally their maggots occur in fetid sores or wounds. Adults of this species look much like house flies, except that they are usually larger and more robust. They are attracted to, and feed on, plant juices, rotting fruits, and insect-excreted honeydew. Females oviposit by scattering their 140 to 200 eggs on the surface of overripe, decaying fruit. They also oviposit on accumulations of dead insects or feces, usually from human sources, and on buried carrion. Early-instar larvae are saprophagous but become predaceous as they mature. Third-instar larvae prey on smaller maggots. This transition from a saprophagous to a predaceous habit has two advantages over species whose larvae remain saprophagous: first, the maturing larvae can store protein resources obtained from their prey to be used by the adult in reproduction; and second, this habit enables this species to exploit a wider range of protein-poor resources as a larval substrate. Larval development varies from two to three weeks. They usually overwinter as pupae.

Hydrotaea species are metallic-colored muscids usually found in association with feces and older carrion. Like *M. stabulans*, their dark, cream-colored larvae (up to 15 mm long) become predaceous on other maggots. In Australia, *H. rostrata* has been implicated as a tertiary invader in myiasis of sheep, but appears to feed only on necrotic tissues.

Tropical Nest Flies

These muscid flies include the genera *Passeromyia*, *Mydaea*, and *Philornis*. Species in these three genera usually parasitize the young of cavity-nesting birds. Although few details are known about these flies, their biologies appear to be similar and are discussed collectively here. Species in the first two genera are widely distributed in the Ethiopian, Oriental, and Australian regions. The latter genus is found in the New World tropics. Currently *Philornis* is the valid generic name for those flies formerly listed in the genus *Neomusca*.

Adults of these tropical nest flies feed on plant juices and wet feces. Gravid females oviposit in bird-nest debris near nesting birds. After the eggs hatch, the maggots crawl to the nestlings, scratch the skin with their mouthhooks, and imbibe blood. As they feed they continue to scratch and penetrate host tissues. In heavy infestations they can penetrate the host body cavity with fatal results for the bird. The larvae develop rapidly, in less than 1 week, before leaving the host to pupate in the nest debris. Post-feeding larvae of *Philornis* species exude a frothy, sticky, salivary spittle that coats the puparium and to which camouflaging debris adheres. In spite of this defensive measure, pupae are subject to attack by parasitic wasps.

Calliphoridae (Blow Flies, Carrion Flies, Floor Maggots, Nest Maggots, Screwworms)

The most generalized of the six families of Oestroidea is the Calliphoridae with over 1,000 species. Among the members of this large family there is a transition from the facultative myiasis habit by a large number of normally saprophagous species to obligatory myiasis by a relatively small number of species (ca. 100). The larvae typically feed on wet, living, or dead flesh. Desiccation is detrimental to both egg and larval survival. The following discussion treats the more important, widely distributed, and common genera of myiasis-causing calliphorids.

Carrion-Associated Blow Flies These are the showy metallic blue-bottle flies, green-bottle flies, and black blow flies (Figure 18.10). They include members of the genera *Calliphora*, *Chrysomya*, *Cynomya*, *Eucalliphora*, *Lucilia* (=*Phaenicia*), *Paralucilia*, *Phormia*, and *Protophormia*, which are commonly associated with dead animal tissues or carrion. The Old World genus *Chrysomya* also includes one species (*C. bezziana*) that causes obligatory myiasis. The importance of this group as primary agents of cutaneous myiasis and impact on animal production cannot be overstated, especially *Lucilia sericata*, *L. cuprina*, and *Phormia regina*. These three species are discussed in the section, "Veterinary Importance." The duration of the life cycle of these flies in carrion differs among species but, in general, roughly one-third of the time is spent as the eggs and larvae, one-third as pupae, and one-third as adults from emergence to mating and oviposition. Life cycles typically take three to four weeks, but are prolonged by cold temperatures.

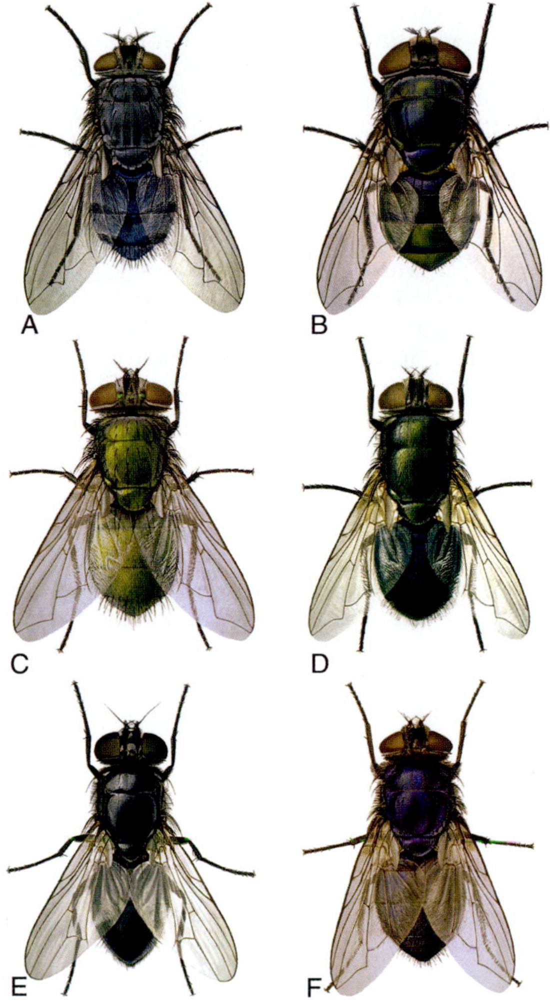

Figure 18.10 Blow flies that cause myiasis, adults (Calliphoridae). (A) *Calliphora vicina*; (B) *Chrysomya megacephala*; (C) *Lucilia* (=*Phaenicia*) *sericata*; (D) *Phormia regina*; (E) *Ophyra leucostoma*; (F) *Protophormia terranovae* (B. Greenberg, 1971; Flies and Disease, Vol. 1, Ecology, Classification and Biotic Associations, Princeton University Press).

These flies are attracted to fetid, purulent open sores and chronic nasopharyngeal or urogenital infections. Heavy larval infestations often result in death of the host, after which the maggots continue to feed on the resulting carrion.

The body form of most of these calliphorid larvae is typical of members of this family (Figure 18.11) and is distinguished by the arrangement and shapes of their spiracles and cephalopharyngeal skeleton. The larvae of a few genera, however, are atypical in possessing numerous girdling bands of fleshy processes (e.g., *Chrysomya* spp.) similar to those of the genus *Fannia*. Like *Muscina stabulans*, the maggots of the hairy maggot blow fly (*Chrysomya rufifacies*) can switch from a myiasis to a predatory role. Postfeeding maggots drop free from the host, or leave the carcass of a dead host to wander in search of pupation sites.

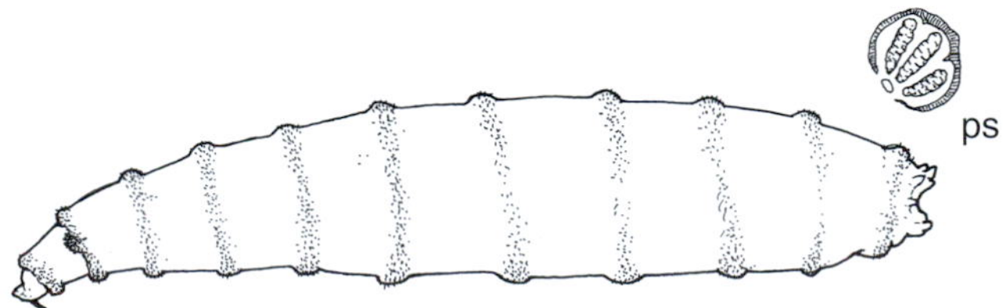

Figure 18.11 Typical blow-fly larva (Calliphoridae), third instar. ps, posterior spiracular plate. (Original by E. P. Catts)

Toad Blow Flies (*Bufolucilia* spp.) These flies include several species that cause primary obligatory myiasis in amphibians. Although they are fairly common in the Old World, *B. silvarum* is the only North American species. The female lays her eggs on the back and flanks of a living toad, a risky activity for any fly. The resultant larvae invade the host's nasal passages or eye sockets where they develop rather quickly, in less than a week. The host dies at about the time of larval maturation, after which some larvae may remain to feed on the dead host. Maggots thus also occur on dead amphibians. The metallic green adults are attracted to other carrion as well.

Screwworms (*Cochliomyia* spp., *Chrysomya* spp.) The New World screwworm (American screwworm) (*Cochliomyia hominivorax*) and the Old World screwworm (*Chrysomya bezziana*) cause obligatory myiasis and can be of major economic importance. Both are primary screwworms with similar biology. Females, when gravid, are attracted to fresh open wounds on any warm-blooded animal. The female quickly oviposits 100 to 200 eggs on the dry perimeter of the chosen site, with each female producing up to 1,000 eggs in her lifetime. Female flies commonly feed at the wound, thus obtaining protein for producing their next egg mass. After a brief incubation period of 10 to 20 hours, the eggs hatch and the larvae begin feeding on the open wound. The maggots develop rapidly to maturity in 4 to 12 days and then drop to the ground to pupate. After about a week as pupae, adults emerge, mate at aggregation sites, and after several days seek a protein meal and new living host. The entire cycle (egg to egg) takes two to three weeks.

Infested wounds range from mere thorn scratches or insect and tick bite marks to gaping lacerations. Infestations of the umbilicus of newborn calves by *C. hominivorax* and *C. bezziana* are common. Livestock husbandry operations such as castrating, dehorning, branding, and shearing also cause wounds subject to invasion. Untreated screwworm cases can be fatal, due to the invasion of host's vital organs, septicemia caused by feeding maggots, or secondary infections. The maggots literally eat the host alive. In areas of mild winters, adults can be active during warm spells. In temperate regions screwworm attacks are restricted to the warm seasons, but in the tropics they are more or less continuous.

Cochliomyia hominivorax (Figure 18.12A) is a major livestock pest, especially to cattle in the Neotropics.

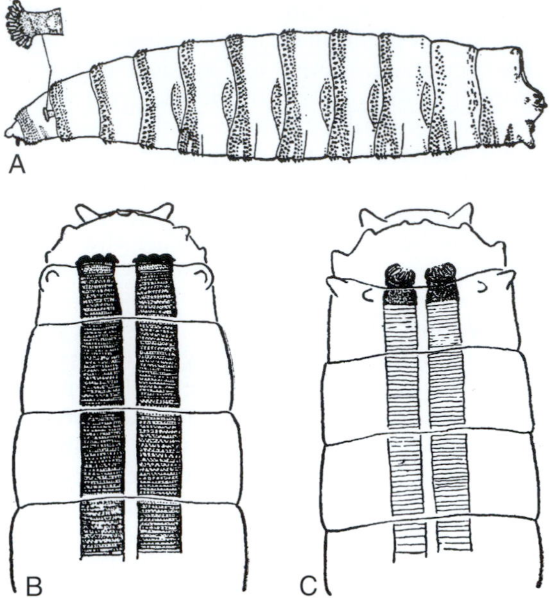

Figure 18.12 New World screwworm larvae (Calliphoridae), third instars. (A–B) Primary screwworm (*Cochliomyia hominivorax*), with enlargement of anterior spiracle and darkened portion of large tracheal trunks from posterior spiracles extending into three or four abdominal segments. (C) Secondary screwworm (*C. macellaria*), with darkened portion of the tracheal trunks restricted to part of one abdominal segment. (Modified from James, 1947)

Although formerly it ranged throughout tropical and temperate regions of the New World, innovative control measures using male sterilization and baiting of females have eliminated it from the Nearctic. *Chrysomya bezziana* is distributed widely in Southeast Asia, New Guinea, and Africa. It is the Afro-Asian counterpart of *C. hominivorax* and also is primarily important as a parasite of livestock. Like the New World screwworm, this species is attracted to open wounds as well as to moist body openings. Generally eggs are deposited in the late afternoon such that their development is completed by the next morning, thereby avoiding lethal exposure to direct sunlight and drying. The rapidly developing maggots consume flesh in localized sites, often penetrating *en masse* deep into the host tissues with fatal consequences.

Certain advantages are gained by these flies attacking only living hosts. There is less interspecies competition than among carrion-exploiting species, and the constant body heat of the host enhances maggot development. Additionally, the host tissues are less acidic and more digestible to the maggots than that of fresh carrion. Finally, and possibly more importantly, they avoid predation by ants and other predators since carcasses tend not to persist for long in the tropics.

Species of secondary screwworms are close relatives of primary screwworms but cause facultative, rather than obligatory, myiasis. They are termed secondary because they often infest wounds after invasion by primary myiasis-causing flies, but also have been observed in the absence of primary screwworm invasion. The most important species are *Cochliomyia macellaria* (Figure 18.12B) and *Chrysomya megacephala*. The latter species has become established in the Neotropics and appears to out-compete the former species. Another related blow fly is *Chrysomya rufifacies*. It is a widely distributed carrion-breeding fly that commonly shifts from being a scavenger in carrion to a predator of other maggots. However, on the island of Maui in Hawaii, maggots of *C. rufifacies* have been known to cause primary myiasis in the umbilicus of newborn calves, with prevalence of infestations as high as 30%.

Nest Blow Flies (*Protocalliphora* spp.) These flies are members of the large genus *Protocalliphora* (90 species), Holarctic flies whose maggots (Figure 18.13) are obligatory parasites of nestling birds. The intermittent, temporary bloodsucking habits of these maggots are similar to those of the tropical muscid nest-maggots. The maggots attach to the host by means of their mouthhooks to feed and then drop free to hide in nest debris while they digest their blood meals. Direct adverse effects on the nestlings are apparent only when maggot numbers are high. There is evidence, however, that their frequent bleeding of the host prolongs nestling development. The longer the time that birds stay in the nest, the more they are at risk of predation. Maggots thus can indirectly influence nestling success. Many species of *Protocalliphora* show high host specificity, indicating a long relationship with their respective hosts. These flies seem to favor cavity-nesting birds. Pupation and oviposition take place in the host nest. How the adult fly locates a specific host's nest is unknown.

Tumbu Fly (*Cordylobia anthropophaga*) The tumbu fly belongs to a small group of African blow flies whose maggots develop in warble-like cysts in hosts ranging from rodents to dogs and people. Rodents are assumed to be their primitive hosts, the flies having become secondarily adapted to other species. Except for their smaller size, these flies have biologies similar to the New World bot flies (Cuterebrinae), in many ways suggesting convergent evolution.

The adult tumbu fly is yellow-brown in color with two rather broad but variable dorsal thoracic stripes. Mature

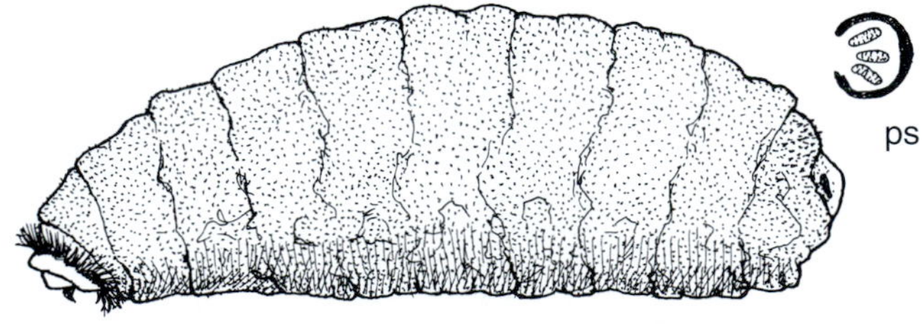

Figure 18.13 Nest blow fly, *Protocalliphora* sp., third-instar larva (Calliphoridae). ps, posterior spiracular plate. (Original by E. P. Catts)

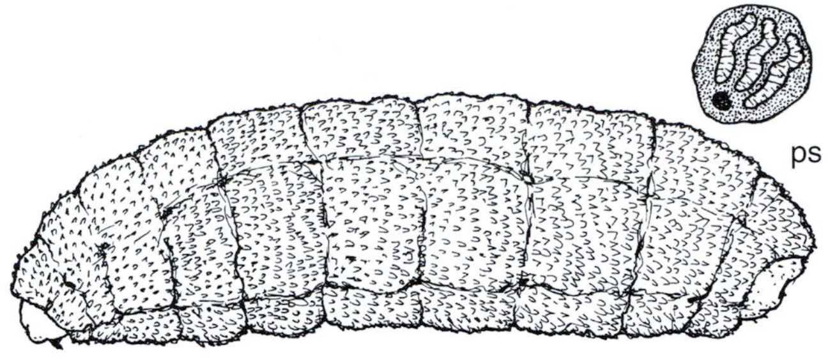

Figure 18.14 Tumbu fly, *Cordylobia anthropophaga*, third-instar larva (Calliphoridae). ps, posterior spiracular plate. (Original by E. P. Catts)

maggots are up to 15 mm in length and densely, but incompletely, covered with small, backwardly-directed, single-toothed spines. The posterior spiracles have a weakly sclerotized peritreme and three sinuous slits (Figure 18.14). Adults feed on decaying fruits, carrion, and feces. Females are shade-loving and deposit eggs singly in dry sand or dirt contaminated with host urine or feces. Females also are attracted to dry, urine-soiled diapers or clothing. They lay up to 500 eggs over their lifespan of two to three weeks. The eggs hatch after several days, following which the first-instar larvae wait in the dry substrate for a host. Contact with a host stimulates the maggots to attach and penetrate the skin. Maggots develop in shallow warbles within or just beneath the skin in about seven to 10 days and drop free to the ground to pupate in surface debris. Adults emerge after another several weeks. Although the tumbu fly invades a wide range of hosts, its successful development varies significantly among different host species. The domestic dog is an important reservoir, but maggots develop best in native rodents.

Congo Floor Maggot (*Auchmeromyia senegalensis*) The Congo floor maggot (formerly *A. luteola*), is one of four or five species in this genus causing obligatory, temporary myiasis similar to that caused by *Protocalliphora* species. They all occur in Africa south of the Sahara, and most are associated with the burrows of larger mammals such as warthogs. *Auchmeromyia senegalensis*, however, appears to prefer humans as hosts. Adults are yellow-brown in color with markings similar to the tumbu fly. They feed on rotting fruits and feces (e.g., human, monkey, pig). Females can lay up to 300 eggs. The maggots (Figure 18.15) spend most of their time hidden in loose dirt and floor debris of native huts. At night they crawl to sleeping hosts, scrape or otherwise break the skin, and suck the oozing blood. After feeding for about 20 minutes the maggots return to hide in debris until the next night. They require two blood meals for each of the three instars and pupate in debris. The life cycle takes about 10 weeks.

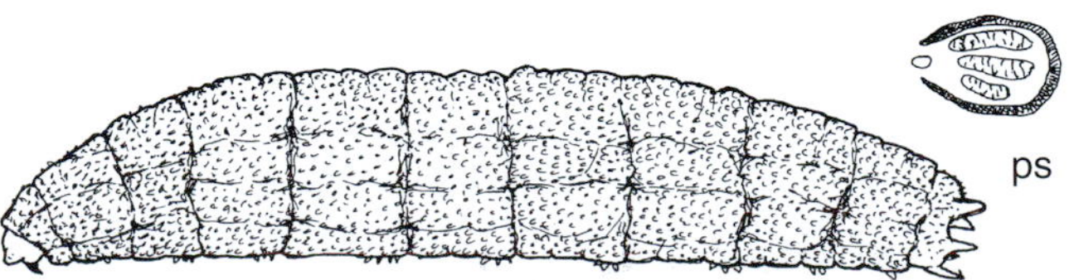

Figure 18.15 Congo floor maggot, *Auchmeromyia senegalensis* (Calliphoridae), third instar. ps, posterior spiracular plate. (Original by E. P. Catts)

Deer and Water Buffalo Skin Maggots (*Boopona* spp.) These myiasis-causing flies include four species of yellow-brown blow flies whose maggots parasitize the skin of the back and feet of cervids and bovids in Eastern Europe, Asia, and the Orient. In the case of cervids they also attack the soft, developing antler buds. Their eggs are attached singly to hairs of the host and require three to five days to develop prior to hatching. These maggots invade the host skin where they develop individually in warble-like boils in about a one-week period. Mature maggots drop from the host to the ground to pupate, and adults emerge in two to three weeks.

Elephant Skin Maggot (*Elepantoloemus indicus*) This species is a small, orange-brown blow fly whose maggots develop only in warble-like boils in the skin of the Asian elephant. Little is known about its biology.

Sarcophagidae (Flesh Flies)

Species of this large, widely distributed family are classified into two subfamilies: the Miltogramminae, which, with few exceptions, are obligatory parasitoids of insects and other arthropods; and the Sarcophaginae, with necrophagous species that include facultative and obligatory parasites causing myiasis. Adults (Figure 18.16) are typically medium to large, black and gray flies with longitudinal thoracic stripes and a checkered, or tessellated, abdominal pattern. All sarcophagid species are larviparous; the gravid female retains the eggs in an expanded, bilobed, uterine pouch until they are ready to hatch. Females produce 30 to 200 larvae, depending on the species involved. Flesh fly larvae (Figure 18.17) are more robust than blow fly larvae and possess paired mandibles in all instars. Their posterior spiracles are recessed in a deep cavity, and the inner slit of each spiracle is parallel to, or slants away from, the ventral mid-body line.

Sarcophaga species usually are associated with carrion or feces, but can cause facultative wounds and accidental gastrointestinal myiasis. About 20 *Sarcophaga* species have been incriminated in cases of gastrointestinal myiasis. One widely distributed species is the red-tailed flesh fly (*S. haemorrhoidalis*), which frequents feces and is attracted indoors by fecal odors. Only a few species in other sarcophagine genera have been recorded as causing myiasis of the gastrointestinal or wound type.

Few flesh flies cause obligatory myiasis. The most widespread and important species are in the miltogrammine genus *Wohlfahrtia*. They include the Old World species *W. magnifica*, and the New World species *W. opaca* and *W. vigil*. These three species have evolved as primary

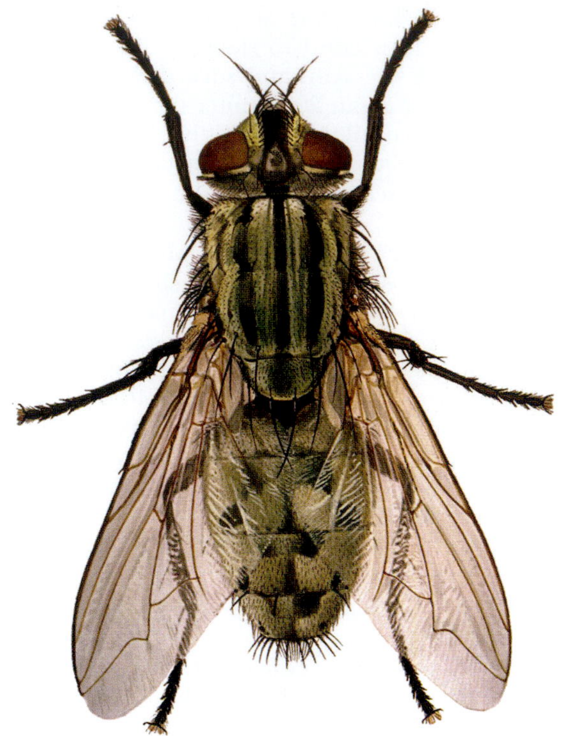

Figure 18.16 Red-tailed flesh fly, *Sarcophaga haemorrhoidalis* (Sarcophagidae), adult (B. Greenberg, 1971; Flies and Disease, Vol. 1, Ecology, Classification and Biotic Associations, Princeton University Press)

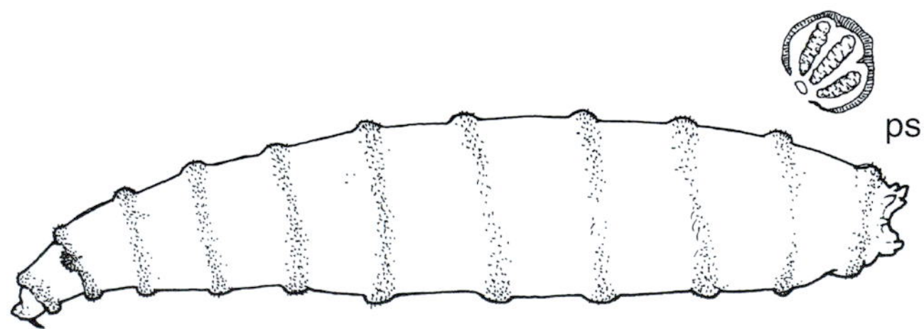

Figure 18.17 Flesh fly, *Sarcophaga* sp. (Sarcophagidae), third-instar larva. ps, posterior spiracular plate. (Original by E. P. Catts)

invaders. Their females larviposit at moist body openings and at fresh wounds or scratches. Larvae can even penetrate thin, unbroken skin. The maggots burrow into the subcutaneous tissue to feed, inducing the formation of a boil-like cyst around groups of larvae with a small, common breathing pore opening to the outside.

Wohlfahrtia adults resemble large house flies with very distinct, longitudinal, thoracic stripes. Unlike other flesh flies, these have the abdomen clearly marked with black spots. A gravid *Wohlfahrtia* female produces 120 to 170 larvae. In a host, the maggots grow rapidly and can cause considerable tissue destruction. After about a week, maggots drop to the ground to pupate and can overwinter in this stage.

The closely related *W. nuba* is not parasitic, but feeds only on necrotic flesh. It has been used successfully in treating ragged, infected wounds similar to the use of *Lucilia sericata*. Most other *Wohlfahrtia* species are scavengers, but all are probably capable of at least facultative myiasis.

A group of four unrelated genera of flesh flies cause facultative, and apparently obligatory, myiasis of certain amphibians and reptiles. In some cases they also attack amphibian and reptilian eggs, killing the developing embryos. These genera are *Anolisimyia*, *Cistudinomyia*, *Eumacronychia*, and *Metoposarcophaga*. Little is known of their biology.

Oestridae (Bot Flies)

Bot flies are the most highly evolved group of obligate myiasis-causing parasites of mammals. They are treated as four distinct subfamilies in the Oestridae. The most primitive are the Cuterebrinae, the New World skin bot flies. Their counterparts are the Hypodermatinae, the Old World skin bot flies. The nose bot flies are in the Oestrinae with their probable center of origin being in Africa. The remaining subfamily is Gasterophilinae, the stomach bot flies, which also appears to have evolved in Africa. All four of these subfamilies were recognized previously as families and are treated as such in earlier literature. The subfamilies of Oestridae can be separated as third-instar larvae by their general appearance, and by the form of their caudal spiracular openings. Bot-fly maggots are thick, robust, grub-like larvae with moderate to heavy spiny armature. As with most other flies discussed in this chapter they pass through three larval instars, and they drop to the ground to pupate. Except for members of the Cuterebrinae, all bot flies overwinter as larvae in the host. The cuterebrines typically overwinter as diapausing pupae free from the host, although there are exceptions in subtropical and tropical regions.

Bot flies differ from other obligatory myiasis producers in several ways. First, the adults do not feed or take in nutrients. Most of them have only rudimentary, nonfunctional mouthparts and are unable to feed. Those few with functional mouthparts and an associated alimentary tract probably only imbibe water to maintain an internal fluid balance. Second, bot flies (with the notable exception of *Dermatobia hominis*) show either a high degree of host specificity or they parasitize only a small group of related hosts. Although some bot fly maggots occasionally occur on atypical hosts, the susceptibility of a host does not necessarily imply the suitability of that host for normal, or successful, bot fly development. Third, bot fly maggots show a marked level of site specificity in a normal host. In abnormal hosts, site specificity can be erratic and can lead to dire results for both host and parasite. Fourth, the site of invasion by the first-instar bot maggot generally is not the site of maggot development. With the exception of *D. hominis*, first-instar maggots of oestrid flies move from the point of invasion to a different site for further development. Interaction between the host and its developing bot fly maggots is generally benign, with the associated pathology and parasite burden being tolerated well by native, coevolved hosts. Bot fly maggots generally cause little injury to their hosts at low to moderate population levels.

Humans are not among the normal hosts for any bot fly species, including the so-called human bot fly (*D. hominis*). However, people may become incidentally infested by bot flies under certain circumstances. In such cases, the associated pathology tends to be more severe than that of their normal hosts.

Burrowing first-instar bot fly larvae occasionally cause paralysis or death of the host. Developing larvae located in warbles at critical sites such as around the eyes and on the feet can increase the risk of predation by interfering with the host's ability to see or escape. Small mammalian hosts encumbered by an ever-enlarging cluster of warbles also may have difficulty in foraging.

Another characteristic of bot flies is that the bee-like adults usually aggregate at specific topographic sites for pairing and copulating. Favored sites are hilltops, cliff faces, steep slopes, prominent rocks or trees, and streambeds. Male flies remain at these sites throughout their brief life, but females leave the sites soon after mating to search for suitable hosts or oviposition sites.

The major importance of bot flies is the economic losses that they cause in livestock operations (e.g., cattle, sheep, goats, reindeer, and horses). Secondary microbial infection of the bot warble is rare because bot fly maggots produce bacteriostatic secretions as they develop. Even after the larva exits the warble, other myiasis-causing flies rarely exploit the empty wound, which normally heals very rapidly. Bot fly maggots cannot complete their development in a dead host. If the host dies, so do its bots.

The evolutionary history of bot flies is not known, but warrants comment. Zumpt (1965) proposed two possible routes for bot fly evolution. One route is through blood-sucking larvae such as nest maggots or floor maggots. The other is through carrion-breeding species and screwworms. Both alternatives seem plausible with regard to skin bots, but they do not explain how the more internally adapted groups such as nose bots and stomach bots originated. The nose bots may have evolved from myiasis-causing flies that were attracted to mucopurulent nasal secretions in hosts suffering from respiratory infections. Stomach bots, on the other hand, may have originated from fly species infesting decaying, fermenting forage, a diet favored by many large herbivores. Interesting discussions of evolution and the oestrid flies and their hosts can be found in treatments of this subject by Papavero (1977) and Pape (2006).

New World Skin Bot Flies (Cuterebrinae)

There are two genera and 58 species in this subfamily of bot flies, all restricted to the western hemisphere. The largest genus is *Cuterebra* (57 spp.), which includes some of the largest bot flies. Some of these robust, thick-bodied bot flies (Figure 18.18) are up to 30 mm in length. Their normal hosts are rodents (e.g., *Microtus*, *Neotoma*, and *Peromyscus* spp.) and lagomorphs (e.g., *Lepus* and *Sylvilagus* spp.). At temperate latitudes cuterebrine maggots show seasonal peaks in prevalence. For example 40% prevalence in *Peromyscus* populations is not unusual during late summer. Non-native rodents and rabbits (e.g., *Mus*, *Rattus*, *Cricetus*, and *Oryctolagus* spp.) also are parasitized, but in these hosts the pathology is more severe and can lead to death of both the host and parasite.

Figure 18.18 Rodent bot, *Cuterebra emasculator*, adult-female (Oestridae, Cuterebrinae); reared from white-footed mouse, *Peromyscus leucopus*. (Photo by Sturgis McKeever)

Cuterebra species oviposit in areas close to the center of host activity (e.g., near nests and lairs or along runs). After about a week, the eggs hatch in response to a sudden increase in temperature, normally indicating a nearby, warm host. First-instar maggots adhere to the host pelage, crawl to natural body orifices of the head, and penetrate the mucosal tissue at such sites as the mouth and nose. After about a week in the pharyngeal areas of the host, the maggots actively burrow through sheets of host connective tissue to a species-specific cutaneous site for maggot development (Figure 18.19). Once there, the maggot cuts an opening through the skin and molts

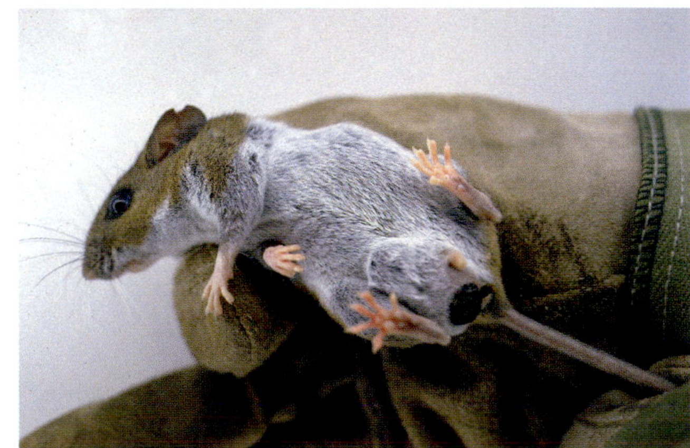

Figure 18.19 Cotton mouse, *Peromyscus gossypinus*, with mature rodent bot, *Cuterebra* sp., or wolf; posterior end of bot, with posterior spiracles exposed, projecting from location at base of host tail. (Photo by Sturgis McKeever)

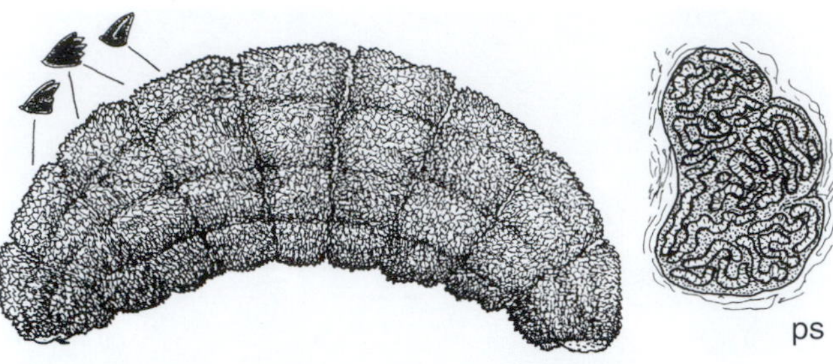

Figure 18.20 Rodent bot, *Cuterebra* sp. (Oestridae, Cuterebrinae), third-instar larva. ps, posterior spiracular plate. (Original by E. P. Catts)

within the newly formed warble. Depending on the bot species involved, maggot development at this site requires three to eight weeks, during which time the much enlarged maggot can increase some 100,000 fold in weight. When mature, the third-instar larva (Figure 18.20) backs out through the warble pore and drops to the ground to pupate. After the bot exits, the collapsed warble heals quickly, usually without secondary infection. In cool, temperate regions it is the pupa that diapauses and overwinters, and there is but one adult flight season per year. Warmer areas probably have two flight seasons where adults are on the wing during late spring and summer. The adult life span is about two weeks.

Figure 18.21 Multiple wolves of squirrel bot (*Cuterebra* sp.) in shoulder area of gray squirrel, *Sciurus carolinensis*. (Courtesy of Department of Pathobiology, Auburn University College of Veterinary Medicine)

Generally there is little economic importance associated with *Cuterebra* species, although they can be a seasonal problem in commercial rabbit operations. Sport hunters often discard bot-infested squirrels (Figure 18.21) and rabbits in the erroneous belief that the carcasses are spoiled by the presence of these maggots. A few *Cuterebra* species, especially *C. fontinella*, have been colonized in the laboratory and used as natural bot-host models for the study of bots affecting livestock.

Tórsalo (*Dermatobia hominis*) The *tórsalo* is a Neotropical species that occurs widely from southern Mexico to Argentina. In Brazil this very important parasite is known locally as *berne*. Although primarily a pest of cattle, it also infests humans, dogs, monkeys, sheep, horses (rarely), and other domestic and wild mammals. This is the only bot fly that frequently parasitizes humans, hence its alternative common name, the human bot fly.

It is a woodland species encountered along forest margins of river valleys and lowlands. It is unusual among cuterebrine flies because of its unique oviposition behavior and means of egg dispersal. Rather than depositing eggs directly on a host, the adult female (Figure 18.22) captures various zoophilic or anthropophilic arthropods, usually dipterans, and glues her eggs in clusters (15–45 eggs) to their abdomen (Figure 18.23). Embryonation requires five to 15 days. These egg carriers, or **porters**, subsequently transport the eggs to a vertebrate host where they hatch while the arthropod feeds. Among the more common porters are day-flying

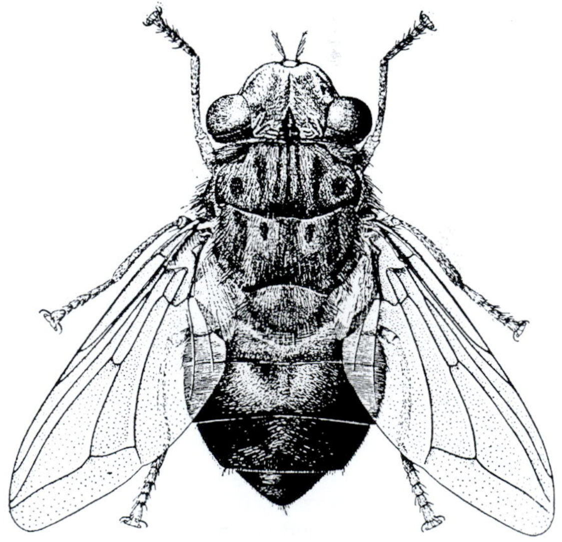

Figure 18.22 Tórsalo, or human bot fly, *Dermatobia hominis* (Oestridae, Cuterebrinae), adult female. (James, 1947)

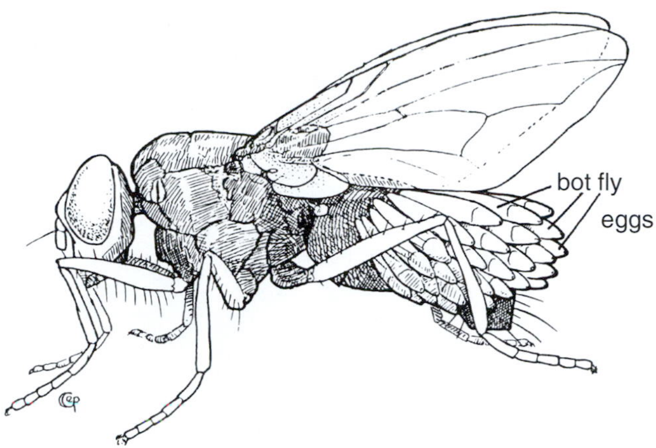

Figure 18.23 A muscid fly porter (*Sarcopromusca arcuata*) to which a tórsalo bot fly (*Dermatobia hominis*) has attached her eggs. (Original by E. P. Catts)

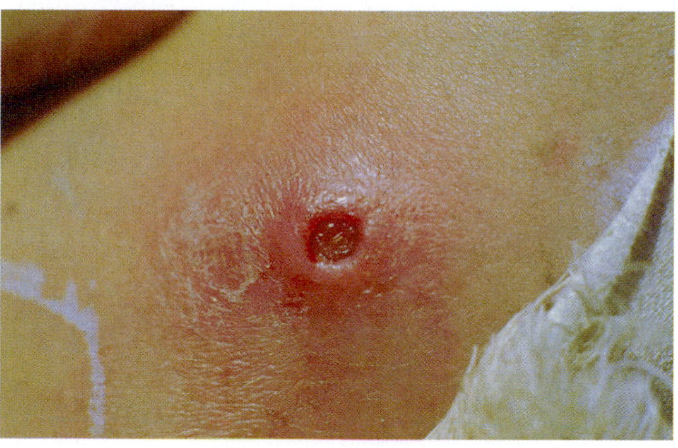

Figure 18.25 Tórsalo, or human bot fly, *Dermatobia hominis*; posterior end of maturing larva visible in wound made by larva in skin on human forearm. Note inflammation and swelling at site of wound. (Photo by Ronald Cave)

mosquitoes (particularly *Psorophora ferox*) and muscid flies (e.g., *Sarcopromusca pruna*, *Stomoxys calcitrans*, *Fannia* spp., and *Synthesiomyia* spp.). The newly emerged larvae enter the skin either through the bite puncture or via hair follicles, soft folds of skin, or areas of moist skin in contact with clothing or bedding. Development occurs at the point of entry, forming a boil-like pocket, or furuncular lesion, where the larva (Figure 18.24) undergoes three instars. This development usually takes five to 10 weeks but sometimes as long as three months or more. During this time the narrower posterior end of the larva is extended into the opening at the skin surface where it exchanges air in respiration (Figure 18.25). After the larva matures, it enlarges the opening and drops to the ground to pupate. The pupal stage lasts 25 to 132 days. The longer period occurs in the colder regions of Brazil and Argentina where the pupae are able to overwinter (Ribeiro, 2007).

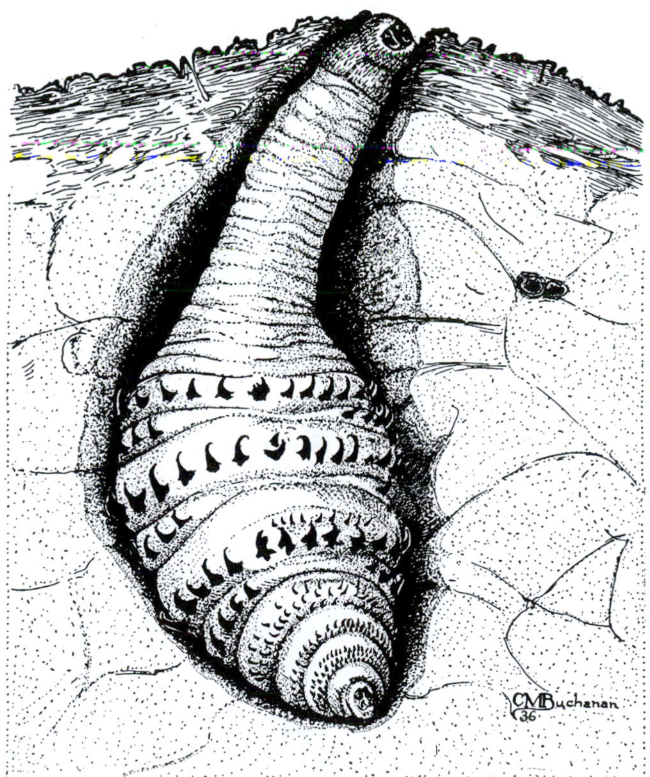

Figure 18.24 Tórsalo, or human bot fly, *Dermatobia hominis* (Oestridae, Cuterebrinae), mature larva in human skin, with posterior spiracles exposed through hole in skin surface. (Craig and Faust, 1940)

Other Cuterebrine Flies The remaining cuterebrine flies are tropical and include only a few species. Most of them parasitize rodents and marsupials, and what is known about their biology has been described by Guimaraes and Papavero (1999) as separate genera. *Cuterebra* (formerly *Allouatamyia*) *baeri* parasitizes howler monkeys (*Alouatta* spp.) in Central and South America and has been implicated in reduced survival of juvenile monkeys. This is the only bot fly specific to a primate host. Warbles of this species usually are located in the cervical and axillary regions of their arboreal hosts.

Old World Skin Bot Flies (Hypodermatinae)

These flies are the Eurasian counterpart of the New World skin bots. There are nine genera and 31 species occurring in rodents, deer, goats, and cattle. All but two species are found only in Asia, Europe, and Africa. The most widespread and important species are in the genus *Hypoderma*, which includes six species, three causing myiasis in cervids, one in yaks, and two in bovids.

Cattle Grubs (*Hypoderma* spp.) The northern cattle grub (*Hypoderma bovis*) and the common cattle grub (*H. lineatum*) (Figure 18.26) are Holarctic in distribution, having been introduced wherever cattle are raised. They are major economic pests of domestic

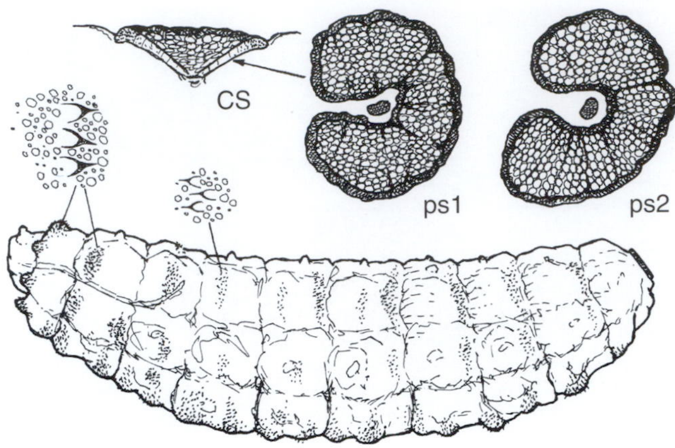

Figure 18.26 Cattle grubs, *Hypoderma* spp. (Oestridae, Hypodermatinae), third instars. cs, cross-section of posterior spiracle of *H. bovis*, showing depressed center that contrasts with flat spiracle of *H. lineatum*; ps1, northern cattle grub (*H. bovis*), third instar; ps2, common cattle grub (*H. lineatum*), third instar. (Original by E. P. Catts)

Figure 18.28 Heel fly, *Hypoderma bovis* (Oestridae, Hypodermatinae), adult female. (Courtesy of Agriculture and Agri-Food Canada)

Figure 18.27 Gadding behavior of calf in response to attack by heel fly (*Hypoderma bovis*), with tail raised and calf kicking up heels of hind legs while running frantically. (Photo by J. Weintraub, Agriculture Canada)

Figure 18.29 Eggs of heel fly, *Hypoderma lineatum* (Oestridae, Hypodermatinae), attached to body hair of cow. (Courtesy of Agriculture Canada)

cattle. Losses include damage to hides and self injury by hosts during headlong flights of panic, or gadding (Figure 18.27), in futile attempts to escape ovipositing flies. Thus the name gad fly often is used to refer to the adults. Adults of *Hypoderma* species also are called heel flies, referring to the defensive behavior of cattle in kicking up their hooves.

The biology of *H. bovis* and *H. lineatum* is very similar. In late spring and early summer, adult females (Figure 18.28) of both species glue their eggs directly to host hairs. *Hypoderma lineatum* usually deposits rows of eggs (Figure 18.29) on the lower body regions of standing or even recumbent hosts, whereas *H. bovis* normally deposits single eggs in the same regions on active hosts. Presence of the latter species is what causes cattle to gad. After an incubation period of three to seven days, eggs hatch and the first-instar larvae crawl to the base of the hairs on which the eggs were glued. They then penetrate the host skin using proteolytic enzymes. A four- to six-month period of migration and overwintering follows as the larvae make their way between sheets of connective tissue and fascial planes within the host, or along nerve pathways for *H. bovis*. During the winter, first-instar larvae of *H. lineatum* can be found in the esophageal submucosa, whereas the larvae of *H. bovis* can be found within the epidural fat along the spinal column. With the onset of spring the larvae of both species leave these "resting" sites and move to the host's back where they cut a hole, the warble pore, and develop through two subsequent larval instars. A boil-like warble develops around the enlarging maggot (Figure 18.30). The cuticle of mature grubs is black in color, and these mature larvae back out of the warble pore and drop to the ground to pupate.

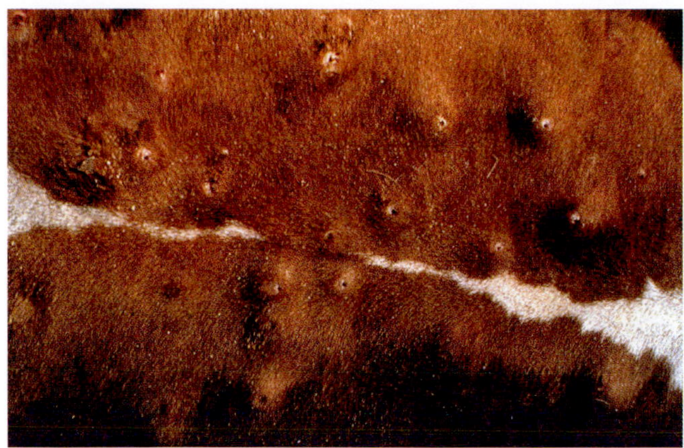

Figure 18.30 Multiple warbles along back of cow, caused by larvae of *Hypoderma bovis* (Oestridae, Hypodermatinae). (Photo by J. Weintraub, Agriculture and Agri-Food Canada, Lethbridge)

As with all bots, it is at this time of dropping from the host that grubs are most vulnerable to predation by birds, rodents, and insectivores. The pupal stage lasts one to three months, depending on ambient temperature. Adult flies live only three to five days and fly quickly after emerging to mate and oviposit throughout late spring and early summer.

Another *Hypoderma* species worthy of note is *H. tarandi* (formerly *Oedemagena tarandi*). In the arctic and subarctic regions, *H. tarandi* causes cutaneous myiasis in the backs of reindeer and caribou, similar to that caused by cattle grubs. Heavier infestations generally occur in yearling fawns rather than in other age groups. Over time infested animals slowly develop partial immunity to these parasites. The other three hypodermatine species (*H. actaeon, H. diana,* and *H. sinensis*) are also Holarctic in distribution and responsible for warbles in cervids and yaks.

Oestromyia spp. (Figure 18.31), *Portschinskia* spp., and *Ochotonia lindneri* commonly cause cutaneous myiasis in wild rodents and lagomorphs in Europe, Asia, and the Far East. The remaining five hypodermatid genera (*Strobiloestrus, Pallasiomyia, Oestroderma, Pavlovskiata,* and *Przhevalskiana*) have been reported from African antelopes and lechwes; Asian antelopes, pika, and gazelles; and Middle Eastern and African sheep, goats, and gazelles, respectively.

Nose Bot Flies (Oestrinae)

This subfamily of oestrid flies includes nine genera and 34 species that parasitize members of the mammalian orders Artiodactyla, Perissodactyla, and Proboscidea (elephants). Most nose bot flies are African or Eurasian in distribution; an exception is the Holarctic genus *Cephenemyia*. The genera *Oestrus* (6 species) and *Rhinoestrus* (11 species) comprise the majority of species in this group. Another species, *Tracheomyia macropi* (Figure 18.32), develops in the trachea of the red Kangaroo in Australia and is the only native bot of that continent.

Nose bot flies differ from other bots in that their eggs develop *in utero*. The first-instar larvae are ejected by the hovering female directly into the muzzle or eye of the host. The larvae crawl down the throat to enter tracheal branches of the lungs, but soon return to the nasal sinuses or pharyngeal region of the host to complete their development. As with other bots in native hosts, there is little pathology at moderate parasite levels. However, purulent mucous exudates associated with an abundance of maggots may lead to respiratory complications or to secondary fly attack. While hosts are under attack by adult nose bots, they stop grazing and attempt to thwart the attack by pushing their muzzles into bushes or clumps of grass. Following development, mature larvae are sneezed from the nostrils of the host, causing some temporary suffering during this time. Occasionally a few larvae may become lodged in the nasal sinuses and can cause the death of their host. After a pupal period of four to six weeks, adults emerge and seek a mate at aggregation sites. Adults generally are univoltine in cold regions and at least bivoltine in tropical and warm temperate areas. This indicates that larval development can be delayed during winter and accelerated during summer. Overwintering takes place within the host, as is the case of most other bot flies.

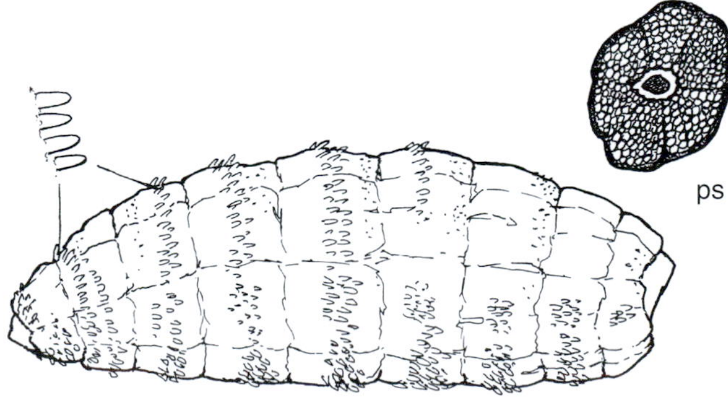

Figure 18.31 *Oestromyia* sp. (Oestridae, Hypodermatinae), third-instar larva; causes myiasis in wild rodents and lagomorphs. ps, posterior spiracular plate. (Original by E. P. Catts)

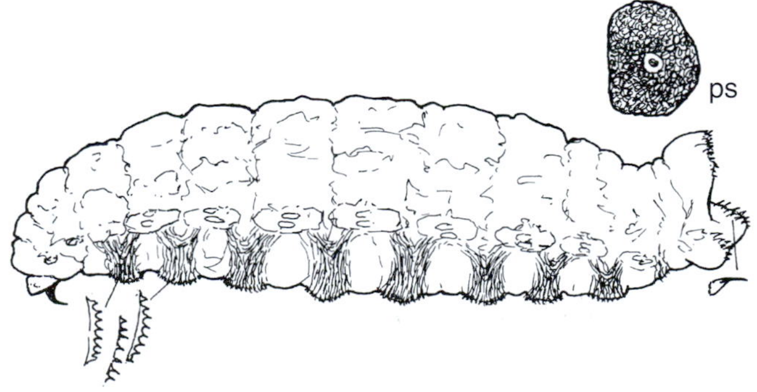

Figure 18.32 Kangaroo throat bot, *Tracheomyia macropi* (Oestridae, Oestrinae), third-instar larva; develops in trachea of the red kangaroo, Australia. ps, posterior spiracular plate. (Original by E. P. Catts)

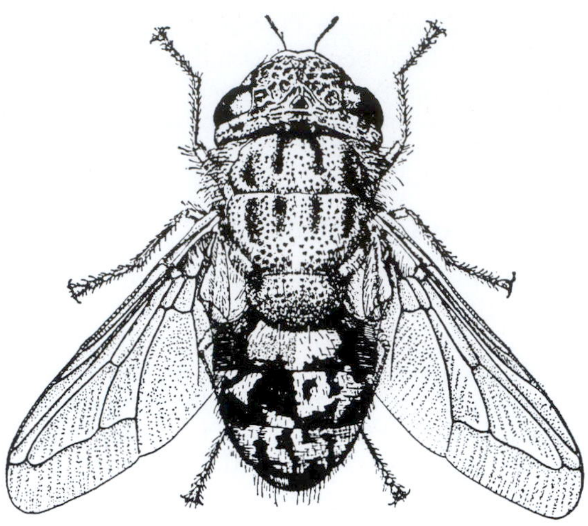

Figure 18.33 Sheep bot fly, *Oestrus ovis* (Oestridae, Oestrinae), female. (James, 1947)

Figure 18.34 Sheep bot, *Oestrus ovis* (Oestridae, Oestrinae), third-instar larva. (Courtesy of P. Scholl)

The most widely distributed and economically important species is the sheep nose bot (*Oestrus ovis*), which parasitizes domestic and wild sheep and goats (Figures 18.33, 18.34). Females produce about 500 progeny, which they larviposit in batches of 10 to 25 at a time. The usual host burden is 12 to 24 maggots, with gradual attrition reducing this number to fewer than 10 survivors by the time the maggots mature. Other *Oestrus* species parasitize antelopes and wild goats in Africa and Asia.

The horse nose bot (*Rhinoestrus pupureus*) is distributed widely throughout Eurasia, Africa, and the Orient. It is most prevalent in Asia where high population levels of these bots in domestic horses have been recorded (>700 maggots in a single host). High parasite loads can cause death of the host. Ocular myiasis in people who live near or handle horses is not uncommon. The general life history of the horse nose bot is like that of *O. ovis. Rhinoestrus* includes 11 species, four in equids and seven others in specific nonequine hosts (e.g., giraffe, antelope, bushpig, hippopotamus, and warthog).

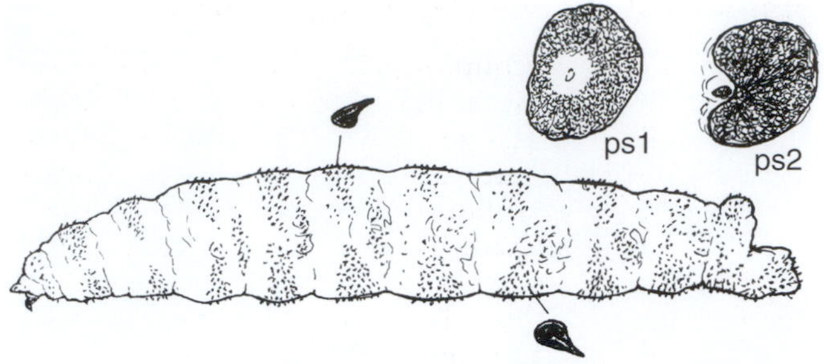

Figure 18.35 Deer nose bot, *Cephenemyia jellisoni* (Oestridae, Oestrinae), third-instar larva. ps, posterior spiracular plate. (Original by E. P. Catts)

Eight species in the genus *Cephenemyia* (Figure 18.35) parasitize cervids (deer). Only one, *C. trompe*, is distributed throughout the northern Holarctic region. The others are confined to either the Nearctic or Palaearctic areas. All are deer nose bots. The four common species in North America are *C. phobifer* (Figure 18.19) to the east of the Continental Divide, and *C. apicata*, *C. jellisoni*, and *C. pratti* to the west. Although *C. trompe* is named the reindeer throat bot, it occurs in deer, moose, and caribou as well as reindeer. The life history of these bots is like that described for other nose bots. Mature maggots usually crowd the retropharyngeal pouches in the throat of their host. On completing their development they crawl to the anterior nasal passages where they are expelled by sneezing and pupate on the ground under surface debris.

The incidental occurrence of several oestrine species in the eyes of atypical hosts suggests that the orbit also may serve as a target for certain larvipositing bot flies. In Africa, first-instar larvae of *Gedoelstia cristata* regularly occur in the eyes of native wildebeest and hartebeest hosts. Later instars are found in the nasal cavities. Ocular invasion by this species in domestic sheep, goats, cattle, and horses produces gross pathological lesions and high levels of mortality. *Cephalopina titillator* (Figure 18.36) is the nose bot of camels and dromedaries from Central Asia to Africa. Its life history is like that of other nose bots.

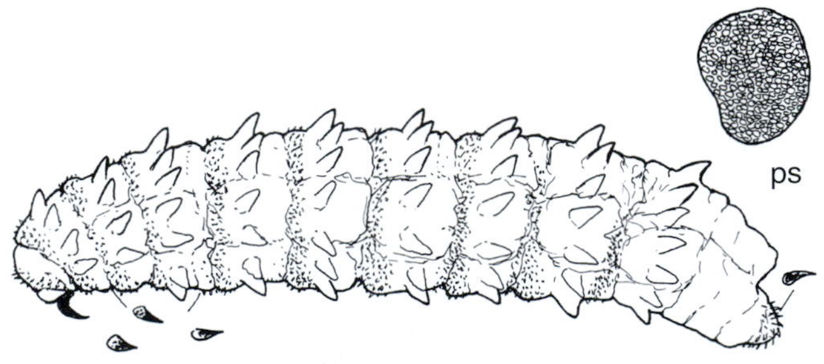

Figure 18.36 Camel nose bot, *Cephalopina titillator* (Oestridae, Oestrinae), third-instar larva. ps, posterior spiracular plate. (Original by E. P. Catts)

Stomach Bot Flies (Gasterophilinae)

Adult flies of this group, represented by 17 species in five genera, resemble honey bees in their general size and color. The largest genus is *Gasterophilus*, the horse stomach bot flies, with nine species, three of which have worldwide distribution (*G. intestinalis*, *G. nasalis*, and *G. haemorrhoidalis*). These parasites are common companions of horses and donkeys wherever these hosts occur. Few horses live to old age without having carried their load of stomach bots along the way.

Following oral entry, developing maggots attach to the gastrointestinal mucosa (Figure 18.37), causing inflammation, sloughing of tissue, and ulcerations. They do not cause warble formations. Burrowing of first-instar larvae in the mouth lining, tongue, and gums can produce pus pockets, loosen teeth, and cause loss of appetite of the host. A gadding response to hovering, ovipositing females can cause self-inflicted injuries as well. However, *G. nigricornis* does not hover but alights on the host's cheek to oviposit, and *G. pecorum* oviposits on leaves and stems of potential host forage.

Most gasterophiline flies glue their eggs to the hair shafts of the host's body. Because of their brief life span of a few days at most, females can deposit all their eggs within a couple of hours if hosts are available and the weather is mild. Their larvae develop in the stomach and intestines of equids, elephants, and rhinoceroses, suggesting a very old coevolutionary relationship to these distantly related mammalian groups. Eggs that are ingested with forage or wetted by self-grooming of the front legs hatch after a brief incubation. Eggs located on the host's head hatch spontaneously. Upon burrowing into the oral tissues for a brief period, first-instar larvae eventually are swallowed and attach to the stomach or intestinal wall. The site of attachment is specific for each fly species. Larvae overwinter in the gastrointestinal tract. After larval development is completed, the mature larvae are expelled with the host feces during the warmer seasons. Pupation occurs in the soil soon after larvae drop from the host, with the pupal stage lasting about three weeks. Adult flies emerge, mate, and quickly resume activity near potential equine hosts. If hosts are not available, the bot flies move to high points to aggregate and mate, when the females initiate a longer-distance search for hosts.

The common horse stomach bot fly (*Gasterophilus intestinalis*) (Figures 18.38, 18.39) is worldwide in distribution and is the predominant species in North America. It prefers to oviposit on the lower forelegs of horses. The two other species in North America are the throat horse bot (*G. nasalis*) and the rarer rectal horse bot (*G. haemorrhoidalis*). The former oviposits on the hairs of the chin and lower jaw and the latter on the hairs of the nose and lips. *Gasterophilus nasalis* is the predominant species in the Southern Hemisphere. The dark-winged horse bot (*G. pecorum*) is the most commonly encountered species in Eurasia and Africa. It is the most pathogenic fly in this genus and can cause host fatalities resulting from constricted swelling of the esophagus due to attached maggots.

Other genera of stomach bots include *Gyrostigma* (3 species) in rhinoceroses and *Cobboldia* (3 species)

Figure 18.37 Common horse stomach bot, *Gasterophilus intestinalis* (Oestridae, Gasterophilinae); larvae attached to mucosa and inner surface of stomach of heavily infested horse. (Courtesy of Department of Pathobiology, Auburn University College of Veterinary Medicine)

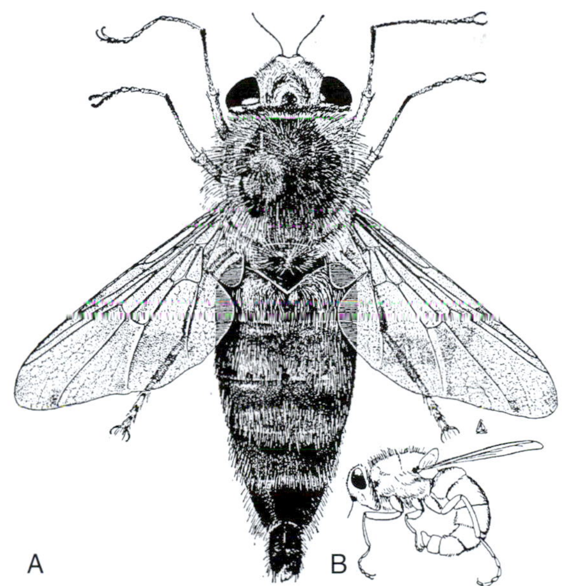

Figure 18.38 Common horse bot fly, *Gasterophilus intestinalis* (Oestridae, Gasterophilinae), adult female. A, Dorsal view; B, Lateral view. (James, 1947)

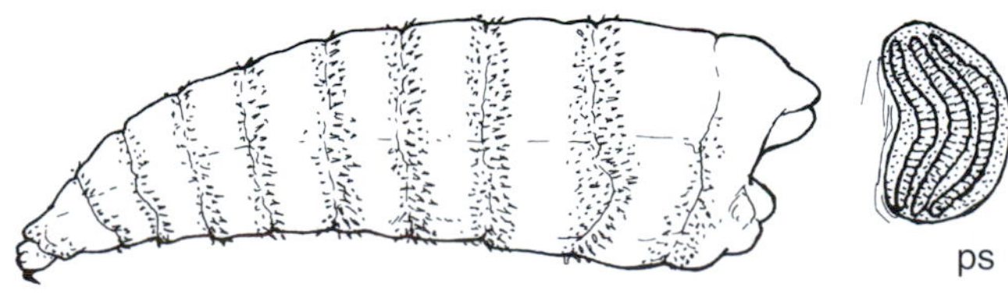

Figure 18.39 Common horse stomach bot, *Gasterophilus intestinalis* (Oestridae, Gasterophilinae), third-instar larva. ps, posterior spiracular plate. (Original by E. P. Catts)

parasitizing African and Indian elephants. Their life histories are similar to that of *Gasterophilus*. Two other obligate-myiasis fly species are associated with the African elephant. Both are cutaneous parasites. *Ruttenia loxodontis* develops in warble-like skin boils on the buttocks and flanks of elephants, whereas *Neocuterebra squamosa* develops in shallow ulcers in the skin crevices of the elephant's feet. Both produce a pseudowarble during their development. The phylogenetic relationship of these unplaced genera to other bot flies is uncertain but recently these genera have been placed with the gasterophilines (Colwell et al., 2006). This placement agrees with Zumpt (1965) who suggested that both species represent early gasterophilines.

Other Oestroid Flies There are three other families of flies included in the Oestroidea in which the species do not parasitize vertebrate animals: Tachinidae, Rhinophoridae, and Mystacinobiidae. All members of the Tachinidae are obligate parasitoids of other insects. None are involved in myiasis. The Rhinophoridae are a small sister group of the tachinids that parasitize isopods. Mystacinobiidae, a sister group of the calliphorids, includes but a single species in New Zealand that is coprophagous on bat guano and phoretic on the bats themselves. None of these families include myiasis-causing species.

PUBLIC HEALTH IMPORTANCE

Myiasis is a relatively uncommon ailment among people worldwide, occurring only seasonally and sporadically in temperate regions and associated with wet seasons in the tropics. The notable exception occurs in South America, where several hundred cases of infestation in humans by the New World screwworm, *Cochliomyia hominivorax*, have been recorded, most often in individuals in a severely debilitated condition (Guimaraes and Papavero 1999). Most human cases of facultative or obligatory myiasis are only temporary, or are aborted because humans are unsuitable hosts (e.g., nose or stomach bots). Maggots normally do not complete their development in people because they usually are interrupted by self-grooming or medical intervention. Infants and infirm or debilitated older age groups are generally at greater risk because of associated difficulties in maintaining a minimal level of personal hygiene. Soiling of bedding or clothing with excrement can result in invasion through urogenital or anal sites. Gastrointestinal myiasis usually results from the ingestion of eggs or maggots with infested foods, commonly causing general malaise, nausea, vomiting, cramps, and diarrhea. Although such cases are seldom serious, adverse effects can be prolonged and may require purging of the alimentary tract to resolve the problem.

Facultative or obligatory wound myiasis is more serious because of the potential for rapid and indiscriminate destruction of tissues. Infirm persons are at particular risk, as are people who work and live in close proximity to livestock operations where carrion and suitable hosts are more available. Myiasis involving other flesh flies and bot flies occasionally occurs in people. The former usually involves infants, and the latter, livestock handlers and field workers. Ocular myiasis is the most common form of human myiasis, for which numerous cases are recorded. It usually is caused by the sheep nose bot when a gravid female larviposits directly into an eye while hovering. As many as 50 first-instar *O. ovis* maggots have been removed from the eye following the strike of a single fly. Although the larvae do not actually penetrate the eye tissues, they can cause extreme irritation as they crawl about on the eyeball. They are best removed by flushing the eye with fluids or by gentle suction.

An aberrant type of cutaneous myiasis in humans is caused by the first-instar larvae of *Gasterophilus* and *Hypoderma* species. In such cases, larvae hatching from eggs laid on a person penetrate the skin and wander in the epidermis, producing visible, sinuous, inflamed tracks accompanied by irritation and pruritus. This is called cutaneous larval migrans. If cases involving horse bot flies (*Gasterophilus* species) are not treated, the larvae eventually die. However cattle grubs (*Hypoderma* species) can cause much more severe pathology in humans, sometimes with fatal results. The larvae are able to penetrate and burrow into deeper tissues with the potential for serious consequences because they are not well adapted to human hosts. More cases involve *H. bovis* than *H. lineatum* and tend to be more severe in children. Both have been reported as causing ocular myiasis and skin warbles, the latter usually in the neck and shoulders. Several human cases involving prolonged but temporary paralysis have been reported.

New World bot flies also parasitize humans. More than 60 cases of human cutaneous myiasis caused by *Cuterebra* species have been documented in North America. Clinical signs appear about 10 to 14 days after the bot fly eggs have been deposited on the skin in outdoor situations. The individual's body heat stimulates the eggs to hatch, following which the first-instar, after contacting the host, enter the skin through cuts, abrasions, or natural body openings. Larval development usually progresses to the second and, occasionally, the third instar. The danger is that in the meantime the wandering first-instar larvae may invade deeper tissues, leading to dire consequences if they should penetrate the cranial cavity, eyes, or vital organs.

The tórsalo, or human bot fly, causes cutaneous myiasis in people throughout its range in Central and South America. Lesions generally occur on the exposed extremities, but also may involve the scalp, forehead, external ears, eyelids, lower back, and thighs. Skin lesions at the site of larval development are relatively minor during the early stages. As the larva grows, however, the site becomes intensely itchy and exudes serosanguineous fluids. About a week before emergence of the mature larva, the lesion becomes tender, moderately swollen, and generally more painful due to the spines of the larva irritating the skin as it moves or rotates within. Upon emergence of the larva the lesion usually heals spontaneously in about a week. If the larva dies before emergence or the wound becomes infected, serious complications can result. These include foul discharges and fetid odors that may attract other skin-invading parasites (e.g., screwworms), leading to cases of secondary myiasis. Occasionally, *D. hominis* infestations of the head have resulted in penetration of brain tissue, particularly in infants and young children, causing convulsions and death (Rossi and Zucoloto, 1972).

Larvae of *D. hominis* can be extracted by physically squeezing them out or killing them by injection with an anesthetic and surgically removing them. They also can be induced to exit the skin by blocking off their air supply with strips of pork fat (bacon therapy), lard, petroleum jelly, soft beeswax, and other materials applied to the lesion. In attempts to reach air, the larvae penetrate the material and are extracted when it is removed.

Clinical Use of Maggots

A few myiasis-causing flies feed only on necrotic tissue as they develop. They are species such as *Lucilia sericata* that are usually associated with carrion. In hospitals the larvae may be deliberately used in the treatment of deep wounds and sores in situations in which surgery is impractical. This approach has been used by people since ancient times for treating deep, chronic, or extensive surface wounds. Anthropological evidence shows that such widely located cultures as the Mayans of Central America, the aboriginals of Australia, and the hill people of Myanmar (formerly North Burma) all used maggots for such wound therapy. Later, with the advent of explosive artillery shells, bombs, and grenades in warfare, battlefield injuries became much more difficult to treat using conventional surgery. During the Napoleonic War, American Civil War, and World War I, military physicians observed that wounds inadvertently infested with maggots healed quickly with minimal scarring. This led to the deliberate use of maggots, hatched from surface-sterilized eggs, to remove necrotic tissue in treating chronic wounds, bed sores, severe burns, and bone infections. The procedure was called maggot therapy, and gained wide use. **Maggot therapy** was a common procedure in medicine until the general use of antibiotics replaced the practice in the 1940s. Now, following a 50-year hiatus, the clinical use of surgical maggots in treating such maladies is being practiced again in cases where antibiotics are ineffective and where surgery is either impractical or is refused by the patient.

Three species of blow flies commonly have been used in treating wounds: *Lucilia illustris*, *L. sericata*, and *Phormia regina*, with *L. sericata* being the species of choice. Care must be taken to assure that maggots from the population selected for colonization feed only on dead tissues. Once the laboratory colony is established, eggs are sterilized in an antiseptic bath (e.g., sodium hypochlorite, formalin, or hydrogen peroxide) and held for hatching. First-instar maggots are transferred to the patient's wound where they are confined with a sterile screen-dressing fastened to the surrounding healthy tissue with adhesive.

The maggots act as microsurgeons, removing necrotic tissue along with debris and associated microorganisms. In moving about the wound while feeding, maggots stimulate the formation of desirable granulation tissue and secrete calcium carbonate, ammonia, and allantonin (a substance that promotes wound healing). Together they produce a more alkaline milieu less conducive to bacterial growth. The maggots develop at a predictable rate and can be removed and counted as they mature. Healing of maggot-treated wounds is very rapid with less scarring than conventional surgery. There is no need to use anesthesia during this process because the active maggots cause little discomfort except for occasional tingling sensations. Another advantage is the much lower cost of the entire procedure even when including the cost of maintaining a fly colony.

VETERINARY IMPORTANCE

Livestock and wildlife are at greater risk of attack by myiasis-causing flies than are humans. This is a result of greater exposure and the tendency to have more untreated open wounds as sites for fly exploitation. The seasonal incidence of myiasis in nonhuman hosts increases during calving, foaling, and lambing when young animals are less resistant in their behavioral and humoral response to maggot invasion. The availability of placental carrion and the exposed umbilicus of newborn animals enhance related fly activity. Open wounds associated with shearing, branding, dehorning, and castrating operations also increase the vulnerability of animals to attack. Blow flies also oviposit on fecal- or urine-soiled pelage. The exploitation of scrapes, scratches, and other skin wounds by myiasis-causing carrion flies, flesh flies, and screwworms can have a major economic impact on livestock operations throughout the world, especially in tropical and warm temperate latitudes.

The sheep blowflies, *Lucilia cuprina* (=*Phaenicia pallescens, Phaenicia cuprina*) and *Lucilia sericata* (=*Phaenicia sericata*) are very important causes of myiasis in sheep in Australia, South Africa, New Zealand, and Great Britain. Such an infestation commonly is referred to as blowfly strike. These flies especially attack Merino and other breeds, which have numerous skin folds and crusted wool that retain excrement, causing fleece rot. Common sites of attack are the anal and genital regions of sheep. Prophylactic measures include keeping these body regions shorn, or surgically removing skin in these areas to limit fouling, termed **Mule's operation** or **mulesing**. *Lucilia sericata* generally invades only fresh carrion at warm latitudes where it is not associated with myiasis. However, in Australia, a race of this species is known to cause secondary facultative myiasis in sheep. This race of fly also exhibits a greater developmental success when infesting live sheep than when infesting dead sheep as carrion. Annual economic losses to blow-fly myiasis in Australia and other sheep-raising countries are very substantial. Pissle strike is another form of myiasis in Australia in which blow flies infest the urethral orifice of a ram's penis, attracted by the surrounding dense, urine-soaked fleece. Sheep strike in the northern Holarctic Region is most often associated with two important and closely related calliphorids, *Phormia regina* and *Protophormia terraenovae* (Wall and Shearer, 1997).

In terms of gross pathology, mortality, and economic losses, the most important myiasis-causing flies worldwide are the screwworms. Untreated screwworm cases nearly always lead to death of the host. The adult flies produced in one infested animal serve as a source for subsequent cases. Fatalities are due to the invasion of vital organs, septicemia resulting from large maggot-infested wounds, or predation of the debilitated host. Prior to the development and implementation of screwworm control measures, annual losses attributed to screwworms amounted to several hundred million dollars annually in North America alone. Because wild vertebrates also are subject to screwworm attack, they constitute important reservoirs of these parasites for domestic species. Any open wound, no matter how small, is a potential site for oviposition by primary screwworms. The treatment of wounds, whether to prevent or remove screwworms, is a costly procedure in time and labor, especially when it involves livestock on open range.

The tórsalo is an important pest of domestic cattle throughout tropical areas of Central and South America. Horses and other equids, however, are seldom bothered. Clusters of warbles can occur anywhere on the host, but only rarely are they secondarily invaded by other myiasis-causing flies. Even then they invade only after the premature death of the larva inside the furuncle. The most important injury is hide damage, causing significant economic losses to the very important leather industry of South America. Heavily infested cattle may become weak and emaciated and have difficulty walking because of *D. hominis* larvae infesting their legs and feet. Animals can become seriously crippled when wounds become infected, especially when retaining dead larvae. Deaths of heavily infested cattle and dogs have occurred in such cases. Certain cattle breeds (e.g., Brahma, Zebu) appear to be differentially resistant to tórsalo infestations and show lower infestation levels than do Herefords and dairy breeds.

Economic losses to cattle grubs, primarily due to hide damage, reduced weight gain, meat trim, and reduced milk production, were estimated to be as high as $100 million annually in North America in the 1980s (Drummond et al., 1988). This figure does not include the cost of controlling cattle grubs. A grub burden of less than 50 larvae per host appears to have little effect on the vitality or weight increase of infested cattle. Economic losses are caused primarily in cattle when open warbles and scars of healed warbles along the host's backline decrease the value of hides used in making leather. The green, or raw, hide constitutes about 10 percent of the total carcass value, and shot-holed hides (warble-scarred) can result in considerable reduction in grower profits. Warbles caused by cattle grubs and tórsalo larvae in the backs of slaughtered animals require excessive trimming of discolored flesh and jellied fat surrounding the warbles, further contributing to decreased profits due to labor costs and loss of meat.

Horses also are commonly infested by *Hypoderma* first-instars, especially *H. bovis*. Grubs in horses, being in an abnormal host, produce abscesses when the larvae make their way to the back and begin to open their breathing holes. Because of the discomfort caused to the backs of infested horses, the use of these animals as saddle mounts is difficult, if not precluded, during certain times of the year.

Gadding by cattle in response to *Hypoderma* females, either in enclosures or on open range can result in serious injuries and even death of animals. Such injuries in turn are open to invasion by other myiasis-causing flies. Headlong galloping by panic-stricken cattle can bruise udders and, in the case of pregnant cows, cause spontaneous abortion. The accidental death of an animal killed in a fall incurred while fleeing from a bot-fly attack has been blamed at times on large predators subsequently discovered feeding on the remains. In the Neotropics, because of the egg-porter phenomenon, hosts take no evasive action to avoid adult tórsalo flies.

Nose bots in sheep cause relatively low economic losses compared to cattle grubs. Unless the nose bots are very numerous, the value gained by the costly handling of sheep to administer controls is questionable. At times individual sheep may die as a result of unusually large numbers of maggots (e.g., 100–300 per host), on the order of 10 times the usual host burden. Significant weight gains and increased fleece length have been reported in Russia and South Africa as a result of controlling nose bots in lambs. Organophosphate and macrocyclic lactone insecticides, such as the avermectins, are effective against nose bots.

Horse stomach bot flies are ubiquitous parasites of horses throughout the world. Chronic and repeated infestations can result in loosened teeth due to larval tunneling in the gums and in ulcerations of the gastric lining. Eventually this will interfere with forage ingestion and digestion. However, most horses tolerate a maggot burden of 100 or so with no apparent ill effects. Depending on the value of the individual animal, whether a race horse or cow pony, a family pet or a "candidate for the glue factory," the economic loss is widely variable. Different breeds of the domestic horse also react differently to egg laying by the hovering adults of the common stomach bot fly. Thoroughbreds, American saddle horses, and Arabians, for example, are much disturbed by the hovering fly and will take evasive action that can cause self-injury. At the other extreme, Shetland ponies, Morgans, and draft breeds appear to ignore the ovipositing flies. All breeds, however, react in wild panic to ovipositing *G. nasalis* females.

Among the horse stomach bot flies, *G. pecorum* is the most pathogenic. Its maggots often attach *en masse* in the oral cavity and upper alimentary tract. This can result in obstruction caused by the enlarging maggots attached to the soft palate and esophagus and the associated inflammation. Heavy populations hinder swallowing, prevent food passage, and can cause severe digestive problems in the host, sometimes with fatal results.

The possible use of cuterebrine bots as biocontrol agents of rodents has been suggested. However, the associated pathology in native hosts is minimal, and bot populations appear to have little effect on rodent numbers.

PREVENTION AND CONTROL

There are three major approaches for controlling myiasis: avoiding contact between potential hosts and myiasis-causing flies, early treatment of wounds to prevent myiasis, and reduction or elimination of myiasis-fly populations. In the area of human health, there is general reliance on hygiene and medical or surgical intervention. In veterinary cases the most common approach is the use of insecticides, especially systemic compounds that target the parasitic larvae, administered to the host.

Preventing unnecessary or avoidable outdoor exposure of humans during fly seasons is one obvious way to control facultative or obligatory myiasis. Attractive odors associated with urine, feces, vomitus, nasal secretions, and purulent sores invite fly strike. Healthy adults are able to react to flies attempting to feed or oviposit, but infants and elderly individuals often are either unaware or unable to take evasive measures. Myiasis-causing flies sometimes enter buildings when attracted by similar odors. Management of geriatric-care facilities should routinely include measures such as screening, diligent resident hygiene, and treatment of sores to avoid attraction and contact with flies. Securing foods in fly-proof containers, under screens, or under refrigeration is the best means of preventing gastrointestinal myiasis. Foods that have been exposed, even when fly eggs are present, can be made safe by freezing or thorough cooking.

The risk of human myiasis is greater in livestock areas. People working or residing close to livestock operations (e.g., stables, dairies, feedlots, and poultry houses) are most at risk. Flies or bee-like insects hovering around a person or alighting at open sores should be suspect. Immediate evasive responses usually will drive the fly away. Where the tórsalo fly occurs in the Neotropics, insect repellents can be used to reduce the attacks of blood-seeking insects carrying tórsalo eggs.

Human myiasis can be prevented by early treatment of open wounds with an antiseptic salve and protective dressing. Wound dressings should be changed regularly and not be allowed to become excessively soiled by wound exudates. *Dermatobia* larvae in furuncles can be induced to back at least part way out of the warble by covering the pore with a thick smear of petroleum jelly. The protruding maggot then can be grasped with forceps or fingers and slowly extracted. Sometimes surgical removal is necessary; this entails snipping the rim of the warble pore to allow greater ease of extraction. Maggots in eyes, nasal sinuses, or the auditory meatus should be removed by flushing with saline or a diluted antiseptic such as carbolic acid. Gastrointestinal myiasis can be treated with purges or emetics, although several treatments may be necessary to remove all maggots.

Preventive measures for the control of myiasis in livestock, zoo animals, and wildlife mainly involve the removal of carrion that may attract flies, early treatment of open wounds to prevent fly strike, and minimizing injuries during the peak fly season. Husbandry practices that can help to reduce the incidence of myiasis include regular inspection of animals to detect and treat all wounds; deep burial (ca. 1.5 m) or cremation of carrion; limiting of dehorning, castrating, and branding to nonfly seasons; removal of sharp objects and other materials that can cause wounds (e.g., horns, barbwire, thorny shrubs); and effective control of biting flies and ticks. The application of salve-like ointments, or smears, can help to prevent fly strike or to kill maggots that are already present. Smears contain larvicides such as benzol or lindane and repellents such as pine-tar oil. Hydrogen peroxide has been infused directly into *Hypoderma* spp. warbles to antiseptically remove the larvae and avoid the consequences of crushing the larvae inside the warble, which can produce an anaphylactic response.

For many years dipping vats have been used throughout the world to treat livestock for arthropod pests, including cattle grubs, screwworms, and other myiasis-causing flies. This entails forcing animals to swim or otherwise move through a water-filled pit containing insecticides or acaricides. This approach, however, has been replaced largely by other methods such as insecticide-impregnated ear tags. The tags slowly release low levels of chemicals that afford protection for extended periods. Another slow-release device developed for cattle grub control was the insecticide-impregnated ankle band attached around the back legs of cattle, which proved to be difficult to prevent soiling and thus quickly losing efficacy. Systemic insecticides such as chlorinated hydrocarbons, organophosphates, carbamates, and sulfur compounds kill cattle grubs before they reach the back of the host. These materials have been almost entirely replaced by the macrocyclic lactones, including the avermectins and milbemycins. Systemic materials can be administered by injection, applied to the back as pour-ons to be absorbed by the bloodstream, as boluses placed in the animal's stomach to provide slow release of systemic compounds, or in mineral blocks (salt licks) for *ad libitum* consumption by livestock and wildlife species. Whereas systemic organophosphates are effective only against prewarble larvae, macrocyclic lactones are effective against all larval instars.

Other methods used to control myiasis-causing flies that attack livestock are low-volume sprays, oral drenches, power dusters, and self-treatment devices such as suspended dust bags and back rubbers. In the case of "fleece worm" blow flies, the topical application of insecticides including organophosphates, synthetic pyrethroids, insect growth regulators, and spinosads to body areas subject to soiling with maggot secretions is an effective, albeit labor intensive, practice. Typically these applications are made using a high-pressure spray, called **jetting**. The use of biocontrol agents, such as parasitic pteromalid wasps that attack fly pupae, and the physical removal of adult flies by trapping can also play a role in integrated pest management programs for some muscoid species that cause myiasis.

The use of genetically altered flies has met with some success in reducing local populations of certain myiasis-causing calliphorid species. This approach shows promise in Australia where *Lucilia cuprina* (*Phaenicia cuprina*) causes annual losses estimated at $120 million (Australian) due to secondary myiasis in sheep. Radiation treatment has been used to induce genetic translocations in chromosomes of a laboratory strain of *L. cuprina* that makes mutant male flies impotent. The fewer daughters that are produced from mutant-male and wild-female matings are able to reproduce normally. However, the second-generation daughters carry about 85% of the induced mutations. The result has been a reduction of *L. cuprina* populations by about half in the first generation and by increased proportions in subsequent generations.

Screwworm Eradication Program

The idea of using sterilized males to eradicate populations of wild flies was first conceived by E. F. Knipling in the 1930s. It was not until 1954 to 1955, however, that the first successful field test was achieved with the eradication of the primary screwworm *C. hominivorax* on the island of Curaçāo (Netherlands Antilles). This entailed the mass production of sterile males from gamma-irradiated pupae, which were then released to mate with wild females. Eggs resulting from these matings do not hatch, thereby reducing fly populations. Subsequent successes in eradicating *C. hominivorax* were achieved in the southeastern United States in 1959, the rest of the United States by 1966, and in Puerto Rico in 1975. Following the establishment of the joint Mexico-United States Commission on Screwworm Eradication in 1972, a cooperative program was begun in 1975 to eliminate *C. hominivorax* in Mexico as far south as the Isthmus of Tehuantepec. Hundreds of millions of flies were mass-reared weekly in facilities at Mission, Texas and Tuxtla Gutierrez, Chiapas, Mexico for aerial releases in the targeted areas. By 1985 screwworms had been successfully eliminated north of the Isthmus of Tehuantepec in Mexico, advancing to the Guatemalan border by 1991, and the rest of Central America by 2001. A sterile-male buffer zone is being successfully maintained at the Darien Gap, the isthmus at the border between Panama and Colombia, with flies produced at the Tuxtla Gutierrez plant. A new mass-rearing fly facility in Panama is in the final stages and flies will soon be available for release and perhaps extend the eradication effort into Colombia and other South American countries. In the meantime, sterile males have been successfully used in areas where *C. hominivorax* has reinvaded former eradication zones and where it has been introduced for the first time. *Cochliomyia hominivorax* was detected for the first time in Libya in 1988, for example, where it was eradicated within a few months following the release of sterile males in 1991. For further details on the history of the screwworm eradication program, see Meyer (1996).

Another approach to screwworm control has been the development of an attractant called **Swormlure** that simulated the odor of animal wounds (Snow et al., 1982; Cunningham et al., 1992). This attractant was combined as a bait with pesticides in a pelletized form that was applied by aircraft to attract and kill gravid females. Called the Screwworm Adult Suppression System (SWASS), this approach has been used successfully as a complement to sterile-male release efforts by reducing the wild fly populations prior to release of sterilized flies.

Cattle Grub Control

The best approach to area-wide control of cattle grubs is integrated management programs enlisting the cooperation of all producers in the targeted region. The most effective method to date is use of **endectocides**, systemic compounds that kill both internal and external parasites (e.g., macrocyclic lactones). Unlike traditional systemic insecticides that kill only the migrating larvae of cattle grubs, macrocyclic lactones are also effective in killing second- and third-instar larvae after they have formed warbles. The other characteristic of these compounds, especially the avermectins, is their high efficacy against migrating *Hypoderma* first-instar larvae even at dramatically low dosage levels (Drummond, 1985). This high efficacy of microdoses of avermectins has been used with success in some of the European eradication programs (Boulard, 2002). However, because these compounds are effective against both internal and external parasites, there are concerns that reducing the dosage for one target could lead to reduced efficacy, or even resistance, with other target species (Leaning, 1984).

Sterile male-release technology also has shown promise but is limited by inherent logistic problems in the mass rearing of *Hypoderma* species for this purpose. By combining the use of systemic compounds and sterile-male releases, cattle grub populations and the associated economic losses were reduced dramatically in the area covered by the joint United States–Canada Cattle Grub Project initiated in 1982 (Kunz et al., 1984; Scholl et al., 1986), and evaluated by Klein et al. (1990) and Kunz et al. (1990). Likewise, significant reductions in cattle grub problems have been achieved in Great Britain and several European countries (Boulard et al., 1984; Wilson, 1986) using combinations of strict compulsory treatment, movement control, serological monitoring, and in some countries, microdoses of avermectins. Although experimental vaccines against *Hypoderma* species have been developed, they have not been widely field tested. They may, however, play a greater role in the future as an important component of integrated management programs. For further information on the biology and control of cattle grubs, see Scholl (1993) and Colwell et al. (2006).

REFERENCES AND FURTHER READING

Baird, C. R., Podgore, J. K., & Sabrosky, C. W. (1982). *Cuterebra* myiasis in humans: Six new case reports from the United States with a summary of known cases (Diptera: Cuterebridae). *Journal of Medical Entomology, 19,* 263–267.

Baird, J. K., Baird, C. R., & Sabrosky, C. W. (1989). North American cuterebrid myiasis. *Journal of the American Academy of Dermatology, 21,* 763–772.

Baumgartner, D. L. (1988). Review of myiasis (Insecta: Diptera: Calliphoridae, Sarcophagidae) of Nearctic wildlife. *Wildlife Rehabilitation, 7,* 3–46.

Boulard, C. (2002). Durably controlling hypodermosis. *Veterinary Research, 33,* 455–464.

Boulard, C., & Thornberry, H. (1984). *Warble Fly Control in Europe.* Rotterdam/Boston: Balkema.

Brewer, T. F., Wilson, M. E., Gonzalez, E., & Felsenstein, D. (1993). Bacon therapy and furuncular myiasis. *JAMA, 270,* 2087–2088.

Carvalho, C. J. B., & Ribeiro, P. B. (2000). *Chave de identificação das espécies de Calliphoridae (Diptera) do Sul do Brasil. Brazilian Journal of Veterinary Parasitology, 9,* 169–173 (In Portuguese).

Catts, E. P. (1982). Biology of New World bot flies: Cuterebridae. *Annual Review of Entomology, 27,* 313–338.

Catts, E. P. (1994). Sex and the bachelor bot fly. *American Entomologist, 40,* 153–160.

Chernin, E. (1986). Surgical maggots. *Southern Medical Journal, 79,* 1143.

Colwell, D. D., Hall, M. J. R., & Scholl, P. J. (2006). *The Oestrid Flies: Biology, Host-Parasite Relationships, Impact and Management.* Wallingford, UK: CABI Publishing.

Craig, C. F., & Faust, E. C. (1940). *Clinical Parasitology* (2nd ed.). Philadelphia: Lea & Febiger.

Cunningham, E. P., Abusowa, M., Lindquist, D. A., Sidahmed, A. E., & Vargas-Teran, M. (1992). Screwworm eradication programme in North Africa. *Revue d'Elevage et de Médicine Vétérinaire des Pays Tropicaux, 45,* 115–118.

Curran, C. H. (1965). *The Families and Genera of North American Diptera* (2nd ed.). Woodhaven, NY: Tripp.

Dear, J. P. (1985). A revision of the New World Chrysomyini (Diptera: Calliphoridae). *Revista Brasileira de Zoologia, 3,* 109–169.

Drummond, R. O. (1985). Effectiveness of ivermectin for control of arthropod pests of livestock. *Southwest Entomologist, 7,* 34–42.

Drummond, R. O., George, J. E., & Kunz, S. E. (1988). *Control of Arthropod Pests of Livestock: A Review of Technology.* Boca Raton, FL: CRC Press.

Ferrar, P. (1987). *A guide to the breeding habits and immature stages of Diptera Cyclorrhapha. Entomography* (Vol. 8, pp. 83–98). Leiden: Brill.

Furman, D. P., & Catts, E. P. (1982). *Manual of Medical Entomology* (4th ed.). Cambridge, UK: Cambridge University Press.

Graham, O. H. (1985). *Symposium on Eradication of the Screwworm from the United States and Mexico. Miscellaneous Publication No. 62.* College Park, MD: Entomological Society of America.

Greenberg, B. (1971). *Flies and Disease. Vol. 1. Ecology, Classification and Biotic Associations.* Princeton, NJ: Princeton University Press.

Greenberg, B., & Singh, D. (1995). Species identification of calliphorid (Diptera) eggs. *Journal of Medical Entomology, 32,* 21–26.

Guimarães, J. H., & Papavero, N. (1999). *Myiasis in Man and Animals in the Neotropical Region; Bibliographic Database.* São Paulo, Brazil: Pleiade/FAPESP.

Hall, D. G. (1948). *The Blow Flies of North America.* College Park, MD: Thomas Say Foundation, Entomological Society of America.

Hall, M. J. R., & Wall, R. (1995). Myiasis of humans and domestic animals. *Advances in Parasitology, 35,* 257–334.

Hendrix, C. M., King-Jackson, D. A., Wilson, M., Blagburn, B. L., & Lindsay, D. S. (1995). Furunculoid myiasis in a dog caused by *Cordylobia anthropophaga. Journal of the American Veterinary Medical Association, 207,* 1187–1189.

James, M. T. (1947). *The Flies that Cause Myiasis in Man. U.S. Department of Agriculture, Miscellaneous Publication 631, U.S.* Washington, DC: Government Printing Office.

Kettle, D. S. (1995). *Medical and Veterinary Entomology.* Oxford, UK: Oxford University Press.

Klein, K. K., Fleming, C. S., Colwell, D. D., & Scholl, P. J. (1990). Economic analysis of an integrated approach to cattle grub (*Hypoderma* spp.) control. *Canadian Journal of Agricultural Economics, 38,* 159–173.

Kunz, S. E., Drummond, R. O., & Weintraub, J. (1984). A pilot test to study the use of the sterile insect technique for eradication of cattle grubs. *Preventive Veterinary Medicine, 2,* 523–527.

Kunz, S. E., Scholl, P. J., Colwell, D. D., & Weintraub, J. (1990). Use of the sterile insect technique for control of cattle grubs (*Hypoderma bovis* and *Hypoderma lineatum*) (Diptera: Oestridae). *Journal of Medical Entomology, 27,* 523–529.

Lane, R. P., & Crosskey, R. W. (1993). *Medical Insects and Arachnids.* London: Chapman & Hall.

Leaning, W. H. D. (1984). Ivermectin as an antiparasitic agent in cattle. *Modern Veterinary Practice, 65,* 669–672.

Leclercq, M. (1969). *Entomological Parasitology—The Relation Between Entomology and the Medical Sciences.* Permagon, (Chapter 5, Myiases).

Liu, D., & Greenberg, B. (1989). Immature stages of some flies of forensic importance. *Annals of the Entomological Society of America, 82,* 80–93.

McAlpine, J. F. et al. (1981). *Manual of Nearctic Diptera* (Vol. 1). Quebec: Research Branch Agriculture Canada, Canadian Government Printing Center, Monograph 27.

McAlpine, J. F. et al. (1987). *Manual of Nearctic Diptera* (Vol. 2). Quebec: Research Branch Agriculture Canada, Canadian Government Printing Center, Monograph 28.

McAlpine, J. F. et al. (1989). *Manual of Nearctic Diptera* (Vol. 3) Quebec: Research Branch Agriculture Canada, Canadian Government Printing Center, Monograph 32.

Meyer, N. V. (1996). *History of the Mexico-United States Screwworm Eradication Program.* New York: Vantage Press.

Norris, K. R. (1965). The bionomics of blow flies. *Annual Review of Entomology, 10,* 47–68.

OISTROS. *An Irregularly Published Newsletter Dealing Exclusively with Current Research on the Oestroidea.* Stockholm, Sweden: Swedish Museum of Natural History.

Papavero, N. (1977). *The World Oestridae (Diptera).* Junk, The Hague: Mammals and Continental Drift.

Pape, T. (2001). Phylogeny of Oestridae (Insecta: Diptera). *Systematic Entomology, 26,* 133–171.

Pape, T. (2006). Phylogeny and evolution of bot flies. In: D. D. Colwell, M. J. R. Hall, P. J. Scholl (Eds.), *The Oestrid Flies: Biology, Host-Parasite Relationships, Impact and Management.* Wallingford, UK: CABI Publishing (Chapter 3).

Reames, M. K., Christensen, C., & Luce, E. A. (1988). The use of maggots in wound debridement. *Annals of Plastic Surgery, 21,* 388.

Ribeiro, P. B. (2007). *Miíase. Doenças de Ruminates e Equideos* (Vol. 1). Pallotti Santa Maria. (In Portuguese).

Ribeiro, P. B., & Carvalho, C. J. B. (1998). Pictorial key to Calliphoridae genera (Diptera) in Southern Brazil. *Brazilian Journal of Veterinary Parasitology, 7,* 137–140.

Rosen, I. J., & Neuberger, N. (1977). Myiasis *Dermatobia hominis,* Linn: Report of a case and review of literature. *Cutis, 19,* 63–66.

Rossi, M. A., & Zucoloto, S. (1972). Fatal cerebral myiasis caused by the tropical warble fly, *Dermatobia hominis. The American Journal of Tropical Medicine and Hygiene, 22,* 267–269.

Sabrosky, C. W. (1986). *North American Species of Cuterebra, the Rabbit and Rodent Bot Flies (Diptera: Cuterebridae).* College Park, MD: Thomas Say Publications, Entomological Society of America.

Sabrosky, C. W., Bennett, G. F., & Whitworth, T. L. (1990). *Bird Blowflies (Protocalliphora) in North America (Diptera: Calliphoridae).* New York: Random House (Smithsonian Institution Press).

Sancho, E. (1988). *Dermatobia,* the Neotropical warble fly. *Parasitology Today, 4,* 242–246.

Scholl, P. J. (1993). Biology and control of cattle grubs. *Annual Review of Entomology, 39,* 53–70.

Scholl, P. J., Colwell, D. D., Weintraub, J., & Kunz, S. E. (1986). Area-wide systemic insecticide treatment for control of cattle grubs, *Hypoderma* spp. (Diptera: Oestridae): Two approaches. *Journal of Economic Entomology, 79,* 1558–1563.

Scott, H. G. (1963). *Myiasis: Epidemiologic Data on Human Cases (North America North of Mexico: 1952–1962 Inclusive).* Atlanta, GA: U.S Department of Health, Education and Welfare, Centers for Disease Control and Prevention.

Sherman, R. A., & Pechter, E. A. (1988). Maggot therapy: A review of the therapeutic applications of fly larvae in human medicine, especially for treating osteomyelitis. *Medical and Veterinary Entomology, 2,* 225–230.

Slansky, F. (2007). Insect/mammal associations: effects of cuterebrid bot fly parasites on their hosts. *Annual Review of Entomology, 52,* 17–36.

Slansky, F., & Kenyon, L. R. (2003). *Cuterebra* bot fly infestation of rodents and lagomorphs. *Journal of Wildlife Rehabilitation, 26,* 7–16.

Smith, K. G. V. (1986). *A Manual of Forensic Entomology.* London: British Museum (Natural History).

Snow, J. W., Siebenaler, A. J., & Newell, F. G. (1981). *Annotated Bibliography of the Screwworm, Cochliomyia hominivorax (Coquerel).* U.S. Department of Agriculture, ARM-S-14. Agricultural Reviews and Manuals, Southern Series No. 14.

Snow, J. W., Coppedge, J. R., Broce, A. B., Goodenough, J. L., & Brown, H. E. (1982). Swormlure: Development and use in detection and suppression systems for adult screwworm (Diptera: Calliphoridae). *Bulletin of the Entomological Society of America, 28,* 277–284.

Stehr, F. W. (1991). *Immature Insects* (Vol. 2). Dubuque, IA: Kendall/Hunt.

Stone, A., Sabrosky, C. W., Wirth, W. W., Foote, R. H., & Coulson, J. R. (1965). *A Catalogue of Diptera of America North of Mexico. Agricultural Handbook 276.* Washington, DC: US Department of Agriculture.

Wall, R., French, N. P., & Morgan, K. L. (1995). Population suppression for control of the blowfly *Lucilia sericata* and sheep blowfly strike. *Ecological Entomology, 20,* 91–97.

Wall, R., & Shearer, D. (1997). *Veterinary Entomology.* Chapman and Hall.

Wall, R., & Shearer, D. (2009). *Veterinary Ectoparasites: Biology, Pathology & Control* (2nd ed.). Weimar, Tx: Culinary and Hospital Industry Publicatons Services.

Whitworth, T. (2006). Keys to the genera and species of blow flies (Diptera: Calliphoridae) of America north of Mexico. *Proceedings of the Entomological Society of Washington, 108,* 689–725.

Wilson, G. W. C. (1986). Control of warble fly in Great Britain and the European community. *The Veterinary Record, 118,* 653–656.

Wood, D. M. (1987). Oestridae. In J. F. McAlpine (Ed.), *Nearctic Diptera* (Vol. 2, pp. 1147–1158). Monograph #28. Quebec: Research Branch Agriculture Canada, Canadian Government Printing Center.

Zumpt, F. (1965). *Myiasis in Man and Animals in the Old World.* London: Butterworth's.

Chapter

19

Louse Flies, Keds, and Related Flies (Hippoboscoidea)

John E. Lloyd

Members of the families Hippoboscidae, Streblidae, and Nycteribiidae are obligate, blood-feeding ectoparasites. Members of the Hippoboscidae are variously called louse flies, bird flies, feather flies, spider flies, flat flies, tick flies, ked flies, and keds. Most species in this family are restricted to a narrow range of hosts. Approximately three-fourths of the known species are ectoparasites of birds, whereas the remainder occur on a variety of mammals other than bats. Members of the Streblidae, called the streblid bat flies or bat flies, and Nycteribiidae, called the nycteribiid bat flies, spider-like bat flies, or bat flies, are ectoparasites of bats and rarely are encountered except by individuals working with bats. For further information on the Hippoboscoidea, refer to the monographs and other works by Bequaert (1942, 1953–1957), Maa (1963, 1966, 1969, 1971), Maa and Peterson (1987), Theodor and Oldroyd (1964), Wenzel (1987), Wenzel and Peterson (1987), and Dick and Patterson (2006).

Although worldwide in distribution, most species of Hippoboscidae are tropical and subtropical in both the Old and New Worlds. The Paleotropics is richer in hippoboscids than any other region. Some hippoboscids may be temporary summer residents of temperate regions due to the migratory habits of their hosts. A few species (e.g., the "grouse fly" *Ornithomya fringillina*) are restricted to temperate regions.

Members of the Streblidae are largely New World and tropical and subtropical in distribution. Relatively few species occur in the warm temperate zones. Most members of the Nycteribiidae are found in the Old World and occur primarily in the tropical and subtropical regions.

TAXONOMY

The Hippoboscidae, Streblidae, and Nycteribiidae represent three families of the superfamily Hippoboscoidea (order Diptera, suborder Cyclorrhapha). Some authors include the Glossinidae in the Hippoboscoidea. Because of similarities in mechanisms of feeding and reproduction as well as specialized morphological characters, the Hippoboscidae, Streblidae, and Nycteribiidae were considered by Theodor (1964) to be monophyletic and were included in the group Pupipara. A recent molecular study by Peterson et al. (2007) would support the view of Theodor. The name Pupipara, however, is inappropriate for the group because the third-instar larva, not the pupa, is deposited by the female. These insects are therefore larviparous and not pupiparous.

The use of morphology to determine phylogenetic relationships within the Hippoboscoidea has been problematic because of the loss of morphological structures to the parasitic way of life. According to Yeates et al. (2007), molecular analysis of higher-level relationships of Hippoboscoidea support the monophyly of the Hippoboscoidea (including Glossinidae, Hippoboscidae, Streblidae, and Nycteribiidae), a sister-group relationship between the Hippoboscoidea and the remaining Calyptrata, and a sister-group relationship between Glossinidae and the Pupipara. Based on morphology, Streblidae and Nycteribiidae would appear to have common structural characters and a common origin, though Dittmar et al. (2006), using molecular techniques, indicate possible paraphyly of the Streblidae. Further molecular studies should provide valuable insight into the phylogeny of these insects.

There are approximately 19 genera and 150 described species in the family Hippoboscidae. Thirteen genera containing 31 species and two subspecies have been reported from the Nearctic region (Maa and Peterson, 1987). Pfadt and Roberts (1978) have presented a list of the louse flies recorded from the United States with their hosts and distribution, both in the United States and worldwide. Maa (1966, 1969) has divided the Hippoboscidae into three subfamilies. The Ornithomyinae includes most of the hippoboscid species, all but seven parasitic on birds. The Lipopteninae contains approximately 34 species, all parasitic on mammals. The Hippoboscinae contains eight species, of which seven infest mammals and one infests ostriches.

Dick and Patterson (2006) recognize 25 genera and 156 described species of Streblidae in the New World, and six genera and 71 described species in the Old World. It is interesting to note that no taxon—either genus or species—of streblid is represented in both the eastern and western hemispheres. Dick and Patterson (2006) also recognize 12 genera and 275 described species of Nycteribiidae worldwide. Although no species of Nycteribiid is found in both the eastern and western hemisphere, several genera are cosmopolitan in distribution.

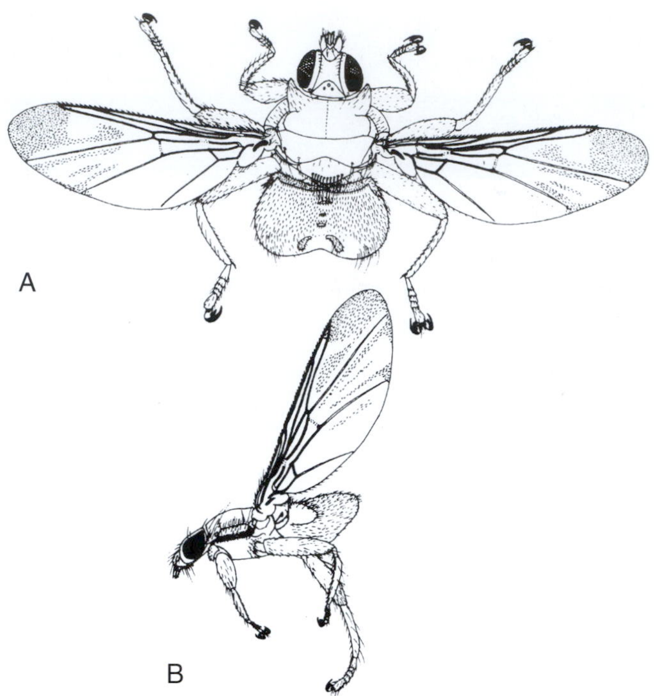

Figure 19.1 *Ornithomya avicularia*, adult female (Hippoboscidae). (A) Dorsal view, with wings spread; (B) lateral view, showing dorso-ventral flattening. (Hutson, 1984)

MORPHOLOGY

All members of the Hippoboscoidea are morphologically adapted for an ectoparasitic existence among the hairs or feathers of their hosts. Certain parts of the exoskeleton have become modified, mainly by fusion and reduction or atrophy, in response to permanent ectoparasitism.

Hippoboscidae

Adults of this family (Figure 19.1) vary in size from 1.5 to 12.0 mm. The body is dorso-ventrally flattened, with the depressed head, thorax, and abdomen giving these insects their louse-like appearance. The mouthparts are directed forward rather than downward. The abdominal integument is soft and flexible, allowing for stretching and distension of the abdomen during larval development within the female and while feeding.

The legs of hippoboscids are generally robust with enlarged femora, flattened tibiae, and short, compact tarsi with one or more basal teeth. The legs tend to be shorter and stouter with heavier tarsal claws in species that infest mammals than in those that infest birds. According to Bequaert (1953), hippoboscids that infest birds have legs that are adapted for scurrying swiftly forward, backward, and sideways amidst the soft feathers. In species parasitizing mammals the legs are built more for grasping and clinging to the skin and the coarse hairs of the pelt.

The compound eyes are generally well developed in those genera with functional wings. The eyes are greatly reduced in genera with small, nonfunctional wings, and those that lose their wings after reaching the host. The sheep ked, *Melophagus ovinus*, which spends its entire life in the wool of sheep, has small compound eyes with relatively few ommatidia. Ocelli are present in several genera of Hippoboscidae, but absent in others. The antennae are small and immovable and are located in deep antennal sockets.

The Hippoboscidae are vessel feeders (solenophages) with both sexes being obligate blood feeders. The proboscis is strongly sclerotized. Its base is partially retracted into a pouch on the ventral side of the head when not in use. At rest, the distal portion is concealed in the palpal sheath. The structure of the proboscis resembles that of blood-sucking muscid flies. The labium is the principal piercing structure. The labrum and hypopharynx lie in a dorsal groove of the labium and together form the food channel. The labella at the tip of the labium are armed with teeth (Figure 19.2).

Adults of most species of Hippoboscidae have relatively long, broad forewings. The hind pair of wings is represented by the halteres, characteristic of dipterans. At rest, the forewings lie flat over each other on the abdomen like closed scissors blades, similar to tsetse flies. In both *Lipoptena* and *Neolipoptena* the newly emerged adult has fully functional wings; this winged adult is referred to as a **volant**. After reaching the host, the wings of these insects break off at the base, leaving a stump. The first blood meal from the

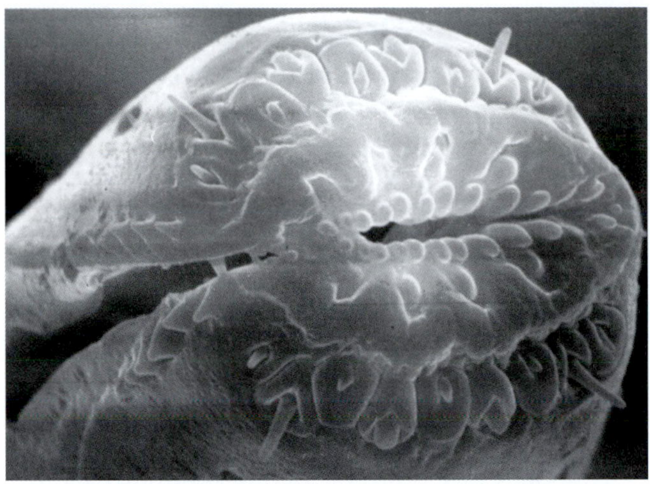

Figure 19.2 Tip of proboscis of the sheep ked, *Melophagus ovinus* (Hippoboscidae), showing labella armed with teeth for cutting into host tissue. (Courtesy of J. E. Lloyd)

host stimulates physiological changes in the fly, including histolysis of flight muscles and growth of leg muscles to accommodate the subsequent parasitic life of the adult.

Both birds and mammals harbor a few species of Hippoboscidae with reduced wings that are not used for flight. Bequaert (1953) noted that at least four genera and 15 species have reduced wings (subapterous). The halteres, however, are not appreciably reduced in these species. The reduced wings are immovable and cover the halteres, which they probably help protect. Both the forewings and halteres are reduced to small stumps and are nonfunctional in the genus *Melophagus*. This genus is possibly the most specialized in the family. The adults of *Melophagus* species emerge from the puparium with rudimentary, nonfunctioning flight muscles that later atrophy.

Streblidae

This family includes a wide variety of forms, some of which exhibit highly specialized adaptations for parasitic life. The adults of most species are 1.5 to 2.5 mm in length; a few Neotropical species may be 0.75 to 5.0 mm (Wenzel and Peterson, 1987). Several genera are flattened dorso-ventrally and superficially resemble hippoboscids or nycteribiids (Figure 19.3). A few are flattened laterally and resemble fleas. Members of the genus *Ascodipteron* demonstrate neosomy similar to some species of fleas. The female burrows into the skin of the host after shedding her wings.

Although most streblids are winged for at least part of their life, and very mobile, there may be different degrees of wing reduction even within a single genus. The head is usually small in the Streblidae and the antennae are inconspicuous. Members of the Streblidae lack ocelli, and the compound eyes are absent or reduced to one or only a few facets.

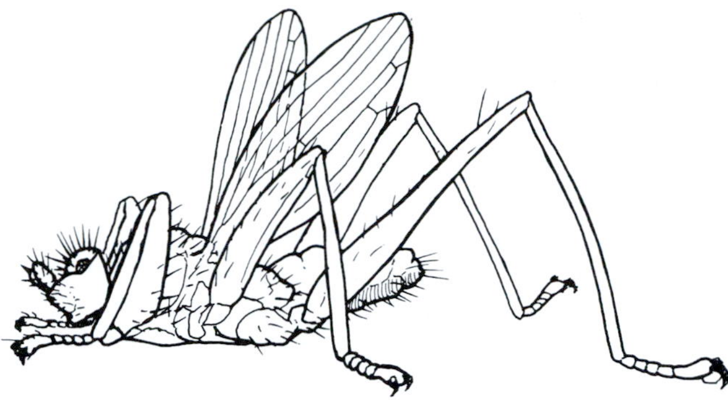

Figure 19.3 Representative adult female bat fly (Streblidae), ectoparasitic on bats. (Furman and Catts, 1982)

Nycteribiidae

Species in this family vary in body length from 1.5 to 5.0 mm. They are an older group than the Streblidae and have become structurally much more modified. The adults are wingless but still possess halteres. Antennae are moderately large in relation to the size of the head and, as in the Hippoboscidae, the antennae of nycteribiids usually are located in antennal pits. The eyes may be absent or are reduced to one, two, or rarely four facets. The marked reduction of the eyes of both Streblidae and Nycteribiidae appears to be associated with their occurrence on bats as nocturnal hosts.

The dorsal plates of the thorax are reduced, and the head and legs of nycteribiids are displaced dorsally. This articular displacement of the legs, together with the complete loss of wings, give the nycteribiids their spider-like appearance (Figure 19.4). The head is narrower than those of the hippoboscids and streblids and, in the resting position, is folded back so that its

Figure 19.4 Spider-like bat fly, *Basilia boardmani* (Nycteribiidae), male, ectoparasite of North American bats. (Courtesy of Jerry F. Butler and Will K. Reeves)

dorsal surface is in contact with the dorsum of the thorax. The head is rotated forward and downward, through about 180 degrees in order to feed.

LIFE HISTORY

Members of the Hippoboscoidea are larviparous. They exhibit a form of viviparity called **adenotrophic viviparity**. A single egg is passed to the uterus where it embryonates and hatches. The egg contains sufficient yolk to nourish the embryo until hatching. The two subsequent larval instars remain in the uterus where they are nourished by a pair of accessory glands or "milk glands" that empty into the uterus. The glands are very similar in structure to those in female tsetse flies, although their secretions are slightly different. Like other insects that feed exclusively on blood, the Hippoboscoidea rely on symbiotic microorganisms to supplement their nutrition. It is interesting that *Bartonella melophagi*, an endosymbiont of the sheep ked, is in the same genus as several pathogenic species. The symbionts of the Hippoboscoidea are passed to the offspring via milk glands in the uterus of the female. Parturition occurs when the larva is fully developed (Figure 19.5) but prior to formation of the puparium (Figure 19.5). The term *prepupa* has been applied to this stage because its structure is similar to that of the third-instar larva. It has ceased to feed, but histolysis of larval organs and formation of the true pupa have not yet started. Shortly after the larva emerges, its integument hardens to form the puparium. In some species, the larva may remain in the uterus until after internal pupal transformations have been initiated. Most puparia are deposited or dropped on the roost substrate, nest, bedding area, wall, or elsewhere in proximity to the host. The sheep ked is unusual in that the puparium is glued by the female to the fleece of the host. The adult fly emerges after a period of several weeks to several months, depending upon the species and temperature. Although detailed life histories of streblids and nycteribiids are described for only a few species, Marshall (1970) and Overal (1980) and have provided detailed life history descriptions, respectively, of *Basilia hispida*, a nycteribiid, and *Megistopoda aranae*, a streblid.

Figure 19.5 Deer ked, *Lipoptena mazamae* (Hippoboscidae); adult, puparium, and third-instar larva on white-tailed deer. (Photo by Nathan D. Burkett-Cadena)

BEHAVIOR AND ECOLOGY

Both sexes of hippoboscoid flies feed as ectoparasites on the blood of birds or mammals. Host specificity varies considerably among different groups. Some are restricted to a single host species. Others are restricted to a genus or to several related genera of hosts, whereas still others are generalists that feed on a relatively wide range of host taxa.

Hippoboscidae occur on 18 orders of birds and five orders of mammals. No species occurs on both birds and mammals. Nor, with the exception of *Hippobosca*, does any genus occur on both bird and mammal hosts. Host specificity is more marked in species parasitic on mammals than in those parasitic on birds. Apterous species and those with reduced wings, or that have lost their wings altogether, tend to be most host-specific. In addition, the more advanced or specialized species tend to be more host-specific.

Members of the Streblidae and Nycteribiidae are exclusively parasites of bats (Order Chiroptera). No species of either family is known to occur naturally on both of the chiropteran suborders, Megachiroptera and Microchiroptera. Host specificity varies widely within the streblid bat flies and nycteribiid bat flies from one to many host species. New World streblids, which tend to be host specific, have become adapted to living and feeding on particular body regions. Individual species are restricted to the wing membranes, head, or trunk. Some bats (e.g., *Phyllostomus hastatus*) commonly harbor three or four species at the same time, with most hosts having at least two species. In temperate regions, at least some bat flies remain physically active on hibernating bats. Some species apparently continue to be reproductively active in spite of the low body temperatures of their hosts.

Nycteribiidae, and most species of Streblidae and Hippoboscidae, deposit their offspring away from their hosts. The fully developed third-stage larva is either dropped to the ground, litter, or nesting material; deposited in a preferred site; or attached to the host or other suitable substrate. Female nycteribiid and streblid flies leave their hosts to deposit larvae in the vicinity of bat roosts. This includes bat-roost surfaces, walls of caves, and branches or leaves of trees.

In the Hippoboscinae, the freshly deposited larva of *Melophagus* is covered with a secretion that hardens upon drying and glues the puparium to the wool fibers of the host. *Neolipoptena* and *Lipoptena*, which shed their wings after reaching the host, also larviposit on the host. These larvae are not fastened to the host and eventually drop to the ground. The duration of the pupal stage of the sheep ked, within the wool of the host is at most 45 days, whereas that of the deer ked on the soil may last several months. Most

Hippobosca species larviposit away from the host in some favored location, as does the pigeon fly *Pseudolynchia canariensis*. Many species of Hippoboscidae that feed on nesting birds larviposit in nesting materials, from which their puparia may be collected.

Winged streblids move readily from one bat to another within a roost. Similarly, winged hippoboscids are very mobile. Newly emerged *Lipoptena* and *Neolipoptena* often swarm in large numbers in search of a host at certain seasons. These volants have functional wings that break off near the base after the host is reached. Once on the host, adult hippoboscids move swiftly among feathers or hair and are difficult to collect. The relatively slow-moving sheep ked is an exception.

COMMON SPECIES OF HIPPOBOSCIDS

A number of louse flies in the genus *Hippobosca* are of particular interest to veterinary entomologists. Most occur in Europe, Africa, and Asia. Occasional introductions have been made into the United States with the importation of zoo animals. With the exception of the ostrich louse fly (*Hippobosca struthionis*), they are parasites of mammals. The sheep ked (*Melophagus ovinus*) is a parasite of sheep and is considered one of the most important insect pests of sheep in many areas of the world.

Figure 19.6 Sheep ked, *Melophagus ovinus* (Hippoboscidae), female, dorsal view. (Courtesy of Cornell University Agricultural Experiment Station)

Sheep Ked (*Melophagus ovinus*)

The sheep ked (Figure 19.6) is a wingless ectoparasite that spends its entire life on domestic sheep. It is worldwide in distribution except in tropical regions where it occurs only in the cooler highlands. It probably was introduced into the United States in the fifteenth century shortly after the European discovery of the New World. Often called the sheep tick by sheep producers, it is found on both range and farm flocks of sheep. The sheep ked is of considerable economic importance and generally is regarded as the most damaging ectoparasite of sheep in North America. A relative of the sheep ked, *Melophagus montanus*, occurs on Dall's sheep (*Ovis dalli*).

Much of what is known about the life history of hippoboscid flies is based on studies of *M. ovinus*. After a period of seven to eight days of feeding and growing in the uterus of the female, the fully developed larva is deposited and cemented to the sheep's wool. Members of the genus *Melophagus* are the only hippoboscids to attach their larvae to the host. The reddish, barrel-shaped puparium (Figure 19.7) is fully formed within 12 hours of parturition. In Wyoming, Swingle (1913) determined that the duration of the pupal stage varied from 19 to 23 days in summer to 20 to 36 days in winter. The variation of this period was attributed to differences in temperature and the distance of the pupa from the skin of the host. Slightly shorter periods have been reported in geographical regions that are warmer than Wyoming. Swingle (1913) further indicated that the duration of the pupal stage is unlikely to be less than 19 days even in the warmest climate, and may increase to 40 to 45 days in the winter.

Figure 19.7 Puparium of sheep ked, *Melophagus ovinus* (Hippoboscidae), adhering to sheep wool. (Courtesy of J. E. Lloyd)

Teneral females of the sheep ked mate within a day of eclosion, although the first larva is not deposited for at least 12 days. This period includes six to seven days for larval maturation. Although one mating provides sufficient sperm for a lifetime, repeated matings usually occur when multiple males are present. A female sheep ked normally lives about four months and produces 10 to 12 larvae during her lifetime. Some, however, can live for six months or more and can produce 15 to 20 larvae. The male life span is slightly shorter, approximately two to three months.

Larviposition by the sheep ked tends to occur on lower body parts, especially under the neck and in the breech area. In unshorn sheep, adults are consistently most numerous in the rib area. Contrary to popular belief, there is no daily or seasonal movement from one location to another on the host. In the spring and summer, sheep keds are more likely to be found on the underside of sheep that recently have been shorn and on young lambs with a very short fleece. On adults they can be found in tufts of longer fleece missed by the shears. Numbers of keds tend to be greater on younger animals.

Sheep keds generally live for only a few days if removed from the host. However, they may live up to five days in wool in the laboratory and, when kept under cool and moist conditions away from the host and fleece, they may live even longer. Their vigor and ability to relocate a host diminish the longer they remain separated from a host animal.

Although sheep keds that become dislodged from their host have the ability to locate a host from the ground, transfer of sheep keds is primarily by animal-to-animal contact. Newborn lambs become infested with keds directly from their mothers soon after birth. Within a flock, transfer occurs when sheep keds move to the tips of the fleece in response to increasing air temperature, and possibly in response to the brisk movement of sheep that accompanies flocking behavior. Air temperature usually must be 21°C (70°F) or above before many sheep keds are observed on the surface of the fleece. At 27 to 58°C (80–90°F) sheep keds are common on the outer wool surface. Thus transfer between animals is more likely, and occurs more rapidly, in summer than in winter.

Like many ectoparasites, populations of *M. ovinus* exhibit annual fluctuations in their numbers. Although minor variations have been reported from different parts of the world, populations of the sheep ked tend to be highest from late winter to early spring, and lowest in summer. In Wyoming ked numbers on ewes tend to increase from September to February. Periods of high populations are extended on rams, pregnant ewes, and undernourished animals, with increases in numbers being prolonged for several weeks in pregnant ewes (Nelson and Qually, 1958). On newborn lambs, which receive only teneral keds from their mothers, numbers of sheep keds increase from their birth in early spring to a couple months later when populations begin their normal decline. Seasonal decline of sheep ked populations is attributed to acquired resistance (Nelson and Bainborough, 1963; Baron and Nelson, 1985). This resistance is apparently caused by a long-lasting, cutaneous, arteriolar vasoconstriction that cuts off much of the capillary blood flow to the upper dermis. Keds are unable to obtain sufficient blood and die of starvation.

Sheep keds feed approximately every 24 to 36 hours, with the feeding time increasing to two-day intervals as the keds become older. The feeding period of an individual ked is typically five to 10 minutes. Feeding is from larger vessels (30–100 μm) near the bases of the wool follicles and often near the apocrine glands or sweat glands associated with primary follicles (Nelson and Petrunia, 1969). Penetration of the dermis by the mouthparts is accomplished by rapid and continuous movement of prestomal teeth on the labellum, followed by movement of the entire haustellum. After piercing a blood vessel, the mouthparts are secured in place by the prestomal teeth, which are everted and serve to anchor the labella to the vessel wall.

Dog Fly (*Hippobosca longipennis*)

The dog fly originally was a parasite of wild carnivores in East Africa. It has since become widely distributed in association with domestic dogs from southern Europe and the Mediterranean region to China. It appears best adapted to warm and arid climates. Up to one-third of dogs in parts of Egypt are infested with this louse fly. It is found mainly on dogs in the Palaearctic region and on wild carnivores in Africa. It has been recorded from members of the families Canidae (dogs, foxes), Viverridae (mongoose, civet), Hyaenidae (hyena), and Felidae (cats).

In 1972, *H. longipennis* was introduced into North America on captive cheetahs from Africa. Subsequently the species has been detected in the United States on cheetahs at wild animal or safari parks in California, Texas, Georgia, and Oregon. Efforts were made by officials in each of the affected states to eradicate this ectoparasite before it escaped from its introduced host to domestic pets, livestock, or wildlife. There is no evidence that this species has become established in the United States or elsewhere in the New World.

Hippobosca equina

This species (Figure 19.8) is normally a parasite of Equidae (horse, donkey, ass) and is a facultative parasite of cattle. Although widespread in the Old World (Europe, northern Africa, western Asia), it does not

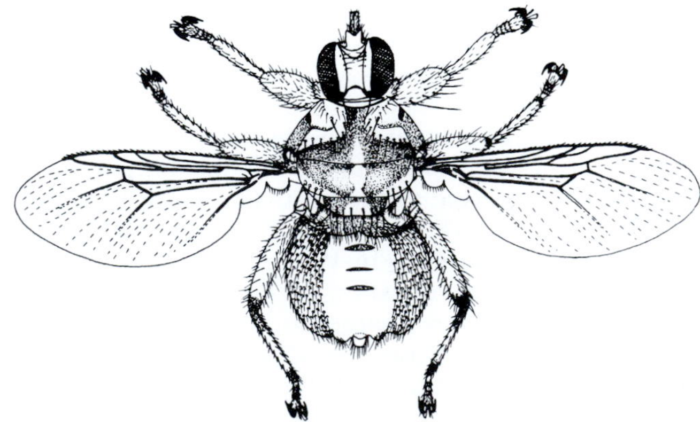

Figure 19.8 Horse ked, *Hippobosca equina* (Hippoboscidae), female, dorsal view. (Hutson, 1984)

occur on wild hosts. The original hosts are unknown. *Hippobosca equina* is a serious and common pest of a wide variety of domestic animals in Egypt (Hafez et al., 1977). It can torment its hosts with painful bites and possibly act as a vector of disease agents, including those that cause piroplasmosis of horses, Q fever, and other types of rickettsioses. In Britain, *H. equina* is called the forest fly.

Hippobosca variegata

Tropical Africa is probably the center of distribution of this species from which it has spread northward to the Mediterranean and eastward into Asia. It is normally a parasite of the domestic horse and its relatives (*Equus* spp.) and cattle (*Bos* spp.). *Hippobosca variegata* is also reported from camels, dromedaries, and water buffalo in Africa and Asia, but these are considered facultative hosts.

Deer Keds (*Lipoptena* and *Neolipoptena* spp.)

Three species of *Lipoptena* and one of *Neolipoptena* are parasites of deer in North America where they are called deer keds. In western North America *L. depressa* and *N. ferrisi* are frequently found on the same host. The wings of deer keds are deciduous. They are fully developed and functional in the newly emerged adult, or volant, although the wing venation is greatly reduced. The wings are shed, probably within 48 hours after the ked reaches a suitable host. Prepupae are deposited while the ked is on the host but eventually fall to the ground.

Lipoptena depressa According to Bequaert (1957) *Lipoptena depressa* is the usual and common parasite of several races of *Odocoileus hemionus*. These include the Rocky Mountain mule deer (*O. h. hemionus*), the Columbian black-tailed deer (*O. h. columbianus*), the California black-tailed deer (*O. h. californicus*), and probably the southern black-tailed deer (*O. h. fulginatus*). It also probably parasitizes the western white-tailed deer (*Odocoileus virginianus leucurus*). Both deer species are efficient breeding hosts for *L. depressa. Lipoptena depressa* is present along the Pacific Coast from southern British Columbia to southern California and as far inland as Alberta, Canada and western South Dakota and Nebraska. Its range includes most of the Rocky Mountain states but apparently not Arizona and New Mexico.

Lipoptena depressa has been divided into two subspecies, *L. d. depressa* and *L. d. pacifica* (Maa, 1969). *Lipoptena d. depressa* is limited in its distribution to the eastern slope of the Rocky Mountain highlands in western Montana, northern Wyoming, southwestern South Dakota, and northwestern Nebraska. The normal host is the Rocky Mountain mule deer. *Lipoptena d. pacifica* is found on the western slopes of the Rocky Mountain lowlands, including British Columbia and the states of Washington, Oregon, Idaho, and California. *Lipoptena d. pacifica* normally breeds on Columbian black-tailed deer and the western subspecies of white-tailed deer.

Lepoptena depressa volants will alight on any moving object. Most leave quickly, however, without dropping the wings if they land on an accidental host such as humans or horses. Westrom and Anderson (1992) found that *L. depressa* was a bivoltine species in California, with volants appearing in peak numbers in October and April. On the host, peak populations of apterous adults occurred in midsummer and early winter following adult flights. Thousands of *L. depressa* may be found on an individual deer, with populations especially heavy in the fall.

Lipoptena cervi This deer ked (Figure 19.9) is a common parasite of the true elk (*Alces alces*), red deer (*Cervus elaphus*), and other species of deer in Europe, Siberia, and northern China. In the United States it was first reported from New Hampshire and Pennsylvania in 1907. It now occurs in New York, New Hampshire, Massachusetts, and Pennsylvania. Bequaert (1942) believed *L. cervi* was introduced into North America by humans with deer from Europe, and the species subsequently spread to white-tailed deer (*Odocoileus virginianus*). It also has been reported from another host native to the United States, the wapiti (*Cervus canadensis*).

Swarms of newly emerged, winged *L. cervi* appear in the fall, but apterous keds may be found on the host year-round. In Denmark, this deer ked was found

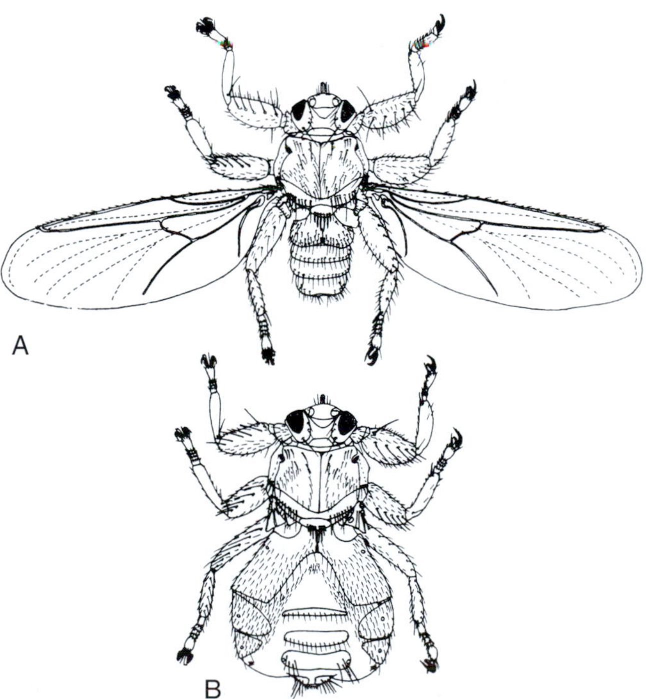

Figure 19.9 Deer ked, *Lipoptena cervi* (Hippoboscidae), females, dorsal view. (A) Winged (alate); (B) wingless (dealate). (Hutson, 1984)

mostly in the fine hair of the neck, anal region, groin, and axilla of the host (Haarløv, 1964). In Russia mass emergence is observed late in August and the first part of September. They are most active on warm, clear afternoons. They are concentrated in low places protected from the wind, in young deciduous forests. The puparia remain on the ground in areas where their hosts are normally found until they emerge in September. Puparia may be particularly numerous at wallows and places where the hosts rub and shed their winter coats.

Lipoptena mazamae The range of this tropical insect extends northward from Argentina through South and Central America, where it is a parasite of brocket deer (*Mazama* spp.), into the United States where it occurs on white-tailed deer in the states bordering the Gulf of Mexico and along the Atlantic coastal states to South Carolina. In surveys of white-tailed deer in southern Texas and Florida, it was the most prevalent ectoparasite (Samuel and Trainer, 1972; Forrester et al., 1996). Volants may be observed in every month from April through November. Numbers on the host appear to be significantly lower in the fall than in the spring or summer. Low populations of this species have been attributed to mortality of the puparia in areas of flooding or high rainfall.

Neolipoptena ferrisi This species occurs in the western United States from Canada to Mexico. Its range includes California, Oregon, Washington, the Rocky Mountain States, and as far east as South Dakota. This deer ked occurs on three races of *Odocoileus hemionus*: the Rocky Mountain mule deer (*O. h. hemionus*), the coastal black-tailed deer (*O. h. columbianus*), and the California black-tailed deer (*O. h. californicus*). Bequaert (1957) considered *O. hemionus* to be the only true breeding host of this insect, and regarded records from western white-tailed deer (*Odocoileus virginianus leucurus*) and pronghorn antelope (*Antilocapra americana*) as accidental occurrences.

In dual infestations, *N. ferrisi* normally is outnumbered by *L. depressa*. *Neolipoptena ferrisi* tends to be collected most frequently from the anterior regions of the body, with the highest population density on the head, whereas *L. depressa* is collected most frequently from the posterior regions of the body, including the tail (Westrom and Anderson, 1992).

Pigeon fly (*Pseudolynchia canariensis*)

The ***pigeon fly*** (Figure 19.10) is a winged hippoboscid that was introduced into North America at least a century ago. The earliest record in North America is 1896 when it was taken on pigeons at Savannah, Georgia (Knab, 1916). Its distribution now is nearly cosmopolitan. This is the only species of hippoboscid that parasitizes domesticated birds. In North America, it is found only on domestic pigeons (*Columba livia*). In the Old World it occurs both on domestic pigeons as well as birds of several other avian orders. Juvenile birds are more frequently attacked. Heavily infested birds become emaciated and susceptible to secondary infections. The pigeon fly can bite humans and can be irritating to individuals handling domestic pigeons.

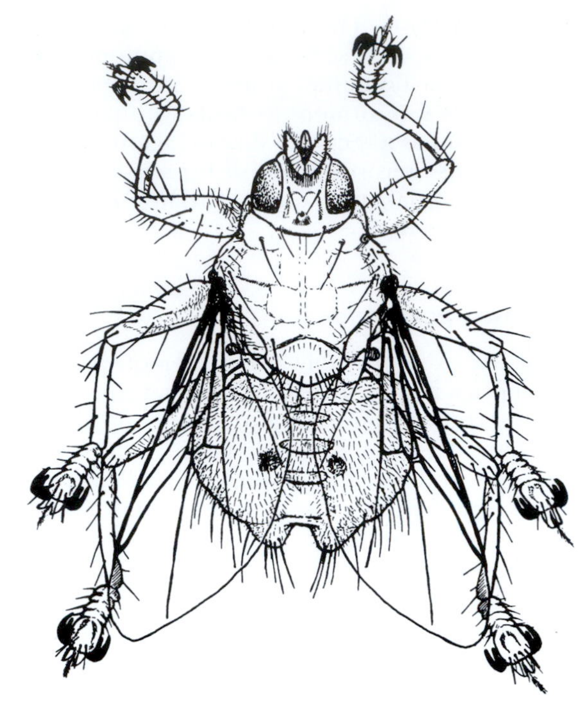

Figure 19.10 Pigeon fly, *Pseudolynchia canariensis* (Hippoboscidae), female, dorsal view. (Furman and Catts, 1982)

PUBLIC HEALTH IMPORTANCE

Humans are not normal hosts of any hippoboscoid species. Occasionally, however, species such as the sheep ked and the pigeon fly bite humans and can be annoying to those routinely handling sheep or domestic pigeons, respectively. Pigeon flies also can be a problem to inhabitants of buildings that have become infested with feral pigeons. Hippoboscids that parasitize nondomesticated animals will bite, and even occasionally feed on a human host. *Olfersia coriacea*, normally a parasite of gallinaceous birds, has been reported attaching to and feeding on the back of the neck of a human in Panama (Harlan and Chaniotis, 1983). Deer keds constitute an annoyance in many areas in Europe because they swarm in large numbers and land on and bite humans, getting into the hair and under clothing.

Human reactions to hippoboscid bites is variable. A redness and swelling at the bite site is reported by some individuals. Others report that the bite is painful, or that there is subsequent irritation at the site of the bite. Individuals have even reported severe pain and swelling that required emergency medical attention. In India *M. ovinus* reportedly causes painful bites to shepherds engaged in shearing (Joseph et al., 1991).

On a personal note, I have been bitten on numerous occasions by the sheep ked and have experienced no associated pain or swelling.

In Europe, feeding on humans by *L. cervi*, resulting in "deer ked dermatitis," has been well documented. In the area of St. Petersburg, Russia, dermatitis was reported among more than 300 individuals bitten by *Lepoptena cervi* during mass flights of volants in August and early September (Chistyakov, 1968). Reunala (1980) reported that the predilection areas were the scalp and upper back. Although the bite of *L. cervi* is barely noticeable, and initially leaves little trace, a hard reddened welt develops within three days. The accompanying itch is intense and typically lasts 14 to 20 days. A pruritic papule may persist for one year. *Bartonella schoenbuchensis*, which colonizes the midgut of *L. cervi* and is transmitted to deer and humans by this ked, is the etiologic agent of deer ked dermatitis (Dehio et al., 2004; Hassler, 2005). *Bartonella schoenbuchensis* was not detected in *Lipoptena mazamae* from Georgia and South Carolina (Reeves et al., 2006). There are no reports of its presence in *L. cervi* in the United States. Rhinoconjunctival allergy to *L. cervi* has also been confirmed in Finland (Laukkanen et al., 2005).

West Nile virus (WNV), which first appeared in the United States in 1999, has been detected in some nonculicine, blood-feeders including *Icosta americana*, a bird-feeding hippoboscid. Gancz et al. (2004) detected WNV RNA in 16 of 18 specimens of *I. americana* collected from owls during a WNV outbreak in Ontario, Canada. Farajollahi et al. (2005) detected WNV RNA from four of 86 specimens of *I. americana* collected from several species of raptors in New Jersey. Since two of the four infected specimens had apparently not blood-fed, and the virus may have been in the hemocoel and tissue, these authors indicted that *I. americana* should be studied further as a potential vector of WNV.

The restriction of streblid and nycteribiid bat flies to bat hosts as well as host specificity among some species of bat flies minimizes the potential for transmission of pathogens to other animals, including humans. Members of both families, however, might be important in maintaining and spreading pathogens among bat populations. Streblids, which apparently will bite humans, are frequent blood-feeders that move readily between bats, and have the potential to quickly spread pathogens from one host animal to another.

Dick and Patterson (2006) suggest it is theoretically possible that bat flies could transmit Ebola virus to humans as both nycteribiid and streblid bat flies are known to infest certain species of Old World fruit bats that may harbor this virus.

VETERINARY IMPORTANCE

Louse flies directly affect their hosts by feeding on blood. Sometimes heavily infested animals become emaciated and susceptible to secondary infections. Juvenile birds and mammals are often more heavily infested with hippoboscids than older animals of the same species. The body conditions of wintering hosts, birds or mammals, may be worsened by infestation of these parasites. Even the streblid bat fly, *Aspidoptera falcate*, has been shown to effect negative weight gains in its bat host, *Megistopoda proxima* (Linhares and Komeno, 2000). In addition to discomfort caused by their biting, louse flies can be annoying to their hosts simply by crawling about on the body. Louse flies also serve as vectors of pathogens and parasites (Table 19.1) and as disseminators of certain ectoparasitic arthropods. These include mammalian trypanosomes and filarial worms, avian trypanosomes, haemosporina blood protozoans, lice, and mites.

Baker (1967) published a review of the role played by the Hippoboscidae as vectors of endoparasites. All known hippoboscid vectors of parasitic protozoans are members of two subfamilies, the Ornithomyinae on birds and the Lipopteninae on mammals. Host-specific hipppoboscid flies are thought to be important vectors for cervid trypanosomes (Rodrigues et al., 2006).

Dipetalonema dracunculoides, a parasitic filarial nematode of dogs and hyenas in the Old World, undergoes cyclical development in the dog fly (*Hippobosca longipennis*), which is thought to be its vector. Hippoboscids may transmit filariae of other mammals, particularly those of camels and lemurs, as well as ostriches and other birds. According to Pfadt and Roberts (1978), the role of hippoboscids as vectors of pathogens is probably much greater than presently known. This may be true of streblids and nycteribiids as well.

The sheep ked transmits *Trypanosoma melophagium*, a nonpathogenic flagellate protozoan of sheep present wherever ked-infested sheep are found. Although it is distributed worldwide, this flagellate protozoan rarely is observed because it is present in relatively small numbers. The trypanosomes are ingested by the keds while feeding on sheep blood. The immature forms of the parasite develop in the posterior midgut of the sheep ked, and infective forms develop in the hindgut and are voided with the feces. They normally do not cross the gut wall into the hemolymph. Flagellates gain entry into the sheep when the keds or their feces are ingested. Nelson (1956) reported mortality of *M.* ovinus as a result of blockage of the posterior midgut by large masses of the crithidial stage of *T. melophagium*. *Lepoptena capreoli*, an ectoparasite of domestic goats and the chamois goat in the Old World, transmits *Trypanosoma theodori* in a similar manner, as does *Ornithomya avicularia*, which transmits *Trypanosoma avium* found in corvid birds. *Pseudolynchia canariensis* and *Stilbometopa impressa* are possible vectors of other avian trypanosomes of pigeons (family Columbidae) and quail (family Phasianidae).

Several hippoboscid flies have been identified as vectors of *Haemoproteus* species, haemosporidian blood parasites that cause bird malarias. The importance of hippoboscids in the natural transmission of most species of *Haemoproteus* is unknown. It is generally assumed that individual species of *Haemoproteus* are transmitted by either hippoboscid flies or *Culicoides* species (Ceratopogonidae), but not both.

Table 19.1 Species of Louse Flies (Hippoboscidae) in the United States, and Selected Species of Veterinary Importance from Other Regions of the World*

Species	Hosts	Geographic Distribution	Parasites or Disease Agents
Subfamily Hippoboscinae			
Hippobosca longipennis	Domestic dogs, hyenas	Southern Europe, northern Africa to China	*Dipetalonema dracunculoides* (filarial nematode)
Subfamily Lipopteninae			
Lipoptena cervi	Deer, elk	Northeastern USA, Europe, and Asia	*Bartonella schoenbuchensis* (bacterium) in Europe and Asia
Lipoptena depressa	White-tailed deer, mule deer	Western USA	*Corynebacterium lipoptenae* (bacterium; symbiote?)
Lipoptena mazamae	White-tailed deer, brocket	Southeastern USA to South America	None
Melophagus ovinus	Domestic sheep	Worldwide	*Trypanosoma malophagium, Rickettsia melophagi* (rickettsia; symbiote?), Bluetongue virus?
Neolipoptena ferrisi	Mule deer	Western USA	None
Subfamily Ornithomyinae			
Icosta albipennis	Egrets, ibises, and other wading birds (Ciconiiformes)	Widespread	None
Icosta americana	Owls, hawks, grouse, turkeys	Widespread	None
Icosta angustifrons	Owls, hawks, falcons	Eastern USA	None
Icosta ardeae boutaurinorum	American bittern	Widespread	None
Icosta hirsuta	Quails, grouse, sage hens	Western USA	*Haemoproteus lophortyx*
Icosta holoptera holoptera	Rails	Eastern and central USA	None
Icosta nigra	Hawks, falcons	Widespread	None
Icosta rufiventris	Hawks, falcons, owls	Eastern and central USA	*Haemoproteus lophortyx*
Microlynchia pusilla	Pigeons, doves, quails, roadrunner	Widespread	*Haemoproteus columbae, H. maccallumi*?, *H. sacharovi*?
Olfersia bisulcata	Vultures	Texas	None
Olfersia fumipennis	Osprey	Widespread	None
Olfersia sordida	Pelicans, cormorants	Widespread	None
Olfersia spinifera	Frigate birds	Florida and Louisiana	None
Ornithoctona erythrocephala	Hawks, pigeons, others	Widespread	None
Ornithoctona fusciventris	Warblers, flycatchers, others	Widespread	None
Ornithoica confluenta	Egrets, ibises, other wading birds	Florida	None
Ornithoica vicina	Owls, sparrows, others	Widespread	None
Ornithomya anchineuria	Hawks, crows, sparrows, others	Widespread	None
Ornithomya bequaerti	Small passerine birds	Widespread	None
Pseudolynchia brunnea	Whippoorwill, nighthawks	Widespread	None
Pseudolynchia canariensis	Domestic pigeons	Widespread	*Haemoproteus columbae, H. maccallumi, Trypanosoma hannai*?
Stilbometopa impresssa	Quails and related game birds	Western USA	*Haemoproteus lophortyx, Trypanosoma* sp.
Stilbometopa podopostyla	Wild pigeons, doves	California	*Trypanosoma avium*

*Included are parasites and disease agents that they reportedly can transmit. Unless otherwise indicated, the parasites and disease agents are all blood protozoans. (Adapted from Baker, 1967, and Pfadt and Roberts, 1978)

Development of *Haemoproteus* in a hippoboscid vector is similar to that of the mosquito-borne malarial parasites in the genus *Plasmodium.* After microgamete production and fertilization of the macrogamete in the hippoboscid midgut, the zygote develops into a motile ookinete, which penetrates and encysts on the outside of the wall of the stomach. The oocyst enlarges and its contents differentiate into sporozoites. The enlarged oocyst bursts to release the sporozoites, some of which enter the salivary glands to be introduced into the next host on which the hippoboscid feeds (Baker, 1967).

The best known *Haemoproteus* species is *H. columbae*, which is parasitic in erythrocytes and visceral endothelial cells of the domestic pigeon (*Columba livia*). It is transmitted by the pigeon fly, *P. canariensis*. Infections can result in anemia and unthriftiness in pigeons, and can cause economic losses to pigeon breeders in the form of nestling mortality. Several other species of *Haemoproteus* are transmitted by hippoboscid flies to a variety of avian hosts (Table 19.1). Proven and presumed vectors of avian haemoproteids include species of *Pseudolynchia*, *Stilbometopa*, *Icosta*, *Ornithomya*, and other hippoboscid genera (Baker, 1967; Pfadt and Roberts, 1978).

A number of nycteribiid species are presumed to be vectors of *Polychromophilus* spp., hemosporidian parasites of bats, primarily in the Old World. Oocysts and sporozoites of *P. murinus* have been found on the midgut and salivary glands, respectively, of the nycteribiid fly *Nycteribia kolenatii*. Developmental stages, mostly sporozoites, of other *Poylchromophilus* spp. have been reported in species of *Nycteribia*, *Penicillida*, and *Basilia* (Garnham, 1973; Gardner and Molyneux, 1988).

Although they are neither mechanical nor biological vectors of any important diseases of sheep, sheep keds have been shown to be capable of transmitting bluetongue virus in experimental studies (Luedke et al., 1965). Such transmission, if it even occurs naturally, is probably only mechanical. In another study sheep keds were unable to transmit *Anaplasma ovis*, the etiologic agent of ovine anaplasmosis, from infected to uninfected sheep (Zaugg and Coan, 1986).

Bartonella are bacteria that infect erythrocytes of vertebrates and are transmitted by blood-sucking arthropods. These bacteria are considered to be emerging pathogens in humans and animals, and it has been proposed that flies of the family Hippoboscidae may be vectors of *Bartonella* to wild and domestic ruminants (Halos et al., 2004).

Feeding by the sheep ked can cause a defect in sheepskins called cockle or rib cockle (Figure 19.11) (Everett et al., 1969; Laidet, 1969). Blemishes appear at the individual bite sites and are presumed to be the result of an allergic reaction to the salivary secretions of the feeding keds. The result is scattered, dense, brownish nodules in the grain layer of sheepskin, which seriously downgrades both grain and suede types of leather. The nodules of dense fibrous material cannot be flattened out and are impenetrable to dyes (Figure 19.12). This defect, especially damaging in garment suede, causes economic losses of several million dollars to the leather industry in the United States each year. When sheep keds are eliminated, the skin recovers from the effect of the bites resulting in usable pelts. The length of time required for recovery by the living animal has not been determined but may be several weeks. A similar defect is caused by the sheep biting louse (*Bovicola ovis*) (Heath, 1994).

Results of studies of weight gains and wool growth of sheep parasitized by sheep keds are equivocal. Several reports in the literature indicate no adverse effects

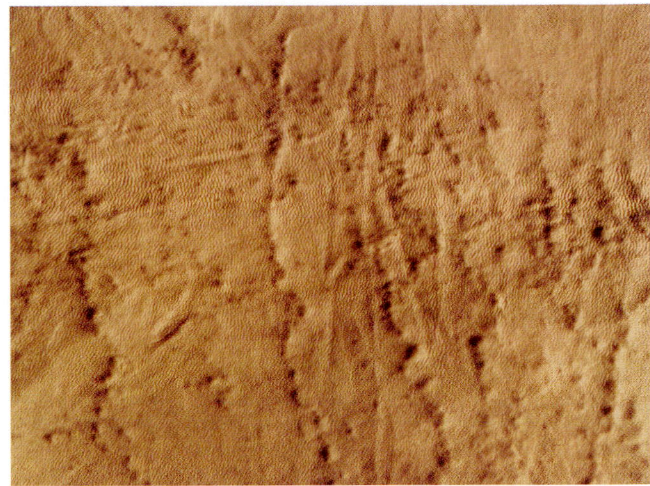

Figure 19.11 Feeding damage (cockle) caused by the sheep ked, *Melophagus ovinus* (Hippoboscidae). Grain side of pickled sheepskin showing pitted surface, or cockle. (Courtesy of J. E. Lloyd)

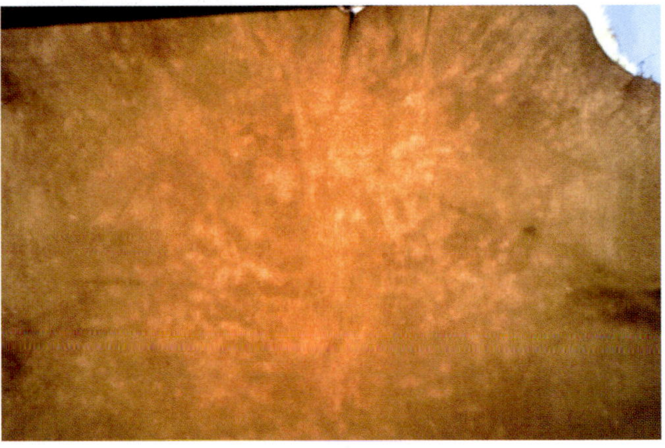

Figure 19.12 Discoloration and mottling of sheep leather due to feeding by sheep ked, *Melophagus ovinus* (Hippoboscidae). (Courtesy of J. E. Lloyd)

due to sheep ked infestations. In a study of the effect of keds on weight gains of feeder lambs in Wyoming, for example, there was no significant difference in gain between ked-infested and uninfested lambs. In that study the number of keds infesting untreated lambs markedly decreased during the period of lamb feeding (Pfadt et al., 1953). In a study in Canada, Nelson and Slen (1968) found that ked-free lambs on various diets gained approximately 1.4 to 3.6 kg (3–8 pounds) more than infested lambs, and that uninfested yearling ewes produced about 11% more wool than infested ones. In a study in New Mexico, a 2% higher dressing percentage was observed in carcasses of uninfested lambs, and carcass weights were significantly heavier (0.9 kg per animal) (Everett et al., 1971). Fleece length was about 8% longer and clean fiber was approximately 7% greater in uninfested lambs, and the difference in clean, dry weight was about 20% in favor of uninfested animals.

Many ranchers in the western United States believe that heavy infestations of sheep ked contribute to the incidence of back loss. This term refers to the death of ewes that roll onto their backs in an apparent attempt to relieve irritation caused by keds. An animal that becomes stuck and remains on its back will eventually suffocate due to the pressure of its internal organs against the diaphragm.

PREVENTION AND CONTROL

Control technology has not been developed for the vast majority of the Hippoboscoidea, and most of these parasites are to be endured. The few species that affect domestic animals and birds may be controlled through treatment of the host with insecticide formulations. The pigeon fly, for example, is controlled by periodic cleaning of the pigeon loft and, as necessary, dusting squabs with an insecticidal dust.

The sheep ked is the only species for which an extensive control technology has been developed. Shearing prior to lambing can reduce sheep ked populations by approximately 75%. Shearing not only removes many pupal and adult keds with the wool but also kills many that are cut with the shears. If ewes are not shorn prior to lambing, substantial numbers of keds can transfer from full-fleeced ewes to their lambs where they are not subjected to the hazards of shearing.

Insecticidal treatment of sheep in the spring following shearing is a common and effective practice. Best results are achieved when ewes are treated following shearing but before lambing. Often, the ewes are treated as they leave the shearing shed. Several traditional treatment methods include whole-body sprays, dusts, and dips. Fall treatments tend to be less effective, possibly due to the greater length of the fleece at that time.

Other methods of treatment are the pour-on and low-volume spray applications. Particularly effective are pyrethroid insecticides, which have become popular because of their ease of application. The pour-on method entails applying a few ounces of insecticide along the backline of each animal (Figure 19.13). The chemical may be poured from a calibrated dipper, or forcefully expelled into the fleece with an application gun. These methods are usually more effective if one waits a few weeks after shearing to permit growth of sufficient wool to retain the liquid formulation.

In the low-volume application, less than an ounce of insecticide is applied to each animal at a pressure of about 50 psi. Large numbers of sheep may be treated in a short period of time by driving the sheep through a spray race equipped with one or more stationary spray nozzles. Animals can be driven through the spray race at approximately one animal per second, providing a treatment method well suited for large range flocks.

Several western states in the United States (Colorado, Idaho, Montana, North Dakota, South Dakota, Utah, and Wyoming) have adopted state-wide, voluntary ked-free programs. These programs involve regular surveillance of sheep, usually at shearing, and prompt treatment for sheep keds when necessary. These programs normally are coordinated by sheep-producer associations within each state. The first statewide ked-free program in the United States was implemented in 1986 by the Wyoming Woolgrowers Association and included 80% of the sheep in that state.

Figure 19.13 Pour-on application of insecticide for control of sheep keds. (Courtesy of J. E. Lloyd)

REFERENCES AND FURTHER READING

Atkinson, C. T. (1991). Vectors, epizootiology, and pathogenicity of avian species of Haemoproteus (Haemosporina: Haemoproteidae). *Bulletin of the Society for Vector Ecology, 16,* 109–126.

Baker, J. R. (1967). A review of the role played by the Hippoboscidae (Diptera) as vectors of endoparasites. *The Journal of Parasitology, 53,* 412–418.

Baron, R. W., & Nelson, W. A. (1985). Aspects of the humoral and cell-mediated immune responses of sheep to the ked *Melophagus ovinus* (Diptera: Hippoboscidae). *Journal of Medical Entomology, 22,* 544–549.

Bequaert, J. C. (1942). A monograph of the Melophaginae, or ked-flies, of sheep, goats, deer and antelopes (Diptera, Hippoboscidae). *Entomological Americana (new series), 22,* 1–220.

Bequaert, J. C. (1952). The Hippoboscidae or louse-flies (Diptera) of mammals and birds. Part 1. Structure, physiology and natural history. *Entomological Americana (new series), 32,* 1–209.

Bequaert, J. C. (1953). The Hippoboscidae or louse-flies (Diptera) of mammals and birds. Part 1. Structure, physiology and natural history. *Entomological Americana (new series), 33,* 211–442.

Bequaert, J. C. (1954). The Hippoboscidae or louse-flies (Diptera) of mammals and birds. Part II. Taxonomy, evolution and revision of American genera and species. *Entomological Americana (new series), 34,* 1–232.

Bequaert, J. C. (1955). The Hippoboscidae or louse-flies (Diptera) of mammals and birds. Part II. Taxonomy, evolution and revision of American genera and species. *Entomological Americana (new series), 35,* 233–416.

Bequaert, J. C. (1957). The Hippoboscidae or louse-flies (Diptera) of mammals and birds. Part II. Taxonomy, evolution and revision of American genera and species. *Entomological Americana (new series), 36,* 417–611.

Chistyakov, A. F. (1968). Skin lesions in people due to bite of *Lipoptena cervi. Vestnik Dermatologii i Venerologii, 42,* 59–62.

Constantine, D. G. (1970). Bats in relation to health, welfare and economy of man. In W. A. Wimsatt (Ed.), Biology of Bats (Vol. II, pp. 319–449). Academic Press.

Dehio, C., Sauder, U., & Hiestand, R. (2004). Isolation of *Bartonella schoenbuchensis* from *Lipoptena cervi,* a blood-sucking arthropod causing deer ked dermatitis. *Journal of Clinical Microbiology, 42,* 5320–5323.

Dick, C. W., & Patterson, B. D. (2006). Bat flies: Obligate ectoparasites of bats. In S. Morand, B. R. Krasnov, R. Poulin (Eds.), Micromammals and Macroparasites, from Evolutionary Ecology to Management (pp. 179–194). Tokyo: Springer-Verlag.

Dittmar, K., Porter, M. L., Murray, S., & Whiting, M. F. (2006). Molecular phylogenetic analysis of nycteribiid and streblid bat flies (Diptera: Brachycera, Calyptratae): Implications for host associations and phylogeographic regions. *Molecular Phylogenetics and Evolution, 38,* 155–170.

Evans, G. O. (1950). Studies on the bionomics of the sheep ked, *Melophagus ovinus* L., in West Wales. *Bulletin of Entomological Research, 40,* 459–478.

Everett, A. L., Roberts, I. H., & Naghski, J. (1971). Reduction in leather value and yields of meat and wool from sheep infested with keds. *Journal of American Leather Chemists Association, 66,* 118–130.

Everett, A. L., Roberts, I. H., Willard, H. J., Apodaca, S. A., Bitcover, E. H., & Naghski, J. (1969). The cause of cockle, a seasonal sheepskin defect, identified by infesting a test flock with keds (*Melophagus ovinus*). *Journal of American Leather Chemists Association, 64,* 460–476.

Farajollahi, A., Crans, W. J., Nickerson, D., Bryant, P., Wolf, B., Glaser, A., et al. (2005). Detection of West Nile virus RNA from the louse fly *Icosta Americana* (Diptera: Hippoboscidae). *Journal of the American Mosquito Control Association, 21,* 474–476.

Forrester, D. J., McLaughlin, G. S., Telford, S. R., Jr., Foster, G. W., & McCown, J. W. (1996). Ectoparasites (Acari, Mallophaga, Anoplura, Diptera) of white-tailed deer, *Odocoileus virginianus,* from southern Florida. *Journal of Medical Entomology, 33,* 96–101.

Furman, D. P., & Catts, E. P. (1982). *Manual of Medical Entomology.* Cambridge University Press.

Gancz, A. Y., Barker, I. K., Lindsay, R., Dibernardo, A., McKeever, K., & Hunter, B. (2004). West Nile virus outbreak in North American owls, Ontario, 2002. *Emerging Infectious Diseases, 10,* 2135–2142.

Gardner, R. A., & Molyneux, D. H. (1988). *Polychromohilus murinus*: A malarial parasite of bats: Life-history and ultrastructural studies. *Parasitology, 96,* 591–605.

Garnham, P. C. C. (1973). The zoogeography of *Polychromophilus* and description of a new species of a gregarine (*Lankestria galliardi*). *Annales de Parasitologie Humaine et Comparée, 48,* 231–242.

Haarløv, N. (1964). Life cycle and distribution pattern of *Lipoptena cervi* (L.) (Dipt., Hippobosc.) on Danish deer. *Oikos, 15,* 93–129.

Hafez, M., Hilali, M., & Fouda, M. (1977). Biological studies on *Hippobosca equina* (L.) (Diptera: Hippoboscidae) infesting domestic animals in Egypt. *Zeitschrift fuer Angewandte Entomologie, 83,* 426–441.

Halos, L., Jamal, T., Maillard, R., Girard, J., Guillot, B., Comel, B., et al. (2004). Role of Hippoboscidae flies as potential vectors of *Bartonella* spp. infecting wild and domestic ruminants. *Applied and Environmental Microbiology, 70,* 6302–6305.

Hare, J. E. (1945). Flying stages of the deer louse fly, *Lipoptena depressa* (Say), in California (Diptera, Hippoboscidae). *Pan-Pacific Entomologist, 21,* 48–57.

Harlan, H. J., & Chaniotis, B. N. (1983). Report of *Olfersia coriacea* (Diptera: Hippoboscidae) feeding on a human in Panama. *The Journal of Parasitology, 69,* 1026.

Hassler, D., Kimmig, P., & Braun, R. (2005). *Bartonella schoenbuchensis* and the deer ked. *Deutsche Medizinische Wochenschrift, 130,* 13–14.

Heath, A. C. G., Cooper, S. M., Cole, D. J. W., & Bishop, D. M. (1994). Evidence for the role of the sheep biting louse *Bovicola ovis* in producing cockle, a sheep pelt defect. *Veterinary Parasitology, 59,* 53–58.

Hutson, A. M. (1984). *Keds, Flat-flies and Bat-flies, Diptera, Hippoboscidae and Nycteribiidae. Handbooks for the Identification of British Insects. Vol. 10, Part 7.* Royal Entomological Society of London.

Jobling, B. (1929). A comparative study of the structure of the head and mouthparts in the Streblidae (Diptera: Pupipara). *Parasitology, 18,* 319–349.

Joseph, S. A., Karunamoorthy, G., Ramachandran, P. K., Sukumaran, D., & Rao, S. S. (1991). *Studies on the Haematophagous Arthropods of Zoonotic Importance in Tamil Nadu. Entomology for Defense Services.* Proceedings of the Symposium held on 12–14 September 1990, 185–192. Gwalior, India: Defense Research & Development Establishment.

Keirans, J. E. (1975). A review of the phoretic relationship between Mallophaga (Phthiraptera: Insecta) and Hippoboscidae (Diptera: Insecta). *Journal of Medical Entomology, 12,* 71–76.

Knab, F. (1916). Four European Diptera established in North America. *Insecutor Inscitiae Menstruus, 4,* 1–4.

Laidet, M. (1969). L'orinine de la noisillure le Melophage. *Technicuir, 4,* 39–50.

Laukkanen, A., Ruoppi, P., & Makinen-Kiljunen, S. (2005). Deer ked-induced occupational allergic rhinoconjunctivitis. *Annals of Allergy, Asthma, and Immunology, 94,* 604–608.

Legg, D. E., Kumar, R., Watson, D. W., & Lloyd, J. E. (1991). Seasonal movement and spatial distribution of the sheep ked (Diptera: Hippoboscidae) on Wyoming lambs. *Journal of Economic Entomology, 84,* 1532–1539.

Lenoble, B. J., & Denlinger, D. L. (1982). The milk gland of the sheep ked, *Melophagus ovinus*: A comparison with *Glossina. Journal of Insect Physiology, 28,* 165–172.

Linhares, A. X., & Komeno, C. A. (2000). *Trichobius joblingi, Aspidoptera falcata,* and *Megistopoda proxima* (Diptera Streblidae) parasitic on *Carolla perspicilllata* and *Sturnira lillum* (Chiroptera: Phyllostomidae) in Southeastern Brazil: Sex ratios, seasonality, host site preference, and effect of parasitism on the host. *The Journal of Parasitology, 86,* 167–170.

Lloyd, J. E. (1985). Arthropod pests of sheep. In R. E. Williams, R. D. Hall, A. B. Broce, P. J. Scholl (Eds.), Livestock Entomology (pp. 253–267). John Wiley.

Lloyd, J. E., Olson, E. J., & Pfadt, R. E. (1978). Low-volume spraying of sheep to control the sheep ked. *Journal of Economic Entomology, 71,* 548–550.

Lloyd, J. E., Pfadt, R. E., & Olson, E. J. (1982). Sheep ked control with pour-on applications of organophosphorus insecticides. *Journal of Economic Entomology, 75,* 5–6.

Luedke, A. J., Jochim, M. M., & Bowne, J. G. (1965). Preliminary bluetongue transmission with the sheep ked *Melophagus ovinus* (L.). *Canadian Journal of Comparative Medicine and Veterinary Science, 29,* 229–231.

Maa, T. C. (1963). Genera and species of Hippoboscidae (Diptera): Types, snyonymy, habitats and natural groupings. *Pacific Insects Monographs, 6,* 1–186.

Maa, T. C. (1966). Studies in Hippoboscidae (Diptera). *Pacific Insects Monographs, 10,* 1–148.

Maa, T. C. (1969). Studies in Hippoboscidae (Diptera). Part 2. *Pacific Insects Monographs, 20,* 1–312.

Maa, T. C. (1971). Studies in batflies (Diptera: Streblidae, Nycteribiidae). Part I. *Pacific Insects Monographs, 28,* 1–248.

Maa, T. C., & Peterson, B. V. (1987). Hippoboscidae. In J. F. McAlpine, B. V. Peterson, G. E. Shewell, H. J. Teskey, J. R. Vockeroth, D. M. Wood (coordinators), Manual of Nearctic Diptera (*Vol. 2.,* pp. 1271–1281). Research Branch, Agriculture Canada, Monograph 28.

Marshall, A. G. (1970). The life cycle of *Basilia hispida* Theodor 1967 (Diptera: Nycteribiidae) in Malaysia. *Parasitology, 61,* 1–18.

Marshall, A. G. (1981). *The Ecology of Parasitic Insects.* London: Academic Press.

Nelson, W. A. (1956). Mortality in the sheep ked, *Melophagus ovinus* (L.) caused by *Trypanosoma melophagium. Nature, 178,* 750.

Nelson, W. A. (1958). Transfer of sheep keds, *Melophagus ovinus* (L.), from ewes to their lambs. *Nature, 181,* 56.

Nelson, W. A., & Bainborough, A. R. (1963). Development in sheep of resistance to the ked *Melophagus vinus* (L.). III. Histopathology of sheep skin as a clue to the nature of resistance. *Experimental Parasitology, 13,* 118–127.

Nelson, W. A., & Petrunia, D. M. (1969). *Melophagus ovinus*: Feeding mechanism on transilluminated mouse ear. *Experimental Parasitology, 26,* 308–313.

Nelson, W. A., & Qually, M. C. (1958). Annual cycles in numbers of the sheep ked, *Melophagus ovinus* (L.). *Canadian Journal of Animal Science, 38,* 194–199.

Nelson, W. A., & Slen, S. B. (1968). Weight gains and wool growth in sheep infested with the sheep ked *Melophagus ovinus. Experimental Parasitology, 22,* 223–226.

Overal, W. L. (1980). Host-relations of the bat fly, *Megistopoda aranea* (Diptera: Streblidae) in Panama. *University of Kansas Science Bulletin, 52,* 1–20.

Peterson, B. V., & Wenzel, R. L. (1987). Nycteribiidae. In J. F. McAlpine, B. V. Peterson, G. E. Shewell, H. J. Teskey, J. R. Vockeroth, D. M. Wood (coordinators), Manual of Nearctic Diptera (*Vol. 2,* pp. 1283–1291). Research Branch, Agriculture Canada, Monograph 28.

Peterson, F. T., Meler, R., Kutty, S. N., & Wiegmann, B. M. (2007). The phylogeny and evolution of host choice in the Hippoboscoidea (Diptera) as reconstructed using four molecular markers. *Molecular Phylogenetics and Evolution, 45,* 111–122.

Pfadt, R. E., Lloyd, J. E., & Spackman, E. W. (1973). *Control of insect and related pests of sheep.* University of Wyoming, *Agricultural Experiment Station Bulletin.*

Pfadt, R. E., Paules, L. H., & DeFoliart, G. R. (1953). Effect of the sheep ked on weight gains of feeder lambs. *Journal of Economic Entomology, 46,* 95–99.

Pfadt, R. E., & Roberts, I. H. (1978). Louse flies (family Hippoboscidae). In R. A. Bram (Ed.), *Surveillance and Collection of Arthropods* of Veterinary Importance (pp. 60–71). *US Department of Agriculture, Agricultural Handbook No. 518.*

Philips, J. R., & Fain, A. (1991). Acarine symbionts of louseflies (Diptera: Hippoboscidae). *Acarologia, 32,* 377–384.

Reeves, W. K., Nelder, M. P., Cobb, K. D., & Dasch, G. A. (2006). *Bartonella* spp. in deer keds, *Lipoptena mazamae* (Diptera: Hippoboscidae), from Georgia and South Carolina, USA. *Journal of Wildlife Diseases, 42,* 391–396.

Reunala, T., Rantanen, T., Vuojolahti, P., & Hackman, W. (1980). Deer ked (*Lipoptena cervi* L.) causes chronic dermatitis in man. *Duodecim, 96,* 897–902.

Rodrigues, A. C., Paiva, F., Campaner, M., Stevens, J. R., Noyes, H. A., & Teixeira, M. M. G. (2006). Phylogeny of *Trypanosoma* (*Megatrypanum*) *theileri* and related trypanosomes reveals lineages of isolates associated with artiodactyl hosts diverging on SSU and ITS ribosomal sequences. *Parasitology, 132,* 215–224.

Samuel, W. M., & Trainer, D. O. (1972). *Lipoptena mazamae* Rodani, 1878 (Diptera: Hippoboscidae) on white-tailed deer in southern Texas. *Journal of Medical Entomology, 9,* 104–106.

Schlein, Y. (1970). A comparative study of the thoracic skeleton and musculature of the Pupipara and the Glossinidae (Diptera). *Parasitology, 60,* 327–373.

Strickman, D., Lloyd, J. E., & Kumar, R. (1984). Relocation of hosts by the sheep ked (Diptera: Hippoboscidae). *Journal of Economic Entomology, 77,* 437–439.

Swingle, L. D. (1913). The life-history of the sheep-tick Melophagus ovinus. *University of Wyoming, Agricultural Experiment Station Bulletin no.* 99.

Theodor, O. (1964). On the relationships between the families of the Pupipara. Proc. 1st Congr. Parasit., Rome, Italy, 999–1000.

Theodor, O. (1967). An Illustrated Catalogue of the Rothschild Collection of Nycteribiidae (Diptera) in the British Museum (Natural History) with Keys and Short Descriptions for the Identification of Subfamilies Genera, Species and Subspecies. London: British Museum (Natural History), Publ. 655.

Theodor, O. (1968). New species and new records of Nycteribiidae from the Ethiopian, Oriental, and Pacific regions. *Parasitology, 58,* 247–276.

Theodor, O. (1975). *Fauna Palaestina, Insecta I: Diptera Pupipara.* Publ. Israel Acad. Sci. Human., Sect. Sciences. Jerusalem: The Jerusalem Post Press.

Theodor, O., & Oldroyd, H. (1964). Hippoboscidae. In E. Lindner (Ed.), *Die fliegen der Palearktischen Region* (Vol. 65, pp. 1–70).

Wenzel, R. L., & Peterson, B. V. (1987). Streblidae. In J. F. McAlpine, B. V. Peterson, G. E. Shewell, H. J. Teskey, J. R. Vockeroth, D. M. Wood (coordinators), *Manual of Nearctic Diptera* (Vol. 2, pp. 1293–1301). Research Branch, Agriculture Canada, Monograph 28.

Wenzel, R. L., Tipton, V. J., & Kiewlicz, A. (1966). The streblid batflies of Panama. In R. L. Wenzel, V. J. Tipton (Eds.), *Ectoparasites of Panama* (pp. 405–675).Chicago: Field Museum, Natural History.

Westrom, D. R., & Anderson, J. R. (1992). The distribution and seasonal abundance of deer keds (Diptera: Hippoboscidae) on Columbian black-tailed deer (*Odocoileus hemionus columbianus*) in northern California. *Bulletin of the Society for Vector Ecology, 17,* 57–69.

Yeates, D. K., Wiegmann, B. M., Courtney, G. W., Meier, R., Lambkin, C., & Pape, T. (2007). Phylogeny and systematics of Diptera: Two decades of progress and prospects. *Zootaxa, 1668,* 565–590.

Zaugg, J. L., & Coan, M. E. (1986). Test of the sheep ked *Melophagus ovinus* (L.) as a vector of *Anaplasma ovis* Lestoquard. *American Journal of Veterinary Research, 47,* 1060–1062.

Chapter

20

Moths and Butterflies (Lepidoptera)

Gary R. Mullen

Many moths and butterflies are recognized as economic pests of row crops, fruit and shade trees, ornamental shrubs, and other plantings on which their larvae feed. Others are household pests that infest cereals, grains, and other stored products, or attack woolen fabrics and other materials of animal origin. Adult moths also can be a nuisance because of their attraction to lights, often entering homes at night. Butterflies, however, are seldom pests. The adults are generally viewed as colorfully attractive insects that are a pleasure to see in flight or visiting flowers for nectar. However, a number of lepidopteran species, notably moths, can cause significant health problems for humans and other animals.

In most cases of a medical-veterinary nature, it is the larval stage that is involved. The larvae of many species are armed with toxic setae or spines that, on contact with skin, can cause a stinging or burning sensation. Under certain circumstances, stinging caterpillars may be ingested by domestic animals resulting in gastrointestinal problems. This most commonly occurs when animals such as cattle and horses graze on infested forage. Adults can also contribute to health problems. The inhalation of wing scales and body hairs of adult moths can induce allergic reactions, and contact with the silk of certain species can cause allergic responses in sensitized individuals. The most unusual lepidopterans from a medical-veterinary perspective are species that, as adults, feed on animal fluids. Some moths feed about the eyes, whereas others are attracted to wounds and in some cases can actually penetrate the skin of humans and other animals to feed directly on blood.

The term **lepidopterism** is applied to adverse reactions of humans and other animals to adult moths or butterflies. Reactions to larvae are called **erucism** (L. *eruca*, caterpillar). Most cases involve **urticaria** (L. *urtica*, nettle), a vascular reaction of the skin in the form of papules or wheals, caused by contact with the specialized defensive spines or setae of certain species. On rare occasions lepidopterous larvae invade animal tissues causing **scoleciasis** (G. *scolec*, worm).

An excellent overview of the Lepidoptera including morphology, developmental stages, behavior, and higher taxonomic classification is provided by Scoble (1992). For information specifically on Lepidoptera-related problems of medical-veterinary significance, the following reviews are recommended: Delgado (1978), Southcott (1978, 1983), Kawamoto and Kumada (1984), and Wirtz (1984).

TAXONOMY

Classification of the Lepidoptera above the superfamily level is subject to debate among taxonomists. The simple separation of the Order into two groups, moths and butterflies, is no longer phylogenetically appropriate. Nonetheless, for the purpose of this chapter, the terms "moths" and "butterflies" are used in the generally accepted context for the species discussed. Virtually all species of medical-veterinary importance are members of the following four superfamilies: Bombycoidea, Noctuoidea, Papilionoidea, and Zygaenoidea. For details on the higher classification of the Lepidoptera, see Kristensen (1984), Nielsen and Common (1991), and Scoble (1992).

Among the more than 100 lepidopteran families, 14 include species that as larvae cause health-related problems. Twelve of these families are moths and two are butterflies. They represent more than 60 genera and 100 species worldwide. Members of at least nine families and 42 genera are known to cause medically related problems as larvae in North America. The families most commonly encountered as problems are Limacodidae, Megalopygidae, and Saturniidae. Other important families in various parts of the world are the Arctiidae, Lasiocampidae, and Lymantriidae.

Six families of Lepidoptera that include species in which adults feed on animal wounds and various body secretions are Geometridae, Noctuidae, Notodontidae, Pyralidae, Sphingidae, and Thyatiridae. The species most commonly observed feeding on animals are *Lobocraspis griseifusa* (Noctuidae), *Hypochrosis* species (Geometridae), *Filodes* and *Microstega* species (Pyralidae) in Southeast Asia. In Africa, *Arcyophora* species (Noctuidae) are more important. The only species that are known to be capable of piercing vertebrate skin are members of the noctuid genus *Calyptra*, in Southeast Asia.

MORPHOLOGY

Adult moths and butterflies are easily recognized by their scale-covered wings, wing venation, and long, coiled proboscis or feeding tube (Figure 20.1). The proboscis serves primarily as a means of imbibing fluids such as nectar, fruit juices, honeydew, and water. While feeding, fluid is drawn through a channel formed by the tightly interlocked pair of elongate maxillae (galeae), each of which bears a longitudinal groove along its inner surface (Figure 20.1B). From there it passes into the alimentary tract by the contraction of pharyngeal muscles. In species that as adults feed on wounds and body fluids of animals, the only significant differences in their feeding mechanism are external modifications, particularly near the tip of the proboscis, to facilitate the rasping or piercing of tissues (Figure 20.2).

The larvae of moths and butterflies are called caterpillars. The larva is typically cylindrical with a well-developed head capsule, three pairs of thoracic legs, and five pairs of fleshy, unsegmented prolegs, one pair each on abdominal segments 3 through 6 and 10 (Figure 20.3). Such larvae are called **eruciform**. Prolegs usually bear tiny hooks called **crochets**, which aid them in clinging to various surfaces while moving about. The mouthparts are adapted for chewing plant material on which the larvae feed. The general body form of larvae is highly modified in some families to the extent that they may not be recognized as lepidopterous larvae by the nonspecialist. Examples include some of the medically important taxa such as puss caterpillars and hag moths.

Caterpillars that cause dermatitis on contact with vertebrate skin are protected by specialized hairs and spines. In some cases these structures cause simple mechanical injury or irritation when they penetrate

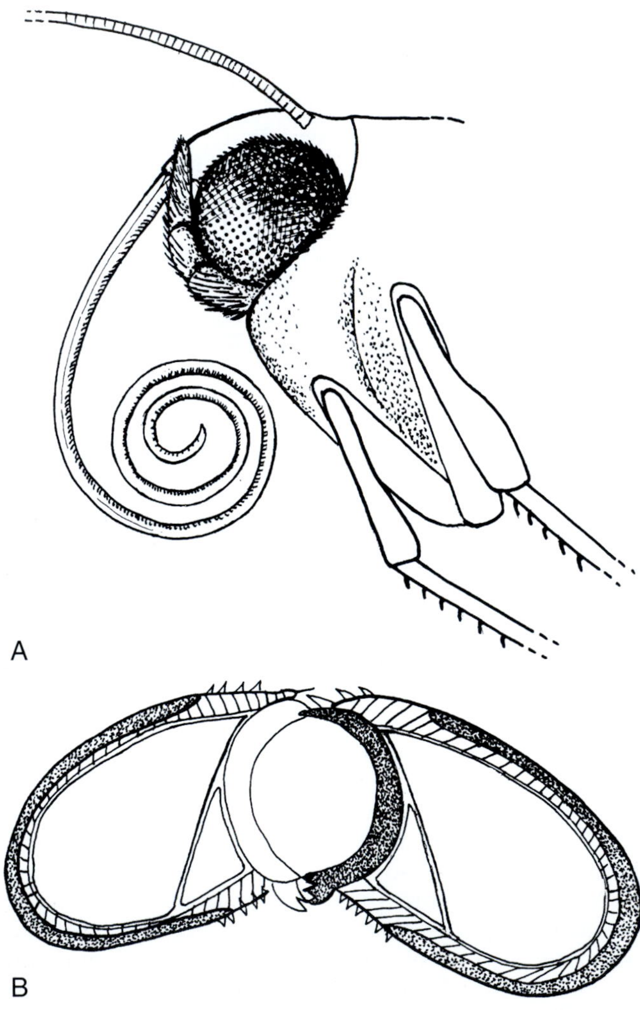

Figure 20.1 Anterior portion of adult moth, showing location and structure of proboscis. (A) Proboscis coiled beneath head when not feeding (original by Margo Duncan); (B) cross-section midway along length of proboscis, showing central food channel formed by the interlocked pair of maxillary galeae. (redrawn from Scoble, 1992)

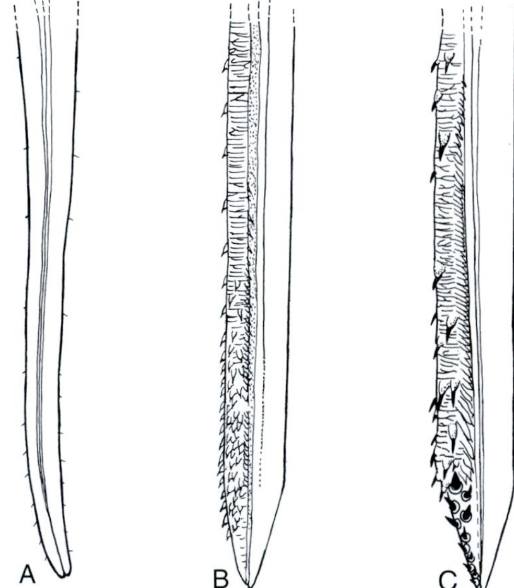

Figure 20.2 Modifications of the distal tip of the proboscis in fruit-piercing and skin-piercing moths. (A) Unmodified proboscis of nectar-sucking moth, *Autographa gamma* (Noctuidae); (B) fruit-piercing moth, *Scolyopteryx libatrix* (Noctuidae); (C) skin-piercing, blood-sucking moth, *Calyptra eustrigata* (Noctuidae). (Redrawn from Bänziger, 1971)

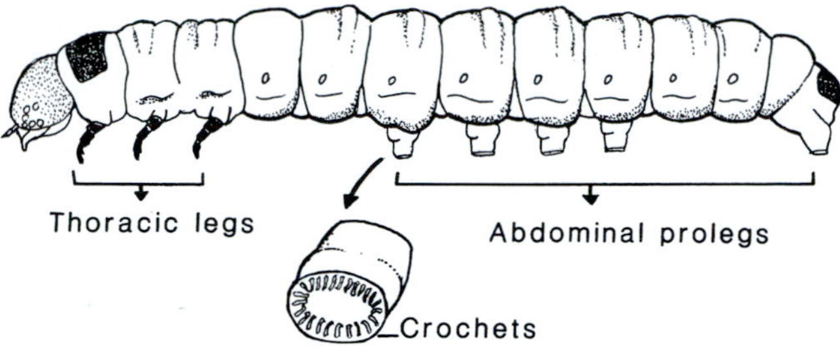

Figure 20.3 External morphology of typical lepidopteran larva, with enlargement of abdominal proleg showing tiny hooks, or crochets. (Redrawn from Romoser, 1981)

the skin. In other cases, they have associated poison glands that secrete toxic substances that elicit varying degrees of inflammation and swelling at the contact site. The location, numbers, and types of urticating hairs and spines vary significantly among different families and genera (Figure 20.4).

Setae and spines are produced by specialized **trichogen cells**, literally "hair-forming" cells. These cells secrete multiple layers of cuticle to form the wall structure, differentiated from the surrounding epidermis.

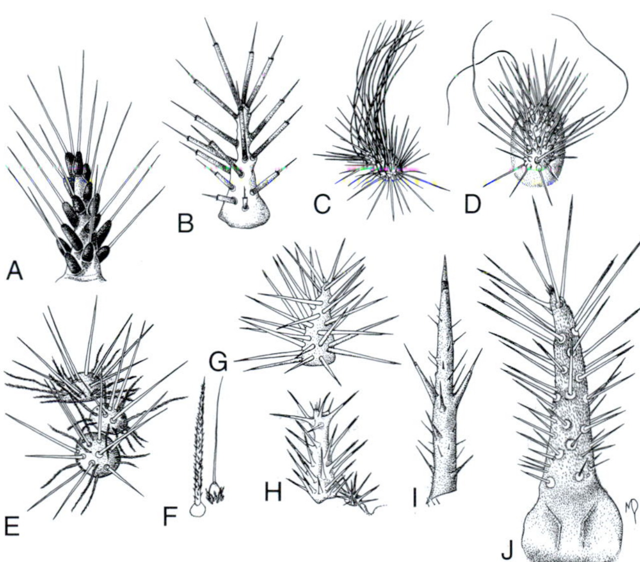

Figure 20.4 Urticating hairs and spines of North American caterpillars. (A) Pandora caterpillar, *Coloradia pandora* (Saturniidae); (B) Buck moth, *Hemileuca maia* (Saturniidae); (C) Puss caterpillar, *Megalopyge opercularis* (Megalopygidae); (D) White flannel moth, *Norape ovina* (Megalopygidae); (E) Smeared dagger moth, *Acronicta oblinita* (Noctuidae); (F) Hag moth, *Phobetron pithecium* (Limacodidae); (G) Io moth, *Automeris io* (Saturniidae); (H) Spiny oak-slug caterpillar, *Euclea delphinii* (Limacodidae); (I) Mourning cloak butterfly, *Nymphalis antiopa* (Nymphalidae); (J) Saddleback caterpillar, *Sibine stimulea* (Limacodidae). (Original by Margo Duncan)

Each spine typically articulates with a socket formed by a **tormogen cell**, or "socket-forming" cell. Setae, or hairs, generally are formed by one or a few trichogen cells, whereas the larger, more robust spines are multicellular in origin and are produced by many trichogen cells. These setae and spines usually are innervated and associated with supporting cells and, depending on the taxon, may or may not have toxin-secreting cells.

The specialized setae and spines of urticating lepidopterans are highly variable in structure. The following categories are based on Kawamoto and Kumada's (1984) modification of a classification proposed by Kano (1967, 1979), which includes types found in adults as well as larvae. According to their classification, there are two major groups of urticating structures, spicule hairs and spine hairs. **Spicule hairs** are detachable setae that are readily rubbed off or can become airborne to cause dermatitis on contact with animal skin. These hairs are easily shed and often are incorporated into the silk of the pupal cocoon. Although some of them cause only mechanical injury, others contain toxins that can affect vertebrates on contact. **Spine hairs**, on the other hand, cause urticaria only when the caterpillar makes direct contact with the skin. They often have associated poison glands and must be innervated in order for the toxins to be released. The following seven types of spicule hairs and four types of spine hairs are recognized.

Spicule Hairs

Type 1 (*Euproctis*-type) These are very small hairs (length 50–200 μm; diameter up to 5 μm), each of which has a pointed end that articulates with its own socket. The distal end has multiple, small barbs. Upon detaching, the pointed tip of the hair penetrates the skin. They occur as small clusters of three to 15 hairs each in cup-shaped papillae (Figure 20.5A) on various parts of the caterpillar. Type 1 hairs are characteristic of the brown-tail moth (*Euproctis chrysorrhoea*) and other *Euproctis* species in the family Lymantriidae. The number of these papillae can be extremely large, as in *Euproctis similis* larvae with an estimated 600,000 spicule hairs and *E. subflava* larvae with more than 6 million.

Type 2 (*Thaumetopoea*-type) These spicule hairs are similar to Type 1 in size and shape but are pointed at both ends. They are inserted point-downward into individual cup-like sockets and occur only in third-instar and older larvae. They are typical of the family Thaumetopoeidae. The processionary caterpillar (*Thaumetopoea processionea*) is estimated to have over 630,000 of these specialized defensive hairs.

Type 3 (*Dendrolimus*-type) These are relatively long (0.5–1.0 mm), slender spicule hairs, with blunt proximal ends and sharply pointed distal tips (Figure 20.5B). They

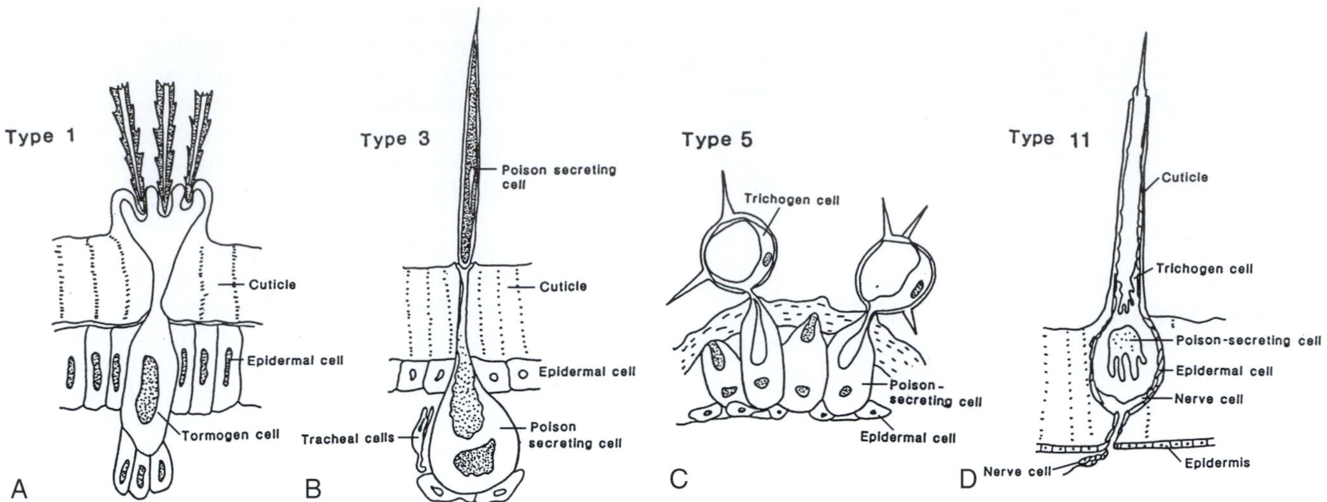

Figure 20.5 Representative types of spicule hairs and spine hairs in urticating caterpillars. (A) Type 1 spicule hair (*Euproctis*, Lymantriidae); (B) Type 3 spicule hair (*Dendrolimus,* Lasiocampidae); (C) Type 5 spicule hair (star-like, Limacodidae); (D) Type 11 spine hair (*Latoia*, Megalopygidae). (Redrawn and modified from Kawamoto and Kumada, 1984)

are loosely articulated with individual sockets and are easily broken off. In addition to mechanical injury to the skin, they can cause localized reactions upon discharge of toxin when the spicule wall is broken. These urticating hairs occur in adults of the Lasiocampidae, notably on the mesothorax and metathorax of the genus *Dendrolimus.*

Type 4 (*Latoia*-type) These spicule hairs are relatively short (0.5–1.0 mm) and stout with a pointed distal tip and three to four basal barbs. They are multicellular in origin, typically having a large poison-secreting cell surrounded by supporting cells. They are characteristic of slug caterpillars of the genus *Latoia,* family Limacodidae, where they occur on the ninth segment of the last larval instar.

Type 5 (Star-like hair) These highly specialized urticating hairs are very small, compact, rhomboid-shaped structures (Figure 20.5C), each produced by a single trichogen cell. Projecting from the outer wall are tiny, pointed spikes or prickles that cause a nettling sensation on contact with skin. These spicule hairs are characteristic of many limacodid larvae, occurring primarily in clusters on the lateral tubercles.

Type 6 (Brush hair) These long (ca. 4.0 mm), slender spicule hairs occur in dense brush-like clusters appearing as tufts on the abdominal segments. The basal end of each hair is sharply pointed with many barbed hooks. They are easily detached and commonly incorporated with the silk as larvae spin their cocoons. These urticating hairs are common in the Lymantriidae, including *Calliteara pundibunda* and the Douglas-fir tussock moth *Orgyia pseudotsugata.* They also occur in the arctiid genus *Premolis* and the Australian anthelid known as the hairymarry caterpillar, *Anthela nicothoe.*

Type 7 (Moth spicules) These spicule hairs occur only in adult moths. They are similar to Type 2 hairs from which they differ primarily in their sockets being located in papillae. The spicule hairs themselves are very small (170 μm long, 3–5 μm diameter) with sharp tips and tiny barbs on the upper one-third of the shaft. They are characteristic of the genus *Hylesia,* family Saturniidae, and adults of several genera of the Thaumetopoeidae such as *Anaphe, Epanaphe, Epicoma,* and *Gazalina.*

Spine Hairs

Type 8 (Primitive type) These are structurally similar to normal body hairs with sharply pointed tips and varying numbers of tiny barbs projecting from their surface. Dermatitis usually is limited to mechanical injury by these hairs because most species lack associated poison-producing cells. Exceptions occur in the genera *Chalcosia* and *Erasmia* (Zygaeneidae) in which poison glands are located at the base of the hairs. Type 8 spine hairs are characteristic of larvae of most species of Arctiidae, Noctuidae, and of such butterflies as Nymphalidae, which are known to cause urticaria.

Type 9 (Simple poisonous spines) These are simple hairs similar to Type 8 spine hairs except that each possesses a poison-secreting cell. Relatively little is known about the morphology of these spines despite their wide occurrence among lepidopterous larvae. They are probably the most common type of urticating spine in caterpillars and cause more severe skin reactions than simple mechanical injury.

Type 10 (*Balataea*-type) These are toxic spines characterized by a bulbous base containing the poison reservoir. The spines are multicellular in origin involving several trichogen cells, the poison cell, supporting cells, and a nerve cell. Toxin secreted by microvilli of the associated poison cell passes into the reservoir via a small duct. Apparently there is no opening at the tip of the spine; the toxin is released only when the tip of the spine breaks off or when the spines are broken upon penetrating the skin. Type 10 spines typically are found in caterpillars of the zygaeneid genera *Balataea* and *Illiberis*, the arctiid genera *Eilema* and *Lithosia*, and first-instar larvae of the gypsy moth *Lymantria dispar* (Lymantriidae).

Type 11 (*Latoia*-type) These spines are relatively short, stout, and conical with a bulb-like base containing a large poison gland (Figure 20.5D). There is no opening at the pointed tip. The structure of Type 11 spine hairs is more complex than the others, involving not only multiple trichogen cells and poison-secreting cells, but also tracheal and nerve cells. The distinguishing feature is the presence of a diaphragm against which the poison cell rests. When mechanically stimulated on contact with skin, pressure of hemolymph acting on the diaphragm causes toxin to be ejected from the tip of each broken spine. The result can be some of the most severe dermal reactions associated with urticating caterpillars. Type 11 spine hairs occur most commonly in the Limacodidae, best represented by *Latoia* larvae. Other limacodid genera with this type of spine are *Microleon*, *Monema*, *Parasa*, and *Scopelodes*. Similar urticarial spines are found in larvae of *Automerisio* and *Dirphia* (Saturniidae), *Catocala* (Noctuidae), and the flannel-moth genera *Doratifera* and *Megalopyge* (Megalopygidae).

LIFE HISTORY

Moths and butterflies undergo holometabolous development. Adults typically deposit their eggs, either singly or in clusters, on host plants that serve as food for the larvae. Soon after eggs hatch the larvae begin feeding. They grow rapidly, molting three to 10 times, depending on the species, as they consume increasing quantities of foliage or other plant material. When time for pupation nears, most species leave the host plant, crawling or descending on a strand of silk to the ground where they seek protected recesses in which to pupate. Others remain on the host plant. The larvae of most moths spin a protective silken cocoon in which pupation takes place. Cocoons may be attached to twigs and leaves or constructed under tree bark, under rocks or ground debris, in the soil or litter, and in other suitable sites. Larvae of moths that do not spin cocoons generally pupate in protected recesses or chambers that they excavate in the soil. The larvae of butterflies do not spin cocoons. Instead they transform into a naked pupa, or chrysalis, which usually hangs exposed on the host plant. Although most chrysalides are green, brown, or otherwise cryptically camouflaged, some are attractively colored or ornamented. Most lepidopterans produce one generation per year, with a few taking two or more years to complete their development. Overwintering usually occurs in the egg or pupal stage.

BEHAVIOR AND ECOLOGY

Only a few groups of moths have become adapted for feeding directly on secretions and body fluids of live animals. This behavior apparently is derived from the commonly observed habit of adult lepidopterans feeding on animal excretory products such as fecal material, urine-contaminated substrates, and animal secretions like saliva and nasal mucous smeared on vegetation or other objects. Imbibing fluids from wounds and the eyes of host animals is a behavioral modification requiring relatively minor morphological changes. In those species that feed on eye secretions, the proboscis is frequently moved over the sensitive eye region to irritate the conjunctiva and increase the flow of tears on which the moths feed. At least one genus (*Calyptra*) includes species that have the ability to actually pierce the skin to feed directly on blood. It has developed unique modifications of the proboscis, notably hooks and erectile barbs (Figure 20.2C). They are movable by internal hemolymph pressure that, aided by special proboscis motions, enables the proboscis to penetrate vertebrate tissues.

Moths that are attracted to animals are called **zoophilous**. They have been observed feeding on the following animal fluids: lachrymal secretions from eyes, blood and skin exudates associated with wounds, nasal secretions, saliva, perspiration, urine, and droplets of blood extruded from the anus of mosquitoes feeding on host animals. Because lepidopteran adults, with few exceptions, lack proteinases and therefore cannot digest proteins from these fluids, this specialized feeding behavior is believed to serve primarily as a means of obtaining salts. The feeding time for most zoophilous species is usually a few minutes.

Species that feed about the eyes on tears from the lachrymal glands are called **lachryphagous**. They also are called tear drinkers and eye-frequenting moths. Lachryphagy is the most striking zoophilous behavior, as documented by Bänziger in Southeast Asia. Since the first observation in 1904 of a moth feeding on the eyes of a horse in Paraguay (Shannon, 1928), lachryphagous moths have been reported attacking a wide range of wild and domestic animals, particularly ungulates and elephants. Zebu and water buffalo tend to be particularly favored hosts; other common hosts include horses, mules, tapirs, rhinoceroses, kangaroos, deer, and humans. It appears that many species exhibit a fairly high degree of host specificity at the order level.

Most lachryphagous species settle on the host to feed; others tend to hover or continually flutter their wings, remaining partially airborne while feeding.

The proboscis is used to feed on tears flowing from the eyes. In some cases, it is used to irritate the conjunctiva or cornea and to feed directly on the eye tissue itself. Many lachryphagous moths are capable of slipping the proboscis between the closed eyelids of sleeping or dozing animals. Some even continue to feed when the host animal tightly closes the eyelids or blinks as a defensive response to the associated irritation. Other moths may irritate eye tissues, particularly the sensitive inner surface of the lids, with their tarsal claws while they attempt to feed.

Some adult moths are attracted to open wounds where they imbibe blood and other host fluids. In most cases the feeding behavior is similar to drinking water and other fluids from the surface of mud, fresh animal feces, or honeydew. In other cases the moths actually probe the wound, penetrating damaged tissues to feed on fresh blood. They are variously called **wound-feeding**, hematophagous, and **blood-sucking moths**. Only a few taxa of skin-piercing moths are capable of piercing intact skin in order to feed.

URTICATING CATERPILLARS

The three more important lepidopteran families with stinging caterpillars are Limacodidae, Megalopygidae, and Saturniidae. Other families include Lymantriidae, Arctiidae, Lasiocampidae, Noctuidae, Thaumetopoeidae, Nymphalidae, and Morphoidae. With the exception of the latter two families, all are moths.

Megalopygidae

Members of this family are called **flannel moths**, referring to the densely hairy adults and larvae. They occur in the Palearctic and Nearctic regions, but especially in South America and the West Indies where the urticating larvae are known as **tataranas** (meaning "like fire"), **cuy machucuy**, and **fire caterpillars**. All megalopygid larvae are protected by poisonous spines concealed beneath the more conspicuous long, fine hairs. These are generally Type 11 spine hairs that cause a nettling sensation on contact with skin. Some species can cause particularly severe reactions and present occupational hazards for tree-plantation workers in parts of South America.

Among the 11 species in North America, the southern flannel moth (*Megalopyge opercularis*) is the most commonly encountered by humans. The larva is called a puss caterpillar, referring to its very hairy appearance (Figure 20.6). The dense, fine, silky hairs vary in color from tan to dark brown or charcoal gray with one or more pairs of small, dorso-lateral patches of white setae. The hairs at the posterior end form a tail-like tuft, whereas the head is concealed beneath the mouse-like pelage. It occurs primarily in the southeastern and south-central United States where the larvae feed on oaks, hackberry, persimmon, apple, orange, almond, pecan, roses, and other trees and shrubs.

Figure 20.6 Puss caterpillar, *Megalopyge opercularis* (Megalopygidae). (Photo by Sturgis McKeever)

The short, toxic spines are arranged in radiating clusters (Figure 20.4C) on three pairs of elevated, longitudinal ridges along the mid-dorsum and sides of the body. The tips of these spines break off on contact with the skin, releasing toxin from the bulbous cavity at their base. All instars are capable of stinging.

The puss caterpillar causes the most painful and severe reactions among urticating species in the United States. Reactions typically include an initial burning sensation, commonly followed by numbness and occasional localized swelling, nausea, and vomiting. Reddened blotches or mottling develop at the contact site, often associated with a glistening appearance as cell fluids are released at the skin surface. Edema may occur, especially if the wrist or lower arm is involved; in such cases, the entire limb from hand to shoulder may become swollen. This often is accompanied by inflammation of lymphatic vessels and dull, throbbing aches involving the axillary nodes, which may persist for 12 hours or more. Stings on the neck can be particularly severe. Occasionally, when populations of *M. opercularis* are unusually high, large numbers of people may be affected. Two such instances have occurred in Texas. One involved hundreds of children that resulted in the closing of public schools (Bishopp, 1923); the other was a widespread outbreak in which over 2,100 cases were reported (Keegan, 1963).

Other megalopygid species that as larvae cause urticaria in North America are the crinkled or black-waved flannel moth (*Lagoa crispata*), yellow flannel moth (*Lagoa pyxidifera*) and white flannel moth (*Norape ovina*) (Figure 20.7).

Limacodidae (Cochlidiidae, Eucleidae)

This is a large, mostly tropical and subtropical family of moths that occurs widely throughout the Neotropical, Ethiopian, Indo-Australian, and Palearctic regions. Approximately 50 species occur in North America. The larvae are called **slug caterpillars** or **nettle grubs**, referring to their unusual shape and the fact that most species have stinging spicule hairs. The larvae are usually

Figure 20.7 White-flannel moth caterpillar, *Norape ovina* (Megalopygidae). (Photo by Sturgis McKeever)

somewhat flattened or slug-like, with a small retractable head, short thoracic legs, and reduced abdominal prolegs that are modified as suckers. They move in a gliding motion, suggestive of slugs. The poisonous setae are usually Type 4 or Type 5 spicule hairs, often in the form of star-like clusters of prickles borne on cone-shaped protuberances; the spines break off easily on contact to cause a nettling sensation.

Six species of urticating slug caterpillars occur in North America. The most commonly encountered is the saddleback caterpillar (*Sibine stimulea*). It is easily recognized by a dorsal, brown oval spot with a white border, in turn surrounded by a green area suggesting a saddle and saddle blanket (Figure 20.8). Urticating hairs are borne on two pairs of large, dark brown, fleshy protuberances (Figure 20.4J), one pair at each end, and on smaller prominences along the sides. In addition, two pairs of rounded lobes bearing specialized, deciduous setae called **caltrops**, which cause irritation to skin, are located at the caudal end. The stinging reaction consists of a burning sensation and erythematous lesion, which is usually much less severe than that of the puss caterpillar. The saddleback caterpillar is found on oaks, elms, dogwoods, linden, corn, ixora, asters, blueberries, grapes, and a number of fruit trees such as apple, citrus, pear, plum, and banana.

Larvae of the hag moth (*Phobetron pithecium*) are sometimes called monkey slugs. Their unkempt, hag-like appearance is attributed to their lateral fleshy processes of variable lengths that are covered with fine, brown or grayish, plumose hairs (Figure 20.9). The relatively few tuberculate stinging hairs (Figure 20.4F) are located at the tips of the processes and laterally on each segment. Contrary to some reports in the literature, the urticarial reaction to hag moth larvae is usually mild. The larvae feed on ashes, birches, hickories, oaks, chestnut, willows, apple, and persimmon.

Another urticating species is the spiny oak caterpillar (*Euclea delphinii*) in the eastern United States (Figure 20.10). The yellow-green larvae feed on a variety of woody plants including beech, cherry, maple, oak, redbud, sycamore, and willow. In addition to urticating spines on the dorsal and lateral processes (Figure 20.4H), they possess a pair of caudal patches of densely clustered, brown, deciduous setae (caltrops), which can be shed defensively, causing skin irritation. These specialized setae also are incorporated into the silk used in spinning the cocoon, providing protection for the developing pupa.

Two closely related limacodid species with urticating larvae are the stinging rose caterpillar (*Parasa indetermina*) (Figure 20.11) and the smaller parasa (*Parasa chloris*) (Figure 20.12). *Parasa indetermina* is found on oaks, hickories, maples, poplars, apple, dogwood, and roses, whereas *P. chloris* occurs most commonly on apple, dogwood, elms, and oaks. Less commonly encountered are the stinging caterpillars of Nason's slug moth (*Natada nasoni*) (Figure 20.13) and *Adoneta spinuloides*. *Natada nasoni* feeds on beeches, hickories, hornbeam, chestnut, and oaks, whereas *A. spinuloides* feeds on beeches, birches, linden, willows, plums, and

Figure 20.8 Saddleback caterpillar, Sibine stimulea (Limacodidae). (Photo by Sturgis McKeever)

Figure 20.9 Hag moth caterpillar, *Phobetron pithecium* (Limacodidae). (Photo by Sturgis McKeever)

Figure 20.10 Spiny oak-slug caterpillar, *Euclea delphinii* (Limacodidae). (Photo by Sturgis McKeever)

Figure 20.13 Nason's slug moth caterpillar, *Nadada nasoni* (Limacodidae). (Photo by Jerry F. Butler)

Figure 20.11 Stinging rose caterpillar, *Parasa indetermina* (Limacodidae). (Photo by Gary R. Mullen)

other trees and shrubs. *Isa textula* can cause a slight urticaria in some individuals; it is found on cherry, maples, and oaks.

Slug caterpillars are common pests among agricultural workers, particularly in Latin America and Southeast Asia. They represent significant occupational hazards for workers in banana and rubber plantations, groves of coconut and oil palms, and other tree crops in the tropics.

Saturniidae

The members of this family are called **giant silk moths**, and include some of the largest and most colorful of all adult moths. Two North American species in particular have urticating caterpillars, the io moth and the buck moth. The io moth (*Automeris io*) occurs throughout the eastern United States and Canada. The larvae feed on a wide range of host plants; these include trees such as oaks, willows, maples, birches, and elms, in addition to other plants like corn, clover, and ixora. The caterpillar is pale green and fairly stout with lateral stripes varying in color from yellow to reddish or maroon (Figure 20.14). The stinging spines are usually

Figure 20.12 Smaller parasa moth caterpillar, *Parasa chloris* (Limacodidae). (Photo by Marc E. Epstein)

Figure 20.14 Io moth caterpillar, *Automeris io* (Saturniidae). (Photo by Sturgis McKeever)

yellow with black tips and are borne on fleshy tubercles along the back and sides (Figure 20.4G). The sharply pointed tips break off easily on contact, allowing the toxin to penetrate the skin. Other *Automeris* species cause urticaria in South America, especially in Brazil and Peru.

The buck moth (*Hemileuca maia*) caterpillar is dark, sometimes almost black, with conspicuous black spines borne on dorsal and lateral tubercles (Figures 20.4B and 20.15). Contact with the skin causes an immediate nettling sensation, often followed by local puffiness. The histamine-induced edema resulting from punctures of the skin by individual spines commonly coalesces to form pronounced wheals. This species occurs in the central and eastern United States where it feeds most commonly on oaks. The adults are unusual in being active fliers during the daytime. Other North American *Hemileuca* species with urticating caterpillars are the New England buck moth (*H. lucina*), Nevada buck moth (*H. nevadensis*), and *H. oliviae*.

In South America, two species of saturniids are involved in urticarial cases: *Lonomia achelous* in Venezuela and other northern countries, and *L. obliqua* (Figure 20.16), particularly in southern Brazil. The larvae are gregarious, feeding at night in trees and moving to the trunk or lower branches during the day. Their toxin contains a potent proteolytic enzyme that breaks down fibrinogen in human blood, interfering with its ability to clot. Dermal contact with *Lonomia* larvae causes an immediate burning sensation, often followed within hours by generalized discomfort, weakness, and headache. This is accompanied by hemorrhaging of capillaries near the skin surface (ecchymosis). In severe or untreated cases, there may be profuse bleeding from the nose, ears, intestinal tract, and vagina within two to 10 days after onset of symptoms. Fatalities have been documented, involving cerebral hemorrhages and kidney failure. An antivenin has been developed for *L. obliqua*.

Other saturniid species that as larvae cause urticaria are *Dirphia multicolor* and *D. sabina* in Brazil and Peru.

Figure 20.15 Buck moth caterpillar, *Hemileuca maia* (Saturniidae). (Photo by Lacy L. Hyche)

Figure 20.16 *Lonomia obliqua* (Saturniidae), larval aggregation on tree trunk, Brazil. (Photo by Germano Woehl, Jr.)

Lymantriidae

Members of this family are called **tussock moths**. The common name refers to the larvae, which are typically hairy with prominent tufts of hairs, or tussocks, on the back. The family occurs throughout the world but mainly in the Nearctic, Palearctic, and Indo-Australian regions. A relatively few species possess urticating hairs. When present they occur as modified, simple hairs arising from cuticular cups clustered in groups at the base of the tussocks. The two most common species that cause urticaria in North America are the brown-tail moth and the gypsy moth. Both are introduced species from Europe that have become serious forest pests in the northeastern United States and the maritime provinces of Canada.

The brown-tail moth (*Euproctis chrysorrhoea*) is so-called because of the brownish tip of the abdomen that contrasts sharply with the otherwise white body of the adult. It was introduced from Europe into Massachusetts in 1897, where it became a major defoliator of New England woodlands. The larvae feed on many species of trees and shrubs, especially members of the rose family. Host plants include apple, pear, plum, cherry, hawthorns, oaks, willows, bayberry, and roses. The larva has a light-brown head and dark-brown, almost black, body with broken lines on either side; two prominent dorsal red spots are located near the posterior end. Numerous tubercles on the back and sides bear long, barbed setae with shorter, brown setae in between. Barbed, stinging setae are borne on tubercles amidst the longer body hairs. These specialized hairs break off easily and cause a nettling sensation when they contact skin. They also are incorporated into the silken cocoon that the larva spins, thereby protecting the pupa from potential enemies. The adult female possesses similarly specialized hairs on her abdomen, which she uses to cover her egg masses while ovipositing. Thus all developmental stages of the brown-tail moth are provided with hairs that can cause urticaria in humans and other animals.

The gypsy moth (*Lymantria dispar*) is similar in many respects to the brown-tail moth. Following its introduction to the United States from Europe about 1868, in an unsuccessful effort to use this species for developing a silk industry in Massachusetts, the gypsy moth escaped and became established in New England. It since has spread widely throughout much of northeastern and Great Lakes area of the United States. The larvae feed on a wide range of deciduous trees, causing extensive damage when their populations are high. They prefer oaks but also attack apple, basswood, alder, birches, boxelder, poplars, willows, hazelnut, mountain-ash, sumac, witch-hazel, and roses. The larvae are quite hairy, with a pair of blue tubercles on each of the thoracic and first two abdominal segments; a pair of red tubercles are present on the next six abdominal segments (Figure 20.17).

Gypsy moth larvae possess two types of defensive setae. One causes irritation to the skin primarily due to mechanical damage by tiny projections on the long, slender shaft of each seta. The other type is represented by shorter, smoothly tapered setae that arise from a ball-in-socket joint; they are connected with poison glands that apparently produce histamine. Reactions to these stinging hairs vary from mild to moderately severe pruritus, with accompanying erythema and papule formation. The onset of discomfort usually is noticed within eight to 12 hours after contact, often becoming more pronounced one or two days later. Most cases resolve in a few days or up to two weeks. Delayed hypersensitivity reactions sometimes result in irritation to the eyes, inflammation of the nasal passages, and shortness of breath. This is especially common in the case of airborne hairs of adult gypsy moths, or contact with clothes hanging on outdoor lines when this moth is locally abundant. A major infestation of the gypsy moth in the northeastern United States in 1981 resulted in thousands of cases of pruritic dermatitis being reported that year. Like the brown-tail moth, female gypsy moths cover their egg masses with specialized body hairs that can cause urticaria upon contact with skin.

Other lymantriid species that cause urticaria in North America are the whitemarked tussock moth (*Orgyia leucostigma*) and the yellow-tailed moth or mulberry tussock moth (*Euproctis similis*). Wind-dispersed hairs of *E. similis* resulted in an estimated 500,000 human cases of pruritic dermatitis in Shanghai, China, in 1981. A similar outbreak involving the Oriental tussock moth (*Euproctis flava*) affected more than 200,000 people in Japan in 1955. The airborne setae are believed to have originated from larval hairs woven into the cocoons that adhered to the adult moths as they emerged. The pale tussock moth (*Dasychira pudibunda*) has been reported as the cause of "hop dermatitis" in Europe.

Figure 20.17 Gypsy moth caterpillar, *Lymantria dispar* (Lymantriidae). (Courtesy of U.S. Department of Agriculture, Forest Service)

Arctiidae

The adults of this family are known as **tiger moths**. The larvae usually are covered with fairly dense hairs of varying colors arising from raised warts, in contrast to the normally bare, shiny head. Included in this group is the familiar "wooly bear" caterpillar of the Isabella tiger moth (*Pyrrharctia isabella*), which, like most arctiids, does not possess urticating hairs or spines. In larvae that cause skin irritation, Type 8 spicule hairs are borne on dorsal tufts, partially concealed by the longer body hairs. Members of the following six genera in North America include urticating caterpillars: *Adolia, Callimorpha, Euchaetes, Halysidota, Lophocampa,* and *Parasemia.* Most of these species are relatively uncommon and only occasionally involved in urticarial cases. However, *Lophocampa caryae* larvae, a common species on hickory, pecan, and walnut trees in the United States, has been responsible for a significant number of cases of mild urticaria, particularly in children. The milkweed moth (*Euchaetes egle*) is perhaps the best known. Its larvae are common on milkweed and are distinguished by dense tufts of black, yellow, and white hairs. Larvae of the hickory halysidota (*Halysidota caryae*), as the name implies, are found on hickories.

Although the family is cosmopolitan, the largest diversity of arctiid moths occurs in the Neotropical and Oriental regions. Where abundant, they can cause occupational erucism among field workers, as in the case of *Premolis semirrufa* in South America.

Lasiocampidae

Members of this family are commonly called **tent caterpillars** or **lappet moths**. The larvae are usually very hairy, often colorful with longitudinal stripes. In the case of lappet moths, the larvae are somewhat flattened with hair-covered, fleshy lobes **(lappets)** on the sides of each segment. They are typically gregarious, forming communal silken webs or "tents" in trees for protection from natural enemies. Their specialized defensive hairs cause only mild, transient discomfort on contact with skin. Urticating hairs are used for strengthening and protecting cocoons and represent another source of contact for humans.

Lasiocampid larvae of only a few North American species reportedly possess urticating hairs. These include the eastern tent caterpillar (*Malacosoma americanum*) commonly found on apple and cherry trees;

and two lappet moths of the genus *Tolype*: the large tolype (*T. velleda*) on apple, oak, ash, elm, birch, plum, and other trees, and the small tolype (*T. notialis*) on conifers.

Several reports in the older literature refer to the larvae of *Bombyx* species (family Bombycidae) causing urticaria in humans. These include species in Great Britain (Sharp, 1885; Jenkyns, 1886; Long, 1886) and in India, Sri Lanka, and Africa (Castellani and Chalmers, 1913). The species reported in Great Britain as *Bombyx rubi* and *B. quercus* are now recognized as the lasiocampids *Macrothylacia rubi* and *Lasiocampus quercus*. Both are said to cause a nettling sensation and small white blisters in some individuals who handle them. The likelihood is that other old reports of bombycid larvae causing urticaria also refer to lasiocampid species.

Noctuidae

This is the largest family of Lepidoptera, with over 20,000 species worldwide. The adults are known as **owlet moths** or simply noctuids. Only a relatively few species are known to possess stinging hairs or spines. Spine hairs are Type 8 with sharp tips that break off and penetrate the skin; the spines may be branched and sometimes form brushes. Urticating species are primarily members of the subfamilies Acronictinae (dagger moths) and Catocalinae (underwings). The larvae of dagger moths are commonly covered with tufts of long hairs, superficially resembling the Arctiidae.

In North America only larvae of the genera *Acronicta* and *Catocala* cause urticaria. The smeared dagger moth (*Acronicta oblinita*), the larva of which is known as the smartweed caterpillar, is a pest of apple and other fruit trees; in addition it is found on elms, oaks, pines, willows, cotton, corn, clover, smartweed, strawberry, and grasses. The cottonwood dagger moth (*Acronicta lepusculina*) feeds on cottonwoods, poplars, aspens, birches, and willows. Larvae of several species of underwing moths (*Catocala* spp.) are protected by stinging hairs.

Nolidae

Larvae of the gum-leaf skeletonizer (*Uraba lugens*), a pest of *Eucalyptus* forests in Australia, causes classic urticaria with associated itching and wheals. This species was introduced to New Zealand in 1992, where it is now established in the Auckland region.

Thaumetopoeidae

The larvae of this family are best known as **processionary caterpillars**. They typically live in communal webs that they leave at night to feed on foliage. When moving, they crawl one behind the other, forming rows or columns, and marching in an orderly fashion. They occur primarily in the Palearctic, Oriental, and Ethiopian regions where they feed on pines and oaks, and less commonly on cedars and walnut. No processionary caterpillars occur in North America. The urticating hairs of their caterpillars are Type 2 spicule hairs similar to those found in the caudal tufts of adult females. Toxin is drawn by capillary action into the tip of the seta from tiny glands in the epidermis. In adults, contractions of the abdomen are sufficient to release these hairs from their sockets. Pruritic dermatitis and urticaria can result from contact with the larvae, airborne setae from the adults, or contact with egg masses in which the barbed hairs from the adult female have been incorporated as a protective covering.

The oak processionary (*Thaumetopoea processionea*) is widely distributed in Europe where it commonly causes urticaria. The larva is covered with long whitish hairs arising from reddish warts, contrasted against a blue-gray coloration above the line of the spiracles and a greenish gray below; velvety black, dorsal patches occur on most of the abdominal segments. The larvae feed primarily on oaks and sometimes on walnut. Other processionary caterpillars that cause discomforting rashes include *Anaphe infracta* in Europe, *Thaumetopoea wilkinsoni*, and several *Thaumetopoea* species in Africa and Madagascar.

Nymphalidae

The only nymphalid butterfly in North America in which the larva possesses urticating hairs is the mourning cloak (*Nymphalis antiopa*). It is an introduced species from Europe that occurs throughout the eastern United States and Canada. The larvae are velvety black, speckled with tiny white dots, with a row of mid-dorsal red spots and several rows of long, branched spines (Figure 20.18) that are capable of piercing skin. The urticating structures are typical Type 8 spine hairs. Caterpillars are found on elm, hackberry, poplar, willow, rose, and other common host plants.

Morphidae

This family includes the showy, brightly iridescent-blue **morpho butterflies** that occur only in the Neotropics. The larvae of at least seven species are known to cause

Figure 20.18 Mourning Cloak caterpillars, *Nymphalis antiopa* (Nymphalidae). (Photo by Sturgis McKeever)

urticaria: *Morpho achillaena, M. anaxibia, M. cypri, M. hercules, M. laertes, M. menelaus,* and *M. rhetenor* (Rotberg, 1971). Most encounters involve accidentally brushing against the larvae feeding on plants in the families Leguminosae and Menispermaceae. Cases are relatively few. They can occur any time of the year but are seen most commonly during the summer when larval populations are highest. Little is known about the nature of the urticating structures.

LACHRYPHAGOUS MOTHS

More than 100 species of zoophilous moths have been observed feeding on lachrymal secretions (Figures 20.19 and 20.20), primarily in Thailand, Malaysia, and other parts of Southeast Asia (see Bänziger references). Most of these moths are members of the Geometridae, Pyralidae, and Notodontidae, with a few species of Noctuidae, Sphingidae, and Thyatiridae.

Geometridae

Members of only a few of the 2,700 genera of geometrid moths are reportedly zoophilous. Nonetheless this family includes the largest number of lachryphagous taxa, with more than 50 species in Southeast Asia. As in other zoophilous moths, except some Noctuidae, only the males are attracted to animals. Most of them feed on mammalian body fluids that either drop to the ground or are smeared on vegetation. They have been observed primarily in association with water buffalo, but other ungulates and elephants also are frequently visited. A few species have been reported imbibing droplets of blood extruded by mosquitoes as they feed on host animals. As a group they do not commonly frequent the eyes; however, *Hypochrosis hyadaria, H. flavifusata, Godonela eleonora,* and to a lesser extent other *Hypochrosis, Godonela, Scopula, Problepsis,* and *Zythos* species are among the more frequent tear drinkers. The only lachryphagous species that has been reported in the United States is the *pectinate euchlaena* or *forked euchlaena* (*Euchlaena pectinaria*) observed feeding on eye secretions of a horse in Arkansas (Selman, 1972).

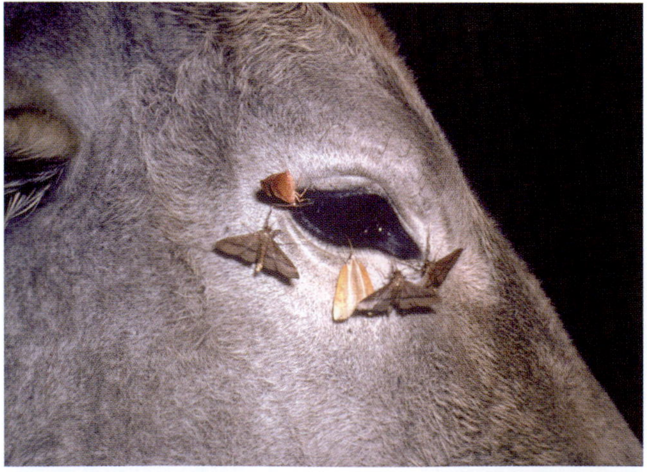

Figure 20.19 Three species of moths feeding on eye secretions of zebu: *Hypochrosis irrorata* (Geometridae), *Filodes mirificalis* (Pyralidae), and *Lobocraspis griseifusa* (Noctuidae). (Photo by H. Bänziger)

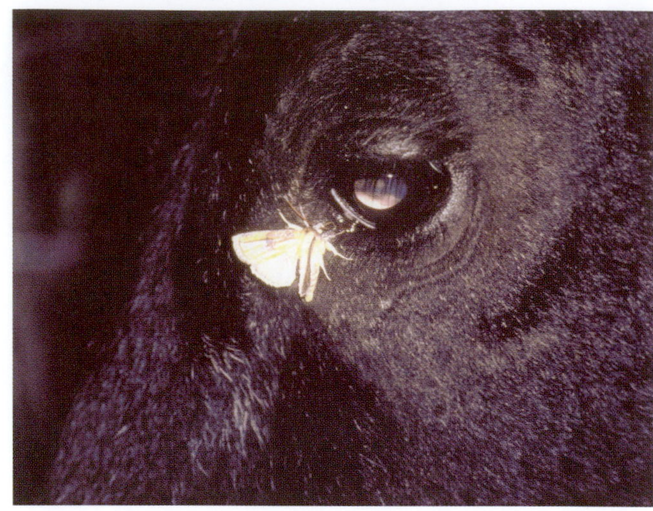

Figure 20.20 Lachryphagous moth, *Chaeopsestis ludovicae* (Thyatiridae), feeding at eye of zebu. (Photo by H. Bänziger)

Pyralidae

Pyralid moths are second only to the Geometridae in the number of species known to feed on lachrymal secretions. Members of the following genera are zoophilous and to various extents lachryphagous: *Botyodes, Epipagis, Hemiscopis, Lamprophaia, Pagyda, Pyrausta,* and *Thliptoceras. Microstega homoculorum, Filodes mirificalis,* and *Paliga damastesalis* are among the more common visitors of human eyes, and *Thliptoceras* and *Hemiscopis* species tend to sip human perspiration. Typically, however, they have been observed feeding on lachrymal and skin secretions of ungulates and elephants.

Notodontidae

Adult males of at least eight species of the genera *Tarsolepis, Togarishachia,* and *Pydnella* are lachryphagous. Elephants appear to be their preferred hosts; however, these moths feed on a wide range of other large mammals in Southeast Asia including water buffalo, zebu, tapir, rhinoceros, deer, and humans. Although they feed primarily on tears, they also have been seen imbibing saliva from around the mouth. They are persistent feeders, some causing only mild discomfort to their hosts, whereas others are very irritating.

Noctuidae

Although only a few species among the more than 3,800 genera of noctuid moths are zoophilous, they are behaviorally the most advanced in terms of lachryphagy, and locally can be the most frequent tear drinkers. The highly flexible proboscis is swept back and forth across the eye to induce tearing as the moth feeds. The extra length of the proboscis allows these moths to feed between the eyelids of dozing animals and reduces the risk of being dislodged by eyelid movements of wakeful hosts. Both males and females of *Arcyophora* and *Lobocraspis* species are lachryphagous and are the only known tear drinkers capable of digesting proteins contained in lachrymal fluids.

Sphingidae

Rhagastis olivacea in Thailand is the only sphingid moth confirmed as being lachryphagous. It feeds while hovering about the eyes of horses, mules, and humans. It also has been observed inserting its proboscis between the lips and into the nostrils of humans to feed on saliva and nasal secretions; the latter has been described as causing a tickling sensation. Only mild discomfort is experienced when they feed on eyes.

Thyatiridae

This is a relatively small family with 70 described genera worldwide. Only a few species in the genera *Chaeopsestis* and *Neotogaria* in Thailand and China are known to be zoophilous. They tend to be avid tear drinkers on zebu, horses, and mules, although they also feed on wounds. *Chaeopsestis ludovicae* is the only thyatirid known to feed on humans (Figure 20.21). It has been observed imbibing perspiration on human skin and clothes, in addition to nasal mucous and saliva of human hosts. This moth is an aggressive feeder, and can cause considerable pain due to irritation of the conjunctiva and inner surface of the eyelid with its tarsal claws. The discomfort has been likened to a grain of sand being rubbed between the eye and eyelid. Adding to the annoyance is its persistence in fluttering about the eyes and repeated attempts to feed even when the eyelids are tightly closed.

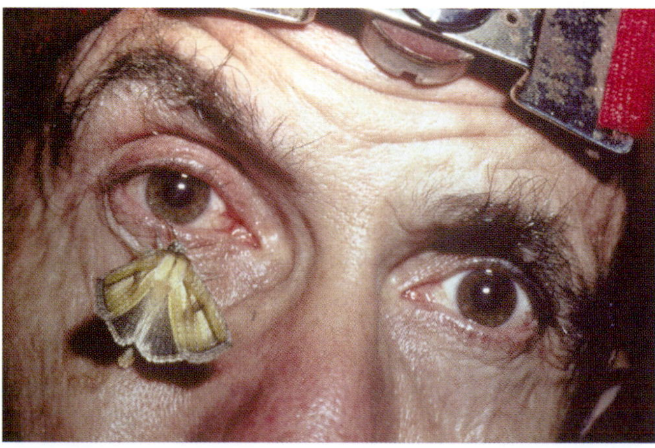

Figure 20.21 Tear-drinking moth, *Chaeopsestis ludovicae* (Thyatiridae), feeding from human eye with tip of proboscis just inside lower eyelid. (Photo by H. Bänziger)

WOUND-FEEDING AND SKIN-PIERCING MOTHS

The only lepidopterans known to be capable of piercing animal skin are members of the noctuid genus *Calyptra* in Southeast Asia. Like many geometrids (Figures 20.22 and 20.23) and other zoophilous moths, they tend to be attracted to wounds, open sores, cuts, scratches, scabs, and other skin lesions. However, although these other moths imbibe only exposed wound fluids, *Calyptra* spp. are capable of piercing the underlying tissue to feed on blood (Figure 20.24). Only the males are hematophagous. Females are believed to feed almost exclusively on fruits, and are able to pierce the outer layers of ripening fruit to

Figure 20.22 Two wound-feeding moths, *Hypochrosis pyrrhophaeata* and *Zythos* sp. (Geometridae), feeding at site of host injury. (Photo by H. Bänziger)

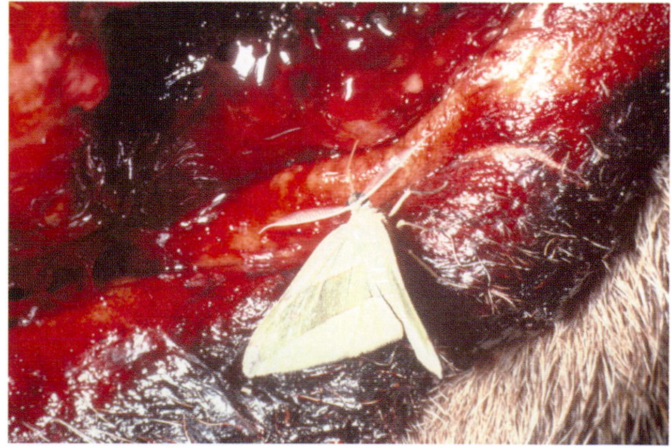

Figure 20.23 *Hypochrosis hyadaria* (Geometridae) feeding at open wound of zebu. (Photo by H. Bänziger)

Figure 20.24 Skin-piercing moth, *Calyptra parva* (Noctuidae), feeding on human. (Photo by H. Bänziger)

reach the sugar-rich juices within. Other moths closely related to *Calyptra* are fruit piercers, suggesting that blood-feeding is a relatively recent development in this group, derived from fruit-piercing behavior.

A number of *Calyptra* species has been observed piercing mammalian skin under natural conditions. Five *Calyptra* species are known to feed on humans: *C. bicolor*, *C. fasciata*, *C. ophideroides*, *C. parva* (Figure 20.24), and *C. pseudobicolor*. *Calyptra* species also have been observed piercing the skin of elephants, water buffalo, zebu, Malayan tapir, rhinoceros, deer, antelope, mules, and pigs. The feeding times typically range from three to 15 minutes. The reaction to the proboscis penetrating the skin varies from being barely felt to causing locally intense pain accompanied by a burning sensation. The latter has been attributed to saliva that is introduced as the moth feeds, whereas the amount of pain is believed to depend on the number of pain receptors that are encountered by the piercing proboscis. Other associated reactions include localized swelling that may persist for several hours, slight numbness or itching, pressure sensitivity at the bite site, and mild induration the following day.

For further details on the biology and behavior of lachryphagous and wound-feeding moths, see references by Bänziger.

PUBLIC HEALTH IMPORTANCE

The severity of reactions to urticating caterpillars is highly variable, depending on the species involved, degree of contact, and nature of the toxin. Toxic components commonly include histamine-like or histamine-releasing substances that cause edema and wheal formation at the site of contact; proteolytic enzymes and esterases; peptides and other substances that can increase vascular permeability, destroy blood cells, or cause local necrosis of tissues; and globulins with immunologic properties.

Many cases involve mild, localized dermatitis in the form of a nettling sensation or minor irritation with transient puffiness or redness at the contact site. In more severe encounters, individuals often experience an intense burning sensation with associated wheal formation and persistent erythema. Other cases may include localized numbness, formation of vesicles, nausea, vomiting, or fever. In cases such as those involving the saddleback caterpillar and puss caterpillar, the affected skin tends to glisten after several minutes as fluids from the damaged dermal cells appear on the skin surface. Without treatment, the burning sensation usually subsides within 30 minutes to an hour, but may persist much longer. Radiating pains and lymphadenopathy may occur in more severe cases involving the limbs. The pain often extends proximally to the axillary or inguinal lymph nodes, sometimes with associated inflammation of the lymphatic vessels, which may be visible as reddened traces on the skin surface. Some cases result in dull, throbbing aches in the lymph nodes that can persist as long as 12 to 24 hours. The contact site may remain discomforting and sensitive to touch for several days thereafter as the skin heals. The pattern of actual contact with the urticating spines may be evident for days, sometimes weeks.

Occupational erucism is a common occurrence in tropical countries, notably in South America and Southeast Asia, where field workers are exposed to stinging caterpillars. In addition to causing temporary discomfort and annoyance affecting primarily the arms and hands, chronic contact with some species (e.g., the arctiid *Premolis semirrufa* in Brazil) can result in persistent swelling and fibrous lesions of the joints of the hands and fingers. Workers in saw mills may experience dermatitis as a result of contact with urticating hairs in egg masses on the trunks of trees being cut for timber or stored at mill sites.

Reactions to adult moths usually occur as dermal irritation induced by airborne setae on contact with the skin, a condition called **moth dermatitis**. Most of the reported cases involve *Hylesia* species (Saturniidae). The response is similar to that caused by urticating larvae. The most commonly affected parts of the body are the face, neck, and upper limbs. As the setae are rubbed into the skin, they may release histamine-like substances that cause erythema and pruritus. The problem is further aggravated and spread by scratching and sweating, or by clothing and bedding contaminated with the irritating hairs. In severe cases, symptoms may persist for several days or longer, sometimes resulting in fever, insomnia, malaise, nausea, vomiting, or muscle spasms.

When airborne body setae or wing scales of moths are drawn into the respiratory tract, they can cause mechanical irritation and inflammation of the nasal passages, pharynx, and trachea. Individuals who become sensitized to these insect parts may develop **inhalation allergies** upon subsequent exposure. Most reported cases involve *Hylesia* species in South America and the Caribbean region. Thousands of microscopic setae from the lateral and ventral areas at the end of the female abdomen become airborne during outbreaks of these moths. Clouds of setae may be released into the air when the moths are attracted to lights at night and bump against windows and outdoor light fixtures. Several such instances involving thousands of human cases of moth dermatitis and inhalation allergies have been reported in Peru (Allard and Allard, 1958) and Brazil (Gusmão et al., 1961).

Another health-related problem is **silk-induced allergies**. This results from contact with the silk of certain moths, with sericin being the main allergen in the silk of *Bombyx mori* and *B. mandarina*. This can be an occupational problem among individuals involved in processing natural silk, resulting in skin irritation and conjunctivitis. Similar allergic responses are caused by contact with silk clothing and silk-filled bed quilts. In addition, inhalation of moth scales by workers in the silk industry can result in allergies, leading to coughing, allergic rhinitis, and asthma. Certain proteins present in silkworm pupae have been reported to cause sensitivity reactions in the sericulture regions of China where eating silkworm pupae is common.

The major health concern from lachryphagous moths feeding on humans is the irritation of eye tissues caused by microscopic lesions as the tip of the proboscis abrades the eyeball or inner surface of the eyelids. The most discomforting cases are caused by the tarsal claws of species that secure themselves to the eyelids while feeding. Actual damage to the eyes is usually minor, however, and generally heals without becoming infected. The risk of mechanically transmitted agents is greater in the case of ocular abrasions by the tarsal claws than by the proboscis, simply because the tarsi are more apt to become contaminated with infectious agents. They come in contact with a wider array of animal substances and potential pathogens. Nonetheless, there is no evidence that any human pathogen is transmitted by eye-frequenting moths.

The potential for mechanical transmission of pathogens is greatest in the case of skin-piercing moths. The relatively large size of these moths provides a greater surface area for picking up contaminants, and the deeper skin punctures and significant blood flow around the bite wound can be contributing factors. The fact that their bites are more painful than lachryphagous species increases the chances of interrupted feedings and attacks on more than one animal, enhancing prospects for transfer of pathogens. However, there is no documented evidence that any disease agent is transmitted to humans by these moths.

Treatment of urticarial cases includes the mechanical removal of urticating spines or hairs and the application of substances to alleviate the symptoms. The broken spines or hairs can be removed from the skin with fine-tipped forceps or by lightly applying sticky tape to the affected skin surface and lifting them away as the tape is removed. It is important not to rub the spines into the skin in the process of trying to remove them. They also may be removed by submerging the affected areas in warm water to float off the hairs, gently washing the skin surface with running water, or showering. Since components of the venom are generally water-soluble, water helps to dilute and destroy the toxins, thereby reducing their potency.

Other measures taken to relieve discomfort include the local application of ice packs, calamine lotion, and antihistamines. Soaking with warm bicarbonate of soda or solutions of household ammonia are generally ineffective, although relief has been reported in some cases. Mild analgesics, such as salicylic acid, do not usually provide much relief, either for the localized pain or accompanying headaches.

VETERINARY IMPORTANCE

Under certain conditions, lepidopterous species can cause veterinary problems. These usually involve the ingestion of urticating caterpillars or the caterpillars of non-urticating species that contain toxins in their body fluids. The result is irritation and inflammation of the lining of the digestive tract, variously called erucic stomatitis, erucic gastroenteritis and erucic gastroenterocolitis. Reactions following ingestion may be immediate or delayed and range from being mild to fatal. Although this type of erucism is most common in grazing cattle, horses, and other herbivores, it also has been reported causing severe stomatitis in dogs and cats. Canine and feline cases have resulted from ingestion of caterpillars and leaves contaminated with urticating hairs. However, severe oral lesions and tongue necrosis have been documented in dogs in the eastern Mediterranean following ingestion of larvae of the pine processionary moth (*Thaumetopoena wilkinsoni*) (Bruchim et al., 2005). In the United States, *Hemileuca maia* and other *Hemileuca* species have been abundant enough to pose serious threats to cattle (Caffrey, 1918). In other cases, cattle have died following ingestion of the larvae or pupae of cabbage butterflies (Pieridae) containing poisonous body fluids that can cause severe enteritis (Delgado, 1978).

Cases of urticaria in domestic animals are seldom reported but occasionally have been observed in horses. Other minor concerns are the suspected involvement of adult moths in transmitting the bacterial agent of bovine infectious keratitis, or pinkeye, in Uganda (Guilbride et al., 1959) and some moths serving as intermediate hosts for the rat tapeworm *Hymenolepis diminuta* (Belding, 1964). There also are reports of adult moths feeding on blood of chickens in coastal Ecuador where they are known by the local inhabitants as *chupa-gallina* (chicken suckers).

Wound-feeding and skin-piercing moths are not known to transmit any animal pathogens. However, their feeding can cause considerable irritation and discomfort, especially when associated moth populations are high and their attacks persistent. Most of the lachryphagous moths cause relatively mild discomfort and are considered more of an annoyance than a significant health problem.

Cases of scoleciasis are occasionally encountered by veterinarians. Three cases have been reported, for example, involving larvae of the Indian meal moth (*Plodia interpunctella*) that penetrated the skin of a wild-caught parakeet and two domestic cats. In the parakeet case a larva made its way to the ventral surface of the brain, where it was recovered as a fourth instar. In the cat cases larvae were removed from subcutaneous tissues of the ear and neck. Both cats were in the same household, separate from the parakeet case, where the source of the moth larvae evidently was infested bird seed (Pinckney et al., 2001).

Caterpillar-induced Equine Abortion

The caterpillars of certain moths have been implicated as the cause of equine abortions and stillbirths in North America and Australia. Although uncommon, high incidences of fetal losses have occurred in localized areas in years when populations of caterpillars are unusually large, leading to horses ingesting caterpillars in fodder or forage. Fragments of barbed setae, which serve a defensive function in the caterpillars, penetrate the intestinal wall and enter the bloodstream. From there they can be carried to various tissues and organs, including the developing fetuses of mares. Evidence indicates that normally nonpathogenic bacteria in the gastrointestinal tract (e.g., alpha *Streptococcus* and *Actinobacillus* spp.) are transported by the setal fragments and can infect reproductive tissues, causing inflammation of the amnion, placenta, and umbilical cord. Fetal pneumonia and death also can occur. In the United States, clinical cases are known as **mare reproductive loss syndrome** (MRLS), associated with species of Lasiocampidae (*Malacosoma*) and Notodontidae (*Datana*). In Australia, the syndrome is called **equine amnionitis and fetal loss** (EAFL), and involves certain caterpillars of the families Thaumetopoeidae (*Ochrogaster*) and Lymantriide (*Euproctis, Leptocneria*). The abortigenic factor that initiates the syndrome in each case remains uncertain but may be a toxin in the integument and setae of the caterpillars.

The largest epidemic of equine abortion in North America occurred in 2001 and 2002, when more than 1500 cases were reported on horse farms in Kentucky. Thoroughbreds were the most affected, with estimated economic losses of $330 to $500 million during the two-year period (Sebastian et al., 2006). The cause was attributed to ingestion of the eastern tent caterpillar (*Malacosoma americanum*) (Figure 20.25)

Figure 20.25 Eastern tent moth caterpillar, *Malacosoma americanum* (Lasiocampidae); upon ingestion by mares can cause equine abortion and stillbirth. (Photo by Lyle Buss)

Figure 20.26 Walnut caterpillar, *Datana integerrima* (Notodontidae); can cause abortion in horses following ingestion by mares. (Photo by James Castner)

crawling on the ground after feeding on nearby heavily infested trees. A similar syndrome occurred in 2005 in Florida, in which cases of equine abortion occurred in mares following ingestion of the walnut caterpillar (*Datana integerrima*) (Figure 20.26), following severe defoliation of hickory trees (Juglandaceae, *Carya*) (John F. Roberts, personal communication).

Fetal losses have been reported in thoroughbred and quarterhorse broodmares in Australia, involving caterpillars of the following three species: processionary caterpillar (*Ochrogaster lunifer*), white cedar moth (*Leptocneria reducta*), and mistletoe brown-tail moth (*Euproctis edwardsi*). Whereas *O. lunifer* feeds on acacia and eucalypt trees, the respective host plants of *L. reducta and E. edwardsi* are reflected by their common names.

PREVENTION AND CONTROL

The greatest problems are presented for individuals working under conditions in which urticating caterpillars commonly pose an occupational hazard. The best line of defense is to recognize, and thereby avoid contact with, those species that cause urticaria and other health-related problems. Protective clothing in the form of long-sleeved shirts, pants, gloves, and suitable headwear can greatly reduce the risks involved. In the case of workers exposed to airborne setae, the use of protective eyeglasses, masks, or respirators is recommended.

To reduce the risk of exposure to *Hylesia* and other moths, lights, burning candles, and fires should be extinguished in the evening to discourage the attraction of moths. Light fixtures, windowpanes, window and door frames, and other surfaces with which the moths may come in contact should be wiped clean with a damp cloth. Clothing and bed linen should be washed daily during periods of flight activity by the adult moths. This not only helps to remove urticating setae but also destroys the water-soluble toxins. Other approaches to reducing the problem include the application of insecticides to kill *Hylesia* adults. It is important to immediately remove the dead moths in order to eliminate them as a source of more setae. Desensitization of individuals with a series of injections of moth extracts is also a consideration under certain circumstances. The resultant immunity, however, is not permanent.

No practical preventive measures are recommended for protecting animals from lachryphagous, wound-feeding, or hematophagous moths.

REFERENCES AND FURTHER READING

Allard, R. F., & Allard, H. A. (1958). Venomous moths and butterflies. *Journal of the Washington Academy of Sciences, 48*, 20–21.

Amarant, T., Burkhart, W., LeVine, H., Arocha-Pinango, C. L., & Parikh, I. (1991). Isolation and complete amino acid sequence of two fibrinolytic proteinases from the toxic Saturnid caterpillar *Lonomia achelous. Biochimica et Biophysica Acta, 1079*, 214–221.

Baerg, W. J. (1924). On the life history and the poison apparatus of the white flannel moth, *Lagoa crispata* Packard. *Annals of the Entomological Society of America, 17*, 403–415.

Bänziger, H. (1968). Preliminary observations on a skin-piercing blood-sucking moth (*Calyptra eustrigata*) (Hmps.) (Lep., Noctuidae)) in Malaya. *Bulletin of Entomological Research, 58*, 159–163.

Bänziger, H. (1969). The extraordinary case of the blood-sucking moth. *Animals Magazine (London)*, 135–137.

Bänziger, H. (1971). Bloodsucking moths of Malaya. *Fauna, 1*, 3–16.

Bänziger, H. (1976). In search of the blood-sucker. *Wildlife Magazine (London)*, 366–369.

Bänziger, H. (1980). Skin-piercing blood-sucking moths. III: Feeding act and piercing mechanism of *Calyptra eustrigata* (Hmps.) (Lep., Noctuidae). *Mitt Schweiz Entomol Ges., 53*, 127–142.

Bänziger, H. (1986). Skin-piercing blood-sucking moths. IV: Biological studies on adults of 4 *Calyptra* species and 2 subspecies (Lep., Noctuidae). *Mitt Schweiz Entomol Ges., 59*, 111–138.

Bänziger, H. (1987). Description of new moths which settle on man and animals in S. E. Asia (genera *Thliptoceras, Hemiscopis, Toxobotys*, Pyarlaidae, Lepid.). *Revue suisse de Zoologie; annals de la Societe zoologique Suisse et du Museum d'historie naturelle de Geneve, 94*, 671–681.

Bänziger, H. (1988a). Lachryphagous Lepidoptera recorded for the first time in Indonesia (Sumatra) and Papua New Guinea. *Heteroc Sumatra, 2*, 133–144.

Bänziger, H. (1988b). The heaviest tear drinkers: Ecology and systematics of new and unusual notodontid moths. *Natural History Bulletin of the Siam Society, 36*, 17–53.

Bänziger, H. (1988c). Unsuspected tear drinking and anthropophily in Thyatirid moths, with similar notes on sphingids. *Natural History Bulletin of the Siam Society, 36*, 117–133.

Bänziger, H. (1989a). A persistent tear drinker: Notodontid moth *Poncetia lacrimisaddicta* sp. n., with notes on its significance to conservation. *Natural History Bulletin of the Siam Society, 37*, 31–46.

Bänziger, H. (1989b). Skin-piercing blood-sucking moths. V. Attacks on man by 5 *Calyptra* spp. (Lepidoptera, Noctuidae) in S and SE Asia. *Mitt Schweiz Ent Ges, 62*, 215–233.

Bänziger, H. (1992). Remarkable new cases of moths drinking human tears in Thailand (Lepidoptera: Thyatiridae, Sphingidae, Notodontidae). *Natural History Bulletin of the Siam Society, 40*, 91–102.

Bänziger, H. (1995). *Microstega homoculorum* sp. n.—The most frequently observed lachryphagous moth of man (Lepidoptera, Pyralidae: Pyraustinae). *Revue suisse de Zoologie; annals de la Societe zoologique Suisse et du Museum d'historie naturelle de Geneve, 102*, 265–276.

Bänziger, H., & Büttiker, W. (1969). Records of eye-frequenting Lepidoptera from man. *Journal of Medical Entomology, 6*, 53–58.

Bänziger, H., & Fletcher, D. S. (1988). Description of five new lachryphagous and zoophilous *Semiothisa* moths from SE Asia, with five new synonymies (Lepid., Geometridae). *Revue suisse de Zoologie; annals de la Societe zoologique Suisse et du Museum d'historie naturelle de Geneve, 95*, 933–952.

Belding, D. L. (1964). Order Lepidoptera. In Textbook of Parasitology. New York: Meredith Publishing (pp. 825–826).

Bettini, S. (Ed.). (1978). Arthropod Venoms. Berlin: Springer-Verlag.

Bishopp, F. C. (1923). *The Puss Caterpillar and the Effects of Its Sting on Man. U.S. Dept. Agric., Dept. Circular* 288.

Bruchim, Y., Ranen, E., Saragusty, J., & Aroch, I. (2005). Severer tongue necrosis associated with the pine processionary moth (*Thaumetopoena wilkinsoni*) ingestion in three dogs. *Toxicon: official journal of the International Society on Toxinology, 45*, 443–447.

Bucherl, W., Buckley, E. E., & Deulofeu, V. (Eds.). (1971). *Venomous Animals and Their Venoms. Vol. 3*, New York: Venomous Insects. Academic Press.

Büttiker, W. (1967). First records of eye-frequenting Lepidoptera from India. *Revue suisse de Zoologie; annals de la Societe zoologique Suisse et du Museum d'historie naturelle de Geneve, 74*, 389–398.

Büttiker, W., & Bezuidenhout, J. D. (1974). First records of eye-frequenting Lepidoptera from South West Africa. *Journal of the Entomological Society of South Africa, 37*, 73–78.

Caffrey, D. J. (1918). Notes on the poisonous urticating spines of *Hemileuca oliviae* larvae. *Journal of Economic Entomology, 11*, 363–367.

Carrijo-Carvalho, L. C., & Chudzinski-Tavassi, A. M. (2007). The venom of the *Lonomia* caterpillar: An overview. *Toxicon: Official Journal of the International Society on Toxinology, 49*, 741–757.

Castellani, A., & Chalmers, A. J. (1913). *Manual of Tropical Medicine.* Tindall and Cox, London: Baillière.

Chan, K., Lee, A., Onell, R., Etches, W., Nahirniak, S., Bagshaw, S. M., et al. (2008). Caterpillar-induced bleeding syndrome in a returning traveler. *The Canadian Medical Association Journal, 179*, 158–161.

Cheverton, R. L. (1936). Irritation caused by contact with the processionary caterpillar (larva of *Thaumetopoea wilkinsoni* Tams and its nest). *Transactions of the Royal Society for Tropical Medicine and Hygiene, 29*, 555–557.

Cock, M. J. W., Godfray, H. C. J., & Holloway, J. D. (1987). *Slug and Nettle Caterpillars: The Biology, Taxonomy and Control of the Limacodidae of Economic Importance on Palms in Southeast Asia.* C.A.B. International.

Davidson, F. F. (1967). Biology of laboratory-reared *Megalopyge opercularis* Sm. & Abb. Morphology and histology of the stinging mechanism of the larvae. *Texas Journal of Science, 19*, 258–274.

Delgado, A. (1978). *Venoms of Lepidoptera.* In S. Bettini (Ed.), *Arthropod Venoms* (pp. 555–611). Berlin: Springer-Verlag.

Derraik, J. (2006). Erucism in New Zealand: Exposure to gum leaf skeletonizer (Uraba lugens) caterpillars in the differential diagnosis of contact dermatitis in the Auckland region. *New Zealand Medical Journal, 119*, 1241, 1242–2143.

Duarte, A. C., Crusius, P. S., Pires, C. A. L., Schilling, M. A., & Fan, H. W. (1996). Intracerebral haemorrhage after contact with *Lonomia* caterpillars. *Lancet, 348*, 1033.

Epstein, M. E. (1996). *Revision and phylogeny of the limacodid-group families, and evolutionary studies on slug caterpillars (Lepidoptera: Zygaenoidea). Smithsonian Contributions to Zoology No. 582.*

Foot, N. C. (1922). Pathology of the dermatitis caused by *Megalopyge opercularis*, a Texan caterpillar. *The Journal of Experimental Medicine, 35*, 737–753.

Gilmer, P. M. (1925). A comparative study of the poison apparatus of certain lepidopterous larvae. *Annals of the Entomological Society of America, 18*, 203–239.

Gilmer, P. M. (1928). The poison and poison apparatus of the white-marked tussock moth *Hemerocampa leucostigma* Smith and Abbot. *The Journal of Parasitology, 10*, 80–86.

Guilbride, P. D. L., Barber, L., & Kalikwani, A. M. (1959). Bovine infectious keratitis suspected moth-borne outbreak in Uganda. *Bulletin of Epizootiology and Disease in Africa, 7*, 149–154.

Gusmão, H. H., Forattini, O. P., & Rotberg, A. (1961). Dermatite provocada por lepidopteros do gênero Hylesia. *Revista do Instituto de Medicina Tropical de São Paulo, 3*, 114–120.

Ishizaki, T., & Nagaisa, R. (1956). Clinical studies on dermatitis due to *Euproctis flava* (Report 1). *Japanese Journal of Zoology, 7*, 113.

Jenkyns, M. S. (1886). Urtication by *Bombyx rubi. Entomologist, 19*, 42.

Jones, D. L., & Miller, J. H. (1959). Pathology of the dermatitis produced by the urticating caterpillar, *Automeris io. American Medical Association, Archives of Dermatology, 79*, 81–85.

Kagan, S. L. (1990). Inhalant allergy to arthropods: insects, arachnids, and crustaceans. *Clinical Reviews in Allergy, 8*, 99–125.

Kano, R. (1967). Venomous Lepidoptera. *Japan J Sanit Zool, 18*, 170–171.

Kano, R. (1979). Lepidoptera (butterflies and moths). In M. Sasa, H. Takahasi, R. Kano, H. Tanaka (Eds.), *Animals of Medical Importance in the Nansei Islands in Japan* (pp. 117–119). Japan: Shinjuku Shobo.

Katzenellengogen, I. (1955). Caterpillar dermatitis as an occupational disease. *Dermatologica, 111,* 99–106.

Kawamoto, F., & Kumada, N. (1984). *Biology and venoms of Lepidoptera.* In A. J. Tu (Ed.), *Handbook of Natural Toxins* (pp. 291–330). New York: Dekker.

Keegan, H. L. (1963). Caterpillars and moths as public health problems. In H. L. Keegan, W. V. Macfarlane (Eds.), *Venomous and Poisonous Animals and Plants of the Pacific Region* (pp. 165–170). Elmsford, NY: Pergamon Press.

Kristensen, N. P. (1984). Studies on the morphology and systematics of primitive Lepidoptera (Insecta). *Steenstrupia, 10,* 141–191.

Kuspis, D. A., Rawlins, J. E., & Krenzelok, E. P. (2001). Human exposures to stinging caterpillar: Lophocampa caryae exposures. *The American Journal of Emergency Medicine, 19,* 396–398.

Lian, Y., & Liu, Z. (2006). Advances in silkwork pupa allergy and their allergens. *Journal of Tropical Medicine (Guangzhou), 6,* 224–226.

Long, F. R. J. (1886). Urtication by larvae of *Bombyx rubi. Entomologist, 19,* 45.

Lucas, T. A. (1942). Poisoning by *Megalopyge opercularis* ('Puss caterpillar'). *Journal of the American Medical Association, 119,* 877–880.

Marshall, G. A. K., Jack, R. W., & Neave, S. A. (1915). *A noctuid feeding on the moisture from the eyes of mules. Proc. Ent. Soc,* 117–119.

McGovern, J. P., Barkin, G. B., McElhenney, T. R., & Wende, R. (1961). *Megalopyge opercularis.* Observations of its life history, of its sting in man, and report of an epidemic. *Journal of the American Medical Association, 175,* 1155.

McMillan, C. W., & Durcell, W. R. (1964). Health hazard from caterpillars. *New England Journal of Medicine, 271,* 147–149.

Mills, R. G. (1923). Observations on a series of cases of dermatitis caused by a liparid moth *Euproctis flava* Bremer. *China Medical Journal, 37,* 351–371.

Nielsen, E. S., & Common, I. F. (1991). Lepidoptera (moths and butterflies). In I. D. Naumann (Ed.), *The Insects of Australia* (2nd ed., Vol. 2., pp. 817–915). Chapter 41. London: Melbourne University Press, Carlton, Victoria and University College of London Press.

Neuedorf, F. (2007). *Caterpillars are aborting our mares.* Horse, (Dec/Jan), 56–58.

Neustater, B. R., Stollman, N. H., & Manten, H. D. (1996). Sting of the puss caterpillar: An unusual cause of abdominal pain. *Southern Medical Journal, 89,* 826–827.

Perlman, F., Press, E., Googins, J., Malley, A., & Poareo, H. (1976). Tussockosis: Reactions to Douglas fir tussock moth. *Annals of Allergy, 36,* 302–307.

Pesce, H., & Delgado, A. (1971). Poisoning from adult moths and caterpillars. In W. Brucherl, et al.(Eds.), *Venomous Animals and Their Venoms* (Vol. 3, pp. 119–156). New York: Academic Press.

Picarelli, Z., & Valle, J. R. (1971). Pharmacological studies of caterpillar venoms. In W. Bucherl, et al.(Eds.), *Venomous Arthropods and Their Venoms* (Vol. 3, pp. 103–118). New York: Academic Press.

Pinckney, R. D., Kanton, K., Foster, C. N., Steinberg, H., & Pellitteri, P. (2001). Infestation of a bird and two cats by larvae of *Ploida interpunetella* (Lepidoptera: Pyralidae). *Journal of Medical Entomology, 38,* 725–727.

Rodriguez-Morales, A. J., Arria, M., Rojas-Mirabel, J., Borges, E., Benitez, J. A., Herrera, M., et al. (2005). Lepidopterism due to exposure to the moth *Hylesia metabus* in northeasern Venezuela. *The American Journal of Tropical Medicine and Hygiene, 73,* 991–993.

Rotberg, A. (1971). Lepidopterism in Brazil. In W. Bucherl, et al. (Eds.), *Venomous Animals and Their Venoms* (Vol. 3, pp. 157–168). New York: Academic Press.

Rothschild, M., Reichstein, T., von Euw, J., Aplin, R., & Harman, R. R. M. (1970). Toxic Lepidoptera. *Toxicon: Official Journal of the International Society on Toxinology, 8,* 293–299.

Scoble, M. J. (1992). *The Lepidoptera: Form, Function and Diversity.* London: Oxford University Press.

Sebastian, M. M., Bernard, W. V., & Fitzgerald, T. D. (2006). *Mare reproductive loss syndrome.* Compendium: Equine Edition (Spring).

Selman, C. L. (1972). Observation of an eye-frequenting geometrid in the United States. *Journal of Medical Entomology, 9,* 276.

Shama, S. K., Etkind, P. H., Odell, T. M., Canada, A. T., Finn, A. M., & Soter, N. A. (1982). Gypsy-moth-caterpillar dermatitis. *New England Journal of Medicine, 30,* 1300–1301.

Shannon, R. C. (1928). Zoophilous moths. *Science, London, 68,* 461–462.

Sharp, H. (1885). Urtication by larvae of *Bombyx rubi. Entomologist, 18,* 324.

Southcott, R. V. (1978). Lepidopterism in the Australian Region. *Rec Adelaide Children's Hospital, 2,* 87–173.

Southcott, R. V. (1983). Lepidoptera and skin infestations. In Parish (Eds.), *Cutaneous Infestations of Man and Animal* (pp. 304–343). New York: Praeger.

Sterling, P. H. (1993). Brown-tail: The invisible itch. *Antenna, 7,* 110–113.

Tu, A. T. (Ed.). (1984). Insect poisons, allergens and other invertebrate venoms. In Handbook of Natural Toxins (Vol. 2). New York: Dekker.

Webb, B. A., Dahlman, W. E., DeBorde, D. L., Weer, S. N., Williams, C., Donahue, N. M., et al. (2004). Eastern tent caterpillars (*Malacosoma americanum*) cause mare reproductive loss syndrome. *Journal of Insect Physiology, 50,* 185–193.

Wen, C. M., Ye, S. T., Zhou, L. X., & Yu, Y. (1990). Silk-induced asthma in children: A report of 64 cases. *Annals of Allergy, 65,* 375–378.

Wirtz, R. A. (1984). Allergic and toxic reactions to non-stinging arthropods. *Annual Review of Entomology, 29,* 47–69.

Zaias, N., Ioannides, G., & Taplin, D. (1969). Dermatitis from contact with moth (Genus *Hylesia*). *Journal of the American Medical Association, 207,* 525.

Chapter

21

Ants, Wasps, and Bees (Hymenoptera)

Hal C. Reed . Peter J. Landolt

A number of species of ants, bees, and wasps are pestiferous and problematic as a stinging hazard to humans. Ants, bees, and wasps generally are abundant throughout much of the world, and make up a significant proportion of all insects (Hölldobler and Wilson, 1990). They are placed in the insect order Hymenoptera, along with numerous nonpest species and also very beneficial insects such as pollinating bees and the parasitoid wasps that serve as natural biological control agents for many insect pests. Also, ants and wasps are among the foremost predators in regulating insect populations in forest communities and agroecosystems.

The Hymenoptera of most concern to human health are those that can use their sting apparatus as either an offensive or defensive weapon. The majority of species of Hymenoptera are not able to sting, but those that can include ants, bees, and certain types of wasps. Most of these stinging species are solitary, nonaggressive, and use their sting and venom primarily to subdue prey. When these solitary species do sting, they usually cause only moderate discomfort to humans and the pain is of short duration. Of far greater concern are the serious stings that can be inflicted by ants, bees, and wasps that are social and live in colonies. Often they possess defensive or aggressive behaviors and their venoms contain chemicals that cause intense pain and that serve as effective deterrents against vertebrate predators. Generally, this chapter summarizes information on the biology and pest status of social species of ants, bees, and wasps, but we include a number of solitary species that are commonly encountered and may be confused with more problematic species.

TAXONOMY

Worldwide, there are more than 115,000 described species in the order Hymenoptera. The Hymenoptera is divided into two major suborders, the Symphyta, or sawflies, which are plant feeders, and the Apocrita, which normally feed on other arthropods, but include the bees. In the Apocrita, the abdomen is narrowly joined to the thorax ("wasp waist"), whereas in the Symphyta the abdomen and thorax are broadly joined. The Apocrita are further divided into the Terebrantia (Parasitica), which use their ovipositor for egg laying, and the Aculeata that have the ovipositor modified as a sting. The 50,000 to 60,000 species of aculeates in the world are classified into eight superfamilies. The work by Goulet and Huber (1993) provides for identification of all families of Hymenoptera. The most important groups that cause health-related problems are members of the Formicoidea, Apoidea, and Vespoidea. It is within these three superfamilies that the social species of ants, bees, and wasps are placed.

The Formicoidea, or ants, are divided into 11 subfamilies (Hölldobler and Wilson, 1990) (Table 21.1). There are about 8,800 described species of ants, although it is estimated that there may be as many as 16,000 to 20,000 species worldwide. Most ants possess a stinger, the exceptions being for members of the subfamilies Formicinae and Dolichoderinae. Many of those that lack stingers squirt various caustic chemicals onto antagonists for defensive purposes. Excellent sources of information on the taxonomy and biology of this diverse group are Hölldobler and Wilson (1990), Bolton (1994), and Fisher and Cover (2007). Klotz, et al. (2008) provide information on the identification, biology, and management of urban ant species in North America and Europe.

Some taxonomists place all the bees, totaling more than 20,000 species, together in the one single family, Apidae. However, we have followed bee taxonomy as presented by Michener (1974, 2007) in which bees are regarded as members of the superfamily Apoidea with nine families. Most of these families consist primarily of solitary or communal species that are rarely a stinging hazard to humans. Most of the stinging bees are social species in the family Apidae, such as the ubiquitous honey bees (*Apis* spp.) and bumble bees (*Bombus* spp.) Some species of sweat bees (*Halictidae*) and carpenter bees (*Anthophoridae*) represent minor stinging threats. For further information on the phylogeny and

Table 21.1 Subfamilies, Common Names, and Distribution of Ants (Formicidae) Based on Hölldobler and Wilson (1990)*

Subfamily	Common Name(s)	Distribution	Sting
Ponerinae	Ponerine ants	Mostly tropical, worldwide	Present
Myrmeciinae	Bull-dog ants	Mostly Australia	Present
Pseudomyrmecinae	Acacia ants	Tropical Asia, Africa, Americas	Present
Myrmicinae	Harvester ants, leaf cutting ants, fire ants, pavement ants, pharaoh ants, others	Worldwide	Present in about half of species
Dorylinae (Old World)	Driver ants, safari ants	Asia, Africa	Present
Ecitoninae (New World)	Army ants, legionary ants	Mostly tropical, Americas	Present
Dolichoderinae	Argentine ants and others	Worldwide	Absent
Formicinae	Thatching ants, carpenter ants, weaver ants, others	Worldwide	Absent

*Omitted are the rare subfamilies Nothomyrmeciinae, Leptanillinae, and Aneuretinae.

classification of bees, see Alexander and Michener (1995), Michener et al. (1994), and Michener (2007).

Social wasps of the family Vespidae, represented by about 860 species worldwide, are the most important stinging wasps. A few species in other families of the Vespoidea, including some sphecid wasps (Sphecidae), velvet ants (Mutillidae), and spider wasps (Pompilidae) also occasionally cause stinging problems. A general reference of Hymenoptera (e.g., Gauld and Bolton, 1988; Goulet and Huber, 1993), supplemented with general textbooks on insects, should suffice for most readers wishing to identify solitary wasps to family or lower taxa. For identifying vespids in the northeastern Nearctic region, Buck et al. (2008) should be consulted. Most social vespids in North America are members of two subfamilies: Vespinae, the hornets and yellowjackets, and Polistinae, the paper wasps (Table 21.2). The Vespinae includes the hornets, *Vespa* spp. found principally in Europe, Asia, and North Africa; *Provespa* spp. in Southeast Asia; and the yellowjackets *Dolichovespula* spp. and *Vespula* spp. of temperate regions around the world. The Polistinae are further divided into four tribes: the Ropalidiini, mostly in the tropics of Africa and Asia; the Epiponini, mostly in tropical Asia, Africa, and South America; the Mischocyttarini in North, Central, and South America; and the Polistini, including the cosmopolitan paper wasps, *Polistes* spp. (Akre and Reed, 1984).

Table 21.2 Subfamilies, Species, Common Names, and Principal Nest Sites of Selected Vespid Wasps in North America

Species	Common Name	Nest Site
Subfamily Polistinae		
Polistes		
P. apachus	None	Aerial
P. aurifer	Golden paper wasp	Aerial
P. annularis	Spanish Jack	Aerial, river and lake shores
P. carolina	Red wasp	Aerial, concealed
P. exclamans	Guinea wasp	Aerial, especially around man-made structures
P. dominulus	European paper wasp	Aerial, especially around man-made structures
P. metricus	None	Aerial, concealed
P. perplexus	Red wasp	Aerial, concealed
Subfamily Vespinae		
Vespa		
V. crabro	European hornet	Hollow trees, attics
Dolichovespula		
D. arenaria	Aerial yellowjacket	Aerial, trees, structures
D. maculata	Baldfaced hornet	Aerial, trees
Vespula		
V. flavopilosa	Hybrid yellowjacket	Subterranean
V. germanica	German yellowjacket	Subterranean, voids in structures
V. maculifrons	Eastern yellowjacket	Subterranean
V. pensylvanica	Western yellowjacket	Subterranean
V. squamosa	Southern yellowjacket	Subterranean
V. sulphurea	California yellowjacket	Subterranean
V. vulgaris	Common yellowjacket	Subterranean

Representatives of a third subfamily of social wasps, the Stenogastrinae, are found in much of Southeast Asia.

MORPHOLOGY

Members of the Hymenoptera range in body length from 0.1 mm for parasitic wasps to over 50 mm for some of the predaceous wasps. The integument of hymenopterans is usually heavily sclerotized, and the pleural sclerites of the thorax are highly modified and fused for strength. Wings are usually well developed in most bees and wasps and in the reproductive, or sexual, forms of ants. However, wings are absent in the nonreproductive worker caste in ants.

In wasps, the first abdominal segment is fused with the thorax as a **propodeum**. The second abdominal segment forms the **petiole** or narrow constriction between the thorax and the more enlarged remaining part of the abdomen called the **gaster**. The prominent petiole of many wasps has given rise to the term "wasp waist" and "wasp-waisted."

Although bees resemble wasps in many features, they generally have more hairs (setae) on the body. Most bees have branched hairs on the legs or gaster and broad hind legs that enable them to gather and transport pollen to their nest. The honey bees and other bees in the family Apidae possess dense rows of branched hairs on the hind legs called corbicula, or pollen baskets, that can store large amounts of pollen.

All ants are social and evolved from a primitive wasp-like ancestor. They are all readily identified by having a narrow abdominal petiole, which consists of one or two segments and bears a dorsal lobe. Worker ants have geniculate (elbowed) antennae with the first segment very long.

Many ant species, like many other social insects, have distinct castes of queens, sterile female workers, and males. The size variation among ant species is extremely great, ranging from the little black ant *Monomorium minimum* with workers 1.5 to 2.0 mm long to the comparatively giant hunting ponerine ant *Dinoponera grandis*, which may be up to 34 mm in length. Some ants such as the primitive, tropical bullet ant *Paraponera clavata* exhibit few, if any, morphological or size differences between the queen and the workers; it is monomorphic, or represented by only one form. However, most ants are polymorphic and are easily separated into castes. Some species of ants exhibit marked worker polymorphism with extremely large individuals, usually with large heads, called **majors**. They may also have intermediate size individuals, and **minors** or small workers. The minors usually constitute most of the **nurse workers** that take care of the brood, while the majors, or **soldiers**, respond quickly to disturbances to defend the colony. Soldiers of army ants have disproportionately large heads that house large adductor muscles that operate the equally large, ice-tong-shaped mandibles. The great differences in size and morphology among the various castes within some species can cause problems in identification, unless the castes are associated with each other and recognized as all belonging to the same species.

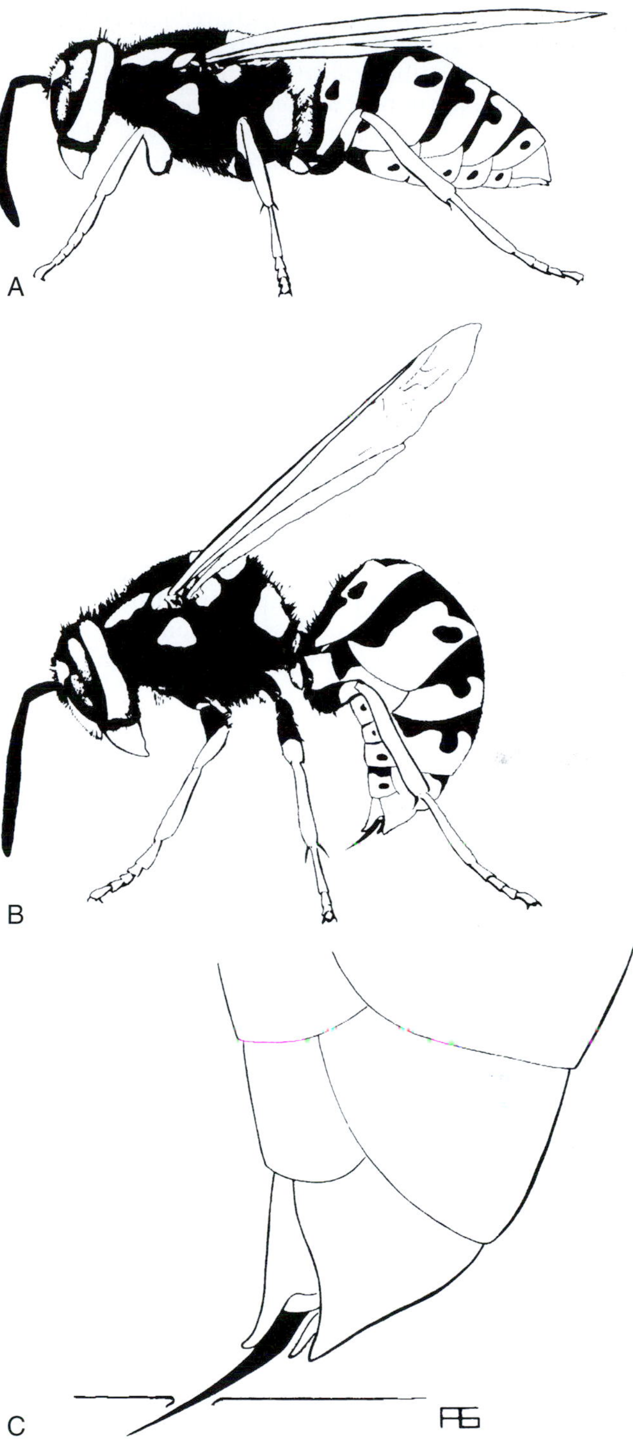

Figure 21.1 Sting action of a yellowjacket (Vespidae) worker. (A) Resting position; (B) sting position with abdomen flexed and lancets of sting extruded; (C) sting at tip of abdomen piercing skin. (Akre et al., 1981)

The sting apparatus is a critically important morphological trait of the aculeate Hymenoptera (Figure 21.1). The basic hymenopteran sting in wasps, bees, and ants is simply a modification of the female egg-laying structure, or ovipositor. Males lack such a structure and cannot sting. In the minority of ants that do not possess a sting, the components of the ovipositor have been reduced or lost; these ants spray defensive secretions from the tip of the gaster into wounds caused by the mandibles.

The visible portion of a typical ovipositor consists of three pairs of elongate structures, or valves, which are used to insert the eggs into a substrate such as plant tissue or soil. One pair of valves often serves as a sheath and is not a piercing structure; the other two pairs form a hollow shaft, which pierces the substrate by a back and forth sawing motion, with one pair held in position by the other (Figure 21.2). The eggs pass down through the shaft in many Hymenoptera, except the stinging species, or Aculeata. During oviposition by aculeate hymenopterans, the sting apparatus is flexed up and out of the way and the eggs are passed from the genital opening at the base of the ovipositor. There are usually two accessory glands in the female that secrete substances to form egg coverings or to glue the eggs to the substrate.

Although cartoonists delight in portraying the sting of wasps and bees as a constantly protruding spike, the sting is part of a complex apparatus usually hidden within a cavity at the end of the abdomen (Figure 21.3). The differences in sting morphology among the various

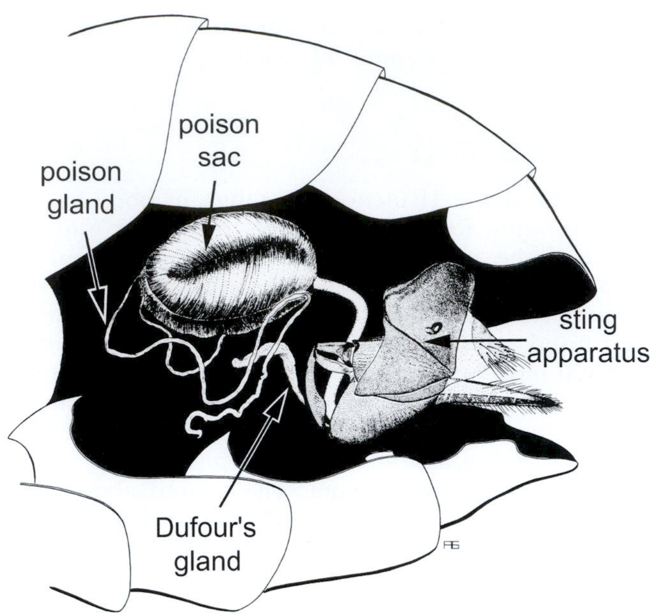

Figure 21.3 Poison gland and associated structures of the sting apparatus of a yellowjacket (Vespidae) worker. (Akre et al., 1981)

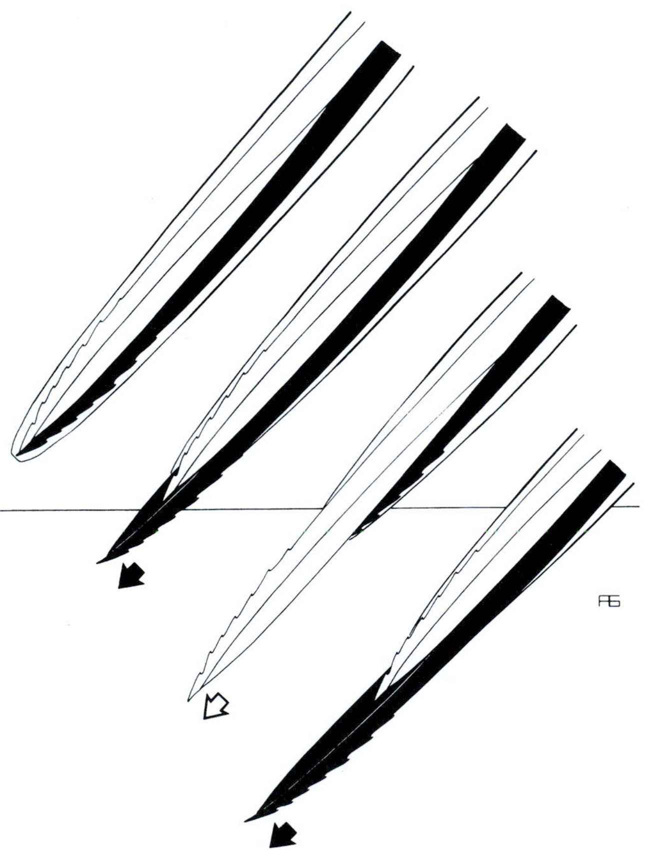

Figure 21.2 Mechanism by which the pair of lancets of hymenopterans penetrate the skin when a victim is stung. The lancets slide alternately back and forth as their serrated tips quickly work their way into the tissue. The opposing surfaces of the lancets are concave, forming a tubular channel through which venom is injected into the wound. (Akre et al., 1981)

aculeate groups are minor. The principal components of the sting apparatus are a pair of long, slender lancets that are encompassed by a single stylet. The ventral edges of the stylet function as a guide rail along which each lancet can slide freely. These structures converge at the base and form a channel through which the venom flows. The distal tip of the lancets is armed with barbs in some species to aid their penetration. When in the retracted position the sting apparatus is covered by two membranous sting sheaths. Penetration by the lancets and stylet into a victim is accomplished by contraction of muscles connected to large sclerotized plates articulating with the base of the stylets and lancets.

The accessory glands of the female reproductive tract have become modified for specialized purposes in stinging hymenopterans (Figure 21.3). One has become a poison gland which produces venom. Another, called Dufour's gland, produces lubricants and coatings for the eggs, linings for brood chambers, and pheromones in some species. The venom actually is produced in the poison gland, which consists of two slender, elongate tubules that empty their products into a prominent, sometimes muscular, reservoir, or poison sac. This reservoir stores the venom until the insect stings, at which time the venom is ejected through a narrow tubule into the base of the sting apparatus. The morphology of the glands associated with the venom apparatus in stinging Hymenoptera is reviewed by van Marle and Piek (1986).

LIFE HISTORY

Hymenopterans undergo holometabolous development with egg, larval, pupal, and adult stages. They exhibit **facultative arrhenotoky**, in which males are produced parthenogenetically, enabling these insects to control the sex of their offspring. Males are haploid and develop from unfertilized eggs whereas the diploid females develop from fertilized eggs. Colonies of many social hymenopterans consist of sterile female workers with a single queen (i.e., monogyny) or multiple queens (i.e., polygyny).

In temperate climates, reproductive males and females of social species of ants, bees, and wasps are produced once per year, typically in late summer and early fall. These reproductives usually leave the colony to find and select mates. Males typically die soon after the mating season and do not contribute to colony activities, such as foraging, brood care, nest maintenance, or defense. An inseminated queen usually will overwinter outside of a colony and start a new colony in the spring, as in the case of yellowjackets, paper wasps, many ants, and bumble bees. However, great variation in colony founding modes is seen among the social species. Honey bee colonies, for example, reproduce when a mated queen leaves the parent colony with a group of workers to begin a new hive, a process known as **swarming**. Colonies of honey bees and many ant species exist for more than a year as perennial colonies, whereas most temperate social wasps, some ants, and bumblebees typically have annual colonies. Development of the queen among social insects is a result of controlled nutrition and greater space, as in cell size, given to a larva during development. These activities are mediated by pheromones that influence the behavior of the workers toward the developing brood. For example, honey bee larvae destined to become queens are reared in large queen cells and fed a special diet of royal jelly, secretions from head glands of nurse bees, throughout their development. Rearing of a new queen usually occurs when the queen is aging, her pheromone secretions are waning, or she has died. New queens also are produced in large, healthy colonies of some species (e.g., *Apis* spp.).

The life histories of social species of Hymenoptera are more diverse in tropical latitudes. Colonies may be started by a single queen, multiple queens, or a swarm of one or more queens along with a retinue of worker females. Colony reproduction may be by individual mated female foundresses, by multiple mated females, by swarms of queens and workers, or by queen replacement. Colony formation, and the later production of new queens and males, may be somewhat synchronized with a wet/dry seasonal cycle, or may appear to be asynchronous. Colony duration may be a fraction of a year, multiple years, or potentially unlimited as in *Eciton* army ants.

BEHAVIOR AND ECOLOGY

Behavior among hymenopteran species is extremely diverse. It varies from individuals that are solitary, with little or no interaction among nesting females, to species that are highly social. Many species have intermediate levels of social behavior in which the parent cares for its offspring and may even share a common nesting area. **Eusocial** species exhibit a reproductive division of labor, a worker caste caring for the young, and overlapping generations in which the offspring assist the parents in rearing the brood. Ants, honey bees, bumble bees, and vespid wasps are eusocial species and typically are found in colonies and often within nest structures.

Stinging behavior is one of the intriguing aspects of social hymenopterans. Visual cues, vibrations, chemicals, and context all play important roles in eliciting a stinging response. Stinging can occur with the disturbance of a nest, in response to swatting, touching, or squeezing of a worker or queen, and in response to alarm chemicals released by other workers. Dark, moving objects are readily attacked by honey bees and vespid wasps when defending their colonies. Vibrations of the ground or other substrate such as a branch that contacts to the nest will stimulate alarm and attack in most species. Alarm pheromones that alert and recruit nestmates to defend the colony are a common strategy among many ants, bees, and wasps, particularly those with large colonies. For example, the honey bee produces several volatile compounds from the sting apparatus and one from the mandibular glands that together alert the colony and help to focus the attack of hive mates so that numerous bees might quickly deter a potential predator. Disturbed workers may also attack when they sense certain chemicals on the attacked individual. Honey bee workers, for example, may sting people when they are applying enamel paints containing **isoamyl acetate**, a natural component of the honey bee alarm pheromone.

Foraging wasps and bees are less prone to sting than individuals near or on the nest. However, foragers may approach and inspect humans, especially if one contains sweet-smelling odors, and if irritated, will sting to defend themselves. Typical stinging behavior of social hymenopterans is illustrated by an attacking yellowjacket (Figure 21.1). The insect grips the skin firmly with its legs and sometimes with its mandibles, exposes the sting apparatus, then plunges the tip of the interlocked lancets and stylet into the skin with a downward thrust of the abdomen. Simultaneously, the contraction of the muscles surrounding the poison sac forces the venom into the sting bulb and through the channel formed by lancets and shaft, much like a hypodermic needle. In order to penetrate, the lancets are moved forward in alternate strokes, each sliding on its track against the stylet shaft (Figure 21.2). The tips of the lancets are equipped with tiny barbs in bees, wasps, and a few ants to facilitate penetration; they literally saw through the victim's flesh as each lancet in turn is thrust forward and anchored in place by the barbs.

Sting autotomy is behavior in which the anchored sting is left embedded in the skin when the insect pulls away (Figure 21.4). It is best known in honey bees, but also occurs in some tropical epiponine social wasps and in harvester ants (*Pogonomyrmex* spp.). Although sting autotomy results in evisceration and eventual death of the insect, the sting with the attached poison gland reservoir continues to pump venom into the wound. Stinger loss is due to the presence of large, recurved barbs on the lancets and to lines of weakness in the structure of the sting apparatus, causing it to readily detach as the

Figure 21.4 Honey bee worker (Apidae) tearing sting apparatus from abdomen as it pulls away, leaving the sting embedded in the skin. (Photo by Justin Schmidt)

insect pulls away after stinging. The barbs on a yellowjacket or paper wasp sting are much smaller and less recurved than those of the honey bee and does not normally become anchored in a person's flesh. The wasp usually can quickly withdraw with an upward pull of its abdomen and sting again. However, a few species will sometimes lose their stingers when they attack human skin (e.g., *Vespula maculifrons*), and workers of most species of yellowjackets will leave their stingers behind when they sting into thick leather gloves.

Aculeate hymenopterans feed on a variety of prey and plant sources. Most food foraging is for larval food that consists of nectar and pollen among bees and masarine wasps, seeds in harvester ants, plant material in some ants, or various arthropod prey in most wasps and many ants. Some yellowjackets will also scavenge for carrion with which to feed the larvae. The adult food may consist of similar materials, but in a liquefied state. Adults of many species of wasps, ants, and bees actively forage also for sweet materials that they may ingest or, in the case of social species, return to the colony. Such foraging may include nectar from flowers or extrafloral nectaries of plants, fruits, saps, and insect honeydews. In eusocial species, liquids are mutually exchanged among workers and between larvae and workers in a process called **trophallaxis**.

The use of chemical communication is very important in maintaining colony cohesion and activities in the societies of most social Hymenoptera. Alarm pheromones often are used to recruit workers to defend the nest and to focus attacks. Many species of ants lay down trail pheromones to food sites to recruit additional foragers, thereby increasing the efficiency of food retrieval. Ants, bees, and wasps produce queen pheromones that limit reproduction by workers and stimulate colony activities such as foraging, construction, and brood care. In some species, queen pheromones may convey information about the status of the colony through trophallactic exchanges of chemical substances among colony members. Additional information on social insect pheromones can be found in Vander Meer, et al. (1997).

HYMENOPTERA VENOMS

Venoms of social hymenopterans are complex biochemical mixtures that paralyze prey, induce pain in vertebrate predators, or act as toxicants (see reference in Piek, 1986b). Probably less than a hundred of these venom compounds have been identified, and many more remain to be discovered and characterized. There are three general categories of venom compounds: (1) small, nonproteinaceous molecules with molecular weights less than 300 Daltons (Da); (2) peptides with molecular weights of 1,500 to 4,000 Da; and (3) larger proteins and enzymes with molecular weights over 10,000 Da. Compounds of the first category include histamines, serotonin and various catecholamines that induce itching, immediate pain, redness, and changes in capillary permeability. The second category is peptides such as hemolysins, which destroy red blood cells and cause pain, neurotoxins, and other pain-inducing compounds such as kinins. The third category of larger proteins and enzymes generally do not cause pain, but aid in the spread and activity of other venom components. An example is hyaluronidase, which facilitates the spread of toxic components through the tissues. Less commonly encountered are phospholipases, which are toxic, disrupt the cell membranes, and cause release of pain-inducing agents (Schmidt, 1986a, 1986c, 1992).

Venoms of solitary hymenopterans such as sphecid and pompilid wasps are designed to cause paralysis in insects, spiders, and other arthropods on which they prey. These venoms directly affect the nervous system and cause a general decline in the rate of metabolic processes. The purpose of these venoms is not to cause the death of the prey but to incapacitate it as food for the larvae. Common components of the venoms of various solitary wasps are histamines, polyamines, and substances such as bradykinins, which cause smooth muscles to contract. Some of their venoms also contain large amounts of acetylcholine as in the case of the sphecid wasp *Philanthus triangulum*. The venoms of solitary wasps generally produce only momentary, slight pain in humans.

Ant Venoms

Ant venoms serve a variety of functions including defense, prey capture, aggregation, trail marking, alarm, and repelling intruders. Only the components of ant venoms that are toxic to vertebrate animals are discussed here. The toxins normally are injected via the sting; however, some ants lack a sting and spray formic acid at their attackers (e.g., many formicine ants). Formic acid is a very effective deterrent, especially if sprayed into the eyes or applied directly into wounds made with the ant's mandibles. Schmidt (1986b) provides an excellent overview of the functions and chemistry of ant venoms.

Venoms of the majority of stinging ants are predominately composed of proteinaceous mixtures. Fire ant venoms, however, largely consist of alkaloids (95%) with only a small proteinaceous component (0.1–1%).

The alkaloids cause most of the local sting reactions whereas the proteins contain active allergenic antigens. The alkaloids are methyl-n-alkylpiperidines called **solenopsins**, and a **piperidine**. The alkaloids are cytotoxic, hemolytic, fungicidal, insecticidal, and bactericidal. The characteristic dermal necrosis that becomes evident at the sting site is due to these alkaloids.

Protein-rich ant venoms are found in most subfamilies of ants, including the Ponerinae, Myrmeciinae, Pseudomyrmecinae, Ecitoninae, and some of the Myrmicinae (Table 21.1). These venoms have not been well investigated because of difficulty in obtaining sufficient quantities of pure venom for analysis. The only studies available are those of primitive ants in the genus *Myrmecia* (Myrmeciinae) and the highly evolved *Myrmica* and *Pogonomyrmex* (Myrmicinae). Harvester ants have a proteinaceous venom with high amounts of phospholipase A and B, hyaluronidase, a potent hemolysin called barbatolysin, and histamines (Schmidt, 1986b). Most ant venoms contain only small amounts of these materials. Other enzymes that have been identified in ant venoms include acid phosphatase, alkaline phosphatase, phosphodiesterase, lipase, esterase, and a nonspecific protease. The primary function of these compounds is to cause pain, either directly or through tissue destruction.

Vespid Venoms

Vespid venoms are biochemically complex and are designed to cause pain (Nakajima, 1986). Their venom typically produces immediate pain, local swelling, and erythema caused by an increase in the permeability of blood vessels at the sting site. The pain often continues for several hours, whereas itching at the sting site may persist for several days. Vespid venoms also cause the contraction of smooth muscles, reduced blood pressure, and the release of histamine and other biogenic amines. Hemolysis induced by lytic peptides and phospholipases may cause kidney damage. There is usually additional damage to surrounding tissues from the products of histolysis. Vespid venoms contain biologically active amines such as serotonin, histamines, tyramine, and catecholamines, all of which tend to produce pain. Acetylcholine has been reported to occur in the venoms of some *Vespa* species. However, the primary pain-causing substances are kinins. In addition, the venoms contain mast-cell degranulating peptides called **mastoparans**, which cause the release of histamines. Venoms also contain enzymes that can act as specific allergens and, in some species, neurotoxic compounds. The immediate pain caused by a vespid sting is principally due to serotonin and kinins. Venoms of some vespids contain alarm pheromones that function to alert nestmates to an intruder and focus stinging attacks.

Honey Bee Venom

The venom of honey bees is a complex mixture of proteins, peptides, and small organic molecules (Banks and Shipolini, 1986; Schmidt, 1992). The most dangerous components for humans are phospholipases and hyaluronidase. Individuals can become sensitized to these materials and subsequently even die from a serious allergic reaction. Bee venom contains large quantities of a potent membrane-disrupting material called **melittin**, which makes membranes extremely susceptible to attack by phospholipases. Melittin is also a cardiotoxin, causes pain, increases capillary blood flow and cell permeability, triggers lysis of red blood cells, and enhances the spread of toxins. The effects of melittin, phospholipase, and a mast cell degranulating peptide cause the release of histamine and serotonin from red blood cells and mast cells. Although the components of honey bee venom that cause pain are very different from those in vespid venom, the end results are very similar. Some components of honey bee venom regulate and/or decrease inflammatory responses in some individuals. This perhaps explains why bee venom therapy has been useful in the treatment of certain forms of arthritis. Another component of honey bee venom, a neurotoxin called **apamin**, seems to cause more effects on insects than humans.

ANTS

Ants are ubiquitous, occurring throughout most of the world, including most oceanic islands. Most major taxa of ants have species that occur worldwide. However, some groups and subgroups are restricted to specific areas. The **acacia ants** (pseudomyrmecines) and ponerines, for example, occur primarily in the tropics, and **bull-dog ants** are found only in Australia. The primarily tropical ecitonine **army ants** are neotropical, with a limited distribution in the United States, and two species occurring as far north as Iowa.

The ants of significant medical-veterinary concern in the United States are the **fire ants** (*Solenopsis* and *Wasmannia* spp.) and **harvester ants** (*Pogonomyrmex* spp.). Most other North American ants rarely sting people, or they are so small that they are incapable of piercing human skin. The **carpenter ants** (*Camponotus* spp.), which are commonly destructive pests of wooden structures, lack a sting, as do members of the subfamilies Formicinae and Dolichoderinae. Other stinging ants such as ponerines (Ponerinae) and army ants (Ecitoninae) are a concern in tropical regions. However, there is one native ponerine species occurring in the southeastern United States, *Odontomachus haematoda*, that can cause painful stings to humans. *Odontomachus haematoda* is peculiar in that it possesses elongated mandibles that are held open at 180 degrees, and snap shut quickly to impale prey or enemies on their sharp teeth. It also uses its mandibles to snip and to jump away when threatened. The **Asian needle ant** (*Pachycondyla chinensis*), also called the **Chinese needle ant**, can cause painful stings. This species was introduced to the southeastern United States sometime before 1932 and has been reported as a particularly pestiferous, stinging ant (Nelder et al., 2006). Other pestiferous ponerine ants that may on occasion be a hazard are *Hypoponera punctatissim, Pseudomyrmex ejectus,* the pavement ant, and pharaoh's ant.

The life cycle of ants is highly varied. In some species, such as the army ants whose queens are wingless throughout their life, colony initiation occurs by budding, a process whereby a colony divides into two colonies. In contrast, many ant species have winged reproductives and colony initiation is typically by a single winged queen. At certain times of the year mature colonies produce an abundance of winged males and queens that leave the nest en masse on a **nuptial flight**. After mating, the males die and the inseminated queens lose their wings before searching for a suitable nest site. The queen lays eggs in the new nest site and feeds the developing larvae from her food reserves stored as fat and flight muscle. The emerging brood becomes the workers and takes over nest maintenance, foraging, and nursing activities. The small colony grows slowly at first and may take a few years to become mature and produce its own reproductives. A single queen is the rule in some ant species, whereas multiple queens occur in others. A few species, like the imported fire ants, have both monogynous and polygynous colonies.

Ants nest in a variety of situations. In the case of army ants, there is no physical nest, but only a bivouac formed from the ants themselves holding on to one another by their legs to form a large mass. Carpenter ants excavate wood to form cavities for their nest, whereas other ants make aerial nests of **carton**, a material formed from soil or plant fibers and saliva. Some *Formica* species build large nests up to one meter high consisting of a mound of small twigs, hence the name "thatching ants." Most ants establish their nests in soil where they excavate extensive galleries and tunnels. Soil is an ideal nesting material as it moderates temperature extremes, holds moisture, and can be easily shaped by the ants into brood or food-holding chambers. Ants often have preferences for specific types of soil.

Ants can have tremendously populous colonies. Primitive ants tend to have only a few hundred workers (e.g., some ponerines); other ants may have up to 10,000 workers in their colonies (e.g., harvester ants). The enormous populations of 1 to 2 million for New World army ants and 22 million for Old World driver ants (Dorylinae) are impressive. However, colonies of these species are small compared to megacolonies of some *Formica* species, which may contain 300 million workers and nearly a 1 million queens occupying an area of a few square kilometers! Such polygynous colonies are usually very tolerant of non-nestmate ants, and it is very difficult to determine whether such colonies are separate units or indeed one giant ant colony.

Fire Ants (*Solenopsis* species)

Most ant stinging problems in North America are due to the two species of imported fire ants in the southern United States: *Solenopsis invicta* (Figure 21.5) and *S. richteri* (Figure 21.6). Less frequent stinging problems are caused by the two native fire ants in the southern United States, *Solenopsis geminata and S. xyloni*, and the little fire ant *Wasmannia auropunctata*.

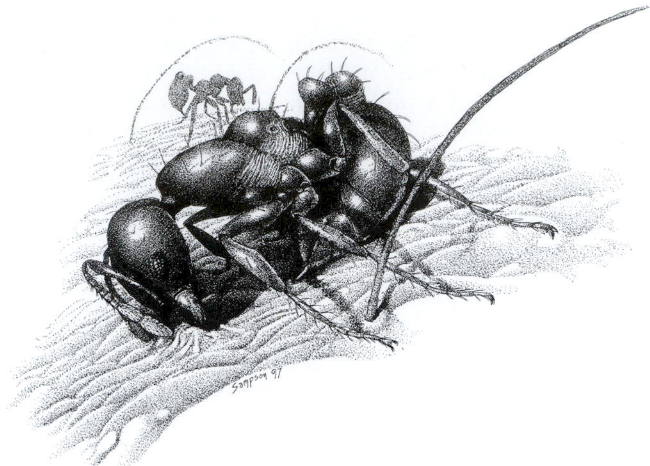

Figure 21.5 Red imported fire ant (*Solenopsis invicta*) worker, stinging skin of human. The ant seizes the skin between its mandibles to provide leverage as it flexes the tip of its abdomen forward to penetrate the skin with its sting. (Original by Blair Sampson)

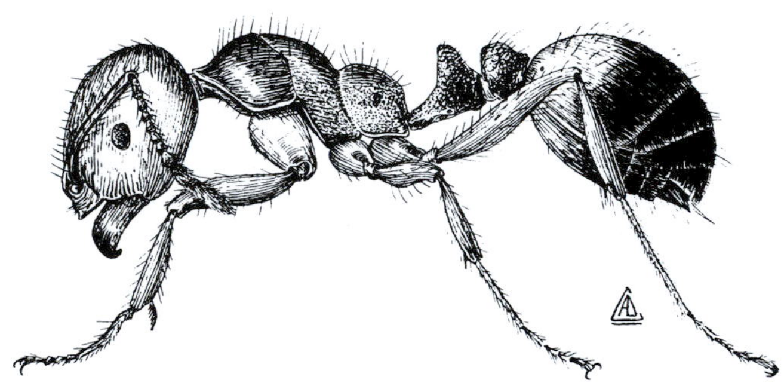

Figure 21.6 Black imported fire ant (*Solenopsis richteri*), worker. (Courtesy of US Department of Agriculture)

The **black imported fire ant** (*S. richteri*) was introduced from South America into the United States at Mobile, Alabama, in the early 1900s, followed by the introduction of the red imported fire ant (*S. invicta*) at the same port in the late 1930s. Since that time, few other stinging insects have created more controversy, generated more research, or received more publicity than these two species of ants. They now inhabit a major portion of 12 southern states (Figure 21.7), with *S. invicta* occupying 95% of the infested area; *S. richteri* occurs only in parts of Mississippi, Alabama, Tennessee, and Georgia. Although *S. invicta* and S. *richteri* are reproductively isolated in their native South America, they sometimes hybridize in the southeastern United States, which can complicate identification and surveillance for these two species. The distributions of these ants probably have reached their northernmost limits where they can survive in the central and eastern United States. There is, however, a possibility that these ants will spread in the western states as far north as Washington. Climatic conditions on the West Coast are suitable, and the westward spread of these ants has

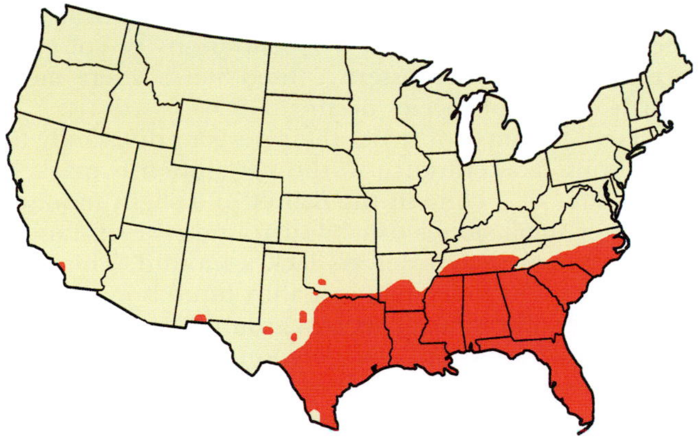

Figure 21.7 Geographic distribution of the **red imported fire ant** (*Solenopsis invicta*) in the continental United States. Note its very restricted occurrence in New Mexico and California. It also occurs in Puerto Rico. (Based on data from the US Department of Agriculture/Animal and Plant Health Inspection Service, 2008)

Figure 21.8 Typical fire ant mound in piedmont region of southeastern United States. (Photo by G. R. Mullen)

been impeded only by the arid regions of western Texas. The **red imported fire ant** was reported in Nevada in 1993 and in Arizona in 1994. More recently, it has invaded California and Oklahoma.

Fire ants are omnivores and opportunistic feeders. They feed mostly on insects and other arthropods. Fire ants also may feed on seeds of some plants and can affect local plant assemblages by transporting viable seeds of other plant species. They also feed on germinating plants, as well as on fruits and roots, and can cause further damage by girdling tree seedlings.

Fire ants are soil nesters, with most colonies being initiated by a single inseminated queen after a nuptial flight during April or August. A queen makes a burrow 3 to 12 cm deep and within 24 hours lays her first eggs in a chamber at the end of the burrow. As many as 2,500 colonies can be initiated per hectare, but few of these incipient colonies survive the next winter. Colony growth is rapid and often produces more than 10,000 workers within a year. Some colonies contain as many as tens of thousands to hundreds of thousands of workers within a few years. Polygynous colonies are fairly common. Brood may be produced year-round in the southernmost US distributions of the ants, but brood production ceases during the winter months north of 30°N latitude. The size and texture of the above-ground mound is variable depending on soil type, moisture, and vegetation (Figure 21.8). Mounds in sandy areas are generally low, whereas those occurring in clay soil may be up to 1 meter tall by 1 meter in diameter. Large colonies often construct several interconnected mounds.

Fire ants quickly respond to disturbances of their nests and attack intruders in force. When a colony is disturbed, the ants swarm over the intruder until the first worker stings and alarm pheromones are released. This triggers stinging behavior in other workers. The workers grasp the skin with their mandibles (Figure 21.5) and sometimes spin in a circle from this attachment, stinging the entire time. The common name "fire ants" reflects the burning sensation caused by their stings.

Fire ants occur in rural, urban, and suburban areas where they constitute a serious stinging problem. In addition to damaging crops, they often sting farm workers as they harvest crops by hand. Large sun-hardened mounds can damage farm equipment or make it nearly impossible to harvest the crops by machine. High densities of colonies in agricultural areas can result in land devaluation. The impact of the red imported fire ant on recreation is decidedly negative as tourists tend to avoid areas with heavy infestations. Colonies are common in urban yards and parks and can be a serious stinging problem to small children and pets. Colonies are even known to nest in traffic lights, air conditioners, and other electrical equipment. Foraging ants will enter houses such that stinging incidents can even occur inside homes.

These introduced pests are out-competing and displacing many native ant species and may even be adversely affecting plant and invertebrate communities. At the same time these ants have had some beneficial effects. They are general predators of a wide variety of crop-damaging insects and of ticks and flies detrimental to livestock and game animals.

The **southern fire ant** (*Solenopsis xyloni*) ranges from California to South Carolina and Florida. This ant usually nests in soil, although it also will nest under stones, in the woodwork of houses, and in masonry. Outdoor nests are marked by excavated soil deposited in irregular piles around the entrances. These ants can girdle nursery stock and other plants and will burrow into plant buds, potato tubers, and strawberries.

The **fire ant** (*Solenopsis geminata*) ranges from Costa Rica to Texas and South Carolina to Florida. Until the imported fire ants were introduced, this ant was the most common and serious ant pest in Florida. Its nests are built in the soil, producing mounds up to 20 cm high. The biology of this species is similar to *S. invicta*. Like *S. xyloni*, it represents only a minor problem and has been largely displaced by *S. invicta* in the United States as *S. invicta* has extended its range.

The **little fire ant** (*Wasmannia auropunctata*) is only 1.5 mm long. It nests in soil, in decayed wood, under stones, in cavities in plants, at the bases of trees, and in houses (Smith, 1965). The limits of a nest are hard to determine, suggesting that nests have satellites and that colonies are polygynous. This ant occurs only in Florida and cannot survive in cooler areas of the United States. Unlike the other fire ants, workers of this ant are not aggressive and sting only when trapped in clothing or similar situations. Unfortunately, these ants will nest in houses and infest clothing, beds, and food. Laborers sometimes refuse to work in cropland where these ants are abundant.

For reviews of fire ants in the United States, see Vander Meer et al. (1990), Taber (2000), and Tschinkel(2006).

Harvester Ants (*Pogonomyrmex* species)

As their common name implies, these ants regularly use seeds as part of their diet. In addition, *Pogonomyrmex* workers scavenge for dead arthropods. Although there are a number of genera in the subfamilies Ponerinae, Myrmicinae, and Formicinae that comprise the harvester ants, only *Pogonomyrmex* species are of concern as a stinging threat in North America. Seven species of *Pogonomyrmex* in North America might constitute a stinging hazard (Cole, 1968). The more common ones are the **western harvester ant** (*P. occidentalis*), the **red harvester ant** (*P. barbatus*), the **California harvester ant** (*P. californicus*), and the **Florida harvester ant** (*P. badius*).

Pogonomyrmex workers (Figure 21.9) are large, up to 10 mm in length. Most are light red or brown, although the gaster of some species may be dark brown to black. These ants are identified by the presence of a psammophore, a fringe of hairs on the underside of the head. These "beards" are used in nest excavation to push material from the nest much like the blade of a bulldozer. Harvester ants are usually slow-moving and cannot walk up slippery vertical surfaces such as glass. Dense concentrations of colonies are common in the western United States where most North American species occur.

Figure 21.9 Harvester ant (*Pogonomyrmex* sp.), workers gathering food. (Courtesy of Roger D. Akre)

Harvester ants construct their nests in dry, sandy to hard soils. The entrance to the nests often is marked by a crater or a cone in the center of a slight mound. A pile of small stones usually surrounds the entrance. Some species in hot deserts lack a mound. The nest can be 1 to 10 m in diameter with tunnels extending down to 5 m or more. The area around the nest is usually completely devoid of vegetation. Colonies of some species have up to 10,000 workers. Individual colonies often survive for 14 to 50 years, reaching maximum densities of 80 or more nests per hectare. Foraging trails from individual nests may extend out 60 m. Where nest densities are high, large expanses of ground may have little vegetation. Because of the habit of harvesting seeds and reducing vegetation, they can damage rangeland used for cattle grazing and sometimes become significant pests locally (MacKay, 1990). At the same time they are beneficial in aerating the soil, providing enrichment, and promoting new plants sprouting from discarded seeds. Nests invariably occur in sunny locations; if nests become shaded by vegetation or human activity, the ants generally move.

Harvester ants sting readily and can inflict intense pain. The incidence of stings is low, however, because their relatively large size and conspicuous nests cause most people to avoid them. Also, their colony numbers are relatively small compared to some other ant species.

Pavement Ant (*Tetramorium caespitum*)

This ant (Figure 21.10) was introduced to North America from Europe. At one time it was especially common only along the Atlantic seacoast, but now it is very common throughout North America. Nests usually are located in soil, but also commonly occur in houses. The pavement ant is omnivorous, being particularly fond of meats and fatty substances. Although it can cause damage to cultivated plants, it is more of a nuisance than a stinging problem for homeowners. This small ant (3–3.5 mm long) usually is not capable of

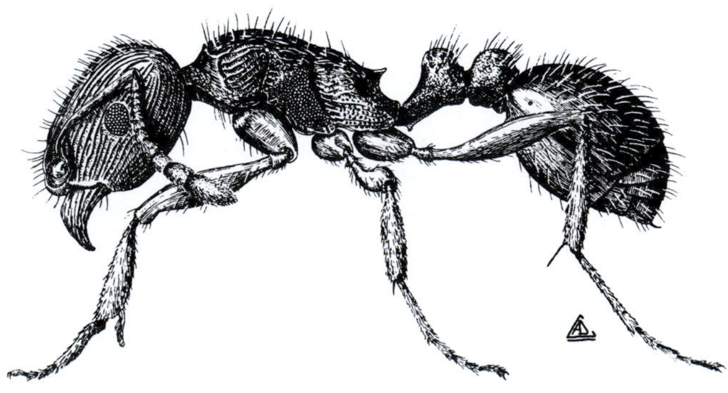

Figure 21.10 Pavement ant (*Tetramorium caespitum*), worker. (Courtesy of US Department of Agriculture)

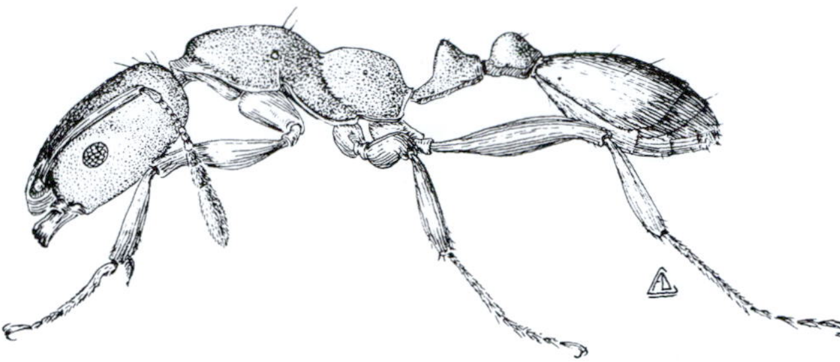

Figure 21.11 Pharaoh ant (*Monomorium pharaonis*), worker. (Courtesy of US Department of Agriculture)

penetrating human skin with its sting. Nevertheless, stings have been reported to cause a skin rash in children, but only rarely have they stimulated serious allergic reactions.

Pharaoh's Ant (*Monomorium pharaonis*)

This tiny ant (Figure 21.11) is only about 2 mm long and nests in every conceivable habitat: in soil, in houses, between sheets of paper or linen, and in trunks of clothing, to mention only a few. The pharaoh's ant probably is native to Africa and has been widely disseminated by commerce so that it is now found throughout most of the world. It occurs in nearly all cities in the United States. It forms huge, polygynous colonies of more than a million workers and produces brood year-round. This ant is omnivorous, feeding on sugary materials, dead insects, breads, and many other food stuffs; and, like the pavement ant, it prefers meats and fats. It is among the more difficult of all ants to control.

In addition to being a concern to human health because it stings, the pharoah's ant has been known to infest surgical dressings and intravenous units in hospitals, and will attack the delicate tissues of newborn babies, especially the eyelids and navel. Because this ant has been found in bedpans, drains, and washbasins, it can come into contact with disease organisms and has been implicated in the transmission of pathogenic organisms in some hospitals. In tropical regions other ant species also have been implicated as vectors of pathogens in hospitals (Fowler et al., 1993). Edwards (1986) summarizes information on the biology and pest status of pharoah's ant.

WASPS

The term "wasp" encompasses a diverse assemblage of hymenopterous groups including solitary-nesting potter and digger wasps, parasitic wasps, and social wasps such as yellowjackets and paper wasps. Parasitic wasps generally oviposit on or into arthropod hosts that remain in place. Generally, the host is much larger than the parasitic wasp, referred to as a parasitoid, with the potential for many wasps to develop from an individual host. These are not a significant stinging problem for people. Solitary-nesting wasps (called here "solitary wasps") and social wasps have a well-developed stinger and will use it defensively.

Solitary Wasps

Most solitary-nesting wasps hunt various arthropods as prey to feed their larvae. Their venom is designed primarily to paralyze their prey and not to cause pain in vertebrates. The female wasp may lay an egg on the immobilized prey, and others provision a cell of a nest with multiple prey items. Since the prey is still alive, it is essentially preserved until it can be consumed by the developing larva. Only a few solitary wasps are a serious stinging threat to people.

Mutillidae Among the nonsocial wasps perhaps the best known for their stings are the brightly-colored velvet ants, or mutillids (Figure 21.12). The sting of large species produces a painful and intense burning sensation and may cause substantial swelling and redness at the sting site. The wingless females may be seen walking over the ground during mid summer, especially in open sandy areas, searching soil nests of bees and wasps to invade and parasitize. One large red and black mutillid, *Dasymutilla occidentalis* (Figure 21.13), is called the cow killer because of its particularly painful sting. For further information on this poorly known group of solitary wasps, including keys to species in the southeastern United States, see Manley (1991).

Pompilidae The metallic blue or black spider wasps in this family may be seen flying low over the ground in search of spiders, which they attack, sting, and carry to a burrow in the soil where they oviposit on their prey. Some spider wasps (e.g., *Pepsis* spp.) (Figure 21.13) can deliver painful stings.

Figure 21.12 A brightly colored velvet ant, *Dasymutilla occidentalis* (Mutillidae), commonly known as "cow killer" because of the intense pain its sting causes. (Photo by Elton J. Hansens)

Figure 21.13 Spider wasp, *Pepsis* sp. (Pompilidae). (Photo by Roger D. Akre)

Sphecidae This family is a diverse and large group of solitary wasps, most of which are not likely to be a stinging threat (Bohart and Menke, 1976; O'Neill, 2001). Sphecid wasps like **mud daubers** (*Sceliphron* and *Chalybion* spp.) and cicada killers (*Sphecius* spp.) commonly nest on human dwellings or in disturbed soil sites. Although they appear threatening, stings to humans are relatively infrequent. Mud daubers, such as the black and yellow mud daubers (*Sceliphron* spp.) (Figure 21.14), commonly build their mud nests in attics, carports, and porches and provision them with spiders. These wasps are often evident at water and around nesting sites as they make frequent trips carrying mud to build their nests. They will rarely sting during these flights; in fact, it is difficult to induce mud daubers and other solitary wasps to sting humans or animals even by disturbing them. When stinging does occur, it usually involves the wasp being trapped inside clothing, being stepped on, or otherwise being pressed against the victim's skin or clothing. Occasionally these solitary wasps sting careless collectors.

The **cicada killers** (*Sphecius speciosus* and other species) are large wasps marked with yellow bands on the gaster

Figure 21.14 Mud dauber, *Sceliphron* sp. (Sphecidae), constructing nest. (Courtesy of Roger D. Akre)

Figure 21.15 Cicada killer, *Sphecius speciosus* (Sphecidae). (Courtesy of the Oklahoma State University Entomology Department)

(Figure 21.15). They capture cicadas and carry their stung prey to burrows they have excavated in soil. The female digs a fairly large burrow with a distinct soil mound surrounding the entrance. They often nest in bare soil areas of lawns, gardens, or flower beds around human dwellings. The female will not sting unless handled, but can inflict a mildly painful sting to people walking barefoot near their nest sites. The males frequently fly around the nesting sites patrolling their territories. During their flights they will approach other insects, birds, and even people that enter their territories. Although they can appear intimidating, these individuals, like all hymenopteran males, cannot sting.

Social Wasps (Vespidae)

In contrast to the relatively innocuous solitary wasps, social wasps generally are defensive and will sting readily. This usually occurs near their nests, but also at foraging sites. Most vespid wasps are unique, and easily recognized, in that they fold their wings longitudinally when at rest. In North America, most social wasps are the yellowjackets and hornets (vespines) and the paper wasps (*Polistes* spp.).

Yellowjacket and hornet nests consist of horizontal, rounded paper combs attached one below the other, with a multilayered paper envelope (Akre et al., 1981; Ross and Matthews, 1991). Some species build nests in exposed aerial locations (Figure 21.16), whereas others build them in subterranean or enclosed and protected sites (Figure 21.17). A few species construct their nests in either type of situation. Colony size ranges from fewer than a hundred to several thousand individuals. Colonies are typically annual and are initiated by a single inseminated queen. Queens are produced late in the season and survive the winter in protected locations. During early spring (April–June), the queen begins to construct a nest, lays eggs, and

Figure 21.16 Aerial paper nest of bald faced hornet, *Dolichovespula maculata* (Vespidae) in tree. (Courtesy of Washington State University)

then forages for arthropods to feed the developing larvae. After the first group of workers emerge, they assume most colony functions of foraging for food, fiber, and water, and taking care of the larvae. The queen no longer leaves the nest after this time but assumes her primary function of laying eggs.

Perennial yellowjacket colonies sometimes occur when the new queens that emerge in the fall mate and then rejoin an active colony. This is most likely to occur in warmer climates and when the foundress queen has died or is losing her influence over the colony. These polygynous colonies can become perennial and contain tens of thousands of workers. Large perennial colonies of *Vespula vulgaris*, *V. pensylvanica*, *V. germanica*, *V. squamosa*, and *V. maculifrons* have been reported in the United States, as well as in other countries, usually in subtropical areas such as Florida or in moderate climates like southwestern California. Such nests are dangerous to destroy even by well-trained people.

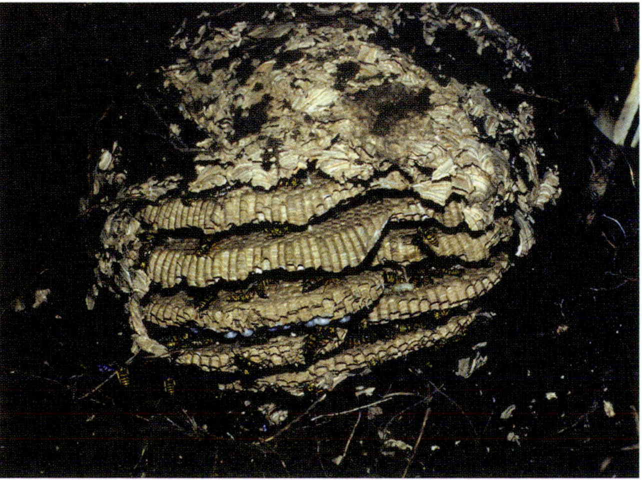

Figure 21.17 Exposed underground nest of eastern yellowjacket, *Vespula maculifrons* (Vespidae). (Photo by Hal C. Reed)

Paper wasp nests are single paper combs with no enclosing envelope. They may be constructed in open situations such as under eaves, or in enclosed places. Paper wasp nests are generally smaller than vespine nests, commonly from 4 cm up to 20 cm in diameter, and numbers of wasps per colony range from several to 100 or so. These nests are founded in the spring by one or more females that were produced the previous summer and overwintered. The offspring of these nests are female workers early in the season, with new queens and males produced in late summer.

Yellowjackets (*Dolichovespula* and *Vespula*) "Yellowjacket" is an American term that is used for all species of wasps in the genera *Dolichovespula* and *Vespula*. The name refers to the yellow and black patterns of most species (as illustrated in Buck et al., 2008). However, some species are black and white, like the baldfaced hornet, *Dolichovespula maculata*. There are 19 species of yellowjackets in North America (6 *Dolichovespula* spp., 13 *Vespula* spp.). Yellowjackets occur throughout all of the United States, including Alaska and Hawaii, which has two introduced species. The greatest diversity of yellowjacket species is found in the northern areas of the United States and southern Canada (Akre et al., 1981; Buck et al., 2008).

Two species of *Dolichovespula* are considered hazardous in North America: the **aerial yellowjacket** (*D. arenaria*) and the **baldfaced hornet** (*D. maculata*). Both species occur primarily in forested areas and are distributed throughout much of North America; however, the aerial yellowjacket does not occur as far south as the baldfaced hornet (Akre et al., 1981). Their large aerial, spherical, or egg-shaped nests are familiar sights in trees and shrubs (Figure 21.16), and are particularly obvious following leaf fall in autumn. The aerial yellowjacket occasionally builds nests at or just below ground level. In urban areas nests are often built under the eaves of houses or on nearby bushes and trees where disturbances of the nests are common. When nests are constructed above a doorway, the vibrations from opening and closing the door can excite the wasps to sting. Both species will attack and sting when the nest is disturbed. They also are capable of squirting venom from the sting into the face and eyes. Fortunately, the average colony size is relatively small by vespine standards, consisting of fewer than 400 workers.

Vespula species are typically subterranean nesters (Figure 21.17). Often the queen initiates the nest in an old rodent burrow or in other cavities in soil, logs, or trees. Some species, however, commonly establish colonies in wall voids of buildings and other enclosed spaces. The paper nests of most yellowjackets are made of gray carton, but those of *V. flavopilosa*, *V. maculifrons*, and *V. vulgaris* consist of tan, fragile carton. Colony cycles of some species, particularly those related to *V. vulgaris*, extend later into the fall (September–December) than those of other species, particularly those related to *V. rufa*. Colonies of those yellowjacket species that are active into the fall construct nests that become quite large and may possess up to 5,000 workers (Table 21.3). Yellowjackets appear to be more aggressive and likely to

Table 21.3 Comparison of Colony Parameters and Foraging Behavior of Yellowjackets in North America

Yellowjacket	Distribution	Foraging Behavior	Colony Size	Colony Decline
Vespula				
germanica	Transcontinental	Predators and scavengers	500–5,000 workers	Late Sept. to early Dec.
maculifrons	Eastern		500–15,000 cells	
pensylvanica	Western		Perennial colonies	
vulgaris	Transcontinental		100,000+ workers	
flavopilosa	Eastern		1,000,000 cells	
Vespula				
atropilosa	Continental	Strictly predators	75–400 workers	Late Aug. to Sept.
acadica	Continental		500–2,500 cells	
consobrina	Continental		No perennial colonies	
Dolichovespula				
arenaria	Transcontinental	Strictly predators	100–700 workers	Early July to Sept.
maculata	Transcontinental		500–4,500 cells	
			No perennial colonies	
Vespula				
squamosa	Southern USA	Predators and some scavengers	500–4,000 workers	Late Aug. to Nov.
sulphurea	California		500–10,000 cells	
			Some perennial colonies	

Colony decline is defined as the period when reproductives emerge.

sting late in the season. This may be related to defense of the colony when queens and males are produced. Additionally, yellowjackets may be more persistent in seeking and scavenging food materials late in the season, bringing them into more frequent contact with people at that time and leading to more stinging episodes.

Workers of several species of *Vespula*, particularly the eastern, western, German, and southern yellowjackets, readily scavenge at garbage or picnic sites (Table 21.3). Scavenging yellowjackets forage for protein-rich foods (e.g., insects, vertebrate flesh, processed meats) and for carbohydrates from a wide variety of sources (e.g., fruits, fruit juices, tree sap, soft drinks, beer, and sweets). This scavenging behavior brings these species into frequent contact with people. These yellowjackets can be problematic for fishermen and hunters when they arrive at freshly captured fish or killed game. They are a problem at outdoor events where food is present and consumed, such as fairs, campgrounds, and outdoor eateries. The same species typically have longer colony durations, larger colony and nest sizes, and are most abundant late in the season. Other yellowjacket species, such as *V. atropilosa* and *V. vidua*, lack this scavenging habit and prey on insects for much of their food, and therefore are less often a stinging nuisance. These species typically have shorter duration colonies, smaller colony and nest sizes, and do not become so abundant. When their nests are disturbed, they will sting, but because their colonies are small the result is not as serious as attacks from larger colonies.

The **eastern yellowjacket** (*V. maculifrons*) and the **hybrid yellowjacket** (*V. flavopilosa*), are major picnic and campsite pests in late summer and fall in the eastern United States. Workers of both species will feed on fresh and decaying fruits. These two species are responsible for many stings to humans and their pets during outdoor recreational activities, especially during years of extremely high yellowjacket populations.

The **western yellowjacket** (*V. pensylvanica*) is the dominant native yellowjacket in dry forests of the western United States. It attains high population densities in some years. This commonly results in increased encounters with humans around homes, picnic areas, and camp grounds. During peak populations of *V. pensylvanica*, recreational areas such as resorts, hunting camps, and parks have been closed. These "wasp years" may coincide with a high incidence of forest fires, posing a serious stinging problem for fire fighters and smoke jumpers. Yellowjacket stings in some years also cause a significant loss of worker time by US Forest Service personnel in the western states. Very large perennial colonies of this species have been reported in Hawaii, where they may disrupt both agriculture and tourism.

The **German yellowjacket** (*V. germanica*) is a species native to Europe that has shown a remarkable propensity for becoming established in temperate areas of the world. It now occurs in many other countries including New Zealand, Australia, South Africa, Argentina, Chile, Canada, and the United States. Colonies tend to be large, especially in warmer areas where perennial polygynous colonies may contain thousands of workers. The German yellowjacket became firmly established in the United States by the late 1960s. The biotype in North America may nest inside structures, unlike the European biotype that typically nests underground. Colonies in buildings benefit from the associated heat and protection and tend to persist very late into the fall and winter months, resulting in larger colonies. At that time of year, workers may chew through walls and emerge inside buildings. The German yellowjacket is well established across much of eastern and western North America.

Early in the colony cycle, German yellowjacket workers scavenge for protein- and carbohydrate-rich foods. This early-season scavenging may give these wasps an advantage over other yellowjacket competitors that exclusively spend time capturing insect prey. This species has become the dominant yellowjacket in some of the areas it has invaded. German yellowjackets are a significant pest species due to their scavenging habits, selection of nesting sites in structures, long colony duration, large colony size, and high population densities.

The **common yellowjacket** (*V. vulgaris*) is native to Europe, temperate areas of Asia, and across North America. It has been introduced into New Zealand and Australia. *Vespula vulgaris* workers are a significant stinging problem, and their pestiferous nature is no doubt related to their scavenging behavior and great numbers where they are prevalent. In North America, they occur principally in moist forest habitats. Perennial colonies of *V. vulgaris* consisting of 50,000 to 100,000 workers have been found in southern California. In more temperate climates, colony sizes are much smaller, typically fewer than 1000 workers.

The **southern yellowjacket** (*V. squamosa*) is widely distributed from the mid-Atlantic states of the United States to the Midwest and south into Central America. In the northern part of its range, such as in north Georgia, it occurs in colonies with a single queen and an annual cycle. Farther south, it produces perennial colonies with numbers of egg-laying queens. Some of these colonies consist of more than 100,000 workers. Because of the large size of some of these colonies, they can be deadly hazardous to humans and animals. One such nest reported from Parrish, Florida, was 3.6 m high by 1.8 m in diameter, built on a broken tree stump. These wasps produce an alarm pheromone that causes stinging attacks on intruders. Workers of some *V. squamosa* colonies are scavengers and are common pests at picnics. A yellowjacket species that is closely related to *V. squamosa*, the California yellowjacket (*V. sulphurea*), also scavenges and has been reported as a nuisance for picnickers. It occurs from southern Oregon through California into northwestern Mexico.

Hornets (*Vespa* species) There is only one true hornet that occurs in North America, the **European hornet** (*Vespa crabro*). This species was accidentally introduced from Europe into the New York area during the mid 1800s. Today it can be found throughout much of the eastern United States and west into Missouri. This is the largest social wasp in North America (body length >20 mm) with contrasting brown and deep yellow bands on the gaster. Although *Vespa* hornets are important stinging hazards in Asia, *V. crabro* is seldom responsible for stinging incidents in North America. This is primarily because *V. crabro* is relatively uncommon and usually nests in the hollows of trees in areas away from human activities. Brown envelope and carton distinguish their nests from the more common gray nests of *D. maculata*. Despite the large size of the workers and the nest, this species is less aggressive than most other North American vespines. A typical colony consists of 200 to 400 workers at peak populations, but large colonies can reach 1,000. Colonies have a long seasonal cycle lasting from early spring into late August through November.

Paper Wasps (*Polistes* species) Paper wasps are the most common stinging wasps encountered by humans in the southern United States, especially during the summer months (Gillaspy, 1986). Some paper wasps, such as *P. exclamans* and the introduced *P. dominulus,* are improperly called "yellowjackets" due to their alternating bands of yellow and dark-brown or black markings on their gaster. Paper wasps are usually longer and have more slender bodies than yellowjackets. They can be most certainly distinguished from yellowjackets by the shape of the anterior-most segment of the gaster, which slopes gradually to the wasp waist in *Polistes,* whereas it is truncated abruptly in the vespines. *Polistes* wasps construct paper nests consisting of a single comb with no envelope (Figure 21.18). Several species tend to nest near or on buildings and as such come in frequent contact with people. Their nests can be found under roof eaves and window sills; around door frames; and inside garages, storage buildings, clothesline poles, bird houses, wall voids and attics. More natural nesting sites include trees, shrubs, and cliff overhangs. Some species, such as *P. exclamans,* prefer nesting in exposed sites on structures, whereas others, like the **red wasp**, (*P. perplexus*), usually nest in concealed sites such as wall voids and attics.

Most colonies are small, with fewer than 100 adults and 100 to 200 cells. Larger nests consist of about 400 cells. The annual colonies are initiated by a single foundress or a group of foundresses that compete for

Figure 21.18 Paper wasp, *Polistes* sp. (Vespidae), with queen and daughter guarding nest. Note eggs and developing larvae in unsealed cells. (Photo by Gary R. Mullen)

reproductive dominance. After the workers emerge, one foundress usually becomes the primary reproductive or queen. Unlike vespine wasps, no queen-worker size dimorphism occurs in paper wasps. Paper wasps forage for caterpillars and other insects and thus are excellent natural control agents of many crop pests. Colonies have been propagated and transplanted into fields to decrease pest populations.

In the fall, queens and males of some species of paper wasp may swarm in large numbers at the top of tall towers and buildings where they may create concerns (Reed and Landolt, 1991). However, many of these wasps are males that cannot sting, whereas the females are involved primarily in mating and aggregation and only rarely sting at these sites. Later in the fall and winter the females aggregate in large numbers at hibernation sites such as the attics of houses, apartments buildings, and barns. Contact with humans is common at this time, resulting in stings. Stinging episodes also occur in the spring when the aggregating wasps come out of hibernation. Unlike yellowjackets, paper wasps do not scavenge. Most stinging occurs when the nests are disturbed.

Polistes annularis is known as **Spanish jack** to fishermen, boaters, and river dwellers in the US Gulf Coast states because of the occurrence of its nests in shrubs, trees, and other structures along streams and lakes. People are commonly stung when they inadvertently disturb a nest. This large wasp produces the largest colonies of paper wasps in North America and can often inflict multiple stings. Several other species such as *Polistes metricus*, *P. fuscatus*, *P. aurifer*, *P. exclamans*, *P. apachus*, and the introduced and now invasive European species *P. dominulus* also commonly are encountered by people, resulting in stings (Table 21.2). This latter species has become the most dominant and ubiquitous species in some areas (e.g., Ohio) of the United States.

BEES

Solitary Bees

Most bees are solitary or, at most, communal or semisocial. Some species of the family Halictidae, which include the sweat bees, are in fact true social bees. A few of the solitary communal and social species form dense nesting aggregations, usually in soil, where they excavate cells in which to rear their larvae. Larvae of all bee species feed on pollen and nectar provisioned in these cells. Nesting activity is quite variable, ranging from a single nest to thousands of individual nests concentrated in a given area. Some bees make nests in wood cavities such as the hollow stems of trees and shrubs, burrows of wood-boring beetles, and artificial cavities such as the hollow pipes of wind chimes and keyholes in doors. Few solitary bees are stinging hazards to humans.

Halictidae Some of these are known as **sweat bees** (Figure 21.19) because of a propensity to land on people to imbibe perspiration. A person may then be stung when swatting the bee. The stings usually are considered minor irritations and simply a nuisance.

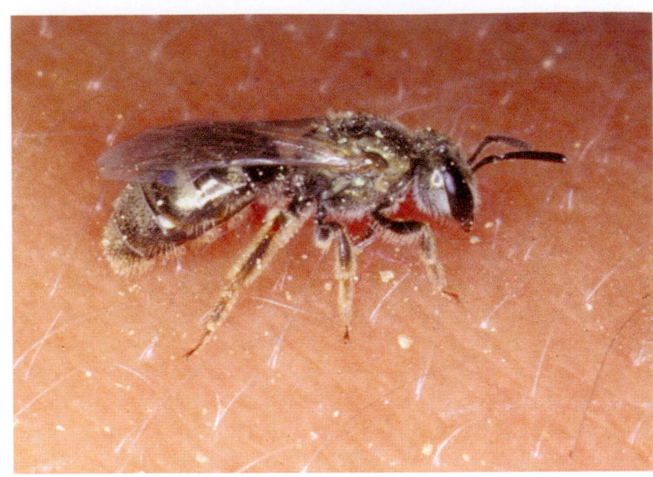

Figure 21.19 Sweat bee, *Lasioglossum zephyrus* (Halictidae), on human arm. (Photo by Gary R. Mullen)

Anthophoridae The **carpenter bees** are similar in size and general appearance to bumble bees. However, they lack the fuzzy appearance and yellow coloration typical of bumble bees and the dorsum of the gaster is mostly shiny black (Figure 21.20). These bees often nest in wood around human dwellings where they bore round holes in window sills, eaves, railings, fence posts, and other wooden structures (Figure 21.21). Female carpenter bees rarely sting, and when they do the pain is relatively mild. Males often are seen flying around nesting sites, may make a loud buzzing noise, and may appear threatening; however, like all male hymenopterans, they cannot sting. The common carpenter bee in eastern North America is *Xylocopa virginica*; the common species in western North America are *X. californica* and *X. varipuncta*.

Figure 21.20 Carpenter bee, *Xylocopa virginica* (Anthophoridae), male. Note that abdomen is mostly black. (Photo by Gary R. Mullen)

Figure 21.21 Circular entrance to nest of carpenter bee, *Xylocopa virinica* (Anthophoridae), in cedar wood of house eave. (Photo by G. R. Mullen)

Social Bees

Apidae Most bee-sting cases are due to social bees in the family Apidae, which in North America include bumble bees (subfamily Bombinae) and honey bees (subfamily Apinae). The stingless bees (subfamily Meliponinae) are exclusively tropical and are found in the Old and New World. It often is assumed that these bees pose no hazard to humans since they are stingless. However, they do swarm out of the nest to attack intruders by biting, especially around the eyes, nose, and ears. One species is known as the **fire bee** (*Oxytrigona tatairas*) because of the caustic defensive secretions produced by its mandibular glands that cause intense burning pain when applied on the skin.

Bumble Bees (*Bombus* species) These large, hairy, yellow and black bees (Figure 21.22) pose little stinging hazard for most people, common as they are around flowers. Most of the 400 species are nonaggressive, even when the nest is disturbed. However, a few species can be very aggressive and persistent in their stinging attacks when their nests are threatened. Bumble bee workers can sting an intruder repeatedly.

Bumble bees generally occur in the more temperate areas of North America where their colony cycle is very similar to that of yellowjackets. Colonies are initiated by an inseminated queen early in the spring, followed by the production of a few hundred workers during the summer. Most nests are established in abandoned rodent burrows or old rodent nests under debris or objects on the ground. They are also known to build nests in attics and wall voids of houses. Nests consist of wax pouches or empty cocoons for storing pollen and honey, and wax chambers for rearing brood (Figure 21.23). Workers forage for pollen and nectar to feed the larvae. These bees are important natural pollinators and also pollinate many agricultural crops, especially cranberries and raspberries.

Honey Bees (*Apis* species) Honey bees (Figures 21.24 and 21.25) are native to Europe, western Asia, and Africa. Four of them are common species that occur from Southeast Asia to Africa and in Europe: the **common honey bee** *(Apis mellifera)*, **giant honey bee** *(Apis dorsata)*, **Asian honey bee** *(A. cerana)*, and **dwarf honey bee** *(A. florea)*. Only the common honey bee occurs in North America. Several subspecies and races of *Apis mellifera* were domesticated by people in Europe and have been introduced into nearly every country in the world. This species is an invaluable pollinator of native and commercial plants, especially fruits and vegetables. Populations of *A. mellifera* have experienced sharp declines in Europe, Asia, North America, and elsewhere in recent years, due to a phenomenon called **colony collapse disorder (CCD)**. Although many theories abound, currently no definitive cause (or causes) for CCD has been identified.

Figure 21.22 Bumble bee, *Bombus* sp. (Apidae), foraging worker. Note conspicuous yellow hairs on abdominal segments. (Photo by Takumasa Kondo)

Figure 21.23 Exposed subterranean nest of bumble bee, *Bombus* sp. (Apidae), showing both open and sealed cells, and surrounding nesting material. (Photo by R. D. Akre)

Figure 21.24 Honey bee, *Apis mellifera* (Apidae), worker foraging on flower. (Courtesy of *California Agriculture*, University of California)

Honey bees build large nests, called **hives**, consisting of several wax combs of hexagonal cells (Figure 21.25) in which they rear the brood and store pollen and nectar. Feral or "wild" honey bee colonies construct nests in cavities of hollow trees and in attics and wall voids of human dwellings. These colonies typically have 15,000 to 30,000 bees, whereas commercial hives are usually larger with 30,000 to 50,000 bees. Despite the fact that commercial breeding of honey bees has dampened the defensive tendencies of many races, honey bee colonies are yet capable of attacking and stinging intruders in large numbers.

The perennial colonies of honey bees usually survive the winter, and in late spring through early summer reproduce by **swarming**. Swarms may be seen resting in exposed sites such as trees, shrubs, and under eaves of buildings while they are seeking a suitable cavity in which to establish a new hive. Although these swarms are less defensive than an established colony because they lack brood, stored pollen, and honey, it is best not to approach or disturb them.

Of particular concern in the United States is the northward movement of the Africanized honey bee (*Apis mellifera scutellata*), an aggressive subspecies of the common honey bee that has spread through tropical regions of South and Central America. It is most abundant in tropical humid areas of Africa but extends into arid regions of South Africa. This honey bee was introduced into Brazil in 1956 in an effort to improve the beekeeping industry in Latin America. Captive bees escaped in 1957 when queen excluders were removed from some of the hives. They spread south to Argentina and north through South America, reaching Panama in 1982. The first Africanized honey bee colony trapped in the United States was in 1990 in Texas. Colonies of Africanized honey bees have been found in southern Texas, New Mexico, Oklahoma, Arizona, California, and Puerto Rico, with additional recent records from Nevada, Louisiana, Arkansas, Alabama, and Florida.

This bee quickly establishes itself in new areas. When foraging or nesting conditions become restrictive, the bees leave their hives in a process called **absconding**, and relocate to new nesting sites. Other subspecies of the common honey bee become established more slowly in new areas and are less likely to abscond when conditions change.

The Africanized honey bee has an increased propensity for mass stinging attacks of both people and animals (Figure 21.26). Detailed, quantitative studies have shown that these bees are much more alert than other subspecies to movement, vibration, and their own pheromones that mediate alarm and attack responses. The alarm compounds are not released in greater quantity by Africanized bees, but rather the threshold of response of the bees to a perceived threat is much lower. Nor is their venom any more potent

Figure 21.25 Honey bees (*Apis mellifera*) on surface of comb in hive. European honey bee (center), surrounded by Africanized honey bee workers. (Courtesy of Entomological Society of America)

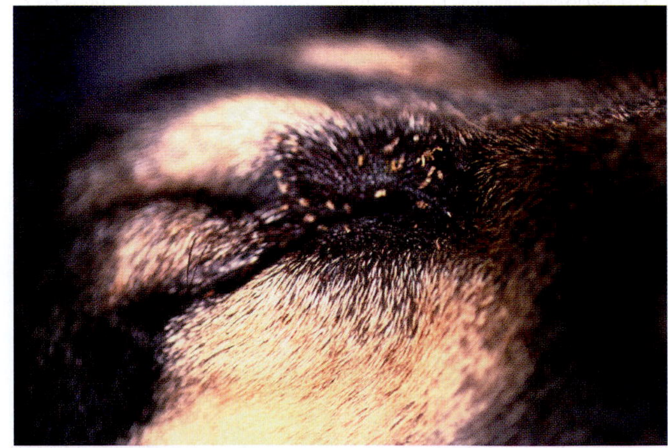

Figure 21.26 Head of German shepherd dog in fatal case of attack by Africanized honey bees (*Apis mellifera*). Note the concentration of bee stings about the eye. (Photo by Justin Schmidt)

than that of other *A. mellifera* subspecies; in fact, the composition is nearly identical.

Aside from their behavioral differences, individual Africanized honey bee workers are indistinguishable from workers of the European subspecies of *A. mellifera* (Figure 21.25). It requires a specialist to conduct DNA identification or to examine and measure specific morphological (e.g., wings) and biochemical features of several bees to determine their identity. Although the Africanized honey bee stings much more readily than the European honey bee, so far deaths of animals and humans due to *A. mellifera scutellata* are not common. About 350 human deaths in Venezuela between 1975 and 1988 were attributed to Africanized honey bees (Winston, 1992). Since they reached Mexico in 1986, the bees have reportedly killed several dozen people. Annual fatalities due to bee stings in Texas have not increased since this bee arrived there in 1990.

Stinging problems due to the Africanized honey bee are actually less of a concern than the potential disruption to beekeeping and agriculture. The bees are more difficult to manage and store less honey than their European counterparts. These traits, coupled with their extremely aggressive nature, poses concerns for beekeeping in North America, which relies upon organized transport of many managed hives for pollination services. However, the impact can be lessened by modifying hive management practices (e.g., requeening with docile European queens), thereby altering their genetic makeup by interbreeding with European strains.

PUBLIC HEALTH IMPORTANCE

Most encounters with venomous arthropods involve stings from ants, wasps, and bees. Probably several million people are stung annually by these insects, most of whom do not require professional medical treatment. Some data on sting frequency are available on fire ants. Studies in the southeastern United States have indicated that 30 to 60% of people are stung by fire ants each year, but only 1 to 5% of these cases require medical treatment (Lofgren, 1986; Schmidt, 1986b). This rate is higher than that of all other hymenopterans. Stinging hymenopterans also pose a hazard during natural catastrophes such as forest fires and hurricanes. Insect stings, primarily due to yellowjackets, were the single most common cause of nonfatal injuries following Hurricane Hugo that hit the southeastern coast of the United States in 1993 (Brewer et al., 1994). Hymenoptera stings represent a major proportion of the annual 900,000 emergency-department visits due to noncanine bite and sting injuries in the United States (O'Neil et al., 2007).

A total of 40 to 50 deaths due to Hymenoptera stings usually are reported in the United States each year (Schmidt, 1986c). However, it is suggested that these figures are very conservative and that a more accurate figure is probably closer to 200 (Akre and MacDonald, 1986). More people than this die each year as a result of allergic reactions to penicillin or from lightning strikes (Camazine, 1988). Honey bees are believed to cause about half of the annually reported human deaths due to hymenopteran stings in the United States (Schmidt, 1992). This figure may be somewhat misleading because the general public and medical profession do not reliably differentiate between honey bees and other types of bees and wasps. Some deaths attributed to bees probably are due to stings from yellowjackets and paper wasps.

The stings of most social hymenopterans cause intense pain to humans, with reactions to various species differing primarily in intensity or duration. Comparative scales for ranking the severity of pain caused by aculeate hymenopterans have been proposed by Starr (1985) and Schmidt (1986c, 1990). The intensity of pain caused by most social bees and wasps is similar, and only the responses to stings of a few species such as the fire and harvester ants are diagnostic. Also, there exists great variation in responses among individuals to stings by the same species. The location of the sting on the body (e.g., finger vs. neck) also may influence individual reactions.

Although most ant stings are painful, those of *Paraponera* and *Pogonomyrmex* species are especially noteworthy. The large ponerine *Paraponera clavata* of Central and South America injects venom that produces intense, debilitating pain lasting several hours. This sting is considered to be the most painful of all Hymenoptera (Schmidt, 1986b, 1990). The affected area can expand 20 to 30 cm within an hour of the sting. The pain induced by stings of *P. clavata* serves as a comparative standard for stings delivered by other stinging species (Starr, 1985). Although less severe than that from *Paraponera*, the pain caused by stings from harvester ants (*Pogonomyrmex* spp.) is extremely intense and can last up to several hours. It has been likened to "turning a screw" into the flesh around the sting site, and also causing a sensation that has been described as "chilling." The sting is unique in that it induces piloerection (elevation of the hairs) and sweating around the sting site for four to eight hours (Schmidt, 1986c). The venoms of some harvester ants are more toxic to mice than any other tested insect venom and eight to 10 times more toxic than honey bee venom; they are almost as toxic per unit volume as the most venomous snakes (Schmidt and Blum, 1978).

Victims of fire ant stings usually experience a temporary burning sensation and discomfort with some swelling around the sting site. In most cases a characteristic vesicle, 3–5 mm in diameter, containing a clear fluid develops within six to 24 hours at each sting site. The fluid becomes cloudy, forming a sterile white pustule (Figures 21.27 and 21.28). These pustules usually disappear within a few days. They are replaced by a discolored lesion, resulting from tissue necrosis caused by alkaloids in the venom that can persists for weeks or months. These lesions can be intensely pruritic and are subject to secondary bacterial infections when they are scratched or otherwise broken.

Reactions to insect stings have been variously classified (Schmidt, 1992; Reisman, 1994a), but can be divided into four categories: local, large local, systemic, and toxic reactions (Table 21.4). Local reactions are

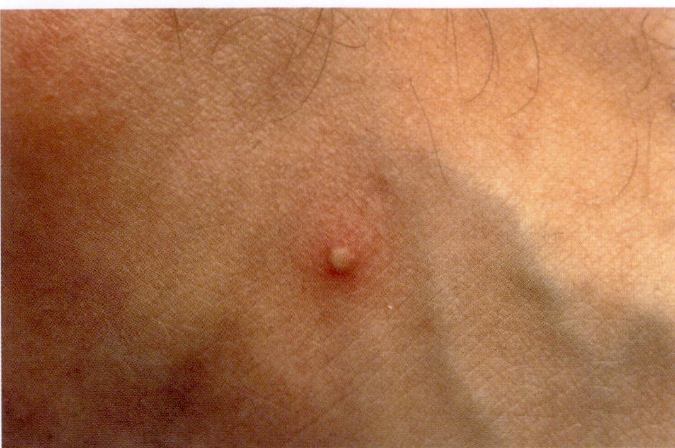

Figure 21.27 Fire ant sting by *Solenopsis invicta* on human ankle, showing characteristic sterile pustule formed at sting site, 24 hours after sting. (Courtesy of M. Horton)

normal for most people and include immediate pain and/or burning at the sting site, development of a flare and wheal, redness (Figure 21.29), and swelling that is limited to the sting site. Later, itching at the sting site usually occurs. A large local reaction involves painful swelling at the site and even the associated extremity (Figure 21.30); this reaction is not considered serious. These are normal responses that are not life-threatening. They may be accompanied by systemic, cutaneous reactions (e.g., hives) on parts of the body other than where the sting occurs.

Systemic reactions are more generalized responses that induce reactions away from the sting site. The most serious of these are allergic reactions, which can be fatal. Allergic reactions typically occur after a second or subsequent sting by the same or closely related species. The first sting causes the body to produce antibodies to specific proteins in the venoms. Later when this hypersensitive individual is stung again, the immune system overreacts to the presence of the same foreign venom proteins. These stinging episodes sometimes result in **anaphylaxis**, a sudden drop of blood pressure and respiratory distress triggered by this immunological response. In some cases this can result in **anaphylactic shock** and death. Such lethal reactions have been reported for stings of vespines, paper wasps, honey bees, and fire ants.

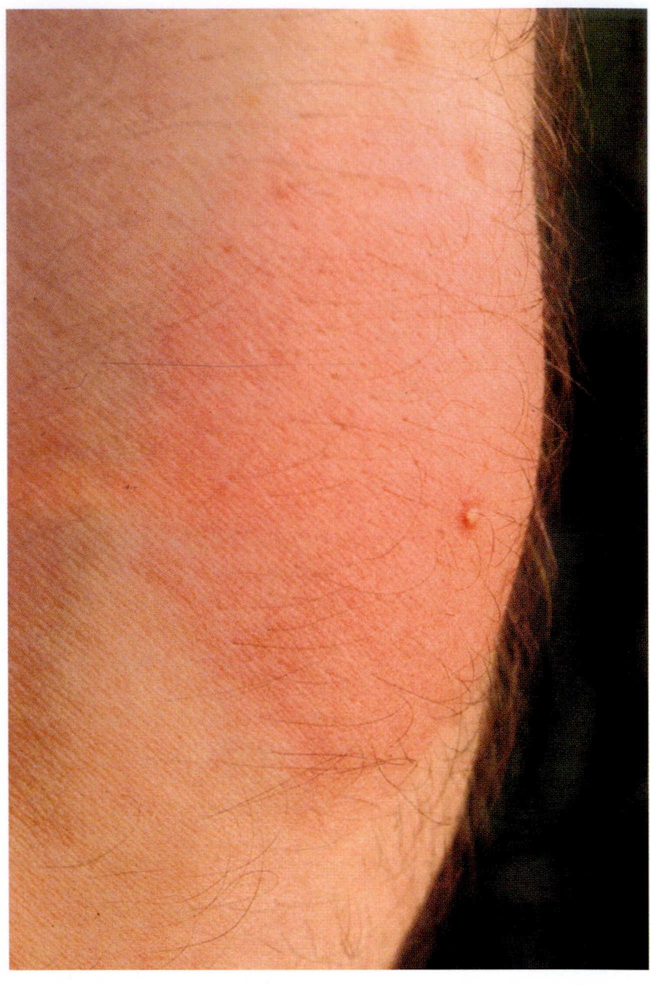

Figure 21.28 Fire ant sting by *Solenopsis invicta* on human forearm, with localized swelling and inflammation, 15 hours following sting. (Photo by Gary R. Mullen)

Deaths from hymenopterous stings are usually due to respiratory failure (70%). Other reported causes of mortality are anaphylaxis (15%), cardiovascular collapse (9%), and neurological complications (6%) (Schmidt, 1986b). Death from stings, when it occurs, is usually very rapid, often within 20 minutes of the sting. Nearly 60% of deaths occur in less than one hour following envenomization. Therefore it is imperative in cases of severe allergic responses that treatment be administered as soon as

Table 21.4 Descriptions of the Four General Types of Reactions to Hymenopteran Stings

Type	Duration	Response
Local	About 2 hrs	Redness, itching, swelling, pain, wheal forms at sting site.
Large local	2–5 days	Painful swelling ($\geq$5 cm) of sting site or entire extremity. General malaise or ill feeling, palpitation of heart, elevated blood pressure.
Systemic	10 min–3 weeks	Reactions to areas beyond the sting site. Normal symptoms plus hives, respiratory distress, laryngeal swelling, gastrointestinal distress, hypotension, cutaneous uticaria, widespread edema, anaphylactic shock, sometimes leading to death.
Toxic	Hours–months	Destruction of muscle and red blood cells, blood pressure drop, kidney failure.

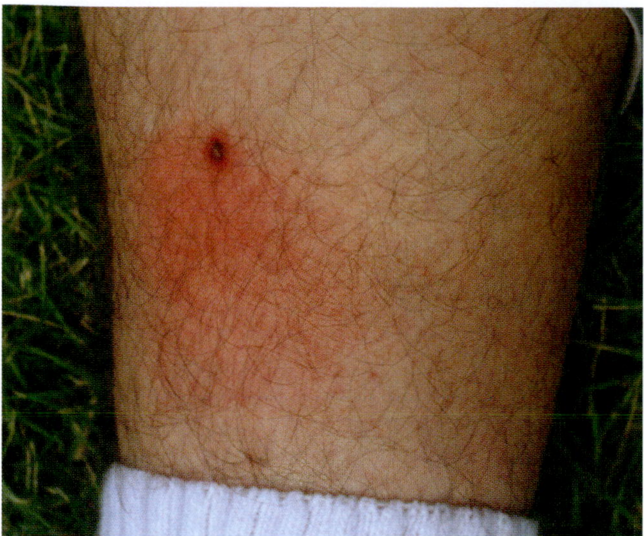

Figure 21.29 Reaction to sting of the southern yellowjacket (*Vespula squamosa*) on lower leg, with reddening and localized hemorrhaging at sting site. (Photo by Gary R. Mullen)

possible after the sting occurs. A few deaths attributed to toxic reactions have been reported in people and animals receiving hundreds to thousands of stings from the Africanized honey bee or yellowjackets. The victims in these cases usually are either young children or people restrained in some way in close proximity to a disturbed colony.

Although relatively few people are killed by hymenopteran stings, many people suffer some degree of sting hypersensitivity. Estimates of the incidence of hypersensitivity reactions, from mild to severe systemic responses, varies from 0.15 to 5% in the United States (Schmidt, 1992). An incidence of 2% in this case represents about 6 million people. Fewer than 1% of fire ant sting victims experience anaphylactic shock when stung. Vespid wasps cause twice as many allergic reactions as honey bees. However, the importance of the species involved differs with the region. In the United States, for example, yellowjackets cause more serious reactions than do honey bees in the Pacific Northwest and Washington, DC areas. Paper wasps are most important in Texas and the southwestern states, whereas fire ants cause more allergic reactions and deaths than do honey bees in the southeastern states (Camazine, 1988). Hypersensitive people often suffer great physical stress and illness; others may experience apprehension and other psychological effects by not knowing when or how serious a subsequent sting might be (Reisman, 1994b).

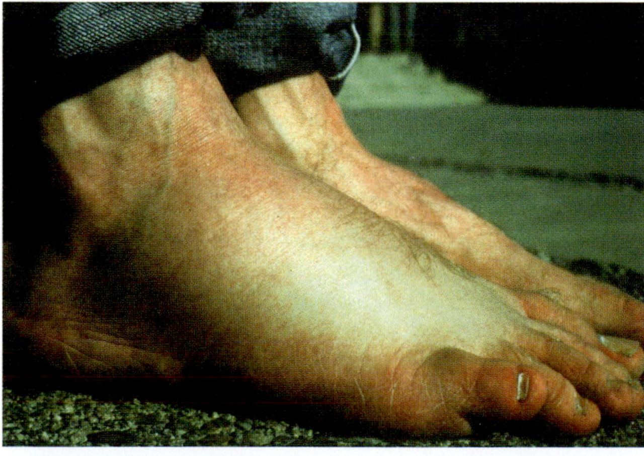

Figure 21.30 Swelling of human foot as a result of yellowjacket (*Vespula* sp.) sting. (Courtesy of Roger D. Akre)

A unique type of allergy has been associated with honey bees. Family members of beekeepers can develop an intense allergy to constituents of honey bee venom while laundering venom-impregnated garments. Most of these people do not realize they are allergic because they have had little or no direct contact with the bees themselves.

Several approaches can be taken in the treatment of sting victims. Embedded stings should be removed as soon as possible to lessen the sting reaction. Conventional advice on sting removal has emphasized that the sting apparatus should be scraped off, not pinched off, as the latter was thought to squeeze more venom into the sting site. Experimental evidence on honey bee stings emphasizes the importance of immediate removal of the embedded sting apparatus in order to minimize the sting reaction (Visscher et al., 1996).

The sting site should be washed with soap and water to minimize the possibility of secondary infection. Ice packs or cool compresses, topical lidocaine, corticosteroid lotions, and antihistamines are recommended treatments for typical local sting reactions (Fitzgerald and Flood, 2006). These treatments reduce the amount of venom absorbed, pain, and local swelling. The more immediately they are applied, the more effective they are in reducing the intensity of the reaction. The use of meat tenderizers, which generally contain a proteolytic enzyme (e.g., papain), is questionable, with no published data from controlled experiments to support their effectiveness. Other home remedies such as baking soda, wet salt, moistened tobacco, 10% ammonia solution, wet aspirin, and commercial sting relievers have been suggested to help to reduce swelling and pain when applied to the sting sites (Weathersby, 1984); however, again no clinical experimental evidence supports these claims.

People with suspected allergic responses can be tested by means of a skin test involving subcutaneous injection of diluted venom, by a radioallergosorbent test (RAST) that requires the taking of a blood sample to combine with the venom to detect reactions, or by a blood test using a histamine leukocyte release assay (Reisman, 1994b; Goddard, 2007). All tests should be done in an allergy clinic where immediate medical treatment is available should adverse reactions to the tests occur.

If a serious systemic reaction occurs, a physician's help should be sought immediately. The prompt injection of epinephrine (adrenaline) is the initial step to combat a life-threatening anaphylactic reaction. Most physicians recommend that persons with demonstrated hypersensitivity wear an identification tag and carry a small emergency sting kit containing antihistamines and a syringe of epinephrine. Such kits are available with a doctor's prescription for a modest cost. Hypersensitive people at high risk of a fatal reaction should consider immunotherapy. These individuals can be desensitized by a series of injections using attenuated doses of the appropriate venom. Such treatment gradually builds up the individual's tolerance to the venom and helps to prevent subsequent systemic reactions. Desensitization programs can be expensive and sometimes require many years.

Additional information about hymenopteran venoms, clinical aspects of stings, allergic reactions, and treatment of sting victims is provided by Piek (1986b), Schmidt (1992), Charpin et al. (1994), Levine and Lockey (1995), Meier (1995), Mosbech (1995), Fitzgerald and Flood (2007), and Golden (2007).

VETERINARY IMPORTANCE

Comparatively little is known about the effects of hymenopteran stings on farm and game animals except for stings by fire ants. Imported fire ants *S. invicta* and *S. richteri* commonly attack and kill newborn game animals such as rabbits, deer, quail, and other ground-nesting birds (Lofgren, 1986). Even the native fire ant in the United States, *S. xyloni*, is known to attack and kill newly hatched poultry. Deaths have been reported only rarely, however, among other newborn farm animals. A few accounts of animal deaths have been attributed to stings of Africanized honey bee, usually involving dogs (Oliveria et al., 2007) (Figure 21.26) or livestock either restrained or enclosed near a hive. The German yellowjacket can injure the teats of milking cows by biting (Braverman et al., 1991), and the resulting lesions may be associated with outbreaks of mastitis in Israel (Shwimmer et al., 1995). Yellowjackets also have been observed cutting flesh from wounds on horses.

Few published data are available on allergic sting reactions in nonhuman animals. Severe systemic, allergic reactions to stings of bees and wasps have been reported in dogs. Fire ant stings in dogs do not develop pustules so characteristic of humans and there have been no reports of anaphylaxis in dogs due to fire ant stings (Rakich et al., 1993). Some dogs, when stung on the face by bees or wasps, develop enormous swelling of the head that may persist for a few days to a week.

Some ant species are intermediate hosts for parasitic helminths of vertebrates. *Formica* species are intermediate hosts for the lancet fluke (*Dicrocoelium dendriticum*), which infests the bile ducts of cattle, sheep, pigs, goats, horses, dogs, and occasionally humans. Pavement ant workers and *Pheidole* ants also serve as intermediate hosts of the poultry tapeworms *Raillietina tetragona* and *R. echinobothrida* (Harwood and James, 1979; Olsen, 1974).

PREVENTION AND CONTROL

Avoidance of stinging insects is the best approach to preventing envenomization. People at risk should avoid areas with high concentrations of ants, bees, or yellowjackets, and be very observant and alert when working in areas where colonies are likely to occur. Colonies of harvester ants can be easily avoided because the ants are slow moving and their nest mounds in the center of an area devoid of vegetation are very noticeable.

Certain colors of clothing should be avoided. Foraging yellowjackets and bees are attracted to yellow. Dark-colored objects may be more attractive to attacking wasps. It is advisable to wear long pants, preferably with the legs tucked into boots, and to wear a long-sleeved shirt to protect the arms. Individuals at risk should avoid using perfumes, hair sprays, sweet-smelling lotions, aftershave lotions, hand and body lotions, and even certain suntan lotions. Some of the odors emanating from these materials are attractive to wasps and bees. Paints containing isoamyl acetate, a component of the honey bee alarm pheromone, should be used with care around honey bees. These materials should not be applied outdoors when and where stinging bees and wasps are likely to be attracted to them.

Repellents commonly used to deter mosquitoes and other biting insects are not effective against attacking ants, bees, and wasps. However, studies have suggested that lotions containing butyl, isopropyl and octyl palmitates, and methyl myristate may repel foraging yellowjackets from human food sources (Henderson et al., 1993).

Care should be taken not to provoke colonies by approaching too closely or creating substrate vibrations that might disturb the nest. Similarly, one should avoid using machines such as power tools and lawnmowers near nests or opening and shutting doors and windows in the vicinity of a nest. Once a colony is aroused, it is best to slowly back out of the area rather than to run. Wasps and bees defending their nests tend to attack nearby moving objects. However, once a person is stung, it is best to leave the area as quickly as possible before other nestmates are recruited to attack in response to alarm pheromone.

In cases of potential mass stinging attacks, such as by the Africanized honey bee or large colonies of yellowjackets, it is best to run away from the nest. The nose and eyes should be covered without blocking vision to avoid stings directed to one's head. Running, perhaps a quarter to half mile, will gradually leave the attacking wasps or bees behind. If you remain in the area, the potential exists to receive many dozens to hundreds of stings.

Insecticides, especially aerosol sprays containing a quick knock-down and a long-lasting pesticide can be used to kill colonies of both yellowjackets and paper wasps. Although these aerosols can propel toxicants up to 6–7 m and are designed primarily for treating aerial nests, they work equally well on colonies nesting underground. Nest destruction should be conducted at night when wasps are least active, and most if not all the individuals are in the nest. If insecticides are applied during the daytime, the nest entrance should not be blocked after treatment so that returning wasps can readily enter the nest and contact the toxicant. Long-lasting residual insecticides will kill workers emerging from capped cells after the initial treatment. Some aerosol-type sprays do not contain a long-lasting toxicant and are not effective on later-emerging adults. Controlling any nest, especially large colonies of paper wasps and yellowjackets, can be difficult and dangerous, so it is often best left to experienced pest control operators. Protective clothing such as a bee suit and veil should be worn when destroying colonies.

Only under certain circumstances is there good reason to control honey bees. When necessary, local beekeepers may be willing to collect swarms near houses or other sites of human activity. Colonies of honey bees nesting in the wall voids of houses probably should be destroyed. They can be killed with the combination of a quick-knockdown insecticide directed into the nest entrance, followed by an application of insecticidal dust into the voids through openings such as electrical plate outlets or holes drilled into the wall. Destruction of well established honey bee colonies in walls of buildings can result in the stored honey dripping through the walls or ceiling. In such cases, the colony and honey combs should be physically removed after the bees are killed or otherwise removed.

Control of some social insects is best accomplished by the use of baits that incorporate a slow-acting insecticide. Foragers are attracted to the bait, return to the colony with the toxic bait, and spread the material throughout the colony by trophallaxis. Several toxicants that show promise for control are classified as insect growth regulators (IGRs), but several other slow-acting toxicants with direct poisoning action are also effective. Although ants in heavily used areas such as picnic grounds can be killed by direct application of insecticides to the nest, it is best to use some type of carbohydrate or fatty baits that contains a slow-acting poison. Baits are the control of choice for pharaoh's ants and pavement ants since treatments with insecticidal sprays in buildings tend to cause the ants to disperse, making them more difficult to control. Traps baited with pheromones are being used to attract honey bee swarms to monitor the northward movement of the Africanized honey bees.

Control measures for fire ants include the direct application of insecticides to the nest, the use of various formulations of toxic baits and IGRs, and the use of biological control agents such as pathogens and parasitoids (Collins, 1992; Drees and Vinson, 1993). The key to effective control is persistence in treating the new colonies as they appear. Unfortunately, despite extensive efforts, control of fire ants has not been very successful except in limited areas.

The control of scavenging yellowjackets may be accomplished by a trapping system with an appropriate attractant or attractive bait. All scavenging *Vespula* species are attracted to meat baits. A simple meat trap with fish or ham suspended over a pan of detergent water can be used to drown thousands of attracted wasps. Workers cut off pieces of meat that are too heavy for them to carry and drop down into the water as they try to fly away. Each day, the drowned wasps should be removed and the detergent water and the meat should be replaced.

Because meat baits spoil and lose their attractiveness quickly, potential synthetic attractants have been investigated. The western yellowjacket, as well as the California and southern yellowjackets, can be lured into traps containing the synthetic attractants hexyl butyrate, heptyl butyrate, or octyl butyrate. Two co-attractants, isobutanol and acetic acid, isolated from fermenting molasses, have been shown to be effective baits for nearly all scavenging yellowjacket species in North America; for example, eastern, western, southern, common, and German yellowjackets, as well as baldfaced hornets, European hornets, and some species of paper wasps (Landolt, 1998; Landolt et al., 1999).

More information on avoidance and control measures for the various stinging hymenopterans can be found in Akre and MacDonald (1986) for yellowjackets, Drees and Vinson (1993) and Lofgren (1986) for fire ants, and Vinson (1986) for social insects in general.

REFERENCES AND FURTHER READING

Agostinucci, W., Cardoni, A. A., & Rosenberg, R. (1981). Effect of papain on bee venom toxicity. *Toxicon, 19*, 851–855.

Akre, R. D. (1982). Social wasps. In H. R. Hermann (Ed.), *Social Insects* (Vol. 4 pp. 1–105). New York: Academic.

Akre, R. D., Greene, A., MacDonald, J. F., Landolt, P. J., & Davis, H. G. (1981). The Yellowjackets of America North of Mexico. *US Department of Agriculture Handbook*, No. 552.

Akre, R. D., & MacDonald, J. F. (1986). Biology, economic importance, and control of yellowjackets. In S. B. Vinson (Ed.), *Economic Impact and Control of Social Insects* (pp. 353–412). New York: Praeger.

Akre, R. D., & Reed, H. C. (1984). Biology and distribution of social Hymenoptera. In A. T. Tu (Ed.), *Insect Poisons, Allergens, and Other Invertebrate Venoms: Handbook of Natural Toxins* (Vol. 2, pp. 3–47). New York: Marcel Dekker.

Alexander, B., & Michener, C. D. (1995). Phylogenetic studies of the families of short-tongued bees (Hymenoptera: Apoidea). *University of Kansas Science Bulletin, 55*, 377–424.

Banks, B. E. C., & Shipolini, R. A. (1986). Chemistry and pharmacology of honey bee venom. In T. Piek (Ed.), *Venoms of the Hymenoptera: Biochemical, Pharmacological and Behavioural Aspects* (pp. 329–416). San Diego, CA: Academic.

Blum, M. S. (1984). Poisonous ants and their venoms. In A. T. Tu (Ed.), *Insect Poisons, Allergens, and Other Invertebrate Venoms: Handbook of Natural Toxins* (Vol. 2, pp. 225–242). New York: Marcel Dekker.

Bohart, R. M., & Menke, S. (1976). *Sphecid Wasps of the World: A Generic Revision*. Berkeley: Univ. California Press.

Bolton, B. (1994). *Identification Guide to the Ant Genera of the World*. Harvard Univ. Press.

Braverman, Y., Marcusfeld, O., Adler, H., & Yakobson, B. (1991). Yellowjacket wasps can damage cow's teats by biting. *Medical and Veterinary Entomology, 5*, 129–130.

Brewer, R. D., Morris, P. D., & Cole, T. B. (1994). Hurricane-related emergency department visits in an inland area: An analysis of the public health impact of Hurricane Hugo in North Carolina. *Annals of Emergency Medicine, 23*, 731–736.

Buck, M. C., Marshall, S. A., & Cheung, D. K. B. (2008). Identification atlas of the Vespidae (Hymenoptera, Aculeata) of the northeastern Nearctic region. *Canadian Journal of Arthropod Identification, 5*, 1–138.

Camazine, S. (1988). Hymenoptera stings: Reactions, mechanisms, and medical treatment. *Bulletin of the Entomological Society of America, 34*, 17–21.

Charpin, D., Birnbaum, J., & Vervloet, D. (1994). Epidemiology of Hymenoptera allergy. *Clinical and Experimental Allergy, 24*, 1010–1015.

Cole, A. C. (1968). *Pogonomyrmex Harvester Ants: A Study of the Genus in North America.* Knoxville: University of Tennessee Press.

Collins, H. (1992). Control of imported fire ants: a review of current knowledge. *USDA Technical Bulletin* No. 1807.

Drees, B. M., & Vinson, S. B. (1993). Fire ants and their management. *Texas AgriLife Extension Service B-1536.*

Edwards, J. P. (1986). The biology, economic importance and control of the Pharaoh's ant, *Monomorium pharaonis* (L.). In S. B. Vinson (Ed.), *Economic Impact and Control of Social Insects.* New York: Praeger.

Fisher, B. L., & Cover, S. P. (2007). *Ants of North America: A Guide to Genera.* Berkeley: University of California Press.

Fitzgerald, K. T., & Flood, A. A. (2006). Hymenoptera stings. *Clinical Techniques in Small Animal Practice, 21,* 194–204.

Fowler, H. G., Bueno, O. C., Sadatsune, T., & Montelli, A. C. (1993). Ants as potential vectors of pathogens in hospitals in the state of São Paulo, Brazil. *Insect Science and Its Application, 14,* 367–370.

Gauld, I. D., & Bolton, B. (1988). *The Hymenoptera.* Oxford University Press.

Gillaspy, J. E. (1986). *Polistes* wasps: Biology and impact on man. In Vinson, S. B. (Ed.), *Economic Impact and Control of Social Insects* (pp. 332–352). New York: Praeger.

Goddard, J. (2007). *Physician's Guide to Arthropods of Medical Importance* (5th ed.). Boca Raton: CRC.

Golden, D. B. K. (2007). Insect sting anaphylaxis. *Immunology and Allergy Clinics of North America, 27,* 261–272.

Goulet, H., & Huber, J. T. (1993). *Hymenoptera of the World: An Identification Guide to Families. Agriculture Canada.* Ottawa: Research Branch.

Graham, J. M. (Ed.). (1992). *The Hive and the Honey Bee.* Hamilton, IL: Dadant.

Harwood, R. F., & James, M. T. (1979). *Entomology in Human and Animal Health* (7th ed.). New York: MacMillan Co.

Henderson, G., Blouin, D. C., & Jeanne, R. L. (1993). Yellowjacket (Hymenoptera: Vespidae) repellency by natural products of paper wasps and Avon's Skin-So-Soft[8]. *Journal of Entomological Science, 28,* 387–392.

Hoffman, D. R. (1984). Insect venom allergy, immunology, and immunotherapy. In Tu, A. T. (Ed.), *Insect Poisons, Allergens, and Other Invertebrate Venoms. Handbook of Natural Toxins* (Vol. 1, pp. 187–223). New York: Marcel Dekker.

Hölldobler, B., & Wilson, E. O. (1990). *The Ants.* Cambridge: Belknap/Harvard Univ. Press.

Klotz, J., Hansen, L., Pospischil, R., & Rust, M. (2008). *Urban Ants of North America and Europe: Identification, Biology, and Management.* Cornell University Press.

Landolt, P. J. (1998). Chemical attractants for trapping yellowjackets *Vespula germanica* and *Vespula pensylvanica* (Hymenoptera: Vespidae). *Environmental Entomology, 27,* 1229–1234.

Landolt, P. J., Reed, H. C., Aldrich, J. R., Antonelli, A. L., & Dickey, C. (1999). Social wasps (Hymenoptera: Vespidae) trapped with acetic acid and isobutanol. *Florida Entomologist, 82,* 609–614.

Levine, M. I., & Lockey, R. F. (Eds.). (1995). *Monograph on Insect Allergy.* Pittsburgh: Dave Lambert Assoc.

Lofgren, C. S. (1986). The economic importance and control of imported fire ants in the United States. In S. B. Vinson (Ed.), *Economic Impact and Control of Social Insects* (pp. 227–256). New York: Praeger.

MacKay, W. P. (1990). The biology and economic impact of *Pogonomyrmex* harvester ants. In R. K. Vander Meer, K. Jaffe, & A. Cedeno (Eds.), *Applied Myrmecology: A World Perspective* (pp. 533–543). Boulder, CO: Westview.

Manley, D. G. (1991). The *velvet ants (Hymenoptera: Mutillidae) of South Carolina.* South Carolina Agric. Expt. Sta., Tech. Bull. 1100

Marle, J. van, & Piek, T., (1986). Morphology of the venom appa ratus. In T. Piek (Ed.), *Venoms of the Hymenoptera: Biochemical, Pharmacological and Behavioural Aspects* (pp. 17–44). San Diego: Academic.

Meier, J. (1995). Biology and distribution of hymenopterans of medical importance, their venom apparatus and venom composition. In J. Meier, & J. White (Eds.), *Handbook of Clinical Toxicology of Animal Venoms and Poisons* (pp. 331–348). Boca Rotan, FL: CRC Press.

Michener, C. D. (1974). *The Social Behavior of the Bees.* Cambridge, MA: Belknap/Harvard University Press.

Michener, C. D. (2007). *The Bees of the World.* Cambridge, MA: Belknap Press.

Michener, C. D., McGinley, R. J., & Danforth, B. N. (1994). *The Bee Genera of North and Central America (Hymenoptera Apoidea).* Washington, DC: Smithsonian Institution Press.

Mosbech, H. (1995). Clinical toxicology of hymenopteran stings. In J. Meier, & J. White (Eds.), *Handbook of Clinical Toxicology of Animal Venoms and Poisons* (pp. 349–359). Boca Rotan, FL: CRC Press.

Nakajima, T. (1984). Biochemistry of vespid venoms. In A. T. Tu (Ed.), *Insect Poisons, Allergens, and Other Invertebrate Venoms. Handbook of Natural Toxins* (Vol. 2, pp. 109–133). New York: Marcel Dekker.

Nakajima, T. (1986). Pharmacological biochemistry of vespid venoms. In T. Piek (Ed.), *Venoms of the Hymenoptera: Biochemical, Pharmacological and Behavioural Aspects* (pp. 309–327). San Diego: Academic.

Nelder, M. P., Paysen, E. S., Zungoli, P. A., & Benson, E. P. (2006). Emergence of the introduced ant *Pachycondyla chinensis* (Formicidae: Ponerinae) as a public health threat in the Southeastern United States. *Journal of Medical Entomology, 43,* 1094–1098.

Oi, D. H. (2008). Pharaoh ants and fire ants. In X. Bonnefoy, H. Kampen, & K. Sweeney (Eds.), *Public Health Significance of Urban Pests* (pp. 175–207). Geneva: World Health Organization.

Oliveira, E. C., Pedroso, R. M. O., Meirelles, A. E. W. B., Pescador, C. A., Gouvea, A. S., & Driemeier, D. (2007). Pathological findings in dogs after multiple Africanized bee stings. *Toxicon, 49,* 1214–1218.

Olsen, O. W. (1974). *Animal Parasites,* Their Life Cycles and Ecology. New York: Dover.

O'Neil, M. E., Mack, K. A., & Gilchrist, J. (2007). Epidemiology of non-canine bite and sting injuries treated in U.S. emergency departments, 2001–2004. *Public Health Reports, 122,* 764–775.

O'Neill, K. M. (2001). *Solitary Wasps: Behavior and Natural History.* Geneva: Cornell University.

Piek, T. (1984). Pharmacology of Hymenoptera venoms. In A. T. Tu (Ed.), *Insect Poisons, Allergens, and Other Invertebrate Venoms. Handbook of Natural Toxins* (Vol. 2, pp. 135–185). New York: Marcel Dekker.

Piek, T. (1986a). Venoms of bumble bees and carpenter bees. In T. Piek (Ed.), *Venoms of the Hymenoptera: Biochemical, Pharmacological and Behavioural Aspects* (pp. 417–424). San Diego: Academic.

Piek, T. (Ed.). (1986b). *Venoms of the Hymenoptera: Biochemical, Pharmacological and Behavioural Aspects.* San Diego: Academic.

Rakich, P. M., Latimer, K. S., Mispagel, M. E., & Steffens, W. L. (1993). Clinical and histological characterization of cutaneous reactions to stings of the imported fire ant (*Solenopsis invicta*) in dogs. *Veterinary Pathology, 30,* 555–559.

Reed, H. C., & Landolt, P. J. (1991). Swarming of paper wasp (Hymenoptera: Vespidae) sexuals at towers in Florida. *Annals of the Entomological Society of America, 84,* 628–635.

Reisman, R. E. (1994a). Insect stings. *The New England Journal of Medicine, 331,* 523–527.

Reisman, R. E. (1994b). Venom hypersensitivity. *The Journal of Allergy and Clinical Immunology, 94,* 651–658.

Rhoades, R. B. Stafford, C. T., James, F.K., Jr. (1989). Survey of fatal anaphylactic reactions to imported fire ant stings. *The Journal of Allergy and Clinical Immunology, 84,* 159–162.

Ross, K. G., & Matthews, R. W. (Eds.). (1991). *The Social Biology of Wasps.* Ithaca, NY: Comstock Publ. Assoc.

Schmidt, J. O. (1986a). Allergy to Hymenoptera venoms. In T. Piek (Ed.), *Venoms of the Hymenoptera: Biochemical, Pharmacological and Behavioural Aspects* (pp. 509–546). San Diego: Academic Press.

Schmidt, J. O. (1986b). Chemistry, pharmacology, and chemical ecology of ant venoms. In T. Piek (Ed.), *Venoms of the Hymenoptera: Biochemical, Pharmacological and Behavioural Aspects* (pp. 425–508). San Diego: Academic Press.

Schmidt, J. O. (1986c). Hymenoptera envenomation. In G. W. Frankie, & C. S. Koehler (Eds.), *Urban Entomology: Interdisciplinary Perspectives* (pp. 187–220). New York: Praeger.

Schmidt, J. O. (1990). Africanized and European honey bee venoms: Implications for beekeepers and the public. *Am Bee J, 130,* 810–811.

Schmidt, J. O. (1990). Hymenoptera venoms: Striving toward the ultimate defense against vertebrates. In D. L. Evans, & J. O. Schmidt (Eds.), *Insect Defenses: Adaptive Mechanisms and Strategies of Prey and Predator* (pp. 390–395). Albany, NY: State University of New York Press.

Schmidt, J. O. (1992). Allergy to venomous insects. In J. Graham (Ed.), *The Hive and the Honey bee.* Hamilton, IL: Dadant and Sons.

Schmidt, J. O., & Blum, M. S. (1978). A harvester ant venom: Chemistry and pharmacology. *Science, 200,* 164–166.

Schmidt, J. O., Menke, G. C., Chen, T. M., & Pinnas, J. L. (1984). Demonstration of cross-allergenicity among harvester ant venoms using RAST and RAST inhibition. *The Journal of Allergy and Clinical Immunology, 73*(1, part 2), 158.

Shipolini, R. A. (1984). Biochemistry of bee venom. In A. T. Tu (Ed.), *Insect Poisons, Allergens, and Other Invertebrate Venoms. Handbook of Natural Toxins* (Vol. 2, pp. 49–85). New York: Marcel Dekker.

Schumacher, M. J., & Egen, N. B. (1995). Significance of Africanized bees for public health. *Archives of Internal Medicine, 155,* 2038–2043.

Shwimmer, A., Shpigel, N. Y., Yeruham, I., & Saren, A. (1995). Epidemiological and bacteriological aspects of mastitis associated with yellowjacket wasp teat lesions in Israeli dairy cows. Proc. Third IDF Internat. Mastitis Seminar 100–102.

Smith, M. R. (1965). *House-infesting ants of the eastern United States: their recognition, biology, and economic importance. United States Department of Agriculture Animal Research Services Technical Bulletin 1326.*

Spivak, M., Fletcher, D. J. C., & Breed, M. D. (Eds.). (1991). *The 'African' Honey Bee.* Boulder, CO: Westview.

Starr, C. K. (1985). A simple pain scale for field comparison of Hymenoptera stings. *Journal of Entomological Science, 20,* 225–232.

Taber, S. W. (2000). *Fire Ants.* College Station, TX: Texas A&M University Press.

Tschinkel, W. R. (2006). *The Fire Ants.* Cambridge, MA: Harvard University Press.

Tu, A. T. (1984). Insect poisons, allergens, and other invertebrate venoms. In A. T. Tu (Ed.), *Handbook of Natural Toxins* (Vol. 2, pp. 49–85). New York: Marcel Dekker.

Ulloa-Chacon, P., & Cherix, D. (1990). The little fire ant *Wasmannia auropunctata* (R.) (Hymenoptera: Formicidae). In R. K. Vander Meer, K. Jaffe, & A. Cedeno (Eds.), *Applied Myrmecology: A World Perspective* (pp. 281–289). Boulder, CO: Westview.

Vander Meer, R. K., Jaffe, K., & Cedeno, A. (Eds.). (1990). *Applied Myrmecology: A World Perspective.* Boulder, CO: Westview.

Vander Meer, R. K., Breed, M. K., Winston, M. L., & Espelie, K. E. (1997). *Pheromone Communication in Social Insects.* Boulder, CO: Westview.

Vinson, S. B. (Ed.). (1986). Economic Impact and Control of Social Insects. New York: Praeger.

Vinson, S. B. (1994). Impact of the invasion of Solenopsis invicta (Buren) on native food webs. In D. F. Williams (Ed.), *Exotic ants: Biology, Impact, and Control of Introduced Species* (pp. 240–258). Boulder, Colorado: Westview.

Vinson, S. B., & Greenberg, L. (1986). The biology, physiology, and ecology of imported fire ants. In S. B. Vinson (Ed.), *Economic Impact and Control of Social Insects* (pp. 193–226). New York: Praeger.

Visscher, P. K., Vetter, R. S., & Camazine, S. (1996). Removing bee stings. *Lancet, 348,* 301–302.

Weathersby, A. B. (1984). Wet salt for envenomization. *Journal of the Georgia Entomological Society, 19,* 1–6.

Winston, M. L. (1992). *Killer Bees: The Africanized Honey Bee in the Americas.* Cambridge: Harvard Univ. Press.

Chapter

22

Scorpions (Scorpiones)

Gary R. Mullen . Scott A. Stockwell

Scorpions represent an ancient group of arachnids that is believed to be descended from now-extinct eurypterids living in estuaries and coastal lagoons. Except for their smaller size, scorpions are notably similar in appearance to their marine ancestors, which first crawled onto land as air-breathing arachnids during the late Devonian and early Carboniferous 325 to 350 million years ago. Throughout recorded history scorpions have intrigued human cultures, being revered and attributed special powers by some, feared as sinister and ominous by others. Scorpion images appear as religious symbols, on seals, magical tablets, amulets, and boundary stones, and in the origin stories of ancient civilizations such as the Chaldeans and Egyptians. They also figured prominently in Greek mythology and as one of the 12 constellations or signs of the Zodiac.

Despite the many superstitions and misconceptions about scorpions that persist to this day, their reputation as venomous arthropods is generally overstated. Most scorpions are not aggressive and inflict only minor, transient pain and discomfort when they do sting, typically to defend themselves when threatened. There are, however, 40 to 50 species worldwide that pose significant health problems. About 25 species are considered to be capable of causing human deaths. Most of them occur in the Tropics and Subtropics or in arid regions of temperate zones.

TAXONOMY

Scorpion taxonomy continues to be unstable. In his review of scorpion classification, Sissom (1990) listed 1,077 species, 117 genera, and nine families. Two of those families (Chactidae and Vaejovidae) could not be distinguished in the identification key provided. During the 1990s, many new families were proposed that helped to create a clearer picture of phylogenetic relationships among the genera of scorpions. In their catalog of the scorpions of the world, Fet et al. (2000) listed 16 families, 154 genera, and 1,252 species of extant scorpions. A phylogenetic study by Soleglad and Fet (2003) resulted in a widely debated revision of the higher classification of scorpions, which seemed to increase taxonomic instability rather than improve it. This arrangement was questioned by Prendini and Wheeler (2005), who challenged most of the changes made by Soleglad and Fet (2003) and essentially returned scorpion taxonomy to its pre-2003 state. The subsequent classification follows that of Prendini and Wheeler (2005), with slight modifications of the included genera. For a listing of valid genera, see the *Conspectus Genericus Scorpionorum* of Dupré (2007).

At least 98 species belonging to 12 genera and five families are known to occur in the continental United States. A single species (*Paruroctonus boreus*) occurs in the extreme southern portions of British Columbia, Alberta, and Saskatchewan, Canada. Mexico is home to at least 178 species spread among 22 genera and seven families. The catalog by Fet et al. (2000) provides the most up-to-date picture of scorpion taxonomy and is an invaluable resource. Keys to most of the genera of scorpions of the world are provided by Sissom (1990). A key to the North American families and genera is provided by Stockwell (1992). For species identification of the North American fauna, see the following regional works: Arizona (Stahnke, 1940), Nevada (Gertsch and Allred, 1965), Utah (Johnson and Allred, 1972), Idaho (Anderson, 1975), California (Hjelle, 1972; Williams, 1976), Baja California, Mexico (Williams, 1980), and Florida (Muma, 1967).

Buthidae

This is the largest and most widespread scorpion family, with 79 currently recognized genera and over 700 valid species. Buthids are found throughout the world with their greatest diversity in the Old World, especially the Afrotropical Region and the southern

Table 22.1 Dangerous Species of Scorpions Based on the Toxicity of Their Venoms

Species	Lethal Dose* (LD_{50})	Geographic Occurrence
Leiurus quinquestriatus	0.25	Turkey, Israel, Egypt, Algeria, Libya, Sudan
Androctonus mauretanicus	0.31	Morocco
Androctonus australis	0.32	Morocco, Algeria, Libya, Tunisia, Egypt
Androctonus crassicauda	0.40	Turkey, Israel, Iraq, Arabian Peninsula
Tityus serrulatus	0.43	Brazil
Centruroides limpidus	0.69	Mexico
Androctonus amoreuxi	0.75	Middle East
Buthus occitanus	0.90	Morocco, Algeria, Jordan, southern Europe
Centruroides exilicauda	1.12	United States, northern Mexico
Parabuthus transvaalicus	4.25	Southern Africa

*The lethal dose is expressed as mg/kg of venom required to kill 50% of mice (LD_{50}) following subcutaneous injection. The lower the LD_{50}, the more potent the venom. All the scorpions listed are members of the family Buthidae. (Compiled from multiple sources.)

Palaearctic Region. Most of the scorpions that are dangerously venomous to humans and other animals belong to this family (Table 22.1). The important genera in this respect are *Androctonus* and *Leiurus* in northern Africa and western Asia, *Hottentotta* in Asia and India, *Parabuthus* in southern Africa, *Centruroides* in North America, and *Tityus* in South America. Numerous other genera may contain members of minor medical importance. Members of this scorpion family commonly are encountered by people and pets, making their status as venomous pests all the more important. The only buthid genus that naturally occurs in North America is *Centruroides*. Members of this genus are crevice dwellers that commonly enter homes. Five species occur in the United States. *Centruroides hentzi* is found throughout Florida and adjoining portions of Alabama and Georgia. *Centruroides guanensis* is common on the islands of the Bahamas and Cuba, but is also found in the southernmost part of the Florida Peninsula and the Florida Keys. *Centruroides gracilis* is native to Central America and Mexico, but has been introduced to many other tropical areas including Florida. *Centruroides vittatus* (Figure 22.1) is the most widespread species of scorpion in the United States, with a range that extends from the Rio Grande River in the west to the Mississippi River in the east. This species has been collected as far north as southernmost Nebraska. It is found throughout Texas and Oklahoma, and in adjoining parts of New Mexico, Colorado, Kansas, Illinois, Missouri, Arkansas, and Louisiana. *Centruroides exilicauda* (formerly and occasionally still known as *C. sculpturatus*) is found in most of Arizona, as well as adjoining parts of New Mexico, Utah, Nevada, and California. All species of *Centruroides* can deliver extremely painful stings that may be accompanied by systemic symptoms. However, only *C. exilicauda* is considered dangerous, and then, only to small children. The only other buthid occurring in the United States is the cosmotropical species *Isometrus maculatus*, an Asian species that has been introduced to tropical port cities around the world. This species is common on the island of Oahu in Hawaii.

Figure 22.1 *Centruroides vittatus* (Buthidae), the most common scorpion in the United States. (Photo by S. A. Stockwell)

Microcharmidae

This family contains three genera of small, tropical, forest-dwelling scorpions from Africa and Madagascar. Traditionally placed in the family Buthidae, their inclusion together in a separate family bears reevaluation.

Pseudochactidae

This recently described family contains two monotypic genera; one from central Asia (Uzbekistan and Tajikistan) and the other from Laos. Members exhibit an unusual trichobothrial pattern similar to that of the Buthidae.

Chaerlidae

This family is represented by the single genus, *Chaerilus*, with 23 described species, none of which are dangerously venomous. These scorpions are unique in many ways, but they share some characteristics with the Buthidae. They occur in the Old World in India,

Sri Lanka, Nepal, Bangladesh, Myanmar, Malaysia, Singapore, and many islands of the Philippines and Indonesia. None is known to be dangerous.

Chactidae

This family contains nine genera and approximately 150 species. Most of the chactid species are found in South America (Colombia, Venezuela, Trinidad and Tobago Guyana, French Guiana, Suriname, Ecuador, Peru, and Brazil). A few species are found as far north as Panama and Costa Rica. One peculiar species, *Nullibrotheas alleni*, is endemic to Baja California Sur (Mexico) and is the only North American representative of the family. None of these is considered dangerous to humans.

Euscorpiidae

This small family shares many characteristics with the Chactidae, of which it was once considered a subfamily. Recent work on this family has resulted in a large number of new species. Additionally, the members of the family Scorpiopidae were transferred to this family (Soleglad and Sissom, 2001), along with the genus *Chactopsis*, which was formerly placed in the family Chactidae. There are currently 11 genera and 77 species. One genus, *Euscorpius*, with 17 species, is found throughout southern (Mediterranean) Europe and has been variously divided into four subgenera and numerous subspecies. The six genera of the former Scorpiopidae (now Scorpioninae) may be found at relatively high altitudes throughout their range. They are native to parts of Afghanistan, Pakistan, India, Sikkim, Nepal, China, Bangladesh, Myanmar, Thailand, Laos, Vietnam, Malaysia, Indonesia, and Bhutan. None of them is regarded as dangerous. *Chactopsis* is native to South America (Peru, Venezuela, Brazil). The other genera in this family are found in eastern Mexico and Guatemala. *Megacormus* and *Plesiochactas* are closely related epigean forms. The two species of *Troglocormus* are troglobitic, having lost the median eyes, and are found only in caves.

Superstitioniidae

This small family previously was regarded as a subfamily in the Chactidae. They share no characters with the Chactidae, but appear to be related to the Vaejovidae and Iuridae. Superstitioniidae contains four genera and nine species. They are generally small in size. At 9 mm total length, *Typhlochactas mitchelli* is the smallest species of scorpion in the world. The one exception in the family is *Alacran tartarus*, which measures up to 70 mm in length. Most of the species tend to be troglobitic in form, lacking eyes and pigmentation. *Superstitionia donensis* is found in the southwestern United States and is the only species with a full complement of eyes. The other species are all found in Mexico in caves or leaf litter. These scorpions are too small or inaccessible to be dangerous to humans. *Alacran* is found only in caves, at depths of up to 800 meters.

Troglotayosicidae

The two monotypic genera in this family were formerly placed in the Superstitioniidae. Both are poorly known and may not be closely related. Both species are troglobites, possessing lateral eyes, but no median eyes. *Troglotayosicus vachoni*, known from a single individual from a cave in Ecuador, shares many characteristics with the Superstitioniidae. *Belisarius xambeui*, from caves in the eastern Pyrenees of Spain and France, was traditionally grouped with *Euscorpius*, with which it bears a superficial resemblance. The trichobothrial pattern of this species, however, is very primitive and much closer to that of the Vaejovidae, Superstitioniidae, and Iuridae.

Iuridae

This interesting but small group of scorpions spans three continents. None of them is considered dangerous. There are six genera and approximately 22 species. The genera *Iurus* and *Calchas*, each with one species, are closely related, relatively large, and found in Turkey and Greece (including Samos, Crete, and other islands). The genera *Caraboctonus* (one species) and *Hadruroides* (nine species) are found in western South America. *Caraboctonus* is found in Chile and Peru; *Hadruroides* is distributed through Ecuador (including the Galapagos Islands), Bolivia, Peru, and Chile. There are six species in the North American genus *Hadrurus*. Additionally, two species formally contained in *Hadrurus* recently were placed in their own genus, *Hoffmannihadrurus* (Fet et al., 2004).

Members of these genera are desert-dwelling burrowers found from south-central Mexico to the western United States. They are the largest scorpion species found in North America. Three species occur in the United States in desert areas of Oregon, Idaho, California, Nevada, Utah, Colorado, and Arizona.

A close relative of *Hadrurus*, *Anuroctonus* contains two species and is distributed from Baja California, Mexico through California, Nevada, Utah, and Idaho. *Anuroctonus* constructs permanent burrows in a variety of habitats throughout its range and is associated with canyons, ravines, and hillsides.

Vaejovidae

This family is comprised of 12 described genera, although the validity of some of them have been questioned, and approximately 159 species that are restricted mostly to North America. Species of Vaejovidae are found in every conceivable habitat in nearly every state of Mexico and much of the United States, especially the west (Washington, Oregon, Idaho, Montana, Wyoming, California, Nevada, Utah, Colorado, North Dakota, South Dakota, Nebraska, Arizona, New Mexico, and Texas). One species ranges south into Guatemala, and another species can be found in Canada.

Vaejovis carolinianus occurs in wooded, mountainous areas of the eastern United States (parts of Kentucky, Tennessee, Virginia, North Carolina, South Carolina,

Georgia, Alabama, Mississippi, and Louisiana). It is a small (about 2.4 cm), dark scorpion that readily enters homes throughout its range. Other species also may occur indoors, such as members of the genera *Pseudouroctonus* and *Uroctonus* that commonly enter homes in California. No members of this family pose any appreciable health threat, but stings from these species are more likely to cause minor localized discoloration, swelling, and necrosis than are the more painful stings of buthid scorpions.

Bothriuridae

This family of 14 genera and 135 species exhibits a Gondwanan distribution. One genus, *Cercophonius*, with seven described species, is distributed throughout Australia, but is also found on New Caledonia and in northern India. The primitive genus *Lisposoma*, with three species, is endemic to Namibia. The remaining genera are distributed throughout western and southern South America (Ecuador, Peru, Brazil, Bolivia, Paraguay, Chile, Argentina, Uruguay).

None of the bothriurids is considered medically important.

Liochelidae

Formerly known as Ischnuridae, and considered a subfamily of the Scorpionidae, the liochelids now are regarded as a separate family. They range in size from small to very large and typically have a flattened body shape. The claws are massive in comparison to the body, but the metasoma is unusually thin, and feeble-looking. Sometimes the metasoma is so short that it cannot reach to the front of the animal. Though capable of burrowing, most of these species are associated with crevice habitats in rocky areas, on trees, under debris, on man-made structures such as stone walls and wooden bridges, and so on. The most impressive of the family is *Hadogenes troglodytes* from South Africa, the males of which attain a body length up to 21 cm.

The family contains 10 genera and 68 species. Representatives are widely distributed throughout the tropics. In the Caribbean they are found in Haiti and Dominican Republic; in Central America, in Panama and Cocos Island (Costa Rica); in South America, in Peru, Colombia, Venezuela, French Guiana, Brazil. In central and southern Africa, they are found in Cameroon, Democratic Republic of Congo, Gabon, Congo, Malawi, Uganda, Ethiopia, Kenya, Tanzania, Angola, Namibia, Botswana, Zimbabwe, Mozambique, South Africa, Lesotho, Swaziland, Mauritius, Round Island, Seychelles, Zanzibar, Madagascar. In Asia, they are found in China, Korea, Japan, India, Aru Islands, Bangladesh, Myanmar, Thailand, Cambodia, Laos, Vietnam, Malaysia, Indonesia, Philippines, Papua New Guinea; in Oceania, in Federated States of Micronesia, Fiji, French Polynesia, Key Islands, Kiribati, Mariana Islands, Marshal Islands, New Caledonia, Palau, Ponape, Tuvalu, Samoa, Solomon Islands, Tonga, Vanuatu; and of course, in Australia.

As a group, they are considered relatively harmless. In the case of stings by the giant *Hadogenes* scorpions, the effect may be so slight as to be barely felt. Other species, such as *Opisthacanthus lepturus* in Panama, can deliver a sting that causes soreness in joints as well as mild, localized discoloration, swelling, and necrosis.

Heteroscorpionidae

This is a tiny family with one genus (*Heteroscorpion*) and five species that are endemic to Madagascar. This genus formerly was placed among the Ischnuridae, with which they share many characteristics. Their medical importance is not known.

Hemiscorpiidae

Members of this family superficially resemble members of the Liochelidae, being flattened and possessing a thin, delicate postabdomen. They also are crevice dwellers like many of the liochelids. In the past, they have been placed variously in the Liochelidae or the Scorpionidae. There is a total of nine species in this group. All are considered to belong to a single genus, *Hemiscorpius*, although some authors still recognize the validity of the monotypic genus *Habibiella*. These species are found in Somalia, Eritrea, Saudi Arabia, Yemen, Oman, United Arab Emirates, Iraq, Iran, and Pakistan. *Hemiscorpius lepturus* in Iran is the only nonbuthid species of scorpion reported to cause significant mortality in humans.

The venom contains a potent cytotoxin that causes severe tissue damage and necrosis near the sting site, as well as severe systemic symptoms. The medical importance of the other species of *Hemiscorpius* is not known; however, it is safe to assume that they have a venom that, at the very least, causes soreness in joints as well as mild, localized discoloration, swelling, and necrosis, as in their close relatives, the Liochelidae and Scorpionidae.

Urodacidae

This family, often considered a subfamily of Scorpionidae, contains a single genus (*Urodacus*) endemic to Australia. There are 20 described species. None is known to be dangerous.

Diplocentridae

This is another comparatively small family with eight genera and about 103 described species that occur primarily in the New World. Exceptions are the genera *Nebo,* found in Syria, Jordan, Lebanon, Israel, Egypt (Sinai), Saudi Arabia, Yemen, and Oman; and two species of *Heteronebo,* both known from the island of Abdel-Kuri (Yemen). Oddly, the other 15 species of *Heteronebo* are found on various islands in the Caribbean, along with the genera *Oiclus*, *Cazierius*, and most of the species in the genus *Didymocentrus.* The latter genus also is represented in Honduras, El Salvador, Nicaragua, and Costa Rica. Another genus, *Tarsoporosus*, is found in Venezuela and Colombia. The genus *Bioculus* is endemic to Baja California Sur (Mexico), and its associated islands. Widespread through Honduras, Guatemala, Belize, and Mexico, *Diplocentrus*, with approximately 47 described species, is the largest genus in the family. Five species occur in the southern parts of Arizona, New Mexico, and Texas.

Diplocentrids generally are not considered dangerous. However, stings from the Middle Eastern species *Nebo hierichonticus* may cause mild, local hemorrhages and slight necrosis.

Scorpionidae

As the oldest recognized scorpion family, this group once included all known scorpions. Over the years, its scope has been reduced. Currently, the family is compact and homogenous, with four genera and 114 species, and includes some of the world's largest and most formidable-looking scorpions. All are heavy-bodied with large, powerful pedipalps. Some members of the Asian genus *Heterometrus* reach lengths of 16 cm or more. *Pandinus imperator* from West Africa often is cited as one of the largest scorpions, occasionally attaining a body length of 18 cm and weighing up to 32 g as nongravid females (Figure 22.2). This large, black scorpion commonly is sold in pet stores.

The 19 subspecies of *Scorpio maurus* are distributed across northern Africa (Senegal, Mauritania, Morocco, Algeria, Tunisia, Libya, Egypt) and the Middle East (Turkey, Lebanon, Syria, Iraq, Kuwait, Iran, Israel, Jordan, Saudi Arabia, Qatar, Yemen). The genus *Pandinus* (22 species) is distributed across central Africa, (Senegal, Gambia, Guinea-Bissau, Guinea, Ivory Coast, Burkina Faso, Ghana, Togo, Nigeria, Cameroon, Equatorial Guinea, Gabon, Congo, Democratic Republic of Congo, Sudan, Eritrea, Ethiopia, Somalia, Kenya, Tanzania, Malawi, Zimbabwe, Mozambique). The genus also is represented on the nearby coasts of Saudi Arabia and Yemen. *Opistophthalmus*, with its 60 species, is found throughout southern Africa (Tanzania, Angola, Zambia, Zimbabwe, Mozambique, Namibia, Botswana, Lesotho, South Africa). The genus *Heterometrus* (32 species) is distributed through India, Sri Lanka, Bangladesh, Myanmar, Thailand, Cambodia, Laos, Vietnam, Malaysia, Singapore, Indonesia, Brunei, and the Philippines.

Though normally not considered dangerous, most species can deliver stings that can cause mild to severe localized discoloration, swelling, and tissue damage. The stings of many species of *Heterometrus* can cause serious localized hemorrhaging and blistering.

MORPHOLOGY

The scorpion body (Figure 22.3) is divided into two major parts, the **prosoma** (cephalothorax) and the **opisthosoma** (abdomen). The opisthosoma is segmented and further divided into the **mesosoma** (preabdomen) and the more slender, tail-like **metasoma** (postabdomen). The metasoma bears at its posterior end a stinging structure called the **telson**.

The dorsal aspect of the prosoma consists of a single sclerotized plate called the **carapace**. It is marked by various furrows, grooves, depressions, and keels that indicate internal apodemes and other surfaces for the attachment of muscles associated with the legs and other structures. A pair of median eyes is located on an ocular tubercle along the midline of the carapace. Two additional groups of smaller lateral eyes

Figure 22.2 Emperor scorpion, *Pandinus imperator* (Scorpionidae). A relatively harmless West African scorpion, despite its large size and intimidating appearance; commonly sold as pets. (Photo by João P. Burini)

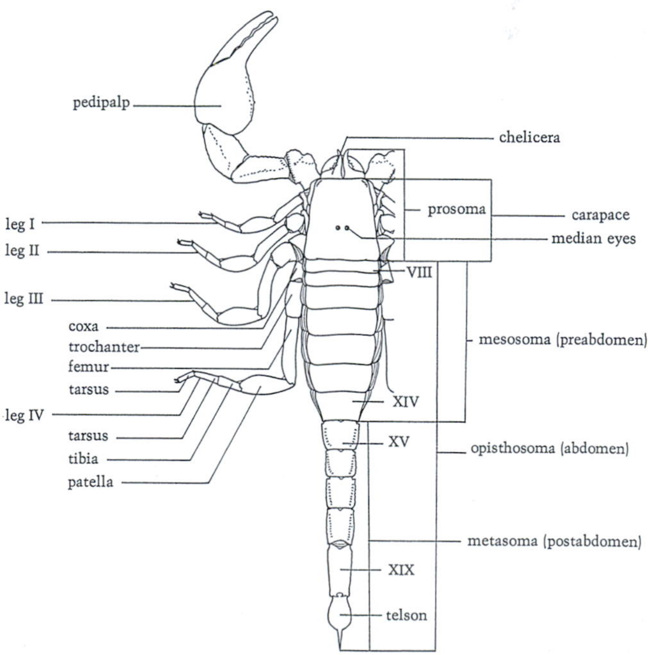

Figure 22.3 Scorpion morphology, adult, dorsal view. (Keegan, 1980, with permission of University Press of Mississippi)

are situated at the anterolateral margins of the carapace. There may be as many as five pairs of lateral eyes, whereas eyes may be lacking altogether in some cave-dwelling species. The ventral aspect of the prosoma consists of a posteromedian **sternum** and the broad coxae of the legs. The sternum is basically pentagonal in shape but may appear to be more triangular in some taxa.

The mesosoma is divided dorsally into seven apparent segments, each bearing a tergite. Ventrally there are five sternites, the first four of which bear a pair of spiracles (Figure 22.4). The genital aperture is located anteriorly between the coxae of the fourth pair of legs and is covered by a pair of small plates called the **genital opercula**. The genital opercula are commonly fused in females, but not in males. Just behind the sternum is a pair of appendages unique to scorpions called **pectines** (sing., pecten) (Figure 22.5). These structures function primarily as mechanoreceptors that can sense the nature of the substrate and apparently aid in detecting substrate vibrations. Although quite variable when all scorpion genera are considered, the pecten typically consists of three anterior marginal lamellae, the middle lamellae, a row of triangular fulcra, and a posterior series of fleshy lamellae called **pectinal teeth**. The ventral surface of each pectinal tooth is covered with mechanoreceptors in the form of tiny sensory pegs visible only at high magnification. Up to 1,200 sensory pegs per pectinal tooth have been reported in *Leiurus quinquestriatus*.

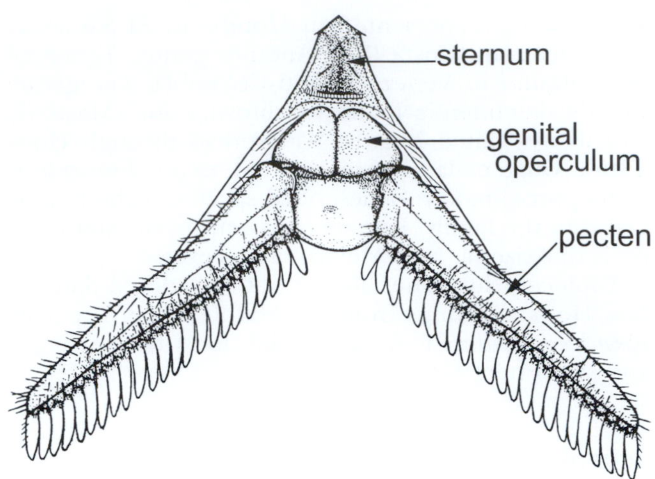

Figure 22.5 Sternum, genital opercula, and pectines of *Centruroides vittatus* (Buthidae); located on venter of scorpion at level of third and fourth pairs of legs (Keegan, 1980).

The metasoma or "tail" is divided into five segments, plus the telson (Figure 22.6). The segments are well sclerotized and may bear longitudinal ridges or **keels** along their dorsal, lateral, and ventral surfaces (Figure 22.6). The nature and location of these keels can serve as important taxonomic characters. The telson consists of a bulbous base, called the **vesicle** or **ampulla**, and a curved, sharply pointed terminal spine, the **aculeus**. Just below the aculeus the telson also may

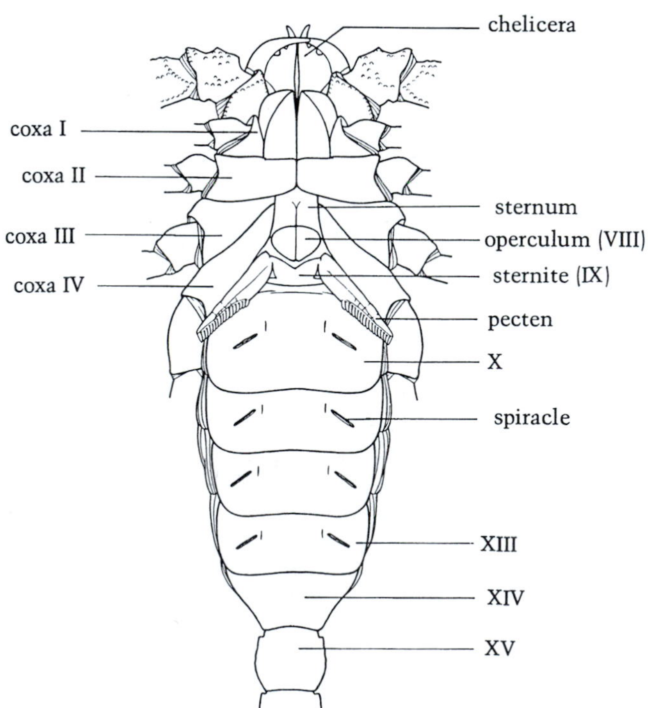

Figure 22.4 Scorpion morphology, adult, ventral view; with legs beyond coxae and most of metasoma (postabdomen) not shown. (Keegan, 1980, with permission of University Press of Mississippi)

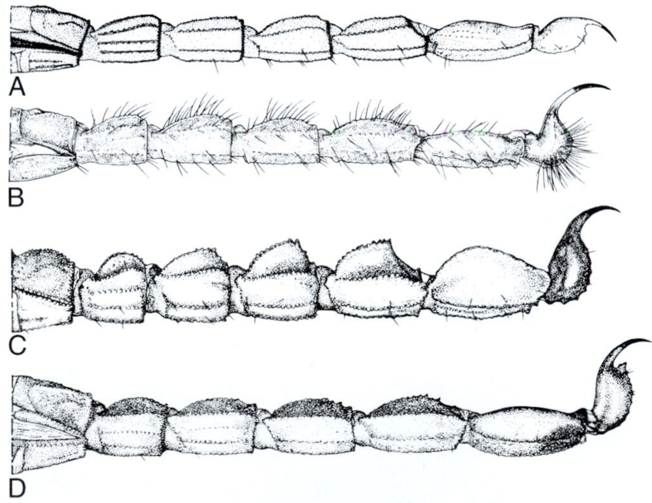

Figure 22.6 Metasoma (postabdomen, including terminal telson) of scorpions, showing various morphological features. (A) *Centruroides vittatus* (Buthidae), with relatively slender segments and inconspicuous keels; (B) *Leiurus quinquestriatus* (Buthidae), with slender segments and numerous, long sensory hairs; (C) *Androctonus australis* (Buthidae), with enlarged, robust segments and prominent keels; (D) *Tityus serrulatus* (Buthidae), with well-developed keels and distinct subaculear tubercle on telson. (Adapted from Keegan, 1980, with permission of University Press of Mississippi)

bear a small, median **subaculear tubercle** or accessory spine (Figure 22.6D). The vescicle contains a pair of **venom glands** and associated musculature. The venom glands may be simple and sac-like or more complexly folded with pouch-like extensions that greatly increase the surface area of the secretory epithelium. The venom is discharged by contraction of the muscles surrounding the glands, which compress the glands against the vesicle wall. The venom is forced out through the pair of venom ducts that open near the tip of the aculeus.

The prosomal appendages of scorpions are a pair of chelicerae (Figure 22.7B), a pair of pincer-like pedipalps (Figure 22.7A), and four pairs of walking legs. Each chelicera consists of three segments. The terminal, third segment serves as a movable finger that opposes the second segment, or hand (*manus*), bearing an anterior apophysis called the fixed finger. Both fingers are armed with teeth that facilitate the grasping and tearing of food. The pedipalps are six-segmented, each consisting of a coxa, trochanter, femur, patella (genu), tibia, and tarsus. The distal-most segment (tarsus) forms a movable finger, which opposes the fixed finger of the tibia, or hand. The various numbers and arrangements of keels, tubercles, denticles, granules, and trichobothria (sensory hairs) on the hand, patella, and femur of the pedipalp are important taxonomically. The walking legs consist of the same six segments as the pedipalps, plus a seventh terminal segment, the **pretarsus**, which bears a pair of lateral claws and a single, small median claw. The tarsus is divided into two tarsomeres. The presence of tibial spurs and pedal spurs on the tarsomeres can be helpful in identifying certain groups of scorpions.

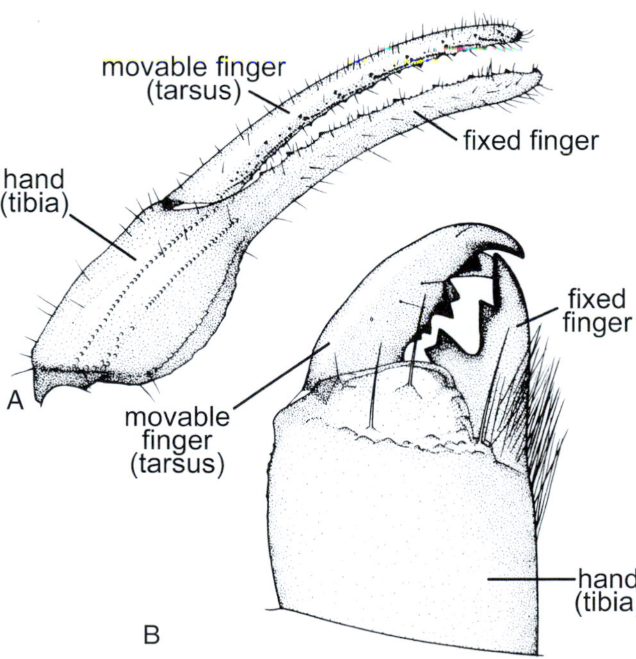

Figure 22.7 *Centruroides vittatus* (Buthidae). (A) Chelate (pincer-like) tibia and tarsus of pedipalp, used for seizing prey; (B) Chelicera, showing the movable tarsus and fixed finger of the tibia, used to crush and tear prey. (Adapted from Keegan, 1980, with permission of University Press of Mississippi)

Although scorpions are sexually dimorphic, it generally takes a specialist to reliably distinguish males and females. There are no uniform gender-specific, external morphological characters for determining the sex that apply to all scorpions. Instead, determinations are based on comparative morphological differences that often require familiarity with gender-related traits within conspecifics. In general, however, males are smaller and less robust compared to females of the same species. Males tend to be relatively more slender than females, but there are many exceptions. In some genera, the pedipalpal chelae of males are longer and more slender, whereas in other genera they are shorter and thicker. The males of some species have special depressions for accepting the female's pedipalpal fingers during mating. Usually, the metasoma of males is longer and more slender than that of females; however, it is sometimes shorter and thicker, being different with each species. The pectines are often strikingly different, with the males having larger pectines with more pectinal teeth. In some cases the males can be recognized by the presence of a pair of genital papillae protruding posteriorly from beneath the genital operculum.

LIFE HISTORY

Although scorpions as a group differ in details relating to their biology and behavior, they follow a similar life-history pattern. Females are viviparous, giving birth to young following a developmental period that varies from two to 18 months, depending on the species. Scorpions are unusual among arachnids in this respect, the only other known case being one family of mites, the Pyemotidae. The average brood size for all scorpions is about 26, but it can range from as few as one up to 105. The sex ratio at birth is about equal, even though this ratio later shifts significantly toward females after reaching maturity. Two species, *Tityus serrulatus* and *Liocheles australasiae,* are reported to be parthenogenic and produce only female offspring.

Following their birth, the newborns immediately crawl onto their mother's back where they remain through the first instar without feeding. If removed during this period, they die without successfully molting. A major contributing factor seems to be their dependence on obtaining water from their mother's cuticle; they are prone to desiccation due to the permeability of their cuticle at this critical time. In most cases, dispersal of the young occurs shortly after they molt to the second instar, usually within three to 14 days after birth. In other cases, the young may remain in the burrow with their mother, where she cares for them and may even feed them captured prey. This type of social behavior appears to be common in the scorpionoids and has been described in other taxa (e.g., *Euscorpius*).

Most species molt five or six times before becoming adults, although the number of molts varies from four to nine, depending on the species. Males generally undergo fewer molts than females of the same species.

No molting occurs in the adults. Maturity is reached in as few as six months in some of the smaller species (e.g., *Centruroides* spp.) but may take as many as three to seven years in some of the largest species (e.g., *Pandinus* spp.). Buthid scorpions develop the fastest, averaging about 18 months, whereas the mean developmental time for all other scorpions is about three years. The longevity of most scorpions is probably about two to five years. It seems likely that scorpions in temperate environments may take longer to mature (one molt per year) and thus live longer overall due to short growing seasons. Scorpions may live still longer in captivity. The longest reliable report is eight years for *Pandinus gambiensis.*

BEHAVIOR AND ECOLOGY

Scorpions are well adapted for surviving in a wide range of habitats, including deserts, grasslands, savannas, and both temperate and tropical forests. In addition, they are found from intertidal zones at sea level to snow-covered mountains at elevations of over 5,500 meters, and in cave systems at depths of more than 800 meters. They can tolerate highly varied environmental conditions, including extremes of temperature, both heat and cold, complete emersion in water for hours, and prolonged periods of drought and starvation. In large part, these adaptations are due to behavioral thermoregulation, low metabolic rates, and high efficiency in conserving water. To moderate their body temperatures, scorpions are typically nocturnal, retreating to the protection of burrows and other sheltered sites during the daytime. They experience minimal water loss via their cuticle, spiracles and booklungs, while excreting nitrogenous wastes in almost insoluble forms such as guanine, xanthine, and uric acid. Similarly, their feces are extremely dry.

Most scorpions live on or very near the ground, where they typically are found under objects, in forest litter, or excavated burrows. The major exception is the large and important family Buthidae in which the species are often excellent climbers. They commonly are found under the bark of trees, in the tops of palms and other plants, and crevices of rocky cliffs. Upon entering houses, these species are likely to be seen on the walls and even ceilings, not infrequently gaining access to the upper floors of multistory buildings. Such climbers include some of the most venomous scorpions, notably members of the genus *Centruroides.* Even some of the common vaejovid scorpions, such as *Vaejovis carolinianus* in the southeastern United States, are excellent climbers. They frequently enter homes where they may be seen on walls or clinging upside-down on ceilings.

Scorpions feed on a variety of prey, notably soft-bodied insects and arachnids. Heavily sclerotized insects and other invertebrates such as certain isopods often are rejected. Common prey items include spiders, solpugids, other scorpions, millipedes, centipedes, gastropods, and other invertebrates. The larger scorpions also will attack and feed on small vertebrates such as lizards, snakes, and rodents. Owing to poor vision, scorpions depend primarily on their sensory hairs and their ability to sense ground vibrations as a means of detecting, locating, and recognizing suitable prey. Using mechanoreceptors on their tarsi, they can detect potential prey up to 15 cm away. Some arboreal scorpions even can capture flying insects that approach close enough for them to detect via air movements with the trichobothria on their pedipalps.

Scorpions with large, robust pedipalps can often subdue their prey with little or no use of their venom. Smaller species with weaker, more slender pedipalps are far more dependent on stinging their prey upon seizing it with their chelate pincers. The thrust with their stinger is usually carefully delivered to penetrate the softer areas of the prey's integument between sclerites or other hard body parts. After locating a suitable site, sufficient venom is injected to immobilize the prey, following which the sting is withdrawn. This is in strong contrast to the defensive strikes directed toward threatening enemies in which the telson lashes forward to sting its target, inject the venom, and be quickly withdrawn.

Once a scorpion has captured a prey item, it crushes it with the coxal bases of its pedipalps and the first two pairs of legs while tearing at it with its chelicerae. Digestive juices from specialized glands in the gut flow through a channel formed by the coxae of the pedipalps and anterior legs. Following extra-oral digestion, the semidigested food is drawn into the mouth assisted by the chelicerae; undigested parts are trapped by setae in the preoral cavity and expelled. The feeding process is slow, taking as long as 2.5 hours to consume an item such as a blow fly. Owing to the efficiency in storing digested food in the hepatopancreas, a well-fed scorpion can survive for months without further feeding.

When it comes time to seek a mate, the female recognizes conspecific males by a behavior called **juddering**, in which the male displays a series of shaking movements, rocking back and forth with his pectines spread out and quivering. The resulting vibrations are communicated via the substrate to the female. There is also evidence to suggest that pheromones may be involved in sex recognition in at least some species. Having located one another, the male initiates courtship by grasping the female's pedipalps with his own and guiding her through a complex courtship behavior in the form of a mating dance, or **promenade**. During the promenade they may engage in cheliceral massages or kissing, in which the male grasps the female's chelicerae with his chelicerae and gently kneads them, apparently serving to suppress her aggressiveness. In many species the male actually stings the female, usually in the tibial joint of the pedipalp, where the inserted sting may be held for three to 20 min or longer. This appears to reduce her aggression and render her more docile during the courtship.

Throughout the promenade the male uses his pectines to sweep back and forth across the ground, sensing the substrate to determine if it is suitable for depositing a spermatophore. Upon finding an acceptable site, he extrudes a complexly structured spermatophore from his genital aperture and attaches it to the substrate in an upright position, with the sticky basal plate firmly anchoring it in place. He then guides the female over the spermatophore to make contact with her genital valves. As the spermatophore bends under pressure from her mesosoma, the sperm is released directly into her genital tract. Contact with the spermatophore may last anywhere from a few seconds to several minutes. Once insemination is completed, the male abruptly disengages from the female and effects an immediate escape, lest he be attacked and eaten by his no longer receptive mate.

When the gravid female is ready to give birth, she assumes a position known as **stilting**, in which she raises the anterior part of her body and forms a birth basket with her pedipalps and first two pairs of legs. While she maintains this posture, the young emerge one at a time from the genital aperture and drop into the birth basket. From there they clamber onto their mother's back. The birth time for an individual varies from one minute up to about an hour, with the total birth process lasting from less than 12 hours to as long as three days.

More than 150 species of predators have been reported to feed on scorpions. Among the most common are birds and lizards, followed by various mammals, frogs, toads, and snakes. Several invertebrates also are natural enemies, including spiders, solpugids, ants, centipedes, and other scorpions. In many cases scorpions are resistant to their own venoms; however, they readily fall prey to attacks by other species. Scorpions comprise a significant part of the diet of burrowing owls, elf owls, and grasshopper mice. They also are hosts for mermithid nematodes and the ectoparasitic larvae of certain mites.

PUBLIC HEALTH IMPORTANCE

Scorpion sting cases can be categorized as two general types: those involving only localized, transitory symptoms usually lasting from a few minutes to several hours, and those involving systemic reactions. Localized responses are characterized by immediate pain followed by moderate swelling at the sting site, often likened to the sting of a wasp or bee. In some cases, the sting may result in a raised, reddened, indurated lesion, even in the case of relatively harmless scorpions (e.g., *Vaejovis carolinianus*). In cases involving cytolytic toxins (e.g., scorpionids and liochelids), swelling may persist up to 72 hours, followed by development of hemorrhages and blood-filled blisters near the sting site. Sloughing of skin may occur, but this varies greatly in severity. Other localized effects include goose flesh, sweating, and muscle spasms near the sting site. In cases of buthid stings, pain usually radiates from the site of the sting up the affected limb. The pain tends to concentrate in the joints, especially the armpits and groin, and often crosses from one side of the body to the other.

In cases of systemic reactions, the clinical signs and symptoms are highly variable, ranging from mild to life-threatening. Systemic reactions commonly are mild and are not necessarily indicators of a serious problem. Often there is no appreciable swelling or discoloration of the skin at the sting site. An intense aching and burning sensation may spread to adjacent tissues, which in turn often throb, sometimes becoming numb. The acute pain at the sting site turns into a chronic, dull pain accompanied by a feeling of numbness around the edge of the sting site, which may persist for one to several days. Numbness in the face, mouth, and throat is fairly common. Muscles may become spasmatic, resulting in muscular twitching, slurred speech, difficulty swallowing, tightness or cramps in the chest and back, rapid heartbeat, and nausea. Often these systemic responses persist less than an hour after the sting and are not considered serious.

In more severe systemic reactions, neurologic effects can lead to profuse sweating and salivation, restlessness, extreme nervousness, respiratory and cardiovascular problems, mental confusion, and convulsions. As the clinical symptoms indicate, the principal components of the venom of dangerous scorpions are neurotoxins. These toxins act on the autonomic, sympathetic, and neuromuscular systems, causing the wide range of systemic reactions reported in sting victims. They act by disrupting the voltage-sensitive sodium and potassium channels of nerves, which in turn causes neural depolarization, prolonged action potentials, repetitive firing, and uncontrolled release of vasodilators and neurotransmitters that affect virtually every major organ system. The effect on neurotransmitters results in depletive release of catecholamines (e.g., adrenaline, noradrenaline) that can severely damage the heart and other organs.

The most commonly reported cause of death in scorpion sting cases is cardiac failure. In other cases, respiratory failure may be the cause, especially in patients with upper respiratory infections or related problems. Death usually occurs several days after envenomation. If symptoms subside during the first two to 12 hours following a sting, the prognosis for recovery is generally good. Mortality rates are quite variable, depending on the species and amount of venom injected. The rates are much higher among children than adults. For further details on the clinical toxicology and symptoms of scorpion stings, see Dehesa-Davila et al. (1995) and Ismail (1995).

Scorpion venom is a very complex mixture of substances that differs significantly among the various taxa, within families and among genera. Differences also occur in different geographic populations of the same species and even within the same population. The toxins are low-molecular-weight proteins that are among the most powerful toxins known. They are comparable in some species to the neurotoxins of certain deadly snakes. Two recognized types of neurotoxins are **α-scorpion toxin**, characteristic of the genera *Androctonus* (Figure 22.8), *Leiurus* (Figure 22.9), and *Buthus*; and **ß-scorpion toxin**, characteristic of *Centruroides*. *Tityus* species appear to have both types. The effects of envenomation by any given scorpion species

Figure 22.8 *Androctonus australis* (Buthidae), adult. A highly toxic scorpion found in North Africa. (Photo by S. A. Stockwell)

Figure 22.9 *Leiurus* sp. (Buthidae). A member of this genus, *L. quinqestriatus* in North Africa and the eastern Mediterranean region, is regarded as one of the most dangerous scorpions in the world. (Photo by S. A. Stockwell)

can differ significantly, owing to a wide range of contributing factors. These include the quantity of venom injected and the age, size, and general health of the victim.

The sting of most scorpions usually requires no special treatment, although the application of ice to the sting site helps to relieve local pain. Incisions, as used in cases of poisonous snake bites, should never be made. Nor is the use of most drugs recommended in uncomplicated cases because antihistamines, steroids, analgesics, and sedatives usually have little or no effect. In more severe cases involving systemic reactions, medical attention should be sought immediately.

Substances that have been found to be effective in treating scorpion stings are atropine to counter effects on the parasympathetic system, calcium gluconate given intravenously to relieve muscle spasms, and sodium phenobarbital administered intravenously to prevent convulsions. Insulin also has been reported to be beneficial in treating cases in India. Morphine and demerol generally are not recommended as pain relievers because of their tendency to act as synergists, increasing the toxicity of certain venoms (e.g., *Centruroides exilicauda*).

Antivenins are recommended where available. However, caution in the use of antivenins must be noted since many antivenins are of poor quality and often are administered at doses far below the level required to produce effective results. They may adequately neutralize the larger venom peptides but not necessarily the more important small-molecular-weight toxins. To be effective, they must be administered within the first hour following the sting. Even then, the cessation of systemic symptoms within about an hour may not necessarily reflect the use of the antivenin, since this is the same period of time that symptoms often subside without treatment of any kind. Another limitation is the fact that most scorpion antivenins are very specific, often for a single species, and are produced only on a limited regional basis (Theakston and Warrell, 1991; Lucas and Meier, 1995).

Scorpions of Medical Importance

Some of the more dangerous species of scorpions for which toxicity data have been reported are shown in Table 22.1. Based on mammalian toxicity, *Leiurus quinquestriatus*, *Androctonus australis*, and *A. mauretanicus* generally are recognized as the most venomous. *Leiurus quinquestriatus* is known in the Middle East and Sahel region of northern Africa as the yellow scorpion, and in the Sudan as the Omdurman scorpion. The *Androctonus* species are commonly called fat-tailed scorpions, referring to the marked thickness and width of their postabdominal segments. *Androctonus australis*, which occurs primarily in arid mountainous regions, causes more deaths than any other species in North Africa. Based on the numbers of cases and fatalities, it is probably the most deadly scorpion worldwide. *Buthus occitanus*, a widely distributed species in the Middle East and North Africa, is the only medically important species in southern Europe. Its toxicity varies markedly in different parts of its range, apparently reflecting different subspecies. The Indian red scorpion (*Hottentotta tamulus*) is the most medically important scorpion on the Indian subcontinent.

The most important venomous genera in the New World are *Centruroides* and *Tityus*. *Centruroides* species occur primarily in Mexico, Central America, and the West Indies. They often are called bark scorpions due to their habit of hiding under loose tree bark or in crevices of dead logs and trees. They commonly are found around domestic settings in piles of wood, stones, bricks, and discarded debris. Stings are likely to occur when they are disturbed in their hiding places or when they enter homes at night in search of prey. *Centruroides exilicauda* (formerly *C. sculpturatus*) often is cited as the most dangerous scorpion in the United States. Deaths from its sting, however, are rare. In fact, the toxicity and effects of its venom are very similar to that of the striped scorpion *Centruroides vittatus*. The latter is the most widely distributed of all American scorpions, occurring in the southern United States from the Mississippi River west to New Mexico, as far north as Kansas and Missouri, and well south into northeastern Mexico.

Among the more than two dozen *Centruroides* species and subspecies in Mexico, the following taxa are of particular concern because of the seriousness of sting cases: *C. elegans*, *C. exilicauda*, *C. infamatus*, *C. noxius*, and *C. suffusus*. All are closely related to one another (Exilicauda group). The most notorious are *C. suffusus* and *C. limpidus*, both of which are capable of causing human deaths. Despite its small size, usually less than 5 cm, *C. noxius* is considered very venomous.

Tityus species are similar to *Centruroides* species in size, general appearance, and behavior. As a group of over 100 species, they occur throughout South America and the Caribbean Basin. In most places where they occur, all *Tityus* species are considered dangerous. The most venomous is the Brazilian species *T. serrulatus*, which is common in urban areas and readily enters homes. Second only to *T. serrulatus* in its medical importance is another house-infesting scorpion in Brazil, *T. bahiensis* (Figure 22.10). Related *Tityus* species that also are highly venomous include *T. cambridgei*, a forest-dwelling species in the Amazon Basin and northern South America, *T. trinitatis*, which can be a serious problem in coconut groves and cane fields in Venezuela and Trinidad, and *T. trivittatus*, a house-infesting scorpion in Argentina.

Figure 22.10 *Tityus bahiensis* (Buthidae). A dangerous scorpion in Brazil that commonly enters homes. (Photo by João P. Burini)

VETERINARY IMPORTANCE

There is little evidence to indicate that scorpions pose any significant threat to domestic animals, including cats and dogs. One can only assume that stinging encounters do occur, perhaps even commonly, but that the effects are seldom serious enough to draw attention of the owners.

PREVENTION AND CONTROL

Pesticides are not generally recommended for controlling scorpions indoors or for preventing their entering homes. Instead, appropriate measures can be taken to "scorpion-proof" buildings or otherwise significantly reduce the prospects of them entering homes. Entry can be discouraged by raising the floor level at least 20 cm above the ground. A single step to reach the threshold is better than multiple steps and should be separated from the wall of the structure by a gap of 6 cm or more. The installation of a horizontal row of glazed ceramic tiles on the vertical surfaces of steps and around the entire perimeter of a building also can provide a barrier that scorpions cannot readily climb. Smooth exterior wall surfaces, such as planed cement, further impede their climbing ability. Worn weather stripping around doors and windows should be replaced, and potential entry sites around water pipes and electrical conduits in foundations should be sealed. To prevent scorpions from gaining access to the roofs of structures, a row of ceramic tiles can be applied to the outer walls just below the roof line.

Scorpions can be discouraged from frequenting the immediate vicinity of homes by trimming plantings that touch buildings and removing piles of firewood, lumber, bricks, and other materials that serve as harborage. The use of coarse bark mulches around plants near the foundation of buildings should be avoided for the same reason.

In areas where climbing scorpions commonly infest homes, measures can be taken to reduce the risk of envenomation. A sheet of muslin or other suitable cloth can be suspended from the ceiling over the sleeping quarters to catch any scorpions that might drop from overhead structures. Mosquito netting over beds affords similar protection. Regularly shaking out clothing and footwear before putting them on is highly recommended.

In temperate regions, the greatest number of complaints of scorpions entering homes is often seasonal, most commonly in the early spring and late fall. Heavy or frequent rains in the spring can saturate the soil and ground litter around building foundations, driving scorpions indoors as they search for drier sites. With the onset of colder weather in the late fall, scorpions are similarly apt to find their way indoors while seeking warmer temperatures. Another circumstance that contributes to scorpion problems is the construction of new homes or subdivisions in previously undisturbed woodlands and other habitats where scorpions are abundant. The clearing of such areas and the associated disturbance of ground litter often causes displaced scorpions to wander extensively. In the process they frequently find their way inside nearby homes. Sealing or blocking possible access sites is the only practical means of preventing their entry.

REFERENCES AND FURTHER READING

Anderson, R. C. (1975). Scorpions of Idaho. *Tebiwa, 18,* 1–17.

Balozet, L. (1971). Scorpionism in the old world. In W. Bücherl, & E. E. Buckely (Eds.), *Venomous Animals and their Venoms* (Vol. 3, pp. 349–371). New York: Academic Press.

Bettini, S. (Ed.). (1978). *Arthropod Venoms.* Berlin: Springer-Verlag.

Briggs, D. E. G. (1987). Palaeontology: Scorpions take to the water. *Nature, 326,* 645–646.

Bücherl, W. (1978). Systematics, distribution, biology, venomous apparatus, etc. of Tityinae; venom collection, toxicity, human accidents and treatment of stings. In S. Bettini (Ed.), *Arthropod Venoms* (pp. 371–378). Berlin: Springer-Verlag.

Couraud, F., & Jover, E. (1984). Mechanisms of action of scorpion toxins. In A. T. Tu (Ed.), *Handbook of Natural Toxins* (Vol. 2, pp. 659–678). New York: Dekker.

Cloudsley-Thompson, J. L. (1990). Scorpions in mythology, folklore, and history. In G. A. Polis (Ed.), *The Biology of Scorpions* (pp. 462–485). Stanford, CA: Stanford University Press.

Cloudsley-Thompson, J. L. (1993). Spiders and scorpions (Araneae and Scorpiones). In R. P. Lane, R. W. Crosskey (Eds.), *Medical Insects and Arachnids* (pp. 659–682). Chapman & Hall.

Coddington, J. A., Larcher, S. F., & Cokendolpher, J. C. (1990). The systematic status of Arachnida, exclusive of Acari, in North America north of Mexico. In M. Kosztarab, & C. W. Schaefer (Eds.), *Systematics of the North American Insects and Arachnids: Status and Needs* (pp. 5–20). Virginia Agricultural Experiment Station Information Series 90–1. Blacksburg: Virginia Polytechnic Institute and State University.

Dehesa-Davila, M., Alagon, A. C., & Possani, L. D. (1995). Clinical toxicology of scorpion stings. In J. Meier, & J. White (Eds.), *Handbook of Clinical Toxicology of Animal Venoms and Poisons* (pp. 221–238). Boca Raton., FL: CRC Press.

Diniz, C. R. (1978). Chemical and pharmacological aspects of Tityinae venoms. In S. Bettini (Ed.), *Arthropod Venoms* (pp. 379–394). Berlin: Springer-Verlag.

Dupré, G. (2007). Conspectus Genericus Scorpionorum 1758–2006 (Arachnida: Scorpiones). *Euscorpius, 50,* 1–31.

El-Asmar, M. F. (1984). Metabolic effect of scorpion venom. In A. T. Tu (Ed.), *Handbook of Natural Toxins* (Vol. 2, pp. 551–576). New York: Dekker.

El-Ayeb, M., & Delori, P. (1984). Immunology and immunochemistry of scorpion neurotoxins. In A. T. Tu (Ed.), *Handbook of Natural Toxins* (Vol. 2, pp. 607–638). New York: Dekker.

Efrati, P. (1978). Symptomatology and treatment of Buthinae stings. In S. Bettini (Ed.), *Arthropod Venoms* (pp. 312–316). Berlin: Springer-Verlag.

Ennik, F. (1972). A short review of scorpion biology, management of stings, and control. *California Vector Views, 19,* 69–80.

Fet, V., & Selden, P. A. (Eds.). (2001). Scorpions. In *memoriam Gary A. Polis.* Burnham Beeches, Bucks: British Arachnological Society.

Fet, V., Sissom, W. D., Lowe, G., & Braunwalder, M. E. (2000). *Catalog of the scorpions of the world (1758–1998).* New York: The New York Entomological Society.

Fet, V., Soleglad, M. E., Neff, D. P. A., & Stathi, I. (2004). Tarsal armature in the superfamily Iuroidea (Scorpiones: Iurida). *Revista Ibérica de Arachnología, 10,* 17–40.

Gertsch, W. J., & Allred, D. M. (1965). Scorpions of the Nevada test site. *Brigham Young University Bulletin, Biological Services, 6*(4), 1–15.

Gertsch, W. J., & Soleglad, M. E. (1972). Studies of North American scorpions of the genera *Uroctonus* and *Vejovis* (Scorpionida, Vejovidae). *Bulletin of the American Museum of Natural History, 148,* 551–608.

Goyffon, M., & Kovoor, J. (1978). Chactoid venoms. In S. Bettini (Ed.), *Arthropod Venoms* (pp. 395–418). Berlin: Springer-Verlag.

Gueron, M., & Ovsychcher, I. (1984). Cardiovascular effects of scorpion venoms. In A. T. Tu (Ed.), *Handbook of Natural Toxins* (Vol. 2, pp. 639–658). New York: Dekker.

Hassan, F. (1984). Production of scorpion antivenin. In A. T. Tu (Ed.), *Handbook of Natural Toxins* (Vol. 2, pp. 577–606). New York: Dekker.

Hjelle, J. T. (1972). Scorpions of the northern California coast ranges (Arachnida: Scorpionida). *Occasional Papers of the California Academy of Science, 92,* 1–59.

Ismail, M. (1995). The scorpion envenoming syndrome. *Toxicon: Official Journal of the International Society on Toxinology, 33,* 825–858.

Johnson, J. D., & Allred, D. M. (1972). Scorpions of Utah. *The Great Basin Naturalist, 32,* 154–170.

Kaestner, A. (1968). Order scorpiones, scorpions. In *Invertebrate Zoology, Vol. II. Arthropod Relatives, Chelicerata and Myriapoda* (pp. 101–114). New York: Interscience.

Keegan, H. L. (1980). *Scorpions of Medical Importance.* Jackson: Univ, Press of Mississippi.

Lucas, S. M., & Meier, J. (1995). Biology and distribution of scorpions of medical importance. In J. Meier & J. White (Eds.), *Handbook of Clinical Toxicology of Animal Venoms and Poisons* (pp. 205–219). Boca Raton, FL: CRC Press.

Muma, M. H. (1967). *Scorpions, whip scorpions and wind scorpions of Florida. In: Arthropods of Florida and Neighboring Land Areas* (Vol. 4). Gainesville, FL: Florida Dept. Agric.

Polis, G. A. (Ed.). (1990). The Biology of Scorpions. Stanford, CA: Stanford Univ. Press.

Possani, L. D. (1984). Structure of scorpion toxins. In A. T. Tu (Ed.), *Handbook of Natural Toxins* (Vol. 2, pp. 513–550). New York: Dekker.

Prendini, L., & Wheeler, W. C. (2005). Scorpion higher phylogeny and classification, taxonomic anarchy, and standards for peer review in online publishing. *Cladistics, 21,* 446–494.

Rankin, W., & Walls, J. G. (1994). *Tarantulas and Scorpions: Their Care in Captivity.* Neptune City, NJ: T.F.H. Publications, Inc.

Shulov, A., & Levy, G. (1978). Systematics and biology of Buthinae. In S. Bettini (Ed.), *Arthropod Venoms* (pp. 309–312). Berlin: Springer-Verlag.

Simard, J. M., & Watt, D. D. (1990). Venoms and toxins. In G. A. Polis (Ed.), *The Biology of Scorpions* (pp. 414–444). Stanford, CA: Stanford Univ. Press.

Sissom, W. D. (1990). Systematics, biogeography, and paleontology. In G. A. Polis (Ed.), *The Biology of Scorpions* (pp. 64–160). Stanford, CA: Stanford Univ. Press.

Soleglad, M. E., & Sissom, W. D. (2001). Phylogeny of the family Euscorpiidae Laurie, 1896: A major revision. In V. Fet & P. A. Selden (Eds.), *Scorpions 2001. In memoriam Gary A. Polis* (pp. 25–111). Burnham Beeches, Bucks: British Arachnological Society.

Soleglad, M. E., & Fet, V. (2003). High-level systematics and phylogeny of the extant scorpions (Scorpiones: Orthosterni). *Euscorpius, 11,* 1–175.

Stahnke, H. L. (1940). The scorpions of Arizona. *Iowa State College Journal of Science, 15,* 101–103.

Stahnke, H. L. (1978). *The genus Centruoides* (Buthidae) and its venom. In S. Bettini (Ed.), *Arthropod Venoms* (pp. 279–308). Berlin: Springer-Verlag.

Stockwell, S. A. (1992). Systematic observations on North American Scorpionida with a key and checklist of the families and genera. *Journal of Medical Entomology, 29,* 407–422.

Theakston, R. D. G., & Warrell, D. A. (1991). Antivenoms: A list of hyperimmune sera currently available for the treatment of envenoming by bites and stings. *Toxicon: Official Journal of the International Society on Toxinology, 29,* 1419–1470.

Tu, A. T. (Ed.). (1984). Insect poisons, allergens, and other invertebrate venoms. In *Handbook of Natural Toxins* (Vol. 2). New York: Dekker.

Wainschel, J., Russell, F. E., & Gertsch, W. S. (1974). Bites of spiders and other arthropods. In H. F. Conn (Ed.), *Current Therapy* (pp. 865–867). Philadelphia: W. B. Saunders Company.

Whittemore, F. W., & Keegan, H. L. (1963). Medically important scorpions in the Pacific area. In H. L. Keegan & M. V. Macfarlane (Eds.), *Venomous and Poisonous Animals and Noxious Plants of the Pacific Region* (pp. 107–110). New York: Permagon Press.

Williams, S. C. (1969). Birth activities of some North American scorpions. *Proceedings of the California Academy of Sciences Series, 4*(37), 1–24.

Williams, S. C. (1976). The scorpion fauna of California. *Bulletin of the Society for Vector Ecology, 3*, 1–4.

Williams, S. C. (1980). Scorpions of Baja California, Mexico and adjacent islands. *Occasional Papers of the California Academy of Science, 135*, 1–127.

Zlotkin, E., Miranda, F., & Rochat, H. (1978). Chemistry and pharmacology of Buthinae scorpion venoms. In S. Bettini (Ed.), *Arthropod Venoms* (pp. 317–370). Berlin: Springer-Verlag.

Chapter

23

Solpugids (Solifugae)

Gary R. Mullen

Solpugids are usually yellow or brownish in color and rather hairy. The body length varies from 1 to 7 cm, with the largest species having a leg span up to 12 cm. The prosoma and opisthosoma are broadly joined, with the latter being visibly segmented (Figure 23.1). The most prominent structures are the greatly enlarged, powerful pair of chelicerae that are used to seize, crush, and tear apart food. With the exception of *Rhagodes nigrocinctus* in India, solpugids generally are believed to lack distinct venom glands, and to rely primarily on their size and strength to overpower prey. The pedipalps are long and leg-like, each ending in an eversible adhesive organ, rather than claws. The first pair of legs is modified as slender tactile organs that are held outstretched as the solpugid moves about. Unique, mallet-shaped structures call **racquet organs** (malleoli) are borne on the underside of the fourth pair of legs in both sexes. They are innervated and function in chemoreception while probing various substrates, presumably to detect chemical cues associated with food and potential mates.

Figure 23.1 Solpugid in desert of southwestern United States. Despite their greatly enlarged pair of chelicerae and formidable appearance, solpugids lack venom glands and are generally harmless to humans. (Photo by Debbie R. Folkerts)

The order Solifugae includes 12 families, approximately 150 genera, and over 900 species worldwide (Punzo, 1998). They occur most commonly in tropical and subtropical deserts in Africa, the Middle East, western Asia, and the Americas. In Africa they also are found in grasslands and forests. They occur in the United States and southern Europe but not in Australia or New Zealand. The two major families in North America are the Ammotrechidae and Eremobatidae, together represented by 11 genera and approximately 120 species. Most of them occur in the western half of the United States. The exception is *Ammotrechella stimpsoni*, which is found under the bark of termite-infested tree stumps in Florida. For a comprehensive treatment of solpugids, including keys to the families and genera worldwide, see Punzo (1998). For further information on solpugids in the United States, see Muma (1951).

Members of this group are known variously as **solpugids**, **sun-spiders**, **wind-spiders**, **wind-scorpions**, **camel-spiders**, **barrel-spiders**, **false-spiders**, and **romans**. Local names in the United States include **bulldozer-spiders** in the Big Bend area of Texas and **sand puppies** in Wyoming. They also are known by the British terms **jerrymander** and **jerrymunglum**. In Mexico they are called *mata venado* ("deer killer") in the mistaken belief that they can kill large animals. In southern Africa solpugids are called **hair-cutters** and **beard-cutters**. According to local lore, females are attracted to the hair of sleeping humans and other animals, which they clip with their chelicerae and carry to their burrows or other retreats to line their nests in preparation for egg laying.

Despite their common names, they do not bear a close resemblance to either spiders or scorpions, although they occur primarily in arid habitats where these other arachnids are found. They are typically nocturnal, hiding during the day under stones and in crevices, or burrowing into loose soil. The name "sun-spider" refers to some species that are active during the daytime. The name "wind-scorpion" reflects their peculiar, rapid movement as they run about the surface of desert sands hunting prey; they give the appearance of being blown across the sand and have been likened by some to tumbleweeds. The name "camel-spider" refers to the arch-shaped plate on the dorsum of the prosoma of many species. Solpugids feed primarily on insects, spiders, and scorpions. The larger species, however, are known to attack and kill small lizards, mice, and birds.

Solpugids will readily attack humans and other animals when provoked. Despite their formidable appearance and aggressive posturing, their bites usually are not serious. However, the larger species can inflict severe wounds with their powerful chelicerae. In one case involving United States military personnel in the Persian Gulf, an individual was bitten on the lip and required 10 stitches to close the wound (Conlon, 1991). The greatest concern is usually preventing secondary infections, which can lead to painful swellings, necrosis of tissues surrounding the bite site, and gangrene.

REFERENCES AND FURTHER READING

Aruchami, M., & Sundara-Rajulu, G. (1978). An investigation on the poison glands and the nature of the venom of *Rhagodes nigrocinctus* (Solifugae: Arachnida). *National Academy of Science Letters (India), 1,* 191–192.

Cloudsley-Thompson, J. L. (1958). *Spiders, Scorpions, Centipedes and Mites.* New York: Pergamon.

Cloudsley-Thompson, J. L. (1992). Solifugae and keeping them in captivity. In J. E. Cooper, P. Pearce-Kelly, & D. L. Williams (Eds.), *Arachnida. Symposium on Spiders and Their Allies, London (1987)* (pp. 52–56). Keighley: Chiron Publications.

Conlon, J. M. (1991). Vectors & war. Part 2 Desert storm. *Wing Beats Florida Mosquito Control Association, 22,* 16–20.

Harvey, M. S. (2003). *Catalogue of the smaller arachnid orders of the world: Amblypygi, Uropygi, Schizomida, Palpigradi, Ricinulei and Solifugae.* Collingwood, Australia: CSIRO Publishing.

Hickin, N. E. (1984). Solifugae. In N. E. Hickin (Ed.), *Pest Animals in Buildings* (pp. 85–86). London: Godwin.

Maury, E. A. (1982). Solifugos de Colombia y Venezuela (Solifugae, Ammotrechidae). *Journal of Arachnology 10,* 123–143.

Muma, M. H. (1951). The arachnid order Solpugida in the United States. *Bulletin of the American Museum of Natural History, 97,* 35–141.

Muma, M. H. (1970). A synoptic review of North American, Central American, and West Indian Solpugida (Arthropoda: Arachnida). *Arthropods of Florida and Neighboring Land Areas* (Vol. 5, pp. 1–62). Gainesville: Florida Department of Agriculture and Consumer Services.

Muma, M. H. (1976). *A review of solpugid families with a annotated list of Western Hemisphere solpugids* Western (Vol 2, pp. 1–33.). New Mexico University, Silver City: Publication of the Office of Research.

Muma, M. H., & Muma, K. E. (1988). *The arachnid order Solpugida in the United States* (Supplement 1, a biological review, pp. 35). Silver City, New Mexico: Southwest Offset.

Punzo, F. (1998). *The Biology of Camel-Spiders (Arachnida, Solifugae).* Dordrecht/Norwell, MA: Kluwer Academic.

Wharton, R.A., 1981. Namibian Solifugae (Arachnida). Cimbebasia Memoir, 5, 1–87.

Chapter

24

Spiders (Araneae)

Gary R. Mullen . Richard S. Vetter

All spiders, except the Symphytognathidae and Uloboridae, possess venom glands that are used to subdue captured prey. When threatened, however, spiders will often defend themselves by biting, thereby injecting those same toxins into vertebrate skin. In most cases, the venom produces only mild, localized reactions that do not warrant medical attention. Other spiders have much more potent venoms that can cause severe reactions in bite victims, occasionally resulting in deaths. Only about 60 species of spiders worldwide are considered to have significant medical importance. Among them are only a few genera that are dangerous to humans. Most occur in the subtropics and tropics. A few tropical species, however, have extended their ranges into temperate regions, particularly those with Mediterranean-like climates.

Envenomation by spiders is called **araneism** after Araneae, the arachnid order to which spiders belong. Separate names, however, commonly are given to bites or syndromes associated with the more dangerous spider genera, each of which is generally characterized by typical clinical signs and symptoms. Examples include atraxism (*Atrax* spp.), latrodectism (*Latrodectus* spp.), loxoscelism (*Loxosceles* spp.), and phoneutriism (*Phoneutria* spp.). There also are cases in which individuals develop an abnormal fear of spiders such that the mere sight of one can cause panic or hysteria, a condition called **arachnophobia**, or more specifically **araneophobia**. This should not be confused with the unfortunate disdain that many people have for spiders, often reflecting their upbringing and misconceptions about spiders in general.

TAXONOMY

Approximately 3,000 genera and 40,000 species of spiders have been described worldwide. In North America alone there are 68 families, 569 genera, and 3,700 species (Ubick et al., 2005). Among the more than 100 families of spiders, about 20 families include species that reportedly cause medical concerns when they bite humans and other animals. The five most important families are the Dipluridae, Hexathelidae, Theraphosidae, Sicariidae (formerly Loxoscelidae), and Theridiidae. See Bettini and Brignoli (1978) and Ori (1984) for additional venomous spiders, representing more than 60 genera worldwide.

The order Araneae is divided into two suborders, the Mesothelae and Opisthothelae. The Mesothelae include the single family Liphistiidae, a small group of primitive spiders in Southeast Asia and the Indo-Malaysian region. The Opisthothelae are comprised of two groups, the Mygalomorphae (tarantula-like spiders) and Araneomorphae (all other spiders). The mygalomorphs are more primitive, represented by trap-door spiders, funnel-web spiders, and tarantulas. They include the largest spiders. The araneomorphs are a very diverse group that include wandering spiders and those familiar to most people by the diversity of silken webs that they produce.

An on-line taxonomic catalogue of the world spider fauna, which is routinely updated, can be found in Platnick (2007). For identification keys to the families and genera of mygalomorphs, see Raven (1985), and for North American families and genera of spiders, see Kaston (1978), Roth (1993), and Ubick et al. (2005).

The following is a synopsis of the major families of spiders of medical-veterinary importance.

Mygalomorph Spiders

Actinopodidae This small group is closely related to the typical trap-door spiders of the family Ctenizidae, which they resemble both morphologically and behaviorally. They construct vertical, silk-lined burrows in the soil, the opening at the surface of which is covered by a hinged "trap door." Their venom is weakly neurotoxic to vertebrates and causes no necrosis. Of the three recognized genera, only *Actinopus* in Central and South America has been reported biting humans, producing only local pain and transient muscle contractions.

Barychelidae Members of this family are closely related to the Theraphosidae, or tarantulas, and are restricted largely to southern Africa and Australia.

Idiommata blackwalli occurs widely throughout Australia in dry areas where it constructs silk-lined burrows provided with a saucer-shaped door that fits tightly into the opening at the ground surface. Its bite is painful, causing local redness and edema in humans. Most encounters occur when wandering males enter homes during the late summer and early fall, or when individuals are dislodged from their burrows in new suburban areas when people rake their yards.

Dipluridae Known as sheet-web or funnel-web-building tarantulas, diplurids construct burrows in the ground in a wide range of habitats. The family includes 19 genera. They are particularly abundant in the southern hemisphere and Australian region. The venom of *Trechona* species is especially toxic to humans and has been reported to cause human deaths in South America. A species of particular importance is *T. venosa*, which occurs in tropical forests and coastal areas where they are encountered on vegetation and along trails. They are very aggressive and, if disturbed, will readily bite.

Hexathelidae Formerly included in the Dipluridae, some members of this family are regarded as among the most toxic of spiders for humans. Among the 11 recognized genera, *Atrax*, *Hadronyche*, and *Macrothele* are the most dangerous. They are known as funnel-web spiders, referring to the extensive silk about the entrance, and extending into their shallow silk-lined burrows in the ground, among rocks, or in stumps and rot holes of trees. The most serious bites are caused by six species, most notably *A. robustus* in Australia.

Theraphosidae This is the largest mygalomorph family, with 84 recognized genera. They are best known for their large size and hairy appearance, familiar to most people as tarantulas. As a group they are primarily tropical and subtropical, occurring widely throughout both the Old World and New World where they are variously known as bird spiders, bird-eating spiders, and monkey or baboon spiders. In North America they extend into the southwestern United States but do not naturally occur east of the Mississippi River. Although the bite of most species is relatively harmless, several genera can cause severe envenomation, particularly in South America where the genera dangerous to humans are *Acanthoscurria*, *Pamphobeteus*, *Phormictopus*, and *Sericopelma*. On the Indian subcontinent, the genus *Poecilotheria* has venom that causes a reaction similar to widow venom, perhaps making it the most dangerous tarantula.

Araneomorph Spiders

Agelenidae Agelenid spiders are called funnel weavers, not to be confused with the mygalomorph funnel-web spiders (Dipluridae and Hexathelidae). They typically build horizontal sheet webs with a tubular retreat or "funnel" leading into a protected recess. When their webs are constructed in vegetation, they are often called grass spiders. A few species occur in homes, especially in basements and cellars, where their chances of encounters with humans are greatest. In North America, the only species that has drawn medical attention is the hobo spider, *Tegenaria agrestis*, In Europe, *Agelena labyrinthica* has been reported to bite humans.

Amaurobiidae Amaurobiid spiders are common in forest ground litter, under stones and debris, in rotting logs, and in cavities of standing trees. They are related to agelenid spiders, often associated with irregular, sheet-like silk webbing extending into protected recesses. A person bitten on the finger by the European species *Coelotes obesus* is said to have experienced intense pain and localized paralysis that persisted for a few hours (Maretic and Lebez, 1979).

Araneidae This is one of the largest families of spiders, familiar to most people because of the symmetrical spiral-like webs that they construct for snaring flying insects. Known as orbweavers, they commonly are found around homes and other dwellings where they take advantage of artificial lights to catch prey at night. They seldom bite humans or other vertebrates and usually cause only minor, temporary discomfort when they do. Even the North American black and yellow garden spider, *Argiope aurantia*, produces only localized pain, redness, and edema on the rare occasions in which it has been known to bite. This is a large and colorful species that often attracts the attention of homeowners in the late summer and fall. *Argiope lobata* in Europe causes a similar, mild reaction. The only araneids that have been reported as a medical concern are members of the genus *Mastophora*, known as bolas spiders, and in Peru, Bolivia, and Chile, as *podadoras*. The bite of some South American species is said to cause localized, necrotic skin lesions, with general pain, edema, and hematuria (Bucherl, 1971). However, based on the rarity of encounters and the lack of documented cases in the literature, there is reason to question the significance of bolas spiders as a cause of bites to humans.

Corinnidae Formerly this family was included in the Clubionidae. Some *Trachelas* species (e.g., *T. volutus*) have been reported to bite humans in the United States, causing a stinging sensation and localized erythema and swelling.

Ctenidae Members of this family are wandering spiders that do not build webs. Most are of moderate size (1.5–2.5 cm in body length) and occur primarily in ground litter and low vegetation where they hunt prey. They resemble wolf spiders (Lycosidae) in both their general appearance and behavior. The taxa of greatest medical concern are the South American species *Phoneutria nigriventer* and *P. keyserlingi* from coastal Brazilian forests. These are relatively large species (>3 cm body length) that occasionally cause severe envenomation; most bites in human adults result in only mild effects. In Central America, several species of *Cupiennius* have been involved in bites, with pain lasting only about 30 minutes (Barth, 2001). Both *Phoneutria* and *Cupiennius* species have been found in international cargo, most commonly bananas. For information on identification of these spiders and medical aspects of their bites, see Vetter and Hillebrecht (2008).

Dysderidae Although there are over 200 *Dysdera* species in the world, only *D. crocata* has become established worldwide via commerce. It has a dark cinnamon-colored cephalothorax, grey-tan abdomen, and very large fangs that it presents when threatened. Its bites are painful, probably due to fang penetration, not venom toxicity, with mild effects that dissipate within an hour.

Gnaphosidae These wandering spiders are commonly found under stones, in rolled leaves, and ground debris, and are known as ground spiders. The bites of most gnaphosids are relatively harmless. *Herpyllus ecclesiasticus* and *Scotophaeus blackwalli* (formerly *H. blackwalli*) have been reported to cause painful bites in the United States, usually upon entering homes at night.

Lamponidae Members of this family are similar to gnaphosids in their habitats and behavior. The white-tailed spider (*Lampona cylindrata/murina* complex) has been associated with necrotic skin lesions for 20 years in Australia. Although bites have resulted in localized inflammation, blistering, ulcerations, and necrosis at the wound site (Sutherland, 1987), there is reason to believe that these conditions were the result of secondary infections and not the spider venom. A total of 130 verified white-tailed spider bites in humans resulted in only mild effects and no necrosis (Isbister and Gray, 2003b).

Lycosidae Commonly known as wolf spiders, lycosids represent a highly successful family of hunting spiders that are noted for their relatively large size (up to 4 cm), and hairy appearance. Their posterior median and posterior lateral eyes are greatly enlarged and aid them visually in capturing prey. Members of the genus *Lycosa* possess cytotoxic venoms that can cause painful bites. Included in this genus are the so-called "tarantulas" of Europe, such as *L. tarentula* of tarantism fame. Although the bites of many wolf spiders are painful, they generally cause only temporary, local discomfort.

Miturgidae Previously the Clubionidae were a large, diverse family known as the sac spiders, referring to the silken tubular retreats that they typically make in rolled-up leaves, other ground litter, and under bark and stones. This family has been divided into many smaller families of which only members of the genus *Cheiracanthium* typically are involved in envenomations. Although bites can be painful, most result in cases of minor medical importance. They are nocturnal, vagrant spiders that commonly are found hunting prey on plants and incidentally enter houses and other buildings.

Oxyopidae Members of this largely tropical and subtropical family are called lynx spiders. They are active hunters that rely on their keen eyesight, speed, and agility to capture prey while climbing in foliage. Although the family generally is not regarded as being medically important, females of the green lynx spider (*Peucetia viridans*) are known to forcibly expel venom from their fangs as a defensive response, especially when guarding their egg sacs. Droplets can be squirted up to 20 cm, and on contact with human eyes can cause impaired vision and moderately severe conjunctivitis (Tinkham, 1946; Fink, 1984).

Pisauridae This family is closely related to the wolf spiders, which they strongly resemble. They occur most frequently near water, where members of the genus *Dolomedes* are adept at moving about on the water surface to capture prey, hence their common name, fishing spiders. Spiders of the genus *Pisaurina* are known as nursery-web spiders because of the habit of females suspending their egg sacs in a protective silken "nursery" in vegetation and guarding the resultant spiderlings until they disperse. Because of their large size (body length up to 4 cm or more) and powerful chelicerae, they can bite if handled, causing local pain and transient swelling. The bite of the European species *Dolomedes fimbriatus* is reported to cause a reaction similar to that of the amaurobiid *Coelotes obesus*.

Salticidae This is the largest family of spiders, with over 400 genera and over 4000 species widely distributed throughout the world. They are known as jumping spiders because of their habit of stalking and pouncing on prey or jumping to escape when threatened. The anterior median eyes are complex and greatly enlarged, providing them with the keenest vision of all spiders. Some of the larger species can be aggressive and inflict painful bites when handled or pressed against the skin. The venom of at least some species contains cytotoxins that cause necrotic lesions at the puncture site, often being slow to heal. The bite of *Phidippus johnsoni* can cause a dull, throbbing pain that may persist for a few hours, in addition to swelling, tenderness, and itching that may last for one to four days following the bite (Russell, 1970).

Segestriidae Members of this relatively small family live in silken retreats under stones and bark or in crevices of wood and rocks. They are active nocturnal hunters that may enter homes or construct their retreats in and around human dwellings. Despite their large, well-developed chelicerae, they are not very aggressive, rarely bite, and are not considered to be very toxic. Nonetheless a few cases of human bites by the European species *Segestria florentina* reportedly have involved local pain, redness, and swelling, and occasionally nausea and vertigo.

Sicariidae (including the former Loxoscelidae) The sicariids are a small group of relatively primitive araneomorphs. Included in this family are the recluse spiders in the genus *Loxosceles*, which can cause severe necrosis in a minority of cases. Approximately 85 species of *Loxosceles* have been described in the Americas, and at least 100 worldwide. They are generally similar in appearance and are difficult to recognize from one another by the nonspecialist. They typically are found in ground litter and under bark or stones; a few species occur in caves, and some are decidedly synanthropic, living in close association with humans. The genus *Sicarius* in Africa also has been shown to be highly toxic in laboratory studies, although its behavior (e.g., burying itself in sand under rocks) and remote distribution limits its encounter with humans.

Theridiidae Members of this large family are called cobweb weavers or comb-footed spiders. The latter refers to a row of serrated bristles on the hind tarsus, which is used to comb the silk from the spinnerets during construction of their irregular webs or wrapping prey. The only genus considered particularly toxic to humans and domestic animals is *Latrodectus*, which includes the widows or shoe-button spiders. Another genus that is less toxic to humans is *Steatoda*. Several species in South America, including *S. andina* (formerly placed in *Lithyphantes*) and the cirari of Bolivia, Chile, and Paraguay, are said to cause serious envenomation (Southcott, 1984). Although venom of the Mediterranean species *Steatoda paykulliana* has been shown to be neurotoxic to guinea pigs, this spider has not been reported to bite humans. The cosmopolitan species *S. grossa*, known as the false black widow, causes only a local bite reaction with mild effects resembling *Latrodectus* bites (Isbister and Gray, 2003a). *Steatoda* spiders have been misidentified as black widows by both patients and physicians. In one case of mistaken identity, black widow spider antivenin was administered for a *Steatoda* bite, which appeared to ameliorate envenomation symptoms.

Thomisidae Members of this family are called crab spiders because of their generally flattened appearance, laterigrade legs, and crab-like gait. They often are cryptically colored and ambush their prey from camouflaged sites such as tree bark and flower heads. As a group they are considered harmless to humans and other animals. Some species of *Misumenoides*, however, have been suspected of causing relatively minor bites in humans (Hickin, 1984).

Zoridae The only zorid spider that has drawn medical attention is *Elassoctenus harpax* of Western Australia, which reportedly can inflict a painful bite. Cases have not been well documented, however, leaving the question of typical bite reactions uncertain.

MORPHOLOGY

The body of a spider is divided into two regions, the anterior **cephalothorax** (prosoma), which represents a fusion of the head and thoracic segments, and the abdomen (opisthosoma) (Figure 24.1). The cephalothorax bears the chelicerae, pedipalps, eyes, and legs. The **chelicerae** (sing. chelicera) are paired structures used to seize prey or to bite defensively when threatened (Figure 24.2). They also serve other functions in different groups of spiders such as digging by burrowing species, transporting prey, and carrying eggs sacs by some spiders. Each chelicera consists of

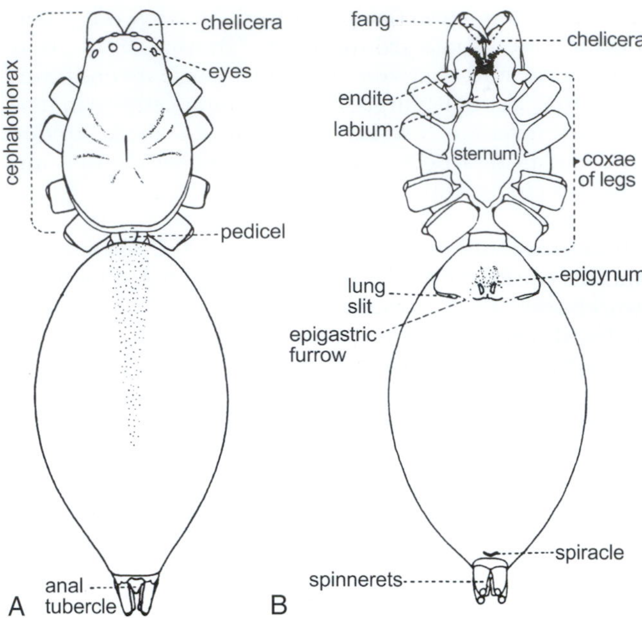

Figure 24.1 Morphology of representative spider. (A) Dorsal view; (B) ventral view. (Modified from Kaston, 1978)

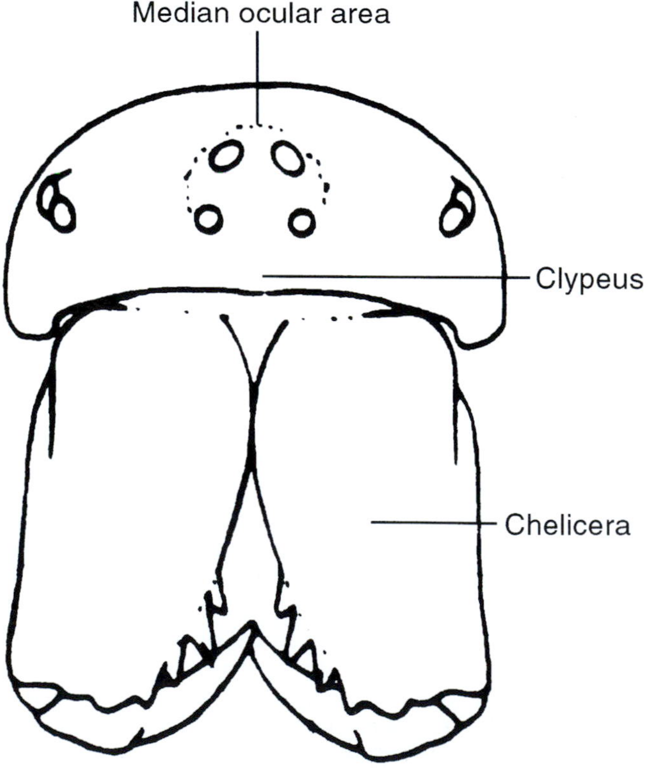

Figure 24.2 Head and chelicerae of representative spider, frontal view. (Modified from Kaston, 1978)

two parts, a stout basal portion and a movable fang. The fang rests in a groove, and is extended when the spider bites or otherwise attempts to grasp something. Near the tip of each fang is a tiny opening to the venom duct, which leads from the **venom gland**; the latter is located in the basal part of the chelicera and usually extends back into the cephalothorax in araneomorphs. In mygalomorphs, the venom glands are restricted to the chelicerae. The mouth is located just behind the bases of the chelicerae. The chelicerae of mygalomorphs are oriented parallel to the long axis of the body and move parallel to one another in a vertical plane. These spiders strike downward when seizing prey. The chelicerae of araneomorphs are oriented perpendicular to the long axis of the body and are opposed to one another, moving together in a pincher-like motion.

Most spiders have eight eyes located anteriorly on the cephalothorax (Figure 24.2). They are simple eyes (ocelli), usually arranged in two rows. Some spiders lack one or more pairs of eyes, as in *Loxosceles* species, which have only six eyes. In other species, such as wolf spiders and jumping spiders, some eyes may be enlarged, reflecting their greater visual acuity. The size and arrangement of the eyes often serve as valuable taxonomic characters.

The **palps** (pedipalps) are a pair of six-segmented appendages that arise immediately behind the mouth. They are primarily tactile structures used in sensing the substrate, perceiving contact stimuli from conspecifics, and both detecting and manipulating prey. Whereas the palps of immature spiders and adult females tend to resemble legs, the palps of adult males are modified as copulatory organs. In such cases the terminal segment (palp tarsus) is enlarged with a ventral, bowl-shaped cavity enclosing a complex of specialized structures formed from the pretarsus that serve as an intromittent organ for inseminating females. Adult males usually can be recognized by the swollen terminus of their palps and their often smaller body size relative to females of the same species.

Spiders have four pairs of legs, each with seven segments: coxa, trochanter, femur, **patella**, tibia, **metatarsus**, and tarsus (Figure 24.3). Spiders that run about on the ground and other substrates without building a trapping web typically have only two tarsal claws on each leg. Many of these hunting spiders possess dense tufts of hairs **(scopulae)** directly beneath the pair of claws or along

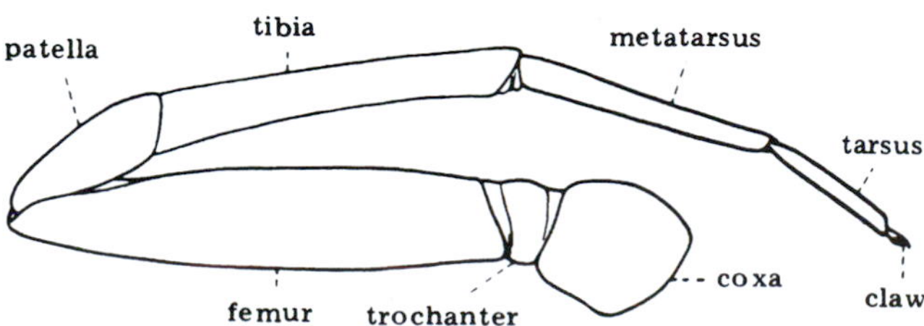

Figure 24.3 Leg of spider showing the seven leg segments. Spiders possess a patella and metatarsus not found in insects. (Kaston, 1978)

the ventral side of the tarsus and metatarsus. They provide physical adhesion to facilitate climbing on smooth surfaces and grasping prey. The scopulae are especially prominent in tarantulas. Three tarsal claws are characteristic of web-building spiders. The single median claw on each leg is used to hold onto silken threads by those spiders that hang suspended in their webs.

The abdomen is connected with the cephalothorax by a narrow pedicel, which provides great flexibility and movement between the two body regions. One or two pairs of slit-like openings to the **book lungs**, the principal respiratory organs in spiders, are located ventrally on the second and third abdominal segments. The more "primitive" spiders tend to retain two pairs of book lungs, whereas most spiders have only one pair. The second pair of book lungs in some spiders is modified to form tubular tracheae that open via a spiracle, or pair of spiracles, on the third abdominal segment. Most spiders, however, have only a single spiracle located in front of the spinnerets.

The genital opening of both sexes is located ventrally on the second abdominal segment between the book lungs. In females of the more "advanced" spiders, a sclerotized copulatory structure called the **epigynum** is located just in front of the genital opening and leads to the **spermathecae**, where sperm is stored after mating. The presence of the epigynum is helpful in distinguishing adult females from immatures and males and in distinguishing species.

Located at the posterior end of the abdomen are the **spinnerets**, through which silk from several types of internal silk glands are extruded via small spigots. Most spiders have three pairs of spinnerets, the size and relative lengths of which provide useful taxonomic characters. In some groups of araneomorph spiders, a sieve-like plate of minute spigots called the **cribellum** is present in front of the anterior pair of spinnerets. These taxa, called cribellate spiders, also possess a row of specialized setae on the metatarsi of the fourth pair of legs called a **calamistrum**. The calamistrum is used to comb silk from the cribellum by rhythmic movements of the hind legs. Most araneomorph spiders lack a cribellum and calamistrum and are referred to as ecribellate spiders.

LIFE HISTORY

Most reproductive activity of spiders occurs during the spring and early summer, when mating takes place and eggs are deposited. Some species, however, are reproductively active in the late summer and fall. Embryonic development commonly takes about two weeks, during which time one or two prelarvae and a larva are formed within individual eggs contained in a silken egg sac. The larva subsequently molts and emerges from the egg sac as a nymph. The young nymph, or **spiderling**, is an miniature of the adult spider with functional spinnerets and poison glands. The nymphs undergo two to 12 molts as juveniles, depending on spider species, before reaching sexual maturity as adults.

In temperate regions, juvenile spiders are present throughout the summer months, with overwintering typically occurring as late-instar nymphs. However, if egg sacs are constructed by females late in the year (e.g., some orb-weaving spiders), the resultant spiderlings remain inside the egg sac until spring. In other cases, spiders overwinter as mature females. The life cycle for most spiders is one year, following which the adults die. Males are shorter lived than females and usually die soon after mating. Some of the more "primitive" spiders live much longer; purse-web spiders (Atypidae) may live seven years, whereas the larger tarantulas (Theraphosidae) may live 20 to 30 years, especially in captivity.

BEHAVIOR AND ECOLOGY

Mating behavior in spiders varies greatly from one group to another and in some cases involves complex courtship displays. Prior to mating, the male typically deposits a droplet of sperm from the genital pore onto a small, silken platform called a sperm web. The droplet is then drawn into the copulatory organ at the tip of each male palp in a process called sperm induction. Sperm is stored in the palp until mating takes place. Following acceptance of a male's advance by a female, copulation is accomplished by indirect insemination in which the male transfers the sperm from his palps to the spermathecal ducts on the underside of the female's abdomen. Although the males of many species die soon after mating, others refill their palps with sperm and may mate with other females one or more times thereafter. Contrary to popular belief, most males either walk away or hurriedly retreat without being attacked and eaten by the female.

Oviposition typically occurs within a few weeks after mating. The female spins a silken sheet onto which she extrudes fertilized eggs from her genital opening, forming a mass in which the individual eggs are cemented together. She then protects the eggs by spinning multiple layers of silk to form the egg sac. More than one egg sac may be produced, as commonly observed in black widow spiders. Although some species of spiders remain with the eggs to protect them until they hatch, most spiders provide no maternal care and leave the eggs to hatch unguarded.

Spiders have been very successful in exploiting a wide range of ecological habitats, including islands. This is accomplished, in part, by their specialized dispersal behavior called **ballooning**. Spiderlings, and even adults of many small spiders, are carried aloft on silken threads by which they may be conveyed by wind currents over considerable distances. To become airborne the spider orients into the wind, lifts the tip of its abdomen, and extrudes silk from the spinnerets. When the amount of silk is sufficient to provide enough buoyancy to support its body weight, the spider releases its grip on the substrate and floats away.

Spiders can be categorized in three major groups based on their general behavior and feeding strategies: burrowers, vagrants or wanderers, and web builders. Burrowing spiders usually excavate their burrows in the soil of suitable sites where they remain more or less permanently throughout their lives. They typically capture prey that comes within reach of the burrow entrance, or make short excursions to capture food and return to the safety of the burrow. Among this group are many of the ground-dwelling tarantulas, trap-door spiders, and burrowing wolf spiders. Vagrant spiders tend to wander extensively and may or may not regularly return to a given location where they have constructed a retreat. They do not produce a silken web for capturing food, but instead hunt or ambush their prey. Examples include many wolf spiders, jumping spiders, sac spiders, gnaphosids, and ctenids. These spiders commonly enter homes during their hunting forays and occasionally bite humans. The brown recluse spider also falls in this group. It differs from the others, however, by actually living indoors rather than wandering inside incidentally, and it makes a flimsy web. Web-building spiders construct various silken structures to detect and capture potential prey. The webs may be sheet-like, as in diplurids and agelenids, or more complex trapping webs like those of comb-footed spiders and orb weavers.

Burrowing and vagrant spiders usually rely on their physical size and strength to capture food, together with sufficiently potent venoms to subdue their prey quickly. The size of acceptable prey items is often positively correlated with the size of the spider itself. Web-building spiders, on the other hand, tend to rely on the use of silk to ensnare or immobilize their prey. They are able to utilize a wider range of prey items, often capturing and feeding on insects and other arthropods much larger than themselves. Those that construct aerial webs also have access to a wider variety of flying insects than do ground-dwelling spiders. The venom of web-building spiders is typically less potent than that of nonweb-building species, accounting in part for the fact that bites of even the larger web builders are usually quite harmless.

PUBLIC HEALTH IMPORTANCE

Recognizing that virtually all spiders possess venom glands, it is not surprising that when they are threatened, species with chelicerae large enough to pierce the skin will bite humans and other animals. In most cases the reaction is minor, usually limited to mild, localized pain and slight to moderate swelling at the bite site. The severity of the reaction is dependent on the species of spider, its size, and the amount of venom injected.

The **venom** varies greatly among different spider taxa in terms of chemical composition and its effects on different animals upon injection. Components include a wide range of proteases, esterases, polyamines, free amino acids, histamine, and specific toxic compounds characteristic of individual groups or species. Whereas some venoms are primarily cytolytic, causing the destruction of cells and tissues with which they come in contact, others act as neurotoxins or disrupt normal blood functions. For details on the biochemistry and pharmacology of spider venoms, see Bucherl (1971), Bettini (1978), Duchen and Gomez (1984), Geren and Odell (1984), and Rash and Hodgson (2002).

Most problems warranting medical attention are not due directly to bites but to secondary infections. In other cases involving the more toxic species, reactions can be much more severe, occasionally causing deaths. See the following sources for information on toxic spiders in different regions of the world: North America (Wong et al., 1987), South America (Lucas, 1988), Europe (Maretic and Lebez, 1979), South Africa (Newlands and Atkinson, 1988), and Australia (Southcott, 1976; Sutherland, 1990); also Isbister and White (2004) and Vetter and Isbister (2008). For a review of antivenins, see Isbister et al. (2003). The following accounts address spider problems of particular medical importance.

Tarantism

The term **tarantism** has special significance from a medical viewpoint. It refers to a condition in which individuals allegedly bitten by a tarantula experience a range of symptoms including tremors, hyperactivity, difficulty breathing, muscular rigidity and priapism (painful penile erection in males), sweating, and uncontrolled crying. In its most extreme form it can lead to fainting spells, delirium, and convulsions. Although descriptions of this syndrome can be traced back as early as Aristotle's writings in the fourth century BC, it was most prevalent in Europe during the Middle Ages. It is believed to have been named after Taranto, Italy, where an epidemic of tarantism occurred in 1370. From there the phenomenon spread throughout Italy to present-day Croatia, Spain, and other parts of the Mediterranean region. The only cure was thought to be prolonged and vigorous dancing to special, lively music to induce profuse sweating and eventual collapse from shear exhaustion. Municipalities sometimes hired musicians to play in shifts for three or four days at a time as victims danced themselves into frenzies, seeking relief from their affliction. Not uncommonly this led to mass hysteria among local residents and shameless exhibitionism on the part of some individuals. Some have linked this choreomania to Saint Vitus' dance, a nervous disease with involuntary jerking motions.

The bite of the European wolf spider *Lycosa tarentula,* commonly called the "tarantula," traditionally has been blamed as the cause of tarantism. The reason for this connection is uncertain; in fact this species seldom comes in contact with people. Its bite causes only mild pain and slight swelling at the bite site, and none of the neurological effects characteristic of tarantism victims. Convincing evidence suggests that the spider involved was actually a *Latrodectus* species. Even as late as the 1950s, spiders of this genus were called *tarantola* in southern Italy, and cases involving their bites were noted in medical records as a tarantola bite or tarantolism. Today tarantism is regarded as largely a psychosomatic response to real or imagined spider bites rooted in legend, ignorance, or superstition linked to cases of latrodectism. Because latrodectism victims tend to move incessantly in response to the neurotoxin, there may be some truth to the notion that dancing helped to ameliorate the effects in envenomation cases (Maretic and Lebez, 1979).

Tarantulism

Tarantulas (Theraphosidae) (Figure 24.4) are typically ground dwellers living in silk-lined burrows. They often leave their burrows at night to hunt prey; at such times may enter homes and other shelters or otherwise come in contact with people. During the mating season, males are more likely to be encountered as they wander in search of females; they are particularly aggressive at this time and are easily provoked. Other circumstances that contribute to human encounters are disturbances of their burrows, land development, and flooding or other natural disasters that tend to displace them. About a dozen genera of tarantulas are considered toxic enough to humans to require medical treatment of bite victims. They are found primarily in the tropics of South America, Africa, and Australia, where most of the serious cases of human envenomation occur.

Despite their large size, powerful fangs, and intimidating appearance (Figure 24.5), most tarantulas are not very toxic. Bite reactions vary from almost painless to moderately or intensely painful with reddening about the puncture site. The sensation is commonly likened to that of a bee sting, except that the pain is not immediate and develops more slowly. The pain subsides gradually, seldom persisting for more than 30 minutes. This may be accompanied by a burning sensation, localized swelling, and tightening of the muscles near the bite wound. Secondary infections can be avoided by cleansing the wound and applying a topical antibiotic.

Figure 24.4 Tarantula, *Vitalius sorocabae* (Theraphosidae), Brazil. (Photo by João P. Burini)

Figure 24.5 Tarantula, ventral view; showing large body size, exposed pair of chelicerae, and extended palps. (Photo by João P. Burini)

Figure 24.6 Tarantula (Theraphosidae), with bald patch on dorsal aspect of abdomen where urticating hairs have been defensively flicked off by the hind legs. (Photo by Nathan Burkett-Cadena)

In cases of more dangerous tarantulas, neurotoxic components in the venom can cause severe, sometimes life-threatening reactions. These toxins are designed to act quickly in subduing vertebrate prey such as frogs, lizards, and birds on which some species feed. When injected into a human bite wound, the venom can cause not only intense pain but also muscle spasms, edema, inflammation of lymphatic vessels, and systemic reactions that can lead to shock and vascular collapse. Other effects reported in laboratory animals include local necrosis, hemoglobin in the urine, and jaundice, indicating the presence of necrotoxic and haemolytic components in the venom of some species.

Members of the genus *Harpactirella* are among those sometimes known as baboon spiders. They occur in South Africa, where they live in silk-lined tunnels under logs, stones, and other debris. They are aggressive hunters, frequently entering homes and animal shelters during their wanderings. The bite of *H. lightfooti* causes an immediate burning pain followed by paleness, vomiting, and severe systemic reactions that can lead to shock and collapse. The bite is not fatal, with recovery usually occurring within 24 hours. None of the tarantulas in the United States are considered to be very toxic. However, the venom of the Texas brown tarantula, *Aphonopelma hentzi,* contains a necrotoxin that has been shown to damage myocardial tissues in mice.

Many tarantulas possess tiny (0.2–1.2 mm long), specialized **urticating hairs** on their abdomen, which are readily detached when stroked with their hind legs (Figures 24.6 and 24.7). Only New World theraphosids in the subfamilies Aviculariinae, Ischnocolinae, and Theraphosinae are known to possess such hairs; the latter includes the genera *Brachypelma* and *Aphonopelma*, representatives of which are commonly sold as pets. These hairs are armed with spines and barbs that can penetrate vertebrate skin and other sensitive tissues (i.e., eyes, nasal cavity) with which they come in contact. Some species have up to 10,000 of these hairs/mm^2, totaling well over 1 million urticating hairs per individual.

Four basic types of hairs are described by Cooke et al. (1972), although there are additional unusual types in a few other species. Type I hairs enter the skin at a shallow angle, do not penetrate very deeply, and cause only a mild reaction. This is the only type found in species in the United States. Type II hairs are not flicked off but are incorporated into the silk lining the tarantula's retreat, eggs sacs, or silk mats used during molting. Type III hairs penetrate the skin up to 2 mm, causing a persistent urticaria and inflammation that may last for two to three weeks; this type is characteristic of many Mexican, Caribbean, Central American, and South American species. Type IV hairs cause inflammation of the respiratory tract in small mammals, although little is known regarding their effects on humans. A given species may have more than one type of urticating hair. Bald patches on the back of the abdomen (Figure 24.6)

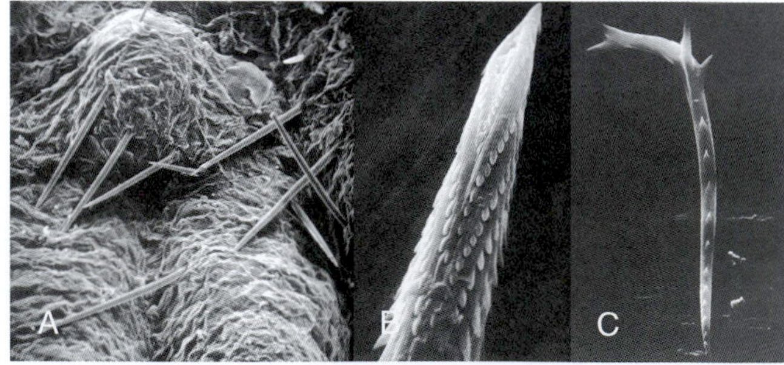

Figure 24.7 Urticating hairs of tarantulas (Theraphosidae). (A) Hairs of *Avicularia surinamensis* embedded in skin of young mouse; (B) basal tip of Type II hair of *A. surinamensis*, with backward-directed barbs that help to work the hair into skin; (C) Type IV hair of unidentified tarantula from Chile that causes inflammation of respiratory passages of small mammals. (From Cooke et al., 1972, courtesy of the American Museum of Natural History)

are usually evidence that a tarantula has defended itself by dislodging these specialized hairs. They are replaced with each molt, even in adult females that continue to molt after reaching maturity.

When threatened, the tarantula flicks a cloud of these hairs at the offender, usually rodents and other small mammals that try to attack them in their burrows. In addition to causing irritation to the skin (Figure 24.7), they can cause severe inflammation of the eyes, mouth, and respiratory passages, serving as an effective deterrent against predators. The effects are solely mechanical and do not involve chemical substances. In the case of certain tarantulas (e.g., *Megaphobema* and *Theraphosa* spp.), the abdominal hairs are incorporated with silk into the egg sacs or the silk mats on which they molt as a defensive barrier to attack by potential predators and parasites.

Humans experience similar reactions to that of other animals when handling or provoking certain tarantulas, including species sold as pets. Common symptoms are urticarial dermatitis, mild edema, and vascular dilation. When the setae come in contact with the eyes, they cause an immediate burning or stinging sensation followed by intense pruritus, lachrymation, swelling of the eyelids, and corneal abrasions. The problem is exacerbated by the natural inclination of the victim to rub the affected areas. Corneal lesions may still be evident six to nine months after the encounter as the embedded hairs are gradually resorbed. The damage is not permanent, and full visual acuity is gradually restored.

Atraxism

The genera *Atrax* and *Hadronyche* occur only in the eastern half of Australia, predominantly in the coastal region, including Tasmania, where its members are known as funnel-web or tube-web spiders. They construct their silken retreats in rock crevices or ground burrows. Six species are considered highly toxic to humans and can cause severe envenomation symptoms (Isbister et al., 2005). The venom of males is more dangerous than that of females.

The **Sydney funnel-web spider**, *A. robustus* (Figure 24.8), the more commonly encountered species, is restricted largely to areas within a 160 km radius of Sydney (New South Wales). It is a large species with males, the larger of the sexes, measuring about 25 cm in body length. They construct their tubular webs under logs, amidst rocks, and in various ground debris. They often are found in suburban gardens where they are attracted to damp, well-watered sites with abundant ground litter. Most bites by this species occur during the summer when roaming males are most likely to enter homes. Wandering males are extremely aggressive, readily attack when provoked, and account for the majority of cases of human envenomation. The bite produces immediate pain and a wide range of neurological symptoms. These include agitation, anxiety, hypertension, generalized muscular twitching, and irregular heart beat, in addition to pulmonary edema and intravascular coagulation. Deaths have been reported, notably in children; no deaths have occurred since the development of antivenin. The principal toxic component of the venom has been named **atraxotoxin**. It acts by stimulating the release of acetylcholine at motor end plates and throughout the autonomic nervous system. The effects are usually transient and reversible, causing no permanent damage.

Figure 24.8 Sydney funnel-web spider, *Atrax robustus* (Hexathelidae), Australia. Note the raised chelicerae and threatening posture. (Photo by Julian White ©)

Hadronyche formidabilis is called the North Coast funnel-web spider or the tree funnel-web spider. It occurs in the rain forests of southeastern Queensland and northern New South Wales. The effects of its bite are similar to that of *A. robustus*, although the toxicity of its venom is greater. It is less commonly encountered, however, with relatively few bite cases having been reported. An antivenin directed at the male toxins is available for treating both *A. robustus* and *H. formidabilis* bites. Other species of concern are *H. infensa* (Figure 24.9), *H. versuta*, and *H. cerberea*, along the southeastern coast of Australia.

Figure 24.9 Funnel-web spider, *Hadronyche infensa* (Hexathelidae), male, Australia. (Photo by Robert Raven)

Phoneutriism

The genus *Phoneutria* is widely distributed throughout South America where members are known as wandering spiders, armed spiders, and banana spiders. They are large aggressive spiders that actively hunt at night, feeding on both invertebrate and vertebrate prey. The venom is highly neurotoxic to humans and acts on both the central and peripheral nervous systems, producing a characteristic syndrome. The most dangerous species are *P. nigriventer* (Figure 24.10.) and *P. keyserlingi,* which occur throughout Brazil, Uruguay, and Argentina, where cases of human envenomation are quite common. It is a large species, with females attaining body lengths up to 5 cm. Its venom contains a number of pharmacologically active compounds including histamine and serotonin, in addition to neurotoxin. Its bite is extremely painful and causes a number of symptoms such as salivation, sweating, muscular spasms, painful penile erection, and visual problems. Deaths are rare and usually are attributed to respiratory failure. In one study of several hundred bite victims, 96% of 10- to 70-year-olds experienced only mild envenomation effects (Bucaretchi et al., 2000).

Cheiracanthism

Only a few members of the families formerly belonging to the Clubionidae, or **sac spiders** (Figure 24.11), pose health concerns. Most are *Cheiracanthium* species (Miturgidae), which as a group occur primarily in the Eastern Hemisphere; only two species are known from North America (*C. inclusum* and *C. mildei*). Envenomation usually occurs when they are trapped against the skin after crawling into clothing or footwear or when an individual rolls over onto one while sleeping. The bite typically causes immediate local pain, redness, and formation of a wheal, similar to the sting of a wasp or bee. In some cases the victim may experience a persistent loss of sensitivity involving the musculature and nerves at the bite site. In more severe cases, systemic responses may include mild fever, nausea, and loss of appetite. No deaths have been attributed to *Cheiracanthium* spiders.

Figure 24.10 *Phoneutria nigriventer* (Ctenidae), female on tree trunk, Brazil. (Photo by João P. Burini)

Figure 24.11 Sac spider, *Clubiona obesa* (Clubionidae), female. (Gertsch, 1979)

Cheiracanthium punctorium is the species most commonly involved in cases of cheiracanthism in Europe. Its bite causes a painful burning sensation and associated swelling that may persist for several days. Other reported symptoms include chills, general muscular aches, and tenderness of regional lymph glands. Envenomation by *C. japonicum* in Japan is similar to that by *C. punctorium* but also may involve local petechiae (purplish, hemorrhagic spots on the skin), nausea, vomiting, and rarely shock. Reactions to one of the two species found in North America, *C. inclusum,* are painful but relatively mild compared to *C. punctorium* and *C. japonicum.* Although *C. mildei* has been reported to sometimes cause severe necrosis similar to a recluse spider bite, a study of verified bites demonstrated only mild effects and no necrosis (Vetter et al., 2006). Several of the *Cheiracanthium* species of medical importance are fairly recent introductions from other parts of the world. These include *C. mildei* introduced to the United States from Europe and *C. mordax* introduced to Hawaii and Fiji from Australia.

Tegenarism

Some agelenid spiders in the genus *Tegenaria* occur in close association with humans. Upon entering homes, they may become established in basements and other relatively dark, damp locations where they construct sheet webs with a funnel-like retreat characteristic of the Agelenidae. In Europe, these species are called house spiders.

One species implicated as medically important in North America is the **hobo spider**, *Tegenaria agrestis* (Figure 24.12), which was introduced from Europe to the Pacific Coast of the United States. It was first

Figure 24.12 Hobo spider, *Tegenaria agrestis* (Agelenidae), female at entrance to funnel-like retreat of sheet web. (Akre and Catts, 1992, courtesy of Washington State University Cooperative Extension)

Figure 24.13 Brown recluse spider, *Loxosceles reclusa* (Sicariidae), female, dorsal view. Note the dark, violin-shaped marking on the cephalothorax and arrangement of the six eyes in three groups of two eyes each. (Photo by Gary R. Mullen)

reported at Seattle in 1930 but did not become common in the Pacific Northwest until the 1960s. The hobo spider is now well established in British Columbia, Washington, and Oregon, with the eastern and southern edge of its distribution extending to Montana, Wyoming, Colorado, and northern Utah. It occurs in funnel-like webs (Figure 24.12), particularly in basements and cellars, window wells of homes, and crawl spaces; around house foundations, in wood piles, under rocks and wood used in landscaping, and other suitable sites at ground level. The males tend to wander at night in search of females, at which time they enter homes and are more likely to be encountered than females. In Europe it is considered harmless. It was not until the 1980s that *T. agrestis* was believed to cause bites in the form of necrotic skin lesions in the Pacific Northwest of the United States. However, the ability of this species to cause dermal necrosis is now in question and has been seriously challenged (Binford, 2001; Vetter and Isbister, 2004).

Loxoscelism

The clinical syndrome called loxoscelism is caused by the bite of *Loxosceles* species known as fiddle-back, brown recluse, or violin spiders. It is also called **necrotic arachnidism** because of cytolytic components of the venom, which cause necrosis of tissues around the bite wound. The common names of these spiders refer to a usually distinct fiddle- or violin-shaped marking on the dorsum of the cephalothorax, the neck of which is directed posteriorly. The base of the "violin" encompasses the eyes and is darkly contrasted against the lighter, general body color in several but not all species (Figure 24.13). The eyes are distinctive among spiders in that there are only six, rather than the usual eight, and that they are arranged in three groups of two eyes each (Figure 24.13). The combination of a violin-shaped marking and this eye pattern distinguishes *Loxosceles* from all other spider genera. The body color and legs are usually light, tawny brown but may be dark brown or even grayish in some populations. *Loxosceles* species quite closely resemble one another, usually requiring a specialist to make species determinations. The legs are relatively long and slender, making them agile spiders that can move quickly. They are primarily nocturnal hunters, either catching prey that comes in contact with their irregular webs or actively wandering from the security of their silken retreats to capture food items. They do not wrap their prey but rely on their potent venom to quickly subdue it.

Approximately 100 species of *Loxosceles* have been described, with over 80 of them being found in the Americas (Gertsch and Ennik, 1983). The other species occur primarily in Europe and Africa. Thirteen *Loxosceles* species are found in the United States (Figure 24.14), including *L. arizonica* (Arizona), *L. blanda* (Texas), *L. deserta* (California, Arizona), *L. devia* (Texas), *L. laeta* (Massachusetts,

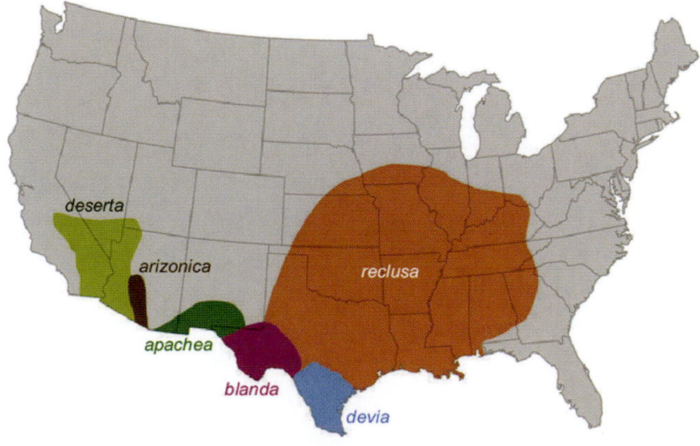

Figure 24.14 Approximate distribution of the more common *Loxosceles* species in the continental United States. (Reprinted with permission from the *Annual Review of Entomology*, Volume 53 © 2008 by Annual Reviews)

California), *L. reclusa* (southeastern except Florida, south-central, and midwestern states), and *L. rufescens* (very disjunct, localized sites in many states; often in a single building). *Loxosceles laeta* and *L. rufescens* are introduced species from South America and the Mediterranean region, respectively. *Loxosceles rufescens* is the most widely distributed member of the genus. It is endemic in southern Europe and northern Africa from which it has spread to northern Europe and parts of the Middle East, and has been introduced to Australia, Japan, Madagascar, and North America. The most important species are *L. reclusa* in North America, *L. laeta* in South America, and *L. rufescens* in the Mediterranean. These species have been introduced into nonnative areas through commerce, although their prevalence is very spotty. A review of aspects relating to *Loxosceles* envenomations can be found in Vetter (2008).

Brown Recluse Spider (*Loxosceles reclusa*) The brown recluse spider (Figures 24.13 and 24.15) occurs primarily in the southeastern and central United States. Records outside this area are believed to represent scattered but not well established populations, unlike *L. rufescens*. It typically is found indoors in warm, dry, undisturbed areas such as closets, attics, basements, storage areas, utility rooms, heated garages, lofts of feed mills, storerooms of broiler houses, and heated warehouses. It also is found hiding in cabinets and furniture, behind baseboards, door facings and wall hangings, and in crevices and corners of rooms. Particularly common sites to find them are in old boxes and accumulations of materials that have not been disturbed for some time.

When mature, *Loxosceles reclusa* females are 7 to 12 mm in body length but may look much larger because of their long legs. The males are slightly smaller (mean 8 mm), have longer legs than the females, and are easily recognized by their bulbous pedipalps. Mating occurs from February to October but most commonly in June and July. The inseminated female produces one to five egg sacs, each containing 20 to 50 eggs. The egg sacs are white, about 17 mm in diameter, flattened on the underside and convex above, and are constructed in the spider's silken retreat. The spiderlings usually emerge in three to seven weeks and remain in the web with the female until after the first or second molt. Development is relatively slow, requiring seven to eight months under favorable conditions. During this time they molt another six or seven times, undergoing eight instars before becoming adults. Adults commonly live up to 2.5 years and have been known to survive five to 10 years under laboratory conditions.

Figure 24.15 Brown recluse spider, *Loxosceles reclusa* (Sicariidae), female. This spider is typically tawny-brown in color, with relatively long legs. (Photo by Sturgis McKeever)

The web of *L. reclusa* is constructed in poorly lighted, undisturbed, out-of-the-way places where the spider spends most of its time. It is a rather irregular, nondescript tangle of silken strands that continues to grow in thickness as new silk is laid down. Freshly deposited silk is sticky but soon becomes covered with dust, contributing to the unkempt appearance of the webbing. In addition to being a retreat, the silk serves to detect the presence of potential prey. When food is scarce, *L. reclusa* will leave the web at night to roam in search of prey. It is under such circumstances that they are most likely to come in contact with humans.

Although being nonaggressive and very retiring as its common name implies, the brown recluse spider will bite if provoked. Most encounters occur either at night when a person rolls onto one of them in bed or when putting on clothes or footwear into which the spider has crawled. Often the victim is not aware of the bite until two or three hours later, whereas in other cases it may be immediately felt as a stinging sensation. This usually is followed by intense local pain with the formation of a small blister at the bite site. The area around the bite becomes reddened and swollen as the venom seeps into the surrounding tissues, making it very sensitive to touch. The extent of the skin area involved is usually evident within six to 12 hours. The venom is highly cytotoxic, killing any cells it contacts. Within 24 hours the involved skin tissue turns dusky or purplish as the blood supply and oxygen to the affected area are cut off. The result is necrosis of tissues and formation of an ulcer (Figure 24.16) within three or four days. Histological evidence indicates acute injury to the blood vessels and infiltration by white blood cells at the site. Phospholipase, a major component of the venom, induces this white-blood-cell response while also causing platelets in the blood to aggregate and the liberation of inflammatory substances that contribute to development of the skin lesion.

The extent of tissue damage is largely dependent on the amount of venom injected at the time of the bite. Small doses can elicit very little response such that many, if not most, *L. reclusa* bites go virtually unnoticed or do not result in ulcerations. A high dose, on the other hand, can result in destruction not only of the skin but also the underlying muscles into which the venom seeps due to gravity. The irregular shape of the skin lesion itself also reflects the effect of gravity, most evident in bites on the arms and legs. Some of the more severe cases result when the bite occurs in areas associated with fat tissue. The enzyme sphingomyelinase D in the venom readily destroys lipid cells, causing saponification and

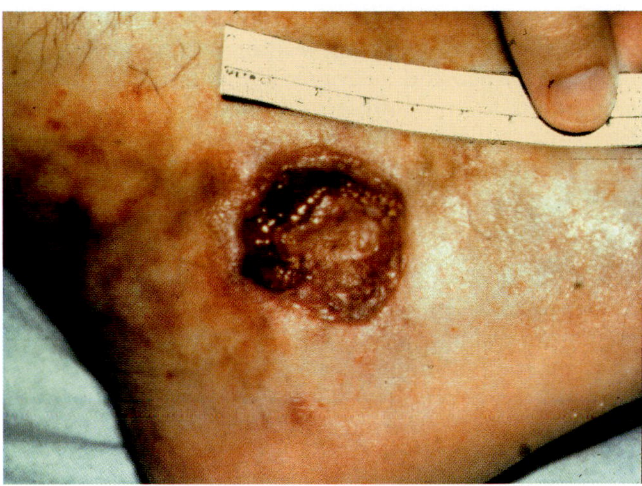

Figure 24.16 Slow-healing skin lesion just below ankle caused by bite of the brown recluse spider, *Loxosceles reclusa*; four months following envenomation, just before the patient underwent a skin graph to repair the tissue damage. (Courtesy of Kevin T. Humphreys, Huntsville, AL)

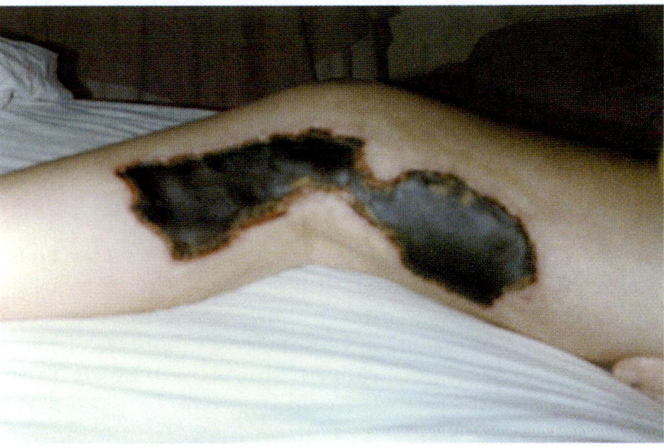

Figure 24.17 Severe case of envenomation by brown recluse spider, *Loxoxsceles reclusa*, on inner surface of leg of 19-year-old woman. The bite occurred at night while she was sleeping in bed. The large, black eschar denotes extent of tissue damage, as evident three months after the bite when picture was taken. (Courtesy of Carolyn Grissom, Shelbyville, TN)

extensive damage to the vasculature. This can cause severe tissue damage to the eyelids and face and to "baby fat" of infants. Because there is a bias toward reporting the more dramatic and graphic bite cases, people tend to mistakenly assume that most recluse bites lead to severe, disfiguring scarring.

Healing occurs very slowly, often requiring two to three months or more. The edges of the wound become thickened and raised as the central area begins to undergo scar formation. The necrotic tissue gradually sloughs away, often exposing underlying muscle. As the wound heals from beneath, a black scab-like **eschar** develops over the damaged area, protecting it during the healing process (Figure 24.17). Throughout this period it is important to keep the wound clean to avoid infections that can significantly prolong the healing time. The end result is typically a sunken scar varying in size from about 2 cm up to 10 cm or more.

A very small percentage of victims of brown recluse spider bites experience systemic reactions, usually within 24 to 48 hours after envenomation. These may include fever, malaise, nausea, vomiting, joint pains, and a generalized pruritic rash. Occasionally the systemic symptoms can be even more serious in the form of hemolysis, intravascular coagulation, and renal failure.

Treatment of *L. reclusa* bite victims remains controversial, with no single approach having been accepted by the medical community. Treatments that have been used with varying success include cleansing the wound with hydrogen peroxide and applying hyperbaric oxygen and burn creams to prevent secondary infection and promote healing. Corticosteroids injected directly into the lesion have been used, but to be effective must be done within a few hours following the bite. In many cases of mild envenomation, however, recluse bites heal well even without medical intervention (Swanson and Vetter, 2005).

Still another approach has been prompt surgical excision of the affected skin, particularly in severe cases, in an effort to remove the venom before it can do further damage. Excising affected tissues, however, often can result in more damage than simply allowing the affected area to heal without surgical intervention. Excision therefore is recommended only after the wound has stopped enlarging and the healing process can begin. Skin grafting and other forms of reconstructive surgery may be required in severe cases to repair extensive tissue damage that otherwise can lead to permanent, disfiguring scars.

In cases of systemic reactions, the anticoagulant heparin can be administered to reduce the risk of intravascular coagulation. Aggressive therapy to counter hemolysis and the use of dialysis in cases of renal failure also may be required. Antivenins are available for treatment of *L. laeta* and other *Loxosceles* species in South America; however, despite the development of antivenin for treatment of *L. reclusa* bites in North America, its production has not proved to be commercially feasible.

Cases of focal skin lesions and necrotic "bite" wounds often are misdiagnosed as brown recluse bites. In some cases other spiders may be involved; more commonly, however, the cause is unknown and is not spider-related at all. Examples of the latter include secondary infections of arthropod bites and stings (e.g., ticks, fire ants), skin abscesses and ulcerous lesions, and slow-healing wounds of diabetic patients. In recent years, an increasing number of cases of skin infections and soft-tissue injury due to methicillin-resistant *Staphylococcus aureus* (MRSA) is being misdiagnosed as loxoscelism. Even when a spider is involved in a suspected case, all too often the specimen is not recovered for identification; and when it is, typically the species is not confirmed by a spider specialist. The situation is further compounded by the tendency of physicians and the general public to blame "bites" of unknown origin on spiders, with no supporting evidence to corroborate it. The result is that the magnitude of misdiagnosed brown recluse bites is difficult to determine and no doubt varies significantly, depending on the geographic area. To date no reliable diagnostic test for recluse envenomation has been developed, although one test, still in the experimental stage, is showing promise.

South American Violin Spider (*Loxosceles laeta*) *Loxosceles laeta* is the largest species in the genus and poses a significant health concern in Central America and South America. It closely resembles *L. reclusa*, from which it is generally distinguished by its more reddish coloration and the fourth pair of legs of the female being longer than the others. In addition to its common names South American violin or brown recluse spider, it is called *araña de los rincones*, or the corner spider, because of its occurrence indoors in the corners of rooms. This spider has been introduced to several parts of North America and Europe where isolated local populations have become established, for example, in museums in New England and Finland. In 1960 an infestation was discovered on the Harvard University campus, where it was believed to have been present for some 20 years. Established populations of *L. laeta* have been documented at several locations in southern California, where they have been known to occur since the 1960s. They have not spread much since that time. At more northern locales, they tend to occur exclusively indoors, whereas in southern California they occasionally are found in sheltered places outdoors.

Females produce multiple egg sacs, each containing about 50 eggs, which are deposited in a dense, cottony part of the web usually at floor level. The number of egg sacs per female varies significantly and may be as high as 15 under laboratory conditions. In natural settings, females produce an average of three to seven egg sacs following a single mating. Most eggs are produced during the spring and summer (October–January) in South America. The developmental time from egg hatch to adult ranges from as short as six to eight months to a year or more. The adults are relatively long-lived, with mated and unmated females surviving about three and four years, respectively. Males live only about half as long as the females. Like other *Loxosceles* species, both sexes of *L. laeta* are able to survive prolonged periods without food and water, reportedly up to two years for some females. *Loxosceles laeta* often produces extensive webbing that is particularly noticeable in corners of rooms and along floor-level runways that they follow at the base of walls. These are composed of multilayers of coarse silk, the amount of which reflects the degree of spider activity and duration of the infestation.

For many years before the cause was determined in 1947, skin lesions resulting from the bite of *L. laeta* in South America were known as gangrenous spot syndrome. The bite reaction is similar to that of *L. reclusa*, producing a necrotic lesion that heals slowly. However, it is more often accompanied by systemic effects that can be life-threatening. Such cases are referred to as **viscerocutaneous loxoscelism**, in which the lungs, kidneys, liver, and central nervous system may be damaged. The venom causes severe inflammatory, cytotoxic, necrotic, and degenerative changes in tissues, leading to fever, jaundice, blood or hemoglobin in the urine, and sensorial involvement.

Latrodectism

The term latrodectism, also known as **neuromyopathic araneism**, is a syndrome caused by the bite of any of several *Latrodectus* species. The venom of these spiders contains potent neurotoxins that cause generalized pain, nausea, vomiting, faintness, dizziness, perspiration, and neuromuscular involvement in the form of muscle weakness, stiffness, cramps, tremors, loss of coordination, numbness or prickling sensations, paralysis, disturbed speech, and difficulty breathing. The main toxic fraction of the venom of *L. mactans* is a protein called α-latrotoxin, which acts on the motor nerve endings at the neuromuscular junctions. It causes depletion of the synaptic vesicles and the selective release of neurotransmitters that cause contraction of voluntary muscles. The autonomic nervous system also is affected. In severe cases the victims typically experience painful abdominal and leg cramps, profuse sweating, lachrymation, and spasms of the jaw muscles that distort the face and cause grimacing. Although symptoms often appear within 10 to 60 minutes, the syndrome may take several hours to develop; symptoms usually persist for 20 to 48 hours. A diagnostic sign of latrodectism is sweating at the bite site. This may be accompanied by localized swelling and redness, increased blood pressure, and the development of various types of rashes, either generalized or limited to the bite area. The fatality rate is relatively low (5%) even in untreated cases.

Prompt treatment of *Latrodectus* bite victims significantly reduces the severity of symptoms and promotes recovery. Antivenin can be very effective as a treatment, especially in cases involving *L. mactans*. Nonetheless, North American physicians, in general, are somewhat reticent to administer antivenin due to possible allergic reaction to the horse-serum base. Considering the efficacy of the antivenin and quickness of pain relief, however, physicians in other parts of the world (e.g., Australia) are more inclined to use it, being vigilant to watch for anaphylaxis. Antiquated remedies include hot baths, tourniquets, and the consumption of whiskey. Calcium gluconate, a commonly recommended treatment, is now considered ineffective (Clark et al., 1992). Current remedy, other than antivenin, entails the use of opiate and nonopiate analgesics and benzodiazepines, although their utility is still questionable (Vetter and Isbister, 2008). Children are more likely to have serious reactions to *Latrodectus* bites and should receive medical attention as quickly as possible.

Latrodectus taxonomy continues to be subject to many changes, with the number of recognized species differing significantly, depending on the author (e.g., 16 species by Levi, 1959, compared to 30 species by Garb et al., 2004). This reflects, in large part, differences of opinion as to which taxa represent species versus subspecies. The following currently recognized species, for example, previously were considered to be subspecies of *Latrodectus mactans*: *L. hasselti*, *L. menadovi*, and *L. tredecimguttatus*. The worldwide distribution of *Latrodectus* species is shown in Figure 24.18.

As a group *Latrodectus* species are medium-sized spiders, seldom more than 1.3 cm in body length, with globose abdomens and relatively long legs. They are generally recognized by their shiny black color and a red or orange hourglass marking on the underside of the abdomen. Some species, however, are more drab in appearance or may be colorfully patterned as in the case of *L. bishopi*. They are known by a variety of common names in different parts of the world. In North America they are usually called black widow spiders and, less commonly, hourglass spiders or button spiders. The term "widow" is derived from the misconception that the females invariably devour the male after mating, whereas the term "button" refers to the resemblance of the shiny black, round abdomen of the female to the buttons on old-fashioned shoes. Other common names include shoe-button spiders in South Africa; jockey in Arabia; *karakurt* (black wolf) in Russia; night stinger or katipo in Australia and New Zealand; *la malmignatte* in the Mediterranean region (notably Italy and Corsica); *araña capulina*, *chintatlahua*, and *viuda negra* in Mexico; *culrouge* and *veintecuatro horas* in the West Indies; *lucacha* in Peru; *mico* in Bolivia; *guina* and *pallu* in Chile; and *araña del lino* (flax spider), *araña del trigo* (wheat spider), and *araña rastrojera* (stubblefield spider) in Argentina and other parts of South America.

Latrodectus species are shy, retiring spiders that construct their tangled webs of coarse silk in dark, undisturbed places, usually close to the ground. They are especially common under logs and stones, in abandoned animal burrows, crevices in protected earthen banks, and various materials stacked on the ground. Some species, however, build large, irregular aerial webs in shrubs and other vegetation, often up to a meter or more above the ground. The adults are primarily nocturnal, spending most of the day in the security of their silken retreats in protected recesses adjoining the web. Prey consists primarily of medium to large-sized insects and other arthropods that stumble into the web.

Following mating, the female constructs one or more egg sacs, which she suspends in the web. Each sac typically contains 200 to 250 eggs. The total number of egg sacs produced by a given female varies considerably among species, with up to 10 for *L. mactans* and 20 for *L. hesperus*. The egg sacs are usually spherical or pyriform, white, yellow, or grayish, with a tough, tightly woven outer covering. The eggs hatch in 14 to 30 days. The spiderlings undergo their first molt within the sac three to four days after hatching, and then emerge via one or more tiny holes that they cut through the silken layers. The spiderlings remain in their mother's web for several weeks before dispersing. They undergo four to nine molts, depending on the species and sex. Males reach maturity in two to five months, whereas females require somewhat longer, usually 3.5 to eight months.

Spiderlings of *Latrodectus* species look quite different from the adult females. Young spiderlings are pale colored with light and dark stripes on the abdomen and legs. They often exhibit patterns of white, yellow, and red bands and spots that are gradually lost in the females as they mature. The males, however, tend to retain the color pattern of the immatures, including the abdominal markings and leg bands. Upon reaching sexual maturity, the males leave their webs to wander in search of a female. Mating takes place in the female's web. Contrary to popular belief, females of most *Latrodectus* species do not kill and devour the males following insemination any more frequently than do most other spiders. In fact, in the case of *L. bishopi* the adult male and female actually live together in the same web.

Five *Latrodectus* species occur in North America: *L. bishopi*, *L. geometricus*, *L. hesperus*, *L. mactans*, and *L. variolus*. *Latrodectus mactans* is the most toxic and widespread of these species, causing most of the cases

Figure 24.18 Generalized distribution of *Latrodectus* species. Not shown is the brown widow (*L. geometricus*), which occurs in multiple locations around the world. (Reprinted with permission from the *Annual Review of Entomology*, Volume 53 © 2008 by Annual Reviews)

of latrodectism requiring medical attention. *Latrodectus tredecimguttatus* is the most important species in Europe, whereas *L. curacaviensis*, *L. hasselti*, and *L. katipo* are important species in South America, Australia, and New Zealand, respectively.

Southern Black Widow (*Latrodectus mactans*) This is the most notorious of all toxic spiders because of its potent venom, widespread occurrence, and likelihood of coming in contact with humans and domestic animals. It is found throughout eastern North America from southern New England to eastern Mexico. The adult female typifies the general description of *Latrodectus* in being shiny black with a prominent red hourglass on the underside of the abdomen (Figure 24.19). It usually has a small red spot just above the spinnerets and occasionally a median series of additional red spots extending anteriorly onto the abdominal dorsum. It is found outdoors in various protected places under rocks, logs, boards and other ground debris, and frequently near buildings. In North America, *L. mactans* also occurs indoors in barns, wood sheds, garages, and various other unheated storage areas, similar to *L. hesperus*. Before the days of indoor plumbing, the widespread use of outdoor privies contributed significantly to the number of human cases of envenomation, especially in males bitten on the genitalia, by black widow spiders in webs under the toilet seats.

Although the bite of *L. mactans* may go unnoticed in some cases, it is more commonly felt as a pin prick or an immediate sharp, burning pain with little or no swelling. The pain spreads from the bite site to regional lymph nodes and other parts of the body, usually reaching its maximum intensity in one to three hours. Thereafter the pain may be continuous or intermittent, lasting up to 48 hours. The accompanying muscular spasms and cramps, especially in the abdomen and legs, can lead to tightness of the chest, board-like rigidity of abdominal muscles, and complete prostration. In extreme cases, complications may occur in the form of shock, leukocytosis, and lesions of the liver, spleen, and kidney evidenced by blood and elevated protein levels in the urine. The death rate can be as high as 4 to 5% in untreated cases, the highest of all *Latrodectus* species.

Figure 24.19 Southern black widow spider, *Latrodectus mactans* (Theridiidae), female, ventral view, showing characteristic red hourglass marking. (Gertsch, 1979)

Northern Black Widow (*Latrodectus variolus*) The northern black widow closely resembles *L. mactans* in both size and general appearance. It is usually distinguished, however, by the ventral hourglass being divided into two transverse bands and a row of prominent red spots along the dorsal midline of the abdomen. Its geographic range largely overlaps that of *L. mactans* in North America, occurring widely throughout the eastern United States as far south as western Florida and eastern Texas. It is more common in the northern states, extending into southeastern Canada. Unlike *L. mactans*, it seldom is found in buildings, preferring outdoor situations such as old stumps, piles of dead tree branches, hollow logs, abandoned animal burrows, cavities in rock walls, and under debris. It is common in wooded areas where it constructs a large tangled web in shrubs and tree branches, sometimes as high as 6 meters above the ground.

Western Black Widow (*Latrodectus hesperus*) This species also is very similar in appearance to *L. mactans*. The abdominal dorsum is typically all black, only infrequently having red markings. The ventral, red hourglass is well defined and usually complete. It occurs from Oklahoma, Kansas and central Texas into the western United States, including the high elevations of Colorado and adjacent Mexico. It also is found in Israel as an introduced species. The western black widow utilizes a wider range of habitats than *L. mactans*. It most commonly constructs its web near the ground, in animal burrows, or under various objects. However it also is found above the ground in shrubs and trees, and in such places as grape arbors and bird nests. It is well adapted to semi-arid and arid habitats where it can be found in soil crevices and various desert plants such as agaves and cacti. It is commonly found around human habitation, often in high densities.

Brown Widow (*Latrodectus geometricus*) As its common name implies, *L. geometricus* females are brownish rather than black and have a highly variable color pattern, giving it a mottled appearance (Figure 24.20). Typically there are three pairs of irregular-shaped spots along the dorsal midline of the abdomen. These vary from simple white spots with black borders to multicolored bull's-eye spots marked with white, yellow, orange, reddish brown, tan, gray, or aqua. The hourglass is generally dull orange, complete, and commonly bordered with yellow. The legs are characterized by brown and black bands. Occasionally individuals are nearly black, more closely resembling other widow spiders.

Although the brown widow occurs in many places around the globe, it is thought to have originated in Africa or possibly in South America. Introductions to the United States have led to established populations

Figure 24.20 Brown widow spider, *Latrodectus geometricus* (Theridiidae), female, dorsolateral view. (Photo by Sturgis McKeever)

Figure 24.21 Red widow spider, *Latrodectus bishopi* (Theridiidae), female, ventral view. This widow spider is unusual in lacking the red hourglass marking; instead the marking typically is reduced to a transverse bar or single, triangular spot. (From Short and Castner, 1992, courtesy of University of Florida-IFAS)

in Texas, Louisiana, Mississippi, Alabama, Florida, Georgia, South Carolina, southern California, and Hawaii. This spider is found in a wide range of domestic settings in or near buildings. It constructs a relatively small tangled retreat in corners and chinks of brick and cement walls, building foundations and fences, under overhangs of steps, and along street curbs. In addition to urban areas, *L. geometricus* is common in South America along ocean beaches where it seems to prefer to construct its web in low, running plants such as the morning-glory *Ipomoea biloba* (Convolvulaceae) above the high tide mark. It is even more shy than other widow spiders, is not aggressive, and rarely has been recorded biting humans. In verified bites involving humans, symptoms typically are limited to pain during fang penetration and a red dermal mark. Brown-widow bites do not typically result in the severe symptoms commonly associated with latrodectism.

Red Widow (*Latrodectus bishopi*) The cephalothorax and legs of the red widow are orange or reddish, contrasted with a black abdomen that has red or orange spots with yellow borders (Figure 24.21). The ventral hourglass is often pale and usually reduced to a transverse bar or a single triangular spot. The red widow is known to occur only in the sand-pine scrub habitat of peninsular Florida, where it commonly builds its web in palmettos (Aracaceae) Although little is known about the toxicity of *L. bishopi* venom, bites by this species are rare and apparently of little, or no, medical concern.

European Black Widow (*Latrodectus tredecimguttatus*) *Latrodectus tredecimguttatus* is the most common widow spider in Europe. It occurs primarily in the Mediterranean region, extending eastward into Eurasia. In Italy and Corsica it is called **la malmignatte** and in Russia, **karakurt** ("black wolf"). It occurs exclusively outdoors where it constructs its web in a variety of herbaceous and shrubby vegetation, including cultivated crops and arbors. The European black widow is characterized by 13 dorsal red spots and absence of an hourglass. It poses a significant occupational hazard for farmers and field workers who are commonly bitten while harvesting or threshing wheat, handling hay, picking fruit, or working in vineyards.

Araña Del Trigo (*Latrodectus curacaviensis*) In South America, *L. curacaviensis* is encountered frequently by humans in buildings, garages, and privies. It also is found in cultivated crops such as wheat, accounting for its common name *araña del trigo*, or wheat spider. It closely resembles *L. mactans*, with which it is easily confused. Like *L. mactans*, its bite can cause serious illness and death.

Australian Redback Spider (*Latrodectus hasselti*) Females of this Australian species are 2 to 3 cm long in leg length. They are black with a prominent red stripe along the dorsal midline of the abdomen (Figure 24.22), hence the common name red back. The immatures of other *Latrodectus* species (e.g., *L. mactans*) sometimes have a dorsal red stripe, which has led to misidentifications as *L. hasselti*. The Australian red-back spider occurs in sheltered, usually dry sites such as hollows of trees and under logs and rocks. It often builds its retreat and web around building foundations and in ventilator gratings, trash cans, and gas-meter boxes. Although this spider is not aggressive, the female causes a painful bite that can lead to systemic envenomation and fatalities (ca. 5%) if untreated. In typical cases, the initial bite is relatively painless, comparable to a pin prick. Thereafter the pain intensifies from a few minutes to a half hour, often accompanied by localized perspiration and edema, and nausea and vomiting. Localized sweating at the bite site is an important diagnostic sign (Wiener, 1961). In untreated systemic cases, recovery usually is protracted and may take three to four weeks. The severity of red-back spider bites was dramatically reduced in Australia with the introduction of an

Figure 24.22 Australian redback spider, *Latrodectus hasselti* (Theridiidae), female, postero-ventral view, showing characteristic red hour-glass marking and wide, reddish stripe along dorsal midline of abdomen. (Photo by Julian White©)

antivenin against *L. hasselti* in 1956; since that time, no fatalities have been recorded.

Katipo Spider (*Latrodectus katipo*) This spider is primarily a coastal species found high on ocean beaches and in river beds, under driftwood, or at the base of vegetation. Although found primarily in New Zealand, where it is known as the New Zealand redback, *L. katipo* also is said to occur in the Caribbean basin on coastal beaches of Jamaica (Southcott, 1976). It is being displaced from its highly restricted distribution in New Zealand by another theridiid spider, *Steatoda capensis.*

VETERINARY IMPORTANCE

Under certain circumstances spiders can pose health threats to household pets, livestock, and other domestic animals. Most cases occur either in stables or in pasture situations where animals are grazing. In the latter situation, high density of a toxic species can cause significant veterinary concerns. Although the evidence of spider bites in nonhuman animals often is circumstantial, enough cases have been documented to show that theridiid and theraphosid spiders are the more common causes of serious spider bites of veterinary importance worldwide.

Most cases of envenomation by spiders in grasslands and pastures are caused by *Latrodectus* species. They usually occur in localized areas where populations of certain toxic taxa are high. Outbreaks involving *L. tredecimguttatus* (published as *L. erebus*) were reported in the steppes of southern Russia in the 1830s. Grazing animals were severely bitten, causing some to stampede due to the pain and to run until they dropped. Fatalities as high as 12% in sheep, 17% in horses, and 33% in camels were reported (Motchoulsky and Becker, 1855). Notable outbreaks of latrodectism affecting cattle and agricultural workers also occurred in Spain in the 1830s and 1840s, and caused deaths in horses, sheep, and other livestock in Chile in the 1870s. For other examples of latrodectism involving goats, sheep, cattle, and horses in Europe, South Africa, and Indonesia, see Maretic and Lebez (1979). Latrodectism does not seem to present a significant problem for livestock in North America or South America.

Cases of *Latrodectus* envenomation in dogs and cats can be serious, with onset of clinical signs usually during the first eight hours following a bite. Moderate to severe cases typically result in extreme pain, abdominal rigidity without tenderness, hypertension, and, in the case of cats, paralysis. Cats are particularly sensitive to the venom, with a significantly higher mortality rate than in dogs. The primary treatment for cases of *Latrodectus* envenomation in veterinary practices is administration of specific antivenin (Peterson, 2006b). Extracts of *Latrodectus* eggs and spiderlings have been shown to induce a potentially fatal hypotension in cats, indicating significantly different toxins in the immature and adult spiders (Ebeling, 1978). The egg toxin has been described by Buffkin et al. (1971).

Brown recluse spiders, notably *Loxosceles reclusa*, cause necrotic skin lesions in household pets and livestock, similar to that in humans. Tissue damage is attributed primarily to rapid blood coagulation and occlusion of capillaries at the bite site, with resultant ischemia. Systemic signs occur in only a small percentage of cases, with the most prevalent signs being hemolytic anemia and significant hemoglobinuria. Systemic cases are potentially fatal and may require intravenous fluid therapy to maintain adequate hydration and to protect renal function (Peterson, 2006a).

Envenomation by *Badumna robusta*, a member of the family Desidae, has been reported as a problem in stables in Queensland, Australia. Known as the black house spider or window spider, *B. robusta* is nocturnal and has been known to bite horses while they are bedded down for the night. It can inflict a painful bite with associated inflammation and swelling, usually about the head and neck (Southcott, 1976).

Tarantulas have been implicated in several reported cases of envenomation of pastured animals in Central America in recent years. *Aphonopelma seemani* and *Sphaerobothria hoffmani*, two theraphosid species that construct deep burrows in pasture soils, are believed to be the cause of necrotic lesions in cattle and horses in Costa Rica (Herrero and Bolanos, 1982). These and other theraphosids in the Americas are suspected as the cause of vesicular lesions on the muzzle, hooves, and udder of cattle, horses, and swine.

PREVENTION AND CONTROL

The best way to avoid being bitten by spiders is not to handle them or allow direct contact while working in areas where they are likely to occur. Gloves can be worn to protect the hands while gardening, potting plants, stacking or handling firewood, moving rocks and other ground materials, or involvement in various other outdoor activities. Spiders can be discouraged from constructing webs around windows and under eaves of buildings by regular removal of their webs with a broom and turning off unnecessary lights at night that attract insect prey. Lumber, trash, garden materials, or other items piled next to buildings should be removed, and cluttered areas in basements, attics, and outbuildings that can serve as protected sites for spiders should be eliminated. Windows and doors should be tightly closed and adequately screened to prevent wandering spiders from entering homes during their activity periods. Regular removal of cobwebs and vacuuming the corners of rooms, window frames and sills, baseboards, and the underside of furniture help to prevent spiders from establishing themselves indoors. The elimination of household insects on which indoor spiders depend for food also helps to reduce infestations.

If necessary, pesticides registered for indoor control of spiders can be used. They are available in the form of sprays, dusts, aerosols, and fogs that help to reach spiders that may escape sweeping and vacuuming by retreating into cracks and crevices. Products with residual activity can be applied to outdoor surfaces such as under eaves, crawl spaces under houses, and around decks and patios.

To avoid bites by black widow spiders and related *Latrodectus* species, precautions should be taken to minimize contact with them in wood piles, under rocks and logs, in outbuildings, and other sheltered outdoor sites. Avoid putting unprotected hands into recesses at ground level, such as water-meter encasements and accumulations of trash and other debris that can serve as retreats. Special precaution should be taken when using outdoor privies and portajohns to avoid contact with webs of black widow spiders on the underside of toilet seats.

In the case of *Loxosceles laeta*, *L. reclusa*, and *L. rufescens*, which occur indoors, appropriate measures can be taken to minimize the risk of being bitten in infested premises. Visually inspect clothes closets, water-heater closets, utility rooms, basements, attics, and other storage areas, killing and removing any specimens that are found. Inspect dry, relatively undisturbed hiding places such as old boxes, wooden cabinets, behind wall hangings, and under beds and other furniture. Shake out clothing that has been hanging unused for some time to avoid encounters with spiders in sleeves and trouser legs. Inspect shoes and other footwear before putting them on, and visually check towels or shake them before use. Where infants or small children are involved, cribs or beds should be pulled away from the wall, and the bed clothes should be kept from reaching the floor.

Established *Loxosceles* populations can be difficult to eliminate without the use of pesticides to reach individuals that are not accessible to sweeping and vacuum cleaning. Sprays of appropriate materials can be applied in attics, crawl spaces, along baseboards, in corners of rooms, under furniture, behind cabinets, and in other out-of-the-way places to kill the spiders. Treating any webs that are found with pesticide dusts generally helps to reduce infestations more quickly than applying materials to other surfaces. Sticky traps can be used to catch *Loxosceles* species and to monitor their activity before and after treatments.

REFERENCES AND FURTHER READING

Akre, R. D., & Myhre, E. A. (1991). Biology and medical importance of the aggressive house spider, *Tegenaria agrestis*, in the Pacific Northwest (Arachnida: Araneae: Agelenidae). *Melanderia, 47*, 1–30.

Anonymous, (1996). Necrotic arachnidism Pacific Northwest, 1988–1996. *Morbidity and Mortality Weekly Report, 45*, 433–436.

Barth, F. G. (2001). *A Spider's Work: Senses and Behavior.* Berlin: Springer-Verlag.

Bettini, S. (Ed.). (1978). *Arthropod Venoms. Handbook of Experimental Pharmacology* (S. Bettini, Ed., Vol. 48). New York: Springer-Verlag.

Bettini, S., & Brignoli, P. M. (1978). Review of the spider families, with notes on the lesser-known poisonous families. In S. Bettini (Ed.), *Arthropod Venoms. Handbook of Experimental Pharmacology* (Vol. 48, pp. 103–120). New York: Springer-Verlag.

Bettini, S., & Maroli, M. (1978). Venoms of Theridiidae, genus *Latrodectus.* In S. Bettini (Ed.), *Arthropod Venoms. Handbook of Experimental Pharmacology* (Vol. 48, pp. 149–184). New York: Springer-Verlag.

Binford, G. J. (2001). An analysis of geographic and intersexual chemical variation in venoms of the spider *Tegenaria agrestis* (Agelenidae). *Toxicon, 39*, 955–968.

Bonnet, P. (1945–1961). Bibliographia Araneorum. *Toulouse, 1*, 1–832, 2, 1–5058.

Breene, R. G. (1996). Arachnid common names in North America. *Newsletter of the British Arachnological Society, 77*, 1–4.

Brignoli, P. M. (1983). *A Catalogue of the Araneae Described Between 1940 and 1981.* Dover, NH: Manchester University Press.

Bucharetchi, F., Deus Reinaldo, C. R., Hyslop, S., Madureira, P. R., Capitani, E. M., & Vieira, R. J. (2000). A clinico-epidemiological study of bites by spiders of the genus *Phoneutria. Revista do Instituto de Medicina Tropical de Sao Paulo 42*, 17–21.

Bucherl, W. (1971). Spiders. In W. Bucherl, & E. Buckley (Eds.), *Venomous Animals and Their Venoms* (Vol. 3, pp. 197–277). New York: Academic Press.

Buffkin, D. C., Russell, F. E., & Deshmukh, A. (1971). Preliminary studies on the toxicity of black widow spider eggs. *Toxicon, 9*, 393–402.

Clark, R. F., Wethern-Kestner, S., Vance, M. V., & Gerkin, R. (1992). Clinical presentation and treatment of black widow spider envenomation: A review of 163 cases. *Annals of Emergency Medicine, 21*, 782–787.

Coddington, J. A., Larcher, S. F., & Cokendolpher, J. C. (1990). The systematic status of Arachnida, exclusive of Acari, in North America north of Mexico. In M. Kosztarab & C. W. Schaefer (Eds.), *Systematics of North American Insects and Arachnids: Status and Needs. Virginia Agricultural Experiment Station Information Services 90–1* (pp. 5–20). Blacksburg, VA: Virginia Polytechnic Institute and State University.

Comstock, J. H. (1948). *The Spider Book.* Cornell University Press. Ithaca, NY.

Cooke, J. A. L., Roth, V., & Miller, F. H. (1972). The urticating hairs of theraphosid spiders. *American Museum Novitates*, No. 2498.

Duchen, L. W., & Gomez, S. (1984). Pharmacology of spider venoms. In A. T. Tu (Ed.), *Handbook of Natural Toxins* (Vol. 2, pp. 483–512). New York: Marcel Dekker.

Ebeling, W. (1978). *Urban Entomology.* Berkeley: University of California.

Fink, L. S. (1984). Venom spitting by the green lynx spider, *Peucetia viridans* (Araneae, Oxyopidae). *Journal of Arachnology, 12*, 372–373.

Foelix, R. F. (1996). *Biology of Spiders* (2nd ed.). Stuttgart: Oxford University Press. *Biologie der Spinnen*, Georg Thieme Verlag.

Garb, J. E., Gonzalez, A., & Gillespie, R. G. (2004). The black widow spider genus *Latrodectus* (Araneae: Theridiidae): Phylogeny, biogeography, and invasion history. *Molecular and Phylogenetic Evolution, 31*, 1127–1142.

Geren, C. R., & Odell, G. V. (1984). The biochemistry of spider venoms. In A. T. Tu (Ed.), *Handbook of Natural Toxins* (Vol. 2, pp. 44–481). New York: Marcel Dekker.

Gertsch, W. J. (1979). *American Spiders* (2nd ed.). New York: Van Nostrand Reinhold Co.

Gertsch, W. J., & Ennik, F. (1983). The spider genus *Loxosceles* in North America, Central America and the West Indies (Araneae,

Loxoscelidae). *Bulletin of the American Museum of Natural History, 175*, 263–360.

Gorham, J. R., & Rheney, T. B. (1968). Envenomation by the spiders *Chiracanthium inclusum* and *Argiope aurantia*. *Journal American Medical Association, 206*, 1958–1962.

Gray, M. R. (1988). Aspects of the systematics of the Australian funnel web spiders (Araneae: Hexathelidae: Atracinae) based upon morphological and electrophoretic data. In A. D. Austin, & N. W. Heather (Eds.), *Australian Entomology* (pp. 113–125). Miscellaneous Publication No. 5. Brisbane: Australian Entomological Society.

Gray, M. R., & Sutherland, S. K. (1978). Venoms of Dipluridae. In S. Bettini (Ed.), *Arthropod Venoms. Handbook of Experimental Pharmacology* (Vol. 48, pp. 121–148). New York: Springer-Verlag.

Herrero, M. V., & Bolanos, R. (1982). Life-history and tunnels of 2 horse-biting spiders from Costa Rica (Araneae: Theraphosidae). Preliminary observations (in Spanish). *Brenesia, 19/20*, 319–324.

Hickin, N. E. (1984). *Pest Animals in Buildings*. London: George Godwin.

Isbister, G. K., Graudins, A., White, J., & Warrell, D. (2003). Antivenom treatment in arachnidism. *Journal of Toxicology-Clinical Toxicology, 41*, 291–300.

Isbister, G. K., & Gray, M. R. (2003a). Effects of envenoming by comb-footed spiders of the genera *Steatoda* and *Achaearanea* (Family Theridiidae: Araneae) in Australia. *Journal of Toxicology-Clinical Toxicology, 41*, 809–819.

Isbister, G. K., & Gray, M. R. (2003b). White-tail spider bite: A prospective study of 130 definite bites by *Lampona* species. *Medical Journal of Australia, 179*, 199–202.

Isbister, G. K., Gray, M. R., Balit, C. R., Raven, R. J., Stokes, B. J., Porges, K. et al. (2005). Funnel-web spider bite: A systematic review of recorded clinical cases. *Medical Journal of Australia, 182*, 407–411.

Isbister, G. K., & White, J. (2004). Clinical consequences of spider bites: Recent advances in our understanding. *Toxicon, 43*, 477–492.

Kaston, B. J. (1970). Comparative biology of American black widow spiders. *Transactions of the San Diego Society of Natural History, 16*, 33–82.

Kaston, B. J. (1978). *How to Know the Spiders* (3rd ed.). Dubuque, IA: William. C. Brown Co.

Levi, H. W. (1959). The spider genus *Latrodectus* (Araneae: Theridiidae). *Transactions of the American Microscopical Society, 78*, 7–43.

Lucas, S. (1988). Spiders in Brazil. *Toxicon, 26*, 759–772.

Maretic, A., & Lebez, D. (1979). *Araneism with Special Reference to Europe*. Belgrade: Nolit Publishing House.

McCrone, J. D., & Levi, H. W. (1964). North American widow spiders of the *Latrodectus curacaviensis* group (Araneae: Theridiidae). *Psyche, 71*, 12–27.

Millikan, L. E. (1984). Biology of spiders. In W. B. Nutting (Ed.), *Mammalian Diseases and Arachnids* (Vol. I, pp. 59–81). Boca Raton, FL: CRC Press.

Motchoulsky, V. N., & Becker, V. (1855). Bulletin de Moscou; as cited by A. Maretic, & D. Lebez (1979). *Araneism with Special Reference to Europe*. Belgrade: Nolit Publishing House.

Newlands, G., & Atkinson, P. (1988). Review of southern African spiders of medical importance, with notes on the signs and symptoms of envenomation. *South African Medical Journal = Suid-Afrikaanse Tydskrif vir Geneeskunde, 73*, 235–239.

Ori, M. (1984). Biology of and poisoning by spiders. In A. T. Tu (Ed.), *Handbook of Natural Toxins* (Vol. 2, pp. 397–440). New York: Marcel Dekker.

Peterson, M. E. (2006a). Brown spider envenomation. *Clinical Techniques in Small Animal Practice, 21*, 191–193.

Peterson, M. E. (2006b). Black widow spider envenomation. *Clinical Techniques in Small Animal Practice, 21*, 187–190.

Platnick, N. I. (1989). *Advances in Spider Taxonomy 1981–1987. A Supplement to Brignoli's A Catalogue of the Araneae Described Between 1940 and 1981*. UK: Manchester University Press.

Platnick, N. I. (1993). *Advances in Spider Taxonomy 1988–1991. With Synonymies and Transfers 1940–1980* (P. Merrett, Ed.). New York: New York Entomological Society.

Platnick, N. I. (2008). *The World Spider Catalog, Version 9.0*. American Museum of Natural History, online at http://research.amnh.org/entomology/spiders/catalog/index.html.

Preston-Mafham, R., & Preston-Mafham, K. (1984). *Spiders of the World*. New York: Facts on File Publications.

Rash, L. D., & Hodgson, W. C. (2002). Pharmacology and biochemistry of spider venoms. *Toxicon, 40*, 225–254.

Raven, R. J. (1985). The spider infraorder Mygalomorphae (Araneae): Cladistics and systematics. *Bulletin of the American Museum of Natural History, 182*, 1–180.

Roewer, C. F. (1942–1954). *Katalog der Araneae*. Bremen: Natura.

Roth, V. D. (1993). *Spider Genera of North America with Keys to Families and Genera and a Guide to Literature* (3rd ed.). Gainesville, FL: American Arachnological Society.

Russell, F. E. (1970). Bites by the spider *Phidippus formosus*. Case history. *Toxicon, 8*, 193–194.

Schenberg, S., & Pereira Lima, F. A. (1978). Venoms of Ctenidae. In S. Bettini (Ed.), *Arthropod Venoms. Handbook of Experimental Pharmacology* (Vol. 48, pp. 217–246). New York: Springer-Verlag.

Schenone, H., & Suarez, G. (1978). Venoms of Scytodidae, genus *Loxosceles*. In S. Bettini (Ed.), *Arthropod Venoms. Handbook of Experimental Pharmacology* (Vol. 48, pp. 247–275). New York: Springer-Verlag.

Short, D. E., & Castner, J. L. (1992). *Venomous Spiders of Florida. Leaflet No. SP 104. Institute of Food and Agricultural Sciences*. Gainesville: University of Florida.

Southcott, R. V. (1976). Arachnidism and allied syndromes in the Australian regions. *Records of Adelaide Children's Hospital, 1*, 97–187.

Southcott, R. V. (1984). Diseases and arachnids in the tropics. In W. B. Nutting (Ed.), Mammalian Diseases and Arachnids (Vol. 2, pp. 15–56). Boca Raton, FL: CRC Press.

Spielman, A., & Levi, H. W. (1970). Probable envenomation by *Chiracanthium mildei*; a spider found in houses. *The American Journal of Tropical Medicine and Hygiene, 19*, 729–732.

Sutherland, S. K. (1987). Watch out, Miss Muffet! *The Medical Journal of Australia, 147*(11–12), 531.

Sutherland, S. K. (1990). Treatment of arachnid poisoning in Australia. *Australian Family Physician, 19*(1), *17*, 50–55, 57–61, 62.

Swanson, D. L., & Vetter, R. S. (2005). Bites of brown recluse spiders and suspected necrotic arachnidism. *The New England Journal of Medicine, 352*, 700–707.

Taylor, S. P., & Greve, J. H. (1984). Suspected case of loxoscelism (spider-bite) in a dog. *Iowa State University Veterinarian, 47*, 84–86.

Tinkham, E. R. (1946). A poison-squirting spider. *Bulletin of the United States Army Medical Department, 5*, 361–362.

Tu, A. T. (1984). *Handbook of Natural Toxins. Vol. 2. Insect Poisons, Allergens, and Other Invertebrate Venoms*. New York: Marcel Dekker.

Ubick, D., Paquin, P., Cushing, P. E., & Roth, V. (Eds.). (2005). *Spiders of North America: An Identification Manual*. Gainesville, FL: American Arachnological Society.

Vest, D. K. (1987). Necrotic arachnidism in the Northwest United States and its probable relationship to *Tegenaria agrestis* (Walckenaer) spiders. *Toxicon, 25*, 175–184.

Vetter, R. S. (2008). Spiders of the genus *Loxosceles* (Araneae: Sicariidae): A review of biological, medical and psychological aspects regarding envenomations. *Journal of Arachnology, 36*, 150–163.

Vetter, R. S., & Hillebrecht, S. (2008). On distinguishing two often-misidentified genera (*Cupiennius, Phoneutria*) (Araneae: Ctenidae) of large spiders found in Central and South American cargo shipments. *American Entomologist, 54*, 82–87.

Vetter, R. S., & Isbister, G. K. (2004). Do hobo spider bites cause dermonecrotic injuries. *Annals of Emergency Medicine, 44*, 605–607.

Vetter, R. S., & Isbister, G. K. (2008). Medical aspects of spider bites. *Annual Review of Entomology, 53*, 409–429.

Vetter, R. S., Isbister, G. K., Bush, S. P., & Boutin, L. J. (2006). Verified bites by yellow sac spiders (Genus *Cheiracanthium*) in the United States and Australia: Where is the necrosis? *American Journal of Tropical Medicine and Hygiene, 74*, 1043–1048.

White, J., Cardoso, J. L., & Hui, W. F. (1995). Clinical toxicology of spider bites. In J. Meier & J. White (Eds.), *Handbook of Clinical Toxicology of Animal Venoms and Poisons*. Boca Raton, FL: CRC Press.

White, J. D., & Hirst Hender, E. (1989). 36 cases of bites by spiders, including the white-tailed spider, *Lampona cylindrata*. *The Medical Journal of Australia, 150*, 401–403.

Wiener, S. (1961). Red back spider bite in Australia: An analysis of 167 cases. *The Medical Journal of Australia, 2*, 44–49.

Wong, R. C., Hughes, S. E., & Voorhees, J. J. (1987). Spider bites. *Archives of Dermatology, 123*, 98–104.

Chapter

25

Mites (Acari)

Gary R. Mullen ▪ Barry M. OConnor

More than 250 species of mites are recognized as the cause of health-related problems for humans and domestic animals. Types of problems include: (1) temporary irritation of the skin due to bites or feeding on host skin, fur, and feathers; (2) persistent dermatitis in response to mites invading the skin or hair follicles; (3) mite-induced allergies; (4) transmission of pathogenic microbial agents and metazoan parasites; (5) serving as intermediate hosts of parasites, notably tapeworms; (6) invasion of respiratory passages, ear canals, and occasionally internal organs; (7) an abnormal fear of mites, or **acarophobia**; and (8) **delusory acariosis**, a psychological condition in which individuals are convinced that they are being attacked by mites when, in fact, no mites are involved. The general term for infestations of animals by mites is called **acarinism**, whereas any disease condition caused by mites is **acariasis** (acarinosis).

For an introduction to mites in general, see Krantz (1978), Woolley (1988), Evans (1992), and Krantz and Walter (2009). For major works dealing specifically with taxa of medical-veterinary importance, the following sources are suggested: Hirst (1922), Baker et al. (1956), Strandtmann and Wharton (1958), Sweatman (1971), Yunker (1973), Nutting (1984), and Baker (1999).

TAXONOMY

Based on the classification scheme described by Evans (1992), mites comprise the arachnid Subclass Acari, which in turn is divided into two major groups, the Anactinotrichida and Actinotrichida. These are further subdivided into seven orders (Table 25.1). The use of alternative names for these orders and the designation of the orders as suborders by various authors causes understandable confusion for those not familiar with mite classification. For discussions of the higher classification of mites see van der Hammen (1972), Krantz (1978), Kethley (1982), OConnor (1984), and Evans (1992).

Members of the orders Ixodida, Mesostigmata, Prostigmata, Astigmata, and Oribatida are the cause of animal-health problems. Most of them are represented by the 50 families listed in Table 25.2, which are covered in this chapter. Not included in the list are the many families of ectoparasitic mites found on wild mammals, birds, and reptiles, most of which cause little or no significant harm to their hosts. Given the number and diversity of taxa involved, there is no single source to which one can turn for identification of mites of medical-veterinary importance. Reliable species determinations usually require the preparation of slide-mounted specimens for microscopic examination and the assistance of an acarologist who has access to the appropriate taxonomic literature. A reference that the nonspecialist can use for identifying some of the more common mite pests of public-health and veterinary importance is the *CDC Pictorial keys: arthropods, reptiles, birds and mammals of public health significance* (Pratt and Stojanovich, 1969).

MORPHOLOGY

The basic body plan of mites is shown in Figure 25.1. The body is divided into two major regions, the anterior **gnathosoma**, bearing the pedipalps and chelicerae, and the **idiosoma**, the remainder of the body bearing the legs and eyes (when present). The **pedipalps** (or palps, palpi) are typically five-segmented but may be greatly reduced and highly modified in different groups of mites. The pedipalps are primarily sensory appendages equipped with chemical and tactile sensors that assist mites in finding food and perceiving environmental cues. In some groups they may be modified as raptorial structures for capturing prey or as attachment devices to facilitate clinging to hosts. The mouthparts consist primarily of a pair of **chelicerae**, each of which is typically three-segmented and terminates in a chela, or pincer. The **chela** is composed of a fixed digit and a movable digit designed for seizing or grasping. In the case of certain parasitic mites the chelicerae are highly modified as long, slender structures for piercing skin to feed on blood and other host tissues. In some groups, structures associated with the mouthparts may be modified as attachment devices to help secure them to their hosts (e.g., chiggers).

Table 25.1 Higher Classification of Mites

Class Arachnida
Subclass Acari

Superorder Anactinotrichida (Parasitiformes)*

Orders:	
	Notostigmata (Opilioacarida)
	Holothyrida (Tetrastigmata)
	Ixodida (Metastigmata)
	Mesostigmata (Gamasida)

Superorder Actinotrichida (Acariformes)

Orders:	
	Prostigmata (Actinedida + Tarsonemida)
	Astigmata (Acaridida)
	Oribatida (Cryptostigmata, Oribatei)

*Alternative names for the respective orders are shown in parentheses.
Based on Evans, 1992

Table 25.2 Orders and Families of Mites That Include Species of Medical-Veterinary Importance or Interest*

Order Ixodida	Cytoditidae
Argasidae	Dermationidae
Ixodidae	Dermoglyphidae
Order Mesostigmata	Echimyopodidae
Ascidae	Epidermoptidae
Dermanyssidae	Gastronyssidae
Entonyssidae	Listrophoridae
Halarachnidae	Lamninosioptidae
Laelapidae	Myocoptidae
Macronyssidae	Pneumocoptidae
Rhinonyssidae	Proctophyllidae
Spinturnicidae	Psoroptidae
Order Prostigmata	Pteryolichidae
Cheyletidae	Pyroglyphidae
Demodicidae	Rhyhncoptidae
Ereynetidae	Sarcoptidae
Harpirhynchidae	Syringobiidae
Myobiidae	Turbinoptidae
Pyemotidae	Order Oribatida
Psorergatidae	Glycyphagidae
Syringophilidae	Ceratozetidae
Tarsonemidae	Hypoderatidae
Trombiculidae	Galumnidae
Order Astigmata	Knemidokoptidae[a]
Acaridae	Oribatulidae
Analgidae	Lemurnyssidae
Atopomelidae	Scheloribatidae
Carpoglyphidae	

*Families are listed alphabetically under each order.
[a]Knemidokoptidae included within Epidermoptidae by some authors (Mironov et al., 2005)

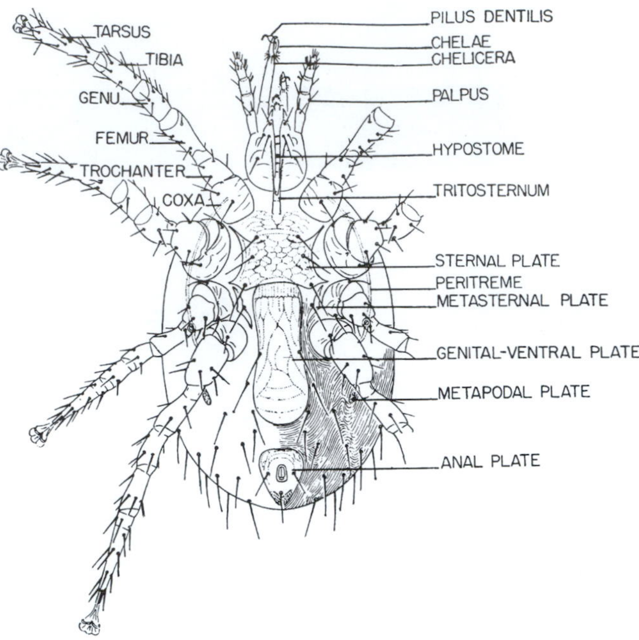

Figure 25.1 Generalized morphology of adult female mite (Mesostigmata, Laelapidae, *Androlaelaps fahrenholzi*), ventral view. (Baker et al., 1956)

The idiosoma can be divided into several regions. The anterior part bearing the legs is the **podosoma**. The posterior section behind the legs is the **opisthosoma**. Other regions include the **propodosoma**, that portion of the idiosoma bearing the first and second pairs of legs, and the **hysterosoma**, extending from just behind the second pair of legs to the posterior end of the body. The designation of these body regions is helpful to morphologists and taxonomists in locating specific setae and other structures. The size and arrangement of sclerotized plates, the **chaetotaxy**, and the nature and location of sensory structures on the idiosoma serve as important taxonomic characters.

Mites typically have four pairs of legs as nymphs and adults, but only three pairs as larvae. The legs are divided into the following segments: coxa, trochanter, femur, genu, tibia, tarsus, and pretarsus. The pretarsus commonly bears a pair of claws, a single median empodium, and in certain groups a membranous pulvillus. These structures are highly variable among different groups of mites and aid in movement or clinging to various surfaces, including hosts.

The respiratory systems of mites often include tracheal ducts that supplement the exchange of oxygen, carbon dioxide, and other gases across the body surface. The presence of spiracular openings associated with the tracheal ducts, commonly denoted by a sclerotized plate, or **stigma** (pl., stigmata), and their location on the body provide important taxonomic characters for recognizing the acarine suborders. In the Prostigmata, for example, the stigmata are located at the anterior margin of the idiosoma. In the Mesostigmata they usually are located dorso-laterally to the third or fourth pairs of legs, whereas in the Oribatida (Cryptostigmata) typically they are hidden, opening ventro-laterally near the bases of the second and third pairs of legs. Tracheal systems and spiracular openings are lacking in the Astigmata.

Reproductive structures are very diverse among mites, providing important characters for distinguishing the sexes and identifying taxa. Sperm transfer may be direct (e.g., insemination by transfer of sperm via the male aedeagus to the sperm storage organ, or spermatheca of the female), or indirect (e.g., transfer of sperm via the male chelicerae to the female genital opening).

A few points should be mentioned regarding the internal morphology of mites that are pertinent to species of medical-veterinary importance. The digestive system handles primarily liquefied food that has been pre-orally digested by enzymes secreted in the saliva. The paired salivary glands typically are located in the anterior portion of the idiosoma and open via ducts into the mouth region of the gnathosoma. In addition to digestive enzymes, these glands secrete anticoagulants in hematophagous mites. In certain groups (e.g., chiggers) they may produce cementing substances to help anchor the mouthparts in host skin. Also important in certain acarine groups are **coxal glands**. They are derived from excretory structures and serve primarily in osmoregulation. Waste products, in the form of guanine, are excreted by one or two pairs of long, slender **Malpighian tubules** that open into the alimentary tract just anterior to the hindgut.

LIFE HISTORY

The basic developmental stages in the life history of mites are the egg, prelarva, larva, **protonymph, deutonymph, tritonymph,** and **adult**. Depending on the taxonomic group, one or more stages may be suppressed resulting in a wide range of life-history patterns (e.g., chiggers, Figure 25.2). Eggs may be deposited externally or retained in the uterus until hatching. The prelarva is a nonfeeding, quiescent stage that may or may not have legs, mouthparts, or other distinct external features. The **larva** is typically an active form, which molts to produce a nymph. The nymphs usually resemble the adults of a given taxon except for their smaller size, pattern of sclerotization, and chaetotaxy. The deutonymph of certain astigmatid mites is noteworthy in that it is highly modified morphologically as a nonfeeding stage adapted for surviving adverse environmental conditions. Such deutonymphs are called **hypopi** (sing., hypopus) or **hypopodes**. They often have specialized clasping structures such as anal or ventral suckers (Figure 25.3), which enable them to adhere to phoretic hosts that aid in carrying them to more favorable sites to continue their development. Certain of these deutonymphs may become parasites in hair follicles or subcutaneous tissues of mammals and birds.

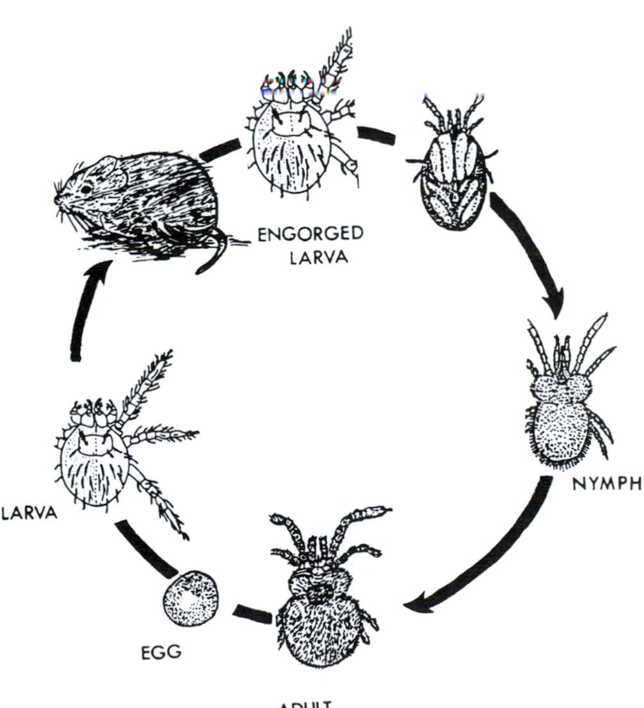

Figure 25.2 Developmental stages in life cycle of chiggers (Trombiculidae). The protonymph and tritonymph are inactive stages passed within the cuticle of the engorged larva and deutonymph, respectively. (Original by Rebecca L. Nims)

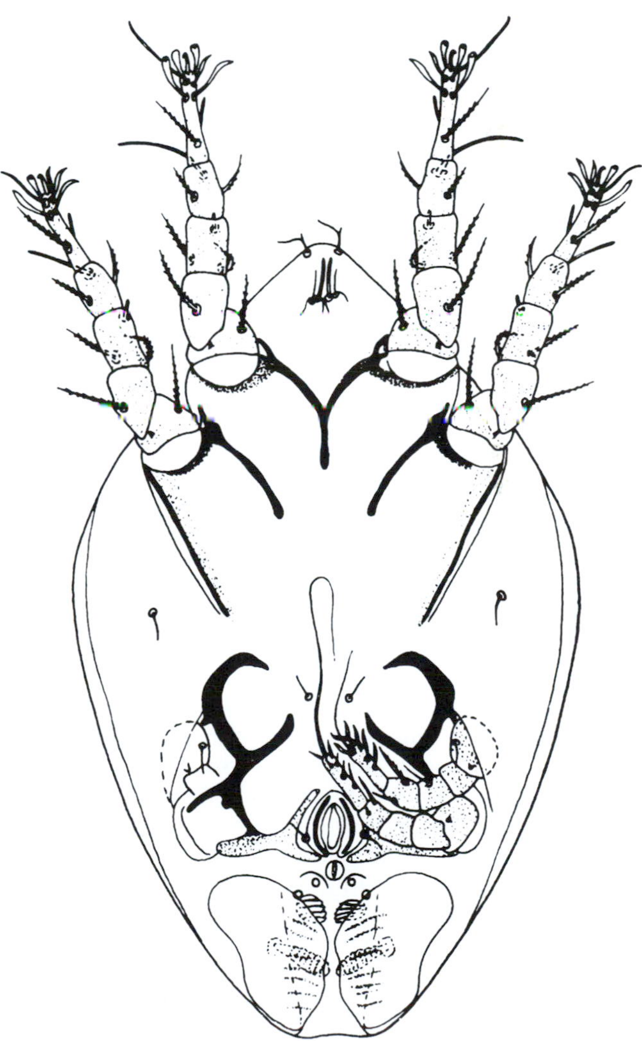

Figure 25.3 Hypopus (deutonymph) of *Glycyphagus hypudaei* (Glycyphagidae), ventral view. This is a specialized, typically phoretic, nonfeeding stage that lacks functional mouthparts. (Whitaker, 1982; original Fain, 1969)

The developmental times from egg to adult and the number of generations per year are too variable to make meaningful generalizations. It is therefore important to understand the developmental biology and life-history patterns of individual groups and species. For further information on the development and life history of mites, see Krantz (1978), Woolley (1988), Schuster and Murphy (1991), Houck (1994), and Krantz and Walter (2009).

BEHAVIOR AND ECOLOGY

Because of the diversity of behavioral and ecological aspects of mites, no attempt is made to discuss them here. Instead they are addressed, where relevant, in the accounts of individual groups and species of mites that follow. For an overview of feeding, mating and reproduction, oviposition, and dispersal behavior of mites, see Woolley (1988) and Evans (1992).

PUBLIC HEALTH IMPORTANCE

Mites can adversely affect human health in many ways. They can infest homes, including carpets, mattresses and bedding, clothing, stored food products, and household pets. Usually they remain unnoticed unless individuals in the household become sensitized and develop various allergies upon subsequent exposure to these mites. Other mites that normally parasitize nonhuman hosts can cause dermatitis in humans when they bite the skin in efforts to feed on blood or other tissues. Most commonly involved in such cases are mite associates of rodents and birds that infest the premises. Such problems typically occur when the natural hosts have died or departed, forcing the mites to seek an alternative food source. A similar situation occurs outdoors when the parasitic larval stage of trombiculid mites, known as chiggers or redbugs, attempt to feed on human skin. Humans are not their normal hosts and often experience intense local skin reactions where these mites attach.

Mites also can pose occupational hazards for farmers, field hands, mill workers, warehouse operators, and others who handle mite-infested materials like straw, hay, and grains. The mites involved normally feed on fungi, plant materials, or various arthropods; however, on contact with humans, they can pierce the skin, sometimes causing severe dermatitis. Other mites actually invade human skin, either burrowing through cutaneous tissues (e.g., scabies mites) or infesting the hair follicles and associated dermal glands (follicle mites). Infestations of these mites can cause persistent, sometimes severe, forms of dermatitis.

In addition to the temporary discomfort or annoyance that mites can cause, some mites are responsible for more serious or chronic medical problems. A number of species may be inhaled or ingested, causing infestations of the respiratory tract and digestive system. Mites even are reported to have been recovered from bile of patients suffering from chronic cholecystitis (inflammation of the gallbladder) and occasionally from the urinogenital tract.

The most widely recognized mite problems affecting human health are respiratory allergies caused by mites infesting house dust. In sensitized individuals, this can lead to chronic respiratory stress, bronchitis, and asthma. There are very few human diseases, however, that involve pathogens transmitted by mites. The most important is tsutsugamushi disease (scrub typhus or chigger-borne rickettsiosis), which occurs primarily in southeastern Asia, Australia, and Pacific islands. The only significant mite-borne disease in the New World is rickettsialpox, reported in the northeastern United States.

For convenience of discussion, problems of a public-health nature caused by mites can be grouped into the following categories: mite-induced dermatitis, respiratory allergies, storage-mite allergies, internal acariasis, mite-borne human diseases, and acarophobia or delusory parasitosis. The mites involved represent at least 18 families in the three suborders Mesostigmata, Prostigmata, and Astigmata.

Mite-Induced Dermatitis

Species in approximately 14 families of mites are known to cause dermatitis in humans. In many cases these represent encounters with species that infest stored products. In other cases, they are ectoparasites of other animals, notably rodents, nesting birds, and poultry. Species in only two families (Demodicidae and Sarcoptidae) utilize humans as their normal hosts. Skin reactions to the feeding or burrowing of these mites range from minor, localized irritation at individual bite sites to severe dermal responses in individuals who become sensitized to specific mite antigens. Still others are free-living, predatory mites that may bite on contact with human skin.

Ascidae The only ascid species thought to be involved in a human case of dermatitis is *Proctolaelaps pygmaeus*, reported in New Zealand by Andrews and Ramsay (1982). This species is probably cosmopolitan and represents an incidental case. Its bites can cause red, papular lesions where the mite pierces the skin.

Dermanyssidae Members of both recognized genera of dermanyssid mites have been reported biting humans: *Dermanyssus* and *Liponyssoides*. They are ectoparasites primarily on wild and domestic rodents and birds. These mites feed on blood of their hosts by piercing the skin with their long, slender, extrusible chelicerae with highly reduced chelae at their tips. Dermanyssids spend most of their time in the nests of their hosts, crawling onto the animals primarily to feed. When they come in contact with human skin, they are prone to bite, typically causing an erythematous papule at each puncture site, often accompanied by intense itching.

Chicken Mite (*Dermanyssus gallinae*) Also called the red poultry mite, this cosmopolitan species (Figure 25.4) is the most common dermanyssid mite that bites people. It parasitizes a very broad range of hosts. This mite is

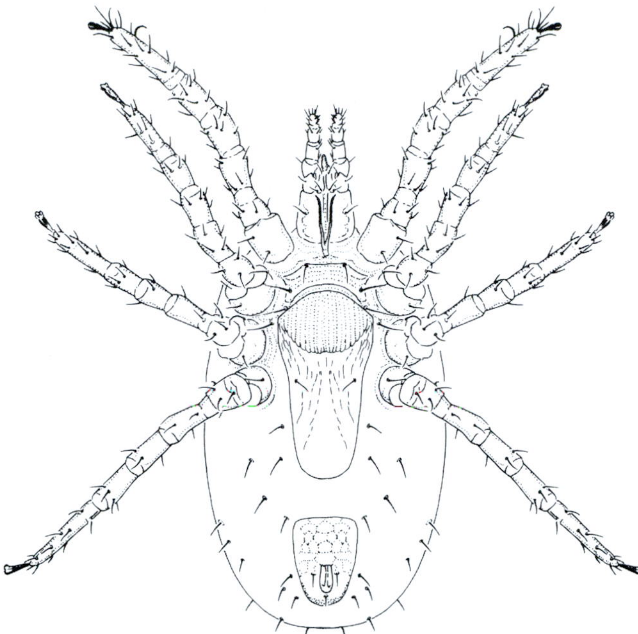

Figure 25.4 Chicken mite, *Dermanyssus gallinae* (Dermanyssidae), female, ventral view. (Modified from Gorham, 1991, courtesy of the US Department of Agriculture)

especially a problem in the Palaearctic region, and in the United States, where most cases occur in poultry houses or around buildings where pigeons, house sparrows, or starlings are nesting. The mites live in nesting materials, where they spend most of their time, moving onto the birds to feed on blood at night. Consequently, workers in poultry operations seldom experience a biting problem while working during the daytime, even when the houses are heavily infested. However, individuals who enter infested buildings at night may be readily bitten. Occasionally, pet canaries and parakeets also serve as sources of human infestations.

The term pigeon mite refers to *D. gallinae* when it infests pigeons or rock doves. The mites frequently enter buildings from pigeon roosts or nests. This tends to happen in the late spring and early summer months when the young pigeons fledge and the nests are abandoned, forcing the mites to seek alternative hosts. Although most human bites occur at night, bites may occur during the daytime when buildings are darkened.

A number of cases have been reported in hospitals and other institutional settings where employees and patients have been bitten by *D. gallinae*. The sources of the problem generally can be traced to nesting birds, notably pigeons, on window sills, ledges, under eaves, in air-intake ducts, or air-conditioners mounted on the outside walls. The mites enter rooms around windows and doors, through crevices and cracks, or via ventilation ducts and air-conditioning systems. In other situations, they may drop onto individuals from roosting or nesting birds in ceilings, or from overhead sites on porches and walkways near buildings. In such cases, close inspection may reveal mites crawling on clothing, furniture, or bed linens, particularly at night when the mites are active.

Human infestations with *D. gallinae* have been variously called chicken tick rash, bird mite disease, psora dermanyssica, pseudogale and gamasidosis. The term fowl mite dermatitis is likewise used, but also can be applied to skin reactions caused by other avian mites that attack people.

Most bites tend to occur on the arms and chest protected by clothing, rather than on exposed skin such as the hands and face. Only in exceptional cases do bites occur in the axillary and pubic areas. The bites are usually painful and typically result in red maculopapular skin lesions on the upper portions of the body and extremities. While being bitten, close examination will reveal the mite as a tiny red speck at the center of the papule. Occasionally the bites produce vesicles, urticarial plaques, and diffuse erythema, with dermatographia frequently being seen. In multiple-bite cases, a pruritic rash may develop and persist until the source of the infestation has been eliminated. Itching tends to be most intense at night. The problem usually is resolved by treatment with antihistamines or topically applied steroids, combined with moving individuals from affected areas.

St. Louis encephalitis (SLE), eastern equine encephalitis (EEE), and western equine encephalitis (WEE) viruses have been isolated from *D. gallinae* infesting wild birds. However, conflicting evidence has been reported regarding the ability of *D. gallinae* to transmit any of these viruses among birds or to humans.

American Bird Mite (*Dermanyssus americanus*) This mite is very closely related to *D. gallinae* but only rarely has been reported biting humans. It can cause acute, generalized, eczematous dermatitis, which is easily misdiagnosed as other skin disorders unless the presence of mites is confirmed. WEE virus has been isolated from this mite infesting nests of the house sparrow but its significance in transmission or maintenance of WEE virus is unknown.

Dermanyssus hirundinis This hematophagous mite is a common ectoparasite of certain birds, especially swallows (*Hirundo* spp.) and the house wren (*Troglodytes aedon*) in North America and Europe. It is not unusual for hundreds or thousands of this mite to infest individual nestling birds. At least one case has been documented in Europe of *D. hirudinis* biting a human and causing urticarial dermatitis (Dietrich and Horstmann, 1983).

House Mouse Mite (*Liponyssoides sanguineus*) This mite (Figure 25.5), referred to in the earlier literature as *Allodermanyssus sanguineus*, is an ectoparasite of domestic and wild rodents. It commonly parasitizes mice, including the house mouse (*Mus musculus*) in the United States and the spiny mouse (*Acomys* spp.) in North Africa. It occurs less commonly on rats (*Rattus* spp.), voles (*Microtus* spp.), and other rodents in localized areas of eastern North America, Europe,

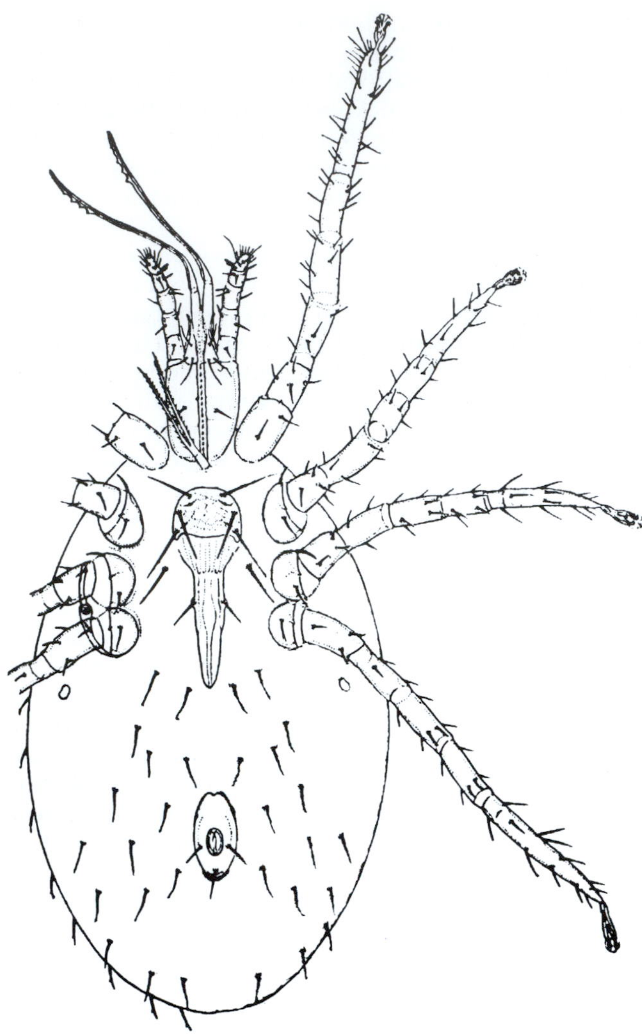

Figure 25.5 House mouse mite, *Liponyssoides sanguineus* (Macronyssidae), female, ventral view. Note the pair of long, attenuated, extruded chelicerae with serrated tips for piercing skin to feed on blood. (Modified from Baker et al., 1956)

Asia, and Africa. It is primarily of interest to medical entomologists because of its role as the vector of *Rickettsia akari*, the etiologic agent of rickettsialpox in humans.

Like most other dermanyssid mites, the house mouse mite lives in nesting materials where it spends most of its time, crawling onto host animals to feed. Its life cycle and behavior are similar in many respects to *D. gallinae*. Females oviposit in rodent nests, or along rodent runways, two to five days after feeding on host blood. The eggs hatch in four or five days to produce larvae that do not feed but instead molt to protonymphs about three days later. The protonymphal stage lasts four to five days, during which time the mite takes a bloodmeal, usually engorging in less than an hour, and then molts to the deutonymph. The deutonymph lives about six to 10 days and requires a bloodmeal before transforming to the adult. The developmental time from egg to adult normally takes two to three weeks. After feeding, blood-engorged females leave the rodent host and can be found in the nests and runways, along the walls of infested premises, and especially in warmer areas of buildings such as furnace and incinerator rooms.

Macronyssidae Macronyssid mites are blood-feeding ectoparasites on reptiles, birds, and mammals. Five species account for most of the cases of medical interest. Three of these are *Ornithonyssus* species infesting rodents or birds; *Chiroptonyssus*, which is parasitic on bats; and *Ophionyssus*, which is parasitic on snakes.

Tropical Rat Mite (*Ornithonyssus bacoti*) This cosmopolitan mite (Figure 25.6) is a parasite of rats, particularly the black rat (*Rattus rattus*), and other rodents in both tropical and temperate regions. In cooler climates this mite occurs only indoors and in nests of wild rodents. Occasionally it also infests carnivores, birds, and humans. When rats are killed or abandon their nests or runways, the mites are left behind and will readily crawl or drop onto humans and other passing animals. Rodents killed by household cats and left near human dwellings also can serve as a source of infestation. The mites are active and can move some distance from their source to enter nearby buildings.

Human bite cases involving the tropical rat mite usually occur in rodent-infested buildings. Using their long, slender chelicerae, they probe the skin in an effort to feed on blood. In some cases they produce a prickling sensation at the bite sites, whereas in other

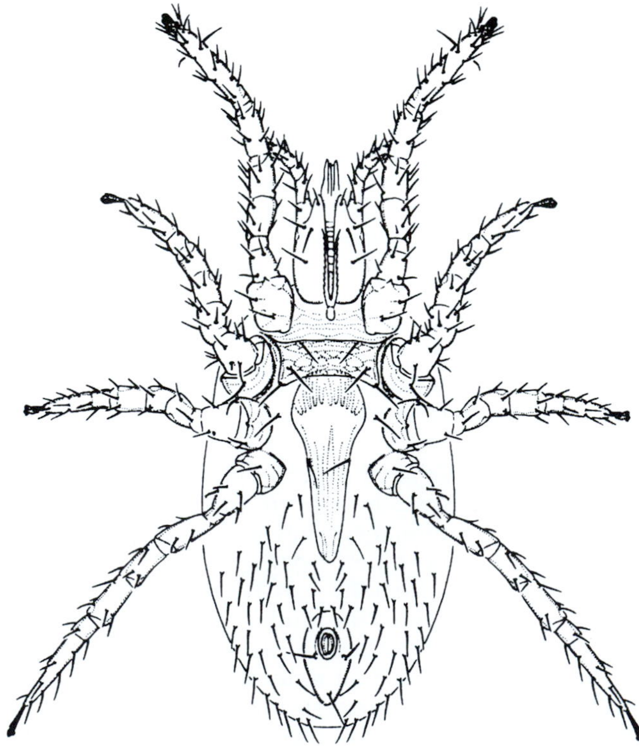

Figure 25.6 Tropical rat mite, *Ornithonyssus bacoti* (Macronyssidae), female, ventral view. (Modified from Gorham 1991, courtesy of the US Department of Agriculture)

cases the bite is painful. Multiple bites often are clustered and subsequently develop into a pruritic, erythematous, papular rash. This may be accompanied by localized swelling and occasional vesicle formation. Although they will bite almost any part of the body, they tend to bite where the clothing is tight; e.g., neck, shoulders, and waist. *Ornithonyssus bacoti* is visible to the unaided eye and may be seen crawling on the clothing or skin, floors, walls, and other structural surfaces.

This mite has been shown experimentally to be capable of being infected with, or transmitting several human pathogens, including those that cause murine typhus, rickettsialpox, plague, tularemia, and Coxsackie virus disease. However, their importance in the epidemiology of these diseases is regarded as negligible. On the other hand, evidence supports the possibility that *O. bacoti* may serve as both a vector and reservoir of Hantaan virus, the causative agent of Korean hemorrhagic fever (epidemic hemorrhagic fever) of humans in Asia.

Tropical Fowl Mite (*Ornithonyssus bursa*) As its common name implies, the tropical fowl mite (Figure 25.7) is distributed widely throughout subtropical and tropical parts of the world, where it parasitizes various domestic and wild birds. It occurs in the eastern and southern United States, Hawaii, Central America, Colombia, South Africa, India, China, and Australia. In addition to being a poultry pest, notably attacking chickens, it parasitizes pigeons, house sparrows, grackles, and other wild avian hosts. Human bite cases can usually be traced to nesting birds under eaves, on ledges, and in other building structures. The mites spend most of their time in the nest, moving onto the host to feed.

Human bites typically occur when young birds leave the nest or the nest is otherwise abandoned, compelling the mites to wander in search of alternative hosts. Wild birds carrying the mites commonly infest poultry operations, leading to workers being bitten when they come in contact with infested, commercially produced birds or their nest materials. Though less common than cases involving the tropical rat mite, human bites due to the tropical fowl mite can be equally discomforting. The latter tend to be more sharply painful and result in more persistent itching. Although WEE virus has been isolated from *O. bursa* in house sparrow nests, there is no evidence that this mite actually transmits the virus. No other human pathogen has been associated with this mite.

Northern Fowl Mite (*Ornithonyssus sylviarum*) The northern fowl mite (Figure 25.8) is widely distributed throughout temperate regions of the world as a parasite of domestic fowl and wild birds. Although regarded as a major pest of chickens, it occasionally bites people. Most human cases result when poultry workers handle infested birds or when the mite enters buildings from nearby bird nests. Although bite cases may occur year-round in commercial poultry operations, most cases in homes and other work places occur about the time young birds fledge and the adults vacate their nests. The bite reaction is similar to that of *O. bursa*, producing red papular skin lesions, often accompanied by intense itching.

The viruses that cause western equine encephalitis and St. Louis encephalitis both have been detected in *O. sylviarum* from nests of wild birds in North America. In the case of WEE virus, it has been shown to persist in avian hosts and to be transmitted by bite to other birds. Newcastle disease virus has been detected in *O. sylviarum* following its feeding on infected chickens, but the virus does not establish persistent infection in the mite. The tropical fowl mite does not appear to play a significant role in the natural transmission of these or other arboviruses affecting humans.

Free-tailed Bat Mite (*Chiroptonyssus robustipes*) This mite is a common blood-feeding ectoparasite on the Brazilian free-tailed bat (*Tadarida braziliensis*) roosting or nesting in walls and attics of buildings throughout its range in the southern United States, Mexico, and

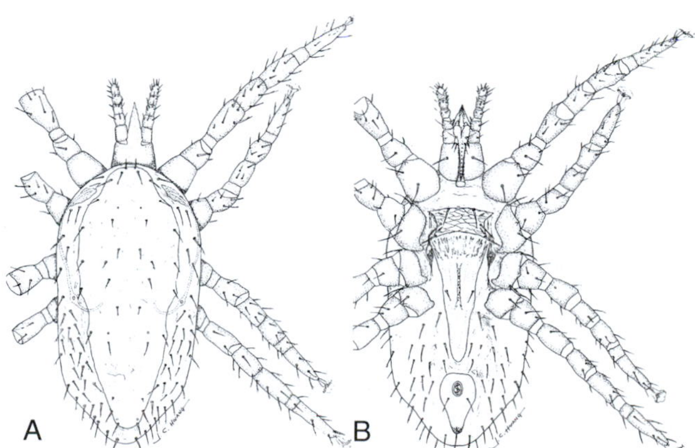

Figure 25.7 Tropical fowl mite, *Ornithonyssus bursa* (Macronyssidae), female. (A) Dorsal view; (B) ventral view. (Strandtmann and Wharton, 1958)

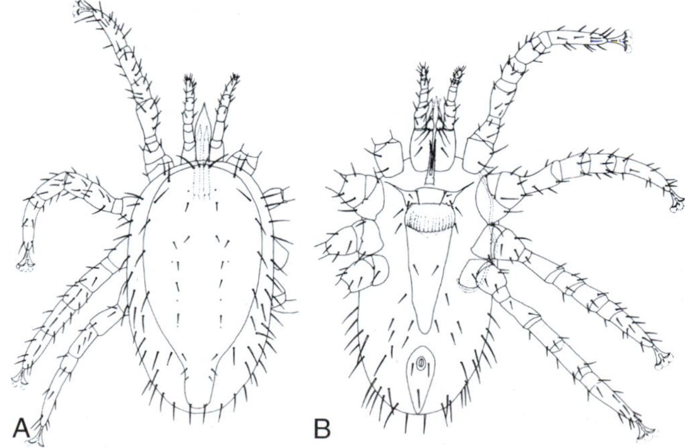

Figure 25.8 Northern fowl mite, *Ornithonyssus sylviarum* (Macronyssidae), female. (A) Dorsal view; (B) ventral view. (Modified from Strandtmann and Wharton, 1958)

Central America. Occasionally it is found in low numbers on other molossid bats roosting with *T. braziliensis*. Only the protonymphs and adult mites feed on blood and other tissue fluids of their bat hosts. Human bite cases are uncommon and usually involve zoologists handling infested bats or otherwise working with mite-infested bat colonies. The mites do not readily bite and are primarily a nuisance as they actively crawl about on the skin and clothing. On rare occasions *C. robustipes* has been reported to invade the living quarters of homes where it has bitten the occupants. One such case involved an 18-month-old boy in California who was bitten repeatedly on the face and abdomen, causing a persistent dermatitis. The problem was not resolved until it was discovered that a wall infestation of Brazilian free-tailed bats was the source of mites that were biting the child each time he was bathed in a bathroom sink (Keh, 1974).

Snake Mite (*Ophionyssus natricis*) This mite (Figure 25.9) is a common pest of captive snakes and only rarely has been reported biting people. Human bites occur primarily in reptile houses at zoological parks, affecting personnel who handle infested snakes. A well-documented case involved several members of a family in a household where a python was kept as a pet (Schultz, 1975). The family members had experienced skin lesions in the form of a papular rash on the forearms and other parts of the body over a five-month period before the source of the problem was identified. Mites were observed to be attached to the skin while attempting to feed and also were found in a chair frequented by the snake. Humans do not serve as suitable hosts for this mite. The mites tend to become immobile with their legs curled underneath the body within a few minutes after they begin feeding on human blood; often they do not recover. They do not transmit any known human pathogens.

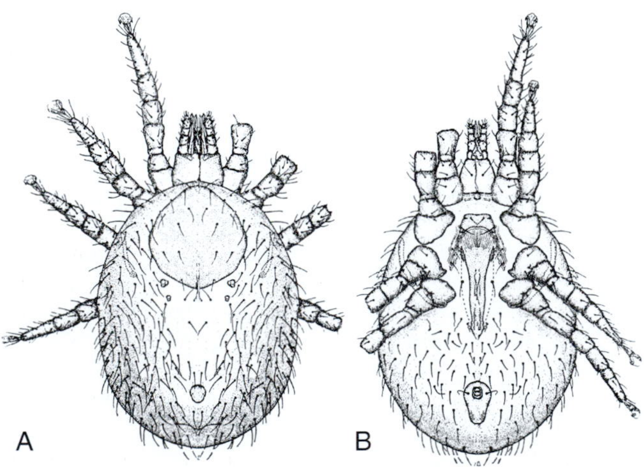

Figure 25.9 Snake mite, *Ophionyssus natricis* (Macronyssidae), female. A, Dorsal view; (B) ventral view. (Modified from Strandtmann and Wharton, 1958)

Laelapidae Members of this family include both free-living and parasitic species, often associated with rodents and other nest-building mammals. The only significant laelapid species that may affect human health is the spiny rat mite. Occasionally other species cause temporary discomfort to humans, as reported in possible cases of *Haemogamasus pontiger* causing dermatitis in England (Theiler and Downes, 1973).

The laelapid *Haemagamasus liponyssoides* is an obligate blood-feeder on wild rodents that has the potential for transmitting human disease agents, even though it has not been reported to bite people (Furman, 1959). Other rodent-associated laelapid mites may play a role in the transmission of Hantaan virus, the causative agent of Korean hemorrhagic fever, based on isolation of this human pathogen from *Laelaps jettmari* in Korea (Traub et al., 1954).

Spiny Rat Mite (*Laelaps echidninus*) The spiny rat mite is a common hematophagous ectoparasite of domestic rats throughout the tropical and temperate zones. Although it is capable of biologically transmitting disease agents, such as the agent of murine typhus among wild rodents, its potential role as a vector of human pathogens remains uncertain. Junin virus, which causes Argentinean hemorrhagic fever, has been isolated from *L. echidninus* and associated rodent hosts in South America (Parodi et al., 1959; Theiler and Downes, 1973).

Trombiculidae Larvae of members of the family Trombiculidae are called **chiggers**. This is the only parasitic stage in the life cycle of trombiculid mites. As a group, they feed on a wide variety of vertebrate hosts including amphibians, reptiles, birds, and mammals, with humans serving only as accidental hosts.

The life history of trombiculid mites includes the following sequence of stages: egg, deutovum, larva, nymphochrysalis, nymph, imagochrysalis, and adult (Figure 25.2). The eggs typically are laid in soil or ground debris. After about six days, the egg shell splits to expose an inactive prelarval stage, the deutovum. After another six days, the active six-legged larva (i.e., chigger) is produced (Figure 25.10). After successfully attaching to a suitable host, the larva generally feeds three to five days on the host before dropping to the ground to form an inactive transitional stage, the nymphochrysalis. This stage in turn develops into the active eight-legged nymph. The nymph subsequently undergoes development as another quiescent stage, the imagochrysalis, to produce the eight-legged adult. The nymph and adult are free-living predators that feed on small arthropods (e.g., collembolans) and their eggs. The duration of the life cycle requires two to 12 months or longer, depending on the species and environmental conditions. In temperate areas, there may be one to three generations per year, whereas in tropical regions generations may be continuous throughout the year.

Although trombiculid larvae usually cause little or no apparent harm to their normal hosts, they often cause dermatitis when they attach to and attempt to feed on

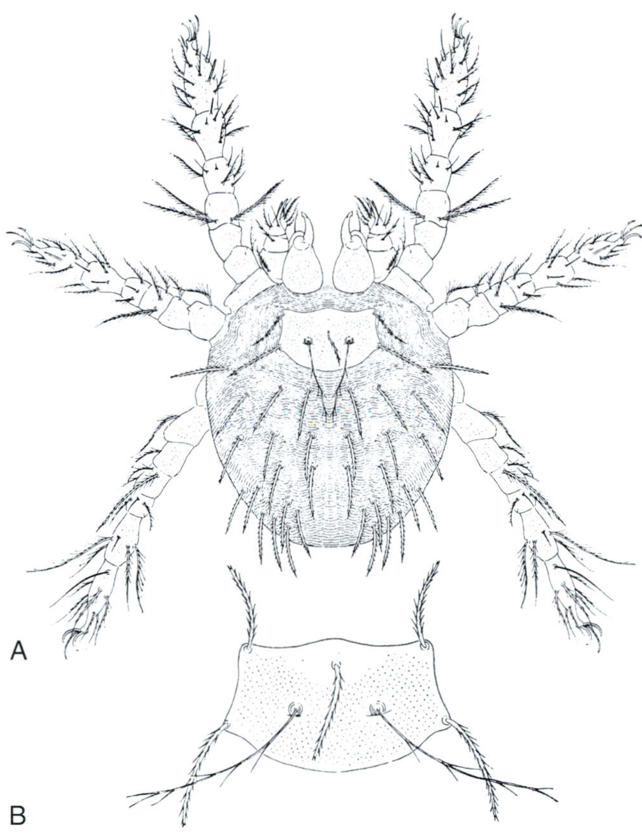

Figure 25.10 Larval stage (chigger) of the harvest mite, *Trombicula autumnalis* (Trombiculidae), of Europe. (A) Dorsal view; (B) scutum (dorsal plate), showing characteristic arrangement of setae. (Modified from Baker et al., 1956)

humans and other atypical hosts. Such an infestation by trombiculid larvae is called chigger dermatitis, or **trombiculosis** (trombidiosis of the older literature).

Chiggers are just large enough (150–300μ) to be visible to the unaided eye. They are yellowish, orange, or red, and can be seen on close inspection at the center of the skin lesions they induce. Unfortunately, chiggers often are encountered in large numbers, resulting in multiple bites (Figure 25.11). Given their preference for attaching where clothing fits snugly against the skin, the bites tend to be concentrated about the ankles and lower legs, waist, and along the elastic borders of undergarments.

Contrary to popular belief, chiggers do not burrow into the skin of their hosts. Instead they attach by piercing the epidermis with their chelicerae and feed externally. Because of their small size and tiny mouthparts, chiggers usually attach where the skin is thin or soft. A preferred site is the opening to hair follicles. There they insert their mouthparts to feed on the thin epidermal lining. In humans, this results in inflammation at the point of attachment and localized swelling of the skin around the chigger, giving the mistaken impression that it has burrowed into the skin. Their food consists primarily of partially digested skin cells and lymph broken down by saliva introduced at the attachment site. They do not feed on blood. Feeding

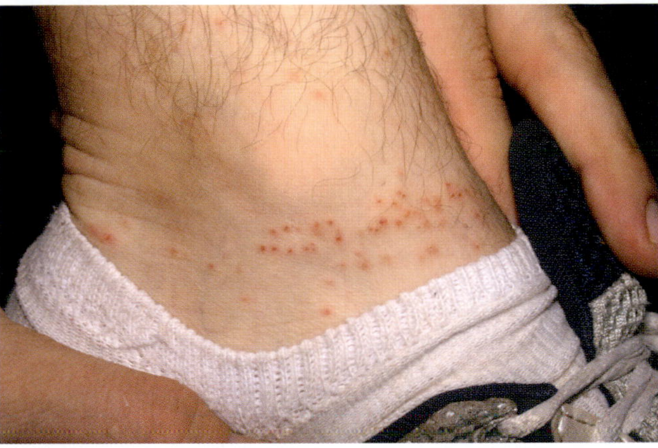

Figure 25.11 Multiple chigger-bite lesions on human foot, with top of ankle sock pulled down to reveal the location of the attached mites where the fabric fits snugly against the skin. (Photo by Nathan D. Burkett-Cadena)

is facilitated by formation of a feeding tube, or stylostome, produced by the interaction of the saliva with surrounding host tissue.

With the exception of *Leptotrombidium* species, chiggers often do not survive more than one or two days on human hosts, due to the adverse host reaction they cause and injury or removal due to scratching. By then, however, the damage is already done, typically producing a discrete, persistent, itching, reddened papule at each attachment site. The lesions persist for several days but may take several weeks to heal if they become secondarily infected. The recovery time can be significantly shortened by prompt treatment to kill the chiggers when they are first detected, generally within three to six hours following attachment; application of a topical medication to alleviate itching and prevent infection; and avoidance of scratching or otherwise excoriating the skin.

More than 50 species of trombiculid mites have been recorded attacking humans. Of this number, about 20 species are considered to be medically important, either due to the dermatitis they cause or due to their role in transmission of disease agents. Four species of particular interest are *Trombicula alfreddugesi* in North America and South America, *T. autumnalis* in Europe, and *Leptotrombidium akamushi* and *L. deliense* in the Orient.

Trombicula alfreddugesi This is the most common and widespread trombiculid mite in the Western Hemisphere, occurring from Canada to Argentina and in the West Indies. In North America, it and related species that attack humans are known as **red bugs**, especially in the southeastern United States. In Mexico it is called tlalzahuatl, and in Mexico and other parts of Latin America as coloradilla and bicho colorado. *Trombicula alfreddugesi* is parasitic as larvae on a variety of amphibians, reptiles, birds, and mammals. It is particularly common in areas of secondary growth, such as shrub and brush thickets, and blackberry and bramble (*Rubus* spp.) patches; along margins of swamps; and ecotones between woodlands and open fields or grasslands. The larvae are present in late summer and early fall in the more temperate parts of its range, and throughout the year in the tropics and subtropics, including southern Florida. Although it is the most common cause of chigger dermatitis in the New World, it is not involved in the natural transmission of any human disease agent.

Trombicula autumnalis Known as the harvest mite, this is the most common chigger that attacks humans in Europe and the British Isles (Figure 25.10). Other names include aoutat and lepte autumnal. As both its common and scientific names suggest, *T. autumnalis* is particularly annoying during harvest time in late summer and fall. The larvae are present from July to the onset of winter, usually reaching peak populations in early September. They tend to be most active on warm, sunny days in grasslands, cultivated grain fields, brush lands, and thickets. The widespread occurrence of this mite reflects its wide variety of natural hosts, especially mammals and certain ground-dwelling birds. Rabbits are a particularly common host. Other hosts include voles, wood mice, hedgehogs, squirrels, cattle, sheep, goats, horses, dogs, cats, pheasants, partridges, chickens, and other domestic fowl.

Other *Trombicula* Species *Trombicula splendens* occurs in the eastern United States from the Gulf Coast north to Massachusetts, Minnesota, and Ontario, Canada. It is especially common in the southeastern United States where it is second only to *T. alfreddugesi* as the cause of trombiculosis in humans. Although it occurs in drier habitats with *T. alfreddugesi*, it is especially abundant in moist habitats such as swamps, bogs, and low-lying areas with rotting stumps and fallen trees. The larva is parasitic on amphibians, reptiles, birds, and mammals, but seems to prefer snakes and turtles as natural hosts. The seasonal occurrence of *T. splendens* is similar to that of *T. alfreddugesi.* Another *Trombicula* species that causes chigger dermatitis of humans in the United States is *T. lipovskyi.* It is restricted to moist areas, generally characterized by an abundance of decaying logs and stumps bordering swamps and streams. It occurs from Alabama and Tennessee west to Arkansas, Oklahoma, and Kansas. Reptiles, rodents, and birds serve as hosts.

***Leptotrombidium* spp.** Several members of the genus *Leptotrombidium* serve as vectors of *Orientia tsutsugamushi,* the causative agent of tsustugamushi disease. They occur widely throughout Southeast Asia and the southwestern Pacific islands. As a group, the larvae of medically important *Leptotrombidium* species are parasitic primarily on ground-dwelling rodents (e.g., *Rattus, Microtus,* and *Apodemus* spp.). Other hosts include insectivores, marsupials, cattle, dogs, and cats. They occur in forests, second-growth areas along the margins of woodlands, in river valleys, and in abandoned agricultural fields where populations of rodents flourish. The principal vectors of the tsutsugamushi disease agent are *L. deliense* in Southeast Asia, the southwestern Pacific islands, and northern Australia; *L. akamushi* in Japan; *L. arenicola* and *L. fletcheri* in the Pacific islands; and *P. pallidum, P. pavlovskyi,* and *P. scutellare* in more restricted regions of the Asian mainland, Japan, and Malaysia (Table 25.3).

Table 25.3 Major Chigger Vectors of *Orientia tsutsugamushi*, the Causative Agent of Tsutsugamushi Disease in Humans

Trombiculid Species	Geographic Occurrence
Leptotrombidium akamushi	Japan
L. arenicola	Malaya, Indonesia, Thailand
L. deliense	Southeast Asia, China, southwestern Pacific islands, northern Australia, Pakistan
L. fletcheri	Malaysia, New Guinea, Philippines, Indonesia, Melanesia
L. pallidum	Japan, Korea, Primorye region of Russia
L. pavlovskyi	Primorye region of Russia
L. scutellare	Japan, China, Thailand, Malaysia, Fugian

from Kawamura et al., 1995

Stored-Products Mites

Members of several families of mites that infest unprocessed and processed plant materials can cause human dermatitis and other health-related problems. Most cases involve people handling infested materials such as grains, flour, hay, straw, dried fruits, and vegetables. Others involve processed materials of animal origin such as meats, hides, cheeses, dried milk, and other dairy products. Such mite infestations are the cause of occupational acarine dermatitis in farm hands, granary operators, warehouse workers, and other personnel.

The stored-products mites responsible for most human cases of acarine dermatitis are members of the families Acaridae, Pyemotidae, and Cheyletidae.

Acaridae and Other Astigmata Acarid mites infest a wide range of stored materials such as grains, milk products, dried fruits, straw, and animal hides in both households and commercial storage facilities. They also are common contaminants of culture media in which insects and other invertebrates are reared; bedding materials for mice, guinea pigs, hamsters, and other vertebrates; and in animal feed and animal-holding cages in pet stores and zoos. Their numbers can build rapidly, especially when the infested materials

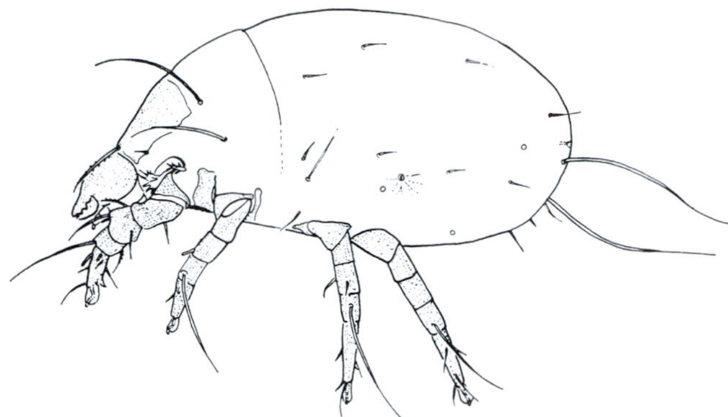

Figure 25.12 *Acarus siro* (Acaridae), female, lateral view. (Modified from Hughes, 1976)

are damp enough to support the growth of fungi on which they typically feed. Dermatitis occurs when the mites pierce the skin in attempts to feed or obtain moisture. The reaction in some cases also may involve contact allergens.

The most important acarid mite in stored products is *Acarus siro* (Figure 25.12), a species found throughout most of the world. It is particularly a pest of processed cereal products (e.g., flour), rather than whole grains or hay. The females are 350 to 650 μm in length, with a colorless body and yellow-to-brown gnathosoma and appendages. It can develop at temperatures of 24 to 32°C and a relative humidity greater than 60%. This mite tends to congregate where the relative humidity is 80 to 85%, at which its reproductive rate is highest. The amount of damage it causes to grains is related directly to the moisture content; the germ is attacked only when the water content is 14% or higher. The dermatitis experienced by food handlers on contact with *A. siro* is commonly known as **grocer's itch**. Other names for dermatitis caused by *A. siro* and related mites are baker's itch, dried-fruit-mite dermatitis (*Carpoglyphus lactis*, Carpoglyphidae, Figure 25.13), wheat pollard itch (*Suidasia nesbitti*, Suidasiidae), and vanillism, reflecting the product or commodity involved.

A mite closely related to *A. siro* that also causes human dermatitis is *Acarus farris* (Figure 25.14). It is a widespread species that has been reported to cause skin irritation to farm workers handling infested bales of hay in England (Hughes, 1976).

Another common acarid mite in stored products is the cosmopolitan *Tyrophagus putrescentiae*. It is particularly a problem in foods with a high protein and fat content such as hams, cheeses, nuts, seeds, dried eggs, and fish meal. This mite feeds primarily on fungi (e.g., *Aspergillus*, *Eurotium*, *Penicillium*) that tend to thrive on foods stored at warm temperatures (>30°C) and relatively high humidities (>85%). Under such conditions it can complete its development from one generation to the next in two to three weeks. It can be a pest in mycology laboratories where it often contaminates fungal cultures. The term mold mite is commonly used to refer to *T. putrescentiae* and a closely related species, *T. longior* (Figure 25.15).

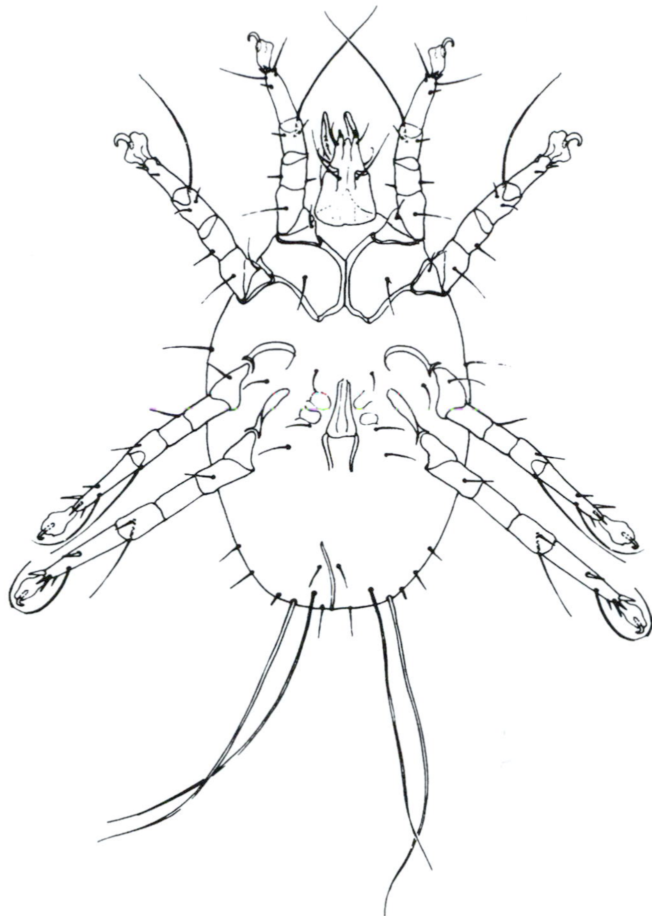

Figure 25.13 *Carpoglyphus lactis* (Carpoglyphidae), male, ventral view. (Modified from Hughes, 1976)

In the tropics, *T. putrescentiae* causes a dermatosis called **copra itch** among workers handling copra, dried coconut kernels from which coconut oil is extracted. In Italy, human cases of cutaneous and respiratory allergies have been attributed to this mite among workers handling raw hams; the mite apparently thrives in the white dust (*ruffino*) that covers hams during the seasoning process (Ottoboni et al., 1989). *Tyrophagus putrescentiae* occurs throughout much of the world, where it is found in a wide range of situations, including grasslands, soil, old hay, mushrooms, and the nest of bees and ducks. This mite was reported as the cause of human dermatitis in a butcher's shop in Austria where it was breeding in molds growing on bacon in a poorly ventilated room (Czarnecki and Kraus, 1978).

A few other acarid mites are known to cause human dermatitis. One is *Tyrolichus casei* reported by Henschel (1929). This is a cosmopolitan species commonly found in stored foods, grains, flour, cheeses, dogmeal, old honey combs, and insect collections (Hughes, 1976). Another is *Suidasia nesbitti*, which occurs in Europe, Africa, North America, and the West Indies. Although it is particularly associated with wheat pollards and bran in England (Hughes, 1976), it also has been reported in rice, whale meat infested with dermestid beetles, dried bird skins, and milking machinery.

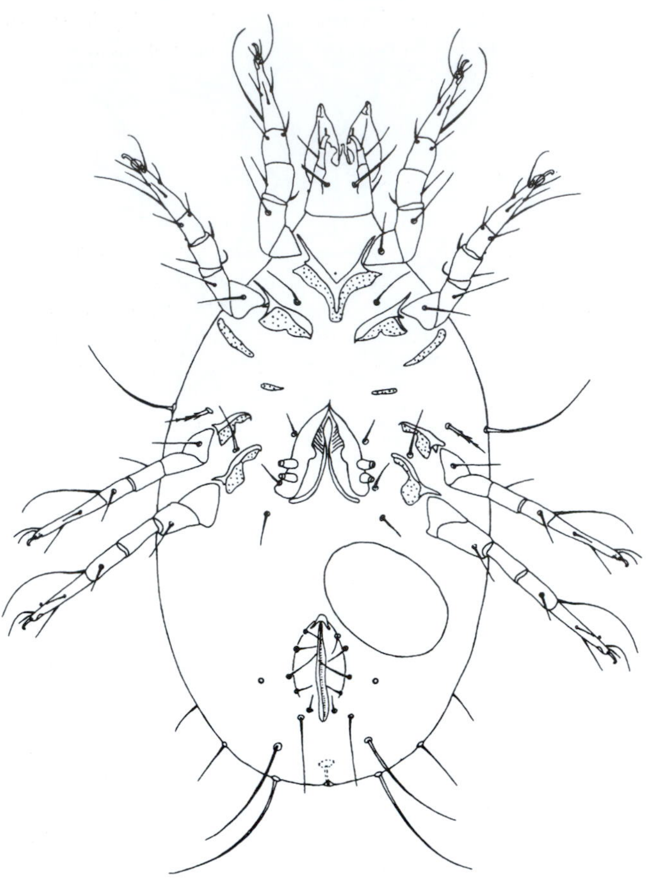

Figure 25.14 *Acarus farris* (Acaridae), female, with single egg, ventral view. (Modified from Hughes, 1976)

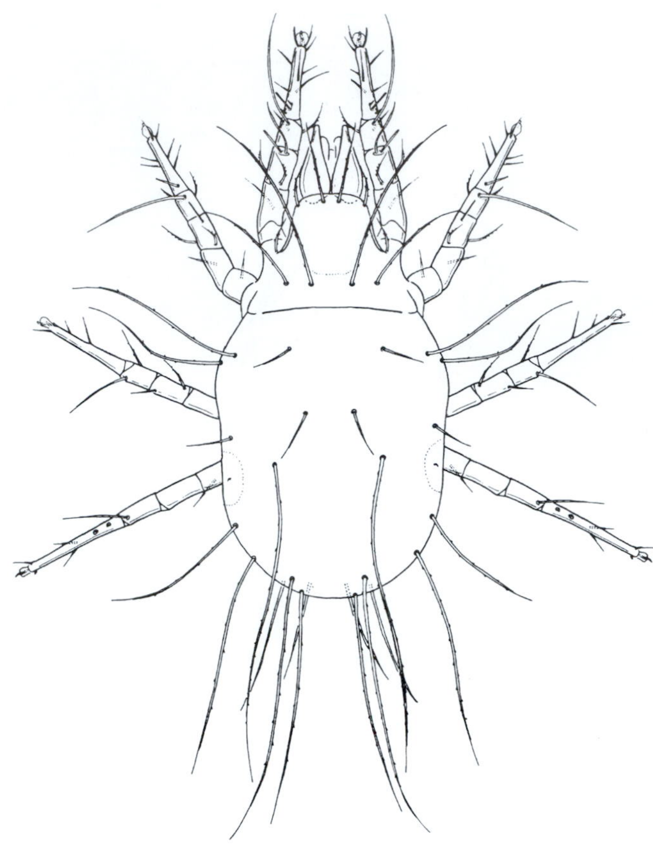

Figure 25.15 *Tyrophagus longior* (Acaridae), male, dorsal view. (Modified from Hughes, 1976)

Pyemotidae Pyemotid mites are ectoparasites of insects that typically attack the larval stage of moths, beetles, and hymenopterans. A few species commonly occur in dried, insect-infested plant products such as hay, straw, and grains. On contact with humans and other animals, these mites cause intense itching when they pierce the skin with their stylet-like chelicerae and inject a toxin produced in their salivary glands. It is a very potent neurotoxin that they use to immobilize their insect prey, enabling them to paralyze insects 150,000 times their size (Tomalski and Miller, 1991).

The most important species affecting humans is *Pyemotes tritici* (Figure 25.16). It is variously known as the **straw itch mite**, **hay itch mite**, and **grain itch mite**, depending on the plant material with which it is associated. Exposure to *P. tritici* represents an occupational hazard for agricultural workers, sales and stock personnel in farm supply stores, and other individuals in the arts-and-crafts field who handle wheat, hay, and straw. People handling infested materials usually develop multiple skin lesions in the form of papules or papulovescicles, accompanied by intense itching. Each bite site typically consists of a minute white wheal with a central erythematous area where a tiny vesicle forms. During the early stages, the mite often is visible as a tiny white speck where the vesicle is located. Although lesions can occur on any exposed part of the body, they usually appear on the back, abdomen, and forearms, where contact with infested materials typically takes place. Lesions seldom occur on the face or hands. Heavily infested or sensitized individuals may experience other symptoms, including headache, fever, nausea, vomiting, diarrhea, and asthma (Southcott, 1984). Less commonly reported are chills, fever, malaise, and anorexia (Betz et al., 1982).

Two other species of *Pyemotes* reportedly cause human dermatitis. Workers in France developed erythematous lesions and complained of intensely itchy papules after handling dried everlasting flowers (*Helichrysium angustifolium*) infested with *P. zwoelferi* imported from Yugoslavia (Le Fichoux et al., 1980). People working in a feed-mixing shed of a pig farrowing house developed a papular rash after contact with grain infested by *P. herfsi* in former Czechoslovakia, whereas workers handling commercial cultures of mealworms (*Tenebrio molitor*) infested with *P. herfsi* developed wheals in former West Germany. *Pyemotes herfsi* also has caused outbreaks of dermatitis in the central United States, where the mite is a parasite of an oak leaf gall midge, *Contarinia* sp. (Diptera: Cecidomyiidae) (Broce et al., 2006).

Cheyletidae Cheyletid mites are mostly free-living predators that commonly feed on other mites and small arthropods in stored products. Occasionally they cause pruritic dermatitis in people handling infested grains

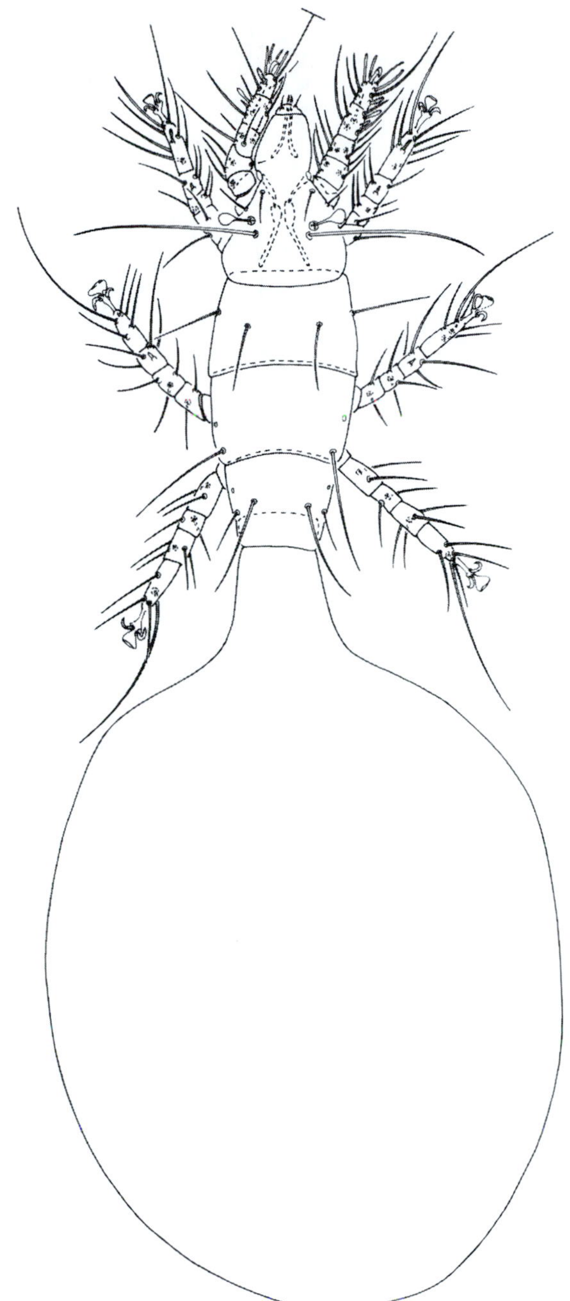

Figure 25.16 Straw itch mite, *Pyemotes tritici* (Pyemotidae), gravid female, dorsal view. (Gorham, 1991; courtesy of the US Department of Agriculture)

and other dried plant materials. The most common cheyletid found in stored products is *Cheyletus eruditus*. This cosmopolitan species has been used commercially as a biological control agent to reduce the numbers of grain mites, notably *Acarus siro* and *Lepidoglyphus destructor*, in granaries and agricultural warehouses. Severe pruritus was reported in a worker at a wholesale florist shop handling fern wreaths imported to the United States from the Philippines (Shelley et al., 1985). The mite involved was apparently *Cheyletus malaccensis*, a species previously shown to cause itching papules in humans (Yashikawa et al., 1983). A second cheyletid mite, *Cheyletomorpha lepidopterorum* also may have been involved.

Skin-Invading Mites

Representatives of only two families of mites typically invade human skin or associated dermal structures and glands. They are the Demodicidae, or follicle mites, and the Sarcoptidae, or scabies mites. Whereas only a relatively small number of humans infested with follicle mites develop clinical problems, most individuals who become infested with the human scabies mite experience an annoying, often severe, dermatitis.

Demodicidae Members of this family are called **follicle mites**. They are extremely tiny, elongate, annulate mites with very short, stout, three-segmented legs (Figure 25.17). They lack body setae and possess a pair of tiny, needle-like chelicerae that are used to pierce dermal cells on which the mites feed. Their minute size and strong reduction of most of the external features represent adaptations for living in the close confines of hair follicles and associated ducts and glands.

Two species of *Demodex* infest humans. *Demodex folliculorum* (Figure 25.17) occurs primarily in hair follicles, whereas *Demodex brevis* generally is found in the sebaceous glands (sweat glands) that open via ducts into the hair follicles. Both species may infest the same host, appearing together in samples taken from a given individual. Adults of the two species closely resemble one another but can be distinguished based on the general body shape and relative size of the males and females. *Demodex folliculorum* females have elongate bodies that are gently tapered from the podosoma to the slender, rounded caudal end. *Demodex brevis* females have bodies that are usually widened posterior to the podosoma and terminate in a more broadly rounded caudum. The eggs are also distinctive, being spindle-shaped in *D. folliculorum* and oval in *D. brevis*.

The entire life cycle of *D. folliculorum* (Figure 25.18) and *D. brevis* is spent on their human host. The mites feed by piercing host cells with their styletiform chelicerae and drawing the cell contents into the esophagus with a pumping action of the pharynx. They are highly host-specific and can survive only on humans. Transfer of mites from one individual to another is presumed to occur primarily between mothers and infants during the intimacy associated with facial contact and nursing. Adults of both sexes are transferred readily between hosts at these times. The fact that 90 to 100% of all humans apparently harbor follicle mites attests to the success with which such transfers are accomplished.

Human follicle mites tend to occur primarily in the regions of the forehead, eyelids, and nose. They also can occur in the eyebrows, the Meibomian glands of the eyelids, perioral mucosa, ear canal, chest, nipples, and other parts of the body (Nutting et al., 1989). In most cases they cause no apparent harm and go virtually unnoticed. Only under unusual circumstances,

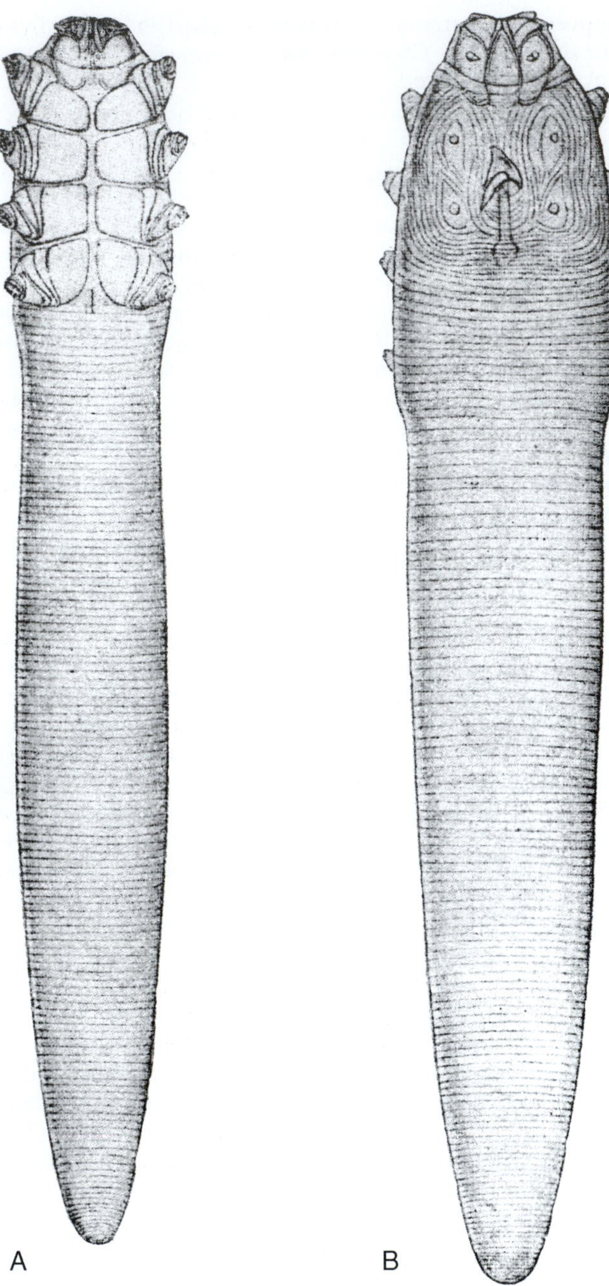

Figure 25.17 Human follicle mite, *Demodex folliculorum* (Demodicidae), female. (A) Female, ventral view; (B) male, dorsal view. (Hirst, 1922)

which remain largely unexplained, do they cause clinical problems that warrant medical attention. Such cases involving dermal reactions to *Demodex* mites are called **demodicosis**. It does not appear that any specific pathogens are involved, although secondary bacterial infections can aggravate the condition.

In addition to differences in one's body chemistry and immunological responses, certain hormones affect population levels of *Demodex* mites and the development of demodicosis. Populations tend to build as the host matures, leveling off in the middle age groups. Substances such as diethylstilbestrol tend to

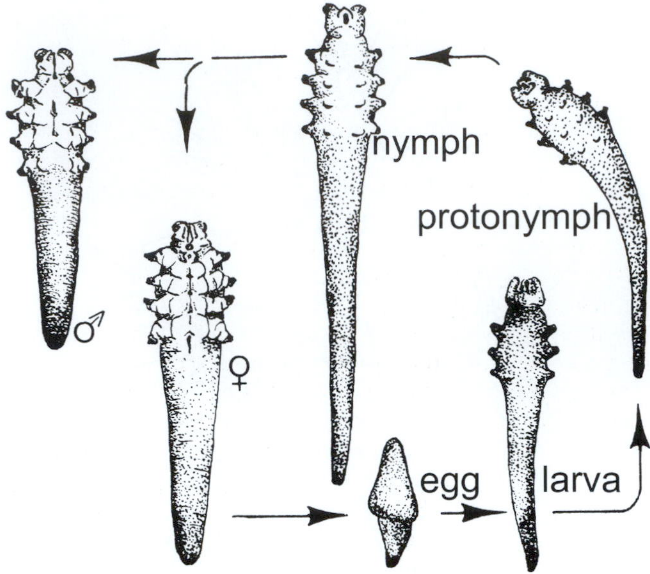

Figure 25.18 Life cycle of human follicle mite, *Demodex folliculorum* (Demodicidae). (Modified from Nutting, 1984)

inhibit mite populations, whereas progesterone and testosterone may promote an increase. The long-term use of topically applied corticosteroids has been correlated with an increased incidence of demodicosis, suggesting a possible link between hormonal levels and the development of inflammatory reactions induced by follicle mites. Cases of human demodicosis can be categorized in five clinical forms: demodex folliculitis, demodex blepharitis, pityriasis folliculorum, demodex granuloma, and human demodectic mange.

Demodex folliculitis occurs most commonly on the face, but also on the forearms and chest. It typically causes rosacea-like skin lesions, initially appearing as red follicular papules and tiny pustules. This is the most difficult *Demodex* infestation to diagnose because it is almost indistinguishable clinically from other skin problems such as acne cosmetica, corticosteroid telangiectasia (dilated blood vessels within the skin that have a tortuous appearance), and rosacea. In some cases it may complicate or aggravate preexisting skin conditions. Confirmation of demodicosis is therefore dependent on demonstrating the presence of the mites, all stages of which can be found in the pustule contents.

Demodex blepharitis, also known as ocular demodicosis, is an inflammation of the hair follicles of the eyelids associated with high populations of demodicid mites. The patient's eyelids typically itch or burn, become reddened, and often are characterized by accumulations of waxy or gelatinous debris at the base of the eyelashes. The presence of mites can usually be detected by plucking affected lashes and examining them under a microscope.

Pityriasis folliculorum is an uncommon form of demodicosis that is clinically recognized as dry, scaly skin with brownish or grayish hyperpigmentation and associated pruritus. The condition often is intensified by scratching or shaving, resulting in erythema, excoriation, mottling, and what has been described as a nutmeg-like roughening of the affected skin. It usually occurs on the face and neck, in both young and older adults.

Demodex granuloma results when follicle mites rupture out of blocked hair follicles into subcutaneous tissues. There they can elicit a response by lymphocytes and histiocytes to form granulomas.

Human demodectic mange is the term applied to the remaining form of human demodicosis in which transient infestations of humans by *Demodex* spp. are contracted from other host species. The most common source is dogs and usually involves intimate contact, such as sleeping with a pet. Patches of papules and vesicles develop, accompanied by a burning or itching sensation. Commonly affected areas include the chin, neck, chest, forearms, stomach, and thighs, reflecting the skin surfaces most likely to come in contact when handling household pets.

A positive diagnosis of demodicosis is made by confirming the presence of large numbers of *Demodex* mites directly associated with the affected areas of skin. The mites can be seen by microscopic examination of skin scrapings, pustule contents, cellular debris from hair follicles, and plucking eyelashes in the case of demodex blepharitis. Adhesive cellophane tape, applied to infested areas of the skin, can be used to recover demodicid mites near the follicular orifices or moving on the surface of the skin. A follicular biopsy also can be helpful, in which a quick-setting cyanoacrylate polymer is used to extract the contents of sebaceous follicles (Mills and Kligman, 1983). Various stages of the mites, including eggs, are usually evident.

Cases of human demodicosis can be effectively treated by daily washing of the affected skin with mild alkaline or sulfur soap, followed by application of a mild sulfur lotion sold for this purpose. Other compounds such as gamma benzene hexachloride (lindane), metronidazole, and physostigmine ophthalmic ointment in blepharitis cases, also are effective. When properly treated, cases often are resolved in two to three weeks, but may take as long as two months. This is not to say that the mites are eliminated; their numbers simply are reduced to lower levels that do not cause pathogenesis. Regular daily washing of the face and eyelids with alkaline soap helps to suppress *Demodex* populations and reduce the risk of developing demodicosis. The use of mascara also seems to retard mite increases. On the other hand, the regular use of medicated creams, skin moisturizers, and topical applications of corticosteroids tend to promote *Demodex* numbers, leading to heavier infestations and increased prospects of related skin problems.

For further information on *Demodex* species and their medical importance, see Desch and Nutting (1972), Nutting (1976a, 1976b, 1976c), English and Nutting (1981), Rufli and Mumculoglu (1981), Franklin and Underwood (1986), and Burns (1992).

Sarcoptidae The only mites in this family that commonly infest humans are members of the genus *Sarcoptes*, generally referred to as scabies mites. They represent a taxonomic complex of varieties or physiological types of the single species *Sarcoptes scabiei.*

Human Scabies Mite (*Sarcoptes scabiei*) The form that typically infests people is called the human scabies mite, or human itch mite, *S. scabiei* var. *hominis*. This mite is cosmopolitan in distribution and infests human populations of all races as an obligate parasite that lives in the skin. The adults are small (females 350–450 μ, males 180–240 μ in length) and rounded in shape, with tiny pointed, triangular spines on their dorsal surface that assist them in burrowing (Figure 25.19). These spines are more numerous and conspicuous in females than in males. The legs are short, with Legs I and II of the female and Legs I to III of the male each bearing a terminal sucker. The two hind pairs of legs of the female and the last pair of legs in the male lack a sucker, and instead terminate in long setae or bristles.

The adult mites can crawl quite rapidly on the surface of the skin, with females traveling up to 2.5 cm/minute. Upon finding a suitable site, the female uses her chelicerae and first two pairs of legs to burrow into the skin, disappearing beneath the surface in about an hour. There she waits in this temporary pit, or shallow burrow, for a wandering male to find her, following which mating takes place. The fertilized female then emerges on the skin surface and searches for a site in which to excavate a permanent burrow. She penetrates the skin once again and makes her way down through the stratum corneum, or horny layer of the skin, to its

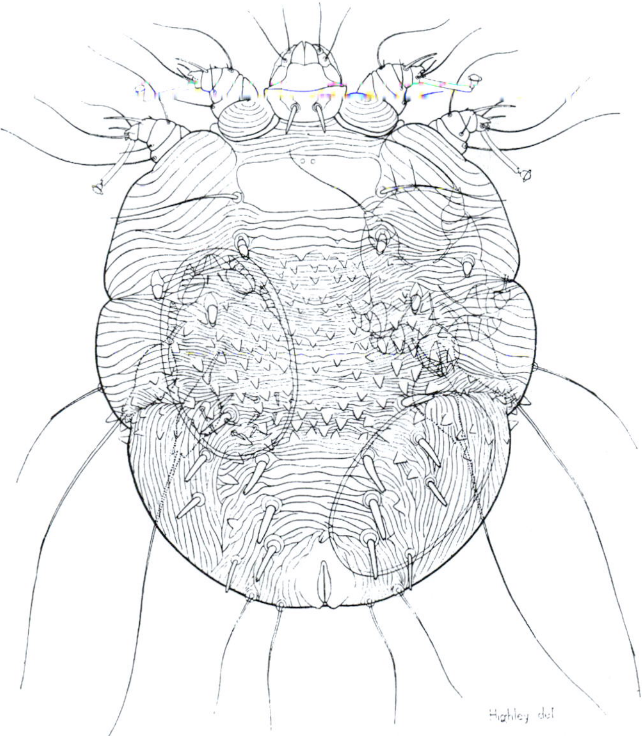

Figure 25.19 Human scabies mite, *Sarcoptes scabiei* (Sarcoptidae), female, with two developing eggs, dorsal view. (Hirst, 1922)

lower boundary with the underlying stratum granulosum. There she excavates a horizontal burrow within the stratum corneum where she will spend the rest of her life, commonly 30 days or more. During this time she continues to extend the length of her burrow by 0.5 mm/day or more, commonly reaching a total length of 1 cm or more. As viewed from the skin surface, fresh burrows appear as tiny grayish, sinuous lines with the adult female discernible as a whitish speck at the end of the tunnel.

Within a few hours, the female begins laying eggs in the burrow, producing two or three each day thereafter. The eggs hatch in three to four days. The resultant larvae often remain in the burrow for about a day before actively crawling out of the burrow onto the surface of the skin. There they excavate shallow burrows in which they molt to nymphs about three days later. The nymphs in turn either remain on the skin surface or dig just beneath the surface where they molt to adults in three or four days. The developmental time from egg to adult typically takes about 10 days for males and 14 days for females.

Although the temporary burrows made by the larvae, nymphs, and virgin females may occur on many parts of the body, the more established burrows made by fertilized females tend to be in very characteristic locations. The most frequent sites are folds of the skin about the wrists and in the sides of, or webbing between, the fingers. Other common sites are the elbows, feet and ankles, axillae, buttocks, penis, scrotum, and breasts. The location of burrows in infants and young children differ somewhat from that of adults, commonly involving the palms, sides and soles of the feet, and areas about the head and neck. In addition to the rash and discomfort directly associated with the burrows, rashes often occur on other parts of the body and do not correspond with the distribution of the adult female mites. These other rashes are believed to be caused, in part, by the shallow burrowing of the immature stages of *S. scabiei* and temporary burrows made by unfertilized females. Unlike adults, children often develop rashes on the face, chest, and back.

The most common means of transmission is by direct contact between individuals when the mites are crawling on the skin surface. However, transmission also can occur via bed linen, clothing, and other fabrics from infested hosts. The mites are able to survive two to three days at room temperatures when the relative humidity is more than 30%. The higher the relative humidity, the higher the survival rate. Larvae of *S. scabiei* can hatch from eggs deposited off the host and infest fomites up to seven days. However, transmission by fomites generally is not of major importance in temperate regions. *Sarcoptes scabiei* from infested horses reportedly has been transmitted to humans via saddle blankets, harnesses, and grooming utensils.

Human Scabies Scabies victims usually experience intense itching, especially at night. The itching typically is out of proportion to the visible signs of the infestation and tends to be aggravated by heat, warm baths, and removal of clothing. The pruritus and rash are attributed to antigens associated with the mite bodies, secretions, and fecal material deposited in the burrows. They stimulate the host's cell-mediated immune response, contributing to the development of acquired immunity to subsequent infestations by *S. scabiei* following initial exposure. This suggests the possibility of vaccines being developed to protect human and other hosts against natural infestations of scabies mites (Arlian et al., 1994a, 1994b). The antigenic nature of the cuticular components of the mites, their secretions, and excretory products helps to explain the persistence of the rash and other clinical signs long after the mites themselves have been killed by acaricidal treatments.

Cases of human scabies occur in a number of clinical forms. The most common type is **papular scabies**, characterized by erythematous papules that erupt as a generalized pruritic rash on various parts of the body. The accompanying itching usually leads to scratching and excoriation of the affected areas, contributing to an eczema-like condition. Vigorous scratched lesions may become secondarily infected with pyogenic, or pus-forming, bacteria, causing an acute, inflammatory, destructive skin condition called pyoderma. In some individuals, tiny vesicles develop in the epidermis in response to burrowing mites. If they become enlarged enough to form macrovesicles, or bullae, they cause what is known as **bullous scabies**. In other cases, the patient may develop urticarial scabies, in which a histamine-like vascular reaction produces wheals or hives that may be intensely itchy and can obscure the primary cause of the problem.

In a small number of individuals, *S. scabiei* may burrow deeper into the skin, penetrating the dermis and inducing infiltration of lymphocytes. This can lead to the formation of firm, reddish brown, pruritic masses and a condition called **nodular scabies**. The nodules tend to occur most commonly at the elbows, axillary region, groin, and male genitalia where they may persist for months or even a year or more despite treatment. Mites seldom are recovered from nodules that are more than a month old. Cases ultimately resolve with or without therapy.

One of the more distinctive, yet rare, clinical types of the disease is **crusted scabies**, also called hyperkeratotic scabies or Norwegian scabies. It is characterized by dry, scaly, or crusted lesions, usually on the hands and feet. Pruritus is typically mild or absent altogether, despite the extremely large numbers of mites, sometimes in the thousands, amidst the overgrowth of keratin tissue in the horny layer of the epidermis. The lack of discomfort and absence of burrows often results in these cases going undiagnosed. This condition is highly contagious and can be spread even on casual contact due to the large numbers of mites involved. Victims thus can serve as silent carriers and often are detected only as a result of clusters of cases of the more common forms of scabies in individuals with whom the source has come in contact, especially in hospitals and other institutional settings. Evidence indicates that the mites even can become airborne along with small scales of skin from the crusted lesions. Crusted scabies generally is associated with immunosuppressed individuals who do not respond normally to infestations of *S. scabiei*, or individuals with nervous disorders that render them insensitive to pain, especially skin sensations. They do not experience the usual itching, and their inclination to scratch is suppressed. Consequently, cases of crusted scabies often are associated with the mentally impaired and physically or immunologically compromised patients.

Despite the high host specificity of the different varieties of *S. scabiei*, many cases have been reported of humans being temporarily infested with scabies mites from other animals. Such cases are referred to as **animal scabies** and human sarcoptic mange. Although these cases usually involve dogs (particularly puppies), sources include livestock such as horses, cattle, sheep, goats, camels, and pigs. Such infestations typically result in localized erythematous papules and pruritus at contact sites. The mites do not form burrows and rarely survive to reproduce. Infestations are self-limiting and usually resolve themselves within a few weeks, provided the source is removed to prevent reinfestation. The absence of burrows and low numbers of mites usually makes it difficult to confirm cases by recovering mites from affected individuals. The diagnosis therefore often is based on demonstrating *S. scabiei* infesting the suspected animals involved.

A diagnosis of scabies can be confirmed by demonstrating the presence of *S. scabiei*. The presence of eggs, immature stages, adults, or fecal material from the burrows are all diagnostic. The presence of burrows in characteristic locations such as the wrists, fingers, elbows, and feet are considered nearly pathognomic (i.e., by themselves virtually confirm the diagnosis). To help in locating burrows, one or two drops of ink can be applied to suspected areas and then wiped off with alcohol after 10 minutes. The ink is retained in the burrows, making them more discernible. Several techniques have been developed to recover mites from scabies patients for microscopic examination and identification. Adult females can be removed from the blind end of their burrows by using a sharp-pointed scalpel blade to pierce the skin and gently pick out the mite. Alternatively, scrapings can be taken by vigorously scraping the affected skin several times with a sterile scalpel blade. The scraping is then transferred to a glass microscope slide for examination. Even in the absence of adult mites, the oval-shaped eggs (ca. $170 \times 190\ \mu$) are often clearly visible, as are the characteristic yellowish brown fecal pellets.

Skin biopsies can be taken and prepared for histological examination. Another method is to place skin scrapings in a small petri dish, or other container, and examine it after 12 to 24 hours for the presence of mites crawling on the bottom. A centrifuge-flotation method also has been used with some success, especially in cases of crusted scabies or when abundant material from affected areas can be collected. The scrapings are placed in 10% potassium hydroxide, or sodium hydroxide, and gently heated. The mixture then is added to a saturated sugar solution in a centrifuge tube and spun until any mites or eggs that are present float to the surface. Drops of the surface fluid can be examined microscopically. Eggs and egg shells have been detected by examining suspected skin scrapings in glycerin preparations using fluorescent microscopy.

The most widely used and effective means of treating scabies cases is the topical application of acaricides to the affected areas of skin. Among the more commonly prescribed acaricides are 1% lindane (gamma benzene hexachloride), crotamiton creams and lotions, sulfur applied directly to the skin or used in baths, 5% flower-of-sulfur suspended in lanolin or petrolatum, benzyl-benzoate emulsions in the form of a lotion or ointment, and tetrahydronaphthalene with copper oleate. It is recommended that these materials be applied after taking a warm, soapy bath. The number of follow-up applications and the prescribed intervals vary depending on the particular product used. Overtreatment can complicate conditions and should be avoided.

In addition to treating known cases and individuals with whom they recently have had contact, fomites should be treated to disrupt possible transmission. Acaricide sprays containing pyrethrins or 5% lindane are commercially available for this purpose. Laundering clothes, bedding, towels, and other fabrics using the hot cycle of a washing machine is usually adequate to kill *S. scabiei*. Hot ironing and placing items in a freezer for a week also is effective in killing them. Clothing and other fomites that cannot be treated (e.g., rugs, couches) should be set aside, if possible, and not touched for two weeks. Any scabies mites that may have been present will have died by then.

For further information on human scabies, see Heilesen (1946), Mellanby (1972), and Orkin et al. (1977).

Human Notoedric Mange Humans occasionally become infested with *Notoedres cati*, a sarcoptid mite that causes notoedric mange in cats. Cases in humans are called human notoedric mange or human notoedric scabies. Following prolonged exposure to infested cats, people may become sensitized to this mite and develop intense pruritus within a few hours of subsequent contact with them. The reaction is induced without the mites actually burrowing. The most common sites of skin lesions are on the hands and legs, reflecting the areas most likely to come in contact with pets. The lesions subside when infested cats either are treated or removed from further contact (Chakrabarti, 1986).

Mite-Induced Allergies

Members of several families of mites can cause allergic responses in humans by either direct contact of mites with the skin, or inhalation of mites or mite parts. The most common sources of allergy-inducing mites are stored products and house dust.

Storage Mites People who handle mite-infested stored products may become sensitized to the mites on subsequent contact, resulting in an immunological response called **storage-mite allergy**. Although the precise nature of the allergens is unknown, these substances include components of both live and dead mites, and material produced in the mite alimentary tract. Sensitized persons may experience either contact dermatitis or respiratory allergy, depending on the type of exposure.

Allergic contact dermatitis results from exposure to mites in grains, dried fruits, flour, and other stored products, causing itching and redness at the contact sites. The families of mites most commonly involved are the Acaridae, Carpoglyphidae, and Glycyphagidae. In addition, what was probably *Dermatophagoides pteronyssus*, but reported as *D. scheremetewski* (Pyroglyphidae), has been associated with cases of feather-pillow dermatitis. Contact with this mite infesting feather pillows is known to cause red papular lesions and pruritus about the scalp, eyes, ears, and nostrils (Traver, 1951; Aylesworth and Baldridge, 1983). A similar allergenic response to *D. farinae* associated with buckwheat-husk pillows has been reported in China (Hong et al., 1987).

Inhalational allergy results when airborne mites and associated allergens are drawn into the respiratory tract. The mucosal membranes lining the nasal and bronchial passages become irritated and inflamed, causing allergic rhinitis and asthma. The mucous membranes lining the eyelids also may be affected, causing conjunctivitis. These responses involve a T-cell-type reaction and both immediate and delayed hypersensitivity. Such reactions to mites present an occupational hazard, especially among farmers and other agricultural workers who handle mite-infested grains and other stored materials. Among the more common storage mites that cause inhalational allergy are *Aleuroglyphus ovatus* and *Tyrophagus putrescentiae* (Acaridae), *Lepidoglyphus destructor* (Echimyopodidae), and *Blomia tropicalis* (Glycyphagidae). For further information on storage-mite allergy, see Cuthbert (1990).

House-Dust Mites A major source of human allergens in the home is house dust and its associated mite fauna. Where humidity is sufficiently high, fungi tend to thrive in accumulated dust, providing food for a variety of house-infesting mites that are primarily saprophages or fungivores. Many of these mites are the same species that infest stored products, nests of rodents and birds, and animal litter. When their populations reach high levels in the home, they can cause acute or chronic allergic reactions commonly known as **house-dust allergy**. The principal allergenic components in house dust are mites and mite feces, rather than the dust material itself.

As many as 10 families and 19 species of mites have been recovered from house dust in a single urban community (Tandon et al., 1988), reflecting the diversity of mites that occur in that microhabitat. The most important taxa that cause human allergy are members of the Pyroglyphidae, notably those belonging to the genera *Dermatophagoides* and *Euroglyphus*. These mites typically comprise 90% or more of the mites found in house dust. The other families of mites commonly associated with house dust are the Acaridae, Glycyphagidae, Cheyletidae, and Echimyopodidae. A member of the latter family, *Blomia tropicalis*, is often the most common house-dust mite in the Neotropics. Many of these same species also infest stored products. Four of the more common storage mites found in house dust are *Acarus siro*, *Tyrophagus putrescentiae*, *Lepidoglyphus destructor*, and *Glycyphagus domesticus* (Figure 25.20) (Wraith et al., 1979).

The most widespread pyroglyphid species that causes house-dust allergy is the **European house-dust mite** (*Dermatophagoides pteronyssinus*), which thrives in floor dust and the surface dust of mattresses. It is regarded as the most frequently encountered house-dust mite, occurring especially in humid coastal areas of western Europe and North America. This was the first mite to be identified as a cause of house-dust allergy in 1966, shortly after the genus *Dermatophagoides* was first linked to house dust and bronchial asthma. The **American house-dust mite**, *D. farinae* (Figure 25.21), tends to be common in drier regions than *D. pteronyssinus*, such as the more continental-type climates of central Europe and the central United States. It is a frequent inhabitant of dried animal meal (e.g., dog biscuits, poultry feed) and coarsely ground wheat. The common name reflects its more common and widespread occurrence in the United States than in Europe and other parts of the

Figure 25.20 *Glycyphagus domesticus* (Glycyphagidae), female with four large, ovoid eggs, ventral view. (Hughes, 1976)

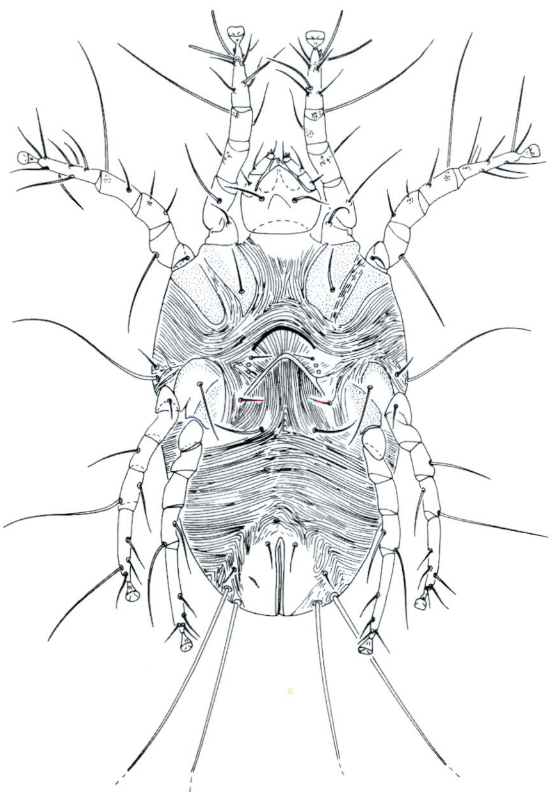

Figure 25.21 American house-dust mite, *Dermatophagoides farinae* (Pyroglyphidae), female, ventral view. (Gorham 1991; courtesy of the US Department of Agriculture)

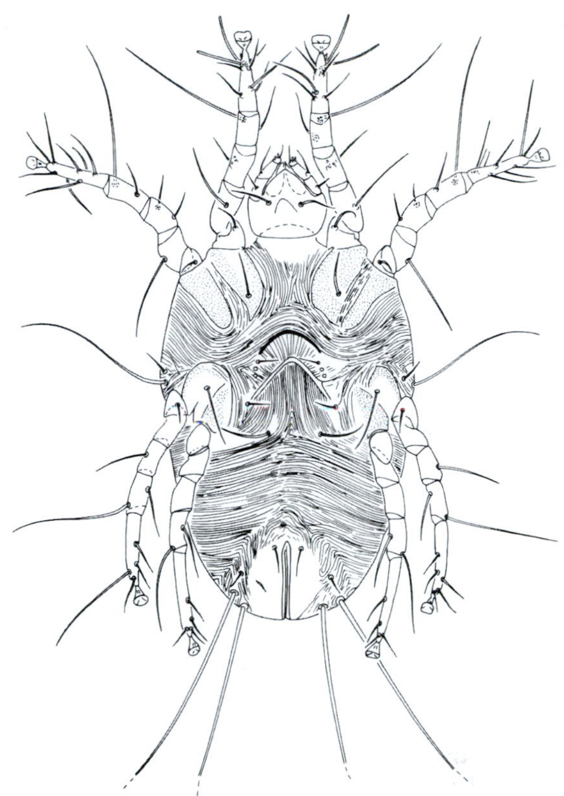

Figure 25.22 House-dust mite, *Euroglyphus maynei* (Pyroglyphidae), female, ventral view. (Gorham 1991; courtesy of the US Department of Agriculture)

world. The third most common mite known to cause house-dust allergy is *Euroglyphus maynei* (Figure 25.22). This is a cosmopolitan species frequently implicated in human allergy cases in Europe and Japan. It typically occurs in damper habitats than that of *D. pteronyssinus*, with which it often is associated.

House-dust mites thrive in environments with relative humidities above 65 to 70%. These mites are dependent on water vapor as their primary source of water, which they extract from the air. They cannot actively survive more than six to 11 days at relative humidities below 50%. They can, however, endure prolonged dry periods by forming desiccation-resistant protonymphs that can survive for months below the critical humidity for the active stages. Their feeding activity, reproductive rate, and amount of fecal material generated are all directly related to humidity levels (Arlian and Hart, 1992). Populations tend to increase beginning in early summer, reach their highest levels in early fall, and remain relatively constant during the winter months, reflecting indoor humidity.

The developmental times of *Dermatophagoides* and *Euroglyphus* species vary with temperature and humidity. Under favorable conditions at room temperature and a relative humidity of 75%, they typically complete a generation in about 30 days. Females do not lay eggs unless fertilized and commonly experience multiple matings. They lay one or two eggs per day during their adult life, usually lasting 30 or more days. Half or more of the mites in dust samples may be represented by eggs and, if overlooked, often leads to underestimates of population sizes when only the nymphs and adults are counted (Colloff and Hart, 1992).

Dermatophagoides and *Euroglyphus* species are saprophages, which, in the home, feed primarily on skin scales shed by humans and indoor pets. Although fungi are commonly found in the alimentary tract of house-dust mites, they presumably are ingested only secondarily and are not a significant nutritional component of their diet. In fact, some fungi, such as *Aspergillus penicillioides*, actually may be detrimental to mite growth and reproduction (Hay et al., 1993). Mattresses, and sleeping quarters in general, provide particularly favorable sites for mites to develop. Human semen has been shown to be a dietary supplement for house-dust mites and can significantly increase the number of eggs a female produces (Colloff et al., 1989).

House-dust mites occur in greatest numbers in the more humid living quarters of homes frequented by the occupants, notably wherever dust accumulates in bedrooms and living rooms. Mattresses are especially suitable, apparently due to the accumulation of human squamal cells and other skin debris. Under optimal conditions, as many as 5,000 mites have been recovered per gram of mattress dust. The type of floor can influence the species and number of mites. In damp houses, carpeted floors contribute to high populations, whereas in drier homes there may be little difference in the numbers of mites in carpeted and noncarpeted flooring. When humidity levels are high, even floors covered with linoleum and other plastic materials will support relatively high numbers of house-dust mites. In general, however, drier floors and carpets support higher populations of *D. farinae* than either *D. pteronysinnus* or *E. maynei. Euroglyphus maynei* has the highest humidity requirements, occurs primarily in mattresses and bedding, and is the least likely to be found in carpets. Although some studies have shown that wool carpets have higher numbers of mites than carpets made from synthetic fibers (e.g., nylon), other studies have shown no significant differences between them.

House-dust mites also may occur in fairly large numbers in other situations in the home. In a survey of household fabrics in Germany, 18% of the mites recovered were found in clothing (e.g., suits) hanging in closets (Elixmann et al., 1991). The same mites also may infest improperly stored food products. In one such case in Alabama, an individual experienced sneezing, intense ocular pruritus, and facial edema within minutes of inhaling a puff of dry pizza-dough mix heavily infested with *D. farinae* (Skoda-Smith et al., 1996).

House-dust mites are now recognized as the source of 20 or more allergens that cause reactions to house dust, especially in children and young adults. The most common clinical manifestation is bronchial asthma, characterized by difficulty breathing, inflammation of the nasal passages, and conjunctivitis. This may be accompanied by atopic eczema in some sensitized individuals. Asthmatic attacks tend to occur at night, especially in poorly ventilated bedrooms with old bedclothes and accumulated mattress and floor dust. Occurrence of symptoms is usually seasonal, reflecting the size of the mite populations. The acuteness of the allergic attacks is directly correlated with the number of mites present.

Dermatophagoides pteronyssinus generally produces the most potent house-dust allergens. However, a portion of individuals of other species, including *D. farinae, E. maynei,* and certain storage mites, can elicit allergic responses as great as that of *D. pteronyssinus.* Each species of mite appears to have its own species-specific antigens and allergens, with differences between those associated with the mite body and feces (scybala). There is significant cross-reactivity among the antigens of different species. This makes it difficult to determine which mite is involved in individual cases, thereby complicating clinical diagnosis and treatment. Diagnostic tests for mite-induced house-dust allergy include skin tests and bronchoprovocation using commercial extracts of individual mite species. Enzyme-linked immunosorbent assays (ELISA) and radioallergosorbent tests (RAST) have been developed to help in the diagnosis of house-dust-mite allergy cases. However, they tend to be less effective than the traditional skin-prick test in identifying people who are only mildly sensitized to mite allergens in house dust.

For further information on dust mites, the allergens they produce, their ecology, and distribution, see Arlian (2002) and Arlian et al. (2002).

Several sampling techniques have been developed to determine if house-dust mites are present in the home and, if so, what species they are and their relative numbers. Most of the techniques entail collecting samples of mattress and floor dust with a vacuum device, and examining the samples microscopically for the presence of mites. Various flotation and staining methods can be used to facilitate the process. Another approach is the use of a guanine test as an indirect means of determining the number of mites present. Guanine is excreted in mite feces and serves as a quantitative index of mite numbers, irrespective of the species. The amount of guanine can be measured using high-performance liquid chromatography (HPLC), providing a simple, rapid method for determining the amount of mite activity in different parts of the home (Quoix et al., 1993).

Internal Acariasis

Under certain exposure conditions, mites may enter natural body orifices, leading to cases of temporary internal acariasis. These cases commonly involve the ingestion of mites with food, and inhalation of airborne mites or mite-contaminated dust via oral or nasal routes. Mites that are swallowed or inhaled can lead to acariasis involving the alimentary tract, whereas mites that are inhaled also can invade the respiratory tract. Cases of mites infesting the urinary tract are rare. For a general discussion of the mites involved, see Ma and Wang (1992).

Pulmonary acariasis, in which mites invade the lungs, occurs most frequently among individuals exposed to mite-infested stored grains and dried herbs. This reportedly can be a serious problem among workers in grain-storage facilities and medicinal-herb warehouses in China (Chen et al., 1990; Li and Li, 1990). Clinical signs and symptoms include cough, expectorated phlegm and blood, difficulty breathing, chest pain, low-grade fever, restlessness, and marked eosinophilia. Pulmonary lesions have been documented on X-ray film as shadows and nodular opacities in lung tissues. The following five families and nine species of mites have been recovered from sputum of affected individuals: Acaridae (*Acarus siro, Tyrophagus putrescentiae, Aleuroglyphus ovatus, Sancassania berlesei* (reported as *Caloglyphus berlesei*), and what was probably *Sancassania mycophaga* (reported as *Carpoglyphus mycophagus*); Pyroglyphidae (*Dermatophagoides farinae, D. pteronyssinus*); Tarsonemidae (*Tarsonemus granarius*); and Cheyletidae (*Cheyletus eruditus*). It is not clear which of these mites cause the more serious problems. Some species, such as *S. berlesei,* can thrive in exceptionally damp food stores covered with a film of water and may be able to survive for some time in the lungs.

A *Carpoglyphus* species (believed to be *C. lactis*) was associated with a case of pulmonary acariasis in Spain (Toboada, 1954), whereas a *Tyrophagus* species and other unidentified mites were recovered from sputum, bronchial washings, and needle-aspirated lung specimens in routine examinations of patients suffering from respiratory ailments (Farley et al., 1989). Mite eggs, larvae, and adults were found in cytology specimens in the latter study, with evidence of their being surrounded by acute inflammatory cells in several cases.

Human cases of **enteric acariasis** occasionally are reported in which mites are found in excreta, suggesting their presence in the digestive tract. In most cases they are acarid mites in the genera *Acarus, Suidasia,* or *Tyrophagus. Suidasia pontifica* (reported as *S. medanensis*) was recovered from feces of a woman and two infants in Mexico (Martinez Maranon and Hoffman, 1976), whereas various stages of an *Acarus* or *Tyrophagus* species, together with eggs, were recovered from bile of a Romanian patient with chronic cholecystitis (Pitariu et al., 1978). It was concluded that the woman probably ingested the mites with her food and that the mites simply aggravated her preexisting cholecystitis by causing inflammation of her digestive tract until the mites were eliminated with the bile. Other cases of enteric acariasis have been reported in children with chronic digestive disorders in Russia (Prisich et al., 1986).

A few cases of **urinary acariasis** have been reported, primarily involving acarid mites in the genus *Tyrophagus.* Two species allegedly recovered from the human urinary tract are *T. putrescentiae* and *T. longior* (Harwood and James, 1979). Many, if not most, of these cases appear to be misleading and probably involve contamination of containers in which urine was collected or examined. A possible exception was the recovery of unidentified mites in urinary samples from several patients in Romania during acute attacks involving inflammation of the kidneys and urinary bladder (Pitariu et al., 1979). Numerous acarid mites and their eggs were observed in the urinary sediments; others were dead and encrusted with salts. Whether or not contamination of samples can be ruled out in these cases is unclear. Other cases of urinary acariasis have been reported in Japan (Harada and Sadaji, 1925) and China (Chen et al., 1992; Ma and Wang, 1992).

Tyrophagus longior (Figure 25.15) occurs primarily in cool temperate regions of Europe where it infests stored grains, hay, and straw; hay stacks in open fields; cucumber plants, tomatoes, and beets; and poultry litter in broiler houses. Cases of digestive and urinary acariasis in humans involving *T. longior* have been reported (Harwood and James, 1979).

Stored-products mites in the genus *Tarsonemus* (family Tarsonemidae) have been reported to be associated with human dermatitis and other skin disorders (Hewitt et al., 1973; Krantz, 1978; Oehlschlaegel et al., 1983) and to invade various organs and body fluids of humans and other animals (Dahl, 1910). The most commonly implicated species is *T. hominis.* It is generally believed that these reports represent contamination of glass slides and other materials used in preparation of tissues for microscopic examinations (Hewitt et al., 1973; Samšiňák et al., 1976).

An unusual case of large numbers of a mite (Histiostomatidae) infesting the external ear canal of a human, actively feeding and reproducing there, has been documented (Al-Arfaj et al., 2007).

MITE-BORNE DISEASES OF HUMANS

Excluding tick-borne diseases, there are only two significant diseases of humans for which mites serve as the principal vectors: rickettsialpox and tsutsugamushi disease.

Rickettsialpox

Rickettsialpox was first recognized in 1946 during an outbreak in New York City (Huebner et al., 1946). Sporadic cases had been reported as early as 1909 in Washington, DC, and other cities along the northeastern seaboard of the United States. Rickettsialpox is a relatively uncommon illness, with only 800 to 900 cases having been reported in the United States. Cases occur primarily in urban areas in crowded living quarters infested with the house mouse (*Mus musculus*) that serves as the major reservoir. The pathogen is transmitted to humans by the bite of the house mouse mite, *Liponyssoides sanguineus* (Dermanyssidae) (Figure 25.5). Other countries in which cases of rickettsialpox have been reported are Russia, Korea, and parts of equatorial and central Africa.

The causative agent of rickettsialpox is *Rickettsia akari.* It is a spotted fever group (SFG) rickettsia, and is morphologically indistinguishable from *R. rickettsii,* the causative agent of Rocky Mountain spotted fever. The intracellular site in which it replicates in human hosts remains unknown.

Rickettsialpox is usually a mild, nonfatal illness that typically begins with the appearance of a nonpruritic, erythematous papule at each infectious bite site, usually within 24 to 48 hours of contact with *L. sanguineus.* Soon thereafter a small vesicle forms at the center of the papule, filling initially with a clear, then cloudy, fluid. The vesicle dries, producing first a crusty lesion and then a brown or black scab, or eschar, in the center of a larger, indurated area 0.5 to 3.0 cm in diameter. These lesions can occur on any part of the body, but usually on the face, trunk, and extremities. They may occur on the palms and soles, and on mucous membranes about the mouth. The latter include the palate and less commonly the general mouth cavity, tongue, and pharynx. Although there usually are only a few, as many as 100 discrete lesions have been reported in some cases.

Systemic symptoms appear about the time that eschars form, nine to 14 days after the initial bites. Fever (usually peaking at 38–40°C), headache, and malaise are characteristic and may be accompanied by muscle aches, especially backaches, drenching sweats, and shaking chills. Less common symptoms are cough, running nose, sore throat, nausea, vomiting, enlarged and tender regional lymph nodes, and abdominal pain. Most cases resolve in six to 10 days without treatment. In some cases, however headache and lassitude may persist for another one to two weeks. Treatment with antibiotics generally alleviates the fever and other symptoms within 48 hours.

Diseases that should be included in the differential diagnosis of rickettsialpox are other members of the SFG rickettsiae, notably Boutonneuse fever, tsutsugamushi disease, Siberian tick typhus, and Queensland tick typhus. They can be distinguished, however, by their geographic occurrence and the clinical nature of the associated skin lesions. The nonrickettsial disease with which rickettsialpox is most commonly confused is chickenpox, caused by a virus. In chickenpox cases, however, the vesicles are not raised on papules, eschars are not formed, and the lesions are much more numerous. The clinical syndrome and a rise in titer of SFG-specific antibodies are generally sufficient to confirm a diagnosis of rickettsialpox. Immunity appears to be complete, perhaps life-long, following recovery from infection. For additional information on the clinical and diagnostic aspects of this disease, see Brettman et al. (1981) and Kass et al. (1994).

Liponyssoides sanguineus, the vector of the rickettsialpox agent, is primarily a parasite on the house mouse. The mite is also found on rats (*Rattus* spp.) and voles, although the role of these and other wild rodents in the ecology of this disease is uncertain. Whereas *L. sanguineus* nymphs generally take a single blood meal, adults move onto and off the host to take several blood meals. Most of the time is spent off the host in nests and runways of mouse-infested areas. Where it occurs in human dwellings, the mite seeks the warmth of furnace rooms and incinerators of old buildings where they may occur in large numbers on the walls and ceilings. Human bites are believed to occur primarily when house mice in apartment buildings become less attractive as hosts, inducing the mites to seek alternate hosts. The occurrence of lymphocytic choriomeningitis in house mice during outbreaks of rickettsialpox in humans has been suggested as a possible factor; such infections cause changes in a mouse's body temperature, perhaps inducing the mites to abandon their natural host (Krinsky, 1983). Starved adults can live seven to eight weeks, whereas blood-fed adults can live nine weeks or longer.

The only other mite reported as a possible vector of *R. akari* is *Ornithonyssus bacoti*, based on experimental transmission studies using laboratory white mice (Philip and Hughes, 1948; Lackman, 1963).

Tsutsugamushi Disease

Tsutsugamushi disease is a mite-borne rickettsiosis of humans that is endemic in eastern and southern Asia, the western Pacific region, along the northern coast of Australia (Queensland and Northern Territory), and the Indian subcontinent. Cases may occur as far west as Afghanistan, Pakistan, and neighboring areas of the former Soviet Union. It is also known as **scrub typhus** and chigger-borne rickettsiosis. The causative agent is *Orientia tsutsugamushi* (formerly *Rickettsia tsutsugamushi*) transmitted by the bite of trombiculid larvae, or chiggers (Figure 25.10).

Tsutsugamushi disease was recognized as early as the fourth century CE when it was described in clinical manuals as an illness associated with mites. It was not until 1930, however, that Japanese workers first isolated and identified the pathogen as a rickettsia. This disease first caught the attention of the western world during World War II when the Allied Forces were severely affected during operations in the Pacific Theater. The number of cases of tsutsugamushi disease exceeded that of direct, wartime casualties among the military forces in that region. Fatality rates as high as 27 to 35% occurred among troops on the islands of Goodenough and Finchhaven in New Guinea (Philip and Kohls, 1945; Philip, 1948). With the advent of effective antibiotics for treatment, the incidence of tsutsugamushi disease decreased dramatically in the region during the late 1940s and 1950s. However, sudden increases have occurred since that time in Japan (ca. 1975), Korea (ca. 1985), and other areas. For comprehensive reviews on this important mite-borne disease, see Traub and Wisseman (1974) and Kawamura et al. (1995).

The causative agent of tsutsugamushi disease is considered to be distinct enough from related rickettsial organisms to be placed in its own genus, *Orientia* (Tamura et al., 1995). Like *Rickettsia* species, it is an obligate intracellular parasite that multiplies in the cytoplasm of host cells. The clinical picture is complicated, however, by a multitude of antigenic variants, or strains, that exhibit various degrees of pathogenicity to humans. Among the better characterized strains are Gillian, Karp, Kato, Kawasaki, Kuroki, and Shimokoski. The relationships among the different strains and their mite vectors remains largely unknown.

The classic form of tsutsugamushi disease varies from a mild to severe illness. It begins with the development of a small papule at the bite site of an infected chigger. The skin reaction varies from hardly noticeable, or mildly itchy, to painful. The latter discomfort is characteristic of bites of the mite *Leptotrombidium akamushi* (Figure 25.23), and has been likened to a tiny thorn that has penetrated the skin and induces pain when it is rubbed. This sensation, called **ira** in endemic areas of Japan, usually appears about 10 to 20 hours after the bite, and is believed to be caused by an inflammatory eruption associated with formation of a feeding tube (**stylostome**) by the attached mite. Bites occur most frequently in the folds of soft skin of the axillary region, upper legs, and abdomen. Other common sites include webs between the fingers, skin behind the knees, genitalia, under breasts, and skin constricted by clothing. The bites become ulcerated and form hard, black scabs (eschars), typically accompanied by fever and a maculopapular rash.

Following an incubation period of about 10 days (range 5–20), symptoms generally include loss of appetite, fever, headache, muscle aches, and general malaise; regional or generalized lymphadenopathy is also

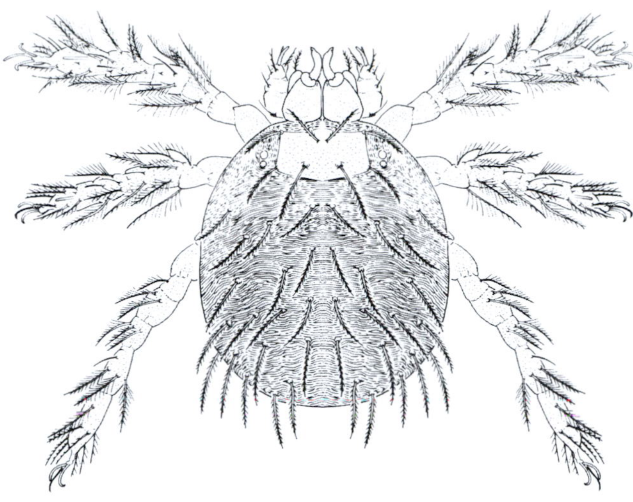

Figure 25.23 *Leptotrombidium akamushi* (Trombiculidae), chigger vector of *Orientia tsutsugamushi*, causative agent of tsutsugamushi disease in Japan. (Modified from Baker et al., 1956)

common. The more virulent strains of *O. tsutsugamushi* can cause hemorrhaging, intravascular coagulation, and other blood disorders as the rickettsiae multiply in epithelial cells of the vascular system. This can lead to microthrombi in the kidneys, lungs, and heart, contributing to fatalities. Mortality varies widely from 3 to 60%, depending on the strain and geographic region. Cases of tsutsugamushi disease can be treated effectively with antibiotics. However, in severe cases with hemorrhagic complications, heparin therapy and platelet transfusion may be necessary. Immunity following recovery is not lasting, such that reinfections are common in endemic areas.

Although cases of tsutsugamushi disease occur throughout the year, they often are seasonal in certain areas, reflecting the activity of local mite vectors. In some parts of Japan, for example, cases known as **Japanese river fever** tend to occur during the warm, summer months along river terraces where larvae of *L. akamushi* are present. Cases in Japan during the autumn and winter months, which may extend into late spring or early summer, are usually associated with two other chigger species, *L. pallidum* and *L. scutellare*. These seasonal differences are reflected in local Japanese names such as **Umayado disease** for the summer form of tsutsugamushi disease in Kagawa Prefecture, and **Shichito fever** for the autumn-winter form in the Hachijo Islands of Izu Shichito. The nonsummer types of tsutsugamushi disease usually are characterized by relatively mild symptoms and low fatality rates. In other subtropical and tropical regions of Southeast Asia and the southwestern Pacific, cases occur independent of seasons, correlated with the presence of chigger vectors that are present year-round (e.g., *L. arenicola*, *L. deliense*, *L. fletcheri*).

More than 40 species of trombiculid mites (13 genera) are known or suspected to be vectors of *O. tsutsugamushi*. The major chigger vectors of this pathogen and their geographic occurrence are shown in Table 22.3. The most important genus is *Leptotrombidium*, represented by approximately 25 vector species, most of which belong to the subgenus *Leptotrombidium*. Two or more vector species are known in each of the following four genera: *Neotrombicula* (6 spp.), *Ascoschoengastia* (2 spp.), *Euschoengastia* (2 spp.), and *Walchia* (2 spp.). Other trombiculid genera that play a role in transmission of *O. tsutsugamushi* to humans are *Acomatacarus*, *Eutrombicula*, *Gahrliepia*, *Leeuwenhoekia*, *Mackiena*, *Neoschoengastia*, *Odontacarus*, and *Shunsennia*.

Chiggers that serve as vectors of tsutsugamushi disease are primarily parasites of wild rodents such as field mice (*Apodemus* spp.), voles (*Microtus* spp.), and rats (*Leopoldamys*, *Maxomys*, *Rattus*, and other genera). They also occur on a wide range of birds (including pheasants, pigeons, and chickens) that are susceptible to infection and can develop at least transient rickettsemia. Although rodents serve as a source of infection of *O. tsutsugamushi* for chiggers that feed on them, it is generally believed that they play only a minor role as reservoirs and as a source of infection for the mites. Instead, the mites themselves serve as the principal natural reservoirs of the disease agent. Certain strains of trombiculid mites of a given species effectively transmit the rickettsia transovarially and/or transstadially; thus the pathogen is transmitted from the adult female to her larval offspring and to each developmental stage that follows. Consequently, *O. tsutsugamushi* is passed from generation to generation of mite and is maintained within local mite populations in endemic areas. Were this not the case, this mite-borne rickettsia could not persist in nature. Because trombiculid mites feed on only a single host, they do not have the opportunity to transmit an acquired pathogen at a subsequent feeding. For further information on the ecology and medical aspects of tsutsugamushi disease, see Kawamura et al. (1995).

Intermediate Hosts of Human Parasites

No metazoan parasites of major health importance to humans involve mites as intermediate hosts. However mites are hosts for a few tapeworms that occasionally infest people. Oribatid mites are intermediate hosts for two *Bertiella* species of anoplocephalid tapeworms. *Bertiella studeri* parasitizes the small intestine of a wide range of Old World primates, including rhesus and cynomolgus monkeys, Japanese macaque, baboons, mandrils, gibbons, orangutan, chimpanzees, and occasionally humans in Asia and Africa. Although several European species of oribatid mites have been shown to support development of *B. studeri* experimentally (Stunkard, 1940; Denegri, 1985), the oribatid species involved in the natural transmission cycle remain unknown. *Bertiella mucronata*, which parasitizes monkeys in South America, develops in the oribatid mites *Dometorina suramerica*, *Scheloribates atahualpensis*, and other species of the genera *Achipteria*, *Galumna*, *Scheloribates*, and *Scutovertex* (Sengbusch, 1977; Denegri, 1985). In the case of both of these tapeworms wild primates are the primary vertebrate hosts, with human cases occurring where infested primates live in close association with people. They cause no apparent lesions or other harm to their hosts.

Occasionally humans may be parasitized by tapeworms of the genus *Mesocestoides* (family Mesocestoididae) that use mammalian carnivores and charadriiform birds as hosts in North America, Europe, Asia, and Africa. Oribatid mites are believed to play a role as intermediate hosts in the relatively complex life cycles of these cestodes.

Delusory Acariasis and Acarophobia

Because of their tiny size and the general lack of knowledge about mites by the general public, mites are often mistakenly blamed as the cause of skin problems or bite-like sensations when the underlying cause is unknown. The term for this is **delusory acariasis**, the imagined notion that mites are biting or infesting the skin when in fact they are not. A rational discussion is unlikely to convince individuals involved otherwise. This is a specific type of the more general phenomenon of **delusory parasitosis**. A typical example is attributing various skin conditions among office workers to "paper mites." Although there are no such creatures, it is difficult to dispel the misconception that such mites are involved. Other imaginary mites are "telephone mites" and "cable mites," blamed as the cause of skin irritation among telephone users and computer operators.

The term **acarophobia** refers to an undue fear of mites that can cause psychological stress. This may develop as the result of an actual experience with mites, or more likely as a consequence of one or more episodes of delusory acariasis.

VETERINARY IMPORTANCE

Mites have been very successful in exploiting vertebrate hosts. Many are ectoparasites of skin, scales, feathers, or fur, whereas others are endoparasites that have invaded body cavities, respiratory passages, and internal tissues and organs. Some mites are vectors of disease agents of domestic and wild animals; still others serve as intermediate hosts for animal parasites, notably tapeworms. Occasionally mites are the cause of allergic reactions of pets and other animals. For overviews of mites of veterinary importance, see Hirst (1922), Baker et al. (1956), Strandtmann and Wharton (1958), Sweatman (1971, 1984), Yunker (1973), Georgi (1980), Whitaker (1982), Nutting (1984), and Pence (1984). For works of a more regional nature, see Domrow (1988, 1991, 1992) for Australia, Mulla and Medina (1980) for South America, and Cosoroabă (1994) for Eurasia.

Mite-Induced Dermatitis

Four families of ectoparasitic mites commonly cause irritation when they bite host animals to feed on blood, lymph, or skin tissues. Three families are mesostigmatid mites: Dermanyssidae, Macronyssidae, and Laelapidae. The fourth is the prostigmatid family Trombiculidae (chiggers).

Dermanyssidae Most dermanyssid mites cause relatively little harm even to heavily infested hosts. *Dermanyssus hirudinis*, for example, has been observed to cause little adverse effects on the survival, growth, or health of house wrens (*Troglodytes aedon*) infested with hundreds or thousands of mites per nestling (Johnson and Albrecht, 1993). Similarly, *D. americanus* and *D. gallinae* seldom cause problems when infesting wild avian hosts. *Dermanyssus gallinae*, however, commonly causes dermatitis in atypical avian hosts and domestic mammals, and can cause severe infestations and economic losses in domestic chickens.

Chicken Mite (*Dermanyssus gallinae*) This mite (Figure 25.4), also known as the red poultry mite and pigeon mite, is an obligate parasite of wild and domestic birds worldwide, including chickens, pigeons, canaries, parakeets, house sparrows, and starlings. Occasionally it infests dogs, cats, horses, cattle, rodents, rabbits, and other mammals. Cases involving domestic animals usually occur in association with poultry houses or infested bird nests.

The chicken mite is especially a problem in poultry operations. It hides by day in crevices and nesting materials, moving onto the birds to feed at night. Skin lesions in chickens are usually inapparent, but may occur as erythematous papules on any part of the body. Chronic or heavy infestations can be debilitating and result in skin irritation, loss of vigor, stunted growth, reduced egg production, anemia, and death due to exsanguination. Newly hatched chicks are particularly vulnerable. Setting hens may be driven from their nests, and susceptibility to disease agents may be significantly increased.

Dogs exposed at night to *D. gallinae* around poultry houses may react adversely to their bites. In addition to developing intense pruritus, they remain awake or howl at night, become less active during the day, and may show signs of depression. Skin lesions include erythematous papules, hyperpigmentation, and scaling, accompanied in some cases by partial loss of hair and slightly enlarged lymph nodes. Removal of an animal from the source of mites usually results in prompt alleviation of symptoms.

Dermanyssus gallinae can develop from egg to adult in as few as five days, completing its life cycle in nine to 10 days. Mating and oviposition occur off the host. The eggs typically are deposited in groups of four to seven eggs, with a given female producing a total of 20 to 24 in her lifetime. The mites usually engorge to repletion at one feeding and lay their eggs at approximately three-day intervals. The eggs hatch in one to two days to produce larvae that do not feed. The protonymphs and deutonymphs, like adults, feed on blood. The adults can endure starvation for several months, enabling them to survive extended periods in abandoned bird nests and unoccupied poultry houses.

Macronyssidae Most members of the Macronyssidae are obligate parasites of vertebrates. As a group they appear to have evolved on bats, from which they have secondarily transferred to other mammals, reptiles, and birds. Although most macronyssid mites cause little or no apparent harm to their bat hosts, *Chiroptonyssus robustipes* has been known to cause the death of heavily infested, captive Brazilian free-tailed bats (*Tadarida brasiliensis*). This mite attaches primarily to the wings where its feeding can result in increased vascularity and edema at the bite sites, enlargement of lymphatic vessels, hyperkeratosis, and excoriation of the stratum corneum (Sweatman, 1971).

A few macronyssid mites have invaded the oral mucosa of pollen-feeding and fruit-feeding bats (Psyllostomatidae). All are members of the genus *Radfordiella* (Figure 25.24). In heavily infested bats they can cause bone damage to the hard palate and destruction of gingival tissues resulting in loss of teeth. The damage is caused by protonymphs. The adult mites presumably are nidicolous and move onto the host only intermittently to feed on parts of the body other than the mouth (Phillips et al., 1969).

Tropical Rat Mite (*Ornithonyssus bacoti*) The tropical rat mite (Figure 25.6) occurs throughout the world, where it parasitizes primarily rodents. It occasionally infests cats, wild carnivores, chickens, and other birds, as well as humans. Most veterinary problems involving this mite occur in laboratory mice, rats, and hamsters. Heavily infested animals may become debilitated and anemic, or experience reduced reproductivity; death can occur in some cases. Infestations usually are recognized by the presence of blood-engorged deutonymphs and adults in the animal bedding, cages, and corners or crevices of cage racks. The tropical rate mite is a vector of *Litomosoides carinii*, a filarial nematode in cotton rats (see the section, "Mite-Borne Diseases").

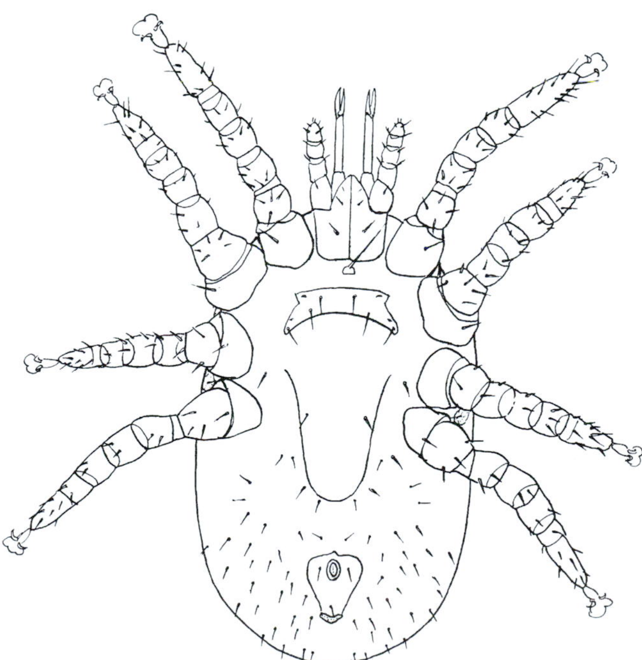

Figure 25.24 *Radfordiella oudemansi* (Macronyssidae), female, ventral view. (Strandtmann and Wharton, 1958)

Blood-fed females lay their eggs in bedding or nest debris of their hosts and in cracks and crevices. The eggs hatch in one to four days, producing larvae that molt to protonymphs about one day later without feeding. The protonymphs feed on blood, molting to deutonymphs in one to two weeks. Deutonymphs molt to adults after one or two days without feeding. Development from egg to adult can be completed in 11 to 16 days under favorable conditions. Adults usually mate within one or two days following emergence and can survive several days to a few weeks without a blood meal. The tropical rat mite is found primarily in rodent nests, moving onto the host animals only to feed.

Tropical Fowl Mite (*Ornithonyssus bursa*) The tropical fowl mite (Figure 25.7) is a common ectoparasite of wild and domestic birds throughout the warmer regions of the world. Domestic or peridomestic birds that may be infested include chickens, ducks, pigeons, starlings, house sparrows, and canaries. Most infestations originate from contact with wild birds or infested nest materials. Heavy infestations in chickens and other domestic fowl can result in anemia, decreased weight gain and egg production, and occasionally death. Newly hatched chicks and young birds are especially vulnerable. Blood-feeding by this mite causes skin irritation that may be intense enough to induce setting hens to leave their nests. Inspection of the plumage will reveal the mites in the down feathers, particularly around or just below the vent. Infested feathers are soiled or dirty in appearance due to the accumulation of mites, exuviae, eggs, and excreta. In the case of young birds, the mite commonly occurs around the eyes and beak.

Ornithonyssus bursa may cause more problems for wild birds than previously suspected. In Denmark, for example, nest infestations of the barn swallow (*Hirundo rustica*) have been shown to decrease reproductive success by reducing clutch sizes, nesting periods, and number of fledglings; and lengthening the time between clutches and the incubation period (Möller, 1990). Barn swallows regularly reuse old nests, but tend to avoid nests infested with this mite.

The tropical fowl mite lays its eggs on the host or in nesting materials where they hatch in two to three days. The larvae do not feed, whereas the protonymphs and adults feed intermittently on host blood. Although relatively few details about its life history have been reported, this mite is believed to develop from egg to adult in six to eight days. In the absence of a host, it can survive about 10 days.

Northern Fowl Mite (*Ornithonyssus sylviarum*) The northern fowl mite (Figure 25.8) is a major pest of chickens and other domestic fowl, particularly in temperate regions of North America and Eurasia. It also is an economic pest of domestic fowl in Australia, New Zealand, and other parts of the world where it has been introduced. In addition to infesting chickens, it is commonly found in the nests of pigeons and various wild birds, and as an incidental pest biting rodents, hamsters, and humans.

The greatest economic impact occurs in chicken houses. Initial infestations usually occur via wild birds or newly acquired chickens already infested with the mite. They can then spread throughout even the largest houses within a few weeks. The northern fowl mite causes problems very similar to those caused by *O. bursa*. These include skin irritation without apparent lesions at the bite sites; matted and grayish feathers, especially around the vent where the mites, their exuviae, and feces are concentrated (Figure 25.25); scaly, scabby, or thickened skin; and general loss of thriftiness. Individual caged birds may have 10,000 or more mites. Heavily infested birds often become anemic, experience decreased weight gains and egg production, and sometimes die. Egg shells may become significantly thickened, and egg production may drop 5 to 15% compared to uninfested birds. The greatest effect on body weight and efficiency of feed conversion generally occurs when hens are infested with mites before they reach full egg production. Although pathogens of poultry, such as viruses that cause Newcastle disease and fowl pox, have been recovered from *O. sylviarum* after feeding on infected chickens, there is little or no evidence that the mite transmits these agents when it bites.

The northern fowl mite spends most of its time on the host. It also occurs, however, in nesting materials and nest debris, roosting areas, cracks and crevices in the floors and walls of chicken houses, where they can be found during the day or night. Oviposition usually occurs on the host about two days following complete engorgement with blood. Eggs typically are laid two to three at a time, but there may be as many as five. The eggs hatch in about 24 hours. The protonymph, the only immature stage that feeds on blood, requires at least two feedings before it molts to produce the deutonymph. The complete life cycle is typically five to seven days, enabling populations to build very rapidly. Survival time without a host is usually three to four days, but may be as long as two to three weeks.

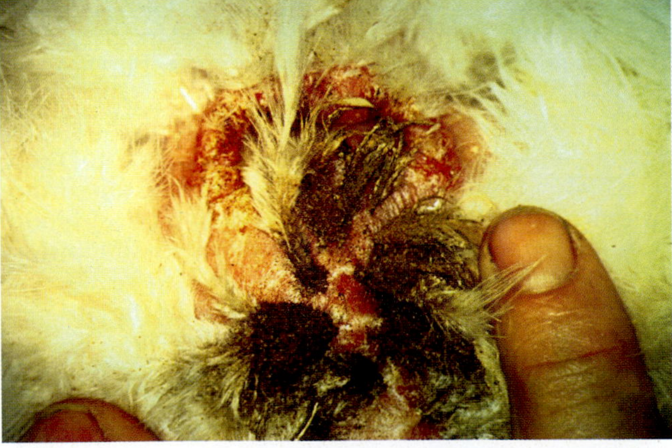

Figure 25.25 Northern fowl mite, *Ornithonyssus sylviarum* (Macronyssidae); heavy infestation of chicken in vent area. (Photo by Jerry F. Butler)

Snake Mite (*Ophionyssus natricis*) This mite (Figure 25.9) is a common parasite on captive snakes where it is usually found around the eyes and under chin scales. It also may be found on captive lizards and other reptiles. The sources of infestation usually are caged snakes in pet stores, zoos, or laboratories. Although *O. natricis* occasionally is found in low numbers (<100/host) on wild snakes, its populations on captive snakes may reach hundreds or thousands per host. Heavy infestations cause listless behavior, loss of appetite, anemia, skin irritation, loss of scales, and in some cases death. Several hundred mites are sufficient to cause severe anemia and other symptoms.

Ophionyssus natricis is a vector of the bacterium *Aeromonas hydrophila* that causes hemorrhagic septicemia in snakes. Infected snakes hemorrhage internally and often die three to four days following infection.

The snake mite lays its eggs in crevices, debris, and on rough surfaces of cages where they hatch in two to four days. The larvae do not feed, whereas the protonymphs, deutonymphs, and adults are obligate parasites that feed exclusively on blood. The life cycle usually is completed in two to three weeks at room temperature, with adult females living five to six weeks. Females typically feed two or three times, depositing about 20 to 25 eggs after each blood meal. See Camin (1953) for further details on the biology and behavior of this mite.

Laelapidae

Laelapid mites are commonly associated with rodents, other nest-building mammals, and bird nests. Those of medical or veterinary concern belong to the subfamilies Laelapinae, Hemogamasinae, and Hirstionyssinae, which include both facultative and obligate parasites that generally cause little or no apparent harm to their hosts. The genera most commonly encountered by veterinarians are *Androlaelaps*, *Haemogamasus*, *Laelaps*, and *Echinonyssus*. *Haemogamasus liponyssoides* is an opportunistic blood-feeder that inhabits nests of rodents and shrews. Its slender, chelate chelicerae are capable of piercing skin to feed on blood, enabling this mite to feed on laboratory mice if they become infested.

Spiny Rat Mite (*Laelaps echidninus*) The spiny rate mite occurs throughout most of the world as an ectoparasite associated primarily with the black rat and Norway rat. Occasionally it is found on *Sigmodon* and *Rattus* species, the house mouse, and other domestic and wild rodents. It is rarely found on laboratory animals. This mite is generally easy to recognize by its large body size (ca. 1 mm long), heavy sclerotization, and long, stout body setae that give it a spiny appearance. It lives primarily in host bedding or nesting materials, moving onto the host at night to feed. Its chelicerae are not capable of piercing intact skin but instead assist the mite in feeding on lachrymal secretions and blood or serous exudates from abraded skin. Rarely does its feeding cause discernible lesions, although injury to the footpads of suckling mice has been reported. The spiny rat mite is a vector of *Hepatozoon muris*, a blood protozoan of rats.

Regular blood meals are required for *L. echidninus* to survive and reproduce. Blood-fed females give birth to live larvae that do not feed. The protonymphs and deutonymphs both apparently feed similarly to adults, completing their development to adults in one to three weeks. The length of their life cycle is variable, requiring at least 16 days. The females can live two to three months; without food, however, they survive only about a week.

Trombiculidae

Although not widely recognized as a problem, larvae of trombiculid mites (chiggers) commonly infest domestic animals. Only in cases of heavy infestation or sensitivity reactions are they likely to be brought to the attention of veterinarians. As in humans, the resultant dermatitis is a response to chiggers that are normally parasitic on other host animals. Only incidentally do they attach to atypical hosts such as cats, dogs, sheep, other livestock, and occasionally to domestic or pet birds. Most cases involve mild pruritus and are likely to go unnoticed. Cases of heavy infestations, however, can result in severe itching with formation of vesicles and crusty or scabby skin lesions, usually about the head and neck. Large numbers of engorged chiggers may be visible as orange patches associated with the lesions. The chiggers typically remain attached only up to two to three days. Treating the lesions with an acaricide and preventing secondary infection usually resolves the problem if the animal is not reinfested.

Some chiggers enter the skin via large hair follicles, crawling down the shaft of the hair, sometimes well beneath the skin surface. An extension of the stylostome, or feeding tube, may extend backward around the mite to form a hyaline capsule. Usually these capsule-forming chiggers are completely intradermal and may cause localized inflammation and edema. In some cases they induce formation of cysts at the base of the hair follicles that can lead to secondary infections and slow-healing lesions. Intradermal chiggers include members of the trombiculid genera *Apollonia*, *Chaldonta*, *Euschoengastia*, *Gahrliepia*, *Guntherana*, *Intercutestrix*, and *Schoutedenichia* (Sweatman, 1971). These mites occur primarily in Southeast Asia, the South Pacific islands, Australia, Africa, and other parts of the Old World. Rodents, shrews, and bandicoots are some of their more common natural hosts.

A few species of chiggers have been identified as the cause of trombiculosis in domestic cats. In the United States, *Walchia americana* is known to cause papules on the face, ears, and thoracic areas of cats, in addition to thickening and crusting of the skin on the abdomen and legs. This is accompanied by hyperkeratosis, eosinophilia, and infiltration of mast cells as evidenced in skin biopsies at the lesion site (Lowenstine et al., 1979). Large numbers of *Eutrombicula alfreddugesi* have been observed as distinct orange patches on the head and ears of a cat in North Carolina, causing inapparent dermatitis (Hardison, 1977). Other chiggers known to infest cats are *Leeuwenhoekia adelaidiae*, *L. australiensis*, *Schoengastia philippinensis*, and *S. westraliensis* in Australia. Natural hosts for these mites include wallabies, grey kangaroos, and wild pigs (Wilson-Hanson and Prescott, 1985). Other cases in cats involving unidentified chigger species have been reported in Australia. Lesions occurred as pin-point erythemas and orange crusts on the ears (pinnae), pruritus, papules, and orange crusty lesions about the eyes and face, conjunctivitis, and ocular discharges. In one case, swelling and irritation of the perineal region with concomitant inability to pass urine was attributed to trombiculid mites (Wilson-Hanson and Prescott, 1985).

Dogs appear to be less commonly bothered by chiggers. Bite reactions are similar to other host animals, with localized redness, pruritus, and development of papules or vesicles at the bite sites. Cases involving heavy infestations may warrant veterinary attention. In Europe, the harvest mite (*Neotrombicula autumnalis*) reportedly has caused nervous symptoms in dogs, including partial paralysis of the limbs and lameness (Prosl et al., 1985).

Virtually all species of livestock are subject to chigger infestations while grazing, walking paths to and from barns, being held in enclosures, or by contact with recently harvested hay or grains infested by these mites. Skin lesions in the form of papules or crusty eruptions can be irritating and lead to self-inflicted skin damage as the host animal rubs and abrades the affected areas. Lesions occur primarily on the lips, muzzle, face, feet, and belly. Pigs have developed a generalized pruritus after feeding on fresh, chigger-infested grains from automatic feeders. Sheep, goats, and cattle are particularly prone to infestations with *Neotrombicula autumnalis* in Europe during the harvest season, causing pruritus, scabs, and loss of hair, particularly about the head and neck. In Australia, sheep have experienced severe dermatitis on the legs and feet due to infestations by *Eurombicula sarcina*, a chigger that normally parasitizes kangaroos. An orf-like condition in sheep, caused by a *Guntheria* species, has been reported during the summer months in South Africa (Otto and Jordaan, 1992).

Domestic birds such as chickens may become parasitized by chiggers (e.g., *Neoschoengastia americana*), leading to itching and dermatitis. In most cases the mites can be found under the wings or around the vent. Reports of anemia in chickens attributed to heavy infestations of chiggers should be treated with skepticism; chiggers feed on dermal tissues and not on blood. Occasionally other captive birds may be affected. A chronic infestation of canaries by an unidentified trombiculid mite has been reported in Australia. The canaries, in a commercial aviary, developed nonpruritic, subcutaneous swellings of the legs and ventral trunk, with acute inflammation and skin necrosis at the sites of mite attachment (Pass and Sue, 1983).

Wild animals generally do not show adverse reactions to chiggers, even when heavily infested. Occasionally, however, they do react severely to bites of certain species that normally parasitize other hosts. Reactions include formation of vesicular or crusty lesions, slow-healing eschars, localized skin discoloration, and some loss of hair. Examinations often reveal orange or red clusters of mites about the head, ears, neck, axillae, or groin. Sometimes infestations of chiggers about the eyes cause ocular lesions in the form of pruritic eyelids and conjunctivitis. Snakes, lizards, skinks, and other reptiles are parasitized by chiggers, most commonly noticed as orange or red clusters of mites on the head and neck. The host seldom shows apparent harm, even in cases of individual snakes infested with several thousand mites.

Fur Mites

Certain families of mites are categorized as fur mites because they are specially adapted for living in the hair coat of mammalian hosts. They often exhibit striking modifications of the palps, legs, and other body structures for grasping or clinging to hair. The five groups of particular veterinary interest are the cheyletoid families Cheyletidae (including the former family Cheyletiellidae) and Myobiidae, and the astigmatid families Listrophoridae, Atopomelidae, and Myocoptidae.

Cheyletidae Parasitic cheyletid mites occur on domestic cats, dogs, and rabbits, as well as many wild mammals and birds. They are nonburrowing mites that live in the pelage of their hosts and feed on lymph and other tissue fluids by piercing the epidermis with their stylet-like chelicerae. The enlarged gnathosoma and pair of large, terminal palpal claws give cheyletiellid mites a characteristic appearance. These structures are used to secure the mites to their hosts and to assist them in inserting their chelicerae. Members of the genus *Cheyletiella* can cause problems that warrant veterinary attention. Although most cases of cheyletiellosis go unnoticed, infestations of these mites can cause eczema-like skin conditions, or **cheyletid mange**, with associated pruritus and hair loss. Three *Cheyletiella* species of veterinary importance are *C. blakei* of cats (Figure 25.26), *C. yasguri* of dogs, and *C. parasitivorax* of rabbits (Figure 25.27). All developmental stages of these mites occur on the host animal. The eggs are glued to hairs but can be dislodged with loose hairs by host grooming. They also can be ingested and passed in the feces. The presence of *Cheyletiella* eggs in cat and dog feces thus serves as evidence of mite infestations even in asymptomatic cases (Fox and Hewes, 1976; McGarry, 1993). Transmission is usually by direct contact with infested animals, including maternal transfer while nursing. Because *Cheyletiella* species can survive up to 10 days or more off a host, animal bedding, household furniture, blankets, and carpets frequented by pets can serve as other sources of these mites. *Cheyletiella* mites are commonly phoretic on cat and dog fleas (*Ctenocephalides* spp.) and also may be transmitted via these ectoparasites.

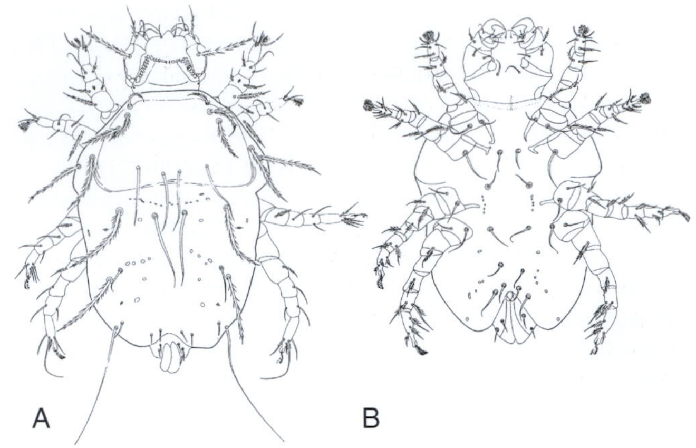

Figure 25.26 *Cheyletiella blakei* (Cheyletiellidae), female. (A) Dorsal view; (B) ventral view. (Modified from Domrow, 1991)

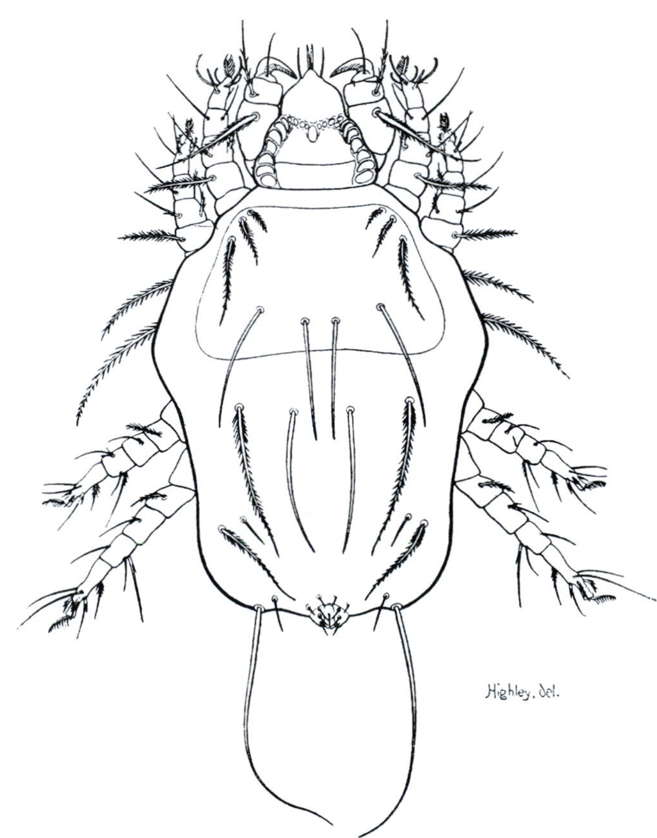

Figure 25.27 Rabbit fur mite, *Cheyletiella parasitivorax* (Cheyletidae), female, dorsal view. (Hirst, 1922)

Cheyletiella blakei usually infests the facial area of cats. Heavy infestations can result in the formation of small, crusty, erythematous papules and loss of hair, accompanied by itching and scratching. Long-haired cats tend to be more commonly infested than short-haired cats and are more likely to be involved in human cases of cheyletiellosis. *Cheyletiella yasguri* parasitizes domestic dogs, particularly in Europe and North America.

It is generally less common than *C. blakei* and only occasionally causes problems warranting veterinary attention. Signs of an infestation are scratching and a mealy or powdery dandruff in the affected areas, commonly the lower back. Heavy infestations can cause scaling, hyperkeratosis and thickening of the skin, erythema, pruritus, and hair loss. Puppies tend to have a higher incidence of *C. yasguri* than do adult dogs and are more likely to exhibit pruritus. This mite can cause dermatitis in humans upon close contact with infested dogs, especially puppies (Figure 25.28). Acaricidal treatments of dogs and their surroundings are effective in controlling *C. yasguri*. This is especially important in kennels, which serve as a common source of mite infestations.

Cheyletiella parasitivorax, the rabbit fur mite, is a common parasite of the European rabbit (*Oryctolagus cuniculus*) in North America, Europe, Asia, Australia, and New Zealand. It occurs most frequently in the posterior-back region of infested hosts, but also may occur on the face, frontal area, and other parts of the body. High mite populations induce the accumulation of epidermal scales and a scurfy appearance, leading in untreated animals to varying degrees of dermatitis, erythema, thickening of the skin, and hair loss. Severe cases may involve serous exudates and hairless patches in which the mites can be found in the disrupted keratin layer amidst epidermal debris. Infestations of *C. parasitivorax* are particularly a problem in commercial rabbit colonies and laboratories where rabbits are closely confined. Wild rabbits seldom exhibit cheyletiellid mange. *Cheyletiella parasitivorax* is capable of transmitting myxomatosis virus among European rabbits in Australia.

For additional information on *Cheyletiella* species and their veterinary importance, see Smiley (1970) and van Bronswijk and De Kreek (1976).

Myobiidae Members of the family Myobiidae are obligate parasites of rodents, bats, insectivores, and certain marsupials. They typically grasp the hairs of their host with their forelegs that are often highly modified for this purpose. The mites move up and down the hair shaft and remain clinging to the hairs as they feed on epidermal fluids. Their chelicerae are long and stylet-like and adapted for puncturing thin epidermal tissues to feed on extracellular fluids. A few species, however, are known to feed on blood (e.g., *Blarinobia simplex* on shrews, and *Eadiea brevihamata* on the shrew-mole). Most species cause little apparent discomfort or harm to their hosts, even when mite populations are high. Exceptions of veterinary interest are a few *Myobia* and *Radfordia* species that commonly infest rats and mice, often causing mild dermatitis and scurfiness in laboratory rodents.

The most widely recognized myobiid is the mouse fur mite, *Myobia musculi* (Figure 25.29). This is a cosmopolitan, ubiquitous species that infests the pelage of both wild and captive house mice (*Mus musculus*). Most of what is known about the development and biology of myobiids is based on this species. Females deposit their eggs singly, gluing them to the bases of hair shafts. The developmental time from egg hatch to adult is about 23 days. All stages, including the larvae, feed on dermal tissue fluids. The host response varies greatly depending on the strain, sex, age, and sensitivity differences of individual mice. Lightly infested hosts are often asymptomatic, or exhibit little adverse reaction. In highly sensitive hosts, however, even a few mites can elicit allergic reactions and severe pathologic responses. Heavy infestations can lead to severe dermatitis, with intense pruritus, hair loss, self-inflicted trauma from scratching, and in some cases

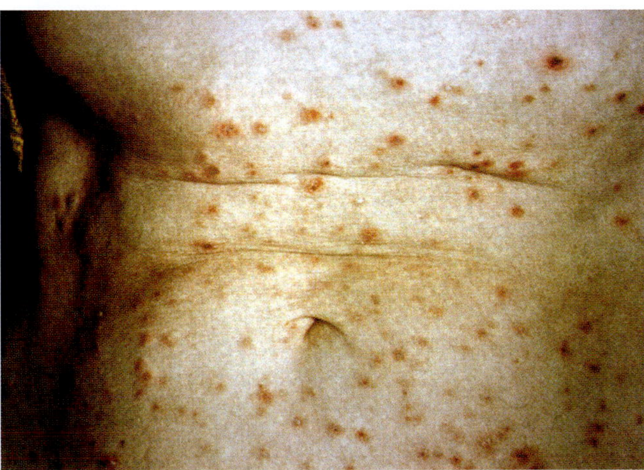

Figure 25.28 Multiple lesions from bites of *Cheyletiella yasguri* (Cheyletidae) on abdomen of woman following contact with infested puppy. (Southcott, 1976)

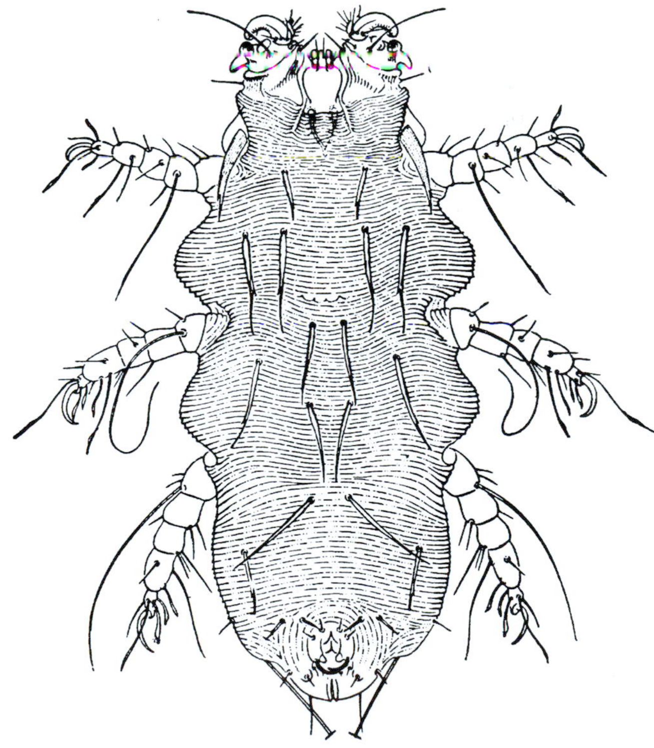

Figure 25.29 Mouse fur mite, *Myobia musculi* (Myobiidae), female, dorsal view. (Baker et al., 1956)

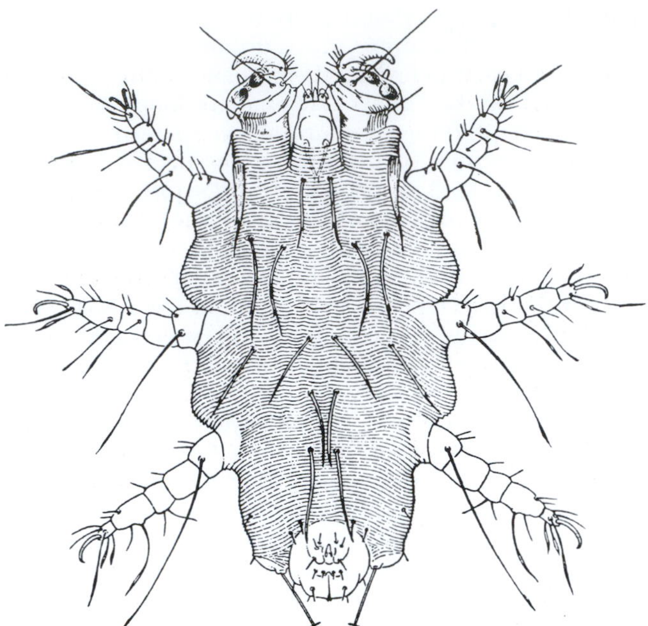

Figure 25.30 *Radfordia ensifera* (Myobiidae), female, dorsal view. (Baker et al., 1956)

death. This especially can be a problem in laboratory mice colonies where infestations are likely to involve virtually all individuals.

Two *Radfordia* species occur worldwide, infesting the fur of wild and laboratory rodents. The more common is *R. ensifera* parasitic on rats (Figure 25.30); the other is *R. affinis* on the house mouse. They closely resemble *Myobia musculi* from which they are distinguished by a pair of tarsal claws (rather than one claw) on the second leg. Although they generally cause little pathologic effect, dermatitis and self-inflicted trauma have been associated with heavy infestations of *R. ensifera* on laboratory rats.

Listrophoridae Listrophorid mites (Figure 25.31) are obligate parasites of rodents, lagomorphs, carnivores, and other mammalian hosts in the Old World and New World. They are well adapted for clasping securely to hair shafts, with appendages modified for this purpose. They attach so firmly that they are difficult to remove and can damage the hairs by bending or crimping them. They feed primarily on sebaceous secretions, usually causing little or no apparent harm even when mite numbers are high. Exceptions occur, however, in rodents and rabbits, especially under conditions of confinement or crowding. Heavy infestations in such cases can result in dermatitis, scratching, and hair loss.

Members of the family are cylindrical to laterally flattened and are distinguished by the anterior coxal fields that are expanded and flattened, with grooved surfaces that serve as attachment organs for grasping host hairs. Rabbits infested with *Leporacarus gibbus* may experience pruritus, hair loss, skin abrasion due to scratching, and occasionally damage to the fur as a result of nibbling. Wild populations of *Rattus* and *Mus* harbor species of *Afrolistrophorus* in tropical

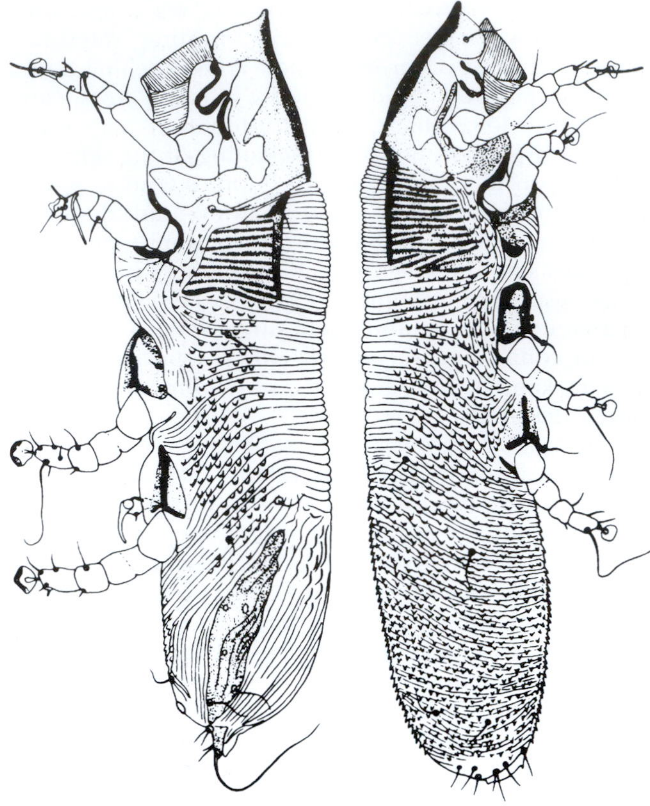

Figure 25.31 *Listrophorus synaptomys* (Listrophoridae), male (left), female (right), lateral views. (Whitaker, 1982; after Fain et al., 1974)

regions, but these have not been reported from laboratory populations. *Lynxacarus radovskyi*, which infests domestic cats, has been reported to cause patches of mange and a scurfy appearance of cats in Hawaii (Tenorio, 1974).

Atopomelidae Atopomelid fur mites are parasites of marsupials, rodents, insectivores, and primates, primarily in the Southern Hemisphere. They attach to hair shafts with their anterior two pairs of legs. They are distinguished from the Listrophoridae by the lack of a projecting tegmen above the gnathosoma, and the striated forecoxal fields do not project around the host's hair.

Guinea pigs are commonly infested with *Chirodiscoides caviae*, particularly in laboratory colonies. This mite usually attaches to hairs on the back but may occur on any part of the body. Attachment to hairs is facilitated by their striated sternal area and legs 1 and 2 that are flattened and curved as clasping structures. Although most infestations in guinea pigs go unnoticed, hair loss and severe pruritus can occur; however, these are more likely the result of *Trixacarus caviae* (Sarcoptidae).

Listrophoroides cucullatus is widely distributed on commensal rats (*Rattus rattus*, *R. norvegicus*), primarily in tropical and subtropical areas. This species rarely is observed in laboratory colonies and is not known to cause damage.

Myocoptidae This family is superficially similar to the Listrophoridae. All stages occur in the pelage of their rodent and marsupial hosts where they attach to hairs with their modified legs 3 and 4. In the female the tibia and tarsus of both pairs of legs fold against the striated inner surfaces of the genu and femur to provide efficient clasping organs. The most common member of this family is *Myocoptes musculinus* (Figure 25.32) that infests wild and captive house mice throughout the world. It attaches its eggs singly to the lower part of the hair shaft and requires about 14 days to complete its life cycle. These mites feed on superficial epidermal tissues, with infestations usually going unnoticed. In laboratory mice, however, conditions often contribute to a build-up of mite numbers that can cause **myocoptic mange**. This is characterized by pruritus, erythema, development of a dull coat, and thinning of the hair due to physical damage by attached mites and scratching. Signs usually are first noticed on the neck and from there spread to the shoulders, back, and other parts of the body. Myocoptic mange generally is more severe in older mice or those with lower resistance.

A few other myocoptic mites occasionally cause mild dermatitis in rodent hosts. These include *Trichoecius romboutsi*, which is known to infest laboratory mice; *T. tenax*, which infests voles (*Microtus* spp.); and *Sciurocoptes sciurinus*, which infests squirrels.

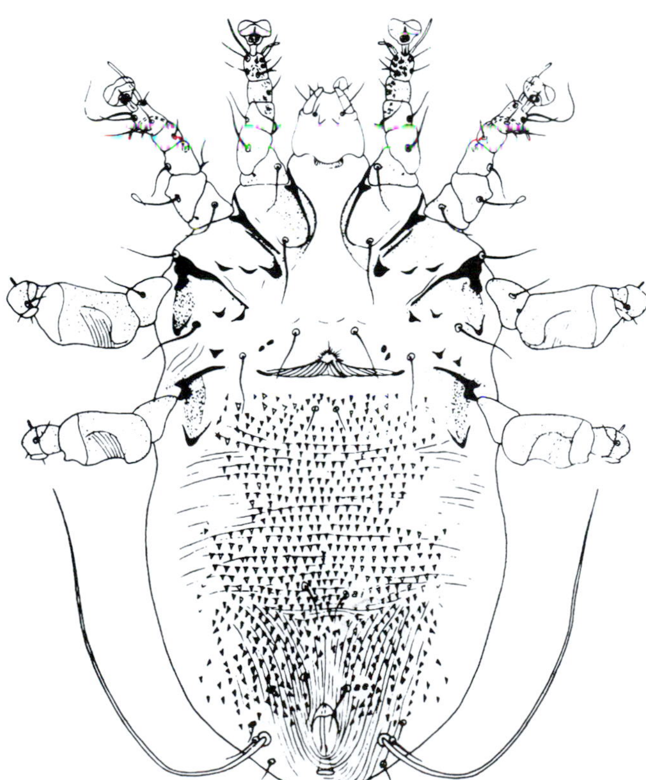

Figure 25.32 *Myocoptes musculinus* (Myocoptidae), female, ventral view. (Modified from Whitaker, 1982; after Fain et al., 1970)

Feather Mites

Mites representing 33 families and approximately 2,000 species live on or in the feathers of birds (Gaud and Atyeo, 1996). Their diversity as a group is reflected in their morphological adaptations for exploiting the many microhabitats that feathers provide. These include different types of feathers (e.g., primary and secondary flight feathers, wing coverts, contour feathers) and their location on the feathers. Mites that live on exposed feather surfaces tend to be more sclerotized, with a reduced setation and prominent terminal body setae. Those that live in more protected sites (e.g., contour feathers or calmus) are generally less sclerotized, often with distinctly modified body forms and pretarsal structures. The majority of feather mites live on the surface of the feathers where they feed primarily as saprophages on skin scales, feather debris, and oily secretions. Diatoms, fungal spores, and other organic materials also serve as food. Among the more common examples of such feather mites are members of the families Analgidae (Figure 25.33), Proctophyllodidae (Figure 25.34), and Pterolichidae (Figure 25.35). The analgid mites *Meginia cubitalis* and *M. ginglymura* and the pterolichid mite *Pterolichus obtusus* occasionally cause dermatitis and reduced thriftiness in chickens. Other mites called **quill mites** live inside the base of feathers (calmus) where they feed on feather tissues or by piercing the calmus wall to feed on host fluids. Two such families are the Syringobiidae (Figure 25.36) and Syringophilidae (Figure 25.37). Like most feather mites, quill mites rarely cause apparent harm to their hosts and generally are not considered to be economically important. Occasionally, however, heavy infestations of *Syringophilus bipectinatus* cause feather loss in chickens.

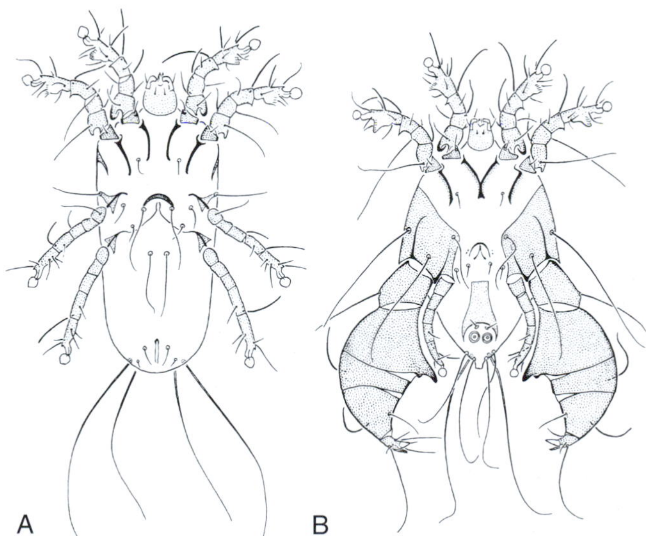

Figure 25.33 *Analges chelopus* (Analgidae), ventral views. (A) Female; (B) male. (Gaud and Atyeo, 1996)

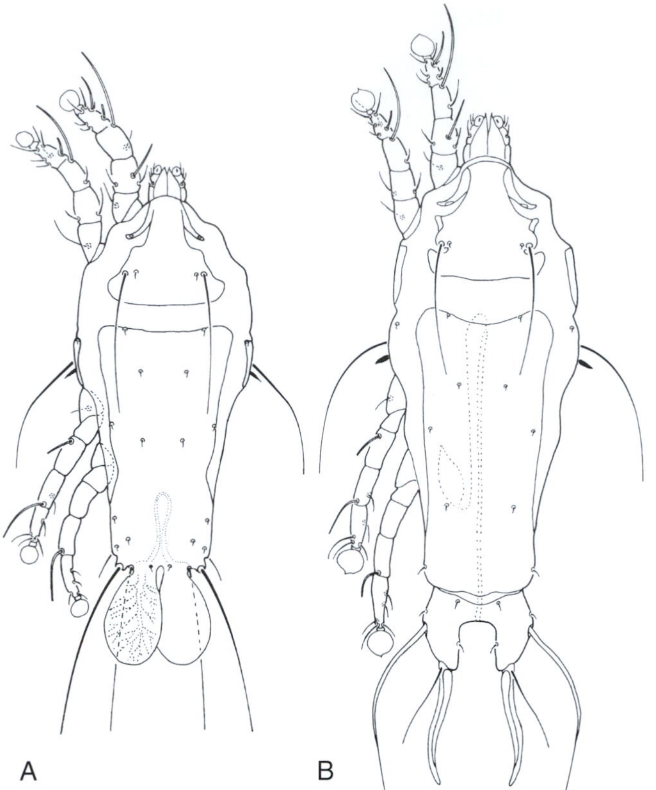

Figure 25.34 *Proctophyllodes glandarinus* (Proctophyllodidae), dorsal views. (A) Male; (B) female. (Gaud and Atyeo, 1996)

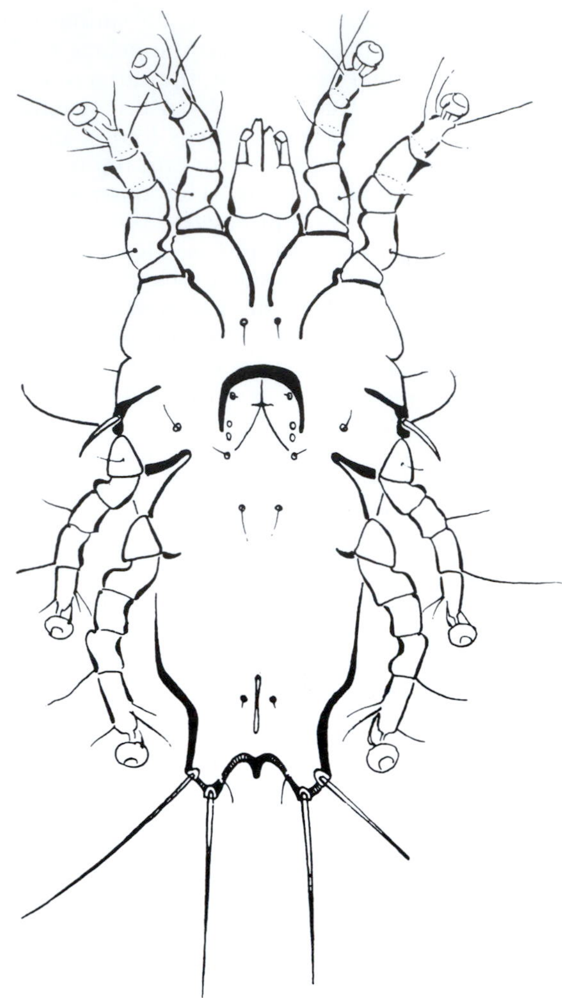

Figure 25.35 *Musophagobius cystodorus* (Pterolichidae), female, ventral view. (Gaud and Atyeo, 1996)

Mange Mites

The following families include species that cause mange in animals, including livestock, poultry, companion animals, and laboratory colonies: Epidermoptidae, Dermationidae, Knemidokoptidae, Laminosioptidae, Demodicidae, Psorergatidae, Sarcoptidae, Psoroptidae, Harpirhynchidae, and Hypoderatidae.

Epidermoptidae and Dermationidae Several species of the Epidermoptidae have been reported to cause discomfort and injury to infested birds. Although commonly referred to as feather mites, they are more appropriately called **avian skin mites**, as reflected in the family name. They generally live on the skin surface or in feather follicles where their feeding can lead to itching, pityriasis (scaly or scabby dermatitis), and various other types of superficial skin lesions. *Epidermoptes bilobatus* (Figure 25.38) commonly infests galliform birds and occasionally causes pityriasis in chickens, whereas *E. odontophori* has been reported to cause mange in African birds. *Microlichus avus* and *M. americanus* are known to produce crateriform skin lesions and severe mange in several avian species. *Myialges* species (e.g., *M. macdonaldi*) may invade the outermost skin layers to produce pityriasis and mange, sometimes severe enough to cause feather loss. Heavy infestations of *Rivoltasia bifurcata* (Dermationidae) on chickens can result in intense itching and pityriasis, especially involving the head. For further information on these and other epidermoptid genera of veterinary interest, see Fain (1965), Krantz (1978), and Krantz and Walter (2009).

Knemidokoptidae Knemidokoptid mites superficially resemble sarcoptids, from which they differ by having short legs without pretarsi or long setae and lacking dorsal triangular setae. They invade the feather follicles and skin of wild and domestic birds worldwide, causing **knemidokoptic mange** in some species. Their life cycle is similar to *Sarcoptes scabiei*. All stages of these mites occur on the host, and transmission is by direct contact with infested birds. There are several species of veterinary importance: *Knemidokoptes mutans* and *Neocnemidocoptes gallinae* infesting poultry, *K. pilae* infesting parakeets, and *K. jamaicensis* infesting passerine birds, including canaries. Based on phylogenetic analyses, some workers have suggested that the Knemidokoptidae be included within the Epidermoptidae.

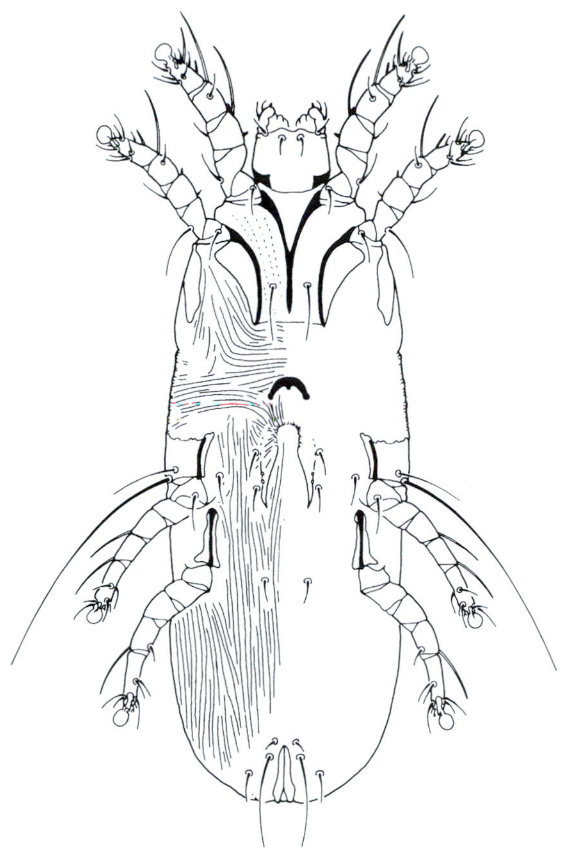

Figure 25.36 *Longipedia tricalcarata* (Syringobiidae), female, ventral view. (Gaud and Atyeo, 1996)

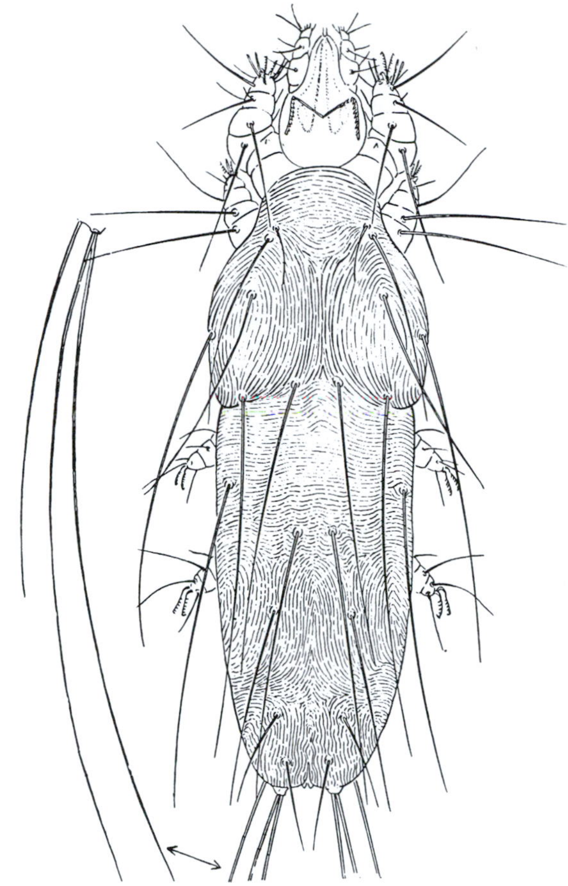

Figure 25.37 *Syringophilus bipectinalis* (Syringophilidae), female, dorsal view. (Baker et al., 1956)

Scaly-leg Mite (*Knemidokoptes mutans*) This mite (Figure 25.39) is a pest of poultry, especially chickens, in North America, Europe, and Africa, and probably occurs worldwide. It burrows beneath the epidermal scales of the legs and feet, causing irritation, inflammation, hyperkeratization, formation of vesicles, and encrustations (Figure 25.40). The crusts may cover entire limbs, hence the term **scaly-leg**, a condition most commonly seen in older birds. In chronic cases, infestations can lead to lameness, deformed legs and feet, and occasionally the loss of digits. The skin of the comb and wattle also may be involved. *Knemidokoptes jamaicensis* causes similar symptoms in wild and cage-reared passerine birds.

Scaly-face Mite (*Knemidokoptes pilae*) This cosmopolitan mite infests captive parakeets, causing crusty lesions primarily about the face, head, and legs. Lesions usually appear initially in the cere (at base of the upper beak) and at the corners of the beak where the mites invade the feather follicles and folds of skin. There they form pouch-like cavities, or pits, and a honeycomb pattern that is discernible on close examination. Early signs include whitish excrescences that may spread to the eyes, forehead, and other parts of the body. These form gray-white to yellow encrustations in chronically infested birds. Lesions of the legs and feet, especially in the early stages, can closely

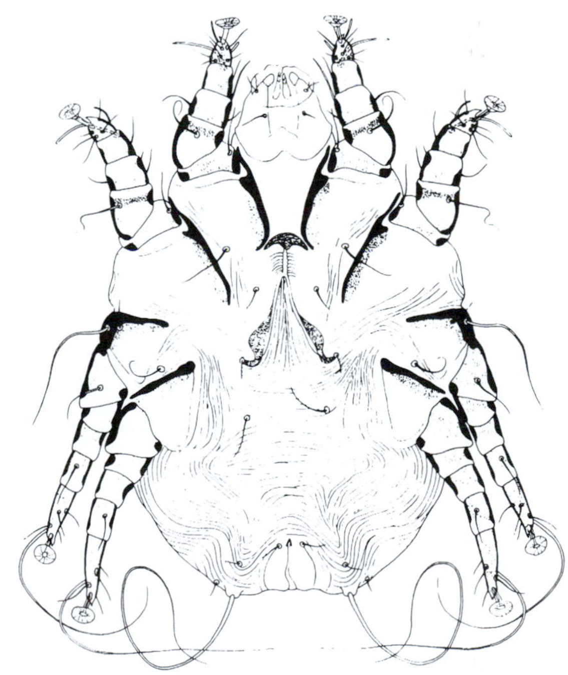

Figure 25.38 *Epidermoptes bilobatus* (Epidermoptidae), female, ventral view. (Gaud and Atyeo, 1996)

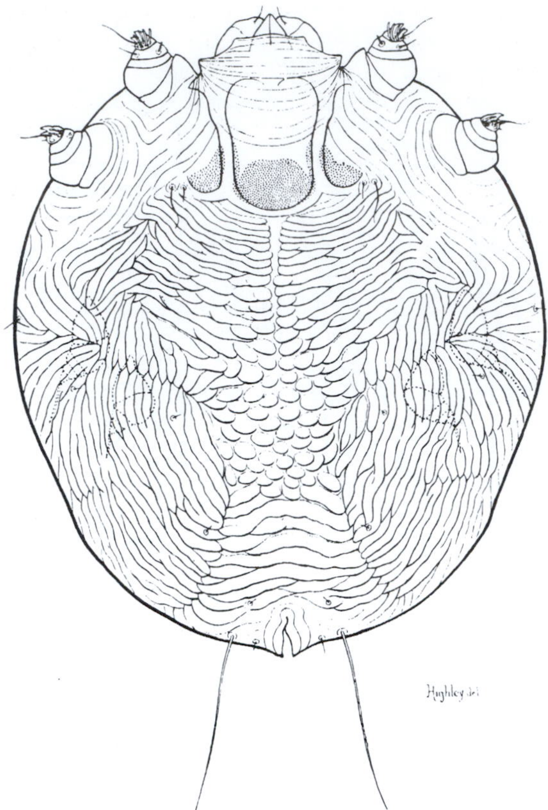

Figure 25.39 Scaly leg mite, *Knemidokoptes mutans* (Knemidokoptidae), female, dorsal view. (Hirst, 1922)

Figure 25.40 Chicken with crusty, raised scales on feet due to infestation with scaly leg mite (*Knemidokoptes mutans*). (Photo by Jerry F. Butler)

resemble those of *K. mutans*. Advanced cases may involve distortions of the beak as the keratinized tissues become overgrown and friable. Even in such cases, there appears to be little or no pruritus, nor rubbing or scratching of the beak.

Depluming Itch Mite (*Neocnemidocoptes gallinae*) Unlike the previous two mites, the depluming mite infests the feathered areas of chickens, burrowing into the epidermis at the base of feathers or into the feather shafts. The parts of the body most commonly affected are the head, neck, back, abdomen, and upper legs. The wings and tail are not usually involved. The mites tend to be confined to the stratum corneum, causing hyperkeratosis, thickening and wrinkling of the skin, and sloughing of keratinous layers. Feathers in the affected areas often break off or fall out, or may be plucked by the bird, accounting for the mite's common name. Severely infested birds may become emaciated and die. A similar species, *Picicnemidocoptes laevis*, may cause similar symptoms in pigeons and doves.

Laminosioptidae The only significant laminosioptid mite of veterinary importance is the fowl cyst mite.

Fowl Cyst Mite (*Laminosioptes cysticola*) This mite (Figure 25.41) occurs worldwide as a parasite of chickens, pheasants, turkeys, geese, pigeons, and other birds. It invades the skin of its avian hosts to form small, yellowish, subcutaneous nodules, or cysts, up to several millimeters in size. The nodules are formed by calcareous deposits produced around the mites after they have died. They are most easily seen in living birds by wetting and parting the breast feathers, and sliding the skin back and forth with the fingertips. Occasionally *L. cysticola* causes heavy infestations that can be fatal. A case was reported in West Virginia (USA) involving a wild turkey with severe neurologic disease (Smith et al., 1997). The affected turkey held its head bent over its

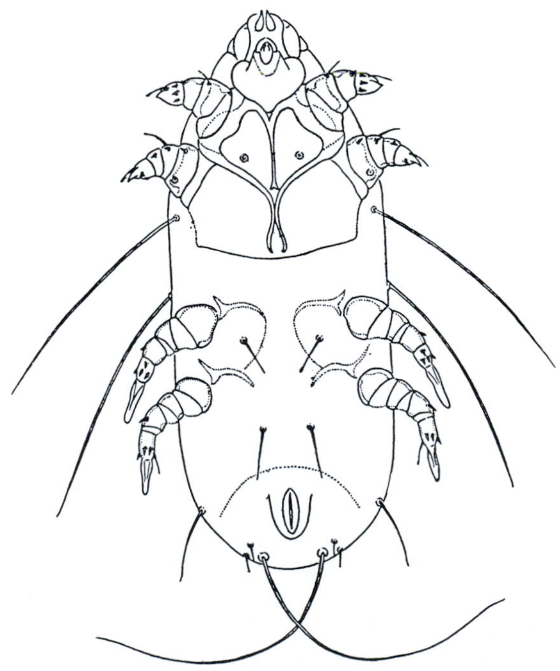

Figure 25.41 *Laminosioptes cysticola* (Laminosioptidae), female, ventral view. (Hirst, 1922)

back when at rest and exhibited circling behavior and falling to one side when it attempted to walk. Histopathologic examination revealed *L. cysticola* mites in enlargements of the wing nerves, inflammation of the brain, and numerous mites in the esophagus, small intestine, and other internal tissues. The fowl cyst mite also has been reported to cause granulomatous pneumonia in dogs (Shaddock and Pakes, 1978). Little is known about the life history or transmission of this mite.

Demodicidae Mites in this family are highly specialized skin parasites that live in the hair follicles and associated glands of domestic and wild mammals. An infestation of demodicid mites is called **demodicosis**, whereas cases with clinical signs are called **demodectic mange**. The mites are very host-specific and typically occur either in hair follicles or dermal glands. These sites include sebaceous glands, modified sebaceous glands (e.g., meibomian, caudal, preputial, and vulval glands), modified sweat glands (e.g., ceruminous glands and submaxillary skin papillae), and mixed sebaceous-sweat glands (e.g., perianal glands). *Opthalmodex* infests lachrymal ducts. A few species burrow into the skin to form epidermal pits, as in *Demodex criceti* of hamsters. Other demodicid species invade oral tissues of their host, infesting the oral epithelium, tongue, and esophagus of the grasshopper mouse (*Onychomys leucogaster*) in western North America and the oral cavities of bats and lemurs in Europe and Africa. In most cases, demodicid mites cause little or no apparent harm to their hosts. In other cases, infestations can lead to varying degrees of dermatitis and other skin problems.

Demodectic mange is common in dogs; livestock such as cattle, goats, sheep, and swine; wild animals such as foxes, other canids, and rabbits; and occasionally laboratory animals such as hamsters, gerbils, guinea pigs, rats, and mice. It is relatively uncommon in cats and horses.

Injury to the host occurs as the mites puncture with their stylet-like chelicerae the epithelial cells lining the hair follicles and glands to feed on the cell contents. In most cases the host response is only mild-to-moderate hypertrophy of the affected epithelia. In other cases, marked hypertrophy and cell destruction may occur. The openings of hair follicles or the ducts of glands may become blocked, leading to the formation of dermal papules and nodules. Damage to the follicles can lead to hair loss, whereas secondary bacterial infections may cause inflammation, pruritus, and the formation of pustules. Lesions often occur first on the face and head, spreading from there to other parts of the body.

Two major clinical forms of demodicosis are recognized, squamous and papulonodular. **Squamous demodicosis** is the more common form, characterized by a dry, scaly dermatitis with itching and loss of hair in the affected areas. Secondary infections often result in the rupture of follicular cells, severe inflammation, and purulent exudates. This can be either a localized or generalized skin condition and occurs in all host groups. **Papulonodular demodicosis** occurs when the hair follicles or gland ducts become obstructed and produce palpable, cyst-like or nodular swellings in the skin, trapping the mites within. The development of demodectic papules and nodules is most commonly seen in cattle, goats, and pigs. These lesions continue to enlarge as the mites multiply, sometimes reaching several thousand mites per lesion, along with accumulated cellular debris and glandular secretions. These nodules may rupture externally leading to secondary infections and abscesses. In other cases they may rupture within the skin, introducing the mites to the circulatory and lymphatic systems. There they can cause thromboses and internal infestations.

Dog Follicle Mite (*Demodex canis*) This mite (Figure 25.42) inhabits the hair follicles and sebaceous glands of dogs throughout the world. It completes its life cycle in three to four weeks, with the eggs and all developmental stages being found in the follicles or glands. Clinical signs are most common in dogs less than a year old, presumably reflecting an immunodeficient state in young animals. Canine demodectic mange (Figure 25.43) usually appears as mildly erythematous patches about the eyes and corners of the mouth, typically associated with hair loss. From there the infestation may spread to the forelegs and trunk as a typical squamous form of demodicosis. Most cases resolve themselves without treatment. In genetically predisposed or immunodepressed animals and cases of secondary infections, the condition can develop into chronically severe, moist, purulent dermatitis

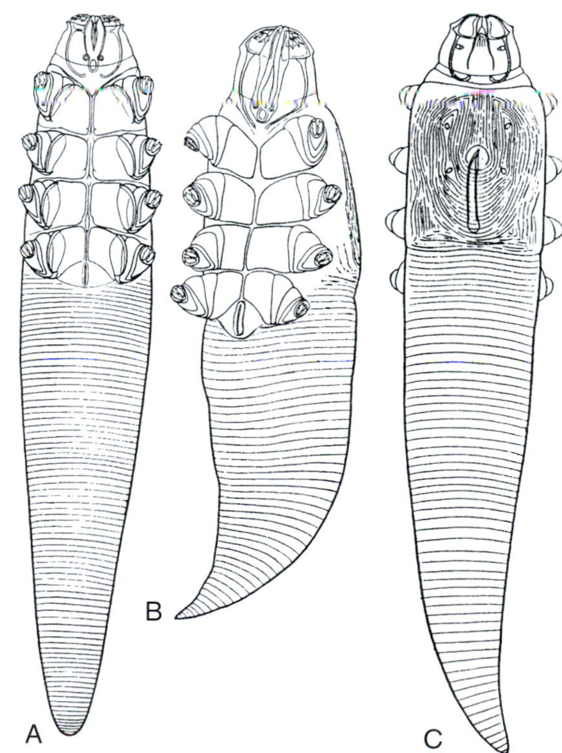

Figure 25.42 Follicle mites (Demodicidae) of domestic animals. (A) *Demodex canis;* (B) *D. phylloides*; (C) *D. cati.* (Hirst, 1922)

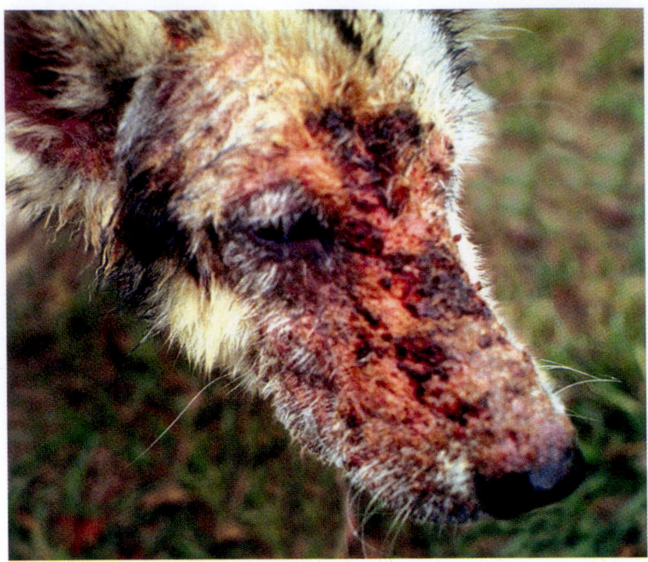

Figure 25.43 Demodectic mange in dog caused by *Demodex canis* (Demodicidae). (Photo by Jerry F. Butler)

known as **pustular demodicosis**. This often is accompanied by an unpleasant odor variously described as rancid or mousy. If this becomes generalized with intense redness and tenderness of the skin, and easily bleeds, it is called **red mange**. Severe cases of red mange also occur in foxes and other wild canids, sometimes leading to deaths in heavily infested animals. Diagnosis of *Demodex* infestations is confirmed by demonstrating the presence of mites in material expressed from hair follicles and in skin scrapings or biopsies of skin lesions. Although most dogs recover from localized infestations even without treatment, other cases may require extended acaricide applications; some are never resolved despite every treatment effort.

Most dogs apparently become infested with *D. canis* as newborn pups, either while nursing or in other intimate contact with their mother or another infested dog. There is little or no evidence to substantiate reports of prenatal transmission of the mite. Nor is the transfer of *D. canis* between mature animals likely, and apparently occurs only in unusual circumstances.

Cat Follicle Mite (*Demodex cati*) Occasionally cats develop lesions attributed to the *Demodex cati* and *D. gatoi*. This is usually the squamous form of demodicosis and occurs on the head or as a generalized condition with varying degrees of pruritus. Cases of feline demodicosis are believed to be associated with underlying immunosuppressive diseases (e.g., feline leukemia, diabetes mellitus).

Cattle Follicle Mite (*Demodex bovis*) Cattle infested with *Demodex bovis* commonly develop a papulonodular form of bovine demodicosis. Adult female mites deposit their eggs in hair follicles where their populations may build to hundreds or thousands of mites per follicle as the follicles become dilated to form dermal papules or cysts. As they enlarge, they can be felt beneath the skin even though they may be difficult to see. Some female mites exit the follicular cysts to invade other hair follicles, thereby spreading the infestation. It is presumably at this time that they also are transferred to other animals by intimate contact. It is postulated that transfer between cattle can occur during copulation.

The lesions tend to be concentrated on the anterior parts of cows, notably the neck, shoulders, and axillary region, but also may occur on the udder. Papular cysts enlarge to form granulomatous nodules when the follicular opening becomes blocked by mite bodies, keratin, and other debris. The occurrence of papulonodular demodicosis in cattle commonly is associated with cows that are stressed by pregnancy or lactation. Individual nodules typically form over the period of a month, then gradually disappear only to be replaced by other developing nodules. Both natural and acquired immunity play a role in reducing mite numbers and associated clinical signs in infested cattle.

The dilated follicles vary in size from that of a pinhead to a chicken egg. Not uncommonly, the larger nodules rupture to produce suppurative sores. The pus-like exudate containing large numbers of *D. bovis* has the consistency of toothpaste and can serve as a means of transfer of mites to other animals. Skin damage resulting from these ruptured nodules can cause defects in raw leather and significant economic losses to the tanning industry in the form of diminished quality of processed cow hides.

A second species of demodicid mite of bovids, *Demodex tauri*, has been reported from hair follicles and ducts of sebaceous glands in eyelids of cattle in Czechoslovakia (Bukva, 1986).

Goat Follicle Mite (*Demodex caprae*) Goats develop dermal papules and nodules similar to those in cattle, when infested with *D. caprae*. Cases of caprine demodicosis occur most commonly in young animals, pregnant does, and dairy goats, the latter presumably reflecting the stress of lactation. Papules usually appear on the face, neck, axillary region, or udder, with a few to several hundred lesions per animal. They are easily palpable in the skin, enlarging to form nodules up to 4 cm in diameter as the mites multiply within. Ruptured nodules tend to suppurate, contributing to transmission of the mites via exudates to other animals. As in dogs, goats have a high incidence of generalized demodicosis that can involve almost any part of the body. If the nodules rupture internally, granulomas develop while phagocytic giant cells of the goat host engulf and destroy the mites. As individual nodules disappear, others are formed. Transmission of *D. caprae* to newborn goats typically occurs within the first day following birth. Other possible means of transfer are parental licking and intimate contact of animals during copulation. Certain breeds of goats (e.g., Saanen) tend to be much more sensitive to demodicosis than are others.

Psorergatidae Psorergatid mites are obligate parasites of mammals that live in superficial layers of the skin of their hosts. Their life histories are similar to those of the Demodicidae. Whereas most species cause no apparent harm to their hosts, a few species cause psorergatid mange in sheep, rodents, and certain primates. Transmission between animals is by direct contact and transfer of adult females that enter hair follicles where they deposit their eggs. All developmental stages occur within the follicles where the mites feed by puncturing follicular cells. Mange-inducing *Psorergates* species cause inflammation and enlargement of the hair follicles to form dermal pouches, or pockets, beneath the stratum corneum. The lesions appear as small, white nodules (up to 2 mm in diameter) on the inner skin surface. The nodules contain all stages of the mite and necrotic debris from the destroyed follicles. In rodents, large lesions may form, with mites lining the inner surface of the lesions.

Sheep Itch Mite (*Psorobia ovis*) The psorergatid mite of most veterinary importance is the sheep itch mite, formerly named *Psorergates ovis.* It infests all breeds of domestic sheep, but especially Merinos. Clinical signs include dry scurfy skin, loss of hair, and sometimes erythema, accompanied by intense pruritus. Infested animals often are restless and bite or rub the affected areas, damaging the skin and wool. Lesions occur most frequently on the neck and shoulders, and gradually may spread to the face, flanks, thighs, and other parts of the body. The spread of *P. ovis* through a flock is slow and is most evident during the winter months.

Mouse Follicle Mite (*Psorergates simplex*) The mouse follicle mite (Figure 25.44) infests wild and laboratory mice in Europe and North America. Its development and life history are similar to those of *P. ovis*. The dermal pouches and resultant white nodules occur most frequently on the head of infested mice, but also may involve the neck, legs, and other parts of the body. Chronic infestations, especially in laboratory mice, can lead to crusty or ulcerous nodules, hypertrophy of skin cells, dermatitis, and hair loss.

Psorergates mites sometimes are associated with cases of murine ear mange, evidenced by pale yellow crusts on the inner and outer ear surfaces. Such cases in *Mus musculus* typically involve *P. muricola.* A third species, *P. hispanicus,* forms lesions on the legs of *M. musculus.*

Other *Psorergates* species that cause lesions in rodents include *P. apodemi, P. dissimilis,* and *P. microti.* Cattle are hosts for *Psorobia bos,* which seldom causes apparent lesions or harm to its bovine hosts. Occasionally, however, it is reported to cause psorergatid mange, as in an infested herd of Bonsmara bulls in South Africa (Oberem and Malan, 1984). *Psorobia cercopitheci* reportedly causes mild dermatitis in mangabey monkeys (Sheldon, 1966) and vervet monkeys (Seier, 1985), whereas unspecified psorergatid species are known to cause dermal cysts and crusty skin in patas monkeys (Raulston, 1972) and macaques (Lee et al., 1981)

Sarcoptidae The family Sarcoptidae includes three important genera that infest domestic and wild animals: *Sarcoptes, Notoedres,* and *Trixacarus.* The adult females burrow into the epidermis of their hosts causing varying degrees of dermatitis with accompanying erythema, pruritus, hair loss, scaling, and dermal encrustations characteristic of sarcoptic mange (Figure 25.45). Lesions may occur on any part of the body but often begin in typical locations depending on the mite species involved.

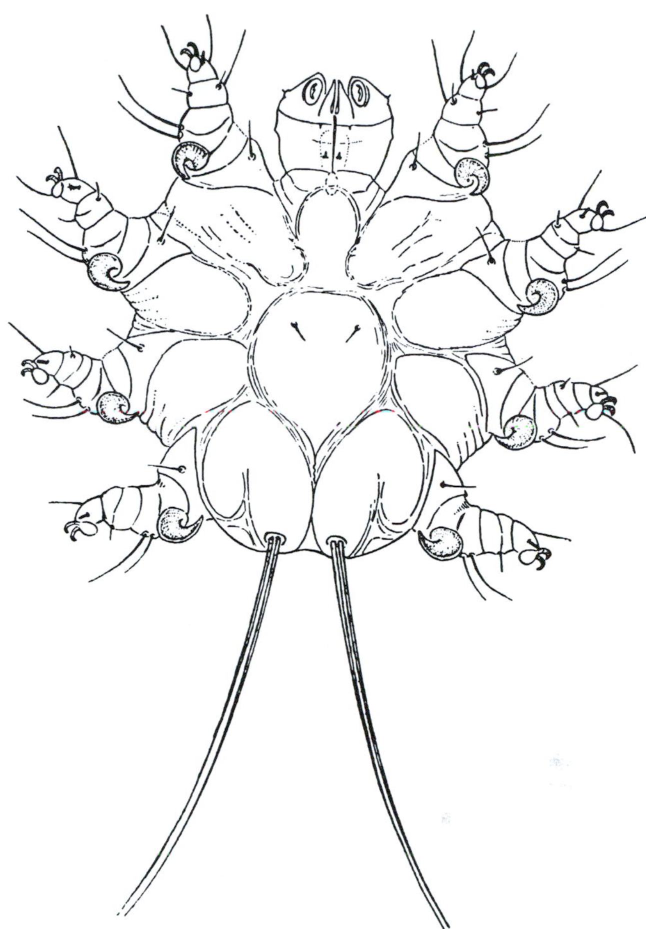

Figure 25.44 Mouse itch mite, *Psorergates simplex* (Psorergatidae), female, ventral view. (Adapted from Baker et al., 1956)

For additional information on sarcoptid taxonomy, phylogenetic relationships, and host associations, see Klompen (1992).

Scabies Mite (*Sarcoptes scabiei*) Most *Sarcoptes* mites that cause mange in animals are morphologically indistinguishable from the mite that causes human scabies, *S. scabiei.* Their life histories are very similar, with all developmental stages (larva, protonymph, tritonymph, adult) living in burrows formed by the adult females in the stratum corneum, stratum lucidum, and upper Malpighian layer of the skin. The female lays up to three eggs per day in the burrow over a period of two to three weeks, following which she dies. Development from egg to adult takes two to three weeks and occurs largely within the burrows. Male protonymphs leave the burrow to establish new epidermal tracts in which they molt to adults. The adult males then mate

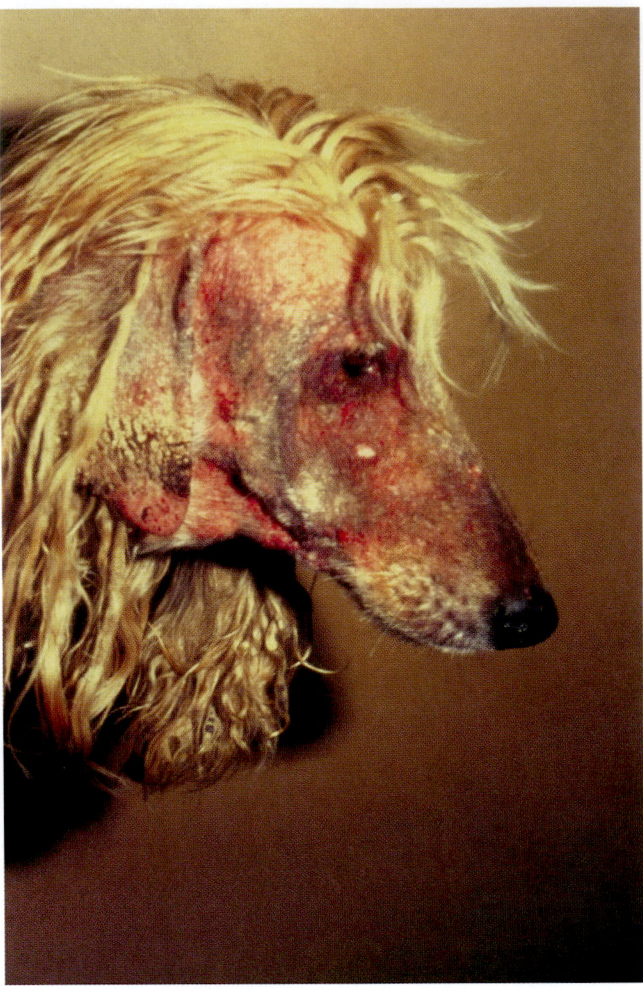

Figure 25.45 Severe case of sarcoptic mange in dog, caused by *Sarcoptes scabiei* var. *canis* (Sarcoptidae). (Courtesy of Department of Pathobiology, Auburn University College of Veterinary Medicine)

with females either on the skin surface or in shallow dermal pits.

Host reactions occur primarily in response to the mites and their fecal deposits in the burrows. This usually occurs three weeks or more after the initial infestation, with the reaction time becoming much shorter (e.g., a few days) after subsequent exposures. Initial lesions can occur anywhere on the body but usually are localized where the hair tends to be thin, most commonly on the head. From there the infestation can spread quickly to cause a more generalized mange. Infestations generally appear as papular eruptions with erythema, pruritus, and hair loss. As it progresses, skin in the affected areas often becomes thickened, crusted with exudates, and secondarily infected following excoriation of the skin due to scratching and rubbing by the host. Scaly areas around the periphery of infested patches often indicate the spread of mites. In extreme cases, severely sensitized animals may experience weight loss, difficulty eating, impaired hearing, blindness, exhaustion, and death.

Burrows seldom are detectable in nonhuman animals. This makes it difficult to recover mites to confirm their identification. Negative skin scrapings therefore are not conclusive. As a result, the diagnosis of scabies in animals is often presumptive, based on clinical signs and positive responses to acaricide treatments.

The various subspecies, or races, of *S. scabiei* tend to be relatively host specific, infesting a range of domestic and wild mammals. Transfer of mites occurs among conspecific hosts by direct contact. Transfer between different host species, when it does occur, often results in only temporary infestations, usually limited to a transient, mild dermatitis. Such a reaction in humans (e.g., from *S. scabiei* of dogs or goats) is called **animal scabies**.

Virtually all domestic animals, except cats and guinea pigs, are subject to infestations of *S. scabiei*. Dogs are the most commonly affected (Figure 25.45). Initial lesions tend to occur on thinly haired parts of the body such as the ear margins, belly, axillary and inguinal regions, elbows, and hocks. If untreated, the infestations generally spread to the head and other parts of the body. In particularly severe cases, dogs may develop thickened, pigmented skin with almost complete hair loss in affected areas of the neck, shoulders, back, trunk, and extremities. The situation often is complicated by secondary infections, self-inflicted trauma in efforts to relieve the itching, emaciation, and sometimes death. Conditions that may be mistaken for canine scabies include seborrhea, eczema, allergic dermatitis, ringworm dermatophytosis, and infestations of other mange mites (notably *Demodex*, *Notoedres*, and *Otodectes* spp.). Detection of mites requires deep skin scrapings, with best results being obtained from the tips of the ears even in the absence of ear lesions.

Initial lesions in farm animals usually are localized in specific body regions. In sheep, goats, and horses this is typically on the face, head, and neck. In cows lesions usually first appear on the underside of the neck, inner thigh, brisket, and tail head. In pigs the lesions tend to be more generalized (Figure 25.46), but are most commonly evident as encrustations and scabs in the ears of chronically infested sows. The highest incidences of *S. scabiei* infestations in farm animals usually occur in the winter months and are attributed to crowding of animals and loss of condition at that time of the year.

In addition to direct transfer between mature animals, *S. scabiei* also is transmitted from mother to offspring at birth. Pruritus has been reported in four-day-old piglets born to infested sows. Maternal antibodies to *S. scabiei* are detectable in neonatal pigs within six hours, reaching their maximum levels 24 to 48 hours after birth (Bornstein and Zakrisson, 1993).

Although cases of sarcoptic mange in goats often resolve themselves without developing severe signs, heavily infested goats may exhibit crusty lesions and extensive hair loss around the muzzle, eyes, and ears; lesions on the inner thighs extending to the hocks, brisket, ventral abdomen, and axillary region; dermal thickening and wrinkling on the scrotum and ears; and dry, scaly skin

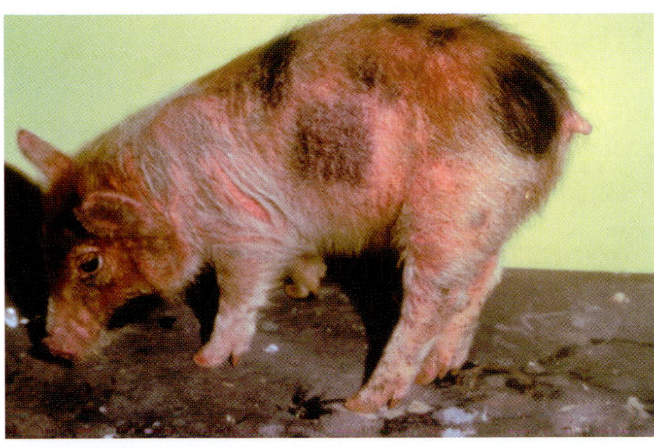

Figure 25.46 Sarcoptic mange in pig, caused by *Sarcoptes scabiei* var. *suis* (Sarcoptidae). Note inflammation of skin and hair loss. (Courtesy of Department of Pathobiology, Auburn University College of Veterinary Medicine)

on all parts of the body, especially in areas of hair loss (Kambarage, 1992). Sarcoptic mange also can cause significant problems in camels (Kumar et al., 1992).

Sarcoptes scabiei may also infest laboratory animals. Canine scabies is common in laboratory dogs obtained from commercial suppliers and animal shelters, with the highest incidence in young dogs and short-haired breeds. Laboratory rabbits usually develop initial lesions on the head, ears, and legs, followed by a more generalized dermatitis with associated erythema, pruritus, scaling of skin, and loss of fur. The situation can be complicated by scratching and self-inflicted injuries within the confines of a cage. Other laboratory animals are infrequently or rarely subject to sarcoptic mange. These include mice, hamsters, and guinea pigs.

In addition to human hosts, *Sarcoptes scabiei* has been reported infesting captive cynomolgus monkeys and simian primates (orangutan, gibbons, and chimpanzees). Lesions are characterized by thickening of the skin of the neck, shoulders, back, and sometimes the lower trunk and extremities. This may be associated with other signs such as skin scales, pruritus, hair loss, and emaciation. For further information on sarcoptic mange in laboratory animals, see Yunker (1973).

Several groups of wild mammals are susceptible to sarcoptic mange by *S. scabiei.* Members of the Canidae and Cervidae are the more commonly infested. Canid hosts are the most severely affected carnivores in North America, notably the red fox (*Vulpes fulva*), gray fox (*Urocyon cinereoargenteus*), wolf (*Canis lupus*), and coyote (*Canis latrans*). The most severe cases have been reported in red foxes during the winter months. In the northeastern United States, eastern Canada, and Russia, epizootic sarcoptic mange has had an economic impact on the pelt industry in some years. Early signs of mite infestations are dry, flaky skin, followed in succeeding weeks and months by crusty lesions, hair loss, eyelid scaling, emaciation, scratching, biting, and even death. Affected areas include the muzzle, neck, shoulders, back, and hind quarters. Successful transfer of *S. scabiei* from red foxes to domestic dogs, gray foxes, and feral dog-coyote hybrids have been demonstrated experimentally (Stone et al., 1972).

Cervid species that are known to develop sarcoptic mange include roe deer (*Capreolus capreolus*) and red deer (*Cervus elaphus*) in Europe, and wapiti (*C. elaphus*) in North America. Lesions in deer occur as encrustations of the outer ear, whereas in wapiti they may appear as moist scabs in the dorsal and lateral thoracic regions. Some cases involve head lesions, impaired vision, blindness, and general debilitation.

Other animals that are known to develop sarcoptic mange are the fisher (*Martes pennanti*), and ferrets (*Mustela* spp.) in North America, Europe, and Asia; llamas (*Lama* spp.) in Central and South America; Thompson's gazelle (*Gazella thompsonii*) and wildebeests (*Connochaetes taurinus*) in Africa; and chamois (*Rupicapra rupicapra*) in Europe and the Caucasus. There is evidence in the case of chamois that mineral imbalances may contribute to the severity of mange cases. Providing mineral blocks or salt licks has been shown to significantly reduce mange in infested animals (Onderscheka et al., 1968). Sarcoptic mange is also a serious problem in Australian wombats, with 35% of the population affected in one study (Hartley and English, 2005). Zoo animals also become infested with *S. scabiei* as reported in the capybara (*Hydrochoerus hydrochoeris*), tapir (*Tapirus* sp.), and camel (*Camelus* sp.) in Poland (Zuchowska, 1991).

Wild primates may develop sarcoptic mange, sometimes in epizootic form. Such epizootics among chimpanzees have resulted in noticeable hair loss. Severe cases in white-headed capuchins attributed to *S. scabiei* have been characterized by abscesses, localized hemorrhages, stratified crusts, weight loss, extreme debilitation, epileptic excitations, and death in some cases (Sweatman, 1971). It is probable, however, that another sarcoptid mite was involved.

Other Sarcoptid Genera

Sarcoptid mites in genera other than *Sarcoptes* also naturally infest primates. These species are very similar morphologically and in the damage they inflict. Typical signs are scabby, encrusted papular lesions, intense pruritus, and hair loss. *Prosarcoptes talapoini* and *P. pitheci* cause mange in guenons (*Cercopithecus* spp.) and baboons (*Papio* spp.), and *P. scanloni* causes mange in macaques (*Macaca* spp.). *Kutzerocoptes gruenbergi* causes a similar condition in capuchin monkeys (*Cebus capucinua*). For further information on sarcoptid mites infesting primates, see Fain (1968), Smiley and OConnor (1980), and Klompen (1992).

The following genera of sarcoptid mites are restricted to bats: *Chirophagoides* (New Zealand), *Chirnyssoides* (Neotropics), *Nycteridocoptes* (Europe, Asia, Africa), *Cynopterocoptes* (Asia), *Rousettocoptes* (Asia), *Tychosarcoptes* (Asia), *Teinocoptes* (Africa, Asia, Australia), and *Chirobia* (Africa, Asia). Most species of *Notoedres* also are restricted to bat hosts.

Whereas some bat hosts exhibit little adverse reaction, others develop pustules, scabby lesions, or small cornified pouches, nodules, or cysts containing the mites.

These lesions can occur on any part of the body including the ears and wing membranes. In some cases the mites burrow into the skin, particularly the anterior part of the body, or excavate shallow epidermal pits in the stratum corneum where the mites can be found. For further details on sarcoptid mites, see Sweatman (1971).

Notoedres Species

The genus *Notoedres* is similar to *Sarcoptes*. These mites differ from *Sarcoptes*, however, by having the anal opening located dorsally, and by lacking dorsal spines. They are skin parasites that typically burrow in the stratum corneum, causing inflammation, crusty lesions, and hair loss known as **notoedric mange**. Common hosts include cats, rats, squirrels, rabbits, and bats. *Notoedres* species have been reported infrequently parasitizing dogs and foxes, civets, lorises (primates), koalas and bandicoots (marsupials), hamsters, and hedgehogs. Most infestations begin on the head causing a condition called **head mange**, commonly observed in cats and rabbits. From there the mites spread to other parts of the body. Severe cases can lead to dehydration, emaciation, and death of the host. Transmission between animals is by direct contact, and only rarely are humans affected. Diagnoses are based on the clinical pattern of lesions and identification of the mites in skin scrapings. *Notoedres* mites are much more readily recovered in skin scrapings than are *Sarcoptes* mites.

Notoedric Cat Mite (*Notoedres cati*) This is the common *Notoedres* mite (Figure 25.47) of domestic cats in North America, Europe, and Africa, although it probably occurs worldwide. It also infests wild cats, laboratory rabbits, and rarely dogs, foxes, other canids, and civets. As the adult female burrows in the skin, she deposits eggs that hatch in three to four days. Development from egg to adult requires six to 10 days. Although *N. cati* typically burrows in the stratum corneum and stratum germativum, it occasionally invades hair follicles and sebaceous glands, causing hyperkeratosis and thickening of the epidermis. Lesions usually appear first on the ears (Figure 25.48), neck, face, and shoulders, but sometimes on the ventral abdomen, legs, and genital area, especially in younger animals. The feet and perineum may become involved due to the cat's sleeping position and grooming behavior. Typical signs are intense pruritus, erythema, skin scaling, grayish-yellow crusts, and loss of hair. As infestations progress, the affected skin becomes thickened, folded, and wrinkled. Scratching to alleviate the itching aggravates the condition by excoriating the skin and causing inflammation. Severe chronic cases can lead to systemic debilitation and death. To distinguish cases from other possible skin problems or mite species, the identity of *N. cati* should be confirmed by examining skin scrapings, usually best taken from the ears.

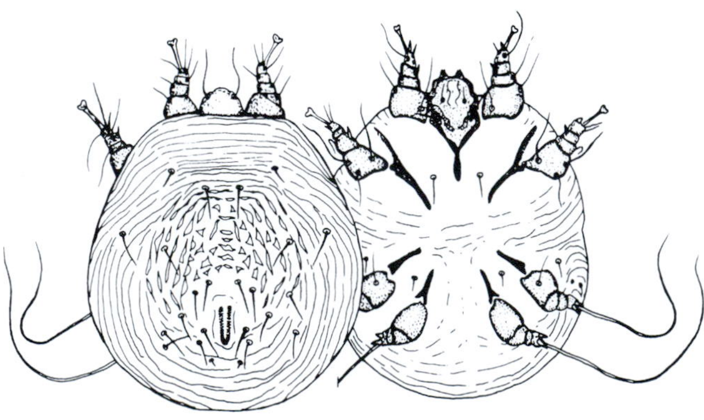

Figure 25.47 *Notoedres cati* (Sarcoptidae), female; dorsal (left) and ventral (right) views. (Nutting, 1984)

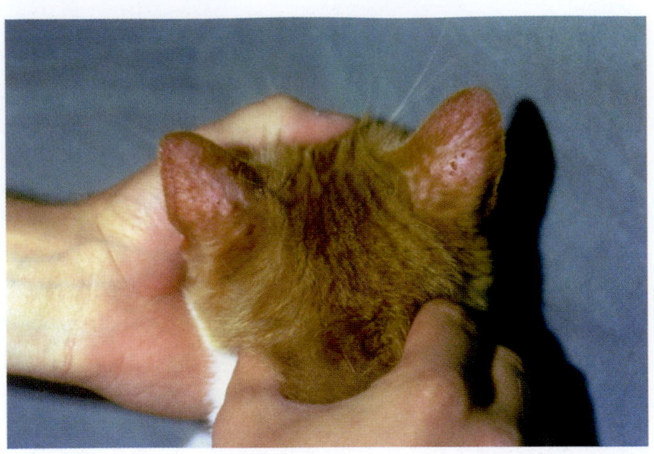

Figure 25.48 Notoedric mange in cat caused by *Notoedres cati* (Sarcoptidae), showing characteristic skin lesions on outer ears. (Courtesy of Department of Pathobiology, Auburn University College of Veterinary Medicine)

Wild cats known to become infested with *N. cati* include the Siberian tiger (*Felis tigris*) and North American bobcat (*Lynx rufus*). A fatal case of notoedric mange in an adult bobcat was reported in Texas. The cat was emaciated, extremely weak, with hair loss about the head, neck, and shoulders, with associated greatly thickened skin and gray encrustations. Bobcat kittens in the same area exhibited similar skin lesions confirmed as notoedric mange (Pence et al., 1982).

Notoedric Squirrel Mite (*Notoedres centrifera*) This mite, formerly known as *N. douglasi*, causes notoedric mange in North American squirrels (*Sciurus* spp.) and porcupines (*Erethizon dorsatum*). Until the first report of infestations in porcupines (Snyder et al., 1991), it was thought to infest only sciurids. Cases in the United States have been reported in the eastern gray squirrel (*S. carolinensis*) in Massachusetts, California gray squirrel (*S. griseus griseus*) in California, and fox squirrel (*S. niger*) in Indiana, Michigan, and West Virginia. Lesions are similar to those of other *Notoedres* species, with thickened and wrinkled skin, scurfy yellowish crusting, and hair loss usually about the head and neck. Other affected areas include the back, torso, limbs, and base of the tail. In some cases multifocal hyperpigmentation, pinpoint nodules, and microabscesses have been reported, in addition to extensive hair loss, dehydration, emaciation, and death. Significant mortality has occurred in epizootics of notoedric mange among California gray squirrels attributed in part to impaired vision and disruption of food-seeking ability of severely affected animals. For further information on *N. centrifera*, see Carlson et al. (1982) and Kazacos et al. (1983).

Notoedric Rat Mite (*Notoderes muris*) This mite infests rodents and is the cause of notoedric ear mange commonly seen in laboratory rats. It is known to parasitize the Norway rat (*Rattus norvegicus*), black rat (*Rattus rattus*), multimammate mouse (*Mastomys natalensis*), and certain wild rodents, marsupials, and hedgehogs. Its distribution is apparently cosmopolitan, albeit sporadic. The mite burrows into the stratum corneum where the eggs are deposited and hatch in four to five days. The entire life cycle is completed in about three weeks. Only occasionally does *N. muris* penetrate the deeper skin tissues. The female lives two to three weeks, laying up to three eggs per day. Although the larvae and nymphs may develop in the parent burrow, they also may move onto the skin surface to excavate pits in the stratum corneum, which in turn become the entrances to new burrows as they continue to develop (Sweatman, 1971). Mating takes place in the burrow, and transmission between hosts typically involves direct transfer of the active immature stages.

Lesions usually appear several weeks after the initial infestation, in the form of wart-like, horny excrescences and yellowish encrustations on the ears, nose, neck, tail, and sometimes the limbs and genitalia. The involvement of the ears and tail are often diagnostic. The skin becomes greatly thickened, and the hairs appear to become shortened and displaced due to proliferation and cornification of epidermal cells in the affected areas. Severe cases can develop erythema, vesicular or papular lesions, serous exudates, and other complications due to secondary bacterial infections.

Related species (*N. musculi, N. oudemansi, N. pseudomuris*) are known from *Rattus* and *Mus* species, as well as from wild rodents and insectivores. These and other *Notoedres* species of rodents show more restricted geographic ranges. The three mentioned are all reported to cause mange similar to that caused by *N. muris*.

***Trixacarus* Species** Two sarcoptid mites in the genus *Trixacarus* can cause severe skin problems called trixacaric mange in guinea pigs and laboratory rats. Like those of *Sarcoptes* and *Notoedres* species, *Trixacarus* females burrow in the upper layers of cornified epithelium where the eggs are deposited and the larvae and nymphs are found. The duration of each developmental stage and the time required to complete the life cycle is unknown. *Trixacarus* species are much smaller than *S. scabiei* (females 140–180 µm versus 400 µm in *S. scabiei*) and can be distinguished from adults of the latter by propodosomal shield reduced to a small circular plate; elongate, sclerotized striae on dorsal and ventral anterior idiosoma; large, weakly sclerotized denticles on dorsum; long, spine-like dorsal setae; and absence of an ambulacral sucker on leg 4 in the male.

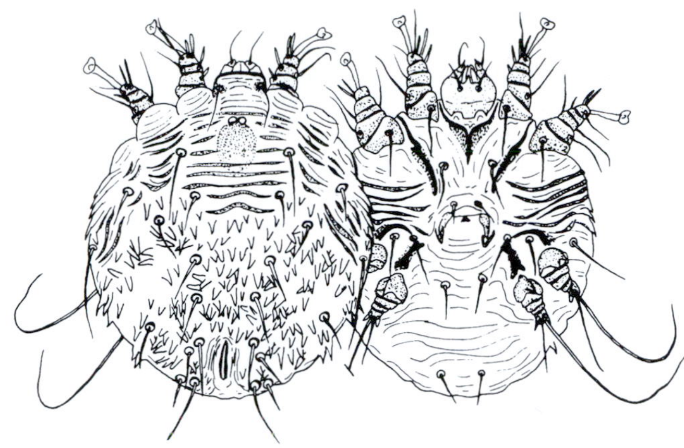

Figure 25.49 *Trixacarus caviae* (Sarcoptidae), female; dorsal (left) and ventral (right) views. (Nutting, 1984)

Trixacarus caviae This mite (Figure 25.49) was first described from a guinea pig colony in England in the early 1970s. Subsequently it has been reported causing mange in laboratory and pet guinea pigs in other parts of Europe and in North America. Lesions begin as an erythematous rash that progresses to a pruritic dermatitis, with thickening and wrinkling of the skin, and induced scratching. Lesions vary from dry and scaly to moist and crusty, with associated hair loss. Affected areas include the head and neck, shoulders, back, sides, lower abdomen, axillary region, and inner thighs. The intense pruritus can lead to self-inflicted trauma, such as frantic running about in cages and blindly striking objects, loss of condition, lethargy, and grand mal seizures. Death commonly occurs within a few weeks or months in heavily infested, untreated animals.

Trixacarus caviae is readily transferred on contact with other guinea pigs, including mothers to neonates, causing rapid spread through colonies. Owners of pet guinea pigs can develop papulovescicular lesions and pruritus by direct contact with infested animals held against the skin. This also is occasionally reported as a problem in people working in animal facilities. For further information on *T. caviae*, see Kummel et al. (1980) and Zenoble and Greve (1980).

Trixacarus diversus This mite is reported to cause severe infestations in the Norway rat, white mice, and hamsters in animal facilities in Europe and in wild field mice (*Calomys musculinus*) in Argentina (Klompen, 1992). Lesions are similar to those described for *T. caviae*, first appearing between the shoulders and from there spreading to the back and sides. Young or weakened animals may die two to three weeks following the appearance of skin problems, whereas untreated adult rats are more likely to die after five to six weeks (Sweatman, 1971).

Rhyncoptidae Rhyncoptid mites are similar to sarcoptid mites but are generally more elongate and are typically restricted to hair follicles. Species of *Audycoptes*, *Saimirioptes*, and most *Rhyncoptes* species occur in monkeys, with other *Rhyncoptes* species in African porcupines, *Caenolestocoptes* in South American marsupials, and *Ursicoptes* species in bears and raccoons. The primate parasites such as *Rhyncoptes grabberi* from rhesus macaques (Klompen, 1989) are generally innocuous, but mange conditions have been reported associated with *Ursicoptes americanus* in black bears in North America (Yunker et al., 1980) and with *U. procyonis* in North American raccoons.

Psoroptidae Mites in the family Psoroptidae are mammalian ectoparasites called **scab mites**. All developmental stages occur on the host. They do not burrow into the epidermis but instead live on the surface of the skin. Some pierce the skin with their chelicerae to feed on lymph, blood, and serous exudates. Others have chelicerae adapted for feeding on sloughed skin scales and other epidermal debris. Feeding injury commonly results in inflammation, pruritus, hair loss, crusting, and scab formation. The host hair often becomes matted and, together with the skin, may be severely damaged due to biting and rubbing by the host against fence posts and other objects. This can result in extensive loss of hair and generalized debilitation, with death occurring in some cases.

Psoroptid mites of veterinary interest parasitize primarily the Artiodactyla, Perissodactyla, and Carnivora. They also are found on certain edentates, carnivores, marsupials, insectivores, and primates. Members of four genera cause problems for domestic and laboratory animals. *Psoroptes* mites infest cattle, sheep, goats, horses, and rabbits; *Chorioptes* mites are primarily a problem on cattle; *Otodectes* mites cause problems for cats and dogs; whereas *Caparinia* mites infest wild and captive hedgehogs.

Psoroptic Scab Mites For many years it generally has been accepted that there are five *Psoroptes* species (*P. cervinus*, *P. cuniculi*, *P. equi*, *P. ovis*, and *P. natalensis*) based on host associations, location on the host, and the length of the outer opisthosomal setae (L_4) of the adult males (Sweatman, 1958b). There is now compelling evidence to indicate that *P. cuniculi* and *P. ovis* are strains, or ecophenotypic variants, of the same species. Evidence to support this view includes: (1) production of viable offspring by reciprocal genetic crosses between *P. cuniculi* and *P. ovis*, (2) presence of both typical and intermediate forms of these two mites associated with body lesions, and (3) the unreliable nature of the L_4 opisthosomal setae as a species-specific character, due to overlapping variations (Wright, et al., 1983, 1984; Bates, 1999). The *cuniculi* form typically occurs in the ears, whereas the *ovis* form is characteristically found on other parts of the body, notably the back and flanks. The *ovis* form causes scabs and other epidermal body lesions, whereas the *cuniculi* form does not cause body lesions but instead tends to move to the ears where it induces aural lesions. For convenience of the following discussion, *P. cuniculi* and *P. ovis* are treated as separate species, while a change in their taxonomic status awaits formal recognition.

Mites in the genus *Psoroptes* (Figure 25.50) cause **psoroptic mange**, a highly contagious form of mange that can spread rapidly by direct transfer of mites

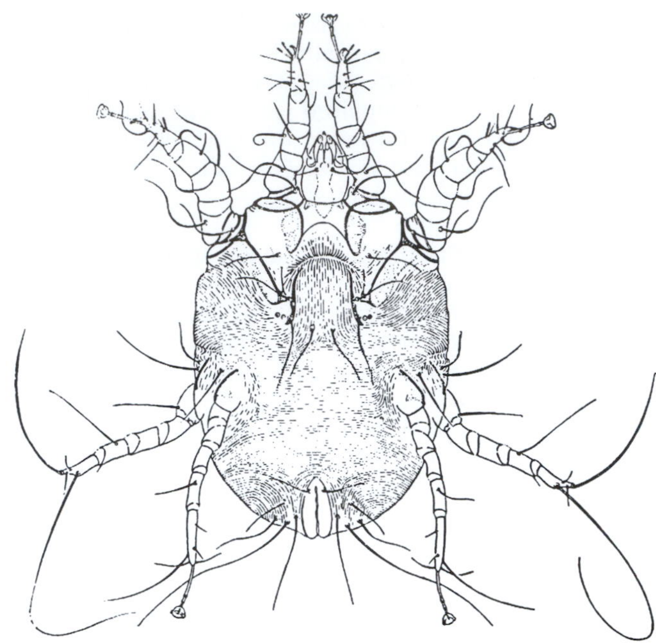

Figure 25.50 *Psoroptes equi* (Psoroptidae), female, ventral view. (Baker et al., 1956)

between animals or indirectly by rubbing against fence posts and other objects. Infestations tend to be more prevalent during the winter months, usually subsiding or disappearing during the warmer seasons. *Psoroptes* mites cause more severe problems than any of the other psoroptid genera, commonly resulting in economic losses in cattle, sheep, and goat operations.

The mouthparts of *Psoroptes* species are adapted for feeding on the surface of the skin rather than piercing the epidermis. The mites abrade the stratum corneum with their chelicerae, ingesting lipids and other dermal substances. The host responds antigenically to the mites by developing localized perivascular dermatitis and edema. Hemoglobin and other blood components are ingested by the mites only as a result of small hemorrhages at the skin surface of abraded sites. Their food primarily consists of digested cells of the stratum corneum and skin exudates.

Species of particular veterinary importance are *P. ovis* of sheep and cattle, *P cervinus* of deer and bighorn sheep, and *P. cuniculi* of rabbits, sheep, and goats. *Psoroptes equi* (Figure 25.50) is a minor pest of horses in England. *Psoroptes natalensis* causes mange in cattle, zebu, Indian water buffalo, and horses. It occurs in South Africa, South America (Brazil, Uruguay), New Zealand, and possibly France (Sweatman 1958b). Other psoroptid mites infest wild animals but seldom cause apparent problems. Exceptions are cases of mange in the African buffalo and in South American primates such as monkeys, marmosets, and tamarins, especially when held in captivity (Sweatman 1971).

Sheep Scab Mite (*Psoroptes ovis*) This mite causes psoroptic mange in sheep and cattle known as **sheep scab** and **cattle scab**. It also is called the psoroptic body mite. Clinical cases of *P. ovis* are misleadingly called *scabies*, a term that is best reserved for mange caused by *Sarcoptes scabiei*. Although more common in domestic stock, *P. ovis* can be transferred from domestic sheep to wild sheep, causing severe infestations and die-offs in bighorn sheep (*Ovis canadensis*) in the western United States.

Psoroptes ovis tends to occur in the more densely haired or wooly parts of sheep, with initial lesions usually appearing on the back and sides. Heavy crusting and scab formations with associated inflammation, hair damage, and depilation are typical in animals that become antigenically sensitized to this mite (Figure 25.51). Heavily infested lambs have been found to have thicker than normal pelts, with matted wool and enlarged lymph glands attributed to *P. ovis* (Cochrane, 1994). Pruritus often induces self-inflicted trauma, with extensive wool loss and skin injury due to licking, biting, and rubbing. Mites move to the periphery of the affected areas, thereby spreading over the body surface. In untreated animals, lesions may develop over major portions of the body, causing extensive loss of wool and reduced weight gains as high as 30% (Kirkwood, 1980). In extremely severe cases, sheep may die. In other cases, sheep heavily infested with *P. ovis* may show no gross evidence of skin damage, yet exhibit adverse effects on wool quality and general body condition.

Transfer of *P. ovis* among sheep primarily involves mites less than two weeks old. Older mites apparently fail to establish infestations even upon successful transfer to a new host. When *P. ovis* is transferred from infested sheep to calves and goats, the mites survive only about a week on the recipient host and do not induce clinical signs, at least not in naive (i.e., previously unexposed) animals (O'Brien et al., 1994).

Figure 25.51 Psoroptic mange, or sheep scab, in sheep caused by *Psoroptes ovis* (Psoroptidae). Note extensive hair loss. (Courtesy of the US Department of Agriculture/Animal Research Service, Kerrville, TX, M&M 8165)

Figure 25.52 Psoroptic mange in calves, caused by *Psoroptes bovis* (Psoroptidae). (Courtesy of the US Department of Agriculture/Animal Research Service, K7268-14)

Infestations of *P. ovis* in cattle cause exudative dermatitis and hair loss similar to that in sheep (Figure 25.52). The severity varies from mild cases to those that can involve virtually the entire body surface. Systemic effects can include mild anemia with hematologic changes such as marked reductions in lymphocytes, neutrophils, and total white blood cells; increases in plasma proteins and fibrinogen; and bone-marrow effects (Stromberg et al., 1986; Stromberg and Guillot, 1987a, 1987b). The degree of these responses is correlated directly with the severity of the associated dermatitis. Infested animals also may experience reduced weight gain, reduced energy conversion rates, and higher maintenance energy requirements (Cole and Guillot, 1987).

Significant differences occur between cattle that have not previously been exposed to *P. ovis* and those that have. In naive animals, skin lesions are slower to appear but progress rapidly. The associated mite populations grow much more quickly, reaching densities 100 to 1,000 times that in previously infested animals. The lower growth rates and lower fecundity in previously exposed cattle is due to both cellular and humoral immune responses involving the development of antibody activity to live *P. ovis* mites infesting the skin. Mites that infest animals with this acquired immunity have much lower ovipositional rates, reflecting decreases in the number of ovigerous females, rather than detrimental effects on egg development (Guillot and Stromberg, 1987). Cattle can become hypersensitive to *P. ovis* antigens, thereby developing severe clinical disease even when infested by relatively modest numbers of mites. This helps to explain why cattle in areas endemic for *P. ovis* generally experience more severe lesions than cattle in nonendemic areas. The severity also is exacerbated by stress caused by stanchioning animals and by extremely cold weather that can contribute to hypothermia and death of infested cattle.

Psoroptes cervinus Psoroptic mange occurs in bighorn sheep, mule deer, elk, and wapiti in the western United States. The identity of the *Psoroptes* species involved, however, remains unclear. Whereas some reports have called them *P. ovis*, others have recognized them as *P. cervinus*. Often these mites are simply referred to as *Psoroptes* sp. because of this taxonomic uncertainty. *Psoroptes* mites collected from bighorn sheep apparently do not establish lasting infestations when transferred to domestic sheep and can be established only with difficulty on cattle (Wright et al., 1981). Comparative studies based on antigenic characterization of *Psoroptes* mites from various hosts further suggests that the mite that infests bighorn sheep and mule deer is different from *P. ovis* on cattle (Boyce and Brown, 1991). For the purpose of the discussion here, the mites causing psoroptic mange in bighorn sheep and cervids are called *P. cervinus*, to distinguish them from their closely related, and possibly conspecific, counterparts on domestic sheep and cattle.

Lesions of psoroptic mange in bighorn sheep occur primarily in the ears or on the face and other parts of the head. The affected areas are characterized by yellowish-white scabs of dried serous exudates and crusty, exfoliated epidermal tissue overlying a reddened and raw epidermis. Other clinical signs are hair loss on the head, neck, and back; droopy ears; and blockage of the outer ear canal with cerumen and exudates. Lesions on other parts of the body generally are less extensive or severe, with mites being recovered primarily from the head and ears.

Psoroptes infestations can be especially severe in desert bighorn sheep (*Ovis canadensis mexicanus*) in the San Andres Mountains of New Mexico (USA). High mortality attributed to psoroptic mange reduced the population of desert bighorn sheep in the San Andres National Wildlife Refuge from more than 200 to about 25 individuals during the 12-year period from 1978 to 1989 (Hoban, 1990). Serologic surveys using immunologic tests for detection of antibodies to *Psoroptes* mites have shown widespread prevalence of these mites in desert bighorn sheep populations in California, where lesions tend to be mild and confined to the ears (Mazet et al., 1992).

Psoroptes cervinus also parasitizes elk (*Cervus elaphus*) and wapiti (*C. canadensis*) in the western United States, where it is called the elk scab mite. Infested elk may develop moist, thick scabs with associated dermatitis and hair loss, especially at the base of the neck and on the dorsal and lateral thorax. Particularly affected are the cows, young males, and calves (Samuel et al., 1991). In wapiti, body lesions occur primarily in the winter months and may involve large areas of the neck, trunk, and upper legs. A wet eczema with overlying scabs and extensive hair loss is similar to that observed in desert bighorn sheep and can be fatal.

Chorioptic Scab Mites Psoroptic mites in the genus *Chorioptes* cause **chorioptic mange** in domestic ungulates, notably cattle, sheep, goats, and horses. The species of greatest importance is *C. bovis*, which infests each of these hosts. Another species, *C. texanus*, infests the ear canals of reindeer in Canada. As a group, chorioptic mites are primarily parasites of herbivores (Artiodactyla, Perissodactyla, and Lagomorpha), including llama, guanaco, alpaca, and rabbits. They feed on sloughed epidermal tissues, sometimes causing irritation and crusty, pruritic lesions that warrant treatment. For further information on *Chorioptes* species and their host associations, see Sweatman (1957, 1958c).

Chorioptes bovis This mite (Figure 25.53) occurs primarily on the legs and feet of its hosts, where all of the developmental stages are likely to be found. Eggs are deposited singly at the rate of one egg per day, and are attached with a sticky substance to the host skin. Adult females usually live for two weeks or more, producing about 14 to 20 eggs during this time. The eggs often are clustered as multiple females oviposit in common sites, or females return on successive days to deposit their eggs. The eggs hatch in four days. The larval and protonymphal stages last three to five days each, whereas the tritonymphal stage takes seven to eight days, with one day of quiescence between each developmental stage. The cycle is completed in about three weeks. Optimum conditions for development are about 35°C and a relative humidity of 80%.

Most animals infested with *C. bovis* do not exhibit noticeable lesions or unusual discomfort due to this mite, even at relatively high mite densities. As a result, infested sheep and cattle often remain asymptomatic, serving as silent carriers and a source of infestation for other animals. Host reactions are induced only

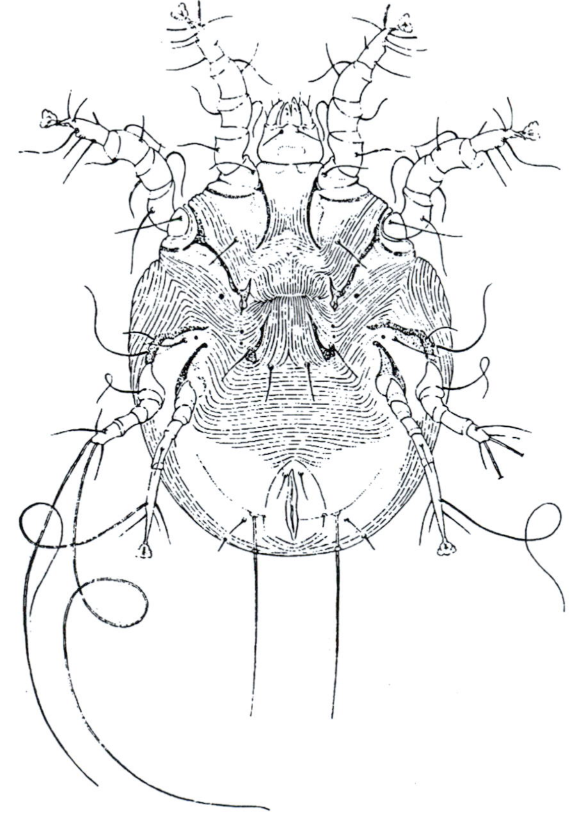

Figure 25.53 *Chorioptes bovis* (Psoroptidae), female, ventral view. (Baker et al., 1956)

when numbers increase to thousands of mites per host, occasionally causing extensive mange and pruritus. Most of the mites are found on the feet, notably the pasterns, regardless of where lesions appear elsewhere on the body. The irritation in sensitized animals can lead to stamping of feet, rubbing and chewing of legs, and other self-inflicted injury. Body lesions in severe cases are characterized by dermal crusting, erythema, and hair loss.

In cattle, *C. bovis* is found more commonly on the hind feet than on the forefeet, and particularly on the pasterns (between fetlock and hoof). The mites move from the feet to other parts of the body to cause mange of the escutcheon, base of the tail, buttocks, and perineum. This has given rise to names such as **foot mange, leg mange,** and **itchy heel** in referring to chorioptic mange of cattle and sheep. Mite populations are highest during the winter and are especially a problem in housed animals. In the spring, their numbers drop sharply and lesions generally disappear when cattle are turned out to pastures.

Although sheep commonly are parasitized by *C. bovis*, the small crusty lesions are hidden beneath the coat and usually go unnoticed. When clinical cases do occur, they are typically in the form of foot mange, affecting the forefeet. The mites occur about the accessory digits and along the coronary border of the outer claws, often in clusters, causing crusting primarily below the accessory digits and in the interdigital spaces. Infestation rates of 30 to 60% in sheep have been reported in the United States, Europe, Australia, and New Zealand. Prevalence of chorioptic mange tends to be highest in rams and generally low in ewes and lambs.

Chorioptes bovis may spread by direct contact of the feet with other parts of the body, notably the upper parts of the hind legs and scrotal area of rams, causing an exudative dermatitis called **scrotal mange**. The scrotal skin develops thick, yellowish, crusty layers as much as 4 cm deep. In severe cases, elevated scrotal temperatures attributed to allergic responses can cause degeneration of the seminiferous tubules, reduced sperm quality, and complete spermatogenic arrest. Testicular weights may become significantly reduced. The effects are reversible, with seminal regeneration and restored sperm production occurring following treatment for mites or spontaneous recovery of infested rams. Prevalence of leg and scrotal mange is usually highest in the fall and winter months, and declines in the spring.

Infestation rates of *C. bovis* tends to be higher in goats than in sheep, with up to 80 to 90% of goats in individual herds being parasitized (Cremers, 1985). As in sheep, the mites occur most commonly on the forefeet of goats, where the largest numbers of mites and lesions usually are associated with the accessory digits and claws. However, they also may occur on the pastern or higher on the foot. Lesions are generally mild and seldom draw attention.

Chorioptic mange due to *C. bovis* occasionally is observed in horses, with Belgian and Frisian breeds being among the more commonly infested. The mites are largely restricted to the pasterns and are most likely to cause foot mange in the horse breeds just mentioned, with long-haired feet (Cremers, 1985). Signs of *C. bovis* infestations in horses include stamping of feet and rubbing one foot against the opposite leg or against some object.

Caparinic Scab Mites Psoroptid mites of the genus *Caparinia* infest hedgehogs and a few other Old World mammals, causing **caparinic mange**. All active stages of these mites feed on sloughed skin cells and epidermal debris, similar to *Chorioptes* species. Although most infestations tend to go unnoticed, high mite populations can cause severely debilitating conditions and even death of the host. Two species that have drawn particular attention in recent years are *Caparinia tripilis* and *C. erinacei*, both of which parasitize wild and captive hedgehogs.

Caparinia tripilis, the Eurasian hedgehog mange mite, infests the Eurasian hedgehog (*Erinaceus europaeus*). It was first recognized in Great Britain in the late 1880s where it is still occasionally reported and has caused the death of at least one captive hedgehog. This mite was introduced to New Zealand on hedgehogs in the nineteenth century, but did not attract attention there until 1955 after hedgehog populations increased dramatically (Brockie, 1974). More recently, *C. tripilis* has been introduced to the United States via breeding colonies of African hedgehogs for sale as pets. An infested colony has been reported in New Mexico (Staley et al., 1994), and the death of a pet hedgehog attributed to *C. tripilis* has been documented in Alabama (Mullen, unpublished). The mites tend to gather in clusters on their host and invade the skin of the head and ears, flanks, and inner surfaces of the legs. The affected skin becomes dry and scaly, may become thickened and folded, and may crack and bleed, leading to secondary infections.

The common association of hedgehog ringworm (*Trichophyton erinacei*) with heavy *C. tripilis* infestations suggests that invasion of the skin by this fungus may contribute to severity of the resultant mange. In severe cases, body spines and hairs may fall out and lesions about the eyes may cause blindness (Brockie, 1974). Heavily infested animals become listless, lose weight, scratch the affected skin, and may abandon their normal nocturnal behavior to become active in the daytime. Male hedgehogs are usually more severely affected than females, with higher mortality occurring in captive than in wild hedgehogs.

Caparinia erinacei, the African hedgehog mange mite, infests the African hedgehog (*Atelerix albiventris*). Unlike *C. tripilis*, this mite does not form clusters on its hosts and exhibits low pathogenicity, occurring most abundantly on the dorsal parts of the body and rarely on the face. It has been reported parasitizing more than 70% of wild hedgehogs in Kenya (Gregory, 1981).

Harpirhynchidae Harpirhynchid mites are typically parasites of birds, in which they invade feather follicles and the skin. Evidence of infestations range from small white cysts or lumps associated with individual feathers to large, irregularly lobed, papilloma-like cysts that can occur on any part of the body. The lesions are typically pale yellow with a dry, granular appearance. Histological preparations of the lesions reveal multiple spaces lined by epidermal cells and packed with large numbers, sometimes thousands, of mites. The most common genus involved is *Harpirhynchus*, which has been reported to cause disfiguring ruffling of feathers and feather loss in a variety of avian hosts (e.g., lorikeets, warblers, eagles) in Australia, the southwestern Pacific region, and North America. Females of the genus *Ophioptes* excavate small, crater-like pits in the body scales of snakes, where they deposit their eggs. The larvae and nymphs lack legs, feed on host tissues, and develop to adults within the pits. Most reported cases have occurred in colubrid snakes in South America and Australia. Except for localized damage to individual scales, these mites do not cause significant harm to their hosts.

Hypoderatidae Hypoderatid mites (previously included in the glycyphagid subfamily Hypodectinae) are parasites of birds, and rarely mammals, in which they develop in subcutaneous fat tissue. These are unusual mites in that they are parasitic as deutonymphs, invading host tissues where they undergo growth and enlargement. The other life stages are nest-inhabiting detritivores, or are rarely nonfeeding. The most common hypoderatid species is *Hypodectes propus*, which parasitizes pigeons and doves in North America, Europe, and Africa. The mites are visible at necropsy as tiny white nodules embedded in fat tissue just beneath the skin. For further information on hypoderatid mites and their host associations, see Pence et al. (1997).

Mite-Induced Allergies

In addition to sensitization of animals to mites that invade the skin (e.g., demodicid and sarcoptid mites), pets and other domestic animals occasionally develop allergic reactions to nonparasitic mites. Reported cases, for example, have involved mites in the family Acaridae, which infest dry pet foods and livestock or poultry feed. Although the ingestion of even heavily mite-infested feed usually causes no apparent harm to animals, some individuals may become sensitized.

Internal Acariasis Approximately 15 families of mites include species that cause internal acariasis in animals of veterinary interest. The most common type is respiratory acariasis, which may involve the nasal passages, nasal sinuses, trachea, bronchi, or pulmonary tissues. Less commonly, mites cause oral, esophageal, gastric, enteric acariasis, and occasionally they invade other internals organs, the body cavity, lymph, and blood.

Species in five families may be found in oral tissues or the alimentary tract of their hosts. Certain *Radfordiella* species (Macronyssidae) invade the oral mucosa of the gums and hard palate of bats, causing erosion of soft tissues and bone. Heavy infestations can result in significant oral lesions, with loss of teeth and sometimes exposure of the maxillary sinus. A few demodicid species invade the tongue, oral epithelium, and esophagus of bats, lemurs, and mice, but rarely cause noticeable problems. Occasionally *Demodex* mites have been recovered from the alimentary tract of dogs, without evidence of penetrating the epithelial lining. *Gastronyssus bakeri* (Gastronyssidae) attaches to the mucosa of the stomach and duodenum of bats, whereas *Paraspinturni globosus* lives in the anal canal of its bat hosts.

Cytodites nudus (Cytoditidae) has been found in the alimentary canal, peritoneum, and body cavity of chickens, and is associated with peritonitis, enteritis, and occasionally deaths. At least some records of *Cytodites* outside the respiratory system could reflect gross dissections of infested hosts. This mite normally is found in the air sacs, but these collapse upon dissection, possibly leading to misleading reports of *Cytodites* in surrounding organs.

In unusual cases, mites have been found in other internal organs. For example, all stages of some *Demodex* species have been found in the liver and spleen (Kirk, 1949), whereas *Cytodites nudus* has been recovered from surface tissues of the heart, liver, and kidneys (Baker et al., 1956). *Demodex* species also have been found alive in lymph nodes, lymphatic vessels, and circulating blood. It is presumed that the mites invade the lymphatic and circulatory systems where extensive destruction of the surrounding dermal tissue occurs in severely infested hosts.

The two most common internal sites of animals that parasitic mites have exploited are the ear canals and respiratory passages.

Ear Mites

Several species of mites infest the ears of domestic and wild animals, causing problems that may warrant veterinary attention. The families most commonly involved are Psoroptidae, Trombiculidae, and Raillietidae.

Psoroptic Ear Mite (*Psoroptes cuniculi*) Although treated here as a separate species, *P. cuniculi* is probably a genetic variant of *P. ovis* (see discussion of psoroptic scab mites). Known as the psoroptic ear mite or ear mange mite of rabbits, *P. cuniculi* causes lesions in laboratory rabbits and commercial rabbit operations (Figure 25.54). Such infestations are referred to as psoroptic otoacariasis. This cosmopolitan mite also infests sheep, goats, and occasionally deer, antelope, and laboratory guinea pigs. Lesions occur primarily in the ears, causing crust formation, malodorous discharges in the external ear canal, and behavioral responses such as scratching the ears, head shaking, loss of equilibrium, and spasmodic contractions of neck muscles (torticollis). In severe cases *P. cuniculi* infestations may spread to other parts of the body, notably the face, neck, and legs.

Psoroptes cuniculi lives its entire life under the margins of scabs formed at infested sites. There the eggs are deposited and hatch in four days. The complete life cycle takes about three weeks. All stages of this nonburrowing mite pierce the stratum corneum to feed on epidermal tissue. Transmission between animals occurs by direct contact.

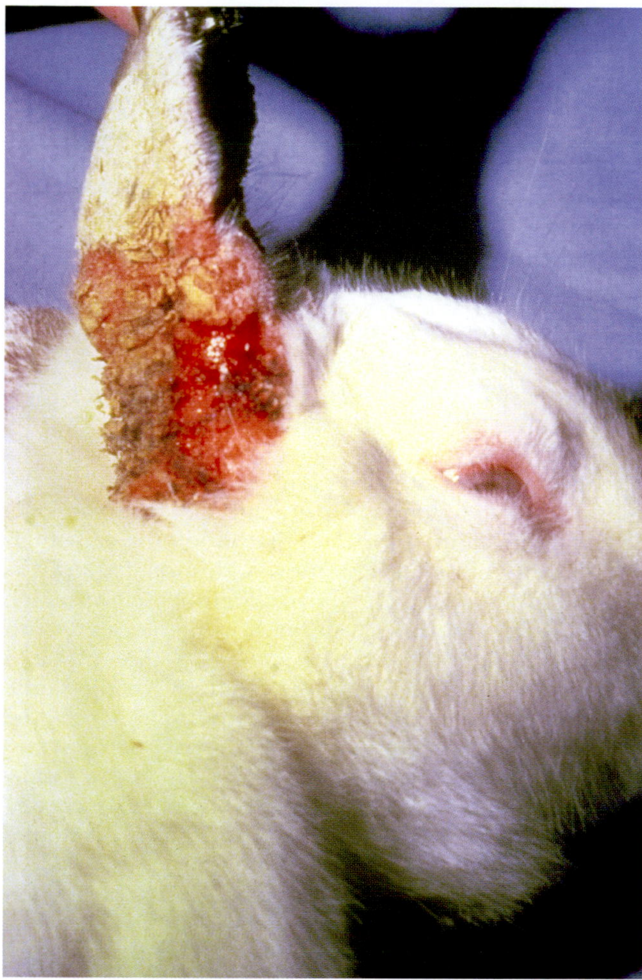

Figure 25.54 Rabbit with ears heavily infested by rabbit ear mite, *Psoroptes cuniculi* (Psoroptidae). Note skin injury, crusty scabs, and bleeding due to rubbing and scratching. (Courtesy of Department of Pathobiology, Auburn University College of Veterinary Medicine)

Infestations of sheep with *P. cuniculi* can cause varying degrees of problems, usually limited to the ears. Lesions in lambs are generally mild and characterized by small, discrete, crusty lesions on the inner surface of the pinnae and around the entrance to the outer auditory canal. Similar lesions may occur at the base of the ears. Severe infestations in older animals can lead to inflammation of the ears (otitis), hematomas, and suppurating abscesses. In other cases sheep may show no clinical signs despite confirmation of mites in the ear canal on otoscopic examination. Survey of sheep flocks in England have shown that the prevalence of *P. cuniculi* is usually higher in lambs than in adults and that up to 60% of some infested flocks are positive for this mite (Morgan, 1992).

Both domestic and wild goats are subject to *P. cuniculi* infestations. Prevalence rates as high as 80 to 90% have been reported in dairy goats, including both kids and adults, in the United States (Williams and Williams, 1978). Goats less than a year old generally exhibit much higher infestation rates than do older animals. Clinical signs of *P. cuniculi* mites in kids often are observed as early as three weeks after birth, reflecting transfer of mites between mother and young. By six weeks of age most kids in infested goat herds are likely to harbor these mites. There is no evidence, however, for cross infections between goats and sheep, even when held in common enclosures (Williams and Williams, 1978). Infestations can cause scaling, crusting, inflammation, and hair loss about the ears; accumulations of wax in the external ear canal; and ear scratching, head shaking, and rubbing of the head and ears against objects in an effort to alleviate the discomfort. Chronic infestations also can lead to anemia and weight loss. Extreme cases are sometimes fatal, with death being preceded by circling, violent fits, and other aberrant behavior. Nondomestic goats reportedly parasitized by *P. cuniculi* include the Nubian mountain goat (*Capra ibex nubiana*) and cross-breeds between domestic and mountain goats, such as Yaez.

Cervid hosts of *P. cuniculi* include both free-ranging and captive white-tailed deer (*Odocoileus virginianus*) and mule deer (*O. hemionus*) in North America. Most reports have come from the southeastern United States where up to 80% of some white-tailed deer populations have been found to be infested. Cases are typically mild, with loss of hair about the ears and base of the antlers, and yellow crusty lesions, accumulated cerumen (ear wax), and serous exudates in the external ear. In more severe infestations, the ear canal may become inflamed and infected, leading to pyogenic bacterial otitis and neurologic disorders (Rollor et al., 1978). In such cases, the infected ear canal and tympanic cavity become filled with mucopurulent exudates, with or without damaging the tympanic membrane. This affects the sensory organs in the inner ear, causing neurologic signs such as excessive salivation, circling behavior, difficulty standing, loss of muscular coordination, and torticollis.

Psoroptes cuniculi is the common ear mange mite of laboratory rabbits that causes psoroptic ear canker. Occasionally wild rabbits and hares (e.g., *Lepus europaeus* in Europe) also are affected. Ear infestations are characterized by loose crusty lesions, excessive cerumen, inflammation, accumulated exudates, and necrotic debris in the external ear. In some heavy infestations, lesions may spread from the ears to other parts of the body, including the face, neck, and genitalia. Often the first signs are behavior such as tilting or shaking of the head, drooping ears, and scratching or self-inflicted trauma to the ears. Inflammation of the middle ear (otitis media) and brown, malodorous discharges in the external auditory canal are common. Cases can become complicated by bacterial infections, causing loss of equilibrium, torticollis and, in some cases, fatal meningitis.

Other animals occasionally infested with *P. cuniculi* are horses, donkeys, mules, antelopes, and guinea pigs. Crusty lesions with accumulated exudates in the ear canal are typical of other hosts. In heavy infestations, mites may spread to other parts of the body (e.g., face, belly, hind legs), causing erythematous and pruritic lesions complicated by secondary bacterial infections. Severe cases of this nature have been reported in the blackbuck antelope (*Antilope cervicapra*) (Wright and Glaze, 1988) and guinea pigs (Yeatts, 1994).

Otodectic Ear Mite (*Otodectes cynotis*) This mite (Figure 25.55) is known as the ear mite or ear canker mite of cats and dogs and as the cause of **otodectic mange**. It occurs worldwide and parasitizes other carnivores such as foxes, ferrets, and raccoons. *Otodectes cynotis* is closely related to *Psoroptes* species, which it resembles in size and general appearance. It can be distinguished from *Psoroptes*, however, by its short, unsegmented tarsal stalks supporting the ambulacral suckers in both sexes; and the greatly reduced hind leg in the female that terminates in two long, whiplike setae.

Otodectes cynotis typically occurs deep in the external ear canal where all the developmental stages are found. Occasionally it secondarily infests other parts of the body including the head, back, tip of the tail, and feet. It does not burrow into the skin, but lives as a surface parasite that may pierce the skin to feed on blood, serum, and lymph. Some workers contend that it feeds more commonly on desquamated epithelial cells and possibly cerumen or other aural exudates. It is believed that development of clinical signs reflects allergic hypersensitivity on the part of the host to antigenic substances introduced while the mites are feeding. This can lead to highly variable responses ranging from asymptomatic or mild cases to severe otitis and convulsive seizures.

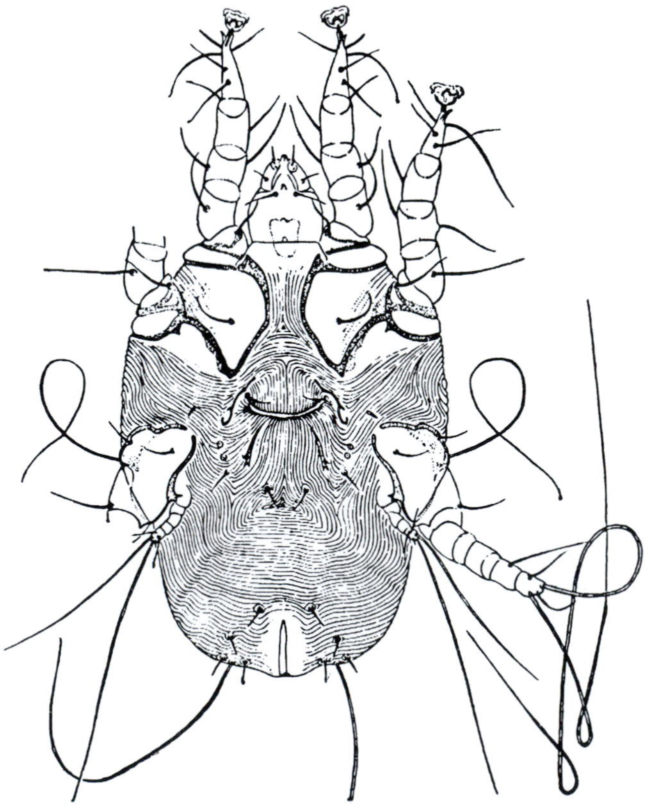

Figure 25.55 Ear mite of cats and dogs, *Otodectes cynotis* (Psoroptidae), female, ventral view. (Baker et al., 1956)

The ear canals of animals infested with *O. cynotis* become excessively moistened with accumulations of cerumen and purulent, brown-black exudates resembling coffee grounds. This is accompanied by inflammation and pruritus, usually involving both ears. As a result of intense itching, infested cats and dogs scratch their ears, shake their head or hold it to one side, and may turn in circles. When the ear canal is massaged, the animal typically responds with pleasurable grunting sounds and by thumping its hind leg on the corresponding side. Severe, untreated cases can lead to emaciation, self-induced trauma, spasms, and convulsions, especially in cats. Diagnosis of *O. cynotis* is confirmed by otoscopic examination and by recovering the mite from aural scrapings.

Cattle and Goat Ear Mites (*Raillietia* spp.) Mites of the genus *Raillietia* (family Halarachnidae) are the only known mesostigmatid species that live in the ear canals of domestic animals. The most widespread species is the cattle ear mite (*R. auris*) (Figure 25.56), which infests dairy and beef cattle in North America, South America, Europe, western Asia, and Australia. Although generally it is considered to be a relatively harmless mite living in cerumen, *R. auris* can cause blockage of the auditory canal by plugs of paste-like wax. Severe cases can result in inflammation of the ear canal, pus formation, ulcerated lesions, and hemorrhaging, with accompanying hearing loss in some animals. Based on studies of *R. flechtmanni* that infests cattle and buffalo (*Bubalus bubalis*) in Brazil, adult mites on pasture vegetation enter the ear canals of grazing animals, where they feed, mate, and oviposit. Larvae, upon hatching from the eggs, leave the host to complete their development as nymphs and adults on pasture (Costa et al., 1992).

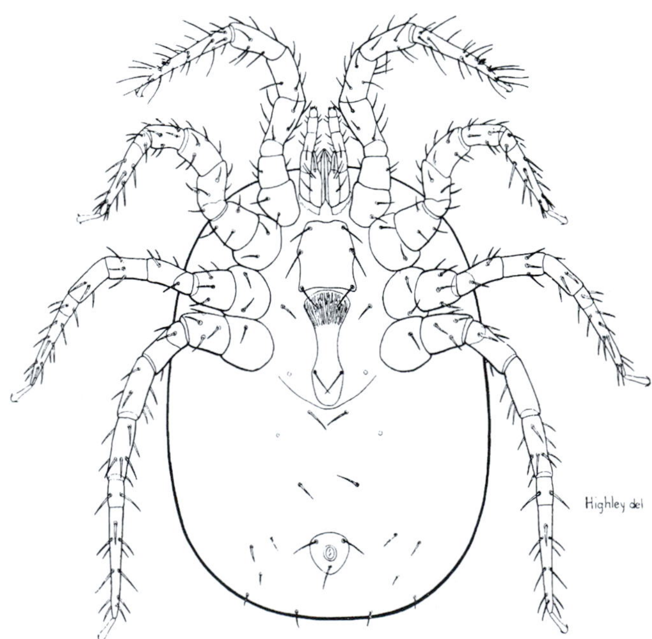

Figure 25.56 *Raillietia auris* (Raillietidae), female, ventral view. (Hirst, 1922)

Other *Raillietia* species include *R. caprae* and *R. manfredi* of goats in Brazil and Australia, respectively; *R. acevedoi* of the Alpine ibex (*Capra ibex*) in Europe and Asia; *Raillietia* species infest the ear canals of the waterbuck (*Kobus ellipsiprymus*) and Uganda kob (*Kobus kob*) in Africa, banteng (*Bos javanicus*) in Indonesia, and wombat (*Vombatus ursinus*) in Australia. *Raillietia caprae* has been implicated in mycoplasma infections of goats (DaMassa et al., 1992).

Histiostomatid Mites Occasionally mites of the family Histiostomatidae, called slime mites, have been found infesting the outer ear canal of large mammals, including elephants, African buffalo, horses, and donkeys. The genera involved are *Auricanoetus*, *Loxanoetus*, and *Otoanoetus*. A related, undescribed species also has been found living and reproducing in the ear canal of a human (Al-Arfaj et al. 2007). Although these mites have been associated with ear infections, they probably play a secondary role, rather than causing the initial infection.

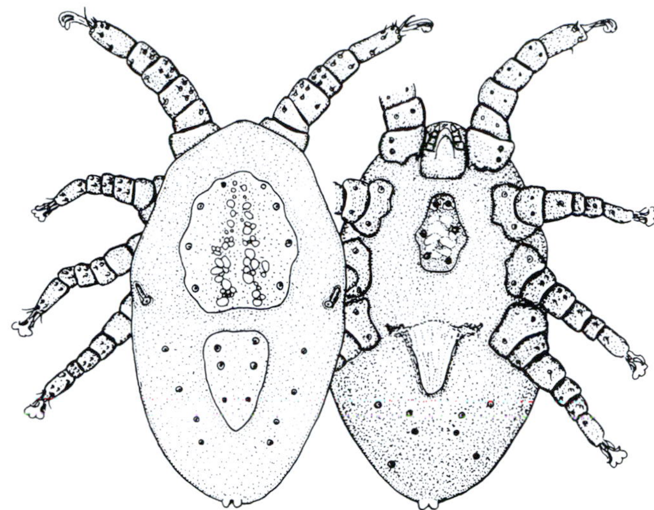

Figure 25.57 *Sternostoma tracheacolum* (Rhinonyssidae), female; dorsal (left) and ventral (right) views (Nutting, 1984).

Respiratory Mites

Representatives of several families of mites are specialized, obligate parasites in the respiratory tracts of reptiles, birds, and mammals. They commonly live in the nasal passages and lungs, causing nasal acariasis and pulmonary acariasis, respectively. Among the taxa of veterinary interest are members of the families Entonyssidae, Rhinonyssidae, Halarachnidae, Ereynetidae, Trombiculidae, Lemurnyssidae, Turbinoptidae, Cytoditidae, Pneumocoptidae, and Gastronyssidae. For keys to species, host lists, and a bibliography for nasal mites of North American birds, see Pence (1975).

Entonyssidae These mites are endoparasites that infest the tracheae and lungs of snakes (Fain, 1961). They rarely seem to cause problems for their hosts, but occasionally induce congestion of the lungs when mite numbers are high.

Rhinonyssidae Rhinonyssid mites are endoparasites in the nasal passages, and occasionally the tracheae, of birds throughout the world. They are common, typically infesting 30 to 50% of birds examined in local surveys (Domrow, 1969; Pence, 1973; Spicer, 1987). Their chelicerae are reduced, membranous structures that are used to imbibe liquid food, including blood, from their hosts as they crawl about on the mucous membranes lining the nasal airways. Feeding is facilitated by the claws on leg 1 that are used in lieu of the chelicerae to tear or otherwise penetrate respiratory tissues. Such injury can cause rhinitis or sinusitis, especially in heavy mite infestations. In most cases, however, infested birds do not experience apparent respiratory problems.

An exception is *Sternostoma tracheacolum* (Figure 25.57) that parasitizes the tracheae, bronchi, parenchymal lung tissue, and air sacs of both wild and captive birds. It does not occur in the nasal cavities. Typical hosts are canaries, parakeets, swallows, and finches. This mite is sometimes called the canary lung mite because of the respiratory problems it causes in captive canaries. Little is known about the behavior of *S. tracheacolum* except that it crawls freely about in the mucous lining of the trachea and bronchi. The mite also may invade the air sacs and lung tissue, where it dies and disintegrates. Its presence causes inflammation and the development of characteristic nodular lesions containing masses of dead mites and purulent, fibrous exudates. Early signs of respiratory distress include listlessness and difficulty breathing.

Heavy infestations can result in tracheitis, bronchitis, hemorrhaging in parenchymal tissue surrounding the terminal bronchi, small foci of bronchial pneumonia, lung congestion, and pneumonitis. As the damage progresses, infested birds may become emaciated and die. Although wild birds are not as severely affected as captive birds, they can develop bronchopneumonia and inflammation of their air sacs.

Halarachnidae Mites in this family are obligate parasites in the respiratory tracts of a variety of mammals. Their hosts include marine mammals, porcupines, squirrels, canids, and nonhuman primates. Halarachnid mites generally are regarded as benign, except for a few species that infest domestic dogs and captive primates. Most species occur in the nasal passages, whereas others may live in sinuses, tracheae, bronchi, or lung tissue. The larvae and adults are the active stages, typically piercing the epithelium with their long chelicerae to feed on lymph and other fluids. Only a few taxa are known to feed on blood. Transmission is presumed to occur by direct transfer of larvae around the host nostrils, or by sneezing and coughing of infested animals. Although most infestations are relatively asymptomatic, others can result in inflammation of the respiratory passages, pulmonary nodules, lung congestion, and host death. Diagnosis of halarachnid

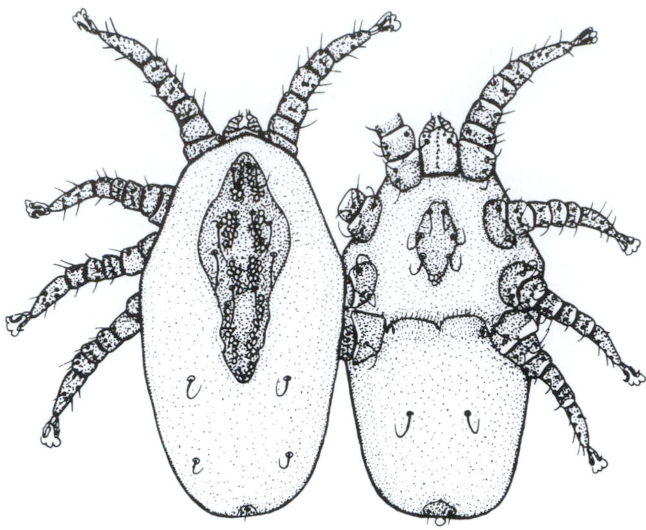

Figure 25.58 Monkey lung mite, *Pneumonyssus simicola* (Halarachnidae), female; dorsal (left) and ventral (right) views. (Nutting, 1984)

infestations usually is based on recovery of larvae in tracheobronchial washings or histological examination of pulmonary tissues. The genera of particular veterinary interest are *Pneumonyssus* of Old World monkeys and apes, and *Halarachne* and *Orthohalarachne* of seals, sea otters, and walruses.

Pneumonyssus simicola, the monkey lung mite (Figure 25.58), is the most common halarachnid mite of primates, infesting the lungs of rhesus, cynomolgus, and macaque monkeys in Africa. It is especially a problem in rhesus colonies, in which up to 100% of the individuals may be infested. Although relatively benign in wild hosts, infestations of *P. simicola* in captive primates in laboratories and zoological parks can cause a wide range of respiratory problems, occasionally proving fatal. The damage results directly from mites attached to the bronchiolar walls piercing the surrounding parenchyma to feed on blood, lymph, and pulmonary epithelial cells. The severity depends largely on the number of mites. Low levels of infestation can cause inflammation of the bronchioles and mild coughing or sneezing. As the number of mites increases, lesions are produced in the form of soft, yellowish nodules containing up to 20 mites each. Other lesions appear as pale spots containing golden-brown, needle-like crystals and dark pigments. The latter are believed to be breakdown products of host blood excreted by the mites. Deaths, in cases of massive infestations, are attributed to congestion of the lungs and alveolar collapse. For additional information on *Pneumonyssus* species, see Hull (1970).

Halarachne species are nasal mites of earless seals (Phocidae) and the Pacific sea otter (*Enhydra lutris*). Captive sea otters may develop heavy infestations of *H. miroungae*, resulting in inflammation of the nasal mucosa, obstruction of nasal passages, destruction of associated bony tissues (turbinates), and pulmonary congestion. More than 3,000 mites have been reported infesting one sea otter that had died. For a key to *Halarachne* species and a review of the genus, see Furman and Murray (1980). Sea lions and fur seals (Otariidae) and walruses (Odobenidae) are hosts for *Orthohalarachne* species. *Orthohalarachne attenuata* infests the nasopharynx, whereas *O. diminuta* infests the lungs. Fur seals often are parasitized simultaneously by both of these mites, with entire populations of seals over three months old being infested. Clinical signs are mucus-filled turbinates, nasal discharges, sneezing, coughing, and impaired respiration. Heavy infestations can lead to alveolar emphysema and a predisposition to more serious ailments that can kill host animals (Kim et al., 1980).

Pneumonyssoides caninum, the dog nasal mite (Figure 25.59), lives in the nasal sinuses of dogs (Figure 25.60) in many parts of the world, including the continental United States and Hawaii, Europe, South Africa, and Australia. Details about its life cycle are largely unknown. This mite is regarded as relatively nonpathogenic, with clinical signs of most infestations being

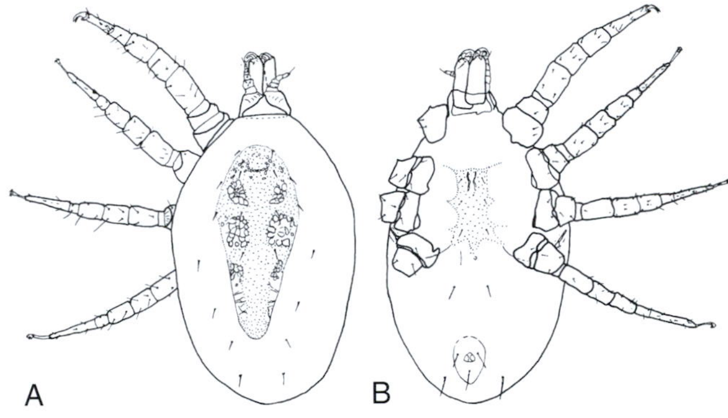

Figure 25.59 Dog nasal mite, *Pneumonyssoides caninum* (Halarachnidae), male. (A) Dorsal view; (B) ventral view. (Strandtmann and Wharton, 1958)

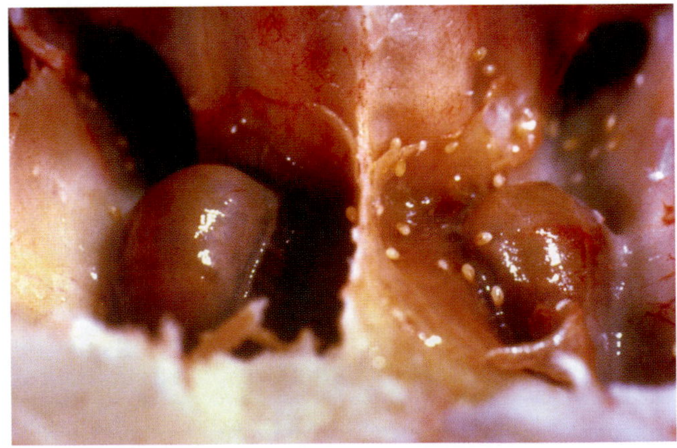

Figure 25.60 Nasal sinuses of dog infested with *Pneumonyssoides caninum* (Halarachnidae); whitish mites seen crawling over tissue surface. (Courtesy of Department of Pathobiology, Auburn University College of Veterinary Medicine)

limited to excessive nasal secretions and hyperemia of the nasal mucosa. More severe cases may involve listlessness, loss of appetite, tearing of the eyes, chronic sneezing, bronchial cough, and rhinitis or sinusitis (Koutz et al., 1954). There also is evidence that *P. caninum* can penetrate host tissues and move beyond the respiratory system to cause lesions in the liver and kidneys (Garlick, 1977).

Ereynetidae Ereynetid mites are primarily free-living detritivores. However, members of two subfamilies are obligate parasites in the respiratory tracts of terrestrial vertebrates. Mites of the subfamily Lawrencarinae infest the nares and nasal passages of amphibians, notably African frogs and toads. Examples are *Lawrencarus eweri* and *Xenopacarus africanus*. Although both feed on tissue fluids, including blood, it is not clear how much harm they cause to their hosts. Mites of the subfamily Speleognathinae live in the mucus-lined nasal passages of a wide range of birds and mammals throughout the world. Occasionally they also invade the lungs. Rarely do they seem to cause harm to their hosts, despite the fact that some species feed on blood (e.g., *Boydaia sturnellus* of meadowlarks, USA). *Speleognathus australis* parasitizes cattle and bison. For additional information on speleognathine mites, see Lawrence (1952), Clark (1960), Baker (1973), Pence (1973, 1975), Fain and Hyland (1975), and Spicer (1987).

Trombiculidae Approximately 20 genera of trombiculid mites are parasitic as larvae (chiggers) in the nasal passages of reptiles, birds, and mammals in both the Old World and New World. Rodents and bats are the most common hosts, parasitized by *Ascoschoengastia, Doloisia, Gahrliepia, Microtrombicula, Schoutedenichia*, and other genera. Other hosts of intranasal chiggers include marsupials (e.g., ring-tailed possum, water opossum, bandicoots), edentates (e.g., armadillos, anteaters), hyraxes (e.g., tree hyrax), lagomorphs (e.g., hares), birds (e.g., sooty tern), felids (e.g., African wild cat), marine iguanas, and sea snakes (Nadchatram 1970). *Vatacarus* species, which parasitize the latter two groups of reptiles, infest not only the nasal fossae, but also the tracheae and lungs of their hosts. The nymphs and adults of intranasal mites are free-living and presumably inhabit the nests, dens, and other sheltered locations of their respective hosts. Virtually nothing is known about possible adverse effects that these chiggers may have on their hosts.

Lemurnyssidae Mites in this psoroptoid family occur as intranasal parasites of lorisid primates (lorises, bush-babies) in Africa and monkeys (Cebidae) in South America (Fain, 1964b). They live in the nasal fossae where they apparently cause little or no harm to their hosts.

Turbinoptidae Members of this family are exclusively intranasal mites of birds and occur in both the Old World and New World. They infest the nasal fossae, without causing apparent harm to their avian hosts. For a review of North American turbinoptid mites, keys to genera and species, and a list of hosts see Pence (1973, 1975). For African and European species, see Fain (1957, 1970); eastern Australia, see Domrow (1969).

Cytoditidae Mites in this family are internal parasites of birds that typically infest the respiratory system, but also may invade the peritoneum and visceral organs. Members of the genus *Cytonyssus* usually are found only in the nasal passages, whereas *Cytodites* species typically inhabit the lungs and air sacs. The species of greatest veterinary importance is *C. nudus*, the air-sac mite of chickens (Figure 25.61), which occurs worldwide. Although low-level infestations do not cause apparent harm, heavy infestations can lead to severe clinical signs and occasionally host death. *Cytodites nudus* is found most commonly in the lining of the air sac, but also may invade the air passages and lungs causing accumulation of mucus in the tracheae and bronchi. Clinical signs include coughing, obstruction of air flow, pulmonary edema, and pneumonia. Infested birds may exhibit weight loss, general weakness, and loss of balance or coordination. In some cases, *C. nudus* invades the body cavity and visceral organs, including the heart, alimentary tract, liver, and kidneys. It shows a particular predilection for the peritoneum. Deaths usually are associated with peritonitis, enteritis, emaciation, and respiratory complications.

Cytodites nudus also has been reported infesting canaries and ruffed grouse, and in one case was reportedly recovered from the peritoneum of a human in

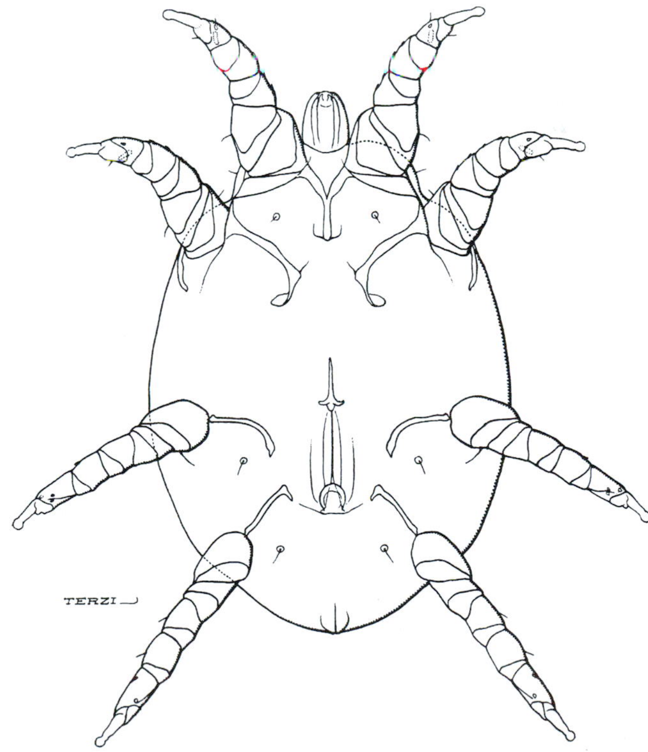

Figure 25.61 Air-sac mite of chickens, *Cytodites nudus* (Cytoditidae), female, ventral view. (Hirst, 1922)

Uganda (Castellani, 1907). Symptoms in canaries are similar to those in chickens but may include bulging eyes and sores or swellings at the corners of the beak (Higby, 1946). Little is known about the life history of this mite. Transmission is presumed to be by coughing.

Pneumocoptidae Members of this family infest the lungs of rodents. *Pneumocoptes jellisoni* and *P. penrosei* are common pulmonary parasites of prairie dogs (*Cynomys* spp.) and mice (*Peromyscus* and *Onychomys* spp.) in the midwestern United States; *P. banksi* parasitizes the California ground squirrel (*Spermophilus beecheyi*), whereas *P. tiollaisi* parasitizes voles (*Clethrionomys* spp.) in Europe. Rarely do these mites cause apparent harm to their hosts. A possible exception is a case involving a captive black-tailed prairie dog (*Cynomys ludovicianus*) at the Philadelphia Zoological Gardens (USA) that died from acute bronchopneumonia. Large numbers of *P. penrosei* were found infesting the lungs at necropsy, with postmortem evidence of emphysema and dilatation of the bronchi and bronchioles (Wiedman, 1916).

Gastronyssidae Gastronyssid mites are primarily intranasal parasites of Old World rodents and bats. Examples include *Opsonyssus* and *Rodhainyssus* species in the nasal fossae of African and Asian bats, and *Sciuracarus paraxeri* in the nasal fossae of South African sun squirrels (Fain, 1956, 1964a, 1967). *Opsonyssus* species also have been reported infesting eye orbits, whereas *Gastronyssus bakeri* lives in the alimentary tract of fruit bats (Pteropodidae). *Yunkeracarus* is common in a diversity of rodents worldwide. Any deleterious effects they may have on their respective hosts remain unknown.

Mite-Borne Diseases

With the exception of ticks and chiggers, only a few species or groups of mites play a significant role in the transmission of disease agents to domestic and wild animals. In many cases the evidence of mite involvement is circumstantial, based primarily on experimental studies. *Dermanyssus gallinae*, for example, has been shown to transmit the viral agents of fowl pox, St. Louis encephalitis, and eastern equine encephalitis among birds, for which mosquitoes are the usual natural vectors. Similarly *Cheyletiella parasitivorax* is capable of experimentally infecting rabbits with the virus causing myxomatosis. Other examples include viruses that cause Newcastle disease of birds involving *Ornithonyssus sylviarum* and western equine encephalitis involving *Dermanyssus americanus*. Perhaps more significant is experimental evidence that *Ornithonyssus bacoti* possibly serves as both a vector and reservoir of Hantaan virus, the causative agent of Korean hemorrhagic fever, which infects mice and humans (Meng et al., 1991).

The importance of mites in the ecology of these diseases should be viewed with caution. The isolation of viruses and other pathogens from naturally infected mites, in the absence of documented evidence of transmission capabilities and natural host associations, should not be construed as evidence that such mites necessarily play a role in the transmission of these disease agents.

With the exception of rickettsial organisms transmitted by ticks and chiggers, bacterial disease agents of animals seldom involve mites as vectors. Some reports, however, suggest the possible role of hay-infesting mites as reservoirs for the causative agents of scrapie in sheep and bovine spongiform encephalopathy (BSE), also known as mad cow disease, in cattle (Wisniewksi et al., 1996). Both diseases are believed to be caused by brain-destroying proteins called prions.

The only significant protozoan disease agents of vertebrates transmitted by mites are blood parasites of the hemogregarine genus *Hepatozoon* that cause **hepatozoonosis**. Their life cycles are similar in many ways to *Plasmodium* species that cause malaria. *Hepatozoon* species parasitize a wide range of hosts, including amphibians (anurans), reptiles, birds, and mammals, in which they undergo development and multiplication in the liver and circulating blood cells. Hematophagous mites ingest infected blood cells while feeding. The *Hepatozoon* parasites are released in the lumen of the mite midgut where they penetrate the gut wall to enter the hemocoel. There they form oocysts containing sporozoites. Vertebrates become infected by ingesting mites with infective oocysts. Among the more common hosts of *Hepatozoon* species that utilize mites as vectors are rodents (e.g., field mice, voles, rats, squirrels), skunks, and lizards. Mites involved are typically mesostigmatid species, including the genera *Haemogamasus*, *Laelaps*, and *Ophionyssus*. Three *Hepatozoon* species are known to infect domestic animals: *H. canis* of dogs, *H. felis* of cats, and *H. muris* of laboratory cotton rats. In addition to mites, many hematophagous insects serve as vectors of other *Hepatozoon* species, including fleas, triatomine bugs, sucking lice, mosquitoes, sand flies, and tsetse flies. For further information on *Hepatozoon* species and their hosts, see Smith (1996).

The only notable nematode parasite of domestic animals transmitted by mites is the filarial worm *Litomosoides sigmodontis* (formerly *L. carinii*), the causative agent of cotton rat **filiariasis**. This nematode occurs in the southeastern United States where it is transmitted among cotton rats (*Sigmodon hispidus*) and squirrels by the tropical rat mite *Ornithonyssus bacoti*. Related *Litomosoides* species occur in South America where their hosts include wood rats (*Neotoma* spp.), marsh rats (*Holochilus* spp.), and the house mouse (*Mus musculus*). Microfilariae of *L. sigmodontis* in host blood are ingested by *O. bacoti*, where they develop to infective third-stage larvae. Rodents become infested when the mite takes a subsequent blood meal. The nematodes become established in the pleural cavity (also the peritoneal cavity in heavy infestations) where they develop to adults. Laboratory-infected cotton rats subjected to chronic reinfection with *L. sigmodontis* tend to lose weight, become feverish, exhibit shallow respiration, and undergo behavioral changes, e.g., sitting in a haunched position with raised fur.

The tropical rat mite (*Ornithonyssus bacoti*) is a vector of *Litomosoides carinii*, a filarial nematode in rats and other wild rodents. The cotton rat (*Sigmodon hispidus*) is a common host in the southern United States. The mites ingest microfilariae with blood from infested rats. The microfilariae penetrate the gut wall and move through the hemocoel to invade the salivary glands, fat-body cells, coxal glands, and glands associated with the female reproductive organs. There they develop to infective third-stage larvae in about two weeks. Infective larvae are introduced to rodent hosts when the mite subsequently feeds, developing to adult worms.

Mites as Intermediate Hosts of Tapeworms

Oribatid mites serve as intermediate hosts for about 27 species (14 genera) of tapeworms in the family Anoplocephalidae. Of this number, approximately 20 species are parasites of domestic animals (Table 25.4). The most important anoplocephalid genus that infests domestic animals is *Moniezia*, represented by two species that parasitize ruminants worldwide. *Moniezia*

Table 25.4 Anoplocephalid Tapeworms of Domestic Animals for Which Oribatid Mites Serve as Intermediate Hosts*

Tapeworm Species	Domestic Hosts	Intermediate Hosts (oribatid genera)	Geographic Occurrence
Anoplocephala perfoliata	Horse, donkey	*Achipteria, Carabodes, Ceratozetes, Eremaeus, Galumna, Hermanniella, Liacarus, Liebstadia, Parachipteria, Platynothrus, Scheloribates, Trichoribates, Urubambates, Zygoribatula*	Cosmopolitan
Anoplocephala magna	Horse, donkey	*Scheloribates*	Cosmopolitan
Avitellina bangaonensis	Cattle, goats	Oribatids and/or psocids?	India
Avitellina centripunctata sensu latu (including *A. goughi, A. lahorea, A. sudanea*)	Sheep (primarily), goat, cattle, buffalo, zebu, camel, other ruminants	*Punctoribates, Scheloribates, Trichoribates* (also psocids, colllembolans)	Europe, Asia, India, Africa
Avitellina chalmersi	Sheep, goat	Oribatids and/or psocids?	India, Africa
Avitellina tatia	Goat	Oribatids and/or psocids?	India
Avitellina woodlandi	Sheep, goat	Oribatids	India
Moniezia autumnalia	Sheep, cattle	Oribatids?	Bulgaria, Tadzhikistan, Russia
Moniezia benedeni	Cattle (primarily), water buffalo, bison, sheep, goat, other ruminants	*Achipteria, Ceratoppia, Ceratozetes, Galumna, Liebstadia, Oribatula, Pergalumna, Platynothrus, Punctoribates, Scheloribates, Spatiodamaemus, Trichoribates, Zygoribatula*	Cosmopolitan
Moniezia expansa	Sheep (primarily), goat, cattle, ibex, gazelle, camel, other ruminants	*Achipteria, Allogalumna [Galumna?], Cepheus, Ceratoppia, Ceratozetes, Eremaeus, Eupelops, Euzetes, Furcoribula, Galumna, Hermanniella, Liacarus, Oribatella, Oribatula, Parachipteria, Peloptulus, Peloribates, Pergalumna, Platynothrus, Protoribates, Punctoribates, Scheloribates, Scutovertex, Spatiodamaeus, Trichoribates, Unguizetes, Xenillus, Zygoribatula*	Cosmopolitan
Moniezia denticulata	Cattle, sheep, goat, others	Oribatids	Cosmopolitan
Moniezia neumani	Sheep	*Punctoribates, Scheloribates, Trichoribates*	?
Paranoplocephala mamillana	Horse	*Achipteria, Allogalumna, Ceratozetes, Galumna, Scheloribates*	?
Stilesia globipunctata	Sheep, goat, cattle, zebu, gazelle, camel, other ruminants	*Africacarus, Allogalumna (Galumna?), Scheloribates, Zygoribatula* (psocids?)	Europe (Spain), Asia Minor (Turkey), Asia, Africa
Stilesia hepatica	Sheep, goat, cattle (rarely); wild ruminants	Oribatids?	Asia, Africa
Stilesia vittata	Camel and dromedary (primarily), sheep, goat	Oribatids?	Ubekistan, India, Africa (eastern and southern)
Thysaniezia giardi (syn. *T. ovilla*)	Sheep, goat (primarily), cattle, buffalo	*Achipteria, Liebstadia, Punctoribates, Scheloribates, Trichoribates, Zygoribatula*	Cosmopolitan
Thysanosoma actinioides	Sheep, goat, cattle, deer, antelope	Oribatids and/or psocids?	North America, South America (western regions)

*The oribatid genera listed include both those that have been found naturally infected and those that have been shown experimentally to support development of cysticercoids.

benedeni is primarily a parasite of cattle, whereas *M. expansa* is primarily a parasite of sheep and goats. These tapeworms are especially prevalent in young host animals less than six to eight months old. Older animals tend to be less susceptible and after two years of age seldom have more than one or a few worms.

Most anoplocephalid tapeworms cause little apparent harm to their hosts even when the parasite burden is high. In some cases, however, they can cause loss of weight gain, unthriftiness, colic, and intestinal blockage that may require veterinary attention. Sheep are the most adversely affected, whereas cattle and horses seldom experience significant health problems due to cestodes associated with oribatid mites. Heavy infestations can lead to severe health problems, especially in Asia, involving weight loss, reduced wool yield, anemia, enteritis, diarrhea, intestinal obstruction, toxemia, convulsions, and death. In the case of *Stilesia hepatica*, which infests the bile ducts of sheep, economic losses can result from condemnation of sheep livers at meat inspection. For further details on the biology and pathogenic consequences of anoplocephalid tapeworms infesting domestic animals, see Graber (1959), Soulsby (1965, 1982), Narsapur (1988), and Kauffmann (1996).

Although tapeworms of the genus *Avitellina* are common parasites of domestic ruminants in Europe, Asia and Africa, they do not appear to significantly harm their hosts. The taxonomy of *Avitellina* remains unstable, with several long-recognized species (e.g., *A. goughi*, *A. lahorea*, *A. sudanea*) being regarded as synonyms of the widespread species *A. centripunctata* by some workers (Raina, 1975), but as valid species by others (Malhotra and Capoor, 1982). At the same time, there is evidence that *A. centripuncta* is represented in Africa as a complex of at least five cryptic species (Ba et al., 1994). This has complicated the interpretation of any meaningful associations of individual oribatid species as intermediate hosts of several *Avitellina* species.

The development and life history of *Moniezia* species is representative of anoplocephalid tapeworms in general. The proglottids of adult tapeworms containing eggs are passed in the host feces, contaminating pasture grasses associated with oribatid mites. The mites feed on the eggs by breaking the shell with their chelicerae and ingesting the developing embryo, or oncosphere. In the mite, the oncosphere penetrates the midgut wall to enter the hemocoel where it slowly develops in about four weeks to the cysticercoid stage (Figure 25.62). The developmental time, however, varies significantly with different species and environmental temperatures, ranging from two to seven months. Infested oribatid mites are consumed with grasses and other forage by ruminants as they graze. The cysticercoids are released as the mites are digested and attach to the wall of the alimentary tract, or bile ducts in some species, where they grow and mature to adult tapeworms in about five to six weeks. Mature tapeworms typically live two to six months before being spontaneously eliminated. During this time they release egg-filled proglottids that are passed in the host feces.

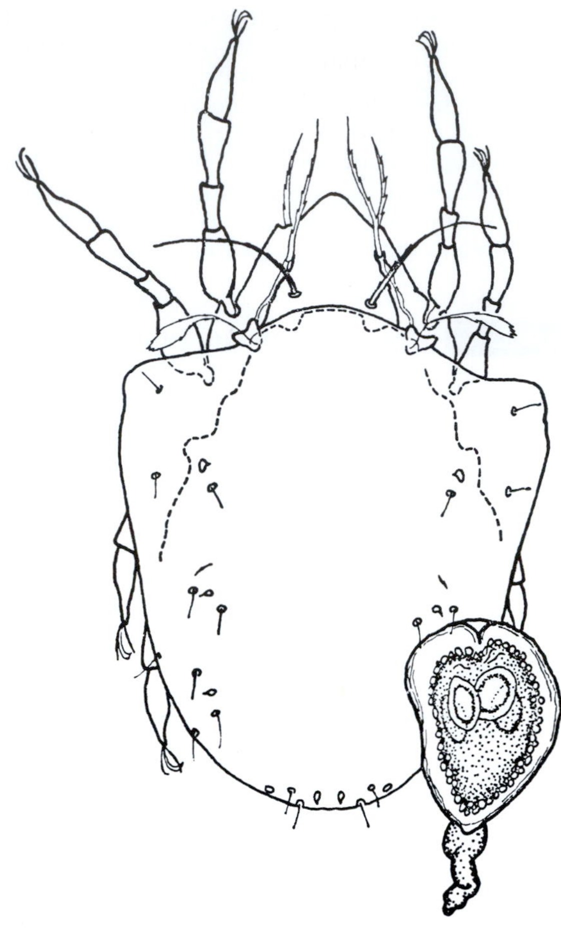

Figure 25.62 Oribatid mite as intermediate host for cysticercoid stage of tapeworms; cysticercoid of tapeworm is depicted at lower right. (Sengbusch, 1977)

More than 125 species of oribatid mites, representing 37 genera and 25 families, have been shown to support development of anoplocephalid tapeworms (Denegri, 1993) (Table 23.5). Of this number, only about 45 oribatid species have been found to be naturally infected with cysticercoids. This indicates that many oribatid species are capable of supporting development of cysticercoids when experimentally infected, but do not necessarily ingest tapeworm eggs under natural conditions. The most important mite families that serve as intermediate hosts are Ceratozetidae, Galumnidae, Oribatulidae, and Scheloribatidae, apparently reflecting their relatively large size and ability to ingest tapeworm eggs. Other contributing factors are the size and structure of the chelicerae, the natural diet of the mites, and their tendency to move upward from the soil into grasses when ruminants are grazing. This combination of traits helps to explain why *Galumna*, *Scheloribates*, and *Zygoribatula* species are the oribatids most commonly infested with cysticercoids of anoplocephalid tapeworms. For further information on oribatid taxa as intermediate hosts for tapeworms, see Sengbusch (1977), Narsapur (1988), and Denegri (1993).

Table 25.5 Families and Genera of Oribatid Mites Known to Support Development of Anoplocephalid Tapeworms That Parsitize Domestic Animals

Achipteriidae	Liacaridae
Achipteria (2)*	*Adoristes* (2)
Parachipteria (2)	*Liacarus* (1)
Astegistidae	*Xenillus* (1)
Furcoribula (1)	Metrioppidae
Camisiidae	*Ceratoppia* (1)
Platynothrus (1)	Mochlozetidae
Carabodidae	*Unguizetes* (1)
Carabodes (1)	Mycobatidae
Cepheus (1)	*Punctoribates* (3)
Ceratozetidae	Oppiidae
Ceratozetes (4)	Oribatellidae
Hypozetes (1)	*Oribatella* (1)
Trichoribates (3)	Oribatulidae
Damaeidae	*Liebstadia* (1)
Spatiodamaeus (1)	*Oribatula* (2)
Epilohmanniidae	*Zygoribatula* (11)
Epilohamminia (1)	Pelopidae
Eremaeidae	*Peloptulus*
Eremaeus (2)	Phenopelopidae
Euzetidae	*Eupelops* (1)
Euzetes (1)	Protoribatidae
Oppiella (1)	*Protoribates* (1)
Galumnidae	Scheloribatidae
Allogalumna	*Scheloribates* (14)
Galumna (15)	*Urubambates* (1)
Pergalumna (2)	Scutoverticidae
Pilogalumna (1)	*Scutovertex* (1)
Haplozetidae	Xylobatidae
Peloribates (2)	*Xylobates* (1)
Hermanniellidae	Family?
Hermanniella (2)	*Africacarus* (1)

*The approximate number of species is indicated in parentheses following each genus.

Based on Allred (1954), Sengbusch (1977), Narsapur (1988), and Denegri (1993)

Heavy tapeworm infestations in ruminants are directly correlated with large oribatid mite populations. Newly seeded and first-year pastures have low mite numbers and result in low infestation rates even in young animals. Older pastures that have been left undisturbed for two or more years support the build-up of oribatid numbers, dramatically increasing the risk of parasitism. To reduce mite populations and avoid this problem, lambs and calves should be started on new pastures, permanent pastures should be cultivated annually, and pastures should be fenced off to prevent livestock from grazing in adjacent rough pasture and woodland.

A few anoplocephalid tapeworms in the genus *Cittotaenia* that utilize oribatid mites as intermediate hosts are parasites of rabbits and hares in North America, Europe, and Asia. Heavy infestations by these tapeworms can cause digestive disturbances, emaciation, and death of hosts. In addition, *C. pusilla* occasionally parasitizes laboratory mice and rats in North America, Europe, and Japan. Its natural hosts include the Norway rat, black rat, house mouse, voles (*Microtus* spp.), and other wild rodents. The grain-infesting mite *Glycyphagus domesticus* (Family Glycyphagidae) reportedly serves as the intermediate host (Allred, 1954).

REFERENCES AND FURTHER READING

Al-Arfaj, A., Mullen, G. R., Rashad, R., Abdel-Hameed, A., OConnor, B. M., Alkhalife, I., et al. (2007). A human case of otoacariasis involving a histiostomatid mite (Acari: Histiostomatidae). *The American Journal of Tropical Medicine and Hygiene, 76*, 967–971.

Allred, D. M. (1954). Mites as intermediate hosts of tapeworms. *Proceedings of the Utah Academy of Sciences, 31*, 44–51.

Andrews, J. R., & Ramsay, G. W. (1982). A case of papular dermatosis in man attributed to an ascid mite (Acari). *Journal of Medical Entomology, 19*, 111–112.

Arlian, L. G. (2002). Arthropod allergens and human health. *Annual Review of Entomology, 47*, 395–433.

Arlian, L. G., & Hart, B. J. (1992). Water balance and humidity requirements of house dust mites. *Experimental & Applied Acarology, 16*, 15–35.

Arlian, L. G., Morgan, M. S., & Neal, J. S. (2002). *Dust mite allergens: Ecology and distribution.* London: Curr Allergy Asthma Rep. BioMed Central Ltd, pp. 2, 5 and 401–411.

Arlian, L. G., Morgan, M. S., Vyszenski-Moher, D. L., & Stemmer, B. L. (1994a). *Sarcoptes scabiei*: The circulating antibody response and inducted immunity to scabies. *Experimental Parasitology, 78*, 37–50.

Arlian, L. G., Rapp, C. M., Vyszenski-Moher, D. L., & Morgan, M. S. (1994b). *Sarcoptes scabiei*: Histopathological changes associated with acquisition and expression of host immunity to scabies. *Experimental Parasitology, 78*, 51–63.

Aylesworth, R., & Baldridge, D. (1983). *Dermatophagoides scheremetewski* and feather pillow dermatitis. *Minnesota Medicine, 66*, 43.

Ba, C. T., Wang, X. Q., Renaud, R., Euzet, L., Marchand, B., & de Meeus, T. (1994). Diversity in the genera *Avitellina* and *Thysaniezia* (Cestoda: Cyclophyllidea): Genetic evidence. *Journal of the Helminthological Society of Washington, 61*, 57–60.

Baker, R. A. (1973). Notes on the internal anatomy, the food requirements and development in the family Ereynetidae (Trombidiformes). *Acarologia, 15*, 43–52.

Baker, E. W., Evans, T. M., Gould, D. J., Hull, W. B., & Keegan, H. L. (1956). *A Manual of Parasitic Mites.* New York: National Pest Control Assoc.

Baker, A. S. (1999). *Mites and ticks of domestic animals. An identification guide and information source.* London: H. M. Stationery Office.

Bates, P. G. (1999). Inter- and intra-specific variation within the genus *Psoroptes* (Acari: Psoroptidae). *Veterinary Parasitology, 83*, 201–217.

Betz, T. G., Davis, B. L., Fournier, P. V., Rawlings, J. A., Elliot, L. B., & Baggett, D. A. (1982). Occupational dermatitis associated with straw itch mites (*Pyemotes ventricosus*). *Journal of the American Medical Association, 247*, 2821–2823.

Borstein, S., & Zakrisson, G. (1993). Clinical picture and antibody response in pigs infected by *Sarcoptes scabiei* var. suis. *Veterinary Dermatology, 4*, 123–131.

Boyce, W. M., & Brown, R. N. (1991). Antigenic characterization of *Psoroptes* spp. (Acari: Psoroptidae) mites from different hosts. *The Journal of Parasitology, 77*, 675–679.

Brettman, L. R., Lewin, S., Holzman, R. S., Goldman, W. D., Marr, J. S., Kechijian, P., et al. (1981). Rickettsialpox: Report of an outbreak and a contemporary review. *Medicine (Baltimore), 60*, 363–372.

Broce, A. B., Zurek, L., Kalisch, J. A., Brown, R., Keith, D. L., Gordon, D., et al. (2006). *Pyemotes herfsi* (Acari: Pyemotidae), a mite new to North America as the cause of bite outbreaks. *Journal of Medical Entomology, 43*, 610–613.

Brockie, R. E. (1974). The hedgehog mange mite, *Caparinia tripilis*, in New Zealand. *New Zealand Veterinary Journal, 2*, 243–247.

Bukva, V. (1986). *Demodex tauri* sp.n. (Acari: Demodicidae), a new parasite of cattle. *Folia Parasitologica, 33*, 363–369.

Bukva, V. (1990). Three species of the hair follicle mites (Acari: Demodicidae) parasitizing sheep, *Ovis aries* L. *Folia Parasitologica, 37*, 81–91.

Burns, D. A. (1992). Follicle mites and their role in disease. *Clinical and Experimental Dermatology, 17,* 152–155.

Camin, J. H. (1953). Observations on the life history and sensory behavior of the snake mite, *Ophionyssus natricis* (Gervais) (Acarina: Macronyssidae). *Special Publications of the Chicago Academy of Sciences, 10,* 75.

Carlson, B. L., Roher, D. P., & Nielsen, S. W. (1982). Notoedric mange in gray squirrels (*Sciurus carolinensis*). *Journal of Wildlife Diseases, 18,* 347–348.

Castellani, A. (1907). Note on an acarid-like parasite found in the omentum of a negro. *Centralbl Bakter I, 43,* 372. [Cytoditidae]

Chakrabarti, A. (1986). Human notoedric scabies from contact with cats infested with *Notoedres cati. International Journal of Dermatology, 25,* 646–648.

Chen, X. B., Fu, C. B., Chen, X. B., & Fu, C. B. (1992). Mites causing pulmonary, intestinal and urinary acariasis. In X. B. Chen, & E. P. Ma (Eds.), *Researches of Acarology in China* (pp. 109–113). Chongqing, China: Chongqing Publishing House.

Chen, X. B., Sun, X., & Hu, S. F. (1990). Clinical manifestation and treatment of pulmonary acariasis. *Chinese Journal of Parasitology & Parasitic Diseases, 8,* 41–44. [in *Chinese*]

Clark, G. M. (1960). Three new nasal mites (Acarina: Speleognathidae) from the grey squirrel, the common grackle, and the meadowlark in the United States. *Proceedings of the Helminthological Society of Washington, 27,* 103–110.

Cochrane, G. (1994). Effects of *Psoroptes ovis* on lamb carcasses. *The Veterinary Record, 134,* 72.

Cole, N. A., & Guillot, F. S. (1987). Influence of *Psoroptes ovis* on the energy metabolism of heifer calves. *Veterinary Parasitology, 23,* 285–295.

Colloff, M. J., ChannaBasavanna, G. P., & Viraktamath, C. A. (1989). Human semen as a dietary supplement for house dust mites (Astigmata: Pyroglyphidae). *Progress in Acarology, 1,* 141–146.

Colloff, M. J., & Hart, B. J. (1992). Age structure and dynamics of house dust mite populations. *Experimental & Applied Acarology, 16,* 49–74.

Cosoroabă, I. (1994). *Acarologie Veterinaria.* Bucharest: Ceres Publishing House. [in *Romanian*]

Costa, A. L., Leite, R. C., Faccini, J. L. H., & DaCosta, A. L. (1992). Preliminary investigations on transmission and life cycle of the ear mites of the genus *Raillietia* Trouessart (Acari: Gamasida) parasites of cattle. *Memorias do Instituto Oswaldo Cruz, 87*(Suppl. I), 97–100.

Cremers, H. J. W. M. (1985). The incidence of *Chorioptes bovis* (Acarina: Psoroptidae) on the feet of horses, sheep, and goats in the Netherlands. *Vet Quart, 7,* 283–289.

Cuthbert, O. D. (1990). Storage mite allergy. *Clinical Reviews in Allergy, 8,* 69–86.

Czarnecki, N., & Kraus, H. (1978). Milbendermatitis durch *Tyrophagus dimidiatus. Zeitschrift fur Hautkrankheiten, 53,* 414–416.

Dahl, F. (1910). Milben als Erzeuger von Zellwucherunge. *Centralbl f Bakter, 53,* 524–533.

DaMassa, A. J., Wakenall, P. S., & Brooks, D. L. (1992). Mycoplasmas of goats and sheep. *Journal of Veterinary Diagnostic Investigation, 4,* 101–113.

Davis, J. W., & Anderson, R. C. (Eds.). (1971). *Parasitic Diseases of Wild Mammals.* Ames: Iowa State Univ. Press.

Denegri, G. M. (1985). Desarrollo experimental de *Bertiella mucronata* Meyner, 1895. (Cestoda: Anoplocephalidae) de origen humano en su huésped intermediario. *Zbt Vet Med B, 32,* 498–504.

Denegri, G. M. (1993). Review of oribatid mites as intermediate hosts of tapeworms of the Anoplocephalidae. *Experimental & Applied Acarology, 17,* 567–580.

Desch, C., & Nutting, W. B. (1972). *Demodex folliculorum* (Simon) and *D. brevis* Akbulatova of man: Redescription and reevaluation. *The Journal of Parasitology, 58,* 169–177.

Dietrich, M., & Horstmann, R. D. (1983). Urtikarielle Dermatitis bei Acariasis durch *Dermanyssus hirundinis. Schwalben-Trugkratze, Pseudoscabies Medizinische Welt, 34,* 595–597.

Domrow, R. (1969). The nasal mites of Queensland birds (Acari: Dermanyssidae, Ereynetidae and Epidermoptidae). *The Linnean Society of New South Wales, 93,* 297–426.

Domrow, R. (1988). Acari Mesostigmata parasitic on Australian vertebrates: An annotated checklist, keys and bibliography. *Invertebrate Taxonomy, 1,* 817–948.

Domrow, R. (1991). Acari Prostigmata (excluding Trombiculidae) parasitic on Australian vertebrates: An annotated checklist, keys and bibliography. *Invertebrate Taxonomy, 4,* 1283–1376.

Domrow, R. (1992). Acari Astigmata (excluding feather mites) parasitic on Australian vertebrates: An annotated checklist, keys and bibliography. *Invertebrate Taxonomy, 6,* 1459–1606.

Elixmann, J. H., Jorde, W., & Schata, M. (1991). Incidence of mites in domestic textiles in Germany. *Allergologie, 14,* 451–460.

English, F. P., & Nutting, W. B. (1981). Demodicosis of ophthalmic concern. *American Journal of Ophthalmology, 91,* 362–372.

Evans, G. O. (1992). *Principles of Acarology.* C.A.B. International.

Fain, A. (1956). Une nouvelle famille d'acariens endoparasites des chauves-souris: Gastronyssidae fam. nov. *Annales de la Societe belge de medecine tropicale, 36,* 87–98.

Fain, A. (1957). Les Acariens des familles Epidermoptidae et Rhinonyssidae parasites des fosses nasales d'oiseaux au Ruanda-Urundi et au Congo Belge. *Annales Du Musée Royal Du Congo Belge, 8*(60), 1–176.

Fain, A. (1961). Les acariens parasites endopulmonaires des serpents (Entonyssidae: Mesostigmata). *Institut Royal des Sciences Naturelles de Belgique, 37,* 1–135.

Fain, A. (1964a). Chaetotaxie et classification des Gastronyssidae avec description d'un nouveau genre parasite nasicole d'un Ecureuil sudafricain (Acarina: Sarcoptiformes). *Revue de zoologie et de botanique africaines, 70,* 40–52.

Fain, A. (1964b). Les Lemurnyssidae parasites nasicoles des Lorisidae africains et des Cebidae sud-américaines. Description d'une espèce nouvelle (Acarina: Sarcoptiformes). *Annales de la Societe Belge de Médecine Tropicale, 44,* 453–458.

Fain, A. (1965). A review of the family Epidermoptidae Trouessart parasitic on the skin of birds. Parts I-II. *Kon Acad Wetensch, Lett Schone Kunsten Belg (Wetensch), 27*(84), 1–176.

Fain, A. (1967). Observations sur les Rodhainyssinae acariens parasites des voies respiratoires des chauves-souris (Gastronyssidae: Sarcoptiformes). *Acta Zoological Pathological Antverp, 44,* 3–35.

Fain, A. (1968). Étude de la variabilite de *Sarcoptes scabiei* avec une revision des Sarcoptidae. *Acta Zoological Pathological Antverp, 47,* 3.

Fain, A. (1970). Novoeaux acarines nasicoles de le famillie Turbinoptidae (Sarcoptiformes). *Bulletin et Annales de la Société royale belge d'Entomologie, 106,* 28–36.

Fain, A., & Hyland, K. E. (1975). Speleognathinae collected from birds in North America (Acarina: Ereynetidae). *Journal of the New York Entomological Society, 83,* 203–208.

Farley, M. L., Mabry, L. C., & Hieger, L. R. (1989). Mites in pulmonary cytology specimens. *Diagnost Cytopathol, 5,* 416–426.

Flynn, R. J. (1973). *Parasites of Laboratory Animals.* Ames: Iowa State Univ. Press.

Fox, J. G., & Hewes, K. (1976). *Cheyletiella* infestation in cats. *Journal of the American Veterinary Medical Association, 169,* 332–333.

Franklin, C. D., & Underwood, J. C. (1986). *Demodex* infestation of oral mucosal sebaceous glands. *Oral Surgery, Oral Medicine, Oral Pathology, 61,* 80–82.

Furman, D. P. (1959). Feeding habits of symbiotic mesostigmatid mites of mammals in relation to pathogen-vector potentials. *The American Journal of Tropical Medicine and Hygiene, 8,* 5.

Furman, D. P., & Dailey, M. D. (1980). The genus *Halarachne* (Acari: Halarachnidae), with the description of a new species from the Hawaiian monk seal. *Journal of Medical Entomology, 17,* 352–359.

Garlick, N. L. (1977). Canine pulmonary acariasis. *Canine Practice, 4,* 42–47.

Gaud, J., & Atyeo, W. T. (1996). Feather mites of the world (Acarina, Astigmata): The supraspecific taxa. Part I. Text. Part II. Illustrations of feather mite taxa. *Annalen Zoologische Wetenschappen-Koninklijk Museum voor Midden-Africa/Musee Royal de l'Afrique Centrale, Tervuren, Belgium, 277*(1), 277(2).

Georgi, J. R., & Whitlock, J. H. (1980). Arachnids. In J. R. Georgi (Ed.), *Parasitology for veterinarians* (3rd ed., pp. 41–67). W. B. Saunders Co., Chapter 3.

Gerson, U., Fain, A., & Smiley, R. L. (1999). Further observations on the Cheyletidae (Acari), with a key to the genera of the Cheyletinae and a list of all known species in the family. *Bulletin de L'Institut Royal des Sciences Naturelles de Belgique Entomologie, 69,* 35–86.

Gorham, J. R. (1991). Insect and Mite Pests in Food: An Illustrated Key. US Department of Agriculture, Agriculture Handbook No. 655, Vol. 2.

Graber, M. (1959). *I.A.C.E.D. Symposium on Helminthiasis in Domestic Animals.* Nairobi, Kenya: C.C.T.A.

Gregory, M. W. (1981). Mites of the hedgehog *Erinaceus albiventris* Wagner in Kenya: Observations on the prevalence and pathogenicity of *Notoedres oudemansi* Fain, *Caparinia erinacei* Fain and *Rodentopus sciuri* Fain. *Parasitology, 82*, 149–157.

Guillot, F. S., & Stromberg, P. C. (1987). Reproductive success of *Psoroptes ovis* (Acari: Psoroptidae) on Hereford calves with a previous infestation of psoroptic mites. *Journal of Medical Entomology, 24*, 416–419.

Harada, S. H., & Sadaji, T. (1925). On a case of mites found in human urine. *Chugai Iji Shimpo, 44*, 859–866. [in *Japanese*]

Hardison, J. L. (1977). A case of *Eutrombicula alfreddugesi* (chiggers) in a cat. *Veterinary Medicine, Small Animal Clinician: VM, SAC, 72*, 47.

Hartley, M., & English, A. (2005). *Sarcoptes scabiei* var. *wombati* infection in the common wombat (*Vombatus ursinus*). *European Journal of Wildlife Research, 51*, 117–121.

Harwood, R. F., & James, M. T. (1979). *Entomology in Human and Animal Health* (7th ed.). New York: Macmillan Co.

Hay, D. B., Hart, B. J., & Douglas, A. E. (1993). Effects of the fungus *Aspergillus penicillioides* on the house dust mite *Dermatophagoides pteronyssinus*: An experimental re-evaluation. *Medical and Veterinary Entomology, 7*, 271–274.

Hay, D. B., Hart, B. J., Pearce, R. B., Kozakiewicz, Z., & Douglas, A. E. (1992). How relevant are house dust mite-fungal interactions in laboratory culture to the natural dust system. *Experimental & Applied Acarology, 16*, 37–47.

Heilesen, B. (1946). *Studies on Acarus Scabiei and Scabies*. Copenhagen: Rosenkilde & Bagger.

Henschel, J. (1929). Reizphysiologische Untersuchung der Käsemilbe *Tyrolichus casei* (Oudemans). *Z Verg Physiol, 9*, 802–837.

Hewitt, M., Barrow, G. I., Miller, D. C., Turk, F., & Turk, S. (1973). Mites in the personal environment and their role in skin disorders. *British Journal of Dermatology, 89*, 401–409.

Higby, W. E. (1946). A new canary plague. *All Pets Magazine*, (December), 8–9. [Cytoditidae]

Hirst, S. (1922). *Mites Injurious to Domestic Animals*. British Museum of Natural History, Econ. Ser. No. 13.

Hoban, P. A. (1990). A review of desert bighorn sheep in the San Andres Mountains, New Mexico. *Desert Bighorn Council Transact, 34*, 14–22.

Hogsette, J. A., Butler, J. F., Miller, W. V., & Hall, R. D. (1988) *Annotated bibliography of the northern fowl mite, Ornithonyssus sylviarum (Canestrini & Fanzago) (Acari: Macronyssidae). Entomological Society of America, Miscellaneous Publications No. 76.*

Hong, C. S., Park, H. S., & Oh, S. H. (1987). *Dermatophagoides farinae*, an important allergenic substance in buckwheat-husk pillows. *Yonsei Medical Journal, 28*, 274–281.

Houck, M. A. (Ed.). (1994). *Mites: Ecological and Evolutionary Analyses of Life-History Patterns*. London: Chapman & Hall.

Huebner, R. J., Jellison, W. L., & Pomerantz, C. (1946). Rickettsialpox—A newly recognized rickettsial disease. IV. Isolation of a rickettsia apparently identical with the causative agent of rickettsialpox from *Allodermanyssus sanguineus*, a rodent mite. *Public Health Report, 61*, 1677–1682.

Hughes, A. M. (1976). *The Mites of Stored Food and Houses* (2nd ed.). Ministry of Agriculture, Fisheries and Food, Tech. Bull. 9. London: Her Majesty's Stationery Office.

Hull, W. B. (1970). Respiratory mite parasites in nonhuman primates. *Laboratory Animal Care, 20*, 402–406.

Johnson, L. S., & Albrecht, D. J. (1993). Effects of haematophagous ectoparasites on nestling house wrens, *Troglodytes aedon*: Who pays the cost of parasitism? *Oikos, 66*, 255–262.

Kambarage, D. M. (1992). Sarcoptic mange infestation in goats. *Bull Anim Hlth Product Africa, 40*, 239–244.

Kass, E. M., Szaniawski, W. K., Levy, H., Leach, J., Srinivasan, K., & Rive, C. (1994). Rickettsialpox in a New York City hospital, 1980–1989. *The New England Journal of Medicine, 331*, 1612–1617.

Kaufmann, J. (1996). *Parasitic Infections of Domestic Animals: A Diagnostic Manual*. Basel: Birkhaser Verlag.

Kawamura, A., Jr., Tanaka, H., & Tamura, A. (Eds.). (1995). *Tsutsugamushi Disease*. Tokyo: University of Tokyo Press.

Kazacos, E. A., Kazacos, K. R., & Demaree, H. A., Jr. (1983). Notoedric mange in two fox squirrels. *Journal of the American Veterinary Medical Association, 183*, 1281–1282.

Keh, B. (1974). Dermatitis caused by the bat mite *Chiroptonyssus robustipes* (Ewing) in California. *Journal of medical entomology, 11*, 498.

Kethley, J. (1982). Acariformes. In S. P. Parker (Ed.), *Synopsis and Classification of Living Organisms* (Vol. 2, p. 117). New York: McGraw-Hill.

Kilpio, O., & Pirila, V. (1952). A new tyroglyphid mite causing dermatitis. *Acta Derm-Ven, 32*, 197–200.

Kim, K. C., Haas, V. L., & Keyes, M. C. (1980). Populations, microhabitat preference and effects of infestations of two species of *Orthohalarachne* (Halarachnidae: Acarina) in the northern fur seal. *Journal of Wildlife Diseases, 16*, 45–51.

Kirk, H. (1949). Demodectic mange. *The Veterinary Record, 61*, 394.

Kirkwood, A. C. (1980). Effect of *Psoroptes ovis* on the weight of sheep. *The Veterinary Record, 107*, 469–470.

Klompen, J. S. H. (1989). Ontogeny of *Rhyncoptes grabberi*, n. sp. (Acari: Astigmata: Rhyncoptidae) associated with *Macaca mulatta*. *Journal of Medical Entomology, 26*, 81–87.

Klompen, J. S. H. (1992). Phylogenetic relationships in the mite family Sarcoptidae (Acari: Astigmata). *Miscellaneous Publications Museum of Zoology, University of Michigan, 180*, 1–154.

Koutz, F. R., Chamberlain, D. M., & Cole, C. R. (1954). *Pneumonyssus caninum* in the nasal cavity and paranasal sinuses. *Journal of the American Veterinary Medical Association, 122*, 106.

Krantz, G. W. (1978). *A Manual of Acarology* (2nd ed.). Corvallis, Oregon: Oregon State University Book Store.

Krantz, G. W., & Walter, D. E. (Eds.). (2009). *A Manual of Acarology*. Lubbock: Texas Tech University Press.

Krinsky, W. L. (1983). Does epizootic lymphocytic choriomeningitis prime the pump for epidemic rickettsialpox? *Reviews of Infectious Diseases, 5*, 1118–1119.

Kumar, D., Raisinghani, P. M., & Manohar, G. S. (1992). Sarcoptic mange in camels: A review. In W. R. Allen, A. J. Higgins, I. G. Mayhew, D. H. Snow, & J. F. Wade (Eds.), *Proceedings of the First International Camel Conference* (pp. 79–82). Newmarket, UK: Dubai. R. W. Publications Ltd.

Kummel, B. A., Estes, S. A., & Arlian, L. G. (1980). *Trixacarus caviae* infestation of guinea pigs. *Journal of the American Veterinary Medical Association, 177*, 903–908.

Lackman, D. B. (1963). A review of information on rickettsialpox in the United States. *Clinical Pediatrics, 2*, 296–301.

Lange, R. E., Sandoval, A. V., & Meleney, W. P. (1980). Psoroptic scabies in bighorn sheep (*Ovis canadensis mexicana*) in New Mexico. *Journal of Wildlife Diseases, 16*, 77–82.

Lawrence, R. F. (1952). A new parasitic mite from the nasal cavities of the South American toad *Bufo regularis* Reuss. *Proceedings of the Zoological Society of London, 121*, 747–752.

Lee, K. J., Lang, C. M., Hughes, H. C., & Hartshorn, R. D. (1981). Psorergatic mange (Acari: Psorergatidae) of the stumptail macaque (*Macaca arctoides*). *Laboratory Animal Science, 31*, 77–79.

Lefer, L. G., & Rosier, R. P. (1988). Presence of a mite in the female genital tract: Some comments. Internat. *Journal of Acarology, 14*, 91–92. [Pyroglyphidae, Dermatophagoides]

Le Fichoux, Rack, Y. G., Motte, P., Dellamonica, P., & Marty, P. (1980). Dermatite prurigineuse due a *Pyemotes zwoelferi* Krczal, 1963, a propos de plusieurs cas dans les alpes-maritimes. *Acta Tropica, 37*, 83–89.

Li, C., & Li, L. (1990). Human pulmonary acariasis in Anhui Province: An epidemiological survey. *Chinese Journal of Parasitology & Parasite Diseases, 8*, 41–44. [in *Chinese*]

Lowenstine, L. J., Carpenter, J. L., & OConnor, B. M. (1979). Trombiculosis in a cat. *Journal of the American Veterinary Medical Association, 175*, 289–292.

Ma, E. P., & Wang, R. S. (1992). Tarsonemid mites. In X. B. Chen & E. P. Ma (Eds.), *Researches of Acarology in China* (pp. 34–37). Chongqing: Chongqing Publishing House.

Malhotra, S. K., & Capoor, V. N. (1982). A new species of *Avitellina* Gough (1911) from Garhwal Hills with a revised key to species of subgenus *Avitellina* Raina (1975). *Proceedings of the Indian Academy of Parasitology, 3*, 12–16.

Martinez Maranon, R., & Hoffman, A. (1976). Tres casos de infestacion del intestino humano por acaros en el sur de Veracruz. *Revista de Investigacion en Salud Publica, 36*, 187–201.

Matthes, H. F. (1994). Investigations of pathogenesis of cattle demodicosis: Sites of predilection, habitat, and dynamics of demodectic nodules. *Veterinary Parasitology, 53*, 283–291.

Mazet, J. A. K., Boyce, W. M., Mellies, J., Gardner, I. A., Clark, R. K., & Jessup, D. A. (1992). Exposure to *Psoroptes* sp. Mites is common among bighorn sheep (*Ovis canadensis*) populations in California. *Journal of Wildlife Diseases, 28*, 542–547.

McGarry, J. W. (1993). Identification of *Cheyletiella* eggs in dog feces. *The Veterinary Record, 132,* 359–360.

Mellanby, K. (1972). *Scabies* (2nd ed.). Hampton, UK: E. W. Classey.

Meng, Y. C., Zhuge, H. X., Lan, M. Y., & Zhon, H. F. (1991). *Experimental study on transmission of hemorrhagic fever with renal syndrome virus by mites.* Ornithonyssus bacoti (Hirst), Proc. VIII International Congress of Acarology, Czechoslovakia: Ceske Budejovice: 1990. Vol. II, pp. 35–39.

Miller, W. H. (1984). Diseases of domestic animals. In W. B. Nutting (Ed.), *Mammalian Diseases and Arachnids* (Vol. 2. pp. 115–126). Boca Raton, FL: CRC Press. Chapter 6.

Mills, O. H., Jr., & Kligman, A. M. (1983). The follicular biopsy. *Dermatologica, 167,* 57–63.

Mironov, S. V., Bochkov, A. V., & Fain, A. (2005). Phylogeny and evolution of parasitism in feather mites of the families Epidermoptidae and Dermationidae (Acari: Analgoidea). *Zoologischer Anezeiger, 243,* 155–179.

Möller, A. P. (1990). Effects of parasitism by a haematophagous mite on reproduction in the barn swallow. *Ecology, 71,* 2345–2357. [*Ornithonyssus bursa*]

Morgan, K. L. (1992). Parasitic otitis in sheep associated with *Psoroptes* infestation: A clinical and epidemiological study. *The Veterinary Record, 130,* 530–532.

Mulla, M., & Medina, M. S. (Eds.). (1980). *Domestic Acari of Colombia: Bionomics, Ecology, and Distribution of Allergenic Mites, Their Role in Allergenic Diseases. (bilingual English/ Spanish).* Bogota: Colciencias.

Nadchatram, M. (1970). A review of intranasal chiggers with descriptions of twelve species from east New Guinea (Acarina: Trombiculidae). *Journal of Medical Entomology, 7,* 1–29.

Naltsas, S., Hodge, S. J., Gataky, G. J., Jr., & Owen, L. G. (1980). Eczematous dermatitis caused by *Dermanyssus americanus. Cutis; Cutaneous Medicine for the Practitioner, 25,* 429–431.

Narsapur, V. S. (1988). Pathogenesis and biology of anoplocephaline cestodes of domestic animals. *Annales de Recherches Veterinaires. Annals of Veterinary research, 19,* 1–17.

Nutting, W. B. (1976a). Hair follicle mites (Acari: Demodicidae) of man. *International Journal of Dermatology, 15,* 79–98.

Nutting, W. B. (1976b). Hair follicle mites (*Demodex* spp.) of medical and veterinary concern. *Cornell Veterinarian, 66,* 214–231.

Nutting, W. B. (1976c). Pathogenesis associated with hair follicle mites (Acari: Demodicidae). *Acarologia, 17,* 493–506.

Nutting, W. B. (Ed.). (1984). *Mammalian Diseases and Arachnids. Vol. I. Pathogen Biology and Clinical Management. Vol. II. Medico-Veterinary Laboratory, and Wildlife Diseases, and Control.* Boca Raton, FL: CRC Press, Inc.

Nutting, W. B., Firda, K. E., & Desch, C. E., Jr. (1989). Topology and histopathology of hair follicle mites (Demodicidae) of man. In G. P. ChannaBassavana & C. A. Viraktamath (Eds.), *Progress in Acarology* (Vol. 1, pp. 113–121). New Delhi: Oxford & IBH Publish. Co.

Oberem, P. T., & Malan, F. S. (1984). A new cause of cattle mange in South Africa: *Psorergates bos* Johnston. *Journal of the South African Veterinary Association, 55,* 121–122.

O'Brien, D. J., Gray, J. S., & O'Reilly, P. F. (1994). Survival and retention of infectivity of the mite *Psoroptes ovis* off the host. *Veterinary Research Communications, 18,* 27–36.

O'Connor, B. M. (1984). Phylogenetic relationships among higher taxa in the Acariformes, with particular reference to the Astigmata. In D. A. Griffiths & C. E. Brown (Eds.), *Acarology VI* (Vol. 1, pp. 19–27). Chichester: Ellis Horwood.

Oehlschlaegel, G., Bayer, F., Disko, R., Fechter, H., & Mahunka, S. (1983). *Tarsonemus hominis* in Hautbindegewebe. *Der Hautarzt; Zeitschrift fur Dermatologie, Venerologie, und verwandte Gebiete, 34,* 632–634.

Onderscheka, K., Kutzer, E., & Richter, H. E. (1968). Die Raeude der Gemse und ihre Bekampfung. II. Zusammenhaenge zischen Ernaehrung und Raeude. *Z Jagdwiss, 14,* 12.

Orkin, M., Maibach, H. I., Parish, L. C., & Schwartzman, R. M. (Eds.). (1977). *Scabies and Pediculosis.* Baltimore: J. B. Lippincott.

Otto, Q. T., & Jordaan, L. C. (1992). An orf-like condition caused by trombiculid mites on sheep in South Africa. *The Onderstepoort Journal of Veterinary Research, 59,* 335–336.

Ottoboni, F., di Loreto, V., Cantoni, A., Lozzia, G. C., Rota, P., Melej, R., et al. (1989). Investigations into allergic diseases among raw ham workers in Langhirano and San Daniele. La difesa antiparassitaria nelle industrie alimentari e la protezion degli alimenti. Atti del 4o simposio 235–241.

Parodi, A. S., Rugiero, H. R., Greenway, D. J., Mettler Martinez, N. A., Boxaca, M., & De la Barerra, J. M. (1959). Aislamiento del virus Junin (F.H.E.) De los acaros de la zona epidemica (Echinolaelaps echidninus, Berlese). *Prensa Medica Argentina, 46,* 2242–2244.

Pass, D. A., & Sue, L. J. (1983). A trombiculid mite infestation of canaries. *Australian Veterinary Journal, 60,* 218–219.

Pence, D. B. (1973). The nasal mites of birds from Louisiana. IX. Synopsis. *Journal of Parasitology, 59,* 881–892.

Pence, D. B. (1975). *Keys species and host list, and bibliography for nasal mites of North American birds (Acarina: Rhinohyssinae, Turbinoptinae, Speleognathinae, and Cytoditidae).* Texas Tech Univ., Mus. Spec. Pub. No. 8.

Pence, D. B. (1984). Diseases of laboratory animals. In W. B. Nutting (Ed.), *Mammalian Diseases and Arachnids* (Vol. 2, pp. 129–187). Boca Raton, FL: CRC Press.

Pence, D. B., Matthews, F. D., & Windberg, L. A. (1982). Notoedric mange in the bobcat, *Felis rufus,* from south Texas. *Journal of Wildlife Diseases, 18,* 47–50.

Pence, D. B., Spalding, M. G., Bergan, J. F., Cole, R. A., Newman, S., & Gray, P. N. (1997). New records of subcutaneous mites (Acari: Hypoderatidae) in birds, with examples of potential host colonization events. *Journal of Medical Entomology, 34,* 411–416.

Philip, C. B. (1948). Tsutsugamushi disease (scrub typhus) in the World War II. *The Journal of Parasitology, 34,* 169–191.

Philip, C. B., & Kohls, G. M. (1945). Studies on Tsutsugamushi disease (scrub typhus, mite-borne typhus) in New Guinea and adjacent islands: Tsutsugamushi disease with high endemicity on a small South Sea island. *American Journal of Hygiene, 42,* 195–202.

Philip, C. B., & Hughes, L. E. (1948). The tropical rat mite, *Liponyssus bacoti,* as an experimental vector of rickettsialpox. *American Journal of Tropical Medicine and Hygiene, 28,* 697–705.

Phillips, C. J., Jones, J. K., & Radovsky, F. J. (1969). Macronyssid mites in oral mucosa of long-nosed bats: Occurrence and associated pathology. *Science, 165,* 1368–1369. [*Radfordiella*]

Pitariu, T., Dinulescu, N., Panaitescu, D., & Silard, R. (1978). Cholangiocholecystitis, an acute attack with acarids in B bile. *Revista de Igiena, Bacteriologie, Virusologie, Parazitologie, Epidemiologie, Pneumoftiziologie, 23,* 189–192. [in *Romanian*]

Pitariu, T. N., Popescu, I. G., & Banescu, O. (1979). Acarids of pathological significance in urine. *Revista de Igiena, Bacteriologie, Virusologie, Parazitologie, Epidemiologie, Pneumoftiziologie, 24,* 55–59. [in *Romanian*]

Pratt, H. D., & Stojanovich, C. J. (1969). Acarina: Illustrated key to some common adult female mites and adult ticks. In: *CDC pictorial keys: Arthropods, reptiles, birds and mammals of public health significance* (pp. 26–44). U.S. Dept. Health and Human Services, Public Health Service, Centers for Disease Control and Prevention.

Prisich, I. I., Dobarskaia, L. I., & Zosimova, A. G. (1986). Acariasis in children with chronic digestive system diseases. *Meditsinskaia Parazitologiia i Parazitarnye Bolezni,* 50–51. [in *Russian*]

Prosl, H., Rabitsch, A., & Brabenetz, J. (1985). Zur Bedeutung der Herbstgrasmilbe-*Neotrombicula autumnalis* (Shaw 1790)-in der Veterinarmedizin: Nervale Symptome bei Hunden nach massiver Infestation. *Tierarztliche Praxis, 13,* 57–64.

Quoix, E., Mao, J., Hoyet, C., & Pauli, G. (1993). Prediction of mite allergen levels by guanine measurements in house-dust samples. *Allergy (Copenhagen), 48,* 306–309.

Rafferty, D. E., & Gray, J. S. (1987). The feeding behaviour of *Psoroptes* spp. mites on rabbits and sheep. *The Journal of Parasitology, 73,* 901–906.

Raina, M. K. (1975). A monograph on the genus *Avitellina* Gough, 1911 (Avitellinidae: Cestoda). *Zool. Jahrb Abteil System Oekol Geogr Tiere, 102,* 508–552.

Raulston, G. L. (1972). Psorergatic mites in patas monkeys. *Laboratory Animal Science, 22,* 107.

Rollor, E. A., Nettles, V. F., Davidson, W. R., & Gerrish, R. R. (1978). Otitis media caused by *Psoroptes cuniculi* in white-tailed deer. *Journal of the American Veterinary Medical Association, 173,* 1242–1243.

Rufli, T., & Mumculoglu, Y. (1981). The hair follicle mites *Demodex folliculorum* and *Demodex brevis*: Biology and medical importance. *A Review Dermatologica, 162,* 1–11.

Samšiňák, K., Pali ka, P., Zítek, K., Mališ, L., & Vobrázková, E. (1976). Are the mites of the genus *Tarsonemus* really parasites of man. *Folia Parasitologia, 23,* 91–93.

Samuel, W. M., Welch, D. A., & Smith, B. L. (1991). Ectoparasites from elk (*Cervus elaphus nelsoni*) from Wyoming. *Journal of Wildlife Diseases, 27*, 446–451.

Schultz, H. (1975). Human infestation by *Ophionyssus natricis* snake mite. *British Journal of Dermatology, 93*, 695–697.

Schuster, R., & Murphy, P. W. (Eds.). (1991). *The Acari: Reproduction, Development and Life-History Strategies.* London: Chapman & Hall.

Scott, D. W. (1979). Canine demodicosis. *Vet Clinics N Amer, 9*, 79–92.

Seier, J. V. (1985). Psorergatic acariasis in vervet monkeys. *Laboratory Animals, 19*, 236–239.

Sengbusch, H. G. (1977). Review of oribatid mite-anoplocephalan tapeworm relationships (Acari; Oribatei: Cestoda; Anoplocephalidae). In D. L. Dindal (Ed.), *Biology of Oribatid Mites* (pp. 87–102). Syracuse, NY: State University of New York, College of Environmental Science and Forestry.

Shaddock, J. W., & Pakes, S. P. (1978). Protozoal and metazoal diseases. In K. Benirschke, F. M. Garner, & T. C. Jones (Eds.), *Pathology of Laboratory Animals* (Vol. 2, p. 1587). Berlin: Springer-Verlag.

Sheldon, W. (1966). Psorergatic mange in the sooty mangabey (*Cercocebus torquates atys*) monkey. *Laboratory Animal Care, 16*, 276.

Shelley, W. B., Shelley, E. D., & Welbourn, W. C. (1985). *Polypodium* fern wreaths (Hagnaya). A new source of occupational mite dermatitis. *Journal of the American Medial Association, 253*, 3137–3138.

Skoda-Smith, S., Mullen, G. R., Oi, F., & Atkinson, T. P. (1996). Angioedema following dust mite exposure presenting as suspected food allergy. *American Academy of Allergy, Asthma and Immunology, 97*, 228. [*Dermatophagoides farinae*]

Smiley, R. L. (1970). A review of the family Cheyletiellidae. *Annals of the Entomological Society of America, 63*, 1056.

Smiley, R. L., & O'Connor, B. M. (1980). Mange in *Macaca arctoides* (Primates: Cercopithecidae) caused by *Cosarcoptes scanloni* (Acari: Sarcoptidae) with possible human involvement and descriptions of the adult male and immature stages. *International Journal of Acarology, 6*, 283–290.

Smith, K. E., Quist, C. F., & Crum, J. M. (1997). Clinical illness in a wild turkey with *Laminosioptes cysticola* infestation of the viscera and peripheral nerves. *Avian Diseases, 41*, 484–489.

Smith, T. G. (1996). The genus *Hepatozoon* (Apicomplexa: Adeleina). *The Journal of Parasitology, 82*, 565–585.

Snyder, D. E., Hamir, A. N., Hanlon, C. A., & Rupprecht, C. E. (1991). Notoedric acariasis in the porcupine (*Erethizon dorsatum*). *Journal of Wildlife Diseases, 27*, 723–726.

Soulsby, E. J. L. (1965). *Textbook of Veterinary Clinical Parasitology.* Philadelphia: F. A. Davis.

Soulsby, E. J. L. (1982). *Helminths, Arthropods and Protozoa of Domesticated Animals* (7th ed.). Philadelphia: Lea & Febiger.

Southcott, R. V. (1984). Diseases and arachnids in the tropics. In W. B. Nutting (Ed.), *Mammalian Diseases and Arachnids* (Vol. 2, pp. 15–56). Boca Raton, FL: CRC Press. Chapter 2.

Spicer, G. S. (1987). Prevalence and host-parasite list of some nasal mites from birds (Acarina: Rhinonyssidae, Speleognathidae). *The Journal of Parasitology, 73*, 259–264.

Staley, E. C., Staley, E. E., & Behr, M. J. (1994). Use of permethrin as a miticide in the African hedgehog (*Atelerix albiventris*). *Veterinary and Human Toxicology, 36*(2), 138.

Stone, W. B., Parks, E., Weber, B. L., & Parks, F. J. (1972). Experimental transfer of sarcoptic mange from red foxes and wild canids to captive wildlife and domestic animals. *NY Fish Game J, 19*, 1–11.

Strandtmann, R. W., & Wharton, G. W. (1958). *A Manual of Mesostigmatid Mites Parasitic on Vertebrates.* Contribution No. 4, Institute of Acarology, College Park: University of Maryland.

Stromberg, P. C., Fisher, W. F., Guillot, F. S., Pruett, J. H., Price, R. E., & Green, R. A. (1986). Systemic pathologic responses in experimental *Psoroptes ovis* infestation of Hereford calves. *American Journal of Veterinary Research, 47*, 1326–1331.

Stromberg, P. C., & Guillot, F. S. (1987a). Hematology in the regressive phase of bovine psoroptic scabies. *Veterinary Pathology, 24*, 371–377.

Stromberg, P. C., & Guillot, F. S. (1987b). Bone marrow response in cattle with chronic dermatitis caused by *Psoroptes ovis. Veterinary Pathology, 24*, 365–370.

Stunkard, H. W. (1940). The morphology and life history of the cestode *Bertiella studeri. American Journal of Tropical Medicine and Hygiene, 20*, 305–333.

Sweatman, G. K. (1957). Life history, non-specificity, and review of the genus *Chorioptes*, a parasitic mite of herbivores. *Canadian Journal of Zoology, 35*, 641.

Sweatman, G. K. (1958a). Biology of *Otodectes cynotis*, the ear canker mite of carnivores. *Canadian Journal of Zoology, 36*, 849.

Sweatman, G. K. (1958b). On the life history and validity of the species in *Psoroptes*, a genus of mange mites. *Canadian Journal of Zoology, 36*, 905–929.

Sweatman, G. K. (1958c). Redescription of *Chorioptes texanus*, a parasitic mite from the ears of reindeer in the Canadian arctic. *Canadian Journal of Zoology, 36*, 525.

Sweatman, G. K. (1971). Mites and pentastomes. In J. W. Davis & R. C. Anderson (Eds.), *Parasitic Diseases of Wild Mammals* (pp. 3–64). Ames: Iowa State Univ. Press. Chapter 1.

Sweatman, G. K. (1984). Diseases of wildlife. In W. B. Nutting (Ed.), *Mammalian Diseases and Arachnids* (Vol. 2, pp. 189–232). Boca Raton, FL: CRC Press. Chapter 8.

Taboada, M. de F. (1954). Pulmonary acariasis in Spain. An illustrative case report. *British Medical Journal*, 4859, 437–438.

Tamura, A., Ohahsi, N., Urakami, H., & Miyamura, S. (1995). Classification of *Rickettsia tsutsugamushi* in a new genus, *Orientia* gen. nov., as *Orientia tsutsugamushi* comb. nov. *International Journal of Systematic Bacteriology, 45*, 589–591.

Tandon, N., Chatterjee, H., Gupta, S. K., & Hati, A. K. (1988). Some observations on house dust mites in relation to naso-bronchial asthma in Calcutta, India. In G. P. ChannaBasavanna & C. A. Viraktamath (Eds.), *Progress in Acarology* (Vol. 1, pp. 163–168). Leiden, Netherlands: E. J. Brill.

Tenorio, J. M. (1974). A new species of *Lynxacarus* (Acarina: Astigmata: Listrophoridae) from *Felis catus* in the Hawaiian Islands. *Journal of Medical Entomology, 11*, 599–604.

Theiler, M., & Downes, W. G. (1973). *The Arthropod-borne Viruses of Vertebrates.* New Haven: Yale University Press.

Tomalski, M. D., & Miller, L. K. (1991). Insect paralysis by baculovirus-mediated expression of a mite neurotoxin gene. *Nature, 352*, 82–85. [*Pyemotes tritici*]

Traub, R., Hertig, M., Lawrence, W. H., & Harris, T. T. (1954). Potential vectors and reservoirs of hemorrhagic fever in Korea. *American Journal of Hygiene, 59*, 291.

Traub, R., & Wisseman, C. L., Jr. (1974). The ecology of chigger-borne rickettsiosis (scrub typhus). *Journal of Medical Entomology, 11*, 237–303.

Traver, J. R. (1951). Unusual scalp dermatitis in humans caused by the mite Dermatophagoides (Acarina: Epidermoptidae). *Proceedings of the Entomological Society of Washington, 53*, 1.

van Bronswijk, J. E. M. H., & Sinha, R. N. (1971). Pyroglyphid mites (Acari) and house dust allergy: a review. *The Journal of Allergy, 47*, 31–52.

van Bronswijk, J. E., & De Kreek, E. J. (1976). *Cheyletiella* (Acari: Cheyletiellidae) of dog, cat and domesticated rabbit, a review. *Journal of Medical Entomology, 13*, 315–327.

van der Hammen, L. (1972). A revised classification of the mites (Arachnidea, Acarida) with diagnoses, a key, and notes on phylogeny. *Zool Meded, 47*, 273–292.

Wharton, G. W., Jr. (1976). House dust mites. *Journal of Medical Entomology, 12*, 577–621.

Wharton, G. W., Jr., & Fuller, H. S. (1952). *A Manual of the Chiggers: The Biology, Classification, Distribution, and Importance to Man of the Larvae of the Family Trombiculidae (Acarina).* Washington, DC: Memoirs of the Entomological Society of Washington, No. 4.

Whitaker, J. O., Jr. (1982). *Ectoparasites of mammals of Indiana.* Indiana *Academic Science Monograph.*

Wiedman, F. D. (1916). *Cytoleichus penrosei*, a new arachnoid parasite found in the diseased lungs of a prairie dog, *Cynomys ludovicianus. The Journal of Parasitology, 3*, 82–89. [Pneumocoptidae]

Williams, J. F., & Williams, C. S. (1978). Psoroptic ear mites in dairy goats. *Journal of the American Veterinary Medical Association, 173*, 1582–1583.

Williams, J. F., & Williams, C. S. (1982). Demodicosis in dairy goats. *Journal of the American Veterinary Medical Association, 180*, 168–169.

Wilson-Hanson, S., & Prescott, C. W. (1985). Trombidiosis in cats. *Australian Veterinary Journal, 62*, 202–203.

Wisniewski, H. M., Sigurdarson, S., Rubenstein, R., Kascsak, R. J., & Carp, R. I. (1996). Mites as vectors of scrapie. *Lancet, 347*, 1114.

Woolley, T. A. (1988). *Acarology: Mites and Human Welfare.* New York: John Wiley & Sons.

Wraith, D. G., Cunnington, A. M., & Seymour, W. M. (1979). The role and allergenic importance of storage mites in house dust and other environments. *Clinical Allergy, 9,* 545–561.

Wright, F. C., Guillot, F. S., & Meleney, W. P. (1981). Transmission of psoroptic mites from bighorn sheep (*Ovis canadensis mexicana*) to domestic sheep, cattle and rabbits. *Journal of Wildlife Diseases, 17,* 381–386.

Wright, F. C., & Glaze, R. L. (1988). Blackbuck antelope (*Antilope cervicapra*), a new host for *Psoroptes cuniculi* (Acari: Psoroptidae). *Journal of Wildlife Diseases, 24,* 168–169.

Wright, F. C., Riner, J. C., & Fisher, W. F. (1984). Comparison of lengths of outer opisthosomal setae of male psoroptic mites collected from various hosts. *The Journal of Parasitology, 70,* 141–143.

Wright, F. C., Riner, J. C., & Guillot, F. S. (1983). Cross-mating studies with *Psoroptes ovis* (Hering) and *Psoroptes cuniculi* Delafond (Acarina: Psoroptidae). *The Journal of Parasitology, 69,* 696–700.

Yashikawa, M., Hanaoka, Y., Yamada, Y., et al. (1983). Experimental proof of itching papules caused by *Cheyletus malaccensis* Oudemans. *Annual Report of Tokyo Metropolitan Research Laboratory of Public Health, 34,* 264–276.

Yeatts, J. W. G. (1994). Rabbit mite infestation. *The Veterinary Record, 134,* 359–360. [*Psoroptes cuniculi*]

Yeruham, I., Rosen, S., & Hadani, A. (1986). Sheep demodicosis (*Demodex ovis* Railliet, 1895) in Israel. *Revue d Elevage et de Medecine Veterinaire des Pays Tropicaux, 39,* 363–365.

Yunker, C. E. (1973). Mites. In R. J. Flynn (Ed.), *Parasites of Laboratory Animals* (pp. 425–492). Ames: Iowa State Univ. Press. Chapter 15.

Yunker, C. E., Binninger, C. E., Keirans, J. E., Beecham, J., & Schlegel, M. (1980). Clinical mange of the black bear, *Ursus americanus,* associated with *Ursicoptes americanus* (Acari: Audycoptidae). *Journal of Wildlife Diseases, 16,* 347–356.

Zahler, M., Hendrikx, W. M. L., Essig, A., Rinder, H., & Gothe, R. (2000). Species of the genus *Psoroptes* (Acari: Psoroptidae): A taxonomic consideration. *Experimental & Applied Acarology, 24,* 213–225.

Zenoble, R. D., & Greve, J. H. (1980). Sarcoptid mite infestation in a colony of guinea pigs. *Journal of the American Veterinary Medical Association, 177,* 903–908.

Zuchowska, E. (1991). Swierzb ssakow w ogrodach zoologiczynch [Scabies in zoo mammals]. *Wiadomosci Parazytologiczne, 37,* 123–125.

Chapter

26

Ticks (Ixodida)

William L. Nicholson ▪ Daniel E. Sonenshine ▪
Robert S. Lane ▪ Gerrit Uilenberg

Ticks are notorious as vectors of human and other animal disease agents. They transmit a greater variety of infectious organisms than any other group of blood-sucking arthropods. Worldwide, they are the most important vectors in the veterinary field and second only to mosquitoes in terms of their public health importance. Ticks transmit numerous protozoan, viral, bacterial (including rickettsial), and fungal pathogens. In humans, tens of thousands of cases of tick-borne diseases caused by these agents occur annually. In addition, the bites of ticks can cause toxic reactions, allergic responses, and even fatal paralysis, and the wounds that they produce can create sites for secondary infections and diminish the value of livestock by damaging their hides. Tick-borne diseases such as babesiosis, anaplasmosis, theileriosis, heartwater, and many others have made it difficult or impossible to raise livestock in many tropical and subtropical regions of the world. Although difficult to measure precisely, the global economic impact of ticks and tick-borne diseases is believed to be in the many billions of dollars (Jongejan and Uilenberg, 2004). The study of ticks has contributed greatly to our ability to understand and control the spread of infectious diseases.

This chapter reviews the remarkable adaptations and behavior of ticks that facilitate their success as blood-feeding parasites and the diverse tick-host pathogen interactions that contribute to their role as vectors of human and other animal-disease agents. For more detailed information about the biology of ticks, refer to Sonenshine (1991, 1993); for information on human tick-borne diseases, refer to Goodman et al. (2005); for more information on veterinary tick-borne diseases, consult Uilenberg (1995) and Jongejan and Uilenberg (2004).

TAXONOMY

Ticks constitute the Suborder Ixodida, of the acarine Order Parasitiformes, and are exclusively parasitic. The Ixodida contain three families, the **Ixodidae**, **Argasidae**, and **Nuttalliellidae** (Table 26.1). The Ixodidae are subdivided into the Prostriata, represented by the single genus *Ixodes* with 241 species, and the Metastriata with 442 species, comprising the remaining 11 genera (Horak et al., 2002). There are 683 recognized species in this family, representing about 80% of all tick species that have been described (Table 26.1). The Argasidae contain four genera and about 183 species (Horak et al., 2002). The **Nuttalliellidae** is a monospecific family, represented by only one species, *Nuttalliella namaqua*.

Hard ticks are grouped into two other taxa. Ticks of the genus *Ixodes* possess an anal groove that extends anteriorly and encloses the anus and are classified as **Prostriata** (also subfamily Ixodinae). In contrast, the ticks of other ixodid genera possess an anal groove located posterior to the anus and are grouped as the **Metastriata** (Table 26.1).

For additional information on taxonomy of the Ixodida, see Filippova (1966, 1967), Roberts (1970), Keirans (1992), Klompen and Oliver (1993), Durden and Keirans (1996), Klompen et al. (2000), Murrell et al. (2000), Walker et al. (2000), Beati and Keirans (2001), Barker and Murrell (2002), and Horak et al. (2002).

The following subsections provide descriptions of some of the more important tick genera and species of particular importance as vectors of human or domestic animal pathogenic agents.

Family Ixodidae (Hard Ticks)

Genus *Ixodes* This is the largest tick genus, with an estimated 241 species. *Ixodes* species are known as the Prostriata, characterized by a distinctive anal groove that encircles the anus anteriorly. They also lack eyes. Males have sclerotized ventral plates which are absent in males of other genera. The genus is worldwide in distribution, including Antarctica. Four species are particularly important as vectors of microbial agents to humans: the blacklegged tick (*Ixodes scapularis*) in

Table 26.1 Families and Genera of Ticks

Family	Subfamily (subgroup)	Genera
Ixodidae	Ixodinae (Prostriata)	*Ixodes*
	Amblyomminae (Metastriata)	*Amblyomma*
	Bothriocrotoninae (Metastriata)	*Bothriocroton*
	Haemaphysalinae (Metastriata)	*Haemaphysalis*
	Hyalomminae (Metastriata)	*Hyalomma*
	Rhipicephalinae (Metastriata)	*Anomalohimalaya, Cosmioma, Dermacentor, Margaropus, Nosoma, Rhipicentor, Rhipicephalus*
Argasidae	Argasinae	*Argas*
	Ornithodorinae	*Ornithodoros, Carios*
	Otobinae	*Otobius*
Nuttalliellidae		*Nuttalliella*

After Horak et al., 2002.

eastern North America; the castor bean tick, or sheep tick (*I. ricinus*) in Europe and western Asia; the taiga tick (*I. persulcatus*) in northeastern Europe and northern Asia; and the western blacklegged tick (*I. pacificus*) in the far western United States. Other, nonhuman-biting *Ixodes* spp. serve as enzootic (maintenance) vectors of important tick-borne disease agents; for example, *I. dentatus* and *I. spinipalpis* (*I. neotomae*) in North America, and *I. ovatus* in northern Asia and Japan.

Genus *Dermacentor* This is one of the more important genera of metastriate ticks, with 33 species. The basis capituli appears rectangular when viewed dorsally. A pair of medially directed spurs occurs on the first pair of coxae. The palps are short and thick. The scutum is almost always ornamented. Most *Dermacentor* spp. are three-host ticks that feed on diverse mammals. Adults attack medium-sized or large mammals, whereas the immatures feed on small mammals and lagomorphs. *Dermacentor* species are found mostly in Europe, Asia, Africa, and North and Central America. In North America, important species are the American dog tick (*D. variabilis*), the Rocky Mountain wood tick (*D. andersoni*), the Pacific Coast tick (*D. occidentalis*), and the winter tick (*D. albipictus*). In Central and South America and some Caribbean islands, an important species is *D. nitens* (designated previously as *Anocentor nitens*). In Europe, two important species are *D. reticulatus* and *D. marginatus.*

Genus *Rhipicephalus* Ticks of the genus *Rhipicephalus* are easily recognized by the hexagonal shape of the basis capituli when viewed dorsally. Important species include the brown dog tick (*R. sanguineus*) and the brown ear tick (*R. appendiculatus*). *Rhipicephalus* ticks mainly parasitize mammals, and rarely are found as larvae or nymphs on birds and reptiles. Representative species are found throughout the world. *Rhipicephalus sanguineus* is cosmopolitan in distribution. Among the more important are the five species of the subgenus *Boophilus*, formerly considered as a separate genus (Murrel and Barker, 2003). Subgenus *Boophilus* ticks are small and lack ornamentation. The basis capituli is short and broad, with rounded lateral margins. These ticks are one-host parasites of ungulates. Subgenus *Boophilus* ticks are found in most tropical and subtropical regions of the world. Important species include the cattle tick (*Rhipicephalus [B] annulatus*) and the **tropical fever tick** or the southern cattle tick (*Rhipicephalus [B.] microplus*). The genus contains 80 described species.

Genus *Haemaphysalis* This is the second largest tick genus, which is recognized by the pronounced lateral projection of palpal segment 2 in most species (including all three North American species), which extends well beyond the basis capituli. These small ticks lack eyes. *Haemaphysalis* species parasitize birds and mammals in most regions of the world. An important species is the rabbit tick *H. leporispalustris*, widespread throughout much of North America. Several species in the Old World are important pests and/or vectors of human or other animal disease agents; e.g., *H. longicornis* in Asia and the Pacific region (including Australia), *H. punctata* in Europe, and *H. spinigera* in India. The genus contains about 168 species.

Genus *Hyalomma* This is a relatively small genus of 32 medium-sized to large Old World species. They are characterized by their elongated palps, which are at least twice as long as wide. The distinct eyes are located in sockets adjacent to the postero-lateral edges of the scutum. *Hyalomma* ticks are unornamented. Most species live in xeric environments where they parasitize small and medium-sized wild mammals and livestock. Some species parasitize birds or reptiles. The distribution of *Hyalomma* species is limited to the Old World, primarily in arid or semi-arid habitats. An important subspecies is *H. marginatum marginatum*, a vector of Crimean-Congo Hemorrhagic Fever (CCHF) virus. Other important species are *H. truncatum* in Africa, *H. asiaticum* in central Asia, and *H. detritum* in Asia and the Mediterranean basin. *Hyalomma detritum* is of major veterinary importance as a vector of the agent of bovine tropical theileriosis.

Genus *Amblyomma* Adults of most species in this genus are medium or large in size. The palps are long with segment 2 at least twice as long as segment 3. The scutum usually is ornamented with varying-colored iridescent patterns. Eyes are present in most species (absent in species formerly assigned to *Aponomma*), and in most species are not situated in sockets. Virtually all terrestrial vertebrates serve as hosts, although amphibians rarely are attacked. The distribution is worldwide, primarily in humid tropical or subtropical regions. Examples of important species include the Gulf Coast tick (*A. maculatum*) and lone star tick (*A. americanum*) (Figure 26.6) in

North America; the tropical bont tick (*A. variegatum*) in Africa and on some islands in the Caribbean Sea; and the bont tick (*A. hebraeum*) in Africa. The genus contains about 129 species (including 20 species formerly assigned to the genus *Aponomma*, now in part a synonym of *Amblyomma*).

The remaining genera of the Ixodidae contain relatively few species, none of which is known to be important in pathogen transmission. These include *Anomalohimalaya*, *Bothriocroton*, *Cosmiomma*, *Nosoma*, *Margaropus*, and *Rhipicentor*. A genus previously designated as *Anocentor* was invalidated and its species transferred to the genus *Dermacentor*. Similarly, the genus *Aponomma* is no longer considered valid and its species were transferred to the genus *Amblyomma* and the new genus *Bothriocroton*.

Family Argasidae (Soft Ticks)

Genus *Argas* *Argas* ticks have a flattened body margin, a lateral sutural line, and a leathery, folded cuticle. The many small integumental folds usually have a button-like appearance, each with a pit on its top. Most species parasitize bats or birds. The genus is worldwide in distribution, mostly in xeric environments or dry caves in otherwise humid environments. Examples of important species are the fowl tick (*A. persicus*) and the pigeon tick (*A. reflexus*). About 57 species have been described.

Genus *Carios* These ticks are similar to those of the genus *Argas* but differ in the structure of their Haller's organ on the tarsus of the foreleg. In most *Carios* spp. the setiform seta is replaced by a second serrate seta. The roof of Haller's organ in both subgenera is solid and lacks perforations. The host range includes mammals (mainly bats) and birds. Approximately 10 species formerly classified in the genus *Antricola*, now a synonym of *Carios*, possess a tuberculated cuticle. The females have a distinctive, scoop-like hypostome; the hypostome is vestigial in the males. These species are parasites of New World bats. Thus far, none has been implicated in the transmission of microbial disease agents. A species formerly classified in the genus *Nothoaspis*, now a synonym of *Carios*, is similar to *Antricola*, but the anterior dorsal surface bears a smooth shield-like structure, the pseudoscutum. The current classification of the genus *Carios* is based on Horak et al. (2002) and Klompen and Oliver (1993). The genus contains about 87 species.

Genus *Ornithodoros* Nymphs and adults have a leathery cuticle with innumerable tiny wrinkles (mammillae) and a rounded body margin; they lack a lateral, sutural line. Mammillae are smaller and more numerous than those found in *Argas*. The host range is diverse and includes reptiles, birds, and mammals. The genus is worldwide in distribution. Examples of important species include the African tampan (*O. moubata*) and the cave tick (*O. tholozani*). In North America, several species (e.g., *O. hermsi*) are important as vectors of relapsing fever spirochetes to humans and other animals. The genus contains about 38 species.

Genus *Otobius* The integument of the nymphs is spinose, whereas that of the adults is granulated. There are just two nymphal instars. The adults do not feed, and the hypostome is vestigial. *Otobius* ticks are found in North America, Africa, and Asia, having been inadvertently introduced onto the latter two continents. The genus contains two species, the spinose ear tick (*O. megnini*) and *O. lagophilus*.

Family Nuttalliellidae

The only known species in this family, *Nuttalliella namaqua*, occurs in eastern and southern Africa. It shares features with both the Argasidae and the Ixodidae but also has several unique morphological traits. This tick has ball-and-socket joints that articulate the leg segments, a small dorsal pseudoscutum, and a highly wrinkled cuticle with numerous pits and elevated rosettes. It has been collected from the nests of rock hyraxes and swallows in South Africa and Tanzania (Keirans et al., 1976). It is rare and of no known medical or veterinary importance.

MORPHOLOGY

External Anatomy

The major external regions of ticks are the capitulum (gnathosoma), idiosoma, and the legs (Figures 26.1, 26.2, and 26.3). The capitulum consists of the basis capituli, which articulates with the body; the segmented palps; the chelicerae; and the toothed hypostome. The capitulum of ixodid ticks is located at the anterior end of the body. Females bear paired clusters of pores, the porose areas, located dorsally on the basis capituli. The porose areas secrete anti-oxidants that inhibit degradation of the waxy compounds in the secretions of Gené's organ, which coat the eggs as they are laid. The chelicerae are located on the dorsal aspect of the capitulum. Their shafts, surrounded by spinose sheaths, lie between the palps and often extend even farther anteriorly than the palps. Each chelicera bears two digits distally. The larger, medial digit can be moved laterally; the smaller outer digit resides in a cavity of the medial digit and moves with it. Both digits have sharp **denticles**. The chelicerae are used to cut host tissues during attachment. The hypostome is a prominent, ventrally located structure that bears rows of recurved teeth on its ventral surface; teeth are absent in some nonfeeding males. A narrow food canal is located on the mid-dorsal surface. The palps consist of four distinct segments. In nymphs and adults of most ixodid species, the small terminal (4th) segment is recessed in a cavity in segment 3 and bears numerous fine setae at its tip.

The capitulum of adult and nymphal argasids is similar. However, it is situated just below an anteriorly protruding body extension, or hood, and is not visible dorsally in nymphs or adults (Figures 26.4 and 26.5). The four palpal segments are about equal in size. Small flaps, the cheeks, occur alongside the capitulum

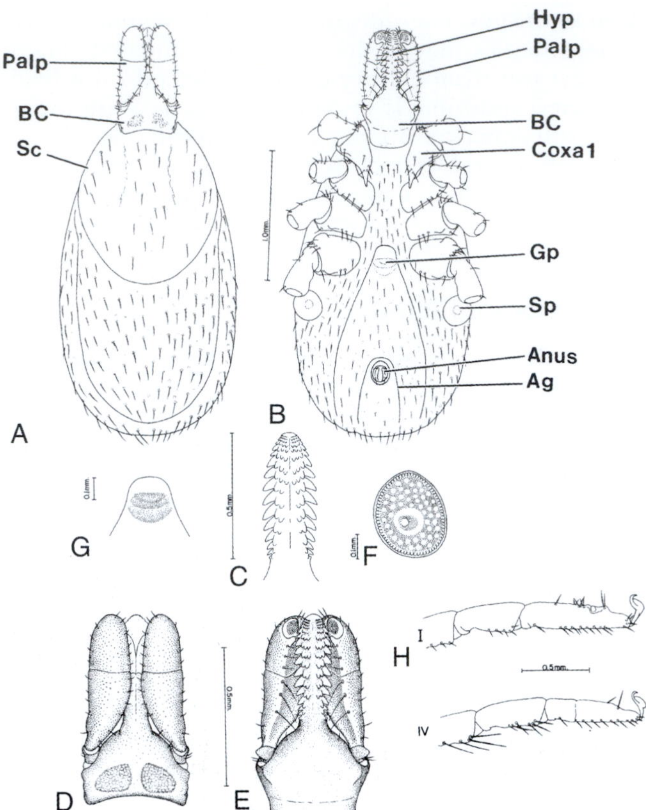

Figure 26.1 External morphology of representative female ixodid tick (*Ixodes pacificus*). (A) Dorsal view; (B) ventral view; (C) hypostome; (D) capitulum, dorsal view; (E) capitulum, ventral view; (F) spiracular plate; (G) genital pore; (H) legs I and IV. Ag, anal groove; BC basis capituli; Gp, genital pore; Hyp, hypostome; Sc, scutum; Sp, spiracle. (Sonenshine 1991, with permission of Oxford University Press)

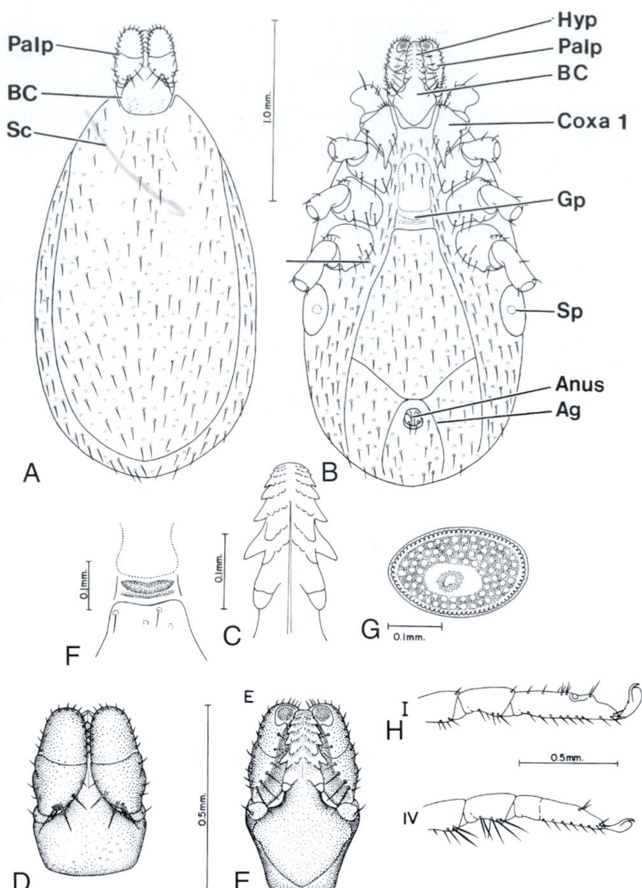

Figure 26.2 External morphology of representative male ixodid tick (*Ixodes pacificus*). (A) Dorsal view; (B) ventral view; (C) hypostome; (D) capitulum, dorsal view; (E) capitulum, ventral view; (F) genital pore; (G) spiracular plate; (H) legs I and IV. Ag, anal groove; BC, basis capituli; Gp, genital pore; Hyp, hypostome; Sc, scutum; Sp, spiracle. (Sonenshine 1991, with permission of Oxford University Press)

in many species and can be folded to cover the delicate mouthparts. In argasid larvae, the mouthparts protrude anteriorly, as in ixodids.

The body, exclusive of the capitulum, is the **idiosoma**. It is divided into two parts, the anterior **podosoma**, which bears the legs and the genital pore; and the posterior **opisthosoma**, the region behind the coxae that bears the spiracles and the anal aperture. The cuticle is relatively tough with sclerotized plates (sclerites) in certain locations. It serves as the site of muscle attachment and protects the animal from desiccation and injury. The cuticle bears numerous sensory setae as well as various pores representing the openings of dermal glands or sensilla.

The legs are jointed and articulate with the body via the **coxae**. Larvae are easily recognized by the presence of only three pairs of legs, whereas nymphs and adults have four pairs of legs. The structure of the legs is similar in the Ixodidae and Argasidae. Each leg is divided into six segments, the coxa, trochanter, femur, patella (genu), tibia, and tarsus. The coxae are inserted ventrally and allow limited rotation in the antero-ventral and dorso-ventral planes. The other segments can be flexed, so that the legs can be either folded against the ventral body surface for protection or extended for walking. A pair of claws and a pad-like pulvillus are present on each tarsus of most species. The pulvillus is absent in argasid nymphs and adults. An odor-detecting sensory apparatus, **Haller's organ** (Figure 26.6), is evident on the dorsal surface of the tarsus of leg I in all stages. This organ consists of an anterior pit and a posterior capsule. Gustatory, thermosensory, and mechanosensory functions also have been associated with this organ. Variations in the structure of Haller's organ are useful for distinguishing genera and species.

Ixodidae

Ixodid ticks, also called hard ticks, are illustrated in Figures 26.1, 26.2, and 26.3. Females have a hard cuticular plate or scutum on the anterior half of the dorsal body surface (Figure 26.1A). In males, the scutum occupies virtually the entire dorsal surface (Figure 26.2A). Elsewhere, the cuticle contains tiny

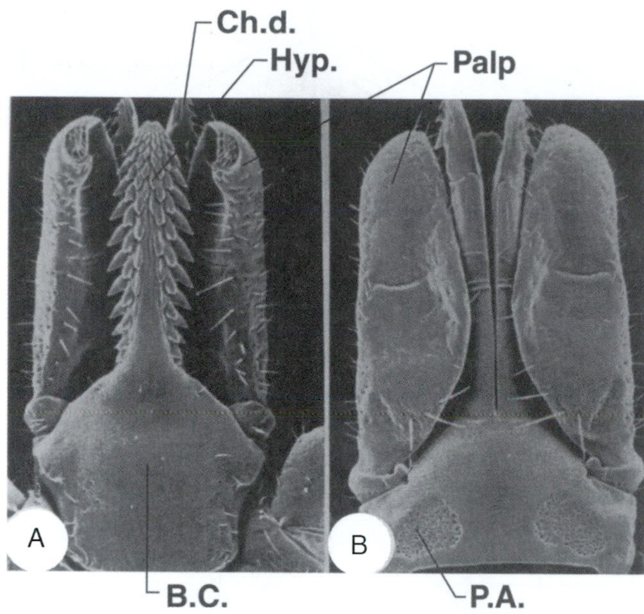

Figure 26.3 Capitulum of a representative ixodid tick (*Ixodes scapularis*), scanning electron micrographs. (A) Ventral view; (B) dorsal view. BC, basis capituli; Ch.d., cheliceral digit; Hyp, hypostome; P.A., porose area. (From Biology of Ticks, Vol. 1 by Daniel E. Sonenshine, 1991; Oxford University Press, Inc. Used by permission)

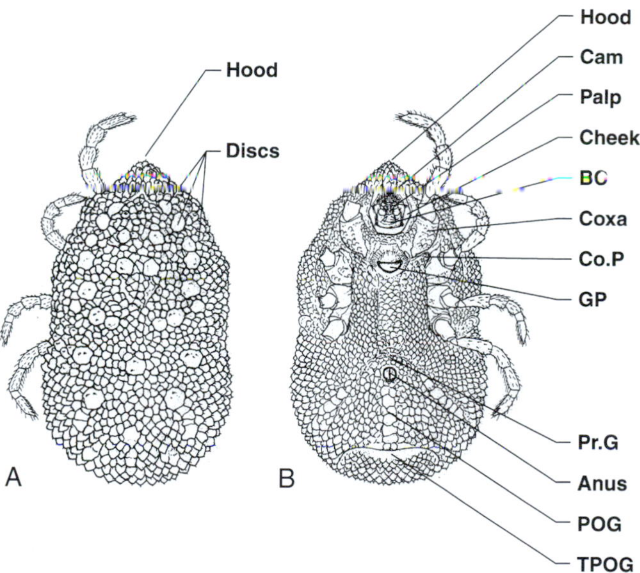

Figure 26.4 External morphology of a generalized argasid tick (*Ornithodorus*). (A) Dorsal view; (B) ventral view. BC, basis capituli; Cam, camerostome; Co.P, coxal pore; GP, genital pore; Pr.G, pre-anal groove; POG, post-anal groove; TPOG, transverse post-anal groove.

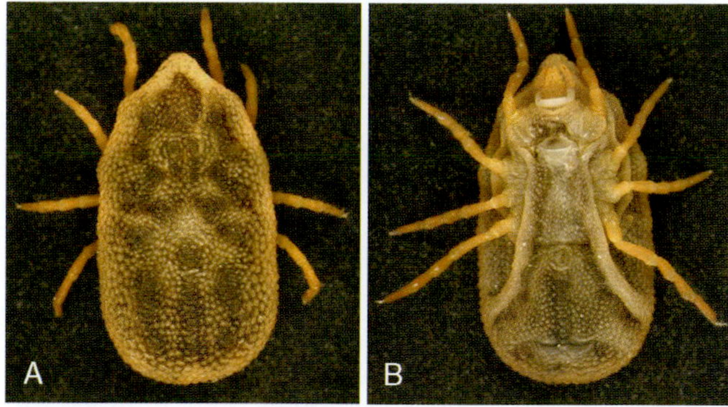

Figure 26.5 A representative argasid tick (*Carios kelleyi*), a parasite of North American bats. (A) Female, dorsal view; (B) female, ventral view.

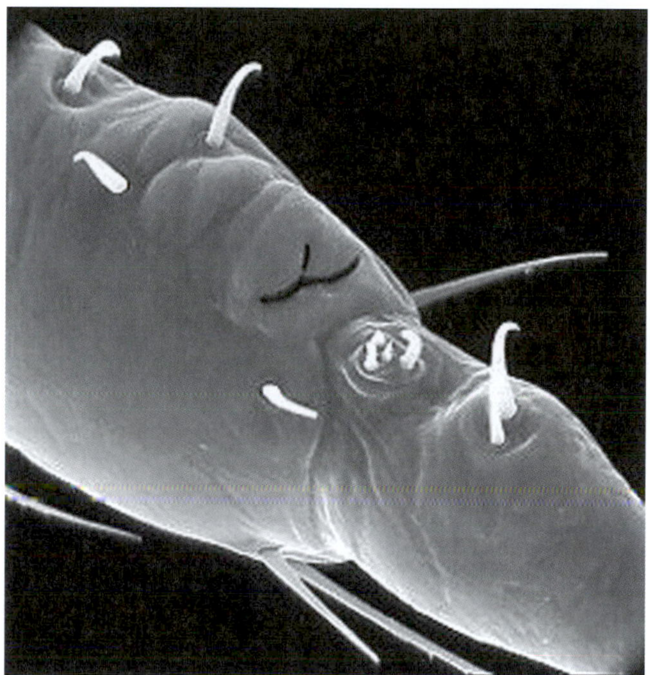

Figure 26.6 Haller's organ of the larval stage of *Amblyomma rotundatum,* a parasite of reptiles and amphibians in the Americas; scanning electron micrograph. (Courtesy of the US National Tick Collection, Georgia Southern University)

surface folds, which give it a fingerprint-like appearance when viewed at high magnification. The body of the female posterior to the scutum expands enormously during feeding as new cuticle is synthesized to accommodate the blood meal. In males, however, the larger scutum limits expansion. The scutum bears setae and tiny pores termed **sensilla auriformia**. The latter are believed to serve as proprioceptive organs. When present, a simple eye occurs along each postero-lateral margin of the scutum.

Immediately posterior to the scutum in the females are paired foveal pores (absent in *Ixodes*) from which a volatile sex pheromone, 2,6-dichlorophenol, is emitted. The dorsal body surface posterior to the scutum, the alloscutum, has innumerable fine folds. In females, a paired protrusible organ, Gené's organ, lies in the dorsal foramen between the scutum and the capitulum (capitular foramen). The ends of this organ protrude during oviposition and apply wax to each egg as it is deposited. In *Ixodes* males, hard sclerotized plates cover the ventral body surface (Figure 26.2B). In females, the genital pore is a U- or V-shaped opening, with prominent marginal folds (Figure 26.1B), but in males it is covered by a movable plate (Figure 26.2B). Other ventral structures include paired spiracular plates behind coxae IV in adults and nymphs (absent in larvae), each with a small ostium that opens to the respiratory system; and the anal aperture, located near the posterior margin. The entire body is covered by numerous setae and the pore-like sensilla auriformia. Larvae possess few setae, although their number and relative placement provide valuable taxonomic characters for generic and subgeneric differentiation (Figure 26.7).

Argasidae

The major external body features of argasid ticks, also known as soft ticks, are illustrated in Figures 26.4 and 26.5. The body margins are rounded in most species. In *Argas*, however, they are flattened and covered by small marginal discs. Eyes, when present, occur on folds lateral to the coxae. A tiny coxal pore, the opening of the duct from the paired coxal glands, occurs bilaterally between the coxae of legs I and II. The spiracular plates, located between coxae III and IV, are relatively small and inconspicuous. In females, the genital pore appears as a horizontal slit surrounded by a prominent fold. In males, the pore is subtriangular or suboval, without a genital apron. There are no foveal pores. Representative argasid ticks are shown in Figures 26.4 and 26.5.

Internal Anatomy

The internal organs of a typical tick are illustrated in Figure 26.8. The organs are bathed in a circulating fluid,

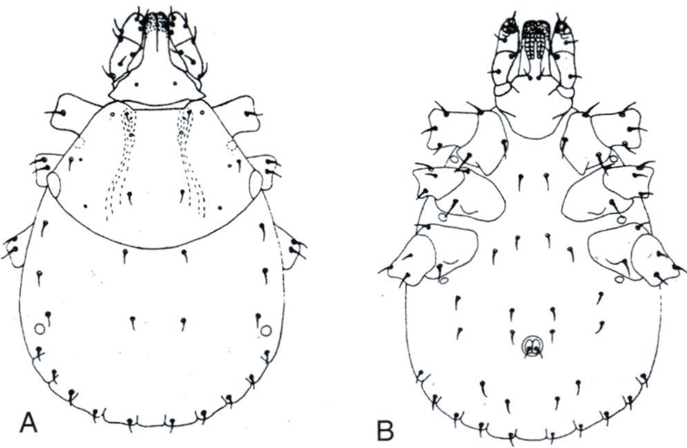

Figure 26.7 Larva of a representative ixodid tick (*Dermacentor variabilis*), with legs beyond coxae and trochanters not shown. (A) Dorsal view; (B) ventral view. (From Clifford, C.M., Anastos, A., and Elbl, A., 1961; The larval ixodid ticks of the eastern United States. Miscellaneous publications of the Entomological Society of America 2, 213–217)

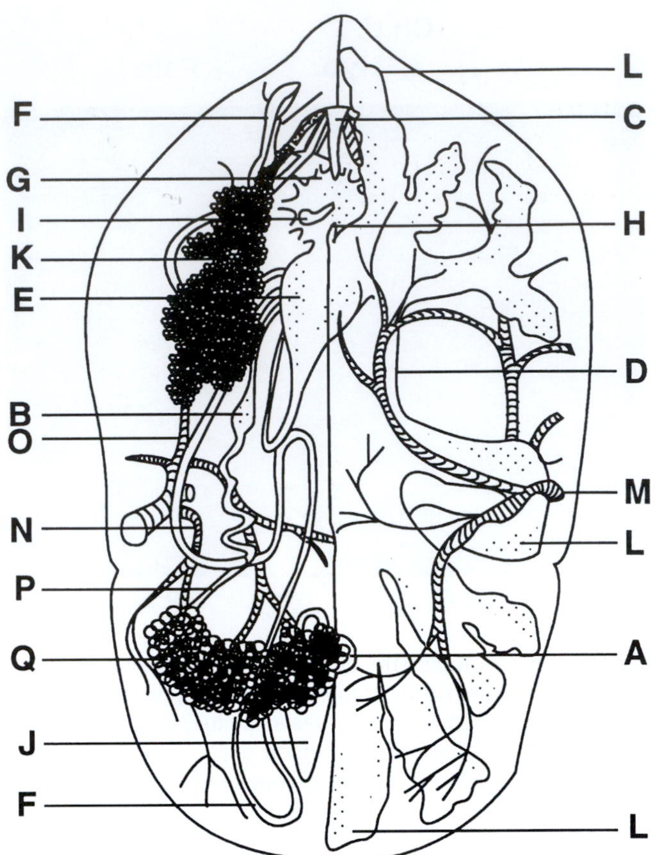

Figure 26.8 Internal anatomy of typical argasid tick, female; midgut shown on right side; midgut removed on left side to reveal underlying organs. A, anus; B, ampulla of oviduct; C, pharynx; D, median region of midgut; E, uterus; F, Malpighian tubule; G, brain (synganglion); H, esophagus; I, accessory gland; J, rectal sac; K, salivary gland; L, midgut diverticula; M, N, O, different regions of tracheal trunks; P, oviduct; Q, ovary. (Balashov 1972, with permission from Copyright Agency of Russia)

the **hemolymph**. The hemolymph is a watery medium rich in salts, amino acids, soluble proteins, and other dissolved substances. In addition, it contains several types of **hemocytes**, the most prominent of which are the phagocytes (plasmatocytes and granulocytes). Four major categories of cell types generally are recognized in tick hemolymph, namely, prohemocytes, nongranular plasmatocytes, granulocytes (type I and II), and spherulocytes (Borovickova and Hypša, 2005). A simple heart, situated middorsally, filters and circulates this vital body fluid. Muscles extend from the dorsal and ventral cuticular surfaces to the inner surfaces of the coxae, chelicerae, and other structures.

The most prominent internal organ is the midgut, a large sac-like structure with numerous lateral diverticula. The shape of the midgut depends on the state of engorgement. In unfed ticks, the diverticula are narrow, tube-like structures. In fed ticks, they enlarge and obscure most of the other organs as they fill with blood. Branches of the tracheal system ramify over the surfaces of the diverticula and surround the other internal organs. Ticks respire through these innumerable tiny air tubes, which open to the exterior via the paired spiracles.

Paired salivary glands are situated anterolaterally. These large glands, which resemble clusters of grapes, are connected via the salivary ducts to the mouthparts. Their salivary secretions empty into the salivarium, located between the chelicerae and the hypostome. Tick saliva contains pharmacologically active compounds that facilitate attachment and suppress host inflammatory responses. The salivary glands eliminate excess water from the blood meal. In ixodid ticks, most water in the blood meal is extracted by specialized salivary-gland cells and excreted into the host as the tick feeds.

Other prominent internal structures are the reproductive organs. In males, these include the testes, the tubular vasa deferentia, the seminal vesicle, and the ejaculatory duct, which is connected to the genital pore. The ejaculatory duct is obscured by the large, multilobed accessory gland, which secretes the components of the spermatophore. In females, the reproductive organs include the ovary, paired oviducts, uterus, vagina, and the seminal receptacle. The ovary is small and inconspicuous in unfed ticks, but expands enormously during feeding and especially after mating. In gravid females, the ovary is distended with large, amber-colored eggs.

In argasid and ixodid ticks, excretion is accomplished by the Malpighian tubules, a pair of long, coiled structures that empty into the rectal sac. Nitrogenous wastes are excreted in the form of guanine. In argasid ticks, paired coxal glands adjacent to the coxae of leg I extract excess water and salts accumulated during feeding, and excrete this watery waste via the coxal pores. Each gland consists of a membranous sac that serves as a filtration chamber and a coiled tubule that selectively reabsorbs small, soluble molecules and ions. Relapsing fever spirochetes may be transmitted to vertebrate hosts via the coxal fluid of certain species of infected ticks.

The central nervous system in ticks is fused to form the **synganglion**, located antero-ventrally above the genital pore. The synganglion, which regulates the function of the structures just described, is the fused central nervous system. Large pedal nerves extend from the synganglion to the legs; smaller nerves innervate the palps, chelicerae, cuticular sensilla, and the internal organs.

LIFE HISTORY

The life cycle includes four stages: the egg, larva, nymph, and adult. Ixodid ticks have only one nymphal instar, whereas argasid ticks have two or more nymphal instars. All ticks feed on blood during some or all stages in their life cycle; that is, they are obligate ectoparasites. Larvae attack hosts, feed, detach, and develop in sheltered microenvironments where they molt to nymphs. Nymphs seek hosts, feed, drop, and molt to adults (except in argasid ticks, which molt into later nymphal instars). Adult ticks seek hosts, feed, and, in the case of engorged ixodid females, drop off to lay their eggs (Figure 26.9).

Figure 26.9 Lone star tick (*Amblyomma americanum*), female, that has just finished depositing an egg mass of about 4,000 eggs. (Photo by Gary R. Mullen)

In contrast to most other hematophagous arthropods, ticks can be remarkably long-lived. Many can survive for one or more years without feeding. Their life cycles vary greatly, with the greatest differences evident between the Ixodidae and Argasidae.

Ixodid Life Cycles

Immature and adult ticks each take a blood meal, except for the nonfeeding males of some species, especially members of the genus *Ixodes*. Following contact with the host, a tick uses its chelicerae to puncture the skin and its hypostome to securely anchor itself. In many species, attachment is known to be reinforced by secretion of cementing substances with the saliva into and around the wound site. Females feed only once. Following mating, females suck blood for several days and swell enormously during the last 24-48 hours of attachment. Replete, mated females drop from their hosts, find a sheltered location, and subsequently oviposit hundreds to thousands of eggs (Figure 26.9). For *D. variabilis*, the average egg production is 5,400. For *Hyalomma impeltatum*, the reported average is nearly 10,700 eggs per female. The greatest number ever recorded was produced by an *Amblyomma variegatum* female that produced over 34,000 eggs. The eggs are deposited in a single, continual mass over many days or weeks. The female dies upon completion of egg laying.

Males swell only slightly during feeding. They usually remain on their hosts, feed repeatedly, and inseminate several females. Mating typically occurs on the host. Certain species of *Ixodes*, however, mate on their hosts, in nests, or in vegetation. Many *Ixodes* males have vestigial hypostomes, and these species invariably mate off the host. Except for *Ixodes* species, males and females require a blood meal to stimulate oogenesis and spermatogenesis. More than 90% of the life cycle is spent off the host. Molting usually occurs in some sheltered microhabitat such as soil and leaf litter, or in host nests. After molting, nymphal and adult ticks must seek another host and feed. When host seeking and feeding occur in all three parasitic stages, the pattern is termed a **three-host life cycle** (Figure 26.10). This is characteristic of more than 90% of ixodid species.

In tropical climates with frequent rainfall, developmental times are relatively short, and several generations may occur each year. In regions with alternating dry and rainy seasons, the life cycle is longer because ticks cease host-seeking during the driest period. In colder temperate or subarctic regions, development is slower, and ticks commonly undergo diapause during the coldest months. As a result, the life cycle may take two or more years. An example of a diapausing species is *Dermacentor variabilis.* Larvae feed on mice or other small mammals, mostly in spring. Fed larvae drop off and molt to nymphs that again attack small mammals. Fed nymphs drop off and molt within a few weeks. If the adults feed and reproduce in the same year, the entire life cycle can be completed within several months. Thus, under favorable conditions in nature, the typical three-host life cycle can be completed in less than one year. In the laboratory, it can be completed even sooner. However, adverse environmental conditions can prolong the life cycle to two or more years. The life cycle of *Ixodes ricinus* may require up to four years in the northern parts of its range in Europe (Hoogstraal, 1985). In Ireland, the life cycle takes three years, with each stage requiring approximately one year before developing to the next (Gray, 1991).

A few ixodid species exhibit a two-host or one-host life cycle (Figure 26.10). For example, in the two-host camel tick (*Hyalomma dromedarii*), fed larvae molt on their hosts, and the unfed nymphs reattach soon after emergence. Following engorgement, the nymphs drop off the host, molt, and feed as adults on a second host. In the one-host cattle tick *R.* (*B.*) *annulatus*, and in other species of the subgenus *Boophilus*, all stages feed and molt on the same host. Mating also occurs on this host. Replete, fertilized females drop off the host and oviposit in soil.

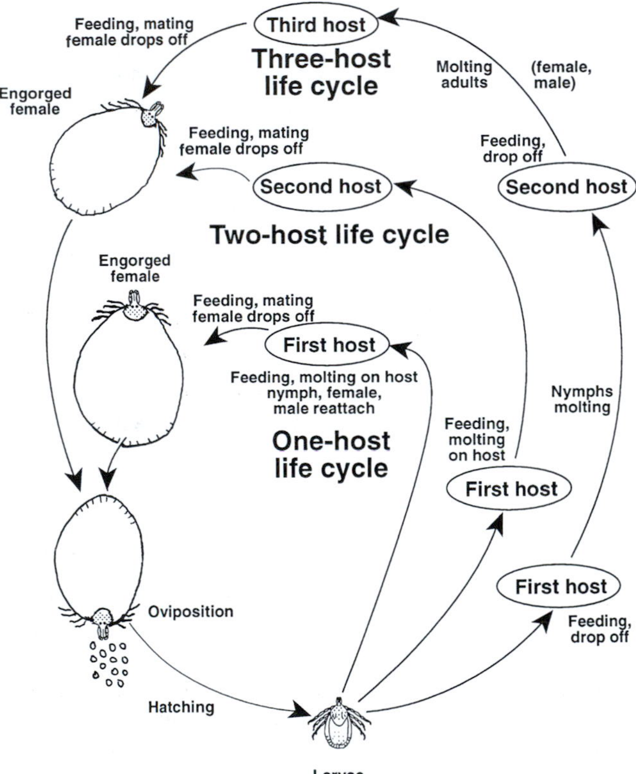

Figure 26.10 Three basic life cycles of ixodid ticks: (1) One-host ticks (inner circle) in which the larva, nymph, and adult all attach to, and develop on, a single host (e.g., *Boophilus annulatus*). (2) Two-host ticks (middle circle) in which larva and nymph feed on one host and the adult attaches and feeds on a second host (e.g., *Hyalomma dromedarii*). (3) Three-host ticks (outer circle) in which larva, nymph, and adult each parasitize a different host (typical of most ixodid ticks). Most argasid ticks have a multihost life cycle involving more than three hosts; with several nymphal instars, each potentially feeding on a different host.

Argasid Life Cycles

In contrast to the ixodids, most argasids have two or more nymphal instars in their life cycle, each of which must consume a blood meal. This pattern is termed the multihost life cycle. Molting occurs off the host in cracks, crevices, or beneath debris in or near the nest. Argasid females take repeated small blood meals and lay small batches of eggs (Figure 26.11), typically less than 500 eggs/batch after each feeding. These are termed **multiple gonotrophic cycles**. The interval between feedings is typically several months. As many as six gonotrophic cycles have been reported in some species. Mating usually occurs off the host. Because of the multiple nymphal instars that may number six or even seven in some species, argasid ticks often live for many years. In addition, these ticks are highly resistant

Figure 26.11 Argasid tick (*Ornithodoros turicata*), female depositing small batches of eggs. (Photo by Jerry F. Butler)

to starvation, which can extend their longevity even further. In some species that feed on migratory bats or birds, diapause serves to delay oviposition or development during the periods when hosts are absent.

The larvae of most *Ornithodoros* species that parasitize bats and birds remain attached to their hosts for many days, just as ixodid ticks do. Following the larval blood meal, they molt twice without additional feeding. Thereafter, the life cycle is similar to the typical argasid pattern. Another species with an unusual life cycle is *Otobius megnini*. This tick exhibits a high degree of host and body-site specificity that regulates its feeding and development. Females do not feed and are autogenous (i.e., oviposit without feeding).

BEHAVIOR AND ECOLOGY

Feeding behavior, even on preferred hosts, is not a uniform process. Blood-feeding begins soon after contact and acceptance of recognition features that determine that the animal is a suitable host. In ixodids, a tick may crawl about the host for several hours in search of a suitable feeding site. Once a site has been selected, the tick cuts into the skin with its cheliceral digits and inserts its hypostome to initiate the attachment process. Shortly after they attach, most ixodid ticks secrete cement during the first one or two days to secure themselves at the wound site. Subsequently, the tick begins salivating into the developing hematoma and sucking blood; salivating and blood sucking alternate, often for extended periods of time for each process. The feeding lesion enlarges as the tick injects anticoagulant and antihemostatic compounds into the wound; recruitment of host leucocytes to the wound site also contributes to tissue lysis and fluid influx around the tick's mouthparts.

Successful blood-feeding depends upon the secretion of an extensive array of antihemostatic, anti-inflammatory, and immunomodulatory proteins and lipids in the tick saliva so as to suppress the host's ability to reject the feeding tick. Of particular importance in tick saliva are antagonists of the intrinsic pathway factor X (Xa) and factor V (Va), which converts prothrombin to thrombin. Thrombin in turn converts plasma fibrinogen to fibrin and clots the blood. Tick saliva blocks blood coagulation by inhibiting factor Xa. In addition, many species also secrete proteins that inhibit thrombin directly or inhibit the conversion of prothrombin to thrombin by inhibiting factor V. Other salivary proteins prevent platelet aggregation, also important for blood coagulation, proteins that bind, antagonize or degrade important host mediators of pain, itching and inflammation, particularly the host's histamine, serotonin and bradykinin.

Our knowledge of the tick's ability to antagonize the host's hemostatic mechanisms is still incomplete. An example is the role of angiogenesis, the formation of new blood vessels that is an essential element in wound healing, and the role of tick saliva in disrupting this process (Francischetti et al., 2008). Much additional information has become or will become available with the publication of cDNA libraries of tick salivary glands (e.g., Ribeiro et al., 2006).

Digestion of the blood meal takes place in the midgut. Erythrocytes and other blood cells are lysed soon after ingestion. Hemoglobin from lysed cells binds to receptors on the midgut epithelial cells and is incorporated by a process known as **receptor-mediated endocytosis** into tiny vacuoles in the digestive cells (Coons et al., 1986). The vacuoles fuse with lysosomes, forming specialized phagolysosomes, wherein hydrolytic enzymes released into these acidic vacuoles carry out digestion of the hemoglobin (Gough and Kemp, 1995; Mendiola et al., 1996). Most of the heme released following hemoglobin digestion is detoxified to hematin, which accumulates in specialized hemosomes (Lara et al., 2003), and the amino acids liberated from the globin moieties are passed into the hemolymph. The latter provides the primary nourishment derived from the tick's blood meal. Although some other proteins, various lipids, and carbohydrates are also digested, most proteins ingested with the blood meal remain in the midgut lumen.

Ixodid ticks feed gradually prior to mating because first they must create new cuticle to accommodate the massive blood meal. Typical attachment periods range from as few as two days for larvae to as long as 13 days for females. When feeding is completed, the weight of blood and other fluids consumed ranges from 11 to 17 times the tick's prefeeding body weight in ixodid larvae and from 60 to 120 times the tick's prefeeding body weight in ixodid females. Measurements of blood volume consumed range from 0.7 ml to as high 8.9 ml per female in some ixodid species. Nymphal and adult argasid ticks attach for only brief periods. This can be as little as 35 to 70 minutes for adults. These ticks do not secrete cement during attachment; instead, they attach solely with the mouthparts, especially the hypostome. Argasid ticks swell to the extent that their cuticles can stretch. New cuticle is not secreted during feeding as it is in ixodid ticks.

In ticks, mating can occur either during feeding or off the host. In the metastriate Ixodidae, mating occurs while the adults are attached and feeding. Unfed adults are sexually immature and require a blood meal to stimulate gametogenesis. Mating usually is regulated by sex pheromones and follows a complex, hierarchical pattern of responses. Feeding females secrete a volatile sex- attractant pheromone, usually 2,6-dichlorophenol, that excites males feeding nearby on the same host. The males detach and seek the females that they recognize by means of the mounting sex pheromone (a mixture of cholesteryl esters) on the female's body surface. The male climbs onto the dorsum of the female, then moves to her ventral surface and searches for her gonopore. Once a male locates the gonopore, he probes the opening with his chelicerae. Spermatophores, containing spermatozoa, are produced in the large accessory gland of the sexually mature male during the mating process. At this time, the spermatophore emerges from the male's genital pore, whereupon the male seizes it with his mouthparts and inserts it into the female's vulva.

Copulation is essential to initiate rapid engorgement by the blood-feeding females.

In most ixodid ticks, the attached females do not fully engorge unless inseminated by a conspecific male, and the ovary remains in the nonvitellogenic state; however, parthenogenesis is known to occur in some species. Full engorgement by the female leads to a remarkable sequence of molecular and physiological changes that result in vitellogenin production, ovarian development, and oviposition. Although our understanding of this complex process is incomplete, much has been learned about several of the most important elements. A signal from the synganglion stimulates greatly increased production of the hormone, 20-hydroxyecdysone (Lomas et al., 1997). In the hard tick, *D. variabilis*, elevated levels of this hormone induce expression of vitellogenin in the fat body (Thompson et al., 2005), followed by secretion into the hemolymph. Elevated 20-hydroxyecdysone levels also induce expression of vitellogenin receptors (VgR) on the developing oocytes (Thompson et al., 2005), resulting in vitellogenic oocytes. The full nucleotide and amino acid sequences of tick vg (Thompson et al., 2007) and VgR (Mitchell et al., 2007) have been described, the first such description for a chelicerate arthropod.

In prostriate Ixodidae and in argasid ticks, the adults become sexually mature soon after the nymphal molt. These ticks usually mate in the nests or in vegetation, although ticks of some prostriate species also may mate on the host. There is evidence that mating in argasid ticks is regulated by one or more sex pheromones, although no specific compounds have been identified to date.

A pre-ovipositional period of up to several weeks precedes egg laying in ixodid ticks. During oviposition, the cuticle of the vulva softens and evaginates as the eggs pass through it, thereby serving as an ovipositor. The emerging eggs are waxed by secretions from Gené's organ. The process of oviposition continues for several weeks. Typically, about 50 to 60% of the female's body weight at the time of drop-off is converted to eggs. The number of eggs deposited is directly proportional to the size of the engorged female. At the completion of oviposition, the spent female dies. Thus, there is only a single gonotrophic cycle among ixodids. In the Argasidae, however, mated females commence oviposition soon after feeding but deposit small clutches containing only a few hundred eggs. Following oviposition, the females remain active and seek hosts again. These ticks feed and lay eggs after each meal. They do not need to mate again. The number of gonotrophic cycles varies but rarely exceeds six.

Most tick species live in forests, savannahs, second-growth areas of scrub and brush, and grassy meadows. Others remain buried in sand or sandy soils, under stones, in crevices, or in the litter, duff, and rotting vegetation at the floor of woods and grasslands. In contrast, almost all argasids and some ixodids, especially males and immatures of several species of the genus *Ixodes*, are nidicolous. They live in the nests, burrows, caves, or other shelters used by their hosts.

Nonnidicolous ticks are active during certain periods of the year when climatic conditions favor development and reproduction. During such periods they engage in host-seeking behavior. In temperate and subpolar regions, seasonal activity is regulated by ambient temperature, changing photoperiod, and incident solar energy. In tropical regions, where day length or temperature vary only slightly throughout the year, tick activity often is controlled by the transition from the dry to the rainy season. Host-seeking ticks exhibit at least two strategies for locating potential targets for their blood meals. Ambush ticks climb onto weeds, grasses, bushes, or other leafy vegetation to wait for passing hosts. When stimulated by the presence of a host, they extend their forelegs anterolaterally in what is called questing behavior (Figure 26.12) and quickly grasp the hair, feathers, or clothing of a passing host. Hunter ticks emerge from their refuges in the soil, sand, or duff when excited by host odors and run rapidly across the ground to attack hosts.

After contacting a potential vertebrate host, the tick must perceive appropriate host-recognition cues that enable it to determine whether to attach and feed, or to drop off and resume host-seeking. Odors, radiant heat, visual images, or vibrations stimulate the tick and enable it to recognize its prospective host. Odors

Figure 26.12 A questing brown dog tick (*Rhipicephalus sanguineus*), female at tip of vegetation. (Courtesy of Rickettsial Zoonoses Branch, Centers for Disease Control and Prevention, USA)

are probably the most important stimuli. Electrophysiological studies have shown that larvae of *Rhipicephalus (B.) microplus* respond to odors from extracts of cattle skin but not to dry air. Human breath also elicits a response, but not as vigorously as that caused by cattle extracts. Among the more important attractants emitted by hosts are carbon dioxide in animal breath, and ammonia in urine and other animal wastes. Other odors that attract ticks are butyric acid and lactic acid, which occur commonly in sweat and other body fluids. Small increases in radiant heat excite ticks and act synergistically with host odors. Visual cues may be important, especially in certain hunter ticks that can discriminate dark shapes against the bright background of the sky. Host-seeking ticks of many species respond to shadows, resulting in extension of their legs to facilitate contact.

Other stimuli that elicit questing behavior include vibrations, sound, and tactile cues. Vibrating the grass stems on which ticks are perched can elicit questing behavior almost immediately. *Rhipicephalus (B.) microplus* larvae respond to sounds in the 80 to 800 Hz range, typical of the frequencies produced by feeding cattle, whereas the sounds produced by barking dogs reportedly attract *Rhipicephalus sanguineus* (Waladde and Rice, 1982). Tactile stimuli perceived when ticks contact their hosts, in combination with short-range attractants such as heat and odor, help to determine the selection of suitable feeding sites. Ticks will not attach to a host unless the appropriate stimuli are received in a particular sequence.

The timing of drop-off from the host offers important ecological advantages. For nonnidicolous ticks, such drop-off rhythms are synchronized with host behavioral patterns. This tends to disperse fed ticks in optimal habitats where they can develop and reproduce. Photoperiod appears to be the dominant exogenous factor affecting drop-off patterns. The daily light:dark cycle induces a regular rhythm of feeding and dropping off. This effect, termed **photoperiodic entrainment**, is partially reversible. In a series of elegant experiments with *H. leporispalustris,* the existence of an endogenous drop-off rhythm entrained by the scotophase or dark period was shown. The rhythm was affected only partially by changing the photoperiodic regime and was maintained even when hosts were held in constant darkness. Detachment may occur while the hosts are inactive in their nests or burrows or, alternatively, it may be coordinated with the period of maximum host activity. In *D. variabilis,* fed larvae and nymphs drop soon after night begins. In contrast, immatures of *H. leporispalustris* drop off during daylight hours when their lagomorph hosts are confined in their forms or warrens.

Ticks exhibit varying degrees of host specificity. More than 85% of argasid and ixodid ticks exhibit relatively strict host specificity. However, the evolutionary significance of this phenomenon is uncertain. More recent studies using molecular methods suggest that most of the existing tick-host-association patterns may be explained as artifacts of biogeography and ecological specificity as well as incomplete sampling (Klomplen et al., 1996). Regardless of how host specificity evolved, it is clear many species are specialists. Examples include certain species of argasids (e.g., some *Carios* spp.), which infest only bats; and the ixodids *Dermacentor albipictus, R. (B.) annulatus,* and *R. (B.) microplus,* which feed only on large ruminants. Many of the nidicolous ticks are highly host-specific. Examples include *Ixodes marxi,* which feeds almost exclusively on squirrels; *Amblyomma tuberculatum,* which as adults, parasitizes the gopher tortoise; and *Argas arboreus,* which attacks herons.

At the opposite extreme are ticks that are opportunistic species with catholic feeding habits. Examples include *Ixodes ricinus* and *I. scapularis.* Larvae and nymphs of these species feed readily on lizards, birds, small mammals, and larger hosts like sheep and humans. Adults feed on larger mammals, especially ungulates, and also attack humans. Over 300 species of vertebrates have been recorded as hosts for *I. ricinus* and over 120 have been recorded for *I. scapularis.*

Host specificity is influenced by evolutionary history, ecological and physiologic factors, and the ability of the ticks to avoid host rejection. Many species belonging to the less specialized and phylogenetically primitive genus *Amblyomma* and all species of the former *Aponomma* (now included in *Amblyomma* or *Bothriocroton*) feed on reptiles or primitive mammals. Ticks adapted to a specific habitat type (e.g., grassland) encounter only those vertebrates adapted to the same habitat. Questing height also is important. Ticks questing on or near the ground are exposed mostly to small animals, whereas those questing higher in the vegetation are more likely to encounter larger animals.

The extent to which different hosts are utilized depends upon host behavior and opportunities for contact, such as foraging range, time of day and time spent foraging, habitats visited, and other factors. White-footed mice, which forage extensively on the ground, acquire numerous *I. scapularis* immatures, whereas flying squirrels, which spend much less time on the ground, are rarely infested. A similar relationship exists among migratory birds. In the United States, ground-feeding birds that forage in habitats shared with cottontail rabbits are heavily infested with immatures of the rabbit tick *H. leporispalustris.* Birds that forage above ground in bushes or trees rarely encounter these ticks. Acceptance of a vertebrate animal also is dependent on physiological factors and the ability of the ticks to recognize it as a host.

Host utilization may be influenced by the ability of ticks to evade or suppress host homeostatic systems and avoid rejection. This was first reported in a landmark paper by Trager (1939), who noted that the feeding success of *D. variabilis* on guinea pigs declined with frequent reexposures. The guinea pig is a South American rodent and an unnatural host for this North American tick. When fed on its natural hosts (e.g., white-footed mice), *D. variabilis* experiences little if any rejection. Similarly, *I. scapularis* saliva contains pharmacologically active compounds that suppress host mediators of edema and inflammation such as anaphylatoxins, bradykinin, and other kinins. These and other salivary components prevent release of host inflammatory agents from leukocytes while enhancing vasodilation, which brings more blood to the mouthparts. *Ixodes scapularis* apparently lacks compounds to suppress histamine. Because histamine-induced edema does not occur in the white-footed mouse, this does not deter tick-feeding on this host. However, histamine-containing basophils are very abundant in guinea pigs, and cutaneous basophil hypersensitivity develops rapidly even after a single feeding by these ticks (Ribeiro, 1989). *Ixodes scapularis* saliva also contains an enzyme that destroys complement, thereby facilitating the survival of pathogens such as *Borrelia burgdorferi* ingested during blood-feeding (Valenzuela et al., 2000).

Ticks occur in many terrestrial habitats ranging from cool, arboreal northern forests to hot, arid deserts. Each species, however, has become adapted to specific types of habitats where generally it is found in greater abundance. Typical habitat associations of nonnidicolous ixodid ticks include forests, meadows and other clearings, grasslands, savannahs, and semi-desert or desert areas. At one end of the spectrum are species that have very limited resistance to desiccation and occur in cool, moist forests (e.g., *I. scapularis* and *I. ricinus*). In the middle are the majority of species that can survive at least brief periods of desiccation during host seeking or development (e.g., *Dermacentor variabilis*, *Amblyomma maculatum*, and *A. americanum*). At the other end of the spectrum are the desiccation-tolerant species adapted to survive in arid steppes, semi-deserts, and other xeric environments (e.g., *Hyalomma asiaticum* and *Ornithodoros savignyi*).

Water balance is a critical determinant of a tick's ability to wait for hosts, sometimes requiring weeks or months. When they begin to desiccate, they retreat to more sheltered, humid microenvironments such as the rotting vegetation in a meadow or damp leaf litter on the forest floor. They secrete a hygroscopic salivary secretion onto their mouthparts that collects atmospheric water (direct sorption). After repeated cycles of secretion and drinking the condensed water, the rehydrated ticks are able to resume host-seeking. Some ticks are able to remain in the questing position for many days without rehydration, whereas others must return to their humid microenvironments each day.

Ixodes scapularis is an example of a tick with very limited desiccation tolerance. Consequently, it is most abundant in dense, humid, forest habitat or in dense shrub-dominant habitats adjacent to large rivers, bays, or the Atlantic Ocean. Another desiccation-intolerant species is *Ixodes ricinus.* This tick is widespread in the British Isles, Continental Europe, and western Asia, where it frequents woodlands, damp meadows, pastures, and ecotones.

Dermacentor variabilis is an example of a species exhibiting greater tolerance to desiccation. It flourishes in the ecotone between secondary growth deciduous forests and lush, grassy meadows, as well as along secondary roads and trails in forested habitats. The dense ecotonal vegetation provides shade, increased moisture, protection from intense solar radiation, and food plants that support the tick's mammalian hosts. This type of environment is ideal for the immature stages of *D. variabilis.* Adults, with their greater resistance to desiccation and greater mobility, venture further afield to quest in sunlit meadows or along roads and trails.

The camel tick, *Hyalomma dromedarii,* is an example of a desiccation-tolerant species. This desert tick is common in the steppes and semi-desert habitats in large areas of northern Africa and the Middle East. Larvae and nymphs generally live in rodent burrows. Adults bury themselves in sand and duff near their hosts, especially around caravansaries and similar locations where camels and other livestock are kept.

Nidicolous ticks living in or near the nests of their hosts are adapted to highly specialized environments. Normally the temperature and relative humidity in a burrow, cave, or similar type of shelter are more uniform throughout the year than in the external macroenvironment. The higher relative humidity in such microenvironments is due in part to the presence of hosts, their wastes, and plant materials used to construct or line their nests. Nidicolous ticks exhibit behavioral patterns that restrict their distribution to these sheltered locations. They avoid bright sunlight and low humidity, the type of conditions prevalent at the entrances of burrows or caves. Confined within these cryptic, restricted locations, nidicolous ticks become active when hosts are present. However, when hosts are absent, they may wait for up to several years for hosts to return, or until they die of starvation.

Seasonal activity refers to the period of the year when ticks actively seek hosts. For example, *D. variabilis* larvae emerge from winter diapause in spring to feed on small mammals, especially mice and voles. Activity accelerates rapidly as increasing numbers of larvae emerge from overwintering sites to attack hosts, culminating in the seasonal peak within a few weeks. Thereafter, activity continues unabated, with larval abundance declining as more individuals find hosts, desiccate, or die of starvation. Nymphal and adult ticks also feed during the warm spring and early summer months. In the southern parts of its range, overwintering *D. variabilis* adults emerge early and soon overlap with those that develop from nymphs fed in the spring. Thus, the seasonal peak for adults occurs in early summer. As a result, most females oviposit in July and August, and the newly hatched larvae enter diapause as day length diminishes. In the southeastern United States, the entire life cycle is completed in one year. Occasionally, a small secondary peak of *D. variabilis* larval activity occurs in the fall. In the northern part of its range, however, tick activity is delayed due to cooler spring temperatures and shorter day lengths. As a result, although larvae and nymphs feed in the late spring and summer, adults emerge too late in the summer to commence questing activity. These adults undergo diapause and emerge the following spring. This pattern of feeding and diapause results in a two-year life cycle.

Ixodes scapularis also exhibits distinct seasonal activity periods. In the northern part of its range, larval activity does not occur until middle or late summer and the nymphs that molt from the fed larvae diapause until the following spring. Nymphs that feed in the spring molt in summer, but the young adults delay host-seeking until fall. This pattern results in a two-year life cycle. In the southernmost part of its range, *I. scapularis* activity occurs earlier in the year. These southern populations may complete their life cycle in just one year.

A few tick species are active during the cooler months of the year, especially fall and winter. Larvae of the winter tick (*Dermacentor albipictus*), a one-host tick that feeds on horses, deer, elk, moose, and other large ungulates, commence host-seeking activity in late summer or early fall. Larvae and nymphs feed and molt on the same hosts and the resulting adults reattach, feed, and mate. Replete females drop off the host and oviposit in the soil. In the northern-most parts of its range, adults usually do not appear until late winter, with peak occurrence in April. Subsequent oviposition and hatching occur in late spring. At this time, larvae undergo diapause, presumably in response to increasing daylight, and do not commence host seeking until after an extended period of declining photoperiod. Development proceeds faster farther south. On stanchioned bovines held in stalls at Kerrville, Texas, the entire process of feeding, molting, and production of engorged females is completed in 21 to 36 days. Engorged females that drop off their hosts in winter do not oviposit until the following spring.

In tropical regions, where day length is nearly uniform throughout the year and where there is no prolonged dry season, the seasonal activity of many tick species (e.g., *Rhipicephalus appendiculatus*) often is influenced by the distribution of rainfall. Farther from the equator, particularly in colder regions in southern Africa, *R. appendiculatus* diapauses during the dry season.

In contrast, most argasid ticks do not exhibit patterns of seasonal activity. This is especially true for ticks infesting the nests or burrows of nonmigratory hosts such as rodents and carnivores. However, nidicolous ticks that parasitize migratory birds and bats tend to delay oviposition so that hatching occurs at about the time the hosts return.

Diapause is an important behavior that enables ticks to survive adverse environmental conditions and conserve energy until conditions improve Diapausing ticks become inactive, reduce their metabolic rates, and do not feed on hosts even when given the opportunity. Newly emerged larvae, freshly molted nymphs, and adults of many species enter diapause before seeking hosts, particularly if they emerge during periods of declining photoperiod. This is termed **host-seeking diapause**. As noted earlier, diapause enables *D. variabilis* larvae and adults to survive the cold winters that occur throughout most of the tick's range. It also determines the length of the life cycle, one year in the south but two years in the northern part of its range. Diapause delays activity of *I. scapularis* nymphs at more northern latitudes so that they do not commence host seeking until spring or early summer. It also may delay adult activity that usually begins in fall, often several months after molting.

Another type of diapause is morphogenetic diapause, in which development or oviposition is delayed. In *Dermacentor marginatus*, oviposition rather than hatching is delayed. Thus, females that feed in spring or early summer lay eggs immediately, but those that feed in late summer or early fall oviposit the following spring. A remarkable example of morphogenetic diapause occurs in certain argasid ticks that inhabit the nests of birds or the roosts of bats. Females of the bat tick *Carios kelleyi* (Figure 26.5) oviposit immediately after feeding in spring. However, those that feed in fall delay oviposition until the following spring. Because the bats migrate to cold caves or caverns far from the tick's normal habitats, this ovipositional delay avoids the risk that larvae will emerge at a time when no hosts are available.

TICK SPECIES OF MEDICAL-VETERINARY IMPORTANCE

The following ticks are important as household pests, species that transmit disease agents to humans or other animals, and species that are injurious to livestock. The more important tick-borne diseases of humans and other animals are listed in Tables 26.2 and 26.3.

The brown dog tick (*Rhipicephalus sanguineus*) (Figures 26.12 and 26.13) is a common household pest throughout most of the world. Its primary host is the dog, which can become heavily infested (Figure 26.14). All life stages can be found on dogs. However, in many areas bordering the Mediterranean Sea, western Asia, and Africa, this tick also feeds readily on a wide range of wildlife (especially small mammals) and also attacks humans. It often infests houses, especially when dogs are kept indoors, which can produce considerable distress when the owners encounter thousands of these ticks. Seasonal activity peaks in summer, although activity peaks can occur throughout the year when ticks inhabit heated homes. This species is the primary vector of *Rickettsia conorii*, which causes Boutonneuse fever in many Mediterranean countries. It recently has been implicated as the vector for *Rickettsia rickettsii* in the southwestern United States. In animals, this tick is a vector of the agents of canine ehrlichiosis (*Ehrlichia canis*) and canine babesiosis (*Babesia vogeli* and a small *B. gibsoni*-like species).

The brown ear tick (*Rhipicephalus appendiculatus*) and related species are major pests of livestock in eastern, central, and southern Africa. Hosts include most domestic ruminants and many wildlife species, to which it attaches predominantly in and around the ears. It is the vector of the agent of East Coast fever, a protozoan disease that afflicts ruminant livestock within its range. Many other species of *Rhipicephalus* are important livestock pests and vectors of pathogens (e.g., *R. bursa* is a vector of the agents of equine,

Table 26.2 Representative Tick-Borne Diseases of Public Health Importance and Associated Characteristics*

Disease	Causative Agent	Primary Tick Vector Species	Affected Host(s)
Human babesiosis	*B. microti, B. divergens, B. duncani* (WA1) *B. venatorum* (EU1)	*Ixodes scapularis, Ixodes ricinus*	Humans, mice, cattle
Tick-borne encephalitis	*Flavivirus*[1]	*I. ricinus, I. persulcatus*	Rodents, insectivores, carnivores, humans, etc.
Kyasanur Forest disease	*Flavivirus*[1]	*Haemaphysalis spinigera*	Monkeys, small mammals, carnivores, birds, cattle, humans
Powassan encephalitis	*Flavivirus*[2]	*Ixodes, Dermacentor,* and *Haemaphysalis* spp.	Rodents, hares, carnivores
Colorado tick fever	*Coltivirus*[2]	*Dermacentor andersoni*	Rodents, carnivores, humans, domestic animals
Crimean-Congo hemorrhagic fever	*Nairovirus*[3]	*Hyalomma m. marginatum, H. m. rufipes,* others	Hares, hedgehogs, small mammals, humans
Rocky Mountain spotted fever	*Rickettsia rickettsii*	*Dermacentor variabilis, D. andersoni,* others	Small mammals, carnivores, humans, rabbits, others
Boutonneuse fever[4]	*Rickettsia conorii*	*R. sanguineus, D. marginatus, D. reticulatus,* others	Small mammals, hedgehogs, dogs, and humans
African tick-bite fever	*Rickettsia africae*	*Amblyomma* spp.	Mammals including humans
Human ehrlichiosis	*Ehrlichia chaffeensis*	*Amblyomma americanum*	Humans, deer
Human ehrlichiosis	*Ehrlichia ewingii*	*Amblyomma americanum*	Dogs, humans
Human anaplasmosis	*Anaplasma phagocytophilum*	*Ixodes scapularis, I. pacificus, I. ricinus, I. persulcatus*	Rodents, deer, humans, dogs
Q Fever	*Coxiella burnetii*	Many tick species	Large domestic livestock, humans
Lyme disease	*Borrelia burgdorferi, B. afzelii, B. garinii,* others	*Ixodes scapularis, I. ricinus, I. pacificus, I. persulcatus,* others	Mammals including humans, birds
Tick-borne relapsing fever	*Borrelia* spp.	*Ornithodoros* spp.	Various mammals including humans
Tularemia	*Francisella tularensis*	Many tick species	Lagomorphs, rodents, carnivores, humans
Tick paralysis	Tick proteins	*I. holocyclus, I. rubicundus, D. variabilis, D. andersoni,* others	Cattle, sheep, dogs, humans, other mammals, birds, others
Tick-bite allergies	Tick proteins	*Argas reflexus, Ornithodoros coriaceus, Ixodes pacificus,* others	Humans

*All Tick-borne Diseases Have Not Been Included.

1. Family Flaviviridae; 2. Family Reoviridae; 3. Family Bunyaviridae; 4. Also known as Mediterranean spotted fever.

bovine, and small ruminant babesiosis and bovine and ovine anaplasmosis in the Mediterranean area). *Rhipicephalus evertsi* and *R. turanicus* are two other species injurious to livestock.

The genus *Haemaphysalis* contains numerous species that attack mammals and birds. In North America, an important species is the widespread rabbit tick (*H. leporispalustris*). Larvae and nymphs attack ground-feeding birds as well as lagomorphs, whereas adults feed only on lagomorphs. Larvae and nymphs are active in the late summer and fall, whereas adults feed in the spring. This species contributes to the maintenance of Rocky Mountain spotted fever among wildlife. In India, an important species of this genus is *H. spinigera*, which occurs in dense forest habitat. Larvae feed on small mammals and ground-feeding birds, but nymphs and adults attack larger animals, including monkeys, cattle, and even humans. Larvae are active during October and November, nymphs from November to June, and adults mostly in July and August. This tick is the principal vector of the virus that causes Kyasanur Forest disease.

In Europe, *H. punctata* may transmit mild forms of the agents of bovine babesiosis (*B. major*) and theileriosis (*T. buffeli*) or babesiosis of small ruminants (*B. motasi*). *Haemaphysalis longicornis* infests cattle and other large mammals in eastern Asia and the Pacific area, transmitting the agent of bovine babesiosis (*B. ovata*) and a more pathogenic version of *T. buffeli* in eastern Asia. Ticks of the *H. leachi* group are vectors of the agent of a severe canine babesiosis (*B. rossi*) in Africa.

The American dog tick (*Dermacentor variabilis*) (Figure 26.15) is a major pest of people and domestic animals throughout much of the eastern and south-central United States as well as some areas of southeastern Canada. Tick populations generally decline west of the Mississippi River basin, although *D. variabilis* may be locally abundant in

Table 26.3 Representative Tick-borne Diseases of Veterinary Importance

Disease	Causative Agent	Primary Tick Vector Species	Affected Host(s)
Bovine babesiosis	*Babesia bigemina* *B. bovis*	*R. (Boophilus) annulatus* *R. (B.) microplus*, others	Cattle, water buffalo
East Coast fever	*Theileria parva*	*Rhipicephalus appendiculatus*	Cattle, Cape buffalo
Tropical theileriosis	*T. annulata*	*Hyalomma* spp.	Cattle, water buffalo
Feline cytauxzoonosis	*Cytauxzoon felis*	*Dermacentor variabilis*	Domestic and wild cats
Louping ill	*Flavivirus*	*Ixodes ricinus*	Sheep, grouse, others
African swine fever	*Iridovirus*	*Ornithodoros porcinus*	Domestic and wild pigs, warthogs
Tick-borne fever	*Anaplasma phagocytophilum*	*I. ricinus, I. scapularis, I. pacificus*	Domestic and wild ruminants, horses, dogs, humans
Canine ehrlichiosis	*Ehrlichia canis* *E. ewingii*	*R. sanguineus, I. ricinus* *A. americanum*, others	Dogs
Heartwater	*Ehrlichia ruminantium*	*Amblyomma hebraeum, A. variegatum*, others	Ruminants
Anaplasmosis	*Anaplasma marginale, A. centrale, A. ovis*	*Dermacentor* spp. *R. (Boophilus)* spp., *Hyalomma* spp., *Rhipicephalus* spp.	Cattle, sheep, other ruminants
Borrelioses	*Borrelia burgdorferi*	*Ixodes scapularis, I. ricinus I. pacificus, I. persulcatus*	Dogs, cats, cattle, others
Avian spirochetosis	*Borrelia anserina*	*Argas persicus*	Turkeys, chickens, birds, other
Epizootic bovine abortion	Probably a delta-proteobacterium	*Ornithodoros coriaceus*	Cattle, deer
Tularemia	*Francisella tularensis*	*D. andersoni*, others	Sheep, horses, rabbits, game birds
Q Fever	*Coxiella burnetii*	Many tick spp.	Most domestic animals
Tick paralysis	Tick proteins	*Ixodes holocyclus, I. brunneus, I. rubicundus, D. variabilis, Argas walkerae Rhipicephalus evertsi, D. andersoni*	Ruminants, other dogs, wild birds, chickens, mammats
Tick toxicoses	Tick proteins	*Ornithodoros savigny, O. lahorensis, A. persicus*	Cattle, sheep, birds
Sweating sickness	Tick proteins	*Hyalomma truncatum*	Cattle, sheep, other ruminants, dogs

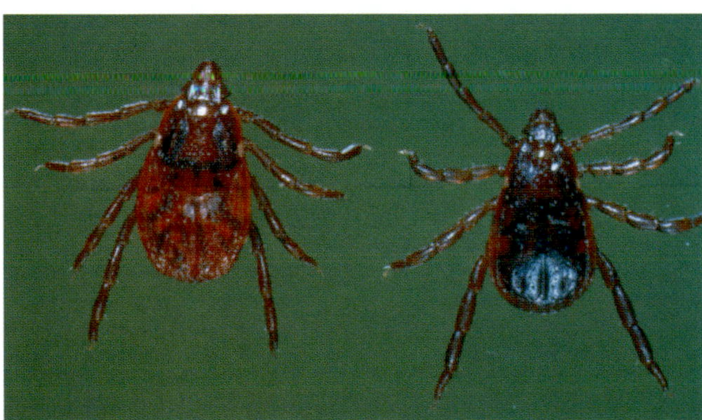

Figure 26.13 Brown dog tick (*Rhipicephalus sanguineus*); female, left; male, right. (Photo by Elton J. Hansens.)

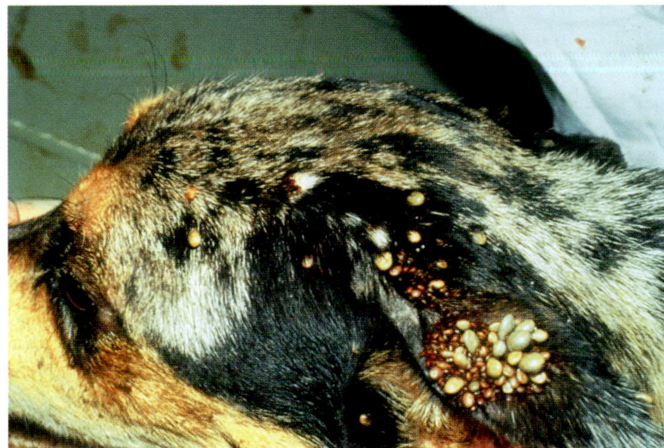

Figure 26.14 Dog heavily infested with the brown dog tick (*Rhipicephalus sanguineus*) about the head, and especially the ears. (Photo by Jerry F. Butler.)

some parts of the midwestern and far western United States. Larvae and nymphs feed on small mammals, but adults attack dogs, other medium-sized mammals, livestock, and humans. Larvae and nymphs are active in late winter and spring, and adults are most abundant in late spring and early summer. This species is the major vector of *Rickettsia rickettsii*, the agent of Rocky Mountain spotted fever, in the eastern United States. It also transmits the agents of tularemia and anaplasmosis and can cause tick paralysis in dogs and humans. In western North America, the closely related Rocky Mountain wood tick (*Dermacentor andersoni*) is an important pest attacking humans, livestock, and wildlife. Adults and nymphs of this tick attack almost any medium- or large-size mammal, whereas larvae attack small mammals. Adults and nymphs are active in late spring and early summer, and larvae are most abundant in the summer. *Dermacentor andersoni* is the primary

Figure 26.15 American dog tick (*Dermacentor variabilis*); female, left; male, right. (Photo by Gary R. Mullen)

Figure 26.16 Western blacklegged tick (*Ixodes pacificus*), female. (Photo by James Gathany, Centers for Disease Control and Prevention, USA)

vector of *R. rickettsii* and Colorado tick fever virus in this region. It also transmits *Anaplasma marginale*, which causes anaplasmosis in domestic ruminants. In the Pacific Northwest, *D. andersoni* is an important cause of tick paralysis.

Dermacentor spp. are also important in Eurasia, and *D. reticulatus* is the main vector of the agents of European canine babesiosis (*Babesia canis*) and equine babesiosis (*B. caballi*); it can also transmit the agent of bovine anaplasmosis. *D. marginatus* is also an importance pest of sheep and is involved in the epidemiology of Q fever.

Dermacentor nitens (until recently placed in a separate genus, *Anocentor*) is an important pest of livestock in tropical Central and South America, and parts of the Caribbean. It infests the ears, and is a vector of the agent of equine babesiosis.

The blacklegged tick (*Ixodes scapularis*) is widespread throughout large areas of the eastern, south-central, and midwestern United States. The immature stages usually feed on small mammals, lizards, and birds whereas adults are most common on white-tailed deer. All stages of *I. scapularis* will bite humans. Nymphal ticks, the stage most likely to transmit Lyme disease spirochetes to people, are active in late spring and early summer. Adults are active in the fall and early spring (and winter in southern latitudes). Larvae are most abundant in the summer. *Ixodes scapularis* is the primary vector of the Lyme disease spirochete *Borrelia burgdorferi*, the protozoan *Babesia microti* that causes human babesiosis, and *Anaplasma phagocytophilum*, the agent of human granulocytic anaplasmosis. In western North America, the western blacklegged tick (*Ixodes pacificus*) (Figure 26.16) is the primary vector of both *B. burgdorferi* and *A. phagocytophilium* to humans. The immatures feed on lizards, ground-frequenting birds, and small mammals, whereas the adult ticks attach to medium- or large-sized mammals (e.g., carnivores, lagomorphs), particularly deer. Like *I. scapularis*, nymphal ticks are most apt to transmit *B. burgdorferi* to people during the spring and early summer. Individuals bitten repeatedly by this tick may develop severe allergic hypersensitivity reactions.

In Europe, the castor bean tick, or sheep tick (*I. ricinus*), is a major pest of livestock and humans. This tick ranges from Ireland, Britain, and Scandinavia across continental Europe to Iran and southward to the Mediterranean Sea. In Britain and Ireland, it is commonly found in poorly maintained, overgrown sheep pastures that contain dense mats of moist, rotting vegetation ideal for tick development and survival. On the European continent, *I. ricinus* abounds in mixed hardwood-pine forests and shrubs but rarely in grassy meadows. Larvae and nymphs attack mostly small mammals, insectivores, birds, and lizards. Adults are found most commonly on sheep, other domestic ruminants, and deer. However, this tick may attack virtually any vertebrate, including humans. Seasonal activity varies greatly in different regions throughout the tick's range. *Ixodes ricinus* transmits the agents of Lyme disease, which, in Europe, include *Borrelia burgdorferi*, *B. garinii*, and *B. afzelius*. In addition, *I. ricinus* is the major vector of the virus that causes Tick-borne encephalitis and of *A. phagocytophilum*. In Ireland, Britain, and some areas of western Europe, *I. ricinus* also transmits the virus that causes louping ill in sheep. It is also a vector of human babesiosis and is of importance to livestock as the vector of *B. divergens*, which causes bovine babesiosis. Further to the east, it is replaced by the aggressive **taiga tick**, *Ixodes persulcatus*, another major vector of human pathogens including borreliae, *A. phagocytophilum*, and tick-borne encephalitits virus. Interestingly, it is the adult female, not the nymph, that serves as the primary vector of zoonotic pathogens to people.

In Australia, an important species is the Australian paralysis tick (*Ixodes holocyclus*). This tick is found along the eastern coast of Queensland and Victoria province. It feeds on most wild mammals, domestic animals, and humans. *Ixodes holocyclus* is notorious as the cause of tick paralysis in Australia. In contrast to other diseases caused by an infectious microbe, tick paralysis is caused by a proteinaceous material, holocyclotoxin in the case of *I. holocyclus*, secreted in the tick's saliva. Even the bite of a single tick may be sufficient to cause a fatal paralysis.

Many species of *Hyalomma* are vectors of *Theileria annulata*, the agent of bovine tropical theileriosis, a major disease of cattle and domestic buffalo in much of Asia, including the Middle East, the Mediterranean basin, parts of southern Europe, and some parts of northern sub-Saharan Africa. The agent of theileriosis virulent to small ruminants, *Theileria recondita*, also is transmitted by *Hyalomma* species. Important vectors include *H. detritum*, *H. anatolicum*, *H. asiaticum*, and *H. lusitanicum*. In the Mediterranean basin and parts of the former Soviet Union (the Crimea and adjacent areas of the former USSR), an important tick is *Hyalomma marginatum*. Larvae and nymphs attack hares, hedgehogs, and birds. Adults attack larger mammals, including domestic ruminants and humans. This tick is one of the most important vectors of Crimean-Congo Hemorrhagic Fever virus. In the adult stage, *Hyalomma* spp. infest particularly the perianal area, the perineum, or the tail switch, where they escape visual detection.

The lone star tick (*Amblyomma americanum*) (Figure 26.17) is one of the most notorious tick pest species in the United States. It is found along the Atlantic coast from New York to Florida and west into Texas and Oklahoma. *Amblyomma americanum* larvae, nymphs, and adults readily attack humans, companion animals, and livestock, as well as wildlife. Virtually any mammal or ground-feeding bird may be infested. It is often abundant in areas with large populations of deer, which serve as the primary hosts for the adult ticks. In the southeastern United States, nymphs and adults emerge from their winter diapause and commence host-seeking activity in late spring. Larvae generally appear in late summer. Seasonal activity may be delayed farther north. *Amblyomma americanum* has been implicated as a vector of the agents that cause human ehrlichiosis (*Ehrlichia chaffeensis and E. ewingii*) and tularemia (*Francisella tularensis*).

Figure 26.17 Lone star tick (*Amblyomma americanum*); female. (Photo by James Gathany, Centers for Disease Control and Prevention, UAS)

Another important species in the United States is the Gulf Coast tick (*Amblyomma maculatum*) (Figure 26.18), which is found in the southeastern and south-central United States and Central America. Larvae and nymphs attack a wide range of birds and mammals, but adults feed largely on ruminants. These ticks feed mainly on the head and ears. *Amblyomma maculatum* can cause severe injury to the skin of cattle and other livestock, often rendering the hides useless from the bites of these ticks or from secondary infections and predisposing to screwworm and severe dermatophilosis. It is an efficient experimental vector of *Ehrlichia ruminantium* that causes heartwater, a major African disease of ruminants, which has been imported into the Caribbean area. This species recently has been shown to transmit *Rickettsia parkeri* to humans.

In Africa, the bont tick (*Amblyomma hebraeum*) (Figure 26.19) and the tropical bont tick (*A. variegatum*) attack livestock as well as wild ruminants. In addition, *A. variegatum* larvae and nymphs will parasitize ground-feeding birds, including herons and other migratory birds, and small mammals. *Amblyomma hebraeum* is restricted to southern Africa, but *A. variegatum* ranges throughout most of sub-Saharan Africa, Madagascar, and several islands in the Caribbean. These ticks are the major vectors of the rickettsia *Ehrlichia* (formerly *Cowdria*) *ruminantium* that causes heartwater in ruminants, whereas *A. variegatum* is also associated with severe forms of ruminant dermatophilosis (see later). They are also vectors of *Rickettsia africae*, the causative agent of African tick bite fever in humans, which also has been inadvertently introduced into the Caribbean.

Arguably the most important livestock tick on a global scale is the one-host southern cattle fever tick, *Rhipicephalus* (*Boophilus*) *microplus*. This species, and

Figure 26.18 Gulf Coast tick (*Amblyomma maculatum*). (A) Female; (B) male. (Photo by James Gathany, Centers for Disease Control and Prevention, USA)

Figure 26.19 Bont tick (*Amblyomma hebraeum*), male. (Courtesy of Jerry F. Butler)

other members of the subgenus *Boophilus*, can cause high infestations in cattle and some other ungulates (Figure 26.20). The southern cattle fever tick originated in southern Asia, but is now also established in Australia, the Pacific area, Mexico, Central and tropical South America, the Antilles, Madagascar, and large parts of eastern and southern Africa, where it has replaced the indigenous tick *B. decoloratus*. When present in large numbers it is the cause of retarded growth and weight loss. It is even more important, however, as the main vector of *Babesia bovis* and *B. bigemina*, agents of bovine babesiosis and of *Anaplasma marginale*, which causes anaplasmosis. Other important species are *B. decoloratus* (although it is not a vector of *Babesia bovis*), present in much of sub-Saharan Africa, and *B. geigyi*, which has replaced it in West Africa.

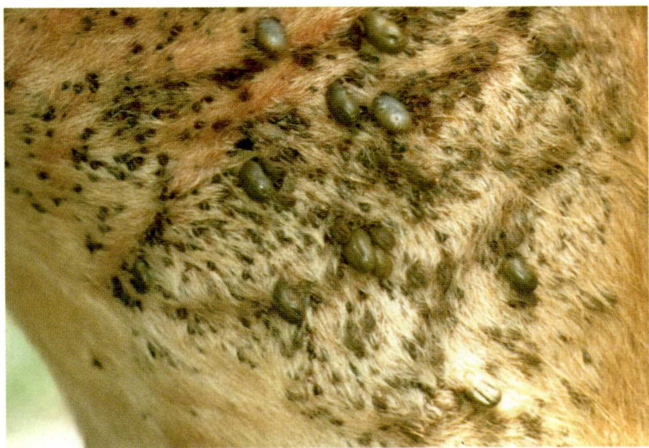

Figure 26.20 Deer heavily infested with ticks, *Rhipicephalus* (*Boophilus*) sp. (Courtesy of US Department of Agriculture, Animal Research Service, Kerrville, TX)

Another species of major importance is the cattle fever tick, *Rhipicephalus (B.) annulatus*, ranging throughout large areas of North and sub-Saharan Africa north of the equator, parts of southern Europe, and western Asia and parts of North America, Central America, and South America. Only intensive surveillance has prevented its reintroduction, as well as that of *B. microplus*, into the United States from tick-infested herds in Mexico. *Boophilus annulatus* is active throughout the year in the tropics. This one-host tick feeds almost exclusively on cattle, but it also infests white-tailed and other species of deer. It is a major pest of cattle, causing reduced weight gains and milk production in heavily infested animals. It is best known for its role in the transmission of the protozoan *Babesia bigemina*, which causes Texas cattle fever, and as a vector of *B. bovis* and *A. marginale*.

Among the Argasidae, the fowl tick (*Argas persicus*) and related species are important parasites of poultry in the Old World. All stages feed on these birds. Populations of this tick can reach enormous numbers in poultry barns and can cause high mortality due to exsanguination. This tick is the vector of the rickettsia *Aegyptianella pullorum*, which causes fowl disease in domestic fowl. In the Mediterranean region, it is a vector of *Borrelia anserina*, the agent of fowl spirochetosis, an important poultry disease. In the New World, the fowl tick is supplemented by a complex of three species (*A. radiatus*, *A. sanchezi*, and *A. miniatus*), in addition to the introduced and now established (but rare) *A. persicus*.

The genus *Ornithodoros* includes several species that live in animal burrows and poorly maintained homes or shelters, where they hide in cracks and crevices of walls, ceilings, and attics. In the western United States, *O. hermsi* often infests mountain cabins and other dwellings. Although rodents that infest dwellings are the principal hosts of *O. hermsi*, humans may be attacked when they enter such dwellings if rodents have been killed or driven out. This tick is notorious as a vector of the relapsing fever spirochete *Borrelia hermsii*.

In eastern and southern Africa, the human-biting African tampan (*Ornithodoros moubata*) and related species, such as *O. porcinus*, coexist with people and animals in mud huts where the tick hides in the walls. These species are the major vectors of the relapsing fever spirochete *Borrelia duttonii*. *Ornithodoros porcinus* may be involved in maintaining the virus of African swine fever between warthogs and ticks. The virus becomes directly contagious when it spreads to domestic pigs, and subsequently has been temporarily introduced to other continents. African swine fever is of major international importance. It has been difficult to eradicate from Spain and Portugal, where it has become established in a local species, *Ornithodoros erraticus*. *Ornithodoros porcinus* also occurs on Madagascar. Another species, *O. sonrai*, is thought to be implicated in the epidemiology of African swine fever in West Africa.

Ornithodoros savignyi is a major pest of camels, other domestic animals, and humans in the drier parts of Africa and southern Asia. Although it is not known to transmit pathogenic organisms, it often occurs in large numbers in the sand of sites where animals and humans congregate (e.g., resting sites, wells) and may be responsible for loss of blood and bites, which remain painful and itch for long periods.

Another important argasid in western North America is the spinose ear tick (*Otobius megnini*). It has become established in India, Madagascar, Kenya, and Turkey. It frequently infests livestock, especially cattle and horses, and most domestic ruminants. *Otobius megnini* also attacks wild ruminants, especially deer, antelope, mountain sheep, and even humans. The larvae and second-stage nymphs feed, whereas the adults do not. This tick feeds in the ears, causing injury to the auditory canal (the nymphs are covered with spines) and secondary infections.

In the central and eastern United States, the bat tick *Carios kelleyi* (Figure 26.5) has been shown to feed occasionally on humans in bat-infested houses. An erythematous skin rash, presumably due to a reaction to the bite, may occur (Gill et al., 2004).

PUBLIC HEALTH IMPORTANCE

Ticks are of public health significance mainly because of the zoonotic disease agents transmitted by them, which include an array of bacterial, viral, and protozoan disease agents (Harwood and James, 1979; Sonenshine, 1993; Goodman et al., 2005). They also are important because their attachments can cause various kinds of dermatoses or skin disorders, such as inflammation, pain, and swelling. Rarely, they invade the auditory canal producing a condition known as otoacariasis. Certain species of ticks may cause a flaccid, ascending, and sometimes fatal paralysis known as tick paralysis. Individuals bitten repeatedly by some ticks may develop allergic or even anaphylactic reactions (Van Wye et al., 1991).

Among the biological factors that contribute to the high vector potential of ticks are their persistent blood-sucking habit, longevity, high reproductive potential, relative freedom from natural enemies, and highly sclerotized bodies that protect them from environmental stresses. Further, the slow feeding behavior of ixodid ticks permits wide dispersal and increases their likelihood of acquiring pathogens during attachment to a host. Transstadial passage of microbial disease agents from larva to nymph or nymph to adult commonly occurs in vector ticks, transovarial transmission of many agents occurs in some ticks, and both phenomena contribute to the maintenance and spread of certain tick-borne agents. Intrastadial transmission of pathogens sometimes occurs when male ixodid ticks, which feed more than once, move from one host individual to another.

Several other biological attributes of ticks also enhance their vector potential. First, pharmacologically active substances present in the saliva of ticks may promote feeding success and aid transmission of microbial agents. For example, the saliva of *I. scapularis* has anti-edema, antihemostatic and immunosuppressive properties. Second, ticks imbibe large quantities of blood during each feeding period. Indeed, certain species may increase their body weight by 100-fold or more. This is actually an underestimate of the amount ingested because feeding ticks concentrate the blood meal by secreting copious amounts of host-derived fluid back into the host. Third, ticks take multiple blood meals during their lifetimes. Those individuals that attain adulthood and that successfully feed as adults feed three (ixodids) or more (argasids) times.

It should be noted that ticks are far more efficient than insects in maintaining microbial agents in their bodies. In ticks, most internal tissues change gradually during development and transstadial survival of pathogens occurs frequently. In holometabolous insects, however, the extensive internal changes that occur during molting seem to have a harmful effect on most microorganisms that cause human disease. Ticks, like most other animals, have a well developed innate immune system for preventing infection. This system includes cellular defenses and humoral elements (Johns et al. 1998; Rudenko et al. 2005). The tick's immune system recognizes and distinguishes invading microbes as foreign by detecting foreign molecular structures, known, collectively, as pathogen-associated molecular patterns (PAMPs). This in turn triggers the various immune responses noted below.

Invading microbes, especially gram-positive bacteria, are recognized and destroyed by phagocytosis, mostly by plasmatocytes or granulocytes (Borovickova and Hypsa 2005; Inoue et al. 2001). Gram-negative bacteria clump together upon contact with the hemolymph, a response presumably mediated by lectins (see below). In this process, known as nodulation, the aggregate is surrounded by degranulating hemocytes and destroyed (Ceraul et al. 2002). In some cases, invading particles are trapped by hemocytes, surrounded and encapsulated (Eggenberger et al. 1990).

Microbial challenge also upregulates humoral expression of antimicrobial peptides (AMPs) and immunoproteins. Among the most important are defensin, typically small (~4 – 6 kilodaltons) cationic peptides that attack gram positive microorganisms (Ceraul et al. 2003; Hynes et al. 2005; Nakajima et al. 2001). Some novel defensins have unusual structures and are active against gram-negative as well as gram-positive bacteria (Lai et al. 2004). Other important humoral agents are the protease inhibitors that inactivate or destroy microbial proteases, lectins that bind to invading bacteria (Grubhoffer et al. 2004), identifying them for phagocytosis (opsonizing activity) or nodulation and factors that induce hemolymph coagulation, entrapping many invading microbes as well as facilitating wound healing.

Blood feeding upregulates expression in the tick's midgut of proteases, protease inhibitors (cystatins and serpins), lectins, GSTs, peroxiredoxins and other oxidative stress reducing proteins that also contribute to the destruction of invading microbes (Anderson et al. 2008). However, pathogenic microbes, e.g., *Borrelia burgdorferi* able to bind to specific receptors (Pau et al. 2004) may avoid destruction and survive to invade the tick's body. For further information about innate immunity in ticks, the reader is referred to Sonenshine and Hynes (2008).

As reviewed by Lane (1994) and Nuttall and Labuda (1994), ticks transmit microbes by several routes, including salivary secretions (e.g., Lyme disease spirochete, Colorado tick fever virus, the agent of heartwater, and spotted fever group rickettsiae), coxal fluids (certain species of relapsing fever spirochetes), regurgitation (e.g., possibly the spirochetes that cause Lyme disease), and via feces (Q fever organisms). A novel type of transmission, saliva-activated transmission, occurs in the case of some tick-borne arboviruses (Jones et al., 1992). In this model, one or more proteins secreted in tick saliva potentiate virus transmission. Moreover, this phenomenon seems to be the mechanism underlying **nonviremic transmission**, whereby arboviruses are transmitted from infected to uninfected ticks feeding simultaneously on a vertebrate host having no or very low levels of viremia (Nuttall and Jones, 1991; Nuttall and Labuda, 1994). Transmission between cofeeding infected and uninfected ticks, which also has been demonstrated for the Lyme disease spirochete, *Borrelia burgdorferi*, is important epidemiologically for two reasons (Randolph et al., 1996). First, some vertebrates that do not develop systemic infections still can serve as competent hosts for infecting vector ticks, and second, it adds yet another transmission route for certain tick-borne pathogens. Although some tick-borne agents may be transmitted by two routes (e.g., transmission of certain relapsing fever spirochetes via coxal fluid secretions and by saliva), only one route is usually significant.

The more important tick-borne diseases of public health concern are summarized in Table 26.2. The causative agent, clinical manifestations, ecology, and epidemiology of each of these diseases are discussed next.

Human Babesiosis

Human babesiosis is an emerging disease caused by several species of protozoans in the genus *Babesia*. This genus also contains species of major veterinary importance, as do the related protozoan genera *Cytauxzoon* and *Theileria*. Species in all three genera belong to the family Babesiidae, order Piroplasmorida, and phylum Apicomplexa. They often are referred to as **piroplasms** because they possess pear-shaped, intraerythrocytic merozoites in the vertebrate host. *Babesia* spp. resemble malarial parasites (*Plasmodium* spp.) and other blood-infecting protozoans, especially regarding their developmental cycles. More information about these important parasites can be found in Kjemtrup and Conrad (2000), Zintl et al. (2003), Schetters and Brown (2006), and Hunfeld et al. (2008).

Wilson and Chowning in the early 1900s were the first to incriminate babesial parasites as a probable cause of human infection among patients with Rocky Mountain spotted fever in the western United States. However, the first definitive case was not described until 1957 in a splenectomized Yugoslavian cattle farmer who died of a babesial infection following an eight-day illness. In the United States, the disease initially was recognized in a Californian in 1968. To date, approximately 60 cases of human babesiosis have been reported from Europe, over 300 cases from the United States, and sporadic cases reported elsewhere. In the United States, human babesiosis occurs principally along the eastern seaboard, especially on Nantucket Island, Massachusetts and Long Island, New York, where the etiologic agent is *Babesia microti*. Other endemic foci of *B. microti* occur in Connecticut, Minnesota, and Wisconsin. The incidence appears to be increasing in Wisconsin, where 72% of the 32 cases reported from 1996 to 2005 occurred during 2004 to 2005. In the far-western United States, the recently described *Babesia duncani* has been identified in nine patients since 1991 (Conrad et al., 2006). Besides the index case, four of the patients had previously had their spleen removed (one died), and two each were blood donors or blood recipients. This piroplasm, formerly designated as the WA1-type *Babesia* in the literature, lies in a distinct clade separable from *Babesia* sensu stricto, *B. microti*, and *Theileria* spp.

At least 70% of the cases in Europe are associated with the cattle piroplasm *B. divergens* (Genchi, 2007). Intriguingly, *Babesia* parasites similar but not identical to *B. divergens* have been detected in three asplenic men in Missouri (1992), Kentucky (2001), and Washington State (2002) (Herwaldt et al., 2004). Serosurveys suggest that a low percentage of Europeans (≤3.4%) from several countries may be infected with *B. microti*, particularly individuals who engage in high-risk outdoor activities like forestry. A new *Babesia* species, *Babesia* sp. EU1 (provisionally named *B. venatorum*), first described in 2003 based on isolates obtained from two asplenic men in Austria and Italy, is an emerging zoonosis (Herwaldt et al., 2003). Phylogenetically, this organism is most closely related to *B. odocoilei*, a parasite of white-tailed deer in North America.

In humans, *Babesia* species may produce a malarial-like disease without the periodicity that often accompanies the human malarias. Following an incubation period of one to four weeks, the clinical course varies according to the etiologic agent and ranges from subclinical infection to a severe disease with sudden onset. Splenectomized persons infected with either *B. divergens* or *B. microti*, or elderly persons infected with *B. microti*, tend to develop severe or sometimes fatal illnesses. Signs and symptoms at onset include fever, chills, profuse sweating, headache, and generalized muscle aches. Joint pain, nausea, vomiting, and prostration may occur. Parasitemia and the resultant clinical course may persist for several months with severe anemia, jaundice, and hemoglobinuria. In many individuals, however, babesiosis is a mild, self-limited disease that requires only supportive therapy. Clindamycin and quinine in combination are the current drugs of choice, but azithromycin and atovaquone in combination are equally effective in treating babesiosis patients with fewer adverse effects.

With few exceptions, *Babesia* species develop entirely within circulating red blood cells. Sporozoites are introduced via the saliva of a *Babesia*-infected tick during feeding. Once they gain entry into the bloodstream, most parasites develop asexually within red blood cells. Occasionally, *Babesia* invade lymphocytes, and only subsequent generations develop in the erythrocytes. Within the host cell, the parasites develop into trophozoites termed **meronts**, which multiply asexually by binary fission to produce merozoites. Some of the merozoites escape from the disintegrating host cells to invade other erythrocytes and continue the cycle. Other merozoites develop into gametocytes called **piroplasms** after entering previously uninfected erythrocytes. The gametocytes remain in an arrested state of development until they are ingested by a feeding tick.

When *Babesia*-infected blood is ingested by ticks, the gametocytes commence development, but the asexual stages are destroyed. The gametocytes escape from the dying host cells and transform into gamete-forming cells called gamonts, which develop structures (rays and spines) that are subsequently used to penetrate cells. Following gametic fusion, the resulting zygotes invade the tick's digestive epithelium, develop into motile kinetes, and migrate to other internal organs. Some other *Babesia* spp., such as *B. divergens* and *Babesia* sp. EU1, invade the female tick's ovaries and are transmitted transovarially to the next generation, whereas others (e.g., *B. microti*) are not passed via the eggs. Instead, immature ticks are infected while feeding on a parasitemic host, the parasites invade the salivary glands, multiply, and are passed transstadially to the next stage.

In the northeastern and upper midwestern United States, *B. microti* is maintained in a transmission cycle involving the black-legged tick (*I. scapularis*) and the white-footed mouse. Meadow voles also are efficient reservoir hosts. Most people who acquire the infection are bitten by nymphal ticks. In the far-western United States the primary tick vector(s) and reservoir host(s) of *B. duncani* have not been identified, but the close similarity of babesial isolates from mule deer with *B. duncani* isolates from humans suggests that large ungulates serve as reservoirs. In Europe, *I. ricinus* is the primary vector of both *B. divergens* and *Babesia* sp. EU1, whereas *Ixodes trianguliceps* transmits *B. microti* among small mammals. The apparent paucity of human *B. microti* infections in Europe may be attributable to the fact that *I. trianguliceps* is a nidicolous tick that seldom attaches to people.

Babesia spp. may be transmitted by two other routes besides tick-feeding: blood transfusion and transplacental. Over 50 cases of transfusion-associated *B. microti* infections have been reported in North America, and 10 cases of neonatal babesiosis have been published. *Babesia* ranks second only to *Plasmodium* among blood transfusion-acquired parasitic infections. Estimates of the percentages of *Babesia*-infected blood products in certain endemic settings in the United States (e.g., Connecticut, Massachusetts) range between 0.17 and 3.7%.

Tick-Borne Encephalitis Complex

First described as Russian spring-summer encephalitis (RSSE) in 1932 from the far-eastern region of the former Soviet Union, Tick-borne Encephalitis (TBE) was recognized after World War II in central Europe, where it was termed Central-European Encephalitis (CEE). RSSE and CEE are now considered to represent a single entity, TBE, which has been classified into three subtypes (European, Siberian, and Far Eastern) that vary in virulence for humans. In 2007, a novel variant of the far-eastern subtype was isolated from the brain of a 15-year old boy in Primorsky District, Russia, who succumbed to the infection. Grard et al. (2007) proposed that TBE viruses be divided into four types—Western, Eastern, Turkish sheep, and Louping ill. TBE is one of at least 12 related, but distinguishable, serotypes of tick-borne flaviviruses (family Flaviviridae) that constitute the TBE complex. It includes such viruses as Louping ill, Kyasanur Forest disease, Omsk hemorrhagic fever, and Powassan encephalitis. Each of these viruses produces a clinically distinctive disease.

TBE is endemic in nearly 30 areas of Europe and northern Asian countries, and the incidence is estimated to be as many as 14,000 cases per year, with about 11,000 of them occurring in Russia. Few arthropod-borne zoonotic agents have received as much scientific scrutiny. In Russia, for instance, researchers published approximately 5,000 to 6,000 articles and 40 to 50 monographs on various aspects of TBE during the first 60 years following its discovery (Korenberg and Kovalevskii, 1999). In that regard, it has been estimated that 20,000 to 30,000 autonomous natural foci, ranging in size from a few square kilometers to several hundred kilometers, exist in Russia.

Illness in humans is accompanied by high, often biphasic, fever and headache, followed soon afterward by inflammation of the brain (encephalitis) and meninges (meningitis). Some patients develop muscle weakness or paralysis, especially in the right shoulder muscles. Case-fatality rates average about 1 to 2% for European strains, 20 to 60% for far-eastern strains, and rarely exceed 6 to 8% for Siberian strains (Charrel et al., 2004). Although Siberian strains are less lethal than far-eastern strains, they nevertheless tend to cause chronic or prolonged infections.

Fortunately, there are several highly effective, safe, and well-tolerated vaccines commercially available against TBE viruses in Europe and Russia. The two highly purified, formalin-inactivated, whole-virus vaccines developed in Europe have an overall efficacy of 99% when used in accordance with the recommended vaccination schedule. Notwithstanding, the number of reported cases of TBE increased an astounding 400% in Europe between 1974 and 2003 due to the complex interaction of various ecological, economic, social, political, and climatic factors (Kunze et al., 2007). A notable exception has been Austria, which has experienced a dramatic decline in clinical cases as a result of increased vaccination coverage, from approximately 6% in 1980 to 88% of the entire population in 2006.

The widespread use of vaccines in Austria from 2000 to 2006 is estimated to have prevented approximately 2,800 cases and 20 deaths from TBE (Heinz et al., 2007). In stark contrast, vaccination coverage in TBE-endemic countries bordering Austria is meager: 11% in the Czech Republic and 13% each in Germany and Switzerland.

Climatic changes may contribute to the geographic expansion or resurgence of some vector-borne diseases. They alone, however, cannot explain the recent upsurge in the incidence of TBE or the pronounced spatio-temporal heterogeneity of the virus in Central Europe and the Baltic Region (Rogers and Randolph, 2006; Randolph and Ŝumilo, 2007). Anthropogenic impacts on the landscape have allowed tick populations to expand and multiply; changes in human behavior may have resulted in a greater degree of contact with virus-laden ticks; and migrating birds can disperse TBE-virus infected *I. ricinus* ticks (Randolph, 2001; Waldenström et al., 2007). In Estonia, Latvia, and Lithuania, environmental changes resulting from political upheaval, and socioeconomic transitional factors following the end of Soviet rule that presumably elevated human contact with infected ticks, have been posited to play an important role in the increased incidence of TBE. Among Latvians, harvesting mushrooms and berries or working in forests have been associated with unemployment, lower incomes, increased forest visitation, and a higher than average risk of being tick-bitten the previous year (Randolph and Ŝumilo, 2007). Weather conditions also may influence the frequency of forest visits and therefore the degree of tick exposure in this region.

The primary vectors of TBE viruses are *Ixodes ricinus* (European subtype) and *I. persulcatus* (Siberian and far-eastern subtypes). Other tick species that have been found infected naturally (*Ixodes arboricola, I. hexagonus, I. trianguliceps*) may amplify viral infection. Although most mammals are susceptible to TBE virus, rodents, especially the bank vole (*Clethrionomys glareolus*), field mice (*Apodemus* spp.), and insectivores are the chief reservoir hosts. Viral amplification and enzootic maintenance occur by means of the seasonally synchronized cofeeding of virus-infected nymphs and large numbers of uninfected larvae during brief periods (2–3 days) of nonviremic infectivity within primary vertebrate hosts (Randolph et al., 1999; Randolph and Ŝumilo, 2007). Thus, transstadially infected nymphs transmit the virus horizontally to uninfected larvae, which molt up to a year later to produce infected nymphs. This nonviremic route of transmission between cofeeding ticks can even occur in rodents that are immune to TBE virus (Labuda et al., 1997). The principal environmental driver for synchronizing the springtime feeding of larvae and nymphs in TBE foci initially was thought to be a rapid rate of cooling in autumn, corrected for mid-summer maximum temperatures (Randolph et al., 2000). More recent evidence suggests that the rate of spring warming, corrected by January minimum temperatures, is more important in synchronizing larval and nymphal feeding activities than is the rate of autumnal cooling (Randolph and Ŝumilio, 2007).

Powassan virus, named after the town in Ontario, Canada, where it was originally isolated from a five-year-old boy who succumbed to the infection, is a *Flavivirus* related to TBE viruses. It was first recognized initially in scattered localities in the United States, Canada, and in eastern parts of the former Soviet Union. The virus causes a disease known as Powassan encephalitis, which is characterized in its acute stage by encephalitis, severe headache, and fever. Nausea, labored breathing, and neurologic disorders, including partial paralysis, occur frequently. Recovered patients may suffer permanent nerve damage. In North America, the disease in humans is rare as only 31 cases were reported from the northeastern United States and Canada from 1958 to July 2001, with an average case-fatality rate of about 10 to 15% (CDC, 2001). Case recognition is increasing in the United States, however, as 9 cases were reported from 5 eastern states from 1999 to 2005. Human cases also have been recorded in Russia (Charrel et al., 2004).

Tick vectors of Powassan virus belong to the genera *Ixodes, Dermacentor,* and *Haemaphysalis.* In the United States, isolates of the virus have been obtained from *Ixodes cookei* and *Ixodes marxi* in the East and *Dermacentor andersoni* and *Ixodes spinipalpis* in the West. *Ixodes cookei* feeds on various wild and domestic animals and rarely on humans. Marmots (woodchucks) are important hosts of *I. cookei* and are excellent reservoirs of the virus. Similarly, the snowshoe hare amplifies populations of vector ticks and the virus. The virus has been isolated twice from naturally infected foxes, a red squirrel, a white-footed mouse and a spotted skunk, but the reservoir competence of these species remains to be determined. Antibodies to the virus have been detected in 38 wild and five domestic mammalian species. *Dermacentor andersoni* is the most important vector in the western United States and Canada. In the former Soviet Union, the virus has been isolated from *Haemaphysalis neumanni, I. persulcatus,* and *Dermacentor silvarum,* and from mosquitoes. *Apodemus* mice and *Microtus* voles are the primary vertebrate hosts in the Eastern Hemisphere (Charrel et al., 2004).

In the northeastern United States, a genotype or subtype of Powassan virus identified during the 1990s is called deer tick virus (DTV). About 3 to 4% of white-footed mice surveyed in Massachusetts, Rhode Island, and Wisconsin were seroreactive against DTV test-antigen, and the virus was detected in host-seeking *Ixodes scapularis* adults from areas that yielded seroreactive mice (Ebel et al., 2000). In the laboratory, 90% of *I. scapularis* larvae acquired DTV from needle-inoculated mice; the efficiency of transstadial passage was 22%; and the resultant nymphs transmitted the infection to naïve mice after having been attached for as few as 15 minutes (Ebel and Kramer, 2004). To date, clinical disease in humans attributable to DTV has not been reported.

Colorado Tick Fever

Colorado Tick Fever (CTF) is caused by a *Coltivirus* in the family Reoviridae. Coltiviruses were divided into two subgroups, A and B, based upon their genetic relatedness. North American and European species were placed in subgroup A and Asian species in subgroup B (Marfin and Campbell, 2005). In 2000, researchers proposed that subgroup B coltiviruses were sufficiently distinct to warrant inclusion in a separate genus, *Seadornavirus*. Subgroup A coltiviruses currently comprise four antigenically related viruses: CTF virus in western North America; Salmon River virus in Idaho; Eyach virus in the Czech Republic, France and Germany; and California hare coltivirus in the United States (Attoui et al., 2005). The latter virus was isolated first from a western gray squirrel in 1965, and a similar if not identical virus was isolated 11 years later from a black-tailed jack rabbit (Lane et al., 1982). Notably, these isolates originated in west-central or northwestern California far outside the distributional ranges of the primary mammalian hosts of CTF virus and of the Rocky Mountain wood tick (*Dermacentor andersoni*), the primary bridging vector to humans. Furthermore, this virus is the only one of the subgroup A coltiviruses that has not been associated definitively with human illness.

Symptoms of CTF usually appear within four days (range, 1–14 days) following the attachment of an infected tick. The disease is characterized by a biphasic fever, chills, headache, generalized musculoskeletal aches, and malaise. Some patients experience eye pain, intolerance of light, chills, sore throat, and nausea. The virus develops in most internal organs and may spread to the brain or bone marrow. Although CTF sometimes is depicted as a mild febrile illness, acutely ill patients usually are bedridden, and as many as 14% require hospitalization (Marfin and Campbell, 2005). Convalescence may be prolonged, with some patients taking several weeks to recover. Case-fatality rates are very low, usually less than 0.2%, and all reported deaths have involved children. Early in the course of disease, CTF may be mistaken for Rocky Mountain spotted fever because up to 12% of CTF patients may develop a maculopapular or petechial rash.

In endemic regions, people engaged in outdoor activities in mountainous or highland areas from about 4000 feet to over 10,000 feet (1219 meters to 3048 meters) are at risk of exposure to virus-infected ticks. In Rocky Mountain National Park, Colorado, natural foci occur on south-facing slopes covered with open stands of pine and shrubs on dry, rocky surfaces. Cases are reported from March to November, but most occur in the spring and early summer when adult and nymphal ticks are active. The distribution of CTF approximates that of *D. andersoni* in western North America. The virus has been isolated from ticks, humans, or both from parts of the United States and Canada. In the United States, 476 (61%) of 777 cases reported to 12 state health departments between 1987 and 2001 were contracted in Colorado, with Utah (n = 122) and Montana (n = 106) ranking second and third (Marfin and Campbell, 2005). Risk factors for the disease include being male, 10 to 49 years of age, and occupational or recreational exposure at higher elevations in the Rocky Mountains or other endemic mountainous areas of the western United States. Transfusion-associated CTF has been reported at least once.

CTF virus is passed efficiently from stage-to-stage in *D. andersoni* ticks, but transovarial passage does not occur. Therefore, the virus is maintained horizontally as host-seeking nymphs, previously infected while feeding as larvae on viremic hosts, attach to and infect susceptible small mammalian hosts. The virus has been isolated from *D. albipictus*, *D. occidentalis*, *D. parumapertus*, *Haemaphysalis leporispalustris*, *Ixodes sculptus*, *I. spinipalpis*, and *Otobius lagophilus*. Larvae and nymphs of *D. andersoni* feed on small mammals, especially ground squirrels, mice, and rabbits. Nymphs, which quest higher in vegetation than larvae, also attack larger mammals like small carnivores and occasionally humans. Important hosts of the immatures include golden-mantled ground squirrels, deer mice, bushy-tailed woodrats, chipmunks, and rabbits. Adults parasitize larger mammals, such as porcupines, elk, deer, antelope, carnivores, and humans. Competent reservoir hosts include the golden-mantled ground squirrel, least chipmunk, deer mouse, bushy-tailed woodrat, and porcupine. Viremia in amplifying hosts may persist for weeks or months, and possibly even longer in hibernating mammals. In the latter case, overwintering hosts may serve as a source of infection for uninfected immature ticks the following spring (Marfin and Campbell, 2005).

Eyach virus, which was isolated for the first time from *I. ricinus* ticks in Germany in 1976, is antigenically related to, but distinct from, CTF virus (Charrel et al., 2004). Additional strains of the virus were isolated from *I. ricinus* and *I. ventalloi* ticks in France in 1981. Serologic surveys demonstrated that Eyach virus occasionally infects people in France and the former Czechoslovakia, and may cause encephalitis and polyradiculoneuritis in some patients. The transmission cycle has not been defined, but the primary reservoir is believed to be the European rabbit (Charrel et al., 2004).

Rocky Mountain Spotted Fever

This disease was first recognized in the Bitterroot Valley of western Montana in 1872. Rocky Mountain spotted fever (RMSF) is widely distributed throughout most of the United States and, to a lesser extent, in Canada and South America. In the United States, the disease was recognized only in the West until the 1930s when cases were detected for the first time in the East. Since 1997, there has been an increase in reported cases classified as RMSF, with about 1000 cases per year since 2002. The annual national incidence has ranged from 1.4 to 3.8 cases per million population. Most cases now occur east of the Mississippi River in the south-central and southeastern states, especially along the Atlantic coast (Figure 26.21). Cases tend to occur in foci in rural areas and suburban communities near major population centers. In the southeastern states, the seasonal peak of reported cases typically occurs in July, coincident with, or shortly after, the period of peak abundance of adult *D. variabilis*. In the

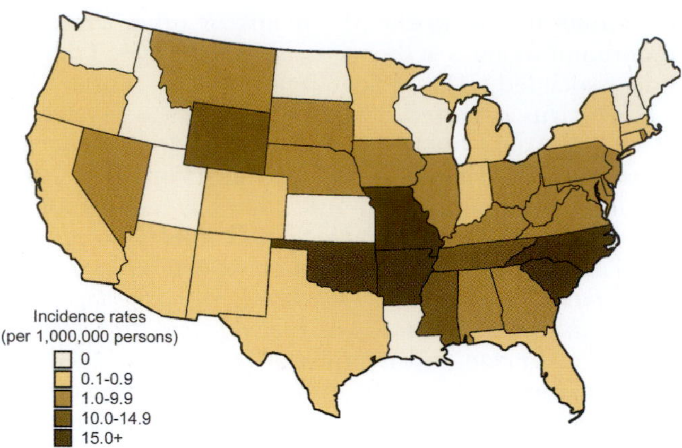

Figure 26.21 Distribution of human cases of Rocky Mountain spotted fever in the United States, shown as annual incidence of cases/million population), 2000–2005. (Courtesy of Rickettsial Zoonoses Branch, Centers for Disease Control and Prevention, USA)

northeastern states, the peak is usually in May or early June, although a bimodal pattern may occur in this region.

RMSF is caused by a rickettsia of the spotted fever group, *Rickettsia rickettsii*, an intracellular bacterium that multiplies freely in the cytoplasm and occasionally in the nuclei of host cells. It can cause a severe disease of the circulatory system with significant mortality in untreated or inappropriately treated cases. Rickettsiae multiply in the endothelial linings of capillaries, smooth muscle of arterioles, and in other blood vessels. After an incubation period of about seven days, patients develop fever, intense headaches, joint pain, muscle aches, nausea, and other symptoms. Although dermatologic features may not appear until later in the disease course, a characteristic maculopapular rash (Figure 26.22) occurs in most patients several days after onset of symptoms. It consists of many tiny pink or reddish spots, some of which may coalesce. The rash first appears on the hands and feet, gradually spreads to cover the entire body, and may persist for a week or longer. This particular pattern of progression is an important clinical feature of RMSF and helps to distinguish it from rashes produced by other vector-borne disease agents, such as epidemic typhus and allergic reactions. However, the extent of the rash indicates further progression of the disease and increasing severity. Severe cases may culminate in delirium or coma. Death can occur at any time during the acute clinical phase as a result of renal failure, clotting within blood vessels, shock, or encephalitis. Currently, 2 to 5% of RMSF patients in the United States may die despite the availability of effective antibiotic therapy. Even treated patients may die, primarily due to delayed treatment.

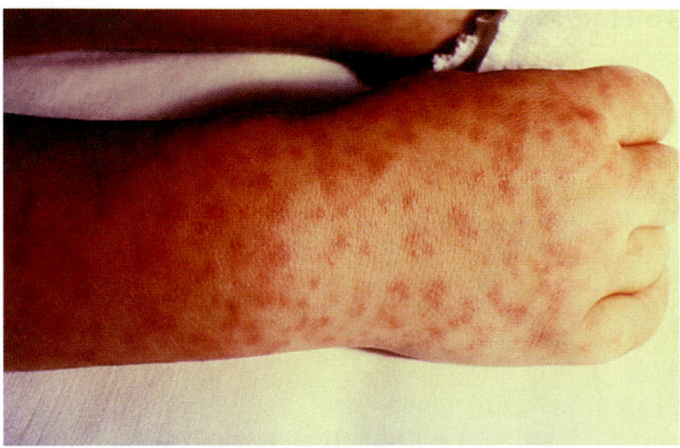

Figure 26.22 Rocky Mountain spotted fever case, with typical maculopapular rash on the arm and hand of a patient infected with *Rickettsia rickettsii*. (Courtesy of US Public Health Image Library, Centers for Disease Control and Prevention)

Rickettsia rickettsii is transmitted by the bite of ixodid ticks. In the United States and Canada, *D. variabilis* (Figure 26.15) and *D. andersoni* are the primary vectors. The elegant, pioneering work of Dr. Howard T. Ricketts at the turn of the twentieth century elucidated the role of *D. andersoni* and its vertebrate hosts in the transmission cycle of *R. rickettsii*. Ricketts detected the agent in wild-caught ticks, and demonstrated experimentally that *D. andersoni* could transmit it to susceptible laboratory and wild rodents by the bite. Further, he showed that *R. rickettsii* is passed transstadially and transmitted transovarially within populations of ticks. His contributions laid the foundation for subsequent studies of tick-borne zoonotic agents in the United States and abroad.

Dermacentor variabilis is abundant throughout eastern North America, but it has a much more limited distribution in the West where it is not known to transmit *R. rickettsii*. In contrast, *D. andersoni* is restricted to western North America. The immature stages of both tick species feed on rodents and other small mammals, and the adults attack larger mammals, including people.

In 2003, cases of RMSF were reported from eastern Arizona, which previously had reported few cases. Subsequent investigation showed a human incidence 300 times greater than anywhere else in the United States. The brown dog tick (*Rhipicephalus sanguineus*) (Figures 26.12 and 26.13) was implicated as the vector (Demma et al., 2005b). The nymphal stage was identified as the likely vector to humans. *Rickettsia rickettsii* infection rates in the ticks reached 10% at one home of a fatal case and was 4% overall. High numbers of ticks were documented in the peridomestic environment, but rarely were found in dwellings. The tick numbers were associated with the lack of control measures and an excessive stray or free-roaming dog population in the affected communities. In retrospect, Parker and coworkers (1933) knew that this species had been shown to transmit this pathogen to humans in Mexico, and conducted studies to demonstrate that *R. sanguineus* is a very efficient vector.

Other tick species that help maintain *R. rickettsii* in nature include *H. leporispalustris*, which feeds on birds and rabbits, and *Ixodes texanus*, which feeds on raccoons. The lone star tick, *A. americanum* (Figure 26.17), has been found infected naturally with spotted fever group rickettsiae that were believed to be nonpathogenic for humans (informally named *R. amblyommii*). However, recent studies have suggested that a mild illness may result from bites of this tick and that serologic evidence can develop after such bites. Although suspected as a vector of RMSF rickettsiae, especially in endemic areas outside the distribution of *D. variabilis*, it is not considered to be a primary vector of *R. rickettsii*. In Central America and South America, various species of *Amblyomma* have been implicated as vectors of *R. rickettsii* or closely related rickettsiae.

Larval and nymphal ticks maintain the infection from year-to-year and infect susceptible rodents when the ticks emerge to feed in the spring. Infected ticks must remain attached for at least 10 hours before transmission can occur; this is known as the reactivation phenomenon. The delay in transmission is due to the fact that *R. rickettsii* seems to be in an avirulent state in unfed ticks. The rickettsiae become virulent only after prolonged attachment of the tick to its host or following ingestion of blood by ticks.

Most humans who contract RMSF are infected by the bite of adult ticks in late spring or summer, although nymphs occasionally transmit the infection. Only about 1 to 3% of adult *Dermacentor* ticks in most foci are infected with spotted fever group rickettsiae, a small proportion of which are *R. rickettsii*. As previously noted, *Rhipicephalus* may show higher prevalences of specific *R. rickettsii* infection. Ticks can be assayed for evidence of rickettsial infection by examination of their hemolymph using immunofluorescence assays (IFA). However, precise estimates of tick-infection prevalences with *R. rickettsii* are complicated by the potential presence of nonpathogenic spotted fever group rickettsiae, such as *Rickettsia montanensis*, *R. bellii*, and *R. rhipicephali*. IFA tests that employ species-specific monoclonal antibodies can resolve these. However, polymerase chain reaction (PCR) assays and nucleotide sequencing have provided more reliable means for determining tick infection prevalences and have largely replaced other methods.

Isolations of *R. rickettsii* have been made from numerous small and medium-sized wild mammals. Species that have been implicated as natural reservoirs include meadow voles and deer mice. In these animals, there are few if any obvious signs of clinical disease during infection.

The period when rickettsiae are present in the blood of reservoir hosts is usually brief, often less than a week. Ticks feeding on infected animals may acquire rickettsiae, which produce generalized infections in tick tissues. In western North America, people normally become infected when they enter tick-infested habitats while engaged in outdoor activities in rural areas. In eastern North America, humans acquire their infections both in rural and peridomestic settings because dogs, which are significant hosts of adult *D. variabilis*, carry infected ticks into the home environment. Dog ownership often is noted as a risk factor in human cases.

Boutonneuse Fever

Boutonneuse fever, also known as Indian tick typhus, Kenya tick typhus, Crimean tick typhus, Marseilles fever, Mediterranean spotted fever, and Mediterranean tick fever, shares many features with Rocky Mountain spotted fever. However, the causative agent, Rickettsia conorii, does not occur in the Americas. It has an extensive range in southern Africa, India, central Asia, the Middle East, Europe, and North Africa.

Patients with Boutonneuse fever develop fever, chills, severe headaches, and a rash. In addition, a button-like (boutonneuse) ulcer called an **eschar** or **tache noir** usually forms at or near the site of tick attachment. The disease is generally milder than Rocky Mountain spotted fever, and most patients recover without antibiotic treatment. However, strains vary in virulence, and one that occurs in Israel has caused severe illness and several deaths. In temperate regions, cases of Boutonneuse fever are most common in late spring and summer, coincident with the seasonal activity of the primary tick vectors.

Rickettsia conorii is transmitted by several species of ixodid ticks in six genera (*Amblyomma, Dermacentor, Haemaphysalis, Hyalomma, Ixodes, Rhipicephalus*). In Europe, *Dermacentor reticulatus*, *D. marginatus*, and *I. ricinus* are important vectors. *Rhipicephalus sanguineus* is the principal vector in southern Europe, the Middle East, and North Africa, especially in countries bordering the Mediterranean Sea. Larvae and nymphs of *R. sanguineus* feed on small mammals, especially rodents and hedgehogs, whereas the adults feed mainly on larger mammals including humans. Lagomorphs, rodents, and possibly birds can serve as reservoir hosts. Dogs are susceptible to infection and transport vector ticks into and around human domiciles. Development of *R. conorii* within populations of ticks is similar to that of *R. rickettsii* in *D. andersoni* and *D. variabilis*.

Separate subspecies of *R. conorii* have been shown to cause human disease in Israel, Sicily, and Portugal (Israeli spotted fever) and areas surrounding the Caspian Sea and in Chad (Astrakhan spotted fever) (Parola et al., 2005).

Other Spotted Fever Group Rickettsiae

African tick bite fever (also called South African tick typhus) is caused by *R. africae*. The clinical features are similar to those of Boutonneuse fever, but multiple eschars are more likely with this infection. This species is responsible for much of the spotted fever group rickettsial infections in sub-Saharan Africa, and has been identified as established in the French West Indies. The pathogen is transmitted by ticks of the genus *Amblyomma*. Additional rickettsioses are summarized by Parola et al. (2005) and will not be discussed at length here. *Rickettsia sibirica* may cause Siberian spotted fever and certain subspecies may cause lymphangitis-associated rickettsiosis or tick-borne lymphadenitis (also known as *Dermacentor*-borne necrosis-erythema-lymphadenopathy). *Rickettsia aeschlimannii* has infected patients in Morocco in South Africa, and *R. helvetica* has been identified in humans from France and Sweden.

Queensland tick typhus is caused by *R. australis*, and is found along the eastern coast of Australia. *Rickettsia honei* is found on Flinders Island near Tasmania (Flinders Island spotted fever), Thailand, and elsewhere. Japanese spotted fever is caused by *R. japonica* in southwestern Japan.

Although the organism had been identified over 60 years earlier, *Rickettsia parkeri* was first reported as a human infection in 2004 (Paddock, 2005). Human infection is characterized by clinical findings similar to, and possibly confused with, RMSF. However, an eschar at the bite site with a maculopapular rash provides evidence of possible *R. parkeri* infection. Several cases of the infection have been confirmed in the United States. These cases were associated with the Gulf Coast tick, *Amblyomma maculatum* (Figure 26.18), in which field studies have demonstrated infection by the rickettsiae.

Human Ehrlichiosis

Ehrlichiae are obligate intracellular organisms that invade the cells of the vertebrate hematopoietic system. Several species are important to human and veterinary health. The various species of the genus *Ehrlichia* grow within cytoplasmic vacuoles of monocytes, granulocytes, lymphocytes, or platelets. Upon infection of a cell, the ehrlichiae divide by binary fission to form microcolonies, known as **morulae**. Cells may contain one or many morulae. Although they are not commonly detected in routine examination of stained peripheral blood smears, presence of morulae is helpful in presumptive diagnosis (Figure 26.23). Specific PCR assays and cell culture provide useful diagnostic methods.

Human ehrlichiosis was first described in Japan in the 1950s. The etiologic agent, *Neorickettsia sennetsu*, is transmitted to humans by an unknown mechanism, but ingestion of infected fish parasites is suspected. In the United States, ehrlichiosis was first recognized as a febrile illness following a tick bite. Investigations of the disease in different parts of the world have found multiple species causing human disease. These organisms were reclassified into the family Anaplasmataceae in 1999, and currently members of three genera are recognized as human pathogens.

Ehrlichia chaffeensis is found primarily in the southeastern and south-central United States. This species primarily invades the monocytic leukocytes (Figure 26.24). The disease, originally called human monocytic ehrlichiosis, manifests as an acute illness with high fever, severe headaches, aching muscles and joints, and other nonspecific signs and symptoms. A rash is not common, but may occur in about 20 to 30% of younger patients. The disease can be mild, although some patients may require hospitalization. Severe cases can occur and may result in death, especially in those with compromised immune systems. Over 1000 cases of *E. chaffeensis* infection have been reported since its recognition, with 358 cases from 2001 to 2002 (Demma et al., 2005a). The primary vector for *E. chaffeensis* is *Amblyomma americanum* (Paddock and Childs, 2003; Childs and Paddock, 2003). This tick species feeds readily on white-tailed deer, which serve as a reservoir for the ehrlichiae and an important host for the tick. Other wild and domestic animals have been identified as potential reservoirs based on serologic, cultural, and molecular studies.

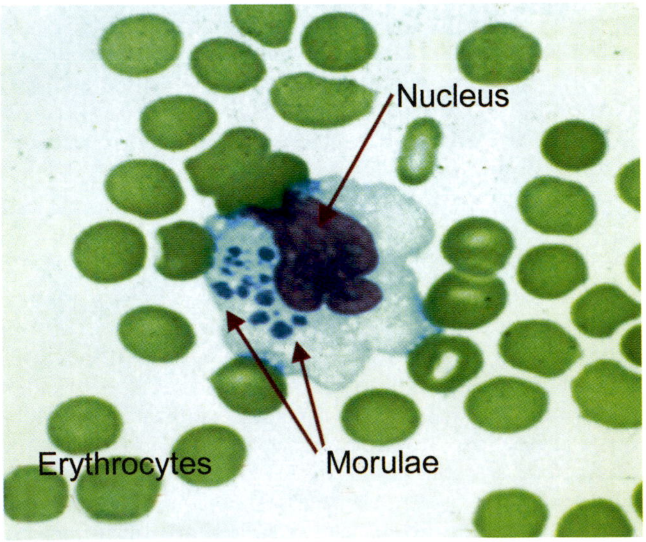

Figure 26.23 Light micrograph of human blood, showing a granulocyte infected with morulae of *Anaplasma phagocytophilum*, the causative agent of human granulocytic anaplasmosis; Wright-Giemsa stain. (Courtesy of Rickettsial Zoonoses Branch, Centers for Disease Control and Prevention, USA)

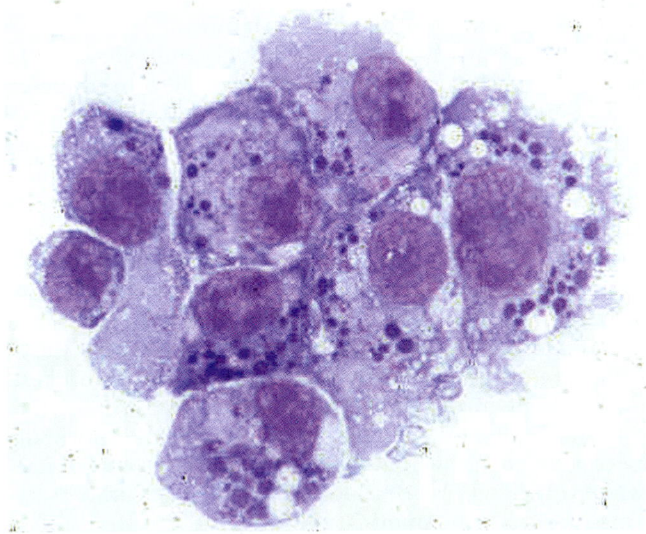

Figure 26.24 Cell culture showing *Ehrlichia chaffeensis*, the causative agent of human monocytic ehrlichiosis, in canine monocytic cells; Wright-Giemsa stain. (Courtesy of Centers for Disease Control and Prevention, USA)

Ehrlichia ewingii was first detected in granulocytes of human patients from Missouri in 1999 (Buller et al., 1999). Since then, additional cases have been reported, primarily in immunocompromised patients. The etiologic agent has been described from several southern states where it was known to be a cause of illness in dogs. Experimental evidence has shown adult *Amblyomma americanum* to be an efficient vector, and recent studies have detected the organism in this tick species. The reservoir for the pathogen is not known, but the organism has been identified in white-tailed deer. Other reservoirs may include wild or domestic canines.

Human Granulocytic Anaplasmosis

In 1994, a new pathogen was identified in Wisconsin and Minnesota. Infected patients were found outside the known geographic range for *E. chaffeensis* and the agent invaded the granulocytes (neutrophils and eosinophils), so the disease was provisionally named human granulocytic ehrlichiosis. Molecular studies later revealed that the organism was identical to the veterinary pathogens, *E. phagocytophila* and *E. equi*. The species were characterized and subsequently combined and reclassified as *Anaplasma phagocytophilum* (Figure 26.23). The disease now is referred to as human granulocytic anaplasmosis (Goodman et al., 2005). The number of cases reported in the United States has increased each year since its recognition, with 789 cases reported in 2001 and 2002 (Demma et al., 2005a).

This pathogen has a wide distribution in the temperate regions of the world, and is found primarily in the northeastern, upper midwestern, and far-western regions in the United States. The primary vector in the eastern United States is *Ixodes scapularis*, with *Peromyscus leucopus* and *P. maniculatus* serving as reservoirs. A wide range of wildlife species have been shown to be infected or are serologically positive, indicating a complex system. *Ixodes pacificus* transmits the pathogen to humans and domestic animals in northern California. Novel cycles of infection involving woodrats (*Neotoma* spp.) have been identified in the western United States, whereby the pathogen is maintained by *I. spinipalpis* among *Neotoma fuscipes* in California and among *N. mexicana* in Colorado (Nicholson et al., 1999; Zeidner et al., 2000). Long-lasting bloodstream infections in woodrats provide infectious blood meals to feeding ticks. Cofeeding *I. pacificus* ticks may then act as bridging vectors to humans or domestic animals.

In Europe, the pathogen is transmitted to humans by the bites of *I. ricinus*. Much work had been done on tick-borne fever, the disease in ruminants. It is now clear that a complex system of reservoirs and other hosts occurs across Europe and Asia. *Ixodes ricinus* is the primary vector in Europe, while *Ixodes persulcatus* is infected in the Far East. The number of human cases has not been as high as that in the United States, but these numbers are increasing as well.

Q Fever

First recognized among livestock handlers in Australia in 1935, Q fever is now known to occur on four other continents (Europe, Asia, Africa, North America) and is probably worldwide in distribution. The etiologic agent, *Coxiella burnetii*, is a bacterium that develops in the phagolysosomes of the cytoplasm of susceptible cells. *Coxiella burnetii* can survive for months or years outside host cells under environmental conditions that are lethal to other bacteria. It can survive in dried tick feces, dried or frozen tissues, soil, and water.

After an incubation period of about 20 days, Q fever is characterized by sudden onset of fever, chills, sweats, diarrhea, sore throat, painful sensitivity to light, muscle pain, and headache. Fever may persist for two weeks and show a biphasic pattern. Fatigue, enlargement of the liver, and inflammation of the lungs, accompanied by a mild cough and chest pain, occur frequently. A rash is usually absent; when present, it appears on the trunk and shoulders. Q fever may become chronic, in which case it causes inflammation of the lining of the heart and its valves. The case fatality rate is less than 1% in acute cases but may rise to 30% in chronic cases.

Transmission by ticks was first reported in 1938. Both argasid and ixodid ticks have been found infected naturally with *C. burnetii.* Subadult ticks infected while feeding upon bacteremic hosts develop a generalized infection in their tissues. Following the transstadial molt, nymphs or adults transmit *C. burnetii* by bite, and females can pass the organism transovarially. Argasid ticks also can disseminate the organism via infectious coxal fluids. Notably, *C. burnetii* can survive in contaminated tick feces for as long as six years, which facilitates spread to humans and domestic animals.

Coxiella burnetii is maintained in enzootic cycles involving domestic animals (e.g., sheep, cattle, goats), wildlife, and their associated ticks. For example, a cycle exists among Australian kangaroos (*Macropus major* and *M. minor*), the marsupial bandicoot (*Isoodon torosus*), and their associated host-specific ticks. Transmission to cattle and humans occurs when wild mammals also are parasitized by the nonspecific *Ixodes holocyclus.* In mammals, infection is usually asymptomatic, but abortions sometimes occur. Small mammals (e.g., *Apodemus, Microtus, Clethrionomys, Arvicola,* and *Pitymys*) living in and around agricultural communities may link the domestic and feral cycles. These animals develop high rickettsemias and shed the organism in their feces for weeks after becoming infected. Dogs, cats, birds, and reptiles also are susceptible to infection and may play a role in maintaining the infection in natural habitats.

Although ticks are important in maintaining the pathogen horizontally and vertically in enzootic cycles, they rarely transmit *C. burnetii* to humans by bite. Instead, persons who handle infected animals or their products, or materials contaminated by tick feces, are at increased risk of acquiring *C. burnetii.* Tick excreta are an important source of infection because they are often highly contaminated and easily aerosolized. However, aerosols emanating from afterbirth membranes and associated fluids, blood, urine, feces, nasopharyngeal discharges, and milk containing high concentrations of the organism constitute the most common means for spreading the infection. As these materials dry, *C. burnetii* can be spread in aerosolized dust and debris present in animal stalls, barns, store rooms, and similar facilities. The most common site of Q fever epidemics is on farms or in farming communities, usually when domestic animals are being handled, such as during wool-shearing, lambing, calving, and slaughtering. Milk and milk products are a particularly important means of disseminating *C. burnetii* to humans; the organism may survive in contaminated milk and butter for up to three months.

Recent studies have shown that *Amblyomma americanum* ticks harbor a *Coxiella burnetii*-like symbiont. The role of this organism in human or other animal disease is undetermined, but the infection has been detected in all populations of this tick that have been sampled (Jasinskas et al., 2007).

Lyme Disease

Lyme disease, also known as Lyme borreliosis, erythema chronicum migrans, Bannwarth's syndrome, and tick-borne meningopolyneuritis, is a tick-borne spirochetosis. It is caused by three genospecies of closely related bacteria: *Borrelia burgdorferi* sensu stricto (s.s.), *B. afzelii,* and *B. garinii.* Twelve additional members of the *B. burgdorferi* sensu lato (s.l.) complex have been described as of 2007, several of which reportedly infect humans only occasionally or rarely (e.g., *B. bissettii, B. lusitaniae, B. spielmanii*). The genome of the B31 type strain of *B. burgdorferi* was sequenced in 1997. North American and European populations of *B. burgdorferi* reportedly belong to genetically distinct populations, and this genospecies may have originated in Europe instead of North America as proposed earlier.

Although the etiologic agent was not discovered until 1981 (Burgdorfer et al., 1982), human cases have been documented in the medical literature dating back to the early nineteenth century. First recognized in the United States as a new form of inflammatory arthritis in the environs of Old Lyme, Connecticut, in the mid-1970s, Lyme disease and related disorders have been reported since then from most states in the United States, and in Canada and various European and Asian countries. Suspected human cases of the disease also have been reported from Africa, Australia, Mexico, and South America.

Currently, Lyme disease is the most commonly reported vector-borne disease in the United States, Europe, and in temperate regions of the Northern Hemisphere generally, where they account for tens of thousands of new cases annually (Steere et al., 2005). In the United States, 248,074 cases were reported to the Centers for Disease Control and Prevention from 1992 to 2006, and the number of cases reported increased 101% during this 14-year period. In Europe, the highest frequencies of the disease occur in central and Scandinavian countries, especially in Austria, Germany, Slovenia, and Sweden, where the incidence has been estimated to be as high as 120 (Slovenia) to 130 (Austria) cases per 100,000 residents (Steere et al., 2005).

In the United States, Lyme disease is most prevalent in the northeastern states, especially in New York, Pennsylvania, New Jersey, and southern New England (Figure 26.25). Other major regional foci occur in the Upper Midwest, especially in Wisconsin and Minnesota, and in northern California. In the northeastern United States, people living in close proximity to forests, or in suburban communities having a mosaic patchwork of wooded areas and homes, have the highest risk of exposure to spirochete-infected ticks. White-tailed deer thrive in these habitats and, consequently, *I. scapularis* abounds. Moreover, infection rates in *I. scapularis* nymphs and adults are high. In one study in New York, 30% of nymphal and 50% of adult ticks were found to be infected with *B. burgdorferi.*

On the other hand, northern Californian cases are most likely to occur in semi-rural or rural settings where *I. pacificus* is abundant. However, infection prevalences in this tick (typically 1–2% in adults, 2–15% in nymphs) are generally much lower than they are in northern populations of *I. scapularis,* and the risk of infection to humans is correspondingly lower. Cases occur throughout the year, but most often are seen in spring and early summer when the nymphs reach peak densities. In contrast, *I. pacificus* adults are primarily active in fall and winter, when temperatures are cool and humidities are high.

Besides greater awareness and increased tick-surveillance activities, several ecological and epidemiological factors have contributed to the current epidemic of Lyme disease in the northeastern United States and Europe. They are the occurrence of an abundant, efficient tick vector on both continents; the presence of numerous natural hosts for the immature and adult stages of the vectors; high rates of spirochetal infection in reservoir populations; anthropogenic changes

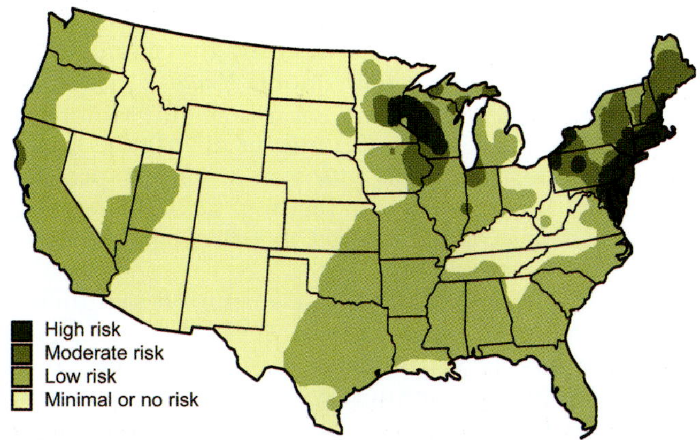

Figure 26.25 Risk of humans acquiring Lyme disease in the United States, based on approximate distributions of four estimated risk categories. (Courtesy of Centers for Disease Control and Prevention, USA)

(e.g., deforestation, reforestation, suburbanization) that favor an increased abundance of amplifying hosts and infected vector ticks; and close proximity of susceptible human populations to populations of tick vectors. Moreover, efficient transspecies transmission among the mammalian hosts of *B. burgdorferi*, a generalist microparasite, seems to have fueled the rapid epidemic spread of Lyme disease in the northeastern United States (Hanincová et al., 2006). In some European countries, the greater species diversity of *B. burgdorferi* s.l. spirochetes pathogenic for humans also may have contributed to the upsurge in reported cases as populations of *I. ricinus* expanded and increased in abundance.

When injected into humans by a feeding tick, *Borrelia* spirochetes multiply and disseminate in the skin. Gradually, they invade the blood stream and may spread throughout the body, often localizing in the bursae of the large joints, in the nervous system, and in the heart. Clinical signs and symptoms usually appear within one to two weeks (range, 3–32 days) following the bite of an infectious tick. Most cases occur during the late spring or summer, coincident with the seasonal activity of the nymphal stages of the primary vectors. *Ixodes persulcatus* is a notable exception; the adult female of this Eurasian tick seems to be the primary life stage that transmits spirochetes to humans.

Early-stage Lyme disease is characterized by nonspecific ("flu-like") symptoms and by an erythematous skin rash, erythema migrans (EM), which is present in 60 to 80% of patients. Erythema migrans is a slowly expanding, usually circular or elliptical, but sometimes triangular- or irregularly shaped lesion that often exhibits bright red outer margins and partial central clearing (Figure 26.26). Most patients have one EM lesion at the site of tick attachment, but 25 to 50% may develop multiple satellite lesions. The rash should not be confused with erythematous skin lesions that develop within minutes to a few hours, and tend to expand rapidly in size, after the attachment of an *Ixodes* tick. Such lesions typically result from allergic hypersensitivity reactions following the injection of tick-salivary proteins.

Untreated patients may manifest no further signs or symptoms of illness, or they may go on to develop late-stage Lyme disease within one to several months. Late manifestations include, either alone or in combination, cardiac, neurologic, arthritic, or further dermatologic abnormalities (e.g., acrodermatitis chronica atrophicans, caused by *B. afzelii*).

The following general account of the remarkably diverse ecology of *B. burgdorferi* s.l.-group spirochetes is by necessity highly selective because of space constraints. The interested reader is referred to pertinent chapters in Gray et al. (2002) and to Piesman and Gern (2004) for greater in-depth coverage of the voluminous literature treating the ecology in different geographic regions.

At least 40 species of ixodid ticks and two species of argasid tick have been found infected naturally with *B. burgdorferi* s.l. spirochetes. In most endemic foci, however, only a single member of the *Ixodes ricinus* complex serves as the primary vector to people. Thus, spirochetes are transmitted to humans by *I. scapularis* (formerly *Ixodes dammini* in part; Oliver et al., 1993) and *I. pacificus*, in eastern and western North America, respectively; by *I. ricinus* in Europe and western Asia; and by *I. persulcatus* in eastern Europe and Asia. Other *Ixodes* ticks that seldom or never attach to humans, and do (e.g., *I. jellisoni*) or do not belong to the *I. ricinus* complex (e.g., *I. spinipalpis* [= *I. neotomae*] in the western United States; *I. dentatus* in the eastern United States; and *I. ovatus* in Japan), may serve as efficient enzootic (maintenance) vectors of *B. burgdorferi* s.l. spirochetes. Ticks in other genera occasionally serve as secondary vectors to people. For example, *A. americanum* in New Jersey has been implicated, even though several experimental studies have shown that it is an incompetent vector of certain isolates of *B. burgdorferi*.

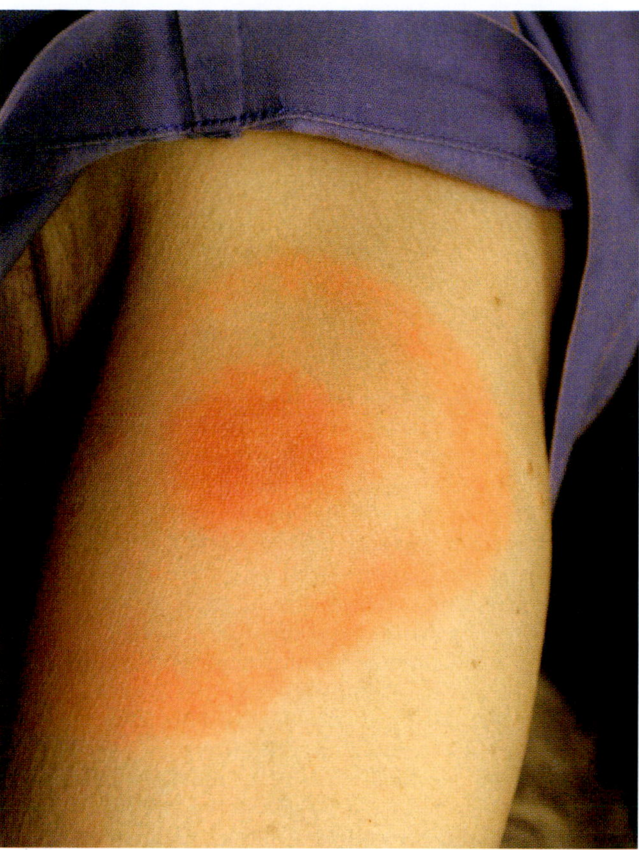

Figure 26.26 Erythema migrans skin lesion on upper arm of human patient, commonly seen in early stage of Lyme disease; caused by the spirochete *Borrelia burgdorferi*. (Courtesy of Centers for Disease Control and Prevention, USA).

In *Ixodes* spp. immatures, development of *B. burgdorferi* s. s. begins with ingestion of an infectious blood meal. *Borrelia burgdorferi* normally develops extracellularly by binary fission in the midgut diverticula of *Ixodes* spp. ticks, although spirochetes have been found occasionally in oocytes of the ovaries and in secretory cells of the salivary glands. Following the transstadial molt and resumption of tick feeding on another host, spirochetes escape from the midgut, enter the hemocoel, and migrate to the salivary glands (Zung et al., 1989). In some ticks, borreliae spread to other organs as well. Thus, spirochetes are maintained within populations of vector ticks by transstadial passage and by replenishment as noninfected ticks feed on infectious hosts. Transovarial transmission of *B. burgdorferi* s.l. has been documented in some of its primary vectors, but this mechanism appears to be inefficient for perpetuating and distributing most genospecies (e.g., *B. burgdorferi* by *I. pacificus* or *I. scapularis*).

In addition to ticks, spirochetes have been detected in mosquitoes, deer flies, and horse flies in the northeastern United States or Europe, where anecdotal accounts suggest that some individuals may acquire spirochetal infections following the bites of blood-sucking insects. Although the overall role of insects in the ecology and epidemiology of *B. burgdorferi* s.l. appears to be minimal, further investigation is warranted.

Wherever the ecology of *B. burgdorferi* s.l. has been studied intensively, typically one or more species of rodents or insectivores, and less often birds or lizards, have been implicated as primary reservoir hosts. Different genospecies may be associated with different vertebrates or even classes of vertebrates. Thus, *B. afzelii* and *B. burgdorferi* are associated predominantly with small mammals, *B. garinii* and *B. valaisiana* primarily with birds, and *B. lusitaniae* with lizards; however, *B. burgdorferi* and *B. garinii* parasitize some birds and small mammals, respectively.

The term **reservoir**, as used herein, pertains to those vertebrates that commonly are infected with a particular borrelial genospecies; that maintain spirochetes within their tissues for prolonged periods, if not for life; that have a significant amount of contact with vector ticks; and that readily infect ticks that feed on them. In North America, competent reservoirs of *B. burgdorferi* s.l. include the white-footed mouse, eastern chipmunk, short-tailed shrew, and masked shrew in the East; and the western gray squirrel, dusky-footed woodrat, deer mouse, and California kangaroo rat in the far-western United States. In Europe, the common shrew, bank vole, wood mouse, yellow-necked field mouse, edible and garden dormice, gray squirrel, red squirrel, hedgehog, and hare exhibit varying degrees of reservoir competence for *B. burgdorferi* s.l., among the approximately 35 host species identified as reservoirs to date (Gern and Humair, 2002). In some geographic regions, a few species of birds (e.g., American robin and song sparrow in the United States; Eurasian blackbird) may play a significant enzootiologic role by providing populations of immature ticks with blood meals, by infecting vector ticks with *B. burgdorferi* s.l. and by transporting infected ticks considerable distances, thereby establishing new foci of infection. Conversely, some vertebrates that are excellent hosts of vector ticks, such as white-tailed deer and western fence lizards in North America, and roe deer in Europe, do not serve as reservoirs (e.g., Telford et al., 1988). They therefore help to reduce tick-infection prevalences and the risk of human exposure to spirochetes; this has been termed **zooprophylaxis**. Nonetheless, these and certain other vertebrate hosts are important tick-maintenance hosts that sustain local or regional tick abundance.

Although it is an inefficient host for *B. burgdorferi* and *B. bissettii* in the far-western United States, the western fence lizard is the predominant maintenance host of *I. pacificus* immatures in many biotopes. Its reservoir incompetence stems from the fact that it contains a heat-labile, spirochete-killing (borreliacidal) factor in its blood. This factor destroys spirochetes in the midgut diverticula of infected nymphs either while they feed, or soon after they have fed, on lizard blood, with the result that, after the transstadial molt, the adult ticks are devoid of spirochetes. Eliminating spirochetal infections from vector ticks like *I. pacificus* may reduce the force of transmission of a zoonotic agent to humans or other animals.

The mechanism responsible for the borreliacidal activity in preimmune sera from the western fence lizard and the southern alligator lizard was demonstrated to reside in proteins comprising the alternative complement pathway (Kuo et al., 2000). In Europe, complement-mediated borreliacidal effects observed for specific combinations of vertebrate-host serum and different genospecies of *B. burgdorferi* s.l. generally coincided with what was known about the reservoir potential of the mammalian and avian species evaluated (Kurtenbach et al., 1998).

To understand the role of lizards, birds or mammals in the community ecology of *B. burgdorferi* s.l. and Lyme disease risk, each vertebrate species must be assessed separately under both field and laboratory conditions before any biologically meaningful conclusions can be reached (e.g., Brisson et al., 2008; LoGiudice et al., 2003; Mather et al., 1988). In that regard, few lizards had been studied intensively until recently, except for the western fence lizard in northern California. Findings stemming from the earlier studies collectively resulted in the belief proffered by some researchers that other lizards similarly might be reservoir-incompetent spirochetal hosts. Recent research, especially since 2005, has demonstrated that although some lizards indeed are nonreservoir hosts, others are clearly not. In the southeastern United States, several lizards reportedly are hosts for *B. burgdorferi* s.l. and serve as a source of spirochete infection for xenodiagnostic ticks (e.g., Clark et al., 2005). Likewise, in several European countries (Germany, Italy, Slovakia) and in Tunisia, four species of lizards have been implicated as primary reservoir hosts of *B. lusitaniae.*

Tick-Borne Relapsing Fever

Tick-borne relapsing fever (TBRF), also known as tick-borne spirochetosis and endemic relapsing fever, has been known since ancient times. Early descriptions confused it with louse-borne relapsing fever. Transmission by ticks was not recognized until the pioneering work of Dutton and Todd (1905), who detected spirochetes in *Ornithodoros moubata* from East Africa. The disease is caused by about 14 species of *Borrelia*, each of which typically is associated with a different species of argasid tick in the genus *Ornithodoros*. Thus, the high vector specificity of the relapsing fever spirochetes stands in marked contrast to the lower vector specificity of the Lyme disease spirochetes, *B. burgdorferi* s.l. The epidemiology of TBRF in the United States, and the possibilities for reemergence of tick- and louse-borne relapsing-fever infections globally, were reviewed recently (Dworkin et al., 2002; Cutler, 2006).

The 14 TBRF spirochetes of known public health importance and their associated vector ticks are *B. duttonii/O. moubata* in central, eastern, and southern Africa; *B. hispanica/O. erraticus* (large variety) in North Africa and southwestern Europe; *B. crocidurae, B. merionesi, B. microti, B. dipodilli/O. erraticus* (small variety) in Africa, the Near East, and Central Asia; *B. persica/O. tholozani* in northeast Africa and Asia; *B. caucasica/O. verrucosus* in the Caucasus Mountains of the former Soviet Union to Iraq; *B. latyschewii/O. tartakovskyi* in central Asia and Iran; *B. hermsii/O. hermsi, B. parkeri/O. parkeri*, and *B. turicatae/O. turicata* in the western or southwestern United States; *B. mazzottii/O. talaje* in the southern United States, Mexico, Central and South America; and *B. venezuelensis/O. rudis* in Central and South America (Burgdorfer and Schwan, 1991). *Borrelia crocidurae* also is transmitted by *O. sonrai* in West Africa.

Several other species of relapsing-fever group spirochetes associated with human-biting ticks have been described for which their public health significance awaits clarification: for example, *B. coriaceae* in the far-western United States; *B. lonestari* in the southern United States; *B. miyamotoi* in Europe, Japan, and the United States; and a novel *Borrelia* sp. from Tanzania that clusters with Nearctic isolates of *B. hermsii* and *B. turicatae*. Interestingly, *B. lonestari* and *B. miyamotoi* have been associated only with *Amblyomma americanum* and four *Ixodes ricinus* complex ticks, respectively. *Borrelia lonestari* has been implicated as the cause of an erythema migrans-like rash in patients diagnosed with an illness known as southern tick-associated rash illness, STARI (James et al., 2001). Failure to detect *B. lonestari* in 31 skin-biopsy specimens taken from 30 Missouri patients with EM-like lesions, or to isolate *B. burgdorferi* from 19 cultures of skin biopsies suggest that these agents are unlikely to be the cause of the rash associated with STARI (Wormser et al., 2005).

Onset of TBRF in humans is characterized by fever, chills, and a throbbing headache, usually without a pronounced rash or an ulcer at the bite site. The episode of fever usually lasts about three to five days, during which time spirochetes are present in the peripheral blood. Symptoms recur after several days without fever. This alternating cycle of febrile and afebrile periods may be repeated two or more times, which extends the illness for several weeks. Other signs or symptoms that often occur are muscle aches, joint pain, abdominal pain, nausea, vomiting, diarrhea, and a petechial rash on the trunk. In the western United States, three severe cases of TBRF in *B. hermsii*-endemic areas were associated with acute respiratory distress syndrome in 2004 and 2005. This indicates that a severe clinical condition might occur more often than previously reported. Additionally, serodiagnosis of *B. hermsii* infection was problematic in the United States for many years because this spirochete shares antigen with the Lyme disease spirochete *B. burgdorferi*. In 1996, however, Schwan and coworkers developed a specific serological test using glycerophosphodiester phosphodiesterase (GlpQ) as antigen that reliably differentiates relapsing fever from Lyme disease infections.

In certain TBRF-endemic areas of Tanzania, the annual incidence of *B. duttonii* infection reportedly is 384 per 1,000 in infants (less than 1 year of age) and 163 per 1,000 in children (less than 5 years of age); the perinatal mortality rate is a staggering 436 per 1,000 (McConnell, 2003). Likewise, in Senegal, the average incidence of *B. crocidurae* infection in all age groups was 11 per 100 person-years from 1990 to 2003 (Vial et al., 2006). In marked contrast, the incidence of TBRF in endemic areas of developed countries (e.g., Israel, United States) is orders of magnitude lower (Sidi et al., 2005).

The relapsing nature of the disease is explained by the antigenic variability of the borreliae. Some spirochetes are able to alter their surface-protein composition, probably through transposition of the genes encoding them. Consequently, new populations of spirochetes emerge in an infected host and multiply before the host can mount an effective antibody response against them.

In their tick vectors, borreliae ingested with blood disseminate to the internal organs, including the salivary glands. Spirochetes are passed transstadially so that once infected, ticks remain so for life. This, together with the long life span of *Ornithodoros* ticks, enhances the likelihood that relapsing fever spirochetes will persist in tick-infested habitats for prolonged periods. Transovarial transmission of relapsing-fever group spirochetes has been demonstrated in *O. coriaceus, O. erraticus, O. hermsi, O. moubata, O. tartakovskyi, O. tholozani, O. turicata, O. sonrai*, and *O. verrucosus*, but not *in O. parkeri, O. rudis*, or *O. talaje* (Burgdorfer and Schwan, 1991). Borreliae also invade the coxal glands of some *Ornithodoros* spp. and can be transmitted to hosts via coxal fluid excreted during or soon after the blood meal.

Primary reservoir hosts of nearly all borreliae transmitted by *Ornithodoros* ticks are rodents. Three notable exceptions are *B. coriaceae, B. lonestari*, and *B. duttonii*. In the United States, Columbian black-tailed deer have been implicated as reservoirs of *B. coriaceae* in the Far West, and white-tailed deer as reservoirs of ixodid-transmitted *B. lonestari* in the South.

The established dogma is that only humans serve as reservoirs of *B. duttonii*, the cause of East African tick-borne or endemic relapsing fever. Recent compelling, and possibly paradigm-shifting field and molecular evidence, however, suggests that *B. duttonii* is a zoonosis in central Tanzania (McCall et al., 2007). Domestic animals associated closely with households (i.e., tembe houses consisting of adobe walls, flat earthen roofs, and soil floors) in a TBRF endemic region were infected with borreliae that shared greatest homology with *B. duttonii.* Nearly half of 122 houses surveyed were infested with *O. moubata* s.l. ticks, and 11% of the chickens and 9% of the pigs tested were PCR positive. Also, a mark-release-recapture experiment revealed that about 3% of the recaptured ticks had moved from their release sites in pigpens into adjoining or nearby houses seven to 25 days postrelease.

Epidemiologically, TBRF is a highly focal disease. In the United States, about 50% of 450 cases identified from 11 western states from 1977 to 2000 originated in only 13 counties (Dworkin et al., 2002). In this region and in Mexico, *O. turicata* inhabits caves and the burrows and nests of rodents, terrapins, and snakes. It also infests homes, slaughterhouses, and pigsties, where it is likely to encounter humans. Another North American vector is *O. hermsi,* which infests cavities or nesting materials in dead trees, fallen logs, rustic cabins, and other dwellings at higher elevations (about 2,000 to 7,000 feet) (610 meters to 2134 meters), where it feeds on various rodents. A *B. hermsii*-like spirochete was detected in a northern spotted owl that had died of an acute septicemic spirochetosis (Thomas et al., 2002). This and other findings suggest that birds may be involved in the transmission cycle of *B. hermsii* in western North America.

People become infected when they are bitten indoors by *O. hermsi* ticks at night, especially if rodent hosts have been driven out or eliminated through house cleaning, rodent trapping, or poisoning activities. In 1973, the largest outbreak of relapsing fever ever recorded in North America was attributed to infection with *Borrelia*-infected *O. hermsi.* This episode involved 62 persons who slept overnight in rustic cabins at the North Rim, Grand Canyon National Park, Arizona (Boyer et al., 1977). More recently, the first multicase outbreak of TBRF from Montana was reported by Schwan et al. (2003). *Borrelia hermsii* was isolated from the blood of two patients, and *O. hermsi* ticks were collected from the cabin where all five patients acquired their infections.

Another vector of relapsing-fever spirochetes in western North America is *O. parkeri,* which typically occurs at lower elevations than *O. hermsi.* However, *B. parkeri* rarely infects humans because *O. parkeri* ticks inhabit the burrows of rodents (e.g., California ground squirrels) and therefore rarely have an opportunity to feed upon people. The few cases of TBRF attributed to *O. parkeri* ticks were contracted in a single Californian county (Dworkin et al., 2002).

Ornithodoros erraticus, a parasite of rats, mice, gerbils, and other small mammals, is an important vector in North Africa, whereas *O. tholozani* is the primary vector in the Middle East, the Balkans, and the southern part of the former Soviet Union. *Ornithodoros tholozani,* an indiscriminant feeder, is common in human shelters, caves, rocky overhangs, and other situations where livestock are housed. In southern Africa, *O. moubata* infests mud and thatch huts, surviving in cracks and crevices in the walls, bed posts, and other locations. While feeding, infected ticks excrete spirochetes in coxal fluid onto the open feeding wound. In such situations, ticks and humans maintain long-term, intimate contact. Recent changes in home construction in some countries, including use of modern building materials and larger homes divided into different rooms, have greatly reduced tick infestations and, concomitantly, the incidence of relapsing fever.

Tularemia

Tularemia was first identified as a distinct clinical entity in Japan in 1837, where it was attributed to infected hares. It was first recognized in the western United States in 1911 by McCoy, who described it as "a plague-like disease of rodents." The organism was detected in tissues of the California ground squirrel. In the United States, it has long been associated with hunters, who acquire the infection while skinning wild rabbits (hence the colloquial name, "rabbit fever"), and with persons who handle infected livestock, especially sheep. Workers in the former Soviet Union recognized this disease among trappers handling European water voles (*Arvicola amphibius*) and established that it was caused by the same bacterium. For excellent overviews of tularemia in North America, see Jellison (1974), Bell (1988), and Eisen (2007).

The causative agent, *Francisella tularensis,* is a pleomorphic, gram-negative, aerobic bacterium. It exists in different forms termed biovars, which are subspecies having special biochemical or physiological properties. Two biovars of *F. tularensis* occur in North America; a highly virulent form associated with rabbits, sheep, and ticks (biovar *F. t. tularensis*), and an apparently waterborne, less virulent form associated with beavers, muskrats, and voles (biovar *F. t. holarctica*). Biovar *F. t. tularensis* is the most common and known from North America and recently Europe. It is fatal in 5 to 7% of untreated patients. Biovar *F. t. holarctica* (*palaearctica*) occurs throughout the Northern Hemisphere and is rarely fatal to humans. A third biovar, *F. t. mediaasiatica,* is found in Central Asia, whereas a fourth biovar, *F. t. novicida,* exists in parts of the United States and possibly in Australia.

Francisella tularensis is transmitted by many blood-feeding arthropods other than ticks, including deer flies, mosquitoes, and fleas. It is also transmitted by handling infected animals, inhalation of contaminated dust, drinking infected water, and eating insufficiently cooked, infected meat. Patients infected with *F. tularensis* experience fever, headache, and nausea, usually accompanied by development of an ulcerated lesion at the site of inoculation. Other clinical manifestations include enlargement of regional lymph nodes, pneumonia, and occasionally a rash. Although seven clinical types have been recognized, the ulceroglandular form accounts for about 80% of cases (Jellison, 1974). The pneumonic form is particularly prone to produce severe illness; if left untreated, it may persist for two to three months and become chronic thereafter.

Many species of ixodid ticks have been found infected naturally with *F. tularensis.* In North America alone, this agent has been detected in, or isolated from, at least 13 species of ixodids in four genera (one *Amblyomma* sp., five *Dermacentor* spp., two *Haemaphysalis* spp., and five *Ixodes* spp.). Ticks are considered by some workers to be reservoirs of *F. tularensis,* or at least part of a multihost reservoir system, together with their primary vertebrate hosts. Transstadial passage of the agent occurs in susceptible ticks, but earlier claims that transovarial transmission also occurs have not been confirmed. Ongoing studies on Martha's Vineyard, Massachusetts have demonstrated a wide diversity of *F. tularensis* genotypes in *D. variabilis,* suggesting long-standing enzootic transmission. All the infected ticks harbored biovar *F. t. tularensis,* yet the infection rate was quite low (<1%) (Goethert et al., 2004).

Although tularemia is mainly an infection of wild lagomorphs and rodents, natural infections have been reported in numerous species of mammals (both domestic and wild) and birds. Fish, frogs, and toads have been found infected occasionally. Skunks and raccoons, both hosts of *D. variabilis,* recently have been identified as important sentinels for enzootic transmission in the eastern United States (Berrada et al., 2006).

Epizootics are spread mainly by water or ticks. Waterborne epizootics occur in western Siberia, where lakes and other fresh water habitats are contaminated by infected dead and dying muskrats and voles. Fur trappers and others who handled carcasses of infected water voles, muskrats, or water contaminated by them, have contracted the disease in large numbers. *Ixodes apronophorus* transmits *F. tularensis* among European water voles, furthering the spread of the disease along shorelines.

In parts of North America, *F. tularensis* is acquired by hunters as they skin freshly killed rabbits. Lagomorphs were identified as the source of infection in over 80% of the cases of tularemia acquired in California between 1927 and 1951; of these, jack rabbits were implicated in 71% of the cases in which a distinction was made between jack rabbits and cottontail rabbits. On the other hand, ticks are much more significant than lagomorphs as a source of human infection in the south-central United States, especially in Arkansas and Missouri, where the lone star tick (*Amblyomma americanum*) is the primary vector (Jellison, 1974). Since 1990, 41% of the tularemia cases in the United States have been reported from this region, with the cases involving *A. americanum* nymphs and adults and *Dermacentor variabilis* adults (Eisen, 2007).

Tick Paralysis

Tick paralysis is a host reaction to compounds secreted in the saliva of feeding ticks. This malady has been reported from North America, Europe, Asia, South Africa, and eastern Australia. It was first reported in Australia in 1824 by William Howell, who described a tick "which buries itself in the flesh and would in the end destroy either man or beast if not removed in time" (cited in Harwood and James, 1979).

The affliction is characterized by a progressive, flaccid, ascending paralysis. In humans, it usually begins in the legs with muscle weakness and loss of motor coordination and sensation. Paralysis gradually extends to the trunk, with loss of coordination in the abdominal muscles, back muscles, and eventually the intercostal muscles of the chest. Paralysis of the latter may lead to death from respiratory failure. During advanced stages, the patient may be unable to sit up or move the arms and legs; chewing, swallowing, and speaking may become difficult. The condition progresses rapidly, and death may ensue within 24 to 48 hours after onset of symptoms. In North America, the case fatality rate is about 10% in the Pacific Northwest; most of those who die are children (Gregson, 1973). In Washington State alone, 33 cases were reported between 1946–1996, mostly in children less than 8 years old (Dworkin et al., 1999). In a one-week period in May 2006, four cases of tick paralysis were reported from Colorado (CDC, 2006). In most North American cases, symptoms abate within hours following detection and removal of the attached tick or ticks, and recovery may be complete within 48 hours. If paralysis has progressed too far, complete recovery may take up to six weeks. In contrast, paralysis induced by the Australian tick, *Ixodes holocyclus,* may worsen after tick removal, and full recovery may take up to several weeks (Grattan-Smith et al., 1997).

The nature of the toxin(s) causing tick paralysis has not been determined for most species of ticks that cause this condition. Intensive research on the salivary components of *I. holocylcus* has revealed the existence of a protein, named **holocyclotoxin**, which produces paralytic symptoms. A different salivary-gland protein has been implicated as the toxin in *D. andersoni* females.

Typically, only female ticks can induce tick paralysis. In southern Africa, however, nymphs of the Karoo paralysis tick (*Ixodes rubicundus*) can cause paralysis in laboratory rabbits, and larvae of the soft tick *Argas walkerae* produce a toxic protein that can paralyze chickens. Female ticks must be attached to the host for several days (usually 4 to 6) before they begin secreting the toxins. For an excellent review of the mechanisms of pathology among the tick paralyses, see Gothe et al. (1999) and Mans et al. (2004).

Forty-six species of ticks in 10 genera reportedly cause tick paralysis in humans, other mammals, and birds. Worldwide, the ticks of greatest concern are *I. holocyclus, I. rubicundus, D. andersoni,* and *D. variabilis.* Recently, *Dermacentor andersoni* populations from various regions in Canada were shown to differ in their paralyzing ability. Apparently this is under genetic control, as selection can increase this ability in the laboratory (Lysyk and Majak, 2003).

Although reported from many regions in North America, tick paralysis in humans and domesticated animals has been documented most often from the Pacific Northwest in the United States and British Columbia in Canada. Cases in humans and other animals occur in spring and early summer, coincident with the activity period of adult *D. andersoni.* In the eastern United States, cases in dogs and humans are caused by *D. variabilis*, whereas in California this tick has been associated only with paralysis in dogs.

In contrast to paralysis produced by *Dermacentor* species or *I. holocyclus*, in which feeding by one female is sufficient to cause this condition, severity of the disease in South Africa is related directly to the number of attached *I. rubicundus*.

In England and France, human paralysis has been attributed infrequently to *I. hexagonus*. In California, several mild cases had been ascribed to *I. pacificus* prior to the 1950s, but these earlier observations were not well documented and no new cases have been reported since then.

Tick-Bite Allergies

The bites of many species of ticks can cause host reactions other than paralysis. These range from minor, localized, inflammatory reactions that subside soon after tick removal to severe, systemic reactions involving skin rash, fever, nausea, vomiting, diarrhea, shock, and death. Severe toxic or allergic reactions may follow the bites of the soft ticks *Argas brumpti*, *A. reflexus*, *Ornithodoros coriaceus*, *O. moubata*, and *O. turicata*. In Europe, severe reactions and even loss of consciousness have occurred following attacks by the pigeon tick, *A. reflexus*, which infests buildings where pigeons roost. This tick is particularly prone to bite people if the birds have been driven away. In the far-western United States, reports that bites of the pajaroello tick (*O. coriaceus*) are more feared than those of rattlesnakes seem exaggerated, but severe allergic reactions have been documented (characterized by edema, pain, erythema, tissue necrosis, ulceration, and prolonged healing). Bites of the bat tick, *Carios kelleyi*, in Iowa induced large erythematous lesions in some individuals (Gill et al., 2004).

Attached *Ixodes pacificus* sometimes causes severe allergic reactions and, in rare cases, anaphylactic shock in persons previously sensitized to its bite (Van Wye et al., 1991). Likewise, individuals bitten by *I. holocyclus* may experience anaphylactic reactions involving tick-specific IgE.

VETERINARY IMPORTANCE

Ticks are of veterinary concern mostly because of the many microbial disease agents they can transmit to livestock, companion animals, and wildlife. These include protozoans, bacteria (including rickettsiae), and arboviruses. Ticks also are important because of the debilitating and sometimes fatal host reactions produced in domesticated livestock and companion animals as a result of the feeding activities of certain species (e.g., tick toxicosis, tick paralysis). Moreover, livestock, as well as wildlife, may suffer from exsanguination, leading to anemia and death. In Oklahoma and Texas, for example, significant mortality in white-tailed deer fawns has been associated with heavy infestations of *Amblyomma americanum*. Livestock breeds in tick-infested areas have been subjected to natural selection and are able to acquire immunity to infestation upon exposure to local tick species. Exotic livestock breeds from tick-free regions, have no such immunity when first imported, and do not acquire a high degree of immunity on exposure.

Livestock and poultry that are heavily infested with ticks may experience economically significant reductions in body weight, milk or egg production, and general unthriftiness, and may on occasion even die of anemia when infested by large numbers of ticks. Some species routinely or incidentally invade the auditory canal of bovines or other mammals, a condition known as otoacariasis, which may be accompanied by serious secondary bacterial infections. Tick bites, especially by species with a long hypostome (e.g., *Amblyomma*) may generate abscesses by secondary bacterial infection and cause the loss of one or more quarters of the udder. Other consequences can be lameness, due to ticks attached in the interdigital space and wounds that attract myiasis-producing flies, causing significant reduction in the value of the hide, fleece, or carcass.

The following section provides an overview of the major tick-borne diseases of veterinary importance.

Piroplasmoses

Piroplasmoses are protozoan diseases caused by *Babesia* (Family Babesiidae) (Figure 26.27) and *Theileria* (Family Theileriidae, which also includes the genus *Cytauxzoon*). Parasites of both genera live and multiply in erythrocytes of the vertebrate host. The term itself refers to the developmental stage of the protozoan in the erythrocytes(i.e.,"piroplasm," literally meaning a pear-shaped structure). This is the stage infective to ticks. In ticks the parasite undergoes a sexual cycle with macro- and microgametes developing in the midgut and ending with infective sporozoites in the salivary glands.

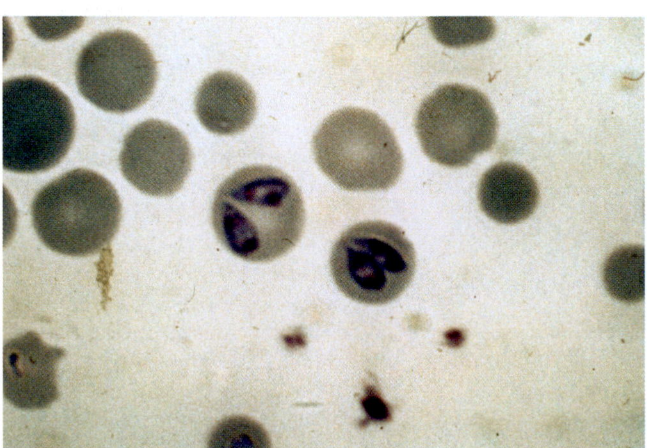

Figure 26.27 *Babesia* sp. in the blood of an infected animal. (Courtesy of Gerrit Uilenberg)

Some piroplasm species were formerly classified as *Babesia*. However, after it was discovered that they undergo schizogony during development in vertebrate hosts and that they are not transmitted transovarially, they had to be excluded from this genus (Uilenberg, 2006). Examples include *Babesia equi* and *B. microti*. *Babesia equi* is now classified as *Theileria equi* (Mehlhorn et al., 1986), whereas, based on phylogenic studies using molecular comparisons, *B. microti* has been placed intermediate between the families Babesiidae and Theileriidae (Kjemtrup et al., 2006).

Babesia and *Theileria* species exhibit basic differences in their developmental biology and ability to be transmitted by infected ticks to their offspring. Whereas *Babesia* (sensu stricto) is transmitted both transstadially (i.e., from one developmental stage of the tick to the next) and transovarially (from mother to offspring via her eggs), *Theileria* is transmitted only transstadially. As a result, *Babesia* persists in infected ticks throughout their development as larvae, nymphs, and adults; unfed larvae of the next generation are already infected when they hatch from the eggs. In contrast, *Theileria* does not persist into the next generation. Unfed tick larvae are never infective. Nymphs and adults become infective only if they were infected in the previous developmental stage.

There are also genus-specific differences between these piroplasms in their vertebrate hosts. *Babesia* develops and multiplies only in erythrocytes (Figure 26.27), whereas *Theileria* develops and multiplies in both erythrocytes and lymphoid cells. The latter first undergoes multiplication (schizogony) in lymphocytes, producing macroschizonts (i.e., schizonts containing large nuclei, also known as Koch's blue bodies), which appear a few days after the onset of symptoms. Later, as infected lymphocytes transform to lymphoblasts, microschizonts appear in the lymphoblasts. Merozoites released from the lymphoblasts then invade host erythrocytes. Piroplasms of both genera multiply in the erythrocytes by budding, with *Babesia* usually producing two daughter cells and *Theileria* usually four.

Babesioses Infections of cattle by *Babesia* species are often severe, especially when involving *B. bovis* or *B. bigemina* (in tropical and subtropical regions) and *B. divergens* (in Europe and parts of North Africa). The severity depends on the susceptibility of the animal (natural resistance) and its immune status (acquired resistance). Animals develop high temperatures, cease feeding, and become anaemic. One of the more characteristic and common symptoms is hemoglobinuria, causing the urine to become red or brownish, hence the name "redwater" for this disease. Icterus (jaundice) often is seen in severe cases. As babesiosis progresses, the animals become lethargic and eventually lapse into a coma and may die. Central nervous symptoms often are seen in infections by *B. bovis*. Considerable variation in severity has been noted in different geographical regions. This is attributed to differences in virulence among the parasites, as well as variation in the susceptibility of different cattle breeds. Nursing animals are protected by passive immunity acquired from antibodies in the colostrum of immune cows and young animals, even born to uninfected dams, are much more resistant than older ones (age-linked resistance, even more important than passive immunity). When all young animals in a population are infected, there is typically little or no clinical disease, particularly in local breeds. This situation is called **endemic stability**. Endemic stability may also occur in theilerial and ehrlichial infections.

The most important vectors of bovine babesiosis are *Rhipicephalus (Boophilus)* species, especially *R. (B.) microplus* and *R. (B.) annulatus*. Although eradicated from the United States in the early decades of the twentieth century, *R. annulatus* and *R. microplus* still occur in Mexico and stringent controls are maintained to prevent their reintroduction. Moreover, white-tailed deer are potential hosts for *R. annulatus* and could serve as a wild reservoir for this vector tick. Relict populations of *R. annulatus* are believed to exist in some areas of Florida (Sonenshine, 1993). *Babesia divergens* is transmitted by the tick *Ixodes ricinus* in Europe and North Africa. In eastern, central, and southern Africa, *R. decoloratus* is an important vector (but only of *B. bigemina*), with *R. geigyi* replacing it in West Africa. Domestic buffalo, or water buffalo, may also contract babesiosis, with *B. orientalis* having been described recently in buffalo in China.

Dogs are often victims of canine babesiosis, caused primarily by *B. canis* (now considered a complex of three species). The brown dog tick (*Rhipicephalus sanguineus*) (Figures 26.12 and 26.13) is a widespread vector. *Dermacentor reticulatus* is an important vector in Europe, and ticks of the *Haemaphysalis leachi* group are known to transmit agents of canine babesiosis in Africa.

Babesiosis also affects other animals, such as small ruminants and horses (*Babesia caballi*) (Schetters and Brown, 2006). In the latter case, this disease is of particular concern regarding the international movement and commerce of horses. Wild mammals commonly are infected, usually with host-specific *Babesia* species.

Immunization against bovine babesiosis is carried out in some countries by injecting attenuated strains of the parasites, produced in donor cattle, with all the inherent risks associated with live vaccines. There also are commercially available inactivated vaccines against canine babesiosis, produced in blood cultures. Their efficacy, however, has not yet been well evaluated, and the existence of several species and/or strains further complicates their practical use.

Theilerioses *Theileria* species infect a wide range of domestic and wild animals, particularly in the Old World. The more important agents of veterinary interest are *T. annulata* and *T. parva* in cattle, *T. equi* in horses, *T. lestoquardi* in sheep, and *Cytauxzoon felis* in domestic cats.

East Coast Fever (ECF) is a disease of cattle and domestic buffalo caused by *Theileria parva*. The disease, which has been known from East Africa since the nineteenth century, is widespread in eastern, central, and southern Africa. Movement of cattle has played a major role in the periodic outbreaks of ECF during the twentieth century. Epizootics with high mortality tend to occur when very susceptible exotic breeds (e.g., taurine breeds, but even Asian zebu breeds) are introduced into areas endemic for *T. parva*. In endemic areas there may be a situation approaching endemic stability in local breeds.

An estimated 25 million cattle are at risk for acquiring ECF. Infected animals develop enlarged lymph glands and, after a few days, develop a high fever, become listless, and stop feeding. This may be followed by diarrhea and mucous discharges from the eyes and nose, and frequently by pulmonary signs, due to edema of the lungs. Mortality may exceed 90% in adult animals but is usually much less in calves.

Theileria parva can also cause another severe illness in cattle, called Corridor disease. This results from transmission of the parasite from wild buffalo, the primary host of *T. parva*, to domestic cattle by ticks. It is believed that classical ECF has evolved from adaptation of the parasite to tick transmission between cattle. Water buffalo are also highly susceptible to ECF, but there are very few of them in endemic regions of Africa.

The primary vector of *T. parva* is *Rhipicephalus appendiculatus*, a tick whose geographical distribution coincides largely with that of ECF throughout much of eastern, central, and southern Africa. Corridor disease also occurs outside the known range of *R. appendiculatus* but within that of another competent vector, *Rhipicephalus zambeziensis*.

Another somewhat milder disease of cattle is tropical theileriosis, caused by *Theileria annulata*. It is transmitted by several *Hyalomma* species (e.g., *H. anatolicum* in Eurasia) and affects both domestic cattle (*Bos* spp.) and the Asian domestic buffalo. Although clinical signs are similar, this disease differs from babesioses in the absence of hemoglobinuria and the less severe anemia that it causes in infected animals.

Tropical theileriosis is arguably of even greater importance than ECF because of its much wider distribution throughout North Africa, northern parts of sub-Saharan Africa, southern Europe, the Middle East, and elsewhere in central Asia, India, and China. In this regard the name of the disease is a misnomer because it also occurs in some temperate regions. Similarly its other name, Mediterranean theileriosis, is misleading, since it occurs outside the Mediterranean basin.

Immunization against tropical theileriosis and East Coast fever of cattle is used in some countries. In the case of tropical theileriosis, live attenuated, schizont-based vaccines, produced in lymphoblastoid cell cultures against *T. annulata*, have provided satisfactory results. However, similar vaccines for ECF have not proved effective. Instead, immunizations entail injection of live, fully virulent sporozoites obtained from ticks, followed by specific treatment to prevent clinical disease. Although homologous immunity is excellent, antigenic diversity complicates the results in the field.

Many other *Theileria* spp. cause infection, and often disease, in other livestock and wildlife in different regions of the world. Like *Babesia caballi*, *Theileria equi* is of significant concern to horse owners and a major obstacle to the international movement of horses. In small ruminants, particularly sheep, *Theileria lestoquardi* can be a serious problem, causing fatalities where it occurs in the Old World. Additionally, recent discoveries of new species of pathogenic *Theileria* in small ruminants have been made in China.

Domestic cats are prone to suffer from a theilerial disease called feline cytauxzoonosis, caused by *Cytauxzoon felis*. It is closely related to *Theileria*, and in the United States is often fatal in untreated animals. *Dermacentor variabilis* is believed to be the primary vector. The bobcat (*Lynx rufus*) may be the primary host of *C. felis* and does not show apparent signs of illness when infected. Other species of *Cytauxzoon* occur in wild cats (Felidae) in various parts of the world (Meinkoth and Kocan, 2005).

Louping Ill

Although known to sheep herders in Scotland for centuries, louping ill (LI) was not recognized as a separate clinical entity until 1913. Its viral etiology was not established until 1931. Louping ill virus causes an economically important disease of sheep and red grouse (a game bird) in England and Scotland. It also occasionally infects cattle, horses, pigs, and humans, often with severe or fatal consequences. The causative agent is a flavivirus that is antigenically similar to other members of the Tick-borne encephalitis complex, the only member of the complex present in the British Isles. Similar diseases occur sporadically in a few other European and Scandinavian countries. The latter are caused by viruses distinguishable from LI virus by nucleotide sequencing. They have been named according to the country in which they were first recognized, such as Spanish sheep encephalomyelitis and Greek goat encephalomyelitis (Gritsun et al., 2003).

Infected sheep lose their appetites and become feverish. On about the fifth day after onset, the fever rises and the animals become uncoordinated and develop tremors. Seriously ill animals walk with an awkward, erratic, "louping" gait, hence the name of the disease. Many sheep die shortly after locomotor signs appear, but others develop a chronic condition that may persist for several weeks. Mortality in different breeds of sheep varies considerably, and reaches 100% in some susceptible flocks. Recovered animals often show signs of permanent neurologic damage. Experimentally infected red grouse experience mortality rates of up to 78%.

Historically, the disease has been most prevalent in areas of unimproved pastures and moorlands where *Ixodes ricinus* is abundant. LI is passed transstadially in ticks but not transovarially. Field and laboratory studies suggest that the vector competence of *I. ricinus* for the virus is low. In enzootic areas of northern Britain, for example, only 0.1 to 0.4% of ticks are infected with the virus.

Although *I. ricinus* parasitizes many vertebrate hosts, tick populations inhabiting sheep rangeland are supported principally by this animal. Small mammals are not abundant on many upland sites grazed by sheep and, when present, tend to support low tick loads. This and other limited evidence suggest that small mammals are not infected with LI virus. Moreover, sheep exposed to LI virus develop high viremias and are infective to ticks for several days. Other vertebrates, both domestic and wild, that may contribute to the maintenance of the virus are cattle, goats, mountain hares, and several species of ground-inhabiting birds, especially red grouse, willow grouse, and ptarmigan. These species serve as hosts for *I. ricinus* and occasionally develop viremias high enough to infect feeding ticks. Two other experimentally proven routes of transmission that may amplify LI virus in natural foci are nonviremic transmission of the virus among infected and uninfected ticks cofeeding on mountain hares, and the ingestion of infected ticks by red grouse during their first season (Gilbert et al., 2004). It has been estimated from field observations that 73 to 98% of viral infections in young-of-year red grouse might occur as a result of eating infected ticks.

African Swine Fever

This disease first was recognized in Kenya in 1921, as a catastrophic illness that killed 99% of infected pigs. Since then, sporadic epidemics of African swine fever have been reported from many African countries south of the Sahara, in Europe, and in the western hemisphere in Cuba, Haiti, the Dominican Republic, and Brazil.

The disease is caused by a large icosahedral DNA virus (family Iridoviridae) that attacks cells of the reticuloendothelial system, especially monocytes. The host range of the virus is limited to domestic pigs, European wild boars, warthogs, and bush pigs, all species of the family Suidae. In its acute form, animals develop fever about three days post-infection, the fever persists for three or four days, the temperature drops, and death ensues. In its subacute form, an irregular fever lasts for three or four weeks, whereupon the animals either recover or die. In its chronic form, animals may survive for long periods before succumbing to a secondary illness, most commonly pneumonia. Chronically infected animals usually experience stunted growth and emaciation, and serve as long-term reservoirs of the virus.

The primary vectors in Africa, and indeed the only proven natural vectors, are *Ornithodoros* spp. of the *moubata* group, particularly *O. porcinus.* These ticks occur in eastern, central, and southern Africa. In these regions there is a sylvatic cycle between ticks and wild Suidae, particularly the warthog, in which the cases of infection show no symptoms. Once the infection has passed to domestic pigs, it spreads as a contagious disease and has been transported as such to various parts of the world. In the Iberian peninsula (Spain and Portugal), it became established in the tick *Ornithodoros erraticus*, which made eradication a difficult and long affair. In West Africa the tick *Ornithodoros sonrai* is suspected of playing a role in the persistence of the virus. In the New World, *O. coriaceus, O. turicata,* and other *Ornithodoros* species have been demonstrated to be competent experimental vectors, raising concern about the potential establishment of African swine fever virus in North America. It is a porcine disease of major global importance, because of the high mortality it causes and the absence of a vaccine, despite numerous attempts to develop one.

Diseases Caused by Members of the Family Anaplasmataceae

The taxonomy of the genus *Ehrlichia* has been changed considerably in recent years (Dumler et al., 2001). Some members have been reclassified in the genus *Anaplasma*, formerly restricted to rickettsias of red cells, some in the genus *Neorickettsia*, while *Cowdria ruminantium* has been reclassified as an *Ehrlichia*. The genera *Anaplasma, Ehrlichia* (including *Cowdria*), *Neorickettsia*, and *Wolbachia* have been united in the family Anaplasmataceae, to which the genus *Aegyptianella* has been added.

Ehrlichioses and Non-Erythrocytic Anaplasmoses The diseases caused by members of the genera *Ehrlichia* and *Anaplasma* have long been known in veterinary medicine. These pathogens infect the leukocytes or platelets of livestock, companion animals, and wildlife. The ehrlichiae grow as distinct microcolonies, or morulae, within the cytoplasm of host cells (Figure 26.23). The disease manifestations caused by these agents can range from asymptomatic to fatal.

Canine ehrlichiosis, due to *E. canis*, was first recognized in 1935. The cosmopolitan distribution of this pathogen corresponds with that of its primary vector, *Rhipicephalus sanguineus. Ehrlichia canis* occurs in mononuclear cells, but it is often difficult to find in blood smears. In the United States, serosurveys of civilian and military dogs have revealed that the disease is present in most states. Dogs infected with *E. canis* develop fever, conjunctivitis, and swelling of various tissues. The disease causes a reduction in the numbers of all blood cells (red, white, and platelets), and thus also is referred to as canine tropical pancytopenia. Infected animals stop eating, lose weight, and frequently appear depressed. Acute infection often is followed by a debilitating chronic phase, accompanied by anemia and sometimes nasal bleeding. The German Shepherd breed is particularly susceptible to acute, severe illness. These animals suffer from low white-blood-cell counts and damage to the lymph glands, bone marrow, and spleen. Animals with severe infection usually die without antibiotic treatment. In other breeds, particularly mongrels, the disease is often milder and ranges from asymptomatic to chronically symptomatic.

Transstadial transmission of *E. canis* occurs in *R. sanguineus*, and transstadially infected nymphal or adult ticks can transmit *E. canis* to susceptible dogs. Transovarial transmission occurs rarely, if at all. Dogs apparently serve as the primary reservoir *of E. canis* because inapparent infections can persist for over five years, and the agent is continually present in chronically infected dogs.

Infection by *E. ewingii* causes a disease in dogs known as canine granulocytic ehrlichiosis. The pathogen grows within neutrophils of infected animals and was once erroneously thought to be an atypical form of *E. canis*. The infection is usually mild and may manifest as polyarthritis. Although its actual distribution may be wider, *E. ewingii* is found primarily in the southern United States. The agent is passed from *A. americanum*, nymphs to adults, and transstadially infected adults can transmit the infection to susceptible dogs (Anziani et al., 1990). Naturally infected lone star ticks have been identified in North Carolina (Wolf et al., 2000). *Ehrlichia ewingii* is morphologically indistinguishable from *A. phagocytophilum* in blood smears.

Heartwater Heartwater is an ehrlichial disease of large ungulates (livestock and game) caused by *Ehrlichia* (formerly *Cowdria*) *ruminantium* (Figure 26.28). The disease occurs primarily in sub-Saharan Africa and neighboring islands (e.g., Madagascar, Reunion Island, Mauritius, the Comoros, and São Tomé). It has been inadvertently transported to the western hemisphere where it now occurs on several Caribbean islands. It is one of "the big four" most important tick-borne cattle diseases in tropical regions (babesioses, theilerioses, anaplasmosis, and heartwater).

Heartwater affects all domestic ruminants, especially cattle, sheep, and goats. Domestic buffaloes are also highly susceptible, but they are almost absent in endemic regions. As in several other tick-borne diseases, local breeds are much less susceptible than exotic ones, and endemic stability may occur, particularly in cattle. However, even local breeds of small ruminants may suffer considerable losses.

Infected animals develop fever, and after a few days central-nervous-system signs. They may become disoriented, and show signs of motor disorder, especially abnormal walking, trembling, and muscle twitching. In particular, cattle may develop profuse diarrhea. As the illness progresses, they develop convulsions and die soon afterward. Dead and dying animals commonly show a massive accumulation of fluids in the membrane surrounding the heart (pericardium) and edema in the lungs and other organs. Surviving animals become immune. Calves are protected by a short period of age-dependent tolerance, and possibly to some extent also by maternal antibodies transferred in milk. As a result, infected calves often develop only mild illness, or none at all. The disease is quite severe in exotic breeds of sheep and goats, Angora goats being probably the most susceptible breed. Introduction of exotic livestock disrupts endemic stability and often leads to epidemics of the disease. Similarly, rapid resurgences of vector-tick populations following drought also can lead to devastating epidemics. Moreover, endemic stability can be disrupted by excessive use of pesticides, which destroys the natural herd immunity that results from constant, low-level challenge by small numbers of infected ticks.

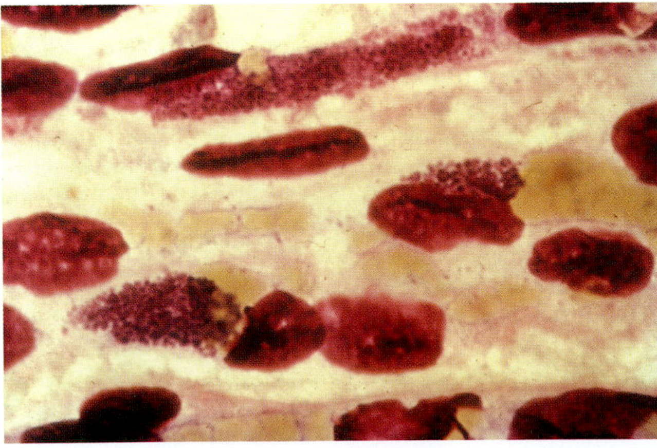

Figure 26.28 *Ehrlichia ruminantium* in brain tissue of a goat that died of heartwater; note the three masses of ehrlichial organisms. (Courtesy of Gerrit Uilenberg)

In ruminants, multiplication of *E. ruminantium* is observed in endothelial cells, which is easiest seen in the capillaries of brain-cortex smears (Figure 26.28). However, its presence in circulating neutrophils in the blood also has been demonstrated.

The agent of heartwater is transmitted by at least 10 African species of *Amblyomma* ticks, most of which have indiscriminant feeding habits. In view of its enormous geographic range and adaptability to varying climatic conditions, the tropical bont tick, *A. variegatum* (Figure 26.19) is the primary vector in most enzootic areas of Africa and the Caribbean. The bont tick, *Amblyomma hebraeum*, which is found in subtropical southern Africa, is also an important vector. Other *Amblyomma* species are important locally as vectors to livestock and wild ungulates. *Ehrlichia ruminantium* is maintained transstadially within populations of ticks, in which there is multiplication in the gut and later in the salivary glands. The pathogen is transmitted by the bite of the tick. Ticks remain infected for long periods, possibly for life. Transovarial transmission has been reported but appears to be exceptional. Male ticks may also be responsible for intrastadial transmission, as they may move from an infected animal to another.

There is considerable strain diversity, which complicates the development of a reliable and safe vaccine. Immunization is currently carried out, particularly in South Africa, by inoculating blood from a sheep reacting to a well-characterized stock. The injection has to be carried out intravenously, and the animals have to be closely monitored in order to treat them in good time. The procedure is risky from several points of view and is labor-intensive and far from 100% effective.

In the eighteenth or nineteenth century, the tick *A. variegatum* and the causal agent of heartwater were introduced into the Caribbean region with cattle from West Africa. The tick is, or has been, present on several islands of the Lesser Antilles, whereas heartwater is known to occur on Guadeloupe, Marie-Galante, and Antigua. It is likely that the tick is spread from island to island by another African immigrant, the cattle egret. Because of the constant threat of invasion of more islands and particularly the American mainland, an eradication program has been set up, but lack of continuity in financing and international coordination has led to intermittent abandonment of this program (Pegram, 2006). Some islands have been freed from the tick, but the constant migration of cattle egrets can easily result in reinvasion. Invasion of the North American or South American mainlands would be disastrous for the livestock industry, not only because of heartwater, but even more so because *A. variegatum* is associated with severe dermatophilosis. Three American species of *Amblyomma* are experimental vectors of heartwater, one of them, *A. maculatum*, being an efficient vector.

For further information about heartwater, see the reviews by Camus et al. (1996) and Mahan (2006).

Granulocytic Anaplasmosis *Anaplasma phagocytophilum* is an important pathogen of livestock, and also a cause of human granulocytic ehrlichiosis, an important zoonosis. In Europe, *Ixodes ricinus* transmits this rickettsia, the causative agent of tick-borne fever (also called pasture disease) to sheep, cattle, and goats. Reduced milk production and abortions can occur in infected animals. The organism infects granulocytes and induces a marked immunosuppression that may predispose animals to secondary infections (such as pyaemia by *Staphylococcus aureus* in lambs) and reduce their antibody response to vaccination against other diseases. *Anaplasma phagocytophilum* also causes infection and disease in horses in the United States and Europe. In the United States it is transmitted by *Ixodes pacificus* and *I. scapularis*. It may also cause a mild illness in dogs, in which it is morphologically indistinguishable from *Ehrlichia ewingii*. *Anaplasma phagocytophilum* appears to be passed transstadially, but not transovarially, in the tick vector.

Other Non-Erythrocytic Anaplasmoses *Anaplasma platys* infects canine platelets, causing their numbers to fluctuate over time. This infection is mild and often is diagnosed as canine cyclic thrombocytopenia. The tick vector(s) of this species have not been determined, although *Rhipicephalus sanguineus* is suspected. This species has been identified in the southern United States, Greece, Taiwan, and Japan.

In many tropical and subtropical regions, cattle and domestic buffalo can be infected with *A. bovis*, which is usually benign, whereas more pathogenic congeneric parasites occur in Central and West Africa, and Brazil. Known vectors belong to the genera *Hyalomma*, *Rhipicephalus*, and *Amblyomma*. *Anaplasma bovis* develops in mononuclear cells, as does the closely related *Ehrlichia ovina* in small ruminants. *Ehrlichia ovina* is known to occur in tropical and subtropical areas of the Old World, but not in the western hemisphere.

Many other *Anaplasma* spp. and *Ehrlichia* spp. have been described, and more are being discovered every year. Most appear not to be important as disease-causing agents, but surprises may lie ahead.

Erythrocytic Anaplasmosis This section is limited to "classical" anaplasmosis, caused by rickettsial agents in red blood cells. Anaplasmosis was first described in South Africa in 1910 by Max Theiler, who also identified and named the primary agent one year later. This parasite, *Anaplasma marginale*, and two related species, sometimes considered to be subspecies or variants of it (*A. centrale* and *A. ovis*), infect red blood cells of cattle and sheep throughout much of the world. Other described species either are not recognized as separate taxa or have no standing in taxonomic nomenclature. Anaplasmosis now is considered one of the most important diseases of livestock. The causative agent, *Anaplasma marginale*, is a pleomorphic, coccoid rickettsia that occurs and multiplies in membrane-bound inclusions called colonies in the cytoplasm of infected erythrocytes.

Disease onset is abrupt following a lengthy incubation period of two to six weeks. Common clinical manifestations include fever of several-days duration, labored breathing, loss of muscle tone in the rumen, constipation, and hemolytic anemia, often followed by jaundice. Animals usually recover when infected with strains of mild virulence, whereas 30 to 50% of those infected with highly virulent strains may die. Moreover, severity of disease and case fatality rates increase with age. Cattle that recover from acute anaplasmosis maintain a persistent, low-level parasitemia that protects them from reinfection; however, they constitute a reservoir of infection in the herd.

Approximately 20 species of ixodid ticks serve as vectors of *Anaplasma* species. The main vectors in subtropical and tropical countries are *Rhipicephalus* (*Boophilus*) ticks: *R.* (*B.*) *microplus*, *R.* (*B.*) *annulatus*, *R.* (*B.*) *decoloratus*, and probably *R.* (*B.*) *geigyi*. *Hyalomma* spp. and other *Rhipicephalus* spp. also play an important role. In North America, the primary vectors are *Dermacentor* species: *D. andersoni*, *D. occidentalis*, and *D. variabilis*. In Europe, *D. reticulatus* and possibly *D. marginatus* appear to play a role. Bloodsucking flies in the family Tabanidae (horse flies and deer flies), and other blood-sucking insects, have been implicated as mechanical vectors. Mechanical transmission can also occur during vaccination and dehorning campaigns, when the same needles, syringes, or other instruments are used on several animals.

Anaplasma marginale undergoes a complex developmental cycle in ticks involving five morphological forms (Kocan et al., 2004). Details of the life cycle have been elucidated in *D. andersoni* and are presumably representative of the parasite's development in its other tick vectors. The genetic and morphologic characteristics of *Anaplasma* species and their development within ticks are similar to those of *Ehrlichia ruminantium* and other *Ehrlichia* species in their tick vectors. *Anaplasma marginale* is passed transstadially, but not transovarially, within ticks. *Dermacentor andersoni* females infected as nymphs begin transmitting the infection by the sixth day of feeding on a susceptible host, whereas male ticks that acquire infection as adults can transmit the pathogen within 24 hours. Male ticks are of considerable importance as vectors because they feed repeatedly. They also readily transfer between hosts that are in close contact, and are therefore capable of transmitting *A. marginale* to multiple hosts (intrastadial transmission). The transfer of ticks between individuals also explains why one-host ticks such as *Rhipicephalus* (*Boophilus*) spp. are vectors of *A. marginale*, which is not transmitted transovarially.

Cases occur frequently on farms located adjacent to tick-infested woodlands. In the eastern United States, the presence of white-tailed deer is considered a risk factor because this cervid is an important host of tick vectors. However, white-tailed deer are not competent reservoirs of *A. marginale* and therefore do not serve as a source of infection for noninfected ticks. In the western United States, mule deer are not only primary hosts but also they are competent reservoirs of *A. marginale*.

Borrelioses

Borrelioses are diseases of birds and mammals caused by spirochetes in the genus *Borrelia* (Figure 26.29). Important tick-borne borrelioses include avian spirochetosis and Lyme disease. Avian spirochetosis is a highly fatal disease of turkeys, pheasants, geese, doves, chickens, and canaries in Europe, Africa, Siberia, Australia, Indonesia, India, and North, Central, and South America. It causes severe losses to the poultry industry in certain countries. Infected birds develop high fever, diarrhea, and become cyanotic. Birds that survive develop a long lasting immunity. *Argas persicus* and related ticks (subgenus *Persicargas*) transmit the etiologic agent, *B. anserina*, via infectious tick feces or by bite. *Borrelia anserina* is related to borreliae in the relapsing fever group. Transstadial passage and transovarial transmission occur, and ticks can remain infective for six months or longer.

Dogs, cats, cattle, horses, and possibly sheep can be infected with certain etiologic agents of Lyme disease (i.e., *B. burgdorferi* s.l.). This disease, or related disorders, have been reported from numerous countries on five continents: Africa, Asia, Europe, North America, and South America. Earlier claims that *B. burgdorferi* is present in Australia have not been substantiated, nor have isolates been obtained in South America. Populations of domestic dogs living in areas highly endemic for *B. burgdorferi* can have seroprevalence rates as high as 90%, even though relatively few seropositive animals manifest overt clinical signs. Indeed, 25 to 50% of apparently healthy dogs in some areas may have significant antibody titers to *B. burgdorferi*. In hyperendemic foci of the northeastern and far-western United States, the most commonly observed clinical manifestations among dogs are lameness, inappetence, fever, and fatigue. Dysfunction of the central nervous system, heart block (secondary to myocarditis), and a renal syndrome have been associated with *B. burgdorferi* infection in some dogs.

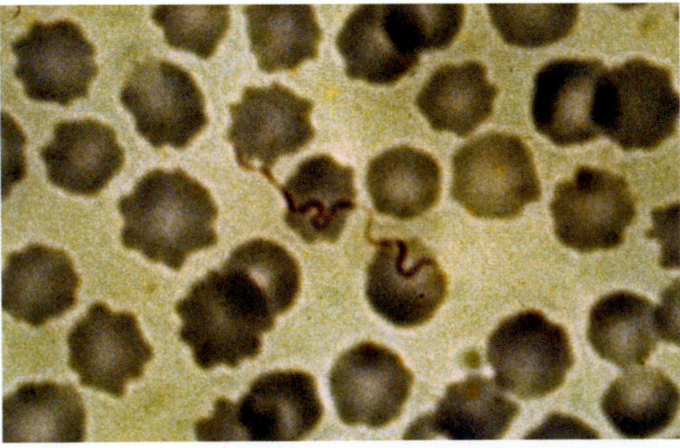

Figure 26.29 A *Borrelia* spirochete in the peripheral blood of an infected cow; Wright-Giemsa stain. (Courtesy of Gerrit Uilenberg)

Cats reportedly are exposed to *Ixodes scapularis* infected with *B. burgdorferi* and can develop elevated antibody titers to this spirochete; however, the clinical significance of these findings is unknown. In northern California, outdoor cats living in rural or semirural settings occasionally are bitten by *I. pacificus* females, but apparently at a frequency much lower than that of dogs.

Lyme disease has been reported in cattle and horses in the northeastern and upper midwestern United States. Antibodies against *B. burgdorferi* have been detected in serum or synovial fluid in cattle exhibiting lameness or arthritis, but a cause-and-effect relationship between exposure to such spirochetes and clinical disease in bovines has not been established. In the Delaware River valley of the eastern United States, serum antibodies against *B. burgdorferi* have been detected in about 10% of horses tested, and in 60% of the animals from one farm. Although arthritis, edema, and dermatitis were observed in some animals, clinical illness attributable to *B. burgdorferi* infection apparently is uncommon in horses.

Tick spirochetosis or bovine borreliosis is a benign disease of cattle, sheep, and horses that occurs in Africa and Australia (Burgdorfer and Schwan, 1991). Infected *Rhipicephalus* (subgenus *Boophilus*) ticks also have been identified along the United States–Mexico border. Infected animals experience one or two attacks of fever, loss of appetite, weight loss, anemia, and weakness. The causative agent, *Borrelia theileri*, is transmitted by *Rhipicephalus* spp. and possibly other genera of ixodid ticks. This spirochete, along with the lone star agent (*B. lonestari*) and *B. miyamotoi*, form a clade that is distinct from, but most closely related to, the relapsing fever-group spirochetes.

Epizootic bovine abortion (EBA), also known as "foothill abortion," is a major disease of rangeland cattle in the far-western United States, particularly California. Since its first recognition as a distinct clinical entity during the 1950s, various microorganisms have been evaluated or proposed as potential etiologic agents. These have included *Borrelia coriaceae*, which occurs in all three parasitic stages of the putative tick vector, *Ornithodoros coriaceus* (Lane et al., 1985). More recently, a novel deltaproteobacterium identified in thymus tissue from EBA-affected fetuses and detected in wild-caught *O. coriaceus* ticks by PCR was implicated as the etiologic agent (King et al., 2005).

Tularemia

Tularemia, caused by the bacterium *Francisella tularensis*, is primarily a disease of rodents and lagomorphs (rabbits and hares) in the Northern Hemisphere.

Although best known for its public health importance, *F. tularensis* also is a veterinary pathogen that can cause devastating epizootics in domestic sheep.

Reliable reports of epizootics in North America date back to 1923, when serious losses occurred in eastern Montana and southern Idaho (Jellison, 1974). These outbreaks shared several features: animals were put on rangeland enzootic for *F. tularensis* in early spring, they grazed in sagebrush areas where *D. andersoni* ticks were abundant, and they became heavily infested with ticks. *Dermacentor andersoni* is a competent vector of virulent strains of *F. tularensis*. As many as 50% of a sheep herd may become sick, and 10% may die within a few days. Diseased animals that survive such outbreaks may lose weight and condition during their illnesses. Although sheep usually are quite resistant to infection with *F. tularensis*, reduced vitality of animals after a long winter, shortage of feed, exposure to early spring storms, and heavy infestations of ticks, predispose flocks to epizootics. Over 14,000 cases of tularemia in sheep were recorded in the United States between 1923 and 1945. In contrast, only about 40 cases were reported in sheep in Canada by 1945.

Epizootics of tularemia among sheep constitute a risk factor for humans. Jellison and Kohls (1955) presented records of 189 human cases of tularemia associated with the sheep industry in the United States. Of these, 66 cases occurred among sheep shearers. Other individuals found to be at particular risk were sheep owners, sheep herders, and housewives of sheep handlers.

Horses infected with *F. tularensis* also can become ill, or die. The agent was isolated from two foals that died of an acute illness in Montana in 1958. Three other horses (one mare, two foals) from the same locality became ill, but recovered and had high agglutination titers to *F. tularensis*. All five horses were infested abundantly with ticks. Notably, one of the veterinarians involved in the study became ill with tularemia.

In North America, upland game birds are sometimes infected with *F. tularensis*. The agent has been isolated from ruffed grouse, sharp-tailed grouse, sage grouse, bobwhite quail, and pheasant. An epizootic in sage grouse was reported from central Montana, in which the birds were heavily parasitized by the ixodid bird tick *Haemaphysalis chordeilis*. Likewise, severe epizootics in domestic turkeys infested with the same tick have been reported. The significance of these sporadic outbreaks is that they occasionally expose hunters and veterinarians to infection with *F. tularensis*.

Q Fever

Coxiella burnetii, the agent that causes Q fever, is widespread in populations of domestic livestock and wildlife. Infection in mammals other than humans, however, is typically benign. For that reason, this disease is discussed primarily in the section on public health importance. Nonetheless, *C. burnetii* is also a veterinary pathogen, which occasionally induces abortion in pregnant cows and ewes. Because infected animals excrete large numbers of *C. burnetii* in their waste and in birthing tissues and fluids, they pose a significant health risk for animal handlers. Infected ticks produce infectious feces that can contaminate the wool of sheep and become aerosolized when handled. However, for the general public the danger of acquiring infection arises from inhalation of bacteria-laden dust as it becomes airborne.

Dermatophilosis

Dermatophilosis, also called cutaneous streptothricosis, is a skin disease of domestic and wild mammals, including occasionally humans, and is caused by the bacterium *Dermatophilus congolensis*. It is especially economically important to cattle and sheep production. In sheep, the disease may be known as **lumpy wool**. The pathogen is transmitted from animal to animal by mobile zoospores. The infection is widespread, probably cosmopolitan, and normally benign. Transmission and the development of severe skin lesions (Figure 26.30) following infection are favored by humidity and heat, and particularly, in a way that is not yet well understood, by the presence of certain *Amblyomma* species, notably the tropical bont tick *A. variegatum*. The role of this tick had long been suspected in Africa, and even earlier in the Caribbean region, where this African tick had been introduced.

The role of *A. variegatum* in this disease was experimentally confirmed by Walker and Lloyd (1993). Although *D. congolensis* is not introduced when the tick feeds, the saliva of attached *A. variegatum* adults influences the pathogenesis of dermatophilosis, causing severe dermatophilosis in ruminants when the tick is present. Dermatophilosis therefore is not a tick-borne disease, but a tick-associated disease.

Host resistance to dermatophilosis varies significantly in different areas, with much higher resistance in local breeds than in exotic stock. The constant high humidity in the Antilles is particularly favorable for the transmission and development of dermatophilosis and also to the multiplication and survival of *A. variegatum*. In contrast, the incidence of dermatophilosis is much lower in drier, less humid parts of Africa where the tick

Figure 26.30 Senepol cattle on the Caribbean island of Nevis, affected by severe dermatophilosis. (Courtesy of Gerrit Uilenberg)

occurs, and where there is a reduction in tick numbers during the long dry season.

Dermatophilosis is considered the most important cattle disease in some of the Caribbean islands, where it presents a greater concern than heartwater. Cattle on many of the islands have been exposed only recently to *A. variegatum* and are particularly susceptible to severe dermatophilosis (Figure 26.30). On the island of Nevis, for example, losses of 75% of cattle to this disease have been reported since *A. variegatum* was introduced, probably about 1977, and where farmers have been forced to abandon their cattle-breeding operations as a result (Hadrill et al., 1990). This is in contrast to the islands of Guadeloupe and Marie-Galante, where the tick was introduced from endemic areas of West Africa two or three centuries ago and "Creole" sheep have become more resistant to infection. No vaccine is currently available, and the only prevention is intensive tick control. This poses a significant concern should *A. variegatum* be introduced to the American mainland.

Tick Paralysis

The first reports of tick paralysis in livestock originated in Australia in 1890 and in British Columbia, Canada, in 1912. This condition is most common in livestock and pets, and causes injury or death to thousands of animals each year. Tick paralysis has been reported from many countries in North America, Europe, Asia, and Africa.

In South Africa, the Karoo paralysis tick (*Ixodes rubicundus*) is responsible for annual losses totaling tens of thousands of sheep and up to 9% of game animals in some areas. Other animals affected include goats, cattle, and species of wild antelope. Induction of paralysis by *I. rubicundus* is related directly to the number of ticks feeding on a host. Stock losses between 1983 and 1986 averaged nearly 29,000 animals per year, of which 91% were sheep. Stock farmers regard tick paralysis as one of the most important problems affecting their operations. The disease occurs in hill rangeland or mountainous terrain covered with a "Karoo" type of vegetation, which is grassy areas interspersed with shrubs or trees. Although *I. rubicundus* parasitizes many wild mammals, only antelopes are known to develop paralysis.

Another tick that paralyzes sheep and goats in Africa, particularly in South Africa, although not as severely or as often as *I. rubicundus*, is *Rhipicephalus evertsi evertsi*. This subspecies has been recognized as a cause of tick paralysis since 1900. The induction, duration, and severity of the paralysis are related to the number of female ticks that have engorged to body weights of 15 to 21 mg. *Hyalomma* ticks have been implicated as causing paresis or paralysis of camels in the Sudan and Somalia.

In North America, three species of *Dermacentor* ticks cause paralysis in companion animals and livestock. In the eastern and western United States, *D. variabilis* is a common cause of tick paralysis in domestic dogs. In the Sierra Nevada foothills of northern California, for example, an average of six cases was seen in two veterinary practices during a one-year investigation (Lane et al., 1984). Dogs were infested with a mean of 32 ticks; 98% were *D. variabilis* adults. Another tick from the same region, *D. occidentalis*, is responsible for occasional cases of tick paralysis in cattle, ponies, and deer, but not dogs.

The most important species of paralysis-inducing tick in North America is *D. andersoni*, which was responsible for paralyzing more than 3,800 sheep and cattle in the Pacific Northwest (British Columbia, Washington, Idaho, and Montana) between 1900 and the early 1970s. Individual outbreaks have involved up to 320 animals with cases occurring most frequently from April to June when adult *D. andersoni* activity is greatest.

In Australia, *Ixodes holocyclus* induces paralysis in dogs and humans. This tick inhabits a narrow zone along the eastern coast of Queensland and Victoria. Drugs administered along with hyperimmune serum to dogs with advanced paralysis improve chances for full recovery. Among several drugs tested, phenoxybenzamine hydrochloride, an alpha-adrenergic blocking agent, has been found to be most effective.

In Europe and Asia, tick paralysis is widely scattered. In Macedonia and Bulgaria, paralysis in sheep, goats, and cattle has been attributed to *Haemaphysalis punctata*, and in Crete, the former Yugoslavia, and the former Soviet Union, livestock are sometimes paralyzed by bites of *I. ricinus*.

Birds are susceptible to tick paralyses also. Larvae of *Argas walkerae* induce paralysis in chickens in South Africa. A toxic fraction isolated from replete larvae of this argasid tick consists of two proteins having "membranophilic" properties and molecular masses of 32 and 60 kilodaltons; extracts containing these proteins induced paralysis in one-day-old chicks. In the southeastern United States, a number of species of wild birds, especially passeriforms, are paralyzed by attached *Ixodes brunneus* females.

General reviews of tick paralysis and tick toxicosis (see below) have been given by Gothe (1999) and Mans et al. (2004).

Tick Toxicoses

Toxic reactions have been associated with the bites of certain species of ticks, notably argasids. In Africa, cattle bitten by *Ornithodoros savignyi* may die of toxicosis in just one day. Sheep attacked by *O. lahorensis* in eastern Europe and in the southern region of the former Soviet Union may tremble, gnash their teeth, exude frothy saliva, experience paralysis, and sometimes perish. In Europe, *Argas persicus* can cause leg weakness in ducks and geese; this condition resembles a true toxicosis and not a paralysis. A toxic illness that affects mainly calves in large areas of central, eastern, and southern Africa is known as sweating sickness. Wild hosts include eland, antelope, and zebra. The illness, which begins four or more days after tick attachment, is characterized by fever, loss of appetite, lachrymation, and salivation, and an eczema-like skin disease (hence the name), but no paralysis. Approximately 75% of afflicted animals die. The active principle in the African tick, *Hyalomma truncatum*, is a salivary gland protein; this is present in females of only certain strains of this tick species. A similar disease has been reported in India and Sri Lanka.

Large numbers of *Rhipicephalus appendiculatus* (the brown ear tick) in southern Africa also have been suspected of causing tick toxicosis in cattle. This may be partly complicated by the transmission of the mildly pathogenic agent *Theileria taurotragi.*

PREVENTION AND CONTROL

Historically, control of ticks and tick-borne diseases almost always was accomplished with pesticides to kill the ticks, and drugs to kill the infectious agents. The cattle tick *R.* (*Boophilus*) *annulatus* was eradicated in the United States by dipping cattle in pesticide solutions, thereby eliminating the deadly Texas cattle fever. Quarantine, pasture rotation, and elimination of deer also were used in the effort to eradicate this vector. It has been estimated that reintroduction of cattle ticks, *Rhipicephalus* (*Boophilus*) spp., could cost the United States cattle industry more than $1 billion/year to achieve eradication again. Costs to the worldwide cattle industry were estimated in 1984 by FAO at more than $7 billion. Damage to other livestock and valuable wildlife by ticks and tick-borne diseases is much more difficult to estimate. Losses due to human illnesses have never been calculated.

Treatment with acaricides still provides the most widely used means to control or prevent tick attacks. Promising alternatives, such as vaccines or pheromone-acaricidal treatments, are being investigated. There are at present two commercial recombinant vaccines available against the cattle tick *R.* (*Boophilus*) *microplus*, based on a so-called concealed antigen, which occurs in the tick gut. Research continues on other antigens and other tick species. These and other novel alternatives are discussed next.

Personal Protection

Preventive measures are the most effective means for protecting persons who enter tick-infested habitats. People should wear boots, socks, long trousers, and light-colored clothing. Trousers should be tucked into the boots, socks drawn over trousers, and the socks taped to form a tight seal. The clothing should be treated with a repellent or acaricide. It is now possible to obtain clothing permanently impregnated with permethrin that remains efficacious for the life of the garment, despite repeated washings (Faulde and Uedelhoven, 2006). A recent study showed that wearing protective clothing was 40% effective in preventing Lyme disease (Vasquez et al., 2008). Exposed skin also should be treated with repellents or acaricides suitable for use on humans. Each person should conduct self-examinations for ticks during and after exposure to tick-infested areas. Early removal of attached ticks is important in minimizing the risk of contracting tick-borne diseases. Ticks should be removed by grasping the capitulum as close to the skin as possible with a pair of fine forceps and gently pulling the tick with a slow, steady force until its mouthparts release their hold. Turning or twisting the tick should be avoided to prevent the hypostome breaking off in the wound.

The most widely used personal protectant is the repellent DEET, available as a lotion or a spray. Applications should be repeated as needed to maintain maximum protection, but should be applied cautiously on children to avoid adverse reactions that occasionally follow overuse. In addition, permethrin is effective when applied to clothing before entering tick-infested habitats. However, permethrin should not be applied to bare skin. Two newer repellents for use in the US include products with either picaridin or oil of eucalyptus. Other repellents are under investigation and appear quite promising.

Acaricides

Acaricides are chemicals used to kill ticks and mites. The term **ixodicides** sometimes is applied to acaricides used against ticks. Acaricides include arsenical preparations, chlorinated hydrocarbons (e.g., DDT and lindane), organophosphorus compounds (e.g., coumaphos), carbamates (e.g., carbaryl), pyrethroids (e.g., permethrin, flumethrin), formamidines (e.g., amitraz), avermectins (e.g., ivermectin), and phenylpyrazoles (e.g., fipronil). The pyrethroids are among the safest and most effective pesticides and are now widely used for tick control. Fipronil is a moderately toxic broad-spectrum phenylpyrazole insecticide, used against ticks and other ectoparasites of pets.

Area applications of acaricides have been shown to be very effective for controlling *Ixodes scapularis* ticks in the domestic environment. A single application in May or April will reduce nymphal tick populations, while a single application in October can control adult ticks. These applications can be very targeted, focusing on lawns, edges of woodland lots, and along borders adjacent to known tick habitats (Stafford 2007). The formulations can vary and may be used as sprays, or granules for ground treatments. Acaricide applications are most effective when combined into an integrated approach, including area acaricide treatments, treatment of animals, habitat management to reduce tick survival, management of host animal abundance, and other efforts.

One way to kill ticks on host animals is to dip livestock and pets in a pesticide bath. When used for cattle, this is termed a cattle dip. Dipping alone is not always effective. Often, ticks hidden in sheltered locations (e.g., between the toes, in the ears, or under the tail) are missed and survive to lay eggs and reestablish the pest population. This is valid for motorized spray-races (facilities where cattle are directed into a chute where pressurized spray is directed from various angles). For intensive tick control, dipping therefore is supplemented by applying an acaricidal cream ("tick-grease") or spray to such sites. Acaricides can be delivered as sprays, using manual or motorized high-pressure sprayers to provide a mist that can reach every part of the animal's body. They can also be delivered as pour-ons or spot-ons. These are formulations in which the acaricide is mixed with surfactants to spread the liquid over the animal's hair coat. Finally, they may be applied as dusts, in which acaricides are mixed with talc and deposited directly onto the animal's fur. The familiar "flea powders" for pets, which are effective against ticks as well as fleas, and the dust bags used for treating cattle are examples of acaricidal dusts.

To achieve long-lasting efficacy, acaricides can be incorporated into plastic or other suitable matrices that provide a slow release of the toxicant over a period of weeks or months. Plastic collars, such as the familiar flea and tick collars, are widely used for control of ticks on cats and dogs. Similarly, acaricide-impregnated plastic ear tags are widely used for control of ear-infesting ticks (e.g., Gulf Coast tick and spinose ear tick) on cattle and other large domestic animals. However, they are much less effective for control of ticks that attach around the groin, udder, and other parts of the hindquarters of these animals. Systemic acaricides offer another means of providing long-lasting and effective tick control. In this case, the toxicant is introduced into the host's blood to kill ticks as they feed on the treated animals. Unfortunately, most acaricides are too toxic to administer to animals systemically. An exception is ivermectin, which can provide excellent control of certain ticks on cattle for two to three months.

Each application method has its advantages and disadvantages. Dips and spray-races are suitable for treating large numbers of animals. The efficacy of manual spraying depends on the person applying it and it can be applied to only a limited number of animals, whereas the application by pour-ons can be expensive. Dip sites where persistent arsenical and chlorinated-hydrocarbon compounds may have been used for many years are often heavily contaminated and can represent a significant environmental hazard.

The development of acaricide resistance by ticks is a continuing concern. Ticks have been found to be resistant to arsenic, cyclodiene pesticides, other chlorinated hydrocarbons, organophosphorus insecticides, pyrethroids, and formamidines. Resistance may occur in one or more species in an area, while other species in the same locality remain acaricide-susceptible. Some strains of cattle ticks in Australia and elsewhere have been found to be resistant to most or all of the acaricides currently in use, including pyrethroids and amitraz. Resistance of cattle ticks to pyrethroids and organophosphorus compounds also has been found in Mexico, and poses concerns regarding the possible reestablishment of these ticks into the United States. Continued research is necessary to discover and develop new pesticide products to overcome resistance to compounds already in use. Future efforts using mass gene sequencing techniques, mining of the *Ixodes scapularis* genome, bioinformatics and RNA-interference to identify potential gene targets may make it possible to minimize acaricide resistance or enhance other tick control strategies (Lees and Bowman, 2007).

Pheromone-Assisted Control

The difficulties and high cost of tick control on animals have stimulated interest in alternatives to the conventional methods just described. Such alternatives help to reduce the use of acaricides. Research with tick pheromones suggests that combinations of pheromones and acaricides can be significantly more effective for controlling ticks than the acaricide alone, because ticks are unlikely to develop resistance to their own pheromones. A pheromone-acaricide combination applied to a single spot on cattle can be effective in killing the Gulf Coast tick. Another promising device is the "tick decoy" in which the sex pheromone 2,6-dichlorophenol and an acaricide are impregnated into plastic beads on the surface of which "mounting" sex pheromone is smeared. Male ticks are attracted to decoys on the animal's hair coat and killed. This also disrupts mating activity, so that any surviving females cannot lay viable eggs. For the livestock-parasitizing bont ticks *Amblyomma hebraeum* and *A. variegatum*, a tail-tag decoy was developed that uses a mixture of tick-specific phenols to attract ticks to specific sites on cattle and kill them when they attach nearby. Field trials with tail-tags have demonstrated promising efficacy for up to three months (Norval et al., 1996). A novel technology for killing *I. scapularis* ticks in their natural habitats was developed by impregnating the components of the tick arrestment pheromone (guanine, xanthine, and haematin) along with permethrin in an oily matrix for dispersal on vegetation. Ticks that encountered these materials assembled on the droplets, imbibed lethal doses, and died. Laboratory and field trials showed that inclusion of the pheromone components increased lethality up to 95% versus only 65% with permethrin alone (Sonenshine et al., 2003).

Passive Treatment

Another way to apply acaricides to animals is by means of self-treating devices. Animals seeking food or nesting materials visit these devices and acquire an acaricide, spreading it over their fur and skin to kill ticks. An example is biodegradable cardboard tubes containing a permethrin-impregnated cotton ball. Mice collect the cotton for nesting material, thereby spreading the pesticide among nest mates. Such tubes have been effective in reducing populations of *Ixodes scapularis* and the occurrence of Lyme disease in some localities, especially in residential communities; however, they have not been effective in other situations. Bait boxes containing the acaracide fipronil were found to be effective in killing *I. scapularis* nymphs and larvae on small mammals, reducing these instars by as much as 68% and 84%, respectively, and the infection rate of white-footed mice with *B. burgdorferi* by as much as 53%. Subsequently, the abundance of *I. scapularis* adults in the targeted area was reduced by 77%. Tick infection with *Anaplasma phagocytophilum* was also significantly reduced (Dolan et al., 2004).

Another example is the self-treating tick applicator for controlling blacklegged ticks (*I. scapularis*) on white-tailed deer (Sonenshine et al., 1996). Animals become coated with oil containing an acaricide as they remove food from the applicator. A similar technique was used in Zimbabwe for treating wild ungulates (Duncan and Monks, 1992). Perhaps the most effective example of this strategy is the "four-poster" self-applicating device for treating white-tailed deer against ticks. Deer attracted to corn or other food bait in the device acquire acaricide from cloth-covered rollers. Field studies showed up to 80% and 99.5% efficacy in controlling *I. scapularis* and *A. americanum* nymphs, respectively (Carroll et al., 2002).

Hormone-Assisted Control

Hormones and insect growth regulators (IGRs) such as methoprene also have been used to disrupt tick development in laboratory experiments. Analogues or mimics of ecdysteroids and juvenile hormone are effective in killing ticks by delaying their development, disrupting oviposition, or killing the larvae when they hatch from eggs deposited by treated females. However, these compounds do not appear to be uniformly effective against all types of ticks.

Vaccines

Anti-tick vaccines have been used successfully. In Australia, a commercial recombinant antigen vaccine has been developed for the control of the cattle tick *Rhipicephalus* (*Boophilus*) *microplus*, based on a so-called concealed antigen (Bm86) in cells of the tick gut. A similar recombinant vaccine, based on the same antigen, has been developed in Cuba. Recent reports suggest that the recombinant Bm86 can reduce tick fecundity by as much as 90%. Although it is possible that antigen-resistant strains of cattle ticks may appear, large-scale vaccination of cattle herds with these recombinant vaccines offers a promising alternative or supplement to acaricides. The vaccines are not only active against *R.* (*B.*) *microplus*, but also against related species.

Research on other antigens and other tick species is in progress. Of special interest is the development of novel combinations using RNAi (see Chapter 27) to silence subolesin and a tick- protective antigen, Rs86 (similar to Bm86) against *R. sanguineus*; the synergistic effect of silencing both genes causes a much greater reduction of tick feeding and oviposition than targeting either one alone (de la Fuente et al., 2006). Another promising vaccine targets tick-cement protein, disrupting attachment success, as well as midgut injury and the tick's ability to transmit pathogens (Labuda et al., 2006).

Management

Management practices provide another means of reducing tick numbers. Zero-grazing—keeping animals confined in stables—minimizes exposure to ticks. Acaricides can be applied directly to vegetation. However, because ticks commonly occur in microhabitats covered by vegetation, leaf litter, soil, and other natural materials, or in the nests, burrows, and other cavities used by their hosts, they often do not come in direct contact with these toxicants. Therefore, to be effective, the acaricides must reach the ticks as vapors or by contact when the ticks move about while seeking hosts. Public opposition to treatment of natural habitats with pesticides has made it unpopular to use this form of tick control except for the most compelling reasons. In recent years, acaricidal treatment of natural areas has been limited largely to military bases or selected recreational areas. Alternatives include habitat modifications such as burning or clearing vegetation or host removal (e.g., removal of deer by hunting and deer-exclusion fences). Burning or clearing vegetation removes the dense cover under which ticks shelter, thereby reducing ground-level humidity as well as exposing them to intense ultraviolet radiation and heat. Such changes can make the habitat unsuitable for tick survival.

Integrated control of ticks also can include the timing of acaricide treatments (e.g., when most engorged females of a particular species drop from their host) and management practices such as rotational grazing; cattle-breed resistance, selected use of acaricides, and predators of ticks.

Tick surveys often are conducted to determine whether or not tick control is warranted and, if so, when it should be implemented. The most common method for sampling ticks is the use of a flag or drag cloth pulled or dragged through the vegetation. Ticks collected on the cloth are counted as the number of a given species per unit of distance dragged (e.g., 100 m), or the number collected per hour of dragging. Although absolute measures of tick population densities cannot be obtained in this manner, the relative abundance of ticks in different areas sampled can be determined. Tick species collected can provide an indication of potential risks of tick-borne diseases in a given area. An alternative to dragging is the use of carbon dioxide traps. Carbon dioxide gas from a block of dry ice or from a compressed gas cylinder is the tick attractant. Ticks adhering to or crawling around the trap are counted after a few hours of operation. When more reliable estimates of tick abundance are required, a mark-and-recapture method can be used. Using this method, the numbers of marked ticks recaptured from a previous sample are compared with the number of unmarked ticks to obtain an estimate of the entire tick population in the area studied.

For tick-infested cattle, horses, mules, and other livestock, a time-tested method is the scratching technique, whereby livestock inspectors pass their hands over different regions of the animal's body to detect attached ticks. A similar technique is used in combination with visual inspection to examine wild animals. For example, investigators can be assigned to deer check stations during hunting season to count all ticks on hunter-killed animals. Another technique for sampling ticks is to trap small and medium-sized wild animals and hold them over trays filled with water or alcohol to catch fed ticks as they detach.

Eradication

In a few cases, tick eradication may be practicable. An example is the Cattle Tick Fever Eradication Program that was initiated in the southern United States in 1907. This program led to the eradication of *R.* (*B.*) *annulatus*, *R.* (*B.*) *microplus*, and Texas Cattle Fever from the United States by 1960. However, reinvasion from Mexico continues to be a constant threat, because of illegal movement of livestock across the US–Mexico border and uncontrollable wild hosts. Attempts to eradicate *R.* (*B.*) *microplus* in other areas (US Virgin Islands, Argentina, Uruguay, Australia, and Papua New Guinea) have been unsuccessful, despite reductions of over 99% of the tick populations in some localities. Acaricide resistance, as well as reinvasion by ticks and their rapid repopulation of areas in the eradication zone, are major contributing factors.

An attempt to eradicate *Amblyomma variegatum*, a vector of the agent of heartwater and associated with bovine dermatophilosis, from islands in the Caribbean has been successful on some islands (Pegram et al., 2004). However, it has not achieved its overall purpose, which was to eradicate the tick from the Western Hemisphere. Although the lack of international collaboration, political will, and funding have been contributing factors, there are also technical reasons that have prevented successful eradication of this tick on some of the islands where it is well established. The tick continues to be a major threat to the American mainland, and the Greater Antilles.

Occasionally eradication has been achieved in the case of exotic species that were recognized soon after their introduction to a new area. An example includes the eradication of the African species *R. evertsi* soon after it was introduced into a wild-animal compound in Florida. Eradication is easiest if exotic species are identified as soon as possible after their introduction. This was addressed following the discovery of *Amblyomma marmoreum* and *A. sparsum*, both vectors of heartwater, on nine different premises in Florida (Burridge et al., 2000).

Enzootic stability, rather than eradication, is the preferred method of controlling tick-borne diseases of livestock in several African countries (Norval et al., 1992).

REFERENCES AND FURTHER READING

Anderson, J. F., Sonenshine, D. E., & Valenzuela, J. (2008). Exploring the mialome of ticks: An annotated catalogue of midgut transcripts from the hard tick *Dermacentor variabilis* (Acari: Ixodidae). *BMC Genomics*.

Anziani, O. S., Ewing, S. A., & Barker, R. W. (1990). Experimental transmission of a granulocytic form of the tribe Ehrlichieae by *Dermacentor variabilis* and *Amblyomma americanum* to dogs. *American Journal of Veterinary Research*, *51*, 929–931.

Attoui, H., Jaafar, F. M., de Micco, P., & de Lamballerie, X. (2005). Coltiviruses and seadornaviruses in North America, Europe, and Asia. *Emerging Infectious Diseases*, *11*, 1673–1679.

Barker, S. C., & Murrell, A. (2002). Phylogeny, evolution and historical zoogeography of ticks: A review of recent progress. *Experimental & Applied Acarology*, *28*, 55–68.

Beati, L., & Keirans, J. E. (2001). Analysis of the systematic relationships among ticks of the genera *Rhipicephalus* and *Boophilus* (Acari: Ixodidae) based on mitochondrial 12S ribosomal DNA gene sequences and morphological characters. *The Journal of Parasitology*, *87*, 32–48.

Bell, J. F. (1988). Tularemia. In J. H. Steele (Ed.), *CRC Handbook Series in Zoonoses. Section A. Bacterial, Rickettsial, and Mycotic Disease* (Vol. 2, pp. 161–193). Boca Raton, FL: CRC Press.

Berrada, Z. L., Goethert, H. K., & Telford, S. R., III (2006). Raccoons and skunks as sentinels for enzootic tularemia. *Emerging Infectious Diseases*, *12*, 1019–1021.

Borovickova, B., & Hypša, V. (2005). Ontogeny of tick hemocytes: A comparative analysis of *Ixodes ricinus* and *Ornithodoros moubata*. *Experimental & Applied Acarology*, *35*, 317–333.

Boyer, K. M., Munford, R. S., Maupin, G. O., Pattison, C. P., Fox, M. D., Barnes, A. M., et al. (1977). Tick-borne relapsing fever: An interstate outbreak originating at Grand Canyon National Park. *American Journal of Epidemiology*, *105*, 469–479.

Brisson, D., DyKhuizen, D. E., & Ostfeld, R. S. (2008). Conspicuous impacts of inconspicuous hosts on the Lyme disease epidemic. *Proceedings of the Royal Society of London, B 275*, 227–235.

Buller, R. S., Arens, M., Hmiel, S. P., Paddock, C. D., Summer, J. W., Rikihisa, Y., et al. (1999). *Ehrlichia ewingii*, a newly recognized agent of human ehrlichiosis. *The New England Journal of Medicine*, *341*, 148–155.

Burgdorfer, W., Barbour, A. G., Hayes, S. F., Benach, J. L., Grunwaldt, E., & Davis, J. P. (1982). Lyme disease—A tick-borne spirochetosis? *Science*, *216*, 1317–1319.

Burgdorfer, W., & Schwan, T. G. (1991). *Borrelia*. In A. Balows, W.J. Hausler, K.L. Herrman Jr., H.D. Isenberg, & H.J. Shadomy (Eds.), Manual of Clinical Microbiology (5th ed., pp. 560–566). Washington, DC: American Society of Microbiology.

Burger, D. B., Crause, J. C., Spickett, A. M., & Neitz, A. W. H. (1991). A comparative study of proteins present in sweating-sickness-inducing and non-inducing strains of *Hyalomma truncatum* ticks. *Experimental & Applied Acarology*, *13*, 59–63.

Burridge, M. J., Simmons, L. A., & Allan, S. A. (2000). Introduction of potential heartwater vectors and other exotic ticks in Florida on imported reptiles. *The Journal of Parasitology*, *86*, 700–704.

Camus, E., & Barreé, N. (1995). Vector situation of tick-borne diseases in the Caribbean Islands. *Veterinary Parasitology*, *57*, 167–176.

Camus, E., Barreé, N., Martinez, D., & Uilenberg, G. (1996). *Heartwater (cowdriosis): A Review* (2nd ed.). Paris: Office International des Épizooties.

Carroll, J. F., Allen, P. C., Hill, D. E., Pound, J. M., Miller, J. A., & George, J. E. (2002). Control of *Ixodes scapularis* and *Amblyomma americanum* through use of the '4-poster' treatment device on deer in Maryland. *Experimental & Applied Acarology*, *28*, 289–296.

Centers for Disease Control and Prevention. (2001). Outbreak of Powassan encephalitis—Maine and Vermont, 1999–2001. *MMWR. Morbidity and Mortality Weekly Report*, *50*, 761–764.

Centers for Disease Control and Prevention. (2006). Cluster of tick paralysis cases—Colorado, 2006. *MMWR. Morbidity and Mortality Weekly Report*, *55*, 933–935.

Ceraul, S. M., Sonenshine, D. E., & Hynes, W. L. (2002). Investigations into the resistance of the tick, *Dermacentor variabilis* (Say) (Acari: Ixodidae) following challenge with the bacterium, *Escherichi coli* (Enterobacteriales: Enterobacteriaceae). *Journal of Medical Entomology*, *39*, 376–383.

Ceraul, S. M., Sonenshine, D. E., Ratzlaff, R. E., & Hynes, W. L. (2003). An arthropod defensin expressed by the hemocytes of the American dog tick, *Dermacentor variabilis* (Acari: Ixodidae). *Insect Biochemistry and Molecular Biology*, *33*, 1099–1103.

Charrel, R. N., Attoui, H., Butenko, A. M., Clegg, J. C., Deubel, V., Frolova, T. V., et al. (2004). Tick-borne virus diseases of human interest in Europe. *Clinical Microbiology and Infection*, *10*, 1040–1055.

Childs, J. E., & Paddock, C. D. (2003). The ascendancy of *Amblyomma americanum* as a vector of pathogens affecting humans in the United States. *Annual Review of Entomology, 48,* 307–337.

Clark, K., Hendricks, A., & Burge, D. (2005). Molecular identification and analysis of *Borrelia burgdorferi* sensu lato in lizards in the southeastern United States. *Applied and Environmental Microbiology, 71,* 2616–2625.

Conrad, P. A., Kjemtrup, A. M., Carreno, R. A., Thomford, J., Wainwright, K., Eberhard, M., et al. (2006). Description of *Babesia duncani* n. sp. (Apicomplexa: Babesiidae) from humans and its differentiation from other piroplasms. *International Journal for Parasitology, 36,* 779–789.

Coons, B., Rosell-Davis, R., & Tarnowski, B. I. (1986). Bloodmeal digestion in ticks. In J. R. Sauer & J. A. Hair (Eds.), *Morphology, Physiology and Behavioral Biology of Ticks* (pp. 248–279). Chichester, UK: Ellis Harwood.

Cutler, S. J. (2006). Possibilities for relapsing fever reemergence. *Emerging Infectious Diseases, 12,* 369–374.

de la Fuente, J., Almazén, C., Naranjo, V., Blouin, E. F., & Kocan, K. M. (2006). Synergistic effect of silencing the expression of tick protective antigens 4D8 and Rs86 in *Rhipicephalus sanguineus* by RNA interference. *Parasitology Research, 99,* 108–113.

Demma, L. J., Holman, R. C., McQuiston, J. H., Krebs, J. W., & Swerdlow, D. L. (2005a). Epidemiology of human ehrlichiosis and anaplasmosis in the United States, 2001–2002. *The American Journal of Tropical Medicine and Hygiene, 73,* 400–409.

Demma, L. J., Traeger, M., Nicholson, W. L., Paddock, C. D., Blau, D., Eremeeva, M., et al. (2005b). Rocky Mountain spotted fever from an unexpected tick vector in Arizona. *The New England Journal of Medicine, 353,* 587–594.

Dumler, J. S., Barbet, A. F., Bekker, C. P. J., Dasch, G. A., Palmer, G. H., Ray, S. C., et al. (2001). Reorganization of genera in the families *Rickettsiaceae* and *Anaplasmataceae* in the order *Rickettsiales*: Unification of some species of *Ehrlichia* with *Anaplasma*. *Cowdria* with *Ehrlichia* and *Ehrlichia* with *Neorickettsia*, descriptions of six new species combinations and designation of *Ehrlichia equi* and 'HGE agent' as subjective synonyms of *Ehrlichia phagocytophila*. *International Journal of Systematic and Evolutionary Microbiology, 51,* 2145–2165.

Duncan, I. M., & Monks, N. (1992). Tick control on eland (*Taurotragus oryx*) and Buffalo (*Syncerus caffer*). *Journal of the South African Veterinary Association, 63,* 7–10.

Durden, L. A., & Keirans, J. E. (1996). Nymphs of the Genus *Ixodes* (Acari: Ixodidae) of the United States: Taxonomy, Identification Key, Distribution, Hosts and Medical/Veterinary Importance. Thomas Say Foundation Monographs. Lanham, MD: ESA.

Dutton, J. E., & Todd, J. L. (1905). The nature of tick fever in the eastern part of the Congo Free State, with notes on the distribution and bionomics of the tick. *British Medical Journal 2,* 1259–1260.

Dworkin, M. S., Shoemaker, P. C., & Anderson, D. E. (1999). Tick paralysis: 33 human cases in Washington state, 1946–1996. *Clinical Infectious Diseases, 29,* 1435–1439.

Dworkin, M. S., Shoemaker, P. C., Fritz, C. L., Dowell, M. E., & Anderson, D. E., Jr. (2002). The epidemiology of tick-borne relapsing fever in the United States. *American Journal of Tropical Medicine and Hygiene, 66,* 753–758.

Ebel, G. D., Campbell, E. N., Goethert, H. K., Spielman, A., & Telford, S. R., III (2000). Enzootic transmission of deer tick virus in New England and Wisconsin sites. *The American Journal of Tropical Medicine and Hygiene, 63,* 36–42.

Ebel, G. D., & Kramer, L. D. (2004). Short report: Duration of tick attachment required for transmission of Powassan virus by deer ticks. *The American Journal of Tropical Medicine and Hygiene, 71,* 268–271.

Eggenberger, L., Lamoreaux, W. R., & Coons, L. B. (1990). Hemocytic encapsulation of implants in the tick *Dermacentor variabilis*. *Experimental & Applied Acarology, 9,* 279–287.

Eisen, L. (2007). A call for renewed research on tick-borne *Francisella tularensis* in the Arkansas-Missouri primary natural focus of tularemia in humans. *Journal of Medical Entomology, 44,* 389–397.

Faulde, M., & Uedelhoven, W. (2006). A new clothing impregnation method for personal protection against ticks and biting insects. *International Journal of Medical Microbiology: IJMM, 296*(Suppl 40), 225–229.

Filippova, N. A. (1966). Argasid ticks (Argasidae). *Fauna SSSR, Paukoobraznye, 4*(3). (In Russian)

Filippova, N. A. (1967). Ixodid ticks of the subfamily Ixodinae. *Fauna SSSR, Paukoobraznye, 4*(4). (In Russian)

Francischetti, I. M. B., Nunes, A. S., Mans, B., & Ribeiro, J. M. C. (2008). The role of saliva in tick feeding. *Frontiers in Bioscience,* (in press).

Genchi, C. (2007). Human babesiosis, an emerging zoonosis. *Parassitologia, 49,* 29–31.

Gern, L., & Humair, P. F. (2002). Ecology of *Borrelia burgdorferi* sensu lato in Europe. In J. Gray, O. Kahl, R. S. Lane, & G. Stanek (Eds.), *Lyme borreliosis: Biology, Epidemiology and Control* (pp. 149–174). New York: CABI Publishing.

Gilbert, L., Jones, L. D., Laurenson, M. K., Gould, E. A., Reid, H. W., & Hudson, P. J. (2004). Ticks need not bite their red grouse hosts to infect them with louping ill virus. *Proceedings of the Royal Society of London, B 271,* S202–S205.

Gill, J. S., Rowley, W. A., Bush, P. J., Viner, J. P., & Gilchrist, M. J. R. (2004). Detection of human blood in the bat tick, *Carios (Ornithodoros) kelleyi* (Acari: Argasidae). *Journal of Medical Entomology, 41,* 1179–1181.

Goodman, J. L., Dennis, D. T., & Sonenshine, D. E. (2005). *Tick-borne diseases of humans.* Washington, DC: ASM Press.

Goethert, H. K., Shani, I., & Telford, S. R., III (2004). Genotypic diversity of *Francisella tularensis* infecting *Dermacentor variabilis* ticks on Martha's Vineyard, Massachusetts. *Journal of Clinical Microbiology, 42,* 4968–4973.

Gothe, R. (1999). *Tick Toxicoses.* Munich: Hieronymus. (In German)

Gough, J. M., & Kemp, D. H. (1995). Acid phosphatase in midgut digestive cells in partially fed females of the cattle tick *Boophilus microplus*. *The Journal of Parasitology, 81,* 341–349.

Grard, G., Moureau, G., Charrel, R. N., Lemasson, J. J., Gonzalez, J. P., Gallian, P., et al. (2007). Genetic characterization of tick-borne flaviviruses: New insights into evolution, pathogenetic determinants and taxonomy. *Virology, 361,* 80–92.

Grattan-Smith, P. J., Morris, J. G., Johnston, H. M., Yiannikas, C., Malik, R., Russell, R., et al. (1997). Clinical and neurophysiological features of tick paralysis. *Brain, 120,* 1975–1987.

Gray, J. S. (1991). The development and seasonal activity of the tick *Ixodes ricinus*: A vector of Lyme borreliosis. *Review of Medical and Veterinary Entomology, 79,* 323–333.

Gray, J., Kahl, O., Lane, R. S., & Stanek, G. (2002). *Lyme borreliosis: Biology, Epidemiology and Control.* New York: CABI Publishing.

Gregson, J. D. (1973). *Tick Paralysis: An Appraisal of Natural and Experimental Data.* Monograph No. 9. Ottawa: Canada Dept. Agric.

Gritsun, T. S., Lashkevich, V. A., & Gould, E. A. (2003). Tick-borne encephalitis. *Antiviral Research, 57,* 129–146.

Grubhoffer, L., Kovar, V., Rudenko, N. (2004). Tick lectins: Structural and functional properties. *Parasitology (Supplement), 129,* S113–S125.

Hadrill, D. J., Boid, R., Jones, T. W., & Bell-Sakyi, L. (1990). Bovine babesiosis on Nevis—Implications for tick control. *The Veterinary Record, 126,* 403–404.

Hanincová, K., Kurtenbach, K., Diuk-Wasser, M., Brei, B., & Fish, D. (2006). Epidemic spread of Lyme borreliosis, northeastern United States. *Emerging Infectious Diseases, 12,* 604–611.

Harwood, R. F., & James, M. T. (1979). *Entomology in Human and Animal Health* (7th ed.). New York: Macmillan Publishing Co.

Heinz, F. X., Holzmann, H., Essl, A., & Kundi, M. (2007). Field effectiveness of vaccination against tick-borne encephalitis. *Vaccine, 25,* 7559–7567.

Herwaldt, B. L., Cacciò, S., Gherlinzoni, F., Aspöck, H., Slemenda, S. B., Piccaluga, P., et al. (2003). Molecular characterization of a non-*Babesia divergens* organism causing zoonotic babesiosis in Europe. *Emerging Infectious Diseases, 9,* 942–948.

Herwaldt, B. L., de Bruyn, G., Pieniazek, N. J., Homer, M., Lofy, K. H., Slemenda, S. B., et al. (2004). *Babesia divergens*-like infection, Washington State. *Emerging Infectious Diseases, 10,* 622–629.

Hoogstraal, H. (1985). Argasid and nuttallielid ticks as parasites and vectors. *Advances in Parasitology, 24,* 135–238.

Horak, I. G., Camicas, J. L., & Keirans, J. E. (2002). The Argasidae, Ixodidae and Nuttallielidae (Acari: Ixodida): A world list of valid tick names. *Experimental & Applied Acarology, 28,* 27–54.

Hunfeld, K.-P., Hildebrandt, A., & Gray, J. S. (2008). Babesiosis: recent insights into an ancient disease. *International Journal for Parasitology. 38,* 1219–1237.

Hynes, W. L., Ceraul, S. M., Todd, S. M., Sequin, K. C., & Sonenshine, D. E. (2005). A defensin-like gene expressed in the black-legged tick, *Ixodes scapularis*. *Medical and Veterinary Entomology, 19*, 339–344.

Inoue, N., Hanada, K., Tsuji, N., Igarashi, I., Nagasawa, H., Mikami, T., et al. (2001). Characterization of phagocytic hemocytes in *Ornithodoros moubata* (Acari: Ixodidae). *Journal of Medical Entomology, 38*, 514–519.

James, A. M., Liveris, D., Wormser, G. P., Schwartz, I., Montecalvo, M. A., & Johnson, B. J. B. (2001). *Borrelia lonestari* infection after a bite by an *Amblyomma americanum* tick. *The Journal of Infectious Diseases, 183*, 1810–1814.

Jasinskas, A., Zhong, J., & Barbour, A. G. (2007). Highly prevalent *Coxiella* sp. bacterium in the tick vector *Amblyomma americanum*. *Applied and Environmental Microbiology, 73*, 334–336.

Jellison, W. L. (1974). *Tularemia in North America 1930–1974*. Missoula, MT: University of Montana Foundation.

Jellison, W. L., & Kohls, G. M. (1955). Tularemia in sheep and sheep industry workers in western United States. *Public Health Monograph, 28*, 1–17.

Johns, R., Sonenshine, D. E., & Hynes, W. L. (1998). Control of bacterial infections in the hard tick *Dermacentor variabilis* (Acari: Ixodidae): Evidence for the existence of antimicrobial proteins in tick hemolymph. *Journal of Medical Entomology, 35*, 458–464.

Jones, L. D., Hodgson, E., Williams, T., Higgs, S., & Nuttall, P. A. (1992). Saliva-activated transmission (SAT) of Thogoto virus: Relationship with vector potential of different haematophagous arthropods. *Medical and Veterinary Entomology, 6*, 261–265.

Jongejan, F., & Uilenberg, G. (2004). The global importance of ticks. *Parasitology, 129*, S3–S14.

Keirans, J. E. (1992). Systematics of the Ixodida (Argasidae, Ixodidae, Nuttalliellidae); an overview and some problems. In B. Fivaz, T. Petney, & I. G. Horak (Eds.), *Tick Vector Biology: Medical and Veterinary Aspects* (pp. 1–21). Berlin: Springer-Verlag.

Keirans, J. E., & Clifford, C. M. (1978). The genus *Ixodes* in the United States: a scanning electron microscope study and key to the adults. *Journal of Medical Entomology*, (Suppl. 2).

Keirans, J. E., Clifford, C. M., Hoogstraal, H., & Easton, E. R. (1976). Discovery of *Nuttalliella namaqua* Bedford (Acarina: Ixodoidea: Nuttalliellidae) in Tanzania and re-description of the female based on scanning electron microscopy. *Annals of the Entomological Society of America, 69*, 926–932.

King, D. P., Chen, C. I., Blanchard, M. T., Aldridge, B. M., Anderson, M., Walker, R., et al. (2005). Molecular identification of a novel deltaproteobacterium as the etiologic agent of epizootic bovine abortion (foothill abortion). *Journal of Clinical Microbiology, 43*, 604–609.

Kjemtrup, A. M., & Conrad, P. A. (2000). Human babesiosis: An emerging tick-borne disease. *International Journal for Parasitology, 30*, 1323–1337.

Kjemtrup, A. M., Wainwright, K., Miller, M., Penzhorn, B. L., & Carreno, R. A. (2006). *Babesia conradae*, sp. nov., a small canine *Babesia* identified in California. *Veterinary Parasitology, 138*, 103–111.

Klompen, J. H. S., Black, W. C., Keirans, J. E., & Norris, D. E. (2000). Systematics and biogeography of hard ticks, a total evidence approach. *Cladistics, 16*, 79–102.

Klompen, J. H. S., Black, W. C., IV, Keirans, J. E., & Oliver, J. H., Jr. (1996). Evolution of ticks. *Annual Review of Entomology, 41*, 141–161.

Klompen, J. H. S., & Oliver, J. H., Jr. (1993). Systematic relationships in the soft ticks (Acari: Ixodida: Argasidae). *Syst Entomol, 18*, 313–331.

Kocan, K. M., de la Fuente, J., Blouin, E. F., & Garcia-Garcia, J. C. (2004). *Anaplasma marginale* (Rickettsiales: Anaplasmataceae): Recent advances in defining host-pathogen adaptations of a tick-borne rickettsia. *Parasitology, 129*, S285–S300.

Korenberg, E. I., & Kovalevskii, Y. V. (1999). Main features of tick-borne encephalitis eco-epidemiology in Russia. *Zentralblatt Bakteriologie, 289*, 525–539.

Kunze, U., & the International Scientific Working Group on Tick-Borne Encephalitis. (2007). Tick-borne encephalitis: From epidemiology to vaccination recommendations in 2007. New issues–best practices. *Wiener medizinische Wochenschrift (1946), 157*, 228–232.

Kuo, M. M., Lane, R. S., & Giclas, P. C. (2000). A comparative study of mammalian and reptilian alternative pathway of complement-mediated killing of the Lyme disease spirochete (*Borrelia burgdorferi*). *The Journal of Parasitology, 86*, 1223–1228.

Kurtenbach, K., Sewell, H. S., Ogden, N. H., Randolph, S. E., & Nuttall, P. A. (1998). Serum complement sensitivity as a key factor in Lyme disease ecology. *Infection and Immunity, 66*, 1248–1251.

Labuda, M., Kozuch, O., Zuffova, E., Elecková, E., Hails, R. S., & Nuttall, P. A. (1997). Tick-borne encephalitis virus transmission between ticks cofeeding on specific immune natural rodent hosts. *Virology, 235*, 138–143.

Labuda, M., Trimnell, A. R., Lickova, M., Kazimirova, M., Davies, G. M., Lissina, O., et al. (2006). An antivector vaccine protects against a lethal vector-borne pathogen. *PLoS Pathogens, 2*, e27.

Lai, R., Lomas, L. O., Jonczy, J., Turner, P. C., & Rees, H. H. (2004). Two novel non-cationic defensin-like antimicrobial peptides from hemolymph of the female tick, *Amblyomma hebraeum*. *The Biochemical Journal, 379*, 681–685.

Lane, R. S. (1994). Competence of ticks as vectors of microbial agents with an emphasis on *Borrelia burgdorferi*. In D. E. Sonenshine & T. N. Mather (Eds.), *Ecological Dynamics of Tick-Borne Zoonoses* (pp. 45–67). New York: Oxford University Press.

Lane, R. S., Burgdorfer, W., Hayes, S. F., & Barbour, A. G. (1985). Isolation of a spirochete from the soft tick, *Ornithodoros coriaceus*: a possible agent of epizootic bovine abortion. *Science, 230*, 85–87.

Lane, R. S., Emmons, R. W., Devlin, V., Dondero, D. V., & Nelson, B. C. (1982). Survey for evidence of Colorado tick fever virus outside of the known endemic area in California. *The American Journal of Tropical Medicine and Hygiene, 31*, 837–843.

Lane, R. S., Peek, J., & Donaghey, P. J. (1984). Tick (Acari: Ixodidae) paralysis in dogs from northern California: Acarological and clinical findings. *Journal of Medical Entomology, 21*, 321–326.

Lara, F. A., Lins, U., Paiva-Silva, G., Almeida, I. C., Braga, C. M., Miguens, F. C., et al. (2003). A new intracellular pathway of haem detoxification in the midgut of the cattle tick *Boophilus microplus*: aggregation inside a specialized organelle, the hemosome. *The Journal of Experimental Biology, 206*, 1707–1715.

Lees, K., & Bowman, A. S. (2007). Tick neurobiology: Recent advances and the post-genomic era. *Invertebrate Neuroscience, 7*, 183–198.

LoGiudice, K., Ostfeld, R. S., Schmidt, K. A., & Keesing, F. (2003). The ecology of infectious disease: Effects of host diversity and community composition on Lyme disease risk. *Proceedings of the National Academy of Sciences, United States of America, 100*, 567–571.

Lomas, L. O., Turner, P. C., & Rees, H. H. (1997). A novel neuropeptide-endocrine interaction controlling ecdysteroid production in ixodid ticks. *Proceedings. Biological Sciences/The Royal Society, 264*, 589–596.

Lysyk, T. J., & Majak, W. (2003). Increasing the paralyzing ability of a laboratory colony of *Dermacentor andersoni* Stiles. *Journal of Medical Entomology, 40*, 185–194.

Mahan, S. M. (2006). Diagnosis and control of heartwater, *Ehrlichia ruminantium*, infection: An update. *CAB Reviews: Perspectives in Agriculture, Veterinary Science, Nutrition and Natural Resources, 1*, (no. 055).

Mans, B. J., Goethe, R., & Neitz, A. W. H. (2004). Biochemical perspectives on paralysis and other forms of toxicoses caused by ticks. *Parasitology, 129*, S95–S111.

Marfin, A. A., & Campbell, G. L. (2005). Colorado tick fever and related *Coltivirus* infections. In J. L. Goodman, D. T. Dennis, & D. E. Sonenshine (Eds.), *Tick-Borne Diseases of Humans* (pp. 143–149). Washington, DC: ASM Press.

Mather, T. N., Wilson, M. L., Moore, S. I., Ribeiro, J. M. C., & Spielman, A. (1989). Comparing the relative potential of rodents as reservoirs of the Lyme disease spirochete (*Borrelia burgdorferi*). *American Journal of Epidemiology. 130*, 143–150.

McCall, P. J., Hume, J. C. C., Motshegwa, K., Pignatelli, P., Talbert, A., & Kisinza, W. (2007). Does tick-borne relapsing fever have an animal reservoir in east Africa. *Vector-Borne and Zoonotic Diseases, 7*, 659–666.

McConnell, J. (2003). Tick-borne relapsing fever under-reported. *The Lancet Infectious Diseases, 3*, 604.

Mehlhorn, H., Raether, W., Schein, E., Weber, M., & Uphoff, M. (1986). Licht- und elektronenmikroskopische Untersuchungen zum Entwicklungszyklus and Einfluss von Pentamidin auf die Morphologie der intraerythrocytären Stadien von *Babesia microti*. *Deutsche Tierärztliche Wochenschrift, 93*, 400–405.

Meinkoth, J. H., & Kocan, A. A. (2005). Feline cytauxzoonosis. *Veterinary Clinics of North America, Small Animal Practice, 35*, 89–101.

Mendiola, J., Alonso, M., Marquetti, M. C., & Finlay, C. (1996). *Boophilus microplus*: Multiple proteolitic activities in the midgut. *Experimental Parasitology, 82*, 27–33.

Mitchell, R. E., 3rd, Ross, E., Osgood, C., Sonenshine, D. E., Donohue, K. V., Khalil, S. M., et al. (2007). Molecular characterization, tissue-specific expression and RNAi knockdown of the first vitellogenin receptor from a tick. *Insect Biochemistry and Molecular Biology, 37*, 375–388.

Murrel, A., & Barker, S. C. (2003). Synonomy of *Boophilus* Curtice, 1891 with *Rhipicephalus* Koch, 1844 (Acari: Ixodidae). *Systematic Parasitology, 56*, 169–172.

Murrel, A., Campbell, N. J., & Barker, S. C. (2000). Phylogenetic analyses of the rhipicephaline ticks indicate that the genus *Rhipicephalus* is paraphyletic. *Molecular Phylogenetics and Evolution, 16*, 1-7.

Nakajima, Y., van der Goes van Naters-Yusui, A., Taylor, D., & Yamakawa, M. (2001). Two isoforms of a member of the arthropod defensin family from the soft tick, *Ornithodoros moubata* (Acari: Argasidae). *Insect Biochemistry and Molecular Biology, 31*, 747–751.

Nicholson, W. L., Castro, M. B., Kramer, V. L., Summer, J. W., & Childs, J. E. (1999). Dusky-footed wood rats (*Neotoma fuscipes*) as reservoirs of granulocytic ehrlichiae (Rickettsiales: Ehrlichieae) in northern California. *Journal of Clinical Microbiology, 37*, 3323–3327.

Norval, R. A. I., Sonenshine, D. E., Allan, S. A., & Burridge, M. J. (1996). Efficacy of pheromone-acaricide impregnated tail-tag decoys for control of bont ticks, *Amblyomma hebraeum* on cattle in Zimbabwe. *Experimental & Applied Acarology, 20*, 31–46.

Nuttall, P. A., & Jones, L. D. (1991). Non–viraemic tick–borne virus transmission: Mechanism and significance. In F. Dusbabek & V. Bukva (Eds.), *Modern Acarology, Vol. 2. Proceedings, 8th Int. Congr. Acarol* (pp. 3–6). The Hague, The Netherlands: Academia Prague and SPB Academic Publishing.

Nuttall, P. A., & Labuda, M. (1994). Tick-borne encephalitis subgroup complex. In D. E. Sonenshine & T. N. Mather (Eds.), *Ecological Dynamics of Tick-borne Zoonoses* (pp. 351–391). New York: Oxford University Press.

Oliver, J. H., Jr., Owsley, M. R., Hutcheson, H. J., James, A. M., Chen, C., Irby, W. S., et al. (1993). Conspecificity of the ticks *Ixodes scapularis* and *I. dammini* (Acari: Ixodidae). *Journal of Medical Entomology, 30*, 54–63.

Paddock, C. D. (2005). *Rickettsia parkeri* as a paradigm for multiple causes of tick-borne spotted fever in the western hemisphere. *Annals of the New York Academy of Sciences, 1061*, 315–326.

Paddock, C. D., & Childs, J. E. (2003). *Ehrlichia chaffeensis*: A prototypical emerging pathogen. *Clinical Microbiology Reviews, 16*, 37–64.

Parker, R. R., Philip, C. B., & Jellison, W. L. (1933). Potentialities of tick transmission in relation to geographical occurrence in the United States. *The American Journal of Tropical Medicine and Hygiene, 13*, 341–379.

Parola, P., Paddock, C. D., & Raoult, D. (2005). Tick-borne rickettsioses around the world: Emerging diseases challenging old concepts. *Clinical Microbiology Reviews, 18*, 719–756.

Pau, U., Li, X., Wang, T., Montgomery, R. R., Ramamoorthy, N., DeSilva, A. M., et al. (2004). TROSPA, an *Ixodes scapularis* receptor for *Borrelia burgdorferi*. *Cell, 119*, 457–468.

Pegram, R. (2006). End of the Caribbean *Amblyomma* Programme. *ICTTD Newsletter*, (no. 30, June), 4–6.

Pegram, R., Indar, L., Eddi, C., & George, J. (2004). The Caribbean *Amblyomma* Program: Some ecological factors affecting its success. *Annals of the New York Academy of Sciences, 1026*, 302–311.

Piesman, J., & Gern, L. (2004). Lyme borreliosis in Europe and North America. *Parasitology, 129*, S191–S220.

Randolph, S. E. (2001). The shifting landscape of tick-borne zoonoses: Tick-borne encephalitis and Lyme borreliosis in Europe. *Philosophical Transactions of the Royal Society of London B 356*, 1045–1056.

Randolph, S. E., Gern, L., & Nuttall, P. A. (1996). Co-feeding ticks: Epidemiological significance for tick-borne pathogen transmission. *Parasitology Today, 12*, 472–479.

Randolph, S. E., Miklisová, D., Lysy, J., Rogers, D. J., & Labuda, M. (1999). Incidence from coincidence: Patterns of tick infestations on rodents facilitate transmission of tick-borne encephalitis virus. *Parasitology, 118*, 177–186.

Randolph, S. E., Green, R. M., Peacey, M. F., & Rogers, D. J. (2000). Seasonal synchrony: The key to tick-borne encephalitis foci identified by satellite data. *Parasitology, 121*, 15–23.

Randolph, S. E., & Ŝumilo, D. (2007). Tick-borne encephalitis in Europe: Dynamics of changing risk. In W. Takken & B. G. J. Knols (Eds.), *Emerging Pests and Vector-borne Diseases in Europe* (pp. 187–206). Wageningen University Publishers.

Ribeiro, J. M. C. (1989). Role of saliva in tick/host interactions. *Experimental & Applied Acarology, 7*, 15–20.

Ribeiro, J. M., Alarcon-Chaidez, F., Francischetti, I. M., Mans, B. J., Mather, T. N., Valenzuela, J. G., et al. (2006). An annotated catalog of salivary gland transcripts from *Ixodes scapularis* ticks. *Insect Biochemistry and Molecular Biology, 36*, 111–129.

Roberts, F. H. S. (1970). *Australian Ticks*. Melbourne: Commonwealth Scientific and Industrial Organization.

Rogers, D. J., & Randolph, S. E. (2006). Climate change and vector-borne diseases. *Advances in Parasitology, 62*, 345–381.

Rudenko, N., Golovchenko, M., Edwards, M. J., & Grubhoffer, L. (2005). Differential expression of *Ixodes ricinus* tick genes induced by blood feeding or *Borrelia burgdorferi* infection. *Journal of Medical Entomology, 42*, 36–41.

Schetters, T. P. M., & Brown, W. C. (Eds.). (2006). Special issue: Babesiosis. *Veterinary Parasitology, 138*, 1–168.

Schwan, T. G., Schrumpf, M. E., Hinnebusch, B. J., Anderson, D. E., Jr., & Konkel, M. E. (1996). GlpQ: An antigen for serological discrimination between relapsing fever and Lyme borreliosis. *Journal of Clinical Microbiology, 34*, 2483–2492.

Schwan, T. G., Policastro, P. F., Miller, Z., Thompson, R. L., Damrow, T., & Keirans, J. E. (2003). Tick-borne relapsing fever caused by *Borrelia hermsii*, Montana. *Emerging Infectious Diseases, 9*, 1151–1154.

Schwan, T. G., Raffel, S. J., Schrumpf, M. E., & Porcella, S. F. (2007). Diversity and distribution of *Borrelia hermsii*. *Emerging Infectious Diseases, 13*, 436–442.

Scott, J. C., Wright, D. J. M., & Cutler, S. J. (2005). Typing African relapsing fever spirochetes. *Emerging Infectious Diseases, 11*, 1722–1729.

Sidi, G., Davidovitch, N., Balicer, R. D., Anis, E., Grotto, I., & Schwartz, E. (2005). Tickborne relapsing fever in Israel. *Emerging Infectious Diseases, 11*, 1784–1786.

Sonenshine, D. E. (1991). *Biology of Ticks* (Vol. I). New York and Oxford: Oxford University Press.

Sonenshine, D. E. (1993). *Biology of Ticks* (Vol. II). New York and Oxford: Oxford University Press.

Sonenshine, D. E., Adams, T., Allan, S. A., McLaughlin, J. R., & Webster, F. X. (2003). Chemical composition of some components of the arrestment pheromone of the black-legged tick, *Ixodes scapularis* (Acari: Ixodidae) and their use in tick control. *Journal of Medical Entomology, 40*, 849–859.

Sonenshine, D. E., Allan, S. A., Norval, R. A. I., & Burridge, M. J. (1996). A self-medicating applicator for control of ticks on deer. *Medical and Veterinary Entomology, 10*, 149–154.

Sonenshine, D. E., & Hynes, W. L. (2008). Molecular characterization and related aspects of the innate immune response in ticks. *Frontiers in Bioscience, 13*, 7046–7063.

Stafford, K. C., III (2007). *Tick Management Handbook: An Integrated Guide for Homeowners, Pest Control Operators, and Public Health Officials for the Prevention of Tick-Associated Diseases* (revised ed.), Connecticut Agricultural Experiment Station Bulletin No. 1010. Available at www.ct.gov/caes

Steere, A. C., Coburn, J., & Glickstein, L. (2005). Lyme borreliosis. In J. L. Goodman, D. T. Dennis, & D. E. Sonenshine (Eds.), *Tick-Borne Diseases of Humans* (pp. 176–206). Washington, DC: ASM Press.

Telford, S. R., III, Mather, T. N., Moore, S. I., Wilson, M. L., & Spielman, A. (1988). Incompetence of deer as reservoirs of the Lyme disease spirochete. *American Journal of Tropical Medicine and Hygiene. 39*, 105–109.

Thomas, N. J., Bunikis, J., Barbour, A. G., & Wolcott, M. J. (2002). Fatal spirochetosis due to a relapsing fever-like *Borrelia* sp. in a northern spotted owl. *Journal of Wildlife Diseases, 38*, 187–193.

Thompson, D. M., Khalil, S. M., Jeffers, L. A., Ananthapadmanaban, U., Sonenshine, D. E., Mitchell, R. D., et al. (2005). In vivo role of 20-hydroxyecdysone in the regulation of the vitellogenin mRNA and egg development in the American dog tick, *Dermacentor variabilis* (Say). *Journal of Insect Physiology, 51*, 1105–1116.

Thompson, D. M., Khalil, S. M., Jeffers, L. A., Sonenshine, D. E., Mitchell, R. D., Osgood, C. J., et al. (2007). Sequence and the developmental and tissue-specific regulation of the first complete vitellogenin messenger RNA from ticks responsible for heme sequestration. *Insect Biochemistry and Molecular Biology, 37*, 363–374.

Trager, W. (1939). Acquired immunity to ticks. *The Journal of Parasitology, 25,* 57–81.

Uilenberg, G. (1995). International collaborative research: Significance of tick-borne hemoparasitic diseases to world animal health. *Veterinary Parasitology, 57,* 19–41.

Uilenberg, G. (2006). *Babesia*—A historical overview. *Veterinary Parasitology, 138,* 3–10.

Van Wye, J. E., Hsu, Y. P., Terr, A. I., Lane, R. S., & Moss, R. B. (1991). Anaphylaxis from a tick bite. *The New England Journal of Medicine, 324,* 777–778.

Valenzuela, J. G., Charlab, R., Mather, T. N., & Ribeiro, J. M. C. (2000). Purification, cloning and expression of a novel salivary anticomplement protein from the tick, *Ixodes scapularis. The Journal of Biological Chemistry, 275,* 18717–18723.

Vázquez, M., Muehlenbein, C., Cartter, M., Hayes, E. B., Ertel, S., & Shapiro, E. D. (2008). Effectiveness of personal protective measures to prevent Lyme disease. *Emerging Infectious Diseases, 14,* 210–216.

Vial, L., Diatta, G., Tall, A., Ba, E. H., Bouganali, H., Durand, P., et al. (2006). Incidence of tick-borne relapsing fever in west Africa: Longitudinal study. *Lancet, 368,* 37–43.

Waladde, S. M., & Rice, M. J. (1982). The sensory basis of tick feeding behavior. In F.D. Obenchain & R. Galun (Eds.), *Physiology of Ticks* (pp. 71–118). Oxford: Pergamon Press.

Waldenström, J., Lundkvist, Å., Falk, K. I., Garpmo, U., Bergström, S., Lindegren, G., et al. (2007). Migrating birds and tickborne encephalitis virus. *Emerging Infectious Diseases, 13,* 1215–1218.

Walker, A. R., & Lloyd, C. M. (1993). Experiments on the relationship between feeding of the tick *Amblyomma variegatum* (Acari: Ixodidae) and dermatophilosis skin disease in sheep. *Journal of Medical Entomology, 30,* 136–143.

Walker, J. B., Kierans, J. E., & Horak, I. G. (2000). *The Genus Rhipicephalus (Acari: Ixodidae): A Guide to the Brown Ticks of the World.* Cambridge: Cambridge University Press.

Wilson, L. B., & Chowning, W. M. (1904). Studies in pyroplasmosis hominis (spotted fever or tick fever of the Rocky Mountains). *Journal of Infectious Diseases. 1,* 31–57.

Wormser, G. P., Masters, E., Liveris, D., Nowakowski, J., Nadelman, R. B., Holmgren, D., et al. (2005). Microbiologic evaluation of patients from Missouri with erythema migrans. *Clinical Infectious Diseases, 40,* 423–428.

Zeidner, N. S., Burkot, T. R., Massung, R., Nicholson, W. L., Dolan, M. C., Rutherford, J. S., et al. (2000). Transmission of the agent of human granulocytic ehrlichiosis by *Ixodes spinipalpis* ticks: Evidence of an enzootic cycle of dual infection with *Borrelia burgdorferi* in northern Colorado. *The Journal of Infectious Diseases, 182,* 616–619.

Zintl, A., Mulcahy, G., Skerrett, H. E., Taylor, S. M., & Gray, J. S. (2003). *Babesia divergens,* a bovine blood parasite of veterinary and zoonotic importance. *Clinical Microbiology Reviews, 16,* 622–636.

Zung, J. L., Lewengrub, S., & Rudzinska, M. A. (1989). Fine structural evidence for the penetration of the Lyme disease spirochete *Borrelia burgdorferi* through the gut and salivary tissues of *Ixodes dammini. Canadian Journal of Zoology, 67,* 1737–1748.

Chapter

27

Molecular Tools Used in Medical and Veterinary Entomology

Dana Nayduch

This section is by no means a comprehensive list of all molecular techniques used in the field of medical and veterinary entomology. Rather, it serves as a resource for the reader to obtain information on, and a brief explanation of, some of the more commonly used molecular techniques and their applications. The section focuses on the description of those techniques and examples of applications currently being used in medical and veterinary entomological research. Molecular tools are used by researchers to address numerous biological questions including, but certainly not limited to: phylogeny and relatedness of arthropods, roles of genes and proteins in vector competence and vector potential, genetic basis of pesticide resistance, and molecular basis of arthropod development. Further, many of these techniques can help in an applied, clinical sense when they are used for the rapid and simple detection of vector-borne pathogens in both arthropods and vertebrates.

CLONING GENES AND GENOMICS

Cloning Genes: Recombinant DNA Technology

"Cloning" genes, as the name implies, involves isolating a gene of interest and making identical copies of it. This is accomplished by harvesting the gene from the host organism, for example an insect, by various techniques (using enzymes to cut the gene out of a stretch of DNA containing the gene, or other techniques to be described in this section) and then inserting it into a *vector*. Vectors, in this context, are usually either circular DNA molecules called **plasmids** or artificial chromosomes (e.g., BACs are bacterial artificial chromosomes) with known DNA sequences, that contain specific sites for cutting, and then inserting (ligating), foreign DNA fragments. When the foreign gene is inserted into the vector, this recombined molecule is referred to as **recombinant DNA**. Methods downstream from this (such as making transgenic organisms and other applications described later) are part of recombinant DNA technology.

One method for propagating (making many copies of) the cloned gene involves putting the recombinant vector back into suitable host organisms, such as bacteria. This process, by which bacteria take up extraneous DNA molecules, is called **transformation**. Transformed cells can be grown in vast quantities, thereby amplifying the inserted gene during each cell division. The recombinant vector can be reisolated from these transformed cells, yielding millions of copies of the cloned gene, which can be cut back out of the vector and used for other purposes, or simply stored as a genetic library. Each gene is housed in a separate bacterial clone (see later). In other cases, the recombinant molecule (e.g., plasmid) is left inside the bacteria, which transcribe and translate the gene into a collectable protein product that may be secreted into culture medium. This technology is used to make many recombinant proteins, such as human insulin used by diabetics. Since bacteria can be manipulated to synthesize introduced genes, microbes can serve as delivery vehicles for these proteins. This is the concept behind **paratransgenesis**.

Paratransgenesis and Controlling Vector-borne Illnesses

As with any host-parasite system, controlling the spread of vector-borne disease involves disrupting the parasite's life cycle. This can be done by treating definitive hosts with medications to kill the parasite (e.g., antimalarial drugs, antibiotics to kill tick-borne bacterial pathogens) or preventing those hosts from becoming infected in the first place (e.g., pesticides to kill infected vectors). However, alternatives to pesticides have been highly sought, since pesticide resistance is on the rise and because these chemicals are harmful to other animals, including beneficial insects. The objective of paratransgenesis is novel but simple: kill the parasite, not in the definitive host, but instead in the vector. Because vectors cannot all be captured and treated with drugs, researchers have devised an ingenious delivery system to use "friendly" microbes to perform the surreptitious delivery of the antiparasitic compound *in vivo*.

Many blood-feeding arthropods harbor symbiotic bacteria that are either commensals or mutuals. In fact, the mutuals are usually obligately symbiotic for both hosts, as they manufacture and provision nutrients to the hosts that have restricted diets (e.g., strict blood-feeders like tsetse flies). Additionally, symbionts are transmitted vertically from mother to offspring so they are pervasive in a population of the host species. The rationale behind paratransgenesis is to genetically engineer symbiotic bacteria by transforming them with a recombinant plasmid containing an antiparasitic gene. The recombinant symbionts can be reintroduced to the vector, and, during their normal biology, synthesize the gene product in proximity to the parasites, killing them *in vivo* and blocking transmission. Candidate genes include those involved in invertebrate immunity (e.g., genes coding for antimicrobial peptides, such as the gene *attacin*, which kills African trypanosomes). This approach to blocking transmission of vector-borne disease seemed promising, since the symbionts frequently reside in proximity to the parasites within the body of the vector and seem to be resistant to the effects of these genes. This second observation is vital in ensuring that the gene persists in the symbiont; if the gene causes any detrimental effect to the symbiont itself, natural selection would hinder its persistence in future generations. Vector biologists continue to assess the feasibility of this option in controlling vector-borne diseases. One system, that of tsetse flies and their symbiotic bacterium, *Sodalis glossinidius*, is outlined in Figure 27.1.

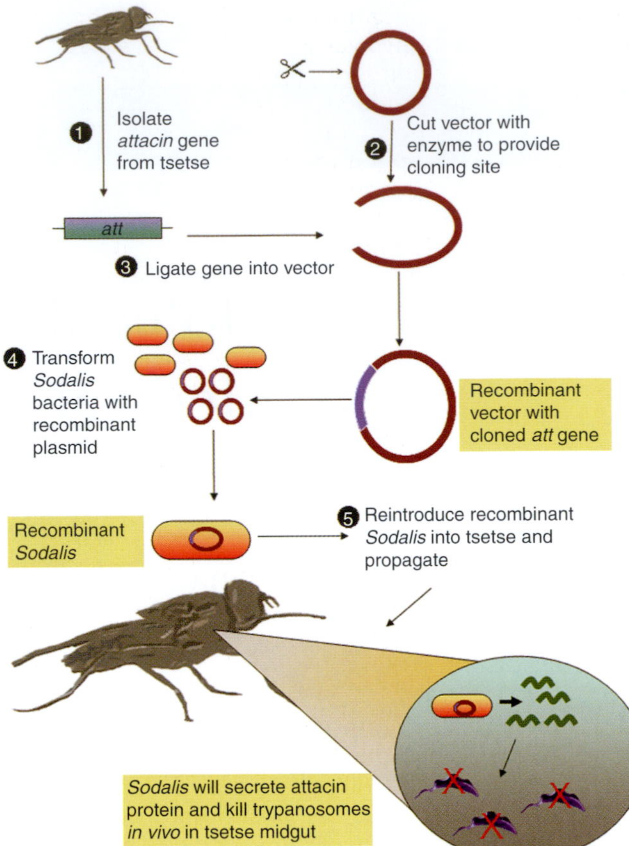

Figure 27.1 Paratransgenesis in tsetse flies. The *attacin* gene, found naturally in tsetse, has been shown to kill trypanosomes *in vitro*. Tsetse harbor symbiotic bacteria, *Sodalis*, which can be transformed with plasmids containing *attacin*. The paratransgenic strategy involves replacing wild-type *Sodalis* with attacin-producing recombinant strains of bacteria, which can kill trypanosomes *in vivo*.

GENOMICS: CATALOGUING AN ORGANISM'S COMPLETE GENETIC SEQUENCE

A genome is a complete collection of an organism's hereditary information and includes both coding and noncoding sequences of DNA (or, in the case of some viruses, RNA). Additionally, for some organisms such as bacteria this consists of both chromosomal DNA and extrachromosomal DNA (e.g., plasmids). Genome sizes vary greatly and, like chromosome number, are no measure of an organism's complexity. For example, the human genome consists of about 3.2×10^9 base pairs (bp), with bases being the nucleotides Adenine (A), Thymine (T), Guanosine (G), or Cytosine (C); and the fruit fly, *Drosophila melanogaster*, is one order of magnitude smaller (1.3×10^8 bp). Conversely, the single-celled protist *Amoeba dubia* has a very large genome (6.7×10^{11} bp).

Library Construction and Genome Assembly

All the DNA from an organism's genome can be harvested and cataloged in a genomic "library." The first step in library construction involves cutting the long stretches of DNA into more manageable pieces and cloning them into suitable vectors. This is analogous to photocopying individual words from a large book and storing each of them in small folders in a filing cabinet. DNA can either be cut specifically with enzymes called **restriction endonucleases** (see RFLP, following) or sheared randomly into fragments and "shotgun" cloned. Shotgun cloning and subsequent sequencing has revolutionized and streamlined the high-throughput method of genomic library construction used today. Advances in sequencing technology and computational bioinformatics have allowed researchers to more easily assemble contiguous stretches of DNA into genes, chromosomes, and eventually whole maps of genomes by piecing together a large library of overlapping short fragments.

There are two techniques commonly used to sequence genomes, the whole-genome shotgun method and the clone contig method. The whole-genome shotgun approach involves sequencing millions of relatively small fragments of genomic DNA, and using bioinformatics (described later) to assemble these fragments into larger, contiguous sequences, or contigs, that can be mapped to the genome using a physical framework of genetic markers. The conventional clone contig method, which is much more labor intensive and incorporates methodology from the shotgun method, is considered to be more accurate and error-free than the whole-genome shotgun method, but is not described here. For the whole-genome shotgun method, first the organism's DNA is isolated from a tissue sample, randomly sheared, and then size-selected to yield two populations of manageable fragments that are typically either 1 to 2 kilobases (1 kb = 1000 bp) or between 10 and 20 kb. These fragments are cloned into appropriate vectors (plasmids or BACs) and transformed into bacteria. Transformed bacteria—each containing a piece of the genome—are grown on culture plates, and naturally amplify recombinant plasmids as they replicate. The library of genes is then stored in these bacterial clones (with necessary overlapping and redundancy).

To begin sequencing the genome, the smaller plasmid inserts are isolated from the bacterial clones and sequenced from each end (~500 bp). Sequencing both ends (mate pairs) of the same fragment provides two valuable pieces of information: (1) the two sequences are oriented in opposite directions, and (2) the distance that these sequences are from each other, extrapolated from the insert size. Computer algorithms analyze the redundancy and overlaps of millions of clones to delineate regions of contiguous sequence. The longer fragments (~10 kb) are sequenced from both ends as well, which confirms relative orientation and provides approximate spacing to assist in the assembly of the smaller insert reads (Figure 27.2). Although gaps in the sequence may remain, especially in areas where many repeats (such as those that occur in eukaryotic introns) exist, this becomes the draft sequence.

A scaffold-like framework of the physical map is used to orient and anchor these contigs. Physical maps are generated ahead of time using molecular methods and contain known landmarks (markers or genes) that assist in assembling contigs in a larger context. The distance between contigs can be inferred from the mate-pair positions in the larger clones. Essentially, by "walking" along the DNA sequence, lining up the overlaps piece by piece along the physical framework, the genome can be assembled in the correct order and orientation (Figure 27.3). Higher order frameworks, containing known markers at defined positions, also are generated in genetic maps, which are constructed using genetic techniques.

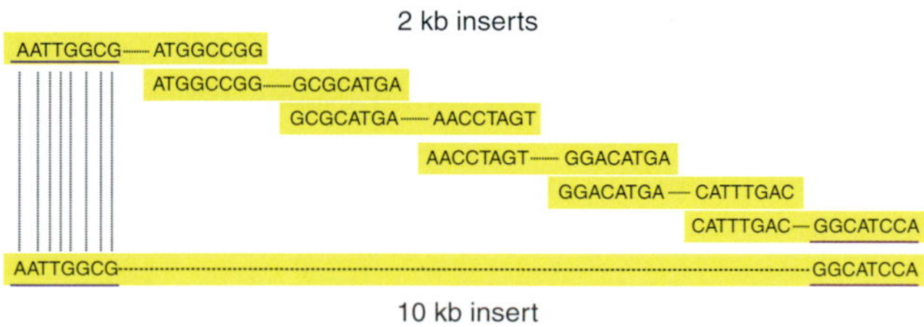

Figure 27.2 Alignment of small contiguous sequences, or contigs, (2 kb) with larger clones (10 kb) in initial genome assembly. The larger clones generally contain markers that are part of the physical map.

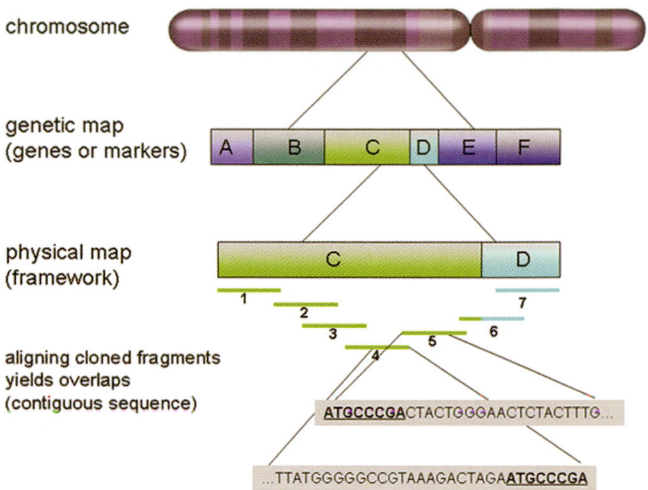

Figure 27.3 Assembly of contiguous sequences in shotgun genomic library construction. Contigs are aligned using overlapping regions (shown in bold). Note that some inserts, such as contig 6, will span gene or marker regions. Assembly, alignment, orientation, and gap filling are assisted by the physical map, which contains landmarks that serve as a framework for reconstruction. The genetic map is generated not by molecular methods, but rather by use of genetic techniques (breeding experiments, pedigrees), and also serves as a framework for hierarchical assembly. By repeating this process (i.e., by sequencing enough clones for sufficient coverage/representation of the genome), eventually the entire chromosome sequence can be constructed. For simplicity, only seven small fragments are shown; in reality, there would be much more redundancy and overlapping.

It is important to note that the whole-genome shotgun approach does not require any prior knowledge of the genome and can be performed in the absence of a genetic or physical map. This is especially the case for simple prokaryotic genomes that do not have long regions of repeated sequences. The original sequence is reconstructed from the insert reads using sequence assembly software that can be used to assign a score for the base quality (Phred, implying "fragment editor"), help assemble sequences (Phrap, implying "fragment assembly program"), and edit assembled sequences (Consed). These tools all are freely available though the University of Washington at www.phrap.org. New methods for sequencing and assembling genomes are evolving rapidly and include cell-free systems such as fiber-optic-based 454 pyrosequencing (www.454.com) and cloning with phi-29 DNA polymerase.

Bioinformatics and databases

Bioinformatics involves using computer algorithms and statistical functions to collect, catalog, and analyze DNA and protein sequences. The major data banks where these sequences are stored are GenBank at the National Institutes of Health, Bethesda, Maryland, and the EMBL Sequence Data Base at the European Bioinformatics Institute in the United Kingdom. Researchers post newly reported sequences in these databases, where they are freely available to the public via the Internet (www.ncbi.nlm.nih.gov and www.ebi.ac.uk/Databases).

New sequences from genome projects can be inputted into these databases to assess putative gene functions by looking for homology to known sequences. Additionally, protein-coding regions can be translated and compared to protein databases, which also are available. Alignment programs such as BLAST (Basic Local Alignment Search Tool) at NCBI (National Center for Biotechnology Information) have wide-ranging capabilities including determining the similarity of arbitrary sequences by aligning them with other sequences in the database. Other programs are used to sort through raw sequence data to identify boundaries between genes and assigning gene functions (genome **annotation**). Others can be used to determine evolutionary relationships between genes across phyla (molecular phylogenetics).

Genomes of Vectors and Vector-borne Pathogens

Genome sequence data and analysis can provide new avenues for vector biologists to study arthropods of medical and veterinary importance and/or the microorganisms they transmit. For example, arthropod and microbial genomics can be used to investigate the phylogenetic relationships between arthropod taxa. Moreover, genome analysis may reveal the genetic basis of vector competence and vector-pathogen interactions, which in turn can lead to novel targets for control. The genomes of numerous arthropod vectors and pests, and vector-borne pathogens have been sequenced or genome projects are underway. Tables 27.1 and 27.2 list those organisms for which genomes have been completely sequenced, or nearly so.

POLYMERASE CHAIN REACTION (PCR)

Although the initial use for **PCR** was to amplify fragments of DNA for subsequent analysis and tests, the technique now has numerous applications. Amplifying DNA from a sample, or template, involves all the key players in cellular DNA replication including (1) nucleotides (building blocks); (2) **primers**, to "prime" the synthesis to begin in certain areas of the template, and which will flank the resulting amplicon on the 5′ and 3′ ends; (3) buffers, to maintain optimal pH and salt concentrations, and provide cofactors for the reaction; and (4) a heat-stable polymerase, the enzyme that will perform the DNA synthesis.

The amplification is indeed a "chain reaction" in that there are sequential rounds of amplification cycles, with each cycle increasing the amount of DNA product (amplicon) exponentially ($1 \rightarrow 2 \rightarrow 4 \rightarrow 8$). After the typical 35 cycles, one molecule of DNA is amplified to more than 34 billion products (Figure 27.4). Each cycle has three temperature-associated steps (Figure 27.5): (1) **denaturation**, where DNA is heated to 94°C, thereby separating the strands of template; (2) **annealing**, in which the primers (short oligonucleotides) anneal, or bind, to the separated strands of DNA template by base-pairing rules; (3) **extension**, during which the polymerase performs its enzymatic reaction and adds the nucleotides to the growing strand by base pairing rules, using the original molecule as a template (at 72°C for most polymerases). The process is performed using an automated cycler, which can heat and cool the tubes containing the reaction mixture (template, primers, buffers, polymerase) in a very short time.

Applications of PCR

PCR is used for a wide range of applications from diagnostics to mutagenesis to phylogenetics. The key component in PCR is in the development of the primers, which can be designed to bind to, and amplify, specific gene regions. For example, PCR can be so precise and sensitive that species- and gene-specific primers can amplify only the product of interest, even though several DNA templates may be present in a mixed sample. The presence of a product of predicted size, based on the distance between the primers on that gene, indicates that the organism's DNA was present in the sample. This technique is used frequently for rapid diagnostic techniques, such as detecting pathogens within total DNA extracted from arthropod vectors (e.g., rickettsia in ticks, *Plasmodium* in mosquitoes).

At the same time, PCR stringency is flexible and temperature-dependent. For example, primers containing an intentional mismatch are used to insert mutations into DNA sequences by performing PCR with a lowered annealing temperature. This allows for the incorporation of that mismatch in the amplicons despite base-pairing rules. The technique also can be used to incorporate other point mutations, such as deletions or insertions. This type of site-directed mutagenesis allows for rapid screening of gene function when the mutated gene is reintroduced into an organism and the new phenotype is expressed.

A brief discussion of some other PCR-based techniques and their applications follows.

Amplified Fragment Length Polymorphism (AFLP) Restriction enzymes (REs) recognize certain DNA sequences and only cut at those locations, or restriction sites. Thus, distances between the sites and the presence or absence of the sites themselves vary between individuals due to mutations such as

Table 27.1 Complete (or Nearly Complete) Genome Projects for Arthropods of Medical and Veterinary Importance

Organism	Significance	Size (Mbp)*	Center
Aedes aegypti	Yellow fever, Dengue, and Chikungunya vector	1380	The Broad Institute; Colorado State Univ.; JCVI
Anopheles gambiae	Malaria vector	278	Celera Genomics; Genoscope
Ixodes scapularis	Vector for Lyme disease, Babesiosis, Anaplasmosis	2100	The Broad Institute; JCVI
Genome nearly complete			
Culex pipiens quinquefasciatus	Vector for numerous viruses (e.g., WNV) and filariasis	540	The Broad Institute; JCVI
Pediculus humanus humanus	Vector for louse-borne relapsing fever, trench fever, and epidemic typhus	112	JCVI

Center: JCVI = J. Craig Venter Institute.
*For diploid organisms, this is the haploid genome size.
Note: Genome projects are currently underway for *Glossina morsitans morsitans, Rhodnius prolixus,* and *Aedes albopictus.*
Source: www.vectorbase.org

Table 27.2 Complete (or Nearly Complete) Genome Projects for Vector-Borne Microorganisms (Bacteria, Protists) and Nematodes

Organism	Disease	Size (Mbp)*	Center
Bacteria			
Anaplasma marginale	Bovine anaplasmosis	1.2	Washington State Univ.; Amplicon Express Inc.; USDA
A. phagocytophilum	Human granulocytic anaplasmosis	1.47	JCVI; Ohio State Univ.
Bartonella bacilliformis	Oroya fever, Carrion's disease	1.4	TIGR; JCVI
Borrelia burgdorferi	Lyme disease	1.52	TIGR; JCVI
Coxiella burnetii	Q fever	1.995–2.25	TIGR; JCVI
Ehrlichia canis	Canine monocytic ehrlichiosis	1.3	Univ. of Texas; DOE Joint Genome Institute
E. chaffeensis	Human monocytic ehrlichiosis	1.18	TIGR; JCVI; Ohio State Univ.
Francisella tularensis	Tularemia	1.89	University of Birmingham, UK; Swedish Defense Research Agency
Rickettsia prowazekii	Epidemic typhus	1.1	Bio Health Base
R. rickettsii	Rocky Mountain Spotted Fever	1.27	Integrated Genomics Inc.; NIAID
R. typhi	Endemic typhus	1.11	Human Genome Sequencing Center, Baylor College of Medicine
Yersinia pestis	Plague	4.5–4.88	JCVI; DOE Joint Genome Institute; Sanger Institute; Univ. of Wisconsin; TIGR
Protists			
Babesia bovis	Bovine babesiosis	8.2	Washington State Univ.; JCVI
Plasmodium falciparum	Human malaria	22.9	Malaria Genome Project Consortium
Plasmodium yoelii	Rodent malaria	23.1	NMRC; JCVI
Leishmania infantum	Visceral leishmaniasis	31.76	Univ. of Glasgow; Imperial College; Sanger Institute
L. major	Cutaneous leishmaniasis	32.8	Seattle Biomedical Research Institute; Sanger Institute
Theileria annulata	Tropical theileriosis	8.35	Univ. of Glasgow; Sanger Institute
T. parva	African east coast fever	8.3	TIGR; International Livestock Research Institute
Trypanosoma brucei	African trypanosomiasis	26	Sanger Institute; JCVI
Trypanosoma cruzi	Chagas' disease	34	TIGR, Seattle Biomedical Research Institute; Uppsala University
Nematode			
Brugia malayi	Filariasis	90	JCVI; Univ. of Pittsburgh

Centers: TIGR = The Institute for Genomic Research; JCVI = J. Craig Venter Institute; NIH = National Institutes of Health; NIAID = National Institute of Allergy and Infectious Disease; DOE = Department of Energy; USDA = United States Department of Agriculture.
*For diploid organisms, this is the haploid genome size.
Sources: www.genonmesonline.org; www.ncbi.nlm.nih.gov

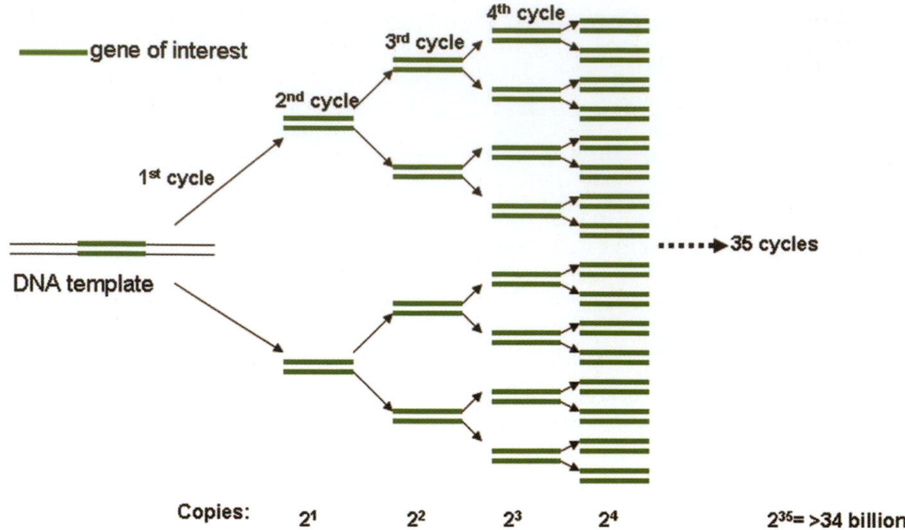

Figure 27.4 Exponential amplification of product from DNA during the course of 35 cycles of PCR.

insertions, deletions, or substitutions. In AFLP, DNA is first cut with REs, and then small DNA adaptors are ligated to the ends of these fragments. Next, the fragments are PCR-amplified using a primer set that is complimentary to the adaptor and restriction site on the fragments. The fragments are size-separated by gel electrophoresis. Organisms with different product lengths (polymorphic DNA, which can be considered different alleles or markers) reveal different banding patterns, or fingerprints, when the fragments are visualized on gels. Organisms with numerous "bands" in common likely share similar alleles and consequently reflect the degree of close relatedness. This technique is used commonly in forensics and to determine paternity. In the field of entomology, AFLP frequently is used to look at genetic relatedness between individuals and thus has been applied to such areas as vector population genetics.

Random Amplification of Polymorphic DNA (RAPD) This technique does not require any specific knowledge of the DNA sequence of the organism. Random primers (8–12 nucleotides long, nt) are used to amplify genomic DNA. The primers either will or will not anneal to the target DNA and result in a product.

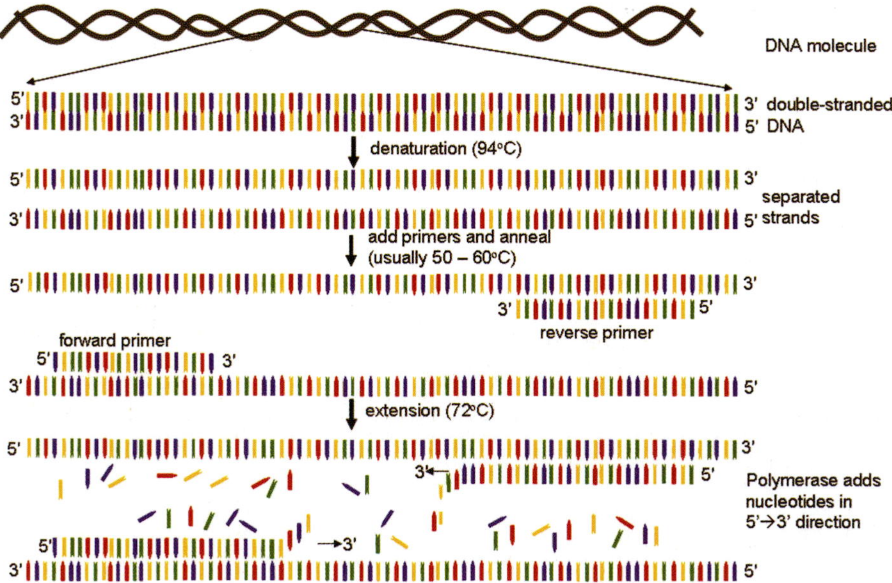

Figure 27.5 Events during one cycle of PCR product amplification from DNA template. Strands of DNA must first be denatured with heat. Primers designed to be complementary to specific sequences within the gene will anneal when temperatures are lowered (usually 50–60°C). Polymerase (not shown) in the primer-binding areas will begin synthesizing DNA in the $5' \rightarrow 3'$ direction.

This is dependent on there being areas on the template that are complementary to the primer sequence. Therefore, if a mutation has occurred in the template DNA of one individual at the site that is complementary to the primer in a second individual, a PCR product of that particular size will be produced from only the second individual. The result is a different banding pattern of amplicons on the gel, or fingerprint. Because fingerprints may be different for individuals in a population or species, RAPDS, like AFLP, can be used to assess genetic relatedness.

Real-time PCR (Quantitative PCR) As the name suggests, Real-time PCR detects the amount of amplicons accumulating during PCR cycles in "real-time." Whereas traditional PCR requires visualization of a product at the endpoint on a gel, real-time PCR utilizes techniques that allow monitoring of products virtually (e.g., on a monitor) during the process of amplification. Accordingly, there are two clear benefits: (1) results can be gained more quickly and with increased sensitivity, and (2) the technique can be used to quantify the starting amount of DNA, and thus the starting amount of organisms providing that template. It is because of this second use that real-time PCR is also referred to as quantitative or Q-PCR.

The mechanism of real-time PCR involves designing a short DNA probe (called a TaqMan® Probe) that will anneal to a specific region of the template that is internal to the forward and reverse PCR primers. This probe can be constructed several ways, but most commonly it is designed to contain a high-energy dye (a fluorescent "reporter") at its 5′ end and a low-energy molecule ("quencher") at the 3′ end. When this probe is intact and excited by a light source, emission by the reporter dye is suppressed by being proximal to the quencher dye. When, or if, DNA polymerase starts to copy DNA during the extension phase, the annealed probe that sits in the polymerization path is cleaved by exonuclease activity of the polymerase and the quencher and reporter are separated. Consequently, the reporter is capable of transmitting the fluorescent signal, which is detected and quantified by the instrument and software. An increase in the reporter signal over time indicates continued (exponential) amplification of product. Further, the amount of reporter signal increase is proportional to the amount of product, which is directly proportional to the amount of starting material (template) for a given sample.

Quantification is performed by specially designed instruments that perform the thermal cycling steps, excite the reporter, and simultaneously monitor fluorescence emission in real-time, during each cycle. DNA product also can be quantified by using SYBR® Green chemistry. This dye binds the minor groove (the smaller of the two grooves along the outer surface of a DNA molecule) in double-stranded DNA. As more double-stranded product is synthesized, SYBR Green dye signal increases. However, SYBR Green dye binds to *any* double-stranded DNA and is therefore less precise than using an internal probe as it does not differentiate between specific and nonspecific PCR products. Nearly exact amounts of DNA (i.e., actual numbers of copies or measurable quantities of product) can be extrapolated from standard curves, which are generated by also performing PCR using the same primer set on a known quantity of template.

Real-time PCR technology is also utilized in quantitative RT-PCR—a technique that is used to monitor gene expression in real-time (described later). This can be a bit confusing since the techniques have the same, or even redundant, acronyms. Although PCR alone can be used to detect the presence of a pathogen in an arthropod vector, Q-PCR is used to further quantify the number of those pathogens in a single arthropod, which is important in assessing vector potential.

Multiplex PCR This variant of PCR utilizes more than one primer pair and therefore results in simultaneous amplification of multiple targets of interest during one reaction. Its application ranges from analyses of gene deletions, mutations, and polymorphisms, to quantitative analysis. Typically, multiplex PCR is used in (1) genotyping, when simultaneous analysis of multiple markers is required, (2) pathogen detection in vectors, and (3) microsatellite analysis in population genetics. Microsatellites are small stretches of repeating units of DNA (1–6 bp in length) and are molecular markers used in population genetics studies since they show polymorphism between individuals. Multiplex assays require tedious and lengthy optimization procedures; however, they can reveal a plethora of information from one reaction.

Nested PCR This technique is used for two main purposes: to increase the amount of product from, and/or to confirm the results of, a primary PCR reaction. To do this, a second PCR is performed with primers internal to the initial primer positions (i.e., those that will anneal within the amplicon, hence the second reaction is "nested" within the first) using a primary PCR product as template. For example, a primary PCR may be performed using *genus*-specific primers and a second (nested) PCR can be performed using *species*-specific primers. Like Q-PCR, this technique is used commonly to detect pathogenic microbes in vectors. A primary PCR can reveal the presence of *Rickettsia* spp. in ticks, whereas a nested PCR can delineate whether a subset of those products is *Rickettsia ricketsii* by using species-specific primers.

ANALYZING GENE EXPRESSION

An animal's phenotype can be related back to the expression of gene products, which typically are proteins. Since the intermediate between DNA and protein is the RNA transcript, expression can be measured by analyzing products of transcription (messenger RNA, mRNA) and translation (protein). Analysis of gene expression is a step beyond genomic sequencing (thereby "functional genomics"), as it allows researchers to look at the biological function of genes. Specific examples of the implementation and application of expression analysis in the study of important arthropods includes:

- Arthropod response to pathogens they vector, which has revealed genes relevant to vector competence
- Genes involved in arthropod development, such as those critical to metamorphosis, which could lead to novel targets for control
- Proteins that promote/enhance an arthropod's ability to feed on hosts, such as salivary-gland proteins that are involved in blood-feeding
- Genome-wide expression analysis of arthropods in response to various stimuli, which helps elucidate entire pathways of gene expression and gene interactions (such stimuli include changing environmental conditions, host immune response, and changes in life history stages/development)
- Defining function of "new" genes discovered in genomic or complementary DNA (cDNA) libraries, as explained later

RNA Analysis

Analysis of gene expression at the RNA level allows the user to determine the transcriptional activation of genes. Many of the following techniques involve the molecular detection of mRNA. When a specific mRNA for a gene is present, that indicates the gene has been "turned on" under those conditions. So, whereas genomics or PCR can search for the presence or absence of genes, synthesis of mRNAs from those genes in those organisms gives more insight into their biological roles and functions.

Complementary DNA (cDNA) libraries Genes are transcribed into RNA, some of which (mRNA) code for proteins that contribute functions and/or phenotypes. Entire catalogs of mRNAs from an arthopod can be stored in cDNA libraries, collectively constituting a **transcriptome**. The transcriptome can be general (a collection of mRNAs from the organism under normal conditions) or specific (a collection of mRNAs when the organism is exposed to a stimulus, such as in response to pathogens or when feeding on a host). cDNA libraries are constructed by first extracting mRNA from an organism's tissues and synthesizing complementary DNA (hence "c"DNA). Fabrication of cDNAs is accomplished by enzymes that reverse-transcribe the mRNA into a DNA sequence, to which a second complementary strand is generated using PCR. These are then cloned into bacteria, where the cDNA library is comprised of inserts that are DNA versions of the original mRNAs. These clones are sequenced and annotated in a way similar to that used for genomic libraries.

Researchers construct cDNA libraries to investigate gene function in important arthropods. Some examples include transcriptomes for (1) midguts of blood-feeding arthropods under starved and fed conditions; (2) the arthropod immune response, including hemocyte and fat body cDNAs from vectors; (3) pesticide-resistant strains of arthropods; and (4) life-history stages of arthropods (e.g., egg, larva, pupa, adult). Function of cDNAs identified from these libraries can be studied by the following methods.

Northern Analysis Northern analysis (of which northern blots are a part) is a method for examining the expression of single genes in organisms or their tissues across various conditions. Northern blots were named as such since the technique was conceived after Southern blots (which were named after their inventor Edwin Southern, hence the capitalization of this term) and involve detection of DNA, rather than RNA. Accordingly, "western" blots (described next), which involve the analysis of protein expression, were also named in reference to Southern. RNA is extracted from the sample, and is denatured (to remain unfolded and linear) and size separated by gel electrophoresis. The RNA is then transferred ("blotted") to a nylon filter, to which it adheres (Northern blot). The RNA is permanently linked to this filter and then exposed to a solution containing a fluorescently or radioactively labeled DNA or RNA probe complementary to the mRNA of interest. These probes frequently are generated by PCR using a known gene or cDNA as template. Probes specifically hybridize by base-pairing rules to the sample mRNA if it is present, even among the mixed population of possibly thousands of mRNAs in the sample. When the blot is exposed to film (autoradiography), a dark band appears in the sample lane where the probe has hybridized. Since the amount of RNA in the sample can be estimated (usually by additionally exposing the blot to a probe for a constitutively expressed "housekeeping" gene), expression can be roughly quantified as upregulation (increased expression) or downregulation (decreased expression) of the gene across experimental treatment groups. Hybridization to the correct size mRNA is confirmed as a standard RNA ladder containing known-sized fragments is also run on the gel.

Northern analysis can be used to compare gene expression in samples subjected to different conditions. For example, suppose a gene involved in the arthropod immune response was identified from a cDNA library. Various samples of a vector's RNA (e.g., harboring or not harboring pathogen, or various species of vector) could each be run in separate lanes on an RNA gel and subjected to Northern analysis using a probe for that immune-response gene. The lane where the probe hybridizes indicates that the gene was expressed under those conditions. Such experiments have been used to determine the role of immune-responsive genes in mosquito-*Plasmodium* and tsetse-trypanosome interactions across a variety of vector species.

Reverse-Transcriptase PCR (RT-PCR) Reverse-transcriptase PCR (not to be confused with real-time PCR, for which the same abbreviation is mistakenly, and confusingly, sometimes used) is used to generate

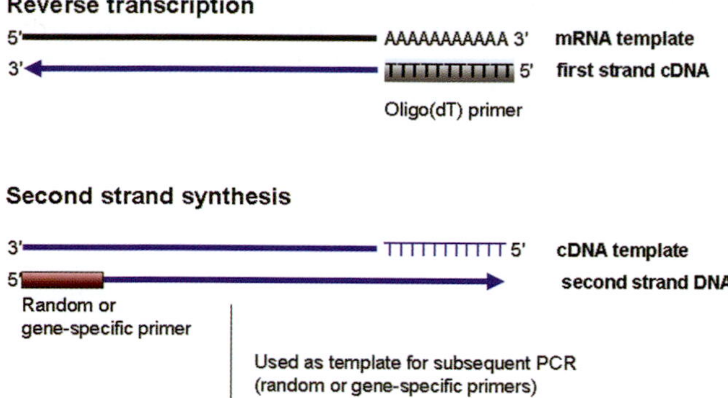

Figure 27.6 RT-PCR: Making dsDNA from mRNA. Reverse transcriptase uses nucleotides (not shown) and oligo(dT) primers to synthesize complementary DNA (cDNA) first strand from mRNA template. Second-strand synthesis is performed using either a random or gene-specific primer and PCR (polymerase, nucleotides not shown). This double-stranded product can be amplified by further cycles of PCR.

cDNA (described earlier) and also is used alone to measure gene expression. RT-PCR uses an enzyme called reverse transcriptase that can synthesize a complementary DNA (cDNA) from an RNA template. Synthesis is performed by providing nucleotides in the reaction mixture (similar to PCR) but is initiated by using a short oligo(dT) primer (a short stretch of repeating T nucleotides) since most eukaryotic mRNAs have a poly(A) tail at their 3′ end. The second step of the reaction involves generating a second strand to the cDNA with either a nonspecific or gene-specific primer using PCR. PCR is used to further amplify the copies of the cDNA (Figure 27.6). The product can be visualized after electrophoresis on a DNA gel, like any PCR product. If gene-specific primers are used, the presence of product on the gel indicates that the gene is being transcribed in the sample. RT-PCR is more sensitive than Northern analysis (mainly because of the PCR step), and methods exist for making it quantitative. In fact, RT-PCR can be combined with real-time PCR technology (sometimes called real-time RT-PCR or qRT-PCR) to sensitively quantify gene expression.

Differential Display This technique is a derivative of RT-PCR that allows for the analysis of the expression of multiple genes simultaneously. In differential display, two additional nucleotides are added to the 3′ end of the oligo(dT) primer, which causes it to bind only to a subset of the mRNAs. For example, if the oligonucleotide TTTTTTTTTTTGA is used as a primer, it will bind and prime mRNAs only where TC precedes the poly(A) tail. The second primer used for the subsequent PCR usually is an arbitrary/random 10-nt short sequence. After amplification, the multiple products produce a complex fingerprint-like series of bands when size-separated and visualized in a long polyacrylamide gel. Unlike microarrays (described next), the identities of the genes that are differentially expressed are not known. However, this type of multiplex gene expression screening is useful in initial comparative studies to reveal how much gene expression varies between organisms, tissues, or cells under different conditions.

Microarrays Microarray technology is used to analyze the global, genome-wide gene expression patterns that occur in an organism or tissue sample. Hence, microarrays are analogous to performing multiple Northern blots on all the mRNAs simultaneously using all possible cDNA probes. DNA microarrays contain thousands of microscopic spots of either ≈1 kb DNA sequences or smaller gene-specific short oligonucleotides (these arrays are called "gene chips") adhered to the surface of small areas on a glass slide. These genes can be selected and spotted onto the microarrays from either genomic or cDNA libraries using specialized robotic techniques. Global expression of the genes represented in the array is assayed by hybridizing fluorescently labeled cDNAs from mRNA preparations isolated from different specimens (e.g., cells or tissues under different conditions). The amount of labeled cDNA that hybridizes at each spot on the array is quantified using a scanning confocal laser. Data for each gene are normalized against "housekeeping" genes using complex computer algorithms. Such analysis allows for the simultaneous examination of all the genes at once—where relative intensity of upregulation or downregulation can be assessed. Additionally, microarrays can be washed and reused numerous times, so that different experimental samples can be applied to the same array for comparative analysis.

Microarray analysis allows researchers to analyze the total transcriptional profile of an organism during specific physiological, environmental, or developmental conditions. Researchers routinely confirm expression (upregulation or downregulation) seen on microarrays by follow-up experiments utilizing Northern or RT-PCR analysis for that particular gene. Genome sequencing and annotation is facilitating the construction of gene-chip microarrays for medically important arthropods. Currently, microarrays are being used to look at genes involved in the physiological response of *Anopheles gambiae* to *Plasmodium* infections. Likewise, gene chips also are being used to look at differential gene expression in pesticide-resistant and susceptible strains of mosquitoes on a large scale.

***In Situ* Hybridization** This technique provides for the evaluation of transcription *in situ*, i.e., the cellular localization of RNA transcripts in a tissue section. The tissue is prepared by embedding in wax and then cutting very thin sections with a microtome. These sections are mounted on a slide, and labeled DNA probes are added (usually a fluorescent or colorimetric label). Probes hybridize to the mRNA transcripts. As a result, the observer can see the cellular location of transcription, and can do comparative analysis either within or between tissue samples.

RNA Interference (RNAi) After genes of interest are identified by these methods, researchers typically proceed to assessing gene function in biological processes. Traditionally, if a researcher wanted to determine the functionality of a gene, a "knockout mutant" would have to be generated. This involved the laborious task of manipulating genes *in vivo* in embryos followed by growing the organism to adulthood and observing the knockout phenotype. RNAi, on the other hand, is a mechanism used to silence genes posttranscriptionally and allows researchers to inject or feed double-stranded RNA (dsRNA) into a whole organism, and look at the resulting phenotype within hours or days. To perform RNAi, a dsRNA is generated that is identical in sequence to the mRNA that is to be silenced. When this dsRNA is introduced *in vivo*, an innate antiviral mechanism (present in most, but not all, metazoan organisms) involving several enzymatic pathways degrades the dsRNA into small fragments called small interfering RNAs (siRNA), which are integral to the RNAi process. Since siRNAs have nucleotide sequences complementary to the targeted mRNA strand, other proteins in the RNAi pathway recognize that sequence, search for mRNAs containing those sequences, and cleave the mRNAs into smaller pieces that are no longer able to be translated into protein. The effect of the gene's silencing can subsequently be determined since the protein is integral to the phenotype.

RNAi has been used in large-scale screens to assess the role of multiple genes and gene families in cellular or organismal processes and in smaller assessments of specific gene functions, such as the insect immune response or in feeding processes, respectively. For example, RNAi silencing of genes expressed in the mosquito salivary gland has revealed proteins that are essential to blood-feeding. Similarly, RNAi has been used to elucidate the role of antimicrobial peptides (immune response molecules) in the refractoriness of tsetse flies to trypanosome infection.

Protein Analysis Expression of genes can vary post-transcriptionally; that is, the mRNA level detected by the earlier techniques may not reflect the amount of protein produced on the translational level. As a consequence, researchers sometimes use additional techniques for detecting expression at the protein level. Many of these techniques employ labeled antibodies that are generated to specifically detect proteins, just as DNA or RNA probes are used to detect nucleic acids. These systems include detection of specific gene products from a sample of total protein, and localization of that product *in situ* (in cells, tissues).

Generation of Antibodies Traditionally, antibodies are generated by immunizing animals (e.g., rodents, rabbits, goats) with either a recombinant protein (synthesized by bacteria containing cDNA) or synthetic peptides (generated based on amino acid sequences deduced from cDNAs). The animal's immune system responds to the injected protein and releases antibodies into the serum. The antiserum that is collected contains a heterogeneous (polyclonal) mixture of antibodies that recognize different parts (epitopes) of the immunogen. If a homogeneous (monoclonal) preparation of antibodies is desired, a clone of B-lymphocytes can be propagated by collecting B-cells from the animal and fusing them with cells derived from an immortal B-cell tumor line. Hybrid cells that have both the ability to make a particular antibody and the ability to multiply indefinitely in culture are selected. These **hybridomas** (hybrids of tumor cells and normal B-cells) are propagated as individual clones, each of which can provide a stable supply of the single type of monoclonal antibody.

Antibodies are labeled in various ways and are used in either direct or indirect detection systems. In direct detection methods, the purified antibody is labeled appropriately with a reporter molecule (e.g., fluorescein and biotin) and then used directly to bind the target protein. In indirect detection systems, the primary antibody is used as an intermediate molecule (not linked directly to a colorimetric label). After binding to the target, a secondary antibody, which is conjugated to a colorimetric reporter (e.g., a fluorochrome, or an enzyme such as horseradish peroxidase or alkaline phosphatase), is added. The secondary antibody is generated to detect the primary antibody and therefore binds to it.

Western Analysis Like northern analysis, western analysis (of which western blotting is a part) allows researchers to look for gene expression (here a protein rather than an RNA product) across samples one at a time. Total proteins are extracted from organisms, denatured with a detergent (such as sodium dodecyl sulfate, SDS), and size-separated by polyacrylamide gel electrophoresis (**SDS-PAGE**). Size fractionated proteins can be visualized by staining the gel with a dye (e. g., Coomassie blue) and quantified. The separated proteins are transferred ("blotted") to a sheet of nitrocellulose and then exposed to a solution containing specific polyclonal or monoclonal antibody designed to bind to the protein of interest (western blot). Molecular-weight markers also are run on the gel, ensuring that the correct size can be confirmed when antibody binds to the protein of interest. Western blotting has been used to identify important salivary-gland proteins used by blood-feeding arthropods such as ticks, sand flies, and mosquitoes.

Immunofluorescence and Immunohistochemistry These techniques allow for the observation and localization of protein expression *in situ*; that is, within tissues or cells. This is analogous to *in situ* hybridization methods used to detect RNA expression in cells. Tissues are either frozen or embedded in wax, and then cut into thin sections and mounted on a slide. The tissue section is exposed to a labeled antibody that binds to the protein of interest. In immunofluorescence, the antibody is conjugated to a fluorescent marker, and special microscopes are used that allow for magnification of the tissue while exposing the section to wavelengths of light that excite the marker and cause it to emit colors of various wavelengths. In immunohistochemistry, the antibody is coupled to an enzyme that converts a colorless substrate into a colored reaction product. In either case, the fluorescent dye or colored product indicates the localization of the protein of interest. This allows one to visualize and compare the cytolocalization of expressed protein in cells compared to surrounding tissues.

Two-dimensional Gels (2D Gels) Two-dimensional gels first separate proteins by isoelectric focusing (by charge in a pH gradient) and then, at right angles to the first separation, by mass (using SDS-PAGE). The result is that the proteins are spread out across the 2D gel by their size and isoelectric point (pI). Each "spot" on a stained gel likely represents a unique protein that can be mapped with x and y coordinates (i.e., mass and pI). Greater resolution is achieved compared to SDS-PAGE alone, since it is unlikely that proteins will be similar in both properties. Like differential display (of RNA, earlier), 2D gels can be compared across different samples to gain information on initial differences in gene expression. A spot present on one gel may not be present on another, or may be present at higher or lower intensity. Computer software can be used to scan gels, perform overlays, and detect these differences for comparative analysis.

DIAGNOSTIC TECHNIQUES

When diagnosing vector-borne diseases, the rapid, reliable, and easy detection of pathogens in both vectors and their hosts is paramount in assessing treatment course and vector control programs. The clear benefit of molecular-based techniques is that they bypass the need for laborious, and sometimes impossible, culture of microbes from hosts. Molecular techniques also are frequently much more sensitive in detecting microbes that may exist in low, unculturable levels in blood or tissues. Additionally, some techniques indirectly assess infection by looking for the host immune response (e.g., antibodies) to the pathogen, thereby bypassing the need to isolate the microbe.

Some researchers have scrutinized the use of molecular techniques alone in detecting pathogens, as Koch's postulates for determining true infection are not satisfied. Essentially, the criticism focuses on the question of whether the presence of a pathogen's DNA or protein in a sample actually qualifies as "infection" or "isolation" per se. Since those molecules are simply components of cells, they may not necessarily reflect the presence of viable organisms that can cause disease. This concern is justified by the fact that hosts can immunologically respond to dead organisms, as well as live ones (the basis of many vaccines). Thus, a positive antibody test (see ELISA, later) may not indicate infection, but rather exposure to the pathogen.

Rapid Detection and Quantification of Pathogens in Hosts and Vectors

Before the advent of molecular-based techniques to detect pathogens in hosts and vectors, microorganisms such as protozoa and bacteria had to be isolated by laboratory culture of host tissues and fluids. Culturing vector-borne microbes can be especially laborious since many of these microorganisms are obligate parasites that require cell-culture techniques (rather than simple media like broths or agar plates). PCR-based detection of microorganisms is clearly more rapid since results can be obtained sometimes in a few hours. This is important when the possibility of an infected human host is considered, because the time between exposure and need for effective treatment can be very short.

Techniques used to rapidly detect microbes include PCR (also nested and multiplex PCR) and real-time PCR (Q-PCR). As described previously, the specificity of these methodologies depends primarily on the design of the primers. Sensitivity can be optimized by using such techniques as nested PCR, which will perform two consecutive amplifications. PCR is so sensitive that it can be used to detect ONE bacterial cell in total DNA extracted from vector or host fluids. Detection of microbial DNA using any of these techniques indicates that the microbe, dead or alive, was present in the sample.

To fully assess vector potential (in arthropod samples) or host infection (in host tissue samples) further tests can be performed to determine whether viable, pathogenic microbes in fact are residing in the host organism. Additionally, PCR detection of microbes in hosts is sometimes confirmed by indirect immune-based methods (see later) to further assess if actual infection exists and persists. Treatment of hosts, however, often is initiated after simply determining exposure, especially in cases involving highly virulent organisms.

Visualization of Pathogens in Hosts and Vectors

Simple light microscopy still is used routinely to visualize larger microbes in vector and host tissues. Examples include viewing protozoa such as trypanosome flagellates or malaria parasites in insect midguts or host blood smears. However, small microbes such as bacteria can all appear the same under the microscope (small cocci or bacilli). Detection of surface or internal proteins (or nucleic acids) with species-specific antibodies (or nucleic acid probes) can help in identifying the species of these organisms in tissue sections. Techniques employing color-tagged antibodies or probes include immunohistochemistry and *in situ* hybridization (described earlier). Detection of the actual organisms using these techniques can be used to support initial screening methods (such as PCR) and to confirm infection, while still bypassing culture techniques. Additionally, infection intensity can be assessed since organisms and their developmental stages can be visualized and enumerated within the specimens.

Immune-based Diagnosis of Host Infection

These techniques involve either direct or indirect detection of pathogens in host body fluids. Enzyme-linked immunosorbent assay (**ELISA**) can be used in a direct or indirect way to detect either host antibodies to pathogens or pathogen antigen in biological samples, such as blood (Figure 27.7). To test for host

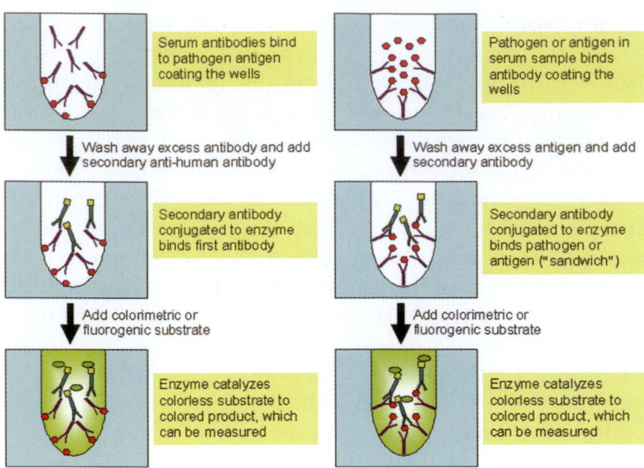

Figure 27.7 Enzyme-linked immunosorbent assay (ELISA) is used to measure host infection indirectly (left) or directly (right). Indirect ELISA measures host exposure to pathogen or antigen by detecting circulating antibodies. In contrast, direct or "sandwich" ELISA is used to detect presence of the pathogen or its products (e.g., toxins, antigens).

antibodies (and indicate exposure) to a pathogen, wells in a polystyrene microtiter plate are coated with microbial antigens, such as surface proteins, and host serum containing antibodies is added to the wells. Antibodies are allowed to recognize and bind the antigen, and any excess is washed away. A second antibody (for example a mouse-derived antihuman antibody) that is linked to an enzyme is then added to the wells. Excess antibody is again washed away, and a colorless substrate is added. This can be a colorimetric substance or a fluorescent substance that, when catalyzed by an enzyme, gives a visible product. Presence of the visible product indicates that the serum sample contains antipathogen antibodies. The amount of circulating antibody that binds can be quantified, and reflects the level of infection.

To test for the presence of pathogen antigen in a serum sample, the reverse is performed in what is called a "sandwich" ELISA. Wells in the microtiter plate are instead coated with antibody that has been raised against the pathogen of interest (e.g., mouse anti-*Rickettsia* surface protein antibodies, known as the **capture antibody**). The serum sample is applied to the plate, and possible pathogens are allowed to bind to the antibody. A second antibody, usually raised against a second antigen or epitope and conjugated to an enzyme, is added. The substrate of that enzyme is added as before, and a colorimetric or fluorescent change is noted. Both tests can be standardized with known quantities of antibody or antigen, respectively. Additionally, equipment known as microplate readers can measure the color or fluorescence intensity, quantify the reaction, and compare it to predetermined cut-off values that truly represent infection.

CONCLUSIONS

Like other disciplines within the biological sciences, the use of molecular tools has advanced the field of medical and veterinary entomology by providing new avenues for researchers to investigate their subjects (e.g., vectors, vertebrate hosts, pathogens). There are many examples of the broad impact of molecular tools: genomic databases can reveal novel targets for pest control existing in an insect's genome; expression analysis and RNAi can help in determining the function of these genes within organisms; paratransgenesis could be promising in manipulating vectors and preventing the spread of devastating diseases; rapid diagnostic techniques aid clinicians in assessing infection and saving lives. Medical and veterinary entomologists are first and foremost biologists; the advent of these molecular tools simply has provided researchers with a novel arsenal of techniques and procedures—many of which were not available just a few decades ago—to use in answering their biological questions.

REFERENCES AND FURTHER READING

Assumpção, T. C., Francischetti, I. M., Andersen, J. F., Schwarz, A., Santana, J. M., & Ribeiro, J. M. (2008). An insight into the sialome of the blood-sucking bug *Triatoma infestans*, a vector of Chagas' disease. *Insect Biochemistry and Molecular Biology, 38*, 213–232.

Boisson, B., Jacques, J., Choumet, V., Martin, E., Xu, J., Vernick, K., et al. (2006). Gene silencing in mosquito salivary glands by RNAi. *FEBS Letters, 580*, 1988–1992.

Brown, T. A. (2002). *Genomes* (2nd ed.). New York and London: Garland Science.

Hutchison, C. A., Smith, H. O., Pfannkoch, C., & Venter, J. C. (2005). Cell-free cloning using phi29 DNA polymerase. *Proceedings of the National Academy of Sciences of the United States of America, 102*, 17332–17336.

Janeway, C.A., Travers, P., Walport, M., & Shlomchik, M. (2001). *Immunobiology* (5th ed.). New York and London: Garland Science.

Kengne, P., Trung, H. D., Baimai, V., Coosemans, M., & Manguin, S. (2001). A multiplex PCR-based method derived from random amplified polymorphic DNA (RAPD) markers for the identification of species of the *Anopheles minimus* group in Southeast Asia. *Insect Molecular Biology, 10*, 427–435.

Lin, T., Oliver, J. H., Jr., & Gao, L. (2003). Comparative analysis of *Borrelia* isolates from southeastern USA based on randomly amplified polymorphic DNA fingerprint and 16S ribosomal gene sequence analyses. *FEMS Microbiology Letters, 228*, 249–257.

Lodish, H., Berk, A., Zipursky, S. L., Matsudaira, P., Baltimore, D., & Darnell, J. E. (2000). *Molecular Cell Biology* (4th ed.). New York: W. H. Freeman & Co.

Margonari, C. S., Fortes-Dias, C. L., & Dias, E.S. (2004). Genetic variability in geographical populations of *Lutzomyia whitmani* elucidated by RAPD-PCR. *Journal of Medical Entomology, 41*, 187–192.

Meister, S., Kanzok, S. M., Zheng, X. L., Luna, C., Li, T. R., Hoa, N. T., et al. (2005). Immune signaling pathways regulating bacterial and malaria parasite infection of the mosquito *Anopheles gambiae*. *Proceedings of the National Academy of Sciences of the United States of America, 102*, 11420–11425.

Moriarity, J. R., Loftis, A. D., & Dasch, G. A. (2005). High-throughput molecular testing of ticks using a liquid-handling robot. *Journal of Medical Entomology, 42*, 1063–1067.

Nayduch, D., & Aksoy, S. A. (2007). Refractoriness in tsetse may be a matter of timing. *Journal of Medical Entomology, 44*, 660–665.

Paupy, C., Orsoni, A., Mousson, L., & Huber, K. (2004). Comparisons of amplified fragment length polymorphism (AFLP), microsatellite, and isoenzyme markers: Population genetics of *Aedes aegypti* (Diptera: Culicidae) from Phnom Penh (Cambodia). *Journal of Medical Entomology, 41*, 664–671.

Pourmand, N., Elahi, E., Davis, R. W., & Ronaghi, M. (2002). Multiplex pyrosequencing. *Nucleic Acids Research, 30*, e31.

Venter, J. C., Adams, M. D., Sutton, G. G., Kerlavage, A. R., Smith, H. O., & Hunkapiller, M. (1998). Shotgun sequencing of the human genome. *Science, 280*, 1540–1542.

Venter, J. C., Smith, H. O., & Hood, L. (1996). A new strategy for genome sequencing. *Nature, 381*, 364–366.

Vontas, J., Blass, C., Koutsos, A. C., David, J. P., Kafatos, F. C., Louis, C., et al. (2005). Gene expression in insecticide resistant and susceptible *Anopheles gambiae* strains constitutively or after insecticide exposure. *Insect Molecular Biology, 14*, 509–521.

Weiss, B. L., Attardo, G. M., Pais, R., Wang, J., & Aksoy, S. A. (2007). Novel strategies targeting pathogen transmission reduction in insect vectors: tsetse-transmitted trypanosomiasis control. *Entomology Research, 37*, 231–237.

Yabsley, M.J., Davidson, W.R., Stallknecht, D.E., Varela, A.S., Swift, P.K., Devos, J.C., Jr., et al. (2005). Evidence of tick-borne organisms in mule deer (*Odocoileus hemionus*) from the western United States. *Vector-Borne and Zoonotic Diseases, 5*, 351–362.

Yan, G., Romero-Severson, J., Walton, M., Chadee, D. D., & Severson, D. W. (1999). Population genetics of the yellow fever mosquito in Trinidad: Comparisons of amplified fragment length polymorphism (AFLP) and restriction fragment length polymorphism (RFLP) markers. *Molecular Ecology, 8*, 951–963.

Yu, J., Ni, P., & Wong, G. K. (2006). Comparing the whole-genome-shotgun and map-based sequences of the rice genome. *Trends in Plant Science, 11*, 387–391.

Appendix

Arthropod-Related Viruses of Medical and Veterinary Importance

Michael J. Turell

Hundreds of different viruses have been associated with arthropods. Although each of these viruses can be classified phylogenetically into various orders, families, genera, and species, it has been useful to characterize them ecologically by their association with arthropods. These viruses can be placed in four ecological groups: (1) arthropod-borne viruses, (2) arthropod-transmitted animal viruses, (3) arthropod viruses, and (4) arthropod-transmitted plant viruses. The first two groups include most of the viruses of medical or veterinary importance. **Arthropod-borne viruses** usually are referred to by the shortened term **arboviruses**. These viruses replicate in both arthropod vectors and in vertebrate hosts and can be transmitted between vertebrate hosts by the arthropod vector (Table A.1). Infections of vertebrates with these viruses may result in clinical disease or death. **Arthropod-transmitted animal viruses** are similar to arboviruses except that they do not replicate in the arthropod vector (Table A.2). Transmission is mechanical rather than biological. **Arthropod viruses** replicate only in an arthropod. Because they do not replicate in a vertebrate host, they do not cause disease in humans or other vertebrates, although infection of the arthropod may be deleterious to that arthropod. **Arthropod-transmitted plant viruses** can be transmitted either mechanically or, in some cases, biologically, by a variety of arthropods, including aphids, leafhoppers, plant bugs, and certain plant mites. Because viruses in the latter two groups do not directly affect vertebrates, they are not included in the following tables.

Arboviruses are often named for (1) the disease that they cause, for example, chikungunya, yellow fever, and bluetongue viruses; (2) the location where the virus or disease was first described, for example, Barmah Forest, West Nile, and Sindbis viruses; or (3) a combination of the two, for example, Crimean-Congo hemorrhagic fever, eastern equine encephalitis, and Colorado tick fever viruses. When the full name of a virus is written out, only proper nouns should be capitalized; for example, western equine encephalitis virus, Venezuelan equine encephalitis virus, and Rift Valley fever virus. Unlike family names for other groups of biological organisms, the current convention is to italicize virus family as well as genus names. A standard two- to four-letter abbreviation has been approved for each of these viruses, as indicated in the tables.

Arboviruses are members of the following five families: *Togaviridae* (genus *Alphavirus*), *Flaviviridae* (genus *Flavivirus*), *Bunyaviridae* (various genera), *Rhabdoviridae* (genus *Vesiculovirus*), and *Reoviridae* (genus *Orbivirus*). Although viruses can contain either RNA or DNA, arboviruses all have nucleic acid in the form of RNA that can be either single- or double-stranded; most, however, are single-stranded. They can be either positive or negative sense. The RNA of a positive-sense virus is infectious if it enters a cell, while a reverse transcriptase is required to transcribe negative-sense RNA to enable it to infect a cell. The viruses can have either a single strand of RNA or be multisegmented. Some viruses, such as the members of the families *Togaviridae* and *Flaviviridae,* contain a single strand of RNA, whereas other viruses, such as the members of the *Bunyaviridae,* contain multiple strands of RNA, each encoding separate genes. If two closely related viruses with multiple strands of RNA infect the same cell, it is possible for the viruses to reassort to produce a new virus that contains genes from each of the two initial viruses. Some arboviruses, such as Japanese encephalitis and the various serotypes of dengue virus, currently cause

Table A.1 Selected Arthropod-Borne Viruses (Arboviruses)*

Virus	Abbreviation	Family	Genus	Serogroup	Animals Affected	Principal Vectors	Disease[a]	Distribution	Comments
African horsesickness	AHS	*Reoviridae*	*Orbivirus*	Bluetongue	Wild/domestic equids horses, mules	*Culicoides* midges *C. imicola*	F	Africa, Europe, India, Middle East	
African swine fever	ASF	*Asfarviridae*	*Asfivirus*	Ungrouped	Swine	*Ornithodoros* ticks	F, H	Sub-Saharan Africa, Eurasia	Also mechanical transmission
Barmah Forest	BF	*Togaviridae*	*Alphavirus*	Barmah Forest	Marsupials, humans	*Culex* mosquitoes *Cx. annulirostris, Ae. vigilax*	F, A	Australia	Similar disease as caused by RR Virus
Bluetongue	BLU	*Reoviridae*	*Orbivirus*	Bluetongue	Wild ruminants, sheep, cattle, goats	*Culicoides* midges *C. sonorensis, C. imicola*	F, V, H	Worldwide	Can be confused with foot and mouth disease
Bunyamwera	BUN	*Bunyaviridae*	*Orthobunyavirus*	Bunyamwera	Rodents	Mosquitoes	F, A	Africa	
Cache Valley	CV	*Bunyaviridae*	*Orthobunyavirus*	Bunyamwera	Ungulates, sheep	Mosquitoes *An. quadrimaculatus*	F	North America	Causes congenital malformations in sheep
Caraparu	CAR	*Bunyaviridae*	*Orthobunyavirus*	Group C	Rodents, humans	Mosquitoes	F	North/South America	
California encephalitis	CE	*Bunyaviridae*	*Orthobunyavirus*	California	Rodents, rabbits, humans	*Aedes* mosquitoes *Ae. dorsalis, Ae. melanimon*	F, N	Western US	
Chandipura	CHP	*Rhabdoviridae*	*Vesiculovirus*	VS	Hedgehogs, humans	Sand flies *P. papatasi*	F, N	India, Africa	
Changuinola	CGL	*Reoviridae*	*Orbivirus*	Changuinola	Sloths, humans	Sand flies *Lutzomyia* spp.	F	Central/ South America	
Chikungunya	CHIK	*Togaviridae*	*Alphavirus*	Semliki Forest	Humans	*Aedes* mosquitoes *Ae. aegypti, Ae. albopictus*	F, A	Africa, Asia	
Colorado tick fever	CTF	*Reoviridae*	*Orbivirus*	Colorado tick fever	Rodents, humans	Ticks *D. andersoni*	F	Western Canada, US, Northern Mexico	200–300 cases/yr in US
Crimean-Congo hemorrhagic fever	CCHF	*Bunyaviridae*	*Nairovirus*	CCHF group	Hares, hedgehogs cattle, humans	*Hyalomma* ticks *H. marginatum, H. asiaticum*	F, H, N	Africa, Asia	
Dengue-1,2,3 & 4	DEN	*Flaviviridae*	*Flavivirus*	Dengue	Humans	*Aedes* (*Stegomyia*) mosquitoes *Ae. aegypti, Ae. albopictus*	F, A, H	Tropics Worldwide	~50 million cases worldwide/year
Eastern equine encephalitis	EEE	*Togaviridae*	*Alphavirus*	EEE	Birds, horses, humans	Mosquitoes *Cs. melanura*	F, N	Eastern Canada to Argentina	

Epizootic hemorrhagic disease	EHD	*Reoviridae*	*Orbivirus*	Bluetongue	Wild ruminants deer, cattle	*Culicoides* midges *C. sonorensis*	F, H	Worldwide	
Highlands J	HJ	*Togaviridae*	*Alphavirus*	WEE	Birds	Mosquitoes *Cs. melanura*	None	Eastern US	Often confused with EEE virus
Itaqui	ITQ	*Bunyaviridae*	*Orthobunyavirus*	Group C	Rodents, humans	Mosquitoes	F	South America	
Jamestown Canyon	JC	*Bunyaviridae*	*Orthobunyavirus*	California	Rodents, humans	Mosquitoes *Aedes* and *Culiseta* spp.	F, N	Southern Canada, Temperate US	Includes SSH virus
Japanese encephalitis	JBE	*Flaviviridae*	*Flavivirus*	Japanese encephalitis	Birds, swine, horses humans	*Culex* mosquitoes *Cx. tritaeniorhynchus*	F, N	Asia	~50,000 cases/year in Asia
Karshi	KSI	*Flaviviridae*	*Flavivirus*	Mammalian tick-borne flavivirus group	Rodents?, humans	Ticks *Ornithodoros* spp.	F, N	Western Asia	
Kunjin	KUN	*Flaviviridae*	*Flavivirus*	Japanese encephalitis	Birds, humans	*Culex* mosquitoes *Cx. annulirostris*	F, N	Australia	Considered a subtype of WNV
Kyasanur Forest disease	KFD	*Flaviviridae*	*Flavivirus*	Mammalian tick-borne flavivirus group	Rodents, monkeys, birds, humans	Ticks, *Haemaphysalis* spp.	F, H, N	Southern Asia	
La Crosse	LAC	*Bunyaviridae*	*Orthobunyavirus*	California	Rodents, humans	*Aedes* mosquitoes *Ae. triseriatus*	F, N	Midwestern-Mid Atlantic states in US	~70 cases/yr in US
Langat	LGT	*Flaviviridae*	*Flavivirus*	Mammalian tick-borne flavivirus group	Rodents	Ticks, both ixodid and argasid	F, N	Asia	
Louping ill	LI	*Flaviviridae*	*Flavivirus*	Mammalian tick-borne flavivirus group	Rodents, sheep, humans	Ticks *I. ricinus*	F, N	United Kingdom	
Mayaro	MAY	*Togaviridae*	*Alphavirus*	Semliki Forest	Primates, rodents, birds, humans	Mosquitoes *Haemagogus* spp.	F, A	South America, Caribbean islands	
Murray Valley encephalitis	MVE	*Flaviviridae*	*Flavivirus*	Japanese encephalitis	Birds, humans	*Culex* mosquitoes *Cx. annulirostris*	F, N	Australia, New Guinea	

Continued

Table A.1 Selected Arthropod-Borne Viruses (Arboviruses)*—Cont'd

Virus	Abbreviation	Family	Genus	Serogroup	Animals Affected	Principal Vectors	Disease[a]	Distribution	Comments
Nairobi sheep disease	NSD	*Bunyaviridae*	*Nairovirus*	NSD group	Rodents, sheep, goats	Ixodid ticks *R. appendiculatus*	F, H	East Africa	
Omsk hemorrhagic fever	OMSK	*Flaviviridae*	*Flavivirus*	Mammalian tick-borne flavivirus group	Rodents, goats, sheep, humans	Ixodid ticks *Dermacentor* and *Ixodes* spp.	F, H	Western Siberia	Also spread from contaminated milk
O'nyong'nyong	ONN	*Togaviridae*	*Alphavirus*	Semliki Forest	Humans	*Anopheles* mosquitoes *An. funestus, An. gambiae*	F, A	Africa	Closely related to CHIKV
Oropouche	ORO	*Bunyaviridae*	*Orthobunyavirus*	Simbu group	Primates, sloths, humans	Mosquitoes? *Culicoides* spp., *C. paraensis*	F, A	South America, Caribbean	
Powassan	POW	*Flaviviridae*	*Flavivirus*	Mammalian tick-borne flavivirus group	Rodents, humans	Ticks *I. cookei*	F, N	Northern US, Canada	
Rift Valley fever	RVF	*Bunyaviridae*	*Phlebovirus*	Rift Valley fever	Cattle, sheep, goats, humans	Mosquitoes *Aedes macintoshi, Culex* spp.	F, H, N	Africa	Major outbreaks every 10–15 years
Rocio	ROC	*Flaviviridae*	*Flavivirus*	Ntaya	Birds, humans	Mosquitoes	F, N	South America	
Ross River	RR	*Togaviridae*	*Alphavirus*	Semliki Forest	Marsupials, humans	Mosquitoes *Aedes* and *Culex* spp.	F, A	Australia, South Pacific	
Russian spring summer encephalitis	RSSE	*Flaviviridae*	*Flavivirus*	Mammalian tick-borne flavivirus group	Rodents, humans	Ticks *I. ricinus*	F, N	Europe, Western Asia	
Sandfly fever Naples	SFN	*Bunyaviridae*	*Phlebovirus*	Phlebotomus fever	Rodents?	Sand flies *Phlebotomus* spp.	F	Southern Europe, India, North Africa	
Sandfly fever Sicilian	SFS	*Bunyaviridae*	*Phlebovirus*	Phlebotomus fever	Rodents?	Sand flies *Phlebotomus* spp.	F	Southern Europe, India, North Africa	
Semliki Forest	SF	*Togaviridae*	*Alphavirus*	Semliki Forest	Rodents, humans	Mosquitoes *Aedes* and *Culex* spp.	F, N	Africa, India-SE Asia	
Sindbis	SIN	*Togaviridae*	*Alphavirus*	WEE	Birds, humans	Mosquitoes *Culex* spp.	F, A	Africa, Europe, Asia, Australia	
St. Louis encephalitis	SLE	*Flaviviridae*	*Flavivirus*	Japanese encephalitis	Passerine birds, humans	*Culex* mosquitoes *Cx. pipiens* complex, *Cx. tarsalis*	F, N	North/South America	
Snowshoe hare	SSH	*Bunyaviridae*	*Orthobunyavirus*	California	Rabbits, humans	Mosquitoes *Aedes* and *Culiseta* spp.	F, N	Northern US and southern Canada	Closely related to La Crosse virus

Tahyna	TAH	*Bunyaviridae*	*Orthobunyavirus*	California	Rabbits, humans	Mosquitoes *Aedes* spp.	F, N	Europe, Asia, Africa	
Tick-borne encephalitis	TBE	*Flaviviridae*	*Flavivirus*	Mammalian tick-borne flavivirus group	Rodents, humans	Ticks *I. ricinus*	F, N	Europe	Western Asia
Toscana	TOS	*Bunyaviridae*	*Phlebovirus*	Phlebotomus fever	Rodents, humans	Sand flies *Phlebotomus* spp.	F, N	Europe	
Venezuelan equine encephalitis	VEE	*Togaviridae*	*Alphavirus*	VEE	Rodents, horses, humans	Mosquitoes *Cx.* (*Melanoconion*) and *Aedes* spp.	F, N	South/Central America	Both epizootic and enzootic strains
Vesicular stomatitis Indiana	VSI	*Rhabdoviridae*	*Vesiculovirus*	VS	Mammals, cattle, swine, horses, humans	Sand flies, black flies? *Lutzomyia* spp.	F, V	South/Central/North America	Can be confused with foot and mouth disease
Vesicular stomatitis New Jersey	VSNJ	*Rhabdoviridae*	*Vesiculovirus*	VS	Mammals, cattle, swine, horses, humans	Sand flies, Black flies *Lutzomyia* spp.	F, V	South/Central/North America	Can be confused with foot and mouth disease
Western equine encephalitis	WEE	*Togaviridae*	*Alphavirus*	WEE	Birds, horses, humans	*Culex* mosquitoes *Cx. tarsalis*	F, N	Western Canada through South America	
West Nile	WNV	*Flaviviridae*	*Flavivirus*	Japanese encephalitis	Birds, horses, humans	Many mosquito species primarily *Culex* (*Culex*) spp.	F, N	Worldwide	Spread to New World in 1999
Yellow fever	YF	*Flaviviridae*	*Flaviviru*	Ungrouped	Primates, humans	*Aedes* (*Stegomyia*) mosquitoes *Ae. aegypti*, *Ae. albopictus*	F, H	Africa, South America	Formerly found in the Central/North America

*Viruses that replicate in arthropods and are dependent on arthropod vectors for transmission.

[a]Disease: F = fever, H = hemorrhagic, N = neurologic (i.e., meningoencephalitis). A = arthritis, V = vesicular rash

Table A.2 Selected Arthropod-Transmitted Animal Viruses*

Virus	Abbreviation	Family	Genus	Animals Affected	Principal Vectors	Disease[a]	Distribution
Bovine leukemia	BL	*Retroviridae*	*Deltaretrovirus*	Cattle	Biting insects	None	Worldwide
Equine infectious anemia	EIA	*Retroviridae*	*Lentivirus*	Horses	Horse flies, deer flies, mosquitoes, lice	F	Americas, Europe, Middle East, Africa
Hog Cholera (Classical swine fever)	CSF	*Flaviviridae*	*Pestivirus*	Swine	Deer flies, stable flies	F, H	Europe, Africa, South/Central America
Myxoma	MYX	*Poxviridae*	*Leporipoxvirus*	Rabbits	Mosquitoes, fleas	V	Europe, Australia, The Americas
Swinepox	SWP	*Poxviridae*	*Suipoxvirus*	Swine	Hog louse, mange mites	F, V	Worldwide

*Viruses that are mechanically transmitted by arthropods, but that do not replicate in the arthropod vector, i.e., not an arbovirus.
[a]Disease: F = fever, H = hemorrhagic, V = vesicular rash

millions of infections in humans and tens of thousands of fatalities each year. Others, such as Rift Valley fever and Venezuelan equine encephalitis viruses can cause outbreaks that can devastate livestock and domestic animal populations as well as cause severe disease in human populations.

Arboviruses and arthropod-transmitted animal viruses can be transmitted by a number of different arthropod groups, ranging from mosquitoes and sand flies to argasid and ixodid ticks. These include viruses that replicate in the arthropod vector as well as those that are merely mechanically transmitted by arthropods. Most of the arboviruses are biologically transmitted; that is, they replicate in their arthropod host prior to being transmitted, by mosquitoes, sand flies, or ticks (Table A.1). In contrast, arthropod-transmitted viruses are transmitted mechanically, without replication in the arthropod, by tabanids, mosquitoes, black flies, lice, and mange mites (Table A.2).

Although there have been reports that some of the hemorrhagic fever viruses, such as Hantaan virus (the causative agent of Korean hemorrhagic fever) have been transmitted by mites, these reports are generally unsubstantiated (Table A.3).

In order to consolidate information on arboviruses, the Subcommittee on Information Exchange (SIE) of the American Committee on Arthropod-borne Viruses (ACAV) published the *International Catalogue of Arboviruses Including Certain Other Viruses of Vertebrates* in 1967. A second edition was published in 1975 and a third edition in 1985. The catalog recently was placed online in electronic format and is maintained by the Centers for Disease Control and Prevention (Fort Collins, Colorado, USA). It can be accessed at http://www2.ncid.cdc.gov/arbocat/index.asp. The number of viruses listed in the catalogue has continued to increase. Whereas the original 1967 edition listed 204 viruses, the second edition in 1975, and the third edition in 1985, listed 359 and 504 viruses, respectively, and the online version as of 2008 lists more than 530 viruses.

The *International Catalogue of Arboviruses* includes viruses believed to be arboviruses as well as those "thought to be of interest to arbovirologists," such as Hantaan, Ebola, Marburg, and Lassa viruses. These latter viruses now are believed to be rodent- or bat-associated, rather than transmitted by arthropods, and therefore are listed separately in Table A.3.

When a potentially new arbovirus is discovered, information about its structure (i.e., enveloped or not, RNA single- or double-stranded, RNA segmented or not, etc.), relationship with other known arboviruses, and isolation history is submitted to two subcommittees of the ACAV: the Subcommittee on Inter-Relationships among Catalogued Arboviruses (SIRACA) and the Subcommittee on the Evaluation of Arthropod-borne Status (SEAS). Based on the findings of SIRACA and SEAS, respectively, the SIE decides if the virus should be included as a new virus in the catalog. If so, a decision is then made as to whether it should be designated as an arbovirus, a rodent-associated virus, or a virus for which additional information is needed before SEAS can make a final decision.

Table A.3 Selected Hemorrhagic Fever Viruses Found in the Catalog of Arboviruses*

Virus	Abbreviation	Family	Genus	Animals Affected	Disease[a]	Distribution	Comments
Ebola	EBO	*Filoviridae*	*Filovirus*	Rodents?, bats, humans, primates	F, H	Africa	Very high case-fatality rate
Hantaan	HTN	*Bunyaviridae*	*Hantavirus*	Rodents, humans	F, H	Asia	Korean hemorrhagic fever
Junin	JUN	*Arenaviridae*	*Arenavirus*	Rodents, humans	F, H	Argentina	Argentinean hemorrhagic fever
Machupo	MAC	*Arenaviridae*	*Arenavirus*	Rodents, humans	F, H	Bolivia	Bolivian hemorrhagic fever
Marburg	MBG	*Filoviridae*	*Filovirus*	Humans	F, H	Africa	Very high case-fatality rate
Prospect Hill	PH	*Bunyaviridae*	*Hantavirus*	Rodents	?	Eastern US	
Puumala	PUU	*Bunyaviridae*	*Hantavirus*	Rodents, humans	F, H	Europe	Nephropathia epidemica
Seoul	SEO	*Bunyaviridae*	*Hantavirus*	Rodents, humans	?	Worldwide	Possible cause of kidney disease/hypertension
Sin nombre	SN	*Bunyaviridae*	*Hantavirus*	Rodents, humans *P. maniculatus*	F, R	North America	Caused an outbreak in southwestern US in 1993

*Now believed to be rodent- or bat-associated viruses rather than arboviruses (that is, not transmitted by an arthropod vector).

[a]Disease: F = fever, H = hemorrhagic, R = respiratory.

REFERENCES AND FURTHER READING

Berge, T. O., Ed. (1975). "*International Catalogue of Arboviruses Including Certain Other Viruses of Vertebrates* (2nd ed.). US Department of Health, Education, and Welfare, DHEW Publication No. (CDC) 75-8301.

International Catalog of Arboviruses Including Certain Other Viruses of Vertebrates (http://www2.ncid.cdc.gov/arbocat/index.asp).

Karabatsos, N., Ed. (1985). "*International Catalogue of Arboviruses Including Certain Other Viruses of Vertebrates* (3rd ed.). American Society of Tropical Medicine and Hygiene, San Antonio, TX.

Monath, T. P., Ed. (1988). "*The Arboviruses: Epidemiology and Ecology.*" Vols. 1-5. CRC Press, Boca Raton, FL.

Taylor, R. M., Ed. (1967). "*The Catalog of Arthropod-borne and Selected Vertebrate Viruses of the World.*" U.S. Public Health Service Publication No. 1760, US Government Printing Office, Washington, DC.

Glossary

The following definitions are provided for words and terms used in this book. They are intended to help you quickly understand terminology with which you may not be familiar, given the breadth of interests and professions of the intended readers.

Abamectin An analogue of the antiparasitic group of compounds called avermectins.

Abatement A reduction in the amount or degree; for example, abatement of mosquito populations, or an abatement district or program.

Abrade (-ed, -sion) To wear away by scraping, rubbing, or friction; for example, skin abrasion.

Abscess Localized accumulation of pus, typically with associated inflammation.

Abscond (-ing) Behavior in bees in which an entire colony abandons an established hive to seek another suitable nest site.

Absolute The actual, or real, numerical value of something; as opposed to a relative number or value; for example, absolute abundance.

Acalyptrate Lacking a calypter; for example, certain muscid flies. Sometimes spelled acalypterate.

Acariasis An infestation with mites; also any disease or other medical condition caused by mites.

Acaricide (-al) A compound or substance that kills mites.

Acarine Referring to mites (Acari).

Acarology The science or study of mites (Acari).

Acarophobia Abnormal fear of mites and ticks; also used in the medical community with reference to a delusion that the skin is infested with mites (not to be confused with acrophobia, a fear of heights).

Acetylcholine A chemical, an ester of choline, that plays a role in transmission of nerve impulses at synapses and myoneural junctions.

Acne cosmetica An acne condition resulting from repeated applications of cosmetics; usually mild and noninflammatory.

Acquired immunity (Acquired resistance) Resistance resulting from previous exposure of the individual in question to an infectious agent or antigen.

Action threshold The level or magnitude at which action should be, or is, initiated; for example, when mosquito population densities reach a particular level (action threshold) such that control measures should be implemented.

Aculeate Insects of the order Hymenoptera that possess a stinger; wasps, bees, and ants.

Acute (1) Sharp or severe; sudden onset and usually of short duration; as opposed to chronic; for example, acute symptoms or illness. (2) A morphological description of a structure that forms a point.

Adenolymphangitis Inflammation of lymphatic glands and vessels.

Adenopathy Swelling or unusual enlargement of lymph nodes.

Adenosine phosphate A compound of adenosine (nucleotide containing adenine and ribose) and one or more phosphoric acid groups; a hydrolytic product of nucleic acids.

Adenosine triphosphate (ATP) A compound of adenosine (nucleotide containing adenine and ribose) and three phosphoric acid groups. Energy of muscle is stored in this compound.

Adenotrophic (viviparity) A form of viviparity in which larvae are fully nourished within the parent female by maternal glands and are larviposited immediately prior to pupation. Occurs in some Diptera such as tsetse flies and some members of the Hippoboscoidea.

Adrenaline See *Epinephrine.*

Adrenergic blocking agent A compound that interferes with the synaptic release of epinephrine (adrenaline) when nerve fibers are stimulated.

Adulticide A chemical compound or substance that kills the adults; for example, insects or mites.

Aedeagus The intromittent organ of male arthropods, used to transfer sperm to female.

Aedine Taxonomic term that refers to mosquitoes of the subfamily Aedinae (e.g., genus *Aedes*).

Aerial In the air, as in flying insects; above the ground; for example, aerial nest of yellowjackets, in trees (see *aerial treatment*).

Aerial treatment Application of insecticides, or other pesticides, from the air, typically by fixed-wing airplanes or helicopter; for example, for mosquito control.

Aerobic Living only where oxygen is present; for example, aerobic bacteria.

Aeropyle A minute respiratory opening in the eggs of some insects; for example, lice.

Aerosol (-ized) A suspension of fine particles in air or other gases.

Aerosol transmission The transfer of an infectious agent, or aerosolized particles, from one individual to another via air; for example, by sneezing or coughing.

Aestivate (ion) To pass the summer, summer months, or other unfavorable periods (except winter) in a dormant state.

Afebrile Without fever.

Afrotropical region Biogeographic area that includes Africa south of the Sahara and the southwestern part of Arabia; also called Ethiopian region.

Age grading A means, or method, for determining the physiological age of an organism; for example, the reproductive status of mosquitoes and other insects (see *physiological age grading*).

Aggregation pheromone A chemical substance produced by an organism that causes individuals of the same species to gather together, or aggregate; for example, cockroaches and some ticks.

Agonistic Aggressive or threatening, as in agonistic behavior.

Ague A term originally used to indicate fever and chills; formerly with particular reference to malaria (pronounced ā′gū).

Alakurt An adult flea belonging to the family Vermipsyllidae, and the genera *Vermipsylla* or *Dorcadia*.

Albendazole A benzimidazole compound used to treat various roundworm and tapeworm infestations; for example, cases of lymphatic filariasis.

Aliphatic compound An organic compound with carbon atoms joined together in a straight, or branched, open chain rather than a ring; for example, hydrocarbons, such as alkanes and alkenes.

Alkaloid An alkaline compound in plants that is typically bitter and has a toxic effect on other organisms; for example, quinine, nicotine, and morphine.

Allantoin A product of purine metabolism that promotes wound healing; for example, a natural antibiotic produced by certain blow-fly larvae.

Allergen Any substance that induces an allergic reaction; in addition to proteins and antigens, includes dusts, fungi, pollen, food components, drugs, and others.

Allergic dermatitis Sensitivity reaction of the skin to an allergen.

Allergic reaction A physiological or dermatological response to an allergen.

Allergy A reaction of body tissues of sensitized individuals to a specific substance or allergen.

Allopatric Occurring in separate geographic areas; usually with reference to the distribution of species.

Alloscutum The dorsal surface of the body (idiosoma) of mites and ticks posterior to the scutum, or dorsal plate.

Alopecia Loss of hair.

Alphavirus An arbovirus belonging to the family *Togaviridae*, genus *Alphavirus*.

Alveolar collapse Compression of the air cells, or alveoli, in the lungs of vertebrates.

Amastigote A cell lacking a flagellum; used to describe a developmental stage of trypanosome protozoans; for example, *Leishmania* and *Trypanosoma* species.

Ambient Surrounding, as in ambient temperature or other environmental conditions.

Ambulacrum (-al) The pretarsus, or terminal structures of the tarsus, in mites and ticks. Consists of the empodium, pulvillus, and tarsal claws, which can be highly modified or reduced.

Ametabolous A type of development in wingless insects in which there is little distinction between the immature stages and adults, other than size and reproductive function. Immatures often are called juveniles; for example, springtails or collembolans.

Amplification The increase of a pathogen in an ecological setting, which can involve the numbers of arthropod vectors or hosts infected and the levels of infection; used particularly with reference to viruses.

Amplification cycle The sequence of organisms involving arthropods and their vertebrate hosts (reservoirs) by which a pathogen multiplies and is spread from one organism to another.

Amplifying host (amplifier host) A host animal, typically a vertebrate, in which a pathogen multiplies, thereby increasing the potential level of infectivity to competent arthropod vectors.

Ampulla A dilatation or sac-like enlargement of a duct or canal; also used to refer to a bulbous enlargement of a structure; for example, basal portion of the sting at the tip of the tail of scorpions.

Anal groove A curved groove that partially surrounds the anal area, anteriorly and laterally, in ticks of the genus *Ixodes*, but is posterior to the anus in other ixodid (hard) ticks.

Anal papilla (-ae) Thin-walled projections of cuticle at the caudal end of some insects, which typically serve an osmoregulatory or respiratory function; for example, larvae of mosquitoes and other aquatic insects.

Anal wing vein(s) Longitudinal, unbranched vein that extends from the base of the insect wing to the outer margin of the wing, below the cubitus.

Analgesic A medication that alleviates pain.

Anaphylactic shock A physiological state of shock (severely depressed vital signs) induced by injection or ingestion of a substance to which the individual has become sensitized (see *anaphylaxis*).

Anaphylatoxin A substance that can trigger anaphylaxis; typically involves release of histamine and other compounds associated with a hypersensitivity reaction.

Anaphylaxis (-tic) A hypersensitivity reaction in which the body responds severely to an injected or ingested protein to which it has been previously exposed (e.g., reaction to bee or wasp sting). an acute antigen-antibody response (see *anaphylactic shock*).

Anaplasmosis A tick-borne disease of cattle, sheep, and other ruminants caused by rickettsial bacteria in the genus *Anaplasma* (e.g., *A. marginale*); often characterized by anemia and jaundice.

Anautogeny (-ous) The inability of females of hematophagous arthropods to produce eggs without taking a blood meal (see *autogeny*).

Anemia (-ic) A deficiency in the oxygen-carrying capacity of the blood due to reduced levels of hemoglobin, typically caused by too few red blood cells.

Angiomatosis A medical condition characterized by multiple angiomas; typically benign lesions or tumors involving lymphatic and blood vessels.

Annulate Formed of ring-like segments, or bands representing a secondarily divided segment.

Annulus (-i) A ring-like structure encircling a joint, segment, or other structure; for example, ring sclerite surrounding the base of an antenna, or annulate terminal part of the antenna, as in horseflies.

Anoplocephalid Referring to taenioid tapeworms in the family Anoplocephalidae.

Anorexia An eating disorder, with loss of appetite or inability to eat, with associated loss of body mass.

Antepygidial Located immediately anterior to the pygidium, or dorsal aspect of the last abdominal segment.

Anterior-station transmission The transfer of a pathogen or parasite from one individual to another via the mouth or mouthparts.

Anthelmintic An agent that kills or expels parasitic intestinal worms (helminths); syn. anthelminthic.

Anthomyiid Refers to members of the dipteran family Anthomyiidae, calyptrate muscoid flies.

Anthroponosis (-es) A disease involving only humans.

Anthropophagic Feeding on humans; for example, anthropophagic insects and other arthropods.

Anthropozoonosis (-es) An arthropod-borne disease that is transmitted primarily among wild animals (zoonotic disease), but for which humans can become incidentally infected with the associated pathogen or parasite; for example, East African trypanosomiasis.

Antibody Any of numerous proteins (immunoglobulins) produced in response to specific foreign antigens, including viruses, bacteria, and protozoans, and capable of reacting specifically with that antigen; an integral part of a body's immune system.

Anticoagulant A substance that prevents, or interferes with, clotting of the blood.

Anticoagulin An anticoagulant, usually referring to substances produced by organisms other than humans that interfere with normal clotting of vertebrate blood; for example, proteins in saliva of blood-sucking arthropods.

Antiedema Counteracting or reducing the amount of fluids that accumulate in body tissues.

Antigen A substance that can stimulate the production of antibodies.

Antigenic complex A composite of different antigenic structures (e.g., molecule, cell, virus, bacterium) with two or more different antigenic groups and specificities.

Antihistamine A compound that counteracts the effect of histamine.

Antimalarial A drug used to prevent or cure cases of malaria; for example, chloroquine; also used as an adjective, as in antimalarial agent or drug.

Antithrombin A substance that inhibits clotting of blood by preventing reaction between thrombin and fibrinogen.

Antivenin A substance that counteracts the effects of venom from bites and stings of insects, snakes, and other organisms; also, animal serum that contains antivenins, used to treat cases of envenomation.

Apex The most distal portion of a structure, or farthest from the base; for example, tip of the insect wing, farthest point from the wing base.

Aphrodisiac A substance or agent that causes sexual arousal or desire.

Apical droplet Small drop of fluid excreted via the anus or specialized glands at the tip of the insect abdomen.

Apocrine gland A coiled, tubular gland that secretes products at the apical end. The latter is then pinched off as a secretion, including cellular and cytoplasmic components; for example, apocrine sweat glands.

Apodeme A hollow invagination of the body wall in arthropods that serves primarily as an attachment site for muscle.

Apolysis The separation of old cuticle from the underlying, newly formed epidermis in the initial stages of molting in arthropods.

Apophysis Solid projections of the body wall of arthropods, both internal and external. In addition to sites for muscle attachment, serve to strengthen the body wall and as both support and protection for various organs.

Aposematic Having a conspicuous structure or color pattern than advertises oneself as a means of defense from potential enemies; for example, the black-and-yellow warning coloration of some wasps and bees.

Apotele A tarsal claw on the palp of certain mites; palptarsal claw.

Apterous Lacking wings.

Apyrase Any of several enzymes that catalyze the hydrolysis of adenosine triphosphate (ATP), releasing phosphate and energy.

Arachnida A large class of arthropods characterized by two body regions and four pairs of jointed legs, with no wings or antennae; includes spiders, mites, ticks, scorpions, and other arachnids.

Arachnology The science and study of arachnids.

Arachnophobia An abnormal, morbid, or pathological fear of arachnids, most commonly involving spiders.

Araneism A medical condition resulting from the bite of a spider; also called arachnidism, a more general term referring to envenomation by spiders, scorpions, and other arachnids.

Araneomorph Spiders in the suborder Araneomorphae, characterized by opposable chelicerae that move in a transverse plane; as distinguished from tarantulas and their relatives in which the fangs move vertically.

Araneophobia An abnormal fear of spiders (see *arachnophobia*).

Arbovirus A shortened name for arthropod-borne virus.

Argasid A member of the tick family Argasidae, or soft ticks.

Arista (-ate) A bristle-like process on the terminal segment of the adults of certain higher Diptera; for example, muscoid flies.

Arolium (-a) A median lobe, or cushion-like pad, between the bases of the tarsal claws in insects.

Arrhenotoky Parthenogenesis in which only males are produced from unfertilized eggs.

Arrhythmia An abnormal rhythm or disturbance of the heartbeat.

Arsenical A preparation or drug containing arsenic.

Arteriole The smaller terminal branches of an artery, particularly those that connect with capillaries.

Arthralgia (-ic) Joint pain.

Arthritis An acute or chronic inflammation of a joint, typically accompanied by pain, swelling, and stiffness.

Arthropod Invertebrates of the phylum Arthropoda, possessing a segmented body, jointed appendages, and a chitinous exoskeleton; includes insects and arachnids.

Arthropod-borne virus A virus that replicates in, and is transmitted by, arthropods.

Artiodactyla An order of hoofed mammals with an even number of functional toes, or digits, on each foot; includes cattle, deer, sheep, goats, and pigs.

Ascites An excessive accumulation of serous fluids in the peritoneal cavity.

Aseptic Free of pathogenic microorganisms or their toxins.

Aseptic meningitis A mild form of meningitis characterized by fever, headache, and stiff neck; usually caused by a virus.

Asexual Lacking evidence of sex organs; involving reproduction without the union of male and female gametes.

Asilid A member of the dipteran family Asilidae, or robber flies.

Asthma (-tic) A chronic respiratory disorder, often caused by allergies. Characterized by recurrent bronchial spasms, coughing, wheezing, and constriction in the chest.

Asthmatic bronchitis Inflammation of the mucous membrane lining the bronchial tubes in individuals with asthma.

Asymptomatic Not showing symptoms.

Atopic Relating to certain allergic conditions of a hereditary nature; for example, atopic dermatitis, eczema, and asthma.

Atrophy A deterioration or reduction in size of tissues or organs of the body as the result of injury, disease, or lack of use.

Attenuate (-ed) To reduce or weaken in strength, amount, or virulence; for example, attenuated virus, attenuated chelicerae of certain mites.

Attractant A substance that lures insects or other animals to an animal or trapping device; can include pheromones and chemical compounds involved in host-seeking behavior.

Augmentative release Repeated releases of commercially reared biological control agents in the form of predators or parasites for the control of pest species.

Aural Relating to the ear or sense of hearing.

Australasian The region of the southern Pacific Ocean that includes Australia, New Zealand, New Guinea, and neighboring islands.

Autoantibody An antibody that acts against tissues, cells, or cell components of the organism that produces it.

Autogeny (-ous) The production and development of eggs without a blood meal in insects and other arthropods that usually require blood in order to do so. Sufficient protein reserves are carried over from the larval and pupal stages to support one or more gonotrophic cycles.

Autolysis Autodigestion resulting from enzymatic digestion of cells, especially when these are degenerate or dead; especially prevalent in cadavers.

Autonomic Occurring involuntarily or spontaneously, independent of external stimuli; for example, autonomic nervous system.

Avermectin A group of anti-parasitic drugs whose analogues include ivermectin and abamectin.

Avirulent Lacking virulence; nonpathogenic.

Axillary Pertaining to the axilla, or arm pit.

Babesiosis (-es) Any of several tick-borne diseases of humans, livestock, and wild animals caused by blood protozoans of the genus *Babesia*.

Bacillus (-i) Rod-shaped, aerobic bacteria of the genus *Bacillus*, or of numerous other genera.

Back rubber A self-treatment device for livestock in which a padded rope saturated with an insecticide or acaricide is suspended between two vertical posts through which animals regularly walk. The toxicant is applied to the head and upper body by direct contact as the animal passes beneath it.

Bacteremia Presence of bacteria in the blood.

Bacteriome An area in or near the gut of some blood-feeding arthropods containing symbiotic microorganisms (typically bacteria).

Bacteriophage A virus that infects and destroys bacterial cells.

Bacteriostatic A chemical or biological agent that inhibits bacterial growth.

Bagasse Fibrous residue from the process of extracting juice from crushed stalks of sugarcane and other plants. Used as a source of cellulose for making paper products; used as a rearing medium for stable flies and horn flies.

Bait A food or other attractive substance containing or associated with a toxicant used to lure insects and other pests as a means of control or extermination.

Bait station A device containing a bait used to attract and kill insects; for example, ants and cockroaches. Also used for attracting rodents and other animals, bringing the latter in contact with insecticides or acaricides that kill certain ectoparasites.

Bandicoot (1) Ratlike marsupials of the family Peramelidae in Australia, Tasmania, New Guinea, and adjacent islands. (2) Any of several large rats in the family Muridae of the genera *Bandicota* and *Nesokia* of southeastern Asia.

Bartonellosis A disease of humans and other mammals caused by bacteria of the genus *Bartonella*.

Basalar flight muscles Muscles in insects that attach to sclerotized plates, or basalar sclerite(s), in the membranous area at the base of the wings. Play important role in flight.

Basis capituli The fused proximal elements of the mouthparts and palps that form the base of the gnathosoma, or capitulum, in ticks.

Basophil A cell, especially white blood cells, with granules that readily stain with methylene blue and other basic dyes.

Bed net A fine-mesh fabric hung tent-like over, or otherwise enclosing, a bed to protect sleeping individuals from biting insects, notably mosquitoes.

Benign Mild; not progressive or malignant.

Benthic Pertaining to the substrate or community of organisms at the bottom of a body of water, such as a lake or ocean.

Benznidazole An antiparasitic drug used to treat cases of Chagas disease.

Berlese funnel A device used to extract insects and other arthropods from soil, ground litter, and other substrates; consists of a funnel over which is suspended an electric light bulb. The light and heat from the bulb slowly dries the substrate, driving the arthropods down the funnel and into a collecting jar, usually containing alcohol.

Bifid Divided into two parts; for example, bifid claw (see *bifurcate*).

Bifurcate (-tion) Divided into two parts or branches; forked or cleft.

Bilirubin The orange or yellow pigment in bile, produced by the degradation of heme in red blood cells. Excessive titers in the blood can impart yellowish color to skin (jaundice).

Bimodal Having two distinct statistical modes.

Binary fission A form of asexual reproduction in which a parent cell divides into two approximately equal parts.

Biogenic Produced by living organisms or biological processes.

Biological control agent A biological organism (e.g., virus, bacterium, fungus, parasitoid) used to reduce the population density of a pest species.

Biological vector A vector in which biological development (cyclodevelopment) of the vector-borne pathogen/parasite under consideration occurs (as opposed to a mechanical vector in which no biological development occurs).

Biopsy Removal of a piece of tissue for microscopic examination or testing.

Biotic potential The ability of an organism to reproduce optimally.

Biovar A term sometimes applied to different strains of certain bacteria, such as those that cause plague (*Yersinia pestis*) or tularemia (*Francisella tularensis*).

Bivoltine Having two generations per year.

Blepharitis Inflammation of the eyelids, involving the hair follicles and glands; characterized by redness, swelling, and crusty exudates.

Blister A local elevation of the skin containing serum or other watery fluid.

Blood meal Blood taken into the midgut of a hematophagous arthropod at a given feeding.

Blood poisoning Septicemia; a systemic disorder caused by pathogenic organisms or their toxins in the blood; characterized by fever, chills, and prostration.

Bolus A soft mass of ingested food in the alimentary tract; a large pill containing a drug, insecticide, parasiticide, or other compound fed to animals, or otherwise inserted in the alimentary tract.

Boric acid A white or colorless crystalline compound of boron and oxygen (H_3BO_3) prepared from borax. When ingested by cockroaches and other insects, damages epithelial cells of the digestive tract, interfering with nutrient absorption.

Borreliacide (-al) A drug or other compound that kills spirochetes of the genus *Borrelia.*

Borrow pit An excavation from which sand or gravel is moved to use as fill at another location, typically a construction site.

Bot Grub-like parasitic larvae of bot flies, family Oestridae.

Bot fly Adults of flies in the family Oestridae.

Botanical A drug, insecticide, or other compound derived from plants.

Bovid A member of the family Bovidae; hoofed, hollow-horned ruminants that include cattle, sheep, goats, buffaloes, and antelopes.

Bovine A member of, or pertaining to, the genus *Bos* (cattle, buffaloes, kudus), subfamily Bovinae.

Brachyceran A member, or pertaining to, the dipteran suborder Brachycera.

Brachypterous Possessing very short or rudimentary wings.

Bradykinin A polypeptide that mediates an inflammatory response; causes dilation and increased permeability of peripheral blood vessels, and contraction of smooth muscle.

Bridge vector A vector that transmits a pathogen or parasite from an enzootic cycle to humans or other animals that are often dead-end hosts.

Brisket The breast, or chest, of a quadruped animal.

Bronchiole (-ar) Small, thin-walled, noncartilaginous branches of the bronchi, the two large branches of the trachea.

Bronchitis Inflammation of the bronchial mucous membranes.

Bronchopneumonia Inflammation of the terminal bronchioles and alveoli in the lungs.

Bronchoprovocation Inhalation of an aerosolized chemical to induce a hypersensitivity response as a means of identifying obstructed airway passages in the lungs of asthmatic patients. Also called bronchial challenge and broncho provocation.

Brucellosis An infectious disease of cattle, sheep, goats, and some rodents caused by bacteria of the genus *Brucella*; for example, Bang's disease of cattle caused by *B. abortus.*

Bti Abbreviation for the bacterium *Bacillus thuringiensis israelensis.*

Bubo An inflamed swollen or enlarged lymph node, typically in the axilla or groin; as in the bubonic form of plague.

Bug In an entomological context, members of the Order Hemiptera, possessing piercing-sucking mouthparts; for example, bed bugs and triatomine bugs; used colloquially to imply any kind of arthropod from Old English bwg, evil spirit.

Bulbous dermatitis Inflammation of the skin, characterized by bulb-shaped lesion(s).

Bulla (-ae) A large blister or dermal vesicle filled with fluid.

Bunch (-ing) A behavioral response of livestock to the presence of biting or pestiferous flies in which the animals cluster tightly together to minimize the exposure of body surfaces to attack.

Bunyavirus A genus of single-stranded RNA virus (mostly arboviruses) in the family Bunyaviridae.

Bursa (-ae) A pouch or sac-like cavity, often filled with fluid; for example, in the vicinity of vertebrate joints where it reduces friction associated with movement of tendons, ligaments, and bones.

Cachexia A general physical wasting and malnutrition usually associated with chronic disease.

Cadaver A dead and decaying human body.

Caecum (-a) A blind pouch-like extension of the alimentary tract; also spelled cecum.

Calabash Any of several gourd-bearing plants, including the annual vine *Lagenaria siceraria*, or bottle gourd, and the tropical American tree *Crescentia cujete.*

Calamistrum A row of spines on the metatarsus of cribellate spiders, used to comb tiny strands of silk as they are extruded from the cribellum.

Calcium gluconate A white water-soluble compound ($CaC_{12}H_{22}O_{14}$), used as a dietary supplement of calcium; in solution is used to counteract the action of black widow venom.

Calliphorid A member of the fly family Calliphoridae, or blow flies.

Callus (-i) A localized thickening and enlargement of the horny layer of skin (vertebrates) or cuticle (arthropods); a callosity.

Calamus The hollow basal part of a bird feather; quill.

Caltrop Specialized defensive setae in caterpillars that readily break off on contact, causing irritation to skin.

Calypter (-ate) An enlarged membranous structure (alula or squama) at the base of the dipteran wing that covers the haltere in some muscoid flies (Calypteratae).

Calyx A cup into which certain structures are set; for example, egg-calyx in insects, the enlarged portion of the oviduct at the opening of the ovarioles in insects, into which eggs are received before entering the vagina.

Camerostome In soft ticks (family Argasidae), the ventral depression, or cavity, in which the mouthparts (capitulum) are situated; more difficult to discern in engorged specimens.

Campestral Pertaining to fields and open country.

Campodeiform A larval form of insects that, at least in their earlier instars, resemble diplurans; includes larvae of Neuroptera, Trichoptera, Strepsiptera, and some Coleoptera.

Canid A member of the family Canidae; includes dogs, coyotes, wolves, foxes, and jackals.

Canopy trap A tent-like structure elevated above the ground and used to capture flying insects; typically uses a large black ball or other visual object to attract certain biting flies (e.g., tabanids).

Canthariasis Invasion of humans and other animals by beetles, as larvae or adults.

Capillary feeder An organism that feeds directly on blood by piercing a capillary with its stylet-like mouthparts.

Capitulum A term for the mouthparts and associated palpal structures of ticks (see *gnathosoma*).

Capuchin Long-tailed monkeys of the genus *Cebus*, family Cebidae, in Central and South America; with a tuft of hair on the head suggesting a monk's cowl, or hood.

Carapace The sclerotized dorsal portion of the prosoma in some arachnids, for example, spiders and scorpions.

Caravansary (-ies) A large hotel or inn in central and western Asia, with a courtyard to accommodate caravans.

Carbamate A salt or ester of carbamic acid, with insecticidal properties.

Cardiomegaly Enlargement of the heart.

Cardiomyopathy A disease or disorder of heart muscle, especially of unknown cause.

Cardiovascular collapse Retraction of the walls of the heart and blood vessels due to loss of blood, lowered blood pressure, anaphylactic reactions, and other causes.

Carrier A person or other animal that harbors an infectious agent but typically shows no apparent symptoms or signs of the disease, and that is capable of transmitting the pathogen to others.

Carrion Dead and decaying animal remains.

Carton Paper made by hymenopteran insects (e.g., social wasps), used to construct their nests.

Caste One of multiple forms in social insects that perform a specialized function in a colony; for example, worker, soldier, and queen.

Cataract Opacity of the lens or capsule of the eye, causing impaired vision or blindness.

Catecholamine Any of a group of amines derived from catechol that act as neurotransmitters and hormones; for example, epinephrine, norepinephrine, and dopamine.

Caudal At or near the posterior end of the body.

Causative agent A biological pathogen or parasite that causes a disease; also sometimes used when referring to a toxin; etiologic agent.

Cell In insect morphology, an enclosed area of an insect wing between, or bounded by, veins named for the vein forming the anterior margin.

Cell-mediated immunity The reaction to antigens by cells (e.g., T cells) rather than antibodies.

Cellular immune response See *Cell-mediated immunity*.

Cellular immunity See *Cell-mediated immunity*.

Cellulitis Inflammation of cells, especially of subcutaneous and connective tissues; characterized by redness, swelling, pain, and sometimes fever.

Cenozoic (Caenozoic) The geologic era from 65 million years ago until the present time; sometimes called "the age of mammals."

Cephalic Pertaining or attached to the head; also, directed toward the head.

Cephalopharyngeal skeleton Internal sclerotized structures in the head region of muscoid fly larvae, serving as muscle attachments for the mouthparts and pharynx; of integumental origin and shed when larva molts.

Cephalothorax A body region in certain arthropods formed by fusion of the head and thorax; for example, arachnids and crustaceans.

Ceratopogonid A member of the dipteran family Ceratopogonidae, or biting midges.

Cercus (-i) A sensory caudal appendage, usually paired, on the tenth abdominal segment of insects; usually slender or filamentous, but highly variable in form and length.

Cerebral Pertaining to the cerebrum, or largest part of the vertebrate brain; as in cerebral malaria.

Cere (-es) A fleshy or wax-like swelling near the base of the upper beak in certain birds (e.g., parrots) through which the nostrils, or nares, open.

Cerumen (-inal, -nous) Soft, yellow-to-brown, wax-like secretions in the external ear canal; earwax.

Cervid A member of the family Cervidae; includes deer, elk, and moose.

Cervix (-cal) Pertaining to the neck.

Cestode A member of the class Cestoda; includes tapeworms.

Chaetotaxy Nomenclature, arrangement, or study of setae and bristles on the integument of insects and other invertebrates.

Chamois A goat-like antelope, *Rupicapra rupicapra,* in the mountainous regions of Europe.

Chaparral An ecological habitat dominated by small-leaved evergreen shrubs, with hot dry summers and cool moist winters; for example, foothills of California.

Chela (-ae, -ate) A pincer-like structure at the tip of a limb in arthropods, such as arachnids and crustaceans, composed of a fixed and movable digit; for example, claw-like palps of scorpions and pseudoscorpions and the chelicerae of many mites.

Chelicera (-ae) The pair of pincer-like or fang-like mouthparts in arachnids, variously modified for grasping or piercing.

Chemoprophylaxis (-actic) The use of chemical agents, drugs, or food supplements to prevent disease.

Chemosensillum (-a) A sense organ that responds to chemical agents or chemical cues.

Chigger The parasitic larval stage of mites in the families Trombiculidae and Leeuwenhoekiidae.

Chironomid A member of the dipteran family Chironomidae, aquatic midges with non-biting adults.

Chitin (-ous) A durable, semitransparent substance, made primarily of nitrogen-containing polysaccharide, which is the principal component of cuticle in arthropods.

Chitin synthesis inhibitor A class of pesticides for which the mode of action is interference with the synthesis of chitin; disrupts normal development during formation of the new cuticle in the molting process.

Chloramphenicol An broad-spectrum antibiotic derived from the soil bacterium *Streptomyces venezuelae,* or produced synthetically.

Chlorpyrifos A broad-spectrum organophosphate insecticide, or acaricide, that acts as an acetylcholinesterase inhibitor.

Cholecystitis Inflammation of the gallbladder.

Choriomeningitis Cerebral meningitis with cellular infiltration of the meninges.

Chorion In arthropods, the membrane(s) surrounding the egg, secreted by follicle cells in the ovary; the outer surface is often sculptured (i.e., with a distinct surface pattern).

Chorionic sculpture (-ing) See *chorion.*

Chromatography (-ic) A chemical analysis in which a mixture of substances is separated into fractions by adsorption on a porous surface; for example, paper chromatography, gel chromatography.

Chronic Persistent, or of long duration; versus acute.

Chrysalis (-ides) The developmental stage in butterflies between the larva and adult (i.e., the butterfly pupa).

Chukar A Eurasian partridge, *Alectoris chukar;* introduced as a game bird in western North America.

Chyluria The presence of fatty globules, or chyle, in the urine; can be caused by blockage of lymphatic vessels by parasites.

Cibarium (-al) A chamber or pocket anterior to the mouth in insects through which fluids are pumped, or otherwise drawn, into the food canal.

Cimicid A member of the insect family Cimicidae; for example, bed bugs and bat bugs.

Circadian rhythm A daily activity cycle in living organisms that follows a rhythmic pattern at intervals of approximately 24 hours; may be physiological or behavioral.

Civet A cat-like mammal of the family Viverridae; in Africa and Asia.

Cladistics An approach to classifying organisms taxonomically based on the branching of descendant lineages from a common ancestor.

Cladistic analysis The quantitative analysis of comparative data used to develop a taxonomic system representing phylogenetic relationships and the evolutionary history of groups of related organisms.

Clasper Variously modified lobe or extensions of abdominal segment 10 in male insects used to grasp females during mating.

Clavate Clubbed, or thickened toward the tip; for example, setae, antennae.

Clavus The oblong or triangular anal region of the forewings, or hemelytra (hemi-elytra), in members of the insect order Hemiptera.

Clinical Based on direct observation of patients and treatment for diseases; for example, clinical cases, signs, and symptoms.

Clubbed See *Clavate.*

Clypeus The front of the insect head between the frons and labrum, usually separated from the latter by a groove.

Coalesce To fuse or grow together; for example, blood vessels, rashes, and other skin lesions.

Coarctate Referring to an insect pupa that is encased in a hardened cuticle of the next-to-last larval instar, or puparium.

Coccobacillus A microorganism, typically a bacterium, that is spherical to ovoid in shape.

Cockle A wrinkle or pucker, as in a fabric; used to refer to blemishes and structural damage of processed sheep skin resulting from feeding by the sheep ked.

Coevolution Biological evolution of two interacting species, with each adapting to changes in the other; for example, between a parasite and its host.

Colic Muscular spasms in any hollow or tubular organ, such as the intestine; severe abdominal pain caused by such spasms.

Collembolan A member of the insect order Collembola, or springtails.

Colostrum A yellowish secretion from the mammary glands of female mammals a few days before and after they give birth; rich in antibodies, other proteins, and minerals.

Colubrid snake A member of the family Colubridae, represented primarily by nonvenomous species including garter snakes, king snakes, and water snakes.

Columbiform Members of the avian order Columbiformes; pigeons and doves.

Coma A state of deep, often prolonged unconsciousness in which an individual does not respond to external stimuli; usually the result of injury or disease.

Comb scale A specialized spicule, or pointed spine, on the lateral aspect of the seventh abdominal segment of culicine and first-instar anopheline mosquito larvae.

Commensal A symbiotic relationship between two species in which one is benefited and the other is generally unaffected.

Communal Organisms living, or performing an activity, together.

Companion animal A domesticated animal with which humans share a special bond or sentimental relationship; a pet, for example, a cat, dog, horse.

Competent [host, reservoir] Providing favorable conditions for the survival or replication of a particular pathogen or parasite.

Complement A group of proteins in blood serum that act as part of the body's immune response; interact with antibodies and other chemical substances to destroy bacteria and other foreign cells.

Complement-fixing antibody An antibody that, in combination with an antigen and complement, inactivates the complement, interfering with the ability to destroy foreign cells.

Complex In classification, referring to a group of closely related organisms or interacting biologically active molecules; for example, taxonomic complex, virus complex, antigenic complex.

Compound eye One of the pair of image-forming structures in many insects, and some crustaceans, composed of large numbers of individual light-sensitive units called ommatidia, visible on the eye surface as facets.

Concave Curved inward, like the inner surface of a sphere.

Confirmed (case) A clinically corroborated infection, infestation, or other condition of humans or animals.

Congenital Relating to a condition present at birth; may be hereditary or a result of environmental influences.

Congestion Accumulation of excessive fluid; for example, blood or other fluids, in tissues or organs.

Congestive heart failure Inability of the heart to maintain adequate circulation of the blood, resulting in distension of the ventricles, congestions in the lungs, and edema in the lower extremities.

Conjunctiva Mucous membrane that lines the inner surface of the eyelid.

Conjunctivitis Inflammation of the conjunctiva of the eye.

Connective tissue Highly vascular tissue derived from embryonic mesoderm that supports and connects various structures of the body; includes elastic and collagenous fibers, mucosal and fat tissues, cartilage and bone.

Connexivum (-al) The lateral margin of the abdomen at the juncture of the dorsal and ventral sclerites in members of the insect order Hemiptera.

Conspecific Belonging to the same species.

Contact dermatitis Inflammation of the skin resulting from contact with an irritating substance or allergen.

Contact pheromone A pheromone that elicits a response after it is touched by a conspecific organism.

Cootie Colloquial term for the human body louse; in a less strict sense, also the human head louse and human pubic louse; from the Malay and Indonesian *kutu*, louse, or other ectoparasites such as ticks and fleas.

Coprophagous (-y) Feeding on excrement, or dung.

Corbicula Pollen basket of bees; modification of the tibia of the hind leg, characterized by a smoothly concave surface with a fringe of hairs along the tibial margin.

Cornified Having undergone cornification, the conversion of squamous epithelial cells into horny materials, such as nails, hair, feather, or scales; keratinized.

Corolla In mosquito eggs deposited as a raft, notably of the genera *Culex* and *Culiseta*; a frill-like collar surrounding the micropyle that helps to keep the egg vertical while floating on the water surface.

Coronary band An area of highly vascular tissue at the base of the hoof in horses, ruminants, and other hoofed animals that secretes the horny wall of the hoof.

Corticosteroid A steroid compound produced by the cortex of the adrenal gland.

Corvid A member of the avian family Corvidae; includes crows, ravens, jays, and magpies.

Cosmopolitan Occurring worldwide, or at least in many parts of the world.

Cospeciation As for coevolution but at the species level.

Costa The thickened anterior margin of the wing in insects, usually referring to the forewing.

Cowdriosis A tick-borne disease of cattle, sheep, and goats caused by the rickettsia *Cowdria ruminatium*; also called heartwater and veldt disease.

Coxa (ae) The basal-most segment of the leg in insects and other arthropods.

Coxal gland An excretory structure in ticks and other mites that opens via a duct on, or near, the coxa; serves primarily in osmoregulation or to excrete noncellular liquid components of a bloodmeal.

Coxal pore The opening of the duct of a coxal gland.

Crack-and-crevice Protected out-of-the-way places indoors that can harbor insect pests; usually used with regard to application of insecticides, for example, crack-and-crevice treatments.

Crateriform Hollowed like a bowl or saucer; crater-like.

Creatinine kinase An enzyme in muscle, brain, and other vertebrate tissues that acts as a catalyst for the conversion of ADP and phosphocreatine into ATP and creatine.

Creeping welt Segmental, annular thickenings of the external abdominal wall of the larvae of certain aquatic or semiaquatic flies (e.g., tabanids) that facilitate locomotion; sometimes used to refer to spiny annular bands of maggots that serve a similar purpose.

Crepuscular Pertaining to the twilight hours at dusk and dawn.

Cretaceous Period The geological period of the Mesozoic Era, 140 to 65 million years ago, that began with the development of flowering plants and modern insects and ended with the extinction of dinosaurs and many other life forms.

Cribellum (-ate) A transverse plate anterior to the spinnerets in certain spiders through which silk is extruded to produce hackled-band threads characteristic of cribellate spiders.

Crop An enlargement of the foregut, usually the posterior part of the esophagus, in which food is stored; also a diverticulum of the foregut in insects that feed on blood and other fluids.

Cross infection The ability of a pathogen or parasite to infect related hosts.

Cross-reactivity For immunological testing, indicates a positive reaction to a pathogen or parasite other than the one under consideration; usually cross-reactivity concerns relatively closely related pathogens, such as the spirochete bacteria that cause Lyme disease and syphilis.

Cryptic infection An infection that is difficult, or sometimes impossible, to detect.

Cryptic species Members of a species complex that are not morphologically distinguishable from one another; can be separated only by nonmorphological criteria such as genetic sequencing and life-history studies.

Ctenidium (-a) A row of short, stiff spines resembling the teeth of a comb, used to cling to hair and feathers; found on the head and thorax (rarely also on the apex of the head or on the abdomen) of some fleas and other ectoparasitic insects; for example, bat bugs (family Polyctenidae) beetles (*Platypsyllus*), and bat flies (families Nycteribiidae and Streblidae).

Culicidology The science and study of mosquitoes.

Culicine Referring to members of the mosquito subfamily Culicinae; includes most mosquito genera other than *Anopheles* and *Toxorhynchites*.

Cuterebrine Referring to members of the subfamily Cuterebrinae, New World skin bots.

Cuticle The noncellular outer layer of the arthropod integument, composed primarily of chitin and protein.

Cuticular hydrocarbon A compound associated with the cuticle, consisting of only carbon and hydrogen.

Cyanosis (-tic) Discoloration of the skin due to reduced hemoglobin in blood; typically bluish, grayish, or purplish.

Cyclical development See *Cyclodevelopment*.

Cyclodevelopment (-al) The sequence of developmental stages in the life cycle of organisms that undergo metamorphosis; for example, *Plasmodium* and filarial nematodes in arthropod hosts.

Cyclodiene An organic compound with a chlorinated methylene group bonded to two carbon atoms of a six-member carbon ring; active ingredient in insecticides such as aldrin, chlordane, and dieldrin.

Cyclorrhapha (-an, -ous) A large group of flies, order Diptera, in the suborder Brachycera, generally characterized as adults by an arista on the third antennal segment; represents a monophyletic group within the infraorder Muscomorpha.

Cynomolgus monkey A primarily arboreal macaque (*Macaca fascicularis*) native to Southeast Asia; also called crab-eating macaque and long-tailed macaque.

Cyst In parasitology a structure formed by and enclosing organisms in a larval or dormant stage; a closed sac or pouch with a defined wall that contains fluid, semifluid, or solid material.

Cysticercus (-i, -oid) The larva of certain tapeworms in invertebrates that consists of a fluid-filled sac with an invaginated scolex (or multiple scolices), in tissues of an intermediate host.

Cytokine Any of several regulatory proteins released by cells of the immune system in the presence of specific antigen that act as intercellular mediators in generating an immune response.

Cytology The study of structure and function of cells.

Cytopathology The study of cellular changes in diseases.

Cytopenia A deficiency, or lack, of cellular elements in circulating blood.

Cytotaxonomy A branch of taxonomy based on cellular structure and function, particularly the structure and number of chromosomes.

Cytotoxin A substance that has a toxic effect on cells.

Dasyurid A member of the marsupial family Dasyuridae found in Australia and New Guinea; includes banded anteaters, pouched mice, and Tasmanian devil.

DDT The chlorinated hydrocarbon compound dichloro-diphenyl-trichloroethane, used as an insecticide; now banned for use in many countries including the United States, because of associated environmental problems.

Dead-end host An animal that harbors a pathogen or parasite but does not serve as a source of infection for transmission to another individual.

DEC The chemical compound diethyl carbamate; used to kill filarial nematodes.

DEET A shortened term for the compound diethyl toluamide, used as an insect, tick, and chigger repellent; chemical names, *N,N*-Diethyl-*m*-toluamide and *N,N*-Diethyl-3-methylbenzamide.

Defervescence Abatement of fever; falling of elevated body temperature.

Definitive host An animal in which a parasite develops to the adult stage and in which sexual reproduction of the parasite takes place.

Delayed (reaction, sensitivity) A nonimmediate immunological or skin response to antigens, including those injected by blood-feeding or skin-infesting arthropods.

Delusional parasitosis See *delusory parasitosis.*

Delusory parasitosis A psychological disorder involving the false belief by an individual that he or she is infested by a parasite or that live organisms, usually mites or insects, are crawling on their skin.

Demodicosis A condition resulting from infestation of animal tissues, usually hair follicles or glands, by mites of the genus *Demodex.*

Dentate Resembling a tooth.

Denticle A small tooth-like projection.

Depilation The removal of hair from the body or animal skins.

Dermatitis Inflammation of the skin; usually with itching and redness.

Dermatographia A condition in which lightly touching or scratching the skin causes raised reddish marks; also a form of urticaria in which pressure to the skin produces wheals.

Dermatomycosis An infection of the skin caused by certain fungi; for example, ringworm.

Dermatophilosis A skin disease caused by the bacterium *Dermatophilus congolensis.*

Dermatophytosis A fungal infection of the skin, especially moist skin covered by clothing; for example, athlete's foot, eczema, and ringworm.

Dermatosis (-es) Any disease of the skin, especially without inflammation.

Dermestid A beetle in the family Dermestidae, or skin beetles.

Desensitization The process of reducing sensitivity or making an individual nonreactive to an antigen or allergen.

Desiccant A substance that absorbs water and is used as a drying agent; for example silica gel and calcium oxide.

Desquamate To shed epithelial cells.

Detritivore An animal that feeds on detritus.

Detritus Partially decomposed, particulate organic matter.

Deutonymph The second of three nymphal stages in the basic life cycle of mites.

Dewlap A fold of loose skin hanging from the throat or neck; for example, in bovines, and the wattle of certain birds.

Diabetes mellitus A carbohydrate-metabolism disorder resulting from inadequate production or utilization of insulin; characterized by elevated glucose levels in the blood and urine.

Diapause A physiological state of quiescence characterized by reduced metabolic activity without growth or development; may be seasonal or when environmental conditions are unfavorable.

Diatomaceous earth A light-colored porous soil or powder with a high silica content, made from the dried cell walls of diatoms; used as an abrasive, adsorbent, filtering, and insect-killing agent.

Diazinon An organophosphate compound used as an insecticide.

Dichoptic In Diptera, having compound eyes noticeably separated from one another along their midline; *cf. holoptic*.

Diel Pertaining to a 24-hour period; for example a daily cycle or activity pattern.

Diel periodicity Occurring at a specific time each day, or at intervals of approximately 24 hours.

Dieldrin A chlorinated hydrocarbon used as an insecticide.

Die-off A sudden, sharp decline in a population or community as a result of a natural cause, such as disease or extreme weather conditions.

Diethylstilbestrol A synthetic, nonsteroidal compound with estrogen-like properties, used in medical treatments and in feed for livestock and poultry.

Diflubenzuron The chemical benzamide that acts as an insect growth regulator; interferes with the formation of insect cuticle and is used as an insecticide.

Dike An embankment or barrier of earth and rock built to prevent flooding or to retain water, for example, for agricultural purposes; also, to construct such embankments.

Dilatation The condition of being stretched or expanded beyond the normal size, as in a tubular structure or orifice; for example, in insects, a dilatation of the stalk of an ovariole where an egg follicle has degenerated.

Dimorphic Occurring in two distinct forms.

Dingo A wild dog, *Canis dingo*, of Australia; often treated as a subspecies, *Canis familiaris dingo*, of the domestic dog.

Diphyletic Derived or descended from two ancestral lines.

Diploid Having two sets of chromosomes, or twice the haploid number; in organisms that reproduce sexually, one pair of chromosomes from each parent.

Dipping vat A large tank of fluid containing an insecticide or acaricide, through which animals are forced to walk or swim as a means of treatment for livestock pests; for example, ticks.

Disease A pathological condition of the body characterized by a group of signs or symptoms.

Disease agent An organism or substance that causes disease; usually refers to a pathogen or parasite; also called the causative agent.

Dispersal The movement of individuals away from a population center; as distinct from dispersion, the spatial distribution of individuals within a given geographic area.

Disseminate (-ed, -ation) To scatter or become distributed over a considerable geographic area, as in the case of a disease agent; also to spread throughout an organ or the body.

Dissemination barrier A physiological mechanism, typically the immune system, that prevents the spread of a pathogen within the body of an infected arthropod vector.

Distal Toward the free end of an appendage, or farthest from the body.

Distotibial Near the distal end of the tibia.

Diuresis Production and passage of abnormally large amounts of urine.

Diverticulum (-a) A blind, tubular sac or pouch branching from a cavity or canal; usually associated with the alimentary tract.

DNA Deoxyribonucleic acid; the main component of chromosomes that carries the genetic information in a cell.

DNA probe A substance or technique used to identify a segment of DNA; uses a known sequence of nucleotide bases from a DNA strand to detect a complementary sequence in a sample.

Domestic Entomologically speaking, found in, or immediately around, the home or other area of human habitation; for example, domestic cockroaches and flies (see *peridomestic*).

Domiciliary Referring to a residence or home.

Doramectin (doramectine) A derivative of ivermectin used as a veterinary drug for treatment of roundworms, grubs, lice, and mites in cattle.

Dormant (-cy) A general term for various inactive states of an organism in which growth and development cease and metabolism is reduced; includes aestivation, quiescence, and diapause.

Doxycycline A broad-spectrum antibiotic derived from tetracycline.

Drag cloth A piece of fabric attached to a horizontal pole that is dragged over vegetation and the ground to collect questing (host-seeking) ticks.

Draught animal Any animal, other than a human, that is used for its physical power to pull or operate equipment, transport goods, or otherwise provide work; also, draft animal.

Dromedary A one-humped camel, *Camelus dromedarius*, native to northern Africa and western Asia; also called the Arabian camel.

Dry ice A solid form of carbon dioxide, commonly used as a cooling agent and as a source of carbon dioxide for various insect traps.

Duff A layer of decomposing leaves, needles, and twigs in a woodland setting, between the ground litter and underlying mineral soil.

Dung Animal excrement, usually referring to vertebrates; feces or manure.

Dung flies Dipterous insects that breed in animal excrement, or dung.

Dung pat A discrete deposit of feces, produced by bovine animals when they defecate.

Dust bag A sack containing insecticidal dust that is suspended across an opening through which livestock pass; insecticide is applied on contact as an animal walks under it.

Dysentery Inflammation of the mucous membranes of the lower intestinal tract, particularly the colon; usually caused by bacteria, protozoa, or other parasitic infections; can result in severe diarrhea with passage of blood and mucus.

Dyspnea Difficult or labored breathing.

Ear tag A plastic strip or other device impregnated with a slow-release insecticide or acaricide fastened to animal's ear; used to control flies and other arthropod pests; also tags with an identification number.

Ecchymosis Seepage of blood from damaged blood vessels into subcutaneous tissue, resulting in purple or black-and-blue discoloration of the skin; for example, as from a bruise.

Ecdysis The last step in the molting process, in which the exoskeleton or outer cuticle of arthropods and other animals is shed.

Ecdysteroid A general term for insect molting hormones; that is, ecdysone and its homologues.

Eclose (-ed, -ion) Hatching of the insect larva from an egg; also used to refer to emergence of an adult insect from the pupal stage.

Ecophenotype (-ic) An ecological form of an organism, with slightly distinct morphology; for example, *Sarcoptes scabiei* on different host species.

Ecotone A transitional zone between two ecological communities or ecosystems.

Ecribellate Referring to spiders that lack a cribellum.

Ectoparasite A parasite that lives on the surface or exterior of another organism.

Eczema An inflammatory skin condition characterized by redness, itching, serous exudates, crusting, and scaling.

Edema An accumulation of an excessive amount of serous or watery fluid in body tissues or cavities.

Edentate Belonging to the Edentata, an order of mammals in Central and South America that have few or no teeth; includes armadillos, sloths, and anteaters.

Epinephrine A hormone produced by the adrenal medulla and other tissues that causes vasoconstriction, stimulates the heart, and relaxes bronchioles; syn. adrenaline.

Efficacious Capable of producing a desired effect or result; effective.

Egg breaker See *Egg burster.*

Egg burster A cuticular projection, typically on the head of an embryo, that is used to break the egg shell when hatching; often evident in the first-instar larva.

Egg raft A floating mass of insect eggs, in which the eggs are oriented vertically side-by-side with the anterior ends downward; for example, characteristic of certain mosquito genera such as *Culex* and *Culiseta.*

Ehrlichiosis (-es) Infection caused by rickettsial bacteria of the genus *Ehrlichia*; many species are transmitted by ticks.

EIA Enzyme immunoassay; an assay in which an enzyme is bound to an antigen or antibody that is used as a label for detecting a specific protein (see *enzyme-linked immunosorbent assay*).

EIA Equine infectious anemia; also known as Coggins of horses; caused by a retrovirus and mechanically transmitted by blood-sucking insects, notably tabanid flies.

Ekbom syndrome See *delusory parasitosis* named after the Swedish neurologist Karl Axel Ekbom, who described this condition in the 1930s; also called Ekbom's syndrome and Ekbom's disease.

Elaeophorosis Parasitism of deer, elk, and moose by filarial nematodes of the genus *Elaeophora*, transmitted by the bite of tabanid flies.

Elateriform A larval form of holometabolous insects resembling that of click beetles (Elateridae), or wireworms; elongate cylindrical body, heavily sclerotized, with short legs.

Elliot's disease An elusive skin condition of uncertain origin so-called in memory of an American by the name of Elliot, who was convinced that fibers were coming out of his skin, with associated itching and burning sensations.

Elephantiasis A chronic, extreme form of human filariasis characterized by greatly enlarged cutaneous and subcutaneous tissues, especially of the legs and scrotum, caused by lymphatic obstruction due to filarial nematodes.

ELISA See *Enzyme-linked immunosorbent assay.*

Elytron (-a) The leathery or hardened forewing of beetles that covers the hindwing.

Emasculate To remove, or render nonfunctional, the testicles of a male animal; castrate.

Embryogenesis The development and growth of an embryo.

Emetic Causing to vomit, or an agent that induces vomiting.

Emphysema A chronic disease of the lungs characterized by abnormal enlargement of the air spaces, or alveoli, and destruction of the alveolar walls.

Empodium (-a) A single, median pad-like or bristle-like structure between the pair of tarsal claws in certain insects and mites.

Emulsifiable concentrate A formulation of insecticide in which a highly concentrated active ingredient in liquefied form is suspended in another liquid, as an emulsion.

Encapsulate (-ion) To surround or encase something, as if in a capsule; for example, the body's defensive response to a parasitic organism, or enclosing a drug or slow-release insecticide in a capsule.

Encephalitis Inflammation of the brain.

Encephalomyelitis Inflammation of the brain and spinal cord.

Encyst (-ed) To enclose or become enclosed in a cyst, a membranous sac filled with fluid, semifluid, or solid material.

Endectocide A systemic compound used to kill both internal and external parasites; for example, in livestock.

Endemic (1) Native to a specific geographic region and not naturally occurring elsewhere, as in endemic species; (2) relating to a pathogen or disease that occurs more or less continuously in a particular locality.

Endemicity Confinement of an organism or disease to a particular geographic area (see *endemic*).

Endocarditis Inflammation of the endocardium, or membranous lining of the heart, particularly of the valves.

Endogenous Derived or originating internally; as opposed to exogenous.

Endoparasite A parasite living within the body of its host.

Endophagic Feeding within a host by certain parasitic insects.

Endophilic (-y) Ecologically associated with humans and their domestic environment; used to refer to insects occurring indoors or under overhangs of human and animal dwellings; for example, certain mosquitoes.

Endothelium (-al) A thin layer of flat epithelial cells lining the blood vessels, lymphatic vessels, heart, and serous cavities.

Endotoxin A toxin produced by certain bacteria that is released only when the bacterial cell breaks down.

Enteric Of or relating to the intestines, or enteron.

Enteritis Inflammation of the intestinal tract, particularly the small intestine.

Enterotoxigenic Producing an enterotoxin, a toxin originating in intestinal contents, usually by bacteria.

Entomology The branch of zoology dealing with the study of insects.

Entomopathogen (-ic) An organism that is pathogenic to insects; for example, certain bacteria and fungi.

Entomophobia An abnormal or morbid fear of insects.

Envenomation The injection of a poisonous substance by bite, sting, spine, or other means; also envenomization and British envenomisation.

Enzootic Relating to a disease involving nonhuman animals in a specific geographic area, where the disease is constantly present but at low incidence.

Enzootic transmission The transfer of a pathogen or parasite among nonhuman animals in an enzootic disease cycle, typically by arthropods.

Enzootic vector An arthropod that plays a role in transmission of a disease agent in an enzootic cycle.

Enzyme-linked immunosorbent assay (ELISA) An immunoassay that uses an enzyme linked to an antigen or antibody as a marker for detection of a specific protein; used in diagnostic tests for infectious agents to identify corresponding antibodies in blood.

Eosinophil A white blood cell with cytoplasmic inclusions or granules that are readily stained by the acid dye eosin; also other cells, microorganisms, or histological elements that are easily stained by eosin.

Eosinophilia An increase in the number of eosinophils in the blood, above normal levels.

Epidemic A widespread outbreak of an infectious or transmissible disease in humans, infecting large numbers of people at the same time; also used as an adjective.

Epidemiology A branch of medical science dealing with the cause, incidence, distribution, ecology, and control of diseases in human populations.

Epidural Located on or outside the dura mater, the outer membrane covering the brain and spinal cord; for example, epidural fat.

Epigean Living or occurring on or near the surface of the ground.

Epimastigote A developmental stage of flagellate protozoans with an undulating membrane, in which the flagellum arises from the kinetoplast and emerges from the anterior end of the organism; for example, trypanosomes; replaces former term "crithidial stage."

Epinephrine See *Adrenaline*.

Epiornitic (epornitic) Outbreak of a disease in a bird population.

Epipharynx A median structure on the posterior, or ventral, surface of the labrum or clypeus overlying the mouth in some insects; has no relation to the pharynx.

Epiponine Pertaining to tropical social wasps belonging to the subfamily Epinoninae of the family Vespidae.

Epizootic An outbreak of a disease in animals other than humans, involving a large number of individuals within a particular geographic area; also used as an adjective, as in epizootic disease; counterpart of epidemic in human disease.

Epizootic transmission The transfer of a pathogen or parasite from one animal to another in the epizootic cycle of a disease.

Epizootiology The study of diseases in nonhuman animals, involving the cause, incidence, distribution, ecology, and control, especially of epizootic diseases; counterpart of epidemiology in human disease.

Epyginum A hardened plate partially covering the genital opening of female spiders.

Equid See *Equine*.

Equine A member of the family Equidae, including horses, zebras, and donkeys; equid; also used as adjective.

Eruciform Caterpillar-like larval form of insects; cylindrical body, with distinct head, and both thoracic legs and abdominal prolegs; characteristic of Lepidoptera, Mecoptera, and some Hymenoptera.

Erucism Urticaria, or other forms of dermatitis, caused by stinging spines of lepidopterous larvae (i.e., caterpillars of moths and butterflies).

Erythema (-atous) Redness of the skin caused by dilatation and congestion of capillaries; a common sign of infection or inflammation.

Erythrocyte (-ic) A red blood cell in vertebrates; transports oxygen and carbon dioxide to and from tissues.

Eschar A dry scab or crust formed on the skin as a result of contact with a caustic or corrosive substance or a burn; often used to describe the black scab associated with brown-recluse spider bites or the characteristic entry lesion for some rickettsial pathogens.

Escutcheon On a cow, the shield-shaped area with a distinct hair pattern between the udders and the genital opening.

Esophagitis Inflammation of the esophagus.

Esterase Any enzyme that hydrolyzes an ester to form an alcohol and an acid.

Estrogen Any of several female hormones, produced primarily by the ovaries, which promote estrus and the developmental growth of typical female sexual characteristics.

Ethiopian The zoogeographical region that includes Africa south of the Tropic of Cancer, Madagascar, and the southern part of the Arabian Peninsula; also called the Afrotropical region.

Etiologic agent A pathogen or substance that causes a disease; a causative agent.

Etiology (-ic) Study of the causes of disease.

Eurypterid A member of the extinct order Eurypterida, marine chelicerate arthropods that resembled scorpions and lived in the Ordovician period to the end of the Palaeozoic era.

Eusocial Behavior in social organisms characterized by cooperation among individuals in rearing young, division of labor, and overlap of generations; for example, ants, social wasps, and bees.

Exarate A type of insect pupa in which the appendages are free, usually not protected by a cocoon; typical of most holometabolous insects other than Diptera and Lepidoptera.

Excoriate (-ed, -ation) To abrade or strip off the epidermis or the coating of any organ of the body by trauma, burns, chemicals, or other causes.

Excrement Waste material, especially feces, passed out of the body following digestion.

Excrescence An abnormal outgrowth or enlargement of a body surface, usually harmless.

Excreta Any waste material excreted from the body; for example, urine, feces, and sweat.

Exflagellation The formation of microgametes (flagellated bodies) from microgametocytes in sporozoans; for example, *Plasmodium* spp. in the mosquito midgut.

Exfoliate (-ed) To remove, come off, or separate as scales, flakes, or layers; for example, exfoliated skin.

Exoerythrocytic Occurring outside red blood cells; for example, multiplication of *Plasmodium* in liver cells.

Exogenous Derived or originating externally; *cf.* endogenous.

Exophilic (-y) Ecologically independent of humans and their domestic environment; found only outdoors.

Exotic From another part of the world; foreign; introduced from abroad but not fully naturalized or acclimatized.

Exotoxin A poisonous substance produced by a microorganism and released into the surrounding medium.

Expectorate (-ed) To expel via the mouth body fluids coughed up from the throat and lungs, including saliva, sputum, and phlegm.

Exsanguinate (-tion) To be drained of blood.

Exserted Projecting beyond the body or over a particular point.

Extrafloral nectar (-y) Nectar that is produced outside the flower, in nectaries generally located on the petiole, mid-rib, or margin of the leaf.

Extraoral digestion The digestion of food material outside the mouth; characteristic of spiders and other predaceous arachnids that cannot ingest solid food.

Extrinsic incubation period The time interval between an arthropod vector acquiring a pathogen or parasite and the ability of the vector to transmit the agent to a susceptible vertebrate host.

Exudate (-ive) A fluid or semisolid material that passes slowly out of a body tissue or its capillaries, due to inflammation or injury.

Exuviae The cast skin, or cuticle, (exoskeleton) of molted insect larvae or nymphs; in the strict entomological sense, used in the plural, referring to multiple shed parts.

Eyespot A rudimentary ocellus.

Facet A small face or surface; the external surface of each individual unit, or ommatidium, that makes up the compound eye in insects.

Facial lunule The crescent-shaped space at the base of the antennae in members of the Cyclorrhapha (Diptera), bounded by the frontal suture; also called frontal lunule.

Facultative May or may not take place; optional, depending on the conditions.

Fanniid A member of the dipterous family of muscoid flies Fanniidae, closely related to Muscidae.

Fascia A sheet or band of fibrous connective tissue supporting, covering, or separating muscles, organs, and other soft body structures.

Fascial plane A sheet of fibrous connective tissue supporting, covering, or separating muscles, organs, and other soft body structures (i.e., fascia); provides means by which cattle grubs can move within a host animal while minimizing direct injury to organs and other tissues.

Fascicle A bundle or tight cluster of elongate structures; in entomology, used to refer to the stylet-like mouthparts of certain insects that collectively form a food canal and mechanism for piercing plant or animal tissues; for example, mouthparts of adult mosquitoes.

Febrile Pertaining or characterized by fever; feverish.

Fecund (-ity) Producing or capable of producing offspring, particularly in large numbers.

Feedlot A tract of land where livestock are fattened for market.

Felid A member of the cat family Felidae.

Femoro-tibial joint The articulation between the femur and tibia.

Femur The leg segment in insects between the coxa and tibia; in arachnids between the coxa and patella (genu).

Fenoxycarb A carbamate juvenile-hormone analog used as an insect growth regulator to control certain insects.

Feral Existing in a wild or untamed state.

Fetid Having an offensive odor.

Fibrilla (-ae) A minute filament or fiber.

Fibrinogen A protein in blood plasma that is converted to fibrin when blood clots.

Fibroma A benign tumor consisting primarily of fibrous tissue.

Fibrosis The formation of excessive fibrous tissue.

Filaria (-ae, -al) See *Filarial nematodes.*

Filarial nematodes Members of the superfamily Filarioidea, parasites of vertebrates as adults and as larvae in mosquitoes, black flies, and other insects; characterized by prelarval microfilaria stage; for example, genera *Wuchereria*, *Brugia*, and *Onchocerca.*

Filariasis A disease caused by infestation of tissues with filarial nematodes.

Filth flies Flies that breed in excrement and other animal wastes, carrion, or garbage; include primarily muscoid flies in the families Muscidae, Calliphoridae, Fanniidae, and Sarcophagidae.

Fistula An abnormal passage from a hollow organ to the surface or from one organ or cavity to another, caused by injury or disease, or for experimental testing and recording.

Flag cloth A piece of fabric attached to a rod in flaglike fashion, for collecting ticks; can be used to probe animal burrows, under fallen trees, and other sites where a drag cloth is not effective.

Flagellomere The individual units, or pseudosegments, that comprise the flagellum of an insect antenna.

Flagellum (-a) The third and apical segment of the basic insect antenna; also a threadlike or whiplike extension of unicellular organisms and other cells that function in locomotion.

Flank The side of an animal between the ribs and hips; also more generally used to refer to the side of livestock animals and other quadrupeds.

Flare A spreading area of redness of the skin surrounding the primary site of an infection or irritation, due to dilatation of arterioles.

Flavivirus A genus of single-stranded RNA virus in the family Flaviviridae, also formerly known as group B arboviruses, transmitted by mosquitoes or ticks.

Float hair A specialized dorsal seta on the abdominal segments of *Anopheles* mosquito larvae that helps to hold the abdomen parallel to the water surface; characterized by flattened, movable, usually horizontal branches radiating from a short stem; also called palmate seta.

Fluke A flatworm of the class Trematoda, including internal and external parasites of vertebrates; with thick integument and one or more suckers or hooks for attaching to its host.

Fly speck A small dark spot or stain made by excrement of a fly; also flyspeck.

Fly strike Cutaneous myiasis, especially those caused by blow flies (family Calliphoridae) in sheep (e.g., Australia).

Focus (-i, -al) In epidemiology, a localized area of disease or infection, or the center from which a disease develops and spreads.

Follicle A crypt or narrow-mouthed, cell-lined depression in the skin from which the hair emerges; also a small spherical group of cells with a central cavity, for example, ovarian follicle.

Fomite Any inanimate object that can be contaminated by an infectious organism and thus is capable of transmitting it from one individual to another; for example, clothing, bedding, towels, doorknobs, toys, cell phones.

Food poisoning An acute gastrointestinal condition caused by food containing natural toxins or contaminated with a toxic chemical; also by bacteria or their toxins, for example, *Salmonella*; characterized by fever, chills, headache, abdominal and muscular pain, nausea, and diarrhea.

Forage To wander in search of food or provisions; also, grass or hay browsed or grazed by horses, cattle, and other livestock; food for domestic animals, or fodder.

Forensic Pertaining to, or used in, courts of law or public discussion; relating to investigations to establish facts or evidence involving legal procedures; for example, forensic entomology and forensic medicine.

Forensic entomology The study of insects and their use involving the courts and legal proceedings; the application of such knowledge and data to forensic investigations.

Foretibia Tibia of the first, or front, leg in insects and other arthropods.

Formicophilia A psychosexual disorder of humans in which an individual experiences erotic arousal and orgasm in response to small creatures (e.g., ants, cockroaches, snails, and other invertebrates) crawling, creeping, or nibbling on the body, especially the genitalia, perianal area, or nipples.

Formulation A substance prepared according to a formula; used to refer to commercial products, such as insecticides, indicating the active ingredient and chemical composition, their respective weight or percentage, and other attributes.

Fossa (-ae) A furrow, cavity, or depression; in insects, mites, and other arthropods often for protecting an appendage that can be drawn into it; for example, antennal fossa and leg fossa; in birds, the nasal fossa containing the nostrils.

Fossula (-ae) A relatively long and narrow depression or groove; in insects, usually on the sides of the head or prothorax.

Foveal pore An opening to a small pit or cuplike depression; for example, opening from which a sex pheromone produced by some ticks is released.

Frons (-tal) The front of the insect head above the clypeus.

Fronto-clypeus (-eal) The front of the insect head composed of the frons and clypeus.

Fulcrum (-a) Any structure that serves as a support to another; for example, fulcra of pectine in scorpions.

Fulminate (-ing) To occur suddenly or intensely.

Fungivore An organism that feeds on molds and other fungi.

Furunculus (-i, -ar) A boil; a deep inflammation of the skin usually resulting in suppuration and necrosis; also furuncle.

Fusiform Cylindrically rounded and tapered from the middle toward each end; spindle-shaped.

Galea (-ae) Outer lobe of the maxilla; highly modified, especially in Diptera and Hymenoptera; forms coiled tongue of Lepidoptera.

Galliform Referring to ground-nesting birds of the order Galliformes; includes grouse, quail, pheasants, chickens, and turkeys; also called gallinaceous.

Gallinaceous See *Galliform.*

Gametocyte A cell from which a gamete develops; for example, oocyte or spermatocyte.

Gamete A mature sexual cell that unites with another cell to form a zygote, and ultimately a new organism; for example, sperm and egg.

Gametogenesis The production and development of gametes.

Gametogony A stage in the sexual cycle of sporozoans in which gametes are formed, often by schizogony.

Ganglion cell A neuron, or nerve cell, having its body outside the central nervous system.

Gangrene Death and decay of body tissue caused by insufficient blood supply, often involving an extremity or a limb; necrotic tissue.

Garbage fly A general term for flies that breed in, or are attracted to, discarded food and other solid wastes.

Gaster In ants and some other hymenopterans the rounded part of the abdomen posterior to the pedicel or petiole.

Gasterophiline Referring to members of the subfamily Gasterophilinae, or horse bots.

Gastric caecum (-a) A pouch or blind outpocketing of the midgut in insects (and some other organisms).

Gastritis Inflammation of the stomach, especially the mucous lining; associated pain, tenderness, nausea, and vomiting.

Gastroenteritis Inflammation of the mucous membrane lining the stomach and intestines; also called enterogastritis.

Gena (-ae, -al) The side of the insect head below the compound eyes.

Genal ctenidium (-a) A comb-like row of strong spines on the anteroventral border of the head in fleas (see *ctenidium*).

Gené's organ In female ticks, a paired extrusible structure located dorsally between the capitulum and scutum, used to apply a coating of wax to eggs as they are deposited.

Geniculate Elbowed, or bent at a joint like a knee; for example, antennae of ants.

Genital chamber In female insects, an invaginated cavity behind the eighth abdominal sternum into which the gonopore and spermathecal duct opens; often forms a pouch-like or tubular vagina or uterus; in males, a ventral invagination behind the ninth sternum containing the intromittent organ (aedeagus).

Genital pore The external opening of the male or female reproductive tract; also called gonopore.

Genital pouch See *Genital chamber.*

Genitalia Organs of the reproductive system, especially the external genital organs.

Genome (-ic) An organism's genetic material, or full DNA sequence of a haploid set of an organism's chromosomes.

Genospecies A group of organisms that can interbreed; all are genotypes within a single species.

Genu In arachnids, the segment of the leg between the femur and tibia; homologous to patella of insects.

Gingival Of or relating to the gums, the firm fleshy tissue overlying the jaw.

Glanders A disease of horses caused by the bacterium *Pseudomonas mallei* and communicable to humans; characterized by swollen lymph glands beneath the jaw and profuse mucous discharge from the nostrils.

Glaucoma An eye disorder characterized by an increase in intraocular fluid pressure that can lead to atrophy of the optic nerve and blindness.

Glycoprotein A macromolecule made up of protein(s) bonded to one or more carbohydrates.

Gnat A nontaxonomic term referring in general to small pesky flies that may or may not bite.

Gnathosoma The anterior-most body region of mites and ticks composed of the mouthparts and palps and their fused bases; also called captitulum.

Gonad A sex gland in animals that produces gametes; for example, ovary and testis.

Gonoactive A blood-fed female arthropod that is progressing through a gonotrophic cycle.

Gonocoxite Basal segment of the external male or female genital appendages on segment 8 or 9 in insects, surrounding the genital opening.

Gonoinactive A blood-fed female arthropod that is not progressing through a gonotrophic cycle.

Gonopod In arthropods, an appendage modified for a reproductive function, such as copulation, intromission, or oviposition; usually associated with genital segments 8 and 9.

Gonopore See *Genital pore.*

Gonostylus (-i) A process on abdominal segment 8 or 9 of insects, in males generally modified to form a clasping organ.

Gonotrophic concordance A situation where a blood-fed female vector follows a predictable gonotrophic cycle.

Gonotrophic cycle The reproductive cycle of blood-feeding, bloodmeal digestion, egg maturation, and oviposition in a vector; most commonly pertaining to mosquitoes and some other dipterans.

Gonotrophic discordance A situation where a blood-fed female vector does not follow a predictable gonotrophic cycle.

Gonotrophic dissociation A situation where a blood-fed female vector does not develop eggs after the blood meal.

Grand mal seizure A sudden attack or convulsion characterized by generalized muscle spasms and loss of consciousness; an epileptic attack.

Granulation tissue Small protuberances that form on the surface of a wound during the healing process; consists of outgrowths of new connective tissue and capillaries.

Granulocyte White blood cells containing granules in the cytoplasm; important in phagocytosis and immunological responses; most numerous of the white cells in humans.

Granuloma (-tous) A mass of granulation tissue resulting from inflammation, infection, or injury; usually associated with skin or lymphoid tissues.

Gravid Carrying eggs or developing young.

Gregarine Sporozoans in the order Gregarinida that are parasitic in the digestive tracts of various invertebrates, including insects, other arthropods, and annelids; produce cysts filled with spores.

Gregarious Tending to form a group with individuals of the same species.

Grub A thick-bodied, typically whitish insect larva with thoracic legs but lacking abdominal prolegs; relatively sluggish or inactive; a term loosely applied to larvae of Coleoptera, Hymenoptera, and certain Diptera (e.g., Oestridae).

Guanine A major excretory product of arachnids; also, one of the constituent purine bases that codes genetic information in DNA and RNA; pairs with cytosine.

Guenon Any of various slender, arboreal African monkeys with long hind legs, long tail, and long hair around the face; members of the genus *Cercopithecus.*

Guinea fowl An African gallinaceous bird, *Numida meleagris*, introduced to many other parts of the world.

Gullet Esophagus.

Gut barrier (midgut barrier) Something in the gut of a particular blood-feeding arthropod that prevents it from becoming infected with a particular pathogen through that inoculation route.

Gut bolus A soft mass of ingested food in the alimentary canal; in vertebrates, usually referring to the bowel or intestines; in invertebrates, may refer to the midgut or hindgut.

Habronemiasis Infestation of horses by parasitic nematodes of the genus *Habronema* (family Spiruridae), which develop in flies of the genera *Musca* and *Stomoxys*; if ingested, can cause inflammation of the stomach; can also cause ulcerated cutaneous lesions, or summer sores.

Haemocoel See *Hemocoel*.

Haemogregarine A member of the sporozoan genus *Haemogregarina* (order Coccidia) that is parasitic in the alimentary tract of invertebrates and the circulatory system of vertebrates; also hemogregarine.

Haemoproteid A sporozoan of the family Haemoproteidae parasitic in birds; for example, *Haemoproteus*.

Haemosporidian A member of the protozoan order Haemosporidia, parasitic at some stage of development in blood cells of vertebrates; includes the families Babesiidae, Haemoproteidae, and Plasmodiidae.

Hair follicle A tubular invagination of the skin from which a hair develops.

Haller's organ A complex sensory apparatus on the dorsal surface of the tarsus of leg 1 in all developmental stages of ticks; includes receptors for taste, odor, temperature, and mechanical stimuli.

Haltere A drumstick-shaped sensory projection homologous to the hindwing of members of the Diptera (and forewing of members of the Strepsiptera); secondarily lost in the sheep ked, *Melophagus ovinus*.

Hammock An elevated tract of forest land surrounded by wetlands, usually applied to areas of freshwater marsh, such as the Everglades, in the southeastern United States.

Hanging groin Folds of atrophic, inelastic skin in the inguinal area (e.g., cases of lymphatic filariasis) typically involving enlarged lymph nodes and pendent tissues with accumulated lymph.

Haploid An organism or cell having only one set of chromosomes, half the number found in somatic cells.

Harborage A sheltered or protected place; used regarding cockroaches, rodents, and other household pests to indicate where they live and hide.

Haustellum (-late) The anterior part of the insect head or basal portion of the mouthparts modified for sucking; for example, the conical projection of sucking lice, and the structures of certain flies with sponging or piercing mouthparts, such as tsetse flies and house flies.

Head capsule The fused sclerites of the insect head that together form a hard exoskeletal case.

Hedgehog Small insectivorous mammals of the genus *Erinaceus*, family Erinaceidae, with erect spiny hairs and the ability to roll into a ball for protection; native to Africa, Asia, and Europe.

Helminth Parasitic worms, usually referring to those living in the intestines of vertebrate animals; includes roundworms, tapeworms, and flukes.

Hematocrit The volume of red blood cells in a sample of blood after centrifugation, expressed as a percentage of the total blood in the sample.

Hematologic Relating to the medical study of blood and blood-forming organs.

Hematoma A localized swelling filled with blood resulting from a break in a blood vessel.

Hematophage (-ous, -y) An organism that feeds on blood; also hemaphage.

Hematophagia (haematophagia) See *Hematophagy*.

Hematopoiesis (-tic) The formation and development of blood cells; *syn.* hemapoiesis.

Heme A deep red, iron-containing, nonprotein component of hemoglobin in vertebrate blood that binds with oxygen.

Hemelytron (-a) The forewing of Hemiptera, in which the basal half is thickened and the distal half is membranous; also hemi-elytron.

Hemelytral pad One of a pair of dorsal case-like projections on the mesothorax of late-instar nymphs of Hemiptera, which give rise to the forewings, or hemelytra, of the adult; also the reduced forewings of adult bed bugs (Cimicidae).

Hemidesmosome A specialized junction that connects the basal surface of epithelial cells to the underlying basement membrane.

Hemimetabolous Pertaining to metamorphosis in insects represented by three developmental stages: egg, nymph (larva), and adult; for example, Blattaria, Phthiraptera, and Hemiptera.

Hemocoel The body cavity of insects and other arthropods filled with hemolymph.

Hemocyte A blood cell.

Hemoglobinuria Presence of hemoglobin in urine.

Hemolymph Circulatory fluid filling the body cavity of arthropods and other invertebrates, analogous to blood and lymph of vertebrates; consists of water, inorganic salts, lipids, amino acids, and sugars.

Hemolysis Destruction of red blood cells releasing hemoglobin.

Hemolysin A substance or agent that causes hemolysis; for example, an antibody or bacterial toxin.

Hemopoiesis See *Hematopoiesis.*

Hemorrhage (-ic) Excessive loss of blood from blood vessels; profuse bleeding.

Hemosporine A member of the order Haemosporida (Haemosporidia); includes the genera *Haemoproteus, Leucocytozoon,* and *Plasmodium.*

Hemostatic Arresting loss of blood due to hemorrhaging.

Hemotoxic A substance that destroys red blood cells.

Hemotropic An entity (often a pathogen or parasite) that attracts phagocytic cells in the blood; *syn.* hematotropic.

Heparin A natural substance found particularly in lung and liver tissues that prevents clotting of blood; also the commercial form.

Hepatitis Inflammation of the liver, usually caused by a virus or toxin.

Hepatopancreas An organ associated with the digestive tract of arthropods, crustaceans, mollusks, and other invertebrates that serves functions similar to the liver and pancreas of mammals.

Hepatosplenomegaly Enlargement of the liver and spleen.

Hibernaculum (-a) A protective site in which an organism overwinters.

Hippoboscid A member of the dipterous family Hippoboscidae; louseflies and keds.

Histamine A compound released primarily by mast cells in allergic reactions that causes dilation and permeability of capillaries, decreased blood pressure, and constriction of bronchial muscles; also the commercial form.

Histiocyte A macrophage cell found in connective tissue.

Histoblast A cell, or group of cells, capable of forming tissues; in dipteran larvae, epithelial cells that give rise to structures other than appendages; for example, respiratory histoblasts of black fly larvae.

Histolysis The breakdown and disintegration of tissues.

HIV Human immunodeficiency virus, which causes acquired immune deficiency syndrome (AIDS) by infecting and killing helper T cells.

Hives An eruption of itching wheals, usually of systemic origin; may be due to hypersensitivity to certain foods, drugs, pathogens, or parasites.

Hock A joint in the hind leg of pigs, cows, and horses; above the fetlock and corresponding to the ankle of humans, although it bends in the opposite direction.

Holarctic Region A zoological area in the Northern Hemisphere encompassing the nontropical parts of Europe and Asia, Africa north of the Sahara, and North America south to the deserts of Mexico; includes the Nearctic and Palearctic regions.

Holocyclotoxin A neurotoxin in the saliva of the Australian tick *Ixodes holocyclus* that causes tick paralysis in cattle and other livestock, humans, dogs, and cats; inhibits release of acetylcholine.

Holometabolous Pertaining to metamorphosis in insects represented by four developmental stages: egg, larva, pupa (chrysalis), and adult; for example, orders Diptera, Siphonaptera, Hymenoptera, Coleoptera, and Lepidoptera.

Holoptic Refers to adult flies in which the pair of eyes are enlarged enough to meet along the dorsal midline of the head; for example, male horse flies; *cf. dichoptic.*

Homeostatic The physiological ability of an organism to maintain internal equilibrium.

Homeothermic Refers to an animal, typically a bird or mammal, capable of maintaining a constant body temperature largely independent of the ambient temperature; endothermic.

Honeydew A sugar-rich secretion deposited on vegetation by aphids, scale insects, and other plant-feeding insects; serves as a natural source of sugar for certain hematophagous flies.

Horizontal transmission The transfer of an infectious agent from one individual to another other than from parent to offspring.

Host preference The species or range of species to which an ectoparasitic or other blood-feeding arthropod, given a choice, is typically attracted and on which it feeds.

Host specificity The degree of selectivity exhibited by ectoparasitic or other blood-feeding arthropods in choosing a host.

Humoral immunity The aspect of immune systems mediated by antibodies produced by lymphocytes (B cells) in bone marrow.

Husbandry The application of scientific principles to agricultural practices and farming, particularly livestock and animal breeding.

Hyaline Glassy or transparent.

Hyaluronidase An enzyme that plays a role in breaking down hyaluronic acid, increasing the permeability of tissues to fluids; also called spreading factor.

Hydramethylnon A chemical compound that acts as a metabolic inhibitor, used as an insecticidal bait for cockroaches and ants.

Hydrocele Accumulation of serous fluid in a body cavity; for example, in scrotum.

Hydroprene An insect growth regulator, used to control cockroaches.

Hyperemia Increased blood flow to an organ or other body part.

Hyperendemic Occurrence of a pathogen or parasite at very high incidence in a host population.

Hypergammaglobulinemia An excess of gamma globulins in the blood, often associated with chronic infectious diseases.

Hyperimmune (-ity) Exhibiting an unusually high degree of immunity in which the body is extremely reactive against a particular antigen.

Hyperkeratosis Proliferation of cells of the cornea, or thickening of the horny layer of the skin.

Hypermetamorphosis (-phic) A type of holometabolous development in insects in which the larval stage is represented by two or more different larval types; for example, blister beetles.

Hyperpigmentation Excess pigmentation, or darkening, usually of the skin.

Hypersensitive (-ity) Responding excessively to an allergen or other foreign agent; abnormally sensitive or allergic.

Hypertension Abnormally elevated blood pressure; arterial disease characterized by chronic high blood pressure.

Hypertrophy An abnormal enlargement of an organ or tissues due to an increase in size of the cells but not their numbers.

Hypnozoite A latent or dormant stage of sporozoan parasites; for example, *Plasmodium* species in liver cells, which contribute to relapses in malaria cases.

Hypoendemic Occurrence of a very low incidence of a pathogen in a host population, with little transmission.

Hypognathous In insects, the mouthparts directed downward or ventrally.

Hypopharynx In insects, a median sensory structure anterior to the labium, usually associated with the salivary ducts; in certain sucking insects, an elongate mouthpart containing the salivary channel.

Hypopleuron The lower part of the external mesothoracic wall immediately above the middle and hind coxae.

Hypopode (-es) See *Hypopus.*

Hypopus (-i) A highly modified, typically nonfeeding form of the deutonymph in astigmatid mites, adapted for dispersal and enduring adverse environmental conditions.

Hypopygium A modification of the ventral aspect of the last abdominal segment(s) in certain insects, notably Diptera and Coleoptera.

Hyposensitization The process of reducing a person's sensitivity to an allergen or other stimulus, usually by injecting progressively larger doses of the allergen involved; desensitization.

Hypostome The median ventral part of the insect head posterior to the mandibles; in mites and ticks, the ventral, basal part of the gnathosoma, greatly enlarged and modified in ticks for host attachment.

Hypotension Low blood pressure.

Hypothermia Abnormally low body temperature.

Hysterosoma In mites, that portion of the body posterior to the second pair of legs.

Ibex Wild goats of the genus *Capra,* with long, ridged, recurved horns; native to the mountainous regions of Eurasia and northern Africa.

Idiosoma The major body region of mites and ticks, excluding the gnathosoma.

IGR See *Insect growth regulator.*

Imidacloprid A chlorinated analog of nicotine used as an insecticide.

Immediate reaction (type I hypersensitivity) Rapid (within minutes) inflammation of the skin, known as wheal and flare, in response to bites by certain arthropods.

Immunity Ability of an organism to resist disease by destroying or inactivating infectious agents or other foreign substances.

Immunocompromised Unable to develop a normal immune response, usually because of malnutrition, disease, or immunosuppressive therapy.

Immunoglobulin (Ig) A large glycoprotein produced by plasma cells in bone marrow and loose connective tissues of vertebrates that function as antibodies in an immune response; five classes: IgA, IgD, IgE, IgG, and IgM.

Immunoglobulin A (IgA) A class of immunoglobulins comprising 10 to 15% of total immunoglobulins; often transferred transplacentally from mother to fetus.

Immunoglobulin D (IgD) A class of immunoglobulins representing less than 0.1% of total immunoglobulins.

Immunoglobulin E (IgE) A class of immunoglobulins produced in the skin, mucous membranes, and lungs that function particularly in allergic reactions, comprising less than 0.01% of total immunoglobulins.

Immunoglobulin G (IgG) A class of immunoglobulins representing the most common antibodies circulating in the blood and lymph (ca 80% of total immunoglobulins); active against invading microorganisms and other foreign agents.

Immunoglobulin M (IgM) A class of immunoglobulins that includes antibodies released into the blood early in the immune response (typically can be detected within three months of an infection), with particular affinity for viruses, comprising 5 to 10% of total immunoglobulins.

Immunomodulator A chemical agent that alters the immune response or functioning of the immune system; may strengthen or suppress the response; for example, stimulation of antibody formation.

Immunosuppression Lowering of the body's normal immune response due to disease, drugs, radiation, or other conditions; for example, HIV infection or side effects of chemotherapy and radiotherapy.

Immunotherapy Treatment to produce immunity to a disease by inducing, suppressing, or enhancing an immune response.

Impetigo Contagious inflammatory skin disease characterized by pustular eruptions and yellow crusts, commonly on the face; caused by staphylococcal and streptococcal bacteria, especially in children.

in copula Linking of a male and female in the act of pairing or mating; commonly used when referring to insects.

Inapparent infection Presence of infection without symptoms; asymptomatic or subclinical infections.

Inappetence Lack of appetite.

Incidence Frequency or extent of an occurrence; for example, infection, disease *cf. prevalence.*

Incidental host An unpredictable and very minor host of a parasite or pathogen.

Incompetent Lacking the ability to play a significant role; for example, incompetent vector, an arthropod that is not susceptible to, or lacks the ability to transmit, a given pathogen or parasite.

Incubation period See *extrinsic* and *intrinsic incubation period.*

Indigenous Native to and occurring naturally in a given area or environment.

Indigenous transmission Transfer of a pathogen or parasite from one host to another in its native country or region.

Indolent ulcer Inactive or painless lesion of the skin or mucous membranes caused by superficial loss of tissue, usually with associated necrosis and inflammation.

Indurate (-ed) (-tion) To become firm or hardened.

Infect (-ed) (-tion) To live in or on a host by a pathogenic microorganism or agent; *cf.* infest.

Infection rate The proportion (sometimes expressed as a percentage) of a population or specified group of individuals infected with a pathogenic microorganism or agent.

Infectious disease Any disease caused by growth and multiplication of pathogenic microorganisms in the body; may or may not be contagious.

Infectious dose A specific quantity of a pathogenic microorganism required to establish infection.

Infective Capable of producing infection; infectious.

Infective stage The developmental form of a pathogen or parasite that invades a vertebrate host; in a strict sense, a misnomer when applied to organisms other than microbes.

Infest (-ed) (-ation) To live as a macroscopic parasite in or on a host, usually implying high enough numbers to be harmful; to parasitize; *cf.* infect.

Inflammation Reaction of tissue to injury, characterized by redness, swelling, tenderness, and pain.

Inguinal Pertaining to, or located in, the groin.

Inoculate (-ion) To inject, or otherwise introduce, a microorganism or other disease agent into a host or culture medium; also, to introduce an antigenic substance, serum, or vaccine into an animal to boost immunity to a specific disease agent.

Insect growth regulator (IGR) Any chemical compound or other substance that modifies, disrupts, or otherwise interferes with normal body development and metamorphosis in insects.

Insectivore An animal or plant that feeds on insects; usually refers to members of the mammalian order Insectivora that include moles, shrews, and hedgehogs.

Installment hatching Hatching of groups of insect eggs within the same batch at different time periods, that is, not simultaneously; often applied to mosquito eggs.

Instar The form of an insect or other arthropod between two successive molts.

Integrated pest management (IPM) The application of a variety of different control techniques against a pest species; for vectors, this may involve, for example, the use of chemical insecticides, insect growth regulators, application of fungal pathogens and/or parasitoids, and management of potential harborage sites.

Integument The outer body surface of an organism, typically applied to arthropods.

Interdigital space The area between fingers and toes, and between the digits of hoofed animals.

Intermediate host An animal in which a multi-host parasite undergoes development but does not become sexually mature.

Interrupted feeding The behavior of arthropods, typically hematophagous insects, in which feeding is disrupted and usually results in subsequent efforts to feed again.

Interstice (-es) A small or narrow space in substrates composed of closely spaced particles such as soil, rocks, sand, and dried mud.

Intranasal Within the nose, or administered via the nose.

Intravascular coagulation Clotting of blood within blood vessels.

Intrinsic incubation period The time period between infection or initial parasitism of a vertebrate host and the onset of symptoms.

Intromittent organ An external, or eversible, genital structure used in copulation; typically refers to the male insect penis or aedeagus (phallosome of some authors).

Inundative release Purposeful release of large numbers of a commercially produced biological control agent into the environment for the purpose of reducing or eliminating a target species; for example, parasitic wasps to control muscoid flies.

Iritis Inflammation of the iris of the eye.

Isoenzyme Chemically distinct but functionally similar enzymes.

Isolate (-tion) In microbiology, to obtain an organism from a sample or the environment in pure culture.

Isopod Member of the crustacean order Isopoda, including pill bugs and sowbugs.

Ivermectin A semisynthetic product derived from the fungus *Streptomyces avermitilis* used as an anthelmintic, insecticide, and acaricide.

Ixodid Member of the tick family Ixodidae (hard ticks).

Jaundice Yellowish discoloration of the skin, whites of the eyes, and mucous membranes caused by abnormally high levels of bile pigments in the blood.

Johnston's organ An auditory structure in the second antennal segment (pedicel) of most adult insects; perceives movement of the antennal flagellum.

Juvenile hormone A hormone in arthropods that promotes larval development and inhibits molting to the adult stage.

Juvenile hormone mimic A chemical compound that simulates the effects of juvenile hormone; can be used to disrupt normal development of immature insects to the adult stage; juvenile hormone analog.

Karyotype A visual array of chromosomes of an organism, typically arranged by size, shape, number, and other characteristics; to classify the chromosomes or prepare a karyotype.

Keratin A tough structural protein found in hair, nails, claws, horns, hooves feathers, and dead outer layers of skin.

Keratitis Inflammation of the cornea; causes watery and painful eyes, blurred vision.

Keratoconjunctivitis Ocular inflammation of the cornea and conjunctiva.

Kinase An enzyme that catalyzes conversion of a proenzyme to an active enzyme; also, an enzyme that catalyzes transfer of a phosphate group from a high-energy phosphate-containing molecule (e.g., ATP, ADP) to a substrate.

Kinetoplast A mass of circular DNA within a large mitochondrion near the base of the flagellum of certain protozoans, for example, trypanosomes.

Kinin A polypeptide that causes contraction of smooth muscle, vasodilation, and altered permeability of capillaries.

K-strategy (-ist) The production of relatively small, constant numbers of offspring in a stable or predictable environment, maintaining a population near the carrying capacity, K; *cf. s-strategy*.

Labellum (-a) Sensory structure at the tip of the labium in mosquitoes, muscoid flies, and some other dipterans; possesses temperature, contact, and chemical receptors.

Labium Posterior-most or ventral-most mouthpart in insects, depending on the orientation of the mouthparts.

Labial palp One of a pair of typically multisegmented appendages of the labium; labial palpus (pl., palps or palpi).

Labral fan A modification of the labrum forming a brush-like structure that can be extended like a fan to trap or filter particles of food from water; for example, black-fly larvae.

Labrum Anterior-most or dorsal-most mouthpart in insects, depending on the orientation of the mouthparts.

Lacerate (-tion) To cut, tear, or rip, leaving irregular or jagged edges.

Lachrymal Of or relating to tears or tear glands; lacrimal.

Lachryphagy (-ous) Feeding on tears from the eyes of vertebrates.

Lacinia (-ae) The inner or medial lobe of the maxilla of insects.

Lacrimation Excessive secretion of tears; lachrymation.

Lactation Secretion or formation of milk by the mammary glands.

Lactic acid A carboxylic acid produced during anaerobic metabolism of glucose, as in muscle tissue during exercise.

Lactophenol A mixture of lactic acid and phenol used to clear small arthropods, especially mites, prior to slide-mounting.

Lacustrine Of or relating to lakes.

Lagomorph A member of the mammalian order Lagomorpha, including rabbits, hares, and pikas.

Lamella (-ae, ate) A thin plate-like or leaf-like structure; as in lamellate antennae of scarab beetles.

Lancet In insects, the first of three blade-like processes that surround the ovipositor.

Lappet A small flap or projecting, lobe-like structure; characteristic of larvae of lappet moths (Lasiocampidae).

Larva (-ae) The immature stage of insects that hatches from the egg and undergoes metamorphosis to the adult; also called nymphs or naiads in certain types of metamorphosis; in mites and ticks, the six-legged stage that hatches from the egg.

Larvicide An insecticide that kills larvae.

Larviparous Bearing or depositing living larvae, rather than eggs.

Larviposit To deposit living larvae, rather than eggs.

Laryngeal Of or relating to the larynx.

Lassitude A feeling or state of weariness, lack of energy; listlessness, languor.

Laterigrade Having a sideways manner of moving, as a crab; used to refer to arthropods that are dorsoventrally flattened with the legs extending laterally in a horizontal plane and move in crab-like fashion, for example, keds.

Latrine A communal toilet, usually in a military area.

LD_{50} The dose of a substance that kills 50% of the treated or targeted organism.

Lechwes An African antelope (*Kobus leche*) that inhabits marshes and wet, grassy plains (pronounced lēchwē).

Leishmaniasis An infection caused by a flagellate protozoan in the genus *Leishmania*.

Lentic Relating to or living in still water, for example, ponds and lakes.

Lepidopterism An affliction caused by direct or indirect contact with hairs, setae, or wing scales of adult moths and butterflies; as distinguished from similar contact with urticating hairs of caterpillars, or erucism.

Lesion Any abnormal structural change in body tissue, usually caused by trauma or infection.

Lethargy (-ic) A state of sluggish inactivity, listlessness, drowsiness, and apathy.

Leucocytozoonosis A disease of birds caused by infection with a protozoan of the genus *Leucocytozoon*.

Leukopenia An abnormally low number of white blood cells circulating in the blood; leucopenia.

Leukemia Cancer of the bone marrow that prevents normal production of red and white blood cells and platelets; results in a proliferation of certain kinds of leukocytes.

Leukocyte A white blood cell.

Leukosis An abnormal proliferation of leukocyte-forming tissues.

Ligula The terminal lobe(s) of the labium in insects.

Lindane An organochlorine insecticide, gamma benzene hexachloride; used to kill lice and scabies mites.

Listeriosis A bacterial disease of domestic and wild animals and sometimes humans, caused by *Listeria monocytogenes*; characterized by fever, meningitis, and encephalitis.

Litter In livestock or poultry operations, material used as bedding or to absorb animals wastes, reduce odors, and facilitate clean-out; for example, sawdust, wood shavings, straw.

Loiasis An infestation with the filarial nematode *Loa loa*, or African eyeworm.

Loin The area in livestock (or humans) situated ventrally on each side of the hipbone and the false ribs.

Loris A common name for Old World primates belonging to the subfamily Lorinae of the family Lorisidae; includes representatives of the genera *Loris* (the slender lorises) and *Nycticebus* (the slow lorises).

Lotic Relating to or living in running water; for example, streams and rivers.

Lufenuron A benzoylurea compound used in veterinary products to control fleas and filarial nematodes; inhibits chitin production in flea larvae.

Lumen The space within a tubular structure or organ, such as a blood vessel or alimentary tract.

Lunule A small crescent-shaped structure between the ptilinal suture and antennal socket; the frontal suture in certain adult flies (Schizophora) (see *facial lunule*).

Lymphadenopathy Pathology of lymph nodes, usually manifesting as chronic node enlargement, often associated with disease.

Lymphedema Swelling, particularly in subcutaneous tissues of the extremities, due to accumulation of lymph resulting from obstruction of lymphatic vessels and lymph nodes.

Lymphocytic choriomeningitis An acute disease caused by an arenavirus and transmitted by rodents; characterized by excessive lymphocytes in cerebrospinal fluid.

Lymphokine A cytokine secreted by helper T cells following stimulation by specific antigens; mediate the immune response by acting on other cells, for example, activating macrophages.

Lysis The dissolution or destruction of cells.

Macaque Short-tailed monkeys of the genus *Macaca*, native primarily to southeast Asia and northern Africa.

Macrogametocyte A female gametocyte that produces a macrogamete.

Macrophage Large cells of the reticuloendothelial system that remove cellular debris and particulate substances, including microorganisms, by phagocytosis.

Macropterous Bearing large, well-developed wings in insects.

Macrotrichium (-ia) Relatively large microscopic hairs; for example, on the wing surface of flies.

Macule (1) A discolored spot or patch of skin not usually elevated, caused by various disease agents; (2) a spot or patch on insect integument, especially on the wings.

Maculopapular Characterized by a skin eruption with both macules and papules.

Maggot Legless, soft-bodied, vermiform fly larva, usually found in decaying matter; typified by larvae of houseflies and blowflies.

Malady Any disease, disorder, or body ailment.

Malaise A vague feeling of physical discomfort, weakness, or uneasiness, often characterizing onset of an illness or disease.

Malathion A dithiophosphorus insecticidal hydrocarbon.

Malignant Dangerous to health, tending toward a progressive, life-threatening condition.

Malodorous Having an unpleasant, offensive odor; foul smelling.

Malpighian tubule Long, slender excretory structures in insects, arachnids, and other terrestrial arthropods, opening into the anterior end of the hindgut.

Mammalophagic Feeding on mammals.

Mammalophily (-ic) Tendency of attraction to mammals by host-seeking arthropods.

Mammilla (-ae) A nipple-like process or protuberance.

Mandible One of a pair of unsegmented mouthparts of insects located between the labrum and maxillae; usually heavily sclerotized for chewing, but highly modified in some insects for piercing-sucking.

Mandibular gland Pheromone-producing gland that opens on the surface of the mandible, found in most Hymenoptera.

Mandibular stylets Insect mandibles that are highly modified as long, slender structures for piercing and sucking.

Mangabey Slender, long-tailed monkeys of the genus *Cercocebus* in forests of central Africa.

Mange A persistent skin condition of mammals caused by parasitic mites; characterized by redness, itching, and hair loss.

Mangrove A tropical or subtropical tree or shrub mostly of the genus *Rhizophora*, with stilt-like prop roots that form dense thickets along shallow tidal areas.

Mansonellosis An infestation of humans by filarial nematodes of the genus *Mansonella*, usually used with reference to *M. ozzardi*.

Manure A mixture of animal excrement and bedding or litter, such as hay or straw; used to fertilize land.

Mark–recapture Technique of marking animals by various methods so that they can be recognized when caught again on subsequent occasions; also called mark–release–recapture.

Marsupial A nonplacental mammal of the order Marsupialia, with a pouch and mammary glands for nurturing the young; includes kangaroos, wombats, bandicoots, and opossums.

Masarine Referring to members of the vespid wasp subfamily Masarinae, found in the western United States.

Mast cell A large granular cell found particularly in connective tissue that releases heparin, histamine, and serotonin in response to allergens, inflammation, or injury.

Mastitis Inflammation of the breast or udder.

Mating plug A physical blockage in the reproductive system of female animals, including some arthropods, caused by a seminal mass introduced by a male during copulation.

Matrone In mosquitoes, a substance in male accessory fluid introduced during copulation that causes the female to become unreceptive to other males.

Mausoleum An above-ground burial chamber or building containing tombs.

Maxilla (-ae) One of the second pair of mouthparts in insects, immediately behind the mandibles.

Maxillary palp Typically one of a pair of segmented, sensory appendages on the maxilla of insects (pl., palps or palpi).

Maxillary sinus One of a pair of air cavities in the upper jaw of vertebrates that opens into the middle passage of the nose.

Maxillary stylets Long slender mouthparts formed by the maxillae; usually associated with piercing-sucking insects.

Mebendazole An anthelmintic compound (methyl-5-benzoyl-2-benzimidazolecarbamate) used to treat roundworm infestations; interferes with carbohydrate metabolism.

Mechanical transmission The transfer of a pathogen or parasite via the external surface of the mouthparts, appendages, or other body parts, without involving biological development (cyclodevelopment) of the organism.

Mechanical vector An arthropod that transmits a pathogen or parasite as a contaminant on the external surface of a body part (see *mechanical transmission*).

Mechanoreceptor A sensory structure that responds to mechanical stimuli such as touch, pressure, tension, stretching, sound. and other vibration.

Mecopteroid A member of the insect superorder Mecopteroidea, which includes scorpion flies (Mecoptera), fleas (Siphonaptera), true flies (Diptera), caddis flies (Trichoptera) and butterflies and moths (Lepidoptera); members of these orders are all endopterygotes that have hypognathous mouthparts.

Mectizan A brand name of ivermectin.

Media (n) Fourth longitudinal vein in the basic insect wing, counting from the anterior margin.

Medicocriminal Pertaining to medically criminal practices, including aspects of forensic entomology.

Medicolegal Pertaining to legal issues in medicine, including aspects of forensic entomology.

Megacolon Abnormal enlargement of the colon, with extreme dilation and hypertrophy.

Megaesophagus Abnormal enlargement and hypertrophy of the lower portion of the esophagus.

Megasyndrome A condition in which various organs in the abdomen are enlarged.

Meibomian gland A sebaceous gland in the eyelid of vertebrates; produces lubricant to prevent eyelids from sticking together; also called tarsal gland (tarsus being the supporting plate of the eyelid, not to be confused with insect tarsus).

Melioidosis An acute infectious bacterial disease caused by *Burkholderia pseudomallei*, primarily affecting rodents in India and Southeast Asia; can be transmitted to humans causing pneumonia, multiple abscesses, and bacteremia.

Melittin A polypeptide and major active component of honey-bee venom that causes localized pain and inflammation; also has antifungal, antibacterial, and anti-inflammatory properties.

Meninges The three membranes that envelop the brain and spinal cord; singular meninx.

Meningitis Inflammation of the membranes surrounding the brain or spinal cord.

Meningoencephalitis Inflammation of the brain and its meninges.

Mentum The lower (distal) part of the insect labrum, usually bearing palps.

Mercaptan A class of sulfur-containing compounds with distinctive and often offensive garlic-like odors; also called thiol.

Mermithid A nematode in the family Mermithidae; parasites of insects and other invertebrates.

Merogony A form of asexual schizogony characteristic of sporozoans; involves division of the nucleus several times before the cytoplasm divides, forming a schizont that further divides to produce merozoites.

Meron The base of the coxa in insects located lateral to, and just posterior to, the point of articulation of the coxa and thorax; may be greatly enlarged in some insects.

Meront A stage in the asexual part of the sporozoan life cycle in which schizogony occurs, producing merozoites; occurs in the vertebrate host.

Merozoite A trophozoite, or vegetative form, in the asexual part of the sporozoan life cycle produced by a mature meront; occurs in the vertebrate host.

Mesad Toward the median plane of the body or body part.

Mesal In a middle line or plane; also mesial.

Mesenteric Referring to the mesentery, or peritoneal tissue that surrounds most of the small intestine and connects it to the abdominal wall.

Mesoendemic Referring to a geographic area in which only modest transmission of a disease agent takes place.

Mesonotal suture A groove along the dorsal surface of the second (middle) thoracic segment of some insects.

Mesonotum The dorsal part of the second (middle) thoracic segment of insects.

Mesostigmatid Referring to mites of the suborder Mesostigmata; Gamasida of some authors.

Mesothorax The second thoracic segment of insects, bearing the middle pair of legs and first pair of wings.

Metabolic inhibitor A substance or agent that slows or stops chemical reactions involved in an organism's metabolism.

Metacyclic In trypanosomes, referring to the developmental form of the protozoan produced in the arthropod that is the infective stage for the vertebrate host.

Metamorphosis A change in form during development of an organism.

Metanotum The dorsal part of the third (most posterior) thoracic segment of insects.

Metastriate Pertaining to hard ticks (family Ixodidae) in which the anal groove extends posterior to the anus; includes all ixodid genera except *Ixodes.*

Metatarsus The tarsus, or last segment, of the second leg in arthropods.

Metathorax The third (most posterior) thoracic segment of insects, bearing the hind legs and hind wings.

Metazoa (-an) A subdivision of the animal kingdom that includes all multicellular animals, with cells differentiated to form tissues and organs.

Methoprene A synthetic insect juvenile hormone used as a pesticide to disrupt normal larval development as a means of control.

Methoxychlor A synthetic organochlorine insecticide that acts both as a neurotoxin and an endocrine disruptor.

Microfilaremia (-ic) The presence of microfilariae in the blood of a vertebrate host.

Microfilaria (-ae) The minute embryonic larva of filarial nematodes produced in the arthropod host and serving as the infective stage of the vertebrate host.

Microgamete The smaller of the two mature sexual cells, usually the male, that unite to produce a zygote in organisms that produce two types of gametes.

Microgametocyte A gametocyte that gives rise to microgametes.

Micropyle In some animals, a minute opening in the membrane covering the ovum through which a spermatozoon can enter.

Microsporidia (-an) Parasitic unicellular fungi, formerly considered to be protozoans, that infect insects, crustaceans, fish, and humans; replicate in host cells by spores.

Microthrombus (-i) A tiny blood clot that can obstruct capillaries and impede blood flow.

Microtine Referring to members of the rodent subfamily Arvicolinae, family Arvicolidae, that includes voles, lemmings, and muskrats; this subfamily sometimes called Microtinae.

Microtrichium (-ia) A minute hair-like structure, typically found on the wings of Diptera.

Microvillus (-i) A microscopic fingerlike or hairlike projection on the surface of an epithelial cell.

Miliary The presence of skin nodules that resemble millet seeds.

Miltogrammine Representatives of the subfamily Miltogramminae within the dipteran family Sarcophagidae.

Minimum infection rate (MIR) A relative measure of infection prevalence based on pools of vectors; number of positive pools/total specimens tested/unit of time × 100 or 1000.

Mode of action How a particular drug or compound, such as an insecticide, works; the specific biochemical interactions that produce the resultant effect; also called mechanism of action.

Molossid Members of the bat family Molossidae, called free-tailed bats and mastiffs.

Molt The act or process of forming a new integument and shedding the old one; in arthropods, shedding the old cuticle; in vertebrates, shedding hair, skin, horn, or feathers; also used as a verb (see *moult* and *ecdysis*).

Moult Verb, to molt; noun, a British variant of molt.

Monocyte A large, circulating white blood cell, produced in bone marrow and the spleen, that engulfs foreign particles and cell debris.

Monograph A treatise or detailed scholarly document on a particular subject, usually in the form of a book.

Monogyny (-ous) In social insects, having only one functioning queen in a colony.

Monophyletic Relating to a taxonomic group of organisms all descended from a single common ancestor.

Monospecific A taxonomic group, such as a family or genus, represented by a single species; in immunology, meaning specific for a single antigen or receptor site on an antigen.

Monotypic Having only one representative; for example, a monotypic genus with only a single species.

Morbidity A state of being diseased; the incidence of a disease in a specific population or geographical area.

Moribund In a dying state; near death.

Morphogenesis (-tic) The formation or development of structural features of an organism, including tissues and organs.

Mortuary A place for keeping dead bodies temporarily prior to burial or cremation; for example, funeral home, morgue.

Morula (-ae) A mass of cells, resembling a bunch of mulberries, resulting from cleavage of a zygote or ovum; characteristic of ehrlichiae inside host cells.

Mosquito coil A flat, spiral device impregnated with an insecticide, usually pyrethrum or a synthetic pyrethroid, which when burned helps to repel or kill mosquitoes and other small flying insects.

Mouth hook One of a pair of claw-like or hook-like sclerites near the oral opening of muscoid fly larvae, which serve the function of mandibles.

Mucocutaneous Of or relating to the skin and mucous membranes.

Mucopurulent Containing mucus and pus.

Multifocal Relating to or arising in many locations.

Multimammate mouse Any of several African species of rodents in the genera *Praomys* or *Mastomys*, family Muridae, so-called because the females possess an unusually large number of teats; also called multimammate rat; *Mastomys natalensis* is often given the name multimammate rat.

Multiplicative transmission (propagative) Transmission of a pathogen or parasite by a vector after asexual reproduction within the vector; the form of the pathogen or parasite transmitted is indistinguishable from the form that was initially ingested by the vector.

Multivoltine Producing three or more broods or generations per year *cf. univoltine, bivoltine.*

Mummification The formation of a desiccated, leathery cadaver (or carrion).

Murine Relating to rodents of the subfamily Murinae of the family Muridae, including Old World and peridomestic mice and rats.

Muscid Of or belonging to the fly family Muscidae.

Muscoid Of or belonging to the dipteran superfamily Muscoidea.

Muscomorpha (-an) An infraorder within the dipteran suborder Brachycera that includes muscoid flies and most brachycerans; adults with short three-segmented antenna and arista; larvae with reduced head capsule; larvae form puparia.

Mutualism (-ist, -istic) An ecological or behavioral association between two species in which both benefit from the relationship.

Muzzle The forward, projecting part of the head of certain animals, including the nose, mouth, and jaws; the snout.

Myalgia Muscular pain or discomfort.

Mycetome Specialized tissues or structures in insects that harbor symbiotic microorganisms; associated with the alimentary tract, fat body, or gonads.

Mycology The biology or scientific study of fungi.

Myelitis Inflammation of the spinal cord or bone marrow.

Myenteric Of or relating to the muscular coat of the intestinal wall.

Mygalomorph Members of the taxonomic group of spiders called Mygalomorphae, represented by tarantulas and their close relatives.

Myiasis Infestation of tissues, wounds, or body cavities of live vertebrate animals by fly larvae.

Myocarditis Inflammation of the myocardium.

Myocardium (-al) The middle muscular layer of the heart wall.

Myofibroblast Cells that give rise to connective tissue (fibroblasts) and that have some structural and functional characteristics of smooth muscle cells.

Myxomatosis An infectious, usually fatal, viral disease of rabbits; characterized by benign skin tumors composed of connective tissue embedded in mucus, called myxomas.

Nagana Highly fatal disease of domestic animals in tropical Africa caused by flagellate protozoans of the genus *Trypanosoma* and transmitted by tsetse and other biting flies.

Naive In reference to animals, not previously having been exposed to a disease agent, or not previously having been used in a scientific experiment; also naïve.

Nasal fossa (-ae) One of the two halves of the nasal cavity, between the roof of the mouth and floor of the cranium.

Nasopharynx (-geal) The part of the pharynx behind the nose and above the soft palate, continuous with the nasal passages.

Natural immunity The normal antipathogen and antiparasitic activity of all animals, both humoral and cellular.

Nearctic Region The biogeographic area of North America characterized by temperate climate, flora, and fauna (New World); together with the Palearctic Region (Old World) comprises the Holarctic Region.

Necrophagous Feeding on carrion, including corpses and other dead animals.

Necrophilous Drawn to, or feeding on, dead animal or human tissue.

Necrosis (-otic) Localized death of cells or tissues due to injury or disease; causes include impaired blood supply, infection, and trauma.

Nectary A glandular organ or structure of plants that secretes nectar; can be within a flower (floral) or outside a flower (extrafloral).

Nematoceran A member of the suborder Nematocera (order Diptera); adults with multisegmented antennae, and larvae with a well-developed head capsule and toothed or brush-like mandibles that move laterally.

Neosomy (-ic) Radical intrastadial growth and subsequent gross body changes of certain ectoparasites or subdermal parasites; for example, female chigoe fleas (*Tunga penetrans*) and females of streblid batflies belonging to the genus *Ascodipteron*.

Neoteny (-ic) The retention of larval or immature structures or traits in adults of a given species.

Neotropical Region The tropical area of the New World extending south from the deserts of Mexico through Central America into South America.

Nettle (-ling) Any of many plants in the genus *Urtica* (family Urticaceae) having stinging hairs that on contact cause skin irritation; also used to refer to other plants that cause similar urticaria.

Neuritis Inflammation of a nerve or group of nerves, characterized by pain, which can lead to loss of function and degeneration of associated muscles.

Neurologic (-cal) Pertaining to the nervous system or associated disorders.

Neuromyopathy (-ic) A disease or disorder affecting nerves and associated muscle tissue.

Neurosis A disorder of the mind and thought processes without evidence of disease or structural change in the nerves or central nervous system.

Neurotoxin (-ic) A toxin that affects nerve cells or tissues.

Neurotropic Having an affinity for, or moving or growing toward, nerve tissue.

Neutrophil (-ic) A phagocytic white blood cell with an abundance of granules in the cytoplasm that are readily stained by neutral dyes.

New World A biogeographical term implying the Americas (North, Central, and South) and the Caribbean islands.

Nidicolous Living in the nest of another animal species; for example, nidicolous insects or mites; *cf. inquilinous*.

Nit The egg of a louse.

Nodule A small, rounded, usually firm or hard mass of body tissue; a node or knot.

Noradrenaline A catecholamine precursor of epinephrine secreted by the adrenal medulla and by nerve endings of the sympathetic nervous system; causes vasoconstriction, increased blood pressure and heart rate, and elevated sugar level in the blood; also called norepinephrine.

Notopleura (-al) A more-or-less triangular area of the insect thorax, notably in Diptera, where the notum and pleuron join above the second pair of legs; may be enlarged as a lobe.

Nuchal (e.g., ligament) Pertaining to the neck or nape; for example, nuchal ligament.

Nullipar (-ous) A female that has not yet produced eggs or young.

Nuptial flight Reproductive behavior among most ants and some bees in which a virgin queen mates with a male, before seeking a suitable site to begin a new colony.

Nymph The immature stage of certain insects that resembles the adult except for its smaller size (e.g., lice) and, in the case of pterous species, the lack of functional wings (e.g., cockroaches).

Obligate Required, necessary, or essential; for example, obligate parasite, a parasite that requires a host in order to complete its development (as opposed to facultative); as in obligate autogeny, obligate hematophagy, and so on.

Obtect Having the antennae, legs, and wings embedded in a secretion that forms a hard cover or protective case, as in obtect pupae characteristic of butterflies and moths.

Occult Concealed, not apparent, obscure; as in an occult infection.

Ocellus (-i) A simple eye in insects, which is sensitive to light and changes in light intensity, but does not form a visual image.

OCP Onchocerciasis Control Programme, in West Africa, under the auspices of the World Health Organization.

Octenol A compound found naturally in bovine breath that is attractive to certain host-seeking flies; used to enhance the attractiveness of various traps for collecting biting flies, such as mosquitoes, biting midges, and tsetse flies.

Ocular point A cuticular projection posterior to the antenna in lice lacking eyes, situated where eyes are located in some other species.

Odonate A member of the insect order Odonata, the dragonflies and damselflies.

Odor plume A mass of air moving downwind from an animal, with higher temperature, humidity, and carbon dioxide levels than that of the ambient conditions; provides indirect clues to the presence of a potential host.

Oedemerid A member of the family Oedemeridae, false blister beetles.

Oestrid A member of the family Oestridae, bot flies and warble flies.

Old World A biogeographical term encompassing Europe, Asia, and Africa.

Oiler In pest control, a self-applicating appliance against which livestock rub and become coated with an oil-based pesticide formulation.

Olfaction Sense of smell, or act of smelling.

Oligochaete Terrestrial and aquatic annelid worms of the class Oligochaeta, with tiny bristles located singly along their length; for example, earthworms.

Oligonucleotide A short nucleic-acid chain with fewer than 20 bases.

Omnivorous Feeding on a varied diet of both plants and animals.

Onchocerciasis An infestation with filarial worms of the genus *Onchocerca*.

Onchocercid Common name for filarial worms of the family Onchocercidae.

Onomatopoetic Formation or use of words that imitate a natural sound associated with the object or action to which it refers.

Oocyst A thick-walled structure that surrounds the developing zygote in the life cycle of certain sporozoan parasites; for example, *Plasmodium* spp.

Oocyte A cell in the ovary of female animals from which an egg, or ovum, develops by meiosis.

Oogenesis The process by which an ovum is formed and develops to maturity.

Ookinete A motile zygote in the life cycle of certain sporozoan parasites; for example, *Plasmodium*, which penetrates the stomach wall of a mosquito and forms an oocyst.

Ootheca (-ae) A case or capsule enclosing the eggs of certain insects; for example, cockroaches.

Operculum (-a) A lid-like part of the insect egg, usually at the anterior end, that opens to allow the insect to emerge, for example, louse egg; also a similar structure in puparia of certain flies that is pushed open to facilitate adult emergence.

Ophthalmia (-ic) Inflammation of the eye, usually involving the conjunctiva.

Opisthosoma That portion of the arachnid body (idiosoma) posterior to the hind pair of legs.

Orbivirus A reovirus of the genus *Orbivirus*; includes the viruses that cause bluetongue disease and Colorado tick fever.

Orf A contagious pustular dermatitis in sheep and goats caused by the orf virus; primarily affects lambs and can be transmitted to humans, causing a pustular lesion.

Organ of Berlese See *Spermalege.*

Organochlorine A chlorinated hydrocarbon, most commonly used to refer to pesticides such as DDT, aldrin, or dieldrin.

Organophosphate An organic compound containing phosphorus; used as an insecticide that acts as a neurotoxin.

Organ of Ribaga See *Spermalege.*

Oriental Region A biogeographical region roughly corresponding to Southeast Asia and some adjacent islands such as Sumatra, Java, and Borneo.

Ornithonosis (es) A disease of birds that occasionally is transmitted to humans.

Ornithophagic Feeding on birds.

Ornithophily (-ic) Attracted to birds.

Orthopteroid An insect superorder including Orthoptera and related insect orders: Phasmatodea, Grylloblattaria, Mantophasmatodea, Mantodea, Blattaria, Isoptera, Dermaptera, and Embiidina (Embioptera).

Ostium (-a) A small opening in a tubular organ or other anatomical structure; for example, insect heart.

Otitis Inflammation of the ear.

Otitis media Inflammation of the middle ear, behind the eardrum.

Otoacariasis Infestation of the ear by mites; also, otacariasis.

Otoscope (-ic) A instrument for visually examining the external ear canal and eardrum.

Ovarian follicle A cell aggregation in the ovary (or ovaries) in which eggs develop and from which they are released.

Ovariole One of the multiple tubes that form the ovary in insects.

Ovate Egg-shaped or oval in outline.

Overwinter To pass and survive the colder months of the year.

Oviduct A tube through which an egg passes from the ovary.

Ovigerous Bearing eggs; gravid.

Oviparous (-ity) Producing eggs that hatch outside the body.

Oviposit (-tion) To deposit or lay eggs.

Ovipositor The egg-laying structure of female insects formed by modifications of the eighth and ninth abdominal segments; also the egg-laying tube through which an egg passes when deposited.

Ovoviviparous (-ity) Producing live young from eggs that hatch within the female body; common in some insects; for example, cockroaches.

Pacific region A biogeographical region that includes land masses in, and adjacent to, the Pacific Ocean.

Paederine Referring to beetles of the subfamily Paederinae, particularly members of the genus *Paederus* and closely related species (family Staphylinidae).

Paleotropics Tropical areas of the Palearctic region.

Palearctic Region A biogeographical region that includes most of Eurasia and Africa north of the Sahara.

Palatal brush A group of hair-like filaments of the palatum (oral surfaces of the labrum and clypeus) in larvae of nematocerous flies; for example, mouthbrushes of mosquito larvae.

Palmate Shaped similar to a hand with the fingers extended or a palm frond; for example, palmate float hairs on abdominal segments of *Anopheles* mosquito larvae.

Palp or Palpus (-i) A segmented appendage associated with the mouthparts of insects, arachnids, and other invertebrates; with sensory receptors for tactile and chemical stimuli.

Palpable Capable of being perceived or felt, especially by touch.

Palpitate (-tion) To beat rapidly, pulsate, or throb; for example, heart palpitation.

Palpomere A subdivision of a palpal segment; not a true segment.

Pancytopenia An abnormal reduction in the number of red blood cells, white blood cells, and platelets in the blood; usually due to disease of the bone marrow.

Pandemic A human disease or other affliction occurring at above the normal incidence over a wide geographic area, such as a country, continent, or the whole world; a widespread or global epidemic; also an adjective.

Papilla (-ae, -ate) A small nipple-like protuberance or projection; also a term applied to more elongate or leaf-like projections of the surface of an organism, as in anal papillae of mosquito and black-fly larvae.

Papilloma A benign epithelial tumor of the skin or mucous membranes.

Papule (-ular) A small, solid, usually inflamed elevation of the skin that does not contain pus.

Papulonodular Referring to a skin condition with both papules and nodules.

Papulovesicle (-ular) A papule that changes to a vesicle or blister.

Paragenital sinus A notch in the posterior margin of the fifth abdominal sclerite of female cimicid bugs that leads via a slit into the sperm-receiving structure.

Paramere A lateral lobe or process at the base of the phallus, or intromittent organ, of male insects.

Parasite An organism that lives in or on another species, the host, and from which it derives nourishment and protection at the expense of the host.

Parasitemia (-ic) Presence of parasites in the blood.

Parasiticide A chemical, other agent, or preparation that kills parasites.

Parasitoid An insect that as a larva develops within the body of another insect, consuming its tissues and eventually killing it; for example, pteromalid wasps and tachinid flies; used as biological-control agents.

Parasitosis (-es) An infestation with parasites, or disease resulting from a parasitic infestation.

Parasympathetic Pertaining to the nerves and ganglia of the autonomic nervous system that arise from the cranial and sacral regions; control involuntary functions such as pupil constriction, dilation of blood vessels, and reduced heart rate.

Parenchyma (-al) The principal tissues of an organ, as distinct from associated connective or supporting tissues.

Parity rate Proportion of parous females per number of females examined.

Parous Having given birth, or in the case of invertebrates having deposited eggs, at least once.

Paroxysm A sudden attack, recurrence, or intensification of disease symptoms.

Parthenogenesis (-tic) A type of reproduction in which the egg develops without fertilization.

Parturition The act or process of giving birth to offspring.

Passeriform Pertaining to birds of the order Passeriformes; passerine.

Passerine See *Passeriform.*

Pastern That part of the equine foot from the fetlock to the top of the hoof.

Patas monkey *Erythrocebus patas*, an African ground-dwelling monkey.

Patella The knee cap of vertebrates; in arachnids, the leg segment between the femur and tibia; genu.

Pathogen A microorganism, virus, fungus, or other agent capable of causing disease.

Pathogenesis The origination and development of a disease.

Pathogenic Capable of producing disease.

Pathologic Pertaining to pathology, the nature and cause of disease; pathological.

Paurometabolous A form of development in insects in which the immature stages and adults resemble one another and both live in the same habitat; for example, cockroaches, lice, and bed bugs.

PCR See *polymerase chain reaction.*

Pecten (-tines) A comb-like structure or row of teeth; for example, on the respiratory tube of mosquito larvae, and, major sensory organ of scorpions.

Pectinal tooth The individual tooth-like structure that forms a pecten.

Pectinate Having closely parallel, tooth-like projections, comb-like; for example, pectinate antennae.

Pedal Pertaining to the leg or foot.

Pedicel The second basal segment of the insect antenna between the scape and flagellum, bearing Johnston's organ; a slender stalk or stem-like structure serving for support.

Pediculicide A chemical or other agent used to kill lice.

Pediculosis An infestation of humans and other animals with lice; from *L. pediculus*, louse; in a restricted sense, refers specifically to human lice.

Pedipalp (-us, -i) One of the second pair of appendages near the mouth of arachnids immediately posterior to the chelicerae; homologous to mandibles of insects.

Pelage Hair, fur, wool, or other soft covering that forms the coat of a mammal.

Peliosis (-es) Any of several blood diseases that cause subcutaneous bleeding marked by purple patches on skin or mucous membranes; purpura.

Penicillin Any of several broad-spectrum antibiotics produced by molds of the genus *Penicillium*; also produced synthetically; most active against gram-positive bacteria.

Pentastomid A member of the Pentastomida, worm-like invertebrates with two pairs of hooks near the mouth; obligate parasites in the respiratory tract of reptiles, birds, and mammals; called tongue worms, due to their resemblance to a vertebrate tongue.

Penultimate Next to the last.

Peptide A compound formed by two or more amino acids linked by the carboxyl group of one amino acid to the amino group of another.

Peracute Very acute or violent.

Perennial Lasting or remaining active on a continual basis from one year to the next.

Perianal gland Glands near or around the anus.

Pericardium (-al) The membranous sac that encloses the heart and the origins of the aorta and other large blood vessels in vertebrate animals.

Peridomestic Found around or in proximity to human dwellings or habitation.

Perineum (-al) The general area between the anus and genital organs of vertebrate animals.

Periodic Recurring at regular or irregular intervals of time.

Periodicity A recurrence at regular intervals of time.

Perioral Surrounding the mouth or involving tissues around the mouth.

Periorbital Involving or located around the orbit of the eye.

Peripheral blood Blood circulating in arteries, veins, and capillaries near the general body surface and in the extremities.

Peripylarian Growth or development of a parasite or pathogen in the hindgut of a vector, as in members of the protozoan subgenus *Viannia* of the genus *Leishmania*; as opposed to suprapylarian, which implies growth or development mainly in the midgut.

Perissodactyla An order of nonruminant grazing ungulates with an odd number of toes on each hoof; includes horses, tapirs, and rhinoceroses.

Peritoneum (-al) The transparent serous membrane lining the abdominal cavity and enclosing most of the viscera.

Peritonitis Inflammation of the peritoneum.

Peritreme A sclerotized area of the body wall in insects surrounding a spiracle; in mites, a sclerotized groove in the body wall leading to a spiracle, or stigma.

Peritrophic Pertaining to a delicate matrix that forms a cylindrical envelope surrounding food in the insect midgut; produced by the midgut wall.

Perivascular Surrounding a vessel, especially a blood vessel.

Permethrin A synthetic pyrethroid compound used as an insecticide, acaricide, and insect repellent; acts as a neurotoxin.

Per oral Via the oral cavity or mouth; for example, per oral infection; per os.

Personal protectant A compound or other substance that, when applied to the skin or clothing, serves to reduce the attractiveness of an individual to annoying or biting insects, mites, or ticks; for example, insect repellents.

Petechia (-ae, -al) Small reddish or purplish spots on the skin or mucous membranes caused by hemorrhaging of capillaries.

Petiole A slender stalk or stem; for example, the constriction between the thorax and abdomen of wasps and ants; see *pedicel*.

Phagocyte (-ic) A cell, such as a macrophage, that engulfs and destroys microorganisms and other foreign bodies in the blood or other body tissues.

Phagolysosome A vesicle formed within a cell by an ingested particle (phagosome) and a lysosome containing hydrolytic enzymes.

Phagostimulant A compound or other substance in plants or animals that induces an organism to feed.

Pharate A term referring to a fully formed insect or mite while still enclosed within the old separated cuticle of the previous developmental stage; an instar between apolysis and ecdysis.

Pharyngeal pump The sucking mechanism of fluid-feeding invertebrates involving the pharyngeal musculature.

Pharyngitis Inflammation of the pharynx, typically in vertebrates.

Pharynx (-geal) In invertebrates, the anterior part of the foregut extending from the mouth to the esophagus.

Pheromone A chemical substance released by an animal that triggers, or otherwise influences, the behavior or physiology of other members of the same species.

Phlebotomine Referring to members of the psychodid subfamily Phlebotominae, commonly called sand flies.

Phlebovirus A genus of viruses of the family Bunyaviridae, including causative agents of sand-fly fever and Rift Valley fever.

Phlegm Thick mucus secreted in the lungs and respiratory passages, caused by infection.

Phobia A persistent, abnormal and irrational fear of, or aversion for, a specific object or situation.

Phoresy (-tic) A commensal relationship between two species in which one is carried on the other, typically as a means of dispersal or escaping adverse environmental conditions.

Phospholipase An enzyme that catalyzes the breakdown of phospholipids, by hydrolysis of ester bonds.

Photoperiod The duration of exposure to light by an organism in a daily 24-hour cycle; also used to refer to the durations of light and darkness during a 24-hour period.

Photophobia An unusual sensitivity to or intolerance of light; an abnormal fear of light.

Phylogeny (-etic) The development of taxonomic groupings and relationships among organisms based on comparative and evolutionary studies; the evolutionary history of groups of organisms.

Physiological age grading Determining the reproductive state of adult organisms, for example, the number of times a female insect has produced eggs or oviposited; based on examination of the reproductive organs.

Phytophagous Feeding on plants.

Piloerection The raising or bristling of hairs due to reflex contraction of tiny muscles at the base of the hair pulling the hair erect; goose bumps or goose flesh.

Pilus dentilus A short, stout seta at the base of the fixed, or unmovable, digit of the chelicerae in gamasid mites.

Pinkeye Acute, contagious conjunctivitis in humans and other animals usually due to infection by bacteria or viruses; bacterial agents include *Hemophilus aegyptius* in humans and *Moraxella bovis* in cattle.

Pinna (-ae) The outer, visibly projecting, cartilaginous structure of the ear in vertebrates.

Pinworm Any of several small nematodes of the family Oxyuridae parasitic in humans, horses, rabbits, and other mammals; infest the intestines and rectum; for example, *Enterobius vermicularis* of humans.

Piroplasm Parasitic sporozoans of the family Babesiidae that infect red blood cells of mammals such as humans, cattle, sheep, and dogs; includes *Babesia* and *Theileria* species.

Piroplasmosis Infection with protozoans of the family Babesiidae, including the tick-borne bovine diseases Texas cattle fever and East Coast fever; for example, babesiosis, theileriosis.

Pissle A penis, especially of rams and bulls.

Pityriasis Any of various skin diseases characterized by the shedding of dry, flaky, or bran-like epidermal scales.

Plaque (1) In pathology, any of various small patches or disk-shaped formations on the skin or surface of mucous tissues; (2) a deposit of fatty material on the inner wall of an artery; (3) an area on a cell culture where cells have been killed by a pathogen being screened (plaque assay).

Plasmatocyte A type of phagocytic cell in the hemolymph of arthropods characterized by relatively large size, irregular outline, and basophilic cytoplasm.

Platelet Round or oval disk-like cytoplasmic bodies in vertebrate blood, lacking a nucleus and hemoglobin; produced in bone marrow and functioning in blood clotting.

Platyform Plate-like in shape.

Pleomorphic Occurrence of two or more structurally distinct forms of a developmental stage in the life cycle of an organism; polymorphism.

Pleural sclerite A small or minor sclerotized plate in the pleural region; pleurite.

Pleuron (-al) The lateral area of any body segment in arthropods, usually referring to the thorax and abdomen; may or may not be hardened, or sclerotized.

Plumose Resembling a plume or feather.

Pneumonia Inflammation of the lung or lungs, usually caused by viruses, bacteria, fungi, or chemical irritants; pneumonitis.

Pneumonic (-ic) Involving the lungs or pneumonia.

Pneumonitis Inflammation of the lung or lungs; pneumonia.

Podosoma The anterior part of the main body region of mites and ticks bearing the legs.

Poikilotherm (-ic) An animal in which the body temperature varies with the temperature of its environment.

Poison Any substance that, when taken into the body, interferes with normal physiological functions; a general term that includes toxins and venoms.

Poliomyelitis Inflammation of the gray matter of the spinal cord and brain stem; a viral disease caused by polioviruses and leading to paralysis, muscular atrophy, and often deformities; polio.

Poll In livestock, the prominent hairy top or back of the head.

Polyamine An organic compound with two or more amino groups.

Polyarthritis Inflammation of multiple joints simultaneously.

Polyctenid A member of the hemipteran family Polyctenidae, ectoparasitic on bats; bat bugs.

Polygenic Pertaining to inheritable characters controlled or caused by interaction of multiple genes.

Polygyny (-ous) In social insects, the presence of more than one functional queen in a colony at the same time; reproductive behavior of animals in which a male mates with more than one female.

Polymerase chain reaction (PCR) A technique for amplifying sequences of DNA involves; separating DNA into two strands and incubating it with oligonucleotide primers and DNA polymerases.

Polymorph (-ic) An organism or structure represented by two or more forms.

Polypeptide A molecular chain of 10 to more than 100 amino acids linked together by peptide bonds.

Polyphaga (-an) The largest suborder of Coleoptera; includes more than 90% of the described species of beetles.

Polytene chromosome A giant chromosome with characteristic dark and light banding patterns, formed by multiple replications of DNA strands that remain tightly together within the cell.

Polytrophic ovariole An ovariole containing a primary oocyte and associated trophocytes, or nutritive cells, all derived from the same female germ cell.

Polyvalent vaccine A vaccine prepared from cultures of two or more strains or species of pathogens, thereby providing protection from more than one pathogenic agent with a single vaccine; multivalent vaccine.

Pool In vector biology, a sample of insects, usually representing a single species, which is prepared for testing to detect evidence of an infectious or other agent.

Pool feeder An ectoparasitic or other blood-feeding organism that lacerates the skin of a host with its mouthparts causing blood or other tissue fluids to accumulate at the bite site, which then are drawn into the oral cavity and alimentary tract; telmophage; *cf. capillary feeder* or *solenophage*.

Porcine Pertaining to or resembling swine or pigs.

Porose area A cluster of tiny pores on the dorsal aspect of the basis capituli of female hard ticks (family Ixodidae), through which are secreted antioxidants that prevent breakdown of waxy compounds protecting the deposited eggs.

Postabdomen The posterior part of the opisthosoma of scorpions that forms the slender, flexible tail terminating in the telson and sting; metasoma.

Posterior-station Referring to the posterior part of the alimentary tract in arthropods; for example, posterior-station transmission of pathogens, that is, pathogens passing from an arthropod host via the anus as a source of infection; stercorarian transmission.

Posterior-station transmission See *Posterior-station.*

Postgena (-al) The portion of the insect head capsule immediately posterior to the gena.

Postgenital Posterior to the genitalia; for example, postgenital segments.

Postmortem Occurring after death.

Postpartum Occurring after childbirth.

Postscutellum Apparent in some insects as the upper, posteriorly produced, part of the metanotum.

Pour-on A liquid formulation of an insecticide or acaricide that typically is applied by pouring it along the midline of animal's back.

Preabdomen In scorpions, the anterior seven segments of the opisthosoma (mesosoma), behind which the opistosoma is modified to form the tail (metasoma); in insects, the relatively unmodified anterior abdominal segments.

Predispose (-ed, -ition) To incline or tend toward something beforehand; to make susceptible, as to a disease.

Prediuresis In hematophagous insects, the separation of liquid components from a bloodmeal in the midgut that are passed directly to the hindgut and excreted as droplets of fluid from the anus, often during feeding.

Prehensile Modified or adapted for seizing, grasping, or holding onto something.

Prelarva (-ae) A nonfeeding developmental stage in certain mites and other arachnids immediately preceding the larval stage.

Prenatal Existing or occurring prior to birth.

Prepupa The last part of the final larval instar in holometabolous insects in which the larva ceases to feed, defecates, and becomes quiescent in preparation for transformation to the pupa.

Preputial gland A sebaceous gland associated with the foreskin of the male penis, the prepuce, and similarly the fold of skin covering the tip of the female clitoris of mammals.

Prestomal teeth Sclerotized teeth arising along the lateral margins of the prestomum, or on the inner wall of the prestomum, of some adult flies; used for scraping food, or lacerating skin in the case of certain hematophagous flies.

Prestomum (-a) A cleft between the two lobes of the labellum just anterior to the opening to the food canal of the adult stage of certain flies.

Presumptive Based on reasonably convincing, albeit unconfirmed, evidence; for example, a presumptive case or presumptive diagnosis.

Pretarsus The distal-most part of the tarsus in insects, arachnids, and other arthropods bearing the claws, empodium, or other terminal appendages.

Prevalence The number of cases of something in a population at a given time; also the percentage of a population affected by a particular cause at a specific time; for example, prevalence of infection or disease; *cf. incidence.*

Priapism A sustained, usually painful erection of the penis not related to sexual arousal.

Primary host A host in which a virus or microorganism most commonly replicates or reproduces; in the case of multicellular parasites, a host in which the organism reaches maturity and reproduces sexually; *syn.* definitive host.

Primary vector An arthropod that plays a major role in transmission of a pathogen or parasite from one vertebrate host to another.

Primer A segment of DNA or RNA that is complementary to a given DNA sequence and that is needed to initiate replication by DNA polymerase; as in PCR.

Privy An outdoor toilet; outhouse.

Proboscidea A mammalian order represented by the single family Elephantidae, elephants, with only three living (extant) species; includes the extinct mastodons and mammoths.

Proboscis In insects and other arthropods, an elongation of the anterior or ventral part of the head composed of the mouthparts and associated structures, typically forming a tube through which food is drawn into the mouth and alimentary tract.

Progesterone A female sex hormone that prepares the uterus for implantation of a fertilized ovum, helps maintain pregnancy, and promotes development of mammary glands.

Proglottid A segment of an adult tapeworm, containing both male and female reproductive organs.

Prognathous With the head directed forward, especially in insects.

Prognosis A prediction of the probable course and outcome of a disease, including the likelihood of recovery.

Proleg A process or appendage of insect larvae that serves the purpose of a leg; usually fleshy and unsegmented; *syn., pseudopod.*

Promastigote A flagellate stage of protozoans in the family Trypanosomatidae, for example, *Leishmania* species; characterized by a single anterior flagellum, without an undulating membrane.

Pronghorn *Antilocapra americana*, family Antilocapridae; an antelope-like ruminant with short forked horns found on the western plains of North America.

Pronotum The dorsal surface of the first thoracic segment of insects.

Propagative transmission Transmission of the same form of a pathogen or parasite by a vector in the form as it was originally ingested from an infected host; occurs after asexual reproduction of the pathogen or parasite in the vector; *syn.*, multiplicative transmission.

Prophylaxis (-actic) Protective treatment or prevention of disease.

Propodeum The first abdominal segment in hymenopterous adults that has become morphologically part of the thorax.

Propodosoma That part of the body of mites and ticks bearing the first two pairs of legs.

Proprioceptor (-ive) A receptor that responds to stimuli originating within the body, such as pressure, stretch, and position.

Prosoma In mites and ticks, the anterior part of the body including the mouthparts (gnathosoma) and leg-bearing region (podosoma); in other arachnids, the combined head and thorax segments, or cephalothorax.

Prosternum (-al) The anterior-most sternal sclerite of insects, between the forelegs; prothoracic sternum.

Prostigmatid Referring to members of the mite suborder Prostigmata.

Prostrate (-tion) Lying on the ground, or other surface, with the belly and/or face downward.

Prostriata (-ate) Hard ticks (family Ixodidae) in which the anal groove curves anteriorly around the anus; includes members of only one genus, *Ixodes.*

Proteolysis (-tic) Breaking down of proteins into simpler compounds.

Prothorax (-ic) The anterior-most thoracic segment of insects, bearing the first pair of legs.

Protist A very diverse group of eukaryote organisms usually referred to as members of the kingdom Protista or Protoctista; includes unicellular or multicellular eukaryotes without highly specialized tissue, that is, distinct from fungi, plants, or animals.

Protonymph The first nymphal stage in the generalized life history of mites and some other arachnids; often a suppressed developmental stage passed within the egg.

Protozoan Single-celled eukaryotic organisms in the kingdom Protista or Protoctista; members of the subkingdom or phylum Protozoa.

Proventriculus (-ar) A typically muscular part of the posterior foregut of insects and other arthropods located just anterior to the constriction leading into the midgut, or ventriculus; in birds, a glandular part of the stomach in which food is partially digested before passing from the crop to the ventriculus, or gizzard.

Pruritus An intense itching sensation.

Psammophore Setae on the ventral surface of the head and mandibles of certain desert-dwelling ants and wasps, forming a basket in which sand or dry soil is moved or transported.

Pseudopenis The term pseudopenis typically is used for the male intromittent organ of lice (order Phthiraptera) and some related insects (see *aedeagus*).

Pseudopod (-ium, -ia) See *Proleg.*

Pseudotrachea (-ae) A structure having the aspect of a trachea, in the labellum of Diptera; pseudotracheae are ringed and ridged grooves used for scraping food and moving liquid material to the mouth by capillary action.

Psoriasis A chronic noncontagious inflammatory skin disease characterized by discrete patches of pink or dull-red lesions, often covered with silvery scales.

Pteridine An organic yellow crystalline compound composed of linked pyrimidine and pyrazine rings includes structural constituents of various animal pigments; useful for age determination in some adult Diptera.

Pteromalid A tiny parasitic wasp of the large chalcidoid family Pteromalidae; includes common parasitoids of the puparia of dung flies.

Ptilinal suture The frontal suture through which the ptilinum is everted during and immediately following emergence of a teneral adult from a puparium.

Ptilinum (-al) A sac-like, inflatable cuticular structure of the head of some adult flies; used to facilitate emergence from the puparium or to burrow through soil to reach the surface.

Pubescence (-ent) A covering of short, closely spaced, fine hairs or setae.

Pubic Referring to the lowermost part of the abdomen between the thighs, overlying the pubes, or pubic bones.

Pulmonary Relating to or affecting the lungs.

Pulvillus (-i) A pad-like structure or lobe of the pretarsus at the base of the tarsal claw.

Pupa (-ae) The quiescent developmental stage between the larva and adult of holometabolous insects; called a chrysalis or chrysalid for members of the Lepidoptera.

Puparium (-ia) The thickened, hardened integument of third-instar larvae of cyclorrhaphous flies within which the pupa and adult are formed.

Pupiparous Referring to an insect in which the larva develops within the body of the parent female and is ready to pupate at the time it is deposited; for example, hippoboscids and tsetse flies.

Purge To evacuate the bowels, or something that causes evacuation of the bowels.

Purpura Hemorrhaging in the skin, mucous membranes, or serosal tissue resulting in purplish spots or patches.

Purpura pulicosa Small, purplish spots on the skin resulting from flea bites and associated capillary leakage.

Purulent Forming, containing, or discharging pus.

Pus A generally viscous yellowish-white fluid resulting from inflammation; consisting of plasma, white blood cells, cellular debris, and necrotic tissue.

Pustule A small inflamed swelling of the skin filled with lymph or pus; a pimple or pus-filled blister.

Putative Commonly believed to be true without conclusive evidence; supposed, reputed.

Putrid (putrefaction) Animal or plant material in an advanced state of decomposition, characterized by a foul odor; rotten.

Pygidial gland Any of various glands that open near the pygidium or anus; common in bees, wasps, ants, and beetles.

Pygidium (-al) The tergum, or dorsal part, of the last abdominal segment of insects and other invertebrates.

Pyloric valve A valve-like fold at the juncture of the midgut and hindgut of insects.

Pylorus The anterior part of the hindgut in insects, usually including the pyloric valve; functionally forms the posterior end of the midgut.

Pyoderma An acute destructive inflammation of the skin characterized by pus-filled lesions.

Pyoderma gangrenosum A chronic skin disease, characterized by purplish nodules, pustules, and spreading ulcers, usually on the trunk.

Pyogenic Producing or relating to formation of pus.

Pyrethrin Either of two esters extracted from the seed cases of chrysanthemum flowers, used as an insecticide.

Pyrethroid A synthetic chemical compound structurally related to natural pyrethrin, widely used in commercial products for insect control (see *permethrin, resmethrin*).

Pyriform Pear-shaped.

Pyriproxyfen A pyridine-based pesticide and juvenile hormone analog that prevents certain insects and other arthropods from developing to adults, thereby precluding reproduction.

Quartan Occurring every fourth day, usually with reference to fever; for example, quartan malaria.

Queen excluder A barrier inside a beehive that prevents queens from entering or exiting the colony while allowing the smaller workers to come and go; typically a perforated plastic or metal sheet or a wire grid.

Quest (-ing) Host-seeking behavior of ticks, which, in most cases, involves a tick crawling upward on vegetation and awaiting contact with a potential host when it passes by; the forelegs are held widely extended, allowing the tick to grasp immediately on contact.

Quiescence (-ent) A general term referring to an inactive, dormant, or latent period in the life or development of an organism; many different levels of physiological activity represented.

Quinone Any of a class of naturally occurring aromatic compounds, characterized by two carbonyl groups in an unsaturated benzene ring; *syn.*, benzoquinone.

R_1 cell Cell of the insect wing bounded anteriorly by the first radial vein, R_1.

Radioallergosorbent test (RAST) An IgE-mediated allergy test; involves incubating serum from a blood sample with a solid-phase antigen and quantifying the amount of allergen-specific IgE using radiolabeled anti-IgE.

Radius (-al) The third longitudinal vein from the anterior margin of the insect wing, behind the costa and subcosta.

Raptorial Modified or adapted for seizing prey.

Receptor A specialized group of nerve endings or a sensory structure for perceiving stimuli; sense organ; particularly concentrated on the antennae, palps, and tarsi of insects.

Recrudesce (-scent, -scence) To assume renewed activity after a dormant or inactive period; for example, recurrent symptoms or relapses of a disease.

Recumbent Lying down; resting or inactive.

Red bug A colloquial name for chigger.

Reflex bleeding The release of hemolymph through the intersegmental membrane of an appendage or other body part of certain insects and other arthropods; usually involves distasteful or toxic chemicals in the hemolymph that serve a defensive function against potential predators.

Refractory Resistant to normal treatment, stimulation, or infection.

Refugium (-a) An area to which an organism can retreat or where it can continue to survive after otherwise being eliminated from the surrounding environment.

Relapse Recurrence of a disease or symptoms after apparent recovery.

Relapsing fever Any of several infectious diseases of humans caused by spirochetes transmitted by lice and ticks, characterized by intermittent attacks of high fever.

Relative abundance A number or index value indicating a comparative estimate of the number of organisms in one area versus another, without knowing the actual or absolute number of individuals present; *cf.* absolute abundance.

Repellent A substance that causes behavioral aversion by an organism; usually a chemical that is distasteful or otherwise offensive, or alters the organism's behavior by blocking or disrupting its sensory receptors.

Replete (-tion) Filled to satiation or satisfaction with food or drink; fully engorged, as in a mosquito or other hematophagous arthropod after taking a bloodmeal.

Reproductive diapause A type of diapause in which reproductive development or activity is temporarily suspended.

Reservoir An organism that supports survival and development of a pathogen or parasite typically without experiencing apparent adverse effects; functions in maintaining pathogens or parasites for an extended period of time and as a source of infection for other organisms in the transmission cycle.

Residue (-ual) In an entomological context, components of a chemical formulation that remain for an extended period of time on a surface following a pesticide application.

Resilin A cuticular protein with elastic properties secreted by epidermal cells of the arthropod integument; functions in storage and release of mechanical energy.

Resource partitioning An ecological concept in which a common finite resource such as food or space is shared by two or more species, each adapted to reduce direct competition among them.

Respiratory horn One of a pair of dorso-lateral appendages on the cephalothorax of larvae of mosquitoes and some other flies with aquatic immature stages; usually open at the distal end to allow for gas exchange at the water surface; *syn.*, respiratory trumpet, air trumpet, trumpet (see *siphon*).

Respiratory trumpet See *Respiratory horn.*

Resting site A location or microhabitat where a flying insect typically is found between periods of flight activity.

Resurge (-ence) To rise again or rebound after a decline or absence; to make a comeback, as in the case of a resurgent disease.

Reticulocyte An immature red blood cell containing a network of granules or filaments.

Reticuloendothelium (-al) Cells and tissues that make up the reticuloendothelial system, part of the immune system consisting primarily of phagocytic cells such as monocytes and macrophages concentrated in the lymph nodes, spleen, and loose connective tissue.

Retinitis Inflammation of the retina of the eye.

Rhinitis Inflammation of the nasal mucosa.

Rhinoconjunctivitis Inflammation of the conjunctiva and adjacent nasal mucosa.

Ribonucleic acid (RNA) A linear polymer of nucleotides found primarily in the cytoplasm, which transfers genetic information from DNA to the cytoplasm, plays a key role in protein synthesis, and controls chemical processes in the cell.

Rickettsemia Presence of rickettsial bacteria in the blood.

Rickettsia (-ae) A member of the genus *Rickettsia*, obligate intracellular proteobacteria in the order Rickettsiales, family Rickettsiaceae, transmitted by fleas, lice, chiggers, and ticks; causative agents of various forms of typhus and of rickettsialpox.

Rickettsialpox A generally mild disease caused by *Rickettsia akari* transmitted from mice to humans by the bite of an infected mite.

Rickettsiosis (-es) Infection with a rickettsia.

Ring gland A ring-like structure of endocrine tissue encircling the aorta of muscomorphan fly larvae, representing fusion of the corpora allata, corpora cardiaca, and prothoracic glands; connected to the brain by a pair of nerves.

Ringworm An eruption of the skin caused by fungi, usually belonging to the genera *Microsporum*, *Trichophyton*, or *Epidermophyton*; can affect any part of the body but more commonly the feet, nails, and scalp.

Riparian Occurring or situated on the bank of a stream, river, or other flowing body of water.

RNA See *Ribonucleic acid.*

Rock pool A natural depression in the rock bed of a stream or river forming isolated pools of water when the water level drops low enough; a temporary breeding site for certain mosquitoes and other aquatic organisms.

Rosacea A chronic skin condition of the face caused by dilation of capillaries and characterized by red or rose-colored acne-like pustular lesions; *syn.*, acne rosacea, acne erythematosa.

Roseola pulicosa A small, slightly swollen, reddish lesion resulting from a flea bite.

Rostrum A general term referring to any snout-like prolongation of the head in arthropods and other invertebrates.

Roundworm A member of the phylum Nematoda, with a cylindrical, unsegmented body usually tapered at both ends; includes parasitic filarial worms, hookworms, and pinworms.

r-strategy (-ist) A reproductive pattern characterized by a high reproductive rate and low survival rate; common in arthropods and other invertebrates; *cf. k-strategy.*

Ruminant Any even-toed; hooved cud-chewing, usually horned mammal of the suborder Ruminantia; with a stomach typically divided into four compartments, the anterior chamber being the rumen; includes cattle, buffalo, sheep, goats, camels, and deer.

Saddle A large sclerite usually covering most of the dorsal and lateral surfaces of the anal segment of larvae of mosquitoes and some other nematocerous flies.

Salicylic acid A white crystalline acid derived from phenol, used in making aspirin and as a topical treatment for certain skin conditions.

Salmonellosis Infection with bacteria of the genus *Salmonella*, caused by consuming contaminated or improperly cooked foods; marked by acute onset of abdominal pain, vomiting, diarrhea, and fever; for example, food poisoning.

Saponify (-ication) To convert into a soap, as when fats or oils chemically react with an alkali; occurs rarely in cadavers and carrion.

Saprophage (-ous, -y) Any vegetable organism (e.g., bacteria, fungi, plants) that lives or feeds on decaying organic matter.

Saprophyte Any organism that feeds on dead plant matter.

Satellite lesion A secondary lesion in a medical condition, in addition to a primary lesion located at the point of injury or infection; for example, secondary erythema migrans lesions in Lyme disease cases.

Sauria (-an) A paraphyletic group of reptiles in the class Sauropsida, order Squamata, which includes lizards, skinks, and dinosaurs; no longer accepted as a taxonomic term.

Savanna A flat grassland with scattered trees, usually in subtropical or tropical regions.

Scabies An infestation of skin by the mite *Sarcoptes scabiei* characterized by dermatitis and intense itching; occurs especially in cattle, sheep, pigs, dogs, and humans.

Scape The basal-most segment of the insect antenna.

Scarabaeid A member of the beetle family Scarabaeidae, or scarab beetles.

Scarabaeiform Grub-shaped; resembling the larvae of scarab beetles.

Scarabiasis An infestation or invasion of living animals by adult beetles.

Scarify To make scratches or small superficial cuts, as in skin or other tissues.

Scavenger An animal that feeds on dead plant or animal matter.

Schizogony Asexual reproduction by multiple fission to produce merozoites, characteristic of sporozoans, for example, malarial parasites; *syn., merogony.*

Sciurid A member of the rodent family Sciuridae; includes ground squirrels, tree squirrels, chipmunks, woodchucks, and marmots.

Sclerite A hardened or sclerotized part of the arthropod integument delineated by membranous areas, sutures, or apodemes.

Sclerotin (-ized) Any of several structural proteins that impart hardness, toughness, and usually darkened coloration to the arthropod integument; formed by cross-linkages of molecules of the protein arthropodin in the cuticle by a tanning process called sclerotization.

Scoleciasis Invasion of living animals by lepidopterous larvae.

Scopula (-ae) A brush or dense tuft of setae.

Scotophase The darkness portion of a 24-hour period of light and darkness.

Scurf (-y, -iness) Scales or flakes of dry skin; for example, dandruff.

Scutellum A small shield-like sclerite; usually referring to a posteromedian projection of the mesonotum of winged insects.

Scutum A shield or shield-like sclerite; in insects, usually referring to the major dorsal part of the mesothoracic or metathoracic notum; in hard ticks (family Ixodidae), represented by a dorsal sclerotized plate of the idiosoma.

Scybalum (-a) In a medical context, a hardened mass of feces; in arthropods, fecal pellets, or scybala; for example, feces produced by scabies mites as they burrow in skin.

Sebaceous gland Oil-secreting glands of the skin, usually opening into hair follicles; produce fatty secretions called sebum.

Seborrhea A disease of the sebaceous glands involving qualitative changes and excessive discharge of sebum, resulting in oily skin and dermal scales or crusts.

Sebum A fatty secretion of the sebaceous glands of the skin.

Secondary host A host of lesser importance than the primary host(s) for a pathogen or parasite.

Secondary infection Infection by a second pathogen after another pathogen is already present.

Secondary vector An arthropod that transmits a pathogen or parasite from one vertebrate host to another but that cannot sustain the organism in a natural cycle without transmission by a primary vector.

Self-limiting disease A disease that tends to run its course little influenced by treatment or with no treatment at all.

Semiaquatic Living in or near water or wet substrates but not completely aquatic.

Seminal bursa An internal pouch in some female insects (e.g., in some mosquitoes) that receives the sperm during copulation.

Seminal vesicle In male insects, a pouch-like or tubular enlargement of the vas deferens in which seminal fluid and spermatozoa are stored.

Sensillum auriformium A sensory organ of arthropods believed to function in proprioception, for example, on scutum of ticks; *pl.*, sensilla auriformia.

Sensillum coeloconicum A sensory organ of arthropods characterized externally by a peg-like or conical projection of the cuticle recessed in a pit; functions in olfaction and thermoregulation; *pl.*, sensilla coeloconica.

Sensilium A sensory patch of integument situated on abdominal tergite 9 or 10 of adult fleas; pygidium of older works.

Sensillum (-a) A simple sensory organ or receptor visible externally on the body or appendages of arthropods; highly variable in both structure and types of stimuli perceived.

Sensitization To render sensitive; to induce acquired sensitivity; immunization, especially with reference to antigens not associated with infection; the induction of acquired sensitivity or allergy.

Sensu lato (s.l.) In the broad sense, from Latin; in taxonomy, referring to a group of broadly related taxa, for example, *Borrelia burgdorferi* s.l.; *cf. sensu stricto.*

Sensu stricto (s.s.) In the narrow sense, from Latin; in taxonomy, referring to a distinct taxon, for example, *Borrelia burgdorferi* s.s.; *cf. sensu lato.*

Sentinel Animals (rarely humans) used to screen for vectors or vector-borne diseases; for example, sentinel chickens or horses from which blood samples are collected on a regular basis and screened for antibodies to zoonotic mosquito-borne arboviruses.

Septicemia (-ic) A systemic infection with a pathogen or its toxin present in the blood; blood poisoning.

Sequela (-ae) A condition resulting from and following a disease, injury, procedure, or treatment.

Seroconversion Production of antibodies in blood serum in response to an antigen, resulting from infection or immunization.

Serogroup A group of serotypes with one or more antigens in common.

Serological survey (serosurvey) Sampling a population to determine the presence, incidence, prevalence, or distribution of a disease based on serological testing of blood or other body tissues.

Serology (-ic, -ical) The scientific study of the properties and reactions of serum, particularly blood serum.

Seropositive Exhibiting a significant level of antibody or other immunologic marker in serum denoting exposure to a given infectious agent, that is, having seroconverted.

Seroprevalence The number of cases or percentage of a population exhibiting the presence of an antibody or other element in serum against a specific pathogen or parasite.

Serosanguineous Consisting or having the nature of blood and serous fluid.

Serosurvey See *Serological survey.*

Serotonin A neurotransmitter in mammals that acts as a vasoconstrictor, stimulates smooth muscle, and regulates certain cyclic processes; involved in neural mechanisms affecting pain perception, mood, and sleep-wake cycles.

Serotype A group of closely related microorganisms characterized by a shared set of specific antigens determined by serologic testing.

Serpentine Snake-like.

Serrate An edge notched like teeth of a saw; saw-like and used to cut or lacerate.

Serum (-a, -ous) A watery fluid from animal tissues, usually used to refer to blood serum, a clear amber fluid resulting when whole blood coagulates or is otherwise separated into its solid and liquid components; contains proteins and antibodies.

Sessile Attached by the base without a stalk or other projecting support; also, firmly attached and not free-moving.

Seta (-ae) A slender, hair-like cuticular projection of an epidermal cell of the arthropod integument.

Sex pheromone A chemical compound produced by an organism used to attract a member of the opposite sex of the same species.

Sexual dimorphism Differences in size, morphology, behavior, or other characters that distinguish males from females of a given species.

Sheep strike Cutaneous myiasis of sheep, typically involving blow flies (family Calliphoridae) (see *fly strike*).

Shigellosis An acute infection of the intestinal tract caused by bacteria of the genus *Shigella*, characterized by diarrhea, abdominal pain, fever, and dehydration.

Shock A potentially fatal physiological condition caused by inadequate oxygen reaching tissues and cells, characterized by pallor, weak pulse, and marked drop in blood pressure and blood flow; collapse of circulatory function caused by blood loss due to injury, hemorrhaging, disease, allergic reactions, and other conditions.

Sibling species Very closely related species that are morphologically indistinguishable from one another and reproductively incompatible; believed to represent recent speciation.

Sigmodontine A member of the mammalian subfamily Sigmodontinae that includes cotton rats (*Sigmodon*), deer mice (*Peromyscus*), wood rats (*Neotoma*), and related New World species.

Sign In a medical context: objective evidence or manifestation of a disease, body dysfunction, or other disorder, excluding impressions or subjective assessments; *cf. symptom.*

Silage Fodder harvested while green and preserved by storing and fermenting in a silo.

Silica gel A gelatinous form of colloidal silica (silicon dioxide) used as a dehumidifying and dehydrating agent.

Silurian Pertaining to the Paleozoic Era of geologic time 425 to 400 million years ago, characterized by the emergence of air-breathing land invertebrates and terrestrial plants (Silurian Period).

Simuliid A member of the dipterous family Simuliidae, black flies.

Sinuous Characterized by curves or turns; winding or linearly wavy.

Sinus Any cavity having a relatively narrow opening.

Sinusitis Inflammation of a sinus, especially the paranasal sinus.

Siphon A respiratory structure formed by a projection of the body wall of aquatic invertebrates bearing a spiracle, or pair of spiracles, at the tip; for example, culicid and psychodid larvae.

Site specificity Pertaining especially to ectoparasites and other blood-feeding arthropods whereby they preferentially, or exclusively, live and/or feed on a particular body region of the host.

Skin biopsy Removal of skin tissue, usually a lesion, for microscopic examination by a pathologist or for diagnostic testing to determine the cause and nature of a lesion or other medical condition.

Skin snip A small sample of skin tissue removed for biopsy.

s. l. See *Sensu latu.*

Social insect An insect that lives in an organized group or colony characterized by overlapping generations, provisioning and care for offspring, and division of labor among the members.

Sodium phenobarbital A short-acting barbiturate in the form of a sodium salt, used medically to treat seizures and as a preoperative sedative or anesthetic.

Solenophage (-y) An insect or other arthropod with mouthparts modified to pierce the skin and feed on blood directly from a blood vessel.

Solitary Living or existing alone, rather than in a group or colony; for example, solitary wasps and bees.

Souma A Sudanese name for nagana, a disease of domestic animals in tropical Africa caused by trypanosomes transmitted by tsetse flies.

Source reduction A general term for reducing or eliminating the source of a problem, such as standing water where mosquitoes breed; involves removal or modification of the environment to make the site unsuitable for producing or harboring pest species.

Sowbug A member of the crustacean class Isopoda in the terrestrial family Oniscidae that feeds on decaying vegetable matter; distinguished from the related pillbugs by a pair of tail-like appendages and the inability to roll into a tight ball.

Spermalege A structure of female bed bugs and other cimicids, separate from the normal reproductive organs, that receives sperm during traumatic insemination; sperm passes from there to the oviduct via the hemocoel also called organ of Berlese and organ of Ribaga.

Spermatheca (-ae) A structure of the female reproductive system of arthropods for receiving and storing spermatozoa from the male.

Spermatogenesis The process of formation and development of spermatozoa.

Spermatogenic arrest The suspension or termination of spermatogenesis as a result of disease, physiological dysfunction, radiation, environmental factors, or other causes.

Spermatophore A capsule or envelope containing spermatozoa and seminal fluid produced by male arthropods and transferred to the female during copulation.

Sphaeromastigote A developmental stage of trypanosome protozoans having a rounded body and an anterior flagellum.

Spine A multicellular, nonarticulated outgrowth of the arthropod integument typically in the shape of a thorn or stiff tapered process with a pointed tip.

Spinneret An external organ via which fluid silk is extruded from glands to produce threads used by insects, spiders, and mites to form cocoons, webs, and other silk structures.

Spiracle A pore or other opening in the arthropod integument connected with a tracheal tube allowing for exchange of gases between the body and surrounding air.

Spiracular plate A sclerotized area of the arthropod integument surrounding a spiracular opening(s).

Spirochete Any of various pathogenic and nonpathogenic spiral-shaped, motile bacteria of the order Spirochaetales; includes *Borrelia* and *Treponema* as pathogens of humans and other animals.

Spirochetosis (-es) An infection caused by spirochetes.

Spirurid A member of the nematode family Spiruridae in which the larva parasitizes insects and the adult parasitizes vertebrates.

Splenomegaly Abnormal enlargement of the spleen.

Sporocyst A walled structure within which sporozoites are formed and develop.

Sporogony Asexual reproduction by multiple fission of a spore or zygote to produce sporozoites, characteristic of sporozoans.

Sporozoan A parasitic spore-forming protozoan of the class Sporozoa; for example, *Babesia, Plasmodium.*

Sporozoite A developmental stage of sporozoans usually produced in the arthropod host and that serves as the infective stage for a vertebrate host.

Sporulation Asexual production and release of spores.

Spot-on A chemical formulation topically applied to the skin of an animal at a localized site where it is absorbed; usually an insecticide or parasiticide.

Spot treat Application of a pesticide to specific, restricted locations where pest species are most likely to encounter it on animals, are absorbed into skin oils and thereby distributed over the body.

Spur A spine-like cuticular structure of the arthropod integument that articulates or attaches at its base with the body wall (or with a leg coxa in hard ticks).

Spurious Resembling something it is not; not genuine or true; for example, spurious wing vein in some insects.

Sputum Saliva mixed with mucus, phlegm, cellular debris, microorganisms and other substances from the respiratory tract expelled by coughing or clearing the throat.

Squama (-ae) A scale or thin plate-like structure; for example, in insects, a scale-like sclerite at the base of the wing or haltere in certain dipterous adults.

s. s. See *sensu stricto.*

Stadium (-ia) The interval of time between two successive molts in the larval or nymphal development of insects and some arachnids.

Stage A specific period of time or step in the development of an organism; in insects, the time between two molts, for example, third-stage larva, or third instar; also used to refer to a major developmental division in a life cycle, for example, larval stage.

Stanchion (-ed, ing) A vertical rod or post used as support; a framework of vertical bars for confining cattle at a feed trough or in a stall.

Steppe A vast, semi-arid plain of grassland, especially in parts of Europe and Asia.

Stercorarian Referring to excrement or dung; in vector biology, specifically referring to transmission of a pathogen or parasite via the feces of the vector; see *Posterior-station.*

Sternite A subdivision of the sternum or a discrete sclerotized plate in the sternal area.

Sternum The ventral sclerotized part of a body segment.

Steroid A class of organic compounds with 17 carbon atoms forming four rings; includes sterols, D vitamins, the sex hormones testosterone and estrogen, adrenal hormones, and plant alkaloids.

Sticky trap A device with adhesive compounds applied to the surface to capture crawling or flying insects (or rodents, etc.).

Stigma (-ata) (1) A spiracle or respiratory opening, usually used when referring to mites; (2) a distinctive sclerotized or pigmented spot on the insect wing.

Stillborn Dead at birth.

Sting A sclerotized apparatus with associated muscles derived from the female reproductive system of some wasps, ants, and bees for delivering venom; any such apparatus associated with venom glands in other insects and arachnids; for example, sting of scorpions; *syn.*, stinger.

Sting autotomy Self-amputation of the sting apparatus in certain hymenopterous insects in which barbed lancets anchor the sting in skin, tearing the sting and associated venom gland from the abdomen as the insect pulls away; for example, honey bees.

Stomatitis Inflammation of the mouth, especially the mucous membranes.

Stool Fecal matter from a single bowel movement.

Stratum corneum The outermost horny layer of the epidermis of vertebrates, consisting chiefly of keratinized dead cells that slough off.

Stratum germinativum The innermost layer of epidermis consisting of a single row of cells that divide to replace other cells of the epidermis as they die and are worn away.

Stratum granulosum A layer of the epidermis between the stratum germinativum and the stratum corneum or stratum lucidum, characterized by deeply staining basophilic granules; functions in strengthening and waterproofing the skin.

Stratum lucidum A translucent layer of epidermal cells underlying the stratum corneum, especially in the thickened skin of the palms and soles; absent in many other epidermal tissues.

Striate (-tion) Marked by long, fine parallel lines, or striae, grooves, or ridges.

Stridulate (-atory) To produce sound by rubbing two body parts together as a means of acoustical communication.

Strike See *Fly strike* and *Sheep strike.*

Stylet A small, rigid, needle-like appendage or other external process in arthropods.

Styletiform Resembling a stylet.

Stylostome A feeding tube produced by chiggers and some other parasitic mites while attached to a host, formed by salivary secretions and through which lysed host tissue is drawn to the mouth.

Stylus (-i) A tapered, nonarticulated process; also, the annulated terminal part of the third antennal segment of some adult flies, for example, tabanids.

Subacute In relation to disease, between acute and chronic but with some acute features.

Subalar Below the wing.

Subclinical Relating to an early stage of a disease before clinical symptoms are apparent.

Subcosta The longitudinal vein near the anterior margin of the insect wing immediately behind and parallel to the costa.

Subcutaneous Beneath the skin.

Subcylindrical More or less cylindrical in shape.

Subfamily A taxonomic subdivision within a family containing one or more closely related genera, named after the type genus and typically ending in -inae.

Subgenital plate A sternal sclerite or process of the eighth abdominal segment in insects, covering, or adjacent to the gonopore.

Subgenus A taxonomic subdivision within a genus, the name of which is capitalized, italicized, and placed in parentheses following the genus name.

Submucosa A layer of loose connective tissue below a mucous membrane.

Suborder A major taxonomic subdivision within an order, containing a group of related superfamilies and families.

Subperiodic (-ity) Occurring at something less than regular intervals of time.

Subscutellum The ventral surface of a transversely infolded postscutellum of adult insects.

Succession Gradual colonization of a newly formed habitat, usually by a series of species assemblages, as in arthropod succession of cadavers or carrion.

Suid A member of the family Suidae, including domestic and wild pigs, hogs, warthog, and babirusa; swine.

Sulfluramid A fluorinated sulfonamide with insecticidal properties, often used in bait traps to control cockroaches, ants, and other insects.

Supportive therapy Treatment based on clinical signs and symptoms, without necessarily knowing or directly addressing the underlying cause of a medical problem.

Suppurative Producing or discharging pus.

Suprageneric In taxonomic classification, anything above the genus level.

Suprapylarian Growth or development of a parasite or pathogen in the midgut of a vector as in members of the protozoan subgenus *Leishmania* of the genus *Leishmania*; as opposed to peripylarian, which implies growth or development in the hindgut.

Supraspecific In taxonomic classification, anything above the species level.

Surra An infectious disease of domestic animals, especially horses, caused by *Trypanosoma evansi* and transmitted by the bite of tabanid flies; characterized by anemia, fever, and emaciation.

Susceptibility Lack of ability to resist an extraneous agent such as a pathogen, parasite, or drug.

Suslik Any of several Eurasian ground squirrels (Sciuridae), especially the European species *Citellus citellus*.

Suspect(ed) case A medical diagnosis based on reasonable but inconclusive evidence or laboratory testing.

Suture (-al, -ed) (1) In arthropods, a seam, seam-like line, or boundary between two sclerotized areas of the arthropod integument; (2) surgically sewn wounds or incisions.

Sweat flies Nonbiting muscid flies that are attracted to, and persistently feed on, animal perspiration and other body secretions; used most commonly with reference to members of the genus *Hydrotaea*.

Sylvatic Occurring in, affecting, or transmitted by wild animals.

Symbiont The smaller of two organisms in a symbiotic relationship between different species, which always benefits from the association.

Symbiosis (-tic) Any close and prolonged physical relationship between individuals of two different species; may be mutualistic, parasitic, or commensal.

Sympatry (-ic) Two or more species with the same or overlapping geographical distributions.

Symptom (-atic) Any perceptible change in the body or its functions indicative of a disease or other disorder; in a more restricted sense, referring to a subjective change, as opposed to an objective change, or sign (see *sign*).

Synanthropic Ecologically associated with humans.

Syndrome A group of signs and symptoms that collectively characterize or indicate a particular disease or abnormal condition.

Synergist (-istic) A compound or other agent that, when combined with another substance, interacts to increase the effectiveness of the latter; a nontoxic substance in a pesticide that increases the potency of the active ingredient.

Synonym One of two or more scientific names for the same taxon; generally applied to a previously used name that has been superseded by the currently accepted name; the relegated (nonvalid) name(s) is (are) called the junior synonym(s).

Systematics The study and classification of organisms and their relationships.

Systemic Pertaining to, or affecting, the entire body or a particular body system, for example, nervous system; translocated throughout the body, typically via the circulatory system.

Tabanid A member of the dipterous family Tabanidae, horse flies and deer flies.

Tachinid A member of the dipterous family Tachinidae, in which the larva is typically parasitic in or on other insects.

Tachycardia Excessively rapid heartbeat, over 100 beats per minute in human adult.

Tachypnea Abnormally rapid breathing or respiration.

Tagma (-ata) A group of successive segments forming a distinct section of an arthropod (especially insect) trunk, such as along the abdomen.

Tail head In ungulates such as cattle and horses, the base of the tail where it joins the rump.

Tapeworm A member of the class Cestoda, flat ribbon-like or tape-like parasites found as adults in the alimentary tract of vertebrate animals; typically develop as immature stages in other vertebrate or arthropod hosts.

Tarsomere A subsegment of the arthropod tarsus.

Tarsus (i) The distal-most apparent segment of the arthropod limb; actually the penultimate segment, usually bearing the claws and associated structures, or pretarsus.

Taxon (-a) A unit of classification for animals, plants, or microorganisms; for example, species, genus, family, order, phylum, kingdom, and so on.

Taxonomy The practice of describing, naming, identifying, and classifying organisms according to a system based on the rules of zoological nomenclature.

T cell Any of several lymphocytes that develop in the thymus and play an important role in the body's immune system by recognizing and destroying infected or malignant cells; *syn.*, *T lymphocyte.*

Tegmen (-ina) The sclerotized, leathery forewing of certain orders of insects, including cockroaches and grasshoppers; protects the more delicate hindwing when folded beneath.

Telmophage (-gy) See *Pool feeder.*

Telson The distal-most part of the insect abdomen, or opisthosoma of certain arachnids, bearing the anus; not a true segment; term applied to the terminal segment of the scorpion tail.

Temephos A nonsystemic organophosphorus insecticide; often applied to lakes, ponds, rivers, or wetlands to control larvae of aquatic Diptera such as those of mosquitoes and black flies; formerly used to control fleas and lice on animals.

Temperate region An area having a moderate climate (i.e., not tropical or subtropical); an area that experiences cooler winter months compared to summer months.

Teneral Pertaining to a recently molted immature or adult arthropod in which the integument has not yet hardened and is still pale compared to the final coloration.

Tentorium The internal skeleton of the insect head.

Tergal gland A gland associated with the tergum of the male cockroach that produces pheromones and other secretions upon which the female feeds during courtship and mating.

Tergite A sclerotized subdivision of the tergum of arthropods.

Tergo-trochanteral Referring to the origin and insertion of arthropod muscle between a tergum and trochanter.

Tergum (-a) The dorsal surface of an arthropod body segment.

Terminalia The posterior-most segments and associated structures of the insect abdomen, including modifications to form the external genitalia.

Terrapin Any of various web-footed turtles of the family Emydidae inhabiting fresh or brackish waters of North America, including Mexico; some turtles from other parts of the world are also sometimes called terrapins.

Territoriality The behavior of an animal in establishing and defending a defined geographic area.

Tertian Recurring at approximately 48-hour intervals, or every other day, or every third day; for example, tertian fever.

Tessellate (-ed) A mosaic composed of little squares, like a checkerboard; for example, color pattern on dorsum of abdomen of adult sarcophagid flies.

Testosterone The male sex hormone of vertebrates secreted primarily by the testes, which stimulates development of male sex organs, secondary sexual traits, and sperm.

Tetracycline A broad-spectrum antibiotic synthesized or derived from soil microorganisms of the genus *Streptomyces.*

Thallus (-i) A vegetative body not differentiated into true roots, stems, or leaves; characteristic of fungi and lichens.

Theileriosis An infection by a protozoan of the genus *Theileria*; for example, East Coast fever of cattle; theileriasis.

Theraphosid A member of the spider family Theraphosidae, or tarantulas.

Thermal fog An insecticidal fog produced by heating the product without degrading the active ingredient.

Thermoregulation The control, adjustment, or maintenance of body temperature within defined limits by physiological or behavioral means.

Thigmotaxis (-actic) Movement of an organism toward or away from a solid object in response to touch or a mechanical stimulus.

Thiocyanate A salt or ester of thiocyanic acid, formed when alkaline cyanides combine with sulfur.

Thorny-headed worm A member of the phylum Acanthocephala, parasitic worms with an eversible proboscis armed with spines used to attach to the gut wall of invertebrate and vertebrate hosts; *syn.*, thorn-headed worms.

Thrombocytopenia An abnormal decrease in the number of platelets in circulating blood.

Thrombosis (-es) the formation of a blood clot, or thrombus, in any part of the circulatory system.

Tibia (-ae) The fourth segment of the insect leg, between the femur and tarsus.

Tibial spur A spur usually located near the distal end of the insect or hard-tick tibia.

Tibio-tarsal claw The terminal grasping structure of an arthropod leg or palp, formed by the tarsal element opposing the tibia; for example, claws on the second and third pairs of legs of sucking lice.

Togavirus A genus of arboviruses belonging to the family Togaviridae.

Tormogen The epidermal cell that forms a cuticular depression at the base of a seta; *syn.*, tormogen cell.

Torticollis Spasmodic contractions of the neck muscles twisting the head to one side with the chin pointing to the other.

Toxemia The presence of a toxin in the blood disseminated from a site of bacterial infection or metabolic toxins resulting from disease or organ failure; *syn., blood poisoning.*

Toxicant A general term for any poison or poisonous agent.

Toxicosis (-es) A pathological condition resulting from a poison or toxin; systemic poisoning.

Toxin A poisonous substance produced by plants or animals.

Toxoplasmosis A disease caused by the protozoan *Toxoplasma gondii*; commonly transmitted to humans from infected cats or cat feces.

Trachea (-eae) (1) In insects and some other arthropods, one of a series of large respiratory tubes that connect externally to the spiracles and internally to the tracheoles; (2) in vertebrates, the windpipe or airway extending from the larynx into the thorax.

Tracheitis Inflammation of the trachea (in vertebrates).

Tracheobronchial Pertaining to both the trachea and bronchi or bronchioles.

Tracheole The smaller, finer tubules of the arthropod tracheal system.

Trachoma A contagious eye disease caused by infection of the conjunctiva and cornea by the bacterium *Chlamydia trachomatis*; characterized by inflammation, granulations, and scarring; a major cause of blindness in Africa and Asia.

Transmissible (-ility) Capable of being carried from one individual to another, as an infectious agent.

Transovarial Relating to the passage of any agent or substance from a female organism to its offspring via the ovary and egg.

Transovarial transmission The passage of a pathogen from a female organism to its offspring by infection of eggs while developing or retained in the ovary.

Transplacental infection The passage of a pathogen from mother to fetus via the placenta.

Transstadial transmission The passage of a pathogen in arthropods from one instar or developmental stage to the next.

Traumatic insemination An unusual form of copulation in bed bugs and other cimicids in which the male aedeagus penetrates the female body wall to deposit sperm directly into a specialized structure (spermalege) of the female reproductive system.

Tree hole Any rotted cavity in a tree; often collects rainwater and serves as a temporary breeding site for mosquitoes, biting midges, and other aquatic insects.

Triatomine Referring to members of the subfamily Triatominae, or kissing bugs (family Reduviidae).

Trichinosis A human disease caused by ingestion of larvae of the parasitic nematode *Trichinella spiralis* by eating raw or insufficiently cooked pork; the larvae can migrate through the intestines to muscles and tissues where they encyst and produce a range of symptoms.

Trichobothrium (-ia) A specialized sensory seta of arthropods characterized by a cuticular depression or socket at the base; mechanoreceptors for tactile stimuli.

Trichogen An epidermal cell that forms a seta; *syn.*, trichogen cell.

Tritonymph The third nymphal stage in the basic life cycle of some mites.

Tritosternum A medioventral, forked structure on the sternite between the first pair of legs of gamasid mites, believed to function in guiding fluids while the mite is feeding.

Trochanter The second, and typically smallest, segment of the insect and arachnid leg.

Troglobite (-tic) A cave-dwelling animal that has become specifically adapted to living in total darkness.

Trombiculid A member of the mite family Trombiculidae, or chiggers.

Trombiculosis An infestation with trombiculid mites (see *chiggers*).

Trombidiosis A synonym in the older literature for trombiculosis; no longer used.

Trophallaxis The direct transfer of food or other fluids between members of a colony of social insects such as ants and bees; may be mouth-to-mouth or anus-to-mouth.

Trophic Relating to nutrition.

Trophozoite A protozoan, notably a member of the class Sporozoa, in the active feeding stage of its life cycle.

Tropicopolitan A biogeographical term implying distribution of an organism throughout all the tropical regions of the world; for example, Tropicopolitan distribution.

Trumpet See *Respiratory horn.*

Trypanosome A flagellate protozoan of the genus *Trypanosoma*, family Trypanosomatidae; parasites in blood and tissues of humans and other animals, usually transmitted by insects.

Trypanosomiasis An infection or disease caused by a trypanosome.

Trypanotolerant Able to endure or resist the adverse effects of infection with trypanosomes.

Trypomastigote Any protozoan of the family Trypanosomatidae that has the typical form of a mature trypanosome, that is, with flagellum and undulating membrane.

Tsutsugamushi A disease of humans caused by the rickettsial organism *Orientia tsutsugamushi* transmitted by chiggers; *syn.*, *tsutsugamushi* fever, tsutsugamushi disease, scrub typhus, chigger-borne rickettsiosis.

Tuberculate (-ed) Covered with small rounded projections, or tubercles.

Tungiasis An infestation of the skin by the flea *Tunga penetrans.*

Turbinate A scroll-like spongy bone of the nasal passages of humans and other vertebrates.

Tylenchid A member of the nematode family Tylenchidae.

Typhoid An infectious enteric disease of humans caused by the bacterium *Salmonella typhi.*

Typhus Any of several forms of infectious disease caused by rickettsial organisms of the genus *Rickettsia* transmitted by lice, fleas, ticks, and chiggers; characterized by high fever and eruption of reddish spots on the skin; sometimes, incorrectly used to imply only louse-borne (epidemic) typhus.

Ulcer (-ate, -ed, -ous, -ation) An open sore or lesion of the skin or mucous membranes with necrosis and formation of pus.

Ultra-low volume Referring to highly concentrated formulations of pesticides that are applied in very small quantities (0.6–4.7 liters per hectare) as microscopic droplets within the size range of 0.1–50 μm in diameter (80% of the droplets must be within the range, 0.1–30 μm).

ULV See *Ultra-low volume.*

Umbilicus (-cal) Depressed scar in middle of abdomen denoting attachment of umbilical cord during fetal development; navel.

Ungulate A hooved mammal; includes members of the two orders Perissodactyla and Artiodactyla.

Univoltine Having one generation per year.

Uric acid A white nitrogenous waste product excreted by insects, most reptiles, and birds.

Urogenital Pertaining to both the urinary and reproductive systems of vertebrates.

Urticaria (-al) A vascular reaction of the skin characterized by eruption of pale wheals and intense itching; caused by, among other things, allergic reaction to insect bites and stinging hairs of caterpillars; *syn.*, hives, nettling rash.

Vaccine (-ation) A suspension of a weakened or killed pathogen, or part of a pathogen, used to inoculate an individual to stimulate the production of antibodies as a means of immunological protection against future infection.

Vaccination Inoculation with a vaccine.

Vasoconstriction Narrowing of the lumen of a blood vessel.

Vasodilatation An enlargement or stretching of a blood vessel, especially small arteries and arterioles; *syn.*, vasodilation.

Vasodilator A nerve, drug, or other agent that dilates a blood vessel.

Vector An organism, usually an arthropod, which transmits a disease agent from an infected to a noninfected animal or plant.

Vector competence The susceptibility of an arthropod species to infection with a pathogen or other parasite and its ability to transmit the acquired organism.

Vector-borne disease A disease involving a vector, usually an arthropod, as the means by which the causative agent is transmitted from one individual to another.

Vectorial capacity A quantitative summary of the basic ecological attributes of a vector relating to its ability to transmit a pathogen or parasite.

Venation See *Wing venation.*

Venereal Pertaining to sexual intercourse or the genitals; for example, venereal disease, venereal transmission.

Venom (-ous) A poison secreted by an animal and usually transmitted by a bite or sting.

Ventriculus The digestive part of the alimentary tract of insects; *syn.*, midgut, stomach.

Venule A small vein; in insects, a branch of a vein in the wing.

Vermiform Resembling a worm; elongate, cylindrical, and tapered or rounded at both ends; for example, vermiform larva.

Vernal Relating to or occurring in the spring.

Vertex The top of the head of an insect between the eyes.

Vertical transmission The passage of a pathogen from parent to offspring.

Vertigo A sensation of moving around in space, objects whirling around the individual, or falling as a result of disturbance to one's equilibrium.

Vesicant A chemical agent that causes blistering.

Vesicle A small sac or bladder-like elevation of the skin containing serous fluid; blister.

Vesicular dermatitis Inflammation of the skin marked by vesicles or small blisters.

Vesicular stomatitis A viral disease of horses, cows, and swine, characterized by erosive blisters in and around the mouth caused by rhabdoviruses of the genus *Vesiculovirus*; can be transmitted by insects during epizootics.

Wallaby Any of various marsupials related to, but generally smaller than, kangaroos and common to Australia, New Guinea, and adjacent islands; includes the genera *Macropus* and *Petrogale*.

Wallow A pool of water or mud where animals slowly and clumsily roll about.

Wapiti A large North American deer, *Cervus canadensis*, with large branched antlers in the male, commonly called American elk.

Warren An area where rabbits live in burrows; a colony of rabbits.

Waste lagoon A natural or artificial shallow body of water in which animal wastes are collected and undergo biodegradation.

Wasting condition A general deterioration of an animal's health characterized by loss of weight and strength, and emaciation; wasting disease.

Wasting disease See *Wasting condition*.

Wattle A fleshy, often wrinkled and brightly colored fold of skin hanging from the throat or neck of certain birds, such as chickens and turkeys.

Welt An unbroken elevation of the skin caused by an allergic reaction or by a lash or blow, as from a stick or whip.

Wettable powder A dry formulation of a pesticide that is mixed with water to form a suspension prior to making an application.

Wheal A small raised mark or swelling of the skin of short duration, as from an insect bite or allergic reaction; usually itches or burns.

Whipworm A slender, whip-shaped nematode of the family Trichuridae, especially *Trichuris trichiura*, which parasitizes the human intestine.

WHO World Health Organization, an agency of the United Nations established in 1948 with headquarters in Geneva, Switzerland.

Wing cell See *Cell*.

Wing venation The complete system and pattern of veins of an insect wing.

Wipe-on A chemical formulation that is topically applied to an animal with a cloth or other fabric, usually as a toxicant or repellent, to protect the animal from attack by biting insects.

Withers The highest part of the back of a horse, cow, sheep, and other livestock, located at the base of the neck between the shoulder blades.

Wool slippage Damage to the fleece of sheep caused by a heavy infestation of biting (chewing) lice, resulting in patches of lost hair.

Xanthine A yellow-white, nitrogenous, crystalline compound produced in the breakdown of purines to uric acid.

Xenodiagnosis A method of diagnosing an arthropod-transmitted disease by allowing an uninfected vector species to feed on the patient, then examining the arthropod for the presence of the infective organism following a suitable incubation period.

Yak An ox, *Bos grunniens*, of the Tibetan highlands of China, with a stout body, short legs, a thick shaggy coat of hair hanging to the ankles, and large upward-curved horns; both wild and domesticated.

Zoogeographic region A large-scale biogeographic division of the earth's surface defined by zoologists on the basis of characteristic flora and fauna, representing long periods of isolation and evolutionary distribution patterns.

Zoonosis (-es, -tic) A disease of wild or domestic animals that can be transmitted to humans.

Zoophagy (-ic) Feeding on live animals.

Zoophily (-ic) Attraction to, or preference for, feeding on animals.

Zooprophylaxis A precaution or preventive measure taken to protect animals from contracting or transmitting a disease agent; for example, immunization, quarantine, screened enclosures.

Zygote The cell formed by the union of two gametes, typically an ovum and sperm, prior to undergoing cleavage; fertilized egg.

Taxonomic Index (to genera, species and subspecies)

Note: Page numbers followed by "*f*" indicate figures and "*t*" indicate tables.

A

Abbreviata caucasica, 53, 54*t*
Acanthocheilonema, 177
Acanthocheilonema reconditum, 123*t*, 132
Acanthoscurria, 414
Acarus, 453
Acarus farris, 443, 444*f*
Acarus siro, 443, 444, 445*f*, 450, 452
Achipteria, 455, 485*t*, 487*t*
Acinetobacter, 50*t*
Acinetobacter baumannii, 74
Acomys, 437
Acronicta, 363
Acronicta lepusculina, 363
Acronicta oblinita, 355*f*, 363
Actinomyces pyogenes, 292
Actinoptera, 318
Actinopus, 413
Actinobacillus, 368
Actornithophilus, 61
Adolia, 362
Adoneta spinuloides, 359
Adoristes, 487*t*
Aedeomyia, 208*t*
Aedes, 14*f*, 33, 208*t*, 209, 215, 217, 218, 222, 226*t*, 228*t*, 558*t*
Aedes aegypti, 2, 23, 32, 182, 208, 209, 210*f*, 215, 215*f*, 216, 219, 220, 222, 226*t*, 229, 230, 230*t*, 231, 232, 232*f*, 234, 250, 252, 253, 547*t*, 558*t*
Aedes aegypti aegypti, 208
Aedes aegypti formosus, 208
Aedes africanus, 209, 226*t*, 229, 230*t*, 231
Aedes albopictus, 32, 33, 208, 209, 222, 226*t*, 229, 230*t*, 233, 233*f*, 234, 558*t*
Aedes bromeliae, 209, 230*t*, 231
Aedes cinereus, 211*f*
Aedes circumluteolus, 249
Aedes cordellieri, 229
Aedes dorsalis, 558*t*
Aedes fortunae, 247*t*
Aedes fulvus pallens, 211*f*
Aedes furcifer, 229, 230*t*, 231, 233
Aedes luteoceophalus, 209, 229, 230*t*, 231, 233
Aedes mcintoshi, 239, 249, 558*t*
Aedes melanimon, 27, 558*t*
Aedes metallicus, 230*t*, 231
Aedes opok, 229, 230*t*, 233
Aedes polynesiensis, 209, 226*t*, 229, 230*t*, 233, 247*t*
Aedes pseudoscutellaris, 209, 230*t*, 233
Aedes rotumae, 230*t*, 233
Aedes scutellaris, 230*t*, 233
Aedes serratus, 176
Aedes tabu, 247*t*
Aedes taylori, 229, 230*t*, 231, 233
Aedes tongae, 247*t*
Aedes triseriatus, 23, 558*t*
Aedes upolensis, 247*t*
Aedes vexans, 222, 235, 237*t*, 239
Aedes vittatus, 231
Aedes vogilax, 558*t*
Aegyptianella pulloum, 510
Aeromonas, 52*t*
Aeromonas hydrophila, 458
Africarus, 487*t*
Afrolistrophorus, 462
Agelena labrinthica, 414
Aglossa, 39*t*
Alacran, 399
Alacran tartarus, 399
Alcaligees faecalis, 52*t*
Alcelaphus buselaphus, 304
Alces alces, 275, 345
Aleochara, 112
Aleuroglyphus ovatus, 450, 452
Allodermanyssus sanguineus, 437
Allogalumna, 485*t*, 487*t*
Allolobophora, 283
Alloxasis, 105
Alouatta, 231, 327
Alphavirus, 223, 224*t*, 225–229, 557, 558*t*
Alphitobius diaperinus, 109, 110*f*, 111
Amara, 110
Amblyomma, 15*f*, 16*f*, 494–495, 494*t*, 503, 506*t*, 517, 525, 531, 533
Amblyomma americanum, 494, 499*f*, 504, 506*t*, 507*t*, 509, 509*f*, 516, 518, 519, 520, 521, 523, 525, 526, 529
Amblyomma hebraeum, 494, 507*t*, 509, 510*f*, 530, 536
Amblyomma maculatum, 494, 503, 504, 509, 509*f*, 518, 530
Amblyomma marmoreum, 538
Amblyomma nuttalli, 499
Amblyomma rotundatum, 497*f*
Amblyomma sparsum, 538
Amblyomma variegatum, 494, 507*t*, 509, 530, 533, 534, 536, 538
Ammotrechella stimpsoni, 411
Analges chelopus, 463*f*
Anaphe, 356
Anaphe infracta, 363
Anaplasma, 529
Anaplasma bovis, 531
Anaplasma centrale, 507*t*, 532
Anaplasma marginale, 271, 272, 506, 507*t*, 509, 510, 531, 532, 547*t*
Anaplasma ovis, 349, 507*t*
Anaplasma phagocytophilum, 506*t*, 507*t*, 508, 518*f*, 519, 531, 537, 547*t*
Anaticola anseris, 60*t*
Anaticola crassicornis, 60*t*
Anaxipha gracilis, 167
Ancyclostoma duodenale, 53
Androctonus, 397
Androctonus amoreuxi, 398*t*
Androctonus australis, 398*t*, 402*f*, 406, 406*f*
Androctonus crassicauda, 398*t*
Androctonus mauretanicus, 398*t*, 406
Androlaelaps, 458
Androlaelaps fahrenholzi, 434*f*
Anisomorpha buprestoides, 7, 7*f*
Anisomorpha ferruginea, 7
Anocentor, 508
Anolisimyia, 324
Anomalohimalaya, 494*t*, 495
Anomalurus, 250
Anopheles, 23, 28, 208, 208*t*, 209, 210*f*, 213*f*, 217, 218, 218*f*, 221, 222, 223, 226*t*, 228*t*, 229, 239, 240, 241, 242–243, 250, 253
Anopheles aconitus, 247*t*
Anopheles albimanus, 242, 242*t*
Anopheles anthropophagus, 242*t*, 247*t*
Anopheles anulipes, 249
Anopheles aquasalis, 247*t*
Anopheles arabiensis, 208, 247*t*
Anopheles atroparvus, 242*t*
Anopheles aztecus, 242*t*
Anopheles balabacentis, 247*t*, 250
Anopheles bancrofti, 242*t*, 247*t*
Anopheles barbirostris, 246, 247*t*
Anopheles bellator, 247*t*
Anopheles bwambae, 247*t*
Anopheles campestris, 242*t*, 247*t*
Anopheles claviger, 242*t*
Anopheles cruzi, 250, 251
Anopheles culcifacies, 242, 242*t*
Anopheles darlingi, 242, 242*t*, 246, 247*t*
Anopheles dirus, 242, 242*t*, 250, 251
Anopheles donaldi, 242*t*, 247*t*

Anopheles earlei, 212*f*
Anopheles elegans, 250
Anopheles farauti, 247*t*
Anopheles flavirostris, 247*t*
Anopheles freeborni, 221, 242*t*
Anopheles funestus, 229, 242, 242*t*, 246, 247*t*, 558*t*
Anopheles gambiae, 208, 212*f*, 215*f*, 216, 220, 242, 243*f*, 245, 247*t*, 251, 252, 547*t*
Anopheles hackeri, 250
Anopheles hermsi, 242, 242*t*
Anopheles hispaniola, 242*t*
Anopheles introlatus, 250
Anopheles koliensis, 247*t*
Anopheles kweiyangensis, 247*t*
Anopheles labranchiae, 242*t*
Anopheles letifer, 242*t*, 247*t*
Anopheles leucosphyrus, 247*t*
Anopheles maculatus, 247*t*, 251
Anopheles maculipennis, 211*f*, 243
Anopheles melas, 247*t*
Anopheles merus, 247*t*
Anopheles messeae, 242*t*
Anopheles minimus, 247*t*
Anopheles multicolor, 242*t*
Anopheles nigerrimus, 242*t*, 247*t*
Anopheles nili, 247*t*
Anopheles pauliani, 247*t*
Anopheles pharoensis, 242*t*
Anopheles philippinensis, 247*t*
Anopheles pseudopunctipennis, 242*t*
Anopheles punctimacula, 242*t*
Anopheles punctipennis, 242*t*
Anopheles punctulatus, 247*t*
Anopheles quadrimaculatus, 211*f*, 221, 242*t*, 558*t*
Anopheles sacharovi, 242*t*
Anopheles sergentii, 242*t*
Anopheles sinensis, 246
Anopheles stephensi, 182, 251
Anopheles subpictus, 247*t*
Anopheles umbrosus, 246
Anopheles vagus, 247*t*
Anopheles whartoni, 242*t*, 247*t*
Anoplocephala magna, 485*t*
Anoplocephala perfoliata, 485*t*
Anthela nicothe, 356
Anthocoris musculus, 83
Anthrenus, 39*t*
Anthrenus flavipes, 108*f*
Anthrenus scophulariae, 108*f*
Anthrenus verbasci, 108*f*
Antilocapra americana, 346
Antilope cervicapra, 479
Antricola, 495
Anuroctonus, 399, 405
Aotus, 231
Apatolestes, 262*t*
Apatolestes actites, 264, 265
Aphodius, 112
Aphonopelma, 420
Aphonopelma hentzi, 420
Aphonopelma seemani, 430
Apis, 371, 375, 377
Apis cerana, 387
Apis dorsata, 387
Apis florea, 387
Apis mellifera, 387, 388*f*
Apis mellifera scutellata, 388, 388*f*, 389
Apodemus, 442, 455
Apodemus, 514, 519
Apodemus sylvaticus, 132
Apollonia, 459
Aponomma, 494, 495, 503
Aproctella, 252
Aprostocetus hagenowii, 55
Arccyophora, 364
Arcyophora, 354
Arenavirus, 563*t*
Argas, 494*t*, 495
Argas arboreus, 503
Argas brumpti, 526
Argas miniatus, 510
Argas persicus, 495, 507*t*, 510, 532, 534
Argas reflexus, 495, 506*t*, 526
Argas sanchezi, 510
Argas walkerae, 507*t*, 525, 534
Argiope aurantia, 414
Argiope lobata, 414
Arilus cristatus, 83
Armigeres, 208*t*, 215, 222
Arvicola, 519
Arvicola amphibius, 524
Ascaris, 53
Ascaris lumbricoides, 53
Ascarops strongylina, 111
Ascodipteron, 341
Ascomatacarus, 455
Ascoschoengastia, 483
Ascoschoengastia indica, 128
Asfivirus, 558*t*
Aspergillus, 109, 443
Aspergillus penicilloides, 451
Aspidoptera falcate, 347
Ateles, 231
Athererus africanus, 250
Atherix, 146*f*
Atrax, 413, 414, 421
Atrax robustus, 414, 421, 421*f*
Attagenus, 39*t*
Atylotus, 262*t*, 269*t*
Atylotus agrestis, 271
Atylotus thoracicus, 264
Auchmeromyia senegalensis, 323, 323*f*
Audycoptes, 473
Auricanoetus, 481
Austenina, 297
Austroconops, 172
Austroconops macmillani, 169
Austrosimulium australense, 195*t*
Austrosimulium pestilens, 192, 195*t*, 203
Austrosimulium ungulatum, 195*t*, 198*t*
Autographa gamma, 354*f*
Automeris, 360
Automeris io, 355*f*, 357, 360, 360*f*
Autoserica castanea, 109
Aviculara surinamensis, 420*f*
Avipoxvirus, 249
Avitellina centripuncta, 485*t*, 486
Avitellina chalmersi, 485*t*
Avitellina goughi, 485*t*, 486
Avitellina lahorea, 485*t*, 486
Avitellina sudanea, 485*t*, 486
Avitellina tatia, 485*t*
Avitellina woodlandi, 485*t*
Ayurakita, 208*t*

B

Babesia, 512–513, 526, 526*f*, 527
Babesia bigema, 2, 507*t*, 509, 510, 527
Babesia bovis, 507*t*, 509, 510, 527, 547*t*
Babesia caballi, 508, 527, 528
Babesia canis, 508, 527
Babesia divergens, 506*t*, 508, 512, 513, 527
Babesia duncani, 513
Babesia equi, 527
Babesia gibsoni, 505
Babesia major, 506
Babesia microti, 506*t*, 508, 512, 513, 527
Babesia orientalis, 527
Babesia rossi, 506
Babesia ventorum, 506*t*
Babesia vogeli, 505
Bacillus, 109
Bacillus anthracis, 223, 269*t*, 270
Bacillus cereus, 52*t*
Bacillus sphaericus, 167, 253
Bacillus subtilis, 51, 52*t*
Bacillus thuringiensis, 167, 203, 253, 273
Baduma robusta, 430
Balataea, 357
Bandicota, 127
Baraimlia fuscipes, 120
Bartonella, 129, 349
Bartonella bacilliformis, 159*t*, 161, 161*f*, 547*t*
Bartonella clarridgeiae, 129
Bartonella elizabethae, 129
Bartonella henselae, 129
Bartonella melophagi, 342
Bartonella quintana, 67, 74, 75*t*
Bartonella schoenbuchensis, 347, 348*t*
Basilia, 349
Basilia hispida, 342
Basilla boardmani, 341*f*
Batrachomyia, 319, 319*f*
Beaveria bassiana, 133
Belisarius xambeui, 399
Bembix texana, 273
Berlinia, 297
Bertiella, 455
Bertiella mucronata, 455
Bertiella studeri, 455
Bioculus, 400
Bironella, 208*t*
Blaberus, 45
Blaberus craniifer, 52*t*
Blaps, 107
Blarinobia simplex, 461
Blastocrithidia, 271
Blatella germanica, 44*f*, 49–50, 49*f*, 51*f*, 52*t*, 54*t*
Blatta lateralis, 46–47, 46*f*, 52*t*
Blatta orientalis, 44*f*, 46, 46*f*, 52*t*, 54*t*
Blattella, 43
Blattella asahinai, 50
Blomia tropicalis, 450
Bodyaia sturnellus, 483
Bombus, 371, 387, 387*f*
Bombyx, 363
Bombyx mandarina, 367
Bombyx mori, 367
Bombyx quercus, 363
Bombyx rubi, 363
Boophilus, 494, 500, 509

Boophilus annulatus, 2, 500*f*, 527
Boophilus decoloratus, 510, 527
Boophilus geigyi, 510, 527
Boophilus microplus, 527
Boopona, 323
Borrelia, 26, 506*t*, 522–524, 532, 532*f*
Borrelia afzelii, 506*t*, 508, 520, 521, 522
Borrelia anserina, 507*t*, 510, 532
Borrelia bissettii, 506*t*, 520, 522
Borrelia burgdorferi, 26, 129, 132, 270, 506*t*, 507*t*, 508, 512, 520–522, 521*f*, 523, 532, 536, 547*t*
Borrelia caucasica, 523
Borrelia coriaceae, 523, 532
Borrelia crocidurae, 523
Borrelia dipodilli, 523
Borrelia duttoni, 129, 510, 523
Borrelia garinii, 506*t*, 508, 520, 522
Borrelia hermsii, 510, 523, 524
Borrelia latyschewi, 523
Borrelia lonestari, 523, 532
Borrelia lusitanae, 520, 522
Borrelia mazzotti, 523
Borrelia merionesi, 523
Borrelia microti, 523
Borrelia miyamatoi, 523, 532
Borrelia parkeri, 523, 524
Borrelia persica, 523
Borrelia recurrentis, 2, 73–74, 75*t*
Borrelia spielmanii, 520
Borrelia theileri, 532
Borrelia turicatae, 523
Borrelia valaisiana, 522
Borrelia venezuelensis, 523
Bos, 528
Bos javanicus, 481
Bothriocroton, 494*t*, 495, 503
Botoydes, 364
Bovicola, 64, 68
Bovicola bovis, 60*t*, 62*f*, 68, 68*f*, 70, 76*f*, 77
Bovicola caprae, 60*t*
Bovicola crassipes, 60*t*
Bovicola equi, 60*t*, 70
Bovicola limbata, 60*t*
Bovicola ocellata, 60*t*
Bovicola ovis, 60*t*, 68, 70, 349
Brachypelma, 420
Brachystegia, 297
Bradysia, 143
Breinlia, 252
Brucella abortis, 129, 132
Brugia malayi, 23, 243, 244–245, 244*f*, 246, 247*t*, 252, 547*t*
Brugia pahangi, 252
Brugia timori, 243, 244–245, 244*f*, 246
Brumptomyia, 154, 155, 158
Bubalus bubalis, 480
Bufolucilia, 311*t*
Bufolucilia silvaraum, 321
Burkholderia mallei, 132
Burkholderia pseudomallei, 132
Buthus, 405
Buthus occitanus, 398*t*, 406

C

Cacobius, 108
Caenolestocoptes, 473
Calchas, 399
Callimorpha, 362
Callipepla californica, 184
Calliphora, 36, 311*t*, 320
Calliphora vicina, 39*t*, 321*f*
Calliphora vomitoria, 39*t*
Calliteara pundibunda, 356
Caloglyphus berlesei, 452
Calomys musculinus, 473
Calyptra, 354, 357, 365, 366
Calyptra bicolor, 366
Calyptra eustrigata, 354*f*
Calyptra fasciata, 366
Calyptra ophideroides, 366
Calyptra parva, 366, 366*f*
Calyptra pseudobicolor, 366
Camelus, 471
Camponotus, 377
Campylobacter jejuni, 52*t*
Canis latrans, 471
Canis lupus, 471
Cannabis sativa, 37
Canthon, 112
Caparinia, 474, 477
Caparinia erinacei, 477
Caparinia tripilis, 477
Capra ibex, 480
Capra ibex nubiana, 479
Capreolus capreolus, 471
Caraboctonus, 399
Carabodes, 485*t*, 487*t*
Carcinops, 112
Cardiofilaria, 252
Carios, 494*t*, 495, 503
Carios kelleyi, 497*f*, 505, 511, 526
Carpoglyphus lactis, 443, 443*f*
Catocala, 356, 357
Cavernicola, 86
Cazierius, 400
Cebus, 231
Cebus capucinua, 471
Cediopsylla simplex, 116*t*, 120
Centruroides, 397, 403, 404, 405, 406
Centruroides elegans, 407
Centruroides exilicauda, 397, 398*t*, 406
Centruroides guanensis, 397
Centruroides hentzi, 397
Centruroides infamatus, 407
Centruroides limpidus, 398*t*, 407
Centruroides noxius, 407
Centruroides sculpturatus, 397
Centruroides suffusus, 407
Centruroides vittatus, 397, 398*f*, 402*f*, 403*f*, 406, 407
Cephalopina titillator, 329*f*, 330
Cephenemyia, 317, 329, 330
Cephenemyia apicata, 330
Cephenemyia jellisoni, 330, 330*f*
Cephenemyia phobifer, 330
Cephenemyia pratti, 330
Cephenemyia trompe, 330
Cephius, 485*t*
Ceratophyllus, 124
Ceratophyllus gallinae, 116*t*, 117*f*, 123, 131
Ceratophyllus niger, 116*t*, 131
Ceratoppia, 485*t*, 487*t*
Ceratozetes, 485*t*, 487*t*
Cercophonius, 400
Cercopithecus, 231, 471
Cercopithecus aethiops, 229
Ceroxys latiusculus, 286
Cervus canadensis, 345, 476
Cervus elaphus, 471, 476
Cetonia, 111
Chactopsis, 399
Chaeopsestis, 365
Chaeopsestis ludovicae, 364*f*, 365, 365*f*
Chaerilus, 398
Chagasia, 208*t*
Chalcosia, 356
Chaldonia, 459
Chalybion, 382
Chaoborus astictopus, 144, 144*f*
Cheiracanthium, 415, 422
Cheiracanthium inclusum, 422
Cheiracanthium japonicum, 422
Cheiracanthium mildei, 413, 422
Cheiracanthium mordax, 422
Cheiracanthium punctorium, 422
Chelopistes meleagridis, 60*t*
Cheyletiella blakei, 460, 460*f*
Cheyletiella parasitivorax, 460, 460*f*, 461, 484
Cheyletiella yasguri, 460, 461*f*
Cheyletomorpha lepidopterorum, 444
Cheyletus eruditus, 444, 452
Cheyletus malaccensis, 444
Chilomastix, 285
Chirnyssoides, 471
Chirobia, 471
Chirodiscoides caviae, 462
Chirophagoides, 471
Chiroptonyssus, 438
Chiroptonyssus robustipes, 439–440, 457
Chlamydia trachomatis, 286
Chlethrionomys, 519
Chlethrionomys glareolus, 514
Chlorotabanus, 262*t*
Chlorotabanus crepuscularis, 266
Choanotaenia infundibulum, 110, 110*f*, 289
Chorioptes, 474, 476–477
Chorioptes bovis, 476–477, 476*f*
Chorioptes texanus, 476
Chrysomya, 311*t*, 320, 321
Chrysomya bezziana, 140*f*, 311*t*, 320, 321
Chrysomya megacephala, 321*f*, 322
Chrysomya rufifacies, 321, 322
Chrysops, 261, 262*t*, 263, 263*f*, 263*t*, 264, 264*f*, 265, 266, 267, 269*t*
Chrysops atlanticus, 269
Chrysops callidus, 262*f*, 264, 265*f*
Chrysops cincticornis, 264, 265*f*
Chrysops dimidiatus, 269*t*, 270
Chrysops discalis, 270
Chrysops fuliginosis, 266
Chrysops silaceus, 269*t*, 270
Cimex, 14*f*, 93, 94
Cimex brevis, 97
Cimex hemipterus, 93, 94, 94*f*, 95, 96
Cimex lectularius, 89, 94, 94*f*, 95, 96, 97
Cimex pilosellus, 97
Cimex pipistrelli, 97
Cimexopsis nyctalis, 97
Cistudinomyia, 324
Citellophilus, 125
Citrobacter, 52*t*
Cittotaenia, 487

Cittotaenia pusilla, 487
Clethrionomys, 116, 484
Clogmia, 153, 154, 155*f*, 156, 157
Clogmia albipunctatus, 158
Clostridium bifermentans, 166
Clostridium novii, 52*t*
Clostridium perfringens, 52*t*
Clubiona obesa, 422*f*
Cnephia, 140*f*
Cnephia ornithophilia, 195*t*, 198*t*, 200
Cnephia pecuarum, 195*t*, 200, 203
Cobboldia, 331
Cochliomyia hominivorax, 311*t*, 321, 322*f*, 332, 336
Cochliomyia macellaria, 39*t*, 309, 311*t*, 322, 322*f*
Coelomomyces, 253
Coelomycidium simuli, 203
Coelotes obesus, 414, 415
Coloradia pandor, 355*f*
Colpocephalum, 64
Coltivirus, 506*t*
Columba livia, 346, 349
Columbicola columbae, 62*f*
Comperia merceti, 55
Conicera tibialis, 146, 147*f*
Conispiculum, 252
Connochaetes taurinus, 471
Conorhinopsylla, 119
Copris, 108
Copris minutus, 108*f*
Coquillettidia, 208, 208*t*, 209, 210, 215, 217, 218, 219, 222
Coquillettidia perturbans, 212*f*, 226, 226*t*, 235
Coquillettidia venezualensis, 176
Cordylobia anthrophaga, 322–323, 323*f*
Corethrella, 144
Corethrella brakeleyi, 144
Corethrella wirthi, 144
Corvus brachyhynchos, 234
Corynebacterium lipoptenae, 348*t*
Cosmiomma, 494*t*, 495
Cowdria ruminantium, 529
Coxiella burnetii, 123*t*, 128, 506*t*, 507*t*, 519–520, 533, 547*t*
Creophilus maxillosus, 39*t*
Cricetomys gambianus, 127
Cricetus, 325
Cryptosporidium, 285
Crypturellis soui, 64
Ctenocephalides, 123*t*, 460
Ctenocephalides canis, 116*t*, 121, 121*f*, 130
Ctenocephalides felis, 14*f*, 16*f*, 116*t*, 117*f*, 118*f*, 121, 121*f*, 123, 128, 130
Ctenophthalmus, 125
Cuclutogaster heterographus, 60*t*, 71
Culex, 20, 23, 32, 33, 208, 208*t*, 209, 210*f*, 213*f*, 215, 217, 218, 218*f*, 222, 223, 234, 236, 250, 253, 558*t*
Culex acossa, 228*t*
Culex annulirostris, 226*t*, 229, 230*t*, 247*t*, 249, 558*t*
Culex australicus, 208
Culex bitaeniorhynchus, 247*t*
Culex cedecei, 228, 228*t*
Culex fuscocephala, 236
Culex gelidus, 230*t*, 236
Culex globocoxitus, 208
Culex molestus, 208, 230*t*, 246, 247*t*
Culex nigripalpus, 226*t*, 230*t*, 236
Culex ocassia, 227
Culex pallens, 208, 247*t*
Culex panocossa, 227, 228*t*
Culex pipiens, 32, 33, 167, 182, 208, 211, 218*f*, 230*t*, 234, 235, 236, 239, 246, 558*t*
Culex pipiens fatigans, 2
Culex pipiens quinqufasciatus, 547*t*
Culex portesi, 228*t*
Culex pseudovishnui, 236
Culex quinquefasciatus, 20, 22, 23, 176, 208, 214*f*, 222, 223, 230*t*, 235, 235*f*, 236, 245, 246, 247, 247*t*
Culex restuans, 210*f*
Culex sitiens, 247*t*
Culex stigmatosoma, 236
Culex taeniopus, 227, 228*t*
Culex tarsalis, 25, 226*t*, 227, 230*t*, 235, 236, 238, 558*t*
Culex theileri, 239
Culex tritaeniorhycus, 230*t*, 235, 236, 559*t*
Culex univittatus, 229, 230*t*, 234
Culex vishnui, 230*t*, 236
Culicinomyces, 253
Culicoides, 14*f*, 169, 170, 170*f*, 171, 173, 176, 179, 180, 185–186, 347, 558*t*
Culicoides actoni, 180
Culicoides adersi, 184*t*
Culicoides arakawae, 184, 184*t*
Culicoides arboricola, 183, 184, 184*t*
Culicoides austeni, 175*t*, 177
Culicoides barbosai, 173, 174, 175*t*, 177
Culicoides baueri, 183
Culicoides biguttatus, 172
Culicoides bolitinos, 180, 182, 183
Culicoides bottimeri, 184
Culicoides brevitarsis, 174*t*, 178, 180, 185
Culicoides chiopterus, 185
Culicoides circumscriptus, 172, 184*t*
Culicoides crepuscularis, 172, 184*t*
Culicoides debilipalpis, 173, 180
Culicoides denningi, 172
Culicoides diabolicus, 174*t*
Culicoides downesi, 184*t*
Culicoides dycei, 174*t*
Culicoides edeni, 183, 184, 184*t*
Culicoides filarifer, 180
Culicoides floridensis, 173
Culicoides fulvithorax, 177
Culicoides fulvus, 174*t*, 180
Culicoides furens, 170*f*, 172, 173, 174, 175*t*, 177, 186
Culicoides grahamii, 175*t*, 177
Culicoides gulbenkiani, 174*t*, 180
Culicoides guttifer, 184*t*
Culicoides guttipennis, 172
Culicoides haematopotus, 172, 183, 184*t*
Culicoides hinmani, 174*t*, 184*t*
Culicoides hollensis, 172, 174
Culicoides hortensis, 177
Culicoides imicola, 170, 174*t*, 178, 180, 182, 183, 558*t*
Culicoides impunctatus, 184, 185
Culicoides inornatipennis, 177
Culicoides insignis, 174*t*, 178, 180, 185
Culicoides knowltoni, 183, 184*t*
Culicoides krameri, 177
Culicoides kumbaensis, 177
Culicoides lahillei, 177
Culicoides magnus, 180
Culicoides marksi, 174*t*
Culicoides melleus, 172, 174
Culicoides milneri, 174*t*, 177
Culicoides mississippiensis, 172, 173, 174
Culicoides nanus, 183
Culicoides niger, 172
Culicoides nubeculosus, 172, 175*t*, 184*t*, 185
Culicoides obsoletus, 170, 174*t*, 175*t*, 180, 185
Culicoides occidentalis, 169
Culicoides oxystoma, 172, 174*t*, 180, 183
Culicoides paraensis, 173, 174, 174*t*, 175*t*, 176, 177, 183
Culicoides peregrinus, 174*t*, 180
Culicoides phlebotomus, 175*t*, 177
Culicoides pulicaris, 180, 185
Culicoides punctatus, 185
Culicoides pungens, 175*t*
Culicoides pycnostictus, 177, 180
Culicoides ravus, 177
Culicoides rutshuruensis, 177
Culicoides scoticus, 180
Culicoides shulzei, 180, 184*t*
Culicoides sonorensis, 158, 169, 172, 173*f*, 178, 179, 180, 181, 183, 185, 186, 558*t*
Culicoides sphagnumensis, 184*t*
Culicoides spinosus, 185
Culicoides stellifer, 172, 173, 180, 185
Culicoides stilobezzioides, 184*t*
Culicoides travisi, 172
Culicoides variipennis, 169, 171, 172, 173, 174*t*, 175*t*, 181, 182, 185, 186
Culicoides venustus, 172, 185
Culicoides victoriae, 185
Culicoides vitshumbiensis, 177
Culicoides wadi, 180
Culicoides wisconensis, 172
Culicoides zuluensis, 180
Culiseta, 208, 208*t*, 209, 215, 217, 222, 559*t*
Culiseta annulata, 239
Culiseta inornata, 216, 220, 237*t*, 238
Culiseta melanura, 22, 226, 226*t*
Culiseta ochroptera, 229
Cupiennius, 414
Cuterebra, 317, 325, 325*f*, 326*f*, 332
Cuterebra baeri, 327
Cuterebra emasculator, 317
Cuterebra fontinella, 325*f*, 326
Cyanocitta cristata, 184
Cyclocephala borealis, 109
Cylindrothorax bisignatus, 105, 105*t*
Cylindrothorax dusalti, 105, 105*t*
Cylindrothorax melanocephalus, 105, 105*t*
Cylindrothorax picticollis, 105, 105*t*
Cylindrothorax ruficollis, 105*t*
Cynomya, 311*t*, 320
Cynomys, 125, 484
Cynomys ludovicianus, 484
Cynopterocoptes, 471
Cyprinus, 253
Cyrnea colini, 54*t*, 55
Cytauxzoon, 512, 526, 527
Cytauxzoon felis, 507*t*, 528

Cytodites, 478, 483
Cytodites nudus, 478, 483, 483*f*
Cytonyssus, 483

D

Damalina, 63, 78
Dasybasis hebes, 252
Dasymutilla occidentalis, 381, 381
Datana, 368
Datana integerrima, 368, 368*f*
Deinocerites, 208, 208*t*, 220, 222
Deinocerites cancer, 216
Deinocerites pseudes, 226*t*, 227
Deltaretrovirus, 562*t*
Demodex, 445, 446, 447, 467, 470, 478
Demodex bovis, 468
Demodex brevis, 445
Demodex canis, 467–468, 467*f*, 468*f*
Demodex caprae, 468
Demodex cati, 468
Demodex criceti, 467
Demodex folliculorum, 445, 446*f*
Demodex tauri, 468
Dendrolimus, 355–356, 356*f*
Deraiophoronema, 252
Dermacentor, 494, 494*t*, 495, 506*t*, 507*t*, 508, 514, 517, 525
Dermacentor albipictus, 494, 503, 505, 515
Dermacentor andersoni, 2, 494, 506, 506*t*, 507*t*, 514, 515, 516, 517, 525, 531, 533, 534, 558*t*
Dermacentor marginatus, 494, 505, 506*t*, 508, 517, 531
Dermacentor nitens, 494, 508
Dermacentor occidentalis, 494, 515, 531, 534
Dermacentor parumapertus, 515
Dermacentor reticulatus, 494, 506*t*, 508, 517, 527
Dermacentor silvarum, 514
Dermacentor variabilis, 494, 498*f*, 499, 500, 502, 503, 504, 505, 506, 506*t*, 507*t*, 508*f*, 516, 517, 525, 528, 531, 534
Dermanyssus, 14, 15*f*, 436
Dermanyssus americanus, 437, 456, 484
Dermanyssus gallinae, 436–437, 437*f*, 438, 456, 484
Dermanyssus hirundinis, 437, 456
Dermatobia hominis, 316, 319, 324, 325, 326–327, 326*f*, 332, 333
Dermatophagoides, 450, 451
Dermatophagoides farinae, 450, 451, 451*f*, 452
Dermatophagoides pteronyssinus, 450, 451, 452
Dermatophagoides scheremetewski, 450
Dermatophilus congolensis, 533
Dermestes, 39*t*
Dermestes maculatus, 111
Diachlorus, 262*t*
Diachlous ferragutus, 262*f*
Dichrocoelium dendriticum, 392
Didelphis virginiana, 127
Didymocentrus, 400
Dipetalogaster maxima, 88
Dipetalonema, 177
Dipetalonema dracunculoides, 347, 348*t*
Dipetalonema reconditum, 78
Dipetalonema spirocauda, 75*t*
Diplicentrus, 400
Diploptera punctata, 45, 52*t*
Dipylidium caninum, 75, 75*t*, 78, 123*t*, 130, 132
Dirofilaria, 252
Dirofilaria carynodes, 252
Dirofilaria immitis, 251, 251*f*, 252*f*
Dirofilaria magnilarvatum, 252
Dirofilaria repens, 252
Dirofilaria roemeri, 252, 272
Dirofilaria scapiceps, 252
Dirofilaria subdermata, 252
Dirofilaria tenuis, 252
Dirofilaria ursi, 198*t*, 202, 252
Dirpha sabina, 361
Dirphia, 357
Dirphia multicolor, 361
Dolichovespula, 372, 383–385
Dolichovespula arenaria, 372*t*, 383, 384*t*
Dolichovespula maculata, 372*t*, 383, 384*t*
Doloisia, 483
Dolomedes fimbriatus, 415
Dometorina suramerica, 455
Doratifera, 357
Dorcadia ioffi, 116*t*, 131
Draschia megastoma, 289
Drosophila, 140*f*, 278*f*, 309, 318–319
Drosophila busckii, 318
Drosophila funebris, 148, 318
Drosophila hydei, 318
Drosophila immigrans, 318
Drosophila melanogaster, 148, 148*f*, 311*t*, 318
Drosophila repleta, 148, 318
Drosophila simulans, 318
Drosophila virilis, 318
Dysdera, 415
Dysdera crocati, 415

E

Eadiea brevihamata, 461
Echidnophaga gallinacea, 116*t*, 122, 122*f*, 131
Echidnophaga larina, 116*t*, 132
Echidnophaga myrmecobii, 116*t*, 131
Echinonyssus, 458
Echinophthirius horridus, 61*t*, 75*t*
Ectemina, 192
Ehrlichia, 518–519, 529–530
Ehrlichia canis, 505, 507*t*, 529, 547*t*
Ehrlichia chaffeensis, 506*t*, 509, 518, 518*f*, 519, 547*t*
Ehrlichia equi, 519
Ehrlichia ewingii, 506*t*, 507*t*, 509, 519, 529, 531
Ehrlichia ovina, 531
Ehrlichia phagocytophilia, 519
Ehrlichia ruminantium, 507*t*, 509, 530, 530*f*, 531
Eilema, 357
Eimeria, 109
Elaeophora schneideri, 269*t*, 272
Elassoctenus harpax, 416
Eleodes, 107
Elepantoloemus indicus, 323
Enhydra lutris, 482
Entamoeba, 149, 285
Enterobacter, 52*t*
Enterobius vermicularis, 53
Enterococcus, 52*t*
Eovbia apicfusca, 105
Epanaphe, 356
Epicauta, 16*f*, 105, 105*f*
Epicauta cinerea, 105*t*
Epicauta fabricii, 109
Epicauta flavicornis, 105*t*
Epicauta funebris, 109
Epicauta hirticornis, 105*t*
Epicauta lemniscata, 109
Epicauta maculata, 104*f*, 105, 105*t*
Epicauta occidentalis, 109
Epicauta pennsylvanica, 104*f*, 105, 105*t*, 109
Epicauta sapphirina, 105*t*
Epicauta temexa, 109
Epicauta tomentosa, 105*t*
Epicauta vestita, 105*t*
Epicauta vittata, 104*f*, 105, 105*t*, 109
Epicoma, 356
Epidermoptes bilobatus, 464, 465*f*
Epidermoptes odontophori, 464
Epilohamminia, 487*t*
Epipagis, 364
Erasmia, 356
Eremaeus, 485*t*, 487*t*
Erethizon dorsatum, 472
Eretmapodites, 208*t*
Eretmapodites chrysogaster, 220
Erinaceus europaeus, 477
Eristalinus, 147
Eristalis, 147, 311*t*
Eristalis arbustorum, 318
Eristalis dimidiata, 318
Eristalis tenax, 147, 147*f*, 318, 318*f*
Erysipelothrix rhusiopathiae, 129
Escherichia, 109
Escherichia coli, 51, 52*t*, 146, 285, 286
Eucalliphora, 311*t*, 320
Eucalyptus, 363
Euchaetes, 362
Euchaetes egle, 362
Euchlaena pectinaria, 364
Euclea delphinii, 355*f*, 359, 360*f*
Euhoploplsyllus glacialis, 116*t*, 119
Eulindana, 75*t*
Eumacronychia, 324
Eumusca, 290
Euoniticellus, 112
Eupelops, 487*t*
Euproctis, 356*f*, 368
Euproctis chrysorrhoea, 355, 361
Euproctis edwardsi, 368
Euproctis flava, 362
Euproctis similis, 355, 362
Euproctis subflava, 355
Euroglyphus, 450, 451
Euroglyphus maynei, 450, 451, 451*f*
Eurotium, 443
Eurycotis floridana, 44, 48, 49*f*
Euschoengastia, 459
Euscorpius, 399, 403
Eutrombicula, 455
Eutrombicula alfreddugesi, 459
Eutrombicula belkini, 38
Eutrombicula sarcina, 459
Euzetes, 487*t*

F

Fannia, 275, 280, 281, 282, 287, 313, 319, 319*f*, 321, 327
Fannia canicularis, 276*t*, 279*f*, 282*f*, 284, 284*f*, 287, 311*t*, 319
Fannia femoralis, 276*t*
Fannia manicata, 319
Fannia scalaris, 39*t*, 276*t*, 284, 319
Fannia thelaziae, 287, 290
Felicola subrostrata, 60*t*, 62*f*, 71, 78
Felis tigris, 472
Ficalbia, 208*t*
Filodes, 354
Filodes mirificalis, 364, 364*f*
Filovirus, 563*t*
Flavivirus, 223, 224*t*, 229, 506*t*, 507*t*, 557, 558*t*
Folyella, 252
Forcipomyia, 169, 170, 172
Forcipomyia townsvillensis, 185
Francisella tularensis, 75*t*, 123*t*, 129, 223, 269*t*, 270, 270*f*, 506*t*, 507*t*, 524–525, 532–533, 547*t*
Frankliniella tritici, 7
Fundulus, 253
Furcoribula, 487*t*

G

Gahrliepia, 455, 459, 483
Galago, 231
Galindomyia, 208*t*
Galumna, 455, 485*t*, 486, 487*t*
Gambusia affinis, 253
Gasterophilus, 317, 331–332
Gasterophilus haemorrhoidalis, 331
Gasterophilus intestinalis, 316, 331, 331*f*
Gasterophilus nasalis, 316, 331
Gasterophilus nigricornis, 331
Gasterophilus pecorum, 331
Gastronyssus bakeri, 478, 484
Gazalina, 356
Gazella thompsonii, 471
Gedoelstia cristata, 330
Geotrupes, 111
Giardia, 285
Glaucomys volans, 73, 128
Gliricola porcelli, 60*t*, 79
Glossina, 86*f*, 297, 298*f*, 299*f*
Glossina austeni, 297, 301
Glossina brevipalpis, 297, 304*t*
Glossina fusca, 297, 304*t*
Glossina fuscipes, 297, 304, 304*t*
Glossina fuscipleuris, 299*f*, 304*t*
Glossina longipalpis, 297, 304*t*
Glossina longipennis, 297
Glossina medicorum, 297, 301
Glossina morsitans, 300*f*, 302*f*, 304, 304*t*
Glossina pallidipes, 297, 304, 304*t*, 307
Glossina palpalis, 2, 297, 304, 304*t*
Glossina swynnertoni, 297, 304, 304*t*
Glossina tabaniformis, 297, 304*t*
Glossina tachinoides, 297, 304, 304*t*
Glossina vanhoofi, 304*t*
Glycyphagus domesticus, 450, 450*f*, 487
Glycyphagus hypudaei, 435*f*
Godonela eleonora, 364
Gongylonema, 54*t*, 55
Gongylonema neoplasticum, 54*t*, 55
Gongylonema pulchrum, 53, 54*t*, 55, 111
Goniocotes gallinae, 60*t*
Goniops, 261, 262*t*, 263*f*
Goniops chrysocoma, 264
Gonioides dissimilis, 60*t*
Gonioides gigas, 60*t*
Guntherana, 459
Guntheria, 459
Gymnopais, 190, 191, 192, 193, 194
Gyropus ovalis, 60*t*, 62*f*, 79
Gyrostigma, 331

H

Habibiella, 400
Habronema microstoma, 291
Habronema muscae, 289
Hadogenes, 400
Hadogenes troglodytes, 400
Hadronyche, 414, 421
Hadronyche cerberea, 421
Hadronyche formidabilis, 421
Hadronyche infensa, 421, 421*f*
Hadronyche versuta, 421
Hadruroides, 399
Hadrurus, 399
Haemagogus, 208, 208*t*, 209, 215, 217, 222, 229, 230*t*, 231, 558*t*
Haemagogus lucifer, 221*f*
Haemaphysalis, 494, 494*t*, 506, 506*t*, 514, 517, 525
Haemaphysalis chordeilis, 533
Haemaphysalis leachi, 527
Haemaphysalis leporispalustris, 494, 503, 506, 506*t*, 515, 516
Haemaphysalis longicornis, 494
Haemaphysalis neumanni, 514
Haemaphysalis punctata, 494, 506, 534
Haemaphysalis spinigera, 494, 506, 506*t*
Haematobia atripalpis, 290
Haematobia irritans, 277*f*, 279*f*, 282*f*, 291*f*, 292
Haematobia irritans exigua, 284, 291–292
Haematobia irritans irritans, 276*t*, 284, 291–292
Haematobosca alcis, 275
Haematomyzus elephantis, 60*t*, 62*f*, 64, 78
Haematopinus, 63, 68
Haematopinus asini, 64, 70
Haematopinus eurysternus, 61*t*, 69, 69*f*, 76*f*
Haematopinus quadripertusus, 61*t*, 69, 70
Haematopinus suis, 61*t*, 63*f*, 70, 75*t*, 77
Haematopinus tuberculatus, 61*t*, 68
Haematopota, 261, 262*t*, 264, 269*t*
Haematopota mericana, 261
Haematosiphon inodorus, 97
Haemodipsus ventricosus, 61*t*, 71, 79
Haemogamasus, 458
Haemogamasus liponyssoides, 440
Haemogamasus pontiger, 440
Haemophilus influenzae, 149
Haemoproteus, 183, 183*f*, 250, 347, 348, 349
Haemoproteus brayi, 183
Haemoproteus columbae, 348*t*, 349
Haemoproteus danilewskyi, 184, 184*t*
Haemoproteus desseri, 184*t*
Haemoproteus fringillae, 184*t*
Haemoproteus kochi, 183
Haemoproteus lophortyx, 184
Haemoproteus maccallumi, 348*t*
Haemoproteus mansoni, 184*t*
Haemoproteus meleagridis, 178, 183–184, 184*t*
Haemoproteus metchnikovi, 272
Haemoproteus nettionis, 184*t*
Haemoproteus sacharovi, 348*t*
Haemoproteus tophortyx, 348*t*
Haemoproteus velans, 184*t*
Hafnia alvei, 52*t*
Halarachne, 481, 482
Halarahne miroungae, 482
Halysidota, 362
Halysidota caryae, 362
Hantavirus, 160, 563*t*
Haplopleura captiosa, 61*t*
Haplopleura pacifica, 61*t*, 71
Harpactirekka lightfooti, 420
Harpactirella, 420
Harpirhyncus, 477–478
Heizmannia, 208*t*
Helichrysium angustifolium, 444
Heliocopris, 112
Helodon decemarticulatus, 198*t*
Hemileuca lucina, 361
Hemileuca maia, 355*f*, 361, 361*f*, 367
Hemileuca nevadensis, 361
Hemileuca oliviae, 361
Hemiscopis, 364
Hemiscorpius, 400
Hemiscorpius lepturus, 400
Hepatocystis, 183, 250
Hepatocystis brayi, 184*t*
Hepatocystis kochi, 184*t*
Hepatozoon, 484
Hepatozoon canis, 484
Hepatozoon erhardovae, 132
Hepatozoon felis, 484
Hepatozoon muris, 458, 484
Hepatozoon pitymysi, 132
Hepatozoon sciuri, 132
Hermanniella, 485*t*, 487*t*
Hermetia illucens, 39*t*, 146, 146*f*, 311*t*, 317, 317*f*
Herpetomonas, 183
Herpyllus ecclesiasticus, 415
Heterodoxus spiniger, 60*t*, 62*f*, 64, 71
Heterometrus, 401
Heteromys, 227
Heteronebo, 400
Heteroscorpion, 400
Hippelates neoproboscideus, 149
Hippobosca, 342, 343–346
Hippobosca equina, 344–345, 344*f*
Hippobosca longipennis, 344, 347, 348*t*
Hippobosca struthionis, 343
Hippobosca variegata, 345
Hirundo, 437
Hister, 39*t*
Hodgesia, 208*t*
Hoffmannihadrurus, 399
Holotrichius innesi, 83
Homo erectus, 66
Homo sapiens, 66
Hoplopleura pacifica, 128
Hoplopsyllus anomalus, 116*t*, 121*f*, 125
Hottentotta, 397
Hottentotta tamulus, 406

Hyalomma, 494, 494*t*, 507*t*, 509, 517, 528, 531, 534, 558*t*
Hyalomma anatolicum, 509, 528
Hyalomma asiaticum, 494, 509, 558*t*
Hyalomma detritum, 494, 509
Hyalomma dromedarii, 182, 500, 500*f*, 504
Hyalomma impeltatum, 499
Hyalomma lusitanicum, 509
Hyalomma marginatum marginatum, 494, 506*t*, 509
Hyalomma marginatum rufipes, 506*t*
Hyalomma truncatum, 494, 507*t*, 534
Hybomitra, 261, 262*t*, 263*f*, 264, 267, 267*f*, 269*t*, 272
Hybomitra bimaculata, 266
Hybomitra hinei, 266
Hybomitra illota, 266
Hybomitra lasiophthalma, 265
Hydrochoerus hydrochoeris, 471
Hydrotaea, 277, 277*f*, 278, 285, 287–288, 292, 293, 319, 320
Hydrotaea aenescens, 276*t*, 285
Hydrotaea ignava, 276*t*, 277*f*, 279*f*, 282*f*, 285
Hydrotaea irritans, 285
Hydrotaea leucostoma, 39*t*
Hydrotaea meteorica, 276*t*, 285, 287
Hydrotaea rostrata, 311*t*, 320
Hydrotaea scambus, 276*t*, 285, 287
Hylemya, 311*t*
Hylesia, 356, 366, 368
Hyleya, 319
Hymenolepis, 53
Hymenolepis diminuta, 104, 110, 123*t*, 130, 132, 367
Hymenolepis nana, 123*t*, 130, 132
Hypochrosis, 354
Hypochrosis flavifusata, 364
Hypochrosis hyadaria, 364, 365*f*
Hypochrosis irrorata, 364*f*
Hypochrosis pyrrhophaeata, 365*f*
Hypodeces propus, 478
Hypoderma, 327–329, 328*f*, 332, 334, 336
Hypoderma actaeon, 329
Hypoderma bovis, 316, 327, 328, 328*f*, 329*f*, 332, 334
Hypoderma diana, 329
Hypoderma lineatum, 316, 327, 328, 328*f*, 332
Hypoderma sinensis, 329
Hypoderma tarandi, 316, 329
Hypoponera punctatissim, 377

I

Icosiella neglecta, 153
Icosta, 349
Icosta albipennis, 348*t*
Icosta americana, 347, 348*t*
Icosta angustifrons, 348*t*
Icosta ardeae boutaurinorum, 348*t*
Icosta hirsuta, 348*t*
Icosta holopter holoptera, 348*t*
Icosta nigra, 348*t*
Icosta rufiventris, 348*t*
Idiommata blackwalli, 413
Illiberis, 357
Intercutestrix, 459
Iridovirus, 507*t*
Isa textula, 359
Isoberlinia, 297
Isodo torosus, 519
Isometrus maculatus, 397
Isostomyia, 208*t*
Iurus, 399
Ixodes, 493–495, 494*t*, 497, 514, 517, 525
Ixodes apronophorus, 525
Ixodes arboricola, 514
Ixodes brunneus, 507*t*, 534
Ixodes cookei, 514, 558*t*
Ixodes dentatus, 493, 521
Ixodes hexagonus, 514, 526
Ixodes holocyclus, 506*t*, 508, 519, 525, 526, 534
Ixodes jellisoni, 521
Ixodes marxi, 503, 514
Ixodes ovatus, 493, 521
Ixodes pacificus, 493, 496*f*, 506*t*, 507*t*, 519, 520, 521, 522, 526, 531, 532
Ixodes persulcatus, 493, 506*t*, 507*t*, 508, 514, 519, 521
Ixodes ricinus, 26, 493, 497*f*, 503, 504, 506*t*, 507*t*, 508, 513, 514, 515, 517, 519, 520, 521, 527, 528, 534, 558*t*
Ixodes rubicundus, 506*t*, 507*t*, 525, 526, 534
Ixodes scapularis, 493, 503, 504, 505, 506*t*, 507*t*, 508, 508*f*, 511, 513, 514, 519, 520, 521, 531, 532, 536, 537, 547*t*
Ixodes sculptus, 515
Ixodes spinipalpis, 493, 514, 515, 519, 521
Ixodes texanus, 516
Ixodes trianguliceps, 513, 514
Ixodes ventalloi, 515

J

Johnbelkinia, 208*t*

K

Klebsiella, 52*t*
Knemidokoptes jamaicensis, 464, 465
Knemidokoptes mutans, 464, 465, 466*f*
Knemidokoptes pilae, 464, 465–466
Kobus ellipisiprymus, 480
Kobus kob, 480
Kutzerocoptes gruenbergi, 471

L

Lachnosterna, 111
Laelaps, 15*f*, 16*f*, 458, 484
Laelaps echidninus, 128, 440, 459
Laelaps jettmari, 440
Lagenidium, 253
Lagoa crispata, 358
Lagoa pyxidifera, 358
Lagothrix, 231
Lama, 471
Lambornella, 253
Laminosioptes cysticola, 438*f*, 466–467
Lampona cylindrata/murina, 415
Lamprophaia, 364
Lasiocampus quercus, 363
Lasiohelia, 169, 174
Lasioglossum zephyrus, 386*f*
Latoia, 356, 356*f*, 357
Latrodectus, 106, 413, 416, 419, 426–430, 427*f*, 431
Latrodectus bishopi, 427, 429, 429*f*
Latrodectus curacaviensis, 427, 429
Latrodectus geometricus, 427, 428–429, 429*f*
Latrodectus hasselti, 426, 427, 429–430, 430*f*
Latrodectus hesperus, 427, 428
Latrodectus kapito, 427, 430
Latrodectus mactans, 426, 427, 428, 428*f*
Latrodectus menadovi, 426
Latrodectus tredecimguttatus, 426, 427, 429, 430
Latrodectus variolus, 427, 428
Lawrencarus eweri, 483
Leeuwenhoekia, 455
Leeuwenhoekia adelaidiae, 459
Leeuwenhoekia australiensis, 459
Leirus, 397, 405
Leirus quinquestriatus, 398*t*, 402*f*, 406, 406*f*
Leishmania, 162–163, 163*f*
Leishmania aethiopica, 159*t*
Leishmania amazonensis, 159*t*, 163, 164
Leishmania archibaldi, 159*t*
Leishmania braziliensis, 159*t*, 164
Leishmania chagasi, 159*t*, 162, 163, 165
Leishmania colombiensis, 159*t*
Leishmania donovani, 159*t*, 163, 165
Leishmania garnhami, 159*t*
Leishmania guyanensis, 159*t*, 163, 164
Leishmania infantum, 159*t*, 162, 163, 165, 547*t*
Leishmania killicki, 159*t*, 162
Leishmania lainsoni, 159*t*
Leishmania major, 159*t*, 164, 547*t*
Leishmania mexicana, 159*t*, 164, 165
Leishmania naiffi, 159*t*
Leishmania panamensis, 159*t*
Leishmania peruviana, 159*t*, 164
Leishmania pifanio, 159*t*
Leishmania shawi, 159*t*
Leishmania tropica, 162, 163, 164
Leishmania venezuelensis, 159*t*
Lentivirus, 562*t*
Leopoldamys, 455
Lepidoglyphus destructor, 444, 450
Leporacarus gibbus, 462
Leporipoxvirus, 562*t*
Leptinillus, 111
Leptinus, 111
Leptocimex, 94
Leptocimex boueti, 93, 94, 95
Leptocneria, 368
Leptocneria reducta, 368
Leptoconops, 169, 170, 171, 171*f*, 172, 173
Leptoconops becquaerti, 173, 174, 175*t*, 177
Leptoconops kerteszi, 173, 174
Leptoconops linleyi, 173, 174
Leptopsylla segnis, 116*t*, 121*f*, 123, 128, 130
Leptotrombidium, 441, 442, 455
Leptotrombidium akamushi, 441, 442, 442*t*, 454, 455, 455*f*
Leptotrombidium arenicola, 442, 442*t*, 455
Leptotrombidium deliense, 441, 442, 455
Leptotrombidium fletcheri, 442, 442*t*, 455
Leptotrombidium pallidum, 442*t*, 455
Leptotrombidium pavlovskyi, 442*t*
Leptotrombidium scutellare, 442*t*, 455

Lepus, 174*t*, 325
Lepus europaeus, 479
Leucocytozoon, 183, 200, 201–202, 250
Leucocytozoon cambournaci, 198*t*
Leucocytozoon caulleryi, 178, 184, 184*t*
Leucocytozoon dubrenili, 198*t*
Leucocytozoon lovti, 198*t*
Leucocytozoon neavei, 198*t*
Leucocytozoon sakharoffi, 198*t*
Leucocytozoon schoutedeni, 198*t*
Leucocytozoon simondi, 198*t*, 201
Leucocytozoon smithi, 198*t*, 201
Leucocytozoon tawaki, 198*t*
Leucocytozoon toddi, 198*t*, 202*f*
Leucocytozoon ziemanni, 198*t*
Leucophaea maderae, 54*t*
Leucotabanus annulatus, 266
Leucotobanus, 262*t*
Leucotobanus annuulatus, 265
Liacarus, 485*t*, 487*t*
Liebstadia, 485*t*, 487*t*
Limatus, 208*t*, 219, 220
Linognathus, 68
Linognathus africanus, 61*t*, 64, 70, 77*f*, 78
Linognathus ovillus, 61*t*
Linognathus pedalis, 61*t*
Linognathus setosus, 61*t*, 63*f*, 64, 71, 78
Linognathus stenopsis, 61*t*
Linognathus vituli, 61*t*, 63*f*, 68, 69*f*
Linshcosteus, 85
Liocheles australasiae, 403
Liohippelates, 137, 149
Liohippelates collusor, 149
Liohippelates flavipes, 149
Liohippelates pallipes, 149
Liohippelates puruanus, 149
Liohippelates pusio, 148*f*, 149
Lipeurus caponis, 60*t*, 71
Liponyssoides, 436
Liponyssoides sanguineus, 437–438, 438*f*, 453, 454
Lipoptena, 340, 342, 343, 345–346
Lipoptena capreoli, 347
Lipoptena cervi, 345–346, 345*f*, 347, 348*t*
Lipoptena mazamae, 346, 347
Liptopena depressa, 345, 346, 348*t*
Liptopena depressa depressa, 345
Liptopena depressa pacifica, 345
Liptopena mazamae, 342*f*, 348*t*
Lisposoma, 400
Listeria monocytogenes, 129, 132
Listrophhoroides cucullatus, 462
Listrophorus synaptomys, 462*f*
Lithosia, 357
Lithyphantes, 416
Litomosoides carinii, 484
Litomosoides sigmondontis, 484
Loa loa, 268, 269, 269*f*, 269*t*, 270
Loa loa papionis, 269
Lobocraspis, 364
Lobocraspis griseifusa, 354, 364*f*
Loiana, 252
Longipedia tricalcarata, 465*f*
Lonomia, 361
Lonomia achelous, 361
Lonomia oblique, 361, 361*f*
Lophocampa, 362
Lophocampa caryae, 362
Loxanoetus, 481
Loxosceles, 413, 415, 423–426, 423*f*
Loxosceles arizonica, 423
Loxosceles blanda, 423
Loxosceles deserta, 423
Loxosceles devia, 423
Loxosceles laeta, 423, 426, 431
Loxosceles reclusa, 423, 424–426, 424*f*, 425*f*, 430, 431
Loxosceles rufescens, 423, 431
Lucilia, 311*t*, 320
Lucilia cuprina, 320, 333, 336
Lucilia illustris, 333
Lucilia sericata, 320, 321*f*, 324, 333
Lutzomyia, 153, 154, 158, 160, 558*t*
Lutzomyia anthophora, 159*t*, 164
Lutzomyia antunesi, 159*t*
Lutzomyia apache, 158
Lutzomyia ayacuchensis, 159*t*
Lutzomyia ayrozai, 159*t*
Lutzomyia camposi, 157
Lutzomyia carpenteri, 157
Lutzomyia carrerai, 159*t*
Lutzomyia christopheri, 159*t*
Lutzomyia columbiana, 159*t*
Lutzomyia cruzi, 159*t*
Lutzomyia dasipodogeton, 160
Lutzomyia davisi, 160
Lutzomyia diabolica, 159*t*, 164
Lutzomyia dysponeta, 157
Lutzomyia evansi, 157, 159*t*
Lutzomyia flaviscutellata, 159*t*
Lutzomyia gomezi, 156, 157, 158, 159*t*
Lutzomyia hartmanni, 159*t*
Lutzomyia intermedia, 159*t*
Lutzomyia lichyi, 156, 159*t*
Lutzomyia longipalpis, 156, 157, 159*t*
Lutzomyia migonei, 159*t*
Lutzomyia nuneztovari, 159*t*
Lutzomyia olmeca, 158, 159*t*
Lutzomyia ovallesi, 159*t*
Lutzomyia panamensis, 156, 157, 158, 159*t*
Lutzomyia paraensis, 159*t*
Lutzomyia pessoai, 159*t*
Lutzomyia pessoana, 156, 157, 158
Lutzomyia reducta, 159*t*
Lutzomyia rorotaensis, 156
Lutzomyia sanguinaria, 157, 158
Lutzomyia shannoni, 158, 166
Lutzomyia spinicrassa, 159*t*
Lutzomyia squamiventris, 159*t*
Lutzomyia townsendii, 159*t*
Lutzomyia trapidoi, 156, 157, 158, 159*t*, 160, 166
Lutzomyia trinidadensis, 156, 159*t*
Lutzomyia triramula, 157
Lutzomyia ubiquitalis, 159*t*, 160
Lutzomyia umbratilis, 159*t*, 160
Lutzomyia verrucarum, 156, 157, 158, 159*t*
Lutzomyia vespertilionis, 157
Lutzomyia vexator, 165
Lutzomyia wellcomei, 158
Lutzomyia ylephiletor, 158, 159*t*, 160, 166
Lutzomyia youngi, 159*t*
Lutzomyia yucumensis, 159*t*
Lycoriella mali, 143
Lycosa, 414–415
Lycosa tarentula, 415, 419
Lyctocoris campestris, 83
Lymantria dispar, 357, 362
Lynx rufus, 472
Lynxacarus radovskyi, 462
Lytta, 105
Lytta phalerata, 105*t*
Lytta vesicatoria, 104*f*, 105, 105*t*

M

Mastophora, 414
Macaca, 471
Mackiena, 455
Macracanthororhyncus, 104
Macracanthororhyncus hirudinaceus, 111
Macracanthororhyncus ingens, 111
Macrodactylus subspinosus, 109
Macropocopris, 111
Macropus major, 519
Macrothele, 414
Macrothylacia rubi, 363
Malacosoma, 368
Malacosoma americanum, 362, 368, 368*f*
Malaya, 208*t*, 220
Mansonella, 176
Mansonella ozzardi, 174, 175*t*, 176–177, 176*f*, 198*t*, 199
Mansonella perstans, 174, 175*t*, 177
Mansonella streptocera, 174, 175*t*, 177–178
Mansonia, 208, 208*t*, 209, 210, 210*f*, 215, 217, 218, 218*f*, 219, 222, 226*t*, 228*t*, 236
Mansonia annulata, 247*t*
Mansonia annulifera, 246
Mansonia bonneae, 246, 247*t*
Mansonia dives, 246, 247*t*
Mansonia indiana, 247*t*
Mansonia titillans, 247*t*
Mansonia uniformis, 246, 247*t*
Maorigoeldia, 208*t*
Margaropus, 494*t*, 495
Martes pennanti, 471
Mastomys natalensis, 125, 473
Mastophorus muris, 54*t*, 55
Mazama, 346
Medicago sativa, 109
Megacormus, 399
Megalopyge, 357
Megalopyge opercularis, 355*f*, 358, 358*f*
Megaphobema, 421
Megarthroglossus, 119
Megaselia, 311*t*
Megaselia scalaris, 146, 147*f*
Meginia cubitalis, 463
Meginia ginglymura, 463
Megistopoda aranae, 342
Megistopoda proxima, 347
Melanoplus sanguinipes, 158
Melaolestes picipes, 83
Melolontha, 111
Melophagus, 341, 342
Melophagus montanus, 343
Melophagus ovinus, 77, 340, 342, 343, 343*f*, 344, 346, 347, 348*t*, 349*f*
Menacanthus, 64
Menacanthus stramineus, 60*t*, 62*f*, 68*f*, 71, 79
Menopon gallinae, 60*t*, 71, 79
Mercomyia, 262*t*
Meromyza americana, 148

Mesocestoides, 456
Metacnephia lyra, 192, 198*t*
Metarhizium, 253
Metarhizium anisopliae, 273
Metoposarcophaga, 324
Microgesta, 354
Microleon, 357
Microlichus americanus, 464
Microlichus avus, 464
Microlynchia pusilla, 348*t*
Microstega homoculorum, 364
Microtrombicula, 483
Microtus, 325, 437, 442, 455, 463, 487, 514, 519
Mimomyia, 208*t*
Misumenoides, 416
Molinema, 252
Monema, 357
Moniezia, 485, 486
Moniezia autumnalia, 485*t*
Moniezia benedeni, 485, 485*t*
Moniezia denticulata, 485*t*
Moniezia expansa, 485, 485*t*
Moniezia neumani, 485*t*
Moniliformis moniliformis, 54*t*, 111
Monoliformis dubius, 5, *t0015*
Monomorium pharaonis, 381, 381*f*
Moraxella bovis, 289
Morganella morganii, 52*t*
Morpho achillaena, 363
Morpho anaxibia, 363
Morpho cypri, 363
Morpho hercules, 363
Morpho laertes, 363
Morpho menelaus, 363
Morpho rhetenor, 363
Mus, 325
Mus musculus, 71, 123, 127, 437, 453, 461, 469
Musca, 278, 290, 320
Musca autumnalis, 276*f*, 276*t*, 277*f*, 278*f*, 283, 286–287, 289–290, 289*f*, 320
Musca biseta, 283
Musca domestica, 39*t*, 276*f*, 276*t*, 277*f*, 279*f*, 282*f*, 285–286, 288–289, 288*f*, 311*t*, 320
Musca sorbens, 283, 286, 286*f*, 320
Musca vetustissima, 112, 283, 286, 289
Muscidifurax, 293
Muscidifurax zaraptor, 293*f*
Muscina, 277, 280, 282, 293, 311*t*, 320
Muscina assimilis, 284, 320
Muscina levida, 276*t*, 284
Muscina pabulorum, 320
Muscina stabulans, 276*t*, 277*f*, 282*f*, 284, 287, 311*t*, 320, 321
Musophagobius cystodorus, 464*f*
Mustela, 471
Myalges, 464
Myalges macdonaldi, 464
Mycobacterum avium, 110
Mycobacterium leprae, 52*t*
Mycoplasma coccoides, 75, 75*t*, 79
Mycoplasma muris, 75, 75*t*, 79
Mycoplasma parvum, 77
Mycoplasma suis, 75, 77
Mycoptes musculinus, 463, 463*f*
Mydaea, 311*t*, 320
Mylabris bifasciata, 105*t*
Mylabris cichorii, 105*t*
Myobia, 461
Myobia musculi, 461, 461*f*, 462
Myodopsylla insignis, 116*t*, 121*f*
Myrmica, 377

N

Nadata nasoni, 359, 360*f*
Nairovirus, 506*t*, 558*t*
Nauphoeta cinerea, 52*t*
Necrobia rufipes, 39*t*, 40
Nebo, 400
Nebo hierichnticus, 401
Necator americanus, 53
Necrophila americana, 39*t*, 40*f*
Nemorhina, 297
Neocnemidocoptes gallinae, 464, 466
Neocuterebra squamosa, 331
Neogaria, 365
Neohaematopinus sciuropteri, 63*f*, 73
Neolipoptena, 340, 342, 343, 345–346
Neolipoptena ferrisi, 346, 348*t*
Neomusca, 311*t*, 320
Neomyia, 319
Neopsylla, 125
Neorickettsia, 529
Neorickettsia sennetsu, 518
Neoschoengastia, 455
Neoschoengastia americana, 459
Neotoma, 92, 325, 519
Neotoma albigula, 164
Neotoma fuscipes, 519
Neotoma mexicana, 519
Neotoma micropus, 164
Neotrombicula, 455, 459
Neottiophilum praeustrum, 311*t*, 318
Neotunga, 119
Nicrophorus, 39*t*
Nocardia, 52*t*
Nopoiulus kochii, 5
Norape ovina, 355*f*, 359*f*
Nosema, 253
Nosema pulicis, 120
Nosopsyllus, 123*t*
Nosopsyllus fasciatus, 116*t*, 121*f*, 122, 125, 128, 130, 131
Nothoaspis, 495
Notoedres, 469, 470, 471, 472–478
Notoedres cati, 472, 472*f*
Notoedres centrifera, 472–473
Notoedres muris, 473
Notoedres musculi, 473
Notoedres oudemansi, 473
Notoedres pseudomuris, 473
Nullibrotheas alleni, 399
Nuttalliella, 494*t*, 495
Nuttalliella namaqua, 493, 495
Nycteribia, 349
Nycteribia kolenatii, 349
Nycteridocoptes, 471
Nymphalis antiopa, 355*f*, 363, 363*f*

O

Ochlerotatus, 208, 208*t*, 209, 215, 217, 218, 222, 223, 226*t*, 228*t*
Ochlerotatus abserratus, 237*t*
Ochlerotatus albifasciatus, 227
Ochlerotatus atlanticus, 237*t*, 238
Ochlerotatus camptorhynchus, 229
Ochlerotatus canadensis, 237*t*, 238
Ochlerotatus caspius, 239
Ochlerotatus communis, 237*t*
Ochlerotatus dorsalis, 222, 226*t*, 227, 238
Ochlerotatus fijiensis, 247*t*
Ochlerotatus fulvus, 237*t*
Ochlerotatus harinasutai, 247*t*
Ochlerotatus infirmatus, 238
Ochlerotatus intrudens, 237*t*
Ochlerotatus melanimon, 226*t*, 227, 237*t*, 238
Ochlerotatus niveus, 230*t*, 233, 246, 247*t*
Ochlerotatus oceanicus, 247*t*
Ochlerotatus poicilius, 247*t*
Ochlerotatus provocans, 237*t*
Ochlerotatus samoanus, 247*t*
Ochlerotatus scapularis, 237, 247*t*
Ochlerotatus sollicitans, 226, 226*t*
Ochlerotatus stimulans, 237*t*
Ochlerotatus taeniorhynchus, 219, 222
Ochlerotatus togoi, 247*t*
Ochlerotatus tormentor, 237*t*, 238
Ochlerotatus triseriatus, 237*t*, 238
Ochlerotatus trivittatus, 237*t*, 238
Ochlerotatus vigilax, 226*t*, 229, 247*t*
Ochotomia lindneri, 329
Ochrogaster, 368
Ochrogaster lunifer, 368
Odocoileus, 78
Odocoileus hemionus, 272, 345, 346, 479
Odocoileus virginianus, 272, 345, 346, 479
Odontacarus, 455
Odontomachus haematoda, 377
Oeciacus, 97
Oeciacus hirundinis, 97
Oeciacus vicarius, 97, 227
Oestroderma, 329
Oestromyia, 329, 329*f*
Oestrus, 329
Oestrus ovis, 330, 330*f*
Oiceoptoma, 39*t*
Oiclus, 400
Olfersia bisulcata, 348*t*
Olfersia cariacea, 346
Olfersia fumipennis, 348*t*
Olfersia sordida, 348*t*
Olfersia spinifera, 348*t*
Oligella urethralis, 52*t*
Omosita, 39*t*
Onchocerca cervicalis, 175*t*, 178, 184, 184*f*, 185
Onchocerca cervipedis, 198*t*, 201
Onchocerca dukei, 198*t*, 201
Onchocerca gibsoni, 175*t*, 185
Onchocerca gutturosa, 175*t*, 185, 198*t*, 201
Onchocerca lienalis, 198*t*, 201
Onchocerca ochengi, 198*t*, 201
Onchocerca ramachandrini, 198*t*, 201
Onchocerca reticulata, 175*t*, 185
Onchocerca skrjabini, 198*t*, 201
Onchocerca sweetae, 175*t*
Onchocerca tarsicola, 198*t*, 201
Onchocerca voluvulus, 178, 196, 196*f*, 197, 197*f*, 198*t*, 199, 204, 270
Onirion, 208*t*
Onthophagus, 108, 111, 112
Onthophagus polyphemi, 108*f*

Onychocerus albitarsis, 104
Onychomys, 484
Onychomys leucogaster, 467
Ophionyssus, 438, 484
Ophionyssus natricis, 440, 440*f*, 458
Ophioptes, 477, 484
Ophthalmodex, 467
Ophyra aenescens, 285
Ophyra leucostoma, 285, 321*f*
Opifex, 208*t*, 220
Opisthacanthus lepturus, 400
Opistophthalmus, 401
Opsonyssus, 484
Orbivirus, 160, 557, 558*t*
Orchpeas howardi, 73, 116*t*, 123*t*, 124, 128
Orgyia leucostigma, 362
Orgyia pseudotsugata, 356
Oribatella, 487*t*
Oribatula, 485*t*, 487*t*
Orientia, 454
Orientia tsutsugamushi, 442, 454, 455
Orius insidiosus, 83
Ornithocoris toledoi, 97
Ornithoctona erythrocephala, 348*t*
Ornithoctona fusciventris, 348*t*
Ornithodoros, 494*t*, 495, 497*f*, 501, 506*t*, 510, 522–524, 558*t*
Ornithodoros coriaceus, 506*t*, 507*t*, 523, 526, 529, 532
Ornithodoros erraticus, 510, 523, 524, 529
Ornithodoros hermsi, 495, 510, 523, 524
Ornithodoros lahorensis, 507*t*, 534
Ornithodoros moubata, 89, 495, 510, 522, 523, 524, 526, 529
Ornithodoros parkeri, 523, 524
Ornithodoros porcinus, 510, 529
Ornithodoros rudis, 523
Ornithodoros savignyi, 507*t*, 510, 534
Ornithodoros sonrai, 510, 523, 529
Ornithodoros talaje, 523
Ornithodoros tartakovskyi, 523
Ornithodoros tholozoni, 495, 523, 524
Ornithodoros turicata, 500*f*, 523, 524, 526, 529
Ornithodoros verrucosus, 523
Ornithoica confluenta, 348*t*
Ornithoica vicina, 348*t*
Ornithomya, 349
Ornithomya anchineuria, 348*t*
Ornithomya avicularia, 340*f*, 347
Ornithomya bequaerti, 348*t*
Ornithomya fringillina, 339
Ornithonyssus, 14, 15*f*, 438
Ornithonyssus bacoti, 128, 438–439, 438*f*, 454, 457, 484
Ornithonyssus bursa, 439, 439*f*, 457
Ornithonyssus sylviarum, 439, 439*f*, 457–458, 458*f*, 484
Oropsylla montana, 116*t*, 121*f*, 125
Orthobunyavirus, 223, 224*t*, 225, 237, 558*t*
Orthohalarachne, 481, 482
Orthohalarachne attenuata, 482
Orthohalarachne diminuata, 482
Orthoperus, 104
Orthopodomyia, 208, 208*t*
Orthopodomyia alba, 216
Oryctolagus, 325
Oryctolagus cuniculus, 130, 249
Orygia pseudotsugata, 356
Oryzomys, 227
Oscinella frit, 148
Oswaldofilaria, 252
Otoanoetus, 481
Otobius, 494*t*, 495
Otobius lagophilus, 495, 515
Otobius megnini, 495, 511
Otodectes, 470, 474
Otodectes cynotis, 480, 480*f*
Ovis canadensis, 474
Ovis canadensis mexicanus, 476
Ovis dalli, 343
Oxacis, 105
Oxycopis, 105
Oxycopis mcdonaldi, 106*f*
Oxylipeurus polytrapezius, 60*t*
Oxyspirura mansoni, 54*t*, 55
Oxyspirura parvorum, 54*t*, 55
Oxytrigona tatairus, 387

P

Pachycondyla chinensis, 377
Paederus, 105*t*, 106, 107*f*
Paederus alternans, 106*t*
Paederus amazonicus, 106*t*
Paederus australis, 106, 106*t*
Paederus brasiliensis, 106, 106*t*
Paederus columbinus, 106, 106*t*
Paederus cruenricollis, 106, 106*t*
Paederus eximius, 106*t*
Paederus ferus, 106*t*
Paederus fuscipes, 106, 106*t*, 109
Paederus islae, 106*t*
Paederus laetus, 106*t*
Paederus melampus, 106*t*
Paederus ornaticornis, 106*t*
Paederus puncticollis, 106*t*
Paederus riparius, 106*t*
Paederus sabaeus, 106, 106*f*, 106*t*
Paederus signaticornis, 106, 106*t*
Paederus tamulus, 106*t*
Pagdya, 364
Paleoleishmania proterus, 163
Paleomyia burmitis, 163
Paliga damastesalis, 364
Pallasiomyia, 329
Pamphlobeteus, 414
Pan troglodytes, 231
Pandinus gambiensis, 403
Pandinus imperator, 401, 401*f*
Panstrongylus chinai, 85*t*
Panstrongylus geniculatus, 83
Panstrongylus hereri, 85*t*
Panstrongylus megistus, 83, 85*t*, 88, 91
Panstrongylus rufotuberculatus, 85*t*
Pantoea, 52*t*
Papio, 231, 471
Papio ursinus, 229
Parabelminus, 86
Parabuthus, 397
Parabuthus transvaalicus, 398*t*
Parachipteria, 485*t*, 487*t*
Parafilaria, 290
Parafilaria bovicola, 290
Parafilaria multipapillosa, 290
Paralucilia, 311*t*, 320
Paraponera, 389
Paraponera clavata, 389
Parasa, 357
Parasa chloris, 359, 360*f*
Parasa indetermina, 359, 360*f*
Paraspinturnix globosus, 478
Parcoblatta, 43, 44, 46
Paruroctonus boreus, 397
Passeromyia, 311*t*, 320
Pavlovskiata, 329
Pedicinus, 61*t*
Pediculus humanus, 14*f*, 16*f*
Pediculus humanus capitis, 61*t*, 62*f*, 66, 67
Pediculus humanus humanus, 61*f*, 61*t*, 63*f*, 66–67, 66*f*, 75*t*, 128, 547*t*
Pelecitus, 252
Pelecitus fulicaeatrae, 75*t*
Peloptulus, 487*t*
Peloribates, 487*t*
Penicillida, 349
Penicillium, 443
Pepsis, 381, 382*f*
Pergalumna, 485*t*, 487*t*
Pericoma, 155*f*
Pericoma funebris, 155
Periplaneta, 14*f*, 43, 45, 51, 55
Periplaneta americana, 44*f*, 47, 47*f*, 52*t*, 54*t*
Periplaneta australasiae, 44*f*, 47–48, 47*f*, 52*t*
Periplaneta brunnea, 44*f*, 48, 48*f*, 54*t*
Periplaneta fuliginosa, 44*f*, 48, 48*f*
Peromyscus, 26, 227, 325, 484
Peromyscus gossypinus, 228, 325*f*
Peromyscus leucopus, 519
Peromyscus manuculatus, 519
Peromyscus polionotus, 127
Pestivirus, 562*t*
Peucetia viridas, 415
Phaenicia coeruleiviridis, 39*t*
Phaenicia sericata, 39*t*, 40*f*
Phanaeus, 112
Pheidole, 392
Pheromermis myopis, 273
Phidippus johnsoni, 415
Philanthus triangulum, 376
Philonthus, 112
Philornis, 320
Phlebotomus, 153, 154, 155*f*, 158, 160, 559*t*
Phlebotomus aculeatus, 159*t*
Phlebotomus alexandri, 159*t*
Phlebotomus ansarii, 159*t*
Phlebotomus argentipes, 156, 157, 166
Phlebotomus ariasi, 157, 159*t*
Phlebotomus bergeroti, 156
Phlebotomus caucasicus, 156, 157, 159*t*, 164
Phlebotomus celiae, 159*t*
Phlebotomus celiae, 159*t*
Phlebotomus chinensis, 159*t*
Phlebotomus duboscqui, 157, 159*t*
Phlebotomus guggisbergi, 159*t*
Phlebotomus kandelakii, 159*t*
Phlebotomus kazeruni, 156
Phlebotomus langeroni, 159*t*
Phlebotomus longicuspis, 159*t*
Phlebotomus longiductus, 159*t*
Phlebotomus longipes, 159*t*
Phlebotomus martini, 159*t*, 167
Phlebotomus neglectus, 159*t*
Phlebotomus orientalis, 157, 159*t*

Phlebotomus papatasi, 154*f*, 156, 157, 158, 159*t*, 160, 166, 167
Phlebotomus pedifer, 159*t*
Phlebotomus perfiliewi, 156, 159*t*, 160, 167
Phlebotomus perniciosus, 157, 159*t*, 160
Phlebotomus rossi, 159*t*
Phlebotomus salehi, 159*t*
Phlebotomus sergenti, 158, 159*t*
Phlebotomus smirnovi, 159*t*
Phlebotomus tobbi, 155, 159*t*
Phlebotomus transcaucasicus, 159*t*
Phlebotomus vansomerenae, 159*t*
Phlebovirus, 160, 223, 224*t*, 237, 239, 558*t*
Phobetron pithecium, 355*f*, 359, 359*f*
Phoneutria, 413, 414, 422
Phoneutria keyserlingi, 414
Phoneutria nigriventer, 414, 422*f*
Phorcotabanus cinereus, 266
Phormia, 320
Phormia regina, 39*t*, 41, 311*t*, 320, 321*f*, 333
Phormictopus, 414
Phortica variegata, 148
Phyllophaga, 111
Phyllotsomus hastatus, 342
Physaloptera caucasica, 111
Physaloptera praeputalis, 54, 54*t*
Physaloptera rara, 54, 54*t*
Physocephalus sexalatus, 111
Piagetiella, 64
Pilogalumna, 487*t*
Piophila, 318*f*
Piophila casei, 39*t*, 311*t*, 318
Piophila vulgaris, 318
Piophilia casei, 147, 148*f*
Pisaurina, 415
Pitymys, 519
Plasmodium, 20, 183, 239, 240–241, 243, 250–251, 348, 484, 512, 513
Plasmodium aegyptensis, 250
Plasmodium berghei, 250
Plasmodium booliati, 250
Plasmodium brasilianum, 250, 251
Plasmodium cathemerium, 250
Plasmodium chabaudi, 250
Plasmodium circumflexum, 250
Plasmodium cynomolgi, 250, 251
Plasmodium elongatum, 250
Plasmodium falciparum, 239, 241, 242, 243, 254, 547*t*
Plasmodium gallinaceum, 250
Plasmodium gonderi, 250
Plasmodium hermansi, 250
Plasmodium incertae, 250
Plasmodium knowlesi, 239, 241, 243, 250, 251
Plasmodium lophurae, 250
Plasmodium malariae, 239, 241, 250
Plasmodium ovale, 239, 241
Plasmodium pitheci, 250
Plasmodium reichenowi, 250
Plasmodium relictum, 20, 250
Plasmodium schwetzi, 250
Plasmodium simium, 250, 251
Plasmodium vinckei, 250
Plasmodium vivax, 27, 239, 240*f*, 241, 242, 250, 254
Plasmodium watteni, 250
Plasmodium yoeli, 250, 547*t*
Platydracus, 39*t*
Platynothrus, 485*t*, 487*t*
Platypsyllus, 111
Plecia nearctica, 143, 143*f*
Plesiochactas, 399
Ploiaria domestica, 167
Pneumocoptes banksi, 484
Pneumocoptes jellisoni, 484
Pneumocoptes penrosei, 484
Pneumocoptes tiollaisi, 484
Pneumonyssoides caninum, 482, 482*f*
Pneumonyssus, 481
Pneumonyssus simicola, 482, 482*f*
Poecilotheria, 414
Pogonomyrmex, 375, 377, 380, 380*f*, 389
Pogonomyrmex badius, 380
Pogonomyrmex barbatus, 380
Pogonomyrmex californicus, 380
Pogonomyrmex occidentalis, 380
Polchromophilus, 349
Polchromophilus murinus, 349
Polistes, 372, 382, 385–386, 385*f*
Polistes annularis, 372*t*, 386
Polistes apachus, 372*t*, 386
Polistes aurifer, 372*t*, 386
Polistes carolina, 372*t*
Polistes dominulus, 372*t*, 385, 386
Polistes exclamans, 372*t*, 385, 386
Polistes fuscatus, 386
Polistes metricus, 372*t*, 386
Polistes perplexus, 372*t*, 385
Pollenia rudis, 283, 286–287
Polyplax serrata, 61*t*, 68, 71, 75*t*, 79
Polyplax spinulosa, 61*t*, 63*f*, 71, 75*t*, 79, 128
Portschinskia, 329
Pranoplocephala amillana, 485*t*
Premolis, 356
Premolis semirrufa, 362, 366
Preochimys, 227
Presbytis cristata, 246
Problepsis, 364
Proctolaelaps pyrgmaeus, 436
Proctophyllodes glandarinus, 464*f*
Prosarcoptes pitheci, 471
Prosarcoptes scanloni, 471
Prosarcoptes talapoini, 471
Prosimulium, 192, 194
Prosimulium magnum, 192
Prosimulium mixtum, 191*f*, 195*t*
Prosimulium ursinum, 193
Prosimulum impostor, 198*t*
Prosthenorchis elegans, 54*t*, 55
Prosthenorchis spirula, 54*t*, 55
Proteus, 51
Proteus mirabilis, 52*t*
Proteus rettgeri, 52*t*
Proteus vulgaris, 52*t*
Protocalliphora, 311*t*, 322, 322*f*, 323
Protophormia, 311*t*, 320
Protophormia terranovae, 321*f*, 333
Protoribates, 487*t*
Protospirura bonnei, 54*t*, 55
Protospirura muricola, 54*t*, 55
Provespa, 372
Przhevalskiana, 329
Psalydolytta fusca, 105*t*
Psalydolytta substrigata, 105*t*
Psammolestes, 86
Psammomys obesus, 167
Pseudolynchia, 349
Pseudolynchia canariensis, 342, 346, 346*f*, 347, 348*t*
Pseudomenopon pilosum, 75*t*
Pseudomonas, 52*t*, 106
Pseudomyrmex ejectus, 377
Pseudouroctonus, 399
Psorergates, 468
Psorergates apodemi, 469
Psorergates bos, 469
Psorergates dissimilis, 469
Psorergates hispanicus, 469
Psorergates microti, 469
Psorergates muricola, 469
Psorergates simplex, 469, 469*f*
Psorobia cercopitheci, 469
Psorobia ovis, 469
Psorophora, 208, 208*t*, 209, 215, 217, 218, 222, 226*t*, 228*t*
Psorophora ferox, 237, 326
Psoroptes, 467, 474, 480
Psoroptes cervinus, 474, 475–476, 475*f*
Psoroptes cuniculi, 474, 478–480, 479*f*
Psoroptes equi, 474, 474*f*
Psoroptes natalensis, 474
Psoroptes ovis, 474–476, 475*f*, 478
Pseudodiamesa, 145*f*
Psychoda, 153, 154, 154*f*, 155*f*, 156, 157, 311*t*
Psychoda albipennis, 157, 158
Psychoda alternata, 156, 157, 158, 166
Psychoda cinerea, 157, 158
Psychoda pacifica, 158
Psychoda phalaenoides, 155
Psychoda severini, 156, 158
Psychodopygus, 154
Pterolichus obtusus, 463
Pterostichus, 110
Pthirus pubis, 16*f*, 61*t*, 63*f*, 67
Pulex irritans, 116*t*, 120, 121*f*, 125, 128, 130, 131
Pulex simulans, 116*t*, 120, 124, 131
Punctoribates, 485*t*, 487*t*
Pycnoscelus surinamensis, 50, 50*f*, 54*t*
Pydnella, 364
Pyemotes, 444
Pyemotes herfsi, 444
Pyemotes tritici, 444, 445*f*
Pyemotes zwoelferi, 444
Pyrausta, 364
Pyrota insulata, 109
Pyrrharctia isabella, 362
Pythium, 143

R

Radfordia, 462
Radfordia affinis, 462
Radfordia ensifera, 462, 462*f*
Radfordiella, 457, 461, 478
Radfordiella oudemansi, 457*f*
Raillietia, 480–481
Raillietia acevedoi, 480
Raillietia auris, 480, 480*f*
Raillietia caprae, 480
Raillietia flechtmanni, 480
Raillietia mafredi, 480
Raillietiella hemidactyli, 54*t*, 55

Raillietina cesticillus, 110
Raillietina echinobothrida, 392
Raillietina tetragona, 392
Rana esculenta, 153
Rasahus biguttatus, 83
Rasahus thoracicus, 83
Rattus, 325, 437, 442, 455, 462
Rattus norvegicus, 71, 124, 127, 462, 473
Rattus rattus, 71, 124, 127, 438, 462, 473
Reduvius, 90
Reduvius personatus, 83, 98
Rhagastis olivacea, 365
Rhagio, 145*f*
Rhagodes nigrocinctus, 411
Rhinoestrus, 329
Rhinoestrus pupureus, 330
Rhipicentor, 494*t*, 495
Rhipicephalus, 494*t*, 505, 507*t*, 517, 531
Rhipicephalus appendiculatus, 494, 505, 507*t*, 528, 535, 558*t*
Rhipicephalus (Boophilus) annulatus, 494, 500, 503, 507*t*, 510, 510*f*, 531, 535, 538
Rhipicephalus (Boophilus) decoloratus, 531
Rhipicephalus (Boophilus) geigyi, 531
Rhipicephalus (Boophilus) microplus, 494, 502, 503, 507*t*, 509, 510*f*, 535, 537, 538
Rhipicephalus bursa, 505
Rhipicephalus evertsi, 505, 507*t*, 538
Rhipicephalus evertsi evertsi, 534
Rhipicephalus sanguineus, 182, 494, 502*f*, 505, 506*t*, 507*f*, 507*t*, 516, 517, 527, 529, 531
Rhipicephalus turanicus, 505
Rhipicephalus zambesiensis, 528
Rhodnius, 86
Rhodnius ecuadoriensis, 85*t*
Rhodnius pallescens, 85*t*
Rhodnius prolixus, 85*f*, 85*t*, 87, 88, 91, 92, 93
Rhombomys opimus, 164, 167
Rhyncoptes, 473
Rhynocoptes grabberi, 473
Rickettsia, 454, 554
Rickettsia aeschlimanni, 517
Rickettsia africae, 506*t*, 509, 517
Rickettsia akari, 437, 453, 454
Rickettsia australis, 518
Rickettsia bellii, 517
Rickettsia conorii, 74, 505, 506*t*, 517
Rickettsia felis, 128
Rickettsia honei, 518
Rickettsia japonica, 518
Rickettsia melophagi, 348*t*
Rickettsia montanensis, 517
Rickettsia parkeri, 509, 518
Rickettsia prowazekii, 67, 72–73, 75*t*, 123*t*, 128, 547*t*
Rickettsia rhipicephali, 517
Rickettsia rickettsii, 2, 74, 453, 505, 506, 506*t*, 516, 516*f*, 517, 547*t*
Rickettsia sibirica, 517
Rickettsia tsutsugasmushi, 454
Rickettsia typhi, 123*t*, 127–128, 547*t*
Rivoltasia bifurcata, 464
Rodhainyssus, 484
Romanomermis culicivorax, 253
Rousettocoptes, 471
Runchomyia, 208*t*
Rupicapra rupicapra, 471
Ruttenia loxodontis, 331

S

Sabethes, 208*t*, 220, 222, 230*t*, 236
Sabethes chloropterus, 231
Saimiri, 231
Saimiricoptes, 473
Salmonella, 51, 52*t*, 285
Salmonella bareilly, 52*t*
Salmonella bovis-mortificans, 51, 52*t*
Salmonella bredeny, 52*t*
Salmonella chester, 109
Salmonella enteritidis, 129, 146
Salmonella newport, 52*t*
Salmonella oranienburg, 52*t*
Salmonella panama, 52*t*
Salmonella paratyphi-B, 52*t*
Salmonella pyogenes, 52*t*
Salmonella typhi, 51, 52*t*, 74
Salmonella typhimurium, 51, 52*t*, 109
Salmonells, 109
Sancassania berlesei, 452
Sancassania mycophagia, 452
Saprinus, 39*t*
Sarconema eurycerca, 75*t*
Sarcophaga, 324, 324*f*
Sarcophaga bullata, 39*t*
Sarcophaga haemorrhoidalis, 39*t*, 324, 324*f*
Sarcopromusca arcuata, 327*f*
Sarcopromusca pruna, 326
Sarcoptes, 447–448, 469, 473
Sarcoptes scabei, 447–448, 447*f*, 448–449, 464, 469–471, 470*f*, 471*f*, 473, 474
Saurositus, 252
Scarabaeus, 111
Scathophaga, 39*t*
Sceliphron, 382, 382*f*
Scheloribates, 455, 485*t*, 486, 487*t*
Scheloribates atahualpensis, 455
Schoengastia philippinensis, 459
Schoengastia westraliensis, 459
Schoutedenichia, 459, 483
Sciuracarus paraxeri, 484
Sciurocoptes sciurinus, 463
Sciurus, 472
Sciurus carolinensis, 326*f*, 472
Sciurus griseus griseus, 472
Sciurus niger, 472
Scoliopteryx libatrix, 354*f*
Scopelodes, 357
Scopula, 364
Scorpio maurus, 401
Scotophaeus blackwalli, 415
Scutovertex, 455, 487*t*
Seadonavirus, 514
Segestria florentina, 415
Sepsis, 39*t*
Sergeia, 183
Sergentomyia, 154, 158, 160
Sergentomyia antennata, 157
Sergentomyia bedfordi, 157
Sergentomyia theodori, 157
Sericopelma, 414
Serretia, 52*t*
Sessinia, 105
Sessinia kanak, 105
Sessinia lineata, 105
Shannoniana, 208*t*
Shigella, 149, 285, 286
Shigella dysenteriae, 52*t*
Shunsennia, 455
Sibine stimulea, 355*f*, 359, 359*f*
Sicarius, 415
Sigmodon, 227
Sigmodon hispidus, 228, 484
Silvius, 261, 262*t*, 272
Simulium, 191, 192, 194
Simulium adersi, 198*t*
Simulium albivirgulatum, 197, 198*t*
Simulium amazonicum, 198*t*
Simulium anatinum, 198*t*, 201*f*
Simulium angustitarse, 198*t*
Simulium annulus, 194, 198*t*
Simulium arakawae, 195*t*, 198*t*
Simulium argentiscutum, 198*t*
Simulium aureum, 198*t*
Simulium bidentatum, 198*t*, 201
Simulium bovis, 198*t*
Simulium buissoni, 195*t*
Simulium callidum, 198*t*
Simulium cholodkovskii, 195*t*
Simulium chutteri, 195*t*
Simulium colombaschense, 203
Simulium congareenarum, 198*t*
Simulium daisense, 198*t*
Simulium damnosum, 189, 191, 192, 193, 194, 197, 198*t*, 199, 201
Simulium decimatum, 195*t*
Simulium decorum, 193, 198*t*
Simulium diguerense, 198*t*
Simulium equinum, 195*t*
Simulium erythrocephalum, 195*t*, 198*t*, 201, 203
Simulium ethiopiense, 198*t*
Simulium exiguum, 177, 197, 198*t*
Simulium fallisi, 198*t*
Simulium guianense, 198*t*
Simulium incrustatum, 195*t*, 198*t*
Simulium jenningsi, 192, 195, 195*t*, 198*t*, 201
Simulium johannseni, 194
Simulium jujuyense, 195*t*
Simulium kilibanum, 198*t*
Simulium konkourense, 198*t*
Simulium kyushuense, 198*t*
Simulium latipes, 198*t*
Simulium leonense, 198*t*
Simulium limbatum, 198*t*
Simulium lineatum, 195*t*
Simulium luggeri, 195*t*, 200
Simulium maculatum, 195*t*
Simulium mengense, 198*t*
Simulium meridionale, 195*t*, 198*t*, 201, 201*f*
Simulium metallicum, 194, 197, 198*t*, 199
Simulium neavei, 197, 198*t*, 199
Simulium nemorale, 191*f*
Simulium nigrogilvum, 195*t*, 201
Simulium noelleri, 191, 192
Simulium notatum, 202
Simulium ochraceum, 194, 195*t*, 197, 198*t*, 199
Simulium oitanum, 198*t*
Simulium ornatum, 195*t*, 198*t*, 203
Simulium oyapockense, 194, 195*t*, 197, 198*t*
Simulium parnassum, 194, 195*t*
Simulium penobscotense, 195*t*
Simulium pertinax, 195*t*
Simulium posticatum, 194, 195*t*
Simulium quadrivittatum, 195*t*, 198*t*

Simulium rasyani, 198*t*
Simulium rendalense, 198*t*
Simulium reptans, 192, 195*t*, 198*t*, 203
Simulium rostratum, 192
Simulium rufocorne, 198*t*
Simulium rugglesi, 194, 195*t*, 198*t*, 201
Simulium sanctipauli, 198*t*
Simulium sanguineum, 195*t*, 198*t*
Simulium sirbanum, 197, 198*t*, 199
Simulium slossonea, 195*t*, 198*t*, 201
Simulium soubrense, 198*t*
Simulium squamosum, 198*t*
Simulium tescorium, 195*t*
Simulium thyolense, 198*t*
Simulium truncatum, 192
Simulium usovae, 198*t*
Simulium vampirum, 192, 193, 195*t*, 200*f*, 203
Simulium venustum, 190*f*, 192, 194, 195*t*, 196*f*, 198*t*, 252
Simulium vernum, 198*t*
Simulium vittatum, 158, 190*f*, 192, 193, 194, 195, 195*t*, 202
Simulium woodi, 198*t*
Simulium yahense, 198*t*
Sinea diadema, 83
Siphunculina, 149
Sitophilus granarius, 104
Skrjabinofilaria, 252
Solenopotes, 68, 69*f*
Solenopotes capillatus, 61*t*, 63*f*, 68, 71*f*
Solenopsis, 377, 378–380
Solenopsis geminata, 378, 379
Solenopsis invicta, 377, 378, 378*f*, 379, 390*f*, 392
Solenopsis richteri, 378, 378*f*, 392
Solenopsis xyloni, 378, 379, 392
Spalangia, 293
Spaniopsis, 145
Spatiodamaemus, 485*t*, 487*t*
Speleognathus australis, 483
Spermophilopsis leptodactylus, 164
Spermophilus beecheyi, 484
Sphaerobothria hoffmani, 430
Sphecius, 372, 382
Sphecius speciosus, 382, 382*f*
Sphingobacterium, 52*t*
Sphingobacterium mizutae, 52*t*
Spilopsyllus cuniculi, 116*t*, 120, 121–122, 123*t*, 130, 131
Spirocerca lupi, 111
Spirura rytipleurites, 54*t*
Splendidofilaria fallisensis, 202
Staphylococcus aureus, 52*t*, 67, 129, 425, 508, 531
Staphylococcus epidermidis, 52*t*
Staribia, 318
Steatoda, 416
Steatoda andina, 416
Steatoda grossa, 416
Steatoda paykulliana, 416
Stegomyia, 209
Stegopterna, 192
Steinernema carpocapsae, 55, 124, 133
Stenoponia tripectinata, 116*t*, 118
Stephanofilaria stilesi, 292
Sternostoma tracheacolum, 481, 481*f*
Stilbometopa, 349
Stilbometopa impressa, 347, 348*t*
Stilbometopa podopostyla, 348*t*
Stilesia globipunctata, 485*t*
Stilesia hepatica, 485*t*
Stilesia vittata, 485*t*
Stomoxys calcitrans, 276*t*, 277*f*, 278*f*, 279*f*, 281*f*, 282*f*, 283–284, 287, 290–291, 291*f*, 326
Stonemyia, 261, 262*t*
Streptococcus, 109, 368
Streptococcus faecalis, 52*t*
Streptococcus pyogenes, 67
Strictia carolina, 273
Strobiloestrus, 329
Suidasia, 453
Suidasia nesbetti, 443
Suidasia pontifica, 453
Suipoxvirus, 562*t*
Suncus murinus, 127
Supella, 43
Supella longipalpa, 44*f*, 49, 49*f*, 54*t*
Superstionia donensis, 399
Suragina, 145
Sycorax silacea, 153
Sylvilagus, 174*t*, 249, 325
Symphoromyia, 145, 145*f*
Synthesiomyia, 326
Syringophilus bipectinalis, 463, 465*f*

T

Tabanus, 140*f*, 261, 262*t*, 263*f*, 263*t*, 264, 264*f*, 267, 267*f*, 269*t*, 272
Tabanus abactor, 264, 266
Tabanus atratus, 261, 265, 266
Tabanus calens, 265
Tabanus conterminus, 273
Tabanus equalis, 266
Tabanus fairchildi, 263
Tabanus fusciostatus, 267
Tabanus imitans, 264*f*
Tabanus marginalis, 140*f*
Tabanus moderator, 266
Tabanus nigrovittatus, 261, 265, 273
Tabanus pallidescens, 266
Tabanus punctifer, 265, 266, 270
Tabanus simulans, 261
Tabanus subsimilis, 264, 265
Tabanus sulcifrons, 264, 267*f*
Tabanus trimaculatus, 262*f*
Tabanus wilsoni, 266
Tadarida braziliensis, 439–440, 457
Taenia saginata, 110
Tanakaius, 208, 208*t*
Tapirus, 471
Tarsolepis, 364
Tarsonemus, 453
Tarsonemus granarius, 452
Tarsonemus hominis, 453
Tarsoporosus, 400
Tegenaria, 422–423
Tegenaria agrestis, 414, 422, 423*f*
Teinocoptes, 471
Telonomus fariai, 93
Tenebrio, 110
Tenebrio molitor, 103, 104, 110, 444, 456
Tenebrio obscurus, 103
Tetrahymena, 253
Tetrameres americana, 54*t*, 55
Tetrameres fissipina, 54*t*, 55
Tetramorium caespitum, 380–381, 380*f*
Tetrapetalonema, 177
Thaumetopoea, 363
Thaumetopoea processionea, 355, 363
Thaumetopoea wilkinsoni, 363, 367
Theileria, 512, 526, 527–528
Theileria annulata, 507*t*, 509, 527, 528, 547*t*
Theileria equi, 527, 528
Theileria lestoquardi, 527, 528
Theileria parva, 507*t*, 527, 528, 547*t*
Theileria recondita, 509
Theileria taurotragi, 535
Thelazia californiensis, 287, 290
Thelazia callipaeda, 148
Thelazia gulosa, 290
Thelazia lacrymalis, 290
Thelazia skrjabini, 290
Theraphosa, 421
Thereva, 311*t*
Thliptoceras, 364
Thyridanthrax, 307
Thysaniezia giardi, 485*t*
Thysanosoma actinoides, 485*t*
Tilapia, 253
Tinea pellionella, 39*t*
Tineola bisselliella, 39*t*
Tipula, 140*f*, 142–143, 143*f*
Tityus, 397, 406, 407
Tityus bahiensis, 407, 407*f*
Tityus cambridgei, 407
Tityus serrulatus, 398*t*, 402*f*, 403, 407
Tityus trinitatis, 407
Tityus trivatatus, 407
Togarishachia, 364
Tolype notialis, 362
Tolype velleda, 362
Topomyia, 208*t*
Toxoplasma gondii, 51
Toxorhynchites, 208, 208*t*, 209, 211, 215, 216, 217, 218, 222, 253
Toxorhynchites, 214*f*, 218*f*
Toxorhynchites brevipalpis, 210*f*, 211*f*
Trachelas volutas, 414
Tracheomyia macropi, 329, 329*f*
Tragelaphus scriptus, 304
Trechona, 414
Trechona venosa, 414
Trentula cubensis, 430
Treponema pertenue, 149
Triatoma, 85, 86
Triatoma barberi, 89
Triatoma brasiliensis, 85*t*
Triatoma carrioni, 85*t*
Triatoma dimidiata, 85*t*, 86*f*
Triatoma gerstaeckeri, 92
Triatoma guasayana, 85*t*
Triatoma infestans, 85*f*, 85*t*, 86, 87, 88, 91, 92, 93
Triatoma lecticularia, 92
Triatoma maculata, 85*t*
Triatoma pallidipennis, 85*t*
Triatoma pantagonica, 85*t*
Triatoma phyllosoma, 85*t*
Triatoma protracta, 88
Triatoma recurva, 92
Triatoma rubida, 92
Triatoma rubrofasciata, 84*f*, 85, 92
Triatoma sanguisuga, 92

Triatoma sordida, 85*t*, 87
Triatoman picturata, 89, 92
Tribolium, 104, 110
Tribolium castaneum, 107, 107*f*
Tribolium confusum, 107, 107*f*
Trichillium, 111
Trichinella spiralis, 132
Trichodectes canis, 60*t*, 62*f*, 64, 71, 75, 75*t*, 78, 130
Trichoecius, 16*f*
Trichoecius romboustsi, 463
Trichoecius tenax, 463
Tricholipeurus parallelus, 62*f*
Trichophaga tapetzella, 39*t*
Trichophyton erinacei, 477
Thrichophyton verrucosum, 75*t*
Trichoprosopon, 208*t*, 215
Trichoprosopon digitatum, 222
Trichoribates, 485*t*, 487*t*
Trichospirura leptostoma, 55
Trichuris trichuria, 53
Trinoton anserinum, 60*t*, 75*t*
Trinoton querquedulae, 60*t*
Tripteroides, 208*t*
Trixacarus, 469, 473
Trixacarus caviae, 462, 473, 473*f*
Trixacarus diversus, 473
Trochilocoetes, 62
Trochoderma glabrum, 103
Trochoderma ornatum, 103
Troglocormus, 399
Troglodytes aedon, 437
Troglotayosicus vachoni, 399
Trogoderma, 108*f*
Trogoderma angustum, 104
Trombicula alfreddugesi, 441–442
Trombicula autumnalis, 441, 441*f*, 442
Trombicula lipovskyi, 442
Trombicula splendens, 442
Trox, 39*t*
Trypanosoma, 144
Trypanosoma avium, 198*t*, 347, 348*t*
Trypanosoma brucei, 271, 302, 304, 304*t*, 305, 306, 547*t*
Trypanosoma confusum, 198*t*, 202
Trypanosoma congolense, 271, 306
Trypanosoma conorhini, 92
Trypanosoma corvi, 202
Trypanosoma cruzi, 85, 88–92, 89*f*, 90*f*, 91*f*, 93, 97, 547*t*
Trypanosoma equinum, 271
Trypanosoma evansi, 269*t*, 271
Trypanosoma gambiense, 302, 303, 304, 304*t*, 306
Trypanosoma grayi, 306
Trypanosoma hannai, 348*t*
Trypanosoma hedricki, 97
Trypanosoma lewisi, 123*t*, 131, 132
Trypanosoma melophagium, 347, 348*t*
Trypanosoma musculi, 132
Trypanosoma myoti, 97
Trypanosoma nabiasi, 123*t*, 132
Trypanosoma numidae, 198*t*
Trypanosoma rabonowitschi, 132
Trypanosoma rangeli, 92
Trypanosoma rhodesiense, 302, 303, 304, 304*t*, 306
Trypanosoma simiae, 304*t*, 306
Trypanosoma suis, 306
Trypanosoma theileri, 271
Trypanosoma uniforme, 304*t*, 306
Trypanosoma vivax, 269*t*, 271, 304*t*, 306
Tunga, 119, 129–130, 131
Tunga monositus, 116*t*, 119
Tunga penetrans, 116*t*, 119, 122, 122*f*, 123*f*, 129–130, 131
Twinnia, 192
Tychosarcoptes, 471
Typhlochactas mitchelli, 399
Tyrolichus casei, 443
Tyrophagus, 452, 453
Tyrophagus longior, 443, 444*f*, 453
Tyrophagus putrescentiae, 443, 450, 452, 453

U

Udaya, 208*t*
Unguizetes, 487*t*
Uraba lugens, 363
Uranotaenia, 208, 208*t*, 215, 222
Uroctonus, 399
Urocyon cinereoargenteus, 471
Urodacus, 400
Uropsylla tasmanica, 116*t*, 119
Ursicoptes, 473
Ursicoptes americanus, 473
Ursicoptes procyonis, 473
Urubambates, 485*t*, 487*t*

V

Vaejovis carolinianus, 399, 404, 405
Vampirolepis nana, 104, 110
Vatacarus, 483
Vermipsylla, 119
Vermipsylla alakurt, 131
Verrallina, 208*t*
Vesiculovirus, 158, 160, 557, 558*t*
Vespa, 372
Vespa crabro, 372*t*, 385
Vespa maculata, 383*f*
Vespula, 372, 383–385, 391*f*
Vespula acadica, 384*t*
Vespula atropilosa, 384, 384*t*
Vespula consobrina, 384*t*
Vespula flavopilosa, 372*t*, 383, 384, 384*t*
Vespula germanica, 372*t*, 383, 384, 384*t*
Vespula maculifrons, 372*t*, 375, 383, 383*f*, 384, 384*t*
Vespula pensylvanica, 372*t*, 383, 384, 384*t*
Vespula rufa, 383
Vespula squamosa, 372*t*, 384*t*, 385, 391*f*
Vespula sulphurea, 372*t*, 384*t*, 385
Vespula vidua, 384
Vespula vulgaris, 372*t*, 383, 384*t*, 385
Vibrio, 52*t*
Vibrio cholerae, 145
Vitalius sorocabae, 419*f*
Vombatus ursinus, 480
Vulpes fulva, 471

W

Walchia americana, 459
Waltonella, 252
Warileya, 154, 158
Wasmannia auropunctata, 378, 380
Wasmannia, 377
Wenzella, 119
Wohlfahrtia, 323, 324
Wohlfahrtia magnifica, 323
Wohlfahrtia nuba, 309, 324
Wohlfahrtia opaca, 323
Wohlfahrtia vigil, 323
Wolbachia, 167, 197, 529
Wuchereria, 252
Wuchereria bancrofti, 2, 19, 20, 22, 25, 243, 244–245, 244*f*, 246, 247*t*
Wuchereria kalimantani, 246
Wyeomyia, 208, 208*t*, 209, 215, 216, 220, 222, 236
Wyeomyia smithii, 215, 216

X

Xenillus, 487*t*
Xenopacarus africanus, 483
Xenopsylla, 123*t*, 124, 125, 127
Xenopsylla astia, 116*t*, 128
Xenopsylla bantorum, 116*t*, 128
Xenopsylla brasiliensis, 116*t*, 128
Xenopsylla cheopis, 116*t*, 121, 121*f*, 124, 125, 128, 130, 131
Xylobates, 487*t*
Xylocopa californica, 386
Xylocopa varipuncta, 386
Xylocopa virginica, 386, 386*f*, 387*f*
Xyloryctes, 111

Y

Yersinia pestis, 52*t*, 120, 123*t*, 124–127, 126*f*, 547*t*
Yersinia pseudotuberculosis, 129
Yunkeracarus, 484

Z

Zeugnomyia, 208*t*
Zygodontomys, 227
Zygodontomys brevicauda, 125
Zygoribatula, 485*t*, 486, 487*t*
Zythos, 364

Subject Index

Note: Page numbers followed by "*f*" indicate figures and "*t*" indicate tables.

A

Aardvark, tsetse flies, 301
Absconding, bees, 388
Acalyptratae, 137
Acanthocephala, 104, 111
Acaraphobia, 433, 528–529
Acari, *see* Mites
Acariasis
  definition, 433
  enteric acariasis, 453
  internal acariasis, 452–453
  pulmonary acariasis, 452
  urinary acariasis, 453
Acaricides, 535–536
Acarinism, 433
Acarophobia, 5
Accidental myiasis, 309
Accumulated degree hours, 41, 42
Actinopodidae, 413
Adenolymphangitis, 245
Adenotrophic viviparity, 342
Adephaga, 101, 102
AFLP, *see* Amplified fragment length polymorphism
African blue louse, 70, 77*f*
African hedgehog mange mite, 477
African horsesickness, 181–183, 181*f*
Africanized honey bee, 388, 388*f*, 389
African sleeping sickness, 302–303, 302*f*
African swine fever, 510, 529
African tampan, 510
Agas, 169, 170
Agelenidae, 414–416
Aggregation pheromones, 45
Agramonte, Aristides, 232
Ague, 239
AHS, *see* African horsesickness
Air trumpet, 210
Air-sac mite, 483*f*
Alakurts, 116
Allergic reactions
  cockroach, 53–54
  equine allergic dermatitis, 185–186, 185*f*
  flea, 131
  hymenopterous sting reactions, 390, 390*t*
  mite-induced allergy, 449–452
  silk, 367
  stages, 4
  sudden death, 37
  tick bite, 526
Alphavirus, mosquito-borne viruses, 225–227, 226*f*, 226*t*, 227–228, 228–229, 228*t*, 229–230
Amaurobiidae, 414
Amblycera, 59–60
American bird mite, 437
American cockroach, 47, 47*f*
American dog tick, 508*f*
American trypanosomiasis, *see* Chagas disease
Ammotrechidae, 411
Amphibians
  biting midges, 173
  chiggers, 440, 441
  flesh flies, 324
  mansonellosis, 199
  mites, 483, 484
  mosquitoes, 220
  Psychodidae, 153
  ticks, 497*f*
  toad blow flies, 321
  triatomines, 86
Amplification, parasites, 21
Amplified fragment length polymorphism, 546–548
Amplifying host, 21
Ampulla, 217
Amastigote, 90*f*
Analgesics, 6
Anaphylactic shock, 390
Anaphylaxis, 390
Anaplasmataceae, 529–532
Anaplasmosis
  canine, 531
  ovine, 531
  public health importance, 519
  veterinary importance, 529–530
Anautogeny, 141, 193, 281, 314
Annealing, DNA, 546
Annotation, genome, 546
Anophelinae, 207, 208*t*
Anoplura, 59–60
Anteater, trypanosomiasis, 92
Antelope
  bots, 329, 330
  deer ked, 346
  ear mite, 478, 479
  skin-piercing moths, 366
  ticks, 511, 516, 534
  trypanosomiasis, 90, 306
Antennal fossae, 116
Anterior-station transmission, 24
Anthocoridae, 93
Anthomyiidae, myiasis, 310, 319
Anthophoridae, 386
Anthrax, tabanid transmission, 270
Anthropogenic changes, 32
Anthropophagic vector, 22
Anticoagulants, 6
Antigenic complex, 178
Antivenins, 406
Ants, *see also* Hymenoptera
  behavior and ecology, 375–376
  fire ants, 377, 378–380, 378*f*, 379*f*
  harvester ant, 377, 380, 380*f*
  life history, 374–375
  morphology, 373–374
  pavement ant, 380–381, 380*f*
  pharoah ant, 381, 381*f*
  prevention and control, 392–393
  public health importance, 389–392
  taxonomy, 371–373, 372*t*
  venoms, 376–377
  veterinary importance, 392
Apamin, 377
Apidae, 371, 387
Apidocere formation, 38
Apinae, 387
Apocrita, 371
Apoidea, 371
Aposematic coloration, 106
Apotele, 16
Apyrase, 6
Arachnophobia, 5
Araña del trigo, 429
Araneae, *see* Spiders
Araneidae, 414
Araneomorphae, 414–416
Arbovirus, 97, 557–564
Arbovirus reaction, 4
Archostemata, 101
Arctiidae, 362
Argasidae
  life cycles, 500–501, 500*f*
  morphology, 498, 498*f*
  taxonomy, 493, 495
Arista, 140
Arthropod-borne viruses, 557–564
Aschiz, 137
Ascidae, 436
Asian cockroach, 50, 50*f*
Asian needle ant, 377, 389, 390*f*
Astigmata, 433
Athericide, 145–146, 146*f*
Atopomelidae, 462
Atratoxin, 421
Attacin, 544
Australian cockroach, 47–48, 47*f*

Australian frog fly, 319, 319*f*
Australian paralysis tick, 508
Australian redback spider, 429–430, 430*f*
Autogeny, 141, 193, 314
Autolysis, 38

B

Babesiosis
  bovine, 527
  canine, 527
  humans and ticks, 512–513
Bald-faced hornet, 383*f*
Ballooning, 418
Bangkok hemorrhagic disease, 184
Barbeiro, 84, 88
Bark lice, 7
Bark scorpion, 406
Barmah Forest virus, 229
Bartonellosis, 161–162, 161*f*
Barychelidae, 413–414
Bat, 497, 497*f*, 501, 505
  Cimicidae, 93
  mites, 438, 439, 457, 461, 471, 478, 483, 484
  myiasis, 310, 332
  Psychodidae, 157
  sarcoptic mange, 471
  ticks
  triatomines, 84, 87
  trypanosomes, 97
  Venezuelan equine encephalomyelitis, 227
  vesicular stomatitis, 158
Bat fly, 341*f*
Bat tick, 511
Bazaar fly, 283, 286, 286*f*
Bed bug
  behavior and ecology, 95–96
  common names, 93
  life history, 95
  morphology, 94–95, 94*f*
  prevention and control, 97–98
  public health importance, 96–97
  taxonomy, 93–94
  veterinary importance, 97
Bees, *see also* Hymenoptera
  behavior and ecology, 375–376
  life history, 374–375
  morphology, 373–374
  prevention and control, 392–393
  public health importance, 389–392
  social bees, 387–389
  solitary bees, 386
  venoms, 376–377
  veterinary importance, 392
Beetles, 101–114
  behavior and ecology, 102–103
  blister beetles, 101, 104–105, 104*f*, 105*f*, 105*t*
  darkling beetles, 101, 107, 107*f*
  dung beetles, 111–112
  false blister beetles, 101, 105–106, 106*f*
  ingestion, 109
  intermediate hosts, 110–111, 110*f*
  lady beetles, 109
  larder beetles, 101, 107, 108*f*
  life history, 102
  morphology, 101–102, 102*f*
  nest associates and ectoparasites, 111
  pathogens, 109–110
  prevention and control, 112
  public health importatance, 103–109, 103*t*
  rove beetles, 101, 106–107, 106*f*, 106*t*
  scarab beetles, 101, 108–109, 108*f*
  taxonomy, 101
  veterinary importance, 109–112
Behavioral defenses, 6
Bembicinae, 273
Bibionidae, 143, 143*f*
Bicho colorado, 441–442
Bicudo, 84
Bioinformatics, 546
Biological transmission, 4
Birds, *see also* Chickens
  black fly attacks, 201, 201*f*
  feather mites, 463
  fleas, 131
  lice, 71–72, 79
  malaria, 250
Bison
  epizootic hemorrhagic disease, 180
  flies, 276*t*, 284
Biting midges, *see* Ceratopogonidae
Biting rate, 28
Blaberidae, 43
Black death, 125
Black flies, 189–206
  behavior and ecology, 192–194
  life history, 191–192
  morphology, 190–191, 190*f*, 191*f*
  nuisance problems, 195, 196*f*
  prevention and control, 203–204
  public health importance, 194–200, 195*t*, 196*f*, 197*f*, 198*t*
  taxonomy, 189–190
  veterinary importance, 200–203, 200*f*, 201*f*
Black fly fever, 195
Black imported fire ant, 378, 378*f*
Black-legged tick, 508, 508*f*, 513
Black widow spiders
  eastern black widow spider, 428
  European black widow spider, 429
  northern black widow spider, 428
  southern black widow spider, 428, 428*f*
Blattaria, 43–58
Blattellidae, 43
Blattidae, 43
Blister beetles, 101, 104–105, 104*f*, 105*f*, 105*t*
Blow flies, *see also* Calliphoriidae
  carrion-associaed flies, 320–321
  myiasis, 320–323, 321*f*
  name origins, 315
  nest blow flies, 322, 322*f*
  postmortem succession, 39, 40*f*, 41*f*
  screwworms, 321–322, 322*f*
  toad blow flies, 321
  tumbu fly, 322–323, 323*f*
Bluetongue disease, 178–179, 179*f*
Body shape, parasitic arthropods, 13
Bombycoidea, 353
Bont tick, 509, 510*f*
Boopiidae, 60
Borrelioses, 532
Bot flies, *see* Oestroidea
Bothuriuridae, 400
Boutonneuse fever, 517
Bovine leukemia virus, 562*t*
Bovine onchocerciasis, 201
Bovine spongiform encephalopathy, 484
Bovine tropical theileriosis, 509
Brachycera, 137
Breakbone fever, 233
Bridge vector, 25
Brill-Zinsser disease, 72
Brownbanded cockroach, 49, 49*f*
Brown cockroach, 48, 48*f*
Brown dog tick, 502*f*
Brown-tail moth, 361
Brown recluse spider, 423*f*, 424–426, 424*f*, 425*f*
Brown widow spider, 428–429, 429*f*
Bruce, David, 2
BSE, *see* Bovine spongiform encephalopathy
Buboes, 126
Bubonic plague, 126, 126*f*
Buck moth, 361, 361*f*
Buffalo fly, 284, 291–292
Bulge-eye, 177
Bulldozer-spider, 411
Bumble bee, 387, 387*f*
Bung-eye, 177
Bunyaviridae, 174*t*, 176, 223, 224*t*, 237*t*, 238–239, 557, 558*t*
Bush chinch, *see* Kissing bugs
Bush fly, 283, 286, 289
Buthidae, 397–398
Butterflies, *see* Lepidoptera

C

Cacodminae, 93
Calabar swellings, 269
Calamistrum, 417
California encephalitis virus, 237*t*, 238
Calliphoridae, myiasis, 138, 139, 140
Calypter, 140, 141*f*
Calyptratae, 137, 140, 141*f*
Camel
  filariasis, 347
  hippoboscids, 345
  mange, 449
  nagana, 305
  spider bites, 430
  stable fly, 430
  trypanosomiasis, 271
Camel nose bot, 330, 329*f*
Camel tick, 171, 504
Campestral plague, 125
Canberra eye, 104
Canine ehrlichiosis, 529–530
Canine trypanosomiasis, 92
Canthariasis, 103
Capture antibody, 554
Carapace, 401
Card Agglutination Trypanosomiasis Test, 306
Carpenter bee, 386, 386*f*, 387*f*
Carroll, James, 232
Castor bean tick, 508
Cat
  bots, 316
  chiggers, 516

cytauxzoonosis, 528
dog fly, 344
ear mites, 480, 480*f*
flas, 120, 121, 122
leishmaniasis, 165
lice, 71, 78
Lyme disease, 532
mites and mange, 438, 442, 449, 454, 456, 460–461, 457, 462, 467, 468, 474, 480, 480*f*
murine typhus, 127
notoedric cat mite, 472, 472*f*
Plague, 125
Q fever, 519
spider bites, 430
tapeworms, 132
theilerioses, 527
trypanosomiasis, 92
Caterpillar-induced equine abortion, 368
Cat flea, 121, 130
Cat follicle mite, 467*f*, 468
CATT, *see* Card Agglutination Trypanosomiasis Test
Cattle
babesiosis, 527
theilerioses, 527–528
heartwater, 530–531
dermatophilosis, 533–534, 533*f*
bluetongue disease, 179*f*
bovine onchocerciasis, 201
ear mites, 480–481
lice, 68–70, 68*f*, 69*f*, 71*f*, 76, 76*f*
pinkeye, 288, 289, 290*f*
psoroptic mange, 475*f*
chorioptic scab mites, 476–477, 476*f*
Tabanid feeding sites, 267, 267*f*
Cattle fever tick, 510
Cattle follicle mite, 468
Cattle grub, 327–329, 328*f*, 334, 336–337
Cattle tail louse, 69
Cat scratch disease, 129
CCD, *see* Colony collapse disorder
cDNA, *see* Complementary DNA
Centipede, 8
Central-European encephalitis, 513
Ceratophyllidae, 115
Ceratopogonidae
behavior and ecology, 172–173, 172*f*, 173*f*
life history, 171–172
morphology, 170, 170*f*, 171*f*
prevention and control, 186
public health importance, 174–178, 174*t*
taxonomy, 169–170
veterinary importance, 178–186, 179*f*, 180*f*, 181*f*, 184*t*
Ceratopopogoninae, 169
Cercopidae, 83
Cestoda, 104
Cuterebrinae, myiasis, 310
Chactidae, 399
Chaerlidae, 398–399
Chagas, Carlos, 2
Chagas disease
acute disease, 90
chronic disease, 90–92
control, 92–93
diagnosis, 91
geographic distribution, 91*f*
kissing-bug vectors, 85*t*, 88–92
trypanosome life cycle, 88, 89*f*
Chagoma, 90
Chamois, sarcoptic mange, 471
Chandipura virus disease, 160
Changuinola virus disease, 160–161
Chaoboridae, 144, 144*f*, 207
Cheyletidae, 444–445, 460–461, 461*f*
Chicken
air-sac mite, 483*f*
poultry leucocytozoonosis, 184, 201–202, 202*f*
Chicken body louse, 68*f*, 71–72, 79
Chicken head louse, 71
Chicken mite, 436, 437*f*, 456
Chiggers, 440–442
Chigoe, 122, 122*f*
Chikungunya virus, 228–229
Chinchorro, 84
Chinese needle ant, 377
Ch'ing yao ch'ung, 106
Chipmunk
Colorado tick fever, 515
La Crosse virus, 23
plague, 125
Chipo, 84
Chirimacho, 84
Chironomidae, 138, 144–145, 145*f*, 207
Chironomoidea, 207
Chloropidae, 148–149, 148*f*, 310, 319, 319*f*
Chorion, 209, 216
Choriothete, 299
Christmas eye, 104
Chrysomyinae, 310, 313
Chrysopsinae, 261
Chupão, 84
Chupon, 84
Chyluria, 245
Cicada killer, 382, 382*f*
Cicadelliae, 83
Cicadidae, 83
Cimicidae, 93
Cimicinae, 93
Classical swine fever virus, 562*t*
Clavus, 83
Cleg, 261
Climate change, 33
Clinical case, 31–32
Cloning, genes, 543–544
Coarctate, 142
Coccinellidae, 109
Cockle, 77, 349, 349*f*
Cockroach
allergy, 53–54
American cockroach, 47, 47*f*
Asian cockroach, 50, 50*f*
Australian cockroach, 47–48, 47*f*
behavior and ecology, 45–46
brown cockroach, 48, 48*f*
brownbanded cockroach, 49, 49*f*
Florida woods cockroach, 48, 49*f*
German cockroach, 49–50, 49*f*
intermediate hosts, 53, 54*t*
life history, 44–45, 44*f*, 45*t*
morphology, 43–44
Oriental cockoach, 46, 46*f*
pathogens, 51–53, 52*t*
prevention and control, 55–56
public health importance, 50–54
smokybrown cockroach, 48, 48*f*
Surinam cockroach, 50, 50*f*
taxonomy, 43
Turkestan cockroach, 46–47, 46*f*
veterinary importance, 54–55
Coffin fly, 147*f*
Coleoptera, 101–114
Colony collapse disorder, 387
Colorado tick fever, 514–515
Complementary DNA, libraries, 550
Conenoses, 84
Congo floor maggot, 323, 323*f*
Connexivum, 85
Copra itch, 443
Corethrellidae, 207
Corinnidae, 414
Corium, 83
Corolla, 209
Coxae, 15, 496
Coyote, sarcoptic mange, 471
Crab louse, *see* Human crab louse
Crane flies, 138, 142–143, 143*f*
Cribellum, 417
Crimean tick typhus, *see* Boutonneuse fever
Crochet, 354
Cryptic species, 189
Ctenidium, 64, 414–415
Ctenophthalmidae, 115
CTF, *see* Colorado tick fever
Culicidae, *see also* Mosquitoes
taxonomy, 207, 208*t*
Culicinae, 207
Cutaneous leishmaniasis, 163–165, 164*f*
Cuterebrinae, myiasis, 325–327
Cyclodevelopmental transmission, 25
Cyclorrhapha, 137
Cytauxzoonosis, 496
Cytoditidae, 483–484

D

Darkling beetles, 101, 107, 107*f*
Darkwinged fungus gnat, 143–144, 144*f*
Darwin, Charles, 90
Dasyheleinae, 169
DDT, *see* Dichloro-diphenyl-trichloroethane
Decomposition
insect species, 39*t*
stages, 38–39
succession, 39–42
Deer
ear mites, 479
epizootic hemorrhagic disease, 180–181, 180*f*, 181*f*
lice, 78
psoroptic mange, 475–476
sarcoptic mange, 471
Deer flies, *see* Tabanidae
Deer keds, 342*f*, 345–346, 345*f*, 347
Deer nose bot, 317, 330*f*
Deer skin maggot, 323
DEET
biting midge control, 186
black fly control, 196

DEET (*Continued*)
flea control, 133
fly control, 150
mosquito control, 252
sand fly control, 166
tick control, 535
Definitive host, 20
Delayed hypersensitivity reaction, 4
Delusory acariosis, 433, 528–529
Delusory parasitosis, 5
Demodecosis, 445, 447
Demodicidae, 445–447, 446*f*, 467
Denaturation, DNA, 546
Dengue
geographic distribution, 232*f*
hemorrhagic fever, 233, 233*f*
shock syndrome, 233
transmission, 233, 233*f*
Depluming itch mite, 466
Dermanyssidae
public health importance, 436–438
veterinary importance, 456, 457, 458
Dermationidae, 464
Dermatitis
flea-bite, 131
mites, 436–442, 442–445, 456–458
Dermatitis linearis, 106
Dermatophilosis, 533–534, 533*f*
Dermestidae, 101, 107, 108*f*
Desensitization reaction, 4
Developmental time, 95
Developmental transmission, 25
Diapause, 217, 222, 282
Dichloro-diphenyl-trichloroethane
flea control, 133
mosquito control, 253
tick control, 535
Differential display, 551
Dioptic, 141
Diplocentridae, 400–401
Dipluridae, 414
Diptera
behavior and ecology, 142
families of minor interest, 142–149
life history, 141–142
morphology, 138–141, 139*f*, 140*f*, 141*f*
public health importance, 138*t*, 149
taxonomy, 137–138, 139*t*
veterinary importance, 138*t*, 150
Disease agent, 20
Dixidae, 207
Dog
African horsesickness, 182
anaplasmosis, 531
babesiosis, 527
borreliosis, 532
bots, 316
chicken mite, 456
ear mites, 480, 480*f*
eastern equine encephalomyelitis, 243
ehrlichiosis, 121
fleas, 120, 121, 122, 131
leishmaniasis, 165
lice, 71, 78
Lyme disease, 475
mites amd mange, 442, 447, 449, 453, 454, 463, 467–468, 472, 480
murine typhus, 127
Q fever, 519
sarcoptic mange, 470*f*
spider bites, 430
stable fly, 283, 290–291
tapeworms, 130, 132
ticks, 505, 534–535
tórsalo, 326–327
trypanosomiasis, 89, 92, 271
tumbu fly, 322–323
tungiasis, 131
walkingstick interactions, 7, 7*f*
Dog flea, 121
Dog fly, 344, 347
Dog follicle mite, 467–468, 467*f*
Dog heartworm, 251, 251*f*, 252*f*
Dog nasal mite, 482, 482*f*
Dolichoderinae, 371
Drone fly, 147, 147*f*
Drosophilidae, 148, 148*f*, 318–319, 318*f*
Dumdum fever, 162
Dung beetles, 111–112
Dung flies, 275, 276*t*
Dysderidae, 415

E

Ear mites, 478–481, 481–484
Earwig, 7
East African trypanosomiasis, 303
East coast fever, 527
Eastern black widow spider, 428
Eastern equine encephalitis, 22, 225–227, 226*f*, 226*t*, 248, 248*f*
Eastern tent caterpillar, 368*f*
Eastern yellowjacket, 384
Ebola virus, 563*t*
ECF, *see* East coast fever
Ectoparasite, 20
EEE, *see* Eastern equine encephalitis
Eggs
cockroach, 43, 44*f*
fleas, 118, 118*f*
lice, 60, 62*f*
mosquito, 209, 210*f*, 216*f*, 222
muscid flies, 276*f*, 287
Tabanidae, 264, 265*f*
EHD, *see* Epizootic hemorrhagic disease
Ehrlichiosis
public health importance, 518–519, 518*f*
veteriary importance, 529–530
EIA, *see* Equine infectious anemia
Elaeophorosis, tabanid transmission, 271, 272*f*
Elephant
African horsesickness, 182
lachryphagous moths, 364, 366
nose bot flies, 329
stomach bot flies, 331
trypanosomiasis, 271
tsetse flies, 301
Elephantiasis, *see* Filariasis
Elephant louse, 78
Elephant skin maggot, 323
ELISA, *see* Enzyme-linked immunosorbent assay
Elk
deer keds, 345
elaeophorosis, 272
Epizootic hemorrahgic disease, 180
mits, 475
ticks, 505, 515
Elytra, 101
Embryogenesis, 45
Emerging disease, 32–33
Emerging infectious diseases, 19
Empodium, 213
Emperor scorpion, 401*f*
Emu
lice, 79
Endemic pemphigus foliaceus, 200
Endemic threshold, 22
Endemic typhus, *see* Murine typhus
Endoparasite, 20
Endophilic vector, 22
Enteric acariasis, 453
Entomological inoculation rate, 28, 31
Entomophobia, 5
Entomyssidae, 481
Envenomation, 3
Enzootic transmission, 25
Enzyme-linked immunosorbent assay, 553, 554*f*
Epidemic polyarthrits, 229
Epidemic threshold, 22
Epidemic typhus
diagnosis, 72
historical perspective, 73
reservoirs, 73
transmission, 72
vector, 72
Epidemiology, vector-borne diseases, 19–34
Epidermoptidae, 464
Epigynum, 417
Epimastigote, 88
Episystem, 178
Epizootic hemorrhagic disease, 180–181, 180*f*, 181*f*
Equine allergic dermatitis, 185–186, 185*f*
Equine infectious anemia, 271, 271*f*, 562*t*
Equine onchocerciasis, 184–185, 184*f*
Equne amnionitis and fetal loss, 368
Eremobatidea, 411
Ereynetidae, 483
Eruciform, 354
Erucism, 353, 366
Erythema migrans, 521
Erythrocytic anaplasmosis, 531–532
Eschar, 517
Eurasian hedgehog mange mite, 477
European black widow spider, 429
European chicken flea, 123
European mouse flea, 123
European rabbit flea, 121–122
Euscorpiidae, 399
Eusocial species, 375
Everglades virus, 228
Exarate, 118
Exflagellation, 240
Exophilic vector, 22
Exophily, 194
Extension, DNA, 546

Extrinsic incubation, 25
Eye gnats, 148–149, 148*f*

F

Facultative egg diapause, 222
Facultative myiasis, 309
Facultative parasite, 20
False blister beetles, 101, 105–106, 106*f*
False stable fly, 284, 287
Fanniidae, myiasis, 310, 319, 319*f*
Fanniinae, 275
Fear, arthropods, 5
Feather mites, 463
Feeding frequency, 28
Femur, 15
Feral species, 45
Ferret, sarcoptic mange, 471
Filariasis, *see also* Dog heartworm; Elaeophorosis; Mansonellosis; Onchocerciasis
  clinical disease, 245–246
  elephantiasis, 245, 245*f*, 246*f*
  geographic distribution, 243–248, 244*f*
  historical perspective, 247–248
  life cycle, 244–245
  mosquito vectors and epidemiology, 246–247, 247*t*
Filth flies, 275, 276*t*
Finlay, Carlos, 2, 232
Fipronil, tick control, 535
Fire ants, 377, 378–380, 378*f*, 379*f*
Five-o, 169
Flaviviridae, 223, 224*t*, 229–230, 230–232, 230*t*, 232–234, 233*f*, 234–235, 236–237, 557, 558*t*
Flea dirt, 118*f*, 119
Flea-borne typhus, *see* Murine typhus
Fleas, 115–136
  allergies, 127–128
  behavior and ecology, 119–120
  cat flea, 121
  chigoe, 122, 122*f*
  dermatitis, 131
  dog flea, 121
  European chicken flea, 123
  European mouse flea, 123
  European rabbit flea, 121–122
  human flea, 120–123
  intermediate hosts, 130, 132
  life history, 118–119
  morphology, 116–118, 116*f*, 117*f*, 118*f*, 121*f*
  murine trypanosomiasis, 131–132
  murine typhus, 127–128
  myxomatosis, 131
  Northern rat flea, 122
  Oriental rat flea, 121
  pathogens, 123*t*, 129, 132
  prevention and control, 132
  plague, 124–127
  sticktight flea, 122, 122*f*
  sylvatic epidemic typhus, 128
  taxonomy, 115, 116*t*
  tungiasis, 122, 122*f*, 131
  veterinary importance, 130–132
Flesh flies, *see* Sarcophagidae
Float, egg, 209
Florida woods cockroach, 48, 49*f*
Fly blown, 315
Fly strike, 315
Flying squirrel
  malaria, 250
  typhus reservoir, 73
Fogo selvagem, 200
Folliculogenesis, 216, 216*f*
Food contamination, arthropods, 5
Forcipomyiinae, 169
Forensic entomology, 35–42
  decomposition stages, 38–39, 39–42, 39*t*
  flies, 137
  historical perspective, 35–36
  homicide and accidental death cases, 36–38
  liability cases, 36
Formicinae, 371
Formicoidea, 371
Formicophilia, 6
Fort Morgan virus, 227
Fossula, 86
Fowl cyst mite, 466–467, 466*f*
Fowl tick, 510
Fowlpox virus, 249
Fox, sarcoptic mange, 471
Free-tailed bat mite, 439–440
Fruit flies, 148, 148*f*
Fulgoroidea, 83
Fur mites, 460–463

G

Gadding behavior, 316, 328*f*
Gametocyte, 183
Gametogony, 240
Garbage flies, 285, 287
Gaster, 373
Gasterophilinae, myiasis, 310, 331–332
Gastronyssidae, 484
Genal ctenidium, 116
Gene
  cloning, 543–544
  expression analysis, 549–553
Genital opercula, 402, 402*f*
Genitalia
  fleas, 117, 117*f*
  lice, 60, 61*f*
  mosquito, 214, 214*f*, 215*f*
  scorpion, 402
Genomics
  bioinformatics, 546
  library construction and genome assembly, 544–545, 545*f*
  sequencing of vectors and pathogens, 546, 547*t*
Genu, 16
Geometridae, 354, 364
Gerbil
  leishmaniasis, 164, 165
  plague, 124, 125
  sand fly fever, 160
German cockroach, 49–50, 49*f*
German yellowjacket, 384, 385
Glossinidae, *see* Tsetse flies
Gnaphosidae, 415
Goat
  ear mites, 479, 480–481
  heartwater, 530–531, 530*f*
  lice, 70, 77
Goat follicle mite, 468
Gonoinactive female, 216
Gonotrophic concordance, 216
Gonotrophic cycle, 20, 193, 216
Gonotrophic discordance, 223
Gorgas, William, 232
Graber's organ, 263
Granulocytic anaplasmosis, 531
Grassi, Govanni Batista, 2, 243
Ground squirrel, *see* Squirrel
Grub, 315
Guanine test, mites, 452
Guinea pig, lice, 79
Gulf Coast tick, 509, 509*f*
Gypsy moth, 362, 362*f*
Gyropidae, 60

H

Haematopinidae, 60
Haemosporidian, 183–184, 183*f*, 184*t*, 382
Hag moth, 359, 359*f*
Halarachnidae, 481–483, 482*f*
Halictidae, 386
Haller's organ, 17, 496, 497*f*
Haltere, 141, 213
Hanging groin, 197
Hantaan virus, 563*t*
Harpirhynchidae, 477–478
Harvester ant, 377, 380, 380*f*
Harvester's keratitis, 104
Harvest mite, 441*f*, 442
Haustellum, 14
HBV, *see* Hepatitis B virus
Heartwater, 530–531
Heartworm, *see* Dog heartworm
Heavy dragoon, 93
Hedgehog
  Chandipura virus disease, 160
  mites and mange, 442, 472, 473, 474, 477
  ticks, 509, 517, 522
Heel fly, 328*f*
Hemelytra, 83, 86
Hemiptera, 83–100
Hemiscorpiidae, 400
Hen flea, 123, 129–130
Hepatitis B virus, 96
Heteroscorpionidae, 400
Hexathelidae, 414
Hibernaculum, 222
Highlands J virus, 227
Hippoboscidae, 141, 339–340, 340–341, 340*f*
Hippoboscoidea, 297, 339–352
  behavior and ecology, 342–343
  deer keds, 342*f*, 345–346, 345*f*
  dog fly, 344
  horse ked, 344–345, 344*f*
  life history, 342
  morphology, 340–342
  pigeon fly, 346, 346*f*
  prevention and control, 350
  public health importance, 346–347

Hippoboscoidea (*Continued*)
  sheep ked, 341*f*, 343–344, 343*f*
  taxonomy, 339–340
  veterinary importance, 347–350, 348*t*, 349*f*
Histiotomatidae, 481
HIV, *see* Human immunodeficiency virus
Hobo's disease, 67
Hobo spider, 422, 423*f*
Hog cholera virus, 562*t*
Hog louse, 64
Holometabolous, 141
Holometabolous development, 102
Holoptic, 141
Homoptera, 83
Homosequential sibling species, 189
Honey bee, 376*f*, 377, 387–389, 388*f*
Hoplopleuridae, 60
Horizontal transmission, 21*f*, 24–25
Horn fly, 284, 291–292, 291*f*
Hornets, 385
Horse bot fly, 331*f*
Horse
  African horsesickness, 181–183, 181*f*
  caterpillar-induced abortion, 368
  chorioptic mange, 477
  eastern equine encephalitis, 225–227, 226*f*, 226*t*, 248, 248*f*
  equine allergic dermatitis, 185–186, 185*f*
  equine infectious anemia, 271, 271*f*
  equine onchocerciasis, 184–185, 184*f*
  lice, 70, 76
  mare reproductive loss syndrome, 368
  theilerioses, 528
  Venezuelan equine encephalitis, 226*f*, 226*t*, 227–228, 228*t*, 249
  western equine encephalitis, 226*f*, 226*t*, 227, 248–249
Horse flies, *see* Tabanidae
Horse ked, 344–345, 344*f*
Horse nose bot, 330
Horse stomach bot, 331*f*, 334
Host accessibility, 22
Host defenses, 6
Host immunity, 20–21
Host selection, 22–23
Host susceptibility, 22
House-dust mite, 451*f*
  allergy, 450–452
  American house dust mite, 450, 451*f*
  European house dust mite, 450
House fly, 282–283, 285–286, 288–289, 288*f*
House mouse mite, 437, 438*f*
Human body louse, 66–67, 66*f*, 72–73
Human bot fly, 326–327, 326*f*, 327*f*, 332
Human crab louse, 67
Human flea, 120–123
Human follicle mite, 445–447, 446*f*
Human head louse, 67
Human immunodeficiency virus, 96, 163
Human notoedric mange, 449
Human scabies, 448–449
Human scabies mite, 447, 447*f*
Hybridoma, 552
Hydrocele, 245
Hymenoptera, 371–396; *see also* Ants; Bees; Wasps
  behavior and ecology, 375–376
  life history, 374–375
  morphology, 373–374
  prevention and control, 392–393
  public health importance, 389–392
  sting reactions, 390, 390*t*
  taxonomy, 371–373, 372*t*
  venoms, 376–377
  veterinary importance, 392
Hyperendemic, 30
Hypersensitivity reaction, 4
Hypnozoite, 241
Hypoderatidae, 478
Hypoendemic, 30
Hypopus, 435
Hypopode, 435
Hypordematinae, myiasis, 310, 327–329, 329*f*
Hypostoma, 190

I

Ibex, epizootic hemorrhagic disease, 180
Idiosoma, 496
Igbo Ora, 229
IgG, *see* Immunoglobulin G
IgM, *see* Immunoglobulin M
IGRs, *see* Insect growth regulators
Immediate-sensitivity reaction, 4
Immunofluorescence microscopy, 552–553
Immunoglobulin G, 21
Immunoglobulin M, 21
Immunohistochemistry, 552–553
Incidental myiasis, 309
Indian red scorpion, 406
Indian tick typhus, *see* Boutonneuse fever
Infection prevalence, 27
Infectious bovine keratoconjunctivitis, 288, 289, 290*f*
Initiation phase, 216
Insect growth regulators
  cockroach control, 56
  flea control, 133
  hymenopteran control, 393
  tick control, 537
In situ hybridization, 551
Installment hatching, 217
Intermediate host, 20
Interseasonal maintenanve, parasites, 26–27
Intrinsic incubation, 25
Io moth, 360*f*
Iquipito, 84
Ira, 454
Ischnocera, 59–60
Ischnopsyllidae, 115
Isoamyl acetate, 375
Iuridae, 399
Ivermectin, 270
Ixodida, *see* Ticks
Ixodidae
  life cycles, 499–500, 500*f*
  morphology, 496–498, 498*f*
  taxonomy, 493–495

J

Jackal
  canine filariasis, 132
  control, 167
Jamestown Canyon virus, 238–239
Japanese encephalitis virus complex
  geographic distribution, 234*f*, 235*f*, 236*f*
  Japanese encephalitis virus, 235–236, 249
  Murray Valley encephalitis virus, 236–237
  St. Louis encephalitis virus, 236, 236*f*
  West Nile virus, 234–235
Jejenes, 169
Jerrymander, 411
Jigger, 122
Johnston's organ, 17, 210
Junin virus, 563*t*
Justinian's plague, 125

K

Kangaroo
  bot flies, 329*f*, 333
  chiggers, 435, 459
  filarial nematodes, 252
  myiasis, 272
  Q fever, 519
  tabanid disease transmission, 272
  tear-drinking moths, 357
Kangaroo rat, Lyme disease, 238
Kangaroo throat bot, 329*f*
Karakurt, 429
Katipo spider, 430
Keds, *see* Hippoboscoidea
Keel, 402
Kenya tick typhus, *see* Boutonneuse fever
Keystone virus, 238
Kilbourne, F. L., 2
Kissing bugs
  behavior and ecology, 86–88
  Chagas disease vectors, 85*t*, 88–92
  life history, 86
  morphology, 85–86, 86*f*
  overview, 84
  prevention and control, 92–93
  public health importance, 88–92
  taxonomy, 85, 85*f*
  veterinary importance, 92
Knemidokoptidae, 464
K-strategist, 142
Kuiki, 169

L

Lachryphagous moths, 357, 364, 365
La Crosse virus, 23, 24*f*, 238
Lady beetles, 109
Laelapidae, 440, 458–459
Lagomorph, *see* Rabbit
La malmignatte, 429
Laminosioptidae, 466
Lamponidae, 415
Lancet fluke, 392
Larder beetles, 101, 107, 108*f*
Larval diapause, 222
Larviparous, 141
Lasiocampidae, 362–363
Latrodectism, 426–430, 427*f*
Laveran, Charles, 243

Lazear, Jesse, 232
Leishmaniasis
  clinical features, 161*f*
  cutaneous leishmaniasis, 163–165, 164*f*
  geographic distribution, 162*f*
  transmission, 162–163
  veterinary importance, 165
  visceral leishmaniasis, 164*f*, 165
Lemurynssidae, 483
Lepidopterism, 353
Lepidoptera, 353–370
  behavior and ecology, 357–358
  hairs, 355–356, 355*f*, 356–357
  lachryphagous moths, 364–365
  life history, 357
  morphology, 341*f*, 354–357
  prevention and control, 368–369
  public health importance, 366–367
  taxonomy, 353–354
  urticating caterpillars, 358–364
  veterinary importance, 367–368
  wound-feeding and skin-piercing moths, 333
Leptoconopinae, 169
Leptopsyllidae, 115
Leucocytozoonosis
  black fly transmission, 201–202, 202*f*
  poultry, 184
Lice
  behavior and ecology, 45–46
  bird lice, 71–72, 79
  cat lice, 71, 78
  cattle lice, 68–70, 68*f*, 69*f*, 71*f*, 76
  deer lice, 78
  dog lice, 71, 78
  horse lice, 70, 76
  human body louse, 66–67, 66*f*
  human crab louse, 67
  human head louse, 67
  laboratory animals, 71, 79
  life history, 44–45
  livestock losses, 76–79
  morphology, 43–44, 44*f*, 46*f*, 47*f*
  pathogens, 72–73, 73–74, 75, 75*t*
  pig lice, 70, 77
  prevention and control, 79–81
  public health importance, 72–75
  sheep and goat lice, 70, 77
  taxonomy, 59–60, 60*t*, 61*t*
  veterinary importance, 75–76
Lignognathidae, 60
Limacodidae, 358–360, 359*f*, 360*f*
Lindane, tick control, 535
Liochelidae, 400
Listrophoridae, 462
Little blue cattle louse, 68, 69*f*, 71*f*
Little fire ant, 380
Little house fly, 284, 284*f*, 287
Lizards
  mites, 458, 460, 484
  scorpions, 404, 405
  spider bites, 420
  ticks, 26, 503, 508, 516, 522
  triatomines, 86
  tsetse flies, 301
Loiasis
  clinical features, 269–270, 269*f*
  control, 270
  tabanid transmission, 269, 270
Lone star tick, 499*f*, 509, 509*f*
Longnosed cattle louse, 68, 69*f*
Louping ill, 528–529
Louse-borne relapsing fever, 73–74
Louse flies, *see* Hippoboscoidea
Love-bug, 143, 143*f*
Loxoscelism, 423–426, 425*f*
Lunule, 140
Lycosidae, 415
Lymantriidae, 361–362
Lyme disease
  public health importance, 520–522, 520*f*
  tabanid transmission, 270
  veterinary importance, 532
Lymphatic filariasis, 243

M

Machupo virus, 563*t*
Mackie, F. P., 2
Macronyssidae, 438–440, 457
Macrophage, 21
Mad cow disease, *see* Bovine spongiform encephalopathy
Maggot
  clinical use, 333
  myiasis, 315
Makunagi, 169
Malaria
  avian, 250
  clinical disease, 241–242
  control, 254
  geographic distribution, 240*f*
  historical perspective, 243
  life cycle of parasite, 240–241, 240*f*
  monkeys, 250–251
  mosquito vectors and epidemiology, 242–243, 242*t*
  overview, 239–243
  reptiles, 250
  rodent, 250
Malignant tertian malaria, 241
Mammalophagic vector, 22
Mammalophily, 194
Mange
  cheyletid mange, 460
  human notoedric mange, 449
  mites, 464–471
  psoroptic mange, 474–476
  sarcoptic mange, 469–471
Mango fly, 261
Manson, Patrick, 2, 243, 247
Mansonellosis
  black fly transmission, 199
  human, 174, 176–178, 179*f*
Marburg virus, 563*t*
March fly, 143, 143*f*, 261
Mare reproductive loss syndrome, 368
Marmoset, wasting disease, 55
Marmots, Powassan encephalitis, 514
Marseilles fever, *see* Boutonneuse fever
Maruin, 169
Mastoparans, 377
Mata venado, 411
Mating plug, 220
Matrone, 220
May fly, 261
Mayaro virus, 229
Mechanical transmission, 24
Median infectious dose, 28
Medical entomology, 1
Medical-veterinary acarology, 1
Medical-veterinary arachnology, 1
Medical-veterinary entomology
  definition, 1
  historical perspective, 2–3
  literature, 1–2
Medicocriminal entomology, 137
Megalopygidae, 358, 358*f*
Mechanical transmission, 4
Melittin, 377
Meloidae, 101, 104–105, 104*f*, 105*f*, 105*t*
Membracidae, 83
Menoponidae, 60
Merogony, 240
Meront, 513
Merozoite, 183
Merutu, 169
Mesembrinellinae, myiasis, 310
Mesoendemic, 30
Mesosoma, 401
Mesostigmata, 433
Mesothelae, 413
Metacyclic forms, 88
Metasoma, 401, 402*f*
Metastriata, 493
Methoprene, 133
Mexican chicken bug, 97
Mexican typhus, *see* Murine typhus
Microarray, 551
Microcharmidae, 398
Microdistribution, 192
Microfilaremia, 244
Microfilariae, 244
Micropyle, 190
Midges, *see* Ceratopogonidae; Chaoboridae; Chironomidae
Migratory vertebrate host, 27
Millipede, 8
Miltogramminae, myiasis, 310
Minors, ants, 373
Mites, 43–58
  acaraphobia, 433, 456
  allergies, 449–452
  behavior and ecology, 436
  delusory acariosis, 433, 456
  dermatitis, 436–442, 442–445, 456–458
  intermediate hosts, 455–456
  internal acariasis, 452–453
  life history, 435–436, 435*f*
  morphology, 433–435, 434*f*
  rickettsialpox, 453–454
  skin-invading mites, 445–449
  stored-products mites, 442–445
  tapeworm intermediate hosts, 455–456, 485–487, 485*t*, 487*t*
  taxonomy, 433, 434*t*
  Tsutsugamushi disease, 442, 442*t*, 454–455
  veterinary importance, 456–487, 471*f*, 485*t*, 487*t*
Miturgidae, 415
Mahogony-flat, *see* Bed bug
Mold mite, 443
Mombinae, 387

Mongoose, dog fly, 344
Monkey lung mite, 482*f*
Monkeys
beetle infestation, 111
blood protozoans, 183
bot flies, 327
cat fleas, 130
Chikungunya virus, 229
Dengue, 232–234
filarial nematodes, 244
malaria, 250–251
mites and mange, 455, 469, 471, 473, 474, 482, 483
Oropouche fever, 176
sarcoptic mange, 471
spider bites, 414
tórsalo, 326
triatomines, 92
vesicular stomatitis, 166
wasting disease, 55
yellow fever, 230–232, 230*t*
Monkey slug, 359
Moose fly, 275
Morbidity, 29
Morgellons disease, 6
Morphoidae, 363–364
Mortality, 29
Morulae, 518
Mosquitoes, 207–260
behavior and ecology, 217–223, 218*f*, 219*f*
bites, 223
filariasis, 243–248, 244*f*, 245*f*, 247*t*, 251, 252
life history, 215–217, 216*f*
malaria, 239–243, 242*t*
morphology, 209–215, 210*f*, 211*f*, 212*f*, 213*f*, 214*f*, 215*f*
prevention and control, 223–248, 252–254
taxonomy, 207–215, 208*t*
viruses, 223–225, 224*t*, 225–227, 226*f*, 226*t*, 227–228, 228–229, 228*t*, 229–230, 230–232, 230*t*, 232–234, 233*f*, 234–235, 236–237, 237*t*, 238–239, 248–249
Moth fly, *see* Psychodinae
Moths, *see* Lepidoptera
Mouse
bot, 325*f*
lice, 71, 79
trypanosomiasis, 92
Mouse follicle mite, 469, 469*f*
Mouse fur mite, 461, 461*f*
Mouthparts
biting midges, 170, 171, 171*f*
black fly, 191
cockroaches, 44
Diptera, 139, 139*f*, 140
fleas, 116
lepidopterans, 354*f*
lice, 62, 63*f*
mosquitoes, 211, 213*f*
muscid flies, 278, 278*f*, 279*f*
parasitic arthropods, 13–15, 14*f*, 15*f*
spiders, 416, 416*f*
Tabanidae, 264
true bugs, 83
tsetse fly, 298–299, 298*f*
Mud dauber, 382*f*
Multiple gonotrophic cycles, 500
Multiplex polymerase chain reaction, 549
Multiplicative transmission, 25
Mummification, 38
Murine trypanosomiasis, 131–132
Murine typhus, 127–129
Murray Valley encephalitis virus, 236–237
Muscidae, 275–296
bazaar fly, 283, 286, 286*f*
behavior and ecology, 280–282
buffalo fly, 284, 291–292
bush fly, 283, 286, 289
cluster fly, 283
face fly, 283, 286–287, 289–290, 289*f*
false stable fly, 284, 287
garbage flies, 285, 287
horn fly, 284, 291–292, 291*f*
house fly, 282–283, 285–286, 288–289, 288*f*
life history, 279–280, 280*t*
little house fly, 284, 284*f*, 287
morphology, 275–279, 276*f*, 277*f*, 278*f*, 279*f*
myiasis, 310, 319–320, 319*f*
prevention and control, 287, 292–294, 293*f*
public health importance, 285–288
stable fly, 281*f*, 283–284, 287, 290–291, 291*f*
sweat flies, 285, 287–288, 292
taxonomy, 275, 276*t*
veterinary importance, 288–292
Muscinae, 275
Muscoidea, 309–338
ecology and behavior, 315–317
life history, 313*f*, 314–315
morphology, 313–314, 313*f*
myiasis types, 309, 310
myiasis myths, 317
prevention and control, 335–337
public health importance, 332–333
taxonomy, 310–313, 311*t*
veterinary importance, 333–335
Muscomorpha, 137, 139, 140
Mutillidae, 381
Mygalomorphae, 413–414
Myiasis, 309–338
ecology and behavior of flies, 315–317
fly families, 317–332
forms, 309, 310
life history of flies, 314–315
morphology of flies, 313–314
myths, 317
prevention and control, 335–337
public health importance, 332–333
taxonomy of flies, 310–313, 311*t*
veterinary importance, 333–335
Myobiidae, 461–462
Myocoptidae, 463
Mystacinobiidae, 310
Myxoma virus, 249, 562*t*
Myxomatosis, 131, 249
Myxophaga, 101

N

Nason's slug moth, 360*f*
Nematocera, 137, 140, 142
Nematoda, 104
Nematodes
biting midge transmission, 175*t*, 176–177, 177–178
Neottiphilidae, 310, 318
Nest blow flies, 322, 322*f*
Nest skipper fly, 318
Nested polymerase chain reaction, 549
Neuromyopathic araneism, 426
Nicolle, Charles, 2
No-no, 169
Noctuidae, 354, 363, 364–365
Noctuoidea, 353
Nolidae, 363
Nonviremic transmission, 25, 512
North Coast funnelweb spider, 421
Northern analysis, 550
Northern black widow spider, 428
Northern fowl mite, 439, 439*f*, 457–458
Northern rat flea, 122
Nose bot flies, 329–330
Notodontidae, 354, 364, 368
Notoedric cat mite, 472, 472*f*
Notoedric rat mite, 473
Notoedric squirrel mite, 472–473
Nott, Joshua, 2
Nukaka, 169
Nulliparity, 20
Nurse worker, ant, 373
Nuttalliellidae, 493, 495
Nycteribiidae, 339–340, 341–342
Nymph, 86, 87*f*
Nymphalidae, 363
Nymphal stage, 95
Nyung noi, 169

O

Obligate egg diapause, 222
Obligate parasite, 20
Obligatory myiasis, 309
Obtect, 142
Occupational hazards, arthropods, 36
Ockelbo, 229
OCP, *see* Onchocerciasis Control Programme
Ocular points, 63
Oedemeridae, 101, 105–106, 106*f*
Oestridae, myiasis, 310, 329*f*
Oestrinae, myiasis, 310, 329–330
Oestroidea, 309–338
Cuterebrinae, 325–327
ecology and behavior, 315–317
Gasterophilinae, 331–332
Hypodermatinae, 327–329, 329–330
life history, 313*f*, 314–315
morphology, 313–314
prevention and control, 335–337
public health importance, 332–333
taxonomy, 310–313, 311*t*
veterinary importance, 333–335
Ombdurman scorpion, 406
Onchocerciasis
bovine onchocerciasis, 201
control, 204
equine onchocerciasis, 184–185, 184*f*
river blindness, 194–200, 196*f*, 197*f*

Onchocerciasis Control Programme, 204
ONN virus, *see* O'nyong-nyong virus
O'nyong-nyong virus, 229
Oocyst, 183
Ookinete, 183
Oothecae, 43
Opisthosoma, 401, 496
Opisthothelae, 413
Oribatida, 433
Organ of Berlese, 95
Organ of Ribaga, 95
Oriental cockoach, 46, 46*f*
Oriental rat flea, 121
Ornithophagic vector, 22
Ornithophily, 194
Ornithonosis, 25
Oropouche fever, 176
Oviparity, 45, 141
Oviposition, 45, 194, 222, 502
Ovoviviparity, 314
Ovoviviparous, 141
Oxyopidae, 415

P

Palatal brushes, 209
Pandemic, 30
Pangoniinae, 15, 261
Paper wasp, 385–386, 385*f*
Papilionoidea, 353
Papulonodular demodicosis, 467
Paragenital sinus, 95
Parakeet
  lice, 79
  mites, 436, 456, 464, 465, 481
Parasimuliinae, 189
Parasitoid, 293
Paratransgenesis, 543–544, 544*f*
Parous female, 20
Pasteur, Louis, 243
Pathogen, 20
Pathogen evolution, 33
Pavement ant, 380–381, 380*f*
PCR, *see* Polymerase chain reaction
Pectinal teeth, 402
Pederin, 106
Pedicinidae, 60
Pediculidae, 60
Pediculicide, 79
Peridomestic species, 45, 86
Peritrophic membrane, 214
Permethrin
  sand-fly control, 166
  tick control, 536
Petiole, 373
Phantom midges, 144, 144*f*
Pharoah ant, 381, 381*f*
Pheromones, 45, 536
Philopteridae, 60
Phlebotominae
  behavior and ecology, 156–157
  life history, 156
  morphology, 155–156
  prevention and control, 166–167
  public health importance, 158–159, 159*t*, 160, 161–162, 162–163, 163–165
  sand fly, 154*f*, 155*f*
  taxonomy, 154
Phoridae, 146–147, 147*f*
Phospholipase, 424
Photoperiod entrainment, ticks, 503
Phthiraptera, 59–82
Phthiridae, 60
Pig
  African swine fever, 529
  beetles, 111
  Chandipura virus, 160
  fleas, 131
  Japanese encephalitis, 235, 236
  lancet fluke, 392
  lice, 70, 77
  louping ill, 528
  mites and mange, 449, 454, 467, 470, 471*f*
  muscid flies, 288
  myiasis, 320, 323
  sand flies, 158
  sarcoptic mange, 471*f*
  simuliotoxicosis, 202
  skin-piercing moths, 366
  Tahnya virus, 239
  tick-borne relapsing fever, 510, 523
  vesicular stomatitis, 166, 202
  vesicular stomatitis, 166, 202
Pigeon
  beetle infestation, 111
  chicken mite, 436
  chiggers, 455
  fowl cyst mite, 441*f*, 445, 466–467
  hypoderatid mites, 478
  louse flies, 349
  northern fowl mite, 458
  ticks, 495
  tropical fowl mite, 439
  trypanosomiasis, 120
Pigeon fly, 346, 346*f*
Pigeon mite, 437
Pigeonpox virus, 249
Pinkeye, 288, 289, 290*f*
Piophilidae, 139, 147, 148, 318, 318*f*
Piperidine, 376
Piroplasmoses, tick transmission, 526–528
Piroplasms, 512, 513
Pisauridae, 415
Pito, 84
Plague
  ecological forms, 125
  flea vectors, 125, 126
  outbreak control, 133
  pandemic history, 125
  pathogen, 124
Plasmid, 543–544
Pneumocoptidae, 484
Pneumonic plague, 126
Podosoma, 496
Poison, 3
Polistinae, 372
Polleniinae, myiasis, 310
Polyctenidae, 93
Polymerase chain reaction
  amplified fragment length polymorphism, 546–548
  applications, 546–549
  multiplex polymerase chain reaction, 549
  nested polymerase chain reaction, 549
  principles, 546–549
  rapid amplification of polymorphic DNA, 548–549
  rapid diagnostics, 553
  real-time polymerase chain reaction, 549
  reverse transciptase-polymerase chain reaction, 550–551
Polyphaga, 101
Polyplacidae, 60
Pompilidae, 381
Pophiliae, 147–148, 148*f*
Porcupine, malaria, 250
Posterior-station transmission, 24
Postgenal cleft, 190
Postinfestation interval, 41, 42
Postmortem interval, 35, 41
Post-trophic phase, 216
Poultry, *see* Chicken: Turkey
Prepupa, 141
Pretarsus, 403
Previtellogenic phase, 216
Primary myiasis, 309
Primates, nonhuman, *see* Monkeys
Pronotal ctenidium, 116
Propagative transmission, 25
Propodeum, 373
Prosimuliinae, 189
Prosoma, 401
Prospect Hill virus, 563*t*
Prostigmata, 433
Prostriata, 493
Pseudochactidae, 398
Psocoptera, 59–60
Psorergatidae, 468–469
Psoroptic mange, 474–476
Psoroptidae, 474
Psychodidae, 153–168, 474
  behavior and ecology, 156–157
  life history, 156
  morphology, 154–156
  moth fly, *see* Psychodinae
  prevention and control, 166–167
  public health importance, 157–158, 159*t*, 160, 161–162, 162–163, 163–165
  sand fly, *see* Phlebotominae
  Syrcoracinae, 153
  taxonomy, 153–154
  veterinary importance, 165–166
Psychodinae
  behavior and ecology, 156
  life history, 156
  morphology, 154–155
  moth fly, 154*f*, 155*f*
  prevention and control, 166
  public health importance, 157–158
  taxonomy, 153
Ptilinum, 142
Pulicidae, 115
Pulmonary acariasis, 452
Pulvilli, 44
Punky, 169
Pupariation, 277
Puparium, 142, 277
Purpura pulicosa, 123
Puss caterpillar, 358*f*
Puumala virus, 563*t*
Pyemotidae, 444
Pygidium, 116

Pygiopsyllidae, 115
Pyralidae, 354, 364
Pyrethrins, 133
Pyriproxyfen, 133

Q

Q fever, tick transmission, 519–520, 533
Quantitative polymerase chain reaction, 549
Queensland tick typhus, 518
Quill mites, 464

R

Rabbit
  California encephalitis, 238
  Colorado tick fever, 514, 515
  demodectic mange, 514, 515
  filarial nematodes, 252
  fleas, 119
  Keystone virus, 222, 231, 238
  lice, 71, 74, 79, 81
  mites, 442, 454, 456, 460, 461, 462, 479
  muscid flies, 287
  myiasis, 315, 317, 325, 326
  myxoma virus, 131, 202, 249
  sarcoptic mange, 472, 528
  snowshoe hare virus, 238
  ticks, 494, 503, 525
  Trivattatus virus, 238
  trypanosomiasis, 132
  tularemia, 270, 524–525, 532–533
  vesicular stomatitis, 158
  western equine encephalomyelitis, 222
Rabbit ear mite, 479, 479*f*
Rabbit fever, *see* Tularemia
Rabbit flea, 121*f*
Rabbit fur mite, 460*f*, 461
Rabbit tick, 506
Raccoon
  fleas, 121, 124, 133
  trypanosomiasis, 92
  vesicular stomatitis, 166
Racquet organ, 411
Radioallergosorbent test, 391
RAPD, *see* Rapid amplification of polymorphic DNA
Rapid amplification of polymorphic DNA, 548–549
RAST, *see* Radioallergosorbent test
Rat
  lice, 71, 79
  murine typhus, 127–128
  notoedric rat mite, 473
Rat typhus, *see* Murine typhus
Rat-tailed maggot, 317–318, 318*f*
Real-time polymerase chain reaction, 549
Receptor-mediated endocytosis, 501
Recombinant DNA, 543–544
Recrudescent typhus, 72
Red bug, 441–442
Red coat, 93
Red imported fire ant, 378–380, 378*f*
Reduviidae, 84
Red widow spider, 429, 429*f*
Reed, Walter, 2, 232
Relapsing fever
  louse-borne, 73–74
  tick-borne, 522–524
Reoviridae, 174*t*, 514, 557, 558*t*
Reproductive diapause, 223
Reptiles, *see also* Lizards; Snakes
  biting midges, 173, 183
  chiggers, 440, 442
  flesh flies, 324
  malaria, 250
  mites, 433, 438, 440, 457, 458, 460, 481, 483, 484
  mosquitoes, 221, 226
  Q fever, 519
  tabanids, 266
  ticks, 494, 495, 497*f*
  triatomines, 86
  tsetse flies, 301
Reservoir host, 21
Resilin, 15, 120
Respiratory mites, 481–484
Respiratory siphon, 210
Respiratory spiracles, 277, 277*f*
Restriction endonuclease, 544
Reticulocyte, 241
Reverse transciptase-polymerase chain reaction, 550–551
Rhabdoviridae, 158, 174*t*, 557, 558*t*
Rhagionidae, 145, 145*f*
Rhiniinae, myiasis, 310
Rhinoceros
  myiasis, 315
  skin-piercing moths, 366
  tear-drinking moths, 357, 364
  tsetse flies, 301
Rhinonyssinae, 481
Rhinophoridae, myiasis, 310
Rhopopsylllidae, 115
Rhynchophthirina, 59–60
Rhyncoptidae, 473–474
Rib cockle, 77, 349, 349*f*
Ricketts, Howard Taylor, 2
Rickettsialpox, 453–454
Rift Valley fever, 224*f*
Rift Valley fever virus, 239, 249
River blindness, 194–200, 196*f*, 197*f*
RMSF, *see* Rocky Mountain spotted fever
RNA analysis
  complementary DNA libraries, 550
  differential display, 551
  in situ hybridization, 551
  microarrays, 551
  Northern analysis, 550
  reverse transcriptase-polymerase chain reaction, 550–551
RNA interference, 552
Rocio virus, 237
Rocky Mountain spotted fever, 515–517, 516*f*
Rodents, *see* Mouse: Rat
Romaña's sign, 90, 90*f*
Root maggot, 319
Roseola pulicosa, 123
Ross, Ronald, 2, 243
Ross River virus, 229
Rostrum, 86
Rove beetles, 101, 106–107, 106*f*, 106*t*
RSSE, *see* Russian spring-summer encephalitis
r-strategist, 142
Rural plague, 125
Russian spring-summer encephalitis, 513
Ruxton, Buck, 36

S

Sac spider, 422*f*
Saddleback caterpillar, 359*f*
St. Louis encephalitis virus, 236, 236*f*
Salmonellosis, 129
Salticidae, 415
Sand flea, 122
Sand fly fever, 160
Sand fly, *see* Phlebotominae
Sand-puppy, 411
Sarcophagidae, myiasis, 310, 323–324, 324*f*
Sarcoptidae, 447–448, 448–449, 469
Saturniidae, 360–361, 361*f*
Scabies
  human, 448–449
  sheep, 449
Scabies mite, 469–471
Scaly-face mite, 465–466
Scaly-leg mite, 465, 466*f*
Scarab beetles, 101, 108–109, 108*f*
Scarabaeidae, 101, 108–109, 108*f*
Scarabiasis, 103
Scatter cockle, 349, 349*f*
Schizogony, 183, 240
Schizophora, 137
Sciaridae, 143–144, 144*f*
Scoleciasis, 353
Scorpionidae, 401
Scorpions
  behavior and ecology, 404–405
  life history, 403–404
  morphology, 401–403, 402*f*, 403*f*
  prevention and control, 407–408
  public health importance, 405–407
  taxonomy, 397–401
  toxicity ranking of species, 398*t*
  veterinary importance, 407
α-Scorpion toxin, 405
β-Scorpion toxin, 405
Screwworm, 315, 321–322, 322*f*, 334, 336
Scuttle flies, 146–147, 147*f*
Secondary myiasis, 309
Segestriidae, 415
Senogastrinae, 372
Sensilium, 17, 116
Sentinel, 31–32
Seoul virus, 563*t*
Septicemic plague, 126
Seroconversion, 30
Seropositive, 30
Shaft louse, 71, 79
Sheep
  African horsesickness, 182
  anaplasmosis, 531
  biting midges, 178
  blow flies, 333
  bluetongue disease, 150, 179*f*
  borreliosis, 532
  bot, 330, 330*f*, 334
  cat fleas, 130
  Chadipura virus, 160

chiggers, 453, 459
chorioptic mange, 476
cockle, 349, 349*f*
domodectic mange, 466
elaeophorosis, 445
epizootic hemorrhagic disease, 180
harvest mite, 442
heartwater, 530–531
lice, 70, 77, 77*f*
louping ill, 528–529
muscid flies, 284
myiasis, 309, 320, 325, 330, 334, 336
nagana, 305
nematodes, 111
psoroptic mites, 474–476, 478–480
Q fever, 508, 533
Rift valley fever, 249
sandfly fever, 160
scabies, 449
simuliotoxicosis, 202
spider bites, 430
tapeworms, 485–487
theileriosis, 527
ticks, 503, 532, 534
tórsalo, 326
trypanosomiasis, 271, 347
tularemia, 524, 532
vesicular stomatitis, 158, 165
Wesselsbron virus, 249
Sheep biting louse, 62*f*, 66, 70, 77
Sheep bot fly, 330*f*
Sheep face louse, 66
Sheep head fly, 285, 292
Sheep itch mite, 469
Sheep ked, 13, 142, 341*f*, 343–344, 343*f*, 347, 349, 349*f*, 350, 350*f*
Sheep nose bot, 330, 334
Sheep scab mite, 474–476, 475*f*
Sheep strike, 150, 333
Sheep tick, 508
Ship typhus, *see* Murine typhus
Shortnosed cattle louse, 69, 69*f*
Sibling species, 189
Sicariidae, 415–416
Silk allergy, 367
Simond, Paul Louis, 2
Simuliidae, *see* Black flies
Simuliinae, 189
Simuliini, 189
Simuliotoxicosis, 202–203
Sindbis virus, 229
Sin nombre virus, 563*t*
Siphonaptera, 115–136
Skipper flies, 147–148, 148*f*, 318, 318*f*
Slender guinea-pig louse, 64, 79
Smith, Theobald, 2
Smokybrown cockroach, 48, 48*f*
Snake mite, 440, 440*f*, 458
Snakes
chiggers, 442
mites, 437, 441, 481, 483
Snipe flies, 145, 145*f*
Snowshoe hare virus, 238
Soldier, ants, 373
Soldier fly, 146, 146*f*, 317, 317*f*
Solenophage, 13
Solenopsins, 376
Soligugae, 411–412, 411*f*
Solpugids, 411–412, 411*f*
Soremuzzle, 178
South American violin spider, 426
Southern black widow spider, 428, 428*f*
Southern cattle fever tick, 509
Southern yellowjacket, 385, 391*f*
Spanish jack, 386
Spermalege, 95
Spermode, 95
Sphaeromastigote, 88
Sphecidae, 273, 382
Sphingidae, 354, 365
Spicule hairs, lepidopterans, 355, 355*f*, 356, 356*f*
Spider wasp, 382*f*
Spider-lick, 106
Spiders, 413–432
araneomorph spiders, 414–416
behavior and ecology, 418
black widow spiders, 428, 429
brown recluse spider, 423*f*, 424–426, 424*f*, 425*f*
cheiracanthism, 422
funnelweb spiders, 421
latrodectism, 426–430, 427*f*
life history, 417–418
loxiscelism, 423–426, 425*f*
morphology, 416–417, 417*f*
mygalomorph spiders, 413–414
phoneutriism, 422
prevention and control, 457
public health importance, 418–430, 420*f*
South American violin spider, 426
tarantulas, 419–421, 420*f*
taxonomy, 413–416
tegenarism, 422–423
veterinary importance, 430
Spine hairs, lepidopterans, 355*f*, 356–357
Spinose ear tick, 511
Spiny oak-slug caterpillar, 360*f*
Spiny rat mite, 440, 458–459
Sporadic epidemic typhus, 128
Sporogony, 240
Sporozoite, 183
Springtail, 7
Squamous demodicosis, 467
Squirrel, *see also* Flying squirrels
blood protozoans, 183
borreliosis, 522
bot, 326*f*
California encephalitis, 229
Colorado tick fever, 515
fleas, 132
Keystone virus, 238
La Crosse encephalitis, 23, 238
leishmaniasis, 164
lice, 64
malaria, 250
mites and mange, 442, 463, 472–473, 481, 483
myiasis, 317
notoedric squirrel mite, 472–473
plague, 125
Powassan encephalitis, 514
ticks, 503, 514
tularemia, 524
western equine encephalitis, 227
Squirrel flea, 124
Stable fly, 281*f*, 283–284, 287, 290–291, 291*f*
Staphylinidae, 101, 106–107, 106*f*, 106*t*
Staphylococcal infection, 129
Stercorarian transmission, 24
Sticktight flea, 122, 122*f*
Stilting, 405
Stinging insects, *see* Hymenoptera: Scorpions
Stinging rose caterpillar, 360*f*
Stomach bot flies, *see* Gasterophilinae
Storage mites, 442–445, 450
Stratiomyidae, 146, 146*f*
Stratiomyidae, myiasis, 317, 317*f*
Stratiomyomorpha, 137
Straw itch mite, 445*f*
Streblidae, 339–340, 341
Strike, 315
Stylostome, 454
Sun-spider, 411
Superstitioniidae, 399
Surinam cockroach, 50, 50*f*
Surra, 271
Surveillance, 29–32
Swallow bugs, 97
Swarming, bees, 375
Sweat bee, 386, 386*f*
Sweat flies, 275, 276*t*, 285, 287–288, 292
Swinepox, 76
Swinepox virus, 562*t*
Sydney funnelweb spider, 421, 421*f*
Sylvatic epidemic typhus, 128
Sylvatic plague, 125
Symphyta, 371
Synanthropic flies, 275
Synganglion, 499
Syrcoracinae, 153
Syrphidae, 147, 147*f*
Syrphidae, myiasis, 317–318, 318*f*

T

Tabanidae, 261–274
behavior and ecology, 266–268, 267*f*
horse flies versus deer flies, 261, 263*t*
life history, 264–266, 265*f*
morphology, 263–264, 263*f*, 264*f*
prevention and control, 272–273
public health importance, 268–270, 269*t*
taxonomy, 261–263, 262*t*
traps, 268, 268*f*, 273
veterinary importance, 270–272
Tabaninae, 261
Tabanomorpha, 137, 139, 140, 261
Tache noir, 517
Tachinidae, myiasis, 310
Tahnya virus, 239
Tangential transmisson, 25
Tapeworm
beetles as intemediate hosts, 110–111, 110*f*
fleas as intermediate hosts, 130, 132
lice as intermediate hosts, 75
mites as intermediate hosts, 455–456, 485–487, 485*t*, 487*t*
Tarantism, 415, 419
Tarantulas, 419*f*, 420*f*
hairs, 420, 420*f*
tarantism, 419
toxicity, 419–421

Tarantulism, 419
Tarsomere, 15
Tarsus, 15
TBE, *see* Tick-borne encephalitis
TBRF, *see* Tick-borne relapsing fever
Tegmen, 44
Telmophage, 13, 194
Tenebrionidae, 101, 107, 107*f*
Terebrantia, 371
Terminalia, 44, 214
Thaumaleidae, 207
Thaumetopoeidae, 363
Theileriosis, tick transmission, 527–528
Theraphosidae, 414, 419–421
Theridiidae, 416
Thomisidae, 416
Thrips, 7
Thyatiridae, 354, 365
Tick-borne encephalitis, 513–514
Tick-borne relapsing fever, 522–524
Tick paralysis, 525–526, 534
Ticks, 493–542
  African swine fever, 529
  allergy, 526
  anaplasmosis, 457, 519, 529–530, 531–532
  babesiosis, 512–513, 527
  behavior and ecology, 501–505
  borrelioses, 532
  Boutonneuse fever, 517
  Colorado tick fever, 514–515
  dermatophilosis, 533–534, 533*f*
  ehrlichiosis, 518–519, 518*f*, 529–530
  hard ticks, 493–495
  heartwater, 530–531
  life history, 499–501
  louping ill, 528–529
  Lyme disease, 520–522, 520*f*, 532
  morphology, 495–499
  piroplasmoses, 526–528
  prevention and control, 535–538
  Q fever, 519–520, 533
  Rocky Mountain spotted fever, 515–517, 516*f*
  soft ticks, 495
  taxonomy, 493–495, 494*t*
  theilerioses, 527–528
  tick paralysis, 525–526, 534
  tick toxicosis, 534–535
  tick-borne encephalitis, 513–514
  tick-borne relapsing fever, 522–524
  tularemia, 524–525, 532–533
  veterinary importance, 507*t*, 526–535
Tick toxicosis, 534–535
Tipulidae, 138, 142–143, 143*f*
Tlalzahuatle, 441–442
Toad blow flies, 321
Togaviridae, 223, 224*t*, 225–229, 557, 558*t*
Tórsalo, 326–327, 326*f*, 327*f*, 332, 334
Townsend, C. H. T., 161
Toxicosis, 3
Toxin, arthropods, 3–4
Tracheal skein, 217
Transcriptome, 550
Transfusion, babesiosis risks, 513
Transgenerational transmission, 23
Transmission rate, 23
Transstadial transmission, 23
Trematode, 111
Trench fever, 74
Triatominae, *see* Kissing bugs
Trichobothria, 17
Trichodectidae, 60
Trivittatus virus, 238
Troglotayosicidae, 399
Trombiculidae, 440–442, 459–460, 483
Trophic phase, 216
Tropical bont tick, 509
Tropical fowl mite, 439, 439*f*, 457
Tropical rat mite, 438, 438*f*, 457, 484
Tropical theileriosis, 528
True bugs, 83–100
True flies, *see* Diptera
Trypanosomiasis, *see* African sleeping sickness: Chagas disease: East African trypanosomiasis: Murine trypanosomiasis: Nagana: Surra: West African trypanosomiasis
Tsetse flies, 297–308
  African sleeping sickness, 302–303, 302*f*
  behavior and ecology, 300–302, 302*f*
  East African trypanosomiasis, 303
  geographic distribution, 297, 298*f*
  life history, 299–300, 300*f*
  morphology, 298–299, 298*f*, 299*f*
  nagana, 305–306, 305*f*
  paratransgenesis, 543–544, 544*f*
  pathogens and vectors, 304*t*
  prevention and control, 306–307
  taxonomy, 297–298
  trypanosome life cycle, 303–305
  West African trypanosomiasis, 303
Tsutsugamushi disease, 442, 442*t*, 454–455, 455*f*
Tularemia
  clinical features, 270, 270*f*
  flea transmssion, 129
  tabanid transmission, 270
  tick transmission, 524–525, 532–533
Tumbler, mosquito, 210
Tumbu fly, 322–323, 323*f*
Tungiasis, 122, 122*f*, 131
Turbinoptidae, 483
Turkestan cockroach, 46–47, 46*f*
Turkey
  beetle infestation, 109
  biting midges, 178, 183
  blood protozoans, 183
  borreliosis, 532
  fleas, 122
  leucocytozoonosis, 202
  malaria, 250
  mites, 466
  nematodes, 55
  tapeworms, 110, 110*f*
  ticks, 533
Turtle
  blood protozons, 272
  chiggers, 442
  tabanid disease transmission, 272
Two-dimensional gel electrophoresis, 553–554
Typhus, *see* Epidemic typhus: Murine typhus: Recrudescent typhus

U

Urban plague, 125
Urban typhus, *see* Murine typhus
Urinary acariasis, 453
Urodacidae, 400
Urticaria, 353
  caterpillars, 358–364

V

Vaccines
  African horsesickness, 182
  babesiosis, 527
  bluetongue disease, 179
  epidemic typhus, 72, 81
  fleas, 133
  human scabies, 448
  leishmaniasis, 167
  malaria, 250
  mosquito-borne diseases, 254
  plague, 127
  polio, 149, 163
  tick-borne encephalitis, 514
  tick control, 535, 537
  western equine encephalomyelitis, 21
  yellow fever, 32, 232
Vaejovidae, 399–400
Vagabond's disease, 67
Vector competence, 28
Vectorial capacity, 28–29
Velvet ant, 361, 372, 381, 381*f*
Venereal transmission, 23
Venezuelan equine encephalitis, 226*f*, 226*t*, 227–228, 228*t*, 249
Venom
  ants, 376–377
  arthropods, 3–4
  bees, 376–377
  latrodectism, 426–430
  loxoscelism, 376–377, 424
  scorpions, 405
  tarantulas, 420
  wasps, 376–377
Vermipsyllidae, 115
Vertical transmission, 21*f*, 23–24
Vesicular stomatitis, 158–159, 165, 166
Vespidae, 372, 382–386
Vespinae, 372
Vespoidea, 371
Vinchuca, 84
Visceral leishmaniasis, 164*f*, 165
Viscerocutaneous loxoscelism, 426
Vitellogenin, 502
Volant, 340

W

Walkingstick, 7
Wallaby, tabanid disease transmission, 272
Wall louse, 93
Walnut caterpillar, 368*f*
Warble, 315, 329*f*

Wasps, *see also* Hymenoptera
behavior and ecology, 375–376
life history, 374–375
morphology, 373–374, 374*f*
prevention and control, 392–393
public health importance, 389–392
social wasps, 382–386
solitary wasps, 381–382
taxonomy, 371–373, 372*t*
venoms, 376–377
veterinary importance, 392
Water buffalo
hippoboscids, 345
horn fly, 284
lachryphagous moths, 357, 364
skin-piercing moths, 366
Water buffalo skin maggot, 323
Wesselsbron virus, 249
West African trypanosomiasis, 303
Western analysis, 552
Western equine encephalitis, 23, 226*f*, 226*t*, 227, 248–249
Western yellowjacket, 384
West Nile virus, 234–235, 235*f*, 347
Whamefly, 261
Wheel bug, 83
Whiplash dermatitis, 106
White-flannel moth, 359*f*
Wind-scorpion, 411
Wing louse, 71
Wings
beetles, 101, 102*f*
black fly, 191
cockroaches, 44
Diptera, 141
kissing bugs, 86
mosquito, 210, 212*f*
muscid flies, 278, 279*f*
parasitic arthropods, 13
true bugs, 83, 84*f*
tsetse fly, 299, 299*f*
Wolf, *see also* Dog
fleas, 121
lice, 64, 70
myiasis, 315, 317
sarcoptic mange, 471
Woodchuck, *see* Marmots
Wriggler, mosquito, 209

X

Xenodiagnosis, 91
Xylophagomorpha, 137

Y

Yellow fever
geographic distribution, 230, 230*f*
transmission, 231, 232
virus, 231
Yellowjacket, 373*f*, 374*f*, 382, 383–385, 383*f*, 384*t*, 391*f*, 393
Yellow scorpion, 406

Z

Zebu, 364*f*, 474, 527
Zinsser, Hans, 72
Zoonosis, 25
Zoophagic vector, 22
Zoophilous, 357
Zoridae, 416
Zygaenoidea, 353